FUNDAMENTALS OF THERMAL-FLUID SCIENCES

MCGRAW-HILL SERIES IN MECHANICAL ENGINEERING

CONSULTING EDITORS

Jack P. Holman, Southern Methodist University
John Lloyd, Michigan State University

Anderson	Computational Fluid Dynamics: The Basics with Applications
Anderson	Modern Compressible Flow: With Historical Perspective
Arora	Introduction to Optimum Design
Borman and Ragland	Combustion Engineering
Çengel	Heat Transfer: A Practical Approach
Çengel	Introduction to Thermodynamics & Heat Transfer
Çengel and Boles	Thermodynamics: An Engineering Approach
Çengel and Turner	Fundamentals of Thermal-Fluid Sciences
Culp	Principles of Energy Conversion
Dieter	Engineering Design: A Materials & Processing Approach
Doebelin	Engineering Experimentation: Planning, Execution, Reporting
Driels	Linear Control Systems Engineering
Edwards and McKee	Fundamentals of Mechanical Component Design
Gibson	Principles of Composite Material Mechanics
Hamrock	Fundamentals of Fluid Film Fabrication
Hamrock	Fundamentals of Machine Elements
Heywood	Internal Combustion Engine Fundamentals
Histand and Alciatore	Introduction to Mechatronics and Measurement Systems
Holman	Experimental Methods for Engineers
Jaluria	Design and Optimization of Thermal Systems
Kays and Crawford	Convective Heat and Mass Transfer
Kelly	Fundamentals of Mechanical Vibrations
Martin	Kinematics and Dynamics of Machines
Mattingly	Elements of Gas Turbine Propulsion
Modest	Radiative Heat Transfer
Norton	Design of Machinery
Oosthuizen and Carscallen	Compressible Fluid Flow
Oosthuizen and Naylor	Introduction to Convective Heat Transfer Analysis
Reddy	An Introduction to Finite Element Method
Rosenberg and Karnopp	Introduction to Physical Systems Dynamics
Schlichting	Boundary Layer Theory
Shames	Mechanics of Fluids
Shigley and Uicker	Theory of Machines and Mechanisms
Stoecker	Design of Thermal Systems
Stoecker and Jones	Refrigeration and Air Conditioning
Turns	An Introduction to Combustion: Concepts and Applications
Ullman	The Mechanical Design Process
Wark	Advanced Thermodynamics for Engineers
Wark and Richards	Thermodynamics
White	Fluid Mechanics
White	Viscous Fluid Flow
Zeid	CAD/CAM Theory and Practice

FUNDAMENTALS OF THERMAL-FLUID SCIENCES

Yunus A. Çengel
Robert H. Turner

both of the Department of Mechanical Engineering
University of Nevada, Reno

Boston   Burr Ridge, IL   Dubuque, IA   Madison, WI   New York   San Francisco   St. Louis
Bangkok   Bogota   Caracas   Kuala Lumpur   Lisbon   London   Madrid   Mexico City
Milan   Montreal   New Delhi   Santiago   Seoul   Singapore   Sydney   Taipei   Toronto

McGraw-Hill Higher Education

A Division of The McGraw-Hill Companies

FUNDAMENTALS OF THERMAL-FLUID SCIENCES
International Edition 2001

Exclusive rights by McGraw-Hill Book Co – Singapore, for manufacture and export. This book cannot be re-exported from the country to which it is sold by McGraw-Hill. The International Edition is not available in North America.

Published by McGraw-Hill, an imprint of The McGraw-Hill Companies, Inc., 1221 Avenue of the Americas, New York, NY 10020. Copyright © 2001, by The McGraw-Hill Companies, Inc. All rights reserved. No part of this publication may be reproduced or distributed in any form or by any means, or stored in a database or retrieval system, without the prior written consent of the McGraw-Hill Companies, Inc., including, but not limited to, in any network or other electronic storage or transmission, or broadcast for distance learning.
Some ancillaries, including electronic and print components, may not be available to customers outside the United States.

10 09 08 07 06 05
20 09 08 07 06 05 04
UPE   BJE

Library of Congress Cataloging-in-Publication Data

Cengel, Yunus A.
   Fundamentals of thermal-fluid sciences / Yunus A.. Cengel, Robert H. Turner.
       p.   cm.
   Includes bibliographical references and index.
   ISBN 0-07-239054-9
     1.   Thermodynamics. 2. Heat—Transmission. 3. Fluid mechanics. I. Turner, Robert H.
II. Title.
TJ265.C42 2001
621.402—dc21   00-025660

www.mhhe.com

When ordering this title, use ISBN 0-07-118152-0

Printed in Singapore

About the Authors

Yunus A. Çengel is Professor of Mechanical Engineering at the University of Nevada, Reno (UNR). He received his Ph.D. in mechanical engineering from North Carolina State University. His research areas are energy conservation, exergy analysis, geothermal energy, desalination, radiation heat transfer, natural convection, and engineering education. He is the director of the Industrial Assessment Center at the University of Nevada, Reno. He has led teams of engineering students to numerous manufacturing plants in northern Nevada and California to conduct industrial assessments and has prepared energy conservation, waste minimization, and productivity enhancement reports for those plants.

Dr. Çengel is the coauthor (with Dr. Michael A. Boles) of the widely adopted textbook, *Thermodynamics: An Engineering Approach,* now in its third edition, published in 1998. He is also the author of *Introduction to Thermodynamics and Heat Transfer,* published in 1997, and *Heat Transfer: A Practical Approach,* published in 1998, all by McGraw-Hill. Some of these books have been translated into Chinese, Japanese, Korean, Spanish, Turkish, Italian, and Greek. Dr. Çengel is the recipient of several outstanding teacher awards and he received the ASEE Meriam/Wiley Distinguished Author Award in 1992 and again in 2000 for excellence in authorship.

Dr. Çengel is a registered Professional Engineer in the State of Nevada and is a member of the American Society of Mechanical Engineers (ASME) and the American Society for Engineering Education (ASEE).

Robert H. Turner is Professor Emeritus of Mechanical Engineering at the University of Nevada, Reno (UNR). He earned a B.S. and M.S. from the University of California at Berkeley, and his Ph.D. from UCLA, all in mechanical engineering. He worked in industry for 18 years, including nine years at Cal Tech's Jet Propulsion Laboratory (JPL). Dr. Turner then joined the University of Nevada in 1983. His research interests include solar and renewable energy applications, thermal sciences, and energy conservation. He established and was the first director of the Industrial Assessment Center at the University of Nevada.

For 20 years Dr. Turner has designed the solar components of many houses. In 1994–95, in a cooperative effort between UNR and Erciyes University in Kayseri, Turkey, he designed and oversaw construction of the fully instrumented Solar Research Laboratory at Erciyes University, featuring 130 square meters of site-integrated solar collectors. His interest in applications has led Dr. Turner to maintain an active consulting practice.

Dr. Turner is a registered Professional Engineer and is a member of the American Society of Mechanical Engineers (ASME) and the American Society of Heating, Refrigeration, and Air Conditioning Engineers (ASHRAE).

Contents

PART IV ▓ APPENDIXES

APPENDIX 1 ▓ PROPERTY TABLES AND CHARTS (SI UNITS)   929

APPLICATIONS CHAPTERS AVAILABLE ON THE WEB:
http://www.mcgraw-hill.com/cengel

Table of Examples

In the future mankind will turn to sciences and learning. It will obtain all its power from sciences. Power and rule will then pass to the hands of sciences and knowledge. Eloquence and beauty of expression, the most brilliant of all arts and sciences, will be the most sought after in all their varieties. Even, in order to make one another accept their ideas and carry out their word, people will find their most effective weapon in eloquent expression, and their most irresistible force in fine oratory.

—Said Nursi, 1930

Preface

OBJECTIVES

This introductory text is intended for use in a comprehensive first course in thermal-fluid sciences for undergraduate engineering students in their sophomore, junior, or senior year with adequate background in calculus and physics, and as a reference book for practicing engineers. The objectives of this text have been

- To cover the *basic principles* of thermodynamics, heat transfer, and fluid mechanics.
- To present a wealth of real-world *engineering applications* to give students a feel of engineering practice.
- To develop an *intuitive understanding* of the subject matter by emphasizing the physics and the physical arguments.

The text contains sufficient material to give instructors considerable flexibility and to accommodate their preferences on the right blend of thermodynamics, heat transfer, and fluid mechanics for their students. By careful selection of topics, an instructor can spend one-third, one-half, or two-thirds of the course on thermodynamics and the rest on selected topics of fluid mechanics and heat transfer.

The text is an abbreviated version of the standard thermodynamics, heat transfer, and fluid mechanics texts, covering topics that the engineering students are most likely to need in their professional lives. The thermodynamics portion of this text is based on the text *Thermodynamics: An Engineering Approach* by Y. A. Çengel and M. A. Boles, and the heat transfer portion is based on *Heat Transfer: A Practical Approach* by Y. A. Çengel, both published by McGraw-Hill. Most chapters are practically independent of each other and can be covered in any order. We recommend covering the first 6 introductory chapters in sequence and then covering topics of interest from the remaining 14 chapters in any order. The text is ideally suited for curriculums that have a common introductory course or a two-course sequence on thermal-fluid sciences.

It is recognized that all topics of thermodynamics, fluid mechanics, and heat transfer cannot be covered adequately in a typical three-semester-hour course, and, therefore, sacrifices must be made from depth if not from the breadth. Selecting the right topics and finding the proper level of depth and breadth are no small challenge for the instructors, and this text is intended to serve as the ground for such selection. Students in a combined thermal-fluids course can gain a basic understanding of energy and energy interactions, various mechanisms of heat transfer, and fundamentals of fluid flow. Such a course can also instill in students the confidence and the background to do further reading of their own and to be able to communicate effectively with specialists in thermal-fluid sciences.

GENERAL APPROACH

The philosophy that contributed to the popularity of our thermodynamics and heat transfer books has remained unchanged in this text. The goal throughout this project has been to offer an engineering textbook that

- Talks directly to the minds of tomorrow's engineers in a simple yet precise manner.
- Encourages *creative thinking* and development of a *deeper understanding* of the subject matter.
- Is *read* by students with *interest* and *enthusiasm* rather than being used as an aid to solve problems.

Special effort is made to touch the curious minds and take them on a pleasant journey in the wonderful world of thermal-fluid sciences and explore the wonders of these exciting subjects.

Yesterday's engineer spent a major portion of his or her time substituting values into the formulas and obtaining numerical results. But all the formula manipulations and number crunching are being left to the computers. Tomorrow's engineer will have to have a clear understanding and a firm grasp of the *basic principles* so that he or she can understand even the most complex problems, formulate them, and interpret the results. A conscious effort is made to lead students in this direction.

CONTENTS

We start by giving a general overview of thermal-fluid sciences and introduce the *basic concepts* and the *conservation of mass principle* in Chapter 1. To maintain utmost flexibility and to accommodate different preferences, we have adopted a modular approach and presented the material under the subcategories of

- Thermodynamics (Chapters 2–8).
- Fluid mechanics (Chapters 9–13).
- Heat transfer (Chapters 14–20).

This way, the book can also be used as a suitable text for courses that do not involve fluid mechanics or heat transfer by ignoring the chapters under those subcategories.

- In Chapter 2 we discuss the basic concepts of thermodynamics such as pressure and temperature. In Chapter 3 we present the properties of pure substances, including specific heats, and illustrate the use of property tables. In Chapter 4 we discuss energy transfer by heat, work, and mass. The *first law of thermodynamics* is introduced in Chapter 5, the *second law* and *entropy* in Chapters 6 and 7, and the *power and refrigeration cycles* in Chapter 8.
- Chapters 9 through 13 deal with fluid mechanics. We introduce the *basic concepts, viscosity,* and *surface tension* in Chapter 9; *fluid statics* and *buoyancy* in Chapter 10; *Bernoulli, energy,* and *momentum equations* in Chapter 11; *flow in pipes* in Chapter 12; and *flow over bodies* in Chapter 13.
- Chapters 14 through 20 deal with heat transfer. We discuss the basic mechanisms of heat transfer in Chapter 14, *steady heat conduction* in Chapter 15, *transient heat conduction* in Chapter 16, *forced convection* in Chapter 17, *natural convection* in Chapter 18, *radiation* in Chapter 19, and *heat exchangers* in Chapter 20. Extensive discussions are given on *thermal insulations* and the *optimum thickness of insulation* because of the widespread use of insulations in industry and the key role they play in any energy conservation project.

Chapters 21 and 22 on *heating and cooling of buildings* and *thermal control of electronic equipment* are placed on the Web to keep the size of the text at a manageable level. These two chapters are available to adopters for downloading.

LEARNING TOOLS

Emphasis on Physics

A distinctive feature of this book is the absence of heavy mathematical and theoretical aspects of subject matter such as the separation of variables and the Navier-Stokes equations. The authors believe that such material has little practical value at early stages, and it is better suited for graduate-level courses. The emphasis in undergraduate education should remain on *developing a sense of underlying physical mechanism* and a *mastery of solving practical problems* an engineer is likely to face in the real world. The absence of such time-consuming and intimidating material should free the instructor to cover more material to give the students a broader background. This should also make the course a more pleasant and worthwhile experience for the students.

Effective Use of Association

An observant mind should have no difficulty understanding engineering sciences. After all, the principles of engineering sciences are based on our *everyday experiences* and *experimental observations*. A more physical, intuitive approach is used throughout this text. Frequently, *parallels are drawn* between the subject matter and students' everyday experiences so that they can relate the subject matter to what they already know. The process of cooking, for example, serves as an excellent vehicle to demonstrate the basic principles of thermodynamics and heat transfer mechanisms.

Self-Instructing

The material in the text is introduced at a level that an average student can follow comfortably. It speaks to the students, not over the students. In fact, it is *self-instructive.* Noting that the principles of science are based on experimental observations, all the derivations in this text are based on physical arguments, and thus they are easy to follow and understand.

Extensive Use of Artwork

Figures are important learning tools that help the students "get the picture." The text makes effective use of graphics. It probably contains more figures and illustrations than any other book in this category. Figures attract attention and stimulate curiosity and interest. Some of the figures in this text are intended to serve as a means of emphasizing some key concepts that would otherwise go unnoticed, or as paragraph summaries.

Chapter Openers and Summaries

Each chapter begins with an overview of the material to be covered and its relation to other chapters. A *summary* is included at the end of each chapter for a quick review of basic concepts and important relations.

Numerous Worked-Out Examples

Each chapter contains several worked-out *examples* that clarify the material and illustrate the use of the basic principles. An *intuitive* and *systematic* approach is used in the solution of the example problems, with particular attention to the proper use of units.

A Wealth of Realistic End-of-Chapter Problems

The *end-of-chapter problems* are grouped under specific topics in the order they are covered to make problem selection easier for both instructors and students. The problems within each group start with concept questions, indicated by "C," to check the students' level of understanding of basic concepts. The problems under *Review Problems* are more comprehensive in nature and are not directly tied to any specific section of a chapter. The problems under the *Computer, Design, and Essay Problems* title are intended to encourage students to make engineering judgments, to conduct independent searches on topics of interest, and to communicate their findings in a professional manner. Several economics- and safety-related problems are incorporated throughout to enhance cost and safety awareness among engineering students. Answers to selected problems are listed immediately following the problem for convenience to the students.

A Systematic Solution Procedure

A well-structured approach is used in problem solving while maintaining an informal conversational style. The problem is first stated and the objectives are identified, and any assumptions made are stated together with their justifications. Numerical values are used together with their units to emphasize that numbers without units are meaningless, and unit manipulations are as important as manipulating the numerical values with a calculator. The significance of the findings are discussed following the solutions. This approach is also used consistently in the solutions presented in the Solutions Manual.

Physically Meaningful Formulas

The physically meaningful forms of the conservation equations instead of formulas are used to foster deeper understanding and to avoid a cookbook approach. For example, the mass and energy balances for *any system* undergoing *any process* are expressed as

Mass balance:	$m_{\text{in}} - m_{\text{out}} = \Delta m_{\text{system}}$
Energy balance:	$\underbrace{E_{\text{in}} - E_{\text{out}}}_{\substack{\text{Net energy transfer} \\ \text{by heat, work, and mass}}} = \underbrace{\Delta E_{\text{system}}}_{\substack{\text{Change in internal, kinetic,} \\ \text{potential, etc., energies}}}$

The relations above reinforce that during an actual process mass and energy are conserved. Students are encouraged to use these forms of balances in early chapters after they specify the system and to simplify them for the particular problem. A more relaxed approach is used in later chapters as students gain mastery.

Relaxed Sign Convention

The use of a formal sign convention for heat and work is abandoned as it often becomes counterproductive. A physically meaningful and engaging approach is adopted for interactions instead of a mechanical approach. Subscripts "in" and "out," rather than the plus and minus signs, are used to indicate the directions of interactions.

A Choice of SI Alone or SI/English Units

In recognition of the fact that English units are still widely used in some industries, both SI and English units are used in this text, with an emphasis on SI. The material in this text can be covered using combined SI/English units or SI units alone, depending on the preference of the instructor. The property tables and charts in the appendixes are presented in both units, except the ones that involve dimensionless quantities. Problems, tables, and charts in English units are designated by "E" after the number for easy recognition, and they can be ignored easily by SI users.

Conversion Factors

Frequently used conversion factors and the physical constants are listed on the inner cover pages of the text for easy reference.

SUPPLEMENTS

The following supplements are available to the adopters of the book.

Instructor's Solutions Manual

Available to instructors only, this manual features detailed solutions prepared using an equation editor complete with illustrations. The solutions are suitable for posting or using as handouts in class. The solutions of problems of the two chapters placed on the Web (Chapters 21 and 22) are also given in the solutions manual.

EES Software

Developed by Sandy Klein and Bill Beckman from the University of Wisconsin–Madison, this software program allows students to solve problems, especially design problems, and to ask "what if" questions. EES (pronounced "ease") is an acronym for Engineering Equation Solver. EES is very easy to master since equations can be entered in any form and in any order. The combination of equation-solving capability and engineering property data makes EES an extremely powerful tool for students.

EES can do optimization, parametric analysis, and linear and nonlinear regression and provides publication-quality plotting capability. Equations can be entered in any form and in any order. EES automatically rearranges the equations to solve them in the most efficient manner. EES is particularly useful for heat transfer problems since most of the property data needed for solving heat transfer problems are provided in the program. For example, the steam tables are implemented such that any thermodynamic property can be obtained from a built-in function call in terms of any two properties. Similar capability is provided for many organic refrigerants, ammonia, methane, carbon dioxide, and many other fluids. Air tables are built-in, as are psychometric functions and JANAF table data for many common gases. Transport properties are also provided for all substances. EES also allows the user to enter property data or functional relationships with lookup tables, with internal functions written with EES, or with externally compiled functions written in Pascal, C, C++, or Fortran.

The EES Software Problems Disk contains EES programs that have been developed to solve some of the problems in this text. Problems solved on the EES problems disk are denoted in the text with a disk symbol. Each program provides detailed comments and on-line help. These programs should help the student master the important concepts without the calculational burden that has been previously required.

Website

Students and instructors can have access to the website http://www.mcgraw-hill.com/cengel where PDF files of applications chapters, additional material, instructional aids, and updates will be available for downloading.

ACKNOWLEDGMENTS

We would like to acknowledge with appreciation the numerous and valuable comments, suggestions, criticisms, and praise of the following academic evaluators:

A. Aziz, *Gonazaga University*
Radu Danescu, *Clemson University*
Ram K. Ganesh, *University of Connecticut*
Larry J. Harvard, Jr., *Louisiana State University*
Mehmet Kanoglu, *Celal Bayar University, Manisa, Turkey*
Robert D. Lotz, *Rensselaer Polytechnic Institute*
D.E. Musielak, *University of Texas at Arlington*
Jay Muthuswamy, *Tennesee State University*
Donald K. Roth, *Gannon University*
C. William Savery, *Portland State University*
Gita Talmage, *Pennsylvania State University*
Yong X. Tao, *Tennessee State University*
Charles S. Tritt, *Milwaukee School of Engineering*
Jay Warner, *Milwaukee School of Engineering*

In particular, we would like to express our gratitude to the following reviewers, many of whom reviewed the manuscript at more than one stage of development:

Pradeep Kumar Bansal, *University of Auckland, New Zealand*
John M. Cimbala, *Pennsylvania State University*
Joseph F. Kmec, *Purdue University*
William E. Lee, III, *University of South Florida*
Frank K. Lu, *University of Texas at Arlington*
T. Terry Ng, *University of Toledo*
Jim A. Nicell, *McGill University, Montreal, Canada*
Arthur E. Ruggles, *University of Tennessee*
Chiang Shih, *FAMU-Florida State University*
Brian E. Thompson, *Rensselaer Polytechnic Institute*

Their suggestions have greatly helped to improve the quality of this text.

We also would like to thank our students at the University of Nevada, Reno, who provided plenty of feedback from students' perspectives. Finally, we would like to express our appreciation to our wives Zehra and Nancy and our children for their continued patience, understanding, and support throughout the preparation of this text.

Yunus A. Çengel
Robert H. Turner

Nomenclature

a	Acceleration, m/s^2	erfc	Complementary error function
A	Area, m^2	E	Total energy, kJ
A_c	Cross-sectional area, m^2	E_b	Blackbody emissive flux
Bi	Biot number	EER	Energy efficiency rating
C	Speed of sound, m/s	f	Friction factor
C	Specific heat, kJ/kg $\cdot$ K	f_λ	Blackbody radiation function
C_D	Drag coefficient	F	Force, N
C_f	Friction coefficient	F_D	Drag force, N
C_L	Lift coefficient	$F_{ij}, F_{i \to j}$	View factor
C_p	Constant pressure specific heat, kJ/kg $\cdot$ K	F_L	Lift force, N
C_v	Constant volume specific heat, kJ/kg $\cdot$ K	g	Gravitational acceleration, m/s^2
COP	Coefficient of performance	G	Incident radiation, W/m^2
COP$_{HP}$	Coefficient of performance of a heat pump	Gr	Grashof number
COP$_R$	Coefficient of performance of a refrigerator	h	Convection heat transfer coefficient, W/m$^2 \cdot$ °C
d, D	Diameter, m	h	Specific enthalpy, $u + Pv$, kJ/kg
D_h	Hydraulic diameter, m	h_c	Thermal contact conductance, W/m$^2 \cdot$ °C
e	Specific total energy, kJ/kg	h_L	Head loss, m

xxxiii

H	Total enthalpy, $U + PV$, kJ	Q_L	Heat transfer with low-temperature body, kJ
I	Electric current, A	r	Compression ratio
J	Radiosity, W/m^2; Bessel function	r_c	Cutoff ratio
k	Specific heat ratio, C_p/C_v	r_p	Pressure ratio
k	Thermal conductivity	R	Gas constant, $kJ/kg \cdot K$
k_{eff}	Effective thermal conductivity, $W/m \cdot °C$	R	Radius, m
k_s	Spring constant	R	Thermal resistance, $°C/W$
ke	Specific kinetic energy, $\mathcal{V}^2/2$, kJ/kg	R_c	Thermal contact resistance, $m^2 \cdot °C/W$
K_{loss}	Loss coefficient	R_f	Fouling factor
KE	Total kinetic energy, $m\mathcal{V}^2/2$, kJ	R_u	Universal gas constant, $kJ/kmol \cdot K$
L	Length; half thickness of a plane wall	R-value	R-value of insulation
L_c	Characteristic or corrected length	Ra	Rayleigh number
L_h	Hydrodynamic entry length	Re	Reynolds number
L_t	Thermal entry length	s	Specific entropy, $kJ/kg \cdot K$
m	Mass, kg	s_{gen}	Specific entropy generation, $kJ/kg \cdot K$
$\dot{m}$	Mass flow rate, kg/s	S	Total entropy, kJ/K
M	Molar mass, kg/kmol	S_{gen}	Total entropy generation, kJ/K
MEP	Mean effective pressure, kPa	t	Time, s
n	Polytropic exponent	t	Thickness, m
N	Number of moles, kmol	T	Temperature, $°C$ or K
NTU	Number of transfer units	$\mathbf{T}$	Torque, $N \cdot m$
Nu	Nusselt number	T_b	Bulk fluid temperature, $°C$
p	Perimeter, m	T_{cr}	Critical temperature, K
pe	Specific potential energy, gz, kJ/kg	T_f	Film temperature, $°C$
P	Pressure, kPa	T_H	Temperature of high-temperature body, K
P_{cr}	Critical pressure, kPa	T_L	Temperature of low-temperature body, K
P_r	Relative pressure	T_R	Reduced temperature
P_R	Reduced pressure	T_s	Surface temperature, $°C$ or K
P_v	Vapor pressure, kPa	u	Specific internal energy, kJ/kg
PE	Total potential energy, mgz, kJ	U	Total internal energy, kJ
Pr	Prandtl number	U	Overall heat transfer coefficient, $W/m^2 \cdot °C$
q	Heat transfer per unit mass, kJ/kg		
Q	Total heat transfer, kJ	v	Specific volume, m^3/kg
$\dot{Q}$	Heat transfer rate, kW	v_{cr}	Critical specific volume, m^3/kg
Q_H	Heat transfer with high-temperature body, kJ	v_r	Relative specific volume

V	Total volume, m^3	
$\dot{V}$	Volume flow rate, m^3/s	
$\mathcal{V}$	Velocity, m/s	
$\mathcal{V}_m$	Mean velocity, m/s	
$\mathcal{V}_\infty$	Free-stream velocity, m/s	
w	Work per unit mass, kJ/kg	
W	Total work, kJ	
$\dot{W}$	Power, kW	
W_{in}	Work input, kJ	
W_{out}	Work output, kJ	
x	Quality	
z	Elevation, m	
Z	Compressibility factor	

Greek Letters

α	Absorptivity
α	Thermal diffusivity, m^2/s
α_s	Solar absorptivity
β	Volume expansivity, 1/K
δ	Characteristic length
δ_v	Velocity boundary layer thickness, m
ΔP	Pressure drop, Pa
ΔT_{lm}	Log mean temperature difference
ε	Emissivity; heat exchanger or fin effectiveness
ε	Roughness size, m
η_{fin}	Fin efficiency
η_{th}	Thermal efficiency
θ	Total energy of a flowing fluid, kJ/kg
μ	Dynamic viscosity, $kg/m \cdot s$ or $N \cdot s/m^2$
ν	Kinematic viscosity $= \mu/\rho$, m^2/s; frequency, 1/s
ρ	Density, kg/m^3
ρ_s	Relative density
σ	Stefan–Boltzmann constant
σ_n	Normal stress, N/m^2

σ_s	Surface tension, N/m	**xxxv**
τ	Shear stress, N/m^2	
τ	Transmissivity; Fourier number	
τ_w	Wall shear stress, N/m^2	
ϕ	Relative humidity	

Subscripts

a	Air
abs	Absolute
act	Actual
atm	Atmospheric
av	Average
b	Boundary; bulk fluid
cond	Conduction
conv	Convection
cr	Critical point
cv	Control volume
e	Exit conditions
f	Saturated liquid
fg	Difference in property between saturated liquid and saturated vapor
g	Saturated vapor
H	High temperature as in T_H and Q_H
i	Inlet conditions
L	Low temperature as in T_L and Q_L
r	Relative
rad	Radiation
s	Surface
surr	Surrounding surfaces
sat	Saturated
sys	System
v	Water vapor
1	Initial or inlet state
2	Final or exit state
∞	Far from a surface; free-flow conditions

Superscripts

˙ (over dot)	Quantity per unit time
‾ (over bar)	Quantity per unit mole
° (circle)	Standard reference state

Introduction and Overview

Many engineering systems involve the transfer, transport, and conversion of energy, and the sciences that deal with these subjects are broadly referred to as *thermal-fluid sciences.* Thermal-fluid sciences are usually studied under the subcategories of *thermodynamics, heat transfer, and fluid mechanics.* We start this chapter with an overview of these sciences, and give some historical background. After reviewing the *unit systems* that will be used, we discuss some basic concepts such as *system, control volume, continuum, intensive and extensive properties,* and *specific volume.* This is followed by a discussion of how engineers solve problems, the importance of modeling, and the proper place of software packages. We then present an intuitive systematic *problem solving technique* that can be used as a model in solving engineering problems. Finally, we discuss the *conservation of mass principle* and apply it to various systems.

1-1 ■ INTRODUCTION TO THERMAL-FLUID SCIENCES

The word *thermal* stems from the Greek word *therme,* which means *heat.* Therefore, thermal sciences can loosely be defined as the sciences that deal with heat. The recognition of different forms of energy and its transformations has forced this definition to be broadened. Today, the physical sciences that deal with energy and the transfer, transport, and conversion of energy are usually referred to as **thermal-fluid sciences** or just **thermal sciences.** Traditionally, the thermal-fluid sciences are studied under the subcategories of thermodynamics, heat transfer, and fluid mechanics. In this book we present the basic principles of these sciences, and apply them to situations that the engineers are likely to encounter in their practice.

The design and analysis of most thermal systems such as power plants, automotive engines, and refrigerators involve all categories of thermal-fluid sciences as well as other sciences (Fig. 1-1). For example, designing the radiator of a car involves the determination of the amount of energy transfer from a knowledge of the properties of the coolant using *thermodynamics,* the determination of the size and shape of the inner tubes and the outer fins using *heat transfer,* and the determination of the size and type of the water pump using *fluid mechanics.* Of course the determination of the materials and the thickness of the tubes requires the use of material science as well as strength of materials. The reason for studying different sciences separately is simply to facilitate learning without being overwhelmed. Once the basic principles are mastered, they can then be synthesized by solving comprehensive real-world practical problems. But first we will present an overview of thermal-fluid sciences.

Application Areas of Thermal-Fluid Sciences

All activities in nature involve some interaction between energy and matter; thus it is hard to imagine an area that does not relate to thermal-fluid sciences in some manner. Therefore, developing a good understanding of basic principles of thermal-fluid sciences has long been an essential part of engineering education.

Thermal-fluid sciences are commonly encountered in many engineering systems and other aspects of life, and one does not need to go very far to see some application areas of them. In fact, one does not need to go anywhere.

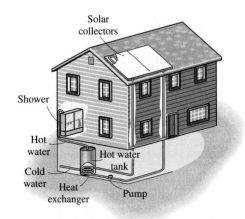

FIGURE 1-1

The design of many engineering systems, such as this solar hot water system, involves all categories of thermal-fluid sciences.

The heart is constantly pumping blood to all parts of the human body, various energy conversions occur in trillions of body cells, and the body heat generated is constantly rejected to the environment. The human comfort is closely tied to the rate of this metabolic heat rejection. We try to control this heat transfer rate by adjusting our clothing to the environmental conditions. Also, any defects in the heart and the circulatory system is a major cause for alarm.

Other applications of thermal sciences are right where one lives. An ordinary house is, in some respects, an exhibition hall filled with wonders of thermal-fluid sciences. Many ordinary household utensils and appliances are designed, in whole or in part, by using the principles of thermal-fluid sciences. Some examples include the electric or gas range, the heating and air-conditioning systems, the refrigerator, the humidifier, the pressure cooker, the water heater, the shower, the iron, the plumbing and sprinkling systems, and even the computer, the TV, and the VCR set. On a larger scale, thermal-fluid sciences play a major part in the design and analysis of automotive engines, rockets, jet engines, and conventional or nuclear power plants, solar collectors, the transportation of water, crude oil, and natural gas, the water distribution systems in cities, and the design of vehicles from ordinary cars to airplanes (Fig. 1-2). The energy-efficient home that you may be living in, for example, is designed on the basis of minimizing heat loss in winter and heat gain in summer. The size, location, and the power input of the fan of your computer is also selected after a thermodynamic, heat transfer, and fluid flow analysis of the computer.

1-2 ■ THERMODYNAMICS

Thermodynamics can be defined as the science of *energy*. Although everybody has a feeling of what energy is, it is difficult to give a precise definition for it. Energy can be viewed as the ability to cause changes.

The name *thermodynamics* stems from the Greek words *therme* (heat) and *dynamis* (power), which is most descriptive of the early efforts to convert

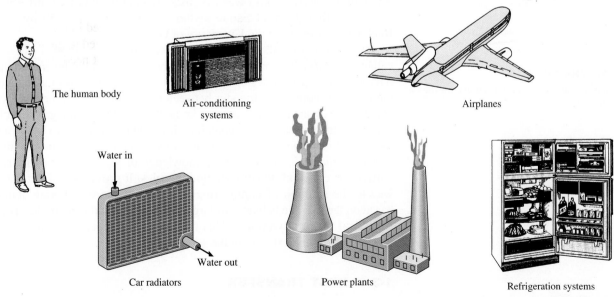

The human body

Air-conditioning systems

Airplanes

Water in

Water out

Car radiators

Power plants

Refrigeration systems

FIGURE 1-2

Some application areas of thermal-fluid sciences.

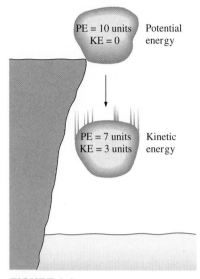

FIGURE 1-3

Energy cannot be created or destroyed; it can only change forms (the first law).

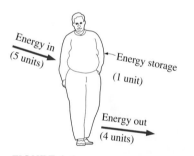

FIGURE 1-4

Conservation of energy principle for the human body.

heat into power. Today the same name is broadly interpreted to include all aspects of energy and energy transformations, including power production, refrigeration, and relationships among the properties of matter.

One of the most fundamental laws of nature is the **conservation of energy principle.** It simply states that during an interaction, energy can change from one form to another but the total amount of energy remains constant. That is, energy cannot be created or destroyed. A rock falling off a cliff, for example, picks up speed as a result of its potential energy being converted to kinetic energy (Fig. 1-3). The conservation of energy principle also forms the backbone of the diet industry: a person who has a greater energy input (food) than energy output (exercise) will gain weight (store energy in the form of fat), and a person who has a smaller energy input than output will lose weight (Fig. 1-4). The change in the energy content of a body or any other system is equal to the difference between the energy input and the energy output, and the energy balance is expressed as $E_{in} - E_{out} = \Delta E$.

The **first law of thermodynamics** is simply an expression of the conservation of energy principle, and it asserts that *energy* is a thermodynamic property. The **second law of thermodynamics** asserts that energy has *quality* as well as *quantity,* and actual processes occur in the direction of decreasing quality of energy. For example, a cup of hot coffee left on a table eventually cools to room temperature, but a cup of cool coffee in the same room never gets hot by itself. The high-temperature energy of the coffee is degraded (transformed into a less useful form at a lower temperature) once it is transferred to the surrounding air.

Although the principles of thermodynamics have been in existence since the creation of the universe, thermodynamics did not emerge as a science until the construction of the first successful atmospheric steam engines in England by Thomas Savery in 1697 and Thomas Newcomen in 1712. These engines were very slow and inefficient, but they opened the way for the development of a new science. The first and second laws of thermodynamics emerged simultaneously in the 1850s, primarily out of the works of William Rankine, Rudolph Clausius, and Lord Kelvin (formerly William Thomson). The term *thermodynamics* was first used in a publication by Lord Kelvin in 1849. The first thermodynamic textbook was written in 1859 by William Rankine, a professor at the University of Glasgow.

It is well known that a substance consists of a large number of particles called *molecules.* The properties of the substance naturally depend on the behavior of these particles. For example, the pressure of a gas in a container is the result of momentum transfer between the molecules and the walls of the container. But one does not need to know the behavior of the gas particles to determine the pressure in the container. It would be sufficient to attach a pressure gage to the container. This macroscopic approach to the study of thermodynamics that does not require a knowledge of the behavior of individual particles is called **classical thermodynamics.** It provides a direct and easy way to the solution of engineering problems. A more elaborate approach, based on the average behavior of large groups of individual particles, is called **statistical thermodynamics.** This microscopic approach is rather involved and is used in this text only in the supporting role.

1-3 ■ HEAT TRANSFER

We all know from experience that a cold canned drink left in a room warms up and a warm canned drink put in a refrigerator cools down. This is accomplished by the transfer of *energy* from the warm medium to the cold one. The energy transfer is always from the higher temperature medium to the lower temperature one, and the energy transfer stops when the two mediums reach the same temperature.

Energy exists in various forms. In heat transfer, we are primarily interested in heat, which is *the form of energy that can be transferred from one system to another as a result of temperature difference.* The science that deals with the determination of the *rates* of such energy transfers is **heat transfer.**

You may be wondering why we need the science of heat transfer. After all, we can determine the amount of heat transfer for any system undergoing any process using a thermodynamic analysis alone. The reason is that thermodynamics is concerned with the *amount* of heat transfer as a system undergoes a process from one equilibrium state to another, and it gives no indication about *how long* the process will take. But in engineering, we are often interested in the *rate* of heat transfer, which is the topic of the science of *heat transfer.* A thermodynamic analysis simply tells us how much heat must be transferred to realize a specified change of state to satisfy the conservation of energy principle.

In practice we are more concerned about the rate of heat transfer (heat transfer per unit time) than we are with the amount of it. For example, we can determine the amount of heat transferred from a thermos bottle as the hot coffee inside cools from 90°C to 80°C by a thermodynamic analysis alone. But a typical user or designer of a thermos is primarily interested in *how long* it will be before the hot coffee inside cools to 80°C, and a thermodynamic analysis cannot answer this question. Determining the rates of heat transfer to or from a system and thus the times of cooling or heating, as well as the variation of the temperature, is the subject of *heat transfer* (Fig. 1-5).

Thermodynamics deals with equilibrium states and changes from one equilibrium state to another. Heat transfer, on the other hand, deals with systems that lack thermal equilibrium, and thus it is a *nonequilibrium* phenomenon. Therefore, the study of heat transfer cannot be based on the principles of thermodynamics alone. However, the laws of thermodynamics lay the framework for the science of heat transfer. The *first law* requires that the rate of energy transfer into a system be equal to the rate of increase of the energy of that system. The *second law* requires that heat be transferred in the direction of decreasing temperature (Fig. 1-6). This is like a car parked on an inclined road must go downhill in the direction of decreasing elevation when its brakes are released. It is also analogous to the electric current flow in the direction of decreasing voltage or the fluid flowing in the direction of decreasing pressure.

The basic requirement for heat transfer is the presence of a *temperature difference.* There can be no net heat transfer between two mediums that are at the same temperature. The temperature difference is the *driving force* for heat transfer; just as the *voltage difference* is the driving force for electric current, and *pressure difference* is the driving force for fluid flow. The rate of heat transfer in a certain direction depends on the magnitude of the *temperature gradient* (the temperature difference per unit length or the rate of change of temperature) in that direction. The larger the temperature gradient, the higher the rate of heat transfer (Fig. 1-7).

Historical Background on Heat

Heat has always been perceived to be something that produces in us a sensation of warmth, and one would think that the nature of heat is one of the first

FIGURE 1-5

We are normally interested in how long it takes for the hot coffee in a thermos to cool to a certain temperature, which cannot be determined from a thermodynamic analysis alone.

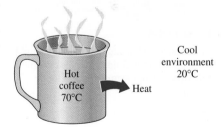

FIGURE 1-6

Heat flows in the direction of decreasing temperature.

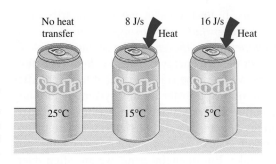

FIGURE 1-7

Temperature difference is the driving force for heat transfer. The larger the temperature difference, the higher is the rate of heat transfer.

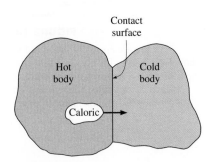

FIGURE 1-8

In the early 19th century, heat was thought to be an invisible fluid called the *caloric* that flowed from warmer bodies to the cooler ones.

things understood by mankind. But it was only in the middle of the nineteenth century that we had a true physical understanding of the nature of heat, thanks to the development at that time of the **kinetic theory,** which treats molecules as tiny balls that are in motion and thus possess kinetic energy. Heat is then defined as the energy associated with the random motion of atoms and molecules. Although it was suggested in the eighteenth and early nineteenth centuries that heat is the manifestation of motion at the molecular level (called the *live force*), the prevailing view of heat until the middle of the nineteenth century was based on the caloric theory proposed by the French chemist Antoine Lavoisier (1743–1794) in 1789. The caloric theory asserts that heat is a fluid-like substance called the **caloric** that is a massless, colorless, odorless, and tasteless substance that can be poured from one body into another (Fig. 1-8). When caloric was added to a body, its temperature increased; and when caloric was removed from a body, its temperature decreased. When a body could not contain any more caloric, much the same way as when a glass of water could not dissolve any more salt or sugar, the body was said to be saturated with caloric. This interpretation gave rise to the terms *saturated liquid* and *saturated vapor* that are still in use today.

The caloric theory came under attack soon after its introduction. It maintained that heat is a substance that could not be created or destroyed. Yet it was known that heat can be generated indefinitely by rubbing one's hands together or rubbing two pieces of wood together. In 1798, the American Benjamin Thompson (Count Rumford) (1753–1814) showed in his papers that heat can be generated continuously through friction. The validity of the caloric theory was also challenged by several others. But it was the careful experiments of the Englishman James P. Joule (1818–1889) published in 1843 that finally convinced the skeptics that heat was not a substance after all, and thus put the caloric theory to rest. Although the caloric theory was totally abandoned in the middle of the nineteenth century, it contributed greatly to the development of thermodynamics and heat transfer.

Heat is transferred by three mechanisms: conduction, convection, and radiation. **Conduction** is the transfer of energy from the more energetic particles of a substance to the adjacent less energetic ones as a result of interaction between particles. **Convection** is the transfer of energy between a solid surface and the adjacent fluid that is in motion, and it involves the combined effects of conduction and fluid motion. **Radiation** is the transfer of energy due to the emission of electromagnetic waves (or photons). The modes of heat transfer are discussed in detail in Chap. 14.

Mechanics is the oldest physical science that deals with both stationary and moving bodies under the influence of forces. The branch of mechanics that deals with bodies at rest is called **statics** while the branch that deals with bodies in motion is called **dynamics.** The subcategory **fluid mechanics** is defined as the science that deals with the behavior of fluids at rest (*fluid statics*) or in motion (*fluid dynamics*), and the interaction of fluids with solids or other fluids at the boundaries. Fluid mechanics is also referred to as **fluid dynamics** by considering fluids at rest as a special case of motion with zero velocity.

Fluid mechanics itself is also divided into several categories. The study of the motion of fluids that are practically incompressible (such as liquids, especially water, and gases at low speeds) is usually referred to as **hydrodynamics.** A subcategory of hydrodynamics is **hydraulics,** which deals with liquid flows in pipes and open channels. **Gas dynamics** deals with flow of fluids that undergo significant density changes, such as the flow of gases through nozzles at high speeds. The category **aerodynamics** deals with the flow of gases (especially air) over bodies such as aircraft, rockets, and automobiles at high or low speeds. Some other specialized categories such as **meteorology, oceanography,** and **hydrology** deal with naturally occurring flows.

You will recall from physics that a substance exists in 3 primary phases: solid, liquid, and gas. A substance in the liquid or gas phase is referred to as a **fluid.** Distinction between a solid and a fluid is made on the basis of their ability to resist an applied shear (or tangential) stress that tends to change the shape of the substance. A solid can resist an applied shear stress by deforming whereas a fluid deforms continuously under the influence of shear stress, no matter how small. You may recall from statics that **stress** is defined as force per unit area, and is determined by dividing the force by the area upon which it acts. The normal component of a force acting on a surface per unit area is called the **normal stress,** and the tangential component of a force acting on a surface per unit area is called **shear stress.** In a fluid, the normal stress is called **pressure** (Fig. 1-9). The supporting walls of a fluid eliminate shear stress, and thus a fluid at rest is at a state of zero shear stress. When the walls are removed or a liquid container is tilted, a shear develops and the liquid splashes or moves to attain a horizontal free surface.

In a liquid, chunks of piled-up molecules can move relative to each other, but the volume remains relatively constant because of the strong cohesive forces between the molecules. As a result, a liquid takes the shape of the container it is in, and it forms a free surface in a larger container in a gravitational field. A gas, on the other hand, does not have a definite volume and it expands until it encounters the walls of the container and fills the entire available space. This is because the gas molecules are widely spaced, and the cohesive forces between them are very small. Unlike liquids, gases cannot form a free surface (Fig. 1-10).

Although solids and fluids are easily distinguished in most cases, this distinction is not so clear in some borderline cases. For example, *asphalt* appears and behaves as a solid since it resists shear stress for short periods of time. But it deforms slowly and behaves like a fluid when these forces are exerted for extended periods of time. Some plastics, lead, and slurry mixtures exhibit similar behavior. Such blurry cases are beyond the scope of this text. The fluids we will deal with in this text will be clearly recognizable as fluids.

Matter is made up of atoms that are widely spaced in the gas phase. Yet it is very convenient to disregard the atomic nature of a substance, and view it as

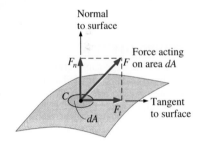

$$\text{Normal stress: } \sigma = \frac{F_n}{dA} = P \text{ (pressure)}$$

$$\text{Shear stress: } \tau = \frac{F_t}{dA}$$

FIGURE 1-9

The normal stress (pressure) and shear (tangential) stress at the surface of a fluid element. Shear stress is zero for fluids at rest.

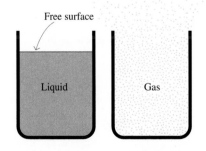

FIGURE 1-10

Unlike a liquid, a gas does not form a free surface, and it expands to fill the entire available space.

a continuous, homogeneous matter with no holes, that is, a **continuum.** The continuum idealization allows us to treat properties as point functions, and to assume the properties to vary continually in space with no jump discontinuities. This idealization is valid as long as the size of the system we deal with is large relative to the space between the molecules. This is the case practically in all problems, except some specialized ones.

To have a sense of the distances involved at the molecular level, consider a container filled with oxygen at atmospheric conditions. The diameter of the oxygen molecule is about 3×10^{-10} m and its mass is 5.3×10^{-26} kg. Also, the *mean free path* of oxygen at 1 atm pressure and 20°C is 6.3×10^{-8} m. That is, an oxygen molecule travels, on average, a distance of 6.3×10^{-10} m (about 200 times of its diameter) before it collides with another molecule. Also, there are about 3×10^{16} molecules of oxygen in the tiny volume of 1 mm^3 at 1 atm pressure and 20°C (Fig. 1-11). The continuum model is applicable as long as the characteristic length of the system (such as its diameter) is much larger than the mean free path of the molecules. At very high vacuums or very high elevations, the mean free path may become large (for example, it is about 0.1 m for atmospheric air at an elevation of 100 km). For such cases the **rarefied gas flow theory** should be used, and the impact of individual molecules should be considered. In this text we will limit our consideration to substances that can be modeled as a continuum.

It is sometimes desirable to study a certain region of a flow, called the **flow field,** to gain a better insight about the behavior of the fluid in that region at the microscopic level and to develop a better intuitive understanding. This is done by viewing the fluid as a pile of tiny spherical particles that are forced to move, and observing the motion of individual particles. The particles that follow the same path form a line, called the **streamline**—just like the cars that travel in the same lane on a highway. The streamlines are tangent to the velocity vector and thus the direction of flow at every point in the flow field, and no fluid particle can cross a streamline. In steady flow, the streamlines do not change with time. Streamlines can be visualized by injecting a light visible substance into the flow—such as smoke into gas flow, and dye or bubbles into a liquid flow (Fig. 1-12).

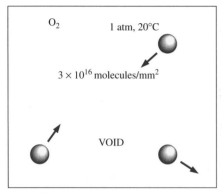

FIGURE 1-11

Despite the large gaps between molecules, a substance can be treated as a continuum because of the very large number of molecules even in the smallest volume.

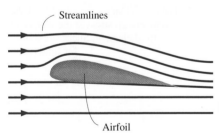

FIGURE 1-12

Streamlines of flow past an airfoil.

1-5 ■ A NOTE ON DIMENSIONS AND UNITS

Any physical quantity can be characterized by **dimensions.** The arbitrary magnitudes assigned to the dimensions are called **units.** Some basic dimensions such as mass m, length L, time t, and temperature T are selected as **primary** or **fundamental dimensions,** while others such as velocity $\mathscr{V}$, energy E, and volume V are expressed in terms of the primary dimensions and are called **secondary dimensions,** or **derived dimensions.**

A number of unit systems have been developed over the years. Despite strong efforts in the scientific and engineering community to unify the world with a single unit system, two sets of units are still in common use today: the **English system,** which is also known as the *United States Customary System* (USCS), and the metric **SI** (from *Le Système International d' Unités*), which is also known as the *International System.* The SI is a simple and logical system based on a decimal relationship between the various units, and it is being used for scientific and engineering work in most of the industrialized nations, including England. The English system, however, has no numerical base, and various units in this system are related to each other rather arbitrarily (12 in. in 1 ft, 16 oz in 1 lb, 4 qt in 1 gal, etc.) which makes it confusing and difficult to learn. The United States is the only industrialized country that has not yet fully converted to the metric system.

The systematic efforts to develop a universally acceptable system of units dates back to 1790 when the French National Assembly charged the French Academy of Sciences to come up with such a unit system. An early version of the metric system was soon developed in France, but it did not find much universal acceptance until 1875 when *The Metric Convection Treaty* was prepared and signed by 17 nations, including the United States. In this international treaty, meter and gram were established as the metric units for length and mass, respectively, and a *General Conference of Weights and Measures* (CGPM) was established that was to meet every six years. In 1960, the CGPM produced the SI, which was based on six fundamental quantities and their units adopted in 1954 at the Tenth General Conference of Weights and Measures: *meter* (m) for length, *kilogram* (kg) for mass, *second* (s) for time, *ampere* (A) for electric current, *degree Kelvin* (°K) for temperature, and *candela* (cd) for luminous intensity (amount of light). In 1971, the CGPM added a seventh fundamental quantity and unit: *mole* (mol) for the amount of matter.

Based on the notational scheme introduced in 1967, the degree symbol was officially dropped from the absolute temperature unit, and all unit names were to be written without capitalization even if they were derived from proper names (Table 1-1). However, the abbreviation of a unit was to be capitalized if the unit was derived from a proper name. For example, the SI unit of force, which is named after Sir Isaac Newton (1647–1723), is *newton* (not Newton), and it is abbreviated as *N*. Also, the full name of a unit may be pluralized, but its abbreviation cannot. For example. the length of an object can be 5 m or 5 meters, *not* 5 ms or 5 meter. Finally, no period is to be used in unit abbreviations unless they appear at the end of a sentence. For example, the proper abbreviation of meter is m (not m.).

The recent move toward the metric system in the United States seems to have started in 1968 when Congress, in response to what was happening in the rest of the world, passed a Metric Study Act. Congress continued to promote a voluntary switch to the metric system by passing the Metric Conversion Act in 1975. A trade bill passed by Congress in 1988 set a September 1992 deadline for all federal agencies to convert to the metric system. But the deadlines were relaxed later with no clear plans for the future.

The industries that are heavily involved in international trade (such as the automotive, soft drink, and liquor industries) have been quick in converting to the metric system for economic reasons (having a single worldwide design, fewer sizes, smaller inventories, etc.). Today, nearly all the cars manufactured in the United States are metric. Most car owners probably do not realize this until they try an inch socket wrench on a metric bolt. Most industries, however, resisted the change, thus slowing down the conversion process.

Presently the United States is a dual-system society, and it will stay that way until the transition to the metric system is completed. This puts an extra burden on today's engineering students, since they are expected to retain their understanding of the English system while learning, thinking, and working in terms of the SI. Given the position of the engineers in the transition period, both unit systems are used in this text, with particular emphasis on SI units.

As pointed out earlier, the SI is based on a decimal relationship between units. The prefixes used to express the multiples of the various units are listed in Table 1-2. They are standard for all units, and the student is encouraged to memorize them because of their widespread use (Fig. 1-13).

TABLE 1-1

The seven fundamental dimensions and their units in SI

Dimension	Unit
Length	meter (m)
Mass	kilogram (kg)
Time	second (s)
Temperature	kelvin (K)
Electric current	ampere (A)
Amount of light	candela (c)
Amount of matter	mole (mol)

TABLE 1-2

Standard prefixes in SI units

Multiple	Prefix
10^{12}	tera, T
10^{9}	giga, G
10^{6}	mega, M
10^{3}	kilo, k
10^{2}	hecto, h
10^{1}	deka, da
10^{-1}	deci, d
10^{-2}	centi, c
10^{-3}	milli, m
10^{-6}	micro, μ
10^{-9}	nano, n
10^{-12}	pico, p

10

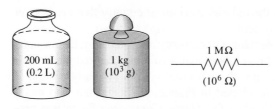

The SI unit prefixes are used in all branches of engineering.

Some SI and English Units

In SI, the units of mass, length, and time are the kilogram (kg), meter (m), and second (s), respectively. The respective units in the English system are the pound-mass (lbm), foot (ft), and second (s). The pound symbol *lb* is actually the abbreviation of *libra,* which was the ancient Roman unit of weight. The English retained this symbol even after the end of the Roman occupation of Britain in 410. The mass and length units in the two systems are related to each other by

$$1 \text{ lbm} = 0.45359 \text{ kg}$$
$$1 \text{ ft} = 0.3048 \text{ m}$$

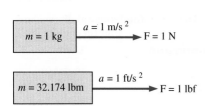

FIGURE 1-14

The definition of the force units.

In the English system, force is usually considered to be one of the primary dimensions and is assigned a nonderived unit. This is a source of confusion and error that necessitates the use of a conversion factor (g_c) in many formulas. To avoid this nuisance, we consider force to be a secondary dimension whose unit is derived from Newton's second law, i.e.,

$$\text{Force} = (\text{Mass})(\text{Acceleration})$$
or
$$F = ma \tag{1-1}$$

In SI, the force unit is the newton (N), and it is defined as the *force required to accelerate a mass of 1 kg at a rate of 1 m/s²*. In the English system, the force unit is the **pound-force** (lbf) and is defined as the *force required to accelerate a mass of 32.174 lbm (1 slug) at a rate of 1 ft/s²* (Fig. 1-14). That is,

$$1 \text{ N} = 1 \text{ kg} \cdot \text{m/s}^2$$
$$1 \text{ lbf} = 32.174 \text{ lbm} \cdot \text{ft/s}^2$$

A force of 1 newton is roughly equivalent to the weight of a small apple ($m = 102$ g), whereas a force of 1 pound-force is roughly equivalent to the weight of 4 medium apples ($m_{\text{total}} = 454$ g), as shown in Fig. 1-15. Another force unit in common use in many European countries is the *kilogram-force* (kgf), which is the weight of 1 kg mass at sea level (1 kgf = 9.807 N).

The term **weight** is often incorrectly used to express mass, particularly by the "weight watchers." Unlike mass, weight W is a *force*. It is the gravitational force applied to a body, and its magnitude is determined from Newton's second law,

$$W = mg \qquad \text{(N)} \tag{1-2}$$

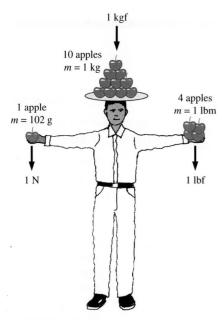

FIGURE 1-15

The relative magnitudes of the force units newton (N), kilogram-force (kgf), and pound-force (lbf).

where m is the mass of the body and g is the local gravitational acceleration (g is 9.807 m/s² or 32.174 ft/s² at sea level and 45° latitude). The ordinary bathroom scale measures the gravitational force acting on a body. The weight of a unit volume of a substance is called the **specific weight** w and is determined from $w = \rho g$, where ρ is density.

The mass of a body will remain the same regardless of its location in the universe. Its weight, however, will change with a change in gravitational acceleration. A body will weigh less on top of a mountain since g decreases with altitude. On the surface of the moon, an astronaut will weigh about one-sixth of what she or he normally weighs on earth (Fig. 1-16).

At sea level a mass of 1 kg will weigh 9.807 N, as illustrated in Fig. 1-17. A mass of 1 lbm, however, will weigh 1 lbf, which misleads people to believe that pound-mass and pound-force can be used interchangeably as pound (lb), which is a major source of error in the English system.

It should be noted that the *gravity force* acting on a mass is due to the *attraction* between the masses, and thus it is proportional to the magnitudes of the masses and inversely proportional to the square of the distance between them. Therefore, the gravitational acceleration g at a location depends on the *local density* of the earth's crust, the *distance* to the center of the earth, and to a lesser extent, the *positions* of the moon and the sun. The value of g varies with location from 9.8295 m/s² at 4500 m below sea level to 7.3218 m/s² at 100,000 m above sea level. However, at altitudes up to 30,000 m, the variation of g from the sea level value of 9.807 m/s² is less than 1 percent. Therefore, for most practical purposes, the gravitational acceleration can be assumed to be *constant* at 9.81 m/s². It is interesting to note that at locations below sea level, the value of g increases with distance from the sea level, reaches a maximum at about 4500 m, and then starts decreasing. (What do you think the value of g will be at the center of the earth?)

The primary cause of confusion between mass and weight is that mass is usually measured *indirectly* by measuring the *gravity force* it exerts. This approach also assumes that the forces exerted by other effects such as air buoyancy and fluid motion are negligible. This is like a car odometer that measures the velocity of a car by measuring the number of revolutions of a wheel and multiplying it by the wheel perimeter. The correct way of measuring mass is to compare it to a known mass. But this is cumbersome, and it is mostly used for calibration and measuring precious metals.

Work, which is a form of energy, can simply be defined as force times distance; therefore, it has the unit "newton-meter (N · m)," which is called a joule (J). That is,

$$1 \text{ J} = 1 \text{ N} \cdot \text{m}$$

A more common unit for energy in SI is the kilojoule (1 kJ = 10³ J). In the English system, the energy unit is the **Btu** (British thermal unit), which is defined as the energy required to raise the temperature of 1 lbm of water at 68°F by 1°F. In the metric system, the amount of energy needed to raise the temperature of 1 g of water at 15°C by 1°C is defined as 1 **calorie** (cal), and 1 cal = 4.1868 J. The magnitudes of the kilojoule and Btu are almost identical (1 Btu = 1.055 kJ).

Dimensional Homogeneity

We all know from grade school that apples and oranges do not add. But we somehow manage to do it (by mistake, of course). In engineering, all

FIGURE 1-16

A body weighing 150 pounds on earth will weigh only 25 pounds on the moon.

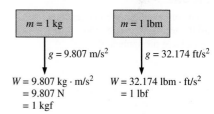

FIGURE 1-17

The weight of a unit mass at sea level.

To be dimensionally homogeneous,
all the terms in an equation
must have the same unit.

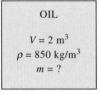

FIGURE 1-19

Schematic for Example 1-2.

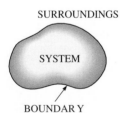

FIGURE 1-20

System, surroundings, and boundary.

equations must be *dimensionally homogeneous.* That is, every term in an equation must have the same unit (Fig. 1-18). If, at some stage of an analysis, we find ourselves in a position to add two quantities that have different units, it is a clear indication that we have made an error at an earlier stage. So checking units can serve as a valuable tool to spot errors.

EXAMPLE 1-1   Spotting Errors from Unit Inconsistencies

While solving a problem, a person ended up with the following equation at some stage:

$$E = 25 \text{ kJ} + 7 \text{ kJ/kg}$$

where E is the total energy and has the unit of kilojoules. Determine the error that may have caused it.

Solution   During an analysis, a relation with inconsistent units is obtained. The probable cause of it is to be determined.

Analysis   The two terms on the right-hand side do not have the same units, and therefore they cannot be added to obtain the total energy. Multiplying the last term by mass will eliminate the kilograms in the denominator, and the whole equation will become dimensionally homogeneous, that is, every term in the equation will have the same unit. Obviously this error was caused by forgetting to multiply the last term by mass at an earlier stage.

We all know from experience that units can give terrible headaches if they are not used carefully in solving a problem. But with some attention and skill, units can be used to our advantage. They can be used to check formulas; they can even be used to derive formulas, as explained in the following example.

EXAMPLE 1-2   Obtaining Formulas from Unit Considerations

A tank is filled with oil whose density is $\rho = 850 \text{ kg/m}^3$. If the volume of the tank is $V = 2 \text{ m}^3$, determine the amount of mass m in the tank.

Solution   The volume of an oil tank is given. The mass of oil is to be determined.

Assumptions   Oil is an incompressible substance and thus its density is constant.

Analysis   A sketch of the system described above is given in Fig. 1-19. Suppose we forgot the formula that relates mass to density and volume. But we know that mass has the unit of kilograms. That is, whatever calculations we do, we should end up with the unit of kilograms. Putting the given information into perspective, we have

$$\rho = 850 \text{ kg/m}^3 \quad \text{and} \quad V = 2 \text{ m}^3$$

It is obvious that we can eliminate m^3 and end up with kg by multiplying these two quantities. Therefore, the formula we are looking for is

$$m = \rho V$$

Thus, $m = (850 \text{ kg/m}^3)(2 \text{ m}^3) = \textbf{1700 kg}$

The student should keep in mind that a formula that is not dimensionally homogeneous is definitely wrong, but a dimensionally homogeneous formula is not necessarily right.

1-6 ▓ CLOSED AND OPEN SYSTEMS

A **thermodynamic system,** or simply a **system,** is defined as a *quantity of matter or a region in space chosen for study.* The mass or region outside the system is called the **surroundings.** The real or imaginary surface that separates the system from its surroundings is called the **boundary.** These terms are illustrated in Fig. 1-20. The boundary of a system can be *fixed* or *movable.* Note that the boundary is the contact surface shared by both the system and the surroundings. Mathematically speaking, the boundary has zero thickness, and thus it can neither contain any mass nor occupy any volume in space.

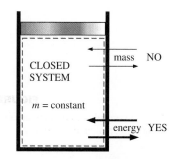

FIGURE 1-21

Mass cannot cross the boundaries of a closed system, but energy can.

Systems may be considered to be *closed* or *open,* depending on whether a fixed mass or a fixed volume in space is chosen for study. A **closed system** (also known as a **control mass**) consists of a fixed amount of mass, and no mass can cross its boundary. That is, no mass can enter or leave a closed system, as shown in Fig. 1-21. But energy, in the form of heat or work, can cross the boundary; and the volume of a closed system does not have to be fixed. If, as a special case, even energy is not allowed to cross the boundary, that system is called an **isolated system.**

Consider the piston-cylinder device shown in Fig. 1-22. Let us say that we would like to find out what happens to the enclosed gas when it is heated. Since we are focusing our attention on the gas, it is our system. The inner surfaces of the piston and the cylinder form the boundary, and since no mass is crossing this boundary, it is a closed system. Notice that energy may cross the boundary, and part of the boundary (the inner surface of the piston, in this case) may move. Everything outside the gas, including the piston and the cylinder, is the surroundings.

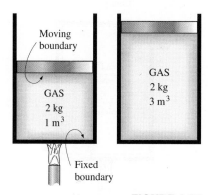

FIGURE 1-22

A closed system with a moving boundary.

An **open system,** or a **control volume,** as it is often called, is a properly selected region in space. It usually encloses a device that involves mass flow such as a compressor, turbine, or nozzle. Flow through these devices is best studied by selecting the region within the device as the control volume. Both mass and energy can cross the boundary of a control volume. This is illustrated in Fig. 1-23.

A large number of engineering problems involve mass flow in and out of a system and, therefore, are modeled as *control volumes.* A water heater, a car radiator, a turbine, and a compressor all involve mass flow and should be analyzed as control volumes (open systems) instead of as control masses (closed systems). In general, *any arbitrary region in space* can be selected as a control volume. There are no concrete rules for the selection of control volumes, but the proper choice certainly makes the analysis much easier. If we were to analyze the flow of air through a nozzle, for example, a good choice for the control volume would be the region within the nozzle.

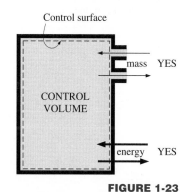

FIGURE 1-23

Both mass and energy can cross the boundaries of a control volume.

The boundaries of a control volume are called a *control surface,* and they can be real or imaginary. In the case of a nozzle, the inner surface of the nozzle forms the real part of the boundary, and the entrance and exit areas form the imaginary part, since there are no physical surfaces there (Fig. 1-24).

A control volume can be fixed in size and shape, as in the case of a nozzle, or it may involve a moving boundary, as shown in Fig. 1-24. Most control volumes, however, have fixed boundaries and thus do not involve any moving

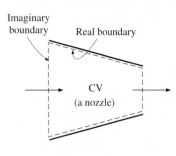

FIGURE 1-24

A control volume may involve fixed,
moving, real, and imaginary boundaries.

(*a*) A control volume with real and
imaginary boundaries

(*b*) A control volume with fixed and
moving boundaries

boundaries. A control volume may also involve heat and work interactions just
as a closed system, in addition to mass interaction.

As an example of an open system, consider the water heater shown in
Fig. 1-25. Let us say that we would like to determine how much heat we must
transfer to the water in the tank in order to supply a steady stream of hot water.
Since hot water will leave the tank and be replaced by cold water, it is not convenient to choose a fixed mass as our system for the analysis. Instead, we can
concentrate our attention on the volume formed by the interior surfaces of the
tank and consider the hot and cold water streams as mass leaving and entering
the control volume. The interior surfaces of the tank form the control surface
for this case, and mass is crossing the control surface at two locations.

In all thermodynamic analyses, the system under study *must* be defined
carefully. In most cases, the system investigated is quite simple and obvious,
and defining the system may seem like a tedious and unnecessary task. In
other cases, however, the system under study may be rather involved, and a
proper choice of the system may greatly simplify the analysis.

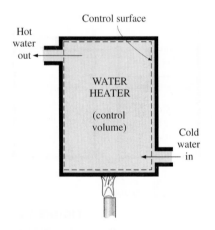

FIGURE 1-25

An open system (a control volume)
with one inlet and one exit.

1-7 ■ PROPERTIES OF A SYSTEM

Any characteristic of a system is called a **property.** Some familiar properties
are pressure P, temperature T, volume V, and mass m. The list can be extended
to include less familiar ones such as viscosity, thermal conductivity, modulus
of elasticity, thermal expansion coefficient, electric resistivity, and even velocity and elevation.

Not all properties are independent, however. Some are defined in terms of
other ones. For example, **density** is defined as *mass per unit volume.*

$$\rho = \frac{m}{V} \qquad (\text{kg/m}^3) \qquad (1\text{-}3)$$

The density of a substance, in general, depends on temperature and pressure.
The density of most gases is proportional to pressure, and inversely proportional to temperature. Liquids and solids, on the other hand, are essentially incompressible substances, and the variation of their density with pressure is
usually negligible. At 20°C, for example, the density of water changes from
998 kg/m³ at 1 atm to 1003 kg/m³ at 100 atm, a change of just 0.5 percent. The

density of liquids and solids depends more strongly on temperature than they do on pressure. At 1 atm, for example, the density of water changes from 998 kg/m³ at 20°C to 975 kg/m³ at 75°C, a change of 2.3 percent, which can still be neglected in most cases.

Sometimes the density of a substance is given relative to the density of a well-known substance. Then it is called **specific gravity,** or **relative density,** and is defined as *the ratio of the density of a substance to the density of some standard substance at a specified temperature* (usually water at 4°C, for which $\rho_{H_2O} = 1000$ kg/m³). That is,

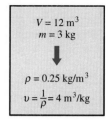

FIGURE 1-26

Density is mass per unit volume; specific volume is volume per unit mass.

$$\rho_s = \frac{\rho}{\rho_{H_2O}} \qquad (1\text{-}4)$$

Note that the specific gravity of a substance is a dimensionless quantity. However, in SI units, the numerical value of the specific gravity of a substance will be exactly equal to its density in g/cm³ or kg/L (or 0.001 times the density in kg/m³) since the density of water at 4°C is 1 g/cm³ = 1 kg/L = 1000 kg/m³. For example, the specific gravity of mercury at 0°C is 13.6. Therefore, its density at 0°C is 13.6 g/cm³ = 13.6 kg/L = 13,600 kg/m³. The specific gravities of some substances at 0°C are 1.0 for water, 0.92 for ice, 2.3 for concrete, 0.3–0.9 for most woods, 1.7–2.0 for bones, 1.05 for blood, 1.025 for sea water, 19.2 for gold, 0.79 for ethyl alcohol, and about 0.7 for gasoline. Note that substances with specific gravities less than 1 are lighter than water, and thus they will float on water.

A more frequently used property in thermodynamics is the **specific volume.** It is the reciprocal of density (Fig. 1-26) and is defined as the *volume per unit mass:*

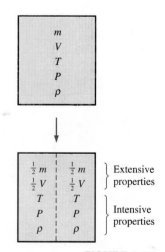

FIGURE 1-27

Criteria to differentiate intensive and extensive properties.

$$v = \frac{V}{m} = \frac{1}{\rho} \qquad (\text{m}^3/\text{kg}) \qquad (1\text{-}5)$$

Properties are considered to be either *intensive* or *extensive*. **Intensive properties** are those that are independent of the size of a system, such as temperature, pressure, and density. **Extensive properties** are those whose values depend on the size—or extent—of the system. Mass m, volume V, and total energy E are some examples of extensive properties. An easy way to determine whether a property is intensive or extensive is to divide the system into two equal parts with a partition, as shown in Fig. 1-27. Each part will have the same value of intensive properties as the original system, but half the value of the extensive properties.

Generally, uppercase letters are used to denote extensive properties (with mass m being a major exception), and lowercase letters are used for intensive properties (with pressure P and temperature T being the obvious exceptions).

Extensive properties per unit mass are called **specific properties.** Some examples of specific properties are specific volume ($v = V/m$) and specific total energy ($e = E/m$).

1-8 ■ SOLVING ENGINEERING PROBLEMS

Thermal equipment such as heat exchangers, boilers, condensers, radiators, heaters, furnaces, refrigerators, and solar collectors are designed primarily on the basis of thermal analysis. Some problems deal with the determination of the size of a system in order to transfer heat at a specified rate for a specified temperature difference, while others deal with the determination of the heat transfer rate for an existing system at a specified temperature difference.

A thermal process or equipment can be studied either *experimentally* (testing and taking measurements) or *analytically* (by analysis or calculations). The experimental approach has the advantage that we deal with the actual physical system, and what we get is what it is, within the limits of experimental error. However, this approach is expensive, time-consuming, and often impractical. Besides, the system we are analyzing may not even exist. For example. the size of a heating system of a building must usually be determined *before* the building is actually built on the basis of the dimensions and specifications given. The analytical approach (including numerical approach) has the advantage that it is fast and inexpensive, but the results obtained are subject to the accuracy of the assumptions and idealizations made in the analysis. In engineering, often a good compromise is reached by reducing the choices to just a few by analysis, and then verifying the findings experimentally and doing some fine-tuning.

Modeling in Engineering

The descriptions of most scientific problems involve relations that relate the changes in some key variables to each other. Usually the smaller the increment chosen in the changing variables, the more general and accurate the description. In the limiting case of infinitesimal or differential changes in variables, we obtain *differential equations* that provide precise mathematical formulations for the physical principles and laws by representing the rates of change as *derivatives*. Therefore, differential equations are used to investigate a wide variety of problems in sciences and engineering (Fig. 1-28). However, most thermal-fluid problems encountered in practice can be solved without resorting to differential equations and the complications associated with them.

The study of physical phenomena involves two important steps. In the first step, all the variables that affect the phenomena are identified, reasonable assumptions and approximations are made, and the interdependence of these variables is studied. The relevant physical laws and principles are invoked, and the problem is formulated mathematically. The equation itself is very instructive as it shows the degree of dependence of some variables on others, and the relative importance of various terms. In the second step, the problem is solved using an appropriate approach, and the results are interpreted.

Many processes that seem to occur in nature randomly and without any order are, in fact, being governed by some visible or not-so-visible physical laws. Whether we notice them or not, these laws are there, governing consistently and predictably what seem to be ordinary events. Most of these laws are well defined and well understood by scientists and engineers. This makes it possible to predict the course of an event before it actually occurs, or to study various aspects of an event mathematically without actually running expensive and time-consuming experiments. This is where the power of analysis lies. Very accurate results to meaningful practical problems can be obtained with relatively little effort by using a suitable and realistic mathematical model. The preparation of such models requires an adequate knowledge of the natural phenomena involved and the relevant laws, as well as a sound judgment. An unrealistic model will obviously give inaccurate and thus unacceptable results.

An analyst working on an engineering problem often finds himself or herself in a position to make a choice between a very accurate but complex

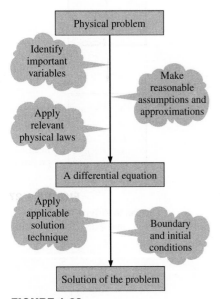

FIGURE 1-28

Mathematical modeling of physical problems.

model, and a simple but not-so-accurate model. The right choice depends on the situation at hand. The right choice is usually the simplest model that yields adequate results. For example, the process of baking potatoes or roasting a round chunk of beef in an oven can be studied analytically in a simple way by modeling the potato or the roast as a spherical solid ball that has the properties of water (Fig. 1-29). The model is quite simple, but the results obtained are sufficiently accurate for most practical purposes. As another example, when we analyze the heat losses from a building in order to select the right size for a heater, we determine the heat losses under anticipated worst conditions and select a furnace that will provide sufficient heat to make up for those losses. Often we tend to choose a larger furnace in anticipation of some future expansion or just to provide a factor of safety. A very simple analysis will be adequate in this case.

When selecting equipment, it is important to consider the actual operating conditions. For example, when purchasing a heat exchanger that will handle hard water, we must consider, that some calcium deposits will form on the heat transfer surfaces over time, causing fouling and thus a gradual decline in performance. The heat exchanger must be selected on the basis of operation under these adverse conditions instead of under new conditions.

Preparing very accurate but complex models is usually not so difficult. But such models are not much use to an analyst if they are very difficult and time-consuming to solve. At the minimum, the model should reflect the essential features of the physical problem it represents. There are many significant real-world problems that can be analyzed with a simple model. But it should always be kept in mind that the results obtained from an analysis are as accurate as the assumptions made in simplifying the problem. Therefore, the solution obtained should not be applied to situations for which the original assumptions do not hold.

A solution that is not quite consistent with the observed nature of the problem indicates that the mathematical model used is too crude. In that case, a more realistic model should be prepared by eliminating one or more of the questionable assumptions. This will result in a more complex problem that, of course, is more difficult to solve. Thus any solution to a problem should be interpreted within the context of how that problem is formulated.

Engineering Software Packages

Perhaps you are wondering why we are about to undertake a painstaking study of the fundamentals of thermal-fluid sciences. After all, almost all such problems we are likely to encounter in practice can be solved using one of several sophisticated software packages readily available in the market today. These software packages not only give the desired numerical results, but also supply the outputs in colorful graphical form for impressive presentations. It is unthinkable to practice engineering today without using some of these packages. This tremendous computing power available to us at the touch of a button is both a blessing and a curse. It certainly enables engineers to solve problems easily and quickly, but it also opens the door for abuses and misinformation. In the hands of poorly educated people, these software packages are as dangerous as sophisticated powerful weapons in the hands of poorly trained soldiers.

Thinking that a person who can use the engineering software packages without proper training on fundamentals can practice engineering is like

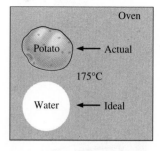

FIGURE 1-29
Modeling is a powerful engineering tool that provides great insight and simplicity at the expense of some accuracy.

thinking that a person who can use a wrench can work as a car mechanic. If it were true that the engineering students do not need all these fundamental courses they are taking because practically everything can be done by computers quickly and easily, then it would also be true that the employers would no longer need high-salaried engineers since any person who knows how to use a word-processing program can also learn how to use those software packages. But the statistics show that the need for engineers is on the rise, not on the decline, despite the availability of these powerful packages.

We should always remember that all the computing power and the engineering software packages available today are just *tools,* and tools have meaning only in the hands of masters. Having the best word-processing program does not make a person a good writer, but it certainly makes the job of a good writer much easier and makes the writer more productive (Fig. 1-30). Hand calculators did not eliminate the need to teach our children how to add or subtract, and the sophisticated medical software packages did not take the place of medical school training. Neither will engineering software packages replace the traditional engineering education. They will simply cause a shift in emphasis in the courses from mathematics to physics. That is, more time will be spent in the classroom discussing the physical aspects of the problems in greater detail, and less time on the mechanics of solution procedures.

All these marvelous and powerful tools available today put an extra burden on today's engineers. They must still have a thorough understanding of the fundamentals, develop a "feel" of the physical phenomena, be able to put the data into proper perspective, and make sound engineering judgments, just like their predecessors. But they must do it much better, and much faster, using more realistic models because of the powerful tools available today. The engineers in the past had to rely on hand calculations, slide rules, and later hand calculators and computers. Today they rely on software packages. The easy access to such power and the possibility of a simple misunderstanding or misinterpretation causing great damage make it more important today than ever to have a solid training in the fundamentals of engineering. In this text we make an extra effort to put the emphasis on developing an intuitive and physical understanding of natural phenomena instead of on the mathematical details of solution procedures.

FIGURE 1-30

An excellent word-processing program does not make a person a good writer; it simply makes a good writer a better and more efficient writer.

1-9 ■ PROBLEM SOLVING TECHNIQUE

The first step in learning any science is to grasp the fundamentals, and to gain a sound knowledge of it. The next step is to master the fundamentals by putting this knowledge to test. This is done by solving significant real-world problems. Solving such problems, especially complicated ones, require a systematic approach. By using a step-by-step approach, an engineer can reduce the solution of a formidable problem into the solution of a series of simple problems (Fig. 1-31). When solving a problem, we recommend that you use the following steps zealously as applicable. This will help you avoid some of the common pitfalls associated with problem solving.

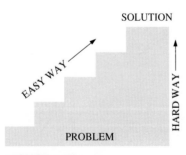

FIGURE 1-31

A step-by-step approach can greatly simplify problem solving.

Step 1: Problem Statement

In your own words, briefly state the problem, the key information given, and the quantities to be found. This is to make sure that you understand the problem and the objectives before you attempt to solve the problem.

Step 2: Schematic

Draw a realistic sketch of the physical system involved, and list the relevant information on the figure. The sketch does not have to be something elaborate, but it should resemble the actual system and show the key features. Indicate any energy and mass interactions with the surroundings. Listing the given information on the sketch helps one to see the entire problem at once. Also, check for properties that remain constant during a process (such as temperature during an isothermal process), and indicate them on the sketch.

Step 3: Assumptions

State any appropriate assumptions made to simplify the problem to make it possible to obtain a solution. Justify the questionable assumptions. Assume reasonable values for missing quantities that are necessary. For example, in the absence of specific data for atmospheric pressure, it can be taken to be 1 atm. But it should be noted in the analysis that the atmospheric pressure decreases with increasing elevation. For example, it drops to 0.83 atm in Denver (elevation 1610 m) (Fig. 1-32).

Step 4: Physical Laws

Apply all the relevant basic physical laws and principles (such as the conservation of energy), and reduce them to their simplest form by utilizing the assumptions made. But the region to which a physical law is applied must be clearly identified first. For example, the heating or cooling of a canned drink is usually analyzed by applying the conservation of energy principle to the entire can.

Step 5: Properties

Determine the unknown properties at known states necessary to solve the problem from property relations or tables. List the properties separately, and indicate their source, if applicable.

Step 6: Calculations

Substitute the known quantities into the simplified relations and perform the calculations to determine the unknowns. Pay particular attention to the units and unit cancellations, and remember that a dimensional quantity without a unit is meaningless. Also, don't give a false implication of high accuracy by copying all the digits from the screen of the calculator—round the results to an appropriate number of significant digits.

Step 7: Reasoning, Verification, and Discussion

Check to make sure that the results obtained are reasonable and intuitive, and verify the validity of the questionable assumptions. Repeat the calculations that resulted in unreasonable values. For example, insulating a water heater that uses $80 worth of natural gas a year cannot result in savings of $200 a year (Fig. 1-33).

Also, point out the significance of the results, and discuss their implications. State the conclusions that can be drawn from the results, and any recommendations that can be made from them. Emphasize the limitations under which the results are applicable, and caution against any possible misunderstandings and using the results in situations where the underlying assumptions do not apply. For example, if you determined that wrapping a water heater with a $20 insulation jacket will reduce the energy cost by $30 a year, indicate that the insulation will pay for itself from the energy it saves in less than a year. But also indicate that the analysis does not consider labor costs, and that this will be the case if you install the insulation yourself.

Given: Air temperature in Denver

To be found: Density of air

Missing information: Atmospheric pressure

Assumption #1: Take P = 1 atm (Inappropriate. Ignores effect of altitude. Will cause more than 15% error.)

Assumption #2: Take P = 0.83 atm (Appropriate. Ignores only minor effects such as weather.)

FIGURE 1-32

The assumptions made while solving an engineering problem must be reasonable and justifiable.

Energy use:	$80/yr
Energy saved by insulation:	$200/yr

IMPOSSIBLE!

FIGURE 1-33

The results obtained from an engineering analysis must be checked for reasonableness.

Keep in mind that you present the solutions to your instructors, and any engineering analysis presented to others is a form of communication. Therefore neatness, organization, completeness, and visual appearance are of utmost importance for maximum effectiveness. Besides, neatness also serves as a great checking tool since it is very easy to spot errors and inconsistencies in a neat work. Carelessness and skipping steps to save time often ends up costing more time and unnecessary anxiety.

The approach described above is used in the solved example problems without explicitly stating each step, as well as in the Solutions Manual of this text. For some problems, some of the steps may not be applicable or necessary. For example, often it is not practical to list the properties separately in thermodynamics problems (Chapters 2–8). However, we cannot overemphasize the importance of a logical and orderly approach to problem solving. Most difficulties encountered while solving a problem are not due to a lack of knowledge; rather, they are due to a lack of coordination. You are strongly encouraged to follow these steps in problem solving until you develop your own approach that works best for you.

A Remark on Significant Digits

In engineering calculations, the information given is not known to more than a certain number of significant digits, usually 3 digits. Consequently, the results obtained cannot possibly be accurate to more significant digits. Reporting results in more significant digits implies greater accuracy than exists, and it should be avoided.

For example, consider a 3.75-L container filled with gasoline whose density is 0.845 kg/L, and try to determine its mass. Probably the first thought that comes to your mind is to multiply the volume and density to obtain 3.16875 kg for the mass, which falsely implies that the mass determined is accurate to 6 significant digits. In reality, however, the mass cannot be more accurate than 3 significant digits since both the volume and the density are accurate to 3 significant digits only. Therefore, the result should be rounded to 3 significant digits, and the mass should be reported to be 3.17 kg instead of what appears in the screen of the calculator. The result 3.16875 kg would be correct only if the volume and density were given to be 3.75000 L and 0.845000 kg/L, respectively. The value 3.75 L implies that we are fairly confident that the volume is accurate within ± 0.01 L, and it cannot be 3.74 or 3.76 L. But the volume can be 3.746, 3.750, 3.753, etc., since they all round to 3.75 L (Fig. 1-34). It is more appropriate to retain all the digits during intermediate calculations, and to do the rounding in the final step since this is what a computer will normally do.

When solving problems, we will assume the given information to be accurate to at least 3 significant digits. Therefore, if the length of a pipe is given to be 40 m, we will assume it to be 40.0 m in order to justify using 3 significant digits in the final results. You should also keep in mind that all experimentally determined values are subject to measurement errors, and such errors will reflect in the results obtained. For example, if the density of a substance has an uncertainty of 2 percent, then the mass determined using this density value will also have an uncertainty of 2 percent.

You should also be aware that we sometimes knowingly introduce small errors in order to avoid the trouble of searching for more accurate data. For example, when dealing with liquid water, we just use the value of 1000 kg/m^3

Given:
    Volume: V = 3.75 L
    Density: ρ = 0.845 kg/L

        (3 significant digits)

Also, $3.75 \times 0.845 = 3.16875$

Find:
    Mass: $m = \rho V = 3.16875$ kg

Rounding to 3 significant digits:

    $m = 3.17$ kg

FIGURE 1-34

A result with more significant digits than that of given data falsely implies more accuracy.

density, which is the density value of pure water at 0°C. Using this value at 75°C will result in an error of 2.5 percent since the density at this temperature is 975 kg/m³. The minerals and impurities in the water will introduce additional error. This being the case, you should have no reservation in rounding the final results to a reasonable number of significant digits. Besides, having a few percent uncertainty in the results of engineering analysis is usually the norm, not the exception.

1-10 ■ CONSERVATION OF MASS PRINCIPLE

The conservation of mass principle is one of the most fundamental principles in nature. We are all familiar with this principle, and it is not difficult to understand. As the saying goes, you cannot have your cake and eat it, too! A person does not have to be an engineer to figure out how much vinegar-and-oil dressing he is going to have if he mixes 100 g of oil with 25 g of vinegar. Even chemical equations are balanced on the basis of the conservation of mass principle. When 16 kg of oxygen reacts with 2 kg of hydrogen, 18 kg of water is formed (Fig. 1-35). In an electrolysis process, the water will separate back to 2 kg of hydrogen and 16 kg of oxygen.

Mass, like energy, is a conserved property, and it cannot be created or destroyed. However, mass m and energy E can be converted to each other according to the famous formula proposed by Einstein:

$$E = mc^2 \tag{1-6}$$

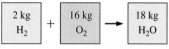

FIGURE 1-35

Mass is conserved even
during chemical reactions.

where c is the speed of light. This equation suggests that the mass of a system will change when its energy changes. However, for all energy interactions encountered in practice, with the exception of nuclear reactions, the change in mass is extremely small and cannot be detected by even the most sensitive devices. For example, when 1 kg of water is formed from oxygen and hydrogen, the amount of energy released is 15,879 kJ, which corresponds to a mass of 1.76×10^{-10} kg. A mass of this magnitude is beyond the accuracy required by practically all engineering calculations and thus can be disregarded.

For *closed systems,* the conservation of mass principle is implicitly used by requiring that the mass of the system remain constant during a process. For *control volumes,* however, mass can cross the boundaries, and so we must keep track of the amount of the mass entering and leaving the control volume (Fig. 1-36).

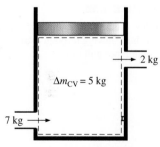

FIGURE 1-36

Conservation of mass principle
for a control volume.

Mass and Volume Flow Rates

The amount of mass flowing through a cross section per unit time is called the **mass flow rate** and is denoted $\dot{m}$. The dot over a symbol is used to indicate a *quantity per unit time.*

A fluid flows in or out of a control volume through pipes (or ducts). The mass flow rate of a fluid flowing in a pipe is proportional to the cross-sectional area A of the pipe, the density ρ, and the velocity $\mathcal{V}$ of the fluid. The mass flow rate through a differential area dA can be expressed as

$$d\dot{m} = \rho \mathcal{V}_n \, dA \tag{1-7}$$

where $\mathcal{V}_n$ is the velocity component normal to dA. The mass flow rate through the entire cross-sectional area of the pipe or duct is obtained by integration:

$$\dot{m} = \int_A \rho \mathcal{V}_n \, dA \quad \text{(kg/s)} \tag{1-8}$$

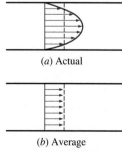

(a) Actual

(b) Average

FIGURE 1-37

Actual and mean velocity profiles for
flow in a pipe (the mass flow rate
is the same for both cases).

In most practical applications, the flow of a fluid through a pipe or duct can be approximated to be *one-dimensional flow,* and thus the properties can be assumed to vary in *one* direction only (the direction of flow). As a result, all properties are *uniform* at any cross section normal to the flow direction, and the properties are assumed to have *bulk average values* over the cross section. But the values of the properties at a cross section *may* change with time unless the flow is steady.

The one-dimensional-flow approximation has little impact on most properties of a fluid flowing in a pipe or duct such as temperature, pressure, and density since these properties usually remain constant over the cross section. But this is not the case for *velocity,* whose value varies from zero at the wall to a maximum at the center because of the viscous effects (friction between fluid layers). Under the one-dimensional-flow assumption, the velocity is assumed to be constant across the entire cross section at some equivalent average value (Fig. 1-37). Then the integration in Eq. 1-8 can be performed for one-dimensional flow to yield

$$\dot{m} = \rho \mathcal{V}_m A \qquad \text{(kg/s)} \qquad (1\text{-}9)$$

where

$$\rho = \text{density of fluid, kg/m}^3 \ (= 1/v)$$
$$\mathcal{V}_m = \text{mean fluid velocity normal to } A, \text{ m/s}$$
$$A = \text{cross-sectional area normal to flow direction, m}^2$$

The volume of the fluid flowing through a cross section per unit time is called the **volume flow rate** $\dot{V}$ (Fig. 1-38) and is given by

$$\dot{V} = \int_A \mathcal{V}_n \, dA = \mathcal{V}_m A \qquad \text{(m}^3\text{/s)} \qquad (1\text{-}10)$$

The mass and volume flow rates are related by

$$\dot{m} = \rho \dot{V} = \frac{\dot{V}}{v} \qquad (1\text{-}11)$$

This relation is analogous to $m = V/v,$ which is the relation between the mass and the volume of a fluid in a container.

For simplicity, we drop the subscript on the mean velocity. Unless otherwise stated, $\mathcal{V}$ denotes the mean velocity in the flow direction. Also, A denotes the cross-sectional area normal to the flow direction.

Conservation of Mass Principle

The **conservation of mass principle** can be expressed as: *net mass transfer to or from a system during a process is equal to the net change (increase or decrease) in the total mass of the system during that process.* That is,

$$\begin{pmatrix} \text{Total mass} \\ \text{entering the system} \end{pmatrix} - \begin{pmatrix} \text{Total mass} \\ \text{leaving the system} \end{pmatrix} = \begin{pmatrix} \text{Net change in mass} \\ \text{within the system} \end{pmatrix}$$

or

$$m_{\text{in}} - m_{\text{out}} = \Delta m_{\text{system}} \qquad \text{(kg)} \qquad (1\text{-}12)$$

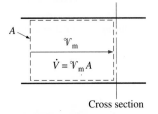

Cross section

FIGURE 1-38

The volume flow rate is the volume of fluid flowing through a cross section per unit time.

where $\Delta m_{\text{system}} = m_{\text{final}} - m_{\text{initial}}$ is the change in the mass of the system during the process (Fig. 1-39). It can also be expressed in the *rate form* as

$$\dot{m}_{\text{in}} - \dot{m}_{\text{out}} = dm_{\text{system}}/dt \qquad \text{(kg/s)} \qquad (1\text{-}13)$$

where $\dot{m}_{\text{in}}$ and $\dot{m}_{\text{out}}$ are the total rates of mass flow into and out of the system and dm_{system}/dt (also denoted by $\Delta\dot{m}_{\text{system}}$ for simplicity) is the rate of change of mass within the system boundaries. The relations above are often referred to as the **mass balance** and are applicable to any kind of system undergoing any kind of process.

The mass balance for a control volume can also be expressed more explicitly as

$$\sum m_i - \sum m_e = (m_2 - m_1)_{\text{system}} \qquad (1\text{-}14)$$

and

$$\sum \dot{m}_i - \sum \dot{m}_e = dm_{\text{system}}/dt \qquad (1\text{-}15)$$

where i = inlet; e = exit; 1 = initial state and 2 = final state of the control volume; and the summation signs are used to emphasize that all the inlets and exits are to be considered.

When the properties at the inlets and the exits as well as within the control volume are not uniform, the mass flow rate can be expressed in the differential form as $d\dot{m} = \rho \mathcal{V}_n\, dA$. Then the general rate form of the mass balance (Eq. 1-15) can be expressed as

$$\sum \int_{A_i} (\rho \mathcal{V}_n\, dA)_i - \sum \int_{A_e} (\rho \mathcal{V}_n\, dA)_e = \frac{d}{dt} \int_V (\rho\, dV)_{CV} \qquad (1\text{-}16)$$

to account for the variation of properties. The integration of $dm_{CV} = \rho\, dV$ on the right-hand side over the volume of the control volume gives the total mass contained within the control volume at time t.

The conservation of mass principle is based on experimental observations and requires every bit of mass to be accounted for during a process. A person who can balance a checkbook (by keeping track of deposits and withdrawals, or simply by observing the "conservation of money" principle) should have no difficulty in applying the conservation of mass principle to engineering systems. The conservation of mass equation is often referred to as the **continuity equation** in fluid mechanics.

Mass Balance for Steady-Flow Processes

During a steady-flow process, the total amount of mass contained within a control volume does not change with time (m_{CV} = constant). Then the conservation of mass principle requires that the total amount of mass entering a control volume equal the total amount of mass leaving it (Fig. 1-40). For a garden hose nozzle, for example, the amount of water entering the nozzle is equal to the amount of water leaving it in steady operation.

When dealing with steady-flow processes, we are not interested in the amount of mass that flows in or out of a device over time; instead, we are interested in the amount of mass flowing per unit time, that is, *the mass flow rate* $\dot{m}$. The **conservation of mass principle** for a general steady-flow system with multiple inlets and exits can be expressed in the rate form as (Fig. 1-41)

$$\begin{pmatrix} \text{Total mass entering CV} \\ \text{per unit time} \end{pmatrix} = \begin{pmatrix} \text{Total mass leaving CV} \\ \text{per unit time} \end{pmatrix}$$

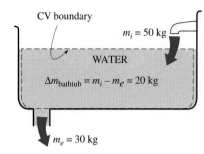

FIGURE 1-39

Conservation of mass principle
for an ordinary bathtub.

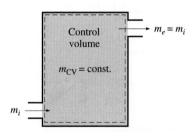

FIGURE 1-40

During a steady-flow process, the amount
of mass entering a control volume
equals the amount of mass leaving.

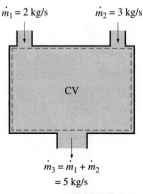

FIGURE 1-41

Conservation of mass principle for a
two-inlet–one-exit steady-flow system.

or

$$\text{Steady Flow:} \qquad \sum \dot{m}_i = \sum \dot{m}_e \qquad \text{(kg/s)} \qquad (1\text{-}17)$$

where the subscript i stands for inlet and e for exit. Many engineering devices such as nozzles, diffusers, turbines, compressors, and pumps involve a single stream (only one inlet and one exit). For these cases, we denote the inlet state by the subscript 1 and the exit state by the subscript 2. We also drop the summation signs. Then Eq. 1-17 reduces, for *single-stream steady-flow systems*, to

$$\text{Steady Flow (single stream):} \quad \dot{m}_1 = \dot{m}_2 \;\rightarrow\; \rho_1 \mathcal{V}_1 A_1 = \rho_2 \mathcal{V}_2 A_2 \qquad (1\text{-}18)$$

Special Case: Incompressible Flow (ρ = constant)

The conservation of mass relations above can be simplified even further when the fluid is incompressible, which is usually the case for liquids, and sometimes the case for gases. Canceling the density from both sides of the steady-flow relations gives

$$\text{Steady Incompressible Flow:} \quad \sum \dot{V}_i = \sum \dot{V}_e \qquad \text{(m}^3\text{/s)} \qquad (1\text{-}19)$$

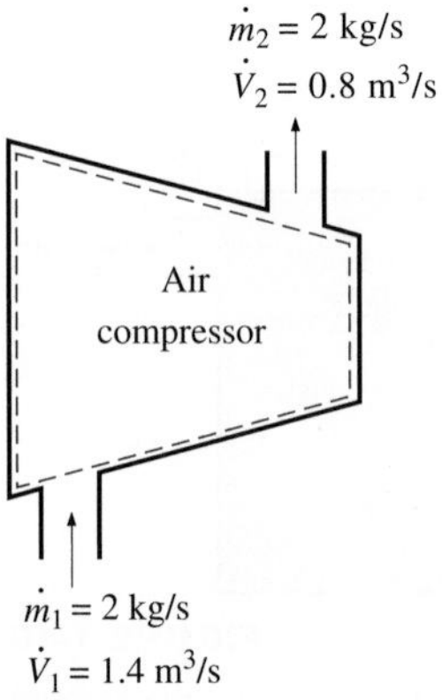

For single-stream steady-flow systems it becomes

$$\begin{array}{l}\text{Steady Incompressible Flow} \\ \textit{(single stream)}:\end{array} \quad \dot{V}_1 = \dot{V}_2 \;\rightarrow\; \mathcal{V}_1 A_1 = \mathcal{V}_2 A_2 \qquad (1\text{-}20)$$

It should always be kept in mind that there is no such thing as a "conservation of volume" principle. Therefore, the volume flow rates into and out of a steady-flow device may be different. The volume flow rate at the exit of an air compressor will be much less than that at the inlet even though the mass flow rate of air through the compressor is constant (Fig. 1-42). This is due to the higher density of air at the compressor exit. For liquid flow, however, the volume flow rates, as well as the mass flow rates, remain constant since liquids are essentially incompressible (constant-density) substances. Water flow through the nozzle of a garden hose is an example for the latter case.

EXAMPLE 1-3   Water Flow through a Garden Hose Nozzle

A garden hose attached with a nozzle is used to fill a 10-gallon bucket. The inner diameter of the hose is 2 cm, and it reduces to 0.8 cm at the nozzle exit (Fig. 1-43). If it takes 50 s to fill the bucket with water, determine (*a*) the volume and mass flow rates of water through the hose, and (*b*) the mean velocity of water at the nozzle exit.

Solution   A garden hose is used to fill water buckets. The volume and mass flow rates of water and the exit velocity are to be determined.

Assumptions   **1** Water is an incompressible substance. **2** Flow through the hose is steady. **3** There is no waste of water by splashing.

Properties   We take the density of water to be 1000 kg/m³ = 1 kg/L.

Analysis   (*a*) Noting that 10 gallons of water are discharged in 50 s, the volume and mass flow rates of water are determined to be

$\dot{m}_2 = 2$ kg/s
$\dot{V}_2 = 0.8$ m³/s

Air
compressor

$\dot{m}_1 = 2$ kg/s
$\dot{V}_1 = 1.4$ m³/s

FIGURE 1-42

During a steady-flow process, volume flow rates are not necessarily conserved.

Garden
hose

Nozzle

Bucket

FIGURE 1-43

Schematic for Example 1-3.

$$\dot{V} = \frac{V}{\Delta t} = \frac{10 \text{ gal}}{50 \text{ s}} \left(\frac{3.7854 \text{ L}}{1 \text{ gal}} \right) = \textbf{0.757 L/s}$$

$$\dot{m} = \rho \dot{V} = (1 \text{ kg/L}) (0.757 \text{ L/s}) = \textbf{0.757 kg/s}$$

(b) The cross-sectional area of the nozzle exit is

$$A_e = \pi r_e^2 = \pi (0.4 \text{ cm})^2 = 0.5027 \text{ cm}^2 = 0.5027 \times 10^{-4} \text{ m}^2$$

The volume flow rate through the hose and the nozzle is constant. Then the velocity of water at the nozzle exit becomes

$$\mathcal{V}_e = \frac{\dot{V}}{A_e} = \frac{0.757 \text{ L/s}}{0.5027 \times 10^{-4} \text{ m}^2} \left(\frac{1 \text{ m}^3}{1000 \text{ L}} \right) = \textbf{15.1 m/s}$$

Discussion   It can be shown that the mean velocity in the hose is 2.4 m/s. Therefore, the nozzle increases the water velocity by over 6 times.

Example 1-4   Discharge of Water from a Tank

A 4-ft-high 3-ft-diameter cylindrical water tank whose top is open to the atmosphere is initially filled with water. Now the discharge plug near the bottom of the tank is pulled out, and a water jet whose diameter is 0.5 in. streams out (Fig. 1-44). The mean velocity of the jet is given by $\mathcal{V} = \sqrt{2gh}$ where h is the height of water in the tank measured from the center of the hole (a variable) and g is the gravitational acceleration. Determine how long it will take for the water level in the tank to drop to 2 ft level from the bottom.

Solution   The plug near the bottom of a water tank is pulled out. The time it will take for half of the water in the tank to empty is to be determined.

Assumptions   **1** Water is an incompressible substance. **2** The distance between the bottom of the tank and the center of the hole is negligible compared to the total water height. **3** The gravitational acceleration is 32.2 ft/s².

Analysis   We take the volume occupied by water as the control volume. The size of the control volume will decrease in this case as the water level drops, and thus this is a variable control volume. (We could also treat this as a fixed control volume which consists of the interior volume of the tank by disregarding the air that replaces the space vacated by the water.) This is obviously an unsteady flow problem since the properties (such as the amount of mass) within the control volume change with time.

The conservation of mass relation for any system undergoing any process is given in the rate form as

$$\dot{m}_{in} - \dot{m}_{out} = \frac{dm_{system}}{dt} \qquad (1)$$

During this process no mass enters the control volume ($\dot{m}_{in} = 0$), and the mass flow rate of discharged water can be expressed as

$$\dot{m}_{out} = (\rho \mathcal{V} A)_{out} = \rho \sqrt{2gh} \, A_{jet} \qquad (2)$$

where $A_{jet} = \pi D_{jet}^2 / 4$ is the cross-sectional area of the jet, which is constant. Noting that the density of water is constant, the mass of water in the tank at any time is

$$m_{system} = \rho V = \rho A_{tank} h \qquad (3)$$

where $A_{tank} = \pi D_{tank}^2 / 4$ is the base area of the cylindrical tank. Substituting Eqs. (2) and (3) into the mass balance relation (1) gives

$$-\rho \sqrt{2gh} \, A_{jet} = \frac{d(\rho A_{tank} h)}{dt} \quad \rightarrow \quad -\rho \sqrt{2gh} \, (\pi D_{jet}^2 / 4) = \frac{\rho (\pi D_{tank}^2 / 4) \, dh}{dt}$$

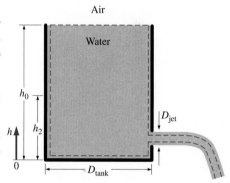

Air

Water

h_0

h h_2

0

D_{jet}

D_{tank}

FIGURE 1-44
Schematic for Example 1-4.

Canceling the densities and other common terms and separating the variables gives

$$dt = -\frac{D_{tank}^2}{D_{jet}^2}\frac{dh}{\sqrt{2gh}}$$

Integrating from $t = 0$ at which $h = h_0$ to $t = t$ at which $h = h_2$ gives

$$\int_0^t dt = -\frac{D_{tank}^2}{D_{jet}^2\sqrt{2g}}\int_{h_0}^{h_2}\frac{dh}{\sqrt{h}} \quad \rightarrow \quad t = \frac{\sqrt{h_0}-\sqrt{h_2}}{\sqrt{g/2}}\left(\frac{D_{tank}}{D_{jet}}\right)^2$$

Substituting, the time of discharge is determined to be

$$t = \frac{\sqrt{4\text{ ft}}-\sqrt{2\text{ ft}}}{\sqrt{(32.2\text{ ft/s}^2)/2}}\left(\frac{3\times12\text{ in.}}{0.5\text{ in.}}\right)^2 = \textbf{12.6 min}$$

Therefore, half of the tank will be emptied in 12.6 min after the discharge hole is unplugged.

Discussion   Using the same relation with $h_2 = 0$ gives $t = 43.1$ min for the discharge of the entire water in the tank. Therefore, emptying the bottom half of the tank will take much longer than emptying the top half. This is due to the decrease in the average discharge velocity of water with decreasing h.

1-11 ■ SUMMARY

In this chapter, some basic concepts of thermal-fluid sciences are introduced and discussed. The physical sciences that deal with energy and the transfer, transport, and conversion of energy are referred to as *thermal-fluid sciences,* and they are studied under the subcategories of thermodynamics, heat transfer, and fluid mechanics.

Thermodynamics is the science that primarily deals with energy. The *first law of thermodynamics* is simply an expression of the conservation of energy principle, and it asserts that *energy* is a thermodynamic property. The *second law of thermodynamics* asserts that energy has *quality* as well as *quantity,* and actual processes occur in the direction of decreasing quality of energy. Determining the rates of heat transfer to or from a system and thus the times of cooling or heating, as well as the variation of the temperature, is the subject of *heat transfer.* The basic requirement for heat transfer is the presence of a *temperature difference.* A substance in the liquid or gas phase is referred to as a *fluid. Fluid mechanics* is the science that deals with the behavior of fluids at rest (*fluid statics*) or in motion (*fluid dynamics*), and the interaction of fluids with solids or other fluids at the boundaries.

A system of fixed mass is called a *closed system,* or *control mass,* and a system that involves mass transfer across its boundaries is called an *open system,* or *control volume.* The mass-dependent properties of a system are called *extensive properties* and the others, *intensive properties. Density* is mass per unit volume, and *specific volume* is volume per unit mass.

When solving a problem, it is recommended that a step-by-step approach be used. Such an approach involves stating the problem, drawing a schematic, making appropriate assumptions, applying the physical laws, listing the relevant properties, making the necessary calculations, and making sure that the results are reasonable.

The *conservation of mass principle* states that the net mass transfer to or from a system during a process is equal to the net change (increase or decrease) in the total mass of the system during that process, and is expressed as

$$m_{in} - m_{out} = \Delta m_{system} \qquad \text{and} \qquad \dot{m}_{in} - \dot{m}_{out} = dm_{system}/dt$$

where $\Delta m_{system} = m_{final} - m_{initial}$ is the change in the mass of the system during the process, $\dot{m}_{in}$ and $\dot{m}_{out}$ are the total rates of mass flow into and out of the system, and dm_{system}/dt is the rate of change of mass within the system boundaries. The relations above are also referred to as the *mass balance* or *continuity equation,* and are applicable to any kind of system undergoing any kind of process.

The amount of mass flowing through a cross section per unit time is called the *mass flow rate,* and is expressed as

$$\dot{m} = \rho \mathcal{V} A$$

where ρ = density of fluid, $\mathcal{V}$ = mean fluid velocity normal to A, and A = cross-sectional area normal to flow direction. The volume of the fluid flowing through a cross section per unit time is called the *volume flow rate* and is expressed as

$$\dot{V} = \mathcal{V} A = \dot{m}/\rho$$

For steady-flow systems, the conservation of mass principle is expressed as

Steady Flow:
$$\sum \dot{m}_i = \sum \dot{m}_e$$

Steady Flow (single stream):
$$\dot{m}_1 = \dot{m}_2 \quad \rightarrow \quad \rho_1 \mathcal{V}_1 A_1 = \rho_2 \mathcal{V}_2 A_2$$

For incompressible fluids, they simplify to

Steady Incompressible Flow:
$$\sum \dot{V}_i = \sum \dot{V}_e$$

Steady Incompressible Flow (single stream):
$$\dot{V}_1 = \dot{V}_2 \quad \rightarrow \quad \dot{V}_1 A_1 = \dot{V}_2 A_2$$

REFERENCES AND SUGGESTED READING

1. American Society for Testing and Materials. *Standards for Metric Practice.* ASTM E 380-79, January 1980.

2. Y. A. Çengel. *Heat Transfer: A Practical Approach.* New York, McGraw-Hill, 1998.

3. Y. A. Çengel and M. A. Boles. *Thermodynamics. An Engineering Approach.* 3rd ed. New York, McGraw-Hill, 1998.

4. R. W. Fox and A. T. McDonald. *Introduction to Fluid Mechanics.* 5th ed. New York, Wiley, 1999.

5. M. C. Potter and D. C. Wiggert. *Mechanics of Fluids.* 2nd ed. Upper Saddle River, Prentice Hall, 1997.

6. J. A. Roberson and C. L. Grove. *Engineering Fluid Mechanics.* 6th ed. New York, Wiley, 1997.

7. F. M. White, *Fluid Mechanics.* 4th ed. New York, McGraw-Hill, 1999.

PROBLEMS*

Thermodynamics, Heat Transfer, and Fluid Mechanics

1-1C   What is the difference between the classical and the statistical approaches to thermodynamics?

1-2C   Why does a bicyclist pick up speed on a downhill road even when he is not pedaling? Does this violate the conservation of energy principle?

1-3C   An office worker claims that a cup of cold coffee on his table warmed up to 80°C by picking up energy from the surrounding air, which is at 25°C. Is there any truth to his claim? Does this process violate any thermodynamic laws?

1-4C   How does the science of heat transfer differ from the science of thermodynamics?

1-5C   What is the driving force for (*a*) heat transfer, (*b*) electric current, and (*c*) fluid flow?

1-6C   Why is heat transfer a nonequilibrium phenomenon?

1-7C   What is the caloric theory? When and why was it abandoned?

1-8C   Can there be any heat transfer between two bodies that are at the same temperature but at different pressures?

1-9C   Define stress, normal stress, shear stress, and pressure.

1-10C   Define flow field and streamline.

Mass, Force, and Acceleration

1-11C   What is the difference between pound-mass and pound-force?

1-12C   What is the difference between kg-mass and kg-force?

1-13C   What is the net force acting on a car cruising at a constant velocity of 70 km/h (*a*) on a level road and (*b*) on an uphill road?

1-14   A 3-kg plastic tank that has a volume of 0.2 m^3 is filled with liquid water. Assuming the density of water is 1000 kg/m^3, determine the weight of the combined system.

1-15   Determine the mass and the weight of the air contained in a room whose dimensions are 6 m $\times$ 6 m $\times$ 8 m. Assume the density of the air is 1.16 kg/m^3.   *Answers:* 334.1 kg, 3277 N

1-16   At 45° latitude, the gravitational acceleration as a function of elevation z above sea level is given by $g = a - bz$, where $a = 9.807$ m/s^2 and $b = 3.32 \times 10^{-6}$ s^{-2}. Determine the height above sea level where the weight of an object will decrease by 1 percent.   *Answer:* 29,539 m

1-17E   A 150-lbm astronaut took his bathroom scale (a spring scale) and a beam scale (compares masses) to the moon where the local gravity is $g = 5.48$ ft/s^2. Determine how much he will weigh (*a*) on the spring scale and (*b*) on the beam scale.   *Answers:* (*a*) 25.5 lbf; (*b*) 150 lbf

---

*Students are encouraged to answer *all* the concept "C" questions.

1-18 The acceleration of high-speed aircraft is sometimes expressed in g's (in multiples of the standard acceleration of gravity). Determine the net upward force, in N, that a 90-kg man would experience in an aircraft whose acceleration is 6 g's.

1-19 A 5-kg rock is thrown upward with a force of 150 N at a location where the local gravitational acceleration is 9.79 m/s^2. Determine the acceleration of the rock, in m/s^2.

1-20 The value of the gravitational acceleration g decreases with elevation from 9.807 m/s^2 at sea level to 9.767 m/s^2 at an altitude of 13,000 m, where large passenger planes cruise. Determine the percent reduction in the weight of an airplane cruising at 13,000 m relative to its weight at sea level.

Systems and Properties

1-21C Most of the energy generated in the engine of a car is rejected to the air by the radiator through the circulating water. Should the radiator be analyzed as a closed system or as an open system? Explain.

1-22C A can of soft drink at room temperature is put into the refrigerator so that it will cool. Would you model the can of soft drink as a closed system or as an open system? Explain.

1-23C What is the difference between intensive and extensive properties?

Solving Engineering Problems

1-24C How do rating problems in heat transfer differ from the sizing problems?

1-25C What is the difference between the analytical and experimental approach to engineering problems? Discuss the advantages and disadvantages of each approach?

1-26C What is the importance of modeling in engineering? How are the mathematical models for engineering processes prepared?

1-27C When modeling an engineering process, how is the right choice made between a simple but crude and a complex but accurate model? Is the complex model necessarily a better choice since it is more accurate?

1-28C How do the differential equations in the study of a physical problem arise?

1-29C What is the value of the engineering software packages in (*a*) engineering education and (*b*) engineering practice?

Conservation of Mass

1-30C Define mass and volume flow rates. How are they related to each other?

1-31C Does the amount of mass entering a control volume have to be equal to the amount of mass leaving during an unsteady-flow process?

1-32C When is the flow through a control volume steady?

1-33C Consider a device with one inlet and one exit. If the volume flow rates at the inlet and at the exit are the same, is the flow through this device necessarily steady? Why?

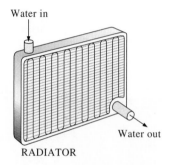

Water in

Water out

RADIATOR

FIGURE P1-21C

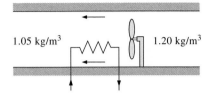

1.05 kg/m³ $\quad$ 1.20 kg/m³

FIGURE P1-36

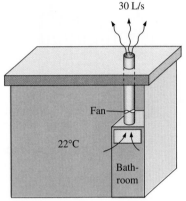

30 L/s

Fan

22°C

Bath-
room

FIGURE P1-39

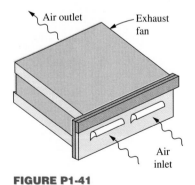

Air outlet $\quad$ Exhaust
fan

Air
inlet

FIGURE P1-41

1-34E A garden hose attached with a nozzle is used to fill a 20-gallon bucket. The inner diameter of the hose is 1 in. and it reduces to 0.5 in. at the nozzle exit. If the mean velocity in the hose is 8 ft/s, determine (*a*) the volume and mass flow rates of water through the hose, (*b*) how long it will take to fill the bucket with water, and (*c*) the mean velocity of water at the nozzle exit.

1-35 Air enters a nozzle steadily at 2.21 kg /m³ and 30 m/s and leaves at 0.762 kg/m³ and 180 m/s. If the inlet area of the nozzle is 80 cm², determine (*a*) the mass flow rate through the nozzle, and (*b*) the exit area of the nozzle. *Answers:* (*a*) 0.5304 kg/s, (*b*) 38.7 cm²

1-36 A hair dryer is basically a duct of constant diameter in which a few layers of electric resistors are placed. A small fan pulls the air in and forces it through the resistors where it is heated. If the density of air is 1.20 kg/m³ at the inlet and 1.05 kg/m³ at the exit, determine the percent increase in the velocity of air as it flows through the dryer.

1-37E Air whose density is 0.078 lbm/ft³ enters the duct of an air-conditioning system at a volume flow rate of 450 ft³/min. If the diameter of the duct is 10 in., determine the velocity of the air at the duct inlet and the mass flow rate of air.

1-38 A 1-m³ rigid tank initially contains air whose density is 1.18 kg/m³. The tank is connected to a high-pressure supply line through a valve. The valve is opened, and air is allowed to enter the tank until the density in the tank rises to 7.20 kg/m³. Determine the mass of air that has entered the tank. *Answer:* 6.02 kg

1-39 The ventilating fan of the bathroom of a building has a volume flow rate of 30 L/s and runs continuously. If the density of air inside is 1.20 kg/m³, determine the mass of air vented out in one day.

1-40E Chickens with an average mass of 4.5 lbm are to be cooled by chilled water in a continuous-flow-type immersion chiller. Chickens are dropped into the chiller at a rate of 500 chickens per hour. Determine the mass flow rate of chickens through the chiller.

1-41 A desktop computer is to be cooled by a fan whose flow rate is 0.34 m³/min. Determine the mass flow rate of air through the fan at an elevation of 3400 m where the air density is 0.7 kg/m³. Also, if the mean velocity of air is not to exceed 110 m/min, determine the diameter of the casing of the fan. *Answers:* 0.238 kg/min, 0.063 cm

Review Problems

1-42 The weight of bodies may change somewhat from one location to another as a result of the variation of the gravitational acceleration g with elevation. Accounting for this variation using the relation in Prob. 1-16, determine the weight of an 80-kg person at sea level ($z = 0$), in Denver ($z = 1610$ m), and on the top of Mount Everest ($z = 8848$ m).

1-43 A $D_0 = 10$ m diameter tank is initially filled with water 2 m above the center of a $D = 10$ cm diameter valve near the bottom. The tank surface is open to the atmosphere, and the tank drains through a $L = 100$ m long pipe connected to the valve. The friction coefficient of the pipe is given to be

$f = 0.015$, and the discharge velocity is expressed as $\mathcal{V} = \sqrt{\dfrac{2gz}{1.5 + fL/D}}$ where z is the water height above the center of the valve. Determine (a) the initial discharge velocity from the tank and (b) the time required to empty the tank. The tank can be considered to be empty when the water level drops to the center of the valve.

1-44 A man goes to a traditional market to buy a steak for dinner. He finds a 12-ounce steak (1 lbm = 16 ounces) for $3.15. He then goes to the adjacent international market and finds a 320-gram steak of identical quality for $2.80. Which steak is a better buy?

1-45 Milk is to be transported from Texas to California for a distance of 2100 km in a 7-m-long, 2-m-external-diameter cylindrical tank. The walls of the tank are constructed of 5-cm-thick urethane insulation sandwiched between two metal sheets of negligible thickness. Determine the amount of milk in the tank in kg and in gallons.

1-46 Underground water is being pumped into a pool whose cross section is 3 m × 4 m while water is discharged through a 5-cm diameter orifice at a constant mean velocity of 5 m/s. If the water level in the pool rises at a rate of 1.5 cm/min, determine the rate at which water is supplied to the pool, in m³/s.

1-47 The reactive force developed by a jet engine to push an airplane forward is called thrust, and the thrust developed by the engine of a Boeing 777 is about 85,000 pounds. Express this thrust in N and kgf.

1-48 The velocity of a liquid flowing in a circular pipe of radius R varies from zero at the wall to a maximum at the pipe center. The velocity distribution in the pipe can be represented as $V(r)$, where r is the radial distance from the pipe center. Based on the definition of mass flow rate $\dot{m}$, obtain a relation for the mean velocity in terms of $V(r)$, R, and r.

1-49 Air at 4.18 kg/m³ enters a nozzle that has an inlet-to-exit area ratio of 2:1 with a velocity of 120 m/s and leaves with a velocity of 380 m/s. Determine the density of air at the exit.   *Answer:* 2.64 kg/m³

1-50 A long roll of 1-m-wide and 0.5-cm-thick 1-Mn manganese steel plate ($\rho = 7854$ kg/m³) coming off a furnace is to be quenched in an oil bath to a specified temperature. If the metal sheet is moving at a steady velocity of 10 m/min, determine the mass flow rate of the steel plate through the oil bath.

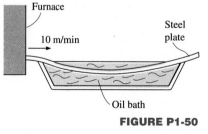

FIGURE P1-50

1-51 The air in a 6 m × 5 m × 4 m hospital room is to be completely replaced by conditioned air every 20 min. If the average air velocity in the circular air duct leading to the room is not to exceed 5 m/s, determine the minimum diameter of the duct.

1-52E It is well-established that indoor air quality (IAQ) has a significant effect on general health and productivity of employees at a workplace. A recent study showed that enhancing IAQ by increasing the building ventilation from 5 cfm (cubic feet per minute) to 20 cfm increased the productivity by 0.25 percent, valued at $90 per person per year, and decreased the respiratory illnesses by 10% for an average annual savings of $39 per person while increasing the annual energy consumption by $6 and the equipment cost by about $4 per person per year (*ASHRAE Journal,* December 1998). For a workplace with 120 employees, determine the net monetary benefit of installing an enhanced IAQ system to the employer per year.

Answer: $14,280/yr

Computer, Design, and Essay Problems

1-53 Write an essay on the various mass- and volume-measurement devices used throughout history. Also, explain the development of the modern units for mass and volume.

1-54 Using a large bucket whose volume is known and measuring the time it takes to fill the bucket with water from a garden hose, determine the mass flow rate and the average velocity of water through the hose.

Thermodynamics

PART I

Basic Concepts of Thermodynamics

Every science has a unique vocabulary associated with it, and thermal-fluid sciences are no exception. Precise definition of basic concepts forms a sound foundation for the development of a science, and prevents possible misunderstandings. We start this chapter with a discussion of some basic concepts such as *state, state postulate, equilibrium, process, energy* and *various forms of energy,* and discuss *temperature* and *temperature scales.* We then present *pressure,* which is the force exerted by a fluid per unit area, and discuss *absolute* and *gage* pressures, the variation of pressure with depth, and pressure measurement devices such as manometers and barometers. Careful study of these concepts is essential for a good understanding of the topics in the following chapters.

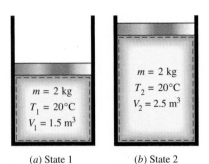

(a) State 1   *(b)* State 2

FIGURE 2-1

A system at two different states.

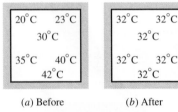

(a) Before   *(b)* After

FIGURE 2-2

A closed system reaching thermal
equilibrium.

FIGURE 2-3

The state of nitrogen is fixed by two
independent, intensive properties.

2-1 ■ STATE AND EQUILIBRIUM

Consider a system not undergoing any change. At this point, all the properties can be measured or calculated throughout the entire system, which gives us a set of properties that completely describes the condition, or the **state,** of the system. At a given state, all the properties of a system have fixed values. If the value of even one property changes, the state will change to a different one. In Fig. 2-1 a system is shown at two different states.

Thermodynamics deals with *equilibrium* states. The word **equilibrium** implies a state of balance. In an equilibrium state there are no unbalanced potentials (or driving forces) within the system. A system in equilibrium experiences no changes when it is isolated from its surroundings.

There are many types of equilibrium, and a system is not in thermodynamic equilibrium unless the conditions of all the relevant types of equilibrium are satisfied. For example, a system is in **thermal equilibrium** if the temperature is the same throughout the entire system, as shown in Fig. 2-2. That is, the system involves no temperature differential, which is the driving force for heat flow. **Mechanical equilibrium** is related to pressure, and a system is in mechanical equilibrium if there is no change in pressure at any point of the system with time. However, the pressure may vary within the system with elevation as a result of gravitational effects. But the higher pressure at a bottom layer is balanced by the extra weight it must carry, and, therefore, there is no imbalance of forces. The variation of pressure as a result of gravity in most thermodynamic systems is relatively small and usually disregarded. If a system involves two phases, it is in **phase equilibrium** when the mass of each phase reaches an equilibrium level and stays there. Finally, a system is in **chemical equilibrium** if its chemical composition does not change with time, that is, no chemical reactions occur. A system will not be in equilibrium unless all the relevant equilibrium criteria are satisfied.

The State Postulate

As noted earlier, the state of a system is described by its properties. But we know from experience that we do not need to specify all the properties in order to fix a state. Once a sufficient number of properties are specified, the rest of the properties assume certain values automatically. That is, specifying a certain number of properties is sufficient to fix a state. The number of properties required to fix the state of a system is given by the **state postulate:**

> *The state of a simple compressible system is completely specified by two independent, intensive properties.*

A system is called a **simple compressible system** in the absence of electrical, magnetic, gravitational, motion, and surface tension effects. These effects are due to external force fields and are negligible for most engineering problems. Otherwise, an additional property needs to be specified for each effect that is significant. If the gravitational effects are to be considered, for example, the elevation z needs to be specified in addition to the two properties necessary to fix the state.

The state postulate requires that the two properties specified be independent to fix the state. Two properties are **independent** if one property can be varied while the other one is held constant. Temperature and specific volume, for example, are always independent properties, and together they can fix the state of a simple compressible system (Fig. 2-3). Temperature and pressure, however, are independent properties for single-phase systems, but are dependent properties for multiphase systems. At sea level ($P = 1$ atm), water boils at 100°C, but on a mountaintop where the pressure is lower, water boils at a lower temperature. That is, $T = f(P)$ during a phase-change process; thus, temperature and pressure are not sufficient to fix the state of a two-phase system. Phase-change processes are discussed in detail in the next chapter.

2-2 ■ PROCESSES AND CYCLES

Any change that a system undergoes from one equilibrium state to another is called a **process,** and the series of states through which a system passes during a process is called the **path** of the process (Fig. 2-4). To describe a process completely, one should specify the initial and final states of the process, as well as the path it follows, and the interactions with the surroundings.

When a process proceeds in such a manner that the system remains infinitesimally close to an equilibrium state at all times, it is called a **quasi-static,** or **quasi-equilibrium, process.** A quasi-equilibrium process can be viewed as a sufficiently slow process that allows the system to adjust itself internally so that properties in one part of the system do not change any faster than those at other parts.

This is illustrated in Fig. 2-5. When a gas in a piston-cylinder device is compressed suddenly, the molecules near the face of the piston will not have enough time to escape and they will have to pile up in a small region in front of the piston, thus creating a high-pressure region there. Because of this pressure difference, the system can no longer be said to be in equilibrium, and this makes the entire process non-quasi-equilibrium. However, if the piston is moved slowly, the molecules will have sufficient time to redistribute and there will not be a molecule pileup in front of the piston. As a result, the pressure inside the cylinder will always be uniform and will rise at the same rate at all locations. Since equilibrium is maintained at all times, this is a quasi-equilibrium process.

It should be pointed out that a quasi-equilibrium process is an idealized process and is not a true representation of an actual process. But many actual processes closely approximate it, and they can be modeled as quasi-equilibrium with negligible error. Engineers are interested in quasi-equilibrium processes for two reasons. First, they are easy to analyze; second, work-producing devices deliver the most work when they operate on quasi-equilibrium processes (Fig. 2-6). Therefore, quasi-equilibrium processes serve as standards to which actual processes can be compared.

Process diagrams plotted by employing thermodynamic properties as co-ordinates are very useful in visualizing the processes. Some common properties that are used as coordinates are temperature T, pressure P, and volume V (or specific volume v). Figure 2-7 shows the P-V diagram of a compression process of a gas.

Note that the process path indicates a series of equilibrium states through which the system passes during a process and has significance for quasi-equilibrium processes only. For non-quasi-equilibrium processes, we are not able to specify the states through which the system passes during the process and so we cannot speak of a process path. A non-quasi-equilibrium process is denoted by a dashed line between the initial and final states instead of a solid line.

The prefix *iso-* is often used to designate a process for which a particular property remains constant. An **isothermal process,** for example, is a process during which the temperature T remains constant; an **isobaric process** is a

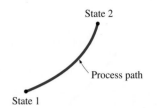

FIGURE 2-4

A process between states 1 and 2 and the process path.

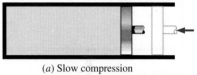

(*a*) Slow compression
(quasi-equilibrium)

(*b*) Very fast compression
(non-quasi-equilibrium)

FIGURE 2-5

Quasi-equilibrium and non-quasi-equilibrium compression processes.

FIGURE 2-6

Work-producing devices operating in a quasi-equilibrium manner deliver the most work.

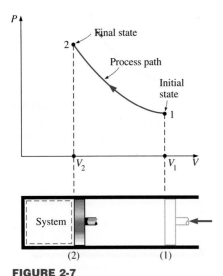

FIGURE 2-7

The *P-V* diagram of a
compression process.

FIGURE 2-8

The macroscopic energy of an object
changes with velocity and elevation.

process during which the pressure *P* remains constant; and an **isochoric** (or **isometric**) **process** is a process during which the specific volume *v* remains constant.

A system is said to have undergone a **cycle** if it returns to its initial state at the end of the process. That is, for a cycle the initial and final states are identical.

2-3 ■ FORMS OF ENERGY

Energy can exist in numerous forms such as thermal, mechanical, kinetic, potential, electric, magnetic, chemical, and nuclear, and their sum constitutes the **total energy** *E* of a system. The total energy of a system on a *unit mass* basis is denoted by *e* and is defined as

$$e = \frac{E}{m} \qquad \text{(kJ/kg)} \tag{2-1}$$

Thermodynamics provides no information about the absolute value of the total energy. It only deals with the *change* of the total energy, which is what matters in engineering problems. Thus the total energy of a system can be assigned a value of zero ($E = 0$) at some convenient reference point. The change in total energy of a system is independent of the reference point selected. The decrease in the potential energy of a falling rock, for example, depends on only the elevation difference and not the reference level selected.

In thermodynamic analysis, it is often helpful to consider the various forms of energy that make up the total energy of a system in two groups: *macroscopic* and *microscopic*. The **macroscopic** forms of energy, on one hand, are those a system possesses as a whole with respect to some outside reference frame, such as kinetic and potential energies (Fig. 2-8). The **microscopic** forms of energy, on the other hand, are those related to the molecular structure of a system and the degree of the molecular activity, and they are independent of outside reference frames. The sum of all the microscopic forms of energy is called the **internal energy** of a system and is denoted by *U*.

The term *energy* was coined in 1807 by Thomas Young, and its use in thermodynamics was proposed in 1852 by Lord Kelvin. The term *internal energy* and its symbol *U* first appeared in the works of Rudolph Clausius and William Rankine in the second half of the nineteenth century, and it eventually replaced the alternative terms *inner work, internal work,* and *intrinsic energy* commonly used at the time.

The macroscopic energy of a system is related to motion and the influence of some external effects such as gravity, magnetism, electricity, and surface tension. The energy that a system possesses as a result of its motion relative to some reference frame is called **kinetic energy** KE. When all parts of a system move with the same velocity, the kinetic energy is expressed as

$$KE = \frac{m\mathcal{V}^2}{2} \qquad \text{(kJ)} \tag{2-2}$$

or, on a unit mass basis,

$$ke = \frac{\mathcal{V}^2}{2} \qquad \text{(kJ/kg)} \tag{2-3}$$

where the script $\mathcal{V}$ denotes the velocity of the system relative to some fixed reference frame. The kinetic energy of a rotating body is given by $\frac{1}{2}I\omega^2$ where *I* is the moment of inertia of the body and ω is the angular velocity.

The energy that a system possesses as a result of its elevation in a gravitational field is called **potential energy** PE and is expressed as

$$PE = mgz \qquad \text{(kJ)} \qquad (2\text{-}4)$$

or, on a unit mass basis,

$$pe = gz \qquad \text{(kJ/kg)} \qquad (2\text{-}5)$$

where g is the gravitational acceleration and z is the elevation of the center of gravity of a system relative to some arbitrarily selected reference plane.

The magnetic, electric, and surface tension effects are significant in some specialized cases only and are usually ignored. In the absence of these effects, the total energy of a system consists of the kinetic, potential, and internal energies and is expressed as

$$E = U + KE + PE = U + \frac{m\mathcal{V}^2}{2} + mgz \qquad \text{(kJ)} \qquad (2\text{-}6)$$

or, on a unit mass basis,

$$e = u + ke + pe = u + \frac{\mathcal{V}^2}{2} + gz \qquad \text{(kJ/kg)} \qquad (2\text{-}7)$$

Most closed systems remain stationary during a process and thus experience no change in their kinetic and potential energies. Closed systems whose velocity and elevation of the center of gravity remain constant during a process are frequently referred to as **stationary systems.** The change in the total energy ΔE of a stationary system is identical to the change in its internal energy ΔU. In this text, a closed system is assumed to be stationary unless it is specifically stated otherwise.

Some Physical Insight to Internal Energy

Internal energy is defined above as the sum of all the *microscopic* forms of energy of a system. It is related to the *molecular structure* and the degree of *molecular activity* and may be viewed as the sum of the *kinetic* and *potential* energies of the molecules.

To have a better understanding of internal energy, let us examine a system at the molecular level. The molecules of a gas move through space with some velocity, and thus possess some kinetic energy. This is known as the *translational energy*. The atoms of polyatomic molecules rotate about an axis, and the energy associated with this rotation is the *rotational kinetic energy*. The atoms of a polyatomic molecule may also vibrate about their common center of mass, and the energy associated with this back-and-forth motion is the *vibrational kinetic energy*. For gases, the kinetic energy is mostly due to translational and rotational motions, with vibrational motion becoming significant at higher temperatures. The electrons in an atom rotate about the nucleus, and thus possess *rotational kinetic energy*. Electrons at outer orbits have larger kinetic energies. Electrons also spin about their axes, and the energy associated with this motion is the *spin energy*. Other particles in the nucleus of an atom also possess spin energy. The portion of the internal energy of a system associated with the kinetic energies of the molecules is called the **sensible energy** (Fig. 2-9). The average velocity and the degree of activity of the molecules are

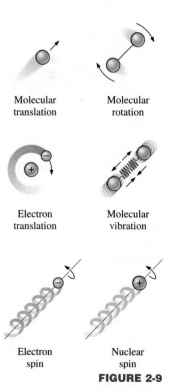

Molecular
translation

Molecular
rotation

Electron
translation

Molecular
vibration

Electron
spin

Nuclear
spin

FIGURE 2-9

The various forms of microscopic energies that make up *sensible* energy.

proportional to the temperature of the gas. Therefore, at higher temperatures, the molecules will possess higher kinetic energies, and as a result the system will have a higher internal energy.

The internal energy is also associated with various *binding forces* between the molecules of a substance, between the atoms within a molecule, and between the particles within an atom and its nucleus. The forces that bind the *molecules* to each other are, as one would expect, strongest in solids and weakest in gases. If sufficient energy is added to the molecules of a solid or liquid, they will overcome these molecular forces and break away, turning the substance into a gas. This is a phase-change process. Because of this added energy, a system in the gas phase is at a higher internal energy level than it is in the solid or the liquid phase. The internal energy associated with the phase of a system is called the **latent energy.** The phase-change process can occur without a change in the chemical composition of a system. Most practical problems fall into this category, and one does not need to pay any attention to the forces binding the atoms in a molecule to each other.

An atom consists of positively charged protons and neutrons bound together by very strong nuclear forces in the nucleus, and electrons orbiting around it. The internal energy associated with the atomic bonds in a molecule is called **chemical energy.** During a chemical reaction, such as a combustion process, some chemical bonds are destroyed while others are formed. As a result, the internal energy changes. The nuclear forces are much larger than the forces that bind the electrons to the nucleus. The tremendous amount of energy associated with the strong bonds within the nucleus of the atom itself is called **nuclear energy** (Fig. 2-10). Obviously, we need not be concerned with nuclear energy in thermodynamics unless, of course, we have a fusion or fission reaction on our hands. A chemical reaction involves changes in the structure of the electrons of the atoms, but a nuclear reaction involves changes in the core or nucleus. Therefore, an atom preserves its identity during a chemical reaction but loses it during a nuclear reaction. Atoms may also possess *electric* and *magnetic dipole-moment energies* when subjected to external electric and magnetic fields due to the twisting of the magnetic dipoles produced by the small electric currents associated with the orbiting electrons.

The forms of energy discussed above, which constitute the total energy of a system, can be *contained* or *stored* in a system, and thus can be viewed as the *static* forms of energy. The forms of energy not stored in a system can be viewed as the *dynamic* forms of energy or as *energy interactions*. The dynamic forms of energy are recognized at the system boundary as they cross it, and they represent the energy gained or lost by a system during a process. The only two forms of energy interactions associated with a closed system are **heat transfer** and **work.** An energy interaction is heat transfer if its driving force is a temperature difference. Otherwise it is work, as explained in Chap. 4. A control volume can also exchange energy via mass transfer since any time mass is transferred into or out of a system, the energy contained in the mass is also transferred with it.

In daily life, we frequently refer to the sensible and latent forms of internal energy as *heat,* and we talk about heat content of bodies. In thermodynamics, however, we usually refer to those forms of energy as **thermal energy** to prevent any confusion with *heat transfer.*

Distinction should be made between the macroscopic kinetic energy of an object as a whole and the microscopic kinetic energies of its molecules that constitute the sensible internal energy of the object (Fig. 2-11). The kinetic

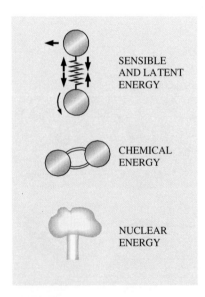

SENSIBLE
AND LATENT
ENERGY

CHEMICAL
ENERGY

NUCLEAR
ENERGY

FIGURE 2-10

The internal energy of a system is the sum of all forms of the microscopic energies.

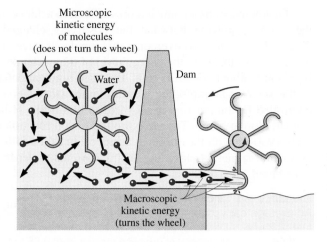

Microscopic
kinetic energy
of molecules
(does not turn the wheel)

Dam

Water

Macroscopic
kinetic energy
(turns the wheel)

Forms of Energy

FIGURE 2-11

The *macroscopic* kinetic energy is an organized form of energy and is much more useful than the disorganized *microscopic* kinetic energies of the molecules.

energy of an object is an *organized* form of energy associated with the orderly motion of all molecules in one direction in a straight path or around an axis. In contrast, the kinetic energies of the molecules are completely *random* and highly *disorganized*. As you will see in later chapters, the organized energy is much more valuable than the disorganized energy, and a major application area of thermodynamics is the conversion of disorganized energy (heat) into organized energy (work). You will also see that the organized energy can be converted to disorganized energy completely, but only a fraction of disorganized energy can be converted to organized energy by specially built devices called *heat engines* (like car engines and power plants). A similar argument can be given for the macroscopic potential energy of an object as a whole and the microscopic potential energies of the molecules.

More on Nuclear Energy

The best known fission reaction involves the split of the uranium atom (the U-235 isotope) into other elements, and is commonly used to generate electricity in nuclear power plants (429 of them in 1990, generating 311,000 MW worldwide), to power nuclear submarines and aircraft carriers, and even to power spacecraft as well as building nuclear bombs. The first nuclear chain reaction was achieved by Enrico Fermi in 1942, and the first large-scale nuclear reactors were built in 1944 for the purpose of producing material for nuclear weapons. When a uranium-235 atom absorbs a neutron and splits during a fission process, it produces a cesium-140 atom, a rubidium-93 atom, 3 neutrons, and 3.2×10^{-11} J of energy. In practical terms, the complete fission of 1 kg of uranium-235 releases 6.73×10^{10} kJ of heat, which is more than the heat released when 3000 tons of coal are burned. Therefore, for the same amount of fuel, a nuclear fission reaction releases several million times more energy than a chemical reaction. The safe disposal of used nuclear fuel, however, remains a concern.

Nuclear energy by fusion is released when two small nuclei combine into a larger one. The huge amount of energy radiated by the sun and the other stars originates from such a fusion process that involves the combination of two hydrogen atoms into a helium atom. When two heavy hydrogen (deuterium) nuclei combine during a fusion process, they produce a helium-3 atom, a free neutron, and 5.1×10^{-13} J of energy (Fig. 2-12).

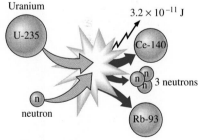

(a) Fission of uranium

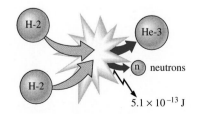

(a) Fusion of hydrogen

FIGURE 2-12

The fission of uranium and the fusion of hydrogen during nuclear reactions, and the release of nuclear energy.

Fusion reactions are much more difficult to achieve in practice because of the strong repulsion between the positively charged nuclei, called the *Coulomb repulsion.* To overcome this repulsive force and to enable the two nuclei to fuse together, the energy level of the nuclei must be raised by heating them to about 100 million °C. But such high temperatures are found only in the stars or in exploding atomic bombs (the A-bomb). In fact, the uncontrolled fusion reaction in a hydrogen bomb (the H-bomb) is initiated by a small atomic bomb. The uncontrolled fusion reaction was achieved in the early 1930s, but all the efforts to achieve controlled fusion by massive lasers, powerful magnetic fields, and electric currents to generate power since then have failed.

EXAMPLE 2-1   A Car Powered by Nuclear Fuel

An average car consumes about 5 L of gasoline a day, and the capacity of the fuel tank of a car is about 50 L. Therefore, a car needs to be refueled once every 10 days. Also, the density of gasoline ranges from 0.68 to 0.78 kg/L, and its lower heating value is about 44,000 kJ/kg (that is, 44,000 kJ of heat is released when 1 kg of gasoline is completely burned). Suppose all the problems associated with the radioactivity and waste disposal of nuclear fuels are resolved, and a car is to be powered by U-235. If a new car comes equipped with 0.1-kg of the nuclear fuel U-235, determine if this car will ever need refueling under average driving conditions (Fig. 2-13).

Nuclear
fuel

FIGURE 2-13
Schematic for Example 2-1.

Solution   A car powered by nuclear energy comes equipped with nuclear fuel. It is to be determined if this car will ever need refueling.

Assumptions   **1** Gasoline is an incompressible substance with an average density of 0.75 kg/L. **2** Nuclear fuel is completely converted to thermal energy.

Analysis   The mass of gasoline used per day by the car is

$$m_{gasoline} = (\rho V)_{gasoline} = (0.75 \text{ kg/L}) (5 \text{ L/day}) = 3.75 \text{ kg/day}$$

Noting that the heating value of gasoline is 44,000 kJ/kg, the energy supplied to the car per day is

$$E = (m_{gasoline}) (\text{Heating value}) = (3.75 \text{ kg/day}) (44,000 \text{ kJ/kg}) = 165,000 \text{ kJ/day}$$

The complete fission of 0.1 kg of uranium-235 releases

$$(6.73 \times 10^{10} \text{ kJ/kg})(0.1 \text{ kg}) = 6.73 \times 10^9 \text{ kJ}$$

of heat, which is sufficient to meet the energy needs of the car for

$$\text{No. of days} = \frac{\text{Energy content of fuel}}{\text{Daily energy use}} = \frac{6.73 \times 10^9 \text{ kJ}}{165,000 \text{ kJ/day}} = \textbf{40,790 days}$$

which is equivalent to about 112 years. Considering that no car will last more than 100 years, this car will never need refueling. It appears that nuclear fuel the size of a cherry is sufficient to power a car during its lifetime.

2-4 ■ TEMPERATURE AND THE ZEROTH LAW OF THERMODYNAMICS

Although we are familiar with temperature as a measure of "hotness" or "coldness," it is not easy to give an exact definition for it. Based on our physiological sensations, we express the level of temperature qualitatively with words like *freezing cold, cold, warm, hot,* and *red-hot.* However, we cannot assign numerical values to temperatures based on our sensations alone.

Furthermore, our senses may be misleading. A metal chair, for example, will feel much colder than a wooden one even when both are at the same temperature.

Fortunately, several properties of materials change with temperature in a *repeatable* and *predictable* way, and this forms the basis for accurate temperature measurement. The commonly used mercury-in-glass thermometer, for example, is based on the expansion of mercury with temperature. Temperature is also measured by using several other temperature-dependent properties.

It is a common experience that a cup of hot coffee left on the table eventually cools off and a cold drink eventually warms up. That is, when a body is brought into contact with another body that is at a different temperature, heat is transferred from the body at higher temperature to the one at lower temperature until both bodies attain the same temperature (Fig. 2-14). At that point, the heat transfer stops, and the two bodies are said to have reached **thermal equilibrium.** The equality of temperature is the only requirement for thermal equilibrium.

The **zeroth law of thermodynamics** states that if two bodies are in thermal equilibrium with a third body, they are also in thermal equilibrium with each other. It may seem silly that such an obvious fact is called one of the basic laws of thermodynamics. However, it cannot be concluded from the other laws of thermodynamics, and it serves as a basis for the validity of temperature measurement. By replacing the third body with a thermometer, the zeroth law can be restated as *two bodies are in thermal equilibrium if both have the same temperature reading even if they are not in contact.*

The zeroth law was first formulated and labeled by R. H. Fowler in 1931. As the name suggests, its value as a fundamental physical principle was recognized more than half a century after the formulation of the first and the second laws of thermodynamics. It was named the zeroth law since it should have preceded the first and the second laws of thermodynamics.

Temperature Scales

Temperature scales enable scientists to use a common basis for temperature measurements, and several have been introduced throughout history. All temperature scales are based on some easily reproducible states such as the freezing and boiling points of water, which are also called the *ice point* and the *steam point,* respectively. A mixture of ice and water that is in equilibrium with air saturated with vapor at 1 atm pressure is said to be at the ice point, and a mixture of liquid water and water vapor (with no air) in equilibrium at 1 atm pressure is said to be at the steam point.

The temperature scales used in the SI and in the English system today are the **Celsius scale** (formerly called the *centigrade scale;* in 1948 it was renamed after the Swedish astronomer A. Celsius, 1701–1744, who devised it) and the **Fahrenheit scale** (named after the German instrument maker G. Fahrenheit, 1686-1736), respectively. On the Celsius scale, the ice and steam points are assigned the values of 0 and 100°C, respectively. The corresponding values on the Fahrenheit scale are 32 and 212°F. These are often referred to as *two-point scales* since temperature values are assigned at two different points.

In thermodynamics, it is very desirable to have a temperature scale that is independent of the properties of any substance or substances. Such a temperature scale is called a **thermodynamic temperature scale,** which is developed in Chap. 6 in conjunction with the second law of thermodynamics. The

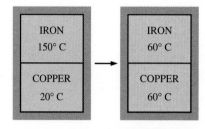

FIGURE 2-14

Two bodies reaching thermal equilibrium after being brought into contact in an isolated enclosure.

thermodynamic temperature scale in the SI is the **Kelvin scale,** named after Lord Kelvin (1824–1907). The temperature unit on this scale is the **kelvin,** which is designated by K (not °K; the degree symbol was officially dropped from kelvin in 1967). The lowest temperature on the Kelvin scale is 0 K. Using nonconventional refrigeration techniques, scientists have approached absolute zero kelvin (they achieved 0.000000002 K in 1989).

The thermodynamic temperature scale in the English system is the **Rankine scale,** named after William Rankine (1820–1872). The temperature unit on this scale is the **rankine,** which is designated by R.

A temperature scale that turns out to be identical to the Kelvin scale is the **ideal gas temperature scale.** The temperatures on this scale are measured using a **constant-volume gas thermometer,** which is basically a rigid vessel filled with a gas, usually hydrogen or helium, at low pressure. This thermometer is based on the principle that *at low pressures, the temperature of a gas is proportional to its pressure at constant volume.* That is, the temperature of a gas of fixed volume varies *linearly* with pressure at sufficiently low pressures. Then the relationship between the temperature and the pressure of the gas in the vessel can be expressed as

$$T = a + bP \qquad (2\text{-}8)$$

where the values of the constants a and b for a gas thermometer are determined experimentally. Once a and b are known, the temperature of a medium can be calculated from the relation above by immersing the rigid vessel of the gas thermometer into the medium and measuring the gas pressure when thermal equilibrium is established between the medium and the gas in the vessel whose volume is held constant.

An ideal gas temperature scale can be developed by measuring the pressures of the gas in the vessel at two reproducible points (such as the ice and the steam points) and assigning suitable values to temperatures at those two points. Considering that only one straight line passes through two fixed points on a plane, these two measurements are sufficient to determine the constants a and b in Eq. 2-8. Then the unknown temperature T of a medium corresponding to a pressure reading P can be determined from that equation by a simple calculation. The values of the constants will be different for each thermometer, depending on the type and the amount of the gas in the vessel, and the temperature values assigned at the two reference points. If the ice and steam points are assigned the values 0 and 100, respectively, then the gas temperature scale will be identical to the Celsius scale. In this case the value of the constant a (which corresponds to an absolute pressure of zero) is determined to be −273.15°C regardless of the type and the amount of the gas in the vessel of the gas thermometer. That is, on a *P-T* diagram, all the straight lines passing through the data points in this case will intersect the temperature axis at −273.15°C when extrapolated, as shown in Fig. 2-15. This is the lowest temperature that can be obtained by a gas thermometer, and thus we can obtain an *absolute gas temperature scale* by assigning a value of zero to the constant a in Eq. 2-8. In that case Eq. 2-8 reduces to $T = bP$, and thus we need to specify the temperature at only *one* point to define an absolute gas temperature scale.

It should be noted that the absolute gas temperature scale is not a thermodynamic temperature scale, since it cannot be used at very low temperatures (due to condensation) and at very high temperatures (due to dissociation and ionization). However, absolute gas temperature is identical to the thermodynamic temperature in the temperature range in which the gas thermometer can

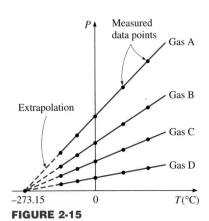

FIGURE 2-15

P versus *T* plots of the experimental data obtained from a constant-volume gas thermometer using four different gases at different (but low) pressures.

be used, and thus we can view the thermodynamic temperature scale at this point as an absolute gas temperature scale that utilizes an "ideal" or "imaginary" gas that always acts as a low-pressure gas regardless of the temperature. If such a gas thermometer existed, it would read zero kelvin at absolute zero pressure, which corresponds to $-273.15°C$ on the Celsius scale (Fig. 2-16).

The Kelvin scale is related to the Celsius scale by

$$T(K) = T(°C) + 273.15 \qquad (2-9)$$

The Rankine scale is related to the Fahrenheit scale by

$$T(R) = T(°F) + 459.67 \qquad (2-10)$$

It is common practice to round the constant in Eq. 2-9 to 273 and that in Eq. 2-10 to 460.

The temperature scales in the two unit systems are related by

$$T(R) = 1.8 \, T(K) \qquad (2-11)$$
$$T(°F) = 1.8 \, T(°C) + 32 \qquad (2-12)$$

A comparison of various temperature scales is given in Fig. 2-17.

At the Tenth Conference on Weights and Measures in 1954, the Celsius scale was redefined in terms of a single fixed point and the absolute temperature scale. The selected single point is the *triple point* of water (the state at which all three phases of water coexist in equilibrium), which is assigned the value 0.01°C. The magnitude of the degree is defined from the absolute temperature scale. As before, the boiling point of water at 1 atm pressure is 100.00°C. Thus the new Celsius scale is essentially the same as the old one.

On the Kelvin scale, the size of the temperature unit *kelvin* is defined as "the fraction 1/273.16 of the thermodynamic temperature of the triple point of water, which is assigned the value of 273.16 K." The ice point on the Celsius and Kelvin scales are 0°C and 273.15 K, respectively.

Note that the magnitudes of each division of 1 K and 1°C are identical (Fig. 2-18). Therefore, when we are dealing with temperature differences ΔT, the temperature interval on both scales is the same. Raising the temperature of a substance by 10°C is the same as raising it by 10 K. That is,

$$\Delta T(K) = \Delta T(°C) \qquad (2-13)$$
$$\Delta T(R) = \Delta T(°F) \qquad (2-14)$$

Some thermodynamic relations involve the temperature T and often the question arises of whether it is in K or °C. If the relation involves temperature differences (such as $a = b\Delta T$), it makes no difference and either can be used. But if the relation involves temperatures only instead of temperature differences (such as $a = bT$) then K must be used. When in doubt, it is always safe to use K because there are virtually no situations in which the use of K is incorrect, but there are many thermodynamic relations that will yield an erroneous result if °C is used.

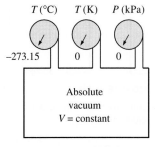

FIGURE 2-16

A constant-volume gas thermometer would read $-273.15°C$ at absolute zero pressure.

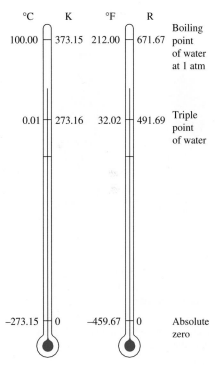

FIGURE 2-17

Comparison of temperature scales.

EXAMPLE 2-2 Expressing Temperature Rise in Different Units
During a heating process, the temperature of a system rises by 10°C. Express this rise in temperature in K, °F, and R.

FIGURE 2-18

Comparison of magnitudes of various temperature units.

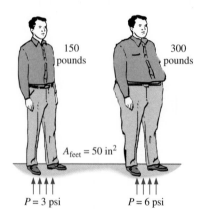

$$P = \sigma_n = \frac{W}{A_{\text{feet}}} = \frac{150 \text{ psi}}{50 \text{ in}^2} = 3 \text{ psi}$$

FIGURE 2-19

The normal stress (or "pressure") on the feet of a chubby person is much greater than that of a slim person.

FIGURE 2-20

A pressure gage open to the atmosphere reads zero.

Solution   The temperature rise of a system is to be expressed in different units.

Analysis   This problem deals with temperature changes, which are identical in Kelvin and Celsius scales. Then from Eq. 2-13,

$$\Delta T(\text{K}) = \Delta T(^\circ\text{C}) = \mathbf{10 \ K}$$

The temperature changes in Fahrenheit and Rankine scales are also identical and are related to the changes in Celsius and Kelvin scales through Eqs. 2-11 and 2-14.

$$\Delta T(\text{R}) = 1.8 \ \Delta T(\text{K}) = (1.8)(10) = \mathbf{18 \ R}$$

and

$$\Delta T(^\circ\text{F}) = \Delta T(\text{R}) = \mathbf{18^\circ F}$$

2-5 ■ PRESSURE

Pressure is defined as *the force exerted by a fluid per unit area.* We speak of pressure only when we deal with a gas or a liquid. The counterpart of pressure in solids is *stress.* Since pressure is defined as force per unit area, it has the unit of newtons per square meter (N/m^2), which is called a **pascal** (Pa). That is,

$$1 \text{ Pa} = 1 \text{ N/m}^2$$

The pressure unit pascal is too small for pressures encountered in practice. Therefore, its multiples *kilopascal* ($1 \text{ kPa} = 10^3 \text{ Pa}$) and *megapascal* ($1 \text{ MPa} = 10^6 \text{ Pa}$) are commonly used. Three other pressure units commonly used in practice, especially in Europe, are *bar, standard atmosphere,* and *kilogram-force per square centimeter:*

$$1 \text{ bar} = 10^5 \text{ Pa} = 0.1 \text{ MPa} = 100 \text{ kPa}$$

$$1 \text{ atm} = 101{,}325 \text{ Pa} = 101.325 \text{ kPa} = 1.01325 \text{ bars}$$

$$1 \text{ kgf/cm}^2 = 9.807 \text{ N/cm}^2 = 9.807 \times 10^4 \text{ N/m}^2 = 9.807 \times 10^4 \text{ Pa}$$

$$= 0.9807 \text{ bar}$$

$$= 0.96788 \text{ atm}$$

Note the pressure units bar, atm, and kgf/cm^2 are almost equivalent to each other. In the English system, the pressure unit is *pound-force per square inch* (lbf/in^2, or psi), and 1 atm = 14.696 psi. The pressure units kgf/cm^2 and lbf/in^2 are also denoted by kg/cm^2 and lb/in^2, respectively, and they are commonly used in tire gages. It can be shown that $1 \text{ kgf/cm}^2 = 14.223$ psi.

Pressure is also used for solids as synonymous to *normal stress,* which is force acting perpendicular to the surface per unit area. For example, a 150-pound person with a total foot imprint area of 50 in^2 will exert a pressure of 150/50 = 3.0 psi. If the person stands on one foot, the pressure will double (Fig. 2-19). If the person gains excessive weight, he or she is likely to encounter foot discomfort because of the increased pressure on the foot (the size of the foot does not change with weight gain). This also explains how a person can walk on fresh snow without sinking by wearing large snowshoes, and how a person cuts with little effort when using a sharp knife.

The actual pressure at a given position is called the **absolute pressure,** and it is measured relative to absolute vacuum (i.e., absolute zero pressure.) Most pressure-measuring devices, however, are calibrated to read zero in the atmosphere (Fig. 2-20), and so they indicate the difference between the absolute pressure and the local atmospheric pressure. This difference is called the **gage pressure.** Pressures below atmospheric pressure are called **vacuum pressures** and are measured by vacuum gages that indicate the difference between the atmospheric pressure and the absolute pressure. Absolute, gage, and vacuum pressures are all positive quantities and are related to each other by

$$P_{\text{gage}} = P_{\text{abs}} - P_{\text{atm}} \qquad \text{(for pressures above } P_{\text{atm}}\text{)} \qquad (2\text{-}15)$$
$$P_{\text{vac}} = P_{\text{atm}} - P_{\text{abs}} \qquad \text{(for pressures below } P_{\text{atm}}\text{)} \qquad (2\text{-}16)$$

This is illustrated in Fig. 2-21.

Like other pressure gages, the gage used to measure the air pressure in an automobile tire reads the gage pressure. Therefore, the common reading of 32 psi (2.25 kgf/cm^2) indicates a pressure of 32 psi above the atmospheric pressure. At a location where the atmospheric pressure is 14.3 psi, for example, the absolute pressure in the tire will be $32 + 14.3 = 46.3$ psi.

In thermodynamic relations and tables, absolute pressure is almost always used. Throughout this text, the pressure P will denote *absolute pressure* unless specified otherwise. Often the letters "a" (for absolute pressure) and "g" (for gage pressure) are added to pressure units (such as psia and psig) to clarify what is meant.

EXAMPLE 2-3   Absolute Pressure of a Vacuum Chamber
A vacuum gage connected to a chamber reads 5.8 psi at a location where the atmospheric pressure is 14.5 psi. Determine the absolute pressure in the chamber.

Solution   The gage pressure of a vacuum chamber is given. The absolute pressure in the chamber is to be determined.

Analysis   The absolute pressure is easily determined from Eq. 2-16 to be
$$P_{\text{abs}} = P_{\text{atm}} - P_{\text{vac}} = 14.5 - 5.8 = 8.7 \text{ psi}$$

Pressure at a Point

Pressure is the *compressive force* per unit area, and it gives the impression of being a vector. However, pressure at any point in a fluid is the same in all directions. That is, it has magnitude but not a specific direction, and thus it is a scalar quantity. This can be demonstrated by considering a small wedge-shaped fluid element of unit length (into the paper) in equilibrium, as shown in

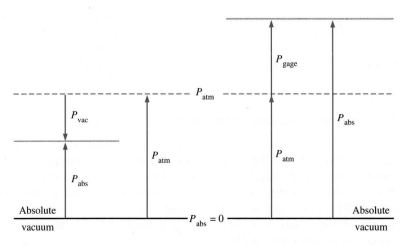

FIGURE 2-21
Absolute, gage, and vacuum pressures.

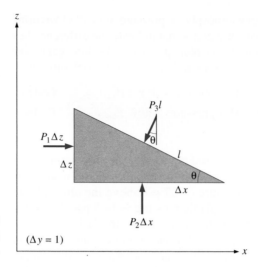

FIGURE 2-22

Forces acting on a wedge-shaped fluid
element in equilibrium.

Fig. 2-22. The mean pressures at the three surfaces are P_1, P_2, and P_3, and the force acting on a surface is the product of mean pressure and the surface area. From Newton's second law, a force balance in the x- and z-directions gives

$$\sum F_x = ma_x = 0: \qquad P_1 \Delta z - P_3\, l \sin\theta = 0 \qquad (2\text{-}17a)$$

$$\sum F_z = ma_z = 0: \qquad P_2 \Delta x - P_3\, l \cos\theta - \frac{1}{2}\rho g\, \Delta x\, \Delta z = 0 \qquad (2\text{-}17b)$$

where ρ is the density and $W = mg = \rho g\, \Delta x\, \Delta z/2$ is the weight of the fluid element. Noting that the wedge is a right triangle, we have $\Delta x = l\cos\theta$ and $\Delta z = l\sin\theta$. Substituting these geometric relations and dividing Eq. 2-17a by Δz and Eq. 2-17b by Δx gives

$$P_1 - P_3 = 0 \qquad (2\text{-}18a)$$

$$P_2 - P_3 - \frac{1}{2}\rho g\, \Delta z = 0 \qquad (2\text{-}18b)$$

The last term in Eq. 2-18b drops out as $\Delta z \to 0$ and the wedge becomes infinitesimal, and thus the fluid element shrinks to a point. Then combining the results of these two relations gives

$$P_1 = P_2 = P_3 = P \qquad (2\text{-}19)$$

regardless of the angle θ. We can repeat the analysis for an element in the xz-plane, and obtain a similar result. Thus we conclude that *the pressure at a point in a fluid has the same magnitude in all directions.* It can be shown in the absence of shear forces that this result is applicable to fluids in motion as well as fluids at rest.

Variation of Pressure with Depth

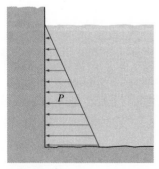

FIGURE 2-23

The pressure of a fluid at rest increases
with depth (as a result of added weight).

It will come as no surprise to you that pressure in a fluid does not change in the horizontal direction. This can be shown easily by considering a thin horizontal layer of fluid, and doing a force balance in any horizontal direction. However, this is not the case in the vertical direction in a gravity field. Pressure in a fluid increases with depth because more fluid rests on deeper layers, and the effect of this "extra weight" on a deeper layer is balanced by an increase in pressure (Fig. 2-23).

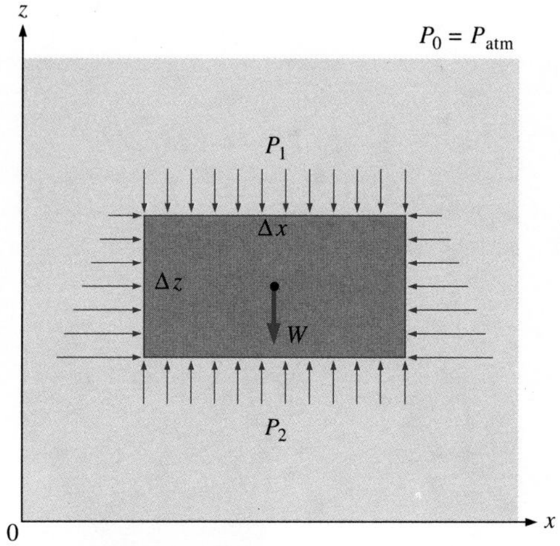

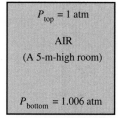

Figure 2-24 shows the free-body diagram with z axis, $P_0 = P_{atm}$ at top, P_1, Δx, Δz, W, P_2 labels.

FIGURE 2-24

Free-body diagram of a rectangular fluid element in equilibrium.

To obtain a relation for the variation of pressure with depth, consider a rectangular fluid element of height Δz, length Δx, and unit depth (into the paper) in equilibrium, as shown in Fig. 2-24. Assuming the density of the fluid ρ to be constant, a force balance in the vertical z-direction gives

$$\sum F_z = ma_z = 0: \qquad P_2 \, \Delta x - P_1 \, \Delta x - \rho g \, \Delta x \, \Delta z = 0 \qquad (2\text{-}20)$$

where $W = mg = \rho g \, \Delta x \, \Delta z$ is the weight of the fluid element. Dividing by Δx and rearranging gives

$$\Delta P = P_2 - P_1 = \rho g \, \Delta z = \gamma \, \Delta z \qquad (2\text{-}21)$$

where $\gamma = \rho g$ is the *specific weight* of the fluid. Thus we conclude that the pressure difference between two points in a constant density fluid is proportional to the vertical distance Δz between the points and the density ρ of the fluid. In other words, pressure in a fluid increases linearly with depth. This is what a diver will experience when diving deeper in a lake. For a given fluid, the vertical distance Δz is sometimes used as a measure of pressure, and it is called the pressure head.

We also conclude from Eq. 2-21 that for small to moderate distances, the variation of pressure with height is negligible for gases because of their low density. The pressure in a tank containing a gas, for example, may be considered to be uniform since the weight of the gas is too small to make a significant difference. Also, the pressure in a room filled with air may be assumed to be constant (Fig. 2-25).

If we take point 1 to be at the free surface of a liquid open to the atmosphere, where the pressure is the atmospheric pressure P_{atm}, then the pressure at a depth h from the free surface becomes

$$P = P_{atm} + \rho g h \qquad \text{or} \qquad P_{gage} = \rho g h \qquad (2\text{-}22)$$

Liquids are essentially incompressible substances, and thus the variation of density with depth is negligible. This is also the case for gases when the elevation change is not very large. The variation of density of liquids or gases with temperature can be significant, however, and may need to be considered when high accuracy is desired. Also, at great depths such as those encountered in oceans, the change in the density of a liquid can be significant because of the compression by the tremendous amount of liquid weight above.

$P_{top} = 1 \text{ atm}$

AIR
(A 5-m-high room)

$P_{bottom} = 1.006 \text{ atm}$

FIGURE 2-25

In a container filled with a gas, the variation of pressure with height is negligible.

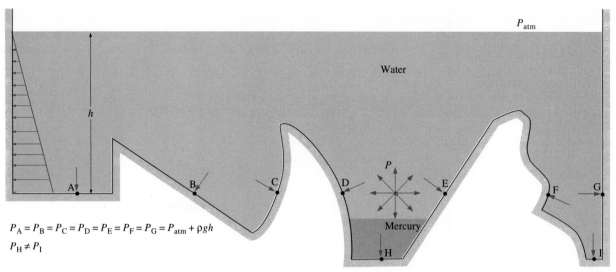

$P_A = P_B = P_C = P_D = P_E = P_F = P_G = P_{atm} + \rho gh$

$P_H \neq P_I$

FIGURE 2-26

The pressure is the same at all points on a horizontal plane in a given fluid regardless of geometry, provided that the points are interconnected by the same fluid.

The gravitational acceleration g varies from 9.807 m/s^2 at sea level to 9.764 m/s^2 at an elevation of 14,000 m where large passenger planes cruise. This is a change of just 0.4% in this extreme case. Therefore, g can be assumed to be constant with negligible error.

For fluids whose density changes significantly with elevation, a relation for the variation of pressure with elevation can be obtained by dividing Eq. 2-20 by $\Delta x \, \Delta z$, and taking the limit as $\Delta z \to 0$. It gives

$$\frac{dP}{dz} = -\rho g \qquad (2\text{-}23)$$

The negative sign is due to our taking the positive z direction to be upward so that dP is negative when dz is positive since pressure decreases in an upward direction. When the variation of density with elevation is known, the pressure difference between points 1 and 2 can be determined by integration to be

$$\Delta P = P_2 - P_1 = -\int_1^2 \rho g \, dz \qquad (2\text{-}24)$$

For constant density and constant gravitational acceleration, this relation reduces to Eq. 2-21, as expected.

Pressure in a fluid is independent of the shape or cross section of the container. It changes with the vertical distance, but remains constant in other directions. Therefore, the pressure is the same at all points on a horizontal plane in a given fluid. This is illustrated in Fig. 2-26. Note that the pressure at points A, B, C, D, E, F, and G are the same since they are at the same depth, and they are interconnected by the same fluid. But the pressures at points H and I are not the same since these two points cannot be interconnected by the same fluid (i.e., we cannot draw a curve from point I to point H while remaining in the same fluid at all times), although they are at the same depth. (Can you tell at which point the pressure is higher?) Also, the pressure force exerted by the fluid is always normal to the surface at the specified points.

A consequence of the pressure in a fluid remaining constant in the horizontal direction is that *the pressure applied to a confined fluid increases the*

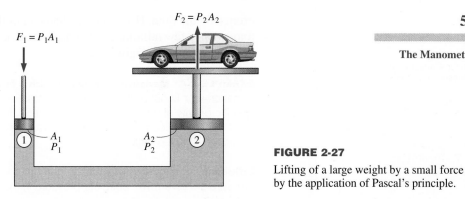

FIGURE 2-27

Lifting of a large weight by a small force by the application of Pascal's principle.

pressure throughout by the same amount. This is called **Pascal's principle,** after Blaise Pascal (1623–1662). Pascal's principle, together with the fact that the pressure force applied by a fluid at a surface is proportional to the surface area, has been the source of important technological innovations. It has resulted in many inventions that impacted many aspects of ordinary life such as hydraulic brakes, hydraulic car jacks, and hydraulic lifts. This is what enables us to lift a car by one arm, as shown in Fig. 2-27. Noting that $P_1 = P_2$ since both pistons are at the same level (the effect of small height differences is negligible, especially at high pressures), the ratio of output force to input force is determined to be

$$P_1 = P_2 \quad \rightarrow \quad \frac{F_1}{A_1} = \frac{F_2}{A_2} \quad \rightarrow \quad \frac{F_2}{F_1} = \frac{A_2}{A_1} \qquad (2\text{-}25)$$

The area ratio A_2/A_1 is called the *ideal mechanical advantage* of the hydraulic lift. Using a hydraulic car jack with a piston area ratio of $A_2/A_1 = 10$, for example, a person can lift a 1000-kg car by applying a force of just 100 kgf (= 908 N).

2-6 ■ THE MANOMETER

We notice from Eq. 2-21 that an elevation change of Δz of a fluid corresponds to a pressure change of $\Delta P/\rho g$, which suggests that a fluid column can be used to measure pressure differences. A device based on this principle is called a **manometer,** and it is commonly used to measure small and moderate pressure differences. A manometer mainly consists of a glass or plastic U-tube containing one or more fluids such as mercury, water, alcohol, or oil. To keep the size of the manometer to a manageable level, heavy fluids such as mercury are used if large pressure differences are anticipated.

Consider the manometer shown in Fig. 2-28 that is used to measure the pressure in the tank. Since the gravitational effects of gases are negligible, the pressure anywhere in the tank and at position 1 has the same value. Furthermore, since pressure in a fluid does not vary in the horizontal direction within a fluid, the pressure at point 2 is the same as the pressure at 1, $P_2 = P_1$.

The differential fluid column of height h is in static equilibrium, and it is open to the atmosphere. Then the pressure at point 2 is determined directly from Eq. 2-22 to be

$$P_2 = P_{\text{atm}} + \rho g h \qquad (2\text{-}26)$$

where ρ is the density of the fluid in the tube. Note that the cross-sectional area of the tube has no effect on the differential height h, and thus the pressure

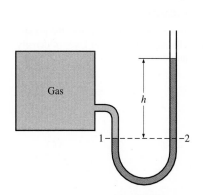

FIGURE 2-28

The basic manometer.

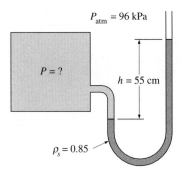

FIGURE 2-29

Sketch for Example 2-4.

exerted by the fluid. However, the diameter of the tube should be large enough (more than a few millimeters) to ensure that the surface tension effect and thus the capillary rise is negligible.

EXAMPLE 2-4 Measuring Pressure with a Manometer

A manometer is used to measure the pressure in a tank. The fluid used has a specific gravity of 0.85, and the manometer column height is 55 cm, as shown in Fig. 2-29. If the local atmospheric pressure is 96 kPa, determine the absolute pressure within the tank.

Solution The reading of a manometer attached to a tank and the atmospheric pressure are given. The absolute pressure in the tank is to be determined.

Assumptions The fluid in the tank is a gas whose density is much lower than the density of oil.

Analysis The density of the fluid is obtained by multiplying its specific gravity by the density of water, which is taken to be 1000 kg/m³:

$$\rho = (\rho_s)(\rho_{H_2O}) = (0.85)(1000 \text{ kg/m}^3) = 850 \text{ kg/m}^3$$

Then from Eq. 2-26,

$$P = P_{atm} + \rho gh$$

$$= 96 \text{ kPa} + (850 \text{ kg/m}^3)(9.81 \text{ m/s}^2)(0.55 \text{ m})\left(\frac{1 \text{ kPa}}{1000 \text{ N/m}^2}\right)$$

$$= \textbf{100.6 kPa}$$

Many engineering problems and some manometers involve multiple immisciple fluids of different densities stacked on top of each other. Such systems can be analyzed easily by remembering that: (1) the pressure change across a fluid column of height h is $\Delta P = \rho gh$, (2) pressure increases downward in a given fluid and decreases upward (i.e., $P_{bottom} > P_{top}$), and (3) two points at the same elevation in a continuous fluid at rest are at the same pressure.

The last principle, also known as *Pascal's law,* allows us to "jump" from one fluid column to the next in manometers without worrying about pressure change as long as we don't jump over a different fluid, and the fluid is at rest. Then the pressure at any point can be determined by starting with a point of known pressure, and adding or subtracting ρgh terms as we advance towards the point of interest. For example, the pressure at the bottom of the tank in Fig. 2-30 can be determined by starting at the free surface where the pressure is P_{atm}, and moving downwards until we reach point 1 at the bottom, and setting the result equal to P_1. It gives

$$P_{atm} + \rho_1 gh_1 + \rho_2 gh_2 + \rho_3 gh_3 = P_1$$

In the special case of all fluids having the same density, the relation above reduces to Eq. 2-26, as expected.

Manometers are particularly well suited to measure pressure drops across a horizontal flow section between two specified points due to the presence of a device such as a valve or heat exchanger or any resistance to flow. This is done by connecting the two legs of the manometer to these two points, as shown in Fig. 2-31. The working fluid can be either a gas or a liquid whose density is ρ_1. The density of the manometer fluid is ρ_2, and the differential fluid height is h.

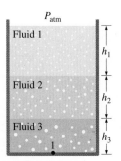

FIGURE 2-30

In stacked-up fluid layers, the pressure change across a fluid layer of density ρ and height h is ρgh.

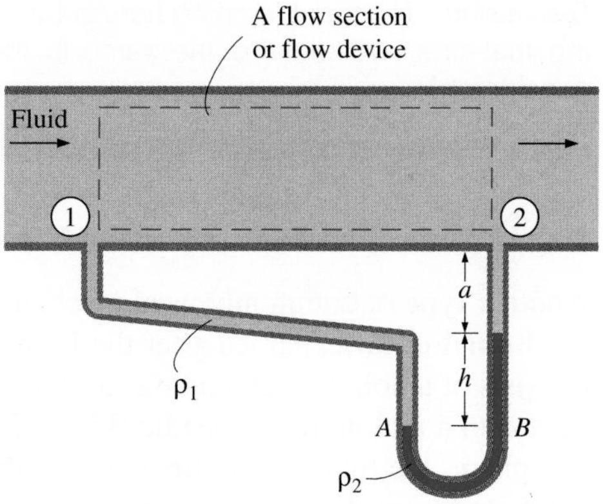

A flow section
or flow device

Fluid

① ②

a

h

ρ_1

A B

ρ_2

FIGURE 2-31
Measuring the pressure drop across a flow section or a flow device by a differential manometer.

A relation for the pressure difference $P_1 - P_2$ can be obtained by starting at point 1 with P_1 and moving along the tube by adding or subtracting the ρgh terms until we reach point 2, and setting the result equal to P_2:

$$P_1 + \rho_1 g\,(a + h) - \rho_2 gh - \rho_1 ga = P_2 \qquad (2\text{-}27)$$

Note that we jumped from point A horizontally to point B and ignored the part underneath since the pressure at both points is the same. Simplifying,

$$P_1 - P_2 = (\rho_2 - \rho_1)\, gh \qquad (2\text{-}28)$$

Note that the distance a has no effect on the result. Also, when the fluid flowing in the pipe is a gas, then $\rho_1 \ll \rho_2$ and the relation above simplifies to $P_1 - P_2 = \rho_2 gh$.

EXAMPLE 2-5 Measuring Pressure with a Multi-Fluid Manometer

The water in a tank is pressurized by air, and the pressure is measured by a multi-fluid manometer as shown in Fig. 2-32. The tank is located on a mountain at an altitude of 1400 m where the atmospheric pressure is 85.6 kPa. Determine the air pressure in the tank if $h_1 = 0.1$ m, $h_2 = 0.2$ m, and $h_3 = 0.35$ m. Take the densities of water, oil, and mercury to be 1000 kg/m³, 850 kg/m³, and 13,600 kg/m³, respectively.

Solution The pressure in a pressurized water tank is measured by a multi-fluid manometer. The air pressure in the tank is to be determined.

Assumption The air pressure in the tank is uniform (i.e., its variation with elevation is negligible due to its low density), and thus we can determine the pressure at the air-water interface.

Analysis Starting with the pressure at point 1 at the air-water interface, and moving along the tube by adding or subtracting the ρgh terms until we reach point 2, and setting the result equal to P_{atm} since the tube is open to the atmosphere gives

$$P_1 + \rho_{water}\, gh_1 + \rho_{oil}\, gh_2 - \rho_{mercury}\, gh_3 = P_{atm}$$

Solving for P_1 and substituting,

$$P_1 = P_{atm} - \rho_{water}\, gh_1 - \rho_{oil}\, gh_2 + \rho_{mercury}\, gh_3$$
$$= P_{atm} + g\,(\rho_{mercury}\, h_3 - \rho_{water}\, h_1 - \rho_{oil}\, h_2)$$
$$= 85.6 \text{ kPa} + (9.81 \text{ m/s}^2)[(13{,}600 \text{ kg/m}^3)\,(0.35 \text{ m}) - (1000 \text{ kg/m}^3)\,(0.1 \text{ m})$$
$$\quad - (850 \text{ kg/m}^3)\,(0.2 \text{ m})]\left(\frac{1 \text{ N}}{1 \text{ kg}\cdot\text{m/s}^2}\right)\left(\frac{1 \text{ kPa}}{1000 \text{ N/m}^2}\right)$$
$$= \mathbf{129.6 \text{ kPa}}$$

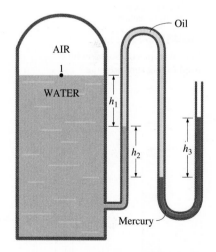

Oil

AIR

1

WATER

h_1

h_2

h_3

Mercury

FIGURE 2-32
Schematic for Example 2-5.

Discussion Note that jumping horizontally from one tube to the next and realizing that pressure remains the same in the same fluid simplifies the analysis considerably.

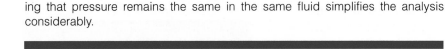

Other Pressure Measurement Devices

Another type of commonly used mechanical pressure measurement device is the **Bourdon tube,** named after the French inventor Eugene Bourdon, which consists of a hollow metal tube bent like a hook whose end is closed and connected to a dial indicator needle (Fig. 2-33). When the tube is open to the atmosphere, the tube is undeflected, and the needle on the dial at this state is calibrated to read zero (gage pressure). When the fluid inside the tube is pressurized, the tube stretches and moves the needle in proportion to the pressure applied.

Electronics have made their way into every aspect of life, including pressure measurement devices. Modem pressure sensors, called **pressure transducers,** are made of semiconductor materials such as silicon and convert the pressure effect to an electrical effect such as a change in voltage, resistance, or capacitance. Pressure transducers are smaller and faster, and they are more sensitive, reliable, and precise than their mechanical counterparts. They can measure pressures from less than a millionth of 1 atm to several thousands of atm.

A wide variety of pressure transducers are available to measure gage, absolute, and differential pressures in a wide range of applications. *Gage pressure transducers* use the atmospheric pressure as a reference by venting the back side of the pressure-sensing diaphragm to the atmosphere, and they give a zero signal output at atmospheric pressure regardless of altitude. The *absolute pressure transducers* are calibrated to have a zero signal output at full vacuum. *Differential pressure transducers* measure the pressure difference between two locations directly instead of using two pressure transducers and taking their difference.

The emergence of an electric potential in a crystalline substance when subjected to mechanical pressure is called the **piezoelectric** (or press-electric) **effect.** This phenomenon, first discovered by brothers Pierre and Jacques Curie in 1880, forms the basis for the widely used **strain-gage** pressure transducers. The sensors of such transducers are made of thin metal wires or foil whose electrical resistance changes when strained under the influence of fluid pressure. The change in the resistance is determined by supplying electric current to the sensor and measuring the corresponding change in voltage drop that is proportional to the applied pressure.

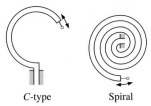

C-type   Spiral

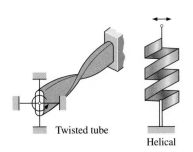

Twisted tube

Helical

Tube cross section

FIGURE 2-33

Various types of Bourdon tubes used to measure pressure.

2-7 ■ BAROMETER AND THE ATMOSPHERIC PRESSURE

The atmospheric pressure is measured by a device called a **barometer;** thus the atmospheric pressure is often referred to as the *barometric pressure.*

As Torricelli discovered a few centuries ago, the atmospheric pressure can be measured by inverting a mercury-filled tube into a mercury container that is open to the atmosphere, as shown in Fig. 2-34. The pressure at point *B* is equal to the atmospheric pressure, and the pressure at *C* can be taken to be zero since there is only mercury vapor above point *C* and the pressure it exerts is negligible. Writing a force balance in the vertical direction gives

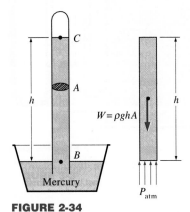

FIGURE 2-34

The basic barometer.

$$P_{atm} = \rho g h \qquad (2\text{-}29)$$

where ρ is the density of mercury, g is the local gravitational acceleration, and h is the height of the mercury column above the free surface. Note that the length and the cross-sectional area of the tube have no effect on the height of the fluid column of a barometer (Fig. 2-35).

A frequently used pressure unit is the *standard atmosphere,* which is defined as the pressure produced by a column of mercury 760 mm in height at 0°C (ρ_{Hg} = 13,595 kg/m³) under standard gravitational acceleration (g = 9.807 m/s²). If water instead of mercury were used to measure the standard atmospheric pressure, a water column of about 10.3 m would be needed. Pressure is sometimes expressed (especially by weather forecasters) in terms of the height of the mercury column. The standard atmospheric pressure, for example, is 760 mmHg (29.92 inHg) at 0°C. The unit mmHg is also called the **torr** in honor of Evangelista Torricelli (1608–1647), who invented the barometer. Therefore, 1 atm = 760 torr, and 1 torr = 133.3 Pa.

The standard atmospheric pressure P_{atm} changes from 101.325 kPa at sea level to 89.88, 79.50, 54.05, 26.5, and 5.53 kPa at altitudes of 1000, 2000, 5000, 10,000, and 20,000 meters, respectively. The standard atmospheric pressure in Denver (elevation = 1610 m), for example, is 83.4 kPa.

Remember that the atmospheric pressure at a location is simply the weight of the air above that location per unit surface area. Therefore, it changes not only with elevation but also with weather conditions.

The decline of atmospheric pressure with elevation has far reaching ramifications in daily life. For example, cooking takes longer at high altitudes since the water boils at a lower temperature at lower atmospheric pressures. Experiencing nose bleeding is a common occurrence at high altitudes since the difference between the blood pressure and the atmospheric pressure is larger in this case, and the delicate walls of veins in the nose are often unable to withstand this extra stress.

For a given temperature, the density of air is lower at high altitudes, and thus a given volume contains less air and less oxygen. So it is no surprise that we tire more easily and experience breathing problems at high altitudes. To compensate for this effect, people living at higher altitudes develop larger lungs and thus larger chests. Similarly, a 2.0-L car engine will act like a 1.7-L car engine at 1500 m altitude (unless it is turbocharged) because of the 15 percent drop in pressure and thus 15 percent drop in the density of air (Fig. 2-36). A fan or compressor will displace 15 percent less air at that altitude for the same volume displacement rate. Therefore, larger cooling fans may need to be selected for operation at high altitudes to assure the specified mass flow rate. The lower pressure and thus lower density also affects lift and drag: airplanes need a longer runway at high altitudes to develop the required lift, and they climb to very high altitudes for cruising for reduced drag and thus better fuel efficiency.

EXAMPLE 2-6   Measuring Atmospheric Pressure with a Barometer

Determine the atmospheric pressure at a location where the barometric reading is 740 mmHg and the gravitational acceleration is g = 9.81 m/s². Assume the temperature of mercury to be 10°C, at which its density is 13,570 kg/m³.

Solution   The barometric reading at a location in height of mercury column is given. The atmospheric pressure is to be determined.

Assumptions   The temperature of mercury is assumed to be 10°C.

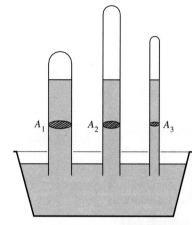

FIGURE 2-35

The length or the cross-sectional area of the tube has no effect on the height of the fluid column of a barometer.

FIGURE 2-36

At high altitudes, a car engine generates less power and a person gets less oxygen because of the lower density of air.

Analysis From Eq. 2-29, the atmospheric pressure is determined to be

$$P_{atm} = \rho g h$$

$$= (13{,}570 \text{ kg/m}^3)\,(9.81 \text{ m/s}^2)\,(0.74 \text{ m})\left(\frac{1 \text{ N}}{1 \text{ kg} \cdot \text{m/s}^2}\right)\left(\frac{1 \text{ kPa}}{1000 \text{ N/m}^2}\right)$$

$$= \textbf{98.5 kPa}$$

EXAMPLE 2-7   Effect of Piston Weight on Pressure in a Cylinder

The piston of a vertical piston-cylinder device containing a gas has a mass of 60 kg and a cross-sectional area of 0.04 m², as shown in Fig. 2-37. The local atmospheric pressure is 0.97 bar, and the gravitational acceleration is 9.81 m/s². (*a*) Determine the pressure inside the cylinder. (*b*) If some heat is transferred to the gas and its volume is doubled, do you expect the pressure inside the cylinder to change?

Solution   A gas is contained in a vertical cylinder with a heavy piston. The pressure inside the cylinder and the effect of volume change on pressure are to be determined.

Assumptions   Friction between the piston and the cylinder is negligible.

Analysis   (*a*) The gas pressure in the piston-cylinder device depends on the atmospheric pressure and the weight of the piston. Drawing the free-body diagram of the piston as shown in Fig. 2-37 and balancing the vertical forces yield

$$PA = P_{atm}A + W$$

Solving for *P* and substituting,

$$P = P_{atm} + \frac{mg}{A}$$

$$= 0.97 \text{ bar} + \frac{(60 \text{ kg})\,(9.81 \text{ m/s}^2)}{0.04 \text{ m}^2}\left(\frac{1 \text{ N}}{1 \text{ kg} \cdot \text{m/s}^2}\right)\left(\frac{1 \text{ bar}}{10^5 \text{ N/m}^2}\right)$$

$$= \textbf{1.117 bars}$$

(*b*) The volume change will have no effect on the free-body diagram drawn in part (*a*), and therefore the pressure inside the cylinder will remain the same.

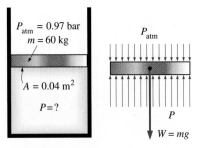

$P_{atm} = 0.97$ bar
$m = 60$ kg

$A = 0.04$ m²

$P = ?$

P_{atm}

P

$W = mg$

FIGURE 2-37

Schematic for Example 2-7, and the free-body diagram of the piston.

2-8 ■ SUMMARY

In this chapter, some basic concepts of thermodynamics are introduced and discussed. Any change from one state to another is called a *process*. A process with identical end states is called a *cycle*. A system is said to be in *thermodynamic equilibrium* if it maintains thermal, mechanical, phase, and chemical equilibrium. During a *quasi-static* or *quasi-equilibrium process,* the system remains practically in equilibrium at all times. The state of a simple, compressible system is completely specified by two independent, intensive properties.

The sum of all forms of energy of a system is called *total energy,* which is considered to consist of internal, kinetic, and potential energies. *Internal energy* represents the molecular energy of a system and may exist in sensible, latent, chemical, and nuclear forms.

The *zeroth law of thermodynamics* states that two bodies are in thermal equilibrium if both have the same temperature reading even if they are not in contact. The temperature scales used in the SI and the English system today are the *Celsius scale* and the *Fahrenheit scale,* respectively. They are related to absolute temperature scales by

$$T(K) = T(°C) + 273.15$$
$$T(R) = T(°F) + 459.67$$

The magnitudes of each division of 1 K and 1°C are identical, and so are the magnitudes of each division of 1 R and 1°F. Therefore,

$$\Delta T(K) = \Delta T(°C) \qquad \text{and} \qquad \Delta T(R) = \Delta T(°F)$$

Force exerted by a fluid per unit area is called *pressure,* and its unit is the *pascal,* 1 Pa = 1 N/m². The pressure relative to absolute vacuum is called the *absolute pressure,* and the difference between the absolute pressure and the local atmospheric pressure is called the *gage pressure.* Pressures below atmospheric pressure are called *vacuum pressures.* The absolute, gage, and vacuum pressures are related by

$$P_{gage} = P_{abs} - P_{atm} \qquad \text{(for pressures above } P_{atm})$$
$$P_{vac} = P_{atm} - P_{abs} \qquad \text{(for pressures below } P_{atm})$$

The pressure at a point in a fluid has the same magnitude in all directions. The variation of pressure with elevation is given by

$$\frac{dP}{dz} = -\rho g$$

where the positive z direction is taken to be upward. When the density of the fluid is constant, the pressure difference across a fluid layer of thickness Δz is

$$\Delta P = P_2 - P_1 = \rho g \Delta z$$

The absolute and gage pressures in a liquid open to the atmosphere at a depth h from the free surface are

$$P = P_{atm} + \rho g h \qquad \text{or} \qquad P_{gage} = \rho g h$$

Small to moderate pressure differences are measured by a *manometer.* The pressure in a fluid remains constant in the horizontal direction. *Pascal's principle* states that the pressure applied to a confined fluid increases the pressure throughout by the same amount. The atmospheric pressure is measured by a *barometer* and is given by

$$P_{atm} = \rho g h$$

where h is the height of the liquid column above the free surface.

REFERENCES AND SUGGESTED READING

1. Y. A. Çengel and M. A. Boles. *Thermodynamics. An Engineering Approach.* 3rd ed. New York, McGraw-Hill, 1998.

2. R. W. Fox and A. T. McDonald. *Introduction to Fluid Mechanics.* 5th ed. New York, Wiley, 1999.

3. B. R. Munson, D. F. Young, and T. Okiishi. *Fundamentals of Fluid Mechanics.* 3rd ed. New York, Wiley, 1998.

4. M. C. Potter and D. C. Wiggert, *Mechanics of Fluids.* 2nd ed. Upper Saddle River, Prentice Hall, 1997.

5. D. C. Giancoli. *Physics*. 3rd ed. Upper Saddle River, Prentice Hall, 1991.

6. J. A. Roberson and C. L. Grove, *Engineering Fluid Mechanics*. 6th ed. New York, Wiley, 1997.

7. *The U.S. Standard Atmosphere*. Washington, D.C., U.S. Government Printing Office, 1976.

8. F. M. White, *Fluid Mechanics*. 4th ed. New York, McGraw-Hill, 1999.

PROBLEMS*

State, Process, Forms of Energy

2-1C Portable electric heaters are commonly used to heat small rooms. Explain the energy transformation involved during this heating process.

2-2C Consider the process of heating water on top of an electric range. What are the forms of energy involved during this process? What are the energy transformations that take place?

2-3C What is the difference between the macroscopic and microscopic forms of energy?

2-4C What is total energy? Identify the different forms of energy that constitute the total energy.

2-5C List the forms of energy that contribute to the internal energy of a system.

2-6C How are heat, internal energy, and thermal energy related to each other?

2-7C For a system to be in thermodynamic equilibrium, do the temperature and the pressure have to be the same everywhere?

2-8C What is a quasi-equilibrium process? What is its importance in engineering?

2-9C Define the isothermal, isobaric, and isochoric processes.

2-10C What is the state postulate?

2-11C Is the state of the air in an isolated room completely specified by the temperature and the pressure? Explain.

2-12 Consider a nuclear power plant that produces 1000 MW of power and has a conversion efficiency of 30 percent (that is, for each unit of fuel energy used, the plant produces 0.3 unit of electrical energy). Assuming continuous operation, determine the amount of nuclear fuel consumed by this plant per year.

2-13 Repeat Prob. 2-12 for a coal power plant that burns coal whose heating value is 28,000 kJ/kg.

---

*Students are encouraged to answer *all* the concept "C" questions.

2-14 When a hydrocarbon fuel is burned, almost all of the carbon in the fuel burns completely to form CO_2 (carbon dioxide), which is the principal gas causing the greenhouse effect and thus global climate change. On average, 0.59 kg of CO_2 is produced for each kWh of electricity generated from a power plant that burns natural gas. A typical new household refrigerator uses about 700 kWh of electricity per year. Determine the amount of CO_2 production that is due to the refrigerators in a city with 200,000 households.

2-15 Repeat Prob. 2-14 assuming the electricity is produced by a power plant that burns coal. The average production of CO_2 in this case is 1.1 kg per kWh.

2-16E Consider a household that uses 8000 kWh of electricity per year and 1500 gallons of fuel oil during a heating season. The average amount of CO_2 produced is 26.4 lbm/gallon of fuel oil and 1.54 lbm/kWh of electricity. If this household reduces its oil and electricity usage by 20 percent as a result of implementing some energy conservation measures, determine the reduction in the amount of CO_2 emissions by that household per year.

2-17 A typical car driven 12,000 miles a year emits to the atmosphere about 11 kg per year of NO_x (nitrogen oxides), which causes smog in major population areas. Natural gas burned in the furnace emits about 4.3 g of NO_x per therm, and the electric power plants emit about 7.1 g of NO_x per kWh of electricity produced. Consider a household that has 2 cars and consumes 9000 kWh of electricity and 1200 therms of natural gas. Determine the amount of NO_x emission to the atmosphere per year for which this household is responsible.

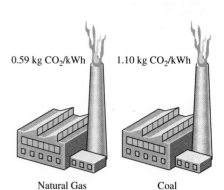

0.59 kg CO_2/kWh 1.10 kg CO_2/kWh

Natural Gas   Coal
Power Plant   Power Plant

FIGURES P2-14 AND P2-15

11 kg NO_x
per year

FIGURE P2-17

Temperature

2-18C What is the zeroth law of thermodynamics?

2-19C What are the ordinary and absolute temperature scales in the SI and the English system?

2-20C Consider an alcohol and a mercury thermometer that read exactly 0°C at the ice point and 100°C at the steam point. The distance between the two points is divided into 100 equal parts in both thermometers. Do you think these thermometers will give exactly the same reading at a temperature of, say, 60°C? Explain.

2-21C The deep body temperature of a healthy person is 37°C. What is it in kelvins? *Answer:* 310 K

2-22E Consider a system whose temperature is 18°C. Express this temperature in R, K, and °F.

2-23 The temperature of a system rises by 30°C during a heating process. Express this rise in temperature in kelvins. *Answer:* 30 K

2-24E The temperature of a system drops by 27°F during a cooling process. Express this drop in temperature in K, R, and °C.

2-25 Consider two closed systems A and B. System A contains 2000 kJ of thermal energy at 20°C whereas system B contains 200 kJ of thermal energy at 50°C. Now the systems are brought into contact with each other. Determine the direction of any heat transfer between the two systems.

Pressure, Manometer, and Barometer

2-26C What is the difference between gage pressure and absolute pressure?

2-27C Explain why some people experience nose bleeding and some others experience shortness of breath at high elevations.

2-28C Someone claims that the absolute pressure in a liquid of constant density doubles when the depth is doubled. Do you agree? Explain.

2-29C A tiny steel cube is suspended in water by a string. If the lengths of the sides of the cube are very small, how would you compare the magnitudes of the pressures on the top, bottom, and side surfaces of the cube?

2-30C Express Pascal's principle, and give a real-world example of it.

2-31C Consider two identical fans, one at sea level and the other on top of a high mountain, running at identical speeds. How would you compare (*a*) the volume flow rates and (*b*) the mass flow rates of these two fans?

2-32 A vacuum gage connected to a chamber reads 24 kPa at a location where the atmospheric pressure is 92 kPa. Determine the absolute pressure in the chamber.

2-33E A manometer is used to measure the air pressure in a tank. The fluid used has a specific gravity of 1.25, and the differential height between the two arms of the manometer is 28 in. If the local atmospheric pressure is 12.7 psia, determine the absolute pressure in the tank for the cases of the manometer arm with the (*a*) higher and (*b*) lower fluid level being attached to the tank.

2-34 The water in a tank is pressurized by air, and the pressure is measured by a multi-fluid manometer as shown in the figure. Determine the gage pressure of air in the tank if $h_1 = 0.2$ m, $h_2 = 0.3$ m, and $h_3 = 0.46$ m. Take the densities of water, oil, and mercury to be 1000 kg/m³, 850 kg/m³, and 13,600 kg/m³, respectively.

2-35 Determine the atmospheric pressure at a location where the barometric reading is 750 mmHg. Take the density of mercury to be 13,600 kg/m³.

2-36 The gage pressure in a liquid at a depth of 3 m is read to be 28 kPa. Determine the gage pressure in the same liquid at a depth of 12 m.

2-37 The absolute pressure in water at a depth of 5 m is read to be 145 kPa. Determine (*a*) the local atmospheric pressure, and (*b*) the absolute pressure at a depth of 5 m in a liquid whose specific gravity is 0.85 at the same location.

2-38E Show that 1 kgf/cm² = 14.223 psi.

2-39E A 200-pound man has a total foot imprint area of 72 in². Determine the pressure this man exerts on the ground if (*a*) he stands on both feet and (*b*) he stands on one foot.

2-40 Consider a 70-kg woman who has a total foot imprint area of 400 cm². She wishes to walk on the snow, but the snow cannot withstand pressures greater than 0.5 kPa. Determine the minimum size of the snow shoes needed (imprint area per shoe) to enable her to walk on the snow without sinking.

2-41 A vacuum gage connected to a tank reads 30 kPa at a location where the barometric reading is 755 mmHg. Determine the absolute pressure in the tank. Take $\rho_{Hg} = 13,590$ kg/m³.   *Answer:* 70.6 kPa

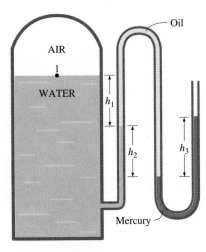

FIGURE P2-34

2-42E A pressure gage connected to a tank reads 50 psi at a location where the barometric reading is 29.1 inHg. Determine the absolute pressure in the tank. Take ρ_{Hg} = 848.4 lbm/ft^3. *Answer:* 64.29 psia

2-43 A pressure gage connected to a tank reads 500 kPa at a location where the atmospheric pressure is 94 kPa. Determine the absolute pressure in the tank.

2-44 The barometer of a mountain hiker reads 930 mbars at the beginning of a hiking trip and 780 mbars at the end. Neglecting the effect of altitude on local gravitational acceleration, determine the vertical distance climbed. Assume an average air density of 1.20 kg/m^3. *Answer:* 1274 m

2-45 The basic barometer can be used to measure the height of a building. If the barometric readings at the top and at the bottom of a building are 730 and 755 mmHg, respectively, determine the height of the building. Assume an average air density of 1.18 kg/m^3.

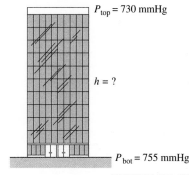

FIGURE P2-45

2-46 Determine the pressure exerted on a diver at 30 m below the free surface of the sea. Assume a barometric pressure of 101 kPa and a specific gravity of 1.03 for seawater. *Answer:* 404.0 kPa

2-47E Determine the pressure exerted on the surface of a submarine cruising 300 ft below the free surface of the sea. Assume that the barometric pressure is 14.7 psia and the specific gravity of seawater is 1.03.

2-48 A gas is contained in a vertical, frictionless piston-cylinder device. The piston has a mass of 4 kg and cross-sectional area of 35 cm^2. A compressed spring above the piston exerts a force of 60 N on the piston. If the atmospheric pressure is 95 kPa, determine the pressure inside the cylinder. *Answer:* 123.4 kPa

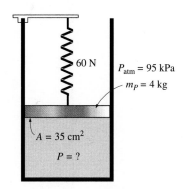

FIGURE P2-48

2-49 Both a gage and a manometer are attached to a gas tank to measure its pressure. If the reading on the pressure gage is 80 kPa, determine the distance between the two fluid levels of the manometer if the fluid is (*a*) mercury (ρ = 13,600 kg/m^3) or (*b*) water (ρ = 1000 kg/m^3).

2-50 A manometer containing oil (ρ = 850 kg/m^3) is attached to a tank filled with air. If the oil-level difference between the two columns is 45 cm and the atmospheric pressure is 98 kPa, determine the absolute pressure of the air in the tank. *Answer:* 101.75 kPa

2-51 A mercury manometer (ρ = 13,600 kg/m^3) is connected to an air duct to measure the pressure inside. The difference in the manometer levels is 15 mm, and the atmospheric pressure is 100 kPa. (*a*) Judging from Fig. P2-51, determine if the pressure in the duct is above or below the atmospheric pressure. (*b*) Determine the absolute pressure in the duct.

2-52 By considering a wedge-shaped fluid element in the *xy*-plane (with unit length in the *z*-direction), show that in the absence of shear forces, the pressure in an accelerating fluid is the same in all directions.

2-53E Blood pressure is usually measured by wrapping a closed air-filled jacket equipped with a pressure gage around the upper arm of a person at the level of the heart. Using a mercury manometer or a stethoscope, the systolic pressure (the maximum pressure when the heart is pumping) and the diastolic pressure (the minimum pressure when the heart is resting) are measured in mmHg. The systolic and diastolic pressures of a healthy person are about

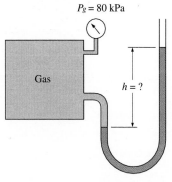

FIGURE P2-49

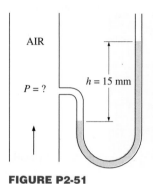

FIGURE P2-51

FIGURE 2-54

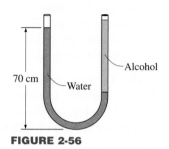

FIGURE 2-56

120 mmHg and 80 mmHg, respectively, and are indicated as 120/80. Express both of these gage pressures in kPa, psi, and meter water column.

2-54 The maximum blood pressure in the upper arm of a healthy person is about 120 mmHg. If a vertical tube open to the atmosphere is connected to the vein in the arm of the person, determine how high the blood will rise in the tube. Take the density of the blood to be 1050 kg/m³.

2-55 Consider a 1.8-m-tall man standing vertically in water and completely submerged in a pool. Determine the difference between the pressures acting at the head and at the toes of this man, in kPa.

2-56 Consider a U-tube whose arms are open to the atmosphere. Now water is poured into the U-tube from one arm, and ethyl alcohol ($\rho = 790$ kg/m³) from the other. One arm contains 70-cm high water while the other arm contains both fluids with a alcohol-to-water height ratio of 6. Determine the height of each fluid in that arm.

2-57 The hydraulic lift in a car repair shop has an output diameter of 30 cm, and is to lift cars up to 2000 kg. Determine the fluid gage pressure that must be maintained in the reservoir.

2-58 Fresh water and seawater flowing in parallel horizontal pipelines are connected to each other by a double U-tube manometer, as shown in the figure. Determine the pressure difference between the two pipelines. Take the

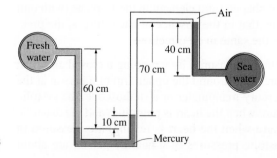

FIGURE P2-58

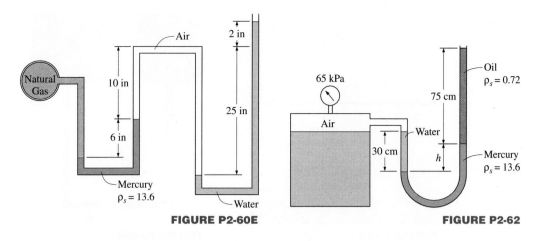

FIGURE P2-60E

FIGURE P2-62

density of seawater at that location to be $\rho = 1035$ kg/m³. Can the air column be ignored in the analysis?

2-59 Repeat Prob. 2-58 by replacing the air with oil whose specific gravity is 0.72.

2-60E The pressure in a natural gas pipeline is measured by the manometer shown in the figure with one of the arms open to the atmosphere where the local atmospheric pressure is 14.2 psia. Determine the absolute pressure in the pipeline.

2-61E Repeat Prob. 2-60E by replacing air by oil with a specific gravity of 0.69.

2-62 The gage pressure of the air in the tank shown in the figure is measured to be 65 kPa. Determine the differential height h of the mercury column.

2-63 Repeat Prob. 2-62 for a gage pressure of 45 kPa.

2-64 The top part of a water tank is divided into two compartments, as shown in the figure. Now a fluid with an unknown density is poured into one side, and the water level rises a certain amount on the other side to compensate for this effect. Based on the final fluid heights shown on the figure, determine the density of the fluid added. Assume the liquid does not mix with water.

2-65 The 500-kg load on the hydraulic lift shown in the figure is to be raised by pouring oil ($\rho = 780$ kg/m³) into a thin tube. Determine how high h should be in order to raise that weight.

2-66E Two oil tanks are connected to each other through a manometer. If the difference between the mercury levels in the two arms is 32 in., determine the pressure difference between the two tanks. The densities of oil and mercury are 45 lbm/ft³ and 848 lbm/ft³, respectively.

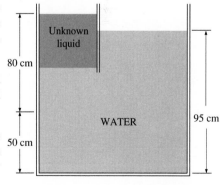

FIGURE P2-64

Review Problems

2-67E The efficiency of a refrigerator increases by 3 percent for each °C rise in the minimum temperature in the device. What is the increase in the efficiency for each (a) K, (b) °F, and (c) R rise in temperature?

2-68E The boiling temperature of water decreases by about 3°C for each 1000 m rise in altitude. What is the decrease in the boiling temperature in (a) K, (b) °F, and (c) R for each 1000 m rise in altitude?

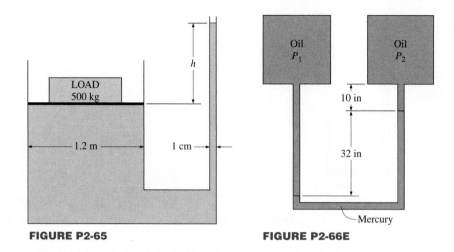

FIGURE P2-65

FIGURE P2-66E

2-69E The average body temperature of a person rises by about 2°C during strenuous exercise. What is the rise in the body temperature in (*a*) K, (*b*) °F, and (*c*) R during strenuous exercise?

2-70E Hyperthermia of 5°C (i.e., 5°C rise above the normal body temperature) is considered fatal. Express this fatal level of hyperthermia in (*a*) K, (*b*) °F, and (*c*) R.

2-71E A house is losing heat at a rate of 3000 kJ/h per °C temperature difference between the indoor and the outdoor temperatures. Express the rate of heat loss from this house per (*a*) K, (*b*) °F, and (*c*) R difference between the indoor and the outdoor temperature.

2-72 The average temperature of the atmosphere in the world is approximated as a function of altitude by the relation

$$T_{atm} = 288.15 - 6.5z$$

where T_{atm} is the temperature of the atmosphere in K and z is the altitude in km with $z = 0$ at sea level. Determine the average temperature of the atmosphere outside an airplane that is cruising at an altitude of 12,000 m.

2-73 Joe Smith, an old-fashioned engineering student, believes that the boiling point of water is best suited for use as the reference point on temperature scales. Unhappy that the boiling point corresponds to some odd number in the current absolute temperature scales, he has proposed a new absolute temperature scale that he calls the Smith scale. The temperature unit on this scale is *smith,* denoted by S, and the boiling point of water on this scale is assigned to be 1000 S. From a thermodynamic point of view, discuss if it is an acceptable temperature scale. Also, determine the ice point of water on the Smith scale and obtain a relation between the Smith and Celsius scales.

2-74 It is well known that cold air feels much colder in windy weather than what the thermometer reading indicates because of the "chilling effect" of the wind. This effect is due to the increase in the convection heat transfer coefficient with increasing air velocities. The *equivalent wind chill temperature* in °F is given by [ASHRAE, *Handbook of Fundamentals* (Atlanta, GA, 1993), p. 8.15]

$$T_{equiv} = 91.4 - (91.4 - T_{ambient})(0.475 - 0.0203\mathcal{V} + 0.304\sqrt{\mathcal{V}})$$

where $\mathcal{V}$ is the wind velocity in mi/h and $T_{ambient}$ is the ambient air temperature in °F in calm air, which is taken to be air with light winds at speeds up to 4 mi/h. The constant 91.4°F in the above equation is the mean skin temperature of a resting person in a comfortable environment. Windy air at temperature $T_{ambient}$ and velocity $\mathcal{V}$ will feel as cold as the calm air at temperature T_{equiv}. Using proper conversion factors, obtain an equivalent relation in SI units where $\mathcal{V}$ is the wind velocity in km/h and $T_{ambient}$ is the ambient air temperature in °C.

Answer: $T_{equiv} = 33.0 - (33.0 - T_{ambient})(0.475 - 0.0126\mathcal{V} + 0.240\sqrt{\mathcal{V}})$

2-75 A 3-m-high rectangular water tank whose cross section is 0.8 m × 1.2 m and whose top is open to the atmosphere is initially filled with water. Now the discharge plug near the bottom of the tank is pulled out, and a water jet whose diameter is 1.2 cm streams out. The mean velocity of the jet is given by $\mathcal{V} = \sqrt{2gh}$ where h is the height of water in the tank measured from the center of the hole and g is the gravitational acceleration. Determine how long it will take for the gage pressure at the bottom of the tank to drop to 15.7 kPa.

2-76 An air-conditioning system requires a 20-m long section of 15-cm diameter duct work to be laid underwater. Determine the upward force the water will exert on the duct. Take the densities of air and water to be 1.3 kg/m³ and 1000 kg/m³, respectively.

2-77 Balloons are often filled with helium gas because it weighs only about one-seventh of what air weighs under identical conditions. The buoyancy force which can be expressed as $F_b = \rho_{air}gV_{balloon}$, will push the balloon upward. If the balloon has a diameter of 10 m and carries two people, 70 kg each, determine the acceleration of the balloon when it is first released. Assume the density of air is $\rho = 1.16$ kg/m³, and neglect the weight of the ropes and the cage. *Answer:* 16.5 m/s²

2-78 Determine the maximum amount of load, in kg, the balloon described in Prob. 2-77 can carry. *Answer:* 520.6 kg

2-79 The pressure in a steam boiler is given to be 75 kgf/cm². Express this pressure in psi, kPa, atm, and bars.

2-80 The basic barometer can be used as an altitude-measuring device in airplanes. The ground control reports a barometric reading of 753 mmHg while the pilot's reading is 690 mmHg. Estimate the altitude of the plane from ground level if the average air density is 1.20 kg/m³.
Answer: 714 m

2-81 The lower half of a 10-m-high cylindrical container is filled with water ($\rho = 1000$ kg/m³) and the upper half with oil that has a specific gravity of 0.85. Determine the pressure difference between the top and bottom of the cylinder. *Answer:* 90.7 kPa

2-82 A vertical, frictionless piston-cylinder device contains a gas at 500 kPa. The atmospheric pressure outside is 100 kPa, and the piston area is 30 cm². Determine the mass of the piston.

2-83 A pressure cooker cooks a lot faster than an ordinary pan by maintaining a higher pressure and temperature inside. The lid of a pressure cooker is

HELIUM
$D = 10$ m
$\rho_{He} = \frac{1}{7}\rho_{air}$

$m = 140$ kg
FIGURE P2-77

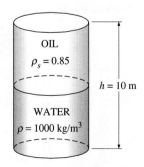

OIL
$\rho_s = 0.85$

WATER
$\rho = 1000$ kg/m³

$h = 10$ m

FIGURE P2-81

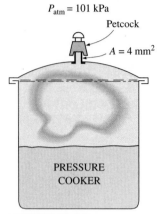

$P_{atm} = 101$ kPa

Petcock

$A = 4$ mm^2

PRESSURE
COOKER

FIGURE P2-83

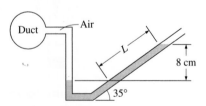

$P_{atm} = 92$ kPa

$h = ?$

water

FIGURE P2-84

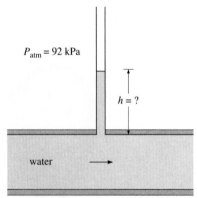

Duct — Air

L

8 cm

35°

FIGURE P2-86

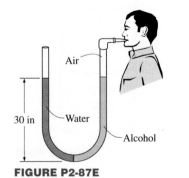

Air

30 in — Water

Alcohol

FIGURE P2-87E

well sealed, and steam can escape only through an opening in the middle of the lid. A separate piece of certain mass, the petcock, sits on top of this opening and prevents steam from escaping until the pressure force overcomes the weight of the petcock. The periodic escape of the steam in this manner prevents any potentially dangerous pressure buildup and keeps the pressure inside at a constant value. Determine the mass of the petcock of a pressure cooker whose operation pressure is 100 kPa gage and has an opening cross-sectional area of 4 mm^2. Assume an atmospheric pressure of 101 kPa, and draw the freebody diagram of the petcock. *Answer: 40.8 g*

2-84 A glass tube is attached to a water pipe as shown in the figure. If the water pressure at the bottom of the tube is 115 kPa and the local atmospheric pressure is 92 kPa, determine how high the water will rise in the tube, in m. Assume $g = 9.8$ m/s^2 at that location and take the density of water to be 1000 kg/m^3.

2-85 The average atmospheric pressure on earth is approximated as a function of altitude by the relation $P_{atm} = 101.325 (1 - 0.02256z)^{5.256}$ where P_{atm} is the atmospheric pressure in kPa and z is the altitude in km with $z = 0$ at sea level. Determine the approximate atmospheric pressures at Atlanta ($z = 306$ m), Denver ($z = 1610$ m), Mexico City ($z = 2309$ m), and the top of Mount Everest ($z = 8848$ m).

2-86 When measuring small pressure differences with a manometer, often one arm of the manometer is inclined to improve the accuracy of reading. (The pressure difference is still proportional to the *vertical* distance, and not the actual length of the fluid along the tube.) The air pressure in a circular duct is to be measured using a manometer whose open arm is inclined 35° from the horizontal, as shown in the figure. The density of the liquid in the manometer is 0.81 kg/L, and the vertical distance between the fluid levels in the two arms of the manometer is 8 cm. Determine the gage pressure of air in the duct, and the length of the fluid column in the inclined arm above the fluid level in the vertical arm.

2-87 Consider a U-tube whose arms are open to the atmosphere. Now equal volumes of water and ethyl alcohol ($\rho = 49.3$ lbm/ft^3) are poured from different arms. A person blows from the alcohol side of the U-tube until the contact surface of the two fluids moves to the bottom of the U-tube, and thus the liquid levels in the two arms are the same. If the fluid height in each arm is 30 in., determine the gage pressure the person exerts on the alcohol by blowing.

2-88 Intravenous infusions are usually driven by gravity by hanging the fluid bottle at sufficient height to counteract the blood pressure in the vein and to force the fluid into the body. The higher the bottle is raised, the higher the flow rate of the fluid will be. (*a*) If it is observed that the fluid and the blood pressures balance each other when the bottle is 1.2 m above the arm level, determine the gage pressure of the blood. (*b*) If the gage pressure of the fluid at the arm level needs to be 20 kPa for sufficient flow rate, determine how high the bottle must be placed. Take the density of the fluid to be 1020 kg/m^3.

2-89 A gasoline line is connected to a pressure gage through a double-U manometer, as shown in the figure. If the reading of the pressure gage is 370 kPa, determine the gage pressure of the gasoline line.

2-90 Repeat Prob. 2-89 for a pressure gage reading of 240 kPa.

2-91E A water pipe is connected to a double-U manometer as shown in the figure at a location where the local atmospheric pressure is 14.2 psia. Determine the absolute pressure at the center of the pipe.

Computer, Design, and Essay Problems

2-92 Write a computer program to express a given temperature in °C, °F, K, and R in terms of the other three units.

2-93 Write an essay on different temperature measurement devices. Explain the operational principle of each device, its advantages and disadvantages, its cost, and its range of applicability. Which device would you recommend for use in the following cases: taking the temperatures of patients in a doctor's office, monitoring the variations of temperature of a car engine block at several locations, and monitoring the temperatures in the furnace of a power plant?

2-94 Write a computer program to express a pressure given in SI units in terms of the heights of the water and mercury columns.

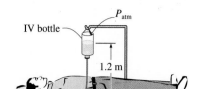

FIGURE P2-88

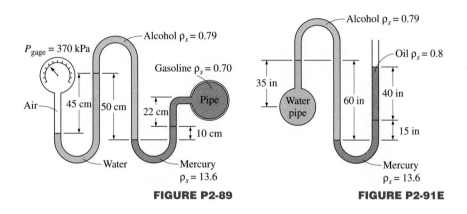

FIGURE P2-89   **FIGURE P2-91E**

Properties of Pure Substances

3

We start this chapter with the introduction of the concept of a *pure substance,* and a discussion of the physics of phase-change processes. We then illustrate the various property diagrams and *P-v-T* surfaces of pure substances. This is followed by discussions of *vapor pressure, phase equilibrium,* and *relative humidity.* After demonstrating the use of the property tables, the hypothetical substance *ideal gas* and the *ideal-gas equation of state* are discussed. The *compressibility factor,* which accounts for the deviation of real gases from ideal-gas behavior, is introduced, and some of the best-known equations of state are presented. Finally, *specific heats* are defined, and relations are obtained for the internal energy and enthalpy of ideal gases in terms of specific heats and temperature. This is also done for solids and liquids, which are approximated as *incompressible substances.*

FIGURE 3-1

Nitrogen and gaseous air are pure substances.

(*a*) H$_2$O   (*b*) AIR

FIGURE 3-2

A mixture of liquid and gaseous water is a pure substance, but a mixture of liquid and gaseous air is not.

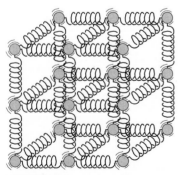

FIGURE 3-3

The molecules in a solid are kept at their positions by the large springlike intermolecular forces.

3-1 ■ PURE SUBSTANCE

A substance that has a fixed chemical composition throughout is called a **pure substance.** Water, nitrogen, helium, and carbon dioxide, for example, are all pure substances.

A pure substance does not have to be of a single chemical element or compound, however. A mixture of various chemical elements or compounds also qualifies as a pure substance as long as the mixture is homogeneous. Air, for example, is a mixture of several gases, but it is often considered to be a pure substance because it has a uniform chemical composition (Fig. 3-1). However, a mixture of oil and water is not a pure substance. Since oil is not soluble in water, it will collect on top of the water, forming two chemically dissimilar regions.

A mixture of two or more phases of a pure substance is still a pure substance as long as the chemical composition of all phases is the same (Fig. 3-2). A mixture of ice and liquid water, for example, is a pure substance because both phases have the same chemical composition. A mixture of liquid air and gaseous air, however, is not a pure substance since the composition of liquid air is different from the composition of gaseous air, and thus the mixture is no longer chemically homogeneous. This is due to different components in air having different condensation temperatures at a specified pressure.

3-2 ■ PHASES OF A PURE SUBSTANCE

We all know from experience that substances exist in different phases. At room temperature and pressure, copper is a solid, mercury is a liquid, and nitrogen is a gas. Under different conditions, each may appear in a different phase. Even though there are three principal phases—solid, liquid, and gas—a substance may have several phases within a principal phase, each with a different molecular structure. Carbon, for example, may exist as graphite or diamond in the solid phase. Helium has two liquid phases; iron has three solid phases. Ice may exist at seven different phases at high pressures. A phase is identified as having a distinct molecular arrangement that is homogeneous throughout and separated from the others by easily identifiable boundary surfaces. The two phases of H$_2$O in iced water represent a good example of this.

When studying phases or phase changes in thermodynamics, one does not need to be concerned with the molecular structure and behavior of different phases. However, it is very helpful to have some understanding of the molecular phenomena involved in each phase, and a brief discussion of phase transformations is given below.

Molecular bonds are strongest in solids and weakest in gases. One reason is that molecules in solids are closely packed together, whereas in gases they are separated by relatively large distances.

The molecules in a **solid** are arranged in a three-dimensional pattern (lattice) that is repeated throughout (Fig. 3-3). Because of the small distances between molecules in a solid, the attractive forces of molecules on each other are large and keep the molecules at fixed positions (Fig. 3-4). Note that the attractive forces between molecules turn to repulsive forces as the distance between the molecules approaches zero, thus preventing the molecules from piling up on top of each other. Even though the molecules in a solid cannot move relative to each other, they continually oscillate about their equilibrium positions. The velocity of the molecules during these oscillations depends on the temperature. At sufficiently high temperatures, the velocity (and thus the momentum) of the molecules may reach a point where the intermolecular forces are partially overcome and groups of molecules break away (Fig. 3-5). This is the beginning of the melting process.

The molecular spacing in the **liquid** phase is not much different from that of the solid phase, except the molecules are no longer at fixed positions relative to each other. In a liquid, chunks of molecules float about each other; however, the molecules maintain an orderly structure within each chunk and retain their original positions with respect to one another. The distances between molecules generally experience a slight increase as a solid turns liquid, with water being a rare exception.

In the **gas** phase, the molecules are far apart from each other, and a molecular order is nonexistent. Gas molecules move about at random, continually colliding with each other and the walls of the container they are in. Particularly at low densities, the intermolecular forces are very small, and collisions are the only mode of interaction between the molecules. Molecules in the gas phase are at a considerably higher energy level than they are in the liquid or solid phases. Therefore, the gas must release a large amount of its energy before it can condense or freeze.

FIGURE 3-4

In a solid, the attractive and repulsive forces between the molecules tend to maintain them at relatively constant distances from each other.

3-3 ■ PHASE-CHANGE PROCESSES OF PURE SUBSTANCES

There are many practical situations where two phases of a pure substance coexist in equilibrium. Water exists as a mixture of liquid and vapor in the boiler and the condenser of a steam power plant. The refrigerant turns from liquid to vapor in the freezer of a refrigerator. Even though many homeowners consider the freezing of water in underground pipes as the most important phase-change process, attention in this section is focused on the liquid and vapor phases and the mixture of these two. As a familiar substance, water will be used to demonstrate the basic principles involved. Remember, however, that all pure substances exhibit the same general behavior.

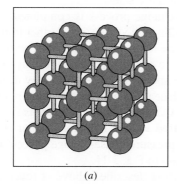
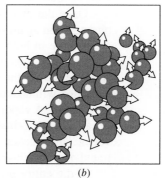
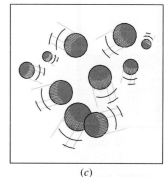

(a)   (b)   (c)

FIGURE 3-5

The arrangement of atoms in different phases: (a) molecules are at relatively fixed positions in a solid, (b) chunks of molecules float about each other in the liquid phase, and (c) molecules move about at random in the gas phase.

FIGURE 3-6

At 1 atm and 20°C, water exists in the liquid phase (*compressed liquid*).

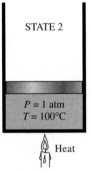

FIGURE 3-7

At 1 atm pressure and 100°C, water exists as a liquid that is ready to vaporize (*saturated liquid*).

FIGURE 3-8

As more heat is transferred, part of the saturated liquid vaporizes (*saturated liquid–vapor mixture*).

Compressed Liquid and Saturated Liquid

Consider a piston-cylinder device containing liquid water at 20°C and 1 atm pressure (state 1, Fig. 3-6). Under these conditions, water exists in the liquid phase, and it is called a **compressed liquid,** or a **subcooled liquid,** meaning that it is *not about to vaporize.* Heat is now transferred to the water until its temperature rises to, say, 40°C. As the temperature rises, the liquid water expands slightly, and so its specific volume increases. To accommodate this expansion, the piston will move up slightly. The pressure in the cylinder remains constant at 1 atm during this process since it depends on the outside barometric pressure and the weight of the piston, both of which are constant. Water is still a compressed liquid at this state since it has not started to vaporize.

As more heat is transferred, the temperature will keep rising until it reaches 100°C (state 2, Fig. 3-7). At this point water is still a liquid, but any heat addition will cause some of the liquid to vaporize. That is, a phase-change process from liquid to vapor is about to take place. A liquid that is *about to vaporize* is called a **saturated liquid.** Therefore, state 2 is a saturated liquid state.

Saturated Vapor and Superheated Vapor

Once boiling starts, the temperature will stop rising until the liquid is completely vaporized. That is, the temperature will remain constant during the entire phase-change process if the pressure is held constant. This can easily be verified by placing a thermometer into boiling water on top of a stove. At sea level ($P = 1$ atm), the thermometer will always read 100°C if the pan is uncovered or covered with a light lid. During a boiling process, the only change we will observe is a large increase in the volume and a steady decline in the liquid level as a result of more liquid turning to vapor.

Midway about the vaporization line (state 3, Fig. 3-8), the cylinder contains equal amounts of liquid and vapor. As we continue transferring heat, the vaporization process will continue until the last drop of liquid is vaporized (state 4, Fig. 3-9). At this point, the entire cylinder is filled with vapor that is on the borderline of the liquid phase. Any heat loss from this vapor will cause some of the vapor to condense (phase change from vapor to liquid). A vapor that is *about to condense* is called a **saturated vapor.** Therefore, state 4 is a saturated vapor state. A substance at states between 2 and 4 is often referred to as a **saturated liquid–vapor mixture** since the *liquid and vapor phases coexist* in equilibrium at these states.

Once the phase-change process is completed, we are back to a single-phase region again (this time vapor), and further transfer of heat will result in an increase in both the temperature and the specific volume (Fig. 3-10). At state 5, the temperature of the vapor is, let us say, 300°C; and if we transfer some heat from the vapor, the temperature may drop somewhat but no condensation will take place as long as the temperature remains above 100°C (for $P = 1$ atm). A vapor that is *not about to condense* (i.e., not a saturated vapor) is called a **superheated vapor.** Therefore, water at state 5 is a superheated vapor. The constant-pressure phase-change process described above is illustrated on a *T-v* diagram in Fig. 3-11.

If the entire process described above is reversed by cooling the water while maintaining the pressure at the same value, the water will go back to state 1, retracing the same path, and in so doing, the amount of heat released will exactly match the amount of heat added during the heating process.

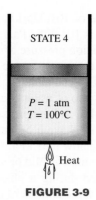

FIGURE 3-9

At 1 atm pressure, the temperature remains constant at 100°C until the last drop of liquid is vaporized (*saturated vapor*).

FIGURE 3-10

As more heat is transferred, the temperature of the vapor starts to rise (*superheated vapor*).

In our daily life, water implies liquid water and steam implies water vapor. In thermodynamics, however, both water and steam usually mean only one thing: H_2O.

Saturation Temperature and Saturation Pressure

It probably came as no surprise to you that the water started to boil at 100°C. Strictly speaking, the statement "water boils at 100°C" is incorrect. The correct statement is "water boils at 100°C at 1 atm pressure." The only reason the water started boiling at 100°C was because we held the pressure constant at 1 atm (101.325 kPa). If the pressure inside the cylinder were raised to 500 kPa by adding weights on top of the piston, the water would start boiling at 151.9°C. That is, *the temperature at which water starts boiling depends on the pressure; therefore, if the pressure is fixed, so is the boiling temperature.*

At a given pressure, the temperature at which a pure substance changes phase is called the **saturation temperature** T_{sat}. Likewise, at a given temperature, the pressure at which a pure substance changes phase is called the

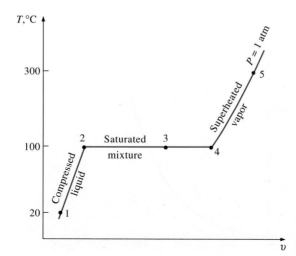

FIGURE 3-11

T-v diagram for the heating process of water at constant pressure.

TABLE 3-1

Saturation (boiling) pressure of water at various temperatures

Temperature, T,°C	Saturation pressure, P_{sat}, kPa
−10	0.26
−5	0.40
0	0.61
5	0.87
10	1.23
15	1.71
20	2.34
25	3.17
30	4.25
40	7.38
50	12.35
100	101.3 (1 atm)
150	475.8
200	1554
250	3973
300	8581

TABLE 3-2

Variation of the standard atmospheric pressure and the boiling (saturation) temperature of water with altitude

Elevation, m	Atmospheric pressure, kPa	Boiling temperature, °C
0	101.33	100.0
1,000	89.55	96.3
2,000	79.50	93.2
5,000	54.05	83.0
10,000	26.50	66.2
20,000	5.53	34.5

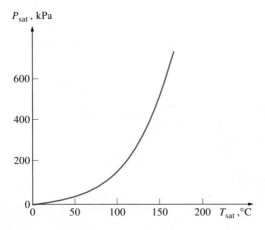

FIGURE 3-12

The liquid–vapor saturation curve of a pure substance (numerical values are for water).

saturation pressure P_{sat}. At a pressure of 101.325 kPa, T_{sat} is 100°C. Conversely, at a temperature of 100°C, P_{sat} is 101.325 kPa.

Saturation tables that list the saturation pressure against the temperature (or the saturation temperature against the pressure) are available for practically all substances. A partial listing of such a table is given in Table 3-1 for water. This table indicates that the pressure of water changing phase (boiling or condensing) at 25°C must be 3.17 kPa, and the pressure of water must be maintained at 3973 kPa (about 40 atm) to have it boil at 250°C. Also, water can be frozen by dropping its pressure below 0.61 kPa.

It takes a large amount of energy to melt a solid or vaporize a liquid. The amount of energy absorbed or released during a phase-change process is called the **latent heat.** More specifically, the amount of energy absorbed during melting is called the **latent heat of fusion** and is equivalent to the amount of energy released during freezing. Similarly, the amount of energy absorbed during vaporization is called the **latent heat of vaporization** and is equivalent to the energy released during condensation. The magnitudes of the latent heats depend on the temperature or pressure at which the phase change is occurring. At 1 atm pressure, the latent heat of fusion of water is 333.7 kJ/kg and the latent heat of vaporization is 2257.1 kJ/kg.

During a phase-change process, pressure and temperature are obviously dependent properties, and there is a definite relation between them, that is, $T_{sat} = f(P_{sat})$. A plot of T_{sat} versus P_{sat}, such as the one given for water in Fig. 3-12, is called a **liquid–vapor saturation curve.** A curve of this kind is characteristic of all pure substances.

It is clear from Fig. 3-12 that T_{sat} increases with P_{sat}. Thus, a substance at higher pressures will boil at higher temperatures. In the kitchen, higher boiling temperatures mean shorter cooking times and energy savings. A beef stew, for example, may take 1 to 2 h to cook in a regular pan that operates at 1 atm pressure, but only 20 to 30 min in a pressure cooker operating at 2 atm absolute pressure (corresponding boiling temperature: 120°C).

The atmospheric pressure, and thus the boiling temperature of water, decreases with elevation. Therefore, it takes longer to cook at higher altitudes than it does at sea level (unless a pressure cooker is used). For example, the standard atmospheric pressure at an elevation of 2000 m is 79.50 kPa, which corresponds to a boiling temperature of 93.2°C as opposed to 100°C at sea level (zero elevation). The variation of the boiling temperature of water with altitude at standard atmospheric conditions is given in Table 3-2. For each

1000 m increase in elevation, the boiling temperature drops by a little over 3°C. Note that the atmospheric pressure at a location, and thus the boiling temperature, changes slightly with the weather conditions. But the corresponding change in the boiling temperature is no more than about 1°C.

Some Consequences of T_{sat} and P_{sat} Dependence

We mentioned earlier that a substance at a specified pressure will boil at the saturation temperature corresponding to that pressure. This phenomenon allows us to control the boiling temperature of a substance by simply controlling the pressure, and it has numerous applications in practice. Below we give some examples. In most cases, the natural drive to achieve phase equilibrium by allowing some liquid to evaporate is at work behind the scenes.

Consider a sealed can of *liquid refrigerant-134a* in a room at 25°C. If the can has been in the room long enough, the temperature of the refrigerant in the can will also be 25°C. Now, if the lid is opened slowly and some refrigerant is allowed to escape, the pressure in the can will start dropping until it reaches the atmospheric pressure. If you are holding the can, you will notice its temperature dropping rapidly, and even ice forming outside the can if the air is humid. A thermometer inserted in the can will register −26°C when the pressure drops to 1 atm, which is the saturation temperature of refrigerant-134a at that pressure. The temperature of the liquid refrigerant will remain at −26°C until the last drop of it vaporizes.

Another aspect of this interesting physical phenomenon is that a liquid cannot vaporize unless it absorbs energy in the amount of the latent heat of vaporization, which is 217 kJ/kg for refrigerant-134a at 1 atm. Therefore, the rate of vaporization of the refrigerant depends on the rate of heat transfer to the can: the larger the rate of heat transfer, the higher the rate of vaporization. The rate of heat transfer to the can and thus the rate of vaporization of the refrigerant can be minimized by insulating the can heavily. In the limiting case of no heat transfer, the refrigerant will remain in the can as a liquid at −26°C indefinitely.

The boiling temperature of *nitrogen* at atmospheric pressure is −196°C (see Table A-3a). This means the temperature of liquid nitrogen exposed to the atmosphere must be −196°C since some nitrogen will be evaporating. The temperature of liquid nitrogen will remain constant at −196°C until it is depleted. For this reason, nitrogen is commonly used in low-temperature scientific studies (such as superconductivity) and cryogenic applications to maintain a test chamber at a constant temperature of −196°C. This is done by placing the test chamber into a liquid nitrogen bath that is open to the atmosphere. Any heat transfer from the environment to the test section is absorbed by the nitrogen, which evaporates isothermally and keeps the test chamber temperature constant at −196°C (Fig. 3-13). The entire test section must be insulated heavily to minimize heat transfer and thus liquid nitrogen consumption. Liquid nitrogen is also used for medical purposes to burn off unsightly spots on the skin. This is done by soaking a cotton swab in liquid nitrogen and wetting the desired area with it. As the nitrogen evaporates, it freezes the affected skin by rapidly absorbing heat from it.

A practical way of cooling leafy vegetables is **vacuum cooling,** which is based on *reducing the pressure* of the sealed cooling chamber to the saturation pressure at the desired low temperature and evaporating some water from the products to be cooled. The heat of vaporization during evaporation is absorbed from the products, which lowers the product temperature. The

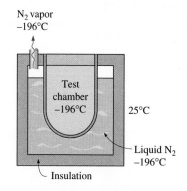

FIGURE 3-13

The temperature of liquid nitrogen exposed to the atmosphere remains constant at −196°C, and thus it maintains the test chamber at −196°C.

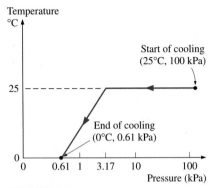

FIGURE 3-14

The variation of the temperature of fruits and vegetables with pressure during vacuum cooling from 25°C to 0°C.

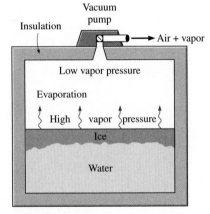

FIGURE 3-15

In 1775, ice was made by evacuating the air in a water tank.

saturation pressure of water at 0°C is 0.61 kPa, and the products can be cooled to 0°C by lowering the pressure to this level. The cooling rate can be increased by lowering the pressure below 0.61 kPa, but this is not desirable because of the danger of freezing and the added cost.

In vacuum cooling, there are two distinct stages. In the first stage, the products at ambient temperature, say at 25°C, are loaded into the flash chamber, and the operation begins. The temperature in the chamber remains constant until the *saturation pressure* is reached, which is 3.17 kPa at 25°C. In the second stage that follows, saturation conditions are maintained inside at progressively *lower pressures* and the corresponding *lower temperatures* until the desired temperature, usually slightly above 0°C, is reached (Fig. 3-14).

Vacuum cooling is usually more expensive than the conventional refrigerated cooling, and its use is limited to applications that result in much faster cooling. Products with large surface area per unit mass and a high tendency to release moisture such as *lettuce* and *spinach* are well-suited for vacuum cooling. Products with low surface area to mass ratio are not suitable, especially those that have relatively impervious peels such as tomatoes and cucumbers. Some products such as mushrooms and green peas can be vacuum cooled successfully by wetting them first.

The vacuum cooling described above becomes **vacuum freezing** if the pressure (actually, the vapor pressure) in the vacuum chamber is dropped below 0.6 kPa, the saturation pressure of water at 0°C. The idea of making ice by using a vacuum pump is nothing new. Dr. William Cullen actually made ice in Scotland in 1775 by evacuating the air in a water tank (Fig. 3-15).

Package icing is commonly used in small-scale cooling applications to remove heat and keep the products cool during transit by taking advantage of the large latent heat of fusion of water, but its use is limited to products that are not harmed by contact with ice. Also, ice provides *moisture* as well as *refrigeration.*

3-4 ■ PROPERTY DIAGRAMS FOR PHASE-CHANGE PROCESSES

The variations of properties during phase-change processes are best studied and understood with the help of property diagrams. Below we develop and discuss the *T-v*, *P-v*, and *P-T* diagrams for pure substances.

1 The *T-v* Diagram

The phase-change process of water at 1 atm pressure was described in detail in the last section and plotted on a *T-v* diagram in Fig. 3-11. Now we repeat this process at different pressures to develop the *T-v* diagram for water.

Let us add weights on top of the piston until the pressure inside the cylinder reaches 1 MPa. At this pressure, water will have a somewhat smaller specific volume than it did at 1 atm pressure. As heat is transferred to the water at this new pressure, the process will follow a path that looks very much like the process path at 1 atm pressure, as shown in Fig. 3-16, but there are some noticeable differences. First, water will start boiling at a much higher temperature (179.9°C) at this pressure. Second, the specific volume of the saturated liquid is larger and the specific volume of the saturated vapor is smaller than the corresponding values at 1 atm pressure. That is, the horizontal line that connects the saturated liquid and saturated vapor states is much shorter.

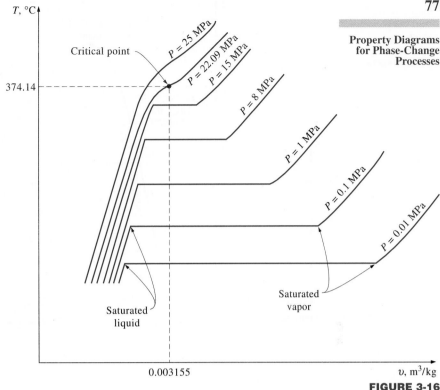

FIGURE 3-16

T-v diagram of constant-pressure phase-change processes of a pure substance at various pressures (numerical values are for water).

As the pressure is increased further, this saturation line will continue to get shorter, as shown in Fig. 3-16, and it will become a point when the pressure reaches 22.09 MPa for the case of water. This point is called the **critical point,** and it may be defined as *the point at which the saturated liquid and saturated vapor states are identical.*

The temperature, pressure, and specific volume of a substance at the critical point are called, respectively, the *critical temperature T_{cr}, critical pressure P_{cr},* and *critical specific volume v_{cr}.* The critical-point properties of water are $P_{cr} = 22.09$ MPa, $T_{cr} = 374.14°C$, and $v_{cr} = 0.003155$ m³/kg. For helium, they are 0.23 MPa, $-267.85°C$, and 0.01444 m³/kg. The critical properties for various substances are given in Table A-1 in the appendix.

At pressures above the critical pressure, there will not be a distinct phase-change process (Fig. 3-17). Instead, the specific volume of the substance will continually increase, and at all times there will be only one phase present. Eventually, it will resemble a vapor, but we can never tell when the change has occurred. Above the critical state, there is no line that separates the compressed liquid region and the superheated vapor region. However, it is customary to refer to the substance as superheated vapor at temperatures above the critical temperature and as compressed liquid at temperatures below the critical temperature.

The saturated liquid states in Fig. 3-16 can be connected by a line called the **saturated liquid line,** and saturated vapor states in the same figure can be connected by another line, called the **saturated vapor line.** These two lines meet at the critical point, forming a dome as shown in Fig. 3-18. All the

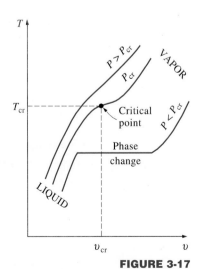

FIGURE 3-17

At supercritical pressures ($P > P_{cr}$), there is no distinct phase-change (boiling) process.

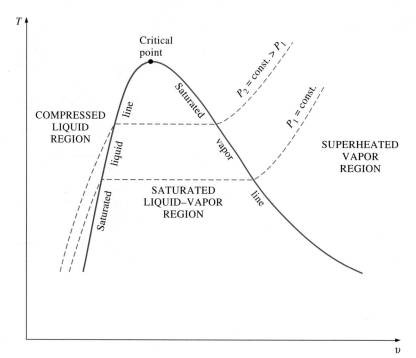

FIGURE 3-18

T-v diagram of a pure substance.

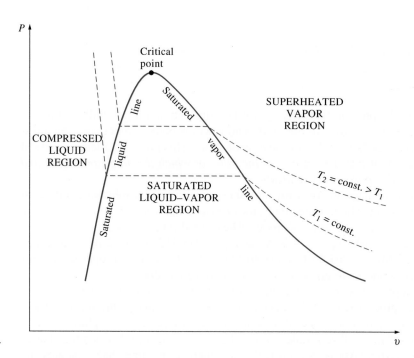

FIGURE 3-19

P-v diagram of a pure substance.

compressed liquid states are located in the region to the left of the saturated liquid line, called the **compressed liquid region.** All the superheated vapor states are located to the right of the saturated vapor line, called the **super-heated vapor region.** In these two regions, the substance exists in a single phase, a liquid or a vapor. All the states that involve both phases in equilibrium are located under the dome, called the **saturated liquid–vapor mixture region,** or the **wet region.**

The general shape of the *P-v* diagram of a pure substance is very much like the *T-v* diagram, but the *T* = constant lines on this diagram have a downward trend, as shown in Fig. 3-19.

Consider again a piston-cylinder device that contains liquid water at 1 MPa and 150°C. Water at this state exists as a compressed liquid. Now the weights on top of the piston are removed one by one so that the pressure inside the cylinder decreases gradually (Fig. 3-20). The water is allowed to exchange heat with the surroundings so its temperature remains constant. As the pressure decreases, the volume of the water will increase slightly. When the pressure reaches the saturation-pressure value at the specified temperature (0.4758 MPa), the water will start to boil. During this vaporization process, both the temperature and the pressure remain constant, but the specific volume increases. Once the last drop of liquid is vaporized, further reduction in pressure results in a further increase in specific volume. Notice that during the phase-change process, we did not remove any weights. Doing so would cause the pressure and therefore the temperature to drop [since $T_{sat} = f(P_{sat})$], and the process would no longer be isothermal.

If the process is repeated for other temperatures, similar paths will be obtained for the phase-change processes. Connecting the saturated liquid and the saturated vapor states by a curve, we obtain the *P-v* diagram of a pure substance, as shown in Fig. 3-19.

Extending the Diagrams to Include the Solid Phase

The two equilibrium diagrams developed so far represent the equilibrium states involving the liquid and the vapor phases only. But these diagrams can easily be extended to include the solid phase as well as the solid–liquid and the solid–vapor saturation regions. The basic principles discussed in conjunction with the liquid–vapor phase-change process apply equally to the solid–liquid and solid–vapor phase-change processes. Most substances contract during a solidification (i.e., freezing) process. Others, like water, expand as they freeze. The *P-v* diagrams for both groups of substances are given in Figs. 3-21 and 3-22. These two diagrams differ only in the solid–liquid saturation region. The *T-v* diagrams look very much like the *P-v* diagrams, especially for substances that contract on freezing.

The fact that water expands upon freezing has vital consequences in nature. If water contracted on freezing as most other substances do, the ice formed would be heavier than the liquid water, and it would settle to the bottom of rivers, lakes, and oceans instead of floating at the top. The sun's rays would never reach these ice layers, and the bottoms of many rivers, lakes, and oceans would be covered with ice year round, seriously disrupting marine life.

We are all familiar with two phases being in equilibrium, but under some conditions all three phases of a pure substance coexist in equilibrium (Fig. 3-23). On *P-v* or *T-v* diagrams, these triple-phase states form a line called the **triple line.** The states on the triple line of a substance have the same pressure and temperature but different specific volumes. The triple line appears as a point on the *P-T* diagrams and, therefore, is often called the **triple point.** The triple-point temperatures and pressures of various substances are given in Table 3-3. For water, the triple-point temperature and pressure are 0.01°C and 0.6113 kPa, respectively. That is, all three phases of water will exist in equilibrium only if the temperature and pressure have precisely these values. No

$P = 1$ MPa
$T = 150$°C

Heat

FIGURE 3-20

The pressure in a piston-cylinder device can be reduced by reducing the weight of the piston.

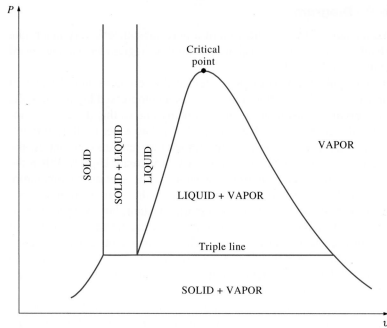

FIGURE 3-21

P-v diagram of a substance that contracts on freezing.

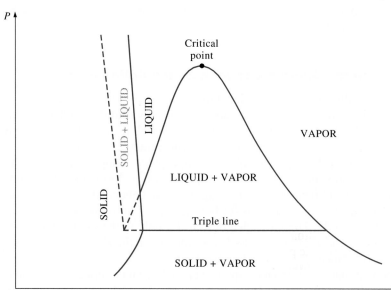

FIGURE 3-22

P-v diagram of a substance that expands on freezing (such as water).

FIGURE 3-23

At triple-point pressure and temperature, a substance exists in three phases in equilibrium.

substance can exist in the liquid phase in stable equilibrium at pressures below the triple-point pressure. The same can be said for temperature for substances that contract on freezing. However, substances at high pressures can exist in the liquid phase at temperatures below the triple-point temperature. For example, water cannot exist in liquid form in equilibrium at atmospheric conditions at temperatures below 0°C, but it can exist as a liquid at −20°C at 200 MPa pressure. Also, ice exists at seven different solid phases at pressures above 100 MPa.

There are two ways a substance can pass from the solid to vapor phase: either it melts first into a liquid and subsequently evaporates, or it evaporates directly without melting first. The latter occurs at pressures below the triple-point value, since a pure substance cannot exist in the liquid phase at those pressures (Fig. 3-24). Passing from the solid phase directly into the vapor

TABLE 3-3

Triple-point temperatures and pressures of various substances

Substance	Formula	T_{tp}, K	P_{tp}, kPa
Acetylene	C_2H_2	192.4	120
Ammonia	NH_3	195.40	6.076
Argon	A	83.81	68.9
Carbon (graphite)	C	3900	10,100
Carbon dioxide	CO_2	216.55	517
Carbon monoxide	CO	68.10	15.37
Deuterium	D_2	18.63	17.1
Ethane	C_2H_6	89.89	8×10^{-4}
Ethylene	C_2H_4	104.0	0.12
Helium 4 (λ point)	He	2.19	5.1
Hydrogen	H_2	13.84	7.04
Hydrogen chloride	HCl	158.96	13.9
Mercury	Hg	234.2	1.65×10^{-7}
Methane	CH_4	90.68	11.7
Neon	Ne	24.57	43.2
Nitric oxide	NO	109.50	21.92
Nitrogen	N_2	63.18	12.6
Nitrous oxide	N_2O	182.34	87.85
Oxygen	O_2	54.36	0.152
Palladium	Pd	1825	3.5×10^{-3}
Platinum	Pt	2045	2.0×10^{-4}
Sulfur dioxide	SO_2	197.69	1.67
Titanium	Ti	1941	5.3×10^{-3}
Uranium hexafluoride	UF_6	337.17	151.7
Water	H_2O	273.16	0.61
Xenon	Xe	161.3	81.5
Zinc	Zn	692.65	0.065

Source: Data from National Bureau of Standards (U.S.) Circ., 500 (1952).

Property Diagrams for Phase-Change Processes

FIGURE 3-24

At low pressures (below the triple-point value), solids evaporate without melting first (*sublimation*).

phase is called **sublimation.** For substances that have a triple-point pressure above the atmospheric pressure such as solid CO_2 (dry ice), sublimation is the only way to change from the solid to vapor phase at atmospheric conditions.

3 The *P-T* Diagram

Figure 3-25 shows the *P-T* diagram of a pure substance. This diagram is often called the **phase diagram** since all three phases are separated from each other by three lines. The sublimation line separates the solid and vapor regions, the vaporization line separates the liquid and vapor regions, and the melting (or fusion) line separates the solid and liquid regions. These three lines meet at the triple point, where all three phases coexist in equilibrium. The vaporization line ends at the critical point because no distinction can be made between liquid and vapor phases above the critical point. Substances that expand and contract on freezing differ only in the melting line on the *P-T* diagram.

The *P-v-T* Surface

In Chap. 2, we indicated that the state of a simple compressible substance is fixed by any two independent, intensive properties. Once the two appropriate

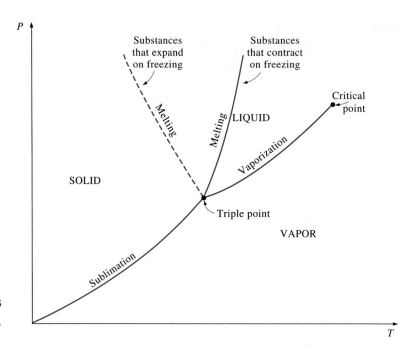

FIGURE 3-25

P-T diagram of pure substances.

properties are fixed, all the other properties become dependent properties. Remembering that any equation with two independent variables in the form $z = z(x, y)$ represents a surface in space, we can represent the *P-v-T* behavior of a substance as a surface in space, as shown in Figs. 3-26 and 3-27. Here *T* and *v* may be viewed as the independent variables (the base) and *P* as the dependent variable (the height).

All the points on the surface represent equilibrium states. All states along the path of a quasi-equilibrium process lie on the *P-v-T* surface since such a process must pass through equilibrium states. The single-phase regions appear

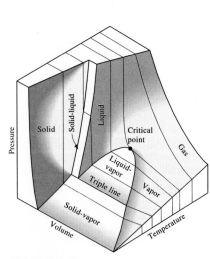

FIGURE 3-26

P-v-T surface of a substance that *contracts* on freezing.

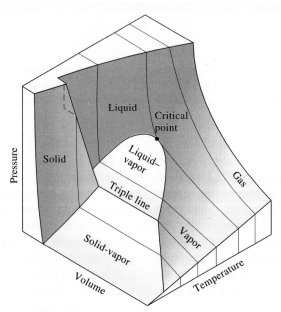

FIGURE 3-27

P-v-T surface of a substance that *expands* on freezing (like water).

as curved surfaces on the P-v-T surface, and the two-phase regions as surfaces perpendicular to the P-T plane. This is expected since the projections of two-phase regions on the P-T plane are lines.

All the two-dimensional diagrams we have discussed so far are merely projections of this three-dimensional surface onto the appropriate planes. A P-v diagram is just a projection of the P-v-T surface on the P-v plane, and a T-v diagram is nothing more than the bird's-eye view of this surface. The P-v-T surfaces present a great deal of information at once, but in a thermodynamic analysis it is more convenient to work with two-dimensional diagrams, such as the P-v and T-v diagrams.

3-5 ■ VAPOR PRESSURE AND PHASE EQUILIBRIUM

The pressure in a gas container is due to the individual molecules striking the wall of the container and exerting a force on it. This force is proportional to the average velocity of the molecules and the number of molecules per unit volume of the container (i.e., molar density). Therefore, the pressure exerted by a gas is a strong function of the density and the temperature of the gas. For a gas mixture, the pressure measured by a sensor such as a transducer is the sum of the pressures exerted by the individual gas species, called the *partial pressure*. It can be shown (see Chap. 12 of Ref. 4) that the partial pressure of a gas in a mixture is proportional to the number of moles (or the mole fraction) of that gas.

Atmospheric air can be viewed as a mixture of dry air (air with zero moisture content) and water vapor (also referred to as moisture), and the atmospheric pressure is the sum of the pressure of dry air P_a and the pressure of water vapor, called the **vapor pressure** P_v (Fig. 3-28). That is,

$$P_{atm} = P_a + P_v \qquad (3\text{-}1)$$

The vapor pressure constitutes a small fraction (usually under 3 percent) of the atmospheric pressure since air is mostly nitrogen and oxygen, and the water molecules constitute a small fraction (usually under 3 percent) of the total molecules in the air. However, the amount of water vapor in the air has a major impact on thermal comfort and many processes such as drying.

Air can hold a certain amount of moisture only, and the ratio of the actual amount of moisture in the air at a given temperature to the maximum amount air can hold at that temperature is called the **relative humidity** ϕ. The relative humidity ranges from 0 for dry air to 100 percent for **saturated air** (air that cannot hold any more moisture). The vapor pressure of saturated air at a given temperature is equal to the saturation pressure of water at that temperature. For example, the vapor pressure of saturated air at 25°C is 3.17 kPa.

The amount of moisture in the air is completely specified by the temperature and the relative humidity, and the vapor pressure is related to relative humidity ϕ by

$$P_v = \phi P_{sat\,@\,T} \qquad (3\text{-}2)$$

where $P_{sat\,@\,T}$ is the saturation pressure of water at the specified temperature. For example, the vapor pressure of air at 25°C and 60 percent relative humidity is

$$P_v = \phi P_{sat\,@\,25°C} = 0.6 \times (3.17 \text{ kPa}) = 1.90 \text{ kPa}$$

$P_{atm} = P_a + P_v$

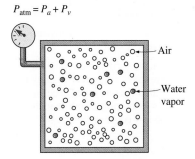

FIGURE 3-28

Atmospheric pressure is the sum of the dry air pressure P_a and the vapor pressure P_v.

The desirable range of relative humidity for thermal comfort is 40 to 60 percent.

Note that the amount of moisture air can hold is proportional to the saturation pressure, which increases with temperature. Therefore, air can hold more moisture at higher temperatures. Dropping the temperature of moist air reduces its moisture capacity and may result in the condensation of some of the moisture in the air as suspended water droplets (fog) or as a liquid film on cold surfaces (dew). So it is no surprise that fog and dew are common occurrences at humid locations especially in the early morning hours when the temperatures are the lowest. Both fog and dew disappear (evaporate) as the air temperature rises shortly after sunrise. You also may have noticed that electronic devices such as camcorders come with warnings against bringing them into moist indoors when the devices are cold to avoid moisture condensation on the sensitive electronics of the devices.

It is a common observation that whenever there is an imbalance of a commodity in a medium, nature tends to redistribute it until a "balance" or "equality" is established. This tendency is often referred to as the *driving force,* which is the mechanism behind many naturally occurring transport phenomena such as heat transfer, fluid flow, electric current, and mass transfer. If we define the amount of a commodity per unit volume as the *concentration* of that commodity, we can say that the flow of a commodity is always in the direction of decreasing concentration, that is, from the region of high concentration to the region of low concentration (Fig. 3-29). The commodity simply creeps away during redistribution, and thus the flow is a *diffusion process.*

We know from experience that a wet T-shirt hanging in an open area eventually dries, a small amount of water left in a glass evaporates, and the aftershave in an open bottle quickly disappears. These and many other similar examples suggest that there is a driving force between the two phases of a substance that forces the mass to transform from one phase to another. The magnitude of this force depends on the relative concentrations of the two phases. A wet T-shirt will dry much faster in dry air than it would in humid air. In fact, it will not dry at all if the relative humidity of the environment is 100 percent and thus the air is saturated. In this case, there will be no transformation from the liquid phase to the vapor phase, and the two phases will be in **phase equilibrium.** For liquid water that is open to the atmosphere, the criterion for phase equilibrium can be expressed as follows: *The vapor pressure in the air must be equal to the saturation pressure of water at the water temperature.* That is (Fig. 3-30),

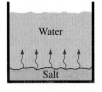

(a) Before   (b) After

FIGURE 3-29

Whenever there is a concentration difference of a physical quantity in a medium, nature tends to equalize things by forcing a flow from the high to the low concentration region.

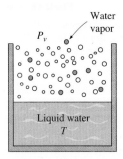

FIGURE 3-30

When open to the atmosphere, water is in phase equilibrium with the vapor in the air if the vapor pressure is equal to the saturation temperature of water.

$$\text{Phase equilibrium criterion for water exposed to air: } P_v = P_{\text{sat @ } T} \qquad (3\text{-}3)$$

Therefore, if the vapor pressure in the air is less than the saturation pressure of water at the water temperature, some liquid will evaporate. The larger the difference between the vapor and saturation pressures, the higher the rate of evaporation. The evaporation will have a cooling effect on water, and thus reduce its temperature. This, in turn, will reduce the saturation pressure of water and thus the rate of evaporation until some kind of quasi-steady operation is reached. This explains why water is usually at a considerably lower temperature than the surrounding air, especially in dry climates. It also suggests that the rate of evaporation of water can be increased by increasing the water temperature and thus the saturation pressure of water.

Note that the air at the water surface will always be saturated because of the direct contact with water, and thus the vapor pressure. Therefore, the vapor pressure at the lake surface will simply be the saturation pressure of water at the temperature of the water at the surface. If the air is not saturated, then the vapor pressure will decrease to the value in the air at some distance from the water surface, and the difference between these two vapor pressures is the driving force for the evaporation of water.

EXAMPLE 3-1   Temperature Drop of a Lake Due to Evaporation

On a summer day, the air temperature over a lake is measured to be 25°C. Determine water temperature of the lake when phase equilibrium conditions are established between the water in the lake and the vapor in the air for relative humidities of 10, 80, and 100 percent for the air (Fig. 3-31).

Solution   The saturation pressure of water at 25°C, from Table 3-1, is 3.17 kPa. Then the vapor pressures at relative humidities of 10, 80, and 100 percent are determined from Eq. 3-2 to be

Relative humidity = 10%: $P_{v1} = \phi_1 P_{sat\,@\,25°C} = 0.1 \times (3.17\ \text{kPa}) = 0.317\ \text{kPa}$

Relative humidity = 80%: $P_{v2} = \phi_2 P_{sat\,@\,25°C} = 0.8 \times (3.17\ \text{kPa}) = 2.536\ \text{kPa}$

Relative humidity = 100%: $P_{v3} = \phi_3 P_{sat\,@\,25°C} = 1.0 \times (3.17\ \text{kPa}) = 3.17\ \text{kPa}$

The saturation temperatures corresponding to these pressures are determined from Table 3-1 by interpolation to be

$$T_1 = -8.0°C \qquad T_2 = 21.2°C \qquad \text{and} \qquad T_3 = 25°C$$

Therefore, water will freeze in the first case even though the surrounding air is hot. In the last case the water temperature will be the same as the surrounding air temperature.

Discussion   You are probably skeptical about the lake freezing when the air is at 25°C, and you are right. The water temperature will drop to −8°C in the limiting case of no heat transfer to the water surface. In practice the water temperature will drop below the air temperature, but it will not drop to −8°C because (1) it is very unlikely for the air over the lake to be so dry (a relative humidity of just 10 percent) and (2) as the water temperature near the surface drops, heat transfer from the air and the lower parts of the water body will tend to make up for this heat loss and prevent the water temperature from dropping too much. The water temperature will stabilize when the heat gain from the surrounding air and the water body equals the heat loss by evaporation, that is, when a *dynamic balance* is established between heat and mass transfer instead of phase equilibrium. If you try this experiment using a shallow layer of water in a well-insulated pan, you can actually freeze the water if the air is really dry and relatively cool.

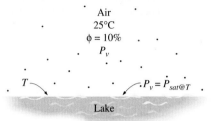

FIGURE 3-31
Schematic for Example 3-1.

The natural tendency of water to evaporate in order to achieve phase equilibrium with the water vapor in the surrounding air forms the basis for the operation of the **evaporative coolers** (also called the *swamp coolers*). In such coolers, hot and dry outdoor air is forced to flow through a wet cloth before entering a building. Some of the water evaporates by absorbing heat from the air, and thus cooling it. Evaporative coolers are commonly used in dry climates and provide effective cooling. They are much cheaper to run than air conditioners since they are inexpensive to buy, and the fan of an evaporative cooler consumes much less power than the compressor of an air conditioner.

Boiling and evaporation are often used interchangeably to indicate *phase change from liquid to vapor.* Although they refer to the same physical process, they differ in some aspects. **Evaporation** occurs at the *liquid–vapor interface*

Boiling   Evaporation

FIGURE 3-32

A liquid-to-vapor phase change process is called *evaporation* if it occurs at a liquid–vapor interface, and *boiling* if it occurs at a solid–liquid interface.

when the vapor pressure is less than the saturation pressure of the liquid at a given temperature. Water in a lake at 20°C, for example, will evaporate to air at 20°C and 60 percent relative humidity since the saturation pressure of water at 20°C is 2.34 kPa, and the vapor pressure of air at 20°C and 60 percent relative humidity is 1.4 kPa. Other examples of evaporation are the drying of clothes, fruits, and vegetables; the evaporation of sweat to cool the human body; and the rejection of waste heat in wet cooling towers. Note that evaporation involves no bubble formation or bubble motion (Fig. 3-32).

Boiling, on the other hand, occurs at the *solid–liquid interface* when a liquid is brought into contact with a surface maintained at a temperature T_s sufficiently above the saturation temperature T_{sat} of the liquid. At 1 atm, for example, liquid water in contact with a solid surface at 110°C will boil since the saturation temperature of water at 1 atm is 100°C. The boiling process is characterized by the rapid motion of *vapor bubbles* that form at the solid–liquid interface, detach from the surface when they reach a certain size, and attempt to rise to the free surface of the liquid. When cooking, we do not say water is boiling unless we see the bubbles rising to the top.

3-6 ■ PROPERTY TABLES

For most substances, the relationships among thermodynamic properties are too complex to be expressed by simple equations. Therefore, properties are frequently presented in the form of tables. Some thermodynamic properties can be measured easily, but others cannot and are calculated by using the relations between them and measurable properties. The results of these measurements and calculations are presented in tables in a convenient format. In the following discussion, the steam tables will be used to demonstrate the use of thermodynamic property tables. Property tables of other substances are used in the same manner.

For each substance, the thermodynamic properties are listed in more than one table. In fact, a separate table is prepared for each region of interest such as the superheated vapor, compressed liquid, and saturated (mixture) regions. Property tables are given in the appendix in both SI and English units. The tables in English units carry the same number as the corresponding tables in SI, followed by an identifier E. Tables A-6 and A-6E, for example, list properties of superheated water vapor, the former in SI and the latter in English units. Before we get into the discussion of property tables, we will define a new property called *enthalpy*.

Enthalpy—A Combination Property

A person looking at the tables carefully will notice two new properties: enthalpy h and entropy s. Entropy is a property associated with the second law of thermodynamics, and we will not use it until it is properly defined in Chap. 7. However, it is appropriate to introduce enthalpy at this point.

In the analysis of certain types of processes, particularly in power generation and refrigeration (Fig. 3-33), we frequently encounter the combination of properties $U + PV$. For the sake of simplicity and convenience, this combination is defined as a new property, **enthalpy,** and given the symbol H:

$$H = U + PV \qquad \text{(kJ)}$$

or, per unit mass,

$$h = u + Pv \qquad \text{(kJ/kg)} \qquad (3\text{-}4)$$

Both the total enthalpy H and specific enthalpy h are simply referred to as enthalpy since the context will clarify which one is meant. Notice that the equations given above are dimensionally homogeneous. That is, the unit of the pressure–volume product may differ from the unit of the internal energy by only a factor (Fig. 3-34). For example, it can be easily shown that $1 \text{ kPa} \cdot \text{m}^3 = 1 \text{ kJ}$. In some tables encountered in practice, the internal energy u is frequently not listed, but it can always be determined from $u = h - Pv$.

The widespread use of the property enthalpy is due to Professor Richard Mollier, who recognized the importance of the group $u + Pv$ in the analysis of steam turbines and in the representation of the properties of steam in tabular and graphical form (as in the famous Mollier chart). Mollier referred to the group $u + Pv$ as *heat contents* and *total heat*. These terms were not quite consistent with the modern thermodynamic terminology and were replaced in the 1930s by the term *enthalpy* (from the Greek word *enthalpien,* which means *to heat*).

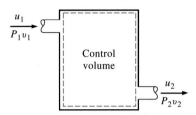

FIGURE 3-33

The combination $u + Pv$ is frequently encountered in the analysis of control volumes.

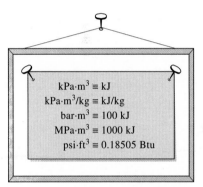

FIGURE 3-34

The product *pressure × volume* has energy units.

1a   Saturated Liquid and Saturated Vapor States

The properties of saturated liquid and saturated vapor for water are listed in Tables A-4 and A-5. Both tables give the same information. The only difference is that in Table A-4 properties are listed under temperature and in Table A-5 under pressure. Therefore, it is more convenient to use Table A-4 when *temperature* is given and Table A-5 when *pressure* is given. The use of Table A-4 is illustrated in Fig. 3-35.

The subscript f is used to denote properties of a saturated liquid, and the subscript g to denote the properties of a saturated vapor. These symbols are commonly used in thermodynamics and originated from German. Another subscript commonly used is fg, which denotes the difference between the saturated vapor and saturated liquid values of the same property. For example,

v_f = specific volume of saturated liquid
v_g = specific volume of saturated vapor
v_{fg} = difference between v_g and v_f (that is, $v_{fg} = v_g - v_f$)

The quantity h_{fg} is called the **enthalpy of vaporization** (or latent heat of vaporization). It represents the amount of energy needed to vaporize a unit mass of saturated liquid at a given temperature or pressure. It decreases as the temperature or pressure increases, and becomes zero at the critical point.

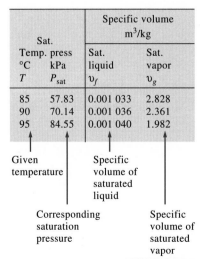

FIGURE 3-35

A partial list of Table A-4.

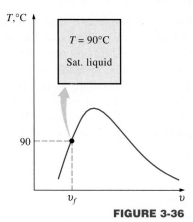

FIGURE 3-36

Schematic and T-v diagram
for Example 3-2.

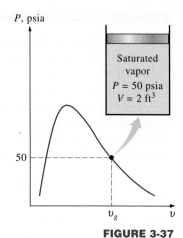

FIGURE 3-37

Schematic and P-v diagram
for Example 3-3.

EXAMPLE 3-2   Finding the Pressure of Saturated Liquid

A rigid tank contains 50 kg of saturated liquid water at 90°C. Determine the pressure in the tank and the volume of the tank.

Solution   The state of the saturated liquid water is shown on a T-v diagram in Fig. 3-36. Since saturation conditions exist in the tank, the pressure must be the saturation pressure at 90°C:

$$P = P_{\text{sat @ 90°C}} = \textbf{70.14 kPa} \qquad \text{(Table A-4)}$$

The specific volume of the saturated liquid at 90°C is

$$v = v_{f\text{@ 90°C}} = 0.001036 \text{ m}^3/\text{kg} \qquad \text{(Table A-4)}$$

Then the total volume of the tank is determined to be

$$V = mv = (50 \text{ kg})(0.001036 \text{ m}^3/\text{kg}) = \textbf{0.0518 m}^3$$

EXAMPLE 3-3   Finding the Temperature of Saturated Vapor

A piston-cylinder device contains 2 ft³ of saturated water vapor at 50-psia pressure. Determine the temperature of the vapor and the mass of the vapor inside the cylinder.

Solution   The state of the saturated water vapor is shown on a P-v diagram in Fig. 3-37. Since the cylinder contains saturated vapor at 50 psia, the temperature inside must be the saturation temperature at this pressure:

$$T = T_{\text{sat @ 50 psia}} = \textbf{281.03°F} \qquad \text{(Table A-5E)}$$

The specific volume of the saturated vapor at 50 psia is

$$v = v_{g\text{@ 50 psia}} = 8.518 \text{ ft}^3/\text{lbm} \qquad \text{(Table A-5E)}$$

Then the mass of water vapor inside the cylinder becomes

$$m = \frac{V}{v} = \frac{2 \text{ ft}^3}{8.518 \text{ ft}^3/\text{lbm}} = 0.235 \text{ lbm}$$

EXAMPLE 3-4 The Volume and Energy Change during Evaporation

A mass of 200 g of saturated liquid water is completely vaporized at a constant pressure of 100 kPa. Determine (a) the volume change and (b) the amount of energy added to the water.

Solution (a) The process described is illustrated on a P-v diagram in Fig. 3-38. The volume change per unit mass during a vaporization process is vfg, which is the difference between vg and vf. Reading these values from Table A-5 at 100 kPa and substituting yield

$$v_{fg} = v_g - v_f = (1.6940 - 0.001043) \text{ m}^3/\text{kg} = 1.6930 \text{ m}^3/\text{kg}$$

Thus, $\Delta V = mv_{fg} = (0.2 \text{ kg})(1.6930 \text{ m}^3/\text{kg}) = \textbf{0.3386 m}^3$

Note that we have considered the first four decimal digits of v_{fg} and disregarded the rest. This is because v_g has significant numbers to the first four decimal places only, and we do not know the numbers in the other decimal places. Copying all the digits from the calculator would mean that we are assuming $v_g = 1.694000$, which is not necessarily the case. It could very well be that $v_g = 1.694038$ since this number, too, would truncate to 1.6940. All the digits in our result (1.6930) are significant. But if we did not truncate the result, we would obtain $v_{fg} = 1.692957$, which falsely implies that our result is accurate to the sixth decimal place.

(b) The amount of energy needed to vaporize the unit mass of a substance at a given pressure is the enthalpy of vaporization at that pressure, which, at 100 kPa, is $h_{fg} = 2258.0 \text{ kJ/kg}$. Thus, the amount of energy added is

$$mh_{fg} = (0.2 \text{ kg})(2258 \text{ kJ/kg}) = \textbf{451.6 kJ}$$

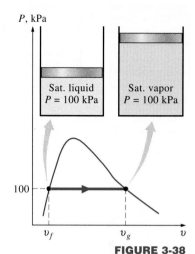

FIGURE 3-38

Schematic and *P-v* diagram for Example 3-4.

1b Saturated Liquid–Vapor Mixture

During a vaporization process, a substance exists as part liquid and part vapor. That is, it is a mixture of saturated liquid and saturated vapor (Fig. 3-39). To analyze this mixture properly, we need to know the proportions of the liquid and vapor phases in the mixture. This is done by defining a new property called the **quality** *x* as the ratio of the mass of vapor to the total mass of the mixture:

$$x = \frac{m_{\text{vapor}}}{m_{\text{total}}} \tag{3-5}$$

where $m_{\text{total}} = m_{\text{liquid}} + m_{\text{vapor}} = m_f + m_g$

Quality has significance for *saturated mixtures* only. It has no meaning in the compressed liquid or superheated vapor regions. Its value is always between 0 and 1. The quality of a system that consists of *saturated liquid* is 0 (or 0 percent), and the quality of a system consisting of *saturated vapor* is 1 (or 100 percent). In saturated mixtures, quality can serve as one of the two independent intensive properties needed to describe a state. Note that *the properties of the saturated liquid are the same whether it exists alone or in a mixture with saturated vapor.* During the vaporization process, only the amount of saturated liquid changes, not its properties. The same can be said about a saturated vapor.

A saturated mixture can be treated as a combination of two subsystems: the saturated liquid and the saturated vapor. However, the amount of mass for

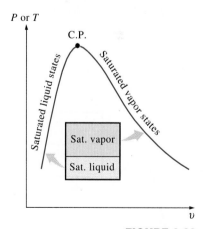

FIGURE 3-39

The relative amounts of liquid and vapor phases in a saturated mixture are specified by the *quality x.*

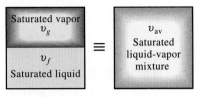

FIGURE 3-40

A two-phase system can be treated as a homogeneous mixture for convenience.

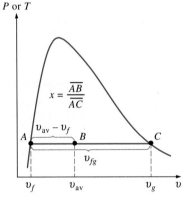

FIGURE 3-41

Quality is related to the horizontal distances on P-v and T-v diagrams.

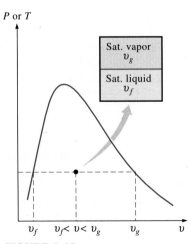

FIGURE 3-42

The v value of a saturated liquid–vapor mixture lies between the v_f and v_g values at the specified T or P.

each phase is usually not known. Therefore, it is often more convenient to imagine that the two phases are mixed very well, forming a homogeneous appearance (Fig. 3-40). Then the properties of this "mixture" will simply be the average properties of the saturated liquid–vapor mixture under consideration. Here is how it is done:

Consider a tank that contains a saturated liquid–vapor mixture. The volume occupied by saturated liquid is V_f, and the volume occupied by saturated vapor is V_g. The total volume V is the sum of these two:

$$V = V_f + V_g$$
$$V = mv \quad \longrightarrow \quad m_t v_{av} = m_f v_f + m_g v_g$$
$$m_f = m_t - m_g \quad \longrightarrow \quad m_t v_{av} = (m_t - m_g)v_f + m_g v_g$$

Dividing by m_t yields

$$v_{av} = (1 - x)v_f + xv_g$$

since $x = m_g/m_t$. This relation can also be expressed as

$$v_{av} = v_f + xv_{fg} \qquad \text{(m}^3\text{/kg)} \qquad (3\text{-}6)$$

where $v_{fg} = v_g - v_f$. Solving for quality, we obtain

$$x = \frac{v_{av} - v_f}{v_{fg}} \qquad (3\text{-}7)$$

Based on this equation, quality can be related to the horizontal distances on a *P-v* or *T-v* diagram (Fig. 3-41). At a given temperature or pressure, the numerator of Eq. 3-7 is the distance between the actual state and the saturated liquid state, and the denominator is the length of the entire horizontal line that connects the saturated liquid and saturated vapor states. A state of 50 percent quality will lie in the middle of this horizontal line.

The analysis given above can be repeated for internal energy and enthalpy with the following results:

$$u_{av} = u_f + xu_{fg} \qquad \text{(kJ/kg)}$$
$$h_{av} = h_f + xh_{fg} \qquad \text{(kJ/kg)}$$

All the results are of the same format, and they can be summarized in a single equation as

$$y_{av} = y_f + xy_{fg} \qquad (3\text{-}8)$$

where y is v, u, or h. The subscript "av" (for "average") is usually dropped for simplicity. The values of the average properties of the mixtures are always *between* the values of the saturated liquid and the saturated vapor properties (Fig. 3-42). That is,

$$y_f \leq y_{av} \leq y_g$$

Finally, all the saturated-mixture states are located under the saturation curve, and to analyze saturated mixtures, all we need are saturated liquid and saturated vapor data (Tables A-4 and A-5 in the case of water).

EXAMPLE 3-5 The Pressure and Volume of a Saturated Mixture

A rigid tank contains 10 kg of water at 90°C. If 8 kg of the water is in the liquid form and the rest is in the vapor form, determine (a) the pressure in the tank and (b) the volume of the tank.

Solution (a) The state of the saturated liquid–vapor mixture is shown in Fig. 3-43. Since the two phases coexist in equilibrium, we have a saturated mixture and the pressure must be the saturation pressure at the given temperature:

$$P = P_{sat @ 90°C} = \textbf{70.14 kPa} \qquad \text{(Table A-4)}$$

(b) At 90°C, v_f and v_g values are $v_f = 0.001036$ m³/kg and $v_g = 2.361$ m³/kg (Table A-4).

One way of finding the volume of the tank is to determine the volume occupied by each phase and then add them:

$$V = V_f + V_g = m_f v_f + m_g v_g$$
$$= (8 \text{ kg})(0.001036 \text{ m}^3/\text{kg}) + (2 \text{ kg})(2.361 \text{ m}^3/\text{kg})$$
$$\cdot = \textbf{4.73 m}^3$$

Another way is to first determine the quality x, then the average specific volume v, and finally the total volume:

$$x = \frac{m_g}{m_t} = \frac{2 \text{ kg}}{10 \text{ kg}} = 0.2$$

$$v = v_f + x v_{fg}$$
$$= 0.001036 \text{ m}^3/\text{kg} + (0.2)[(2.361 - 0.001036) \text{ m}^3/\text{kg}]$$
$$= 0.473 \text{ m}^3/\text{kg}$$

and $\qquad V = mv = (10 \text{ kg})(0.473 \text{ m}^3/\text{kg}) = \textbf{4.73 m}^3$

Discussion The first method appears to be easier in this case since the masses of each phase are given. But in most cases, the masses of each phase are not available, and the second method becomes more convenient.

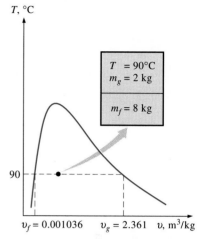

FIGURE 3-43

Schematic and *T-v* diagram for Example 3-5.

EXAMPLE 3-6 The Properties of Saturated Liquid–Vapor Mixture

An 80-L vessel contains 4 kg of refrigerant-134a at a pressure of 160 kPa. Determine (a) the temperature of the refrigerant, (b) the quality, (c) the enthalpy of the refrigerant, and (d) the volume occupied by the vapor phase.

Solution (a) The state of the saturated liquid–vapor mixture is shown in Fig. 3-44. At this point we do not know whether the refrigerant is in the compressed liquid, superheated vapor, or saturated mixture region. This can be determined by comparing a suitable property to the saturated liquid and saturated vapor data. From the information given, we can determine the specific volume:

$$v = \frac{V}{m} = \frac{0.080 \text{ m}^3}{4 \text{ kg}} = 0.02 \text{ m}^3/\text{kg}$$

At 160 kPa, we read

$$\begin{aligned} v_f &= 0.0007435 \text{ m}^3/\text{kg} \\ v_g &= 0.1229 \text{ m}^3/\text{kg} \end{aligned} \qquad \text{(Table A-12)}$$

Obviously, $v_f < v < v_g$, and, therefore, the refrigerant is in the saturated mixture region. Thus, the temperature must be the saturation temperature at the specified pressure:

$$T = T_{sat @ 160kPa} = \textbf{-15.62°C}$$

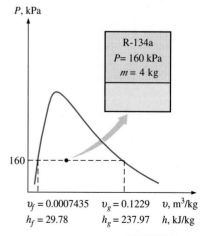

FIGURE 3-44

Schematic and *P-v* diagram for Example 3-6.

(b) Quality can be determined from Eq. 3-7:

$$x = \frac{v - v_f}{v_{fg}} = \frac{0.02 - 0.0007435}{0.1229 - 0.0007435} = \mathbf{0.158}$$

(c) At 160 kPa, we also read from Table A-12 that $h_f = 29.78$ kJ/kg and $h_{fg} = 208.18$ kJ/kg. Then,

$$h = h_f + x h_{fg}$$
$$= 29.78 \text{ kJ/kg} + (0.158)(208.18 \text{ kJ/kg})$$
$$= \mathbf{62.7 \text{ kJ/kg}}$$

(d) The mass of the vapor can be determined from

$$m_g = x m_t = (0.158)(4 \text{ kg}) = 0.632 \text{ kg}$$

and the volume occupied by the vapor phase is

$$V_g = m_g v_g = (0.632 \text{ kg})(0.1229 \text{ m}^3/\text{kg}) = \mathbf{0.0777 \text{ m}^3} \text{ (or 77.7 L)}$$

The rest of the volume (2.3 L) is occupied by the liquid.

Property tables are also available for saturated solid–vapor mixtures. Properties of saturated ice–water vapor mixtures, for example, are listed in Table A-8. Saturated solid–vapor mixtures can be handled just as saturated liquid–vapor mixtures.

2 Superheated Vapor

In the region to the right of the saturated vapor line, a substance exists as superheated vapor. Since the superheated region is a single-phase region (vapor phase only), temperature and pressure are no longer dependent properties and they can conveniently be used as the two independent properties in the tables. The format of the superheated vapor tables is illustrated in Fig. 3-45.

In these tables, the properties are listed versus temperature for selected pressures starting with the saturated vapor data. The saturation temperature is given in parentheses following the pressure value.

Superheated vapor is characterized by

Lower pressures ($P < P_{\text{sat}}$ at a given T)

Higher temperatures ($T > T_{\text{sat}}$ at a given P)

Higher specific volumes ($v > v_g$ at a given P or T)

Higher internal energies ($u > u_g$ at a given P or T)

Higher enthalpies ($h > h_g$ at a given P or T)

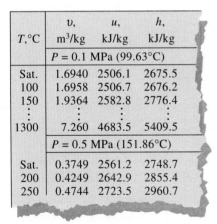

T,°C	v, m³/kg	u, kJ/kg	h, kJ/kg
$P = 0.1$ MPa (99.63°C)			
Sat.	1.6940	2506.1	2675.5
100	1.6958	2506.7	2676.2
150	1.9364	2582.8	2776.4
⋮	⋮	⋮	⋮
1300	7.260	4683.5	5409.5
$P = 0.5$ MPa (151.86°C)			
Sat.	0.3749	2561.2	2748.7
200	0.4249	2642.9	2855.4
250	0.4744	2723.5	2960.7

FIGURE 3-45

A partial listing of Table A-6.

EXAMPLE 3-7 Finding the Internal Energy of Superheated Vapor
Determine the internal energy of water at 20 psia and 400°F.

Solution At 20 psia, the saturation temperature is 227.96°F. Since $T > T_{\text{sat}}$, the water is in the superheated vapor region. Then the internal energy is determined from the superheated vapor table (Table A-6E) to be

$$u = \mathbf{1145.1 \text{ Btu/lbm}}$$

at the given temperature and pressure.

EXAMPLE 3-8 Finding the Temperature of Superheated Vapor

Determine the temperature of water at a state of $P = 0.5$ MPa and $h = 2890$ kJ/kg.

Solution At 0.5 MPa, the enthalpy of saturated water vapor is $h_g = 2748.7$ kJ/kg. Since $h > h_g$, as shown in Fig. 3-46, we again have superheated vapor. Under 0.5 MPa in Table A-6 we read

T, °C	h, **kJ/kg**
200	2855.4
250	2960.7

Obviously, the temperature is between 200 and 250°C. By linear interpolation it is determined to be

$$T = 216.4°C$$

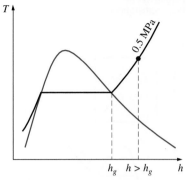

FIGURE 3-46

At a specified P, superheated vapor exists at a higher h than the saturated vapor (Example 3-8).

3 Compressed Liquid

There are not many data for compressed liquid in the literature, and Table A-7 is the only compressed liquid table in this text. The format of Table A-7 is very much like the format of the superheated vapor tables. One reason for the lack of compressed liquid data is the relative independence of compressed liquid properties from pressure. Variation of properties of compressed liquid with pressure is very mild. Increasing the pressure 100 times often causes properties to change less than 1 percent. The property most affected by pressure is enthalpy.

In the absence of compressed liquid data, a general approximation is *to treat compressed liquid as saturated liquid at the given temperature* (Fig. 3-47). This is because the compressed liquid properties depend on temperature more strongly than they do on pressure. Thus,

$$y \cong y_{f @ T}$$

for compressed liquids, where y is v, u, or h. Of these three properties, the property whose value is most sensitive to variations in the pressure is the enthalpy h. Although the above approximation results in negligible error in v and u, the error in h may reach undesirable levels. However, the error in h at very high pressures can be reduced significantly by evaluating it from

$$h \cong h_{f @ T} + v_f(P - P_{sat})$$

instead of taking it to be just h_f. Here P_{sat} is the saturation pressure at the given temperature.

In general, a compressed liquid is characterized by

Higher pressures ($P > P_{sat}$ at a given T)

Lower temperatures ($T < T_{sat}$ at a given P)

Lower specific volumes ($v < v_f$ at a given P or T)

Lower internal energies ($u < u_f$ at a given P or T)

Lower enthalpies ($h < h_f$ at a given P or T)

But these effects are not as pronounced as they are for the superheated vapor.

Given: P and T

$$v \cong v_{f @ T}$$
$$u \cong u_{f @ T}$$
$$h \cong h_{f @ T}$$

FIGURE 3-47

A compressed liquid may be approximated as a saturated liquid at the given temperature.

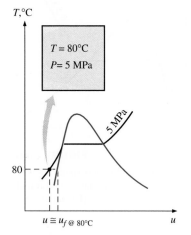

FIGURE 3-48

Schematic and T-u diagram for
Example 3-9.

EXAMPLE 3-9   Approximating Compressed Liquid as Saturated Liquid

Determine the internal energy of compressed liquid water at 80°C and 5 MPa,
using (a) data from the compressed liquid table and (b) saturated liquid data.
What is the error involved in the second case?

Solution   At 80°C, the saturation pressure of water is 47.39 kPa, and since
5 MPa > P_{sat}, we obviously have compressed liquid, as shown in Fig. 3-48.

(a) From the compressed liquid table (Table A-7)

$$\left. \begin{array}{l} P = 5 \text{ MPa} \\ T = 80°C \end{array} \right\} \quad u = \textbf{333.72 kJ/kg}$$

(b) From the saturation table (Table A-4), we read

$$u \cong u_{f@\,80°C} = \textbf{334.86 kJ/kg}$$

The error involved is

$$\frac{334.86 - 333.72}{333.72} \times 100 = \textbf{0.34\%}$$

which is less than 1 percent.

Reference State and Reference Values

The values of u, h, and s cannot be measured directly, and they are calculated
from measurable properties using the relations between thermodynamic properties. However, those relations give the *changes* in properties, not the values
of properties at specified states. Therefore, we need to choose a convenient
reference state and assign a value of *zero* for a convenient property or properties at that state. For water, the state of saturated liquid at 0.01°C is taken as
the reference state, and the internal energy and entropy are assigned zero values at that state. For refrigerant-134a, the state of saturated liquid at −40°C is
taken as the reference state, and the enthalpy and entropy are assigned zero
values at that state. Note that some properties may have negative values as a
result of the reference state chosen.

It should be mentioned that sometimes different tables list different values for some properties at the same state as a result of using a different reference state. However, in thermodynamics we are concerned with the *changes*
in properties, and the reference state chosen is of no consequence in calculations as long as we use values from a single consistent set of tables or charts.

EXAMPLE 3-10   The Use of Steam Tables to Determine Properties

Determine the missing properties and the phase descriptions in the following
table for water:

	T, °C	P, kPa	u, kJ/kg	x	Phase description
(a)		200		0.6	
(b)	125		1600		
(c)		1000	2950		
(d)	75	500			
(e)		850		0.0	

Solution   (a) The quality is given to be $x = 0.6$, which implies that 60 percent
of the mass is in the vapor phase and the remaining 40 percent is in the liquid
phase. Therefore, we have saturated liquid–vapor mixture at a pressure of

200 kPa. Then the temperature must be the saturation temperature at the given pressure:

$$T = T_{sat @ 200kPa} = \textbf{120.23°C} \qquad \text{(Table A-5)}$$

At 200 kPa, we also read from Table A-5 that $u_f = 504.49$ kJ/kg and $u_{fg} = 2025.0$ kJ/kg. Then the average internal energy of the mixture is determined from Eq. 3-8 to be

$$u = u_f + xu_{fg}$$
$$= 504.49 \text{ kJ/kg} + (0.6)(2025.0 \text{ kJ/kg})$$
$$= \textbf{1719.49 kJ/kg}$$

(*b*) This time the temperature and the internal energy are given, but we do not know which table to use to determine the missing properties because we have no clue as to whether we have saturated mixture, compressed liquid, or superheated vapor. To determine the region we are in, we first go to the saturation table (Table A-4) and determine the u_f and u_g values at the given temperature. At 125°C, we read $u_f = 524.74$ kJ/kg and $u_g = 2534.6$ kJ/kg. Next we compare the given u value to these u_f and u_g values, keeping in mind that

if $\quad u < u_f \quad$ we have *compressed liquid*

if $\quad u_f \le u \le u_g \quad$ we have *saturated mixture*

if $\quad u > u_g \quad$ we have *superheated vapor*

In our case the given u value is 1600, which falls between the u_f and u_g values at 125°C. Therefore, we have saturated liquid–vapor mixture. Then the pressure must be the saturation pressure at the given temperature:

$$P = P_{sat @ 125°C} = \textbf{232.1 kPa} \qquad \text{(Table A-4)}$$

The quality is determined from

$$x = \frac{u - u_f}{u_{fg}} = \frac{1600 - 524.74}{2009.9} = \textbf{0.535}$$

The criteria above for determining whether we have compressed liquid, saturated mixture, or superheated vapor can also be used when enthalpy h or specific volume v is given instead of internal energy u, or when pressure is given instead of temperature.

(*c*) This is similar to case (*b*), except pressure is given instead of temperature. Following the argument given above, we read the u_f and u_g values at the specified pressure. At 1 MPa, we have $u_f = 761.68$ kJ/kg and $u_g = 2583.6$ kJ/kg. The specified u value is 2950 kJ/kg, which is greater than the u_g value at 1 MPa. Therefore, we have superheated vapor, and the temperature at this state is determined from the superheated vapor table by interpolation to be

$$T = \textbf{395.6°C} \qquad \text{(Table A-6)}$$

We would leave the quality column blank in this case since quality has no meaning for a superheated vapor.

(*d*) In this case the temperature and pressure are given, but again we cannot tell which table to use to determine the missing properties because we do not know whether we have saturated mixture, compressed liquid, or superheated vapor. To determine the region we are in, we go to the saturation table (Table A-5) and determine the saturation temperature value at the given pressure. At 500 kPa, we have $T_{sat} = 151.86$°C. We then compare the given T value to this T_{sat} value, keeping in mind that

if $\quad T < T_{sat @ given P} \quad$ we have *compressed liquid*

if $\quad T = T_{sat @ given P} \quad$ we have *saturated mixture*

if $\quad T > T_{sat @ given P} \quad$ we have *superheated vapor*

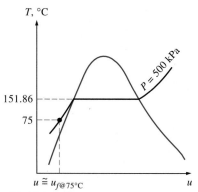

$$u \cong u_{f@\,75°C}$$

FIGURE 3-49

At a given P and T, a pure substance will exist as a compressed liquid if $T < T_{sat\,@\,P}$.

In our case, the given T value is 75°C, which is less than the T_{sat} value at the specified pressure. Therefore, we have compressed liquid (Fig. 3-49), and normally we would determine the internal energy value from the compressed liquid table. But in this case the given pressure is much lower than the lowest pressure value in the compressed liquid table (which is 5 MPa), and therefore we are justified to treat the compressed liquid as saturated liquid at the given temperature (*not* pressure):

$$u \cong u_{f@\,75°C} = \textbf{313.90 kJ/kg} \qquad \text{(Table A-4)}$$

We would leave the quality column blank in this case since quality has no meaning in the compressed liquid region.

(*e*) The quality is given to be $x = 0.0$, and thus we have saturated liquid at the specified pressure of 850 kPa. Then the temperature must be the saturation temperature at the given pressure, and the internal energy must have the saturated liquid value:

$$T = T_{sat\,@\,850kPa} = \textbf{172.96°C} \qquad \text{(Table A-5)}$$
$$u = u_{f@\,850kPa} = \textbf{731.27 kJ/kg} \qquad \text{(Table A-5)}$$

3-7 ■ THE IDEAL-GAS EQUATION OF STATE

One way of reporting property data for pure substances is to list values of properties at various states. The property tables provide very accurate information about the properties, but they are very bulky and vulnerable to typographical errors. A more practical and desirable approach would be to have some simple relations among the properties that are sufficiently general and accurate.

Any equation that relates the pressure, temperature, and specific volume of a substance is called an **equation of state.** Property relations that involve other properties of a substance at equilibrium states are also referred to as equations of state. There are several equations of state, some simple and others very complex. The simplest and best-known equation of state for substances in the gas phase is the ideal-gas equation of state. This equation predicts the *P-v-T* behavior of a gas quite accurately within some properly selected region.

Gas and *vapor* are often used as synonymous words. The vapor phase of a substance is customarily called a *gas* when it is above the critical temperature. *Vapor* usually implies a gas that is not far from a state of condensation.

In 1662, Robert Boyle, an Englishman, observed during his experiments with a vacuum chamber that the pressure of gases is inversely proportional to their volume. In 1802, J. Charles and J. Gay-Lussac, Frenchmen, experimentally determined that at low pressures the volume of a gas is proportional to its temperature. That is,

$$P = R\left(\frac{T}{v}\right)$$

or
$$Pv = RT \qquad (3-9)$$

where the constant of proportionality R is called the **gas constant.** Equation 3-9 is called the **ideal-gas equation of state,** or simply the **ideal-gas relation,** and a gas that obeys this relation is called an **ideal gas.** In this equation, P is the absolute pressure, T is the absolute temperature, and v is the specific volume.

The gas constant R is different for each gas (Fig. 3-50) and is determined from

$$R = \frac{R_u}{M} \qquad (\text{kJ/kg} \cdot \text{K or kPa} \cdot \text{m}^3/\text{kg} \cdot \text{K}) \tag{3-10}$$

where R_u is the **universal gas constant** and M is the molar mass (also called *molecular weight*) of the gas. The constant R_u is the same for all substances, and its value is

$$R_u = \begin{cases} 8.314 \text{ kJ/kmol} \cdot \text{K} \\ 8.314 \text{ kPa} \cdot \text{m}^3/\text{kmol} \cdot \text{K} \\ 0.08314 \text{ bar} \cdot \text{m}^3/\text{kmol} \cdot \text{K} \\ 1.986 \text{ Btu/lbmol} \cdot \text{R} \\ 10.73 \text{ psia} \cdot \text{ft}^3/\text{lbmol} \cdot \text{R} \\ 1545 \text{ ft} \cdot \text{lbf/lbmol} \cdot \text{R} \end{cases} \tag{3-11}$$

Substance	R, kJ/kg·K
Air	0.2870
Helium	2.0769
Argon	0.2081
Nitrogen	0.2968

FIGURE 3-50

Different substances have different gas constants.

The **molar mass** M can simply be defined as *the mass of one mole* (also called a *gram-mole,* abbreviated gmol) *of a substance in grams,* or *the mass of one kmol* (also called a *kilogram-mole,* abbreviated kgmol) *in kilograms.* In English units, it is the mass of 1 lbmol (1 pound-mole = 0.4536 kmol) in lbm (1 pound-mass = 0.4536 kg). Notice that the molar mass of a substance has the same numerical value in both unit systems because of the way it is defined. When we say the molar mass of nitrogen is 28, it simply means the mass of 1 kmol of nitrogen is 28 kg, or the mass of 1 lbmol of nitrogen is 28 lbm. That is, M = 28 kg/kmol = 28 lbm/lbmol. The mass of a system is equal to the product of its molar mass M and the mole number N:

$$m = MN \qquad (\text{kg}) \tag{3-12}$$

Per unit mass	Per unit mole
v, m^3/kg	$\bar{v}$, m^3/kmol
u, kJ/kg	$\bar{u}$, kJ/kmol
h, kJ/kg	$\bar{h}$, kJ/kmol

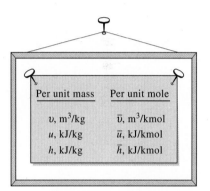

FIGURE 3-51

Properties per unit mole are denoted with a bar on the top.

The values of R and M for several substances are given in Table A-1.

The ideal-gas equation of state can be written in several different forms:

$$V = mv \longrightarrow PV = mRT \tag{3-13}$$

$$mR = (MN)R = NR_u \longrightarrow PV = NR_uT \tag{3-14}$$

$$V = N\bar{v} \longrightarrow P\bar{v} = R_uT \tag{3-15}$$

where $\bar{v}$ is the molar specific volume, that is, the volume per unit mole (in m^3/kmol or ft^3/lbmol). A bar above a property will denote values on a *unit-mole basis* throughout this text (Fig. 3-51).

By writing Eq. 3-13 twice for a fixed mass and simplifying, the properties of an ideal gas at two different states are related to each other by

$$\frac{P_1V_1}{T_1} = \frac{P_2V_2}{T_2} \tag{3-16}$$

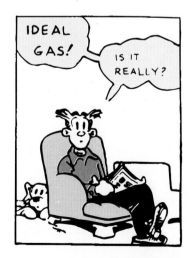

FIGURE 3-52

The ideal-gas relation often is not applicable to real gases; thus, care should be exercised when using it.

An ideal gas is an *imaginary* substance that obeys the relation $Pv = RT$ (Fig. 3-52). It has been experimentally observed that the ideal-gas relation given above closely approximates the P-v-T behavior of real gases at low densities. At low pressures and high temperatures, the density of a gas decreases, and the gas behaves as an ideal gas under these conditions. What constitutes low pressure and high temperature is explained later in this section.

In the range of practical interest, many familiar gases such as air, nitrogen, oxygen, hydrogen, helium, argon, neon, krypton, and even heavier gases

such as carbon dioxide can be treated as ideal gases with negligible error (often less than 1 percent). Dense gases such as water vapor in steam power plants and refrigerant vapor in refrigerators, however, should not be treated as ideal gases. Instead, the property tables should be used for these substances.

EXAMPLE 3-11 Finding the Mass of an Ideal Gas

Determine the mass of the air in a room whose dimensions are 4 m × 5 m × 6 m at 100 kPa and 25°C.

Solution A sketch of the room is given in Fig. 3-53. Air at specified conditions can be treated as an ideal gas. From Table A-1, the gas constant of air is $R = 0.287$ kPa · m³/kg · K, and the absolute temperature is $T = 25°C + 273 = 298$ K. The volume of the room is

$$V = (4 \text{ m})(5 \text{ m})(6 \text{ m}) = 120 \text{ m}^3$$

By substituting these values into Eq. 3-13, the mass of air in the room is determined to be

$$m = \frac{PV}{RT} = \frac{(100 \text{ kPa})(120 \text{ m}^3)}{(0.287 \text{ kPa} \cdot \text{m}^3/\text{kg} \cdot \text{K})(298 \text{ K})} = \textbf{140.3 kg}$$

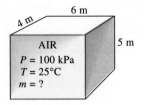

FIGURE 3-53

Schematic for Example 3-11.

Is Water Vapor an Ideal Gas?

This question cannot be answered with a simple yes or no. The error involved in treating water vapor as an ideal gas is calculated and plotted in Fig. 3-54. It is clear from this figure that at pressures below 10 kPa, water vapor can be treated as an ideal gas, regardless of its temperature, with negligible error (less than 0.1 percent). But at higher pressures, the ideal-gas assumption yields unacceptable errors, particularly in the vicinity of the critical point and the saturated vapor line (over 100 percent). Therefore, in air-conditioning applications, the water vapor in the air can be treated as an ideal gas with essentially no error since the pressure of the water vapor is very low. In steam power plant applications, however, the pressures involved are usually very high; therefore, ideal-gas relations should not be used.

Compressibility Factor—
A Measure of Deviation from Ideal-Gas Behavior

The ideal-gas equation is very simple and thus very convenient to use. But, as illustrated in Fig. 3-54, gases deviate from ideal-gas behavior significantly at states near the saturation region and the critical point. This deviation from ideal-gas behavior at a given temperature and pressure can accurately be accounted for by the introduction of a correction factor called the **compressibility factor Z**. It is defined as

$$Z = \frac{Pv}{RT} \tag{3-17}$$

or
$$Pv = ZRT \tag{3-18}$$

It can also be expressed as

$$Z = \frac{v_{\text{actual}}}{v_{\text{ideal}}} \tag{3-19}$$

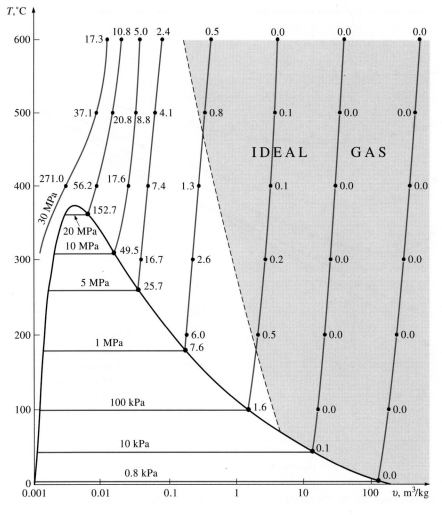

FIGURE 3-54

Percentage of error
($[|v_{table} - v_{ideal}|/v_{table}] \times 100$) involved
in assuming steam to be an ideal gas,
and the region where steam can be
treated as an ideal gas with less
than 1 percent error.

where $v_{ideal} = RT/P$. Obviously, $Z = 1$ for ideal gases. For real gases Z can be greater than or less than unity (Fig. 3-55). The farther away Z is from unity, the more the gas deviates from ideal-gas behavior.

We have repeatedly said that gases follow the ideal-gas equation closely at low pressures and high temperatures. But what exactly constitutes low pressure or high temperature? Is $-100°C$ a low temperature? It definitely is for most substances, but not for air. Air (or nitrogen) can be treated as an ideal gas at this temperature and atmospheric pressure with an error under 1 percent. This is because nitrogen is well over its critical temperature ($-147°C$) and away from the saturation region. But at this temperature and pressure, most substances would exist in the solid phase. Therefore, the pressure or temperature of a substance is high or low relative to its critical temperature or pressure.

Gases behave differently at a given temperature and pressure, but they behave very much the same at temperatures and pressures normalized with respect to their critical temperatures and pressures. The normalization is done as

FIGURE 3-55

The compressibility
factor is unity for ideal gases.

$$P_R = \frac{P}{P_{cr}} \quad \text{and} \quad T_R = \frac{T}{T_{cr}} \quad (3\text{-}20)$$

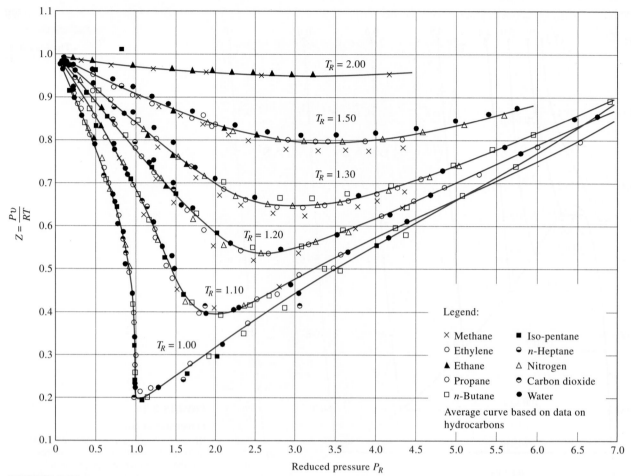

FIGURE 3-56

Comparison of Z factors for various gases. [*Source:* Gour-Jen Su, "Modified Law of Corresponding States," *Ind. Eng. Chem.* (international ed.) 38 (1946), p. 803.]

Legend:

× Methane	■ Iso-pentane
○ Ethylene	◒ n-Heptane
▲ Ethane	△ Nitrogen
○ Propane	◓ Carbon dioxide
□ n-Butane	● Water

Average curve based on data on hydrocarbons

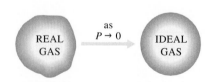

FIGURE 3-57

At very low pressures, all gases approach ideal-gas behavior (regardless of their temperature).

Here P_R is called the **reduced pressure** and T_R the **reduced temperature.** The Z factor for all gases is approximately the same at the same reduced pressure and temperature. This is called the **principle of corresponding states.** In Fig. 3-56, the experimentally determined Z values are plotted against P_R and T_R for several gases. The gases seem to obey the principle of corresponding states reasonably well. By curve-fitting all the data, we obtain the **generalized compressibility chart** that can be used for all gases.

The following observations can be made from the generalized compressibility chart:

1. At very low pressures ($P_R \ll 1$), the gases behave as an ideal gas regardless of temperature (Fig. 3-57),

2. At high temperatures ($T_R > 2$), ideal-gas behavior can be assumed with good accuracy regardless of pressure (except when $P_R \gg 1$).

3. The deviation of a gas from ideal-gas behavior is greatest in the vicinity of the critical point (Fig. 3-58).

The ideal-gas equation of state is very simple, but its range of applicability is limited. It is desirable to have equations of state that represent the P-v-T behavior of substances accurately over a larger region with no limitations. Such equations are naturally more complicated. Several equations have been proposed for this purpose (Fig. 3-59), but we shall discuss only three: the *van der Waals* equation because it is one of the earliest, the *Beattie-Bridgeman* equation of state because it is one of the best known and is reasonably accurate, and the *Benedict-Webb-Rubin* equation because it is one of the more recent and is very accurate.

Van der Waals Equation of State

The van der Waals equation of state was proposed in 1873, and it has two constants that are determined from the behavior of a substance at the critical point. The van der Waals equation of state is given by

$$\left(P + \frac{a}{v^2}\right)(v - b) = RT \tag{3-21}$$

Van der Waals intended to improve the ideal-gas equation of state by including two of the effects not considered in the ideal-gas model: the *intermolecular attraction forces* and the *volume occupied by the molecules themselves*. The term a/v^2 accounts for the intermolecular attraction forces, and b accounts for the volume occupied by the gas molecules. In a room at atmospheric pressure and temperature, the volume actually occupied by molecules is only about one-thousandth of the volume of the room. As the pressure increases, the volume occupied by the molecules becomes an increasingly significant part of the total volume. Van der Waals proposed to correct this by replacing v in the ideal-gas relation with the quantity $v - b$, where b represents the volume occupied by the gas molecules per unit mass.

The determination of the two constants appearing in this equation is based on the observation that the critical isotherm on a P-v diagram has a horizontal inflection point at the critical point (Fig. 3-60). Thus, the first and the second derivatives of P with respect to v at the critical point must be zero. That is,

$$\left(\frac{\partial P}{\partial v}\right)_{T = T_{cr} = \text{const}} = 0 \quad \text{and} \quad \left(\frac{\partial^2 P}{\partial v^2}\right)_{T = T_{cr} = \text{const}} = 0 \tag{3-22}$$

By performing the differentiations and eliminating v_{cr}, the constants a and b are determined to be

$$a = \frac{27R^2 T_{cr}^2}{64P_{cr}} \quad \text{and} \quad b = \frac{RT_{cr}}{8P_{cr}} \tag{3-23}$$

The constants a and b can be determined for any substance from the critical-point data alone (Table A-1).

The accuracy of the van der Waals equation of state is often inadequate, but it can be improved by using the values of a and b that are based on the actual behavior of the gas over a wider range instead of a single point. Despite its limitations, the van der Waals equation of state has a historical value in that it was one of the first attempts to model the behavior of real gases. The van

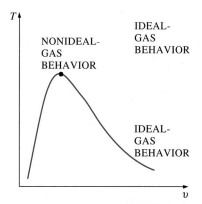

Other Equations of State

FIGURE 3-58

Gases deviate from the ideal-gas behavior most in the neighborhood of the critical point.

van der Waals
Berthelet
Redlich-Kwang
Beattie-Bridgeman
Benedict-Webb-Rubin
Strobridge
Virial

FIGURE 3-59

Several equations of state have been proposed throughout history.

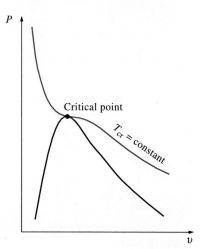

FIGURE 3-60

Critical isotherm of a pure substance has an inflection point at the critical state.

der Waals equation of state can also be expressed on a unit-mole basis by replacing the v in Eq. 3-21 by $\bar{v}$ and the R in Eqs. 3-21 and 3-23 by R_u.

Beattie-Bridgeman Equation of State

The Beattie-Bridgeman equation, proposed in 1928, is an equation of state based on five experimentally determined constants. It was proposed in the form of

$$P = \frac{R_u T}{\bar{v}^2}\left(1 - \frac{c}{\bar{v}T^3}\right)(\bar{v} + B) - \frac{A}{\bar{v}^2} \qquad (3-24)$$

where $\qquad A = A_0\left(1 - \frac{a}{\bar{v}}\right) \qquad$ and $\qquad B = B_0\left(1 - \frac{b}{\bar{v}}\right) \qquad (3-25)$

The constants appearing in the above equation are given in Table 3-4 for various substances. The Beattie-Bridgeman equation is known to be reasonably accurate for densities up to about $0.8\rho_{cr}$, where ρ_{cr} is the density of the substance at the critical point.

Benedict-Webb-Rubin Equation of State

Benedict, Webb, and Rubin extended the Beattie-Bridgeman equation in 1940 by raising the number of constants to eight. It is expressed as

$$P = \frac{R_u T}{\bar{v}} + \left(B_0 R_u T - A_0 - \frac{C_0}{T^2}\right)\frac{1}{\bar{v}^2} + \frac{b R_u T - a}{\bar{v}^3}$$
$$+ \frac{a\alpha}{\bar{v}^6} + \frac{c}{\bar{v}^3 T^2}\left(1 + \frac{\gamma}{\bar{v}^2}\right)e^{-\gamma/\bar{v}^2} \qquad (3-26)$$

The values of the constants appearing in this equation are given in Table 3-4. This equation can handle substances at densities up to about $2.5\rho_{cr}$. In 1962, Strobridge further extended this equation by raising the number of constants to 16 (Fig. 3-61).

Virial Equation of State

The equation of state of a substance can also be expressed in a series form as

$$P = \frac{RT}{v} + \frac{a(T)}{v^2} + \frac{b(T)}{v^3} + \frac{c(T)}{v^4} + \frac{d(T)}{v^5} + \cdots \qquad (3-27)$$

This and similar equations are called the *virial equations of state,* and the coefficients $a(T)$, $b(T)$, $c(T)$, and so on, that are functions of temperature alone are called *virial coefficients.* These coefficients can be determined experimentally or theoretically from statistical mechanics. Obviously, as the pressure approaches zero, all the virial coefficients will vanish and the equation will reduce to the ideal-gas equation of state. The P-v-T behavior of a substance can be represented accurately with the virial equation of state over a wider range by including a sufficient number of terms. All equations of state discussed above are applicable to the gas phase of the substances only, and thus should not be used for liquids or liquid–vapor mixtures.

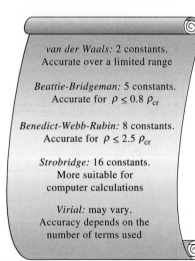

van der Waals: 2 constants.
Accurate over a limited range

Beattie-Bridgeman: 5 constants.
Accurate for $\rho \leq 0.8\,\rho_{cr}$

Benedict-Webb-Rubin: 8 constants.
Accurate for $\rho \leq 2.5\,\rho_{cr}$

Strobridge: 16 constants.
More suitable for
computer calculations

Virial: may vary.
Accuracy depends on the
number of terms used

FIGURE 3-61

Complex equations of state represent the P-v-T behavior of gases more accurately over a wide range.

Complex equations represent the *P-v-T* behavior of substances reasonably well and are very suitable for digital computer applications. For hand calculations, however, it is suggested that the reader use the property tables or the simpler equations of state for convenience. This is particularly true for specific-volume calculations since all the equations above are implicit in v and will require a trial-and-error approach. The accuracy of the van der Waals, Beattie-Bridgeman, and Benedict-Webb-Rubin equations of state is illustrated in Fig. 3-62. It is obvious from this figure that the Benedict-Webb-Rubin equation of state is the most accurate.

EXAMPLE 3-12 Different Methods of Evaluating Gas Pressure

Predict the pressure of nitrogen gas at $T = 175$ K and $v = 0.00375$ m³/kg on the basis of (a) the ideal-gas equation of state, (b) the van der Waals equation of state, (c) the Beattie-Bridgeman equation of state, and (d) the Benedict-Webb-Rubin equation of state. Compare the values obtained to the experimentally determined value of 10,000 kPa.

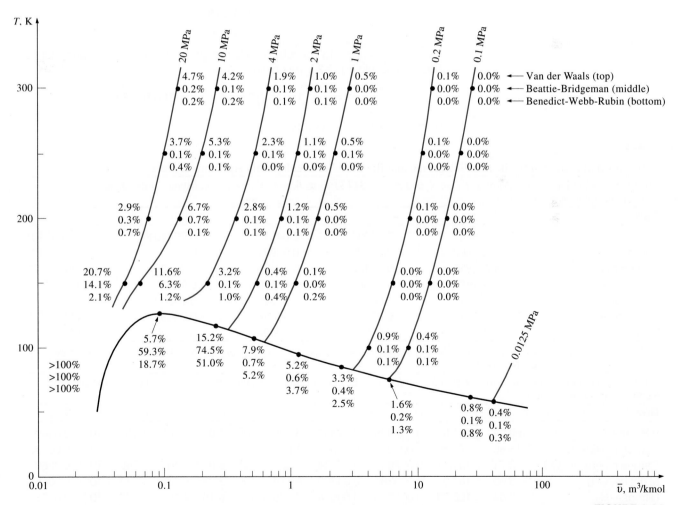

FIGURE 3-62

Percentage of error involved in various equations of state for nitrogen
(% error = [($|v_{\text{table}} - v_{\text{equation}}|$)/$v_{\text{table}}$] × 100).

Solution (a) By using the ideal-gas equation of state, the pressure is found to be

$$P = \frac{RT}{v} = \frac{(0.2968 \text{ kPa} \cdot \text{m}^3/\text{kg} \cdot \text{K})(175 \text{ K})}{0.00375 \text{ m}^3/\text{kg}} = \textbf{13,851 kPa}$$

which is in error by 38.5 percent.

(b) The van der Waals constants for nitrogen are determined from Eq. 3-23 to be

$$a = 0.175 \text{ m}^6 \cdot \text{kPa}/\text{kg}^2$$
$$b = 0.00138 \text{ m}^3/\text{kg}$$

From Eq. 3-21,

$$P = \frac{RT}{v - b} - \frac{a}{v^2} = \textbf{9471 kPa}$$

which is in error by 5.3 percent.

(c) The constants in the Beattie-Bridgeman equation are determined from Table 3-4 to be

$$A = 102.29$$
$$B = 0.05378$$
$$c = 4.2 \times 10^4$$

Also, $\bar{v} = Mv = (28.013 \text{ kg/mol})(0.00375 \text{ m}^3/\text{kg}) = 0.10505 \text{ m}^3/\text{kmol}$. Substituting these values into Eq. 3-24, we obtain

$$P = \frac{R_u T}{\bar{v}^2}\left(1 - \frac{c}{\bar{v}T^3}\right)(\bar{v} + B) - \frac{A}{\bar{v}^2} = \textbf{10,110 kPa}$$

which is in error by 1.1 percent.

TABLE 3-4

Constants that appear in the Beattie-Bridgeman and the Benedict-Webb-Rubin equations of state

(a) When P is in kPa, $\bar{v}$ is in m³/kmol, T is in K, and $R_u = 8.314$ kPa · m³/kmol · K, the five constants in the Beattie-Bridgeman equation are as follows:

Gas	A_0	a	B_0	b	c
Air	131.8441	0.01931	0.04611	−0.001101	4.34×10^4
Argon, Ar	130.7802	0.02328	0.03931	0.0	5.99×10^4
Carbon dioxide, CO_2	507.2836	0.07132	0.10476	0.07235	6.60×10^5
Helium, He	2.1886	0.05984	0.01400	0.0	40
Hydrogen, H_2	20.0117	−0.00506	0.02096	−0.04359	504
Nitrogen, N_2	136.2315	0.02617	0.05046	−0.00691	4.20×10^4
Oxygen, O_2	151.0857	0.02562	0.04624	0.004208	4.80×10^4

Source: Gordon J. Van Wylen and Richard E. Sonntag, *Fundamentals of Classical Thermodynamics,* English/SI Version, 3rd ed. (New York: John Wiley & Sons, 1986), p. 46, table 3.3.

(b) When P is in kPa, $\bar{v}$ is in m³/kmol, T is in K, and $R_u = 8.314$ kPa · m³/kmol · K, the eight constants in the Benedict-Webb-Rubin equation are as follows:

Gas	a	A_0	b	B_0	c	C_0	α	γ
n-Butane, C_4H_{10}	190.68	1021.6	0.039998	0.12436	3.205×10^7	1.006×10^8	1.101×10^{-3}	0.0340
Carbon dioxide, CO_2	13.86	277.30	0.007210	0.04991	1.511×10^6	1.404×10^7	8.470×10^{-5}	0.00539
Carbon monoxide, CO	3.71	135.87	0.002632	0.05454	1.054×10^5	8.673×10^5	1.350×10^{-4}	0.0060
Methane, CH_4	5.00	187.91	0.003380	0.04260	2.578×10^5	2.286×10^6	1.244×10^{-4}	0.0060
Nitrogen, N_2	2.54	106.73	0.002328	0.04074	7.379×10^4	8.164×10^5	1.272×10^{-4}	0.0053

Source: Kenneth Wark, *Thermodynamics,* 4th ed. (New York: McGraw-Hill, 1983), p. 815, table A-21M. Originally published in H. W. Cooper and J. C. Goldfrank, *Hydrocarbon Processing* 46, no. 12 (1967), p. 141.

(d) The constants in the Benedict-Webb-Rubin equation are determined from Table 3-4 to be

$$a = 2.54 \qquad A_0 = 106.73$$
$$b = 0.002328 \qquad B_0 = 0.04074$$
$$c = 7.379 \times 10^4 \qquad C_0 = 8.164 \times 10^5$$
$$\alpha = 1.272 \times 10^{-4} \qquad \gamma = 0.0053$$

Substituting these values into Eq. 3-26, we obtain

$$P = \frac{R_u T}{\bar{v}} + \left(B_0 R_u T - A_0 - \frac{C_0}{T^2}\right)\frac{1}{\bar{v}^2} + \frac{b R_u T - a}{\bar{v}^3} + \frac{a\alpha}{\bar{v}^6} + \frac{c}{\bar{v}^3 T^2}\left(1 + \frac{\gamma}{\bar{v}^2}\right)e^{-\gamma/\bar{v}^2}$$
$$= \mathbf{10{,}009\ kPa}$$

which is in error by only 0.09 percent. Thus, the accuracy of the Benedict-Webb-Rubin equation of state is rather impressive in this case.

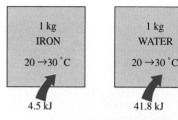

FIGURE 3-63

It takes different amounts of energy to raise the temperature of different substances by the same amount.

3-9 ■ SPECIFIC HEATS

We know from experience that it takes different amounts of energy to raise the temperature of identical masses of different substances by one degree. For example, we need about 4.5 kJ of energy to raise the temperature of 1 kg of iron from 20 to 30°C, whereas it takes about 9 times this energy (41.8 kJ to be exact) to raise the temperature of 1 kg of liquid water by the same amount (Fig. 3-63). Therefore, it is desirable to have a property that will enable us to compare the energy storage capabilities of various substances. This property is the specific heat.

The **specific heat** is defined as *the energy required to raise the temperature of a unit mass of a substance by one degree* (Fig. 3-64). In general, this energy will depend on how the process is executed. In thermodynamics, we are interested in two kinds of specific heats: **specific heat at constant volume** C_v and **specific heat at constant pressure** C_p.

Physically, the specific heat at constant volume C_v can be viewed as *the energy required to raise the temperature of the unit mass of a substance by one degree as the volume is maintained constant.* The energy required to do the same as the pressure is maintained constant is the specific heat at constant pressure C_p. This is illustrated in Fig. 3-65. The specific heat at constant pressure C_p is always greater than C_v because at constant pressure the system is allowed to expand and the energy for this expansion work must also be supplied to the system.

Now we will attempt to express the specific heats in terms of other thermodynamic properties. First, consider a fixed mass in a stationary closed system undergoing a constant-volume process (and thus no expansion or compression work is involved). The conservation of energy principle $e_{in} - e_{out} = \Delta e_{system}$ for this process can be expressed in the differential form as

$$\delta e_{in} - \delta e_{out} = du$$

The left-hand side of this equation represents the net amount of energy transferred to the system in the form of heat and/or work. From the definition of C_v, this energy must be equal to $C_v\, dT$, where dT is the differential change in temperature. Thus,

$$C_v\, dT = du \qquad \text{at constant volume}$$

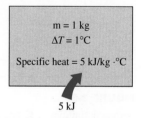

FIGURE 3-64

Specific heat is the energy required to raise the temperature of a unit mass of a substance by one degree in a specified way.

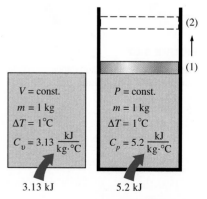

FIGURE 3-65

Constant-volume and constant-pressure specific heats C_v and C_p (values given are for helium gas).

or

$$C_v = \left(\frac{\partial u}{\partial T}\right)_v \qquad (3\text{-}28)$$

Similarly, an expression for the specific heat at constant pressure C_p can be obtained by considering a constant-pressure expansion or compression process. It yields

$$C_p = \left(\frac{\partial h}{\partial T}\right)_p \qquad (3\text{-}29)$$

Equations 3-28 and 3-29 are the defining equations for C_v and C_p, and their interpretation is given in Fig. 3-66.

Note that C_v and C_p are expressed in terms of other properties; thus, they must be properties themselves. Like any other property, the specific heats of a substance depend on the state that, in general, is specified by two independent, intensive properties. That is, the energy required to raise the temperature of a substance by one degree will be different at different temperatures and pressures (Fig. 3-67). But this difference is usually not very large.

A few observations can be made from Eqs. 3-28 and 3-29. First, these equations are *property relations* and as such *are independent of the type of processes.* They are valid for *any* substance undergoing *any* process. The only relevance C_v has to a constant-volume process is that C_v happens to be the energy transferred to a system during a constant-volume process per unit mass per unit degree rise in temperature. This is how the values of C_v are determined. This is also how the name *specific heat at constant volume* originated. Likewise, the energy transferred to a system per unit mass per unit temperature rise during a constant-pressure process happens to be equal to C_p. This is how the values of C_p can be determined and also explains the origin of the name *specific heat at constant pressure.*

Another observation that can be made from Eqs. 3-28 and 3-29 is that C_v is related to the changes in *internal energy* and C_p to the changes in *enthalpy.* In fact, it would be more proper to define C_v as *the change in the internal energy of a substance per unit change in temperature at constant volume.* Likewise, C_p can be defined as *the change in the enthalpy of a substance per unit change in temperature at constant pressure.* In other words, C_v is a measure of the variation of internal energy of a substance with temperature, and C_p is a measure of the variation of enthalpy of a substance with temperature.

Both the internal energy and enthalpy of a substance can be changed by the transfer of *energy* in any form, with heat being only one of them. Therefore, the term *specific energy* is probably more appropriate than the term *specific heat,* which implies that energy is transferred (and stored) in the form of heat.

A common unit for specific heats is kJ/kg · °C or kJ/kg · K. Notice that these two units are *identical* since $\Delta T(°C) = \Delta T(K)$, and 1°C change in temperature is equivalent to a change of 1 K. The specific heats are sometimes given on a *molar basis.* They are then denoted by $\overline{C}_v$ and $\overline{C}_p$ and have the unit kJ/kmol · °C or kJ/kmol · K.

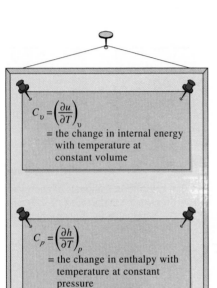

FIGURE 3-66

Formal definitions of C_v and C_p.

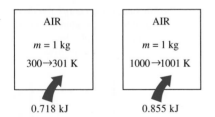

FIGURE 3-67

The specific heat of a substance changes with temperature.

3-10 ■ INTERNAL ENERGY, ENTHALPY, AND SPECIFIC HEATS OF IDEAL GASES

We defined an ideal gas as a gas whose temperature, pressure, and specific volume are related by

$$Pv = RT$$

It has been demonstrated mathematically and experimentally (Joule, 1843) that for an ideal gas the internal energy is a function of the temperature only. That is,

$$u = u(T) \tag{3-30}$$

In his classical experiment, Joule submerged two tanks connected with a pipe and a valve in a water bath, as shown in Fig. 3-68. Initially, one tank contained air at a high pressure and the other tank was evacuated. When thermal equilibrium was attained, he opened the valve to let air pass from one tank to the other until the pressures equalized. Joule observed no change in the temperature of the water bath and assumed that no heat was transferred to or from the air. Since there was also no work done, he concluded that the internal energy of the air did not change even though the volume and the pressure changed. Therefore, he reasoned, the internal energy is a function of temperature only and not a function of pressure or specific volume. (Joule later showed that for gases that deviate significantly from ideal-gas behavior, the internal energy is not a function of temperature alone.)

Using the definition of enthalpy and the equation of state of an ideal gas, we have

$$\left. \begin{array}{c} h = u + Pv \\ Pv = RT \end{array} \right\} \qquad h = u + RT$$

Since R is constant and $u = u(T)$, it follows that the enthalpy of an ideal gas is also a function of temperature only:

$$h = h(T) \tag{3-31}$$

Since u and h depend only on temperature for an ideal gas, the specific heats C_v and C_p also depend, at most, on temperature only. Therefore, *at a given temperature, u, h, C_v, and C_p of an ideal gas will have fixed values regardless of the specific volume or pressure* (Fig. 3-69). Thus, for ideal gases, the partial derivatives in Eqs. 3-28 and 3-29 can be replaced by ordinary derivatives. Then the differential changes in the internal energy and enthalpy of an ideal gas can be expressed as

$$du = C_v(T)\, dT \tag{3-32}$$

and
$$dh = C_p(T)\, dT \tag{3-33}$$

The change in internal energy or enthalpy for an ideal gas during a process from state 1 to state 2 is determined by integrating these equations:

$$\Delta u = u_2 - u_1 = \int_1^2 C_v(T)\, dT \qquad \text{(kJ/kg)} \tag{3-34}$$

and
$$\Delta h = h_2 - h_1 = \int_1^2 C_p(T)\, dT \qquad \text{(kJ/kg)} \tag{3-35}$$

To carry out these integrations, we need to have relations for C_v and C_p as functions of temperature.

At low pressures, all real gases approach ideal-gas behavior, and therefore their specific heats depend on temperature only. The specific heats of real gases at low pressures are called *ideal-gas specific heats,* or *zero-pressure specific heats,* and are often denoted C_{p0} and C_{v0}. Accurate analytical expressions for ideal-gas specific heats, based on direct measurements or calculations from statistical behavior of molecules, are available and are given as

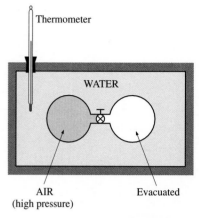

Thermometer

WATER

AIR (high pressure)

Evacuated

FIGURE 3-68

Schematic of the experimental apparatus used by Joule.

$$u = u(T)$$
$$h = h(T)$$
$$C_v = C_v(T)$$
$$C_p = C_p(T)$$

FIGURE 3-69

For ideal gases, u, h, C_v, and C_p vary with temperature only.

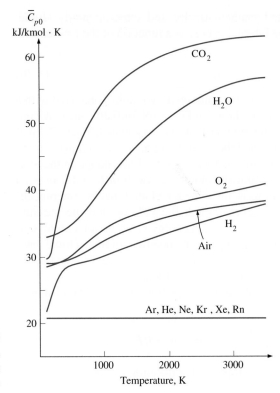

FIGURE 3-70

Ideal-gas constant-pressure
specific heats for some gases
(see Table A-2c for $\overline{C}_p$ equations).

third-degree polynomials in the appendix (Table A-2c) for several gases. A plot of $\overline{C}_{p0}(T)$ data for some common gases is given in Fig. 3-70.

The use of ideal-gas specific heat data is limited to low pressures, but these data can also be used at moderately high pressures with reasonable accuracy as long as the gas does not deviate from ideal-gas behavior significantly.

The integrations in Eqs. 3-34 and 3-35 are straightforward but rather time-consuming and thus impractical. To avoid these laborious calculations, u and h data for a number of gases have been tabulated over small temperature intervals. These tables are obtained by choosing an arbitrary reference point and performing the integrations in Eqs. 3-34 and 3-35 by treating state 1 as the reference state. In the ideal-gas tables given in the appendix, zero kelvin is chosen as the reference state, and both the enthalpy and the internal energy are assigned zero values at that state (Fig. 3-71). The choice of the reference state has no effect on Δu or Δh calculations. The u and h data are given in kJ/kg for air (Table A-17) and usually in kJ/kmol for other gases. The unit kJ/kmol is very convenient in the thermodynamic analysis of chemical reactions, which are not considered in this text.

Some observations can be made from Fig. 3-70. First, the specific heats of gases with complex molecules (molecules with two or more atoms) are higher and increase with temperature. Also, the variation of specific heats with temperature is smooth and may be approximated as linear over small temperature intervals (a few hundred degrees or less). Then the specific heat functions in Eqs. 3-50 and 3-51 can be replaced by the constant average specific heat values. Now the integrations in these equations can be performed, yielding

FIGURE 3-71

In the preparation of ideal-gas tables, 0 K
is chosen as the reference temperature.

AIR		
T, K	u, kJ/kg	h, kJ/kg
0	0	0
.	.	.
.	.	.
.	.	.
300	214.17	300.19
310	221.25	310.24
.	.	.
.	.	.
.	.	.

$$u_2 - u_1 = C_{v,\,av}(T_2 - T_1) \qquad \text{(kJ/kg)} \qquad \text{(3-36)}$$

and

$$h_2 - h_1 = C_{p,\,av}(T_2 - T_1) \qquad \text{(kJ/kg)} \qquad \text{(3-37)}$$

The specific heat values for some common gases are listed as a function of temperature in Table A-2b. The average specific heats $C_{p,\,av}$ and $C_{v,\,av}$ are evaluated from this table at the average temperature $(T_1 + T_2)/2$, as shown in Fig. 3-72. If the final temperature T_2 is not known, the specific heats may be evaluated at T_1 or at anticipated average temperature. Then T_2 can be determined by using these specific heat values. The value of T_2 can be refined, if necessary, by evaluating the specific heats at the new average temperature.

Another way of determining the average specific heats is to evaluate them at T_1 and T_2 and then take their average. Usually both methods give reasonably good results, and one is not necessarily better than the other.

Another observation that can be made from Fig. 3-70 is that the ideal-gas specific heats of *monatomic gases* such as argon, neon, and helium remain constant over the entire temperature range. Thus, Δu and Δh of monatomic gases can easily be evaluated from Eqs. 3-36 and 3-37.

Note that the Δu and Δh relations given above are not restricted to any kind of process. They are valid for all processes. The presence of the constant-volume specific heat C_v in an equation should not lead one to believe that this equation is valid for a constant-volume process only. On the contrary, the relation $\Delta u = C_{v,\,av}\,\Delta T$ is valid for *any* ideal gas undergoing *any* process (Fig. 3-73). A similar argument can be given for C_p and Δh.

To summarize, there are three ways to determine the internal energy and enthalpy changes of ideal gases (Fig. 3-74):

1. By using the tabulated u and h data. This is the easiest and most accurate way when tables are readily available.

2. By using the C_v or C_p relations as a function of temperature and performing the integrations. This is very inconvenient for hand calculations but quite desirable for computerized calculations. The results obtained are very accurate.

3. By using average specific heats. This is very simple and certainly very convenient when property tables are not available. The results obtained are reasonably accurate if the temperature interval is not very large.

Specific-Heat Relations of Ideal Gases

A special relationship between C_p and C_v for ideal gases can be obtained by differentiating the relation $h = u + RT$, which yields

$$dh = du + R\,dT$$

Replacing dh by $C_p\,dT$ and du by $C_v\,dT$ and dividing the resulting expression by dT, we obtain

$$C_p = C_v + R \qquad \text{(kJ/kg} \cdot \text{K)} \qquad \text{(3-38)}$$

This is an important relationship for ideal gases since it enables us to determine C_v from a knowledge of C_p and the gas constant R.

When the specific heats are given on a molar basis, R in the above equation should be replaced by the universal gas constant R_u (Fig. 3-75). That is,

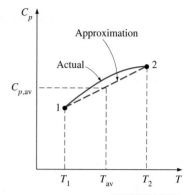

FIGURE 3-72

For small temperature intervals, the specific heats may be assumed to vary linearly with temperature.

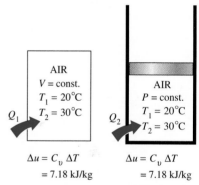

FIGURE 3-73

The relation $\Delta u = C_v\,\Delta T$ is valid for *any* kind of process, constant-volume or not.

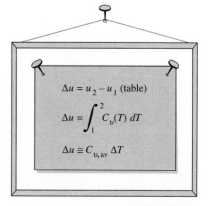

FIGURE 3-74

Three ways of calculating Δu.

FIGURE 3-75

The C_p of an ideal gas can be determined from a knowledge of C_v and R.

AIR at 300 K

$\left.\begin{array}{l} C_v = 0.718 \text{ kJ/kg} \cdot \text{K} \\ R = 0.287 \text{ kJ/kg} \cdot \text{K} \end{array}\right\} C_p = 1.005 \text{ kJ/kg} \cdot \text{K}$

or,

$\left.\begin{array}{l} \overline{C}_v = 20.80 \text{ kJ/kmol} \cdot \text{K} \\ R_u = 8.314 \text{ kJ/kmol} \cdot \text{K} \end{array}\right\} \overline{C}_p = 29.114 \text{ kJ/kmol} \cdot \text{K}$

$$\overline{C}_p = \overline{C}_v + R_u \qquad (\text{kJ/kmol} \cdot \text{K}) \qquad (3\text{-}39)$$

At this point, we introduce another ideal-gas property called the **specific heat ratio** k, defined as

$$k = \frac{C_p}{C_v} \qquad (3\text{-}40)$$

The specific heat ratio also varies with temperature, but this variation is very mild. For monatomic gases, its value is essentially constant at 1.667. Many diatomic gases, including air, have a specific heat ratio of about 1.4 at room temperature.

EXAMPLE 3-13   Evaluation of the Δu of an Ideal Gas

Air at 300 K and 200 kPa is heated at constant pressure to 600 K. Determine the change in internal energy of air per unit mass, using (*a*) data from the air table (Table A-17), (*b*) the functional form of the specific heat (Table A-2*c*), and (*c*) the average specific heat value (Table A-2*b*).

Solution   At specified conditions, air can be considered to be an ideal gas since it is at a high temperature and low pressure relative to its critical-point values. The internal energy change Δu of ideal gases depends on the initial and final temperatures only, and not on the type of process. Thus, the solution given below is valid for any kind of process.

Analysis   (*a*) One way of determining the change in internal energy of air is to read the u values at T_1 and T_2 from Table A-17 and take the difference:

$$u_1 = u_{@ \, 300K} = 214.07 \text{ kJ/kg}$$

$$u_2 = u_{@ \, 600K} = 434.78 \text{ kJ/kg}$$

Thus, $\Delta u = u_2 - u_1 = (434.78 - 214.07) \text{ kJ/kg} = 220.71 \text{ kJ/kg}$

(*b*) The $\overline{C}_p(T)$ of air is given in Table A-2*c* in the form of a third-degree polynomial expressed as

$$\overline{C}_p(T) = a + bT + cT^2 + dT^3$$

where $a = 28.11$, $b = 0.1967 \times 10^{-2}$, $c = 0.4802 \times 10^{-5}$, and $d = -1.966 \times 10^{-9}$. From Eq. 3-39,

$$\overline{C}_v(T) = \overline{C}_p - R_u = (a - R_u) + bT + cT^2 + dT^3$$

From Eq. 3-34,

$$\Delta \overline{u} = \int_1^2 \overline{C}_v(T) \, dT = \int_{T_1}^{T_2} [(a - R_u) + bT + cT^2 + dT^3] \, dT$$

Performing the integration and substituting the values, we obtain

$$\Delta \overline{u} = 6447.15 \text{ kJ/kmol}$$

The change in the internal energy on a unit-mass basis is determined by dividing this value by the molar mass of air (Table A-1):

$$\Delta u = \frac{\Delta \overline{u}}{M} = \frac{6447.15 \text{ kJ/kmol}}{28.97 \text{ kg/kmol}} = \textbf{222.55 kJ/kg}$$

which differs from the exact result by 0.8 percent.

(c) The average value of the constant-volume specific heat $C_{v, av}$ is determined from Table A-2b at the average temperature $(T_1 + T_2)/2 = 450$ K to be

$$C_{v, av} = C_{v @ 450K} = 0.733 \text{ kJ/kg} \cdot \text{K}$$

Thus, $\Delta u = C_{v, av}(T_2 - T_1) = (0.733 \text{ kJ/kg} \cdot \text{K})[(600 - 300) \text{ K}]$
$$= \textbf{219.9 kJ/kg}$$

Discussion   This answer differs from the exact result (220.71 kJ/kg) by only 0.4 percent. This close agreement is not surprising since the assumption that C_v varies linearly with temperature is a reasonable one at temperature intervals of only a few hundred degrees. If we had used the C_v value at $T_1 = 300$ K instead of at T_{av}, the result would be 215.4 kJ/kg, which is in error by about 2 percent. Errors of this magnitude are acceptable for most engineering purposes.

3-11 ■ INTERNAL ENERGY, ENTHALPY, AND SPECIFIC HEATS OF SOLIDS AND LIQUIDS

A substance whose specific volume (or density) is constant is called an **incompressible substance.** The specific volumes of solids and liquids essentially remain constant during a process (Fig. 3-76). Therefore, liquids and solids can be approximated as incompressible substances without sacrificing much in accuracy. The constant-volume assumption should be taken to imply that the energy associated with the volume change, such as the boundary work, is negligible compared with other forms of energy. Otherwise, this assumption would be ridiculous for studying the thermal stresses in solids (caused by volume change with temperature) or analyzing liquid-in-glass thermometers.

It can be mathematically shown that the constant-volume and constant-pressure specific heats are identical for incompressible substances (Fig. 3-77). Therefore, for solids and liquids, the subscripts on C_p and C_v can be dropped, and both specific heats can be represented by a single symbol C. That is,

$$C_p = C_v = C \qquad (3\text{-}41)$$

This result could also be deduced from the physical definitions of constant-volume and constant-pressure specific heats. Specific heat values for several common liquids and solids are given in Table A-3.

Internal Energy Changes

Like those of ideal gases, the specific heats of incompressible substances depend on temperature only. Thus, the partial differentials in the defining equation of C_v (Eq. 3-28) can be replaced by ordinary differentials, which yield

$$du = C_v \, dT = C(T) \, dT \qquad (3\text{-}42)$$

The change in internal energy between states 1 and 2 is then obtained by integration:

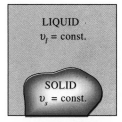

FIGURE 3-76

The specific volumes of incompressible substances remain constant during a process.

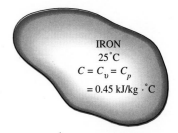

FIGURE 3-77

The C_v and C_p values of incompressible substances are identical and are denoted by C.

$$\Delta u = u_2 - u_1 = \int_1^2 C(T)\, dT \qquad \text{(kJ/kg)} \qquad (3\text{-}43)$$

The variation of specific heat C with temperature should be known before this integration can be carried out. For small temperature intervals, a C value at the average temperature can be used and treated as a constant, yielding

$$\Delta u \cong C_{av}(T_2 - T_1) \qquad \text{(kJ/kg)} \qquad (3\text{-}44)$$

Enthalpy Changes

Using the definition of enthalpy $h = u + Pv$ and noting that $v = \text{constant}$, the differential form of the enthalpy change of incompressible substances can be determined by differentiation to be

$$dh = du + v\, dP + P\, dv^{\,0} = du + v\, dP \qquad (3\text{-}45)$$

Integrating,

$$\Delta h = \Delta u + v\, \Delta P \cong C_{av}\, \Delta T + v\, \Delta P \qquad \text{(kJ)} \qquad (3\text{-}46)$$

For *solids,* the term $v\,\Delta P$ is insignificant and thus $\Delta h = \Delta u \cong C_{av}\,\Delta T$. For *liquids,* two special cases are commonly encountered:

1. *Constant pressure processes,* as in heaters ($\Delta P = 0$): $\Delta h = \Delta u \cong C_{av}\,\Delta T$

2. *Constant temperature processes,* as in pumps ($\Delta T = 0$): $\Delta h = v\,\Delta P$

For a process between states 1 and 2, the last relation can be expressed as $h_2 - h_1 = v(P_2 - P_1)$. By taking state 2 to be the compressed liquid state at a given T and P and state 1 to be the saturated liquid state at the same temperature, the enthalpy of the compressed liquid can be expressed as

$$h_{@\,P,T} \cong h_{f\,@\,T} + v_{f\,@\,T}\,(P - P_{sat}) \qquad (3\text{-}47)$$

where P_{sat} is the saturation pressure at the given temperature. This is an improvement over the assumption that the enthalpy of the compressed liquid could be taken as h_f at the given temperature (that is, $h_{@\,P,T} \cong h_{f\,@\,T}$). However, the contribution of the last term is often very small, so it is usually neglected.

EXAMPLE 3-14   Enthalpy of Compressed Liquid
Determine the enthalpy of liquid water at 100°C and 15 MPa (*a*) by using compressed liquid tables, (*b*) by approximating it as a saturated liquid, and (*c*) by using the correction given by Eq. 3-47.

Solution   At 100°C, the saturation pressure of water is 101.33 kPa, and since $P > P_{sat}$, the water exists as a compressed liquid at the specified state.

Analysis   (*a*) From compressed liquid tables, we read

$$\left.\begin{array}{l} P = 15 \text{ MPa} \\ T = 100°C \end{array}\right\} \quad h = \mathbf{430.28\ kJ/kg} \qquad \text{(Table A-7)}$$

This is the exact value.

(*b*) Approximating the compressed liquid as a saturated liquid at 100°C, as is commonly done, we obtain

$$h \cong h_{f @ 100°C} = \textbf{419.04 kJ/kg}$$

This value is in error by about 2.6 percent.

(*c*) From Eq. 3-47,

$$h_{@ P, T} = h_{f @ T} + v_f(P - P_{sat})$$

$$= (419.04 \text{ kJ/kg}) + (0.001 \text{ m}^3/\text{kg})[(15,000 - 101.33) \text{ kPa}] \left(\frac{1 \text{ kJ}}{1 \text{ kPa} \cdot \text{m}^3}\right)$$

$$= \textbf{434.60 kJ/kg}$$

The correction term reduced the error from 2.6 to about 1 percent. But this improvement in accuracy is often not worth the extra effort involved.

3-12 ■ SUMMARY

A substance that has a fixed chemical composition throughout is called a *pure substance*. A pure substance exists in different phases depending on its energy level. In the liquid phase, a substance that is not about to vaporize is called a *compressed* or *subcooled liquid*. In the gas phase, a substance that is not about to condense is called a *superheated vapor*. During a phase-change process, the temperature and pressure of a pure substance are dependent properties. At a given pressure, a substance changes phase at a fixed temperature, called the *saturation temperature*. Likewise, at a given temperature, the pressure at which a substance changes phase is called the *saturation pressure*. During a boiling process, both the liquid and the vapor phases coexist in equilibrium, and under this condition the liquid is called *saturated liquid* and the vapor *saturated vapor*.

In a saturated liquid–vapor mixture, the mass fraction of the vapor phase is called the *quality* and is defined as

$$x = \frac{m_{vapor}}{m_{total}}$$

The quality may have values between 0 (saturated liquid) and 1 (saturated vapor). It has no meaning in the compressed liquid or superheated vapor regions. In the saturated mixture region, the average value of any intensive property y is determined from

$$y = y_f + x y_{fg}$$

where f stands for saturated liquid and g for saturated vapor.

In the absence of compressed liquid data, a general approximation is to treat a compressed liquid as a saturated liquid at the given *temperature,* that is,

$$y \cong y_{f @ T}$$

where y stands for v, u, or h.

The state beyond which there is no distinct vaporization process is called the *critical point*. At supercritical pressures, a substance gradually and uniformly expands from the liquid to vapor phase. All three phases of a substance coexist in equilibrium at states along the *triple line* characterized by triple-line temperature and pressure. Various properties of some pure substances are listed in the appendix. As can be noticed from these tables, the compressed liquid has lower v, u, and h values than the saturated liquid at the same T or P. Likewise, superheated vapor has higher v, u, and h values than the saturated vapor at the same T or P.

Any relation among the pressure, temperature, and specific volume of a substance is called an *equation of state*. The simplest and best-known equation of state is the *ideal-gas equation of state,* given as

$$Pv = RT$$

where R is the gas constant. Caution should be exercised in using this relation since an ideal gas is a fictitious substance. Real gases exhibit ideal-gas behavior at relatively low pressures and high temperatures.

The deviation from ideal-gas behavior can be properly accounted for by using the *compressibility factor Z*, defined as

$$Z = \frac{Pv}{RT} \qquad \text{or} \qquad Z = \frac{v_{\text{actual}}}{v_{\text{ideal}}}$$

The Z factor is approximately the same for all gases at the same *reduced temperature* and *reduced pressure,* which are defined as

$$T_R = \frac{T}{T_{\text{cr}}} \qquad \text{and} \qquad P_R = \frac{P}{P_{\text{cr}}}$$

where P_{cr} and T_{cr} are the critical pressure and temperature, respectively. This is known as the *principle of corresponding states.*

The P-v-T behavior of substances can be represented more accurately by the more complex equations of state. Three of the best known are

van der Waals:
$$\left(P + \frac{a}{v^2}\right)(v - b) = RT$$

where
$$a = \frac{27R^2T_{\text{cr}}^2}{64P_{\text{cr}}} \qquad \text{and} \qquad b = \frac{RT_{\text{cr}}}{8P_{\text{cr}}}$$

Beattie-Bridgeman:
$$P = \frac{R_uT}{\bar{v}^2}\left(1 - \frac{c}{\bar{v}T^3}\right)(\bar{v} + B) - \frac{A}{\bar{v}^2}$$

where
$$A = A_0\left(1 - \frac{a}{\bar{v}}\right) \qquad \text{and} \qquad B = B_0\left(1 - \frac{b}{\bar{v}}\right)$$

Benedict-Webb-Rubin:
$$P = \frac{R_uT}{\bar{v}} + \left(B_0R_uT - A_0 - \frac{C_0}{T^2}\right)\frac{1}{\bar{v}^2} + \frac{bR_uT - a}{\bar{v}^3} + \frac{a\alpha}{\bar{v}^6}$$
$$+ \frac{c}{\bar{v}^3T^2}\left(1 + \frac{\gamma}{\bar{v}^2}\right)e^{-\gamma/\bar{v}^2}$$

The amount of energy needed to raise the temperature of a unit mass of a substance by one degree is called the *specific heat at constant volume C_v* for a constant-volume process and the *specific heat at constant pressure C_p* for a constant-pressure process. They are defined as

$$C_v = \left(\frac{\partial u}{\partial T}\right)_v \qquad \text{and} \qquad C_p = \left(\frac{\partial h}{\partial T}\right)_p$$

For ideal gases u, h, C_v, and C_p are functions of temperature alone. The Δu and Δh of ideal gases can be expressed as

$$\Delta u = u_2 - u_1 = \int_1^2 C_v(T)\, dT \cong C_{v,\,av}(T_2 - T_1)$$

$$\Delta h = h_2 - h_1 = \int_1^2 C_p(T)\, dT \cong C_{p,\,av}(T_2 - T_1)$$

For ideal gases, C_v and C_p are related by

$$C_p = C_v + R \qquad \text{(kJ/kg} \cdot \text{K)}$$

where R is the gas constant. The *specific heat ratio k* is defined as

$$k = \frac{C_p}{C_v}$$

For *incompressible substances* (liquids and solids), both the constant-pressure and constant-volume specific heats are identical and denoted by C:

$$C_p = C_v = C \qquad \text{(kJ/kg} \cdot \text{K)}$$

The Δu and Δh of incompressible substances are given by

$$\Delta u = \int_1^2 C(T)\, dT \cong C_{av}(T_2 - T_1) \qquad \text{(kJ/kg)}$$
$$\Delta h = \Delta u + v\, \Delta P \qquad\qquad\qquad \text{(kJ/kg)}$$

REFERENCES AND SUGGESTED READING

1. ASHRAE *Handbook of Fundamentals.* SI version. Atlanta, GA: American Society of Heating, Refrigerating, and Air-Conditioning Engineers, Inc., 1993.

2. ASHRAE *Handbook of Refrigeration.* SI version. Atlanta, GA: American Society of Heating, Refrigerating, and Air-Conditioning Engineers, Inc., 1994.

3. A. Bejan. *Advanced Engineering Thermodynamics.* New York: John Wiley & Sons, 1998.

4. Y. A. Çengel and M. A. Boles. *Thermodynamics: An Engineering Approach.* 3rd ed. New York: McGraw-Hill, 1998.

5. K. Wark and D. E. Richards. *Thermodynamics.* 6th ed. New York: McGraw-Hill, 1999.

PROBLEMS*

Pure Substances, Phase-Change Processes, Phase Diagrams

3-1C Is iced water a pure substance? Why?

3-2C What is the difference between saturated liquid and compressed liquid?

3-3C What is the difference between saturated vapor and superheated vapor?

---

*Students are encouraged to answer *all* the concept "C" questions.

3-4C Is there any difference between the properties of saturated vapor at a given temperature and the vapor of a saturated mixture at the same temperature?

3-5C Is there any difference between the properties of saturated liquid at a given temperature and the liquid of a saturated mixture at the same temperature?

3-6C Is it true that water boils at higher temperatures at higher pressures? Explain.

3-7C If the pressure of a substance is increased during a boiling process, will the temperature also increase or will it remain constant? Why?

3-8C Why are the temperature and pressure dependent properties in the saturated mixture region?

3-9C What is the difference between the critical point and the triple point?

3-10C Is it possible to have water vapor at $-10°C$?

3-11C A househusband is cooking beef stew for his family in a pan that is (*a*) uncovered, (*b*) covered with a light lid, and (*c*) covered with a heavy lid. For which case will the cooking time be the shortest? Why?

3-12C How does the boiling process at supercritical pressures differ from the boiling process at subcritical pressures?

Property Tables

3-13C In what kind of pot will a given volume of water boil at a higher temperature: a tall and narrow one or a short and wide one? Explain.

3-14C A perfectly fitting pot and its lid often stick after cooking, and it becomes very difficult to open the lid when the pot cools down. Explain why this happens and what you would do to open the lid.

3-15C It is well known that warm air in a cooler environment rises. Now consider a warm mixture of air and gasoline on top of an open gasoline can. Do you think this gas mixture will rise in a cooler environment?

3-16C In 1775, Dr. William Cullen made ice in Scotland by evacuating the air in a water tank. Explain how that device works, and discuss how the process can be made more efficient.

3-17C Does the amount of heat absorbed as 1 kg of saturated liquid water boils at 100°C have to be equal to the amount of heat released as 1 kg of saturated water vapor condenses at 100°C?

3-18C Does the reference point selected for the properties of a substance have any effect on thermodynamic analysis? Why?

3-19C What is the physical significance of h_{fg}? Can it be obtained from a knowledge of h_f and h_g? How?

3-20C Is it true that it takes more energy to vaporize 1 kg of saturated liquid water at 100°C than it would to vaporize 1 kg of saturated liquid water at 120°C?

3-21C   What is quality? Does it have any meaning in the superheated vapor region?

3-22C   Which process requires more energy: completely vaporizing 1 kg of saturated liquid water at 1 atm pressure or completely vaporizing 1 kg of saturated liquid water at 8 atm pressure?

3-23C   Does h_{fg} change with pressure? How?

3-24C   Can quality be expressed as the ratio of the volume occupied by the vapor phase to the total volume?

3-25C   In the absence of compressed liquid tables, how is the specific volume of a compressed liquid at a given P and T determined?

3-26   Complete the following table for H_2O:

T, °C	P, kPa	v, m³/kg	Phase description
50		4.16	
	200		Saturated vapor
250	400		
110	600		

3-27E   Complete the following table for H_2O:

T, °F	P, psia	u, Btu/lbm	Phase description
250		851	
	20		Saturated liquid
500	120		
400	400		

3-28   Complete the following table for H_2O:

T, °C	P, kPa	h, kJ/kg	x	Phase description
	325		0.4	
160		1682		
	950		0.0	
80	500			
	800	3161.7		

3-29   Complete the following table for refrigerant-134a:

T, °C	P, kPa	v, m³/kg	Phase description
−12	600		
20		0.022	
	320		Saturated vapor
100	600		

3-30   Complete the following table for refrigerant-134a:

T, °C	P, kPa	u, kJ/kg	Phase description
30		120	
−8			Saturated liquid
	400	300	
8	600		

3-31E Complete the following table for refrigerant-134a:

T, °F	P, psia	h, Btu/lbm	x	Phase description
	70	64		
20			0.7	
10	70			
	180	128.77		
110			1.0	

3-32 Complete the following table for H_2O:

T, °C	P, kPa	v, m³/kg	Phase description
125		0.53	
	1000		Saturated liquid
25	750		
500		0.130	

3-33 Complete the following table for H_2O:

T, °C	P, kPa	u, kJ/kg	Phase description
	325	2452	
170			Saturated vapor
190	2000		
	4000	3040	

3-34E The temperature in a pressure cooker during cooking at sea level is measured to be 250°F. Determine the absolute pressure inside the cooker in psia and in atm. Would you modify your answer if the place were at a higher elevation?

3-35E The atmospheric pressure at a location is usually specified at standard conditions, but it changes with the weather conditions. As the weather forecasters frequently state, the atmospheric pressure drops during stormy weather and it rises during clear and sunny days. If the pressure difference between the two extreme conditions is given to be 0.3 in. of mercury, determine how much the boiling temperatures of water will vary as the weather changes from one extreme to the other.

3-36 A person cooks a meal in a 30-cm-diameter pot that is covered with a well-fitting lid and lets the food cool to the room temperature of 20°C. The total mass of the food and the pot is 8 kg. Now the person tries to open the pan by lifting the lid up. Assuming no air has leaked into the pan during cooling, determine if the lid will open or the pan will move up together with the lid.

3-37 Water is to be boiled at sea level in a 30-cm-diameter stainless steel pan placed on top of a 3-kW electric burner. If 60 percent of the heat generated by the burner is transferred to the water during boiling, determine the rate of evaporation of water.

3-38 Repeat Prob. 3-37 for a location at an elevation of 1500 m where the atmospheric pressure is 84.5 kPa and thus the boiling temperature of water is 95°C.

3-39 Water is boiled at 1 atm pressure in a 20-cm-internal-diameter stainless steel pan on an electric range. If it is observed that the water level in the pan drops by 10 cm in 30 min, determine the rate of heat transfer to the pan.

Pressure
cooker
250°F

FIGURE P3-34E

Vapor

60%   40%

3 kW

FIGURE P3-37

3-40 Repeat Prob. 3-39 for a location at 2000-m elevation where the standard atmospheric pressure is 79.5 kPa.

3-41 Saturated steam coming off the turbine of a steam power plant at 30°C condenses on the outside of a 4-cm-outer-diameter, 20-m-long tube at a rate of 45 kg/h. Determine the rate of heat transfer from the steam to the cooling water flowing through the pipe.

3-42 The average atmospheric pressure in Denver (elevation = 1610 m) is 83.4 kPa. Determine the temperature at which water in an uncovered pan will boil in Denver.   *Answer:* 94.4°C.

3-43 Water in a 5-cm-deep pan is observed to boil at 98°C. At what temperature will the water in a 40-cm-deep pan boil? Assume both pans are full of water.

3-44 A cooking pan whose inner diameter is 20 cm is filled with water and covered with a 4-kg lid. If the local atmospheric pressure is 101 kPa, determine the temperature at which the water will start boiling when it is heated.
    Answer: 100.2°C

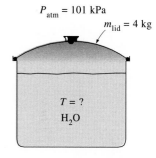

3-45 Water is being heated in a vertical piston-cylinder device. The piston has a mass of 20 kg and a cross-sectional area of 100 cm². If the local atmospheric pressure is 100 kPa, determine the temperature at which the water will start boiling.

3-46 A rigid tank with a volume of 2.5 m³ contains 5 kg of saturated liquid–vapor mixture of water at 75°C. Now the water is slowly heated. Determine the temperature at which the liquid in the tank is completely vaporized. Also, show the process on a *T-v* diagram with respect to saturation lines.
    Answer: 140.7°C

FIGURE P3-44

3-47 A rigid vessel contains 2 kg of refrigerant-134a at 900 kPa and 80°C. Determine the volume of the vessel and the total internal energy.
    Answers: 0.0572 m³, 577.7 kJ

3-48E A 5-ft³ rigid tank contains 5 lbm of water at 20 psia. Determine (*a*) the temperature, (*b*) the total enthalpy, and (*c*) the mass of each phase of water.

3-49 A 0.5-m³ vessel contains 10 kg of refrigerant-134a at −20°C. Determine (*a*) the pressure, (*b*) the total internal energy, and (*c*) the volume occupied by the liquid phase.
    Answers: (*a*) 132.99 kPa, (*b*) 889.5 kJ, (*c*) 0.00487 m³

3-50 A piston-cylinder device contains 0.1 m³ of liquid water and 0.9 m³ of water vapor in equilibrium at 800 kPa. Heat is transferred at constant pressure until the temperature reaches 350°C.
    (*a*) What is the initial temperature of the water?
    (*b*) Determine the total mass of the water.
    (*c*) Calculate the final volume.
    (*d*) Show the process on a *P-v* diagram with respect to saturation lines.

3-51E Superheated water vapor at 180 psia and 500°F is allowed to cool at constant volume until the temperature drops to 250°F. At the final state, determine (*a*) the pressure, (*b*) the quality, and (*c*) the enthalpy. Also, show the process on a *T-v* diagram with respect to saturation lines.
    Answers: (*a*) 29.82 psia, (*b*) 0.219, (*c*) 425.7 Btu/lbm

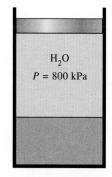

FIGURE P3-50

3-52 A piston-cylinder device initially contains 50 L of liquid water at 25°C and 300 kPa. Heat is added to the water at constant pressure until the entire liquid is vaporized.

(a) What is the mass of the water?
(b) What is the final temperature?
(c) Determine the total enthalpy change.
(d) Show the process on a *T-v* diagram with respect to saturation lines.
Answers: (a) 49.85 kg, (b) 133.55°C, (c) 130,627 kJ

3-53 A 0.5-m³ rigid vessel initially contains saturated liquid–vapor mixture of water at 100°C. The water is now heated until it reaches the critical state. Determine the mass of the liquid water and the volume occupied by the liquid at the initial state. *Answers:* 158.28 kg, 0.165 m³

3-54 Determine the specific volume, internal energy, and enthalpy of compressed liquid water at 100°C and 15 MPa using the saturated liquid approximation. Compare these values to the ones obtained from the compressed liquid tables.

3-55E A 15-ft³ rigid tank contains saturated mixture of refrigerant-134a at 30 psia. If the saturated liquid occupies 10 percent of the volume, determine the quality and the total mass of the refrigerant in the tank.

3-56 A piston-cylinder device contains 0.8 kg of steam at 300°C and 1 MPa. Steam is cooled at constant pressure until one-half of the mass condenses.

(a) Show the process on a *T-v* diagram.
(b) Find the final temperature.
(c) Determine the volume change.

3-57 A rigid tank contains water vapor at 300°C and an unknown pressure. When the tank is cooled to 180°C, the vapor starts condensing. Estimate the initial pressure in the tank. *Answer:* 1.325 MPa

Vapor Pressure and Phase Equilibrium

3-58 Consider a glass of water in a room that is at 20°C and 60 percent relative humidity. If the water temperature is 15°C, determine the vapor pressure (a) at the free surface of the water and (b) at a location in the room far from the glass.

3-59 During a hot summer day at the beach when the air temperature is 30°C, someone claims the vapor pressure in the air to be 5.2 kPa. Is this claim reasonable?

3-60 On a certain day, the temperature and relative humidity of air over a large swimming pool are measured to be 20°C and 40 percent, respectively. Determine the water temperature of the pool when phase equilibrium conditions are established between the water in the pool and the vapor in the air.

3-61 Consider two rooms that are identical except that one is maintained at 30°C and 40 percent relative humidity while the other is maintained at 20°C and 70 percent relative humidity. Noting that the amount of moisture is proportional to the vapor pressure, determine which room contains more moisture.

3-62E A thermos bottle is half-filled with water and is left open to the atmospheric air at 70°F and 35 percent relative humidity. If heat transfer to the water through the thermos walls and the free surface is negligible, determine the temperature of water when phase equilibrium is established.

3-63 During a hot summer day when the air temperature is 35°C and the relative humidity is 70 percent, you buy a supposedly "cold" canned drink from a store. The store owner claims that the temperature of the drink is below 10°C. Yet the drink does not feel so cold and you are skeptical since you notice no condensation forming outside the can. Can the store owner be telling the truth?

Ideal Gas

3-64C Propane and methane are commonly used for heating in winter, and the leakage of these fuels, even for short periods, poses a fire danger for homes. Which gas leakage do you think poses a greater risk for fire? Explain.

3-65C Under what conditions is the ideal-gas assumption suitable for real gases?

3-66C What is the difference between R and R_u? How are these two related?

3-67C What is the difference between mass and molar mass? How are these two related?

3-68C What is the physical significance of the compressibility factor Z?

3-69C What is the principle of corresponding states?

3-70C How are the reduced pressure and reduced temperature defined?

3-71 A spherical balloon with a diameter of 6 m is filled with helium at 20°C and 200 kPa. Determine the mole number and the mass of the helium in the balloon.   *Answers:* 9.28 kmol, 37.15 kg

3-72 The pressure in an automobile tire depends on the temperature of the air in the tire. When the air temperature is 25°C, the pressure gage reads 210 kPa. If the volume of the tire is 0.025 m³, determine the pressure rise in the tire when the air temperature in the tire rises to 50°C. Also, determine the amount of air that must be bled off to restore pressure to its original value at this temperature. Assume the atmospheric pressure to be 100 kPa.

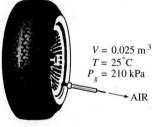

$V = 0.025 \text{ m}^3$
$T = 25°C$
$P_g = 210 \text{ kPa}$

→ AIR

FIGURE P3-72

3-73E The air in an automobile tire with a volume of 0.53 ft³ is at 90°F and 20 psig. Determine the amount of air that must be added to raise the pressure to the recommended value of 30 psig. Assume the atmospheric pressure to be 14.6 psia and the temperature and the volume to remain constant.
    Answer: 0.0260 lbm

3-74 The pressure gage on a 1.2-m³ oxygen tank reads 500 kPa. Determine the amount of oxygen in the tank if the temperature is 24°C and the atmospheric pressure is 97 kPa.

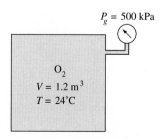

$P_g = 500 \text{ kPa}$

O_2
$V = 1.2 \text{ m}^3$
$T = 24°C$

FIGURE P3-74

3-75E A rigid tank contains 20 lbm of air at 20 psia and 70°F. More air is added to the tank until the pressure and temperature rise to 35 psia and 90°F, respectively. Determine the amount of air added to the tank.
    Answer: 13.73 lbm

3-76 A 800-L rigid tank contains 10 kg of air at 25°C. Determine the reading on the pressure gage if the atmospheric pressure is 97 kPa.

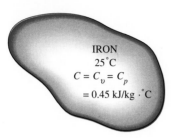

IRON
25°C
$C = C_v = C_p$
$= 0.45$ kJ/kg · °C

FIGURE P3-77

3-77 A 1-m³ tank containing air at 25°C and 500 kPa is connected through a valve to another tank containing 5 kg of air at 35°C and 200 kPa. Now the valve is opened, and the entire system is allowed to reach thermal equilibrium with the surroundings, which are at 20°C. Determine the volume of the second tank and the final equilibrium pressure of air.   *Answer:* 284.1 kPa

Other Equations of State

3-78C What is the physical significance of the two constants that appear in the van der Waals equation of state? On what basis are they determined?

3-79 A 3.27-m³ tank contains 100 kg of nitrogen at 225 K. Determine the pressure in the tank, using (*a*) the ideal-gas equation, (*b*) the van der Waals equation, and (*c*) the Beattie-Bridgeman equation. Compare your results with the actual value of 2000 kPa.

3-80 A 1-m³ tank contains 2.841 kg of steam at 0.6 MPa. Determine the temperature of the steam, using (*a*) the ideal gas equation, (*b*) the van der Waals equation, and (*c*) the steam tables.
    Answers: (*a*) 457.6 K, (*b*) 465.9 K, (*c*) 473 K

3-81E Refrigerant-134a at 100 psia has a specific volume of 0.5388 ft³/lbm. Determine the temperature of the refrigerant based on (*a*) the ideal-gas equation, (*b*) the van der Waals equation, and (*c*) the refrigerant tables.

3-82 Nitrogen at 150 K has a specific volume of 0.041884 m³/kg. Determine the pressure of the nitrogen, using (*a*) the ideal-gas equation and (*b*) the Beattie-Bridgeman equation. Compare your results to the experimental value of 1000 kPa.   *Answers:* (*a*) 1063 kPa, (*b*) 1000.4 kPa

Specific Heats, Δu, and Δh of Ideal Gases

3-83C Is the relation $\Delta U = mC_{v,\ av}\ \Delta T$ restricted to constant-volume processes only, or can it be used for any kind of process of an ideal gas?

3-84C Is the relation $\Delta H = mC_{p,\ av}\ \Delta T$ restricted to constant-pressure processes only, or can it be used for any kind of process of an ideal gas?

3-85C Show that for an ideal gas $\overline{C}_p = \overline{C}_v + R_u$.

3-86C Is the energy required to heat air from 295 to 305 K the same as the energy required to heat it from 345 to 355 K? Assume the pressure remains constant in both cases.

3-87C In the relation $\Delta U = mC_v\ \Delta T$, what is the correct unit of C_v— kJ/kg · °C or kJ/kg · K?

3-88C A fixed mass of an ideal gas is heated from 50 to 80°C at a constant pressure of (*a*) 1 atm and (*b*) 3 atm. For which case do you think the energy required will be greater? Why?

3-89C A fixed mass of an ideal gas is heated from 50 to 80°C at a constant volume of (*a*) 1 m³ and (*b*) 3 m³. For which case do you think the energy required will be greater? Why?

3-90C A fixed mass of an ideal gas is heated from 50 to 80°C (*a*) at constant volume and (*b*) at constant pressure. For which case do you think the energy required will be greater? Why?

3-91 Determine the enthalpy change Δh of nitrogen, in kJ/kg, as it is heated from 600 to 1000 K, using (*a*) the empirical specific heat equation as a function of temperature (Table A-2*c*), (*b*) the C_p value at the average temperature (Table A-2*b*), and (*c*) the C_p value at room temperature (Table A-2*a*).

Answers: (*a*) 447.8 kJ/kg, (*b*) 448.4 kJ/kg, (*c*) 415.6 kJ/kg

3-92E Determine the enthalpy change Δh of oxygen, in Btu/lbm, as it is heated from 800 to 1500 R, using (*a*) the empirical specific heat equation as a function of temperature (Table A-2E*c*), (*b*) the C_p value at the average temperature (Table A-2E*b*), and (*c*) the C_p value at room temperature (Table A-2E*a*).

Answers: (*a*) 170.1 Btu/lbm, (*b*) 178.5 Btu/lbm, (*c*) 153.3 Btu/lbm

3-93 Determine the internal energy change Δu of hydrogen, in kJ/kg, as it is heated from 400 to 1000 K, using (*a*) the empirical specific heat equation as a function of temperature (Table A-2*c*), (*b*) the C_v value at average temperature (Table A-2*b*), and (*c*) the C_v value at room temperature (Table A-2*a*).

Review Problems

3-94 A smoking lounge is to accommodate 15 heavy smokers. The minimum fresh air requirements for smoking lounges are specified to be 30 L/s per person (ASHRAE, *Standard 62,* 1989). Determine the minimum required flow rate of fresh air that needs to be supplied to the lounge, and the diameter of the duct if the air velocity is not to exceed 8 m/s.

3-95 The minimum fresh air requirements of a residential building are specified to be 0.35 air change per hour (ASHRAE, *Standard 62,* 1989). That is, 35 percent of the entire air contained in a residence should be replaced by fresh outdoors air every hour. If the ventilation requirements of a 2.7-m-high, 200-m² residence is to be met entirely by a fan, determine the size of the fan, in L/min, that needs to be installed. Also determine the diameter of the duct if the air velocity is not to exceed 6 m/s.

3-96 The gage pressure of an automobile tire is measured to be 200 kPa before a trip and 220 kPa after the trip at a location where the atmospheric pressure is 90 kPa. Assuming the volume of the tire remains constant at 0.022 m³, determine the percent increase in the absolute temperature of the air in the tire.

3-97 Although balloons have been around since 1783 when the first balloon took to the skies in France, a real breakthrough in ballooning occurred in 1960 with the design of the modern hot-air balloon fueled by inexpensive propane and constructed of lightweight nylon fabric. Over the years, ballooning has become a sport and a hobby for many people around the world. Unlike balloons filled with the light helium gas, hot-air balloons are open to the atmosphere. Therefore, the pressure in the balloon is always the same as the local atmospheric pressure, and the balloon is never in danger of exploding.

Hot-air balloons range from about 15 to 25 m in diameter. The air in the balloon cavity is heated by a propane burner located at the top of the passenger cage. The flames from the burner that shoot into the balloon heat the air in the balloon cavity, raising the air temperature at the top of the balloon from 65°C to over 120°C. The air temperature is maintained at the desired levels by periodically firing the propane burner. The buoyancy force that pushes the

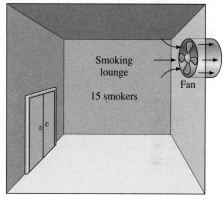

FIGURE P3-94

FIGURE P3-97
A hot-air balloon.

balloon upward is proportional to the density of the cooler air outside the balloon and the volume of the balloon, and can be expressed as

$$F_B = \rho_{\text{cool air}}\, g V_{\text{balloon}}$$

where g is the gravitational acceleration. When air resistance is negligible, the buoyancy force is opposed by (1) the weight of the hot air in the balloon, (2) the weight of the cage, the ropes, and the balloon material, and (3) the weight of the people and other load in the cage. The operator of the balloon can control the height and the vertical motion of the balloon by firing the burner or by letting some hot air in the balloon escape, to be replaced by cooler air. The forward motion of the balloon is provided by the winds.

Consider a 20-m-diameter hot-air balloon that, together with its cage, has a mass of 80 kg when empty. This balloon is hanging still in the air at a location where the atmospheric pressure and temperature are 90 kPa and 15°C, respectively, while carrying three 65-kg people. Determine the average temperature of the air in the balloon. What would your response be if the atmospheric air temperature were 30°C?

3-98 Consider an 18-m-diameter hot-air balloon that, together with its cage, has a mass of 120 kg when empty. The air in the balloon, which is now carrying two 70-kg people, is heated by propane burners at a location where the atmospheric pressure and temperature are 93 kPa and 12°C, respectively. Determine the average temperature of the air in the balloon when the balloon first starts rising. What would your response be if the atmospheric air temperature were 25°C?

3-99E Water in a pressure cooker is observed to boil at 250°F. What is the absolute pressure in the pressure cooker, in psia?

3-100 A rigid tank with a volume of 0.07 m³ contains 1 kg of refrigerant-134a vapor at 400 kPa. The refrigerant is now allowed to cool. Determine the pressure when the refrigerant first starts condensing. Also, show the process on a P-v diagram with respect to saturation lines.

3-101 A 4-L rigid tank contains 2 kg of saturated liquid–vapor mixture of water at 50°C. The water is now slowly heated until it exists in a single phase. At the final state, will the water be in the liquid phase or the vapor phase? What would your answer be if the volume of the tank were 400 L instead of 4 L?

3-102 A 10-kg mass of superheated refrigerant-134a at 0.8 MPa and 40°C is cooled at constant pressure until it exists as a compressed liquid at 20°C.
(a) Show the process on a T-v diagram with respect to saturation lines.
(b) Determine the change in volume.
(c) Find the change in total internal energy.
Answers: (b) −0.261 m³, (c) −1753 kJ

3-103 A 0.5-m³ rigid tank containing hydrogen at 20°C and 600 kPa is connected by a valve to another 0.5-m³ rigid tank that holds hydrogen at 30°C and 150 kPa. Now the valve is opened and the system is allowed to reach thermal equilibrium with the surroundings, which are at 15°C. Determine the final pressure in the tank.

3-104 A 20-m³ tank contains nitrogen at 25°C and 800 kPa. Some nitrogen is allowed to escape until the pressure in the tank drops to 600 kPa. If the temperature at this point is 20°C, determine the amount of nitrogen that has escaped. *Answer:* 42.9 kg

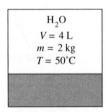

H₂O
$V = 4$ L
$m = 2$ kg
$T = 50°C$

FIGURE P3-101

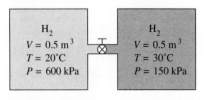

H₂
$V = 0.5$ m³
$T = 20°C$
$P = 600$ kPa

H₂
$V = 0.5$ m³
$T = 30°C$
$P = 150$ kPa

FIGURE P3-103

3-105 Steam at 400°C has a specific volume of 0.02 m³/kg. Determine the pressure of the steam based on (*a*) the ideal-gas equation, and (*b*) the steam tables.

    Answers: (*a*) 15,529 kPa, (*b*) 12,500 kPa

3-106 A tank whose volume is unknown is divided into two parts by a partition. One side of the tank contains 0.01 m³ of refrigerant-134a that is a saturated liquid at 0.8 MPa, while the other side is evacuated. The partition is now removed, and the refrigerant fills the entire tank. If the final state of the refrigerant is 25°C and 200 kPa, determine the volume of the tank.

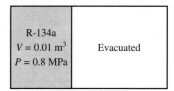

FIGURE P3-106

3-107 Liquid propane is commonly used as a fuel for heating homes, powering vehicles such as forklifts, and filling portable picnic tanks. Consider a propane tank that initially contains 5 L of liquid propane at the environment temperature of 20°C. If a hole develops in the connecting tube of a propane tank and the propane starts to leak out, determine the temperature of propane when the pressure in the tank drops to 1 atm. Also, determine the total amount of heat transfer from the environment to the tank to vaporize the entire propane in the tank.

3-108 Repeat Prob. 3-107 for isobutane.

Computer, Design, and Essay Problems

3-109 Write a computer program to express $T_{sat} = f(P_{sat})$ for steam as a fifth-degree polynomial where the pressure is in kPa and the temperature is in °C. Use tabulated data from Table A-4. What is the accuracy of this equation?

3-110 It is claimed that fruits and vegetables are cooled by 6°C for each percentage point of weight loss as moisture during vacuum cooling. Using calculations, demonstrate if this claim is reasonable.

3-111 Write a computer program to determine the specific volume of a substance at a given temperature and pressure, using the Beattie-Bridgeman equation. Check your program by evaluating the specific volume of refrigerant-134a at 10 different states and comparing them to the tabulated values.

3-112 A solid normally absorbs heat as it melts, but there is a known exception at temperatures close to absolute zero. Find out which solid it is and give a physical explanation for it.

3-113 Numerous equations of state have been proposed throughout history. Write an essay on two equations of state not discussed in this chapter, describe them in sufficient detail, and discuss the accuracy and range of applicability of each equation.

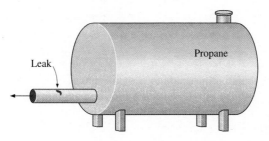

FIGURE P3-107

3-114   It is well known that water freezes at 0°C at atmospheric pressure. The mixture of liquid water and ice at 0°C is said to be at stable equilibrium since it cannot undergo any changes when it is isolated from its surrounding. However, when water is free of impurities and the inner surfaces of the container are smooth, the temperature of water can be lowered to −2°C or even lower without any formation of ice at atmospheric pressure. But at that state even a small disturbance can initiate the formation of ice abruptly, and the water temperature stabilizes at 0°C following this sudden change. The water at −2°C is said to be in a *metastable state*. Write an essay on metastable states and discuss how they differ from stable equilibrium states.

3-115   Using a thermometer, measure the boiling temperature of water and calculate the corresponding saturation pressure. From this information, estimate the altitude of your town and compare it with the actual altitude value.

3-116   Write a computer program to express the variation of specific heat $\overline{C}_p$ of air with temperature as a third-degree polynomial, using the data in Table A-2b. Compare your result with that given in Table A-2c.

3-117   Find out how the specific heats of gases, liquids, and solids are determined in national laboratories. Describe the experimental apparatus and the procedures used.

3-118   Design an experiment complete with instrumentation to determine the specific heats of a gas using a resistance heater. Discuss how the experiment will be conducted, what measurements need to be taken, and how the specific heats will be determined. What are the sources of error in your system? How can you minimize the experimental error?

3-119   Design an experiment complete with instrumentation to determine the specific heats of a liquid using a resistance heater. Discuss how the experiment will be conducted, what measurements need to be taken, and how the specific heats will be determined. What are the sources of error in your system? How can you minimize the experimental error? How would you modify this system to determine the specific heat of a solid?

Energy Transfer by Heat, Work, and Mass

Energy can be transferred to or from a closed system (a fixed mass) in two distinct forms: *heat* and *work*. For control volumes, energy can also be transported by mass. An energy transfer to or from a closed system is *heat* if it is caused by a temperature difference between the system and its surroundings. Otherwise it is *work*, and it is caused by a force acting through a distance. We start this chapter with a discussion of energy transfer by *heat*. We then introduce various forms of *work*, with particular emphasis on the *moving boundary work* or *P dV work* commonly encountered in reciprocating devices such as automotive engines and compressors. We continue with the *flow work*, which is the work associated with forcing a fluid into or out of a control volume, and show that the combination of the internal energy and the flow work gives the property *enthalpy*. Finally, we show that $h + \text{ke} + \text{pe}$ represents the energy entering or leaving a control volume with a fluid stream per unit of its mass.

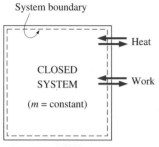

FIGURE 4-1

Energy can cross the boundaries of
a closed system in the form
of heat and work.

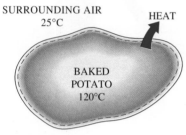

FIGURE 4-2

Heat is transferred from hot bodies
to colder ones by virtue of a
temperature difference.

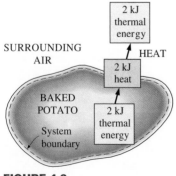

FIGURE 4-3

Energy is recognized as heat only as it
crosses the system boundary.

4-1 ■ HEAT TRANSFER

Energy can cross the boundary of a closed system in two distinct forms: *heat* and *work* (Fig. 4-1). It is important to distinguish between these two forms of energy. Therefore, they will be discussed first, to form a sound basis for the development of the first law of thermodynamics.

We know from experience that a can of cold soda left on a table eventually warms up and that a hot baked potato on the same table cools down (Fig. 4-2). When a body is left in a medium that is at a different temperature, energy transfer takes place between the body and the surrounding medium until thermal equilibrium is established, that is, the body and the medium reach the same temperature. The direction of energy transfer is always from the higher-temperature body to the lower-temperature one. Once the temperature equality is established, energy transfer stops. In the processes described above, energy is said to be transferred in the form of heat.

Heat is defined as *the form of energy that is transferred between two systems (or a system and its surroundings) by virtue of a temperature difference.* That is, an energy interaction is heat only if it takes place because of a temperature difference. Then it follows that there cannot be any heat transfer between two systems that are at the same temperature.

In daily life, we frequently refer to the sensible and latent forms of internal energy as *heat,* and we talk about the heat content of bodies. In thermodynamics, however, we usually refer to those forms of energy as *thermal energy* to prevent any confusion with *heat transfer.*

Several phrases in common use today—such as *heat flow, heat addition, heat rejection, heat absorption, heat removal, heat gain, heat loss, heat storage, heat generation, electrical heating, resistance heating, frictional heating, gas heating, heat of reaction, liberation of heat, specific heat, sensible heat, latent heat, waste heat, body heat, process heat, heat sink,* and *heat source*— are not consistent with the strict thermodynamic meaning of the term *heat,* which limits its use to the *transfer* of thermal energy during a process. However, these phrases are deeply rooted in our vocabulary, and they are used by both ordinary people and scientists without causing any misunderstanding since they are usually interpreted properly instead of being taken literally. (Besides, no acceptable alternatives exist for some of these phrases.) For example, the phrase *body heat* is understood to mean *the thermal energy content* of a body. Likewise, *heat flow* is understood to mean *the transfer of thermal energy,* not the flow of a fluidlike substance called heat, although the latter incorrect interpretation, which is based on the caloric theory, is the origin of this phrase. Also, the transfer of heat into a system is frequently referred to as *heat addition* and the transfer of heat out of a system as *heat rejection.* Perhaps there are thermodynamic reasons for being so reluctant to replace *heat* by *thermal energy*: It takes less time and energy to say, write, and comprehend *heat* than it does *thermal energy.*

Heat is energy in transition. It is recognized only as it crosses the boundary of a system. Consider the hot baked potato one more time. The potato contains energy, but this energy is heat transfer only as it passes through the skin of the potato (the system boundary) to reach the air, as shown in Fig. 4-3. Once in the surroundings, the transferred heat becomes part of the internal energy of the surroundings. Thus, in thermodynamics, the term *heat* simply means *heat transfer.*

A process during which there is no heat transfer is called an **adiabatic process** (Fig. 4-4). The word *adiabatic* comes from the Greek word *adiabatos,* which means *not to be passed.* There are two ways a process can be adiabatic: Either the system is well insulated so that only a negligible amount of heat can pass through the boundary, or both the system and the surroundings are at the same temperature and therefore there is no driving force (temperature difference) for heat transfer. An adiabatic process should not be confused with an isothermal process. Even though there is no heat transfer during an adiabatic process, the energy content and thus the temperature of a system can still be changed by other means such as work.

As a form of energy, heat has energy units, kJ (or Btu) being the most common one. The amount of heat transferred during the process between two states (states 1 and 2) is denoted by Q_{12}, or just Q. Heat transfer *per unit mass* of a system is denoted q and is determined from

$$q = \frac{Q}{m} \qquad \text{(kJ/kg)} \qquad (4\text{-}1)$$

Sometimes it is desirable to know the *rate of heat transfer* (the amount of heat transferred per unit time) instead of the total heat transferred over some time interval (Fig. 4-5). The heat transfer rate is denoted $\dot{Q}$, where the over-dot stands for the time derivative, or "per unit time." The heat transfer rate $\dot{Q}$ has the unit kJ/s, which is equivalent to kW. When $\dot{Q}$ varies with time, the amount of heat transfer during a process is determined by integrating $\dot{Q}$ over the time interval of the process:

$$Q = \int_{t_1}^{t_2} \dot{Q}\; dt \qquad \text{(kJ)} \qquad (4\text{-}2)$$

When $\dot{Q}$ remains constant during a process, the relation above reduces to

$$Q = \dot{Q}\, \Delta t \qquad \text{(kJ)} \qquad (4\text{-}3)$$

where $\Delta t = t_2 - t_1$ is the time interval during which the process occurs.

4-2 ■ ENERGY TRANSFER BY WORK

Work, like heat, is an energy interaction between a system and its surroundings. As mentioned earlier, energy can cross the boundary of a closed system in the form of heat or work. Therefore, *if the energy crossing the boundary of a closed system is not heat, it must be work.* Heat is easy to recognize: Its driving force is a temperature difference between the system and its surroundings. Then we can simply say that an energy interaction that is not caused by a temperature difference between a system and its surroundings is work. More specifically, *work is the energy transfer associated with a force acting through a distance.* A rising piston, a rotating shaft, and an electric wire crossing the system boundaries are all associated with work interactions.

Work is also a form of energy transferred like heat and, therefore, has energy units such as kJ. The work done during a process between states 1 and 2 is denoted by W_{12}, or simply W. The work done *per unit mass* of a system is denoted by w and is expressed as

$$w = \frac{W}{m} \qquad \text{(kJ/kg)} \qquad (4\text{-}4)$$

The work done *per unit time* is called **power** and is denoted $\dot{W}$ (Fig. 4-6). The unit of power is kJ/s, or kW.

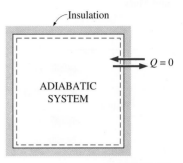

FIGURE 4-4

During an adiabatic process, a system exchanges no heat with its surroundings.

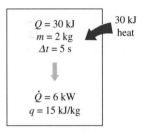

FIGURE 4-5

The relationships among q, Q, and $\dot{Q}$.

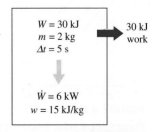

FIGURE 4-6

The relationships among w, W, and $\dot{W}$.

Surroundings

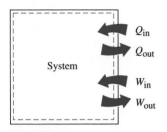

FIGURE 4-7

Specifying the directions of heat and work.

Heat and work are *directional quantities,* and thus the complete description of a heat or work interaction requires the specification of both the *magnitude* and *direction.* One way of doing that is to adopt a sign convention. The generally accepted **formal sign convention** for heat and work interactions is as follows: *heat transfer to a system and work done by a system are positive; heat transfer from a system and work done on a system are negative.* Another way is to use the subscripts *in* and *out* to indicate direction (Fig. 4-7). For example, a work input of 5 kJ can be expressed as $W_{in} = 5$ kJ, while a heat loss of 3 kJ can be expressed as $Q_{out} = 3$ kJ. When the direction of a heat or work interaction is not known, we can simply *assume* a direction for the interaction (using the subscript *in* or *out*) and solve for it. A positive result indicates the assumed direction is right. A negative result, on the other hand, indicates that the direction of the interaction is the opposite of the assumed direction. This is just like assuming a direction for an unknown force when solving a statics problem, and reversing the direction when a negative result is obtained for the force. We will use this *intuitive approach* in this book as it eliminates the need to adopt a formal sign convention and the need to carefully assign negative values to some interactions.

Note that a quantity that is transferred to or from a system during an interaction is not a property since the amount of such a quantity depends on more than just the state of the system. Heat and work are *energy transfer mechanisms* between a system and its surroundings, and there are many similarities between them:

1. Both are recognized at the boundaries of a system as they cross them. That is, both heat and work are *boundary* phenomena.

2. Systems possess energy, but not heat or work.

3. Both are associated with a *process,* not a state. Unlike properties, heat or work has no meaning at a state.

4. Both are *path functions* (i.e., their magnitudes depend on the path followed during a process as well as the end states).

Path functions have **inexact differentials** designated by the symbol δ. Therefore, a differential amount of heat or work is represented by δQ or δW, respectively, instead of dQ or dW. Properties, however, are **point functions** (i.e., they depend on the state only, and not on how a system reaches that state), and they have **exact differentials** designated by the symbol d. A small change in volume, for example, is represented by dV, and the total volume change during a process between states 1 and 2 is

$$\int_1^2 dV = V_2 - V_1 = \Delta V$$

That is, the volume change during process 1-2 is always the volume at state 2 minus the volume at state 1, regardless of the path followed (Fig. 4-8). The total work done during process 1-2, however, is

$$\int_1^2 \delta W = W_{12} \qquad (not \ \Delta W)$$

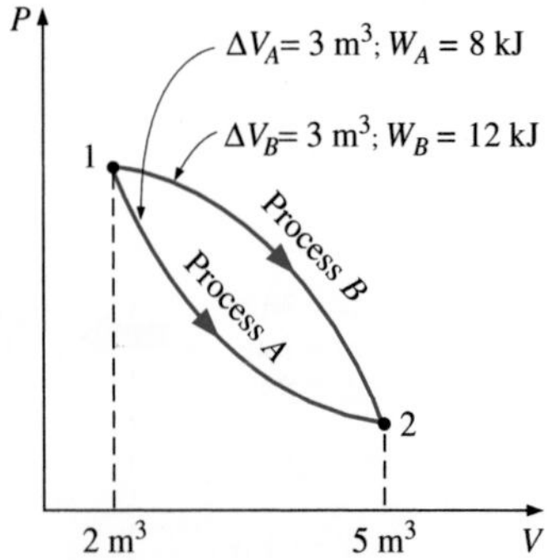

That is, the total work is obtained by following the process path and adding the differential amounts of work (δW) done along the way. The integral of δW *is not* $W_2 - W_1$ (i.e., the work at state 2 minus work at state 1), which is

P

$\Delta V_A = 3$ m³; $W_A = 8$ kJ

$\Delta V_B = 3$ m³; $W_B = 12$ kJ

1

Process B

Process A

2

2 m³   5 m³ V

FIGURE 4-8

Properties are point functions; but heat and work are path functions (their magnitudes depend on the path followed).

meaningless since work is not a property and systems do not possess work at a state.

EXAMPLE 4-1   Burning of a Candle in an Insulated Room

A candle is burning in a well-insulated room. Taking the room (the air plus the candle) as the system, determine (*a*) if there is any heat transfer during this burning process and (*b*) if there is any change in the internal energy of the system.

Solution   (*a*) The interior surfaces of the room form the system boundary, as indicated by the dashed lines in Fig. 4-9. As pointed out earlier, heat is recognized as it crosses the boundaries. Since the room is well insulated, we have an adiabatic system and no heat will pass through the boundaries. Therefore, $Q = 0$ for this process.

(*b*) The internal energy involves energies that exist in various forms (sensible, latent, chemical, nuclear). During the process described above, part of the chemical energy is converted to sensible energy. Since there is no increase or decrease in the total internal energy of the system, $\Delta U = 0$ for this process.

EXAMPLE 4-2   Heating of a Potato in an Oven

A potato initially at room temperature (25°C) is being baked in an oven that is maintained at 200°C, as shown in Fig. 4-10. Is there any heat transfer during this baking process?

Solution   This is not a well-defined problem since the system is not specified. Let us assume that we are observing the potato, which will be our system. Then the skin of the potato may be viewed as the system boundary. Part of the energy in the oven will pass through the skin to the potato. Since the driving force for this energy transfer is a temperature difference, this is a heat transfer process.

EXAMPLE 4-3   Heating of an Oven by Work Transfer

A well-insulated electric oven is being heated through its heating element. If the entire oven, including the heating element, is taken to be the system, determine whether this is a heat or work interaction.

Solution   For this problem, the interior surfaces of the oven form the system boundary, as shown in Fig. 4-11. The energy content of the oven obviously increases during this process, as evidenced by a rise in temperature. This energy transfer to the oven is not caused by a temperature difference between the oven and the surrounding air. Instead, it is caused by *electrons* crossing the system boundary and thus doing work. Therefore, this is a work interaction.

EXAMPLE 4-4   Heating of an Oven by Heat Transfer

Answer the question in Example 4-3 if the system is taken as only the air in the oven without the heating element.

Solution   This time, the system boundary will include the outer surface of the heating element and will not cut through it, as shown in Fig. 4-12. Therefore, no electrons will be crossing the system boundary at any point. Instead, the energy generated in the interior of the heating element will be transferred to the air around it as a result of the temperature difference between the heating element and the air in the oven. Therefore, this is a heat transfer process.

FIGURE 4-9

Schematic for Example 4-1.

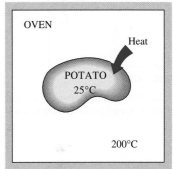

FIGURE 4-10

Schematic for Example 4-2.

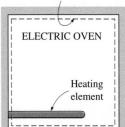

FIGURE 4-11

Schematic for Example 4-3.

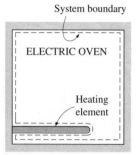

System boundary

FIGURE 4-12

Schematic for Example 4-4.

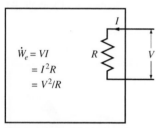

$\dot{W}_e = VI$
$\quad = I^2R$
$\quad = V^2/R$

FIGURE 4-13

Electrical power in terms of resistance R, current I, and potential difference V.

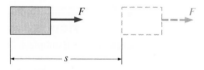

FIGURE 4-14

The work done is proportional to the force applied (F) and the distance traveled (s).

Discussion For both cases, the amount of energy transfer to the air is the same. These two examples show that the same interaction can be heat or work depending on how the system is selected.

Electrical Work

It was pointed out in Example 4-3 that electrons crossing the system boundary do electrical work on the system. In an electric field, electrons in a wire move under the effect of electromotive forces, doing work. When N coulombs of electrons move through a potential difference V, the electrical work done is

$$W_e = VN$$

which can also be expressed in the rate form as

$$\dot{W}_e = VI \qquad \text{(W)} \qquad (4\text{-}5)$$

where $\dot{W}_e$ is the **electrical power** and I is the number of electrons flowing per unit time, that is, the *current* (Fig. 4-13). In general, both V and I vary with time, and the electrical work done during a time interval Δt is expressed as

$$W_e = \int_1^2 VI \, dt \qquad \text{(kJ)} \qquad (4\text{-}6)$$

When both V and I remain constant during the time interval Δt, it reduces to

$$W_e = VI \, \Delta t \qquad \text{(kJ)} \qquad (4\text{-}7)$$

4-3 ■ MECHANICAL FORMS OF WORK

There are several different ways of doing work, each in some way related to a force acting through a distance (Fig. 4-14). In elementary mechanics, the work done by a constant force F on a body displaced a distance s in the direction of the force is given by

$$W = Fs \qquad \text{(kJ)} \qquad (4\text{-}8)$$

If the force F is not constant, the work done is obtained by adding (i.e., integrating) the differential amounts of work,

$$W = \int_1^2 F \, ds \qquad \text{(kJ)} \qquad (4\text{-}9)$$

Obviously one needs to know how the force varies with displacement to perform this integration. Equations 4-8 and 4-9 give only the magnitude of the work. The sign is easily determined from physical considerations: The work done on a system by an external force acting in the direction of motion is negative, and work done by a system against an external force acting in the opposite direction to motion is positive.

There are two requirements for a work interaction between a system and its surroundings to exist: (1) there must be a *force* acting on the boundary, and (2) the boundary must *move*. Therefore, the presence of forces on the boundary without any displacement of the boundary does not constitute a work interaction. Likewise, the displacement of the boundary without any force to oppose or drive this motion (such as the expansion of a gas into an evacuated space) is not a work interaction since no energy is transferred.

In many thermodynamic problems, mechanical work is the only form of work involved. It is associated with the movement of the boundary of a system or with the movement of the entire system as a whole (Fig. 4-15). Some common forms of mechanical work are discussed below.

1 Moving Boundary Work

One form of mechanical work frequently encountered in practice is associated with the expansion or compression of a gas in a piston-cylinder device. During this process, part of the boundary (the inner face of the piston) moves back and forth. Therefore, the expansion and compression work is often called **moving boundary work,** or simply **boundary work** (Fig. 4-16). Some call it the *P dV* work for reasons explained below. Moving boundary work is the primary form of work involved in *automobile engines.* During their expansion, the combustion gases force the piston to move, which in turn forces the crankshaft to rotate.

The moving boundary work associated with real engines or compressors cannot be determined exactly from a thermodynamic analysis alone because the piston usually moves at very high speeds, making it difficult for the gas inside to maintain equilibrium. Then the states through which the system passes during the process cannot be specified, and no process path can be drawn. Work, being a path function, cannot be determined analytically without a knowledge of the path. Therefore, the boundary work in real engines or compressors is determined by direct measurements.

In this section, we analyze the moving boundary work for a *quasi-equilibrium process,* a process during which the system remains in equilibrium at all times. A quasi-equilibrium process, also called a *quasi-static process,* is closely approximated by real engines, especially when the piston moves at low velocities. Under identical conditions, the work output of the engines is found to be a maximum, and the work input to the compressors to be a minimum when quasi-equilibrium processes are used in place of non-quasi-equilibrium processes. Below, the work associated with a moving boundary is evaluated for a quasi-equilibrium process.

Consider the gas enclosed in the piston-cylinder device shown in Fig. 4-17. The initial pressure of the gas is P, the total volume is V, and the cross-sectional area of the piston is A. If the piston is allowed to move a distance ds in a quasi-equilibrium manner, the differential work done during this process is

$$\delta W_b = F\, ds = PA\, ds = P\, dV \qquad (4\text{-}10)$$

That is, the boundary work in the differential form is equal to the product of the absolute pressure P and the differential change in the volume dV of the system. This expression also explains why the moving boundary work is sometimes called the *P dV* work.

Note in Eq. 4-10 that P is the absolute pressure, which is always positive. However, the volume change dV is positive during an expansion process (volume increasing) and negative during a compression process (volume decreasing). Thus, the boundary work is positive during an expansion process and negative during a compression process. Therefore, Eq. 4-10 can be viewed as an expression for boundary work output, $W_{b,\,out}$. A negative result indicates boundary work input (compression).

FIGURE 4-15

If there is no movement, no work is done.

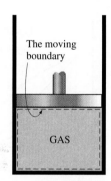

FIGURE 4-16

The work associated with a moving boundary is called *boundary work.*

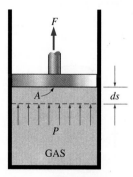

FIGURE 4-17

A gas does a differential amount of work δW_b as it forces the piston to move by a differential amount ds.

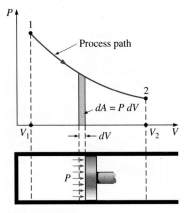

FIGURE 4-18

The area under the process curve on a P-V diagram represents the boundary work.

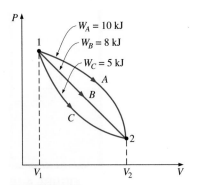

FIGURE 4-19

The boundary work done during a process depends on the path followed as well as the end states.

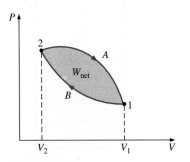

FIGURE 4-20

The net work done during a cycle is the difference between the work done by the system and the work done on the system.

The total boundary work done during the entire process as the piston moves is obtained by adding all the differential works from the initial state to the final state:

$$W_b = \int_1^2 P \, dV \qquad \text{(kJ)} \qquad (4\text{-}11)$$

This integral can be evaluated only if we know the functional relationship between P and V during the process. That is, $P = f(V)$ should be available. Note that $P = f(V)$ is simply the equation of the process path on a P-V diagram.

The quasi-equilibrium expansion process described above is shown on a P-V diagram in Fig. 4-18. On this diagram, the differential area dA is equal to $P \, dV$, which is the differential work. The total area A under the process curve 1-2 is obtained by adding these differential areas:

$$\text{Area} = A = \int_1^2 dA = \int_1^2 P \, dV \qquad (4\text{-}12)$$

A comparison of this equation with Eq. 4-11 reveals that *the area under the process curve on a P-V diagram is equal, in magnitude, to the work done during a quasi-equilibrium expansion or compression process of a closed system.* (On the P-v diagram, it represents the boundary work done per unit mass.)

A gas can follow several different paths as it expands from state 1 to state 2. In general, each path will have a different area underneath it, and since this area represents the magnitude of the work, the work done will be different for each process (Fig. 4-19). This is expected, since work is a path function (i.e., it depends on the path followed as well as the end states). If work were not a path function, no cyclic devices (car engines, power plants) could operate as work-producing devices. The work produced by these devices during one part of the cycle would have to be consumed during another part, and there would be no net work output. The cycle shown in Fig. 4-20 produces a net work output because the work done by the system during the expansion process (area under path A) is greater than the work done on the system during the compression part of the cycle (area under path B), and the difference between these two is the net work done during the cycle (the colored area).

If the relationship between P and V during an expansion or a compression process is given in terms of experimental data instead of in a functional form, obviously we cannot perform the integration analytically. But we can always plot the P-V diagram of the process, using these data points, and calculate the area underneath graphically to determine the work done.

Strictly speaking, the pressure P in Eq. 4-11 is the pressure at the inner surface of the piston. It becomes equal to the pressure of the gas in the cylinder only if the process is quasi-equilibrium and thus the entire gas in the cylinder is at the same pressure at any given time. Equation 4-11 can also be used for non-quasi-equilibrium processes provided that the pressure *at the inner face of the piston* is used for P. (Besides, we cannot speak of the pressure of a *system* during a non-quasi-equilibrium process since properties are defined for equilibrium states.) Therefore, we can generalize the boundary work relation by expressing it as

$$W_b = \int_1^2 P_i \, dV \qquad (4\text{-}13)$$

where P_i is the pressure at the inner face of the piston.

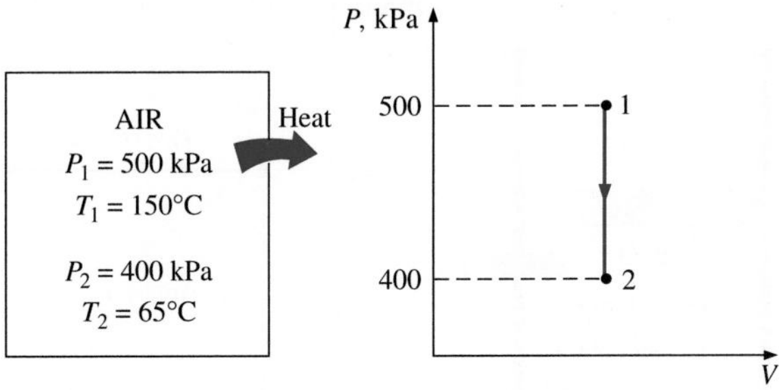

FIGURE 4-21

Schematic and P-V diagram for Example 4-5.

Note that work is a mechanism for energy interaction between a system and its surroundings, and W_b represents the amount of energy transferred from the system during an expansion process (or to the system during a compression process). Therefore, it has to appear somewhere else and we must be able to account for it since energy is conserved. In a car engine, for example, the boundary work done by the expanding hot gases is used to overcome friction between the piston and the cylinder, to push atmospheric air out of the way, and to rotate the crankshaft. Therefore,

$$W_b = W_{\text{friction}} + W_{\text{atm}} + W_{\text{crank}} = \int_1^2 (F_{\text{friction}} + P_{\text{atm}} A + F_{\text{crank}}) \, dx \quad (4\text{-}14)$$

Of course the work used to overcome friction will appear as frictional heat and the energy transmitted through the crankshaft will be transmitted to other components (such as the wheels) to perform certain functions. But note that the energy transferred by the system as work must equal the energy received by the crankshaft, the atmosphere, and the energy used to overcome friction.

The use of the boundary work relation is not limited to the quasi-equilibrium processes of gases only. It can also be used for solids and liquids.

EXAMPLE 4-5 Boundary Work during a Constant-Volume Process

A rigid tank contains air at 500 kPa and 150°C. As a result of heat transfer to the surroundings, the temperature and pressure inside the tank drop to 65°C and 400 kPa, respectively. Determine the boundary work done during this process.

Solution A sketch of the system and the P-V diagram of the process are shown in Fig. 4-21.

Analysis The boundary work can be determined from Eq. 4-11 to be

$$W_b = \int_1^2 P \, dV^{\,0} = 0$$

This is expected since a rigid tank has a constant volume and $dV = 0$ in the above equation. Therefore, there is no boundary work done during this process. That is, the boundary work done during a constant-volume process is always zero. This is also evident from the P-V diagram of the process (the area under the process curve is zero).

EXAMPLE 4-6 Boundary Work during a Constant-Pressure Process

A frictionless piston-cylinder device contains 10 lbm of water vapor at 60 psia and 320°F. Heat is now transferred to the steam until the temperature reaches 400°F. If the piston is not attached to a shaft and its mass is constant, determine the work done by the steam during this process.

Solution A sketch of the system and the P-v diagram of the process are shown in Fig. 4-22.

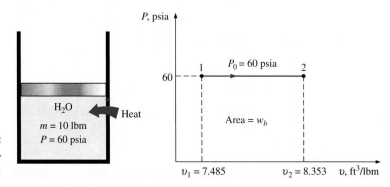

FIGURE 4-22

Schematic and *P-v* diagram for
Example 4-6.

Assumption   The expansion process is quasi-equilibrium.

Analysis   Even though it is not explicitly stated, the pressure of the steam within the cylinder remains constant during this process since both the atmospheric pressure and the weight of the piston remain constant. Therefore, this is a constant-pressure process, and, from Eq. 4-11

$$W_b = \int_1^2 P \, dV = P_0 \int_1^2 dV = P_0(V_2 - V_1) \qquad (4\text{-}15)$$

or
$$W_b = mP_0(v_2 - v_1)$$

since $V = mv$. From the superheated vapor table (Table A-6E), the specific volumes are determined to be $v_1 = 7.485$ ft³/lbm at state 1 (60 psia, 320°F) and $v_2 = 8.353$ ft³/lbm at state 2 (60 psia, 400°F). Substituting these values yields

$$W_b = (10 \text{ lbm})(60 \text{ psia})[(8.353 - 7.485) \text{ ft}^3/\text{lbm}]\left(\frac{1 \text{ Btu}}{5.404 \text{ psia} \cdot \text{ft}^3}\right)$$

$$= \textbf{96.4 Btu}$$

Discussion   The positive sign indicates that the work is done by the system. That is, the steam used 96.4 Btu of its energy to do this work. The magnitude of this work could also be determined by calculating the area under the process curve on the *P-V* diagram, which is simply $P_0 \, \Delta V$ for this case.

EXAMPLE 4-7   Boundary Work during an Isothermal Process
A piston-cylinder device initially contains 0.4 m³ of air at 100 kPa and 80°C. The air is now compressed to 0.1 m³ in such a way that the temperature inside the cylinder remains constant. Determine the work done during this process.

Solution   A sketch of the system and the *P-V* diagram of the process are shown in Fig. 4-23.

Assumptions   **1** The compression process is quasi-equilibrium. **2** At the specified conditions, air can be considered to be an ideal gas since it is at a high temperature and low pressure relative to its critical-point values.

Analysis   For an ideal gas at constant temperature T_0,

$$PV = mRT_0 = C \qquad \text{or} \qquad P = \frac{C}{V}$$

where C is a constant. Substituting this into Eq. 4-11, we have

$$W_b = \int_1^2 P \, dV = \int_1^2 \frac{C}{V} \, dV = C \int_1^2 \frac{dV}{V} = C \ln \frac{V_2}{V_1} = P_1 V_1 \ln \frac{V_2}{V_1} \qquad (4\text{-}16)$$

In the above equation, $P_1 V_1$ can be replaced by $P_2 V_2$ or mRT_0. Also, V_2/V_1 can be replaced by P_1/P_2 for this case since $P_1 V_1 = P_2 V_2$.

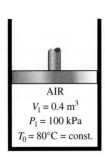

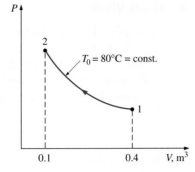

FIGURE 4-23

Schematic and *P-V* diagram for Example 4-7.

Substituting the numerical values into the above equation yields

$$W_b = (100 \text{ kPa})(0.4 \text{ m}^3)\left(\ln \frac{0.1}{0.4}\right)\left(\frac{1 \text{ kJ}}{1 \text{ kPa} \cdot \text{m}^3}\right)$$

$$= -55.45 \text{ kJ}$$

Discussion   The negative sign indicates that this work is done on the system (a work input), which is always the case for compression processes.

Polytropic Process

During expansion and compression processes of real gases, pressure and volume are often related by $PV^n = C$, where n and C are constants. A process of this kind is called a **polytropic process** (Fig. 4-24). Below we develop a general expression for the work done during a polytropic process. The pressure for a polytropic process can be expressed as

$$P = CV^{-n} \qquad (4\text{-}17)$$

Substituting this relation into Eq. 4-11, we obtain

$$W_b = \int_1^2 P \, dV = \int_1^2 CV^{-n} \, dV = C\frac{V_2^{-n+1} - V_1^{-n+1}}{-n + 1} = \frac{P_2V_2 - P_1V_1}{1 - n} \quad (4\text{-}18)$$

since $C = P_1V_1^n = P_2V_2^n$. For an ideal gas ($PV = mRT$), this equation can also be written as

$$W_b = \frac{mR(T_2 - T_1)}{1 - n}, \qquad n \neq 1 \qquad \text{(kJ)} \qquad (4\text{-}19)$$

The special case of $n = 1$ is equivalent to the isothermal process discussed in the previous example.

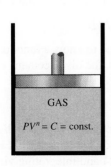

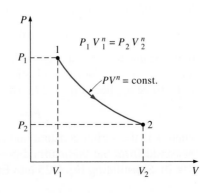

FIGURE 4-24

Schematic and *P-V* diagram for a polytropic process.

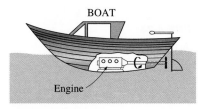

FIGURE 4-25

Energy transmission through rotating shafts is commonly encountered in practice.

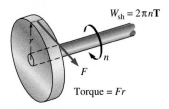

FIGURE 4-26

Shaft work is proportional to the torque applied and the number of revolutions of the shaft.

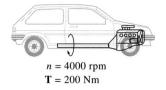

$n = 4000$ rpm
$\mathbf{T} = 200$ Nm

FIGURE 4-27

Schematic for Example 4-8.

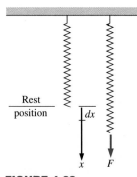

FIGURE 4-28

Elongation of a spring under the influence of a force.

2 Shaft Work

Energy transmission with a rotating shaft is very common in engineering practice (Fig. 4-25). Often the torque $\mathbf{T}$ applied to the shaft is constant, which means that the force F applied is also constant. For a specified constant torque, the work done during n revolutions is determined as follows: A force F acting through a moment arm r generates a torque $\mathbf{T}$ of (Fig. 4-26)

$$\mathbf{T} = Fr \quad \longrightarrow \quad F = \frac{\mathbf{T}}{r} \tag{4-20}$$

This force acts through a distance s, which is related to the radius r by

$$s = (2\pi r)n \tag{4-21}$$

Then the shaft work is determined from

$$W_{\text{sh}} = Fs = \left(\frac{\mathbf{T}}{r}\right)(2\pi rn) = 2\pi n\mathbf{T} \qquad \text{(kJ)} \tag{4-22}$$

The power transmitted through the shaft is the shaft work done per unit time, which can be expressed as

$$\dot{W}_{\text{sh}} = 2\pi\dot{n}\mathbf{T} \qquad \text{(kW)} \tag{4-23}$$

where $\dot{n}$ is the number of revolutions per unit time.

EXAMPLE 4-8   Power Transmission by the Shaft of a Car
Determine the power transmitted through the shaft of a car when the torque applied is 200 N · m and the shaft rotates at a rate of 4000 revolutions per minute (rpm).

Solution   A sketch of the car is given in Fig. 4-27. The shaft power is determined directly from

$$\dot{W}_{\text{sh}} = 2\pi\dot{n}\mathbf{T} = (2\pi)\left(4000 \frac{1}{\text{min}}\right)(200 \text{ N} \cdot \text{m})\left(\frac{1 \text{ min}}{60 \text{ s}}\right)\left(\frac{1 \text{ kJ}}{1000 \text{ N} \cdot \text{m}}\right)$$
$$= \textbf{83.8 kW (or 112.3 hp)}$$

3 Spring Work

It is common knowledge that when a force is applied on a spring, the length of the spring changes (Fig. 4-28). When the length of the spring changes by a differential amount dx under the influence of a force F, the work done is

$$\delta W_{\text{spring}} = F \, dx \tag{4-24}$$

To determine the total spring work, we need to know a functional relationship between F and x. For linear elastic springs, the displacement x is proportional to the force applied (Fig. 4-29). That is,

$$F = kx \qquad \text{(kN)} \tag{4-25}$$

where k is the spring constant and has the unit kN/m. The displacement x is measured from the undisturbed position of the spring (that is, $x = 0$ when $F = 0$). Substituting Eq. 4-25 into Eq. 4-24 and integrating yield

$$W_{spring} = \tfrac{1}{2}k(x_2^2 - x_1^2) \qquad \text{(kJ)} \qquad \text{(4-26)}$$

where x_1 and x_2 are the initial and the final displacements of the spring, respectively, measured from the undisturbed position of the spring.

EXAMPLE 4-9 Expansion of a Gas against a Spring

A piston-cylinder device contains 0.05 m³ of a gas initially at 200 kPa. At this state, a linear spring that has a spring constant of 150 kN/m is touching the piston but exerting no force on it. Now heat is transferred to the gas, causing the piston to rise and to compress the spring until the volume inside the cylinder doubles. If the cross-sectional area of the piston is 0.25 m², determine (a) the final pressure inside the cylinder, (b) the total work done by the gas, and (c) the fraction of this work done against the spring to compress it.

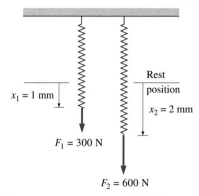

FIGURE 4-29

The displacement of a linear spring doubles when the force is doubled.

Solution A sketch of the system and the *P-V* diagram of the process are shown in Fig. 4-30.

Assumptions **1** The expansion process is quasi-equilibrium. **2** The spring is linear in the range of interest.

Analysis (a) The enclosed volume at the final state is

$$V_2 = 2V_1 = (2)(0.05 \text{ m}^3) = 0.1 \text{ m}^3$$

Then the displacement of the piston (and of the spring) becomes

$$x = \frac{\Delta V}{A} = \frac{(0.1 - 0.05) \text{ m}^3}{0.25 \text{ m}^2} = 0.2 \text{ m}$$

The force applied by the linear spring at the final state is

$$F = kx = (150 \text{ kN/m})(0.2 \text{ m}) = 30 \text{ kN}$$

The additional pressure applied by the spring on the gas at this state is

$$P = \frac{F}{A} = \frac{30 \text{ kN}}{0.25 \text{ m}^2} = 120 \text{ kPa}$$

Without the spring, the pressure of the gas would remain constant at 200 kPa while the piston is rising. But under the effect of the spring, the pressure rises linearly from 200 kPa to

$$200 + 120 = \textbf{320 kPa}$$

at the final state.

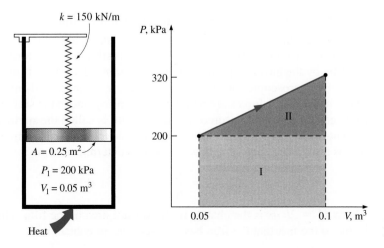

FIGURE 4-30

Schematic and *P-V* diagram for Example 4-9.

(*b*) An easy way of finding the work done is to plot the process on a *P-V* diagram and find the area under the process curve. From Fig. 4-30 the area under the process curve (a trapezoid) is determined to be

$$W = \text{area} = \frac{(200 + 320)\text{ kPa}}{2}\,[(0.1 - 0.05)\text{ m}^3]\left(\frac{1\text{ kJ}}{1\text{ kPa}\cdot\text{m}^3}\right) = \mathbf{13\ kJ}$$

Note that the work is done by the system.

(*c*) The work represented by the rectangular area (region I) is done against the piston and the atmosphere, and the work represented by the triangular area (region II) is done against the spring. Thus,

$$W_{\text{spring}} = \tfrac{1}{2}\,[(320 - 200)\text{ kPa}](0.05\text{ m}^3)\left(\frac{1\text{ kJ}}{1\text{ kPa}\cdot\text{m}^3}\right) = \mathbf{3\ kJ}$$

This result could also be obtained from Eq. 4-26:

$$W_{\text{spring}} = \tfrac{1}{2}k(x_2^2 - x_1^2) = \tfrac{1}{2}(150\text{ kN/m})[(0.2\text{ m})^2 - 0^2]\left(\frac{1\text{ kJ}}{1\text{ kN}\cdot\text{m}}\right) = 3\text{ kJ}$$

4   Other Mechanical Forms of Work

There are many other forms of mechanical work. Below we introduce some of them briefly.

Work Done on Elastic Solid Bars

Solids are often modeled as linear springs because under the action of a force they contract or elongate, as shown in Fig. 4-31, and when the force is lifted, they return to their original lengths, like a spring. This is true as long as the force is in the elastic range, that is, not large enough to cause permanent (plastic) deformations. Therefore, the equations given for a linear spring can also be used for elastic solid bars. Alternately, we can determine the work associated with the expansion or contraction of an elastic solid bar by replacing pressure *P* by its counterpart in solids, *normal stress* $\sigma_n = F/A$, in the boundary work expression:

$$W_{\text{elastic}} = \int_1^2 \sigma_n\,dV = \int_1^2 \sigma_n A\,dx \qquad \text{(kJ)} \qquad (4\text{-}27)$$

where *A* is the cross-sectional area of the bar. Note that the normal stress has pressure units.

Work Associated with the Stretching of a Liquid Film

Consider a liquid film such as soap film suspended on a wire frame (Fig. 4-32). We know from experience that it will take some force to stretch this film by the movable portion of the wire frame. This force is used to overcome the microscopic forces between molecules at the liquid–air interfaces. These microscopic forces are perpendicular to any line in the surface, and the force generated by these forces per unit length is called the **surface tension** σ_s, whose unit is N/m. Therefore, the work associated with the stretching of a film is also called *surface tension work*. It is determined from

$$W_{\text{surface}} = \int_1^2 \sigma_s\,dA \qquad \text{(kJ)} \qquad (4\text{-}28)$$

where $dA = 2b\,dx$ is the change in the surface area of the film. The factor 2 is due to the fact that the film has two surfaces in contact with air. The force

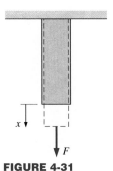

FIGURE 4-31

Solid bars behave as springs under the influence of a force.

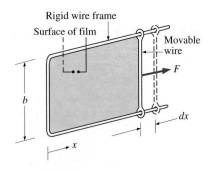

FIGURE 4-32

Stretching a liquid film with a movable wire.

acting on the movable wire as a result of surface tension effects is $F = 2b\sigma_s$ where σ_s is the surface tension force per unit length.

Work Done to Raise or to Accelerate a Body

When a body is raised in a gravitational field, its potential energy increases. Likewise, when a body is accelerated, its kinetic energy increases. The conservation of energy principle requires that an equivalent amount of energy must be transferred to the body being raised or accelerated. Remember that energy can be transferred to a given mass by heat and work, and the energy transferred in this case obviously is not heat since it is not driven by a temperature difference. Therefore, it must be work. Then we conclude that (1) *the work transfer needed to raise a body is equal to the change in the potential energy of the body, and* (2) *the work transfer needed to accelerate a body is equal to the change in the kinetic energy of the body* (Fig. 4-33). Similarly, the potential or kinetic energy of a body represents the work that can be obtained from the body as it is lowered to the reference level or decelerated to zero velocity.

The discussion given above together with the consideration for friction and other losses form the basis for determining the required power rating of motors used to drive devices such as elevators, escalators, conveyor belts, and ski lifts. It also plays a primary role in the design of automotive and aircraft engines, and in the determination of the amount of hydroelectric power that can be produced from a given water reservoir, which is simply the potential energy ofthe water relative to the location of the hydraulic turbine.

EXAMPLE 4-10   Power Needs of a Car to Climb a Hill

Consider a 1200-kg car cruising steadily on a level road at 90 km/h. Now the car starts climbing a hill that is sloped 30° from the horizontal (Fig. 4-34). If the velocity of the car is to remain constant during climbing, determine the additional power that must be delivered by the engine.

Solution   The additional power required is simply the work that needs to be done per unit time to raise the elevation of the car, which is equal to the change in the potential energy of the car per unit time:

$$\dot{W}_g = mg\,\Delta z/\Delta t = mg\,\mathcal{V}_{\text{vertical}}$$

$$= (1200 \text{ kg})(9.81 \text{ m/s}^2)(90 \text{ km/h})(\sin 30°)\left(\frac{1 \text{ m/s}}{3.6 \text{ km/h}}\right)\left(\frac{1 \text{ kJ/kg}}{1000 \text{ m}^2/\text{s}^2}\right)$$

$$= 147 \text{ kJ/s} = \textbf{147 kW} \qquad \text{(or 197 hp)}$$

Therefore, the car engine will have to produce almost 200 hp of additional power while climbing the hill if the car is to maintain its velocity.

EXAMPLE 4-11   Power Needs of a Car to Accelerate

Determine the power required to accelerate a 900-kg car shown in Fig. 4-35 from rest to a velocity of 80 km/h in 20 s on a level road.

Solution   The work needed to accelerate a body is simply the change in the kinetic energy of the body,

$$W_a = \tfrac{1}{2}m(\mathcal{V}_2^2 - \mathcal{V}_1^2) = \tfrac{1}{2}(900 \text{ kg})\left[\left(\frac{80,000 \text{ m}}{3600 \text{ s}}\right)^2 - 0^2\right]\left(\frac{1 \text{ kJ/kg}}{1000 \text{ m}^2/\text{s}^2}\right)$$

$$= 222.2 \text{ kJ}$$

The average power is determined from

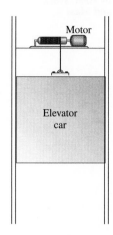

FIGURE 4-33
The energy transferred to a body while being raised is equal to the change in its potential energy.

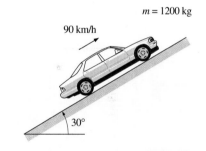

$m = 1200$ kg

90 km/h

30°

FIGURE 4-34
Schematic for Example 4-10.

0 → 80 km/h

$m = 900$ kg

FIGURE 4-35
Schematic for Example 4-11.

$$\dot{W}_a = \frac{W_a}{\Delta t} = \frac{222.2 \text{ kJ}}{20 \text{ s}} = \textbf{11.1 kW} \qquad \text{(or 14.9 hp)}$$

This is in addition to the power required to overcome friction, rolling resistance, and other imperfections.

4-4 ■ NONMECHANICAL FORMS OF WORK

The treatment above represents a fairly comprehensive coverage of mechanical forms of work. But some work modes encountered in practice are not mechanical in nature. However, these nonmechanical work modes can be treated in a similar manner by identifying a *generalized force F* acting in the direction of a *generalized displacement x*. Then the work associated with the differential displacement under the influence of this force is determined from $\delta W = F \, dx$.

Some examples of nonmechanical work modes are **electrical work,** where the generalized force is the *voltage* (the electrical potential) and the generalized displacement is the *electrical charge* as discussed in the Section 4-2; **magnetic work,** where the generalized force is the *magnetic field strength* and the generalized displacement is the total *magnetic dipole moment;* and **electrical polarization work,** where the generalized force is the *electric field strength* and the generalized displacement is the *polarization of the medium* (the sum of the electric dipole rotation moments of the molecules). Detailed consideration of these and other nonmechanical work modes can be found in specialized books on these topics.

4-5 ■ FLOW WORK AND THE ENERGY OF A FLOWING FLUID

Unlike closed systems, control volumes involve mass flow across their boundaries, and some work is required to push the mass into or out of the control volume. This work is known as the **flow work,** or **flow energy,** and is necessary for maintaining a continuous flow through a control volume.

To obtain a relation for flow work, consider a fluid element of volume V as shown in Fig. 4-36. The fluid immediately upstream will force this fluid element to enter the control volume; thus, it can be regarded as an imaginary piston. The fluid element can be chosen to be sufficiently small so that it has uniform properties throughout.

If the fluid pressure is P and the cross-sectional area of the fluid element is A (Fig. 4-37), the force applied on the fluid element by the imaginary piston is

$$F = PA \qquad (4\text{-}29)$$

To push the entire fluid element into the control volume, this force must act through a distance L. Thus, the work done in pushing the fluid element across the boundary (i.e., the flow work) is

$$W_{\text{flow}} = FL = PAL = PV \qquad \text{(kJ)} \qquad (4\text{-}30)$$

The flow work per unit mass is obtained by dividing both sides of this equation by the mass of the fluid element:

$$w_{\text{flow}} = Pv \qquad \text{(kJ/kg)} \qquad (4\text{-}31)$$

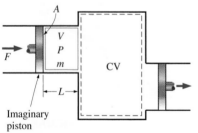

FIGURE 4-36

Schematic for flow work.

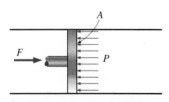

FIGURE 4-37

In the absence of acceleration, the force applied on a fluid by a piston is equal to the force applied on the piston by the fluid.

The flow work relation is the same whether the fluid is pushed into or out of the control volume (Fig. 4-38).

It is interesting that unlike other work quantities, flow work is expressed in terms of properties. In fact, it is the product of two properties of the fluid. For that reason, some people view it as a *combination property* (like enthalpy) and refer to it as *flow energy, convected energy,* or *transport energy* instead of flow work. Others, however, argue rightfully that the product Pv represents energy for flowing fluids only and does not represent any form of energy for nonflow (closed) systems. Therefore, it should be treated as work. This controversy is not likely to end, but it is comforting to know that both arguments yield the same result for the energy equation. In the discussions that follow, we consider the flow energy to be part of the energy of a flowing fluid, since this greatly simplifies the energy analysis of control volumes.

Total Energy of a Flowing Fluid

As we discussed in Chap. 2, the total energy of a simple compressible system consists of three parts: internal, kinetic, and potential energies (Fig. 4-39). On a unit-mass basis, it is expressed as

$$e = u + \text{ke} + \text{pe} = u + \frac{\mathcal{V}^2}{2} + gz \qquad \text{(kJ/kg)} \qquad (4\text{-}32)$$

where $\mathcal{V}$ is the velocity and z is the elevation of the system relative to some external reference point.

The fluid entering or leaving a control volume possesses an additional form of energy—the *flow energy Pv,* as discussed above. Then the total energy of a **flowing fluid** on a unit-mass basis (denoted by θ) becomes

$$\theta = Pv + e = Pv + (u + \text{ke} + \text{pe}) \qquad (4\text{-}33)$$

But the combination $Pv + u$ has been previously defined as the enthalpy h. So the above relation reduces to

$$\theta = h + \text{ke} + \text{pe} = h + \frac{\mathcal{V}^2}{2} + gz \qquad \text{(kJ/kg)} \qquad (4\text{-}34)$$

Professor J. Kestin proposed in 1966 that the term θ be called **methalpy** (from *metaenthalpy,* which means *beyond enthalpy*).

By using the enthalpy instead of the internal energy to represent the energy of a flowing fluid, one does not need to be concerned about the flow work. The energy associated with pushing the fluid into or out of the control

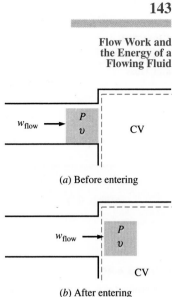

(a) Before entering

(b) After entering

FIGURE 4-38
Flow work is the energy needed to push a fluid into or out of a control volume, and it is equal to Pv.

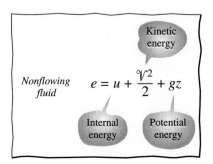

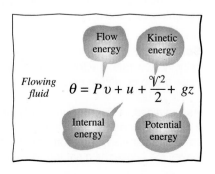

FIGURE 4-39
The total energy consists of three parts for a nonflowing fluid and four parts for a flowing fluid.

volume is automatically taken care of by enthalpy. In fact, this is the main reason for defining the property enthalpy. From now on, the energy of a fluid stream flowing into or out of a control volume is represented by Eq. 4-34, and no reference will be made to flow work or flow energy.

Energy Transport by Mass

Noting that θ is total energy per unit mass, the total energy of a flowing fluid of mass m is simply $m\theta$, provided that the properties of the mass m are uniform. Also, when a fluid stream with uniform properties is flowing at a mass flow rate of $\dot{m}$, the rate of energy flow with that stream is $\dot{m}\theta$ (Fig. 4-40). That is,

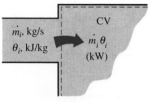

FIGURE 4-40

The product $\dot{m}_i\theta_i$ is the energy transported into the control volume by mass per unit time.

$$
\begin{array}{lll}
\textit{Amount of} & & \\
\textit{Energy Transport:} & E_{\text{mass}} = m\theta = m\left(h + \dfrac{\mathcal{V}^2}{2} + gz\right) & \text{(kJ)} \quad (4\text{-}35)
\end{array}
$$

$$
\begin{array}{lll}
\textit{Rate of} & & \\
\textit{Energy Transport:} & \dot{E}_{\text{mass}} = \dot{m}\theta = \dot{m}\left(h + \dfrac{\mathcal{V}^2}{2} + gz\right) & \text{(kW)} \quad (4\text{-}36)
\end{array}
$$

When the kinetic and potential energies of a fluid stream are negligible, as is usually the case, the relations above simplify to $E_{\text{mass}} = mh$ and $\dot{E}_{\text{mass}} = \dot{m}h$.

In general, the total energy transported by mass into or out of the control volume is not easy to determine since the properties of the mass at each inlet or exit may be changing with time as well as over the cross section. Thus, the only way to determine the energy transport through an opening as a result of mass flow is to consider sufficiently small differential masses δm that have uniform properties and to add their total energies during flow.

Again noting that θ is total energy per unit mass, the total energy of a flowing fluid of mass δm is $\theta\,\delta m$. Then the total energy transported by mass through an inlet or exit ($m_i\theta_i$ and $m_e\theta_e$) is obtained by integration. At an inlet, for example, it becomes

$$
E_{\text{in, mass}} = \int_{m_i} \theta_i\,\delta m_i = \int_{m_i}\left(h_i + \frac{\mathcal{V}_i^2}{2} + gz_i\right)\delta m_i \qquad (4\text{-}37)
$$

Most flows encountered in practice can be approximated as being steady and one-dimensional, and thus the simple relations in Eqs. 4-35 and 4-36 can be used to represent the energy transported by a fluid stream.

EXAMPLE 4-12   Energy Transport by Mass

Steam is leaving a 4-L pressure cooker whose operating pressure is 150 kPa (Fig. 4-41). It is observed that the amount of liquid in the cooker has decreased by 0.6 L in 40 minutes after the steady operating conditions are established, and the cross-sectional area of the exit opening is 8 mm². Determine (a) the mass flow rate of the steam and the exit velocity, (b) the total and flow energies of the steam per unit mass, and (c) the rate at which energy is leaving the cooker by steam.

Solution   Steam is leaving a pressure cooker at a specified pressure. The velocity, flow rate, the total and flow energies, and the rate of energy transfer by mass are to be determined.

Assumptions   **1** The flow is steady, and the initial start-up period is disregarded. **2** The kinetic and potential energies are negligible, and thus they are not considered. **3** Saturation conditions exist within the cooker at all times so that steam leaves the cooker as a saturated vapor at the cooker pressure.

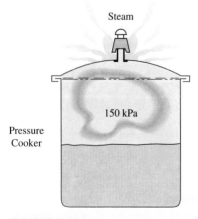

Steam

150 kPa

Pressure
Cooker

FIGURE 4-41

Schematic for Example 4-12.

Properties The properties of saturated liquid water and water vapor at 150 kPa are $v_f = 0.001053$ m³/kg, $v_g = 1.1593$ m³/kg, $u_g = 2519.7$ kJ/kg, and $h_g = 2693.6$ kJ/kg (Table A-5).

Analysis (*a*) Saturation conditions exist in a pressure cooker at all times after the steady operating conditions are established. Therefore, the liquid has the properties of saturated liquid and the existing steam has the properties of saturated vapor at the operating pressure. The amount of liquid that has evaporated, the mass flow rate of the exiting steam, and the exit velocity are

$$m = \frac{\Delta V_{\text{liquid}}}{v_f} = \frac{0.6 \text{ L}}{0.001053 \text{ m}^3/\text{kg}} \left(\frac{1 \text{ m}^3}{1000 \text{ L}} \right) = 0.570 \text{ kg}$$

$$\dot{m} = \frac{m}{\Delta t} = \frac{0.570 \text{ kg}}{40 \text{ min}} = 0.0142 \text{ kg/min} = \mathbf{2.37 \times 10^{-4} \text{ kg/s}}$$

$$\mathcal{V} = \frac{\dot{m}}{\rho_g A_c} = \frac{\dot{m} v_g}{A_c} = \frac{(2.37 \times 10^{-4} \text{ kg/s})(1.1593 \text{ m}^3/\text{kg})}{8 \times 10^{-6} \text{ m}^2} = \mathbf{34.3 \text{ m/s}}$$

(*b*) Noting that $h = u + Pv$ and that the kinetic and potential energies are disregarded, the flow and total energies of the exiting steam are

$$e_{\text{flow}} = Pv = h - u = 2693.6 - 2519.7 = \mathbf{173.9 \text{ kJ/kg}}$$

$$\theta = h + ke + pe \cong h = \mathbf{2693.6 \text{ kJ/kg}}$$

Note that the kinetic energy in this case is $ke = \mathcal{V}^2/2 = (34.3 \text{ m/s})^2/2 = 588 \text{ m}^2/\text{s}^2 = 0.588$ kJ/kg, which is very small compared to enthalpy.

(*c*) The rate at which energy is leaving the cooker by mass is simply the product of the mass flow rate and the total energy of the exiting steam per unit mass,

$$\dot{E}_{\text{mass}} = \dot{m}\theta = (2.37 \times 10^{-4} \text{ kg/s})(2693.6 \text{ kJ/kg}) = 0.638 \text{ kJ/s} = \mathbf{0.638 \text{ kW}}$$

Discussion The numerical value of the energy leaving the cooker with steam alone does not mean much since this value depends on the reference point selected for enthalpy (it could even be negative). The significant quantity is the difference between the enthalpies of the exiting vapor and the liquid inside (which is h_{fg}) since it relates directly to the amount of energy supplied to the cooker, as we will discuss in the next chapter.

4-6 ■ SUMMARY

Energy can cross the boundaries of a closed system in the form of heat or work. For control volumes, energy can also be transported by mass. If the energy transfer is due to a temperature difference between a closed system and its surroundings, it is *heat;* otherwise, it is *work.*

Work is the energy transferred as a force acts on a system through a distance. The most common form of mechanical work is the *boundary work,* which is the work associated with the expansion and compression of substances. On a *P-V* diagram, the area under the process curve represents the boundary work for a quasi-equilibrium process. Various forms of work are expressed as follows:

Electrical work:
$$W_e = VI\,\Delta t$$

Boundary work:

    (1) General
$$W_b = \int_1^2 P\,dV$$

    (2) Isobaric process $\quad W_b = P_0(V_2 - V_1)$
$(P_1 = P_2 = P_0 = \text{constant})$

(3) Polytropic process
$(Pv^n = \text{constant})$
$$W_b = \frac{P_2 V_2 - P_1 V_1}{1 - n} \qquad (n \neq 1)$$

(4) Isothermal process
$(PV = mRT_o = \text{constant})$
$$W_b = P_1 V_1 \ln \frac{V_2}{V_1} = mRT_0 \ln \frac{V_2}{V_1}$$

Shaft work:
$$W_{sh} = 2\pi n\mathbf{T}$$

Spring work:
$$W_{spring} = \frac{1}{2} k_s (x_2^2 - x_1^2)$$

The work required to push a unit mass of fluid into or out of a control volume is called *flow work* or *flow energy,* and is expressed as $w_{flow} = Pv$. In the analysis of control volumes, it is convenient to combine the flow energy and internal energy into *enthalpy.* Then the total energy of a flowing fluid is expressed as

$$\theta = h + ke + pe = h + \frac{\mathcal{V}^2}{2} + gz$$

The total energy transported by a flowing fluid of mass m with uniform properties is $m\theta$. The rate of energy transport by a fluid with a mass flow rate of $\dot{m}$ is $\dot{m}\theta$. When the kinetic and potential energies of a fluid stream are negligible, the amount and rate of energy transport become $E_{mass} = mh$ and $\dot{E}_{mass} = \dot{m}h$, respectively.

REFERENCES AND SUGGESTED READING

1. ASHRAE *Handbook of Fundamentals.* SI version. Atlanta, GA: American Society of Heating, Refrigerating, and Air-Conditioning Engineers, Inc., 1993.

2. A. Bejan. *Advanced Engineering Thermodynamics.* New York: John Wiley & Sons, 1988.

3. Y. A. Çengel and M. A. Boles. *Thermodynamics: An Engineering Approach.* 3rd ed. New York, McGraw-Hill, 1998.

4. K. Wark and D. E. Richards. *Thermodynamics.* 6th ed. New York, McGraw-Hill, 1999.

5. M .J. Moran and H. N. Shapiro. *Fundamentals of Engineering Thermodynamics.* 3rd ed. Wiley, New York, 1996.

PROBLEMS*

Heat Transfer and Work

4-1C In what forms can energy cross the boundaries of a closed system?

4-2C When is the energy crossing the boundaries of a closed system heat and when is it work?

4-3C What is an adiabatic process? What is an adiabatic system?

4-4C A gas in a piston-cylinder device is compressed, and as a result its temperature rises. Is this a heat or work interaction for the gas?

---

*Students are encouraged to answer *all* the concept "C" questions.

4-5C A room is heated by an iron that is left plugged in. Is this a heat or work interaction? Take the entire room, including the iron, as the system.

4-6C A room is heated as a result of solar radiation coming in through the windows. Is this a heat or work interaction for the room?

4-7C An insulated room is heated by burning candles. Is this a heat or work interaction? Take the entire room, including the candles, as the system.

4-8C What are point and path functions? Give some examples.

Boundary Work

4-9C On a P-v diagram, what does the area under the process curve represent?

4-10C Is the boundary work associated with constant-volume systems always zero?

4-11C An ideal gas at a given state expands to a fixed final volume first at constant pressure and then at constant temperature. For which case is the work done greater?

4-12C Show that $1 \text{ kPa} \cdot \text{m}^3 = 1 \text{ kJ}$.

4-13 A mass of 5 kg of saturated water vapor at 200 kPa is heated at constant pressure until the temperature reaches 300°C. Calculate the work done by the steam during this process.   *Answer:* 430.5 kJ

4-14 A frictionless piston-cylinder device initially contains 200 L of saturated liquid refrigerant-134a. The piston is free to move, and its mass is such that it maintains a pressure of 800 kPa on the refrigerant. The refrigerant is now heated until its temperature rises to 50°C. Calculate the work done during this process.   *Answer:* 5227 kJ

4-15E A frictionless piston-cylinder device contains 12 lbm of superheated water vapor at 60 psia and 500°F. Steam is now cooled at constant pressure until 70 percent of it, by mass, condenses. Determine the work done during this process.

4-16 A mass of 1.2 kg of air at 150 kPa and 12°C is contained in a gas-tight, frictionless piston-cylinder device. The air is now compressed to a final pressure of 600 kPa. During the process, heat is transferred from the air such that the temperature inside the cylinder remains constant. Calculate the work input during this process.   *Answer:* 136.1 kJ

4-17 Nitrogen at an initial state of 300 K, 150 kPa, and 0.2 m³ is compressed slowly in an isothermal process to a final pressure of 800 kPa. Determine the work done during this process.

4-18 A gas is compressed from an initial volume of 0.42 m³ to a final volume of 0.12 m³. During the quasi-equilibrium process, the pressure changes with volume according to the relation $P = aV + b$, where $a = -1200 \text{ kPa/m}^3$ and $b = 600 \text{ kPa}$. Calculate the work done during this process (*a*) by plotting the process on a P-V diagram and finding the area under the process curve and (*b*) by performing the necessary integrations.

4-19E During an expansion process, the pressure of a gas changes from 15 to 100 psia according to the relation $P = aV + b$, where $a = 5 \text{ psia/ft}^3$ and

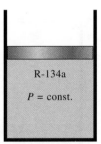

FIGURE P4-14

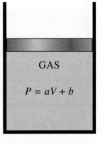

FIGURE P4-18

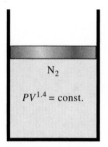

FIGURE P4-21

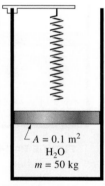

FIGURE P4-25

FIGURE P4-26

b is a constant. If the initial volume of the gas is 7 ft^3, calculate the work done during the process.   *Answer:* 180.9 Btu

4-20   During some actual expansion and compression processes in piston-cylinder devices, the gases have been observed to satisfy the relationship $PV^n = C$, where n and C are constants. Calculate the work done when a gas expands from a state of 150 kPa and 0.03 m^3 to a final volume of 0.2 m^3 for the case of $n = 1.3$.

4-21   A frictionless piston-cylinder device contains 2 kg of nitrogen at 100 kPa and 300 K. Nitrogen is now compressed slowly according to the relation $PV^{1.4} = $ constant until it reaches a final temperature of 360 K. Calculate the work input during this process.   *Answer:* 89.0 kJ

4-22   The equation of state of a gas is given as $\bar{v}\,(P + 10/\bar{v}^2) = R_u T$, where the units of $\bar{v}$ and P are m^3/kmol and kPa, respectively. Now 0.5 kmol of this gas is expanded in a quasi-equilibrium manner from 2 to 4 m^3 at a constant temperature of 300 K. Determine (*a*) the unit of the quantity 10 in the equation and (*b*) the work done during this isothermal expansion process.

4-23   Carbon dioxide contained in a piston-cylinder device is compressed from 0.3 to 0.1 m^3. During the process, the pressure and volume are related by $P = aV^{-2}$, where $a = 8$ kPa · m^6. Calculate the work done on the carbon dioxide during this process.   *Answer:* 53.3 kJ

4-24E   Hydrogen is contained in a piston-cylinder device at 14.7 psia and 15 ft^3. At this state, a linear spring ($F \propto x$) with a spring constant of 15,000 lbf/ft is touching the piston but exerts no force on it. The cross-sectional area of the piston is 3 ft^2. Heat is transferred to the hydrogen, causing it to expand until its volume doubles. Determine (*a*) the final pressure, (*b*) the total work done by the hydrogen, and (*c*) the fraction of this work done against the spring. Also, show the process on a *P-V* diagram.

4-25   A piston-cylinder device contains 50 kg of water at 150 kPa and 25°C. The cross-sectional area of the piston is 0.1 m^2. Heat is now transferred to the water, causing part of it to evaporate and expand. When the volume reaches 0.2 m^3, the piston reaches a linear spring whose spring constant is 100 kN/m. More heat is transferred to the water until the piston rises 20 cm more. Determine (*a*) the final pressure and temperature and (*b*) the work done during this process. Also, show the process on a *P-V* diagram.
    Answers: (*a*) 350 kPa, 138.88°C; (*b*) 27.5 kJ

4-26   A piston-cylinder device with a set of stops contains 10 kg of refrigerant-134a. Initially, 8 kg of the refrigerant is in the liquid form, and the temperature is −8°C. Now heat is transferred slowly to the refrigerant until the piston hits the stops, at which point the volume is 400 L. Determine (*a*) the temperature when the piston first hits the stops and (*b*) the work done during this expansion process. Also, show the process on a *P-V* diagram.
    Answers: (*a*) −8°C, (*b*) 45.6 kJ

4-27   A frictionless piston-cylinder device contains 10 kg of saturated refrigerant-134a vapor at 50°C. The refrigerant is then allowed to expand isothermally by gradually decreasing the pressure in a quasi-equilibrium manner to a final value of 500 kPa. Determine the work done during this expansion process (*a*) by using the experimental specific volume data from the tables and (*b*) by treating the refrigerant vapor as an ideal gas. Also, determine the error involved in the latter case.

4-28 Determine the boundary work done by a gas during an expansion process if the pressure and volume values at various states are measured to be 300 kPa, 1 L; 290 kPa, 1.1 L; 270 kPa, 1.2 L; 250 kPa, 1.4 L; 220 kPa, 1.7 L; and 200 kPa, 2 L.

Other Forms of Mechanical Work

4-29C A car is accelerated from rest to 85 km/h in 10 s. Would the work energy transferred to the car be different if it were accelerated to the same speed in 5 s?

4-30C Lifting a weight to a height of 20 m takes 20 s for one crane and 10 s for another. Is there any difference in the amount of work done on the weight by each crane?

4-31 Determine the work required to accelerate an 800-kg car from rest to 100 km/h on a level road. *Answer:* 308.6 kJ

4-32 Determine the work required to accelerate a 2000-kg car from 20 to 70 km/h on an uphill road with a vertical rise of 40 m.

4-33E Determine the torque applied to the shaft of a car that transmits 450 hp and rotates at a rate of 3000 rpm.

4-34 Determine the work required to deflect a linear spring with a spring constant of 70 kN/m by 20 cm from its rest position.

4-35 The engine of a 1500-kg automobile has a power rating of 75 kW. Determine the time required to accelerate this car from rest to a speed of 85 km/h at full power on a level road. Is your answer realistic?

4-36 A ski lift has a one-way length of 1 km and a vertical rise of 200 m. The chairs are spaced 20 m apart, and each chair can seat three people. The lift is operating at a steady speed of 10 km/h. Neglecting friction and air drag and assuming that the average mass of each loaded chair is 250 kg, determine the power required to operate this ski lift. Also estimate the power required to accelerate this ski lift in 5 s to its operating speed when it is first turned on.

4-37 Determine the power required for a 2000-kg car to climb a 100-m-long uphill road with a slope of 30° (from horizontal) in 10 s (*a*) at a constant velocity, (*b*) from rest to a final velocity of 30 m/s, and (*c*) from 35 m/s to a final velocity of 5 m/s. Disregard friction, air drag, and rolling resistance.
 Answers: (*a*) 98.07 kW, (*b*) 188.07 kW, (*c*) −21.93 kW

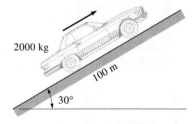

2000 kg

100 m

30°

FIGURE P4-37

4-38 A damaged 1200-kg car is being towed by a truck. Neglecting the friction, air drag, and rolling resistance, determine the extra power required (*a*) for constant velocity on a level road, (*b*) for constant velocity of 50 km/h on a 30° (from horizontal) uphill road, and (*c*) to accelerate on a level road from stop to 90 km/h in 12 s. *Answers:* (*a*) 0, (*b*) 81.7 kW, (*c*) 31.25 kW

Flow Work and Energy Transfer by Mass

4-39C What are the different mechanisms for transferring energy to or from a control volume?

4-40C What is flow energy? Do fluids at rest possess any flow energy?

4-41C How do the energies of a flowing fluid and a fluid at rest compare? Name the specific forms of energy associated with each case.

4-42E Steam is leaving a pressure cooker whose operating pressure is 30 psia. It is observed that the amount of liquid in the cooker has decreased by 0.4 gal in 45 minutes after the steady operating conditions are established, and the cross-sectional area of the exit opening is 0.15 in². Determine (*a*) the mass flow of the steam and the exit velocity, (*b*) the total and flow energies of the steam per unit mass, and (*c*) the rate at which energy is leaving the cooker by steam.

4-43 Refrigerant-134a enters the compressor of a refrigeration system as saturated vapor at 0.14 MPa, and leaves as superheated vapor at 0.8 MPa and 50°C at a rate of 0.04 kg/s. Determine the rates of energy transfers by mass into and out of the compressor. Assume the kinetic and potential energies to be negligible.

4-44 A house is maintained at 1 atm and 24°C, and warm air inside a house is forced to leave the house at a rate of 150 m³/h as a result of outdoor air at 5°C infiltrating into the house through the cracks. Determine the rate of net energy loss of the house due to mass transfer. *Answer:* 0.945 kW

Review Problems

4-45 Consider a vertical elevator whose cabin has a total mass of 800 kg when fully loaded and 150 kg when empty. The weight of the elevator cabin is partially balanced by a 400-kg counterweight that is connected to the top of the cabin by cables that pass through a pulley located on top of the elevator well. Neglecting the weight of the cables and assuming the guide rails and the pulleys to be frictionless, determine (*a*) the power required while the fully loaded cabin is rising at a constant speed of 2 m/s and (*b*) the power required while the empty cabin is descending at a constant speed of 2 m/s.

What would your answer be to (*a*) if no counterweight were used? What would your answer be to (*b*) if a friction force of 1200 N has developed between the cabin and the guide rails?

4-46 A frictionless piston-cylinder device initially contains air at 200 kPa and 0.2 m³. At this state, a linear spring (*F* ∝ *x*) is touching the piston but exerts no force on it. The air is now heated to a final state of 0.5 m³ and 800 kPa. Determine (*a*) the total work done by the air and (*b*) the work done against the spring. Also, show the process on a *P-v* diagram.
Answers: (*a*) 150 kJ, (*b*) 90 kJ

4-47 A mass of 5 kg of saturated liquid–vapor mixture of water is contained in a piston-cylinder device at 100 kPa. Initially, 2 kg of the water is in the liquid phase and the rest is in the vapor phase. Heat is now transferred to the water, and the piston, which is resting on a set of stops, starts moving when the pressure inside reaches 200 kPa. Heat transfer continues until the total volume increases by 20 percent. Determine (*a*) the initial and final temperatures, (*b*) the mass of liquid water when the piston first starts moving, and (*c*) the work done during this process. Also, show the process on a *P-v* diagram.

4-48E A spherical balloon contains 10 lbm of air at 30 psia and 800 R. The balloon material is such that the pressure inside is always proportional to the square of the diameter. Determine the work done when the volume of the balloon doubles as a result of heat transfer. *Answer:* 715.3 Btu

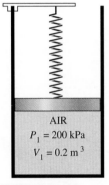

AIR
$P_1 = 200$ kPa
$V_1 = 0.2$ m^3

FIGURE P4-46

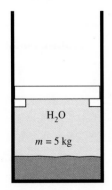

H$_2$O

$m = 5$ kg

FIGURE P4-47

4-49 Design a reciprocating compressor capable of supplying compressed air at 800 kPa at a rate of 15 kg/min. Also specify the size of the electric motor capable of driving this compressor. The compressor is to operate at no more than 2000 rpm (revolutions per minute).

4-50 A considerable fraction of energy loss in residential buildings is due to the cold outdoor air infiltrating through the cracks mostly around the doors and windows of the building. Write an essay on infiltration losses, their cost to homeowners, and the measures to prevent them.

The First Law of Thermodynamics

5

The first law of thermodynamics is simply a statement of the *conservation of energy principle,* and it asserts that *total energy* is a thermodynamic property. In the last chapter, energy transfer to or from a system by heat, work, and mass flow was discussed. In this chapter, the general *energy balance* relation, which is expressed as $E_{in} - E_{out} = \Delta E_{system}$, is developed in a step-by-step manner using an intuitive approach. The energy balance is first used to solve problems that involve heat and work interactions, but not mass flow (i.e., *closed systems*) for general pure substances, ideal gases, and incompressible substances. Then the energy balance is applied to *steady flow systems,* and common steady-flow devices such as nozzles, compressors, turbines, throttling valves, mixers, and heat exchangers are analyzed. Finally, the energy balance is applied to general *unsteady flow processes* such as charging and discharging of vessels.

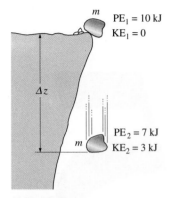

m PE$_1$ = 10 kJ
      KE$_1$ = 0

Δz

                PE$_2$ = 7 kJ
m KE$_2$ = 3 kJ

FIGURE 5-1

Energy cannot be created or destroyed;
it can only change forms.

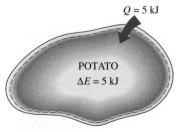

$Q = 5$ kJ

POTATO
$\Delta E = 5$ kJ

FIGURE 5-2

The increase in the energy of a potato
in an oven is equal to the amount of
heat transferred to it.

5-1 ■ THE FIRST LAW OF THERMODYNAMICS

So far, we have considered various forms of energy such as heat Q, work W, and total energy E individually, and no attempt has been made to relate them to each other during a process. The *first law of thermodynamics,* also known as *the conservation of energy principle,* provides a sound basis for studying the relationships among the various forms of energy and energy interactions. Based on experimental observations, the first law of thermodynamics states that *energy can be neither created nor destroyed; it can only change forms.* Therefore, every bit of energy should be accounted for during a process.

We all know that a rock at some elevation possesses some potential energy, and part of this potential energy is converted to kinetic energy as the rock falls (Fig. 5-1). Experimental data show that the decrease in potential energy ($mg\Delta z$) exactly equals the increase in kinetic energy [$m(\mathcal{V}_2^2 - \mathcal{V}_1^2)/2$] when the air resistance is negligible, thus confirming the conservation of energy principle.

Consider a system undergoing a series of *adiabatic* processes from a specified state 1 to another specified state 2. Being adiabatic, these processes obviously cannot involve any heat transfer, but they may involve several kinds of work interactions. Careful measurements during these experiments indicate the following: *For all adiabatic processes between two specified states of a closed system, the net work done is the same regardless of the nature of the closed system and the details of the process.* Considering that there are an infinite number of ways to perform work interactions under adiabatic conditions, the statement above appears to be very powerful, with a potential for far-reaching implications. This statement, which is largely based on the experiments of Joule in the first half of the nineteenth century, cannot be drawn from any other known physical principle and is recognized as a fundamental principle. This principle is called the **first law of thermodynamics** or just the **first law.**

A major consequence of the first law is the existence and the definition of the property *total energy E.* Considering that the net work is the same for all adiabatic processes of a closed system between two specified states, the value of the net work must depend on the end states of the system only, and thus it must correspond to a change in a property of the system. This property is the *total energy.* Note that the first law makes no reference to the value of the total energy of a closed system at a state. It simply states that the *change* in the total energy during an adiabatic process must be equal to the net work done. Therefore, any convenient arbitrary value can be assigned to total energy at a specified state to serve as a reference point.

Implicit in the first law statement is the conservation of energy. Although the essence of the first law is the existence of the property *total energy,* the first law is often viewed as a statement of the *conservation of energy* principle. Below we develop the first law or the conservation of energy relation for closed systems with the help of some familiar examples using intuitive arguments.

First, we consider some processes that involve heat transfer but no work interactions. The potato baked in the oven is a good example for this case (Fig. 5-2). As a result of heat transfer to the potato, the energy of the potato will increase. If we disregard any mass transfer (moisture loss from the potato), the increase in the total energy of the potato becomes equal to the amount of heat transfer. That is, if 5 kJ of heat is transferred to the potato, the energy increase of the potato will also be 5 kJ.

As another example, consider the heating of water in a pan on top of a range (Fig. 5-3). If 15 kJ of heat is transferred to the water from the heating element and 3 kJ of it is lost from the water to the surrounding air, the increase in energy of the water will be equal to the net heat transfer to water, which is 12 kJ.

Now consider a well-insulated (i.e., adiabatic) room heated by an electric heater as our system (Fig. 5-4). As a result of electrical work done, the energy of the system will increase. Since the system is adiabatic and cannot have any heat transfer to or from the surroundings ($Q = 0$), the conservation of energy principle dictates that the electrical work done on the system must equal the increase in energy of the system.

Next, let us replace the electric heater with a paddle wheel (Fig. 5-5). As a result of the stirring process, the energy of the system will increase. Again, since there is no heat interaction between the system and its surroundings ($Q = 0$), the paddle-wheel work done on the system must show up as an increase in the energy of the system.

Many of you have probably noticed that the temperature of air rises when it is compressed (Fig. 5-6). This is because energy is transferred to the air in the form of boundary work. In the absence of any heat transfer ($Q = 0$), the entire boundary work will be stored in the air as part of its total energy. The conservation of energy principle again requires that the increase in the energy of the system be equal to the boundary work done on the system.

We can extend the discussions above to systems that involve various heat and work interactions simultaneously. For example, if a system gains 12 kJ of heat during a process while 6 kJ of work is done on it, the increase in the energy of the system during that process is 18 kJ (Fig. 5-7). That is, the change in the energy of a system during a process is simply equal to the net energy transfer to (or from) the system.

Energy Balance

In the light of the discussions above, the conservation of energy principle may be expressed as follows: *The net change (increase or decrease) in the total energy of the system during a process is equal to the difference between the total energy entering and the total energy leaving the system during that process.* That is, during a process,

$$\begin{pmatrix} \text{Total energy} \\ \text{entering the system} \end{pmatrix} - \begin{pmatrix} \text{Total energy} \\ \text{leaving the system} \end{pmatrix} = \begin{pmatrix} \text{Change in the total} \\ \text{energy of the system} \end{pmatrix}$$

or

$$E_{in} - E_{out} = \Delta E_{system}$$

This relation is often referred to as the **energy balance** and is applicable to any kind of system undergoing any kind of process. The successful use of this relation to solve engineering problems depends on understanding the various forms of energy and recognizing the forms of energy transfer.

Energy Change of a System, ΔE_{system}

The determination of the energy change of a system during a process involves the evaluation of the energy of the system at the beginning and at the end of the process, and taking their difference. That is,

$$\text{Energy change} = \text{Energy at final state} - \text{Energy at initial state}$$

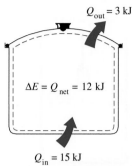

FIGURE 5-3

In the absence of any work interactions, energy change of a system is equal to the net heat transfer.

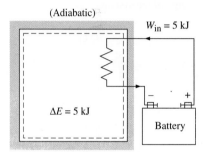

FIGURE 5-4

The work (electrical) done on an adiabatic system is equal to the increase in the energy of the system.

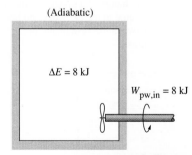

FIGURE 5-5

The work (shaft) done on an adiabatic system is equal to the increase in the energy of the system.

$W_{b,in} = 10$ kJ

$\Delta E = 10$ kJ

(Adiabatic)

FIGURE 5-6

The work (boundary) done on an adiabatic system is equal to the increase in the energy of the system.

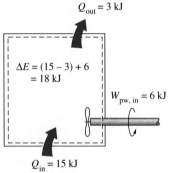

$Q_{out} = 3$ kJ

$\Delta E = (15 - 3) + 6$
$= 18$ kJ

$W_{pw, in} = 6$ kJ

$Q_{in} = 15$ kJ

FIGURE 5-7

The energy change of a system during a process is equal to the *net* work and heat transfer between the system and its surroundings.

Stationary Systems

$z_1 = z_2 \rightarrow \Delta PE = 0$

$\mathcal{V}_1 = \mathcal{V}_2 \rightarrow \Delta KE = 0$

$\Delta E = \Delta U$

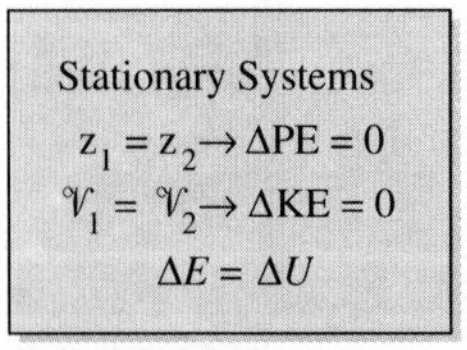

FIGURE 5-8

For stationary systems.
$\Delta KE = \Delta PE = 0$; thus $\Delta E = \Delta U$.

or

$$\Delta E_{system} = E_{final} - E_{initial} = E_2 - E_1 \qquad (5\text{-}1)$$

Note that energy is a property, and the value of a property does not change unless the state of the system changes. Therefore, the energy change of a system is zero if the state of the system does not change during the process. Also, energy can exist in numerous forms such as internal (sensible, latent, chemical, and nuclear), kinetic, potential, electrical, and magnetic, and their sum constitutes the *total energy E* of a system. In the absence of electric, magnetic, and surface tension effects (i.e., for simple compressible systems), the change in the total energy of a system during a process is the sum of the changes in its internal, kinetic, and potential energies and can be expressed as

$$\Delta E = \Delta U + \Delta KE + \Delta PE \qquad (5\text{-}2)$$

where

$$\Delta U = m(u_2 - u_1)$$
$$\Delta KE = \tfrac{1}{2}m(\mathcal{V}_2^2 - \mathcal{V}_1^2)$$
$$\Delta PE = mg(z_2 - z_1)$$

When the initial and final states are specified, the values of the specific internal energies u_1 and u_2 can be determined directly from the property tables or thermodynamic property relations.

Most systems encountered in practice are stationary, that is, they do not involve any changes in their velocity or elevation during a process (Fig. 5-8). Thus, for **stationary systems,** the changes in kinetic and potential energies are zero (that is, $\Delta KE = \Delta PE = 0$), and the total energy change relation above reduces to $\Delta E = \Delta U$ for such systems. Also, the energy of a system during a process will change even if only one form of its energy changes while the other forms of energy remain unchanged.

Mechanisms of Energy Transfer, E_{in} and E_{out}

Energy can be transferred to or from a system in three forms: *heat, work,* and *mass flow.* Energy interactions are recognized at the system boundary as they cross it, and they represent the energy gained or lost by a system during a process. The only two forms of energy interactions associated with a fixed mass or closed system are *heat transfer* and *work.*

1. Heat Transfer, Q   Heat transfer to a system (heat gain) increases the energy of the molecules and thus the internal energy of the system, and heat transfer from a system (heat loss) decreases it since the energy transferred out as heat comes from the energy of the molecules of the system.

2. Work, W   An energy interaction that is not caused by a temperature difference between a system and its surroundings is work. A rising piston, a rotating shaft, and an electrical wire crossing the system boundaries are all associated with work interactions. Work transfer to a system (i.e., work done on a system) increases the energy of the system, and work transfer from a system (i.e., work done by the system) decreases it since the energy transferred out as work comes from the energy contained in the system. Car engines, hydraulic, steam, or gas turbines produce work while compressors, pumps, and mixers consume work.

3. Mass Flow, m Mass flow in and out of the system serves as an additional mechanism of energy transfer. When mass enters a system, the energy of the system increases because mass carries energy with it (in fact, mass is energy). Likewise, when some mass leaves the system, the energy contained within the system decreases because the leaving mass takes out some energy with it. For example, when some hot water is taken out of a water heater and is replaced by the same amount of cold water, the energy content of the hot-water tank (the control volume) decreases as a result of this mass interaction (Fig. 5-9).

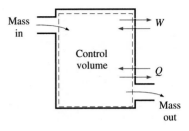

FIGURE 5-9
The energy content of a control volume can be changed by mass flow as well as heat and work interactions.

Noting that energy can be transferred in the forms of heat, work, and mass, and that the net transfer of a quantity is equal to the difference between the amounts transferred in and out, the energy balance can be written more explicitly as

$$E_{\text{in}} - E_{\text{out}} = (Q_{\text{in}} - Q_{\text{out}}) + (W_{\text{in}} - W_{\text{out}}) + (E_{\text{mass, in}} - E_{\text{mass, out}}) = \Delta E_{\text{system}} \tag{5-3}$$

where the subscripts "in" and "out" denote quantities that enter and leave the system, respectively. All six quantities on the right side of the equation represent "amounts," and thus they are *positive* quantities. The direction of any energy transfer is described by the subscripts "in" and "out." Therefore, we do not need to adopt a formal sign convention for heat and work interactions. When heat or work is to be determined and their direction is unknown, we can assume any direction (in or out) for heat or work and solve the problem. A negative result in that case will indicate that the assumed direction is wrong, and it is corrected by reversing the assumed direction. This is just like assuming a direction for an unknown force when solving a problem in statics and reversing the assumed direction when a negative quantity is obtained.

The heat transfer Q is zero for adiabatic systems, the work W is zero for systems that involve no work interactions, and the energy transport with mass E_{mass} is zero for systems that involve no mass flow across their boundaries (i.e., closed systems).

Energy balance for any system undergoing any kind of process can be expressed more compactly as

$$\underbrace{E_{\text{in}} - E_{\text{out}}}_{\substack{\text{Net energy transfer} \\ \text{by heat, work, and mass}}} = \underbrace{\Delta E_{\text{system}}}_{\substack{\text{Change in internal, kinetic,} \\ \text{potential, etc., energies}}} \quad \text{(kJ)} \tag{5-4}$$

or, in the **rate form,** as

$$\underbrace{\dot{E}_{\text{in}} - \dot{E}_{\text{out}}}_{\substack{\text{Rate of net energy transfer} \\ \text{by heat, work, and mass}}} = \underbrace{\Delta \dot{E}_{\text{system}}}_{\substack{\text{Rate of change in internal,} \\ \text{kinetic, potential, etc., energies}}} \quad \text{(kW)} \tag{5-5}$$

For constant rates, the total quantities during a time interval Δt are related to the quantities per unit time as

$$Q = \dot{Q}\,\Delta t, \quad W = \dot{W}\,\Delta t, \quad \text{and} \quad \Delta E = \Delta \dot{E}\,\Delta t \quad \text{(kJ)} \tag{5-6}$$

The energy balance can also be expressed on a **per unit mass** basis as

$$e_{\text{in}} - e_{\text{out}} = \Delta e_{\text{system}} \qquad \text{(kJ/kg)} \qquad (5\text{-}7)$$

which is obtained by dividing all the quantities in Eq. 5-4 by the mass m of the system. Energy balance can also be expressed in the differential form as

$$\delta E_{\text{in}} - \delta E_{\text{out}} = dE_{\text{system}} \qquad \text{or} \qquad \delta e_{\text{in}} - \delta e_{\text{out}} = de_{\text{system}} \qquad (5\text{-}8)$$

For a closed system undergoing a **cycle,** the initial and final states are identical, and thus $\Delta E_{\text{system}} = E_2 - E_1 = 0$. Then the energy balance for a cycle simplifies to $E_{\text{in}} - E_{\text{out}} = 0$ or $E_{\text{in}} = E_{\text{out}}$. Noting that a closed system does not involve any mass flow across its boundaries, the energy balance for a cycle can be expressed in terms of heat and work interactions as

$$W_{\text{net, out}} = Q_{\text{net, in}} \qquad \text{or} \qquad \dot{W}_{\text{net, out}} = \dot{Q}_{\text{net, in}} \qquad \text{(for a cycle)} \quad (5\text{-}9)$$

That is, the net work output during a cycle is equal to net heat input (Fig. 5-10).

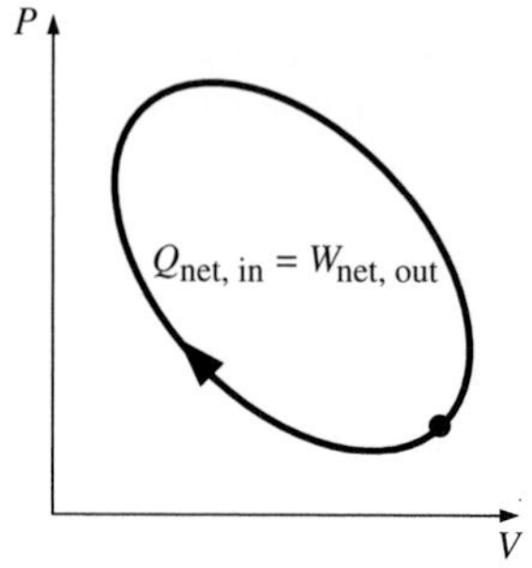

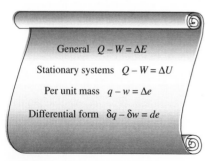

P

$Q_{\text{net, in}} = W_{\text{net, out}}$

V

FIGURE 5-10

For a cycle $\Delta E = 0$, thus $Q_{\text{net, in}} = W_{\text{net, out}}$.

5-2 ■ ENERGY BALANCE FOR CLOSED SYSTEMS

The energy balance (or the first law) relations given above are intuitive in nature and are easy to use when the magnitudes and directions of heat and work transfers are known. But when performing a general analytical study or solving a problem that involves an unknown heat or work interaction, we need to assume a direction for the heat or work interactions. In such cases, it is common practice to assume heat to be transferred *into the system* (heat input) in the amount of Q and work to be done *by the system* (work output) in the amount of W, and then to solve the problem. The energy balance relation in that case for a closed system becomes

$$Q_{\text{net, in}} - W_{\text{net, out}} = \Delta E_{\text{system}} \qquad \text{or} \qquad Q - W = E \qquad (5\text{-}10)$$

where $Q = Q_{\text{net, in}} = Q_{\text{in}} - Q_{\text{out}}$ is the *net heat input* and $W = W_{\text{net, out}} = W_{\text{out}} - W_{\text{in}}$ is the *net work output*. Obtaining a negative quantity for Q or W simply means that the assumed direction for that quantity is wrong and should be reversed. Various forms of this "traditional" first law relation for closed systems are given in Fig. 5-11.

The first law cannot be proven mathematically, but no process in nature is known to have violated the first law, and this should be taken as sufficient proof. Note that if it were possible to prove the first law on the basis of other physical principles, the first law then would be a consequence of those principles instead of being a fundamental physical law itself.

As energy quantities, heat and work are not that different, and you probably wonder why we keep distinguishing them. After all, the change in the energy content of a system is equal to the amount of energy that crosses the system boundaries, and it makes no difference whether the energy crosses the boundary as heat or work. It seems as if the first-law relations would be much simpler if we had just one quantity that we could call *energy interaction* to represent both heat and work. Well, from the first-law point of view, heat and work are not different at all. But from the second-law point of view, heat and work are very different, as is discussed in later chapters.

General $Q - W = \Delta E$

Stationary systems $Q - W = \Delta U$

Per unit mass $q - w = \Delta e$

Differential form $\delta q - \delta w = de$

FIGURE 5-11

Various forms of the first-law relation for closed systems.

EXAMPLE 5-1 Cooling of a Hot Fluid in a Tank

A rigid tank contains a hot fluid that is cooled while being stirred by a paddle wheel. Initially, the internal energy of the fluid is 800 kJ. During the cooling process, the fluid loses 500 kJ of heat, and the paddle wheel does 100 kJ of work on the fluid. Determine the final internal energy of the fluid. Neglect the energy stored in the paddle wheel.

Solution We take the contents of the tank as the *system* (Fig. 5-12). This is a *closed system* since no mass crosses the boundary during the process. We observe that the volume of a rigid tank is constant, and thus there is no boundary work and $v_2 = v_1$. Also, heat is lost from the system and shaft work is done on the system.

Assumptions The tank is stationary and thus the kinetic and potential energy changes are zero, $\Delta KE = \Delta PE = 0$. Therefore, $\Delta E = \Delta U$ and internal energy is the only form of the system's energy that may change during this process.

Analysis Applying the energy balance on the system gives

$$\underbrace{E_{in} - E_{out}}_{\substack{\text{Net energy transfer} \\ \text{by heat, work, and mass}}} = \underbrace{\Delta E_{system}}_{\substack{\text{Change in internal, kinetic,} \\ \text{potential, etc., energies}}}$$

$$W_{pw,\,in} - Q_{out} = \Delta U = U_2 - U_1$$

$$100\ \text{kJ} - 500\ \text{kJ} = U_2 - 800\ \text{kJ}$$

$$U_2 = \textbf{400 kJ}$$

Therefore, the final internal energy of the system is 400 kJ.

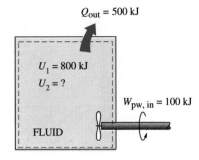

FIGURE 5-12
Schematic for Example 5-1.

EXAMPLE 5-2 Electric Heating of a Gas at Constant Pressure

A piston-cylinder device contains 25 g of saturated water vapor that is maintained at a constant pressure of 300 kPa. A resistance heater within the cylinder is turned on and passes a current of 0.2 A for 5 min from a 120-V source. At the same time, a heat loss of 3.7 kJ occurs. (*a*) Show that for a closed system the boundary work W_b and the change in internal energy ΔU in the first-law relation can be combined into one term, ΔH, for a constant-pressure process. (*b*) Determine the final temperature of the steam.

Solution We take the contents of the cylinder, including the resistance wires, as the *system* (Fig. 5-13). This is a *closed system* since no mass crosses the system boundary during the process. We observe that a piston-cylinder device typically involves a moving boundary and thus boundary work, W_b. The pressure remains constant during the process and thus $P_2 = P_1$. Also, heat is lost from the system and electrical work W_e is done on the system.

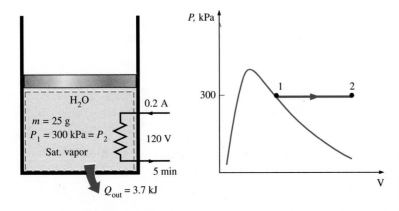

FIGURE 5-13
Schematic and *P-V* diagram for Example 5-2.

Assumptions **1** The tank is stationary and thus the kinetic and potential energy changes are zero, $\Delta KE = \Delta PE = 0$. Therefore, $\Delta E = \Delta U$ and internal energy is the only form of energy of the system that may change during this process. **2** Electrical wires constitute a very small part of the system, and thus the energy change of the wires can be neglected.

Analysis (*a*) This part of the solution involves a general analysis for a closed system undergoing a quasi-equilibrium constant-pressure process, and thus we consider a general closed system. We take the direction of heat transfer Q to be to the system and the work W to be done by the system. We also express the work as the sum of boundary and other forms of work (such as electrical and shaft). Then the energy balance can be expressed as

$$\underbrace{E_{in} - E_{out}}_{\substack{\text{Net energy transfer} \\ \text{by heat, work, and mass}}} = \underbrace{\Delta E_{system}}_{\substack{\text{Change in internal, kinetic,} \\ \text{potential, etc., energies}}}$$

$$Q - W = \Delta U + \cancelto{0}{\Delta KE} + \cancelto{0}{\Delta PE}$$

$$Q - W_{other} - W_b = U_2 - U_1$$

For a constant-pressure process, the boundary work is given as $W_b = P_0(V_2 - V_1)$. Substituting this into the above relation gives

$$Q - W_{other} - P_0(V_2 - V_1) = U_2 - U_1$$

But $\qquad P_0 = P_2 = P_1 \quad \longrightarrow \quad Q - W_{other} = (U_2 + P_2V_2) - (U_1 + P_1V_1)$

Also $H = U + PV$, and thus

$$Q - W_{other} = H_2 - H_1 \qquad \text{(kJ)} \qquad\qquad (5\text{-}11)$$

which is the desired relation (Fig. 5-14). *This equation is very convenient to use in the analysis of closed systems undergoing a constant-pressure quasi-equilibrium process since the boundary work is automatically taken care of by the enthalpy terms, and one no longer needs to determine it separately.*

(*b*) For our case, the only other form of work is the electrical work, which can be determined from

$$W_e = VI\Delta t = (120 \text{ V})(0.2 \text{ A})(300 \text{ s})\left(\frac{1 \text{ kJ/s}}{1000 \text{ VA}}\right) = 7.2 \text{ kJ}$$

State 1: $\left.\begin{array}{l} P_1 = 300 \text{ kPa} \\ \text{sat. vapor} \end{array}\right\} \quad h_1 = h_{g\,@\,300kPa} = 2725.3 \text{ kJ/kg} \qquad \text{(Table A-5)}$

The enthalpy at the final state can be determined directly from Eq. 5-11 by expressing heat transfer from the system and work done on the system as negative quantities (since their directions are opposite to the assumed directions). Alternately, we can use the general energy balance relation with the simplification that the boundary work is considered automatically by replacing ΔU by ΔH for a constant-pressure expansion or compression process:

$$\underbrace{E_{in} - E_{out}}_{\substack{\text{Net energy transfer} \\ \text{by heat, work, and mass}}} = \underbrace{\Delta E_{system}}_{\substack{\text{Change in internal, kinetic,} \\ \text{potential, etc., energies}}}$$

$$W_{e,in} - Q_{out} - W_b = \Delta U$$

$$W_{e,in} - Q_{out} = \Delta H = m(h_2 - h_1) \qquad \text{(since } P = \text{constant)}$$

$$7.2 \text{ kJ} - 3.7 \text{ kJ} = (0.025 \text{ kg})(h_2 - 2725.3) \text{ kJ/kg}$$

$$h_2 = 2865.3 \text{ kJ/kg}$$

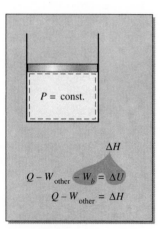

FIGURE 5-14

For a closed system undergoing a quasi-equilibrium, $P = $ constant process, $\Delta U + W_b = \Delta H$.

Now the final state is completely specified since we know both the pressure and the enthalpy. The temperature at this state is

State 2: $\left.\begin{array}{l} P_2 = 300 \text{ kPa} \\ h_2 = 2865.3 \text{ kJ/kg} \end{array}\right\}$ $T_2 = 200°C$ (Table A-6)

Therefore, the steam will be at 200°C at the end of this process.

Discussion   Strictly speaking, the potential energy change of the steam is not zero for this process since the center of gravity of the steam rose somewhat. Assuming an elevation change of 1 m (which is rather unlikely), the change in the potential energy of the steam would be 0.0002 kJ, which is very small compared to the other terms in the first-law relation. Therefore, in problems of this kind, the potential energy term is always neglected.

EXAMPLE 5-3   Unrestrained Expansion of Water into an Evacuated Tank

A rigid tank is divided into two equal parts by a partition. Initially, one side of the tank contains 5 kg of water at 200 kPa and 25°C, and the other side is evacuated. The partition is then removed, and the water expands into the entire tank. The water is allowed to exchange heat with its surroundings until the temperature in the tank returns to the initial value of 25°C. Determine (*a*) the volume of the tank, (*b*) the final pressure, and (*c*) the heat transfer for this process.

Solution   We take the contents of the tank, including the evacuated space, as the *system* (Fig. 5-15). This is a *closed system* since no mass crosses the system boundary during the process. We observe that the water fills the entire tank when the partition is removed (possibly as a liquid–vapor mixture).

Assumptions   **1** The system is stationary and thus the kinetic and potential energy changes are zero, $\Delta KE = \Delta PE = 0$ and $\Delta E = \Delta U$. **2** The direction of heat transfer is to the system (heat gain, Q_{in}). A negative result for Q_{in} will indicate the assumed direction is wrong and thus it is heat loss. **3** The volume of the rigid tank is constant, and thus there is no energy transfer as boundary work. **4** The water temperature remains constant during the process. **5** There is no electrical, shaft, or any other kind of work involved.

Analysis   (*a*) Initially the water in the tank exists as a compressed liquid since its pressure (200 kPa) is greater than the saturation pressure at 25°C (3.169 kPa). Approximating the compressed liquid as a saturated liquid at the given temperature, we find

$$v_1 \cong v_{f@\,25°C} = 0.001003 \text{ m}^3/\text{kg} \cong 0.001 \text{ m}^3/\text{kg} \qquad \text{(Table A-4)}$$

Then the initial volume of the water is

$$V_1 = mv_1 = (5 \text{ kg})(0.001 \text{ m}^3/\text{kg}) = 0.005 \text{ m}^3$$

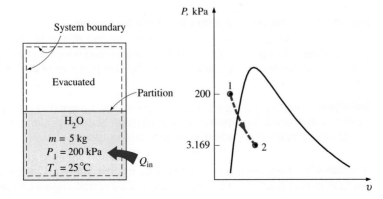

FIGURE 5-15

Schematic and *P-v* diagram for Example 5-3.

The total volume of the tank is twice this amount:

$$V_{tank} = (2)(0.005 \text{ m}^3) = \textbf{0.01 m}^3$$

(b) At the final state, the specific volume of the water is

$$v_2 = \frac{V_2}{m} = \frac{0.01 \text{ m}^3}{5 \text{ kg}} = 0.002 \text{ m}^3/\text{kg}$$

which is twice the initial value of the specific volume. This result is expected since the volume doubles while the amount of mass remains constant.

At 25°C: $v_f = 0.001003 \text{ m}^3/\text{kg}$ and $v_g = 43.36 \text{ m}^3/\text{kg}$ (Table A-4)

Since $v_f < v_2 < v_g$, the water is a saturated liquid–vapor mixture at the final state, and thus the pressure is the saturation pressure at 25°C:

$$P_2 = P_{sat \, @ \, 25°C} = \textbf{3.169 kPa}$$ (Table A-4)

(c) Under stated assumptions and observations, the energy balance on the system can be expressed as

$$\underbrace{E_{in} - E_{out}}_{\substack{\text{Net energy transfer} \\ \text{by heat, work, and mass}}} = \underbrace{\Delta E_{system}}_{\substack{\text{Change in internal, kinetic,} \\ \text{potential, etc., energies}}}$$

$$Q_{in} = \Delta U = m(u_2 - u_1)$$

Notice that even though the water is expanding during this process, the system chosen involves fixed boundaries only (the dashed lines) and therefore the moving boundary work is zero (Fig. 5-16). Then $W = 0$ since the system does not involve any other forms of work. (Can you reach the same conclusion by choosing the water as our system?) Initially,

$$u_1 \cong u_{f @ \, 25°C} = 104.88 \text{ kJ/kg}$$

The quality at the final state is determined from the specific-volume information:

$$x_2 = \frac{v_2 - v_f}{v_{fg}} = \frac{0.002 - 0.001}{43.36 - 0.001} = 2.3 \times 10^{-5}$$

Then $u_2 = u_f + x_2 u_{fg}$
$$= 104.88 \text{ kJ/kg} + (2.3 \times 10^{-5})(2304.9 \text{ kJ/kg})$$
$$= 104.93 \text{ kJ/kg}$$

Substituting yields

$$Q_{in} = (5 \text{ kg})[(104.93 - 104.88) \text{ kJ/kg}] = 0.25 \text{ kJ}$$

Discussion   The positive sign indicates that the assumed direction is correct, and heat is transferred to the water.

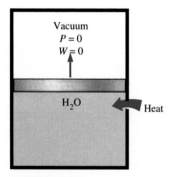

Vacuum
$P = 0$
$W = 0$

H_2O Heat

FIGURE 5-16

Expansion against a vacuum involves no work and thus no energy transfer.

EXAMPLE 5-4   Heating of a Gas in a Tank by Stirring
An insulated rigid tank initially contains 1.5 lbm of helium at 80°F and 50 psia. A paddle wheel with a power rating of 0.02 hp is operated within the tank for 30 min. Determine (a) the final temperature and (b) the final pressure of the helium gas.

Solution   We take the contents of the tank as the *system* (Fig. 5-17). This is a *closed system* since no mass crosses the system boundary during the process. We observe that there is paddle work done on the system.

Assumptions   **1** Helium is an ideal gas since it is at a very high temperature relative to its critical point value of −451°F. **2** Constant specific heats can be used for

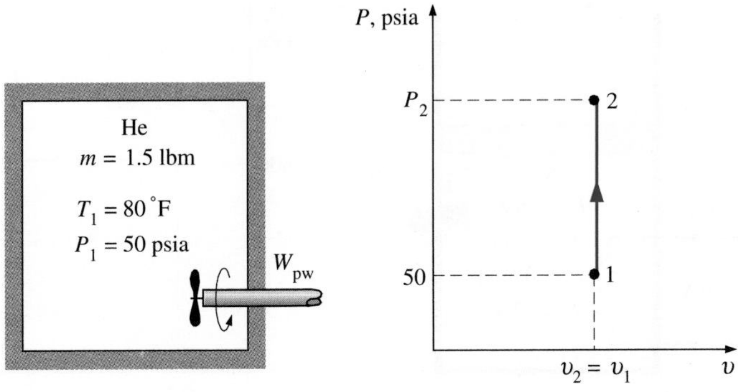

FIGURE 5-17

Schematic and P-v diagram for Example 5-4.

helium. **3** The system is stationary and thus the kinetic and potential energy changes are zero, $\Delta KE = \Delta PE = 0$ and $\Delta E = \Delta U$. **4** The volume of the tank is constant, and thus there is no boundary work and $V_2 = V_1$. **5** The system is adiabatic and thus there is no heat transfer.

Analysis (*a*) The amount of paddle-wheel work done on the system is

$$W_{pw} = \dot{W}_{pw}\,\Delta t = (0.02\text{ hp})(0.5\text{ h})\left(\frac{2545\text{ Btu/h}}{1\text{ hp}}\right) = 25.45\text{ Btu}$$

Under stated assumptions and observations, the energy balance on the system can be expressed as

$$\underbrace{E_{in} - E_{out}}_{\substack{\text{Net energy transfer} \\ \text{by heat, work, and mass}}} = \underbrace{\Delta E_{system}}_{\substack{\text{Change in internal, kinetic,} \\ \text{potential, etc., energies}}}$$

$$W_{pw,\,in} = \Delta U = m(u_2 - u_1) = mC_{v,\,av}(T_2 - T_1)$$

As we pointed out earlier, the ideal-gas specific heats of monatomic gases (helium being one of them) are constant. The C_v value of helium is determined from Table A-2E*a* to be $C_v = 0.753$ Btu/lbm · °F. Substituting this and other known quantities into the above equation, we obtain

$$25.45\text{ Btu} = (1.5\text{ lbm})(0.753\text{ Btu/lbm} \cdot \text{°F})(T_2 - 80\text{°F})$$

$$T_2 = \mathbf{102.5\text{°F}}$$

(*b*) The final pressure is determined from the ideal-gas relation

$$\frac{P_1 V_1}{T_1} = \frac{P_2 V_2}{T_2}$$

where V_1 and V_2 are identical and cancel out. Then the final pressure becomes

$$\frac{50\text{ psia}}{(80 + 460)\text{ R}} = \frac{P_2}{(102.5 + 460)\text{ R}}$$

$$P_2 = \mathbf{52.1\text{ psia}}$$

EXAMPLE 5-5   Heating of a Gas by a Resistance Heater

A piston-cylinder device initially contains 0.5 m³ of nitrogen gas at 400 kPa and 27°C. An electric heater within the device is turned on and is allowed to pass a current of 2 A for 5 min from a 120-V source. Nitrogen expands at constant pressure, and a heat loss of 2800 J occurs during the process. Determine the final temperature of the nitrogen.

Solution   We take the contents of the cylinder as the *system* (Fig. 5-18). This is a *closed system* since no mass crosses the system boundary during the process.

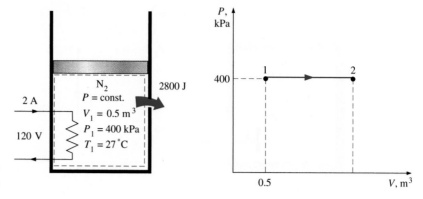

FIGURE 5-18

Schematic and *P-V* diagram for
Example 5-5.

We observe that a piston-cylinder device typically involves a moving boundary
and thus boundary work, W_b. Also, heat is lost from the system and electrical work
W_e is done on the system.

Assumptions   **1** Nitrogen is an ideal gas since it is at a high temperature and low
pressure relative to its critical point values of $-147°C$, and 3.39 MPa. **2** The
system is stationary and thus the kinetic and potential energy changes are
zero, $\Delta KE = \Delta PE = 0$ and $\Delta E = \Delta U$. **3** The pressure remains constant during
the process and thus $P_2 = P_1$. **4** Nitrogen has constant specific heats at room
temperature.

Analysis   First, let us determine the electrical work done on the nitrogen:

$$W_e = VI\,\Delta t = (120\text{ V})(2\text{ A})(5 \times 60\text{ s})\left(\frac{1\text{ kJ/s}}{1000\text{ VA}}\right) = 72\text{ kJ}$$

The mass of nitrogen is determined from the ideal-gas relation:

$$m = \frac{P_1 V_1}{RT_1} = \frac{(400\text{ kPa})(0.5\text{ m}^3)}{(0.297\text{ kPa}\cdot\text{m}^3/\text{kg}\cdot\text{K})(300\text{ K})} = 2.245\text{ kg}$$

Under stated assumptions and observations, the energy balance on the system
can be expressed as

$$\underbrace{E_{\text{in}} - E_{\text{out}}}_{\substack{\text{Net energy transfer} \\ \text{by heat, work, and mass}}} = \underbrace{\Delta E_{\text{system}}}_{\substack{\text{Change in internal, kinetic,} \\ \text{potential, etc., energies}}}$$

$$W_{e,\text{in}} - Q_{\text{out}} - W_b = \Delta U$$

$$W_{e,\text{ in}} - Q_{\text{out}} = \Delta H = m(h_2 - h_1) = mC_p(T_2 - T_1)$$

since $\Delta U + W_b \equiv \Delta H$ for a closed system undergoing a quasi-equilibrium
expansion or compression process at constant pressure. From Table A-2a, $C_p =$
1.039 kJ/kg · K for nitrogen at room temperature. The only unknown quantity in the
above equation is T_2, and it is found to be

$$72\text{ kJ} - 2.8\text{ kJ} = (2.245\text{ kg})(1.039\text{ kJ/kg}\cdot\text{K})(T_2 - 27°C)$$

$$T_2 = \textbf{56.7°C}$$

EXAMPLE 5-6   Heating of a Gas at Constant Pressure

A piston-cylinder device initially contains air at 150 kPa and 27°C. At this state,
the piston is resting on a pair of stops, as shown in Fig. 5-19, and the enclosed
volume is 400 L. The mass of the piston is such that a 350-kPa pressure is required to move it. The air is now heated until its volume has doubled. Determine

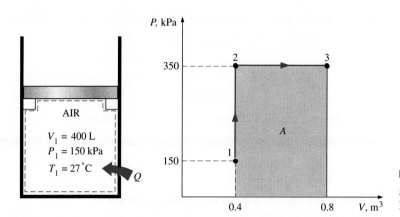

FIGURE 5-19

Schematic and P-V diagram for
Example 5-6.

(*a*) the final temperature, (*b*) the work done by the air, and (*c*) the total heat transferred to the air.

Solution We take the contents of the cylinder as the *system* (Fig. 5-19). This is a *closed system* since no mass crosses the system boundary during the process. We observe that a piston-cylinder device typically involves a moving boundary and thus boundary work, W_b. Also, the boundary work is done by the system, and heat is transferred to the system.

Assumptions **1** Air is an ideal gas since it is at a high temperature and low pressure relative to its critical point values. **2** The system is stationary and thus the kinetic and potential energy changes are zero, $\Delta KE = \Delta PE = 0$ and $\Delta E = \Delta U$. **3** The volume remains constant until the piston starts moving, and the pressure remains constant afterwards. **4** There are no electrical, shaft, or other forms of work involved.

Analysis (*a*) The final temperature can be determined easily by using the ideal-gas relation between states 1 and 3 in the following form:

$$\frac{P_1 V_1}{T_1} = \frac{P_3 V_3}{T_3} \longrightarrow \frac{(150 \text{ kPa})(V_1)}{300 \text{ K}} = \frac{(350 \text{ kPa})(2V_1)}{T_3}$$

$$T_3 = \textbf{1400 K}$$

(*b*) The work done could be determined by integration, but for this case it is much easier to find it from the area under the process curve on a P-V diagram, shown in Fig. 5-19:

$$A = (V_2 - V_1)(P_2) = (0.4 \text{ m}^3)(350 \text{ kPa}) = 140 \text{ m}^3 \cdot \text{kPa}$$

Therefore, $\qquad\qquad W_{13} = \textbf{140 kJ}$

The work is done by the system (to raise the piston and to push the atmospheric air out of the way), and thus it is work output.

(*c*) Under stated assumptions and observations, the energy balance on the system between the initial and final states (process 1-3) can be expressed as

$$\underbrace{E_{\text{in}} - E_{\text{out}}}_{\substack{\text{Net energy transfer} \\ \text{by heat, work, and mass}}} = \underbrace{\Delta E_{\text{system}}}_{\substack{\text{Change in internal, kinetic,} \\ \text{potential, etc., energies}}}$$

$$Q_{\text{in}} - W_{b,\text{ out}} = \Delta U = m(u_3 - u_1)$$

The mass of the system can be determined from the ideal-gas equation of state:

$$m = \frac{P_1 V_1}{RT_1} = \frac{(150 \text{ kPa})(0.4 \text{ m}^3)}{(0.287 \text{ kPa} \cdot \text{m}^3/\text{kg} \cdot \text{K})(300 \text{ K})} = 0.697 \text{ kg}$$

The internal energies are determined from the air table (Table A-17) to be

$$u_1 = u_{@\ 300K} = 214.07\ kJ/kg$$
$$u_3 = u_{@\ 1400K} = 1113.52\ kJ/kg$$

Thus, $Q_{in} - 140\ kJ = (0.697\ kg)[(1113.52 - 214.07)\ kJ/kg]$
$$Q_{in} = \textbf{766.9 kJ}$$

The positive sign verifies that heat is transferred to the system.

EXAMPLE 5-7   Cooling of an Iron Block by Water
A 50-kg iron block at 80°C is dropped into an insulated tank that contains 0.5 m³ of liquid water at 25°C. Determine the temperature when thermal equilibrium is reached.

Solution   We take the entire contents of the tank, water + iron block, as the *system* (Fig. 5-20). This is a *closed system* since no mass crosses the system boundary during the process. We observe that the volume of a rigid tank is constant, and thus there is no boundary work.

Assumptions   **1** Both water and the iron block are incompressible substances. **2** Constant specific heats at room temperature can be used for water and the iron. **3** The system is stationary and thus the kinetic and potential energy changes are zero, $\Delta KE = \Delta PE = 0$ and $\Delta E = \Delta U$. **4** There are no electrical, shaft, or other forms of work involved. **5** The system is well-insulated and thus there is no heat transfer.

Analysis   The energy balance on the system can be expressed as

$$\underbrace{E_{in} - E_{out}}_{\substack{\text{Net energy transfer} \\ \text{by heat, work, and mass}}} = \underbrace{\Delta E_{system}}_{\substack{\text{Change in internal, kinetic,} \\ \text{potential, etc., energies}}}$$

$$0 = \Delta U$$

The total internal energy U is an extensive property, and therefore it can be expressed as the sum of the internal energies of the parts of the system. Then the total internal energy change of the system becomes

$$\Delta U_{sys} = \Delta U_{iron} + \Delta U_{water} = 0$$
$$[mC(T_2 - T_1)]_{iron} + [mC(T_2 - T_1)]_{water} = 0$$

The specific volume of liquid water at or about room temperature can be taken to be 0.001 m³/kg. Then the mass of the water is

$$m_{water} = \frac{V}{v} = \frac{0.5\ m^3}{0.001\ m^3/kg} = 500\ kg$$

The specific heats of iron and liquid water are determined from Table A-3 to be $C_{iron} = 0.45\ kJ/kg \cdot °C$ and $C_{water} = 4.18\ kJ/kg \cdot °C$. Substituting these values into the energy equation, we obtain

$$(50\ kg)(0.45\ kJ/kg \cdot °C)(T_2 - 80°C) + (500\ kg)(4.18\ kJ/kg \cdot °C)(T_2 - 25°C) = 0$$
$$T_2 = \textbf{25.6°C}$$

Therefore, when thermal equilibrium is established, both the water and iron will be at 25.6°C. The small rise in water temperature is due to its large mass and large specific heat.

FIGURE 5-20

Schematic for Example 5-7.

EXAMPLE 5-8 Temperature Rise due to Slapping

If you ever slapped someone or got slapped yourself, you probably remember the burning sensation on your hand or your face. Imagine you had the unfortunate occasion of being slapped by an angry person, which caused the temperature of the affected area of your face to rise by 1.8°C (ouch!). Assuming the slapping hand has a mass of 1.2 kg and about 0.150 kg of the tissue on the face and the hand is affected by the incident, estimate the velocity of the hand just before impact. Take the specific heat of the tissue to be 3.8 kJ/kg · °C.

Solution We will analyze this incident in a professional manner without involving any emotions. First, we identify the system, draw a sketch of it, state our observations about the specifics of the problem, and make appropriate assumptions.

We take the hand and the affected portion of the face as the system (Fig. 5-21). This is a *closed system* since it involves a fixed amount of mass (no mass transfer). We observe that the kinetic energy of the hand decreases during the process, as evidenced by a decrease in velocity from initial value to zero, while the internal energy of the affected area increases, as evidenced by an increase in the temperature. There seems to be no significant energy transfer between the system and its surroundings during this process.

Assumptions **1** The hand is brought to a complete stop after the impact. **2** The face takes the blow well without significant movement. **3** No heat is transferred from the affected area to the surroundings, and thus the process is adiabatic. **4** No work is done on or by the system. **5** The potential energy change is zero, $\Delta PE = 0$ and $\Delta E = \Delta U + \Delta KE$.

Analysis Under the stated assumptions and observations, the energy balance on the system can be expressed as

$$\underbrace{E_{in} - E_{out}}_{\substack{\text{Net energy transfer} \\ \text{by heat, work, and mass}}} = \underbrace{\Delta E_{system}}_{\substack{\text{Change in internal, kinetic,} \\ \text{potential, etc., energies}}}$$

$$0 = \Delta U_{\text{affected tissue}} + \Delta KE_{\text{hand}}$$

$$0 = (mC\,\Delta T)_{\text{affected tissue}} + [m(0 - \mathcal{V}^2)/2]_{\text{hand}}$$

FIGURE 5-21
Schematic for Example 5-8.

That is, the decrease in the kinetic energy of the hand must be equal to the increase in the internal energy of the affected area. Solving for the velocity and substituting the given quantities, the impact velocity of the hand is determined to be

$$\mathcal{V}_{\text{hand}} = \sqrt{\frac{2(mC\,\Delta T)_{\text{affected tissue}}}{m_{\text{hand}}}} = \sqrt{\frac{2(0.15\text{ kg})(3.8\text{ kJ/kg}\cdot\text{°C})(1.8\text{°C})}{1.2\text{ kg}}\left(\frac{1000\text{ m}^2/\text{s}^2}{1\text{ kJ/kg}}\right)}$$

$$= \textbf{41.4 m/s} \text{ (or 149 km/h)}$$

EXAMPLE 5-9 Cooling of Bananas in a Cold Storage Room

A typical one-half carlot capacity banana room contains 18 pallets of bananas. Each pallet consists of 24 boxes, and thus the room stores 432 boxes of bananas. A box holds an average of 19 kg of bananas and is made of 2.3 kg of fiberboard. The specific heats of banana and the fiberboard are 3.55 kJ/kg · °C and 1.7 kJ/kg · °C, respectively. The peak heat of respiration of bananas is 0.3 W/kg. The bananas are cooled at a rate of 0.2°C/h. Disregarding any heat gain through the walls or other surfaces, determine the required rate of heat removal from the banana room.

Solution We take the contents of the banana room as the *system* (Fig. 5-22). This is a *closed system* since it involves a fixed mass. We observe that the heat of respiration can be treated as an energy input to the system.

Banana cooling room

432 boxes
19 kg banana/box
2.3 kg fiberboard/box

Heat of respiration = 0.3 W/kg
Cooling rate = 0.2°C/h

FIGURE 5-22
Schematic for Example 5-9.

Assumptions **1** There is no heat gain through the walls and other surfaces. **2** The energy change of the air in the banana room is negligible. **3** Thermal properties of air, bananas, and boxes are constant. **4** The system is stationary and involves changes in its internal energy only (due to temperature change), and thus $\Delta E = \Delta U$. **5** There is no electrical, shaft, boundary, or any other kind of work involved.

Analysis Under the stated assumptions and observations, the rate form energy balance on the system reduces to

$$\underbrace{\dot{E}_{in} - \dot{E}_{out}}_{\substack{\text{Rate of net energy transfer} \\ \text{by heat, work, and mass}}} = \underbrace{\Delta \dot{E}_{system}}_{\substack{\text{Rate of change in internal, kinetic,} \\ \text{potential, etc., energies}}}$$

$$\dot{E}_{respiration} - \dot{Q}_{out} = \Delta \dot{U}_{banana} + \Delta \dot{U}_{box}$$

Noting that the banana room holds 432 boxes, the total mass of bananas and the boxes is determined to be

$$m_{banana} = \text{(Mass per box)(Number of boxes)} = \text{(19 kg/box)(432 boxes)} = 8208 \text{ kg}$$
$$m_{box} = \text{(Mass per box)(Number of boxes)} = \text{(2.3 kg/box)(432 boxes)} = 993.6 \text{ kg}$$

Noting that the quantity $\Delta T/\Delta t$ is the rate of change in temperature of the products and is given to be $-0.2°C/h$ (a temperature drop).

$$\dot{Q}_{respiration} = m_{banana}\dot{q}_{respiration} = \text{(8208 kg)(0.3 W/kg)} = 2462 \text{ W}$$

$$\Delta \dot{U}_{banana} = (mC\,\Delta T/\Delta t)_{banana} = \text{(8208 kg)(3.55 kJ/kg} \cdot °\text{C)}(-0.2°C/h)$$
$$= -5828 \text{ kJ/h} = -1619 \text{ W} \qquad \text{(since 1 W = 3.6 kJ/h)}$$

$$\Delta \dot{U}_{box} = (mC_p\,\Delta T/\Delta t)_{box} = \text{(993.6 kg)(1.7 kJ/kg} \cdot °\text{C)}(-0.2°C/h) = -338 \text{ kJ/h}$$
$$= -94 \text{ W} \qquad \text{(since 1 W = 3.6 kJ/h)}$$

Substituting,

$$\dot{Q}_{out} = \dot{E}_{respiration} - \Delta \dot{E}_{banana} - \Delta \dot{E}_{box} = 2462 + 1619 + 94 = \textbf{4175 W}$$

Therefore, the refrigeration system must remove heat at a rate of 4175 W from the banana room to achieve the desired cooling rate.

EXAMPLE 5-10   Freezing of Chicken in a Box

A supply of 50 kg of chicken at 6°C contained in a box is to be frozen to $-18°C$ in a freezer. Determine the amount of heat that needs to be removed. The latent heat of the chicken is 247 kJ/kg, and its specific heat is 3.32 kJ/kg · °C above freezing and 1.77 kJ/kg · °C below freezing. The container box is 1.5 kg, and the specific heat of the box material is 1.4 kJ/kg · °C. Also, the freezing temperature of chicken is $-2.8°C$.

Solution We take the chicken and the box they are in as the *system* (Fig. 5-23). This is a *closed system* since it involves a fixed mass (no mass transfer). We observe that the total amount of heat that needs to be removed (the cooling load of the freezer) is the sum of the latent heat and the sensible heats of the chicken before and after freezing, as well as the sensible heat of the box, and is determined below.

Assumptions **1** The energy change of the air in the box is negligible. **2** The thermal properties of fresh and frozen chicken are constant. **3** The entire water content of chicken freezes during the process. **4** The system is stationary and thus the kinetic and potential energy changes are zero, $\Delta KE = \Delta PE = 0$ and $\Delta E = \Delta U$. **5** There is no electrical, shaft, boundary, or any other kind of work involved.

Chicken
$m = 50$ kg
$T_1 = 6°C$
$T_2 = -18°C$
$T_{freezing} = -2.8°C$

Container = 1.5 kg

FIGURE 5-23

Schematic for Example 5-10.

Analysis Under the stated assumptions and observations, the energy balance on the system reduces to

$$\underbrace{E_{in} - E_{out}}_{\substack{\text{Net energy transfer} \\ \text{by heat, work, and mass}}} = \underbrace{\Delta E_{system}}_{\substack{\text{Change in internal, kinetic,} \\ \text{potential, etc., energies}}}$$

$$-Q_{out} = \Delta U_{chicken} + \Delta U_{box}$$

The total amount of heat that needs to be removed is the sum of the latent heat and the sensible heats of the chicken before and after freezing, as well as the sensible heat of the box:

Cooling fresh chicken from 6°C to −2.8°C:

$$\Delta U_{\text{fresh chicken}} = mC\,\Delta T = (50 \text{ kg})(3.32 \text{ kJ/kg} \cdot °C)(-2.8 - 6) \text{ °C} = -1461 \text{ kJ}$$

Freezing chicken at −2.8°C:

$$\Delta U_{\text{freezing}} = mu_{\text{latent}} = (50 \text{ kg})(-247 \text{ kJ/kg}) = -12,350 \text{ kJ}$$

Cooling frozen chicken from −2.8°C to −18°C:

$$\Delta U_{\text{frozen chicken}} = mC\,\Delta T = (50 \text{ kg})(1.77 \text{ kJ/kg} \cdot °C)[-18 - (-2.8)]°C = -1345 \text{ kJ}$$

Cooling the box from 6°C to −18°C:

$$\Delta U_{\text{box}} = (mC\,\Delta T)_{\text{box}} = (1.5 \text{ kg})(1.4 \text{ kJ/kg} \cdot °C)(-18 - 6)°C = -50 \text{ kJ}$$

Therefore, the total amount of heat that needs to be removed is

$$\begin{aligned} Q_{out} &= -\Delta U_{chicken} - \Delta U_{box} \\ &= -(\Delta U_{\text{fresh chicken}} + \Delta U_{\text{freezing}} + \Delta U_{\text{frozen chicken}}) - \Delta U_{box} \\ &= 1461 + 12,350 + 1345 + 50 = \textbf{15,206 kJ} \end{aligned}$$

Discussion Note that most of the cooling load of the refrigeration system (81 percent of it) is due to the removal of the latent heat during the phase-change process. Also note that the cooling load due to the box is negligible (less than 1 percent) and can be ignored in calculations.

5-3 ■ ENERGY BALANCE FOR STEADY-FLOW SYSTEMS

A large number of engineering devices such as turbines, compressors, and nozzles operate for long periods of time under the same conditions, and they are classified as *steady-flow devices.*

Processes involving steady-flow devices can be represented reasonably well by a somewhat idealized process, called the **steady-flow process,** which can be defined as *a process during which a fluid flows through a control volume steadily.* That is, the fluid properties can change from point to point within the control volume, but at any point, they remain constant during the entire process. (Remember, *steady* means *no change with time.*)

During a steady-flow process, no intensive or extensive properties *within the control volume* change with time. Thus, the volume V, the mass m, and the total energy content E of the control volume remain constant (Fig. 5-24). As a result, the boundary work is zero for steady-flow systems (since $V_{CV} = $ constant), and the total mass or energy entering the control volume must be equal to the total mass or energy leaving it (since $m_{CV} = $ constant and $E_{CV} = $ constant). These observations greatly simplify the analysis.

The fluid properties at an inlet or exit remain constant during a steady-flow process. The properties may, however, be different at different inlets and

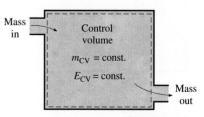

FIGURE 5-24

Under steady-flow conditions, the mass and energy contents of a control volume remain constant.

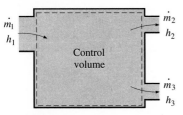

FIGURE 5-25

Under steady-flow conditions, the fluid properties at an inlet or exit remain constant (do not change with time).

exits. They may even vary over the cross section of an inlet or an exit. But all properties, including the velocity and elevation, must remain constant with time at a fixed point at an inlet or exit. It follows that the mass flow rate of the fluid at an opening must remain constant during a steady-flow process (Fig. 5-25). As an added simplification, the fluid properties at an opening are usually considered to be uniform (at some average value) over the cross section. Thus, the fluid properties at an inlet or exit may be specified by the average single values. Also, the *heat* and *work* interactions between a steady-flow system and its surroundings do not change with time. Thus, the power delivered by a system and the rate of heat transfer to or from a system remain constant during a steady-flow process.

Some cyclic devices, such as reciprocating engines or compressors, do not satisfy any of the conditions stated above since the flow at the inlets and the exits will be pulsating and not steady. However, the fluid properties vary with time in a periodic manner, and the flow through these devices can still be analyzed as a steady-flow process by using time-averaged values for the properties and the heat transfer rates through the boundaries.

Steady-flow conditions can be closely approximated by devices that are intended for continuous operation such as turbines, pumps, boilers, condensers, and heat exchangers of steam power plants. The equations that are developed later in this section can be used for these and similar devices once the transient start-up period is completed and a steady operation is established.

Mass Balance for Steady-Flow Systems

The conservation of mass principle for any system undergoing any process was expressed in Chap. 1 as

$$m_{in} - m_{out} = \Delta m_{system} \qquad (5-12)$$

where $\Delta m_{system} = m_{final} - m_{initial}$ is the change in the mass of the system during the process. But during a steady-flow process, the total amount of mass contained within a control volume does not change with time (m_{CV} = constant) and thus $\Delta m_{system} = 0$. Then the conservation of mass principle requires that the total amount of mass entering a steady-flow system equal the total amount of mass leaving it. For a garden hose nozzle, for example, the amount of water entering the nozzle is equal to the amount of water leaving it during steady operation.

When dealing with steady-flow processes, we are not interested in the *amount* of mass that flows in and out of a device over time; instead, we are interested in the amount of mass flowing per unit time, that is, the *mass flow rate* $\dot{m}$. The *mass balance* for a general steady-flow system can be expressed in the rate form as

Mass balance for steady-flow systems: $\quad \dot{m}_{in} = \dot{m}_{out} \qquad$ (kg/s) $\quad$ (5-13)

The mass balance can also be expressed for a steady-flow system with multiple inlets and exits more explicitly as (Fig. 5-26)

Multiple inlets and exits: $\qquad \sum \dot{m}_i = \sum \dot{m}_e \qquad$ (kg/s) $\quad$ (5-14)

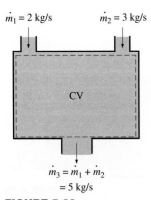

FIGURE 5-26

Conservation of mass principle for a two-inlet–one-exit steady-flow system.

where the subscript i stands for *inlet* and e for *exit,* and the summation signs are used to emphasize that all the inlets and exits are to be considered.

Most engineering devices such as nozzles, diffusers, turbines, compressors, and pumps involve a single stream (one inlet and one exit only). For these cases, we denote the inlet state by the subscript 1 and the exit state by the subscript 2, and drop the summation signs. Then the mass balance for a single-stream steady-flow system becomes

One inlet and one exit: $\quad \dot{m}_1 = \dot{m}_2 \quad$ or $\quad \rho_1 \mathcal{V}_1 A_1 = \rho_2 \mathcal{V}_2 A_2 \quad$ (5-15)

where ρ is density, $\mathcal{V}$ is the average flow velocity in the flow direction, and A is the cross-sectional area normal to the flow direction.

Energy Balance for Steady-Flow Systems

During a steady-flow process, the total energy content of a control volume remains constant (E_{CV} = constant), and thus the change in the total energy of the control volume is zero ($\Delta E_{CV} = 0$). Therefore, the amount of energy entering a control volume in all forms (by heat, work, and mass) must be equal to the amount of energy leaving it. Then the rate form of the general energy balance reduces for a steady-flow process to

$$\underbrace{\dot{E}_{in} - \dot{E}_{out}}_{\substack{\text{Rate of net energy transfer} \\ \text{by heat, work, and mass}}} = \underbrace{\Delta \dot{E}_{system}^{\nearrow 0 \text{ (steady)}}}_{\substack{\text{Rate of change in internal, kinetic,} \\ \text{potential, etc., energies}}} = 0 \qquad (5\text{-}16)$$

or

Energy balance: $\qquad \underbrace{\dot{E}_{in}}_{\substack{\text{Rate of net energy transfer in} \\ \text{by heat, work, and mass}}} = \underbrace{\dot{E}_{out}}_{\substack{\text{Rate of net energy transfer out} \\ \text{by heat, work, and mass}}} \qquad \text{(kW)} \quad (5\text{-}17)$

Noting that energy can be transferred by heat, work, and mass only, the energy balance above for a general steady-flow system can also be written more explicitly as

$$\dot{Q}_{in} + \dot{W}_{in} + \sum \dot{m}_i \theta_i = \dot{Q}_{out} + \dot{W}_{out} + \sum \dot{m}_e \theta_e \qquad (5\text{-}18)$$

or

$$\dot{Q}_{in} + \dot{W}_{in} + \underbrace{\sum \dot{m}_i \left(h_i + \frac{\mathcal{V}_i^2}{2} + gz_i \right)}_{\text{for each inlet}} = \dot{Q}_{out} + \dot{W}_{out} + \underbrace{\sum \dot{m}_e \left(h_e + \frac{\mathcal{V}_e^2}{2} + gz_e \right)}_{\text{for each exit}}$$

$$(5\text{-}19)$$

since the energy of a flowing fluid per unit mass is $\theta = h + ke + pe = h + \mathcal{V}^2/2 + gz$. The energy balance relation for steady-flow systems first appeared in 1859 in a German thermodynamics book written by Gustav Zeuner.

Consider, for example, an ordinary electric hot-water heater under steady operation, as shown in Fig. 5-27. A cold-water stream with a mass flow rate $\dot{m}$ is continuously flowing into the water heater, and a hot-water stream of the same mass flow rate is continuously flowing out of it. The water heater (the control volume) is losing heat to the surrounding air at a rate of $\dot{Q}_{out}$, and the electric heating element is supplying electrical work (heating) to the water

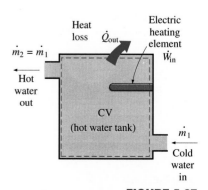

FIGURE 5-27

A water heater in steady operation.

at a rate of $\dot{W}_{in}$. On the basis of the conservation of energy principle, we can say that the water stream will experience an increase in its total energy as it flows through the water heater that is equal to the electric energy supplied to the water minus the heat losses.

The energy balance relation given above is intuitive in nature and is easy to use when the magnitudes and directions of heat and work transfers are known. But when performing a general analytical study or solving a problem that involves an unknown heat or work interaction, we need to assume a direction for the heat or work interactions. In such cases, it is common practice to assume heat to be transferred *into the system* (heat input) at a rate of $\dot{Q}$, and work produced *by the system* (work output) at a rate of $\dot{W}$, and then solve the problem. The first law or energy balance relation in that case for a general steady-flow system becomes

$$\dot{Q} - \dot{W} = \underbrace{\sum \dot{m}_e \left(h_e + \frac{V_e^2}{2} + gz_e \right)}_{\text{for each exit}} - \underbrace{\sum \dot{m}_i \left(h_i + \frac{V_i^2}{2} + gz_i \right)}_{\text{for each inlet}} \quad (5\text{-}20)$$

That is, the rate of heat transfer to a system minus power produced by the system is equal to the net change in the energy of the flow streams. Obtaining a negative quantity for Q or W simply means that the assumed direction for that quantity is wrong and should be reversed.

For single-stream (one-inlet–one-exit) systems, the summations over the inlets and the exits drop out, and the inlet and exit states in this case are denoted by subscripts 1 and 2, respectively, for simplicity. The mass flow rate through the entire control volume remains constant ($\dot{m}_1 = \dot{m}_2$) and is denoted by $\dot{m}$. Then the energy balance for *single-stream steady-flow systems* becomes

$$\dot{Q} - \dot{W} = \dot{m} \left[h_2 - h_1 + \frac{V_2^2 - V_1^2}{2} + g(z_2 - z_1) \right] \quad (5\text{-}21)$$

Dividing the equation above by $\dot{m}$ gives the energy balance on a unit-mass basis as

$$q - w = h_2 - h_1 + \frac{V_2^2 - V_1^2}{2} + g(z_2 - z_1) \quad (5\text{-}22)$$

where $q = \dot{Q}/\dot{m}$ and $w = \dot{W}/\dot{m}$ are the heat transfer and work done per unit mass of the working fluid, respectively.

If the fluid experiences a negligible change in its kinetic and potential energies as it flows through the control volume (that is, $\Delta\text{ke} \cong 0$, $\Delta\text{pe} \cong 0$), then the energy equation for a single-stream steady-flow system reduces further to

$$q - w = h_2 - h_1 \quad (5\text{-}23)$$

The various terms appearing in the above equations are as follows:

$\dot{Q}$ = **rate of heat transfer between the control volume and its surroundings.** When the control volume is losing heat (as in the case of the water heater), $\dot{Q}$ is negative. If the control volume is well insulated (i.e., adiabatic), then $\dot{Q} = 0$.

$\dot{W}$ = **power.** For steady-flow devices, the control volume is constant; thus, there is no boundary work involved. The work required to push mass into and out of the control volume is also taken care of by using enthalpies for the energy of fluid streams instead of internal energies.

Then $\dot{W}$ represents the remaining forms of work done per unit time (Fig. 5-28). Many steady-flow devices, such as turbines, compressors, and pumps, transmit power through a shaft, and $\dot{W}$ simply becomes the shaft power for those devices. If the control surface is crossed by electric wires (as in the case of an electric water heater), $\dot{W}$ will represent the electrical work done per unit time. If neither is present, then $\dot{W} = 0$.

$\Delta h = h_{\text{exit}} - h_{\text{inlet}}$. The enthalpy change of a fluid can easily be determined by reading the enthalpy values at the exit and inlet states from the tables. For ideal gases, it may be approximated by $\Delta h = C_{p,\text{av}}(T_2 - T_1)$. Note that $(\text{kg/s})(\text{kJ/kg}) \equiv \text{kW}$.

$\Delta\text{ke} = (\mathcal{V}_2^2 - \mathcal{V}_1^2)/2$. The unit of kinetic energy is m^2/s^2, which is equivalent to J/kg (Fig. 5-29). The enthalpy is usually given in kJ/kg. To add these two quantities, the kinetic energy should be expressed in kJ/kg. This is easily accomplished by dividing it by 1000.

A velocity of 45 m/s corresponds to a kinetic energy of only 1 kJ/kg, which is a very small value compared with the enthalpy values encountered in practice. Thus, the kinetic energy term at low velocities can be neglected. When a fluid stream enters and leaves a steady-flow device at about the same velocity ($\mathcal{V}_1 \cong \mathcal{V}_2$), the change in the kinetic energy is close to zero regardless of the velocity. Caution should be exercised at high velocities, however, since small changes in velocities may cause significant changes in kinetic energy (Fig. 5-30).

$\Delta\text{pe} = g(z_2 - z_1)$. A similar argument can be given for the potential energy term. A potential energy change of 1 kJ/kg corresponds to an elevation difference of 102 m. The elevation difference between the inlet and exit of most industrial devices such as turbines and compressors is well below this value, and the potential energy term is always neglected for these devices. The only time the potential energy term is significant is when a process involves pumping a fluid to high elevations and we are interested in the required pumping power.

5-4 ■ SOME STEADY-FLOW ENGINEERING DEVICES

Many engineering devices operate essentially under the same conditions for long periods of time. The components of a steam power plant (turbines, compressors, heat exchangers, and pumps), for example, operate nonstop for months before the system is shut down for maintenance (Fig. 5-31). Therefore, these devices can be conveniently analyzed as steady-flow devices.

In this section, some common steady-flow devices are described, and the thermodynamic aspects of the flow through them are analyzed. The conservation of mass and the conservation of energy principles for these devices are illustrated with examples.

1 Nozzles and Diffusers

Nozzles and diffusers are commonly utilized in jet engines, rockets, spacecraft, and even garden hoses. A **nozzle** is a device that *increases the velocity of a fluid* at the expense of pressure. A **diffuser** is a device that *increases the pressure of a fluid* by slowing it down. That is, nozzles and diffusers perform opposite tasks. The cross-sectional area of a nozzle decreases in the flow

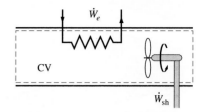

FIGURE 5-28

Under steady operation, shaft work and electrical work are the only forms of work a simple compressible system may involve.

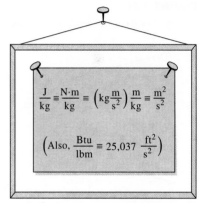

FIGURE 5-29

The units m^2/s^2 and J/kg are equivalent.

$\mathcal{V}_1$ m/s	$\mathcal{V}_2$ m/s	Δke kJ/kg
0	40	1
50	67	1
100	110	1
200	205	1
500	502	1

FIGURE 5-30

At very high velocities, even small changes in velocities may cause significant changes in the kinetic energy of the fluid.

174

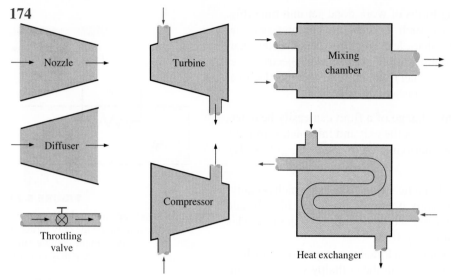

FIGURE 5-31

Steady-flow devices operate steadily for long periods.

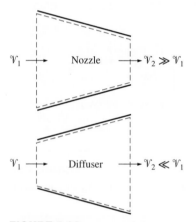

FIGURE 5-32

Nozzles and diffusers are shaped so that they cause large changes in fluid velocities and thus kinetic energies.

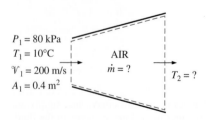

$P_1 = 80$ kPa
$T_1 = 10°C$
$\mathcal{V}_1 = 200$ m/s
$A_1 = 0.4$ m^2

AIR
$\dot{m} = ?$

$T_2 = ?$

FIGURE 5-33

Schematic for Example 5-11.

direction for subsonic flows and increases for supersonic flows. The reverse is true for diffusers.

The rate of heat transfer between the fluid flowing through a nozzle or a diffuser and the surroundings is usually very small ($\dot{Q} \approx 0$) since the fluid has high velocities, and thus it does not spend enough time in the device for any significant heat transfer to take place. Nozzles and diffusers typically involve no work ($\dot{W} = 0$) and any change in potential energy is negligible ($\Delta pe \cong 0$). But nozzles and diffusers usually involve very high velocities, and as a fluid passes through a nozzle or diffuser, it experiences large changes in its velocity (Fig. 5-32). Therefore, the kinetic energy changes must be accounted for in analyzing the flow through these devices ($\Delta ke \neq 0$).

EXAMPLE 5-11   Deceleration of Air in a Diffuser

Air at 10°C and 80 kPa enters the diffuser of a jet engine steadily with a velocity of 200 m/s. The inlet area of the diffuser is 0.4 m². The air leaves the diffuser with a velocity that is very small compared with the inlet velocity. Determine (a) the mass flow rate of the air and (b) the temperature of the air leaving the diffuser.

Solution   We take the *diffuser* as the system (Fig. 5-33). This is a *control volume* since mass crosses the system boundary during the process. We observe that there is only one inlet and one exit and thus $\dot{m}_1 = \dot{m}_2 = \dot{m}$.

Assumptions   **1** This is a steady-flow process since there is no change with time at any point and thus $\Delta m_{CV} = 0$ and $\Delta E_{CV} = 0$. **2** Air is an ideal gas since it is at a high temperature and low pressure relative to its critical point values. **3** The potential energy change is zero, $\Delta pe = 0$. **4** Heat transfer is negligible. **5** Kinetic energy at the diffuser exit is negligible. **6** There are no work interactions.

Analysis   (a) To determine the mass flow rate, we need to find the specific volume of the air first. This is determined from the ideal-gas relation at the inlet conditions:

$$v_1 = \frac{RT_1}{P_1} = \frac{(0.287 \text{ kPa} \cdot \text{m}^3/\text{kg} \cdot \text{K})(283 \text{ K})}{80 \text{ kPa}} = 1.015 \text{ m}^3/\text{kg}$$

Then,

$$\dot{m} = \frac{1}{v_1}\mathscr{V}_1 A_1 = \frac{1}{1.015 \text{ m}^3/\text{kg}}(200 \text{ m/s})(0.4 \text{ m}^2) = \textbf{78.8 kg/s}$$

Since the flow is steady, the mass flow rate through the entire diffuser will remain constant at this value.

(*b*) Under stated assumptions and observations, the energy balance for this steady-flow system can be expressed in the rate form as

$$\underbrace{\dot{E}_\text{in} - \dot{E}_\text{out}}_{\substack{\text{Rate of net energy transfer} \\ \text{by heat, work, and mass}}} = \underbrace{\Delta\dot{E}_\text{system}}_{\substack{\text{Rate of change in internal, kinetic,} \\ \text{potential, etc., energies}}}^{\nearrow 0 \text{ (steady)}} = 0$$

$$\dot{E}_\text{in} = \dot{E}_\text{out}$$

$$\dot{m}\left(h_1 + \frac{\mathscr{V}_1^2}{2}\right) = \dot{m}\left(h_2 + \frac{\mathscr{V}_2^2}{2}\right) \qquad (\text{since } \dot{Q} \cong 0, \dot{W} = 0, \text{ and } \Delta\text{pe} \cong 0)$$

$$h_2 = h_1 - \frac{\mathscr{V}_2^2 - \mathscr{V}_1^2}{2}$$

The exit velocity of a diffuser is usually small compared with the inlet velocity ($\mathscr{V}_2 \ll \mathscr{V}_1$); thus, the kinetic energy at the exit can be neglected. The enthalpy of air at the diffuser inlet is determined from the air table (Table A-17) to be

$$h_1 = h_{@ 283 \text{ K}} = 283.14 \text{ kJ/kg}$$

Substituting, we get

$$h_2 = 283.14 \text{ kJ/kg} - \frac{0 - (200 \text{ m/s})^2}{2}\left(\frac{1 \text{ kJ/kg}}{1000 \text{ m}^2/\text{s}^2}\right)$$

$$= 303.14 \text{ kJ/kg}$$

From Table A-17, the temperature corresponding to this enthalpy value is

$$T_2 = \textbf{303.1 K}$$

which shows that the temperature of the air increased by about 20°C as it was slowed down in the diffuser. The temperature rise of the air is mainly due to the conversion of kinetic energy to internal energy.

EXAMPLE 5-12 Acceleration of Steam in a Nozzle

Steam at 250 psia and 700°F steadily enters a nozzle whose inlet area is 0.2 ft². The mass flow rate of the steam through the nozzle is 10 lbm/s. Steam leaves the nozzle at 200 psia with a velocity of 900 ft/s. The heat losses from the nozzle per unit mass of the steam are estimated to be 1.2 Btu/lbm. Determine (*a*) the inlet velocity and (*b*) the exit temperature of the steam.

Solution We take the *nozzle* as the system (Fig. 5-34). This is a *control volume* since mass crosses the system boundary during the process. We *observe* that there is only one inlet and one exit and thus $\dot{m}_1 = \dot{m}_2 = \dot{m}$.

Assumptions **1** This is a steady-flow process since there is no change with time at any point and thus $\Delta m_\text{CV} = 0$ and $\Delta E_\text{CV} = 0$. **2** There are no work interactions. **3** The potential energy change is zero, $\Delta\text{pe} = 0$.

Analysis (*a*) The specific volume of the steam at the nozzle inlet is

$$\left. \begin{array}{l} P_1 = 250 \text{ psia} \\ T_1 = 700°\text{F} \end{array} \right\} \quad \begin{array}{l} v_1 = 2.688 \text{ ft}^3/\text{lbm} \\ h_1 = 1371.1 \text{ Btu/lbm} \end{array} \qquad \text{(Table A-6E)}$$

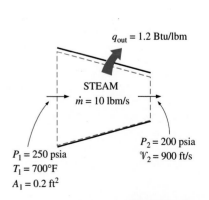

FIGURE 5-34
Schematic for Example 5-12.

Then,

$$\dot{m} = \frac{1}{v_1} \mathcal{V}_1 A_1$$

$$10 \text{ lbm/s} = \frac{1}{2.688 \text{ ft}^3/\text{lbm}} (\mathcal{V}_1)(0.2 \text{ ft}^2)$$

$$\mathcal{V}_1 = \textbf{134.4 ft/s}$$

(*b*) Under stated assumptions and observations, the energy balance for this steady-flow system can be expressed in the rate form as

$$\underbrace{\dot{E}_{\text{in}} - \dot{E}_{\text{out}}}_{\substack{\text{Rate of net energy transfer} \\ \text{by heat, work, and mass}}} = \underbrace{\Delta \dot{E}_{\text{system}}}_{\substack{\text{Rate of change in internal, kinetic,} \\ \text{potential, etc., energies}}}^{\nearrow 0 \text{ (steady)}} = 0$$

$$\dot{E}_{\text{in}} = \dot{E}_{\text{out}}$$

$$\dot{m}\left(h_1 + \frac{\mathcal{V}_1^2}{2}\right) = \dot{Q}_{\text{out}} + \dot{m}\left(h_2 + \frac{\mathcal{V}_2^2}{2}\right) \qquad (\text{since } \dot{W} \cong 0, \text{ and } \Delta\text{pe} \cong 0)$$

Dividing by the mass flow rate $\dot{m}$ and substituting, h_2 is determined to be

$$h_2 = h_1 - q_{\text{out}} - \frac{\mathcal{V}_2^2 - \mathcal{V}_1^2}{2}$$

$$= (1371.1 - 1.2) \text{ Btu/lbm} - \frac{(900 \text{ ft/s})^2 - (134.4 \text{ ft/s})^2}{2}\left(\frac{1 \text{ Btu/lbm}}{25,037 \text{ ft}^2/\text{s}^2}\right)$$

$$= 1354.1 \text{ Btu/lbm}$$

Then,

$$\left.\begin{array}{l} P_2 = 200 \text{ psia} \\ h_2 = 1354.1 \text{ Btu/lbm} \end{array}\right\} \qquad T_2 = \textbf{661.9°F} \qquad (\text{Table A-6E})$$

Therefore, the temperature of steam will drop by 38.1°F as it flows through the nozzle. This drop in temperature is mainly due to the conversion of internal energy to kinetic energy. (The heat loss is too small to cause any significant effect in this case.)

2 Turbines and Compressors

In steam, gas, or hydroelectric power plants, the device that drives the electric generator is the turbine. As the fluid passes through the turbine, work is done against the blades, which are attached to the shaft. As a result, the shaft rotates, and the turbine produces work. The work done in a turbine is positive since it is done by the fluid.

Compressors, as well as pumps and fans, are devices used to increase the pressure of a fluid. Work is supplied to these devices from an external source through a rotating shaft. Therefore, compressors involve work inputs. Even though these three devices function similarly, they do differ in the tasks they perform. A *fan* increases the pressure of a gas slightly and is mainly used to mobilize a gas. A *compressor* is capable of compressing the gas to very high pressures. *Pumps* work very much like compressors except that they handle liquids instead of gases.

Note that turbines produce power output whereas compressors, pumps, and fans require power input. Heat transfer from turbines is usually negligible ($\dot{Q} \approx 0$) since they are typically well insulated. Heat transfer is also negligible for compressors unless there is intentional cooling. Potential energy changes are negligible for all of these devices ($\Delta\text{pe} \cong 0$). The velocities involved in these devices, with the exception of turbines, are usually too low to cause any significant change in the kinetic energy (Δke $\cong$ 0). The fluid velocities encountered in most turbines are very high, and the fluid experiences a significant change in its kinetic energy. However, this change is usually very small relative to the change in enthalpy, and thus it is often disregarded.

EXAMPLE 5-13   Compressing Air by a Compressor

Air at 100 kPa and 280 K is compressed steadily to 600 kPa and 400 K. The mass flow rate of the air is 0.02 kg/s, and a heat loss of 16 kJ/kg occurs during the process. Assuming the changes in kinetic and potential energies are negligible, determine the necessary power input to the compressor.

Solution   We take the *compressor* as the system (Fig. 5-35). This is a *control volume* since mass crosses the system boundary during the process. We observe that there is only one inlet and one exit and thus $\dot{m}_1 = \dot{m}_2 = \dot{m}$. Also, heat is lost from the system and work is supplied to the system.

Assumptions   1 This is a steady-flow process since there is no change with time at any point and thus $\Delta m_{CV} = 0$ and $\Delta E_{CV} = 0$. **2** Air is an ideal gas since it is at a high temperature and low pressure relative to its critical point values. **3** The kinetic and potential energy changes are zero, Δke $= \Delta$pe $= 0$.

Analysis   Under stated assumptions and observations, the energy balance for this steady-flow system can be expressed in the rate form as

$$\underbrace{\dot{E}_{in} - \dot{E}_{out}}_{\substack{\text{Rate of net energy transfer} \\ \text{by heat, work, and mass}}} = \underbrace{\Delta \dot{E}_{system}}_{\substack{\text{Rate of change in internal, kinetic,} \\ \text{potential, etc., energies}}}^{\nearrow^{0 \text{ (steady)}}} = 0$$

$$\dot{E}_{in} = \dot{E}_{out}$$
$$\dot{W}_{in} + \dot{m}h_1 = \dot{Q}_{out} + \dot{m}h_2 \qquad (\text{since } \Delta\text{ke} = \Delta\text{pe} \cong 0)$$
$$\dot{W}_{in} = \dot{m}q_{out} + \dot{m}(h_2 - h_1)$$

The enthalpy of an ideal gas depends on temperature only, and the enthalpies of the air at the specified temperatures are determined from the air table (Table A-17) to be

$$h_1 = h_{@\ 280\ K} = 280.13 \text{ kJ/kg}$$
$$h_2 = h_{@\ 400\ K} = 400.98 \text{ kJ/kg}$$

Substituting, the power input to the compressor is determined to be

$$\dot{W}_{in} = (0.02 \text{ kg/s})(16 \text{ kJ/kg}) + (0.02 \text{ kg/s})(400.98 - 280.13 \text{ kJ/kg})$$
$$= \textbf{2.74 kW}$$

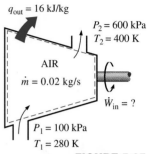

q_{out} = 16 kJ/kg

P_2 = 600 kPa
T_2 = 400 K

AIR

$\dot{m}$ = 0.02 kg/s

$\dot{W}_{in}$ = ?

P_1 = 100 kPa
T_1 = 280 K

FIGURE 5-35

Schematic for Example 5-13.

EXAMPLE 5-14   Power Generation by a Steam Turbine

The power output of an adiabatic steam turbine is 5 MW, and the inlet and the exit conditions of the steam are as indicated in Fig. 5-36.

(*a*) Compare the magnitudes of Δh, Δke, and Δpe.

(*b*) Determine the work done per unit mass of the steam flowing through the turbine.

(*c*) Calculate the mass flow rate of the steam.

Solution   We take the *turbine* as the system (Fig. 5-36). This is a *control volume* since mass crosses the system boundary during the process. We observe that

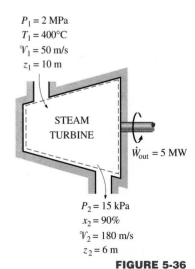

P_1 = 2 MPa
T_1 = 400°C
$\mathcal{V}_1$ = 50 m/s
z_1 = 10 m

STEAM
TURBINE

$\dot{W}_{out}$ = 5 MW

P_2 = 15 kPa
x_2 = 90%
$\mathcal{V}_2$ = 180 m/s
z_2 = 6 m

FIGURE 5-36

Schematic for Example 5-14.

there is only one inlet and one exit and thus $\dot{m}_1 = \dot{m}_2 = \dot{m}$. Also, work is done by the system. The inlet and exit velocities and elevations are given, and thus the kinetic and potential energies are to be considered.

Assumptions **1** This is a steady-flow process since there is no change with time at any point and thus $\Delta m_{CV} = 0$ and $\Delta E_{CV} = 0$. **2** The system is adiabatic and thus there is no heat transfer.

Analysis (*a*) At the inlet, steam is in a superheated vapor state, and its enthalpy is

$$\left.\begin{array}{r} P_1 = 2\text{ MPa} \\ T_1 = 400°C \end{array}\right\} \qquad h_1 = 3247.6\text{ kJ/kg} \qquad \text{(Table A-6)}$$

At the turbine exit, we obviously have a saturated liquid–vapor mixture at 15-kPa pressure. The enthalpy at this state is

$$h_2 = h_f + x_2 h_{fg} = [225.94 + (0.9)(2373.1)]\text{ kJ/kg} = 2361.73\text{ kJ/kg}$$

Then

$$\Delta h = h_2 - h_1 = (2361.73 - 3247.6)\text{ kJ/kg} = \mathbf{-885.87\text{ kJ/kg}}$$

$$\Delta\text{ke} = \frac{\mathcal{V}_2^2 - \mathcal{V}_1^2}{2} = \frac{(180\text{ m/s})^2 - (50\text{ m/s})^2}{2}\left(\frac{1\text{ kJ/kg}}{1000\text{ m}^2/\text{s}^2}\right) = \mathbf{14.95\text{ kJ/kg}}$$

$$\Delta\text{pe} = g(z_2 - z_1) = (9.81\text{ m/s}^2)[(6 - 10)\text{ m}]\left(\frac{1\text{ kJ/kg}}{1000\text{ m}^2/\text{s}^2}\right) = \mathbf{-0.04\text{ kJ/kg}}$$

Two observations can be made from the above results. First, the change in potential energy is insignificant in comparison to the changes in enthalpy and kinetic energy. This is typical for most engineering devices. Second, as a result of low pressure and thus high specific volume, the steam velocity at the turbine exit can be very high. Yet the change in kinetic energy is a small fraction of the change in enthalpy (less than 2 percent in our case) and is therefore often neglected.

(*b*) The energy balance for this steady-flow system can be expressed in the rate form as

$$\underbrace{\dot{E}_{in} - \dot{E}_{out}}_{\substack{\text{Rate of net energy transfer} \\ \text{by heat, work, and mass}}} = \underbrace{\Delta\dot{E}_{system}}_{\substack{\text{Rate of change in internal, kinetic,} \\ \text{potential, etc., energies}}}^{\nearrow 0\text{ (steady)}} = 0$$

$$\dot{E}_{in} = \dot{E}_{out}$$

$$\dot{m}(h_1 + \mathcal{V}_1^2/2 + gz_1) = \dot{W}_{out} + \dot{m}(h_2 + \mathcal{V}_2^2/2 + gz_2) \qquad \text{(since } \dot{Q} = 0\text{)}$$

Dividing by the mass flow rate $\dot{m}$ and substituting, the work done by the turbine per unit mass of the steam is determined to be

$$w_{out} = -\left[(h_2 - h_1) + \frac{\mathcal{V}_2^2 - \mathcal{V}_1^2}{2} + g(z_2 - z_1)\right] = -(\Delta h + \Delta\text{ke} + \Delta\text{pe})$$

$$= -[-885.87 + 14.95 - 0.04]\text{ kJ/kg} = \mathbf{870.96\text{ kJ/kg}}$$

(*c*) The required mass flow rate for a 5-MW power output is

$$\dot{m} = \frac{\dot{W}_{out}}{w_{out}} = \frac{5000\text{ kJ/s}}{870.96\text{ kJ/kg}} = \mathbf{5.74\text{ kg/s}}$$

3 Throttling Valves

Throttling valves are *any kind of flow-restricting devices* that cause a significant pressure drop in the fluid. Some familiar examples are ordinary adjustable valves, capillary tubes, and porous plugs (Fig. 5-37). Unlike turbines, they produce a pressure drop without involving any work. The pressure drop in the fluid is often accompanied by a *large drop in temperature,* and for that reason throttling devices are commonly used in refrigeration and air-conditioning applications. The magnitude of the temperature drop (or, sometimes, the temperature rise) during a throttling process is governed by a property called the *Joule-Thomson coefficient.*

Throttling valves are usually small devices, and the flow through them may be assumed to be adiabatic ($q \cong 0$) since there is neither sufficient time nor large enough area for any effective heat transfer to take place. Also, there is no work done ($w = 0$), and the change in potential energy, if any, is very small ($\Delta pe \cong 0$). Even though the exit velocity is often considerably higher than the inlet velocity, in many cases, the increase in kinetic energy is insignificant ($\Delta ke \cong 0$). Then the conservation of energy equation for this single-stream steady-flow device reduces to

$$h_2 \cong h_1 \qquad \text{(kJ/kg)} \qquad (5\text{-}24)$$

That is, enthalpy values at the inlet and exit of a throttling valve are the same. For this reason, a throttling valve is sometimes called an *isenthalpic device.*

To gain some insight into how throttling affects fluid properties, let us express Eq. 5-24 as follows:

$$u_1 + P_1 v_1 = u_2 + P_2 v_2$$

or   Internal energy + Flow energy = Constant

Thus the final outcome of a throttling process depends on which of the two quantities increases during the process. If the flow energy increases during the process ($P_2 v_2 > P_1 v_1$), it can do so at the expense of the internal energy. As a result, internal energy decreases, which is usually accompanied by a drop in temperature. If the product Pv decreases, the internal energy and the temperature of a fluid will increase during a throttling process. In the case of an ideal gas, $h = h(T)$, and thus the temperature has to remain constant during a throttling process (Fig. 5-38).

EXAMPLE 5-15   Expansion of Refrigerant-134a in a Refrigerator

Refrigerant-134a enters the capillary tube of a refrigerator as saturated liquid at 0.8 MPa and is throttled to a pressure of 0.12 MPa. Determine the quality of the refrigerant at the final state and the temperature drop during this process.

Solution   A capillary tube is a simple flow-restricting device that is commonly used in refrigeration applications to cause a large pressure drop in the refrigerant. Flow through a capillary tube is a throttling process; thus, the enthalpy of the refrigerant remains constant (Fig. 5-39).

At inlet: $P_1 = 0.8$ MPa $\}$ $T_1 = T_{sat\,@\,0.8\,MPa} = 31.33°C$ (Table A-12)
      sat. liquid $h_1 = h_{f\,@\,0.8\,MPa} = 93.42$ kJ/kg

At exit: $P_2 = 0.12$ MPa $\longrightarrow$ $h_f = 21.32$ kJ/kg $T_{sat} = -22.36°C$
     $(h_2 = h_1)$ $h_g = 233.86$ kJ/kg

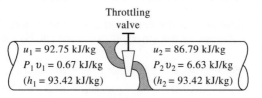

Throttling valve

$u_1 = 92.75$ kJ/kg	$u_2 = 86.79$ kJ/kg
$P_1 v_1 = 0.67$ kJ/kg	$P_2 v_2 = 6.63$ kJ/kg
$(h_1 = 93.42$ kJ/kg$)$	$(h_2 = 93.42$ kJ/kg$)$

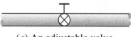

(a) An adjustable valve

(b) A porous plug

(c) A capillary tube

FIGURE 5-37

Throtting valves are devices that cause large pressure drops in the fluid.

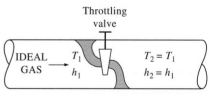

Throttling valve

IDEAL GAS T_1 $T_2 = T_1$
       h_1 $h_2 = h_1$

FIGURE 5-38

The temperature of an ideal gas does not change during a throttling ($h = $ constant) process since $h = h(T)$.

FIGURE 5-39

During a throttling process, the enthalpy (flow energy + internal energy) of a fluid remains constant. But internal and flow energies may be converted to each other.

Obviously $h_f < h_2 < h_g$; thus, the refrigerant exists as a saturated mixture at the exit state. The quality at this state is determined from

$$x_2 = \frac{h_2 - h_f}{h_{fg}} = \frac{93.42 - 21.32}{233.86 - 21.32} = \mathbf{0.339}$$

Since the exit state is a saturated mixture at 0.12 MPa, the exit temperature must be the saturation temperature at this pressure, which is $-22.36°C$. Then the temperature change for this process becomes

$$\Delta T = T_2 - T_1 = (-22.36 - 31.33)°C = \mathbf{-53.69°C}$$

That is, the temperature of the refrigerant drops by 53.69°C during this throttling process. Notice that 33.9 percent of the refrigerant vaporizes during this throttling process, and the energy needed to vaporize this refrigerant is absorbed from the refrigerant itself.

4a Mixture Chambers

In engineering applications, mixing two streams of fluids is not a rare occurrence. The section where the mixing process takes place is commonly referred to as a **mixing chamber.** The mixing chamber does not have to be a distinct "chamber." An ordinary T-elbow or a Y-elbow in a shower, for example, serves as the mixing chamber for the cold- and hot-water streams (Fig. 5-40).

The conservation of mass principle for a mixing chamber requires that the sum of the incoming mass flow rates equal the mass flow rate of the outgoing mixture.

Mixing chambers are usually well insulated ($q \cong 0$) and do not involve any kind of work ($w = 0$). Also, the kinetic and potential energies of the fluid streams are usually negligible (ke $\cong 0$, pe $\cong 0$). Then all there is left in the energy balance is the total energies of the incoming streams and the outgoing mixture. The conservation of energy principle requires that these two equal each other. Therefore, the conservation of energy equation becomes analogous to the conservation of mass equation for this case.

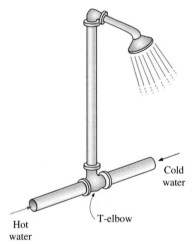

FIGURE 5-40

The T-elbow of an ordinary shower serves as the mixing chamber for the hot- and the cold-water streams.

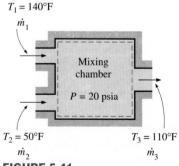

FIGURE 5-41

Schematic for Example 5-16.

EXAMPLE 5-16 Mixing of Hot and Cold Waters in a Shower

Consider an ordinary shower where hot water at 140°F is mixed with cold water at 50°F. If it is desired that a steady stream of warm water at 110°F be supplied, determine the ratio of the mass flow rates of the hot to cold water. Assume the heat losses from the mixing chamber to be negligible and the mixing to take place at a pressure of 20 psia.

Solution We take the *mixing chamber* as the system (Fig. 5-41). This is a *control volume* since mass crosses the system boundary during the process. We observe that there are two inlets and one exit.

Assumptions **1** This is a steady-flow process since there is no change with time at any point and thus $\Delta m_{CV} = 0$ and $\Delta E_{CV} = 0$. **2** The kinetic and potential energies are negligible, ke $\cong$ pe $\cong 0$. **3** Heat losses from the system are negligible and thus $\dot{Q} \cong 0$. **4** There is no work interaction involved.

Analysis Under the stated assumptions and observations, the mass and energy balances for this steady-flow system can be expressed in the rate form as follows:

Mass balance: $\dot{m}_{in} - \dot{m}_{out} = \Delta \dot{m}_{system}^{\,0\,(steady)} = 0$

$$\dot{m}_{in} = \dot{m}_{out} \quad \rightarrow \quad \dot{m}_1 + \dot{m}_2 = \dot{m}_3$$

Energy balance:

$$\underbrace{\dot{E}_{in} - \dot{E}_{out}}_{\substack{\text{Rate of net energy transfer} \\ \text{by heat, work, and mass}}} = \underbrace{\Delta \dot{E}_{system}^{\nearrow 0 \text{ (steady)}}}_{\substack{\text{Rate of change in internal, kinetic,} \\ \text{potential, etc., energies}}} = 0$$

$$\dot{E}_{in} = \dot{E}_{out}$$

$$\dot{m}_1 h_1 + \dot{m}_2 h_2 = \dot{m}_3 h_3 \quad (\text{since } \dot{Q} \cong 0, \dot{W} = 0, \text{ke} \cong \text{pe} \cong 0)$$

Combining the mass and energy balances,

$$\dot{m}_1 h_1 + \dot{m}_2 h_2 = (\dot{m}_1 + \dot{m}_2) h_3$$

Dividing this equation by $\dot{m}_2$ yields

$$y h_1 + h_2 = (y + 1) h_3$$

where $y = \dot{m}_1 / \dot{m}_2$ is the desired mass flow rate ratio.

The saturation temperature of water at 20 psia is 227.96°F. Since the temperatures of all three streams are below this value ($T < T_{sat}$), the water in all three streams exists as a compressed liquid (Fig. 5-42). A compressed liquid can be approximated as a saturated liquid at the given temperature. Thus,

$$h_1 \cong h_{f @ 140°F} = 107.96 \text{ Btu/lbm}$$
$$h_2 \cong h_{f @ 50°F} = 18.06 \text{ Btu/lbm}$$
$$h_3 \cong h_{f @ 110°F} = 78.02 \text{ Btu/lbm}$$

Solving for y and substituting yields

$$y = \frac{h_3 - h_2}{h_1 - h_3} = \frac{78.02 - 18.06}{107.96 - 78.02} = \mathbf{2.0}$$

Thus the mass flow rate of the hot water must be twice the mass flow rate of the cold water for the mixture to leave at 110°F.

4b Heat Exchangers

As the name implies, **heat exchangers** are devices where two moving fluid streams exchange heat without mixing. Heat exchangers are widely used in various industries, and they come in various designs.

The simplest form of a heat exchanger is a *double-tube* (also called *tube-and-shell*) *heat exchanger,* shown in Fig. 5-43. It is composed of two concentric pipes of different diameters. One fluid flows in the inner pipe, and the other in the annular space between the two pipes. Heat is transferred from the hot fluid to the cold one through the wall separating them. Sometimes the inner tube makes a couple of turns inside the shell to increase the heat transfer area, and thus the rate of heat transfer. The mixing chambers discussed earlier are sometimes classified as *direct-contact* heat exchangers.

The conservation of mass principle for a heat exchanger in steady operation requires that the sum of the inbound mass flow rates equal the sum of the outbound mass flow rates. This principle can also be expressed as follows: *Under steady operation, the mass flow rate of each fluid stream flowing through a heat exchanger remains constant.*

Heat exchangers typically involve no work interactions ($w = 0$) and negligible kinetic and potential energy changes ($\Delta\text{ke} \cong 0$, $\Delta\text{pe} \cong 0$) for each fluid stream. The heat transfer rate associated with heat exchangers depends on how the control volume is selected. Heat exchangers are intended for heat transfer between two fluids *within* the device, and the outer shell is usually well insulated to prevent any heat loss to the surrounding medium.

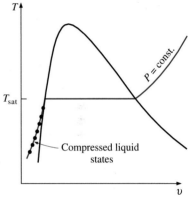

FIGURE 5-42

A substance exists as a compressed liquid at temperatures below the saturation temperatures at the given pressure.

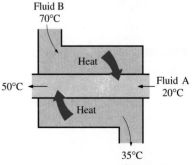

FIGURE 5-43

A heat exchanger can be as simple as two concentric pipes.

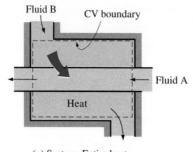

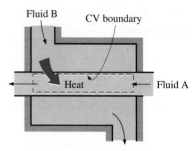

(a) System: Entire heat exchanger ($Q_{CV} = 0$)

(a) System: Fluid A ($Q_{CV} \neq 0$)

FIGURE 5-44

The heat transfer associated with a heat exchanger may be zero or nonzero depending on how the system is selected.

When the entire heat exchanger is selected as the control volume, $\dot{Q}$ becomes zero, since the boundary for this case lies just beneath the insulation and little or no heat crosses the boundary (Fig. 5-44). If, however, only one of the fluids is selected as the control volume, then heat will cross this boundary as it flows from one fluid to the other and $\dot{Q}$ will not be zero. In fact, $\dot{Q}$ in this case will be the rate of heat transfer between the two fluids.

EXAMPLE 5-17   Cooling of Refrigerant-134a by Water

Refrigerant-134a is to be cooled by water in a condenser. The refrigerant enters the condenser with a mass flow rate of 6 kg/min at 1 MPa and 70°C and leaves at 35°C. The cooling water enters at 300 kPa and 15°C and leaves at 25°C. Neglecting any pressure drops, determine (a) the mass flow rate of the cooling water required and (b) the heat transfer rate from the refrigerant to water.

Solution   We take the *entire heat exchanger* as the system (Fig. 5-45). This is a *control volume* since mass crosses the system boundary during the process. In general, there are several possibilities for selecting the control volume for multiple-stream steady-flow devices, and the proper choice depends on the situation at hand. We observe that there are two fluid streams (and thus two inlets and two exits) but no mixing.

Assumptions   **1** This is a steady-flow process since there is no change with time at any point and thus $\Delta m_{CV} = 0$ and $\Delta E_{CV} = 0$. **2** The kinetic and potential energies are negligible, ke $\cong$ pe $\cong$ 0. **3** Heat losses from the system are negligible and thus $\dot{Q} \cong 0$. **4** There is no work interaction.

Analysis   (a) Under the stated assumptions and observations, the mass and energy balances for this steady-flow system can be expressed in the rate form as follows:

Mass balance: $\qquad\qquad\qquad \dot{m}_{in} = \dot{m}_{out}$

for each fluid stream since there is no mixing. Thus,

$$\dot{m}_1 = \dot{m}_2 = \dot{m}_w$$
$$\dot{m}_3 = \dot{m}_4 = \dot{m}_R$$

Energy balance: $\qquad \underbrace{\dot{E}_{in} - \dot{E}_{out}}_{\substack{\text{Rate of net energy transfer} \\ \text{by heat, work, and mass}}} = \underbrace{\Delta \dot{E}_{system}}_{\substack{\text{Rate of change in internal, kinetic,} \\ \text{potential, etc., energies}}}^{\nearrow 0 \text{ (steady)}} = 0$

$$\dot{E}_{in} = \dot{E}_{out}$$

$$\dot{m}_1 h_1 + \dot{m}_3 h_3 = \dot{m}_2 h_2 + \dot{m}_4 h_4 \quad \text{(since } \dot{Q} \cong 0, \dot{W} = 0, \text{ke} \cong \text{pe} \cong 0\text{)}$$

Combining the mass and energy balances and rearranging give

$$\dot{m}_w(h_1 - h_2) = \dot{m}_R(h_4 - h_3)$$

Water
15°C
①

R-134a
③
70°C

④
35°C

②
25°C

FIGURE 5-45

Schematic for Example 5-17.

Now we need to determine the enthalpies at all four states. Water exists as a compressed liquid at both the inlet and the exit since the temperatures at both locations are below the saturation temperature of water at 300 kPa (133.55°C). Approximating the compressed liquid as a saturated liquid at the given temperature, we have

$$h_1 \cong h_{f@\ 15°C} = 62.99 \text{ kJ/kg}$$
$$h_2 \cong h_{f@\ 25°C} = 104.89 \text{ kJ/kg}$$

(Table A-4)

The refrigerant enters the condenser as a superheated vapor and leaves as a compressed liquid at 35°C. From refrigerant-134a tables,

$$\left.\begin{array}{l} P_3 = 1 \text{ MPa} \\ T_3 = 70°C \end{array}\right\} \quad h_3 = 302.34 \text{ kJ/kg} \qquad \text{(Table A-13)}$$

$$\left.\begin{array}{l} P_4 = 1 \text{ MPa} \\ T_4 = 35°C \end{array}\right\} \quad h_4 \cong h_{f@\ 35°C} = 98.78 \text{ kJ/kg} \qquad \text{(Table A-11)}$$

Substituting, we find

$$\dot{m}_w(62.99 - 104.89) \text{ kJ/kg} = (6 \text{ kg/min}) [(98.78 - 302.34) \text{ kJ/kg}]$$
$$\dot{m}_w = \textbf{29.15 kg/min}$$

(*b*) To determine the heat transfer from the refrigerant to the water, we have to choose a control volume whose boundary lies on the path of the heat flow. We can choose the volume occupied by either fluid as our control volume. For no particular reason, we choose the volume occupied by the water. All the assumptions stated earlier apply, except that the heat flow is no longer zero. Then assuming heat to be transferred to water, the energy balance for this single-stream steady-flow system reduces to

$$\underbrace{\dot{E}_{in} - \dot{E}_{out}}_{\substack{\text{Rate of net energy transfer} \\ \text{by heat, work, and mass}}} = \underbrace{\Delta \dot{E}_{system}}_{\substack{\text{Rate of change in internal, kinetic,} \\ \text{potential, etc., energies}}}{}^{\nearrow 0 \text{ (steady)}} = 0$$

$$\dot{E}_{in} = \dot{E}_{out}$$
$$\dot{Q}_{w,\ in} + \dot{m}_w h_1 = \dot{m}_w h_2$$

Rearranging and substituting,

$$\dot{Q}_{w,\ in} = \dot{m}_w(h_2 - h_1) = (29.15 \text{ kg/min})[(104.89 - 62.99) \text{ kJ/kg}]$$
$$= \textbf{1221 kJ/min}$$

Discussion   Had we chosen the volume occupied by the refrigerant as the control volume (Fig. 5-46), we would have obtained the same result for $\dot{Q}_{R,\ out}$ since the heat gained by the water is equal to the heat lost by the refrigerant.

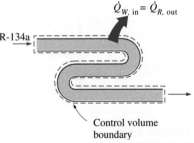

FIGURE 5-46

In a heat exchanger, the heat transfer depends on the choice of the control volume.

5   Pipe and Duct Flow

The transport of liquids or gases in pipes and ducts is of great importance in many engineering applications. Flow through a pipe or a duct usually satisfies the steady-flow conditions and thus can be analyzed as a steady-flow process. This, of course, excludes the transient start-up and shut-down periods. The control volume can be selected to coincide with the interior surfaces of the portion of the pipe or the duct that we are interested in analyzing.

Under normal operating conditions, the amount of heat gained or lost by the fluid may be very significant, particularly if the pipe or duct is long (Fig. 5-47). Sometimes heat transfer is desirable and is the sole purpose of the flow. Water flow through the pipes in the furnace of a power plant, the flow of

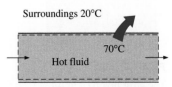

FIGURE 5-47

Heat losses from a hot fluid flowing through an uninsulated pipe or duct to the cooler environment may be very significant.

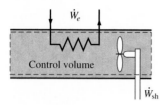

FIGURE 5-48

Pipe or duct flow may involve more than one form of work at the same time.

refrigerant in a freezer, and the flow in heat exchangers are some examples of this case. At other times, heat transfer is undesirable, and the pipes or ducts are insulated to prevent any heat loss or gain, particularly when the temperature difference between the flowing fluid and the surroundings is large. Heat transfer in this case is negligible.

If the control volume involves a heating section (electric wires), a fan, or a pump (shaft), the work interactions should be considered (Fig. 5-48). Of these, fan work is usually small and often neglected in energy analysis.

The velocities involved in pipe and duct flow are relatively low, and the kinetic energy changes are usually insignificant. This is particularly true when the pipe or duct diameter is constant and the heating effects are negligible. Kinetic energy changes may be significant, however, for gas flow in ducts with variable cross-sectional areas especially when the compressibility effects are significant. The potential energy term may also be significant when the fluid undergoes a considerable elevation change as it flows in a pipe or duct.

EXAMPLE 5-18 Electric Heating of Air in a House

The electric heating systems used in many houses consist of a simple duct with resistance wires. Air is heated as it flows over resistance wires. Consider a 15-kW electric heating system. Air enters the heating section at 100 kPa and 17°C with a volume flow rate of 150 m³/min. If heat is lost from the air in the duct to the surroundings at a rate of 200 W, determine the exit temperature of air.

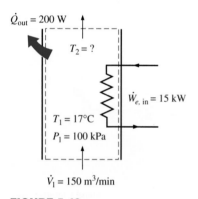

FIGURE 5-49

Schematic for Example 5-18.

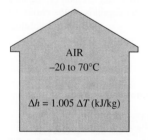

FIGURE 5-50

The error involved in $\Delta h = C_p \Delta T$, where $C_p = 1.005$ kJ/kg · °C, is less than 0.5 percent for air in the temperature range −20 to 70°C.

Solution We take the *heating section portion of the duct* as the system (Fig. 5-49). This is a *control volume* since mass crosses the system boundary during the process. We observe that there is only one inlet and one exit and thus $\dot{m}_1 = \dot{m}_2 = \dot{m}$. Also, heat is lost from the system and electrical work is supplied to the system.

Assumptions **1** This is a steady-flow process since there is no change with time at any point and thus $\Delta m_{CV} = 0$ and $\Delta E_{CV} = 0$. **2** Air is an ideal gas since it is at a high temperature and low pressure relative to its critical point values. **3** The kinetic and potential energy changes are negligible, $\Delta ke \cong \Delta pe \cong 0$. **4** Constant specific heats at room temperature can be used for air.

Analysis At temperatures encountered in heating and air-conditioning applications, Δh can be replaced by $C_p \Delta T$ where $C_p = 1.005$ kJ/kg · °C—the value at room temperature—with negligible error (Fig. 5-50). Then the energy balance for this steady-flow system can be expressed in the rate form as

$$\underbrace{\dot{E}_{in} - \dot{E}_{out}}_{\substack{\text{Rate of net energy transfer} \\ \text{by heat, work, and mass}}} = \underbrace{\Delta \dot{E}_{system}}_{\substack{\text{Rate of change in internal, kinetic,} \\ \text{potential, etc., energies}}}^{\nearrow 0 \text{ (steady)}} = 0$$

$$\dot{E}_{in} = \dot{E}_{out}$$

$$\dot{W}_{e,\,in} + \dot{m}h_1 = \dot{Q}_{out} + \dot{m}h_2 \quad \text{(since } \Delta ke \cong \Delta pe \cong 0\text{)}$$

$$\dot{W}_{e,\,in} - \dot{Q}_{out} = \dot{m}C_p(T_2 - T_1)$$

From the ideal gas relation, the specific volume of air at the inlet of the duct is

$$v_1 = \frac{RT_1}{P_1} = \frac{(0.287 \text{ kPa} \cdot \text{m}^3/\text{kg} \cdot \text{K})(290 \text{ K})}{100 \text{ kPa}} = 0.832 \text{ m}^3/\text{kg}$$

The mass flow rate of the air through the duct is determined from

$$\dot{m} = \frac{\dot{V}_1}{v_1} = \frac{150 \text{ m}^3/\text{min}}{0.832 \text{ m}^3/\text{kg}} \left(\frac{1 \text{ min}}{60 \text{ s}}\right) = 3.0 \text{ kg/s}$$

Substituting the known quantities, the exit temperature of the air is determined to be

$$(15 \text{ kJ/s}) - (0.2 \text{ kJ/s}) = (3 \text{ kg/s})(1.005 \text{ kJ/kg} \cdot °C)(T_2 - 17)°C$$

$$T_2 = \mathbf{21.9°C}$$

5-5 ■ ENERGY BALANCE FOR UNSTEADY-FLOW PROCESSES

During a steady-flow process, no changes occur within the control volume; thus, one does not need to be concerned about what is going on within the boundaries. Not having to worry about any changes within the control volume with time greatly simplifies the analysis.

Many processes of interest, however, involve *changes* within the control volume with time. Such processes are called unsteady-flow, or transient-flow, processes. The steady-flow relations developed earlier are obviously not applicable to these processes. When an unsteady-flow process is analyzed, it is important to keep track of the mass and energy contents of the control volume as well as the energy interactions across the boundary.

Some familiar unsteady-flow processes are the charging of rigid vessels from supply lines (Fig. 5-51), discharging a fluid from a pressurized vessel, driving a gas turbine with pressurized air stored in a large container, inflating tires or balloons, and even cooking with an ordinary pressure cooker.

Unlike steady-flow processes, unsteady-flow processes start and end over some finite time period instead of continuing indefinitely. Therefore in this section, we deal with changes that occur over some time interval Δt instead of with the rate of changes (changes per unit time). An unsteady-flow system, in some respects, is similar to a closed system, except that the mass within the system boundaries does not remain constant during a process.

Another difference between steady- and unsteady-flow systems is that steady-flow systems are fixed in space, size, and shape. Unsteady-flow systems, however, are not (Fig. 5-52). They are usually stationary; that is, they are fixed in space, but they may involve moving boundaries and thus boundary work.

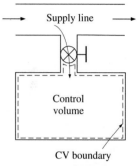

FIGURE 5-51

Charging of a rigid tank from a supply line is an unsteady-flow process since it involves changes within the control volume.

Mass Balance

Unlike the case of steady-flow processes, the amount of mass within the control volume *does* change with time during an unsteady-flow process. The magnitude of change depends on the amounts of mass that enter and leave the control volume during the process. The mass balance for a system undergoing any process was expressed earlier as

Mass balance: $m_{in} - m_{out} = \Delta m_{system}$ (kg)   (5-25)

where $\Delta m_{system} = m_{final} - m_{initial}$ is the change in the mass of the system during the process. The mass balance for a control volume can also be expressed more explicitly as

$$\sum m_i - \sum m_e = (m_2 - m_1)_{system} \quad (5\text{-}26)$$

where i = inlet; e = exit; 1 = initial state and 2 = final state of the control volume; and the summation signs are used to emphasize that all the inlets and

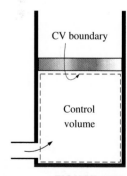

FIGURE 5-52

The shape and size of a control volume may change during an unsteady-flow process.

exits are to be considered. Often one or more terms in the equation above are zero. For example, $m_i = 0$ if no mass enters the control volume during the process, $m_e = 0$ if no mass leaves the control volume during the process, and $m_1 = 0$ if the control volume is initially evacuated.

Energy Balance

The energy content of a control volume changes with time during an unsteady-flow process. The magnitude of change depends on the amount of energy transfer across the system boundaries as heat and work as well as on the amount of energy transported into and out of the control volume by mass during the process. When analyzing an unsteady-flow process, we must keep track of the energy content of the control volume as well as the energies of the incoming and outgoing flow streams.

The general energy balance was given earlier as

$$Energy\ balance: \quad \underbrace{E_{\text{in}} - E_{\text{out}}}_{\substack{\text{Net energy transfer} \\ \text{by heat, work, and mass}}} = \underbrace{\Delta E_{\text{system}}}_{\substack{\text{Changes in internal, kinetic,} \\ \text{potential, etc., energies}}} \qquad (\text{kJ}) \quad (5\text{-}27)$$

The general unsteady-flow process, in general, is difficult to analyze because the properties of the mass at the inlets and exits may change during a process. Most unsteady-flow processes, however, can be represented reasonably well by the **uniform-flow process,** which involves the following idealization: *The fluid flow at any inlet or exit is uniform and steady, and thus the fluid properties do not change with time or position over the cross section of an inlet or exit. If they do, they are averaged and treated as constants for the entire process.*

Note that unlike the steady-flow systems, the state of an unsteady-flow system may change with time, and that the state of the mass leaving the control volume at any instant is the same as the state of the mass in the control volume at that instant. The initial and final properties of the control volume can be determined from the knowledge of the initial and final states, which are completely specified by two independent intensive properties for simple compressible systems.

Then the energy balance for a uniform-flow system can be expressed explicitly as

$$\left(Q_{\text{in}} + W_{\text{in}} + \sum m_i \theta_i\right) - \left(Q_{\text{out}} + W_{\text{out}} + \sum m_e \theta_e\right) = (m_2 e_2 - m_1 e_1)_{\text{system}}$$
$$(5\text{-}28)$$

where $\theta = h + \text{ke} + \text{pe}$ is the energy of a flowing fluid at any inlet or exit per unit mass, and $e = u + \text{ke} + \text{pe}$ is the energy of the non-flowing fluid within the control volume per unit mass. When the kinetic and potential energy changes associated with the control volume and fluid streams are negligible, as is usually the case, the energy balance above simplifies to

$$\left(Q_{\text{in}} + W_{\text{in}} + \sum m_i h_i\right) - \left(Q_{\text{out}} + W_{\text{out}} + \sum m_e h_e\right) = (m_2 u_2 - m_1 u_1)_{\text{system}}$$
$$(5\text{-}29)$$

Note that if no mass enters or leaves the control volume during a process ($m_i = m_e = 0$, and $m_1 = m_2 = m$), this equation reduces to the energy balance relation for closed systems (Fig. 5-53). Also note that an unsteady-flow

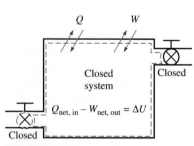

FIGURE 5-53

The energy equation of a uniform-flow system reduces to that of a closed system when all the inlets and exits are closed.

system may involve boundary work as well as electrical and shaft work (Fig. 5-54).

Although both the steady-flow and uniform-flow processes are somewhat idealized, many actual processes can be approximated reasonably well by one of these with satisfactory results. The degree of satisfaction depends upon the desired accuracy and the degree of validity of the assumptions made.

EXAMPLE 5-19   Charging of a Rigid Tank by Steam

A rigid, insulated tank that is initially evacuated is connected through a valve to a supply line that carries steam at 1 MPa and 300°C. Now the valve is opened, and steam is allowed to flow slowly into the tank until the pressure reaches 1 MPa, at which point the valve is closed. Determine the final temperature of the steam in the tank.

Solution   We take the *tank* as the system (Fig. 5-55). This is a *control volume* since mass crosses the system boundary during the process. We observe that this is an unsteady-flow process since changes occur within the control volume. The control volume is initially evacuated and thus $m_1 = 0$ and $m_1 u_1 = 0$. Also, there is one inlet and no exits for mass flow.

Assumptions   **1** This process can be analyzed as a *uniform-flow process* since the properties of the steam entering the control volume remain constant during the entire process. **2** The kinetic and potential energies of the streams are negligible, ke ≅ pe ≅ 0. **3** The tank is stationary and thus its kinetic and potential energy changes are zero; that is, $\Delta KE = \Delta PE = 0$ and $\Delta E_{system} = \Delta U_{system}$. **4** There are no boundary, electrical, or shaft work interactions involved. **5** The tank is well insulated and thus there is no heat transfer.

Analysis   Noting that microscopic energies of flowing and nonflowing fluids are represented by enthalpy h and internal energy u, respectively, the mass and energy balances for this uniform-flow system can be expressed as

Mass balance: $m_{in} - m_{out} = \Delta m_{system}$ → $m_i = m_2 - \cancelto{0}{m_1} = m_2$

Energy balance: $\underbrace{E_{in} - E_{out}}_{\substack{\text{Net energy transfer} \\ \text{by heat, work, and mass}}} = \underbrace{\Delta E_{system}}_{\substack{\text{Change in internal, kinetic,} \\ \text{potential, etc., energies}}}$

$m_i h_i = m_2 u_2$ (since $W = Q = 0$, ke ≅ pe ≅ 0, $m_1 = 0$)

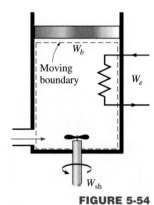

FIGURE 5-54

An unsteady-flow system may involve electrical, shaft, and boundary work all at once.

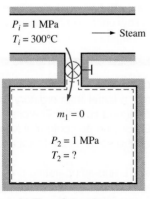

(a) Flow of steam into an evacuated tank

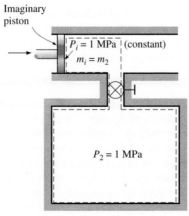

(b) The closed-system equivalence

FIGURE 5-55

Schematic for Example 5-19.

Combining the mass and energy balances gives

$$u_2 = h_i$$

That is, the final internal energy of the steam in the tank is equal to the enthalpy of the steam entering the tank. The enthalpy of the steam at the inlet state is

$$\left.\begin{array}{l} P_i = 1 \text{ MPa} \\ T_i = 300°C \end{array}\right\} \quad h_i = 3051.2 \text{ kJ/kg} \quad \text{(Table A-6)}$$

which is equal to u_2. Since we now know two properties at the final state, it is fixed and the temperature at this state is determined from the same table to be

$$\left.\begin{array}{l} P_2 = 1 \text{ MPa} \\ u_2 = 3051.2 \text{ kJ/kg} \end{array}\right\} \quad T_2 = \mathbf{456.2°C}$$

Discussion   Note that the temperature of the steam in the tank has increased by 156.2°C. This result may be surprising at first, and you may be wondering where the energy to raise the temperature of the steam came from. The answer lies in the enthalpy term $h = u + Pv$. Part of the energy represented by enthalpy is the flow energy Pv, and this flow energy is converted to sensible internal energy once the flow ceases to exist in the control volume, and it shows up as an increase in temperature (Fig. 5-56).

Alternative solution   This problem can also be solved by considering the region within the tank and the mass that is destined to enter the tank as a closed system, as shown in Fig. 5-55b. Since no mass crosses the boundaries, viewing this as a closed system is appropriate.

During the process, the steam upstream (the imaginary piston) will push the enclosed steam in the supply line into the tank at a constant pressure of 1 MPa. Then the boundary work done during this process is

$$W_{b,\text{ in}} = -\int_1^2 P_i \, dV = -P_i(V_2 - V_1) = -P_i[V_{\text{tank}} - (V_{\text{tank}} + V_i)] = P_i V_i$$

where V_i is the volume occupied by the steam before it enters the tank and P_i is the pressure at the moving boundary (the imaginary piston face). The energy balance for the closed system gives

$$\underbrace{E_{\text{in}} - E_{\text{out}}}_{\substack{\text{Net energy transfer} \\ \text{by heat, work, and mass}}} = \underbrace{\Delta E_{\text{system}}}_{\substack{\text{Change in internal, kinetic,} \\ \text{potential, etc., energies}}}$$

$$W_{b,\text{in}} = \Delta U$$
$$m_i P_i v_i = m_2 u_2 - m_i u_i$$
$$u_2 = u_i + P_i v_i = h_i$$

since the initial state of the system is simply the line conditions of the steam. This result is identical to the one obtained with the uniform-flow analysis. Once again, the temperature rise is caused by the so-called flow energy or flow work, which is the energy required to push the substance into the tank.

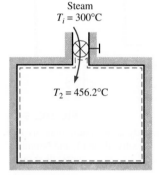

FIGURE 5-56

The temperature of steam rises from 300 to 456.2°C as it enters a tank as a result of flow energy being converted to internal energy.

EXAMPLE 5-20   Cooking with a Pressure Cooker

A pressure cooker is a pan that cooks food much faster than ordinary pans by maintaining a higher pressure and temperature during cooking. The pressure inside the pan is controlled by a pressure regulator (the petcock) that keeps the pressure at a constant level by periodically allowing some steam to escape, thus preventing any excess pressure buildup.

A certain pressure cooker has a volume of 6 L and an operating pressure of 75 kPa gage. Initially, it contains 1 kg of water. Heat is supplied to the pressure cooker at a rate of 500 W for 30 min after the operating pressure is reached. Assuming an atmospheric pressure of 100 kPa. determine (*a*) the temperature at which cooking takes place and (b) the amount of water left in the pressure cooker at the end of the process.

Solution We take the *pressure cooker* as the system (Fig. 5-57). This is a *control volume* since mass crosses the system boundary during the process. We observe that this is an unsteady-flow process since changes occur within the control volume. Also, there is one exit and no inlets for mass flow.

Assumptions **1** This process can be analyzed as a *uniform-flow process* since the properties of the steam leaving the control volume remain constant during the entire cooking process. **2** The kinetic and potential energies of the streams are negligible, ke $\cong$ pe $\cong$ 0. **3** The pressure cooker is stationary and thus its kinetic and potential energy changes are zero; that is, $\Delta KE = \Delta PE = 0$ and $\Delta E_{system} = \Delta U_{system}$. **4** The pressure (and thus temperature) in the pressure cooker remains constant. **5** Steam leaves as a saturated vapor at the cooker pressure. **6** There are no boundary, electrical, or shaft work interactions involved. **7** Heat is transferred to the cooker at a constant rate.

Analysis (a) The absolute pressure within the cooker is

$$P_{abs} = P_{gage} + P_{atm} = 75 + 100 = 175 \text{ kPa}$$

Since saturation conditions exist in the cooker at all times (Fig. 5-58), the cooking temperature must be the saturation temperature corresponding to this pressure. From Table A-5, it is

$$T = T_{sat @ 175 \text{ kPa}} = \mathbf{116.06°C}$$

which is about 16°C higher than the ordinary cooking temperature.

(b) Noting that the microscopic energies of flowing and nonflowing fluids are represented by enthalpy h and internal energy u, respectively, the mass and energy balances for this uniform-flow system can be expressed as

Mass balance:
$$m_{in} - m_{out} = \Delta m_{system} \rightarrow -m_e = (m_2 - m_1)_{CV} \quad \text{or} \quad m_e = (m_1 - m_2)_{CV}$$

Energy balance:
$$\underbrace{E_{in} - E_{out}}_{\substack{\text{Net energy transfer} \\ \text{by heat, work, and mass}}} = \underbrace{\Delta E_{system}}_{\substack{\text{Change in internal, kinetic,} \\ \text{potential, etc., energies}}}$$

$$Q_{in} - m_e h_e = (m_2 u_2 - m_1 u_1)_{CV} \quad \text{(since } W = 0, \text{ ke} \cong \text{pe} \cong 0)$$

Combining the mass and energy balances gives

$$Q_{in} = (m_1 - m_2)h_e + (m_2 u_2 - m_1 u_1)_{CV}$$

The amount of heat transfer during this process is found from

$$Q_{in} = \dot{Q}_{in} \, \Delta t = (0.5 \text{ kJ/s})(30 \times 60 \text{ s}) = 900 \text{ kJ}$$

Steam leaves the pressure cooker as saturated vapor at 175 kPa at all times (Fig. 5-59). Thus,

$$h_e = h_{g @ 175 \text{ kPa}} = 2700.6 \text{ kJ/kg}$$

The initial internal energy is found after the quality is determined:

$$v_1 = \frac{V}{m_1} = \frac{0.006 \text{ m}^3}{1 \text{ kg}} = 0.006 \text{ m}^3/\text{kg}$$

$$x_1 = \frac{v_1 - v_f}{v_{fg}} = \frac{0.006 - 0.001}{1.004 - 0.001} = 0.005$$

Thus,

$$u_1 = u_f + x_1 u_{fg} = 486.8 + (0.005)(2038.1) \text{ kJ/kg} = 497.0 \text{ kJ/kg}$$

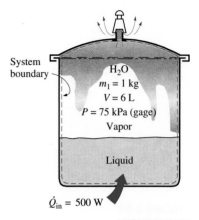

H_2O
$m_1 = 1$ kg
$V = 6$ L
$P = 75$ kPa (gage)
Vapor

Liquid

System boundary

$\dot{Q}_{in} = 500$ W

FIGURE 5-57

Schematic for Example 5-20.

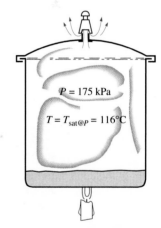

$P = 175$ kPa

$T = T_{sat@P} = 116°C$

FIGURE 5-58

As long as there is liquid in a pressure cooker, the saturation conditions exist and the temperature remains constant at the saturation temperature.

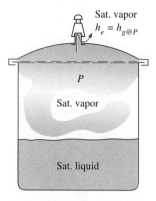

Sat. vapor
$h_e = h_{g @ P}$

P

Sat. vapor

Sat. liquid

FIGURE 5-59

In a pressure cooker, the enthalpy of the exiting steam is $h_{g @ P}$ (enthalpy of the saturated vapor at the given pressure).

and

$$U_1 = m_1 u_1 = (1 \text{ kg})(497 \text{ kJ/kg}) = 497 \text{ kJ}$$

The mass of the system at the final state is $m_2 = V/v_2$. Substituting this into the energy equation yields

$$Q_{in} = \left(m_1 - \frac{V}{v_2}\right)h_e + \left(\frac{V}{v_2}u_2 - m_1 u_1\right)$$

There are two unknowns in this equation, u_2 and v_2. Thus we need to relate them to a single unknown before we can determine these unknowns. Assuming there is still some liquid water left in the cooker at the final state (i.e., saturation conditions exist), v_2 and u_2 can be expressed as

$$v_2 = v_f + x_2 v_{fg} = 0.001 + x_2(1.004 - 0.001) \text{ m}^3/\text{kg}$$
$$u_2 = u_f + x_2 u_{fg} = 486.8 + x_2(2038.1) \text{ kJ/kg}$$

Notice that during a boiling process at constant pressure, the properties of each phase remain constant (only the amounts change). When these expressions are substituted into the above energy equation, x_2 becomes the only unknown, and it is determined to be

$$x_2 = 0.009$$

Thus,

$$v_2 = 0.001 + (0.009)(1.004 - 0.0001) \text{ m}^3/\text{kg} = 0.010 \text{ m}^3/\text{kg}$$

and

$$m_2 = \frac{V}{v_2} = \frac{0.006 \text{ m}^3}{0.01 \text{ m}^3/\text{kg}} = \textbf{0.6 kg}$$

Therefore, after 30 min there is 0.6 kg water (liquid + vapor) left in the pressure cooker.

5-6 ■ SUMMARY

The first law of thermodynamics is essentially an expression of the conservation of energy principle, also called the energy balance. The general mass and energy balances for *any system* undergoing *any process* can be expressed as

$$m_{in} - m_{out} = \Delta m_{system} \qquad \text{(kg)}$$
$$\underbrace{E_{in} - E_{out}}_{\substack{\text{Net energy transfer} \\ \text{by heat, work, and mass}}} = \underbrace{\Delta E_{system}}_{\substack{\text{Changes in internal, kinetic,} \\ \text{potential, etc., energies}}} \qquad \text{(kJ)}$$

They can also be expressed in the *rate form* as

$$\dot{m}_{in} - \dot{m}_{out} = \Delta \dot{m}_{system} \qquad \text{(kg/s)}$$
$$\underbrace{\dot{E}_{in} - \dot{E}_{out}}_{\substack{\text{Rate of net energy transfer} \\ \text{by heat, work, and mass}}} = \underbrace{\Delta \dot{E}_{system}}_{\substack{\text{Rate of change in internal, kinetic,} \\ \text{potential, etc., energies}}} \qquad \text{(kW)}$$

Taking heat transfer *to* the system and work done *by* the system to be positive quantities, the energy balance for a closed system can also be expressed as

$$Q - W = \Delta U + \Delta KE + \Delta PE \qquad \text{(kJ)}$$

$$W = W_{\text{other}} + W_b$$
$$\Delta U = m(u_2 - u_1)$$
$$\Delta \text{KE} = \tfrac{1}{2}m(\mathcal{V}_2^2 - \mathcal{V}_1^2)$$
$$\Delta \text{PE} = mg(z_2 - z_1)$$

For a *constant-pressure process*, $W_b + \Delta U = \Delta H$. Thus,

$$Q - W_{\text{other}} = \Delta H + \Delta \text{KE} - \Delta \text{PE} \qquad \text{(kJ)}$$

Thermodynamic processes involving control volumes can be considered in two groups: steady-flow processes and unsteady-flow processes. During a *steady-flow process*, the fluid flows through the control volume steadily, experiencing no change with time at a fixed position. The mass and energy content of the control volume remain constant during a steady-flow process. Taking heat transfer *to* the system and work done *by* the system to be positive quantities, the conservation of mass and energy equations for steady-flow processes are expressed as

$$\sum \dot{m}_i = \sum \dot{m}_e \qquad \text{(kg/s)}$$

$$\dot{Q} - \dot{W} = \sum \underbrace{\dot{m}_e\left(h_e + \frac{\mathcal{V}_e^2}{2} + gz_e\right)}_{\text{for each exit}} - \sum \underbrace{\dot{m}_i\left(h_i + \frac{\mathcal{V}_i^2}{2} + gz_i\right)}_{\text{for each inlet}} \qquad \text{(kW)}$$

where the subscript i stands for inlet and e for exit. These are the most general forms of the equations for steady-flow processes. For single-stream (one-inlet–one-exit) systems such as nozzles, diffusers, turbines, compressors, and pumps, they simplify to

$$\dot{m}_1 = \dot{m}_2 \qquad \text{(kg/s)}$$

or

$$\frac{1}{v_1}\mathcal{V}_1 A_1 = \frac{1}{v_2}\mathcal{V}_2 A_2$$

and

$$\dot{Q} - \dot{W} = \dot{m}\left[h_2 - h_1 + \frac{\mathcal{V}_2^2 - \mathcal{V}_1^2}{2} + g(z_2 - z_1)\right] \qquad \text{(kW)}$$

In the above relations, subscripts 1 and 2 denote the inlet and exit states, respectively.

Most unsteady-flow processes can be modeled as a *uniform-flow process*, which requires that the fluid flow at any inlet or exit is uniform and steady, and thus the fluid properties do not change with time or position over the cross section of an inlet or exit. If they do, they are averaged and treated as constants for the entire process. The energy balance for a uniform-flow system is expressed explicitly as

$$\left(Q_{\text{in}} + W_{\text{in}} + \sum m_i \theta_i\right) - \left(Q_{\text{out}} + W_{\text{out}} + \sum m_e \theta_e\right) = (m_2 e_2 - m_1 e_1)_{\text{system}}$$

When the kinetic and potential energy changes associated with the control volume and fluid streams are negligible, the energy relation simplifies to

$$\left(Q_{\text{in}} + W_{\text{in}} + \sum m_i h_i\right) - \left(Q_{\text{out}} + W_{\text{out}} + \sum m_e h_e\right) = (m_2 u_2 - m_1 u_1)_{\text{system}}$$

When solving thermodynamic problems, it is recommended that the general form of the energy balance $E_{\text{in}} - E_{\text{out}} = \Delta E_{\text{system}}$ be used for all problems, and simplify it for the particular problem instead of using the specific relations given above for different processes.

REFERENCES AND SUGGESTED READING

1. ASHRAE Handbook of Fundamentals. SI version. Atlanta, GA: American Society of Heating, Refrigerating, and Air-Conditioning Engineers, Inc., 1993.

2. ASHRAE Handbook of Refrigeration. SI version. Atlanta, GA: American Society of Heating, Refrigerating, and Air-Conditioning Engineers, Inc., 1994.

3. A. Bejan. *Advanced Engineering Thermodynamics.* New York: John Wiley & Sons, 1988.

4. Y. A. Çengel. "An Intuitive and Unified Approach to Teaching Thermodynamics." ASME International Mechanical Engineering Congress and Exposition, Atlanta, Georgia, AES-Vol. 36, pp. 251–260, November 17–22, 1996.

5. Y. A. Çengel and M. A. Boles. *Thermodynamics: An Engineering Approach.* 3rd ed. New York: McGraw-Hill, 1998.

6. W. C. Reynolds and H. C. Perkins. *Engineering Thermodynamics.* 2nd ed. New York: McGraw-Hill, 1977.

7. K. Wark and D. E. Richards. *Thermodynamics.* 6th ed. New York: McGraw-Hill, 1999.

PROBLEMS*

Closed-System Energy Balance: General Systems

5-1C For a cycle, is the net work necessarily zero? For what kind of systems will this be the case?

5-2C On a hot summer day, a student turns his fan on when he leaves his room in the morning. When he returns in the evening, will the room be warmer or cooler than the neighboring rooms? Why? Assume all the doors and windows are kept closed.

5-3C Consider two identical rooms, one with a refrigerator in it and the other without one. If all the doors and windows are closed, will the room that contains the refrigerator be cooler or warmer than the other room? Why?

5-4C What are the different mechanisms for transferring energy to or from a control volume?

5-5 Water is being heated in a closed pan on top of a range while being stirred by a paddle wheel. During the process, 30 kJ of heat is transferred to the water, and 5 kJ of heat is lost to the surrounding air. The paddle-wheel work amounts to 500 N · m. Determine the final energy of the system if its initial energy is 10 kJ.
Answer: 35.5 kJ

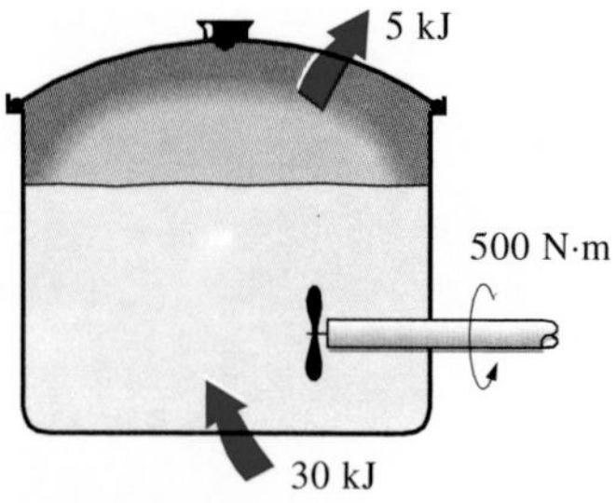

5 kJ

500 N·m

30 kJ

FIGURE P5-5

*Students are encouraged to answer *all* the concept "C" questions.

5-6E A vertical piston-cylinder device contains water and is being heated on top of a range. During the process, 50 Btu of heat is transferred to the water, and heat losses from the side walls amount to 8 Btu. The piston rises as a result of evaporation, and 5 Btu of boundary work is done. Determine the change in the energy of the water for this process.   *Answer:* 37 Btu

5-7 A classroom that normally contains 40 people is to be air-conditioned with window air-conditioning units of 5-kW rating. A person at rest may be assumed to dissipate heat at a rate of about 360 kJ/h. There are 10 light bulbs in the room, each with a rating of 100 W. The rate of heat transfer to the classroom through the walls and the windows is estimated to be 15,000 kJ/h. If the room air is to be maintained at a constant temperature of 21°C, determine the number of window air-conditioning units required.   *Answer:* 2 units

5-8 The lighting requirements of an industrial facility are being met by 700 40-W standard fluorescent lamps. The lamps are close to completing their service life and are to be replaced by their 34-W high-efficiency counterparts that operate on the existing standard ballasts. The standard and high-efficiency fluorescent lamps can be purchased in quantity at a cost of $1.77 and $2.26 each, respectively. The facility operates 2800 hours a year, and all of the lamps are kept on during operating hours. Taking the unit cost of electricity to be $0.08/kWh and the ballast factor to be 1.1 (i.e., ballasts consume 10 percent of the rated power of the lamps), determine how much energy and money will be saved per year as a result of switching to the high-efficiency fluorescent lamps. Also, determine the simple payback period.

5-9 The lighting needs of a storage room are being met by 6 fluorescent light fixtures, each fixture containing four lamps rated at 60 W each. All the lamps are on during operating hours of the facility, which are 6 A.M. to 6 P.M. 365 days a year. The storage room is actually used for an average of three hours a day. If the price of electricity is $0.08/kWh, determine the amount of energy and money that will be saved as a result of installing motion sensors. Also, determine the simple payback period if the purchase price of the sensor is $32 and it takes 1 hour to install it at a cost of $40.

5-10 A university campus has 200 classrooms and 400 faculty offices. The classrooms are equipped with 12 fluorescent light bulbs, each consuming 110 W, including the electricity used by the ballasts. The faculty offices, on average, have half as many light bulbs. The campus is open 240 days a year. The classrooms and faculty offices are not occupied an average of four hours a day, but the lights are kept on. If the unit cost of electricity is $0.075/kWh, determine how much the campus will save a year if the lights in the classrooms and faculty offices are turned off during unoccupied periods.

5-11 The radiator of a steam heating system has a volume of 20 L and is filled with superheated vapor at 300 kPa and 250°C. At this moment both the inlet and exit valves to the radiator are closed. Determine the amount of heat that will be transferred to the room when the steam pressure drops to 100 kPa. Also, show the process on a *P-v* diagram with respect to saturation lines.   *Answer:* 33.4 kJ

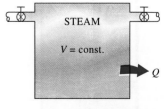

FIGURE P5-11

5-12 A 0.5-m³ rigid tank contains refrigerant-134a initially at 200 kPa and 40 percent quality. Heat is now transferred to the refrigerant until the pressure reaches 800 kPa. Determine (*a*) the mass of the refrigerant in the tank and (*b*) the amount of heat transferred. Also, show the process on a *P-v* diagram with respect to saturation lines.

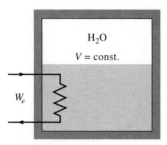

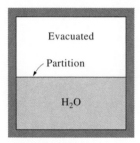

FIGURE P5-14

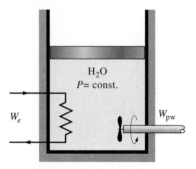

FIGURE P5-15

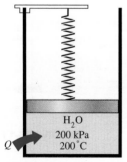

FIGURE P5-18

5-13E A 20-ft³ rigid tank initially contains refrigerant-134a in the saturated vapor form at 120 psia. As a result of heat transfer from the refrigerant, the pressure drops to 30 psia. Show the process on a *P-v* diagram with respect to saturation lines, and determine (*a*) the final temperature, (*b*) the amount of refrigerant that has condensed, and (*c*) the heat transfer.

5-14 A well-insulated rigid tank contains 5 kg of a saturated liquid–vapor mixture of water at l00 kPa. Initially, three-quarters of the mass is in the liquid phase. An electric resistor placed in the tank is connected to a 110-V source, and a current of 8 A flows through the resistor when the switch is turned on. Determine how long it will take to vaporize all the liquid in the tank. Also, show the process on a *T-v* diagram with respect to saturation lines.

5-15 An insulated tank is divided into two parts by a partition. One part of the tank contains 2.5 kg of compressed liquid water at 60°C and 600 kPa while the other part is evacuated. The partition is now removed, and the water expands to fill the entire tank. Determine the final temperature of the water and the volume of the tank for a final pressure of 10 kPa.

5-16 A piston-cylinder device contains 5 kg of refrigerant-134a at 800 kPa and 60°C. The refrigerant is now cooled at constant pressure until it exists as a liquid at 20°C. Determine the amount of heat loss and show the process on a *T-v* diagram with respect to saturation lines.   *Answer:* 1089 kJ

5-17E A piston-cylinder device contains 0.5 lbm of water initially at 120 psia and 2 ft³. Now 200 Btu of heat is transferred to the water while its pressure is held constant. Determine the final temperature of the water. Also, show the process on a *T-v* diagram with respect to saturation lines.

5-18 An insulated piston-cylinder device contains 5 L of saturated liquid water at a constant pressure of 150 kPa. Water is stirred by a paddle wheel while a current of 8 A flows for 45 min through a resistor placed in the water. If one-half of the liquid is evaporated during this constant-pressure process and the paddle-wheel work amounts to 300 kJ, determine the voltage of the source. Also, show the process on a *P-v* diagram with respect to saturation lines.
    Answer: 230.9 V

5-19 A piston-cylinder device contains steam initially at 1 MPa, 350°C, and 1.5 m³. Steam is allowed to cool at constant pressure until it first starts condensing. Show the process on a *T-v* diagram with respect to saturation lines and determine (*a*) the mass of the steam, (*b*) the final temperature, and (*c*) the amount of heat transfer.

5-20 A piston-cylinder device initially contains steam at 200 kPa, 200°C, and 0.5 m³. At this state, a linear spring ($F \propto x$) is touching the piston but exerts no force on it. Heat is now slowly transferred to the steam, causing the pressure and the volume to rise to 500 kPa and 0.6 m³, respectively. Show the process on a *P-v* diagram with respect to saturation lines and determine (*a*) the final temperature, (*b*) the work done by the steam, and (*c*) the total heat transferred.
    Answers: (*a*) 1131°C, (*b*) 35 kJ, (*c*) 807 kJ

5-21 A piston-cylinder device initially contains 0.5 m³ of saturated water vapor at 200 kPa. At this state, the piston is resting on a set of stops, and the mass of the piston is such that a pressure of 300 kPa is required to move it. Heat is now slowly transferred to the steam until the volume doubles. Show the process on a *P-v* diagram with respect to saturation lines and determine (*a*) the final temperature, (*b*) the work done during this process, and (*c*) the total heat transfer.

Answers: (*a*) 878.90°C, (*b*) 150 kJ, (*c*) 875 kJ

Closed-System Energy Balance: Ideal Gases

5-22C   Is it possible to compress an ideal gas isothermally in an adiabatic piston-cylinder device? Explain.

5-23E   A rigid tank contains 20 lbm of air at 50 psia and 80°F. The air is now heated until its pressure doubles. Determine (*a*) the volume of the tank and (*b*) the amount of heat transfer.   *Answers:* (*a*) 80 ft³, (*b*) 2035 Btu

5-24   A 1-m³ rigid tank contains hydrogen at 250 kPa and 500 K. The gas is now cooled until its temperature drops to 300 K. Determine (*a*) the final pressure in the tank and (*b*) the amount of heat transfer.

5-25   A 4-m × 5-m × 6-m room is to be heated by a baseboard resistance heater. It is desired that the resistance heater be able to raise the air temperature in the room from 7 to 23°C within 15 min. Assuming no heat losses from the room and an atmospheric pressure of 100 kPa, determine the required power of the resistance heater. Assume constant specific heats at room temperature.

Answer: 1.91 kW

5-26   A 4-m × 5-m × 7-m room is heated by the radiator of a steam-heating system. The steam radiator transfers heat at a rate of 10,000 kJ/h, and a 100-W fan is used to distribute the warm air in the room. The rate of heat loss from the room is estimated to be about 5000 kJ/h. If the initial temperature of the room air is 10°C, determine how long it will take for the air temperature to rise to 20°C. Assume constant specific heats at room temperature.

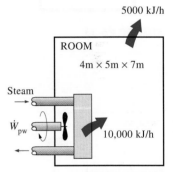

FIGURE P5-26

5-27   A student living in a 4-m × 6-m × 6-m dormitory room turns on her 150-W fan before she leaves the room on a summer day, hoping that the room will be cooler when she comes back in the evening. Assuming all the doors and windows are tightly closed and disregarding any heat transfer through the walls and the windows, determine the temperature in the room when she comes back 10 h later. Use specific heat values at room temperature, and assume the room to be at 100 kPa and 15°C in the morning when she leaves.

Answer: 58.2°C

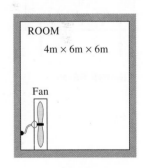

FIGURE P5-27

5-28E   A 10-ft³ tank contains oxygen initially at 14.7 psia and 80°F. A paddle wheel within the tank is rotated until the pressure inside rises to 20 psia. During the process 20 Btu of heat is lost to the surroundings. Determine the paddle-wheel work done. Neglect the energy stored in the paddle wheel.

5-29   An insulated rigid tank is divided into two equal parts by a partition. Initially, one part contains 3 kg of an ideal gas at 800 kPa and 50°C, and the other part is evacuated. The partition is now removed, and the gas expands into the entire tank. Determine the final temperature and pressure in the tank.

5-30   A piston-cylinder device whose piston is resting on top of a set of stops initially contains 0.5 kg of helium gas at 100 kPa and 25°C. The mass of the piston is such that 500 kPa of pressure is required to raise it. How much heat must be transferred to the helium before the piston starts rising?

Answer: 1857 kJ

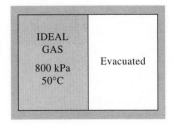

FIGURE P5-29

5-31 An insulated piston-cylinder device contains 100 L of air at 400 kPa and 25°C. A paddle wheel within the cylinder is rotated until 15 kJ of work is done on the air while the pressure is held constant. Determine the final temperature of the air. Neglect the energy stored in the paddle wheel.

5-32E A piston-cylinder device contains 25 ft³ of nitrogen at 50 psia and 700°F. Nitrogen is now allowed to cool at constant pressure until the temperature drops to 140°F. Using specific heats at the average temperature, determine the amount of heat loss.

5-33 A mass of 15 kg of air in a piston-cylinder device is heated from 25 to 77°C by passing current through a resistance heater inside the cylinder. The pressure inside the cylinder is held constant at 300 kPa during the process, and a heat loss of 60 kJ occurs. Determine the electric energy supplied, in kWh.
   Answer: 0.235 kWh

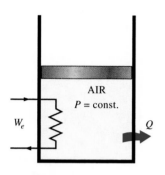

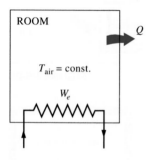

FIGURE P5-33

5-34 An insulated piston-cylinder device initially contains 0.3 m³ of carbon dioxide at 200 kPa and 27°C. An electric switch is turned on, and a 110-V source supplies current to a resistance heater inside the cylinder for a period of 10 min. The pressure is held constant during the process, while the volume is doubled. Determine the current that passes through the resistance heater.

5-35 A piston-cylinder device contains 0.8 kg of nitrogen initially at 100 kPa and 27°C. The nitrogen is now compressed slowly in a polytropic process during which $PV^{1.3}$ = constant until the volume is reduced by one-half. Determine the work done and the heat transfer for this process.

5-36 A room is heated by a baseboard resistance heater. When the heat losses from the room on a winter day amount to 8000 kJ/h, the air temperature in the room remains constant even though the heater operates continuously. Determine the power rating of the heater, in kW.

5-37E A piston-cylinder device contains 3 ft³ of air at 60 psia and 150°F. Heat is transferred to the air in the amount of 40 Btu as the air expands isothermally. Determine the amount of boundary work done during this process.

FIGURE P5-36

5-38 A piston-cylinder device contains 5 kg of argon at 400 kPa and 30°C. During a quasi-equilibrium, isothermal expansion process, 15 kJ of boundary work is done by the system, and 3 kJ of paddle-wheel work is done on the system. Determine the heat transfer for this process.   *Answer:* 12 kJ

5-39 A piston-cylinder device, whose piston is resting on a set of stops initially contains 3 kg of air at 200 kPa and 27°C. The mass of the piston is such that a pressure of 400 kPa is required to move it. Heat is now transferred to the air until its volume doubles. Determine the work done by the air and the total heat transferred to the air during this process. Also show the process on a P-v diagram.
   Answers: 516 kJ, 2674 kJ

5-40 A piston-cylinder device, with a set of stops on the top, initially contains 3 kg of air at 200 kPa and 27°C. Heat is now transferred to the air, and the piston rises until it hits the stops, at which point the volume is twice the initial volume. More heat is transferred until the pressure inside the cylinder also doubles. Determine the work done and the amount of heat transfer for this process. Also, show the process on a P-v diagram.

5-41 In a manufacturing facility, 5-cm-diameter brass balls ($\rho = 8522$ kg/m^3 and $C_p = 0.385$ kJ/kg $\cdot$ °C) initially at 120°C are quenched in a water bath at 50°C for a period of 2 minutes at a rate of 100 balls per minute. If the temperature of the balls after quenching is 74°C, determine the rate at which heat needs to be removed from the water in order to keep its temperature constant at 50°C.

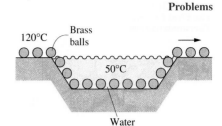

FIGURE P5-41

5-42 Repeat Prob. 5-41 for aluminum balls.

5-43E During a picnic on a hot summer day, all the cold drinks disappeared quickly, and the only available drinks were those at the ambient temperature of 75°F. In an effort to cool a 12-fluid-oz drink in a can, a person grabs the can and starts shaking it in the iced water of the chest at 32°F. Using the properties of water for the drink, determine the mass of ice that will melt by the time the canned drink cools to 45°F.

5-44 Consider a 1000-W iron whose base plate is made of 0.5-cm-thick aluminum alloy 2024-T6 ($\rho = 2770$ kg/m^3 and $C_p = 875$ J/kg $\cdot$ °C). The base plate has a surface area of 0.03 m^2. Initially, the iron is in thermal equilibrium with the ambient air at 22°C. Assuming 85 percent of the heat generated in the resistance wires is transferred to the plate, determine the minimum time needed for the plate temperature to reach 140°C.

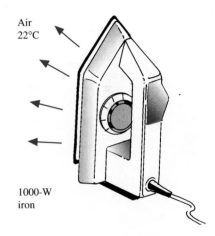

FIGURE P5-44

5-45 Stainless steel ball bearings ($\rho = 8085$ kg/m^3 and $C_p = 0.480$ kJ/kg $\cdot$ °C) having a diameter of 1.2 cm are to be quenched in water at a rate of 1400 per minute. The balls leave the oven at a uniform temperature of 900°C and are exposed to air at 30°C for a while before they are dropped into the water. If the temperature of the balls drops to 850°C prior to quenching, determine the rate of heat transfer from the balls to the air.

5-46 Carbon steel balls ($\rho = 7833$ kg/m^3 and $C_p = 0.465$ kJ/kg $\cdot$ °C) 8 mm in diameter are annealed by heating them first to 900°C in a furnace, and then allowing them to cool slowly to 100°C in ambient air at 35°C. If 2500 balls are to be annealed per hour, determine the total rate of heat transfer from the balls to the ambient air. *Answer: 542 W*

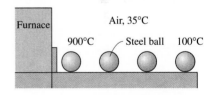

FIGURE P5-46

5-47 An electronic device dissipating 30 W has a mass of 20 g and a specific heat of 850 J/kg $\cdot$ °C. The device is lightly used, and it is on for 5 min and then off for several hours, during which it cools to the ambient temperature of 25°C. Determine the highest possible temperature of the device at the end of the 5-min operating period. What would your answer be if the device were attached to a 0.2-kg aluminum heat sink? Assume the device and the heat sink to be nearly isothermal.

5-48 An ordinary egg can be approximated as a 5.5-cm-diameter sphere. The egg is initially at a uniform temperature of 8°C and is dropped into boiling water at 97°C. Taking the properties of the egg to be $\rho = 1020$ kg/m^3 and $C_p = 3.32$ kJ/kg $\cdot$ °C, determine how much heat is transferred to the egg by the time the average temperature of the egg rises to 70°C.

5-49E In a production facility, 1.2-in.-thick 2-ft × 2-ft square brass plates ($\rho = 532.5$ lbm/ft^3 and $C_p = 0.091$ Btu/lbm $\cdot$ °F) that are initially at a uniform temperature of 75°F are heated by passing them through an oven at 1300°F at a rate of 300 per minute. If the plates remain in the oven until their average

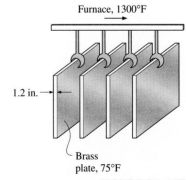

FIGURE P5-49E

temperature rises to 1000°F, determine the rate of heat transfer to the plates in the furnace.

5-50 Long cylindrical steel rods ($\rho = 7833$ kg/m^3 and $C_p = 0.465$ kJ/kg · °C) of 10-cm diameter are heat-treated by drawing them at a velocity of 3 m/min through an oven maintained at 900°C. If the rods enter the oven at 30°C and leave at a mean temperature of 700°C, determine the rate of heat transfer to the rods in the oven.

5-51 Fresh strawberries with a water content of 88 percent (by mass) at 30°C are stored in 0.8-kg boxes made of nylon ($C_p = 1.7$ kJ/kg · °C). Each box contains 23 kg of strawberries, and the strawberries are to be cooled to an average temperature of 4°C at a rate of 60 boxes per hour. Taking the average specific heat of the strawberries to be $C_p = 3.89$ kJ/kg · °C and the average rate of heat of respiration to be 210 mW/kg, determine the rate of heat removal from the strawberries and their boxes, in kJ/h. What would be the percent error involved if the strawberry boxes were ignored in the calculations?

5-52 Lettuce is to be vacuum cooled from the environment temperature of 24°C to a temperature of 2°C in 45 min in a 4-m outer-diameter insulated spherical vacuum chamber whose walls consist of 3-cm-thick urethane insulation sandwiched between thin metal plates. The vacuum chamber contains 5000 kg of lettuce when loaded. Disregarding any heat transfer through the walls of the vacuum chamber, determine (*a*) the final pressure in the vacuum chamber and (*b*) the amount of moisture removed from the lettuce, in kg.
   Answers: (*a*) 0.714 kPa, (*b*) 179 kg

5-53 A supply of 40 kg of shrimp at 8°C contained in a box is to be frozen to −18°C in a freezer. Determine the amount of heat that needs to be removed. The latent heat of the shrimp is 277 kJ/kg and its specific heat is 3.62 kJ/kg · °C above freezing and 1.89 kJ/kg · °C below freezing. The container box is 1.2 kg and is made up of polyethylene, whose specific heat is 2.3 kJ/kg · °C. Also, the freezing temperature of shrimp is −2.2°C.

5-54 Repeat Prob. 5-53 by disregarding the container box.

Steady-Flow Energy Balance: Nozzles and Diffusers

5-55C How is a steady-flow system characterized?

5-56C Can a steady-flow system involve boundary work?

5-57C A diffuser is an adiabatic device that decreases the kinetic energy of the fluid by slowing it down. What happens to this *lost* kinetic energy?

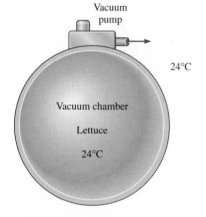

FIGURE P5-52

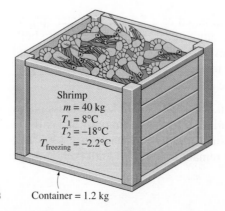

FIGURE P5-53

5-58C The kinetic energy of a fluid increases as it is accelerated in an adiabatic nozzle. Where does this energy come from?

5-59C Is heat transfer to or from the fluid desirable as it flows through a nozzle? How will heat transfer affect the fluid velocity at the nozzle exit?

5-60 Air enters an adiabatic nozzle steadily at 300 kPa, 200°C, and 30 m/s and leaves at 100 kPa and 180 m/s. The inlet area of the nozzle is 80 cm². Determine (a) the mass flow rate through the nozzle, (b) the exit temperature of the air, and (c) the exit area of the nozzle.
     Answers: (a) 0.5304 kg/s, (b) 184.60°C, (c) 38.7 cm²

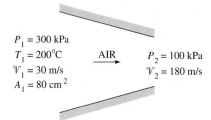

FIGURE P5-60

5-61 Steam at 5 MPa and 500°C enters a nozzle steadily with a velocity of 80 m/s, and it leaves at 2 MPa and 400°C. The inlet area of the nozzle is 50 cm², and heat is being lost at a rate of 90 kJ/s. Determine (a) the mass flow rate of the steam, (b) the exit velocity of the steam, and (c) the exit area of the nozzle.

5-62E Air enters a nozzle steadily at 50 psia, 140°F, and 150 ft/s and leaves at 14.7 psia and 900 ft/s. The heat loss from the nozzle is estimated to be 6.5 Btu/lbm of air flowing. The inlet area of the nozzle is 0.1 ft². Determine (a) the exit temperature of air and (b) the exit area of the nozzle.
     Answers: (a) 508 R, (b) 0.048 ft²

5-63 Steam at 3 MPa and 400°C enters an adiabatic nozzle steadily with a velocity of 40 m/s and leaves at 2.5 MPa and 300 m/s. Determine (a) the exit temperature and (b) the ratio of the inlet to exit area A_1/A_2.

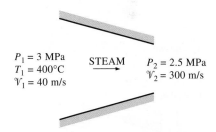

FIGURE P5-63

5-64 Air at 600 kPa and 500 K enters an adiabatic nozzle that has an inlet-to-exit area ratio of 2:1 with a velocity of 120 m/s and leaves with a velocity of 380 m/s. Determine (a) the exit temperature and (b) the exit pressure of the air.   *Answers:* (a) 436.5 K, (b) 330.8 kPa

5-65 Air at 80 kPa and 127°C enters an adiabatic diffuser steadily at a rate of 6000 kg/h and leaves at 100 kPa. The velocity of the airstream is decreased from 230 to 30 m/s as it passes through the diffuser. Find (a) the exit temperature of the air and (b) the exit area of the diffuser.

5-66E Air at 13 psia and 20°F enters an adiabatic diffuser steadily with a velocity of 600 ft/s and leaves with a low velocity at a pressure of 14.5 psia. The exit area of the diffuser is 5 times the inlet area. Determine (a) the exit temperature and (b) the exit velocity of the air.

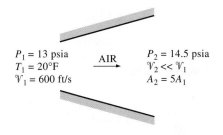

FIGURE P5-66E

Turbines and Compressors

5-67C Consider an adiabatic turbine operating steadily. Does the work output of the turbine have to be equal to the decrease in the energy of the steam flowing through it?

5-68C Consider an air compressor operating steadily. How would you compare the volume flow rates of the air at the compressor inlet and exit?

5-69C Will the temperature of air rise as it is compressed by an adiabatic compressor? Why?

5-70C Somebody proposes the following system to cool a house in the summer: Compress the regular outdoor air, let it cool back to the outdoor temperature, pass it through a turbine, and discharge the cold air leaving the turbine

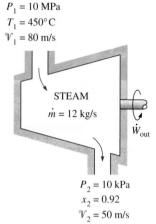

$P_1 = 10$ MPa
$T_1 = 450°C$
$\mathcal{V}_1 = 80$ m/s

STEAM
$\dot{m} = 12$ kg/s

$\dot{W}_{out}$

$P_2 = 10$ kPa
$x_2 = 0.92$
$\mathcal{V}_2 = 50$ m/s

FIGURE P5-71

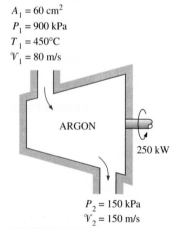

$A_1 = 60$ cm^2
$P_1 = 900$ kPa
$T_1 = 450°C$
$\mathcal{V}_1 = 80$ m/s

ARGON

250 kW

$P_2 = 150$ kPa
$\mathcal{V}_2 = 150$ m/s

FIGURE P5-75

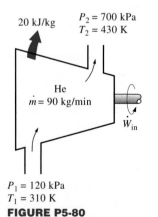

20 kJ/kg
$P_2 = 700$ kPa
$T_2 = 430$ K

He
$\dot{m} = 90$ kg/min

$\dot{W}_{in}$

$P_1 = 120$ kPa
$T_1 = 310$ K
FIGURE P5-80

into the house. From a thermodynamic point of view, is the proposed system sound?

5-71 Steam flows steadily through an adiabatic turbine. The inlet conditions of the steam are 10 MPa, 450°C, and 80 m/s, and the exit conditions are 10 kPa, 92 percent quality, and 50 m/s. The mass flow rate of the steam is 12 kg/s. Determine (*a*) the change in kinetic energy, (*b*) the power output, and (*c*) the turbine inlet area.
Answers: (*a*) -1.95 kJ/kg, (*b*) 10.2 MW, (*c*) 0.00446 m^2

5-72 Steam enters an adiabatic turbine at 10 MPa and 400°C and leaves at 20 kPa with a quality of 90 percent. Neglecting the changes in kinetic and potential energies, determine the mass flow rate required for a power output of 5 MW. *Answer:* 6.919 kg/s

5-73E Steam flows steadily through a turbine at a rate of 45,000 lbm/h, entering at 1000 psia and 900°F and leaving at 5 psia as saturated vapor. If the power generated by the turbine is 4 MW, determine the rate of heat loss from the steam.

5-74 Steam enters an adiabatic turbine at 10 MPa and 500°C at a rate of 3 kg/s and leaves at 20 kPa. If the power output of the turbine is 2 MW, determine the temperature of the steam at the turbine exit. Neglect kinetic energy changes. *Answer:* 110.8°C

5-75 Argon gas enters steadily an adiabatic turbine at 900 kPa and 450°C with a velocity of 80 m/s and leaves at 150 kPa with a velocity of 150 m/s. The inlet area of the turbine is 60 cm^2. If the power output of the turbine is 250 kW, determine the exit temperature of the argon.

5-76E Air flows steadily through an adiabatic turbine, entering at 150 psia, 900°F, and 350 ft/s and leaving at 20 psia, 300°F, and 700 ft/s. The inlet area of the turbine is 0.1 ft^2. Determine (*a*) the mass flow rate of the air and (*b*) the power output of the turbine.

5-77 Refrigerant-134a enters an adiabatic compressor as saturated vapor at $-20°C$ and leaves at 0.7 MPa and 70°C. The mass flow rate of the refrigerant is 1.2 kg/s. Determine (*a*) the power input to the compressor and (*b*) the volume flow rate of the refrigerant at the compressor inlet.

5-78 Air enters the compressor of a gas-turbine plant at ambient conditions of 100 kPa and 25°C with a low velocity and exits at 1 MPa and 347°C with a velocity of 90 m/s. The compressor is cooled at a rate of 1500 kJ/min, and the power input to the compressor is 250 kW. Determine the mass flow rate of air through the compressor.

5-79E Air is compressed from 14.7 psia and 60°F to a pressure of 150 psia while being cooled at a rate of 10 Btu/lbm by circulating water through the compressor casing. The volume flow rate of the air at the inlet conditions is 5000 ft^3/min, and the power input to the compressor is 700 hp. Determine (*a*) the mass flow rate of the air and (*b*) the temperature at the compressor exit.
Answers: (*a*) 6.36 lbm/s, (*b*) 781 R

5-80 Helium is to be compressed from 120 kPa and 310 K to 700 kPa and 430 K. A heat loss of 20 kJ/kg occurs during the compression process. Neglecting kinetic energy changes, determine the power input required for a mass flow rate of 90 kg/min.

Throttling Valves

5-81C Why are throttling devices commonly used in refrigeration and air-conditioning applications?

5-82C During a throttling process, the temperature of a fluid drops from 30 to −20°C. Can this process occur adiabatically?

5-83C Would you expect the temperature of air to drop as it undergoes a steady-flow throttling process?

5-84C Would you expect the temperature of a liquid to change as it is throttled? How?

5-85 Refrigerant-134a is throttled from the saturated liquid state at 800 kPa to a pressure of 140 kPa. Determine the temperature drop during this process and the final specific volume of the refrigerant.
      Answers: 50.1°C, 0.0454 m³/kg

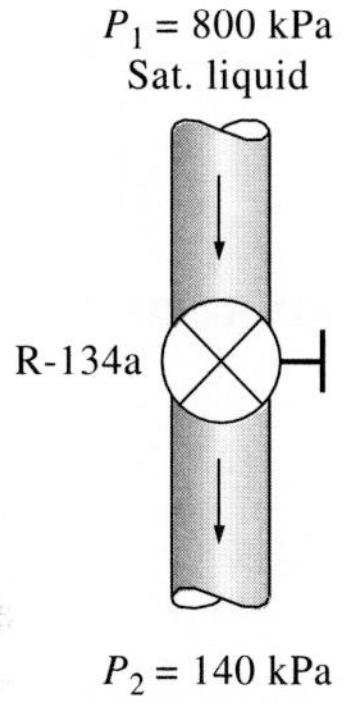

5-86 Refrigerant-134a at 800 kPa and 25°C is throttled to a temperature of −20°C. Determine the pressure and the internal energy of the refrigerant at the final state.   *Answers:* 133 kPa, 78.8 kJ/kg

5-87 A well-insulated valve is used to throttle steam from 8 MPa and 500°C to 6 MPa. Determine the final temperature of the steam.
      Answer: 490.1°C

5-88E Air at 200 psia and 90°F is throttled to the atmospheric pressure of 14.7 psia. Determine the final temperature of the air.

$P_1 = 800$ kPa
Sat. liquid

R-134a

$P_2 = 140$ kPa
FIGURE P5-85

Mixing Chambers and Heat Exchangers

5-89C When two fluid streams are mixed in a mixing chamber, can the mixture temperature be lower than the temperature of both streams? Explain.

5-90C Consider a steady-flow mixing process. Under what conditions will the energy transported into the control volume by the incoming streams be equal to the energy transported out of it by the outgoing stream?

5-91C Consider a steady-flow heat exchanger involving two different fluid streams. Under what conditions will the amount of heat lost by one fluid be equal to the amount of heat gained by the other?

5-92 A hot-water stream at 80°C enters a mixing chamber with a mass flow rate of 0.5 kg/s where it is mixed with a stream of cold water at 20°C. If it is desired that the mixture leave the chamber at 42°C, determine the mass flow rate of the cold-water stream. Assume all the streams are at a pressure of 250 kPa.   *Answer:* 0.864 kg/s

5-93 Liquid water at 300 kPa and 20°C is heated in a chamber by mixing it with superheated steam at 300 kPa and 300°C. Cold water enters the chamber at a rate of 1.8 kg/s. If the mixture leaves the mixing chamber at 60°C, determine the mass flow rate of the superheated steam required.
      Answer: 0.107 kg/s

5-94 In steam power plants, open feedwater heaters are frequently utilized to heat the feedwater by mixing it with steam bled off the turbine at some intermediate stage. Consider an open feedwater heater that operates at a pressure of 800 kPa. Feedwater at 50°C and 800 kPa is to be heated with superheated steam at 200°C and 800 kPa. In an ideal feedwater heater, the

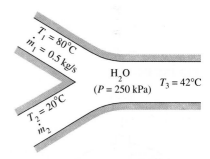

$T_1 = 80°C$
$\dot{m}_1 = 0.5$ kg/s

H_2O
$(P = 250$ kPa$)$ $T_3 = 42°C$

$T_2 = 20°C$
$\dot{m}_2$

FIGURE P5-92

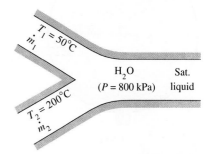

$T_1 = 50°C$
$\dot{m}_1$

H_2O Sat.
$(P = 800$ kPa$)$ liquid

$T_2 = 200°C$
$\dot{m}_2$

FIGURE P5-94

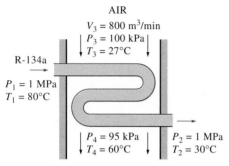

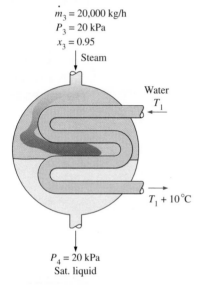

FIGURE P5-97

FIGURE P5-101

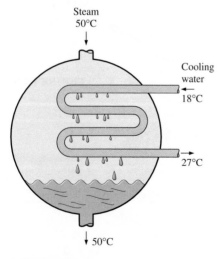

FIGURE P5-102

mixture leaves the heater as saturated liquid at the feedwater pressure. Determine the ratio of the mass flow rates of the feedwater and the superheated vapor for this case.　*Answer:* 4.14

5-95E　Water at 50°F and 50 psia is heated in a chamber by mixing it with saturated water vapor at 50 psia. If both streams enter the mixing chamber at the same mass flow rate, determine the temperature and the quality of the exiting stream.　*Answers:* 281°F, 0.374

5-96　A stream of refrigerant-134a at 1 MPa and 12°C is mixed with another stream at 1 MPa and 60°C. If the mass flow rate of the cold stream is twice that of the hot one, determine the temperature and the quality of the exit stream.

5-97　Refrigerant-134a at 1 MPa and 80°C is to be cooled to 1 MPa and 30°C in a condenser by air. The air enters at 100 kPa and 27°C with a volume flow rate of 800 m³/min and leaves at 95 kPa and 60°C. Determine the mass flow rate of the refrigerant.　*Answer:* 139 kg/min

5-98E　Air enters the evaporator section of a window air conditioner at 14.7 psia and 90°F with a volume flow rate of 200 ft³/min. Refrigerant-134a at 20 psia with a quality of 30 percent enters the evaporator at a rate of 4 lbm/min and leaves as saturated vapor at the same pressure. Determine (*a*) the exit temperature of the air and (*b*) the rate of heat transfer from the air.

5-99　Refrigerant-134a at 800 kPa, 70°C, and 8 kg/min is cooled by water in a condenser until it exists as a saturated liquid at the same pressure. The cooling water enters the condenser at 300 kPa and 15°C and leaves at 30°C at the same pressure. Determine the mass flow rate of the cooling water required to cool the refrigerant.　*Answer:* 27.0 kg/min

5-100E　In a steam heating system, air is heated by being passed over some tubes through which steam flows steadily. Steam enters the heat exchanger at 30 psia and 400°F at a rate of 15 lbm/min and leaves at 25 psia and 212°F. Air enters at 14.7 psia and 80°F and leaves at 130°F. Determine the volume flow rate of air at the inlet.

5-101　Steam enters the condenser of a steam power plant at 20 kPa and a quality of 95 percent with a mass flow rate of 20,000 kg/h. It is to be cooled by water from a nearby river by circulating the water through the tubes within the condenser. To prevent thermal pollution, the river water is not allowed to experience a temperature rise above 10°C. If the steam is to leave the condenser as saturated liquid at 20 kPa, determine the mass flow rate of the cooling water required.　*Answer:* 17,866 kg/min

5-102　Steam is to be condensed in the condenser of a steam power plant at a temperature of 50°C (h_{fg} = 2305 kJ/kg) with cooling water (C_p = 4.20 kJ/kg · °C) from a nearby lake, which enters the tubes of the condenser at 18°C at a rate of 101 kg/s and leaves at 27°C. Determine the rate of condensation of the steam in the condenser.　*Answer:* 1.66 kg/s

5-103　A heat exchanger is to heat water (C_p = 4.18 kJ/kg · °C) from 25°C to 60°C at a rate of 0.2 kg/s. The heating is to be accomplished by geothermal water (C_p = 4.31 kJ/kg · °C) available at 140°C at a mass flow rate of 0.3 kg/s. The inner tube is thin-walled and has a diameter of 0.8 cm. Determine the rate of heat transfer in the heat exchanger and the exit temperature of geothermal water.

5-104 A heat exchanger is to cool ethylene glycol ($C_p = 2.56$ kJ/kg · °C) flowing at a rate of 2 kg/s from 80°C to 40°C by water ($C_p = 4.18$ kJ/kg · °C) that enters at 20°C and leaves at 55°C. Determine (a) the rate of heat transfer and (b) the mass flow rate of water.

5-105 A thin-walled double-pipe counter-flow heat exchanger is to be used to cool oil ($C_p = 2.20$ kJ/kg · °C) from 150°C to 40°C at a rate of 2 kg/s by water ($C_p = 4.18$ kJ/kg · °C) that enters at 22°C at a rate of 1.5 kg/s. Determine the rate of heat transfer in the heat exchanger and the exit temperature of water.

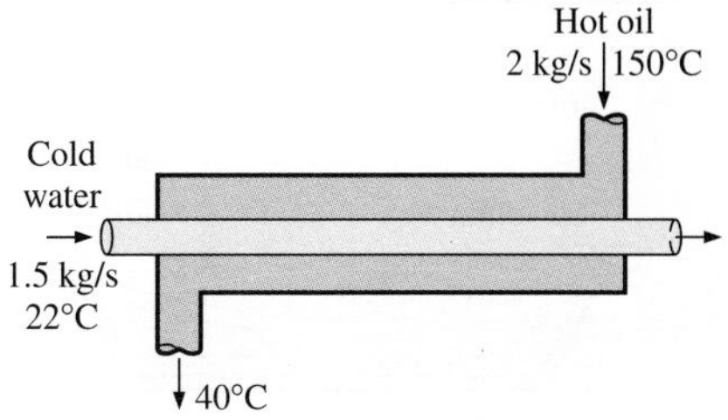

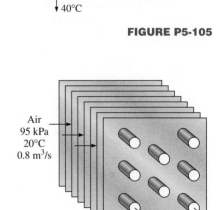

FIGURE P5-105

5-106 Cold water ($C_p = 4.18$ kJ/kg · °C) leading to a shower enters a thin-walled double-pipe counter-flow heat exchanger at 15°C at a rate of 0.25 kg/s and is heated to 45°C by hot water ($C_p = 4.19$ kJ/kg · °C) that enters at 100°C at a rate of 3 kg/s. Determine the rate of heat transfer in the heat exchanger and the exit temperature of the hot water.

5-107 Air ($C_p = 1.005$ kJ/kg · °C) is to be preheated by hot exhaust gases in a cross-flow heat exchanger before it enters the furnace. Air enters the heat exchanger at 95 kPa and 20°C at a rate of 0.8 m³/s. The combustion gases ($C_p = 1.10$ kJ/kg · °C) enter at 180°C at a rate of 1.1 kg/s and leave at 95°C. Determine the rate of heat transfer to the air and its outlet temperature.

5-108 A well-insulated shell-and-tube heat exchanger is used to heat water ($C_p = 4.18$ kJ/kg · °C) in the tubes from 20°C to 70°C at a rate of 4.5 kg/s. Heat is supplied by hot oil ($C_p = 2.30$ kJ/kg · °C) that enters the shell side at 170°C at a rate of 10 kg/s. Determine the rate of heat transfer in the heat exchanger and the exit temperature of oil.

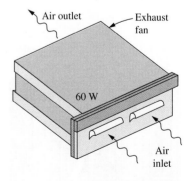

FIGURE P5-107

5-109E Steam is to be condensed on the shell side of a heat exchanger at 90°F ($h_{fg} = 1043$ Btu/lbm). Cooling water ($C_p = 1.0$ Btu/lbm · °F) enters the tubes at 60°F at a rate of 115.3 lbm/s and leaves at 73°F. Assuming the heat exchanger to be well-insulated, determine the rate of heat transfer in the heat exchanger and the rate of condensation of the steam.

Pipe and Duct Flow

5-110 A desktop computer is to be cooled by a fan. The electronic components of the computer consume 60 W of power under full load conditions. The computer is to operate in environments at temperatures up to 45°C and at elevations up to 3400 m where the average atmospheric pressure is 66.63 kPa. The exit temperature of air is not to exceed 60°C to meet the reliability requirements. Also, the average velocity of air is not to exceed 110 m/min at the exit of the computer case where the fan is installed to keep the noise level down. Determine the flow rate of the fan that needs to be installed and the diameter of the casing of the fan.

5-111 Repeat Prob. 5-110 for a computer that consumes 100 W of power.

5-112E Water enters the tubes of a cold plate at 95°F with an average velocity of 60 ft/min and leaves at 105°F. The diameter of the tubes is 0.25 in. Assuming 15 percent of the heat generated is dissipated from the components to the surroundings by convection and radiation, and the remaining 85 percent is removed by the cooling water, determine the amount of heat generated by the electronic devices mounted on the cold plate. *Answer:* 263 W

FIGURE P5-110

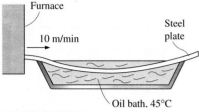

FIGURE P5-115

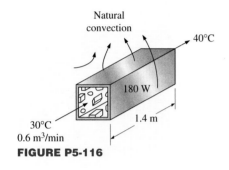

FIGURE P5-116

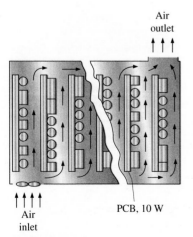

FIGURE P5-120

5-113 A sealed electronic box is to be cooled by tap water flowing through the channels on two of its sides. It is specified that the temperature rise of the water not exceed 4°C. The power dissipation of the box is 2 kW, which is removed entirely by water. If the box operates 24 hours a day, 365 days a year, determine the mass flow rate of water flowing through the box and the amount of cooling water used per year.

5-114 Repeat Prob. 5-113 for a power dissipation of 3 kW.

5-115 A long roll of 2-m-wide and 0.5-cm-thick 1-Mn manganese steel plate ($\rho = 785$ kg/m^3 and $C_p = 0.454$ kJ/kg · °C) coming off a furnace at 820°C is to be quenched in an oil bath at 45°C to a temperature of 51.1°C. If the metal sheet is moving at a steady velocity of 10 m/min, determine the required rate of heat removal from the oil to keep its temperature constant at 45°C.
Answer: 437 kW

5-116 The components of an electronic system dissipating 180 W are located in a 1.4-m-long horizontal duct whose cross section is 20 cm × 20 cm. The components in the duct are cooled by forced air that enters the duct at 30°C and 1 atm at a rate of 0.6 m^3/min and leaves at 40°C. Determine the rate of heat transfer from the outer surfaces of the duct to the ambient.
Answer: 63 W

5-117 Repeat Prob. 5-116 for a circular horizontal duct of diameter 10 cm.

5-118E The hot water needs of a household are to be met by heating water at 55°F to 200°F by a parabolic solar collector at a rate of 4 lbm/s. Water flows through a 1.25-in.-diameter thin aluminum tube whose outer surface is black-anodized in order to maximize its solar absorption ability. The centerline of the tube coincides with the focal line of the collector, and a glass sleeve is placed outside the tube to minimize the heat losses. If solar energy is transferred to water at a net rate of 350 Btu/h per ft length of the tube, determine the required length of the parabolic collector to meet the hot water requirements of this house.

5-119 Consider a hollow-core printed circuit board 12 cm high and 18 cm long, dissipating a total of 20 W. The width of the air gap in the middle of the PCB is 0.25 cm. If the cooling air enters the 12-cm-wide core at 32°C at a rate of 0.8 L/s, determine the average temperature at which the air leaves the hollow core. *Answer:* 53.4°C

5-120 A computer cooled by a fan contains eight PCBs, each dissipating 10 W power. The height of the PCBs is 12 cm and the length is 18 cm. The cooling air is supplied by a 25-W fan mounted at the inlet. If the temperature rise of air as it flows through the case of the computer is not to exceed 10°C, determine (*a*) the flow rate of the air that the fan needs to deliver and (*b*) the fraction of the temperature rise of air that is due to the heat generated by the fan and its motor. *Answers:* (*a*) 0.0104 kg/s, (*b*) 24 percent

5-121 Hot water at 90°C enters a 15-m section of a cast iron pipe whose inner diameter is 4 cm at an average velocity of 0.8 m/s. The outer surface of the pipe is exposed to the cold air at 10°C in a basement. If water leaves the basement at 88°C, determine the rate of heat loss from the water.

5-122 A 5-m × 6-m × 8-m room is to be heated by an electric resistance heater placed in a short duct in the room. Initially, the room is at 15°C, and the local atmospheric pressure is 98 kPa. The room is losing heat steadily to the outside at a rate of 200 kJ/min. A 200-W fan circulates the air steadily through the duct and the electric heater at an average mass flow rate of 50 kg/min. The duct can be assumed to be adiabatic, and there is no air leaking in or out of the room. If it takes 15 min for the room air to reach an average temperature of 25°C, find (a) the power rating of the electric heater and (b) the temperature rise that the air experiences each time it passes through the heater.

5-123 A house has an electric heating system that consists of a 300-W fan and an electric resistance heating element placed in a duct. Air flows steadily through the duct at a rate of 0.6 kg/s and experiences a temperature rise of 5°C. The rate of heat loss from the air in the duct is estimated to be 400 W. Determine the power rating of the electric resistance heating element.
  Answer: 3.12 kW

5-124 A hair dryer is basically a duct in which a few layers of electric resistors are placed. A small fan pulls the air in and forces it through the resistors where it is heated. Air enters a 1200-W hair dryer at 100 kPa and 22°C and leaves at 47°C. The cross-sectional area of the hair dryer at the exit is 60 cm². Neglecting the power consumed by the fan and the heat losses through the walls of the hair dryer, determine (a) the volume flow rate of air at the inlet and (b) the velocity of the air at the exit.
  Answers: (a) 0.0404 m³/s, (b) 7.31 m/s

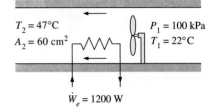

FIGURE P5-124

5-125 The ducts of an air heating system pass through an unheated area. As a result of heat losses, the temperature of the air in the duct drops by 4°C. If the mass flow rate of air is 120 kg/min, determine the rate of heat loss from the air to the cold environment.

5-126E Air enters the duct of an air-conditioning system at 15 psia and 50°F at a volume flow rate of 450 ft³/min. The diameter of the duct is 10 in., and heat is transferred to the air in the duct from the surroundings at a rate of 2 Btu/s. Determine (a) the velocity of the air at the duct inlet and (b) the temperature of the air at the exit.

5-127 Water is heated in an insulated, constant-diameter tube by a 7-kW electric resistance heater. If the water enters the heater steadily at 15°C and leaves at 70°C, determine the mass flow rate of water.

5-128 Steam enters a long, horizontal pipe with an inlet diameter of $D_1 = 12$ cm at 1 MPa and 250°C with a velocity of 2 m/s. Further downstream, the conditions are 800 kPa and 200°C, and the diameter is $D_2 = 10$ cm. Determine (a) the mass flow rate of the steam and (b) the rate of heat transfer.
  Answers: (a) 0.0972 kg/s, (b) 10.04 kJ/s

Energy Balance for Charging and Discharging Processes

5-129 Consider a 5-L evacuated rigid bottle that is surrounded by the atmosphere at 100 kPa and 17°C. A valve at the neck of the bottle is now opened and the atmospheric air is allowed to flow into the bottle. The air trapped in the bottle eventually reaches thermal equilibrium with the atmosphere as a result of heat transfer through the wall of the bottle. The valve remains open during the process so that the trapped air also reaches mechanical equilibrium with the atmosphere. Determine the net heat transfer through the wall of the bottle during this filling process.   *Answer:* $Q_{out} = 0.5$ kJ

5-130 An insulated rigid tank is initially evacuated. A valve is opened, and atmospheric air at 95 kPa and 17°C enters the tank until the pressure in the

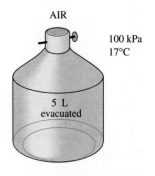

FIGURE P5-129

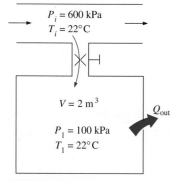

$P_i = 600$ kPa
$T_i = 22°C$

$V = 2$ m³

Q_{out}

$P_1 = 100$ kPa
$T_1 = 22°C$

FIGURE P5-131

tank reaches 95 kPa, at which point the valve is closed. Determine the final temperature of the air in the tank. Assume constant specific heats.
Answer: 406 K

5-131 A 2-m³ rigid tank initially contains air at 100 kPa and 22°C. The tank is connected to a supply line through a valve. Air is flowing in the supply line at 600 kPa and 22°C. The valve is opened, and air is allowed to enter the tank until the pressure in the tank reaches the line pressure, at which point the valve is closed. A thermometer placed in the tank indicates that the air temperature at the final state is 77°C. Determine (*a*) the mass of air that has entered the tank and (*b*) the amount of heat transfer.
Answers: (*a*) 9.58 kg, (*b*) Q_{out} = 339.4 kJ

5-132 A 0.2-m³ rigid tank initially contains refrigerant-134a at 8°C. At this state, 60 percent of the mass is in the vapor phase, and the rest is in the liquid phase. The tank is connected by a valve to a supply line where refrigerant at 1 MPa and 120°C flows steadily. Now the valve is opened slightly, and the refrigerant is allowed to enter the tank. When the pressure in the tank reaches 800 kPa, the entire refrigerant in the tank exists in the vapor phase only. At this point the valve is closed. Determine (*a*) the final temperature in the tank, (*b*) the mass of refrigerant that has entered the tank, and (*c*) the heat transfer between the system and the surroundings.

5-133E A 4-ft³ rigid tank initially contains saturated water vapor at 250°F. The tank is connected by a valve to a supply line that carries steam at 160 psia and 400°F. Now the valve is opened, and steam is allowed to enter the tank. Heat transfer takes place with the surroundings such that the temperature in the tank remains constant at 250°F at all times. The valve is closed when it is observed that one-half of the volume of the tank is occupied by liquid water. Find (*a*) the final pressure in the tank, (*b*) the amount of steam that has entered the tank, and (*c*) the amount of heat transfer.
Answers: (*a*) 29.82 psia, (*b*) 117.5 lbm, (*c*) 117,539 Btu

5-134 A vertical piston-cylinder device initially contains 0.01 m³ of steam at 200°C. The mass of the frictionless piston is such that it maintains a constant pressure of 500 kPa inside. Now steam at 1 MPa and 350°C is allowed to enter the cylinder from a supply line until the volume inside doubles. Neglecting any heat transfer that may have taken place during the process, determine (*a*) the final temperature of the steam in the cylinder and (*b*) the amount of mass that has entered.   *Answers:* (*a*) 262.6°C, (*b*) 0.0176 kg

5-135 An insulated, vertical piston-cylinder device initially contains 10 kg of water, 8 kg of which is in the vapor phase. The mass of the piston is such that it maintains a constant pressure of 300 kPa inside the cylinder. Now steam at 0.5 MPa and 350°C is allowed to enter the cylinder from a supply line until all the liquid in the cylinder has vaporized. Determine (*a*) the final temperature in the cylinder and (*b*) the mass of the steam that has entered.
Answers: (*a*) 133.6°C, (*b*) 9.78 kg

5-136 A 0.1-m³ rigid tank initially contains refrigerant-134a at 1 MPa and 100 percent quality. The tank is connected by a valve to a supply line that carries refrigerant at 1.2 MPa and 30°C. Now the valve is opened, and the refrigerant is allowed to enter the tank. The valve is closed when it is observed that the tank contains saturated liquid at 1.2 MPa. Determine (*a*) the mass of the refrigerant that has entered the tank and (*b*) the amount of heat transfer.
Answers: (*a*) 107.1 kg, (*b*) 1825 kJ

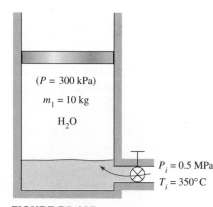

$(P = 300$ kPa$)$

$m_1 = 10$ kg

H₂O

$P_i = 0.5$ MPa
$T_i = 350°C$

FIGURE P5-135

5-137 A 0.3-m³ rigid tank is filled with saturated liquid water at 200°C. A valve at the bottom of the tank is opened, and liquid is withdrawn from the tank. Heat is transferred to the water such that the temperature in the tank remains constant. Determine the amount of heat that must be transferred by the time one-half of the total mass has been withdrawn.

5-138 A 0.1-m³ rigid tank contains saturated refrigerant-134a at 800 kPa. Initially, 40 percent of the volume is occupied by liquid and the rest by vapor. A valve at the bottom of the tank is now opened, and liquid is withdrawn from the tank. Heat is transferred to the refrigerant such that the pressure inside the tank remains constant. The valve is closed when no liquid is left in the tank and vapor starts to come out. Determine the total heat transfer for this process. *Answer:* 267.6 kJ

5-139E A 4-ft³ rigid tank contains saturated refrigerant-134a at 100 psia. Initially, 20 percent of the volume is occupied by liquid and the rest by vapor. A valve at the top of the tank is now opened, and vapor is allowed to escape slowly from the tank. Heat is transferred to the refrigerant such that the pressure inside the tank remains constant. The valve is closed when the last drop of liquid in the tank is vaporized. Determine the total heat transfer for this process.

5-140 A 0.2-m³ rigid tank equipped with a pressure regulator contains steam at 2 MPa and 300°C. The steam in the tank is now heated. The regulator keeps the steam pressure constant by letting out some steam, but the temperature inside rises. Determine the amount of heat transferred when the steam temperature reaches 500°C.

5-141 A 4-L pressure cooker has an operating pressure of 175 kPa. Initially, one-half of the volume is filled with liquid and the other half with vapor. If it is desired that the pressure cooker not run out of liquid water for 1 h, determine the highest rate of heat transfer allowed.

5-142 An insulated 0.08-m³ tank contains helium at 2 MPa and 80°C. A valve is now opened, allowing some helium to escape. The valve is closed when one-half of the initial mass has escaped. Determine the final temperature and pressure in the tank. *Answers:* 225 K, 637 kPa

5-143E An insulated 60-ft³ rigid tank contains air at 75 psia and 120°F. A valve connected to the tank is now opened, and air is allowed to escape until the pressure inside drops to 30 psia. The air temperature during this process is maintained constant by an electric resistance heater placed in the tank. Determine the electrical work done during this process.

5-144 A vertical piston-cylinder device initially contains 0.2 m³ of air at 20°C. The mass of the piston is such that it maintains a constant pressure of 300 kPa inside. Now a valve connected to the cylinder is opened, and air is allowed to escape until the volume inside the cylinder is decreased by one-half. Heat transfer takes place during the process so that the temperature of the air in the cylinder remains constant. Determine (*a*) the amount of air that has left the cylinder and (*b*) the amount of heat transfer. *Answers:* (*a*) 0.357 kg, (*b*) 0

Review Problems

5-145 A mass of 12 kg of saturated refrigerant-134a vapor is contained in a piston-cylinder device at 200 kPa. Now 250 kJ of heat is transferred to the

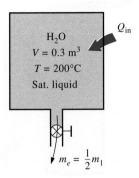

FIGURE P5-137

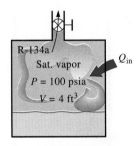

FIGURE P5-139E

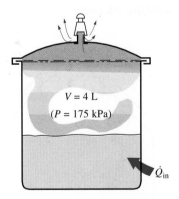

FIGURE P5-141

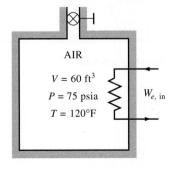

FIGURE P5-143E

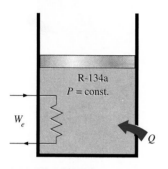

FIGURE P5-145

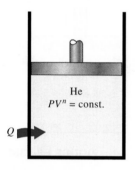

FIGURE P5-147

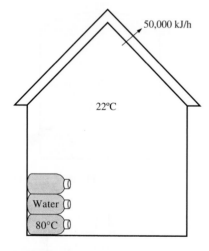

FIGURE P5-149

refrigerant at constant pressure while a 110-V source supplies current to a resistor within the cylinder for 6 min. Determine the current supplied if the final temperature is 70°C. Also, show the process on a *T-v* diagram with respect to the saturation lines. *Answer:* 15.7 A

5-146 A mass of 0.2 kg of saturated refrigerant-134a is contained in a piston-cylinder device at 200 kPa. Initially, 75 percent of the mass is in the liquid phase. Now heat is transferred to the refrigerant at constant pressure until the cylinder contains vapor only. Show the process on a *P-v* diagram with respect to saturation lines. Determine (*a*) the volume occupied by the refrigerant initially, (*b*) the work done, and (*c*) the total heat transfer.

5-147 A piston-cylinder device contains helium gas initially at 150 kPa, 20°C, and 0.5 m³. The helium is now compressed in a polytropic process (PV^n = constant) to 400 kPa and 140°C. Determine the heat loss or gain during this process. *Answer:* 11.2 kJ loss

5-148 A frictionless piston-cylinder device and a rigid tank initially contain 12 kg of an ideal gas each at the same temperature, pressure, and volume. It is desired to raise the temperatures of both systems by 15°C. Determine the amount of extra heat that must be supplied to the gas in the cylinder which is maintained at constant pressure to achieve this result. Assume the molar mass of the gas is 25.

5-149 A passive solar house that is losing heat to the outdoors at an average rate of 50,000 kJ/h is maintained at 22°C at all times during a winter night for 10 h. The house is to be heated by 50 glass containers each containing 20 L of water that is heated to 80°C during the day by absorbing solar energy. A thermostat controlled 15-kW back-up electric resistance heater turns on whenever necessary to keep the house at 22°C. (*a*) How long did the electric heating system run that night? (*b*) How long would the electric heater run that night if the house incorporated no solar heating?
Answers: (*a*) 4.77 h, (*b*) 9.26 h

5-150 A 50-cm-long, 800-W electric resistance heating element whose diameter is 0.5 cm and surface temperature 120°C is immersed in 40 kg of water initially at 20°C. Determine how long it will take for this heater to raise the water temperature to 80°C.

5-151 One ton (1000 kg) of liquid water at 80°C is brought into a well-insulated and well-sealed 4-m × 5-m × 6-m room initially at 22°C and 100 kPa. Assuming constant specific heats for both air and water at room temperature, determine the final equilibrium temperature in the room.
Answer: 78.6°C

5-152 A 4-m × 5-m × 6-m room is to be heated by one ton (1000 kg) of liquid water contained in a tank that is placed in the room. The room is losing heat to the outside at an average rate of 10,000 kJ/h. The room is initially at 20°C and 100 kPa and is maintained at an average temperature of 20°C at all times. If the hot water is to meet the heating requirements of this room for a 24-h period, determine the minimum temperature of the water when it is first brought into the room. Assume constant specific heats for both air and water at room temperature.

5-153 The energy content of a certain food is to be determined in a bomb calorimeter that contains 3 kg of water by burning a 2-g sample of it in the presence of 100 g of air in the reaction chamber. If the water temperature rises by 3.2°C when equilibrium is established, determine the energy content of the food, in kJ/kg, by neglecting the thermal energy stored in the reaction chamber and the energy supplied by the mixer. What is a rough estimate of the error involved in neglecting the thermal energy stored in the reaction chamber?
Answer: 20,083 kJ/kg

5-154 A 68-kg man whose average body temperature is 38°C drinks 1 L of cold water at 3°C in an effort to cool down. Taking the average specific heat of the human body to be 3.6 kJ/kg · °C, determine the drop in the average body temperature of this person under the influence of this cold water.

5-155 A 0.2-L glass of water at 20°C is to be cooled with ice to 5°C. Determine how much ice needs to be added to the water, in grams, if the ice is at (*a*) 0°C and (*b*) −8°C. Also determine how much water would be needed if the cooling is to be done with cold water at 0°C. The melting temperature and the heat of fusion of ice at atmospheric pressure are 0°C and 333.7 kJ/kg, respectively, and the density of water is 1 kg/L.

5-156 In order to cool 1 ton (1000 kg) of water at 20°C in an insulated tank, a person pours 80 kg of ice at −5°C into the water. Determine the final equilibrium temperature in the tank. The melting temperature and the heat of fusion of ice at atmospheric pressure are 0°C and 333.7 kJ/kg, respectively.
Answer: 12.4°C

5-157 An insulated piston-cylinder device initially contains 0.01 m³ of saturated liquid–vapor mixture with a quality of 0.2 at 100°C. Now some ice at 0°C is added to the cylinder. If the cylinder contains saturated liquid at 100°C when thermal equilibrium is established, determine the amount of ice added. The melting temperature and the heat of fusion of ice at atmospheric pressure are 0°C and 333.7 kJ/kg, respectively.

5-158 The early steam engines were driven by the atmospheric pressure acting on the piston fitted into a cylinder filled with saturated steam. A vacuum was created in the cylinder by cooling the cylinder externally with cold water, and thus condensing the steam.

Consider a piston-cylinder device with a piston surface area of 0.1 m² initially filled with 0.05 m³ of saturated water vapor at the atmospheric pressure of 100 kPa. Now cold water is poured outside the cylinder, and the steam inside starts condensing as a result of heat transfer to the cooling water outside. If the piston is stuck at its initial position, determine the friction force acting on the piston and the amount of heat transfer when the temperature inside the cylinder drops to 30°C.

5-159 A 1.6-kg box made of polypropylene (C_p = 1.9 kJ/kg · °C) contains 32 kg of haddock fish with a water content of 83 percent (by mass) at 16°C. The fish is to be frozen to an average temperature of −20°C in 4 h in its box. The specific heat of fish is 3.62 kJ/kg · °C above the freezing temperature of −2.2°C, and 1.89 kJ/kg · °C below the freezing temperature. The latent heat of the fish is 277 kJ/kg. Determine (*a*) the total amount of heat that must be removed from the fish and (*b*) the average rate of heat removal from the fish.

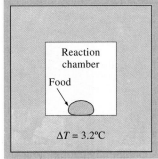

FIGURE P5-153

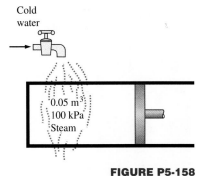

Cold
water

FIGURE P5-158

FIGURE P5-160

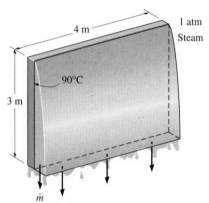

FIGURE P5-164

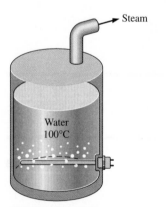

FIGURE P5-165

5-160 Water is boiled at sea level in a coffee maker equipped with an immersion-type electric heating element. The coffee maker contains 1 L of water when full. Once boiling starts, it is observed that half of the water in the coffee maker evaporates in 25 min. Determine the power rating of the electric heating element immersed in water. Also, determine how long it will take for this heater to raise the temperature of 1 L of cold water from 18°C to the boiling temperature.

5-161 In a gas-fired boiler, water is boiled at 150°C by hot gases flowing through a stainless steel pipe submerged in water. If the rate of heat transfer from the hot gases to water is 74 kJ/s, determine the rate of evaporation of water.

5-162 Cold water enters a steam generator at 20°C and leaves as saturated vapor at 100°C. Determine the fraction of heat used in the steam generator to preheat the liquid water from 20°C to the saturation temperature of 100°C.

5-163 Cold water enters a steam generator at 20°C and leaves as saturated vapor at the boiler pressure. At what pressure will the amount of heat needed to preheat the water to saturation temperature be equal to the heat needed to vaporize the liquid at the boiler pressure?

5-164 Saturated steam at 1 atm condenses on a 3-m-high, 4-m-wide vertical plate that is maintained at 90°C by circulating cooling water through the other side. If the rate of heat transfer by condensation to the plate is 180 kJ/s, determine the rate at which the condensate drips off the plate at the bottom and the average heat flux on the plate.

5-165 Water is boiled at 100°C electrically by a 50-cm-long, 2-mm-diameter, 5-kW resistance wire. Determine (a) the rate of evaporation of water and (b) the heat flux at the surface of the wire.

5-166 Consider a well-insulated piston-cylinder device that contains 4 kg of liquid water and 1 kg of water vapor at 120°C and is maintained at constant pressure. Now a 5-kg copper block at 30°C is dropped into the cylinder. Determine the equilibrium temperature inside the cylinder once thermal equilibrium is established, and the mass of the water vapor at the final state.

5-167 The gage pressure of an automobile tire is measured to be 200 kPa before a trip and 220 kPa after the trip at a location where the atmospheric pressure is 90 kPa. Assuming the volume of the tire remains constant and the tire is initially at 25°C, determine the temperature rise of air in the tire during the trip.

5-168 Consider two identical buildings: one in Los Angeles, California, where the atmospheric pressure is 101 kPa and the other in Denver, Colorado, where the atmospheric pressure is 83 kPa. Both buildings are maintained at 21°C, and the infiltration rate for both buildings is 1.2 air changes per hour (ACH). That is, the entire air in the building is replaced completely by the outdoor air 1.2 times per hour on a day when the outdoor temperature at both locations is 10°C. Disregarding latent heat, determine the ratio of the heat losses by infiltration at the two cities.

5-169 The ventilating fan of the bathroom of a building has a volume flow rate of 30 L/s and runs continuously. The building is located in San Francisco, California, where the average winter temperature is 12.2°C, and is maintained at 22°C at all times. The building is heated by electricity whose unit cost is $0.09/kWh. Determine the amount and cost of the heat "vented out" per month in winter.

5-170 Consider a large classroom on a hot summer day with 150 students, each dissipating 60 W of sensible heat. All the lights, with 4.0 kW of rated power, are kept on. The room has no external walls, and thus heat gain through the walls and the roof is negligible. Chilled air is available at 15°C, and the temperature of the return air is not to exceed 25°C. Determine the required flow rate of air, in kg/s, that needs to be supplied to the room to keep the average temperature of the room constant.   *Answer:*   1.29 kg/s

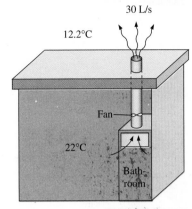

FIGURE P5-169

5-171 A typical full-carlot-capacity banana room contains 36 pallets of bananas. Each pallet consists of 24 boxes, and thus the room stores 864 boxes of bananas. A box holds an average of 19 kg of bananas and is made of 2.3 kg of fiberboard. The specific heats of banana and the fiberboard are 3.55 kJ/kg · °C and 1.7 kJ/kg · °C, respectively. The peak heat of respiration of bananas is 0.3 W/kg. The bananas are cooled at a rate of 0.4°C/h. The rate of heat gain through the walls and other surfaces of the room is estimated to be 1800 kJ/h. If the temperature rise of refrigerated air is not to exceed 2.0°C as it flows thorough the room, determine the minimum flow rate of air needed. Take the density and specific heat of air to be 1.2 kg/m³ and 1.0 kJ/kg · °C, respectively.

5-172 Chickens with an average mass of 2.2 kg and average specific heat of 3.54 kJ/kg · °C are to be cooled by chilled water that enters a continuous-flow-type immersion chiller at 0.5°C. Chickens are dropped into the chiller at a uniform temperature of 15°C at a rate of 500 chickens per hour and are cooled to an average temperature of 3°C before they are taken out. The chiller gains heat from the surroundings at a rate of 200 kJ/h. Determine (*a*) the rate of heat removal from the chickens, in kW, and (*b*) the mass flow rate of water, in kg/s, if the temperature rise of water is not to exceed 2°C.

5-173 In a dairy plant, milk at 4°C is pasteurized continuously at 72°C at a rate of 12 L/s for 24 hours a day and 365 days a year. The milk is heated to the pasteurizing temperature by hot water heated in a natural-gas-fired boiler that has an efficiency of 82 percent. The pasteurized milk is then cooled by cold water at 18°C before it is finally refrigerated back to 4°C. To save energy and money, the plant installs a regenerator that has an effectiveness of 82 percent. If the cost of natural gas is $0.52/therm (1 therm = 105,500 kJ), determine how much energy and money the regenerator will save this company per year.

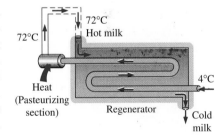

FIGURE P5-173

5-174E A refrigeration system is being designed to cool eggs (ρ = 67.4 lbm/ft³ and C_p = 0.80 Btu/lbm · °F) with an average mass of 0.14 lbm from an initial temperature of 90°F to a final average temperature of 50°F by air at 34°F at a rate of 10,000 eggs per hour. Determine (*a*) the rate of heat removal from the eggs, in Btu/h and (*b*) the required volume flow rate of air, in ft³/h, if the temperature rise of air is not to exceed 10°F.

5-175 The heat of hydration of dough, which is 15 kJ/kg, will raise its temperature to undesirable levels unless some cooling mechanism is utilized. A practical way of absorbing the heat of hydration is to use refrigerated water when kneading the dough. If a recipe calls for mixing 2 kg of flour with 1 kg of water, and the temperature of the city water is 15°C, determine the temperature to which the city water must be cooled before mixing in order for the water to absorb the entire heat of hydration when the water temperature

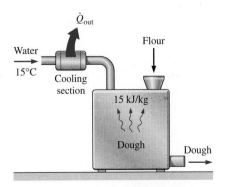

FIGURE P5-175

rises to 15°C. Take the specific heats of the flour and the water to be 1.76 and 4.18 kJ/kg · °C, respectively. *Answer:* 4.2°C

5-176 A glass bottle washing facility uses a well-agitated hot water bath at 55°C that is placed on the ground. The bottles enter at a rate of 800 per minute at an ambient temperature of 20°C and leave at the water temperature. Each bottle has a mass of 150 g and removes 0.2 g of water as it leaves the bath wet. Make-up water is supplied at 15°C. Disregarding any heat losses from the outer surfaces of the bath, determine the rate at which (*a*) water and (*b*) heat must be supplied to maintain steady operation.

5-177 Repeat Prob. 5-176 for a water bath temperature of 50°C.

5-178 Long aluminum wires of diameter 3 mm ($\rho = 2702$ kg/m^3 and $C_p = 0.896$ kJ/kg · °C are extruded at a temperature of 350°C and are cooled to 50°C in atmospheric air at 30°C. If the wire is extruded at a velocity of 10 m/min, determine the rate of heat transfer from the wire to the extrusion room.

5-179 Repeat Prob. 5-178 for a copper wire ($\rho = 8950$ kg/m^3 and $C_p = 0.383$ kJ/kg · °C).

5-180 Steam at 40°C condenses on the outside of a 5-mm-long, 3-cm-diameter thin horizontal copper tube by cooling water that enters the tube at 25°C at an average velocity of 2 m/s and leaves at 35°C. Determine the rate of condensation of steam. *Answer:* 0.0245 kg/s

5-181E The condenser of a steam power plant operates at a pressure of 0.95 psia. The condenser consists of 144 horizontal tubes arranged in a 12 × 12 square array. Steam condenses on the outer surfaces of the tubes whose inner and outer diameters are 1 in. and 1.2 in., respectively. If steam is to be condensed at a rate of 6800 lbm/h and the temperature rise of the cooling water is limited to 8°F, determine (*a*) the rate of heat transfer from the steam to the cooling water and (*b*) the average velocity of the cooling water through the tubes.

5-182 Saturated refrigerant-134a vapor at 30°C is to be condensed as it flows in a 1-cm-diameter horizontal tube at a rate of 0.1 kg/min. Determine the rate of heat transfer from the refrigerant. What would your answer be if the condensed refrigerant is cooled to 20°C?

5-183E The average atmospheric pressure in Spokane, Washington (elevation = 2350 ft), is 13.5 psia, and the average winter temperature is 36.5°F. The pressurization test of a 9-ft-high, 3000-ft^2 older home revealed that the seasonal average infiltration rate of the house is 2.2 air changes per hour (ACH). That is, the entire air in the house is replaced completely 2.2 times per hour by the outdoor air. It is suggested that the infiltration rate of the house can be reduced by half to 1.1 ACH by winterizing the doors and the windows. If the house is heated by natural gas whose unit cost is $0.62/therm and the heating season can be taken to be six months, determine how much the home owner will save from the heating costs per year by this winterization project. Assume the house is maintained at 72°F at all times and the efficiency of the furnace is 0.65. Also assume the latent heat load during the heating season to be negligible.

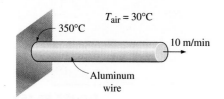

FIGURE P5-178

FIGURE P5-180

5-184 Determine the rate of sensible heat loss from a building due to infiltration if the outdoor air at $-10°C$ and 90 kPa enters the building at a rate of 35 L/s when the indoors is maintained at 22°C.

5-185 The maximum flow rate of standard shower heads is about 3.5 gpm (13.3 L/min) and can be reduced to 2.75 gpm (10.5 L/min) by switching to low-flow shower heads that are equipped with flow controllers. Consider a family of four, with each person taking a five-minute shower every morning. City water at 15°C is heated to 55°C in an electric water heater and tempered to 42°C by cold water at the T-elbow of the shower before being routed to the shower heads. Assuming a constant specific heat of 4.18 kJ/kg · °C for water, determine (*a*) the ratio of the flow rates of the hot and cold water as they enter the T-elbow and (*b*) the amount of electricity that will be saved per year, in kWh, by replacing the standard shower heads by the low-flow ones.

5-186 A fan is powered by a 0.5-hp motor and delivers air at a rate of 130 m³/min. Determine the highest value for the average velocity of air mobilized by the fan. Take the density of air to be 1.18 kg/m³.

5-187 An air-conditioning system requires airflow at the main supply duct at a rate of 180 m³/min. The average velocity of air in the circular duct is not to exceed 10 m/s to avoid excessive vibration and pressure drops. Assuming the fan converts 70 percent of the electrical energy it consumes into kinetic energy of air, determine the size of the electric motor needed to drive the fan and the diameter of the main duct. Take the density of air to be 1.20 kg/m³.

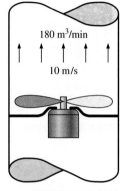

FIGURE P5-187

5-188 Consider an evacuated rigid bottle of volume *V* that is surrounded by the atmosphere at pressure P_0 and temperature T_0. A valve at the neck of the bottle is now opened and the atmospheric air is allowed to flow into the bottle. The air trapped in the bottle eventually reaches thermal equilibrium with the atmosphere as a result of heat transfer through the wall of the bottle. The valve remains open during the process so that the trapped air also reaches mechanical equilibrium with the atmosphere. Determine the net heat transfer through the wall of the bottle during this filling process in terms of the properties of the system and the surrounding atmosphere.

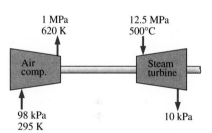

FIGURE P5-189

5-189 An adiabatic air compressor is to be powered by a direct-coupled adiabatic steam turbine that is also driving a generator. Steam enters the turbine at 12.5 MPa and 500°C at a rate of 25 kg/s and exits at 10 kPa and a quality of 0.92. Air enters the compressor at 98 kPa and 295 K at a rate of 10 kg/s and exits at 1 MPa and 620 K. Determine the net power delivered to the generator by the turbine.

5-190 Water flows through a shower head steadily at a rate of 10 L/min. An electric resistance heater placed in the water pipe heats the water from 16°C to 43°C. Taking the density of water to be 1 kg/L, determine the electric power input to the heater, in kW.

In an effort to conserve energy, it is proposed to pass the drained warm water at a temperature of 39°C through a heat exchanger to preheat the incoming cold water. If the heat exchanger has an effectiveness of 0.50 (that is, it recovers only half of the energy that can possibly be transferred from the drained water to incoming cold water), determine the electric power input required in this case. If the price of the electric energy is 8.5 ¢/kWh, determine how much money is saved during a 10-min shower as a result of installing this heat exchanger.

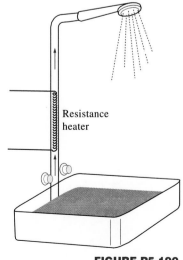

FIGURE P5-190

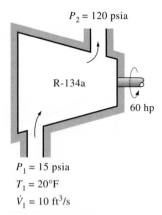

$P_2 = 120$ psia

R-134a

60 hp

$P_1 = 15$ psia
$T_1 = 20°F$
$\dot{V}_1 = 10$ ft³/s

FIGURE P5-192E

5-191 Steam enters a turbine steadily at 10 MPa and 550°C with a velocity of 60 m/s and leaves at 25 kPa with a quality of 95 percent. A heat loss of 30 kJ/kg occurs during the process. The inlet area of the turbine is 150 cm², and the exit area is 1400 cm². Determine (a) the mass flow rate of the steam, (b) the exit velocity, and (c) the power output.

5-192E Refrigerant-134a enters an adiabatic compressor at 15 psia and 20°F with a volume flow rate of 10 ft³/s and leaves at a pressure of 120 psia. The power input to the compressor is 60 hp. Find (a) the mass flow rate of the refrigerant and (b) the exit temperature.

5-193 In large gas-turbine power plants, air is preheated by the exhaust gases in a heat exchanger called the *regenerator* before it enters the combustion chamber. Air enters the regenerator at 1 MPa and 550 K at a mass flow rate of 800 kg/min. Heat is transferred to the air at a rate of 3200 kJ/s. Exhaust gases enter the regenerator at 140 kPa and 800 K and leave at 130 kPa and 600 K. Treating the exhaust gases as air, determine (a) the exit temperature of the air and (b) the mass flow rate of exhaust gases.
Answers: (a) 775 K, (b) 14.9 kg/s

5-194 It is proposed to have a water heater that consists of an insulated pipe of 5-cm diameter and an electric resistor inside. Cold water at 15°C enters the heating section steadily at a rate of 30 L/min. If water is to be heated to 50°C, determine (a) the power rating of the resistance heater and (b) the average velocity of the water in the pipe.

Computer, Design, and Essay Problems

5-195 You are asked to design a heating system for a swimming pool that is 2 m deep, 25 m long, and 25 m wide. Your client desires that the heating system be large enough to raise the water temperature from 20°C to 30°C in 3 h. The rate of heat loss from the water to the air at the outdoor design conditions is determined to be 960 W/m², and the heater must also be able to maintain the pool at 30°C at those conditions. Heat losses to the ground are expected to be small and can be disregarded. The heater considered is a natural gas furnace whose efficiency is 80 percent. What heater size (in Btu/h input) would you recommend to your client?

5-196 A 1982 U.S. Department of Energy article (FS #204) states that a leak of one drip of hot water per second can cost $1.00 per month. Making reasonable assumptions about the drop size and the unit cost of energy, determine if this claim is reasonable.

5-197 Using a thermometer and a tape measure only, explain how you can determine the average velocity of air at the exit of your hair dryer at its highest power setting.

5-198 Design a 1200-W electric hair dryer such that the air temperature and velocity in the dryer will not exceed 50°C and 3 m/s, respectively.

5-199 Design an electric hot water heater for a family of four in your area. The maximum water temperature in the tank and the power consumption are not to exceed 60°C and 4 kW, respectively. There are two showers in the house, and the flow rate of water through each of the shower heads is about 10 L/min. Each family member takes a 5-min shower every morning. Explain why a hot water tank is necessary, and determine the proper size of the tank for this family.

5-200 A manufacturing facility requires saturated steam at 120°C at a rate of 1.2 kg/min. Design an electric steam boiler for this purpose under the following constraints:

- The boiler will be in cylindrical shape with a height-to-diameter ratio of 1.5. The boiler can be horizontal or vertical.
- A commercially available plug-in type electrical heating element made of mechanically polished stainless steel will be used. The diameter of the heater can be between 0.5 cm and 3 cm. Also, the heat flux at the surface of the heater cannot exceed 150 kW/m^2.
- Half of the volume of the boiler should be occupied by steam, and the boiler should be large enough to hold enough water for a 2-h supply of steam. Also, the boiler will be well-insulated.

You are to specify the following: (1) The height and inner diameter of the tank, (2) the length, diameter, power rating, and surface temperature of the electric heating element, and (3) the maximum rate of steam production during short periods (less than 30 min) of overload conditions, and how it can be accomplished.

5-201 Design a scalding unit for slaughtered chicken to loosen their feathers before they are routed to feather-picking machines with a capacity of 1200 chickens per hour under the following conditions:

The unit will be of an immersion type filled with hot water at an average temperature of 53°C at all times. Chicken with an average mass of 2.2 kg and an average temperature of 36°C will be dipped into the tank, held in the water for 1.5 min, and taken out by a slow-moving conveyor. The chicken is expected to leave the tank 15 percent heavier as a result of the water that sticks to its surface. The center-to-center distance between chickens in any direction will be at least 30 cm. The tank can be as wide as 3 m and as high as 60 cm. The water is to be circulated through and heated by a natural gas furnace, but the temperature rise of water will not exceed 5°C as it passes through the furnace. The water loss is to be made up by the city water at an average temperature of 16°C. The walls and the floor of the tank are well-insulated. The unit operates 24 hours a day and 6 days a week. Assuming reasonable values for the average properties, recommend reasonable values for (a) the mass flow rate of the make-up water that must be supplied to the tank, (b) the rate of heat transfer from the water to the chicken, in kW, (c) the size of the heating system in kJ/h, and (d) the operating cost of the scalding unit per month for a unit cost of $0.56/therm of natural gas (1 therm = 105,500 kJ).

The Second Law of Thermodynamics

To this point, we have focused our attention on the first law of thermodynamics, which requires that energy be conserved during a process. In this chapter, we introduce the second law of thermodynamics, which asserts that processes occur in a certain direction and that energy has quality as well as quantity. A process cannot take place unless it satisfies both the first and second laws of thermodynamics. In this chapter, the thermal energy reservoirs, reversible and irreversible processes, heat engines, refrigerators, and heat pumps are introduced first. Various statements of the second law are followed by a discussion of perpetual-motion machines and the absolute thermodynamic temperature scale. The Carnot cycle is introduced next, and the Carnot principles, idealized Carnot heat engines, refrigerators, and heat pumps are examined. Finally, energy conservation associated with the use of household refrigerators is discussed.

FIGURE 6-1

A cup of hot coffee does not get hotter in a cooler room.

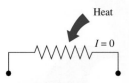

FIGURE 6-2

Transferring heat to a wire will not generate electricity.

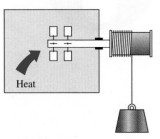

FIGURE 6-3

Transferring heat to a paddle wheel will not cause it to rotate.

FIGURE 6-4

Processes occur in a certain direction, and not in the reverse direction.

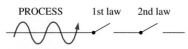

FIGURE 6-5

A process must satisfy both the first and second laws of thermodynamics to proceed.

6-1 ■ INTRODUCTION TO THE SECOND LAW

In the preceding chapter, we applied the *first law of thermodynamics,* or the *conservation of energy principle,* to processes involving closed and open systems. As pointed out repeatedly in that chapter, energy is a conserved property, and no process is known to have taken place in violation of the first law of thermodynamics. Therefore, it is reasonable to conclude that a process must satisfy the first law to occur. However, as explained below, satisfying the first law alone does not ensure that the process will actually take place.

It is common experience that a cup of hot coffee left in a cooler room eventually cools off (Fig. 6-1). This process satisfies the first law of thermodynamics since the amount of energy lost by the coffee is equal to the amount gained by the surrounding air. Now let us consider the reverse process—the hot coffee getting even hotter in a cooler room as a result of heat transfer from the room air. We all know that this process never takes place. Yet, doing so would not violate the first law as long as the amount of energy lost by the air is equal to the amount gained by the coffee.

As another familiar example, consider the heating of a room by the passage of current through an electric resistor (Fig. 6-2). Again, the first law dictates that the amount of electric energy supplied to the resistance wires be equal to the amount of energy transferred to the room air as heat. Now let us attempt to reverse this process. It will come as no surprise that transferring some heat to the wires will not cause an equivalent amount of electric energy to be generated in the wires, even though doing so would not violate the first law.

Finally, consider a paddle-wheel mechanism that is operated by the fall of a mass (Fig. 6-3). The paddle wheel rotates as the mass falls and stirs a fluid within an insulated container. As a result, the potential energy of the mass decreases, and the internal energy of the fluid increases in accordance with the conservation of energy principle. However, the reverse process, raising the mass by transferring heat from the fluid to the paddle wheel, does not occur in nature, although doing so would not violate the first law of thermodynamics.

It is clear from the above arguments that processes proceed in a *certain direction* and not in the reverse direction (Fig. 6-4). The first law places no restriction on the direction of a process, but satisfying the first law does not ensure that that process will actually occur. This inadequacy of the first law to identify whether a process can take place is remedied by introducing another general principle, the *second law of thermodynamics.* We show later in this chapter that the reverse processes discussed above violate the second law of thermodynamics. This violation is easily detected with the help of a property, called *entropy,* defined in the next chapter. *A process will not occur unless it satisfies both the first and the second laws of thermodynamics* (Fig. 6-5).

There are numerous valid statements of the second law of thermodynamics. Two such statements are presented and discussed later in this chapter in relation to some engineering devices that operate on cycles.

The use of the second law of thermodynamics is not limited to identifying the direction of processes, however. The second law also asserts that energy has *quality* as well as quantity. The first law is concerned with the quantity of energy and the transformations of energy from one form to another with no regard to its quality. Preserving the quality of energy is a major concern to engineers, and the second law provides the necessary means to determine the quality as well as the degree of degradation of energy during a process. As discussed later in this chapter, more of high-temperature energy can be converted to work, and thus it has a higher quality than the same amount of energy at a lower temperature.

The second law of thermodynamics is also used in determining the *theoretical limits* for the performance of commonly used engineering systems, such as heat engines and refrigerators, as well as predicting the *degree of completion* of chemical reactions.

6-2 ■ THERMAL ENERGY RESERVOIRS

In the development of the second law of thermodynamics, it is very convenient to have a hypothetical body with a relatively large *thermal energy capacity* (mass × specific heat) that can supply or absorb finite amounts of heat without undergoing any change in temperature. Such a body is called a **thermal energy reservoir,** or just a reservoir. In practice, large bodies of water such as oceans, lakes, and rivers as well as the atmospheric air can be modeled accurately as thermal energy reservoirs because of their large thermal energy storage capabilities or thermal masses (Fig. 6-6). The *atmosphere,* for example, does not warm up as a result of heat losses from residential buildings in winter. Likewise, megajoules of waste energy dumped in large rivers by power plants do not cause any significant change in water temperature.

A *two-phase system* can be modeled as a reservoir also since it can absorb and release large quantities of heat while remaining at constant temperature. Another familiar example of a thermal energy reservoir is the *industrial furnace.* The temperatures of most furnaces are carefully controlled, and they are capable of supplying large quantities of thermal energy as heat in an essentially isothermal manner. Therefore, they can be modeled as reservoirs.

A body does not actually have to be very large to be considered a reservoir. Any physical body whose thermal energy capacity is large relative to the amount of energy it supplies or absorbs can be modeled as one. The air in a room, for example, can be treated as a reservoir in the analysis of the heat dissipation from a TV set in the room, since the amount of heat transfer from the TV set to the room air is not large enough to have a noticeable effect on the room air temperature.

A reservoir that supplies energy in the form of heat is called a **source,** and one that absorbs energy in the form of heat is called a **sink** (Fig. 6-7). Thermal energy reservoirs are often referred to as **heat reservoirs** since they supply or absorb energy in the form of heat.

Heat transfer from industrial sources to the environment is of major concern to environmentalists as well as to engineers. Irresponsible management of waste energy can significantly increase the temperature of portions of the environment, causing what is called *thermal pollution.* If it is not carefully controlled, thermal pollution can seriously disrupt marine life in lakes and rivers. However, by careful design and management, the waste energy dumped into large bodies of water can be used to improve the quality of marine life by keeping the local temperature increases within safe and desirable levels.

6-3 ■ HEAT ENGINES

As pointed out earlier, work can easily be converted to other forms of energy, but converting other forms of energy to work is not that easy. The mechanical work done by the shaft shown in Fig. 6-8, for example, is first converted to the internal energy of the water. This energy may then leave the water as heat. We

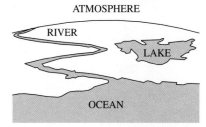

FIGURE 6-6

Bodies with relatively large thermal masses can be modeled as thermal energy reservoirs.

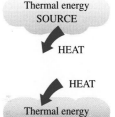

FIGURE 6-7

A source supplies energy in the form of heat, and a sink absorbs it.

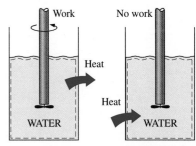

FIGURE 6-8

Work can always be converted to heat directly and completely, but the reverse is not true.

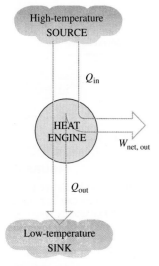

FIGURE 6-9

Part of the heat received by a heat engine is converted to work, while the rest is rejected to a sink.

know from experience that any attempt to reverse this process will fail. That is, transferring heat to the water will not cause the shaft to rotate. From this and other observations, we conclude that work can be converted to heat directly and completely, but converting heat to work requires the use of some special devices. These devices are called **heat engines.**

Heat engines differ considerably from one another, but all can be characterized by the following (Fig. 6-9):

1. They receive heat from a high-temperature source (solar energy, oil furnace, nuclear reactor, etc.).

2. They convert part of this heat to work (usually in the form of a rotating shaft).

3. They reject the remaining waste heat to a low-temperature sink (the atmosphere, rivers, etc.).

4. They operate on a cycle.

Heat engines and other cyclic devices usually involve a fluid to and from which heat is transferred while undergoing a cycle. This fluid is called the **working fluid.**

The term *heat engine* is often used in a broader sense to include work-producing devices that do not operate in a thermodynamic cycle. Engines that involve internal combustion such as gas turbines and car engines fall into this category. These devices operate in a mechanical cycle but not in a thermodynamic cycle since the working fluid (the combustion gases) does not undergo a complete cycle. Instead of being cooled to the initial temperature, the exhaust gases are purged and replaced by fresh air-and-fuel mixture at the end of the cycle.

The work-producing device that best fits into the definition of a heat engine is the *steam power plant,* which is an external-combustion engine. That is, combustion takes place outside the engine, and the thermal energy released during this process is transferred to the steam as heat. The schematic of a basic steam power plant is shown in Fig. 6-10. This is a rather simplified diagram,

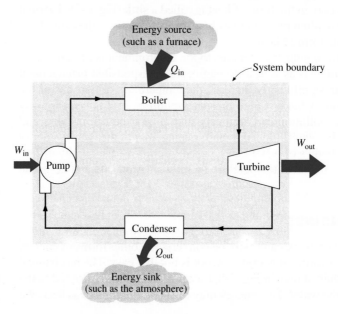

FIGURE 6-10

Schematic of a steam power plant.

and the discussion of actual steam power plants with all their complexities is left to Chap. 8. The various quantities shown on this figure are as follows:

Q_{in} = amount of heat supplied to steam in boiler from a high-temperature source (furnace)

Q_{out} = amount of heat rejected from steam in condenser to a low-temperature sink (the atmosphere, a river, etc.)

W_{out} = amount of work delivered by steam as it expands in turbine

W_{in} = amount of work required to compress water to boiler pressure

Notice that the directions of the heat and work interactions are indicated by the subscripts *in* and *out*. Therefore, all four quantities described above are always *positive*.

The net work output of this power plant is simply the difference between the total work output of the plant and the total work input (Fig. 6-11):

$$W_{net,\,out} = W_{out} - W_{in} \qquad (kJ) \qquad (6-1)$$

The net work can also be determined from the heat transfer data alone. The four components of the steam power plant involve mass flow in and out, and therefore they should be treated as open systems. These components, together with the connecting pipes, however, always contain the same fluid (not counting the steam that may leak out, of course). No mass enters or leaves this combination system, which is indicated by the shaded area on Fig. 6-10; thus, it can be analyzed as a closed system. Recall that for a closed system undergoing a cycle, the change in internal energy ΔU is zero, and therefore the net work output of the system is also equal to the net heat transfer to the system:

$$W_{net,\,out} = Q_{in} - Q_{out} \qquad (kJ) \qquad (6-2)$$

Thermal Efficiency

In Eq. 6-2, Q_{out} represents the magnitude of the energy wasted in order to complete the cycle. But Q_{out} is never zero; thus, the net work output of a heat engine is always less than the amount of heat input. That is, only part of the heat transferred to the heat engine is converted to work. *The fraction of the heat input that is converted to net work output is a measure of the performance of a heat engine and is called the* **thermal efficiency** η_{th} (Fig. 6-12).

Performance or efficiency, in general, can be expressed in terms of the desired output and the required input as (Fig. 6-13)

$$\text{Performance} = \frac{\text{Desired output}}{\text{Required input}} \qquad (6-3)$$

For heat engines, the desired output is the net work output, and the required input is the amount of heat supplied to the working fluid. Then the thermal efficiency of a heat engine can be expressed as

$$\text{Thermal efficiency} = \frac{\text{Net work output}}{\text{Total heat input}} \qquad (6-4)$$

or

$$\eta_{th} = \frac{W_{net,\,out}}{Q_{in}}$$

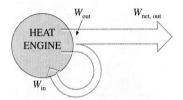

FIGURE 6-11

A portion of the work output of a heat engine is consumed internally to maintain continuous operation.

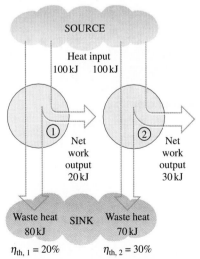

FIGURE 6-12

Some heat engines perform better than others (convert more of the heat they receive to work).

FIGURE 6-13

The definition of performance is not
limited to thermodynamics only.

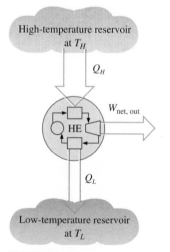

FIGURE 6-14

Schematic of a heat engine.

It can also be expressed as

$$\eta_{th} = 1 - \frac{Q_{out}}{Q_{in}} \qquad (6\text{-}5)$$

since $W_{net,\,out} = Q_{in} - Q_{out}$.

Cyclic devices of practical interest such as heat engines, refrigerators, and heat pumps operate between a high-temperature medium (or reservoir) at temperature T_H and a low-temperature medium (or reservoir) at temperature T_L. To bring uniformity to the treatment of heat engines, refrigerators, and heat pumps, we define the following two quantities:

Q_H = magnitude of heat transfer between the cyclic device and the
high-temperature medium at temperature T_H

Q_L = magnitude of heat transfer between the cyclic device and the
low-temperature medium at temperature T_L

Notice that both Q_L and Q_H are defined as *magnitudes* and therefore are positive quantities. The direction of Q_H and Q_L is easily determined by inspection. Then the net work output and thermal efficiency relations for any heat engine (shown in Fig. 6-14) can also be expressed as

$$W_{net,\,out} = Q_H - Q_L$$

and

$$\eta_{th} = \frac{W_{net,\,out}}{Q_H}$$

or

$$\eta_{th} = 1 - \frac{Q_L}{Q_H} \qquad (6\text{-}6)$$

The thermal efficiency of a heat engine is always less than unity since both Q_L and Q_H are defined as positive quantities.

Thermal efficiency is a measure of how efficiently a heat engine converts the heat that it receives to work. Heat engines are built for the purpose of converting heat to work, and engineers are constantly trying to improve the efficiencies of these devices since increased efficiency means less fuel consumption and thus lower fuel bills and less pollution.

The thermal efficiencies of work-producing devices are relatively low. Ordinary spark-ignition automobile engines have a thermal efficiency of about 25 percent. That is, an automobile engine converts about 25 percent of the chemical energy of the gasoline to mechanical work. This number is as high as 40 percent for diesel engines and large gas-turbine plants and as high as 60 percent for large combined gas-steam power plants. Thus, even with the most efficient heat engines available today, almost one-half of the energy supplied ends up in the rivers, lakes, or the atmosphere as waste or useless energy (Fig. 6-15).

Can We Save Q_{out}?

In a steam power plant, the condenser is the device where large quantities of waste heat is rejected to rivers, lakes, or the atmosphere. Then one may ask, can we not just take the condenser out of the plant and save all that waste energy? The answer to this question is, unfortunately, a firm *no* for the simple reason that without a heat rejection process in a condenser, the cycle cannot be completed. (Cyclic devices such as steam power plants cannot run continuously unless the cycle is completed.) This is demonstrated below with the help of a simple heat engine.

Consider the simple heat engine shown in Fig. 6-16 that is used to lift weights. It consists of a piston-cylinder device with two sets of stops. The working fluid is the gas contained within the cylinder. Initially, the gas temperature is 30°C. The piston, which is loaded with the weights, is resting on top of the lower stops. Now 100 kJ of heat is transferred to the gas in the cylinder from a source at 100°C, causing it to expand and to raise the loaded piston until the piston reaches the upper stops, as shown in the figure. At this point, the load is removed, and the gas temperature is observed to be 90°C.

The work done on the load during this expansion process is equal to the increase in its potential energy, say 15 kJ. Even under ideal conditions (weightless piston, no friction, no heat losses, and quasi-equilibrium expansion), the amount of heat supplied to the gas is greater than the work done since part of the heat supplied is used to raise the temperature of the gas.

Now let us try to answer the following question: *Is it possible to transfer the 85 kJ of excess heat at 90°C back to the reservoir at 100°C for later use?* If it is, then we will have a heat engine that can have a thermal efficiency of 100 percent under ideal conditions. The answer to this question is again *no,* for the very simple reason that heat always flows from a high-temperature medium to a low-temperature one, and never the other way around. Therefore, we cannot cool this gas from 90 to 30°C by transferring heat to a reservoir at 100°C. Instead, we have to bring the system into contact with a low-temperature reservoir, say at 20°C, so that the gas can return to its initial state by rejecting its 85 kJ of excess energy as heat to this reservoir. This energy cannot be recycled, and it is properly called *waste energy.*

We conclude from the above discussion that every heat engine must *waste* some energy by transferring it to a low-temperature reservoir in order to complete the cycle, even under idealized conditions. The requirement that a heat engine exchange heat with at least two reservoirs for continuous operation forms the basis for the Kelvin-Planck expression of the second law of thermodynamics discussed later in this section.

EXAMPLE 6-1   Net Power Production of a Heat Engine

Heat is transferred to a heat engine from a furnace at a rate of 80 MW. If the rate of waste heat rejection to a nearby river is 50 MW, determine the net power output and the thermal efficiency for this heat engine.

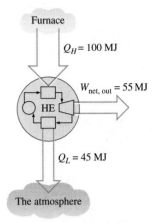

FIGURE 6-15

Even the most efficient heat engines reject almost one-half of the energy they receive as waste heat.

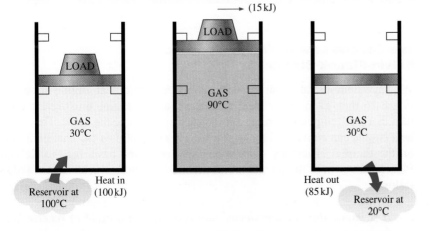

FIGURE 6-16

A heat-engine cycle cannot be completed without rejecting some heat to a low-temperature sink.

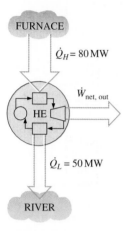

FIGURE 6-17

Schematic for Example 6-1.

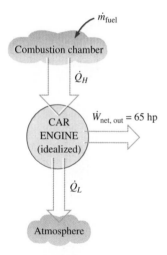

FIGURE 6-18

Schematic for Example 6-2.

Solution A schematic of the heat engine is given in Fig. 6-17. The furnace serves as the high-temperature reservoir for this heat engine and the river as the low-temperature reservoir.

Assumptions Heat losses through the pipes and other components are negligible.

Analysis The given quantities can be expressed in rate form as

$$\dot{Q}_H = 80 \text{ MW and } \dot{Q}_L = 50 \text{ MW}$$

The net power output of this heat engine is

$$\dot{W}_{net,\,out} = \dot{Q}_H - \dot{Q}_L = (80 - 50) \text{ MW} = \textbf{30 MW}$$

Then the thermal efficiency is easily determined to be

$$\eta_{th} = \frac{\dot{W}_{net,\,out}}{\dot{Q}_H} = \frac{30 \text{ MW}}{80 \text{ MW}} = \textbf{0.375} \text{ (or 37.5\%)}$$

That is, the heat engine converts 37.5 percent of the heat it receives to work.

EXAMPLE 6-2   Fuel Consumption Rate of a Car

A car engine with a power output of 65 hp has a thermal efficiency of 24 percent. Determine the fuel consumption rate of this car if the fuel has a heating value of 19,000 Btu/lbm (that is, 19,000 Btu of energy is released for each lbm of fuel burned).

Solution A schematic of the car engine is given in Fig. 6-18. The car engine is powered by converting 24 percent of the chemical energy released during the combustion process to work.

Assumptions The power output of the car is constant.

Analysis The amount of energy input required to produce a power output of 65 hp is determined from the definition of thermal efficiency to be

$$\dot{Q}_H = \frac{\dot{W}_{net,\,out}}{\eta_{th}} = \frac{65 \text{ hp}}{0.24} \left(\frac{2545 \text{ Btu/h}}{1 \text{ hp}} \right) = 689,271 \text{ Btu/h}$$

To supply energy at this rate, the engine must burn fuel at a rate of

$$\dot{m} = \frac{689,271 \text{ Btu/h}}{19,000 \text{ Btu/lbm}} = \textbf{36.3 lbm/h}$$

since 19,000 Btu of thermal energy is released for each lbm of fuel burned.

The Second Law of Thermodynamics: Kelvin-Planck Statement

We have demonstrated earlier with reference to the heat engine shown in Fig. 6-16 that, even under ideal conditions, a heat engine must reject some heat to a low-temperature reservoir in order to complete the cycle. That is, no heat engine can convert all the heat it receives to useful work. This limitation on the thermal efficiency of heat engines forms the basis for the Kelvin-Planck statement of the second law of thermodynamics, which is expressed as follows:

> *It is impossible for any device that operates on a cycle to receive heat from a single reservoir and produce a net amount of work.*

That is, a heat engine must exchange heat with a low-temperature sink as well as a high-temperature source to keep operating. The Kelvin-Planck statement can also be expressed as *no heat engine can have a thermal efficiency of 100 percent* (Fig. 6-19), or as *for a power plant to operate, the working fluid must exchange heat with the environment as well as the furnace.*

Note that the impossibility of having a 100 percent efficient heat engine is not due to friction or other dissipative effects. It is a limitation that applies to both the idealized and the actual heat engines. Later in this chapter, we develop a relation for the maximum thermal efficiency of a heat engine. We also demonstrate that this maximum value depends on the reservoir temperatures only.

6-4 ■ ENERGY CONVERSION EFFICIENCIES

Efficiency is one of the most frequently used terms in thermodynamics, and it indicates how well an energy conversion or transfer process is accomplished. The *thermal efficiency* of a heat engine, for example, is the fraction of the thermal energy a heat engine converts to work. Efficiency is also one of the most frequently misused terms in thermodynamics and a source of misunderstandings. This is because efficiency is often used without being properly defined first. Below we will clarify this further, and define some efficiencies commonly used in practice.

If you are shopping for a water heater, a knowledgeable salesperson will tell you that the efficiency of a conventional electric water heater is about 90 percent (Fig. 6-20). You may find this confusing, since the heating elements of electric water heaters are resistance heaters, and the efficiency of all resistance heaters is 100 percent as they convert all the electrical energy they consume into heat. A knowledgeable salesperson will clarify this by explaining that the heat losses from the hot water tank to the surrounding air amount to 10 percent of the electrical energy consumed, and the **efficiency of a water heater** is defined as the ratio of the *energy delivered to the house by hot water* to the *energy supplied to the water heater*. A clever salesperson may even talk you into buying a more expensive water heater with thicker insulation that has an efficiency of 94 percent. If you are a knowledgeable consumer and have access to natural gas, you will probably purchase a gas water heater whose efficiency is only 55 percent since a gas unit costs about the same as an electric unit to purchase and install, but the annual energy cost of a gas unit will be less than half of that of an electric unit at national average electricity and gas prices.

Perhaps you are wondering how the efficiency for a gas water heater is defined, and why it is much lower than the efficiency of an electric heater. As a general rule, the efficiency of equipment that involves the combustion of a fuel is based on the **heating value of the fuel**, which is *the amount of heat released when a specified amount of fuel (usually a unit mass) at room temperature is completely burned and the combustion products are cooled to the room temperature* (Fig. 6-21). Then the performance of combustion equipment can be characterized by **combustion efficiency,** defined as

$$\eta_{\text{combustion}} = \frac{Q}{\text{HV}} = \frac{\text{Amount of heat released during combustion}}{\text{Heating value of the fuel burned}} \quad (6\text{-}7)$$

A combustion efficiency of 100 percent indicates that the fuel is burned completely and the stack gases leave the combustion chamber at room temperature, and thus the amount of heat released during a combustion process is equal to the heating value of the fuel.

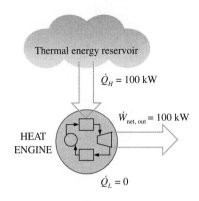

FIGURE 6-19

A heat engine that violates the Kelvin-Planck statement of the second law.

Type	Efficiency
Gas, conventional	55%
Gas, high efficiency	62%
Elect., conventional	90%
Elect., high-efficiency	94%

FIGURE 6-20

Typical efficiencies of conventional and high-efficiency electric and natural gas water heaters.

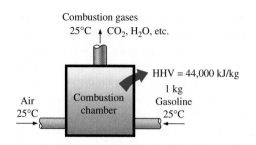

FIGURE 6-21

The definition of the heating value of gasoline.

Most fuels contain hydrogen, which forms water when burned, and the heating value of a fuel will be different, depending on whether the water in combustion products is in the liquid or vapor form. The heating value is called the *lower heating value,* or LHV, when the water leaves as a vapor, and the *higher heating value,* or HHV, when the water in the combustion gases is completely condensed and thus the heat of vaporization is also recovered. The difference between these two heating values is equal to the product of the amount of water and the enthalpy of vaporization of water at room temperature. For example, the lower and higher heating values of gasoline are 44,000 kJ/kg and 47,300 kJ/kg, respectively. An efficiency definition should make it clear whether it is based on the higher or lower heating value of the fuel. Efficiencies of cars and jet engines are normally based on *lower heating values* since water normally leaves as a vapor in the exhaust gases, and it is not practical to try to recuperate the heat of vaporization. Efficiencies of furnaces, on the other hand, are based on *higher heating values.*

The efficiency of space heating systems of residential and commercial buildings is usually expressed in terms of the **annual fuel utilization efficiency,** or **AFUE,** which accounts for the combustion efficiency as well as other losses such as heat losses to unheated areas and start-up and cool-down losses. The AFUE of most new heating systems is close to 85 percent, although the AFUE of some old heating systems is under 60 percent. The AFUE of some new high-efficiency furnaces exceeds 96 percent, but the high cost of such furnaces cannot be justified for locations with mild to moderate winters. Such high efficiencies are achieved by reclaiming most of the heat in the flue gases, condensing the water vapor, and discharging the flue gases at temperatures as low as 38°C (or 100°F) instead of about 200°C (or 400°F) for the conventional models.

For *car engines,* the work output is understood to be the power delivered by the crankshaft. But for power plants, the work output can be the mechanical power at the turbine exit, or the electrical power output of the generator.

A generator is a device that converts mechanical energy to electrical energy, and the effectiveness of a generator is characterized by the **generator efficiency,** which is the ratio of the *electrical power output* to the *mechanical power input.* The *thermal efficiency* of a power plant, which is of primary interest in thermodynamics, is usually defined as the ratio of the shaft work output of the turbine to the heat input to the working fluid. The effects of other factors are incorporated by defining an **overall efficiency** for the power plant as the ratio of the *net electrical power output* to the *rate of fuel energy input.* That is,

$$\eta_{\text{overall}} = \eta_{\text{combustion}}\,\eta_{\text{thermal}}\,\eta_{\text{generator}} = \frac{\dot{W}_{\text{net, electric}}}{\text{HHV} \times \dot{m}_{\text{fuel}}} \qquad (6\text{-}8)$$

The overall efficiencies are about 25–28 percent for gasoline automotive engines, 34–38 percent for diesel engines, and 40–60 percent for large power plants.

Electrical energy is commonly converted to *rotating mechanical energy* by electric motors to drive fans, compressors, robot arms, car starters, and so forth. The effectiveness of this conversion process is characterized by the **motor efficiency** η_{motor}, which is the ratio of the *mechanical energy output* of the motor to the *electrical energy input*. The full-load motor efficiencies range from about 35 percent for small motors to over 96 percent for large high-efficiency motors. The difference between the electrical energy consumed and the mechanical energy delivered is dissipated as waste heat.

We are all familiar with the conversion of electrical energy to *light* by incandescent light bulbs, fluorescent tubes, and high-intensity discharge lamps. The efficiency for the conversion of electricity to light can be defined as the ratio of the energy converted to light to the electrical energy consumed. For example, common incandescent light bulbs convert about 10 percent of the electrical energy they consume to light; the rest of the energy consumed is dissipated as heat, which adds to the cooling load of the air conditioner in summer. However, it is more common to express the effectiveness of this conversion process by **lighting efficacy,** which is defined as the *amount of light output in lumens per W of electricity consumed.*

The efficacy of different lighting systems is given in Table 6-1. Note that a compact fluorescent light bulb produces about four times as much light as an incandescent light bulb per W, and thus a 15-W fluorescent bulb can replace a 60-W incandescent light bulb (Fig. 6-22). Also, a compact fluorescent bulb lasts about 10,000 h, which is 10 times as long as an incandescent bulb, and it plugs directly into the socket of an incandescent lamp. Therefore, despite their higher initial cost, compact fluorescents reduce the lighting costs considerably through reduced electricity consumption. Sodium-filled high-intensity discharge lamps provide the most efficient lighting, but their use is limited to outdoor use because of their yellowish light.

We can also define efficiency for cooking appliances since they convert electrical or chemical energy to heat for cooking. The **efficiency of a cooking appliance** can be defined as the ratio of the *useful energy transferred to the food* to the *energy consumed by the appliance* (Fig. 6-23). Electric ranges are

TABLE 6-1

The efficacy of different lighting systems

Type of lighting	Efficacy, lumens/W
Combustion	
Candle	0.2
Incandescent	
Ordinary	5–20
Halogen	15–25
Fluorescent	
Ordinary	40–60
High output	70–90
Compact	50–80
High-intensity discharge	
Mercury vapor	50–60
Metal halide	55–125
High-pressure sodium	100–150
Low-pressure sodium	up to 200

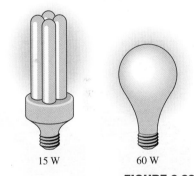

15 W   60 W

FIGURE 6-22

A 15-W compact fluorescent lamp provides as much light as a 60-W incandescent lamp.

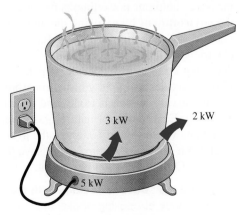

$$\text{Efficiency} = \frac{\text{Energy utilized}}{\text{Energy supplied to appliance}}$$

$$= \frac{3 \text{ kWh}}{5 \text{ kWh}} = 0.60$$

FIGURE 6-23

The efficiency of a cooking appliance represents the fraction of the energy supplied to the appliance that is transferred to the food.

TABLE 6-2

Energy costs of cooking a casserole with different appliances*
[from A. Wilson and J. Morril, *Consumer Guide to Home Energy Savings,*
Washington, D.C.: American Council for an Energy-Efficient Economy, 1996,
p. 192.]

Cooking appliance	Cooking temperature	Cooking time	Energy used	Cost of energy
Electric oven	350°F (177°C)	1 h	2.0 kWh	$0.16
Convection oven (elect.)	325°F (163°C)	45 min	1.39 kWh	$0.11
Gas oven	350°F (177°C)	1 h	0.112 therm	$0.07
Frying pan	420°F (216°C)	1 h	0.9 kWh	$0.07
Toaster oven	425°F (218°C)	50 min	0.95 kWh	$0.08
Crockpot	200°F (93°C)	7 h	0.7 kWh	$0.06
Microwave oven	"High"	15 min	0.36 kWh	$0.03

*Assumes a unit cost of $0.08/kWh for electricity and $0.60/therm for gas.

more efficient than gas ranges, but it is much cheaper to cook with natural gas than with electricity because of the lower unit cost of natural gas (Table 6-2).

The cooking efficiency depends on user habits as well as the individual appliances. Convection and microwave ovens are inherently more efficient than conventional ovens. On average, convection ovens save about *one-third* and microwave ovens save about *two-thirds* of the energy used by conventional ovens. The cooking efficiency can be increased by using the smallest oven for baking, using a pressure cooker, using a crockpot for stews and soups, using the smallest pan that will do the job, using the smaller heating element for small pans on electric ranges, using flat-bottomed pans on electric burners to assure good contact, keeping burner drip pans clean and shiny, defrosting frozen foods in the refrigerator before cooking, avoiding preheating unless it is necessary, keeping the pans covered during cooking, using timers and thermometers to avoid overcooking, using the self-cleaning feature of ovens right after cooking, and keeping inside surfaces of microwave ovens clean.

Using energy-efficient appliances and practicing energy conservation measures help our pocketbooks by reducing our utility bills. It will also help the **environment** by reducing the amount of pollutants emitted to the atmosphere during the combustion of fuel at home or at the power plants where electricity is generated. The combustion of *each therm of natural gas* produces 6.4 kg of carbon dioxide, which causes global climate change; 4.7 g of nitrogen oxides and 0.54 g of hydrocarbons, which cause smog; 2.0 g of carbon monoxide, which is toxic; and 0.030 g of sulfur dioxide, which causes acid rain. Each therm of natural gas saved eliminates the emission of these pollutants while saving $0.60 for the average consumer in the United States. Each kWh of electricity saved will save 0.4 kg of coal and 1.0 kg of CO_2 and 15 g of SO_2 from a coal power plant.

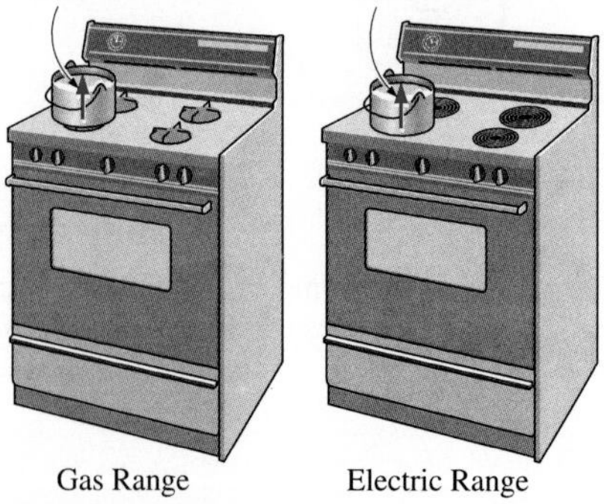

38%   73%

Gas Range   Electric Range

FIGURE 6-24

Schematic of the 73 percent efficient electric heating unit and 38 percent efficient gas burner discussed in Example 6-3.

EXAMPLE 6-3   Cost of Cooking with Electric and Gas Ranges

The efficiency of cooking appliances affects the internal heat gain from them since an inefficient appliance consumes a greater amount of energy for the same task, and the excess energy consumed shows up as heat in the living space. The efficiency of open burners is determined to be 73 percent for electric units and 38 percent for gas units (Fig. 6-24). Consider a 2-kW electric burner at a location where the unit costs of electricity and natural gas are $0.09/kWh and $0.55/therm, respectively. Determine the rate of energy consumption by the burner and the unit cost of utilized energy for both electric and gas burners.

Solution   This example is to demonstrate the economics of electric and gas ranges.

Analysis   The efficiency of the electric heater is given to be 73 percent. Therefore, a burner that consumes 2 kW of electrical energy will supply

$$\dot{Q}_{utilized} = (\text{Energy input}) \times (\text{Efficiency}) = (2 \text{ kW})(0.73) = \textbf{1.46 kW}$$

of useful energy. The unit cost of utilized energy is inversely proportional to the efficiency, and is determined from

$$\text{Cost of utilized energy} = \frac{\text{Cost of energy input}}{\text{Efficiency}} = \frac{\$0.09/\text{kWh}}{0.73} = \textbf{\$0.123/kWh}$$

Noting that the efficiency of a gas burner is 38 percent, the energy input to a gas burner that supplies utilized energy at the same rate (1.46 kW) is

$$\dot{Q}_{input, gas} = \frac{\dot{Q}_{utilized}}{\text{Efficiency}} = \frac{1.46 \text{ kW}}{0.38} = \textbf{3.84 kW} \qquad (= 13{,}100 \text{ Btu/h})$$

since 1 kW = 3412 Btu/h. Therefore, a gas burner should have a rating of at least 13,100 Btu/h to perform as well as the electric unit.

Noting that 1 therm = 29.3 kWh, the unit cost of utilized energy in the case of a gas burner is determined to be

$$\text{Cost of utilized energy} = \frac{\text{Cost of energy input}}{\text{Efficiency}} = \frac{\$0.55/29.3 \text{ kWh}}{0.38} = \textbf{\$0.049/kWh}$$

Discussion   The cost of utilized gas is less than half of the unit cost of utilized electricity. Therefore, despite its higher efficiency, cooking with an electric burner will cost more than twice as much compared to a gas burner in this case. This explains why cost-conscious consumers always ask for gas appliances, and it is not wise to use electricity for heating purposes.

6-5 ▨ REFRIGERATORS AND HEAT PUMPS

We all know from experience that heat flows in the direction of decreasing temperature, that is, from high-temperature mediums to low-temperature ones. This heat transfer process occurs in nature without requiring any devices. The reverse process, however, cannot occur by itself. The transfer of heat from a low-temperature medium to a high-temperature one requires special devices called **refrigerators.**

Refrigerators, like heat engines, are cyclic devices. The working fluid used in the refrigeration cycle is called a **refrigerant.** The most frequently used refrigeration cycle is the *vapor-compression refrigeration cycle,* which involves four main components: a compressor, a condenser, an expansion valve, and an evaporator, as shown in Fig. 6-25.

The refrigerant enters the compressor as a vapor and is compressed to the condenser pressure. It leaves the compressor at a relatively high temperature and cools down and condenses as it flows through the coils of the condenser by rejecting heat to the surrounding medium. It then enters a capillary tube where its pressure and temperature drop drastically due to the throttling effect. The low-temperature refrigerant then enters the evaporator, where it evaporates by absorbing heat from the refrigerated space. The cycle is completed as the refrigerant leaves the evaporator and reenters the compressor.

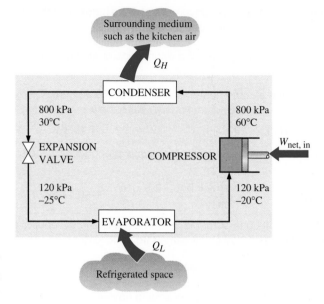

FIGURE 6-25

Basic components of a refrigeration
system and typical operating conditions.

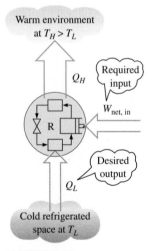

FIGURE 6-26

The objective of a refrigerator is to remove
Q_L from the cooled space.

In a household refrigerator, the freezer compartment where heat is picked up by the refrigerant serves as the evaporator, and the coils behind the refrigerator where heat is dissipated to the kitchen air serve as the condenser.

A refrigerator is shown schematically in Fig. 6-26. Here Q_L is the magnitude of the heat removed from the refrigerated space at temperature T_L, Q_H is the magnitude of the heat rejected to the warm environment at temperature T_H, and $W_{net, in}$ is the net work input to the refrigerator. As discussed before, Q_L and Q_H represent magnitudes and thus are positive quantities.

Coefficient of Performance

The *efficiency* of a refrigerator is expressed in terms of the **coefficient of performance** (COP), denoted by COP_R. The objective of a refrigerator is to remove heat (Q_L) from the refrigerated space. To accomplish this objective, it requires a work input of $W_{net, in}$. Then the COP of a refrigerator can be expressed as

$$COP_R = \frac{\text{Desired output}}{\text{Required input}} = \frac{Q_L}{W_{net, in}} \qquad (6\text{-}9)$$

This relation can also be expressed in rate form by replacing Q_L by $\dot{Q}_L$ and $W_{net, in}$ by $\dot{W}_{net, in}$.

The conservation of energy principle for a cyclic device requires that

$$W_{net, in} = Q_H - Q_L \qquad \text{(kJ)} \qquad (6\text{-}10)$$

Then the COP relation can also be expressed as

$$COP_R = \frac{Q_L}{Q_H - Q_L} = \frac{1}{Q_H/Q_L - 1} \qquad (6\text{-}11)$$

Notice that the value of COP_R can be *greater than unity*. That is, the amount of heat removed from the refrigerated space can be greater than the amount of work input. This is in contrast to the thermal efficiency, which can never be greater than 1. In fact, one reason for expressing the efficiency of a refrigerator by another term—the coefficient of performance—is the desire to avoid the oddity of having efficiencies greater than unity.

Heat Pumps

Another device that transfers heat from a low-temperature medium to a high-temperature one is the **heat pump,** shown schematically in Fig. 6-27. Refrigerators and heat pumps operate on the same cycle but differ in their objectives. The objective of a refrigerator is to maintain the refrigerated space at a low temperature by removing heat from it. Discharging this heat to a higher-temperature medium is merely a necessary part of the operation, not the purpose. The objective of a heat pump, however, is to maintain a heated space at a high temperature. This is accomplished by absorbing heat from a low-temperature source, such as well water or cold outside air in winter, and supplying this heat to the high-temperature medium such as a house (Fig. 6-28).

An ordinary refrigerator that is placed in the window of a house with its door open to the cold outside air in winter will function as a heat pump since it will try to cool the outside by absorbing heat from it and rejecting this heat into the house through the coils behind it (Fig. 6-29).

The measure of performance of a heat pump is also expressed in terms of the **coefficient of performance** COP_{HP}, defined as

$$\text{COP}_{\text{HP}} = \frac{\text{Desired output}}{\text{Required input}} = \frac{Q_H}{W_{\text{net, in}}} \qquad (6\text{-}12)$$

which can also be expressed as

$$\text{COP}_{\text{HP}} = \frac{Q_H}{Q_H - Q_L} = \frac{1}{1 - Q_L/Q_H} \qquad (6\text{-}13)$$

A comparison of Eqs. 6-9 and 6-12 reveals that

$$\text{COP}_{\text{HP}} = \text{COP}_{\text{R}} + 1 \qquad (6\text{-}14)$$

for fixed values of Q_L and Q_H. This relation implies that the coefficient of performance of a heat pump is always greater than unity since COP_{R} is a positive quantity. That is, a heat pump will function, at worst, as a resistance heater, supplying as much energy to the house as it consumes. In reality, however, part of Q_H is lost to the outside air through piping and other devices, and COP_{HP} may drop below unity when the outside air temperature is too low. When this happens, the system usually switches to a resistance heating mode. Most heat pumps in operation today have a seasonally averaged COP of 2 to 3.

Most existing heat pumps use the cold outside air as the heat source in winter, and they are referred to as *air-source heat pumps.* The COP of such heat pumps is currently about 3.0 at design conditions. Air-source heat pumps are not appropriate for cold climates since their efficiency drops considerably when temperatures are below the freezing point. In such cases, geothermal (also called ground-source) heat pumps that use the ground as the heat source can be used. Geothermal heat pumps require the burial of pipes in the ground 1 to 2 m deep. Such heat pumps are more expensive to install, but they are also more efficient (up to 45 percent more efficient than air-source heat pumps). The COP of ground-source heat pumps currently approaches 4.0.

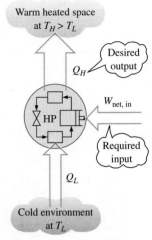

FIGURE 6-27

The objective of a heat pump is to supply heat Q_H into the warmer space.

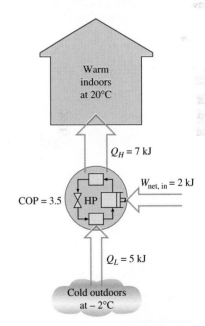

FIGURE 6-28

The work supplied to a heat pump is used to extract energy from the cold outdoors and carry it into the warm indoors.

FIGURE 6-29

When installed backwards, an air
conditioner will function as a heat pump.

Air conditioners are basically refrigerators whose refrigerated space
is a room or a building instead of the food compartment. A window air-
conditioning unit cools a room by absorbing heat from the room air and discharging it to the outside. The same air-conditioning unit can be used as a heat
pump in winter by installing it backwards. In this mode, the unit will pick up
heat from the cold outside and deliver it to the room. Air-conditioning systems
that are equipped with proper controls and a reversing valve operate as air
conditioners in summer and as heat pumps in winter.

The performance of refrigerators and air conditioners in the United States
is often expressed in terms of the **Energy Efficiency Rating** (EER), which is
the amount of heat removed from the cooled space in Btu's for 1 Wh (watt-
hour) of electricity consumed. Considering that 1 kWh = 3412 Btu and thus
1 Wh = 3.412 Btu, a unit that removes 1 kWh of heat from the cooled space
for each kWh of electricity it consumes (COP = 1) will have an EER of
3.412. Therefore, the relation between EER and COP is

$$\text{EER} = 3.412 \, \text{COP}_R$$

Most air conditioners have an EER between 8 and 12 (a COP of 2.3
to 3.5). A high-efficiency heat pump recently manufactured by the Trane
Company using a reciprocating variable-speed compressor is reported to have
a COP of 3.3 in the heating mode and an EER of 16.9 (COP of 5.0) in the air-
conditioning mode. Variable-speed compressors and fans allow the unit to operate at maximum efficiency for varying heating/cooling needs and weather
conditions as determined by a microprocessor. In the air-conditioning mode,
for example, they operate at higher speeds on hot days and at lower speeds on
cooler days, enhancing both efficiency and comfort.

The EER or COP of a refrigerator decreases with decreasing refrigeration
temperature. Therefore, it is not economical to refrigerate to a lower temperature than needed. The COPs of refrigerators are in the range of 2.5–3.0 for cutting and preparation rooms; 2.3–2.6 for meat, deli, dairy, and produce; 1.2–1.5
for frozen foods; and 1.0–1.2 for ice cream units. Note that the COP of freezers is about half of the COP of meat refrigerators, and thus it will cost twice as
much to cool the meat products with refrigerated air that is cold enough to
cool frozen foods. It is good energy conservation practice to use separate refrigeration systems to meet different refrigeration needs.

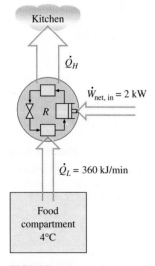

FIGURE 6-30

Schematic for Example 6-4.

EXAMPLE 6-4   Heat Rejection by a Refrigerator

The food compartment of a refrigerator, shown in Fig. 6-30, is maintained at 4°C
by removing heat from it at a rate of 360 kJ/min. If the required power input to the
refrigerator is 2 kW, determine (*a*) the coefficient of performance of the refrigerator and (*b*) the rate of heat rejection to the room that houses the refrigerator.

Solution   The power consumption of a refrigerator is given. The COP and the
rate of heat rejection are to be determined.

Assumptions   Steady operating conditions exist.

Analysis   (*a*) The coefficient of performance of a refrigerator is defined by
Eq. 6-9, which can be expressed in rate form as

$$\text{COP}_R = \frac{\dot{Q}_L}{\dot{W}_{\text{net, in}}} = \frac{360 \text{ kJ/min}}{2 \text{ kW}} \left(\frac{1 \text{ kW}}{60 \text{ kJ/min}} \right) = \mathbf{3}$$

That is, 3 kJ of heat is removed from the refrigerated space for each kJ of work
supplied.

(*b*) The rate at which heat is discharged to the room that houses the refrigerator is determined from the conservation of energy relation for cyclic devices (Eq. 6-10), expressed in rate form as

$$\dot{Q}_H = \dot{Q}_L + \dot{W}_{net,\,in} = 360\ \text{kJ/min} + (2\ \text{kW})\left(\frac{60\ \text{kJ/min}}{1\ \text{kW}}\right) = \textbf{480 kJ/min}$$

Discussion   Notice that both the energy removed from the refrigerated space as heat and the energy supplied to the refrigerator as electrical work eventually show up in the room air and become part of the internal energy of the air. This demonstrates that energy can change from one form to another, can move from one place to another, but is never destroyed during a process.

EXAMPLE 6-5   Heating a House by a Heat Pump

A heat pump is used to meet the heating requirements of a house and maintain it at 20°C. On a day when the outdoor air temperature drops to −2°C, the house is estimated to lose heat at a rate of 80,000 kJ/h. If the heat pump under these conditions has a COP of 2.5, determine (*a*) the power consumed by the heat pump and (*b*) the rate at which heat is absorbed from the cold outdoor air.

Solution   The COP of a heat pump is given. The power consumption and the rate of heat absorption are to be determined.

Assumptions   Steady operating conditions exist.

Analysis   (*a*) The power consumed by this heat pump, shown in Fig. 6-31, can be determined from the definition of the coefficient of performance of a heat pump (Eq. 6-12), expressed in the rate form as

$$\dot{W}_{net,\,in} = \frac{\dot{Q}_H}{\text{COP}_{HP}} = \frac{80{,}000\ \text{kJ/h}}{2.5} = \textbf{32,000 kJ/h (or 8.9 kW)}$$

(*b*) The house is losing heat at a rate of 80,000 kJ/h. If the house is to be maintained at a constant temperature of 20°C, the heat pump must deliver heat to the house at the same rate, that is, at a rate of 80,000 kJ/h. Then the rate of heat transfer from the outdoor air is determined from the conservation of energy principle for a cyclic device (Eq. 6-10):

$$\dot{Q}_L = \dot{Q}_H - \dot{W}_{net,\,in} = (80{,}000 - 32{,}000)\ \text{kJ/h} = \textbf{48,000 kJ/h}$$

Discussion   Note that 48,000 of the 80,000 kJ/h heat delivered to the house is actually extracted from the cold outdoor air. Therefore, we are paying only for the 32,000-kJ/h energy that is supplied as electrical work to the heat pump. If we were to use an electric resistance heater instead, we would have to supply the entire 80,000 kJ/h to the resistance heater as electric energy. This would mean a heating bill that is 2.5 times higher. This explains the popularity of heat pumps as heating systems and why they are preferred to simple electric resistance heaters despite their considerably higher initial cost.

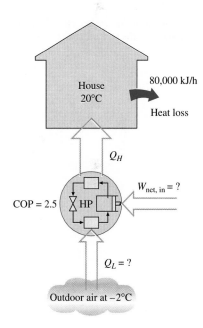

FIGURE 6-31

Schematic for Example 6-5.

The Second Law of Thermodynamics: Clausius Statement

There are two classical statements of the second law—the Kelvin-Planck statement, which is related to heat engines and discussed in the preceding section, and the Clausius statement, which is related to refrigerators or heat pumps. The Clausius statement is expressed as follows:

It is impossible to construct a device that operates in a cycle and produces no effect other than the transfer of heat from a lower-temperature body to a higher-temperature body.

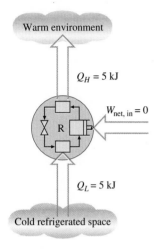

FIGURE 6-32

A refrigerator that violates the Clausius
statement of the second law.

It is common knowledge that heat does not, of its own volition, flow from a cold medium to a warmer one. The Clausius statement does not imply that a cyclic device that transfers heat from a cold medium to a warmer one is impossible to construct. In fact, this is precisely what a common household refrigerator does. It simply states that a refrigerator will not operate unless its compressor is driven by an external power source, such as an electric motor (Fig. 6-32). This way, the net effect on the surroundings involves the consumption of some energy in the form of work, in addition to the transfer of heat from a colder body to a warmer one. That is, it leaves a trace in the surroundings. Therefore, a household refrigerator is in complete compliance with the Clausius statement of the second law.

Both the Kelvin-Planck and the Clausius statements of the second law are negative statements, and a negative statement cannot be proved. Like any other physical law, the second law of thermodynamics is based on experimental observations. To date, no experiment has been conducted that contradicts the second law, and this should be taken as sufficient evidence of its validity.

Equivalence of the Two Statements

The Kelvin-Planck and the Clausius statements are equivalent in their consequences, and either statement can be used as the expression of the second law of thermodynamics. Any device that violates the Kelvin-Planck statement also violates the Clausius statement, and vice versa. This can be demonstrated as follows:

Consider the heat-engine-refrigerator combination shown in Fig. 6-33a, operating between the same two reservoirs. The heat engine is assumed to have, in violation of the Kelvin-Planck statement, a thermal efficiency of 100 percent, and therefore it converts all the heat Q_H it receives to work W. This work is now supplied to a refrigerator that removes heat in the amount of Q_L from the low-temperature reservoir and rejects heat in the amount of $Q_L + Q_H$ to the high-temperature reservoir. During this process, the high-temperature reservoir receives a net amount of heat Q_L (the difference

FIGURE 6-33
Proof that the violation of the Kelvin-
Planck statement leads to the violation of
the Clausius statement.

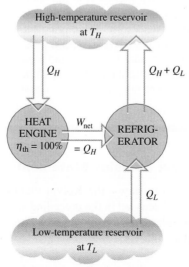

(a) A refrigerator which is powered
by a 100% efficient heat engine

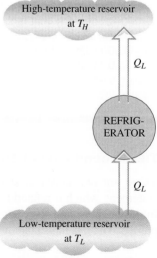

(b) The equivalent refrigerator

between $Q_L + Q_H$ and Q_H). Thus, the combination of these two devices can be viewed as a refrigerator, as shown in Fig. 6-33b, that transfers heat in an amount of Q_L from a cooler body to a warmer one without requiring any input from outside. This is clearly a violation of the Clausius statement. Therefore, a violation of the Kelvin-Planck statement results in the violation of the Clausius statement.

It can also be shown in a similar manner that a violation of the Clausius statement leads to the violation of the Kelvin-Planck statement. Therefore, the Clausius and the Kelvin-Planck statements are two equivalent expressions of the second law of thermodynamics.

6-6 ▩ PERPETUAL-MOTION MACHINES

We have repeatedly stated that a process cannot take place unless it satisfies both the first and second laws of thermodynamics. Any device that violates either law is called a **perpetual-motion machine,** and despite numerous attempts, no perpetual-motion machine is known to have worked. But this has not stopped inventors from trying to create new ones.

A device that violates the first law of thermodynamics (by *creating* energy) is called a **perpetual-motion machine of the first kind** (PMM1), and a device that violates the second law of thermodynamics is called a **perpetual-motion machine of the second kind** (PMM2).

Consider the steam power plant shown in Fig. 6-34. It is proposed to heat the steam by resistance heaters placed inside the boiler, instead of by the energy supplied from fossil or nuclear fuels. Part of the electricity generated by the plant is to be used to power the resistors as well as the pump. The rest of the electric energy is to be supplied to the electric network as the net work output. The inventor claims that once the system is started, this power plant will produce electricity indefinitely without requiring any energy input from the outside.

Well, here is an invention that could solve the world's energy problem—if it works, of course. A careful examination of this invention reveals that the system enclosed by the shaded area is continuously supplying energy to the outside at a rate of $\dot{Q}_{out} + \dot{W}_{net, out}$ without receiving any energy. That is, this system is creating energy at a rate of $\dot{Q}_{out} + \dot{W}_{net, out}$, which is clearly a violation of the first law. Therefore, this wonderful device is nothing more than a PMM1 and does not warrant any further consideration.

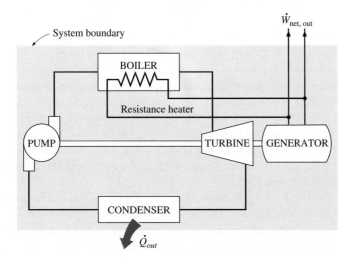

FIGURE 6-34

A perpetual-motion machine that violates the first law of thermodynamics (PMM1).

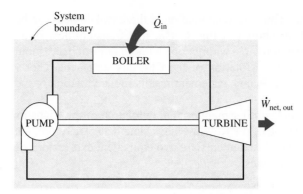

FIGURE 6-35

A perpetual-motion machine
that violates the second law of
thermodynamics (PMM2).

Now let us consider another novel idea by the same inventor. Convinced that energy cannot be created, the inventor suggests the following modification that will greatly improve the thermal efficiency of that power plant without violating the first law. Aware that more than one-half of the heat transferred to the steam in the furnace is discarded in the condenser to the environment, the inventor suggests getting rid of this wasteful component and sending the steam to the pump as soon as it leaves the turbine, as shown in Fig. 6-35. This way, all the heat transferred to the steam in the boiler will be converted to work, and thus the power plant will have a theoretical efficiency of 100 percent. The inventor realizes that some heat losses and friction between the moving components are unavoidable and that these effects will hurt the efficiency somewhat, but still expects the efficiency to be no less than 80 percent (as opposed to 40 percent in most actual power plants) for a carefully designed system.

Well, the possibility of doubling the efficiency would certainly be very tempting to plant managers and, if not properly trained, they would probably give this idea a chance, since intuitively they see nothing wrong with it. A student of thermodynamics, however, will immediately label this device as a PMM2, since it works on a cycle and does a net amount of work while exchanging heat with a single reservoir (the furnace) only. It satisfies the first law but violates the second law, and therefore it will not work.

Countless perpetual-motion machines have been proposed throughout history, and many more are being proposed. Some proposers have even gone so far as to patent their inventions, only to find out that what they actually have in their hands is a worthless piece of paper.

Some perpetual-motion machine inventors were very successful in fund raising. For example, a Philadelphia carpenter named J. W. Kelly collected millions of dollars between 1874 and 1898 from investors in his *hydro-pneumatic-pulsating-vacu-engine,* which supposedly could push a railroad train 3000 miles on one liter of water. Of course, it never did. After his death in 1898, the investigators discovered that the demonstration machine was powered by a hidden motor. Recently a group of investors was set to invest $2.5 million into a mysterious *energy augmentor,* which multiplied whatever power it took in, but their lawyer wanted an expert opinion first. Confronted by the scientists, the "inventor" fled the scene without even attempting to run his demo machine.

Tired of applications for perpetual-motion machines, the U.S. Patent Office decreed in 1918 that it would no longer even consider any perpetual-motion applications. But several such patent applications were still filed, and some made it through the patent office undetected. Some applicants whose patent applications were denied sought legal action. For example, in 1982 the U.S. Patent Office dismissed as just another perpetual-motion machine a huge device that involves several hundred kilograms of rotating magnets and kilometers of copper wire that is supposed to be generating more electricity than it is consuming from a battery pack. But the inventor challenged the decision, and in 1985 the National Bureau of Standards finally tested the machine just to certify that it is battery-operated. But it did not convince the inventor that his machine will not work.

The proposers of perpetual-motion machines generally have innovative minds, but they usually lack formal engineering training, which is very unfortunate. No one is immune from being deceived by an innovative perpetual-motion machine. But, as the saying goes, if something sounds too good to be true, it probably is.

6-7 ■ REVERSIBLE AND IRREVERSIBLE PROCESSES

The second law of thermodynamics states that no heat engine can have an efficiency of 100 percent. Then one may ask, What is the highest efficiency that a heat engine can possibly have? Before we can answer this question, we need to define an idealized process first, which is called the *reversible process.*

The processes that were discussed in Sec. 6-1 occurred in a certain direction. Once having taken place, these processes cannot reverse themselves spontaneously and restore the system to its initial state. For this reason, they are classified as *irreversible processes.* Once a cup of hot coffee cools, it will not heat up retrieving the heat it lost from the surroundings. If it could, the surroundings, as well as the system (coffee), would be restored to their original condition, and this would be a reversible process.

A **reversible process** is defined as a *process that can be reversed without leaving any trace on the surroundings* (Fig. 6-36). That is, both the system *and* the surroundings are returned to their initial states at the end of the reverse process. This is possible only if the net heat *and* net work exchange between the system and the surroundings is zero for the combined (original and reverse) process. Processes that are not reversible are called **irreversible processes.**

It should be pointed out that a system can be restored to its initial state following a process, regardless of whether the process is reversible or irreversible. But for reversible processes, this restoration is made without leaving any net change on the surroundings, whereas for irreversible processes, the surroundings usually do some work on the system and therefore will not return to their original state.

Reversible processes actually do not occur in nature. They are merely *idealizations* of actual processes. Reversible processes can be approximated by actual devices, but they can never be achieved. That is, all the processes occurring in nature are irreversible. You may be wondering, then, *why* we are bothering with such fictitious processes. There are two reasons. First, they are easy to analyze, since a system passes through a series of equilibrium states during a reversible process; second, they serve as idealized models to which actual processes can be compared.

In daily life, the concepts of Mr. Right and Ms. Right are also idealizations, just like the concept of a reversible (perfect) process. People who insist on finding Mr. or Ms. Right to settle down are bound to remain Mr. or Ms. Single for the rest of their lives. The possibility of finding the perfect prospective mate is no higher than the possibility of finding a perfect (reversible)

(a) Frictionless pendulum

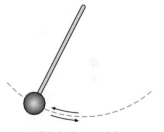

(b) Quasi-equilibrium expansion and compression of a gas

FIGURE 6-36

Two familiar reversible processes.

FIGURE 6-37

Reversible processes deliver the most and consume the least work.

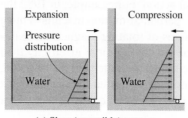

(a) Slow (reversible) process

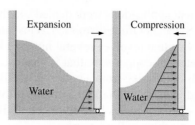

(b) Fast (irreversible) process

process. Likewise, a person who insists on perfection in friends is bound to have no friends.

Engineers are interested in reversible processes because work-producing devices such as car engines and gas or steam turbines *deliver the most work,* and work-consuming devices such as compressors, fans, and pumps *require the least work* when reversible processes are used instead of irreversible ones (Fig. 6-37).

Reversible processes can be viewed as *theoretical limits* for the corresponding irreversible ones. Some processes are more irreversible than others. We may never be able to have a reversible process, but we may certainly approach it. The more closely we approximate a reversible process, the more work delivered by a work-producing device or the less work required by a work-consuming device.

The concept of reversible processes leads to the definition of the **second-law efficiency** for actual processes, which is the degree of approximation to the corresponding reversible processes. This enables us to compare the performance of different devices that are designed to do the same task on the basis of their efficiencies. The better the design, the lower the irreversibilities and the higher the second-law efficiency.

Irreversibilities

The factors that cause a process to be irreversible are called **irreversibilities.** They include friction, unrestrained expansion, mixing of two gases, heat transfer across a finite temperature difference, electric resistance, inelastic deformation of solids, and chemical reactions. The presence of any of these effects renders a process irreversible. A reversible process involves none of these. Some of the frequently encountered irreversibilities are discussed briefly below.

Friction is a familiar form of irreversibility associated with bodies in motion. When two bodies in contact are forced to move relative to each other (a piston in a cylinder, for example, as shown in Fig. 6-38), a friction force that opposes the motion develops at the interface of these two bodies, and some work is needed to overcome this friction force. The energy supplied as work is eventually converted to heat during the process and is transferred to the bodies in contact, as evidenced by a temperature rise at the interface. When the direction of the motion is reversed, the bodies will be restored to their original position, but the interface will not cool, and heat will not be converted back to work. Instead, more of the work will be converted to heat while overcoming the friction forces that also oppose the reverse motion. Since the system (the moving bodies) and the surroundings cannot be returned to their original states, this process is irreversible. Therefore, any process that

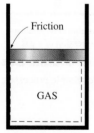

FIGURE 6-38

Friction renders a process irreversible.

involves friction is irreversible. The larger the friction forces involved, the more irreversible the process is.

Friction does not always involve two solid bodies in contact. It is also encountered between a fluid and solid and even between the layers of a fluid moving at different velocities. A considerable fraction of the power produced by a car engine is used to overcome the friction (the drag force) between the air and the external surfaces of the car, and it eventually becomes part of the internal energy of the air. It is not possible to reverse this process and recover that lost power, even though doing so would not violate the conservation of energy principle.

Another example of irreversibility is the **unrestrained expansion of a gas** separated from a vacuum by a membrane, as shown in Fig. 6-39. When the membrane is ruptured, the gas fills the entire tank. The only way to restore the system to its original state is to compress it to its initial volume, while transferring heat from the gas until it reaches its initial temperature. From the conservation of energy considerations, it can easily be shown that the amount of heat transferred from the gas equals the amount of work done on the gas by the surroundings. The restoration of the surroundings involves conversion of this heat completely to work, which would violate the second law. Therefore, unrestrained expansion of a gas is an irreversible process.

A third form of irreversibility familiar to us all is **heat transfer** through a finite temperature difference. Consider a can of cold soda left in a warm room (Fig. 6-40). Heat will flow from the warmer room air to the cooler soda. The only way this process can be reversed and the soda restored to its original temperature is to provide refrigeration, which requires some work input. At the end of the reverse process, the soda will be restored to its initial state, but the surroundings will not be. The internal energy of the surroundings will increase by an amount equal in magnitude to the work supplied to the refrigerator. The restoration of the surroundings to the initial state can be done only by converting this excess internal energy completely to work, which is impossible to do without violating the second law. Since only the system, not both the system and the surroundings, can be restored to its initial condition, heat transfer through a finite temperature difference is an irreversible process.

Heat transfer can occur only when there is a temperature difference between a system and its surroundings. Therefore, it is physically impossible to have a reversible heat transfer process. But a heat transfer process becomes less and less irreversible as the temperature difference between the two bodies approaches zero. Then heat transfer through a differential temperature difference dT can be considered to be reversible. As dT approaches zero, the process can be reversed in direction (at least theoretically) without requiring any refrigeration. Notice that reversible heat transfer is a conceptual process and cannot be duplicated in the real world.

The smaller the temperature difference between two bodies, the smaller the heat transfer rate will be. Any significant heat transfer through a small temperature difference will require a very large surface area and a very long time. Therefore, even though approaching reversible heat transfer is desirable from a thermodynamic point of view, it is impractical and not economically feasible.

Internally and Externally Reversible Processes

A process is an interaction between a system and its surroundings, and a reversible process involves no irreversibilities associated with either of them.

(a) Fast compression

(b) Fast expansion

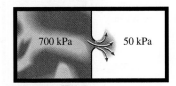

700 kPa   50 kPa

(c) Unrestrained expansion

FIGURE 6-39

Irreversible compression and expansion processes.

20°C   SODA

Heat   20°C

5°C

(a) An irreversible heat transfer process

20°C   SODA

Heat   5°C

2°C

(b) An impossible heat transfer process

FIGURE 6-40

(a) Heat transfer through a temperature difference is irreversible, and (b) the reverse process is impossible.

A process is called **internally reversible** if no irreversibilities occur within the boundaries of the system during the process. During an internally reversible process, a system proceeds through a series of equilibrium states, and when the process is reversed, the system passes through exactly the same equilibrium states while returning to its initial state. That is, the paths of the forward and reverse processes coincide for an internally reversible process. The quasi-equilibrium process discussed earlier is an example of an internally reversible process.

A process is called **externally reversible** if no irreversibilities occur outside the system boundaries during the process. Heat transfer between a reservoir and a system is an externally reversible process if the surface of contact between the system and the reservoir is at the temperature of the reservoir.

A process is called **totally reversible,** or simply **reversible,** if it involves no irreversibilities within the system or its surroundings (Fig. 6-41). A totally reversible process involves no heat transfer through a finite temperature difference, no non-quasi-equilibrium changes, and no friction or other dissipative effects.

As an example, consider the transfer of heat to two identical systems that are undergoing a constant-pressure (thus constant-temperature) phase-change process, as shown in Fig. 6-42. Both processes are internally reversible, since both take place isothermally and both pass through exactly the same equilibrium states. The first process shown is externally reversible also, since heat transfer for this process takes place through an infinitesimal temperature difference dT. The second process, however, is externally irreversible, since it involves heat transfer through a finite temperature difference ΔT.

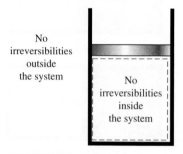

FIGURE 6-41

A reversible process involves no internal and external irreversibilities.

6-8 ▪ THE CARNOT CYCLE

We mentioned earlier that heat engines are cyclic devices and that the working fluid of a heat engine returns to its initial state at the end of each cycle. Work is done by the working fluid during one part of the cycle and on the working fluid during another part. The difference between these two is the net work delivered by the heat engine. The efficiency of a heat-engine cycle greatly depends on how the individual processes that make up the cycle are executed. The net work, thus the cycle efficiency, can be maximized by using processes that require the least amount of work and deliver the most, that is, by using *reversible processes.* Therefore, it is no surprise that the most

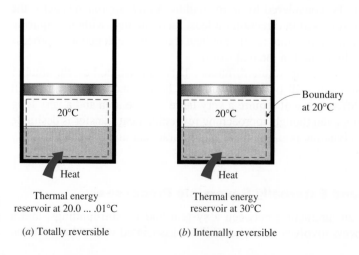

FIGURE 6-42

Totally and internally reversible heat transfer processes.

Thermal energy reservoir at 20.0 ... .01°C

(a) Totally reversible

Thermal energy reservoir at 30°C

(b) Internally reversible

efficient cycles are reversible cycles, that is, cycles that consist entirely of reversible processes.

Reversible cycles cannot be achieved in practice because the irreversibilities associated with each process cannot be eliminated. However, reversible cycles provide upper limits on the performance of real cycles. Heat engines and refrigerators that work on reversible cycles serve as models to which actual heat engines and refrigerators can be compared. Reversible cycles also serve as starting points in the development of actual cycles and are modified as needed to meet certain requirements.

Probably the best known reversible cycle is the **Carnot cycle,** first proposed in 1824 by French engineer Sadi Carnot. The theoretical heat engine that operates on the Carnot cycle is called the **Carnot heat engine.** The Carnot cycle is composed of four reversible processes—two isothermal and two adiabatic—and it can be executed either in a closed or a steady-flow system.

Consider a closed system that consists of a gas contained in an adiabatic piston-cylinder device, as shown in Fig. 6-43. The insulation of the cylinder head is such that it may be removed to bring the cylinder into contact with reservoirs to provide heat transfer. The four reversible processes that make up the Carnot cycle are as follows:

Reversible Isothermal Expansion   (process 1-2, T_H = constant). Initially (state 1), the temperature of the gas is T_H and the cylinder head is in close contact with a source at temperature T_H. The gas is allowed to expand slowly, doing work on the surroundings. As the gas expands, the temperature of the gas tends to decrease. But as soon as the temperature drops by an infinitesimal amount dT, some heat flows from the reservoir into the gas, raising the gas temperature to T_H. Thus, the gas temperature is kept constant at T_H. Since the temperature difference between the gas and the reservoir never exceeds a differential amount dT, this is a reversible heat transfer process. It continues until the piston reaches position 2. The amount of total heat transferred to the gas during this process is Q_H.

Reversible Adiabatic Expansion   (process 2-3, temperature drops from T_H to T_L). At state 2, the reservoir that was in contact with the cylinder head is removed and replaced by insulation so that the system becomes adiabatic. The gas continues to expand slowly, doing work on the surroundings until its temperature drops from T_H to T_L (state 3). The piston is assumed to be frictionless and the process to be quasi-equilibrium, so the process is reversible as well as adiabatic.

Reversible Isothermal Compression   (process 3-4, T_L = constant). At state 3, the insulation at the cylinder head is removed, and the cylinder is brought into contact with a sink at temperature T_L. Now the piston is pushed inward by an external force, doing work on the gas. As the gas is compressed, its temperature tends to rise. But as soon as it rises by an infinitesimal amount dT, heat flows from the gas to the sink, causing the gas temperature to drop to T_L. Thus, the gas temperature is maintained constant at T_L. Since the temperature difference between the gas and the sink never exceeds a differential amount dT, this is a reversible heat transfer process. It continues until the piston reaches state 4. The amount of heat rejected from the gas during this process is Q_L.

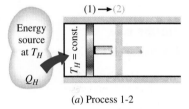

(a) Process 1-2

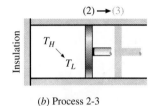

(b) Process 2-3

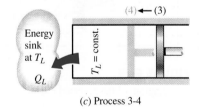

(c) Process 3-4

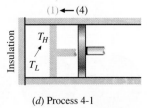

(d) Process 4-1

FIGURE 6-43

Execution of the Carnot cycle in a closed system.

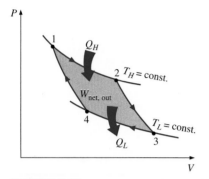

FIGURE 6-44

P-V diagram of the Carnot cycle.

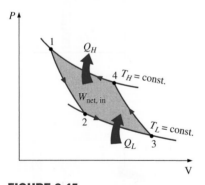

FIGURE 6-45

P-V diagram of the reversed Carnot cycle.

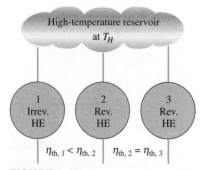

FIGURE 6-46

The Carnot principles.

Reversible Adiabatic Compression (process 4-1, temperature rises from T_L to T_H). State 4 is such that when the low-temperature reservoir is removed, the insulation is put back on the cylinder head, and the gas is compressed in a reversible manner, the gas returns to its initial state (state 1). The temperature rises from T_L to T_H during this reversible adiabatic compression process, which completes the cycle.

The *P-V* diagram of this cycle is shown in Fig. 6-44. Remembering that on a *P-V* diagram the area under the process curve represents the boundary work for quasi-equilibrium (internally reversible) processes, we see that the area under curve 1-2-3 is the work done by the gas during the expansion part of the cycle, and the area under curve 3-4-1 is the work done on the gas during the compression part of the cycle. The area enclosed by the path of the cycle (area 1-2-3-4-1) is the difference between these two and represents the net work done during the cycle.

Notice that if we acted stingily and compressed the gas at state 3 adiabatically instead of isothermally in an effort *to save Q_L*, we would end up back at state 2, retracing the process path 3-2. By doing so we would save Q_L, but we would not be able to obtain any net work output from this engine. This illustrates once more the necessity of a heat engine exchanging heat with at least two reservoirs at different temperatures to operate in a cycle and produce a net amount of work.

The Carnot cycle can also be executed in a steady-flow system. It is discussed in Chap. 8 in conjunction with other power cycles.

Being a reversible cycle, the Carnot cycle is the most efficient cycle operating between two specified temperature limits. Even though the Carnot cycle cannot be achieved in reality, the efficiency of actual cycles can be improved by attempting to approximate the Carnot cycle more closely.

The Reversed Carnot Cycle

The Carnot heat-engine cycle described above is a totally reversible cycle. Therefore, all the processes that comprise it can be *reversed,* in which case it becomes the **Carnot refrigeration cycle.** This time, the cycle remains exactly the same, except that the directions of any heat and work interactions are reversed: Heat in the amount of Q_L is absorbed from the low-temperature reservoir, heat in the amount of Q_H is rejected to a high-temperature reservoir, and a work input of $W_{net, in}$ is required to accomplish all this.

The *P-V* diagram of the reversed Carnot cycle is the same as the one given for the Carnot cycle, except that the directions of the processes are reversed, as shown in Fig. 6-45.

6-9 ■ THE CARNOT PRINCIPLES

The second law of thermodynamics puts limits on the operation of cyclic devices as expressed by the Kelvin-Planck and Clausius statements. A heat engine cannot operate by exchanging heat with a single reservoir, and a refrigerator cannot operate without a net work input from an external source.

We can draw valuable conclusions from these statements. Two conclusions pertain to the thermal efficiency of reversible and irreversible (i.e., actual) heat engines, and they are known as the **Carnot principles** (Fig. 6-46), expressed as follows:

1. *The efficiency of an irreversible heat engine is always less than the efficiency of a reversible one operating between the same two reservoirs.*

2. *The efficiencies of all reversible heat engines operating between the same two reservoirs are the same.*

These two statements can be proved by demonstrating that the violation of either statement results in the violation of the second law of thermodynamics.

To prove the first statement, consider two heat engines operating between the same reservoirs, as shown in Fig. 6-47. One engine is reversible and the other is irreversible. Now each engine is supplied with the same amount of heat Q_H. The amount of work produced by the reversible heat engine is W_{rev}, and the amount produced by the irreversible one is W_{irrev}.

In violation of the first Carnot principle, we assume that the irreversible heat engine is more efficient than the reversible one (that is, $\eta_{th,\,irrev} > \eta_{th,\,rev}$) and thus delivers more work than the reversible one. Now let the reversible heat engine be reversed and operate as a refrigerator. This refrigerator will receive a work input of W_{rev} and reject heat to the high-temperature reservoir. Since the refrigerator is rejecting heat in the amount of Q_H to the high-temperature reservoir and the irreversible heat engine is receiving the same amount of heat from this reservoir, the net heat exchange for this reservoir is zero. Thus, it could be eliminated by having the refrigerator discharge Q_H directly into the irreversible heat engine.

Now considering the refrigerator and the irreversible engine together, we have an engine that produces a net work in the amount of $W_{irrev} - W_{rev}$ while exchanging heat with a single reservoir—a violation of the Kelvin-Planck statement of the second law. Therefore, our initial assumption that $\eta_{th,\,irrev} > \eta_{th,\,rev}$ is incorrect. Then we conclude that no heat engine can be more efficient than a reversible heat engine operating between the same reservoirs.

The second Carnot principle can also be proved in a similar manner. This time, let us replace the irreversible engine by another reversible engine that is

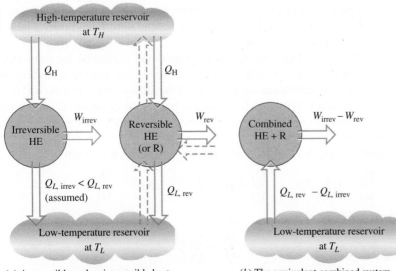

(a) A reversible and an irreversible heat engine operating between the same two reservoirs (the reversible heat engine is then reversed to run as a refrigerator)

(b) The equivalent combined system

FIGURE 6-47

Proof of the first Carnot principle.

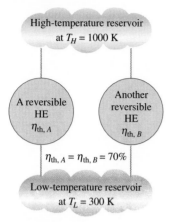

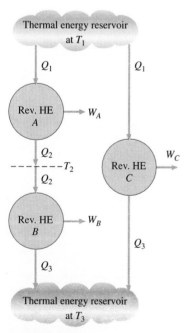

FIGURE 6-48

All reversible heat engines operating between the same two reservoirs have the same efficiency (the second Carnot principle).

FIGURE 6-49

The arrangement of heat engines used to develop the thermodynamic temperature scale.

more efficient and thus delivers more work than the first reversible engine. By following through the same reasoning as above, we will end up having an engine that produces a new amount of work while exchanging heat with a single reservoir, which is a violation of the second law. Therefore we conclude that no reversible heat engine can be more efficient than a reversible one operating between the same two reservoirs, regardless of how the cycle is completed or the kind of working fluid used.

6-10 ■ THE THERMODYNAMIC TEMPERATURE SCALE

A temperature scale that is independent of the properties of the substances that are used to measure temperature is called a **thermodynamic temperature scale.** Such a temperature scale offers great conveniences in thermodynamic calculations, and its derivation is given below using some reversible heat engines.

The second Carnot principle discussed in Sec. 6-9 states that all reversible heat engines have the same thermal efficiency when operating between the same two reservoirs (Fig. 6-48). That is, the efficiency of a reversible engine is independent of the working fluid employed and its properties, the way the cycle is executed, or the type of reversible engine used. Since energy reservoirs are characterized by their temperatures, the thermal efficiency of reversible heat engines is a function of the reservoir temperatures only. That is,

$$\eta_{\text{th, rev}} = g(T_H, T_L)$$

or
$$\frac{Q_H}{Q_L} = f(T_H, T_L) \tag{6-15}$$

since $\eta_{\text{th}} = 1 - Q_L/Q_H$. In these relations T_H and T_L are the temperatures of the high- and low-temperature reservoirs, respectively.

The functional form of $f(T_H, T_L)$ can be developed with the help of the three reversible heat engines shown in Fig. 6-49. Engines A and C are supplied with the same amount of heat Q_1 from the high-temperature reservoir at T_1. Engine C rejects Q_3 to the low-temperature reservoir at T_3. Engine B receives the heat Q_2 rejected by engine A at temperature T_2 and rejects heat in the amount of Q_3 to a reservoir at T_3.

The amounts of heat rejected by engines B and C must be the same since engines A and B can be combined into one reversible engine operating between the same reservoirs as engine C and thus the combined engine will have the same efficiency as engine C. Since the heat input to engine C is the same as the heat input to the combined engines A and B, both systems must reject the same amount of heat.

Applying Eq. 6-15 to all three engines separately, we obtain

$$\frac{Q_1}{Q_2} = f(T_1, T_2), \qquad \frac{Q_2}{Q_3} = f(T_2, T_3), \qquad \text{and} \qquad \frac{Q_1}{Q_3} = f(T_1, T_3)$$

Now consider the identity

$$\frac{Q_1}{Q_3} = \frac{Q_1}{Q_2}\frac{Q_2}{Q_3}$$

which corresponds to

$$f(T_1, T_3) = f(T_1, T_2) \cdot f(T_2, T_3)$$

A careful examination of this equation reveals that the left-hand side is a function of T_1 and T_3, and therefore the right-hand side must also be a function of T_1 and T_3 only, and not T_2. That is, the value of the product on the right-hand side of this equation is independent of the value of T_2. This condition will be satisfied only if the function f has the following form:

$$f(T_1, T_2) = \frac{\phi(T_1)}{\phi(T_2)} \quad \text{and} \quad f(T_2, T_3) = \frac{\phi(T_2)}{\phi(T_3)}$$

so that $\phi(T_2)$ will cancel from the product of $f(T_1, T_2)$ and $f(T_2, T_3)$, yielding

$$\frac{Q_1}{Q_3} = f(T_1, T_3) = \frac{\phi(T_1)}{\phi(T_3)} \tag{6-16}$$

This relation is much more specific than Eq. 6-15 for the functional form of Q_1/Q_3 in terms of T_1 and T_3.

For a reversible heat engine operating between two reservoirs at temperatures T_H and T_L, Eq. 6-16 can be written as

$$\frac{Q_H}{Q_L} = \frac{\phi(T_H)}{\phi(T_L)} \tag{6-17}$$

This is the only requirement that the second law places on the ratio of heat flows to and from the reversible heat engines. Several functions $\phi(T)$ will satisfy this equation, and the choice is completely arbitrary. Lord Kelvin first proposed taking $\phi(T) = T$ to define a thermodynamic temperature scale as (Fig. 6-50)

$$\left(\frac{Q_H}{Q_L}\right)_{\text{rev}} = \frac{T_H}{T_L} \tag{6-18}$$

This temperature scale is called the **Kelvin scale,** and the temperatures on this scale are called **absolute temperatures.** On the Kelvin scale, the temperature ratios depend on the ratios of heat transfer between a reversible heat engine and the reservoirs and are independent of the physical properties of any substance. On this scale, temperatures vary between zero and infinity.

The thermodynamic temperature scale is not completely defined by Eq. 6-18 since it gives us only a ratio of absolute temperatures. We also need to know the magnitude of a kelvin. At the International Conference on Weights and Measures held in 1954, the triple point of water (the state at which all three phases of water exist in equilibrium) was assigned the value 273.16 K (Fig. 6-51). The *magnitude of a kelvin* is defined as 1/273.16 of the temperature interval between absolute zero and the triple-point temperature of water. The magnitudes of temperature units on the Kelvin and Celsius scales are identical (1 K $\equiv$ 1°C). The temperatures on these two scales differ by a constant 273.15:

$$T(°C) = T(K) - 273.15 \tag{6-19}$$

Even though the thermodynamic temperature scale is defined with the help of the reversible heat engines, it is not possible, nor is it practical, to actually operate such an engine to determine numerical values on the absolute temperature scale. Absolute temperatures can be measured accurately by other means, such as the constant-volume ideal-gas thermometer discussed in Chap. 2 together with extrapolation techniques. The validity of Eq. 6-18 can be demonstrated from physical considerations for a reversible cycle using an ideal gas as the working fluid.

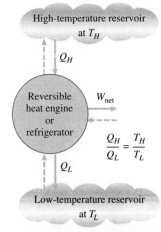

FIGURE 6-50

For reversible cycles, the heat transfer ratio Q_H/Q_L can be replaced by the absolute temperature ratio T_H/T_L.

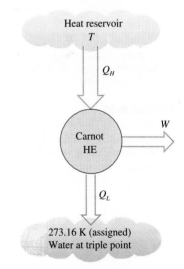

FIGURE 6-51

A conceptual experimental setup to determine thermodynamic temperatures on the Kelvin scale by measuring heat transfers Q_H and Q_L.

6-11 ■ THE CARNOT HEAT ENGINE

The hypothetical heat engine that operates on the reversible Carnot cycle is called the **Carnot heat engine.** The thermal efficiency of any heat engine, reversible or irreversible, is given by Eq. 6-6 as

$$\eta_{th} = 1 - \frac{Q_L}{Q_H}$$

where Q_H is heat transferred to the heat engine from a high-temperature reservoir at T_H, and Q_L is heat rejected to a low-temperature reservoir at T_L. For reversible heat engines, the heat transfer ratio in the above relation can be replaced by the ratio of the absolute temperatures of the two reservoirs, as given by Eq. 6-18. Then the efficiency of a Carnot engine, or any reversible heat engine, becomes

$$\eta_{th, rev} = 1 - \frac{T_L}{T_H} \qquad (6\text{-}20)$$

This relation is often referred to as the **Carnot efficiency,** since the Carnot heat engine is the best known reversible engine. *This is the highest efficiency a heat engine operating between the two thermal energy reservoirs at temperatures T_L and T_H can have* (Fig. 6-52). All irreversible (i.e., actual) heat engines operating between these temperature limits (T_L and T_H) will have lower efficiencies. An actual heat engine cannot reach this maximum theoretical efficiency value because it is impossible to completely eliminate all the irreversibilities associated with the actual cycle.

Note that T_L and T_H in Eq. 6-20 are *absolute temperatures.* Using °C or °F for temperatures in this relation will give results grossly in error.

The thermal efficiencies of actual and reversible heat engines operating between the same temperature limits compare as follows (Fig. 6-53):

$$\eta_{th} \begin{cases} < \eta_{th, rev} & \text{irreversible heat engine} \\ = \eta_{th, rev} & \text{reversible heat engine} \\ > \eta_{th, rev} & \text{impossible heat engine} \end{cases} \qquad (6\text{-}21)$$

Most work-producing devices (heat engines) in operation today have efficiencies under 40 percent, which appear low relative to 100 percent. However, when the performance of actual heat engines is assessed, the efficiencies should not be compared to 100 percent; instead, they should be compared to

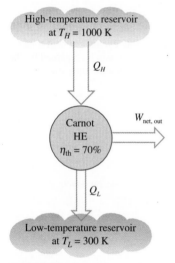

FIGURE 6-52

The Carnot heat engine is the most efficient of all heat engines operating between the same high- and low-temperature reservoirs.

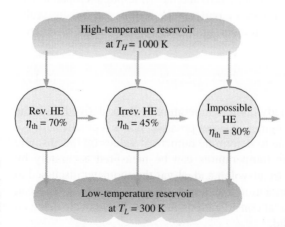

FIGURE 6-53

No heat engine can have a higher efficiency than a reversible heat engine operating between the same high- and low-temperature reservoirs.

the efficiency of a reversible heat engine operating between the same temperature limits—because this is the true theoretical upper limit for the efficiency, not 100 percent.

The maximum efficiency of a steam power plant operating between $T_H = 750$ K and $T_L = 300$ K is 60 percent, as determined from Eq. 6-20. Compared with this value, an actual efficiency of 40 percent does not seem so bad, even though there is still plenty of room for improvement.

It is obvious from Eq. 6-20 that the efficiency of a Carnot heat engine increases as T_H is increased, or as T_L is decreased. This is to be expected since as T_L decreases, so does the amount of heat rejected, and as T_L approaches zero, the Carnot efficiency approaches unity. This is also true for actual heat engines. *The thermal efficiency of actual heat engines can be maximized by supplying heat to the engine at the highest possible temperature* (limited by material strength) *and rejecting heat from the engine at the lowest possible temperature* (limited by the temperature of the cooling medium such as rivers, lakes, or the atmosphere).

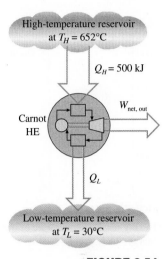

FIGURE 6-54
Schematic for Example 6-6.

EXAMPLE 6-6   Analysis of a Carnot Heat Engine

A Carnot heat engine, shown in Fig. 6-54, receives 500 kJ of heat per cycle from a high-temperature source at 652°C and rejects heat to a low-temperature sink at 30°C. Determine (*a*) the thermal efficiency of this Carnot engine and (*b*) the amount of heat rejected to the sink per cycle.

Solution   The heat supplied to a Carnot heat engine is given. The thermal efficiency and the heat rejected are to be determined.

Analysis   (*a*) The Carnot heat engine is a reversible heat engine, and so its efficiency can be determined from Eq. 6-20 to be

$$\eta_{th,\,C} = \eta_{th,\,rev} = 1 - \frac{T_L}{T_H} = 1 - \frac{(30 + 273)\text{ K}}{(652 + 273)\text{ K}} = \textbf{0.672}$$

That is, this Carnot heat engine converts 67.2 percent of the heat it receives to work.

(*b*) The amount of heat rejected Q_L by this reversible heat engine is easily determined from Eq. 6-18 to be

$$Q_{L,\,rev} = \frac{T_L}{T_H} Q_{H,\,rev} = \frac{(30 + 273)\text{ K}}{(652 + 273)\text{ K}} (500\text{ kJ}) = \textbf{163.8 kJ}$$

Discussion   Note that this Carnot heat engine rejects to a low-temperature sink 163.8 kJ of the 500 kJ of heat it receives during each cycle.

The Quality of Energy

The Carnot heat engine in Example 6-6 receives heat from a source at 925 K and converts 67.2 percent of it to work while rejecting the rest (32.8 percent) to a sink at 303 K. Now let us examine how the thermal efficiency varies with the source temperature when the sink temperature is held constant.

The thermal efficiency of a Carnot heat engine that rejects heat to a sink at 303 K is evaluated at various source temperatures using Eq. 6-20 and is listed in Fig. 6-55. Clearly the thermal efficiency decreases as the source temperature is lowered. When heat is supplied to the heat engine at 500 instead of

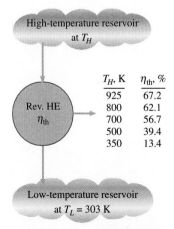

T_H, K	η_{th}, %
925	67.2
800	62.1
700	56.7
500	39.4
350	13.4

FIGURE 6-55
The fraction of heat that can be converted to work as a function of source temperature (for $T_L = 303$ K).

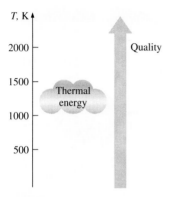

FIGURE 6-56

The higher the temperature of the thermal energy, the higher its quality.

925 K, for example, the thermal efficiency drops from 67.2 to 39.4 percent. That is, the fraction of heat that can be converted to work drops to 39.4 percent when the temperature of the source drops to 500 K. When the source temperature is 350 K, this fraction becomes a mere 13.4 percent.

These efficiency values show that energy has **quality** as well as quantity. It is clear from the thermal efficiency values in Fig. 6-55 that *more of the high-temperature thermal energy can be converted to work. Therefore, the higher the temperature, the higher the quality of the energy* (Fig. 6-56).

Large quantities of solar energy, for example, can be stored in large bodies of water called *solar ponds* at about 350 K. This stored energy can then be supplied to a heat engine to produce work (electricity). However, the efficiency of solar pond power plants is very low (under 5 percent) because of the low quality of the energy stored in the source, and the construction and maintenance costs are relatively high. Therefore, they are not competitive even though the energy supply of such plants is free. The temperature (and thus the quality) of the solar energy stored could be raised by utilizing concentrating collectors, but the equipment cost in that case becomes very high.

Work is a more valuable form of energy than heat since 100 percent of work can be converted to heat, but only a fraction of heat can be converted to work. When heat is transferred from a high-temperature body to a lower-temperature one, it is degraded since less of it now can be converted to work. For example, if 100 kJ of heat is transferred from a body at 1000 K to a body at 300 K, at the end we will have 100 kJ of thermal energy stored at 300 K, which has no practical value. But if this conversion is made through a heat engine, up to $1 - 300/1000 = 70$ percent of it could be converted to work, which is a more valuable form of energy. Thus 70 kJ of work potential is wasted as a result of this heat transfer, and energy is degraded.

Quantity versus Quality in Daily Life

At times of energy crisis, we are bombarded with speeches and articles on how to "conserve" energy. Yet we all know that the *quantity* of energy is already conserved. What is not conserved is the *quality* of energy, or the work potential of energy. Wasting energy is synonymous to converting it to a less useful form. One unit of high-quality energy can be more valuable than three units of lower-quality energy. For example, a finite amount of heat energy at high temperature is more attractive to power plant engineers than a vast amount of heat energy at low temperature, such as the energy stored in the upper layers of the oceans at tropical climates.

As part of our culture, we seem to be fascinated by quantity, and little attention is given to quality. But quantity alone cannot give the whole picture, and we need to consider quality as well. That is, we need to look at something from both the first- and second-law points of view when evaluating some thing, even in nontechnical areas. Below we present some ordinary events and show their relevance to the second law of thermodynamics.

Consider two students Andy and Wendy. Andy has 10 friends who never miss his parties and are always around during fun times. But they seem to be busy when Andy needs their help. Wendy, on the other hand, has five friends. But they are never too busy for her, and she can count on them at times of need. Let us now try to answer the question, *Who has more friends?* From the first law point of view, which considers quantity only, it is obvious that Andy has more friends. But from the second-law point of view, which considers quality as well, there is no doubt that Wendy is the one with more friends.

Another example with which most people will identify is the multibillion-dollar diet industry, which is primarily based on the first law of thermodynamics. But considering that 90 percent of the people who lose weight gain it back quickly, with interest, suggests that the first law alone does not give the whole picture. This is also confirmed by recent work that shows that calories that come from fat are more likely to be stored as fat than the calories that come from carbohydrates and protein. A Stanford study found that body weight was related to fat calories consumed and not calories per se. A Harvard study found no correlation between calories eaten and degree of obesity. A major Cornell University survey involving 6500 people in nearly all provinces of China found that the Chinese eat more—gram for gram, calorie for calorie—than Americans do, but they weigh less, with less body fat. Studies indicate that the metabolism rates and hormone levels change noticeably in the mid 30s. Some researchers concluded that prolonged dieting teaches a body to survive on fewer calories, making it more *fuel efficient.* This probably explains why the dieters gain more weight than they lost once they go back to their normal eating levels.

People who seem to be eating whatever they want, whenever they want, are living proof that the calorie-counting technique (the first law) leaves many questions on dieting unanswered. Obviously, more research focused on the second-law effects of dieting is needed before we can fully understand the weight-gain and weight-loss process.

It is tempting to judge things on the basis of their *quantity* instead of their *quality* since assessing quality is much more difficult than assessing quantity. However, assessments made on the basis of quantity only (the first law) may be grossly inadequate and misleading.

6-12 ■ THE CARNOT REFRIGERATOR AND HEAT PUMP

A refrigerator or a heat pump that operates on the reversed Carnot cycle is called a **Carnot refrigerator,** or a **Carnot heat pump.** The coefficient of performance of any refrigerator or heat pump, reversible or irreversible, is given by Eqs. 6-11 and 6-13 as

$$COP_R = \frac{1}{Q_H/Q_L - 1} \quad \text{and} \quad COP_{HP} = \frac{1}{1 - Q_L/Q_H}$$

where Q_L is the amount of heat absorbed from the low-temperature medium and Q_H is the amount of heat rejected to the high-temperature medium. The COPs of all reversible (such as Carnot) refrigerators or heat pumps can be determined by replacing the heat transfer ratios in the above relations by the ratios of the absolute temperatures of the high- and low-temperature media, as expressed by Eq. 6-18. Then the COP relations for reversible refrigerators and heat pumps become

$$COP_{R, \text{rev}} = \frac{1}{T_H/T_L - 1} \tag{6-22}$$

and

$$COP_{HP, \text{rev}} = \frac{1}{1 - T_L/T_H} \tag{6-23}$$

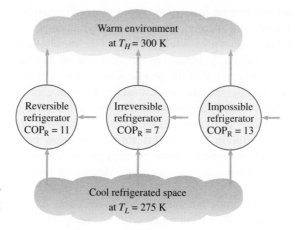

FIGURE 6-57

No refrigerator can have a higher COP
than a reversible refrigerator operating
between the same temperature limits.

*These are the highest coefficients of performance that a refrigerator or a heat
pump operating between the temperature limits of T_L and T_H can have.* All
actual refrigerators or heat pumps operating between these temperature limits
(T_L and T_H) will have lower coefficients of performance (Fig. 6-57).

The coefficients of performance of actual and reversible (such as Carnot)
refrigerators operating between the same temperature limits can be compared
as follows:

$$\text{COP}_R \begin{cases} < \text{COP}_{R,\,rev} & \text{irreversible refrigerator} \\ = \text{COP}_{R,\,rev} & \text{reversible refrigerator} \\ > \text{COP}_{R,\,rev} & \text{impossible refrigerator} \end{cases} \quad (6\text{-}24)$$

A similar relation can be obtained for heat pumps by replacing all values
of COP_R in Eq. 6-24 by COP_{HP}.

The COP of a reversible refrigerator or heat pump is the maximum theoretical value for the specified temperature limits. Actual refrigerators or heat
pumps may approach these values as their designs are improved, but they can
never reach them.

As a final note, the COPs of both the refrigerators and the heat pumps decrease as T_L decreases. That is, it requires more work to absorb heat from
lower-temperature media. As the temperature of the refrigerated space approaches zero, the amount of work required to produce a finite amount of refrigeration approaches infinity and COP_R approaches zero.

EXAMPLE 6-7   A Questionable Claim for a Refrigerator

An inventor claims to have developed a refrigerator that maintains the refrigerated
space at 35°F while operating in a room where the temperature is 75°F and that
has a COP of 13.5. Is this claim reasonable?

Solution   An extraordinary claim made for the performance of a refrigerator is
to be evaluated.

Assumptions   Steady operating conditions exist.

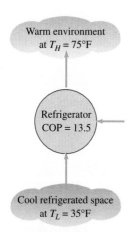

Warm environment
at $T_H = 75°F$

Refrigerator
COP = 13.5

Cool refrigerated space
at $T_L = 35°F$

FIGURE 6-58
Schematic for Example 6-7.

Analysis   The performance of this refrigerator (shown in Fig. 6-58) can be evaluated by comparing it with a Carnot or any other reversible refrigerator operating
between the same temperature limits:

$$\text{COP}_{R,\,max} = \text{COP}_{R,\,rev} = \frac{1}{T_H/T_L - 1}$$

$$= \frac{1}{(75 + 460\ R)/(35 + 460\ R) - 1} = 12.4$$

This is the highest COP a refrigerator can have when removing heat from a cool medium at 35°F to a warmer medium at 75°F. Since the COP claimed by the inventor is above this maximum value, the claim is *false*.

EXAMPLE 6-8 Heating a House by a Carnot Heat Pump

A heat pump is to be used to heat a house during the winter, as shown in Fig. 6-59. The house is to be maintained at 21°C at all times. The house is estimated to be losing heat at a rate of 135,000 kJ/h when the outside temperature drops to −5°C. Determine the minimum power required to drive this heat pump.

Solution A heat pump maintains a house at a fixed temperature. The required minimum power input to the heat pump is to be determined.

Assumptions Steady operating conditions exist.

Analysis The heat pump must supply heat to the house at a rate of Q_H = 135,000 kJ/h = 37.5 kW. The power requirements will be minimum if a reversible heat pump is used to do the job. The COP of a reversible heat pump operating between the house (T_H = 21 + 273 = 294 K) and the outside air (T_L = −5 + 273 = 268 K) is

$$\text{COP}_{\text{HP, rev}} = \frac{1}{1 - T_L/T_H} = \frac{1}{1 - (268 \text{ K}/294 \text{ K})} = 11.3$$

Then the required power input to this reversible heat pump is determined from the definition of the COP, Eq. 6-12:

$$\dot{W}_{\text{net, in}} = \frac{\dot{Q}_H}{\text{COP}_{\text{HP}}} = \frac{37.5 \text{ kW}}{11.3} = \textbf{3.32 kW}$$

Discussion This heat pump can meet the heating requirements of this house by consuming electric power at a rate of 3.32 kW only. If this house were to be heated by electric resistance heaters instead, the power consumption would jump up 11.3 times to 37.5 kW. This is because in resistance heaters the electric energy is converted to heat at a one-to-one ratio. With a heat pump, however, energy is absorbed from the outside and carried to the inside using a refrigeration cycle that consumes only 3.32 kW. Notice that the heat pump does not create energy. It merely transports it from one medium (the cold outdoors) to another (the warm indoors).

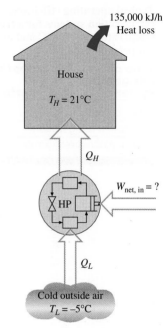

FIGURE 6-59

Schematic for Example 6-8.

6-13 ■ HOUSEHOLD REFRIGERATORS

Refrigerators to preserve perishable foods have long been one of the essential appliances in a household. They have proven to be highly durable and reliable, providing satisfactory service for over 15 years. A typical household refrigerator is actually a combination refrigerator-freezer since it has a freezer compartment to make ice and to store frozen food.

Today's refrigerators use much less energy as a result of using *smaller* and *higher-efficiency* motors and compressors, *better insulation materials, larger coil surface areas,* and *better door seals* (Fig. 6-60). At an average electricity rate of 8.3 cents per kWh, an average refrigerator costs about $72 a year to run, which is half the annual operating cost of a refrigerator 20 years ago. Replacing a 20-year-old, 18-ft³ refrigerator with a new energy-efficient model will save over 1000 kWh of electricity per year. For the environment, this means a reduction of over 1 ton of CO_2, which causes global climate change, and over 10 kg of SO_2, which causes acid rain.

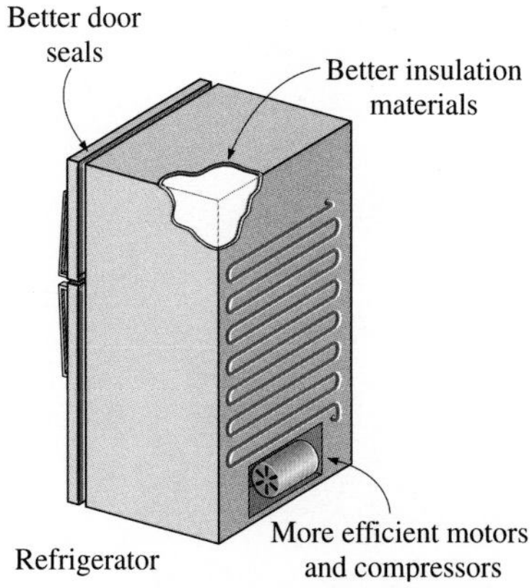

Better door
seals

Better insulation
materials

More efficient motors
and compressors

Refrigerator

FIGURE 6-60

Today's refrigerators are much more efficient because of the improvements in technology and manufacturing.

TABLE 6-3

Typical operating efficiencies of some refrigeration systems for a freezer temperature of −18°C and ambient temperature of 32°C.

Type of refrigeration system	Coefficient of performance
Vapor-compression	1.3
Absorption refrigeration	0.4
Thermoelectric refrigeration	0.1

Despite the improvements made in several areas during the past 100 years in household refrigerators, the basic *vapor-compression refrigeration cycle* has remained unchanged. The alternative *absorption refrigeration* and *thermoelectric refrigeration* systems are currently more expensive and less efficient, and they have found limited use in some specialized applications (Table 6-3).

A household refrigerator is designed to maintain the freezer section at −18°C (0°F) and the refrigerator section at 3°C (37°F). Lower freezer temperatures increase energy consumption without improving the storage life of frozen foods significantly. Different temperatures for the storage of specific foods can be maintained in the refrigerator section by using *special-purpose* compartments.

Practically all full-size refrigerators have a large *air-tight* drawer for leafy vegetables and fresh fruits to seal in moisture and to protect them from the drying effect of cool air circulating in the refrigerator. A covered *egg compartment* in the lid extends the life of eggs by slowing down the moisture loss from the eggs. It is common for refrigerators to have a special warmer compartment for *butter* in the door to maintain butter at spreading temperature. The compartment also isolates butter and prevents it from absorbing *odors* and *tastes* from other food items. Some upscale models have a temperature-controlled *meat compartment* maintained at −0.5°C (31°F), which keeps meat at the lowest safe temperature without freezing it, and thus extending its storage life. The more expensive models come with an automatic *icemaker* located in the freezer section that is connected to the water line, as well as automatic ice and chilled-water dispensers. A typical icemaker can produce 2 to 3 kg of ice per day and store 3 to 5 kg of ice in a removable ice storage container.

Household refrigerators consume from about 90 W to 600 W of electrical energy when running and are designed to perform satisfactorily in environments at up to 43°C (110°F). Refrigerators run intermittently, as you may have noticed, running about 30 percent of the time under normal use in a house at 25°C (77°F).

For specified external dimensions, a refrigerator is desired to have *maximum* food storage volume, *minimum* energy consumption, and the *lowest* possible cost to the consumer. The total food storage volume has been increased over the years without an increase in the external dimensions by using thinner but more effective insulation and minimizing the space occupied by the compressor and the condenser. Switching from the fiberglass insulation (thermal conductivity $k = 0.032$–0.040 W/m · °C) to expanded-in-place urethane foam insulation ($k = 0.019$ W/m · °C) made it possible to reduce the wall thickness of the refrigerator by almost half, from about 90 mm to 48 mm for the freezer section and from about 70 mm to 40 mm for the refrigerator section. The rigidity and bonding action of the foam also provide additional structural support. However, the entire shell of the refrigerator must be carefully sealed to prevent any water leakage or moisture migration into the insulation since moisture degrades the effectiveness of insulation.

The size of the compressor and the other components of a refrigeration system are determined on the basis of the anticipated heat load (or refrigeration load), which is the rate of heat flow into the refrigerator. The heat load consists of the *predictable part,* such as heat transfer through the walls and door gaskets of the refrigerator, fan motors, and defrost heaters (Fig. 6-61), and the *unpredictable part,* which depends on the user habits such as opening the door, making ice, and loading the refrigerator. The amount of *energy*

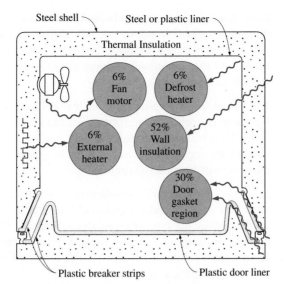

Steel shell — Steel or plastic liner —
Thermal Insulation

6%
Fan
motor

6%
Defrost
heater

6%
External
heater

52%
Wall
insulation

30%
Door
gasket
region

Plastic breaker strips   Plastic door liner

FIGURE 6-61

The cross section of a refrigerator showing the relative magnitudes of various effects that constitute the predictable heat load (from ASHRAE *Handbook of Refrigeration,* Chap. 48, Fig. 2).

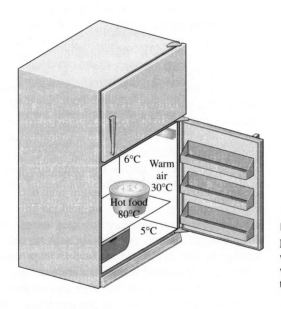

6°C   Warm
air
30°C

Hot food
80°C

5°C

FIGURE 6-62

Putting hot foods into the refrigerator without cooling them first not only wastes energy but also could spoil the foods nearby.

consumed by the refrigerator can be minimized by practicing good *conservation measures* as discussed below.

1. *Open the refrigerator door the fewest times possible* for the shortest duration possible. Each time the refrigerator door is opened, the cool air inside is replaced by the warmer air outside, which needs to be cooled. Keeping the refrigerator or freezer full will save energy by reducing the amount of cold air that can escape each time the door is opened.

2. *Cool the hot foods* to room temperature first before putting them into the refrigerator. Moving a hot pan from the oven directly into the refrigerator not only wastes energy by making the refrigerator work longer, but it also causes the nearby perishable foods to spoil by creating a warm environment in its immediate surroundings (Fig. 6-62).

3. *Clean the condenser coils* behind the refrigerator. The dust and grime that collect on the coils act as insulation that slows down heat dissipation through them. Cleaning the coils a couple of times a year with a damp cloth or a vacuum cleaner will improve cooling ability of the refrigerator while cutting down the power consumption by a few percent. Sometimes a fan is used to force-cool the condensers of large or built-in refrigerators, and the strong air motion keeps the coils clean.

4. *Check the door gasket* for air leaks. This can be done by placing a flashlight into the refrigerator, turning off the kitchen lights, and looking for light leaks. Heat transfer through the door gasket region accounts for almost one-third of the regular heat load of the refrigerators, and thus any defective door gaskets must be repaired immediately.

5. *Avoid unnecessarily low temperature settings.* The recommended temperatures for freezers and refrigerators are −18°C (0°F) and 3°C (37°F), respectively. Setting the freezer temperature below −18°C adds significantly to the energy consumption but does not add much to the storage life of frozen foods. Keeping temperatures 6°C (or 10°F) below recommended levels can increase the energy use by as much as 25 percent.

6. *Avoid excessive ice build-up* on the interior surfaces of the evaporator. The ice layer on the surface acts as insulation and slows down heat transfer from the freezer section to the refrigerant. The refrigerator should be defrosted by manually turning off the temperature control switch when the ice thickness exceeds a few millimeters.

Defrosting is done automatically in no-frost refrigerators by supplying heat to the evaporator by a 300-W to 1000-W resistance heater or by hot refrigerant gas, periodically for short periods. The water is then drained to a pan outside where it is evaporated using the heat dissipated by the condenser. The no-frost evaporators are basically finned tubes subjected to air flow circulated by a fan. Practically all the frost collects on fins, which are the coldest surfaces, leaving the exposed surfaces of the freezer section and the frozen food frost-free.

7. *Use the power-saver switch* that controls the heating coils and prevents condensation on the outside surfaces in humid environments. The low-wattage heaters are used to raise the temperature of the outer surfaces of the refrigerator at critical locations above the dew point in order to avoid water droplets forming on the surfaces and sliding down. Condensation is most likely to occur in summer in hot and humid climates in homes without air-conditioning. The moisture formation on the surfaces is undesirable since it may cause the painted finish of the outer surface to deteriorate and it may wet the kitchen floor. About 10 percent of the total energy consumed by the refrigerator can be saved by turning this heater off and keeping it off unless there is visible condensation on the outer surfaces.

8. *Do not block the air flow passages* to and from the condenser coils of the refrigerator. The heat dissipated by the condenser to the air is carried away by air that enters through the bottom and sides of the refrigerator and leaves through the top. Any blockage of this natural convection air circulation path by large objects such as several cereal boxes on top of the refrigerator will degrade the performance of the condenser and thus the refrigerator (Fig. 6-63).

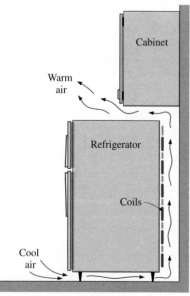

Cabinet

Warm
air

Refrigerator

Coils

Cool
air

FIGURE 6-63

The condenser coils of a refrigerator must be cleaned periodically, and the airflow passages must not be blocked to maintain high performance.

These and other commonsense conservation measures will result in a reduction in the energy and maintenance costs of a refrigerator as well as an extended trouble-free life of the device.

EXAMPLE 6-9 Malfunction of a Refrigerator Light Switch

The interior lighting of refrigerators is provided by incandescent lamps whose switches are actuated by the opening of the refrigerator door. Consider a refrigerator whose 40-W light bulb remains on continuously as a result of a malfunction of the switch (Fig. 6-64). If the refrigerator has a coefficient of performance of 1.3 and the cost of electricity is 8 cents per kWh, determine the increase in the energy consumption of the refrigerator and its cost per year if the switch is not fixed.

Solution The light bulb consumes 40 W of power when it is on, and thus adds 40 W to the heat load of the refrigerator.

Assumptions The life of the light bulb is more than 1 year.

Analysis Noting that the coefficient of performance of the refrigerator is 1.3, the power consumed by the refrigerator to remove the heat generated by the light bulb is determined from

$$\dot{W}_{\text{refrig}} = \frac{\dot{Q}_{\text{refrig}}}{\text{COP}_R} = \frac{40\ \text{W}}{1.3} = 30.8\ \text{W}$$

Therefore, the total additional power consumed by the refrigerator is

$$\dot{W}_{\text{total, additional}} = \dot{W}_{\text{light}} + \dot{W}_{\text{refrig}} = 40 + 30.8 = 70.8\ \text{W}$$

The total number of hours in a year is

$$\text{Annual hours} = (365\ \text{days/yr})(24\ \text{h/day}) = 8760\ \text{h/yr}$$

Assuming the refrigerator is opened 20 times a day for an average of 30 s, the light would normally be on for

$$\text{Normal operating hours} = (20\ \text{times/day})(30\ \text{s/time})(1\ \text{h/3600 s})(365\ \text{days/yr})$$
$$= 61\ \text{h/yr}$$

Then the additional hours the light remains on as a result of the malfunction becomes

$$\text{Additional operating hours} = \text{Annual hours} - \text{Normal operating hours}$$
$$= 8760 - 61 = 8699\ \text{h/yr}$$

Therefore, the additional electric power consumption and its cost per year are

$$\text{Additional power consumption} = \dot{W}_{\text{total, additional}} \times (\text{Additional operating hours})$$
$$= (0.0708\ \text{kW})(8699\ \text{h/yr}) = \mathbf{616\ kWh/yr}$$

and

$$\text{Additional power cost} = (\text{Additional power consumption})(\text{Unit cost})$$
$$= (616\ \text{kWh/yr})(\$0.08/\text{kWh}) = \mathbf{\$49.3/yr}$$

Discussion Note that not repairing the switch will cost the homeowner about $50 a year. This is alarming when we consider that at $0.08/kWh, a typical refrigerator consumes about $70 worth of electricity a year.

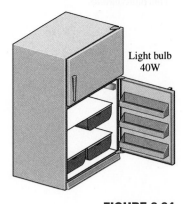

Light bulb
40W

FIGURE 6-64
Schematic for Example 6-9.

6-14 ■ SUMMARY

The *second law of thermodynamics* states that processes occur in a certain direction, not in any direction. A process will not occur unless it satisfies both the first and the second laws of thermodynamics. Bodies that can absorb or reject finite amounts of heat isothermally are called *thermal energy reservoirs* or *heat reservoirs*.

Work can be converted to heat directly, but heat can be converted to work only by some devices called heat engines. The *thermal efficiency* of a heat engine is defined as

$$\eta_{th} = \frac{W_{net,\,out}}{Q_H} = 1 - \frac{Q_L}{Q_H}$$

where $W_{net,\,out}$ is the net work output of the heat engine, Q_H is the amount of heat supplied to the engine, and Q_L is the amount of heat rejected by the engine.

Refrigerators and heat pumps are devices that absorb heat from low-temperature media and reject it to higher-temperature ones. The performance of a refrigerator or a heat pump is expressed in terms of the *coefficient of performance,* which is defined as

$$COP_R = \frac{Q_L}{W_{net,\,in}} = \frac{1}{Q_H/Q_L - 1}$$

$$COP_{HP} = \frac{Q_H}{W_{net,\,in}} = \frac{1}{1 - Q_L/Q_H}$$

The *Kelvin-Planck statement* of the second law of thermodynamics states that no heat engine can produce a net amount of work while exchanging heat with a single reservoir only. The *Clausius statement* of the second law states that no device can transfer heat from a cooler body to a warmer one without leaving an effect on the surroundings.

Any device that violates the first or the second law of thermodynamics is called a *perpetual-motion machine.*

A process is said to be *reversible* if both the system and the surroundings can be restored to their original conditions. Any other process is *irreversible.* The effects such as friction, non-quasi-equilibrium expansion or compression, and heat transfer through a finite temperature difference render a process irreversible and are called *irreversibilities.*

The *Carnot cycle* is a reversible cycle that is composed of four reversible processes, two isothermal and two adiabatic. The *Carnot principles* state that the thermal efficiencies of all reversible heat engines operating between the same two reservoirs are the same, and that no heat engine is more efficient than a reversible one operating between the same two reservoirs. These statements form the basis for establishing a *thermodynamic temperature scale* related to the heat transfers between a reversible device and the high- and low-temperature reservoirs by

$$\left(\frac{Q_H}{Q_L}\right)_{rev} = \frac{T_H}{T_L}$$

Therefore, the Q_H/Q_L ratio can be replaced by T_H/T_L for reversible devices, where T_H and T_L are the absolute temperatures of the high- and low-temperature reservoirs, respectively.

A heat engine that operates on the reversible Carnot cycle is called a *Carnot heat engine.* The thermal efficiency of a Carnot heat engine, as well as all other reversible heat engines, is given by

$$\eta_{th,\,rev} = 1 - \frac{T_L}{T_H}$$

This is the maximum efficiency a heat engine operating between two reservoirs at temperatures T_H and T_L can have.

The COPs of reversible refrigerators and heat pumps are given in a similar manner as

$$COP_{R, rev} = \frac{1}{T_H/T_L - 1}$$

and

$$COP_{HP, rev} = \frac{1}{1 - T_L/T_H}$$

Again, these are the highest COPs a refrigerator or a heat pump operating between the temperature limits of T_H and T_L can have.

REFERENCES AND SUGGESTED READING

1. W. Z. Black and J. G. Hartley. *Thermodynamics.* New York: Harper & Row, 1985.

2. Y. A. Çengel and M. A. Boles. *Thermodynamics: An Engineering Approach.* 3rd ed. New York: McGraw-Hill, 1998.

3. D. Stewart. "Wheels Go Round and Round, but Always Run Down." November 1986, *Smithsonian,* pp. 193–208.

4. K. Wark and D. E. Richards. *Thermodynamics.* 6th ed. New York: McGraw-Hill, 1999.

PROBLEMS*

Introduction to the Second Law and Heat Reservoirs

6-1C A mechanic claims to have developed a car engine that runs on water instead of gasoline. What is your response to this claim?

6-2C Describe an imaginary process that satisfies the first law but violates the second law of thermodynamics.

6-3C Describe an imaginary process that satisfies the second law but violates the first law of thermodynamics.

6-4C Describe an imaginary process that violates both the first and the second laws of thermodynamics.

6-5C An experimentalist claims to have raised the temperature of a small amount of water to 150°C by transferring heat from high-pressure steam at 120°C. Is this a reasonable claim? Why? Assume no refrigerator or heat pump is used in the process.

6-6C Consider the energy dissipated by a computer in a room. What is a suitable choice for a thermal energy reservoir?

6-7C What is a thermal energy reservoir? Give some examples.

6-8C Consider the process of baking potatoes in a conventional oven. Can the hot air in the oven be treated as a thermal energy reservoir? Explain.

6-9C Consider the energy generated by a TV set. What is a suitable choice for a thermal energy reservoir?

*Students are encouraged to answer all the concept "C" questions.

Heat Engines and Thermal Efficiency

6-10C   Is it possible for a heat engine to operate without rejecting any waste heat to a low-temperature reservoir? Explain.

6-11C   What are the characteristics of all heat engines?

6-12C   Describe two ways to determine the net work output of a heat engine.

6-13C   Consider the process of baking potatoes in a conventional oven. How would you define the efficiency of the oven for this baking process?

6-14C   Consider a pan of water being heated (*a*) by placing it on an electric range and (*b*) by placing a heating element in the water. Which method is a more efficient way of heating water? Explain.

6-15C   Which is a more efficient way of converting electricity to light—using a light bulb or using a fluorescent tube?

6-16C   Baseboard heaters are basically electric resistance heaters and are frequently used in space heating. A homeowner claims that her 5-year-old baseboard heaters have a conversion efficiency of 100 percent. Is this claim in violation of any thermodynamic laws? Explain.

6-17C   What is the Kelvin-Planck expression of the second law of thermodynamics?

6-18C   Does a heat engine that has a thermal efficiency of 100 percent necessarily violate (*a*) the first law and (*b*) the second law of thermodynamics? Explain.

6-19C   In the absence of any friction and other irreversibilities, can a heat engine have an efficiency of 100 percent? Explain.

6-20C   Are the efficiencies of all the work-producing devices, including the hydroelectric power plants, limited by the Kelvin-Planck statement of the second law? Explain.

6-21   An 800-MW steam power plant, which is cooled by a nearby river, has a thermal efficiency of 40 percent. Determine the rate of heat transfer to the river water. Will the actual heat transfer rate be higher or lower than this value? Why?

6-22   A steam power plant receives heat from a furnace at a rate of 280 GJ/h. Heat losses to the surrounding air from the steam as it passes through the pipes and other components are estimated to be about 8 GJ/h. If the waste heat is transferred to the cooling water at a rate of 145 GJ/h, determine (*a*) net power output and (*b*) the thermal efficiency of this power plant.
Answers: (*a*) 35.3 MW, (*b*) 45.4 percent

6-23E   A car engine with a power output of 95 hp has a thermal efficiency of 28 percent. Determine the rate of fuel consumption if the heating value of the fuel is 19,000 Btu/lbm.

6-24   A steam power plant with a power output of 150 MW consumes coal at a rate of 60 tons/h. If the heating value of the coal is 30,000 kJ/kg, determine the thermal efficiency of this plant (1 ton $\equiv$ 1000 kg).
Answer: 30.0 percent

6-25 An automobile engine consumes fuel at a rate of 20 L/h and delivers 60 kW of power to the wheels. If the fuel has a heating value of 44,000 kJ/kg and a density of 0.8 g/cm³, determine the efficiency of this engine.

Answer: 30.7 percent

6-26E Solar energy stored in large bodies of water, called solar ponds, is being used to generate electricity. If such a solar power plant has an efficiency of 4 percent and a net power output of 300 kW, determine the average value of the required solar energy collection rate, in Btu/h.

6-27 The United States produces about 55 percent of its electricity from coal, and the efficiency of conventional coal-fired power plants is about 34 percent. The combined power rating of the power plants in the United States is about 500,000 MW. Assuming steady power generation at 400 MW, determine the amount of heat rejected by the coal-fired power plants in the United States per year, in kJ.

6-28 The Department of Energy projects that between the years 1995 and 2010, the United States will need to build new power plants to generate an additional 150,000 MW of electricity to meet the increasing demand for electric power. One possibility is to build coal-fired power plants, which cost $1300 per kW to construct and have an efficiency of 34 percent. Another possibility is to use the clean-burning Integrated Gasification Combined Cycle (IGCC) plants where the coal is subjected to heat and pressure to gasify it while removing sulfur and particulate matter from it. The gaseous coal is then burned in a gas turbine, and part of the waste heat from the exhaust gases is recovered to generate steam for the steam turbine. Currently the construction of IGCC plants costs about $1500 per kW, but their efficiency is about 45 percent. The average heating value of the coal is about 28,000,000 kJ per ton (that is, 28,000,000 kJ of heat is released when 1 ton of coal is burned). If the IGCC plant is to recover its cost difference from fuel savings in five years, determine what the price of coal should be in $ per ton.

6-29 Repeat Prob. 6-28 for a simple payback period of three years instead of five years.

6-30 Wind energy has been used since 4000 BC to power sailboats, grind grain, pump water for farms, and, more recently, generate electricity. In the United States alone, more than 6 million small windmills, most of them under 5 hp, have been used since the 1850s to pump water. Small windmills have been used to generate electricity since 1900, but the development of modern wind turbines occurred only recently in response to the energy crises in the early 1970s. The cost of wind power has dropped an order of magnitude from about $0.50/kWh in the early 1980s to about $0.05/kWh in the mid-1990s, which is about the price of electricity generated at coal-fired power plants. Areas with an average wind speed of 6 m/s (or 14 mph) are potential sites for economical wind power generation. Commercial wind turbines generate from 100 kW to 3.2 MW of electric power each at peak design conditions. The blade span (or rotor) diameter of the 3.2 MW wind turbine built by Boeing Engineering is 320 ft (97.5 m). The rotation speed of rotors of wind turbines is usually under 40 rpm (under 20 rpm for large turbines). Altamont Pass in California is the world's largest windfarm with 15,000 modern wind turbines. This farm and two others in California produced 2.8 billion kWh of electricity in 1991, which is enough power to meet the electricity needs of San Francisco. Many wind turbines currently in operation have just two blades. This is

FIGURE P6-30

because at tip speeds of 100 to 200 mph, the efficiency of the two bladed turbine approaches the theoretical maximum, and the increase in the efficiency by adding a third or fourth blade is so little that they do not justify the added cost and weight.

Consider a wind turbine with an 80-m-diameter rotor that is rotating at 20 rpm under steady winds at an average velocity of 30 km/h. Assuming the turbine has an efficiency of 35 percent (i.e., it converts 35 percent of the kinetic energy of the wind to electricity), determine (*a*) the power produced, in kW; (*b*) the tip speed of the blade, in km/h; and (*c*) the revenue generated by the wind turbine per year if the electric power produced is sold to the utility at $0.06/kWh.

6-31   Repeat Prob. 6-30 for an average wind velocity of 25 km/h.

6-32E   An Ocean Thermal Energy Conversion (OTEC) power plant built in Hawaii in 1987 was designed to operate between the temperature limits of 86°F at the ocean surface and 41°F at a depth of 2100 ft. About 13,300 gpm of cold sea water was to be pumped from deep ocean through a 40-in-diameter pipe to serve as the cooling medium or heat sink. If the cooling water experiences a temperature rise of 6°C and the thermal efficiency is 3 percent, determine the amount of power generated.

Energy Conversion Efficiencies

6-33   Consider a 3-kW hooded electric open burner in an area where the unit costs of electricity and natural gas are $0.07/kWh and $0.60/therm, respectively. The efficiency of open burners can be taken to be 73 percent for electric burners and 38 percent for gas burners. Determine the rate of energy consumption by the burner and the unit cost of utilized energy for both electric and gas burners.

6-34   A 75-hp motor that has an efficiency of 91.0 percent is worn out and is replaced by a high-efficiency 75-hp motor that has an efficiency of 95.4 percent. Determine the reduction in the heat gain of the room due to higher efficiency under full-load conditions.

6-35   A 60-hp electric car is powered by an electric motor mounted in the engine compartment. If the motor has an average efficiency of 88 percent, determine the rate of heat supply by the motor to the engine compartment at full load.

6-36   A 75-hp motor that has an efficiency of 91.0 percent is worn out and is to be replaced by a high-efficiency motor that has an efficiency of 95.4 percent. The motor operates 4368 hours a year at a load factor of 0.75. Taking the cost of electricity to be $0.08/kWh, determine the amount of energy and money saved as a result of installing the high-efficiency motor instead of the standard motor. Also, determine the simple payback period if the purchase prices of the standard and high-efficiency motors are $5449 and $5520, respectively.

6-37E   The steam requirements of a manufacturing facility are being met by a boiler whose rated heat input is 3.6×10^6 Btu/h. The combustion efficiency of the boiler is measured to be 0.7 by a hand-held flue gas analyzer. After tuning up the boiler, the combustion efficiency rises to 0.8. The boiler operates

1500 hours a year intermittently. Taking the unit cost of energy to be $4.35/$10^6$ Btu, determine the annual energy and cost savings as a result of tuning up the boiler.

6-38 The space heating of a facility is accomplished by natural gas heaters that are 80 percent efficient. The compressed air needs of the facility are met by a large liquid-cooled compressor. The coolant of the compressor is cooled by air in a liquid-to-air heat exchanger whose airflow section is 1.0 m high and 1.0 m wide. During typical operation, the air is heated from 20°C to 52°C as it flows through the heat exchanger. The average velocity of air on the inlet side is measured to be 3 m/s. The compressor operates 20 hours a day and 5 days a week throughout the year. Taking the heating season to be 6 months (26 weeks) and the cost of the natural gas to be $0.50/therm (1 therm = 105,500 kJ), determine how much money will be saved by diverting the compressor waste heat into the facility during the heating season.

6-39 An exercise room has 8 weight-lifting machines that have no motors and 4 treadmills each equipped with a 2.5-hp motor. The motors operate at an average load factor of 0.7, at which their efficiency is 0.77. During peak evening hours, all 12 pieces of exercising equipment are used continuously, and there are also two people doing light exercises while waiting in line for one piece of the equipment. Determine the rate of heat gain of the exercise room from people and the equipment at peak load conditions.

6-40 Consider a classroom for 40 students and one instructor, each generating heat at a rate of 100 W. Lighting is provided by 18 fluorescent light bulbs, 40 W each, and the ballasts consume an additional 10 percent. Determine the rate of internal heat generation in this classroom when it is fully occupied.

6-41 A room is cooled by circulating chilled water through a heat exchanger located in a room. The air is circulated through the heat exchanger by a 0.25-hp fan. Typical efficiency of small electric motors driving 0.25-hp equipment is 54 percent. Determine the rate of heat supply by the fan–motor assembly to the room.

Refrigerators and Heat Pumps

6-42C What is the difference between a refrigerator and a heat pump?

6-43C What is the difference between a refrigerator and an air conditioner?

6-44C In a refrigerator, heat is transferred from a lower-temperature medium (the refrigerated space) to a higher-temperature one (the kitchen air). Is this a violation of the second law of thermodynamics? Explain.

6-45C A heat pump is a device that absorbs energy from the cold outdoor air and transfers it to the warmer indoors. Is this a violation of the second law of thermodynamics? Explain.

6-46C Define the coefficient of performance of a refrigerator in words. Can it be greater than unity?

6-47C Define the coefficient of performance of a heat pump in words. Can it be greater than unity?

6-48C A heat pump that is used to heat a house has a COP of 2.5. That is, the heat pump delivers 2.5 kWh of energy to the house for each 1 kWh of

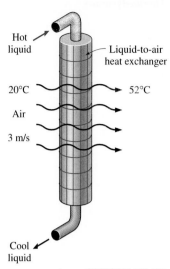

FIGURE P6-38

electricity it consumes. Is this a violation of the first law of thermodynamics? Explain.

6-49C A refrigerator has a COP of 1.5. That is, the refrigerator removes 1.5 kWh of energy from the refrigerated space for each 1 kWh of electricity it consumes. Is this a violation of the first law of thermodynamics? Explain.

6-50C What is the Clausius expression of the second law of thermodynamics?

6-51C Show that the Kelvin-Planck and the Clausius expressions of the second law are equivalent.

6-52 A household refrigerator with a COP of 1.8 removes heat from the refrigerated space at a rate of 90 kJ/min. Determine (*a*) the electric power consumed by the refrigerator and (*b*) the rate of heat transfer to the kitchen air.   *Answers:* (*a*) 0.83 kW, (*b*) 140 kJ/min

6-53 An air conditioner removes heat steadily from a house at a rate of 750 kJ/min while drawing electric power at a rate of 6 kW. Determine (*a*) the COP of this air conditioner and (*b*) the rate of heat transfer to the outside air.
    Answers: (*a*) 2.08, (*b*) 1110 kJ/min

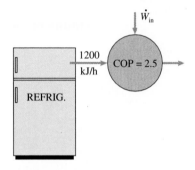

FIGURE P6-54

6-54 A household refrigerator runs one-fourth of the time and removes heat from the food compartment at an average rate of 1200 kJ/h. If the COP of the refrigerator is 2.5, determine the power the refrigerator draws when running.

6-55E Water enters an ice machine at 55°F and leaves as ice at 25°F. If the COP of the ice machine is 2.4 during this operation, determine the required power input for an ice production rate of 20 lbm/h. (169 Btu of energy needs to be removed from each lbm of water at 55°F to turn it into ice at 25°F.)

6-56 A household refrigerator that has a power input of 450 W and a COP of 2.5 is to cool five large watermelons, 10 kg each, to 8°C. If the watermelons are initially at 20°C, determine how long it will take for the refrigerator to cool them. The watermelons can be treated as water whose specific heat is 4.2 kJ/kg · °C. Is your answer realistic or optimistic? Explain.
    Answer: 2240 s

6-57 When a man returns to his well-sealed house on a summer day, he finds that the house is at 32°C. He turns on the air conditioner, which cools the entire house to 20°C in 15 min. If the COP of the air-conditioning system is 2.5, determine the power drawn by the air conditioner. Assume the entire mass within the house is equivalent to 800 kg of air for which $C_v = 0.72$ kJ/kg · °C and $C_p = 1.0$ kJ/kg · °C.

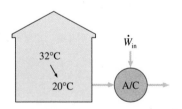

FIGURE P6-57

6-58 Determine the COP of a refrigerator that removes heat from the food compartment at a rate of 8000 kJ/h for each kW of power it consumes. Also, determine the rate of heat rejection to the outside air.

6-59 Determine the COP of a heat pump that supplies energy to a house at a rate of 8000 kJ/h for each kW of electric power it draws. Also, determine the rate of energy absorption from the outdoor air.   *Answers:* 2.22, 4400 kJ/h

6-60 A house that was heated by electric resistance heaters consumed 1200 kWh of electric energy in a winter month. If this house were heated instead by a heat pump that has an average COP of 2.4, determine how much money the homeowner would have saved that month. Assume a price of 8.5¢/kWh for electricity.

6-61E A heat pump with a COP of 2.5 supplies energy to a house at a rate of 60,000 Btu/h. Determine (*a*) the electric power drawn by the heat pump and (*b*) the rate of heat removal from the outside air.
  Answers: (*a*) 9.43 hp, (*b*) 36,000 Btu/h

6-62 A heat pump used to heat a house runs about one-third of the time. The house is losing heat at an average rate of 15,000 kJ/h. If the COP of the heat pump is 3.5, determine the power the heat pump draws when running.

6-63 A heat pump is used to maintain a house at a constant temperature of 23°C. The house is losing heat to the outside air through the walls and the windows at a rate of 60,000 kJ/h while the energy generated within the house from people, lights, and appliances amounts to 4000 kJ/h. For a COP of 2.5, determine the required power input to the heat pump.   *Answer:* 6.22 kW

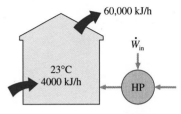

FIGURE P6-63

6-64 Consider an office room that is being cooled adequately by a 12,000 Btu/h window air conditioner. Now it is decided to convert this room into a computer room by installing several computers, terminals, and printers with a total rated power of 3.5 kW. The facility has several 4000 Btu/h air conditioners in storage that can be installed to meet the additional cooling requirements. Assuming a usage factor of 0.4 (i.e., only 40 percent of the rated power will be consumed at any given time) and additional occupancy of four people, each generating heat at a rate of 100 W, determine how many of these air conditioners need to be installed to the room.

6-65 Consider a building whose annual air-conditioning load is estimated to be 120,000 kWh in an area where the unit cost of electricity is $0.10/kWh. Two air conditioners are considered for the building. Air conditioner A has a seasonal average COP of 3.2 and costs $5500 to purchase and install. Air conditioner B has a seasonal average COP of 5.0 and costs $7000 to purchase and install. All else being equal, determine which air conditioner is a better buy.

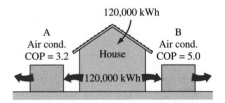

FIGURE P6-65

Perpetual-Motion Machines

6-66C An inventor claims to have developed a resistance heater that supplies 1.2 kWh of energy to a room for each kWh of electricity it consumes. Is this a reasonable claim, or has the inventor developed a perpetual-motion machine? Explain.

6-67C It is common knowledge that the temperature of air rises as it is compressed. An inventor thought about using this high-temperature air to heat buildings. He used a compressor driven by an electric motor. The inventor claims that the compressed hot-air system is 25 percent more efficient than a resistance heating system that provides an equivalent amount of heating. Is this claim valid, or is this just another perpetual-motion machine? Explain.

Reversible and Irreversible Processes

6-68C A cold canned drink is left in a warmer room where its temperature rises as a result of heat transfer. Is this a reversible process? Explain.

6-69C A hot baked potato is left on a table where it cools to the room temperature. Is this a reversible or an irreversible process? Explain.

6-70C Why are engineers interested in reversible processes even though they can never be achieved?

6-71C   Air is compressed from 20°C and 100 kPa to 300°C and 800 kPa first in a reversible manner and then in an irreversible manner. Which case do you think will require more work input?

6-72C   Why does a non-quasi-equilibrium compression process require a larger work input than the corresponding quasi-equilibrium one?

6-73C   Why does a non-quasi-equilibrium expansion process deliver less work than the corresponding quasi-equilibrium one?

6-74C   How do you distinguish between internal and external irreversibilities?

6-75C   Is a reversible expansion or compression process necessarily quasi-equilibrium? Is a quasi-equilibrium expansion or compression process necessarily reversible? Explain.

The Carnot Cycle and Carnot Principles

6-76C   What are the four processes that make up the Carnot cycle?

6-77C   What are the two statements known as the Carnot principles?

6-78C   Somebody claims to have developed a new reversible heat-engine cycle that has a higher theoretical efficiency than the Carnot cycle operating between the same temperature limits. How do you evaluate this claim?

6-79C   Somebody claims to have developed a new reversible heat-engine cycle that has the same theoretical efficiency as the Carnot cycle operating between the same temperature limits. Is this a reasonable claim?

6-80C   Is it possible to develop (a) an actual and (b) a reversible heat-engine cycle that is more efficient than a Carnot cycle operating between the same temperature limits? Explain.

Carnot Heat Engines

6-81C   Is there any way to increase the efficiency of a Carnot heat engine other than by increasing T_H or decreasing T_L?

6-82C   Consider two actual power plants operating with solar energy. Energy is supplied to one plant from a solar pond at 80°C and to the other from concentrating collectors that raise the water temperature to 600°C. Which of these power plants will have a higher efficiency? Explain.

6-83   A Carnot heat engine operates between a source at 1000 K and a sink at 300 K. If the heat engine is supplied with heat at a rate of 800 kJ/min, determine (a) the thermal efficiency and (b) the power output of this heat engine.   *Answers:* (a) 70 percent, (b) 9.33 kW

6-84   A Carnot heat engine receives 500 kJ of heat from a source of unknown temperature and rejects 200 kJ of it to a sink at 17°C. Determine (a) the temperature of the source and (b) the thermal efficiency of the heat engine.

6-85   A heat engine operates between a source at 550°C and a sink at 25°C. If heat is supplied to the heat engine at a steady rate of 1200 kJ/min, determine the maximum power output of this heat engine.

6-86E A heat engine is operating on a Carnot cycle and has a thermal efficiency of 55 percent. The waste heat from this engine is rejected to a nearby lake at 60°F at a rate of 800 Btu/min. Determine (*a*) the power output of the engine and (*b*) the temperature of the source.
Answers: (*a*) 23.1 hp, (*b*) 1155.6 R

6-87 In tropical climates, the water near the surface of the ocean remains warm throughout the year as a result of solar energy absorption. In the deeper parts of the ocean, however, the water remains at a relatively low temperature since the sun's rays cannot penetrate very far. It is proposed to take advantage of this temperature difference and construct a power plant that will absorb heat from the warm water near the surface and reject the waste heat to the cold water a few hundred meters below. Determine the maximum thermal efficiency of such a plant if the water temperatures at the two respective locations are 24 and 4°C.

6-88 An innovative way of power generation involves the utilization of geothermal energy—the energy of hot water that exists naturally underground—as the heat source. If a supply of hot water at 140°C is discovered at a location where the environmental temperature is 20°C, determine the maximum thermal efficiency a geothermal power plant built at that location can have.
Answer: 29.1 percent

6-89 An inventor claims to have developed a heat engine that receives 800 kJ of heat from a source at 400 K and produces 250 kJ of net work while rejecting the waste heat to a sink at 300 K. Is this a reasonable claim? Why?

6-90E An experimentalist claims that, based on his measurements, a heat engine receives 300 Btu of heat from a source of 900 R, converts 160 Btu of it to work, and rejects the rest as waste heat to a sink at 540 R. Are these measurements reasonable? Why?

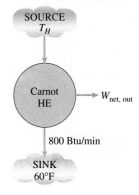

FIGURE P6-86E

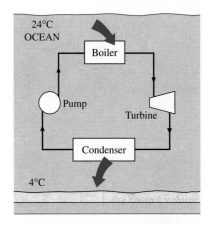

FIGURE P6-87

Carnot Refrigerators and Heat Pumps

6-91C How can we increase the COP of a Carnot refrigerator?

6-92C What is the highest COP that a refrigerator operating between temperature levels T_L and T_H can have?

6-93C In an effort to conserve energy in a heat-engine cycle, somebody suggests incorporating a refrigerator that will absorb some of the waste energy Q_L and transfer it to the energy source of the heat engine. Is this a smart idea? Explain.

6-94C It is well established that the thermal efficiency of a heat engine increases as the temperature at which heat is rejected from the heat engine T_L decreases. In an effort to increase the efficiency of a power plant, somebody suggests refrigerating the cooling water before it enters the condenser, where heat rejection takes place. Would you be in favor of this idea? Why?

6-95C It is well known that the thermal efficiency of heat engines increases as the temperature of the energy source increases. In an attempt to improve the efficiency of a power plant, somebody suggests transferring heat from the available energy source to a higher-temperature medium by a heat pump before energy is supplied to the power plant. What do you think of this suggestion? Explain.

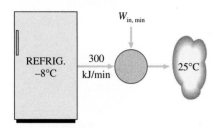

FIGURE P6-97

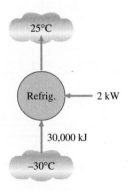

FIGURE P6-102

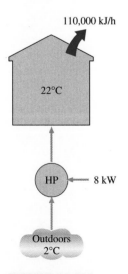

FIGURE P6-105

6-96 A Carnot refrigerator operates in a room in which the temperature is 25°C and consumes 2 kW of power when operating. If the food compartment of the refrigerator is to be maintained at 3°C, determine the rate of heat removal from the food compartment.

6-97 A refrigerator is to remove heat from the cooled space at a rate of 300 kJ/min to maintain its temperature at −8°C. If the air surrounding the refrigerator is at 25°C, determine the minimum power input required for this refrigerator.   *Answer:* 0.623 kW

6-98 An air-conditioning system operating on the reversed Carnot cycle is required to transfer heat from a house at a rate of 750 kJ/min, to maintain its temperature at 20°C. If the outdoor air temperature is 35°C, determine the power required to operate this air-conditioning system.   *Answer:* 0.64 kW

6-99E An air-conditioning system is used to maintain a house at 70°F when the temperature outside is 90°F. If this air-conditioning system draws 5 hp of power when operating, determine the maximum rate of heat removal from the house that it can provide.

6-100 A Carnot refrigerator operates in a room in which the temperature is 25°C. The refrigerator consumes 500 W of power when operating and has a COP of 4.5. Determine (*a*) the rate of heat removal from the refrigerated space and (*b*) the temperature of the refrigerated space.
    Answers: (*a*) 135 kJ/min, (*b*) −29.2°C

6-101 An inventor claims to have developed a refrigeration system that removes heat from the closed region at −5°C and transfers it to the surrounding air at 22°C while maintaining a COP of 8.2. Is this claim reasonable? Why?

6-102 During an experiment conducted in a room at 25°C, a laboratory assistant measures that a refrigerator that draws 2 kW of power has removed 30,000 kJ of heat from the refrigerated space, which is maintained at −30°C. The running time of the refrigerator during the experiment was 20 min. Determine if these measurements are reasonable.

6-103E An air-conditioning system is used to maintain a house at 75°F when the temperature outside is 95°F. The house is gaining heat through the walls and the windows at a rate of 750 Btu/min, and the heat generation rate within the house from people, lights, and appliances amounts to 150 Btu/min. Determine the minimum power input required for this air-conditioning system.   *Answer:* 0.79 hp

6-104 A heat pump is used to heat a house and maintain it at 20°C. On a winter day when the outdoor air temperature is −5°C, the house is estimated to lose heat at a rate of 75,000 kJ/h. Determine the minimum power required to operate this heat pump.

6-105 A heat pump is used to maintain a house at 22°C by extracting heat from the outside air on a day when the outside air temperature is 2°C. The house is estimated to lose heat at a rate of 110,000 kJ/h, and the heat pump consumes 8 kW of electric power when running. Is this heat pump powerful enough to do the job?

6-106 The structure of a house is such that it loses heat at a rate of 5400 kJ/h per °C difference between the indoors and outdoors. A heat pump that requires a power input of 6 kW is used to maintain this house at 21°C. Determine the lowest outdoors temperature for which the heat pump can meet the heating requirements of this house.   *Answer:* −13.3°C

6-107 The performance of a heat pump degrades (i.e., its COP decreases) as the temperature of the heat source decreases. This makes using heat pumps at locations with severe weather conditions unattractive. Consider a house that is heated and maintained at 20°C by a heat pump during the winter. What is the maximum COP for this heat pump if heat is extracted from the outdoor air at (*a*) 10°C, (*b*) −5°C, and (*c*) −30°C?

6-108E A heat pump is to be used for heating a house in winter. The house is to be maintained at 78°F at all times. When the temperature outdoors drops to 25°F, the heat losses from the house are estimated to be 80,000 Btu/h. Determine the minimum power required to run this heat pump if heat is extracted from (*a*) the outdoor air at 25°F and (*b*) the well water at 50°F.

6-109 A Carnot heat pump is to be used to heat a house and maintain it at 20°C in winter. On a day when the average outdoor temperature remains at about 2°C, the house is estimated to lose heat at a rate of 82,000 kJ/h. If the heat pump consumes 8 kW of power while operating, determine (*a*) how long the heat pump ran on that day; (*b*) the total heating costs, assuming an average price of 8.5¢/kWh for electricity; and (*c*) the heating cost for the same day if resistance heating is used instead of a heat pump.   *Answers:* (*a*) 4.19 h, (*b*) $2.85, (*c*) $46.47

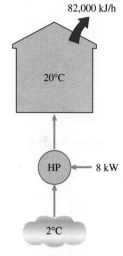

FIGURE P6-109

6-110 A Carnot heat engine receives heat from a reservoir at 900°C at a rate of 800 kJ/min and rejects the waste heat to the ambient air at 27°C. The entire work output of the heat engine is used to drive a refrigerator that removes heat from the refrigerated space at −5°C and transfers it to the same ambient air at 27°C. Determine (*a*) the maximum rate of heat removal from the refrigerated space and (*b*) the total rate of heat rejection to the ambient air.
    Answers: (*a*) 4982 kJ/min, (*b*) 5782 kJ

6-111E A Carnot heat engine receives heat from a reservoir at 1700°F at a rate of 700 Btu/min and rejects the waste heat to the ambient air at 80°F. The entire work output of the heat engine is used to drive a refrigerator that removes heat from the refrigerated space at 20°F and transfers it to the same ambient air at 80°F. Determine (*a*) the maximum rate of heat removal from the refrigerated space and (*b*) the total rate of heat rejection to the ambient air.
    Answers: (*a*) 4200 Btu/min, (*b*) 3850 Btu/min

Household Refrigerators

6-112C Someone proposes that the refrigeration system of a supermarket be overdesigned so that the entire air-conditioning needs of the store can be met by refrigerated air without installing any air-conditioning system. What do you think of this proposal?

6-113C Someone proposes that the entire refrigerator/freezer requirements of a store be met using a large freezer that supplies sufficient cold air at −20°C instead of installing separate refrigerators and freezers. What do you think of this proposal?

6-114C Explain how you can reduce the energy consumption of your household refrigerator.

6-115C Why is it important to clean the condenser coils of a household refrigerator a few times a year? Also, why is it important not to block airflow through the condenser coils?

Refrigerator

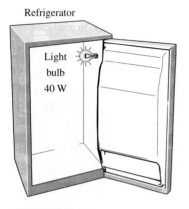

FIGURE P6-118

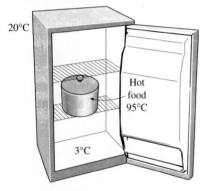

FIGURE P6-119

6-116C   Why are today's refrigerators much more efficient than those built in the past?

6-117   The "Energy Guide" label of a refrigerator states that the refrigerator will consume $83 worth of electricity per year under normal use if the cost of electricity is $0.08/kWh. If the electricity consumed by the light bulb is negligible and the refrigerator consumes 300 W when running, determine the fraction of the time the refrigerator will run.

6-118   The interior lighting of refrigerators is usually provided by incandescent lamps whose switches are actuated by the opening of the refrigerator door. Consider a refrigerator whose 40-W light bulb remains on about 60 h per year. It is proposed to replace the light bulb by an energy-efficient bulb that consumes only 18 W but costs $25 to purchase and install. If the refrigerator has a coefficient of performance of 1.3 and the cost of electricity is 8 cents per kWh, determine if the energy savings of the proposed light bulb justify its cost.

6-119   It is commonly recommended that hot foods be cooled first to room temperature by simply waiting a while before they are put into the refrigerator to save energy. Despite this commonsense recommendation, a person keeps cooking a large pan of stew twice a week and putting the pan into the refrigerator while it is still hot, thinking that the money saved is probably too little. But he says he can be convinced if you can show that the money saved is significant. The average mass of the pan and its contents is 5 kg. The average temperature of the kitchen is 20°C, and the average temperature of the food is 95°C when it is taken off the stove. The refrigerated space is maintained at 3°C, and the average specific heat of the food and the pan can be taken to be 3.9 kJ/kg · °C. If the refrigerator has a coefficient of performance of 1.2 and the cost of electricity is 10 cents per kWh, determine how much this person will save a year by waiting for the food to cool to room temperature before putting it into the refrigerator.

6-120   It is often stated that the refrigerator door should be opened as few times as possible for the shortest duration of time to save energy. Consider a household refrigerator whose interior volume is 0.9 m^3 and average internal temperature is 4°C. At any given time, one-third of the refrigerated space is occupied by food items, and the remaining 0.6 m^3 is filled with air. The average temperature and pressure in the kitchen are 20°C and 95 kPa, respectively. Also, the moisture contents of the air in the kitchen and the refrigerator are 0.010 and 0.004 kg per kg of air, respectively, and thus 0.006 kg of water vapor is condensed and removed for each kg of air that enters. The refrigerator door is opened an average of 8 times a day, and each time half of the air volume in the refrigerator is replaced by the warmer kitchen air. If the refrigerator has a coefficient of performance of 1.4 and the cost of electricity is 7.5 cents per kWh, determine the cost of the energy wasted per year as a result of opening the refrigerator door. What would your answer be if the kitchen air were very dry and thus a negligible amount of water vapor condensed in the refrigerator?

Review Problems

6-121   Consider a Carnot heat-engine cycle executed in a steady-flow system using steam as the working fluid. The cycle has a thermal efficiency of 30 percent, and steam changes from saturated liquid to saturated vapor at 300°C during the heat addition process. If the mass flow rate of the steam is 5 kg/s, determine the net power output of this engine, in kW.

6-122 A heat pump with a COP of 2.4 is used to heat a house. When running, the heat pump consumes 8 kW of electric power. If the house is losing heat to the outside at an average rate of 40,000 kJ/h and the temperature of the house is 3°C when the heat pump is turned on, determine how long it will take for the temperature in the house to rise to 22°C. Assume the house is well sealed (i.e., no air leaks) and take the entire mass within the house (air, furniture, etc.) to be equivalent to 2000 kg of air.

6-123 An old gas turbine has an efficiency of 17 percent and develops a power output of 6000 kW. Determine the fuel consumption rate of this gas turbine, in L/min, if the fuel has a heating value of 46,000 kJ/kg and a density of 0.8 g/cm³.

6-124 Show that $COP_{HP} = COP_R + 1$ when both the heat pump and the refrigerator have the same Q_L and Q_H values.

6-125 An air-conditioning system is used to maintain a house at a constant temperature of 20°C. The house is gaining heat from outdoors at a rate of 20,000 kJ/h, and the heat generated in the house from the people, lights, and appliances amounts to 8000 kJ/h. For a COP of 2.5, determine the required power input to this air-conditioning system. *Answer:* 3.11 kW

6-126 Consider a Carnot heat-engine cycle executed in a closed system using 0.01 kg of refrigerant-134a as the working fluid. The cycle has a thermal efficiency of 15 percent, and the refrigerant-134a changes from saturated liquid to saturated vapor at 70°C during the heat addition process. Determine the net work output of this engine, in kJ.

6-127 A heat pump with a COP of 3.2 is used to heat a house. When running, the heat pump consumes power at a rate of 5 kW. If the temperature in the house is 7°C when the heat pump is turned on, how long will it take for the heat pump to raise the temperature of the house to 22°C? Is this answer realistic or optimistic? Explain. Assume the entire mass within the house (air, furniture, etc.) is equivalent to 1500 kg of air for which $C_v = 0.72$ kJ/kg · °C and $C_p = 1.0$ kJ/kg · °C. *Answer:* 1012 s

6-128 A promising method of power generation involves collecting and storing solar energy in large artificial lakes a few meters deep, called solar ponds. Solar energy is absorbed by all parts of the pond, and the water temperature rises everywhere. The top part of the pond, however, loses to the atmosphere much of the heat it absorbs, and as a result, its temperature drops. This cool water serves as insulation for the bottom part of the pond and helps trap the energy there. Usually, salt is planted at the bottom of the pond to prevent the rise of this hot water to the top. A power plant that uses an organic fluid, such as alcohol, as the working fluid can be operated between the top and the bottom portions of the pond. If the water temperature is 35°C near the surface and 80°C near the bottom of the pond, determine the maximum thermal efficiency that this power plant can have. Is it realistic to use 35 and 80°C for temperatures in the calculations? Explain. *Answer:* 12.7 percent

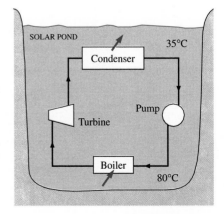

FIGURE P6-128

6-129 Consider a Carnot heat-engine cycle executed in a closed system using 0.0103 kg of steam as the working fluid. It is known that the maximum absolute temperature in the cycle is twice the minimum absolute temperature, and the net work output of the cycle is 25 kJ. If the steam changes from saturated vapor to saturated liquid during the heat rejection process, determine the temperature of the steam during the heat rejection process, in °C.

6-130 Consider a Carnot refrigeration cycle executed in a closed system in the saturated liquid–vapor mixture region using 0.96 kg of refrigerant-134a as the working fluid. It is known that the maximum absolute temperature in the cycle is 1.2 times the minimum absolute temperature, and the net work input to the cycle is 22 kJ. If the refrigerant changes from saturated liquid to saturated vapor during the heat rejection process, determine the minimum pressure in the cycle, in kPa.

6-131 Consider two Carnot heat engines operating in series. The first engine receives heat from the reservoir at 1200 K and rejects the waste heat to another reservoir at temperature T. The second engine receives this energy rejected by the first one, converts some of it to work, and rejects the rest to a reservoir at 300 K. If the thermal efficiencies of both engines are the same, determine the temperature T.   *Answer:* 600 K

6-132 The COP of a refrigerator decreases as the temperature of the refrigerated space is decreased. That is, removing heat from a medium at a very low temperature will require a large work input. Determine the minimum work input required to remove 1 kJ of heat from liquid helium at 3 K when the outside temperature is 300 K.   *Answer:* 99 kJ

6-133E A Carnot heat pump is used to heat and maintain a residential building at 75°F. An energy analysis of the house reveals that it loses heat at a rate of 2500 Btu/h per °F temperature difference between the indoors and the outdoors. For an outdoor temperature of 35°F, determine (*a*) the coefficient of performance and (*b*) the required power input to the heat pump.
    Answers: (*a*) 13.4, (*b*) 2.93 hp

6-134 A Carnot heat engine receives heat at 750 K and rejects the waste heat to the environment at 300 K. The entire work output of the heat engine is used to drive a Carnot refrigerator that removes heat from the cooled space at −15°C at a rate of 400 kJ/min and rejects it to the same environment at 300 K. Determine (*a*) the rate of heat supplied to the heat engine and (*b*) the total rate of heat rejection to the environment.

6-135 A heat engine operates between two reservoirs at 800 and 20°C. One-half of the work output of the heat engine is used to drive a Carnot heat pump that removes heat from the cold surroundings at 2°C and transfers it to a house maintained at 22°C. If the house is losing heat at a rate of 95,000 kJ/h, determine the minimum rate of heat supply to the heat engine required to keep the house at 22°C.

6-136 Consider a Carnot refrigeration cycle executed in a closed system in the saturated liquid–vapor mixture region using 0.8 kg of refrigerant-134a as the working fluid. The maximum and the minimum temperatures in the cycle are 20°C and −10°C, respectively. It is known that the refrigerant is saturated liquid at the end of the heat rejection process, and the net work input to the cycle is 12 kJ. Determine the fraction of the mass of the refrigerant that vaporizes during the heat addition process, and the pressure at the end of the heat rejection process.

6-137 Consider a Carnot heat-pump cycle executed in a steady-flow system in the saturated liquid–vapor mixture region using refrigerant-134a flowing at a rate of 0.264 kg/s as the working fluid. It is known that the maximum absolute temperature in the cycle is 1.15 times the minimum absolute temperature, and the net power input to the cycle is 5 kW. If the refrigerant changes from saturated vapor to saturated liquid during the heat rejection process, determine the ratio of the maximum to minimum pressures in the cycle.

6-138 A Carnot heat engine is operating between a source at T_H and a sink at T_L. If it is desired to double the thermal efficiency of this engine, what should the new source temperature be? Assume the sink temperature is held constant.

6-139 When discussing Carnot engines, it is assumed that the engine is in thermal equilibrium with the source and the sink during the heat addition and heat rejection processes, respectively. That is, it is assumed that $T_H^* = T_H$ and $T_L^* = T_L$ so that there is no external irreversibility. In that case, the thermal efficiency of the Carnot engine is $\eta_C = 1 - T_L/T_H$.

In reality, however, we must maintain a reasonable temperature difference between the two heat transfer media in order to have an acceptable heat transfer rate through a finite heat exchanger surface area. The heat transfer rates in that case can be expressed as

$$\dot{Q}_H = (hA)_H(T_H - T_H^*)$$
$$\dot{Q}_L = (hA)_L(T_L^* - T_L)$$

where h and A are the heat transfer coefficient and heat transfer surface area, respectively. When the values of h, A, T_H, and T_L are fixed, show that the power output will be a maximum when

$$\frac{T_L^*}{T_H^*} = \left(\frac{T_L}{T_H}\right)^{1/2}$$

Also, show that the maximum net power output in this case is

$$\dot{W}_{C,\,\text{max}} = \frac{(hA)_H T_H}{1 + (hA)_H/(hA)_L}\left[1 - \left(\frac{T_L}{T_H}\right)^{1/2}\right]^2$$

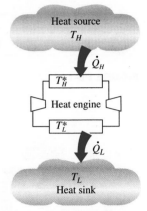

FIGURE P6-139

6-140 Consider a homeowner who is replacing his 20-year-old natural gas furnace that has an efficiency of 60 percent. The homeowner is considering a conventional furnace that has an efficiency of 82 percent and costs $1600 and a high-efficiency furnace that has an efficiency of 95 percent and costs $2700. The home owner would like to buy the high-efficiency furnace if the savings from the natural gas pay for the additional cost in less than 8 years. If the home owner presently pays $1100 a year for heating, determine if he should buy the conventional or high-efficiency model.

6-141 Replacing incandescent lights with energy-efficient fluorescent lights can reduce the lighting energy consumption to one-fourth of what it was before. The energy consumed by the lamps is eventually converted to heat, and thus switching to energy-efficient lighting also reduces the cooling load in summer but increases the heating load in winter. Consider a building that is heated by a natural gas furnace with an efficiency of 80 percent and cooled by an air conditioner with a COP of 3.5. If electricity costs $0.08/kWh and natural gas costs $0.70/therm, determine if efficient lighting will increase or decrease the total energy cost of the building (*a*) in summer and (*b*) in winter.

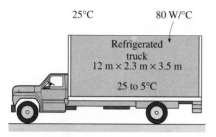

25°C   80 W/°C

Refrigerated
truck
12 m × 2.3 m × 3.5 m
25 to 5°C

FIGURE P6-142

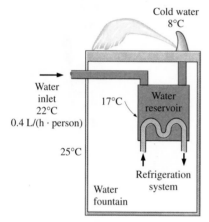

Cold water
8°C

Water
inlet
22°C
0.4 L/(h · person)

17°C   Water
reservoir

25°C

Refrigeration
system

Water
fountain

FIGURE P6-144

Water
heater

FIGURE P6-147

6-142 The cargo space of a refrigerated truck whose inner dimensions are 12 m × 2.3 m × 3.5 m is to be precooled from 25°C to an average temperature of 5°C. The construction of the truck is such that a transmission heat gain occurs at a rate of 80 W/°C. If the ambient temperature is 25°C, determine how long it will take for a system with a refrigeration capacity of 8 kW to precool this truck.

6-143 A refrigeration system is to cool bread loaves with an average mass of 450 g from 22°C to −10°C at a rate of 500 loaves per hour by refrigerated air at −30°C. Taking the average specific and latent heats of bread to be 2.93 kJ/kg · °C and 109.3 kJ/kg, respectively, determine (a) the rate of heat removal from the breads, in kJ/h; (b) the required volume flow rate of air, in m³/h, if the temperature rise of air is not to exceed 8°C; and (c) the size of the compressor of the refrigeration system, in kW, for a COP of 1.2 for the refrigeration system.

6-144 The drinking water needs of a production facility with 20 employees is to be met by a bobbler type water fountain. The refrigerated water fountain is to cool water from 22°C to 8°C and supply cold water at a rate of 0.4 L per hour per person. The outer diameter of the water reservoir where water is cooled and stored is 20 cm, and its height is 25 cm. The outer surface temperature of the reservoir is 17°C, and heat is transferred to the reservoir from the surroundings at 25°C with a heat transfer coefficient of 10 W/m² · °C. If the COP of the refrigeration system is 2.9, determine the size of the compressor, in W, that will be suitable for the refrigeration system of this water cooler.

6-145 The "Energy Guide" label on a washing machine indicates that the washer will use $85 worth of hot water per year if the water is heated by an electric water heater at an electricity rate of $0.082/kWh. If the water is heated from 15°C to 55°C, determine how many liters of hot water an average family uses per week. Disregard the electricity consumed by the washer, and take the efficiency of the electric water heater to be 90 percent.

6-146 The "Energy Guide" label on a washing machine indicates that the washer will use $33 worth of hot water if the water is heated by a gas water heater at a natural gas rate of $0.605/therm. If the water is heated from 60°F to 130°F, determine how many gallons of hot water an average family uses per week. Disregard the electricity consumed by the washer, and take the efficiency of the gas water heater to be 55 percent.

6-147 A typical electric water heater has an efficiency of 90 percent and costs $390 a year to operate at a unit cost of electricity of $0.08/kWh. A typical heat pump–powered water heater has a COP of 2.2 but costs about $800 more to install. Determine how many years it will take for the heat pump water heater to pay for its cost differential from the energy it saves.

6-148E The energy contents, unit costs, and typical conversion efficiencies of various energy sources for use in water heaters are given as follows: 1025 Btu/ft³, $0.0060/ft³, and 55 percent for natural gas; 138,700 Btu/gal, $1.15 gal, and 55 percent for heating oil; and 1 kWh/kWh $0.084/kWh, and 90 percent for resistance electric heaters, respectively. Determine the lowest-cost energy source for water heaters.

6-149E A homeowner is considering the following heating systems for heating his house. Electric resistance heating with $0.09/kWh and 1 kWh = 3412 Btu, gas heating with $0.062/therm and 1 therm = 105,500 kJ, and oil heating with $1.25/gal and 1 gal of oil = 138,500 kJ. Assuming efficiencies of 100 percent for the electric furnace and 87 percent for the gas and oil furnaces, determine the heating system with the lowest energy cost.

6-150 A homeowner is trying to decide between a high-efficiency natural gas furnace with an efficiency of 97 percent and a ground-source heat pump with a COP of 3.7. The unit costs of electricity and natural gas are $0.092/kWh and $0.71/therm (1 therm = 105,500 kJ). Determine which system will have a lower energy cost.

6-151 The maximum flow rate of a standard shower head is about 3.5 gpm (13.3 L/min) and can be reduced to 2.75 gpm (10.5 L/min) by switching to a low-flow shower head that is equipped with flow controllers. Consider a family of four, with each person taking a 6-minute shower every morning. City water at 15°C is heated to 55°C in an oil water heater whose efficiency is 65 percent and then tempered to 42°C by cold water at the T-elbow of the shower before being routed to the shower head. The price of heating oil is $1.20/gal and its heating value is 146,300 kJ/gal. Assuming a constant specific heat of 4.18 kJ/kg · °C for water, determine the amount of oil and money saved per year by replacing the standard shower heads by the low-flow ones.

6-152 A typical household pays about $1200 a year on energy bills, and the U.S. Department of Energy estimates that 46 percent of this energy is used for heating and cooling, 15 percent for heating water, 15 percent for refrigerating and freezing, and the remaining 24 percent for lighting, cooking, and running other appliances. The heating and cooling costs of a poorly insulated house can be reduced by up to 30 percent by adding adequate insulation. If the cost of insulation is $200, determine how long it will take for the insulation to pay for itself from the energy it saves.

6-153 The kitchen, bath, and other ventilation fans in a house should be used sparingly since these fans can discharge a houseful of warmed or cooled air in just one hour. Consider a 200 m² house whose ceiling height is 2.4 m. The house is heated by a 96 percent efficient gas heater and is maintained at 22°C and 92 kPa. If the unit cost of natural gas is $0.60/therm (1 therm = 105,500 kJ), determine the cost of energy "vented out" by the fans in 1 h. Assume the average outdoor temperature during the heating season to be 5°C.

6-154 Repeat Prob. 6-153 for the air-conditioning cost in a dry climate for an outdoor temperature of 32°C. Assume the COP of the air-conditioning system to be 3.2, and the unit cost of electricity to be $0.10/kWh.

6-155 The U.S. Department of Energy estimates that up to 10 percent of the energy use of a house can be saved by caulking and weatherstripping doors and windows to reduce air leaks at a cost of about $50 for materials for an average home with 12 windows and 2 doors. Caulking and weatherstripping every gas-heated home properly would save enough energy to heat about 4 million homes. The savings can be increased by installing storm windows. Determine how long it will take for the caulking and weatherstripping to pay for itself from the energy they save for a house whose annual energy use is $1100.

6-156 The U.S. Department of Energy estimates that 570,000 barrels of oil would be saved per day if every household in the United States lowered the thermostat setting in winter by 6°F (3.3°C). Assuming the average heating season to be 180 days and the cost of oil to be $20/barrel, determine how much money would be saved per year.

Computer, Design, and Essay Problems

6-157 Write a computer program to determine the maximum work that can be extracted from a pond containing 10^5 kg of water at 350 K when the temperature of the surroundings is 300 K. Notice that the temperature of water in the pond will be gradually decreasing as energy is extracted from it; therefore, the efficiency of the engine will be decreasing. Use temperature intervals of (*a*) 5 K, (*b*) 2 K, and (*c*) 1 K until the pond temperature drops to 300 K. Also solve this problem exactly by integration, and compare the results.

6-158 Find out the prices of heating oil, natural gas, and electricity in your area, and determine the cost of each per kWh of energy supplied to the house as heat. Go through your utility bills and determine how much money you spent for heating last January. Also determine how much your January heating bill would be for each of the heating systems if you had the latest and most efficient system installed.

6-159 Prepare a report on the heating systems available in your area for residential buildings. Discuss the advantages and disadvantages of each system and compare their initial and operating costs. What are the important factors in the selection of a heating system? Give some guidelines. Identify the conditions under which each heating system would be the best choice in your area.

6-160 The performance of a cyclic device is defined as the ratio of the desired output to the required input, and this definition can be extended to nontechnical fields. For example, your performance in this course can be viewed as the grade you earn relative to the effort you put in. If you have been investing a lot of time in this course and your grades do not reflect it, you are performing poorly. In that case, perhaps you should try to find out the underlying cause and how to correct the problem. Give three other definitions of performance from nontechnical fields and discuss them.

6-161 Devise a Carnot heat engine using steady-flow components, and describe how the Carnot cycle is executed in that engine. What happens when the directions of heat and work interactions are reversed?

6-162 When was the concept of the heat pump conceived and by whom? When was the first heat pump built, and when were the heat pumps first mass-produced?

6-163 Your neighbor lives in a 2500-square-foot (about 250 m²) older house heated by natural gas. The current gas heater was installed in the early 1970s and has an efficiency (called the Annual Fuel Utilization Efficiency rating, or AFUE) of 65 percent. It is time to replace the furnace, and the neighbor is trying to decide between a conventional furnace that has an efficiency of 80 percent and costs $1500 and a high-efficiency furnace that has an efficiency of 95 percent and costs $2500. Your neighbor offered to pay you $100 if you help him make the right decision. Considering the weather data, typical heating loads, and the price of natural gas in your area, make a recommendation to your neighbor based on a convincing economic analysis.

6-164 Using a thermometer, measure the temperature of the main food compartment of your refrigerator, and check if it is between 1 and 4°C. Also, measure the temperature of the freezer compartment, and check if it is at the recommended value of −18°C.

6-165 Using a timer (or watch) and a thermometer, conduct the following experiment to determine the rate of heat gain of your refrigerator. First make sure that the door of the refrigerator is not opened for at least a few hours so that steady operating conditions are established. Start the timer when the refrigerator stops running and measure the time Δt_1 it stays off before it kicks in. Then measure the time Δt_2 it stays on. Noting that the heat removed during Δt_2 is equal to the heat gain of the refrigerator during $\Delta t_1 + \Delta t_2$ and using the power consumed by the refrigerator when it is running, determine the average rate of heat gain for your refrigerator, in W. Take the COP (coefficient of performance) of your refrigerator to be 1.3 if it is not available.

6-166 Design a hydrocooling unit that can cool fruits and vegetables from 30°C to 5°C at a rate of 20,000 kg/h under the following conditions:

The unit will be of flood type, which will cool the products as they are conveyed into the channel filled with water. The products will be dropped into the channel filled with water at one end and be picked up at the other end. The channel can be as wide as 3 m and as high as 90 cm. The water is to be circulated and cooled by the evaporator section of a refrigeration system. The refrigerant temperature inside the coils is to be −2°C, and the water temperature is not to drop below 1°C and not to exceed 6°C.

Assuming reasonable values for the average product density, specific heat, and porosity (the fraction of air volume in a box), recommend reasonable values for (*a*) the water velocity through the channel and (*b*) the refrigeration capacity of the refrigeration system.

Entropy

CHAPTER 7

In Chap. 6, we introduced the second law of thermodynamics and applied it to cycles and cyclic devices. In this chapter, we apply the second law to processes. The first law of thermodynamics deals with the property *energy* and the conservation of it. The second law leads to the definition of a new property called *entropy*. Entropy is a somewhat abstract property, and it is difficult to give a physical description of it. Entropy is best understood and appreciated by studying its uses in commonly encountered engineering processes, and this is what we intend to do.

This chapter starts with a discussion of the Clausius inequality, which forms the basis for the definition of entropy, and continues with the increase of entropy principle. Unlike energy, entropy is a nonconserved property, and there is no such thing as a *conservation of entropy principle*. Next, the entropy changes that take place during processes for pure substances, incompressible substances, and ideal gases are discussed, and a special class of idealized processes, called *isentropic processes,* are examined. Finally, the reversible steady-flow work and the isentropic efficiencies of various engineering devices such as turbines and compressors are considered.

7-1 ■ ENTROPY

The second law of thermodynamics often leads to expressions that involve inequalities. An irreversible (i.e., actual) heat engine, for example, is less efficient than a reversible one operating between the same two thermal energy reservoirs. Likewise, an irreversible refrigerator or a heat pump has a lower coefficient of performance (COP) than a reversible one operating between the same temperature limits. Another important inequality that has major consequences in thermodynamics is the **Clausius inequality.** It was first stated by the German physicist R. J. E. Clausius (1822–1888), one of the founders of thermodynamics, and is expressed as

$$\oint \frac{\delta Q}{T} \leq 0$$

That is, *the cyclic integral of $\delta Q/T$ is always less than or equal to zero.* This inequality is valid for all cycles, reversible or irreversible. The symbol $\oint$ (integral symbol with a circle in the middle) is used to indicate that the integration is to be performed over the entire cycle. Any heat transfer to or from a system can be considered to consist of differential amounts of heat transfer. Then the cyclic integral of $\delta Q/T$ can be viewed as the sum of all these differential amounts of heat transfer divided by the absolute temperature at the boundary.

To demonstrate the validity of the Clausius inequality, consider a system connected to a thermal energy reservoir at a constant absolute temperature of T_R through a *reversible* cyclic device (Fig. 7-1). The cyclic device receives heat δQ_R from the reservoir and supplies heat δQ to the system whose absolute temperature at that part of the boundary is T (a variable) while producing work δW_{rev}. The system produces work δW_{sys} as a result of this heat transfer. Applying the energy balance to the combined system identified by dashed lines yields

$$\delta W_C = \delta Q_R - dE_C$$

where δW_C is the total work of the combined system ($\delta W_{rev} + \delta W_{sys}$) and dE_C is the change in the total energy of the combined system. Considering that the cyclic device is a *reversible* one, we have (Eq. 6-18)

$$\frac{\delta Q_R}{T_R} = \frac{\delta Q}{T}$$

where the sign of δQ is determined with respect to the system (positive if *to* the system and negative if *from* the system) and the sign of δQ_R is determined with respect to the reversible cyclic device. Eliminating δQ_R from the two relations above yields

$$\delta W_C = T_R \frac{\delta Q}{T} - dE_C$$

We now let the system undergo a cycle while the cyclic device undergoes an integral number of cycles. Then the relation above becomes

$$W_C = T_R \oint \frac{\delta Q}{T}$$

since the cyclic integral of energy (the net change in the energy, which is a property, during a cycle) is zero. Here W_C is the cyclic integral of δW_C, and it represents the net work for the combined cycle.

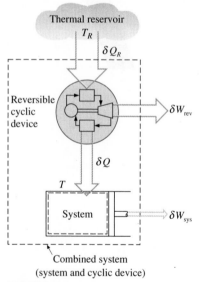

Thermal reservoir
T_R

δQ_R

Reversible cyclic device

δW_{rev}

δQ

T

System

δW_{sys}

Combined system
(system and cyclic device)

FIGURE 7-1

The system considered in the development of the Clausius inequality.

It appears that the combined system is exchanging heat with a single thermal energy reservoir while involving (producing or consuming) work W_C during a cycle. On the basis of the Kelvin-Planck statement of the second law, which states that *no system can produce a net amount of work while operating in a cycle and exchanging heat with a single thermal energy reservoir,* we reason that W_C cannot be a work output, and thus it cannot be a positive quantity. Considering that T_R is an absolute temperature and thus a positive quantity, we must have

$$\oint \frac{\delta Q}{T} \leq 0 \qquad (7\text{-}1)$$

which is the *Clausius inequality.* This inequality is valid for all thermodynamic cycles, reversible or irreversible, including the refrigeration cycles.

If no irreversibilities occur within the system as well as the reversible cyclic device, then the cycle undergone by the combined system will be internally reversible. As such, it can be reversed. In the reversed cycle case, all the quantities will have the same magnitude but the opposite sign. Therefore, the work W_C, which could not be a positive quantity in the regular case, cannot be a negative quantity in the reversed case. Then it follows that $W_{C,\text{ int rev}} = 0$ since it cannot be a positive or negative quantity, and therefore

$$\oint \left(\frac{\delta Q}{T}\right)_{\text{int rev}} = 0 \qquad (7\text{-}2)$$

for internally reversible cycles. Thus we conclude that *the equality in the Clausius inequality holds for totally or just internally reversible cycles and the inequality for the irreversible ones.*

To develop a relation for the definition of entropy, let us examine Eq. 7-2 more closely. Here we have a quantity whose cyclic integral is zero. Let us think for a moment what kind of quantities can have this characteristic. We know that the cyclic integral of *work* is not zero. (It is a good thing that it is not. Otherwise, heat engines that work on a cycle such as steam power plants would produce zero net work.) Neither is the cyclic integral of heat.

Now consider the volume occupied by a gas in a piston-cylinder device undergoing a cycle, as shown in Fig. 7-2. When the piston returns to its initial position at the end of a cycle, the volume of the gas also returns to its initial value. Thus the net change in volume during a cycle is zero. This is also expressed as

$$\oint dV = 0 \qquad (7\text{-}3)$$

That is, the cyclic integral of volume (or any other property) is zero. Conversely, a quantity whose cyclic integral is zero depends on the *state* only and not the process path, and thus it is a property. Therefore, the quantity $(\delta Q/T)_{\text{int rev}}$ must represent a property in the differential form.

Clausius realized in 1865 that he had discovered a new thermodynamic property, and he chose to name this property **entropy.** It is designated S and is defined as

$$dS = \left(\frac{\delta Q}{T}\right)_{\text{int rev}} \qquad (\text{kJ/K}) \qquad (7\text{-}4)$$

Entropy is an extensive property of a system and sometimes is referred to as *total entropy.* Entropy per unit mass, designated s, is an intensive property and

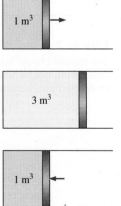

$$\oint dV = \Delta V_{\text{cycle}} = 0$$

FIGURE 7-2

The net change in volume (a property during a cycle is always zero).

has the unit kJ/kg · K. The term *entropy* is generally used to refer to both total entropy and entropy per unit mass since the context usually clarifies which one is meant.

The entropy change of a system during a process can be determined by integrating Eq. 7-4 between the initial and the final states:

$$\Delta S = S_2 - S_1 = \int_1^2 \left(\frac{\delta Q}{T}\right)_{\text{int rev}} \quad \text{(kJ/K)} \quad (7\text{-}5)$$

Notice that we have actually defined the *change* in entropy instead of entropy itself, just as we defined the change in energy instead of the energy itself when we developed the first-law relation. Absolute values of entropy are determined on the basis of the third law of thermodynamics, which is discussed later in this chapter. Engineers are usually concerned with the *changes* in entropy. Therefore, the entropy of a substance can be assigned a zero value at some arbitrarily selected reference state, and the entropy values at other states can be determined from Eq. 7-5 by choosing state 1 to be the reference state ($S = 0$) and state 2 to be the state at which entropy is to be determined.

To perform the integration in Eq. 7-5, one needs to know the relation between Q and T during a process. This relation is often not available, and the integral in Eq. 7-5 can be performed for a few cases only. For the majority of cases we have to rely on tabulated data for entropy.

Note that entropy is a property, and like all other properties, it has fixed values at fixed states. Therefore, the entropy change ΔS between two specified states is the same no matter what path, reversible or irreversible, is followed during a process (Fig. 7-3).

Also note that the integral of $\delta Q/T$ will give us the value of entropy change *only if* the integration is carried out along an *internally reversible* path between the two states. The integral of $\delta Q/T$ along an irreversible path is not a property, and in general, different values will be obtained when the integration is carried out along different irreversible paths. Therefore, even for irreversible processes, the entropy change should be determined by carrying out this integration along some convenient *imaginary* internally reversible path between the specified states.

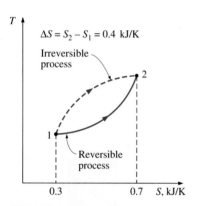

$\Delta S = S_2 - S_1 = 0.4$ kJ/K

FIGURE 7-3

The entropy change between two specified states is the same whether the process is reversible or irreversible.

A Special Case: Internally Reversible Isothermal Heat Transfer Processes

We pointed out in Chap. 6 that isothermal heat transfer processes are internally reversible. Therefore, the entropy change of a system during an internally reversible isothermal heat transfer process can be determined by performing the integration in Eq. 7-5:

$$\Delta S = \int_1^2 \left(\frac{\delta Q}{T}\right)_{\text{int rev}} = \int_1^2 \left(\frac{\delta Q}{T_0}\right)_{\text{int rev}} = \frac{1}{T_0}\int_1^2 (\delta Q)_{\text{int rev}}$$

which reduces to

$$\Delta S = \frac{Q}{T_0} \quad \text{(kJ/K)} \quad (7\text{-}6)$$

where T_0 is the constant absolute temperature of the system and Q is the heat transfer for the internally reversible process. Equation 7-6 is particularly useful for determining the entropy changes of thermal energy reservoirs that can absorb or supply heat indefinitely at a constant temperature.

Notice that the entropy change of a system during an internally reversible isothermal process can be positive or negative, depending on the direction of heat transfer. Heat transfer to a system will increase the entropy of a system, whereas heat transfer from a system will decrease it. In fact, losing heat is the only way the entropy of a system can be decreased.

EXAMPLE 7-1   Entropy Change during an Isothermal Process

A piston-cylinder device contains a liquid–vapor mixture of water at 300 K. During a constant pressure process, 750 kJ of heat is transferred to the water. As a result, part of the liquid in the cylinder vaporizes. Determine the entropy change of the water during this process.

Solution   We take the *entire water* (liquid + vapor) in the cylinder as the system (Fig. 7-4). This is a *closed system* since no mass crosses the system boundary during the process. We note that the temperature of the system remains constant at 300 K during this process since the temperature of a pure substance remains constant at the saturation value during a phase-change process at constant pressure.

Assumptions   No irreversibilities occur within the system boundaries during the process.

Analysis   The system undergoes an internally reversible, isothermal process, and thus its entropy change can be determined directly from Eq. 7-6 to be

$$\Delta S_{\text{sys, isothermal}} = \frac{Q}{T_{\text{sys}}} = \frac{750 \text{ kJ}}{300 \text{ K}} = \textbf{2.5 kJ/K}$$

Discussion   Note that the entropy change of the system is positive, as expected, since heat transfer is *to* the system.

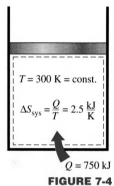

FIGURE 7-4

Schematic for Example 7-1.

7-2 ▓ THE INCREASE OF ENTROPY PRINCIPLE

Consider a cycle that is made up of two processes: process 1-2, which is arbitrary (reversible or irreversible), and process 2-1, which is internally reversible, as shown in Fig. 7-5. From the Clausius inequality,

$$\oint \frac{\delta Q}{T} \leq 0$$

or

$$\int_1^2 \frac{\delta Q}{T} + \int_2^1 \left(\frac{\delta Q}{T}\right)_{\text{int rev}} \leq 0$$

The second integral in the above relation is readily recognized as the entropy change $S_1 - S_2$. Therefore,

$$\int_1^2 \frac{\delta Q}{T} + S_1 - S_2 \leq 0$$

which can be rearranged as

$$S_2 - S_1 \geq \int_1^2 \frac{\delta Q}{T} \tag{7-7}$$

Equation 7-7 can be viewed as a mathematical statement of the second law of thermodynamics for a fixed mass. It can also be expressed in differential form as

$$dS \geq \frac{\delta Q}{T} \tag{7-8}$$

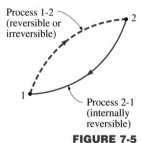

FIGURE 7-5

A cycle composed of a reversible and an irreversible process.

where the equality holds for an internally reversible process and the inequality for an irreversible process. We may conclude from these equations that *the entropy change of a closed system during an irreversible process is greater than the integral of* $\delta Q/T$ *evaluated for that process. In the limiting case of a reversible process, these two quantities become equal.* We again emphasize that T in the above relations is the *absolute temperature* at the *boundary* where the differential heat δQ is transferred between the system and the surroundings.

The quantity $\Delta S = S_2 - S_1$ represents the *entropy change* of the system. For a reversible process, it becomes equal to $\int_1^2 \delta Q/T$, which represents the *entropy transfer* with heat.

The inequality sign in the relations above is a constant reminder that the entropy change of a closed system during an irreversible process is always greater than the entropy transfer. That is, some entropy is *generated* or *created* during an irreversible process, and this generation is due entirely to the presence of irreversibilities. The entropy generated during a process is called **entropy generation** and is denoted by S_{gen}. Noting that the difference between the entropy change of a closed system and the entropy transfer is equal to entropy generation, Eq. 7-7 can be rewritten as an equality as

$$\Delta S_{sys} = S_2 - S_1 = \int_1^2 \frac{\delta Q}{T} + S_{gen} \qquad (7\text{-}9)$$

Note that the entropy generation S_{gen} is always a *positive* quantity or zero. Its value depends on the process, and thus it is *not* a property of the system. Also, in the absence of any entropy transfer, the entropy change of a system is equal to the entropy generation.

Equation 7-7 has far-reaching implications in thermodynamics. For an isolated system (or simply an adiabatic closed system), the heat transfer is zero, and Eq. 7-7 reduces to

$$\Delta S_{isolated} \geq 0 \qquad (7\text{-}10)$$

This equation can be expressed as *the entropy of an isolated system during a process always increases or, in the limiting case of a reversible process, remains constant.* In other words, it *never* decreases. This is known as the **increase of entropy principle.** Note that in the absence of any heat transfer, entropy change is due to irreversibilities only, and their effect is always to increase entropy.

Entropy is an extensive property, and thus the total entropy of a system is equal to the sum of the entropies of the parts of the system. An isolated system may consist of any number of subsystems (Fig. 7-6). A system and its surroundings, for example, constitute an isolated system since both can be enclosed by a sufficiently large arbitrary boundary across which there is no heat, work, or mass transfer (Fig. 7-7). Therefore, a system and its surroundings can be viewed as the two subsystems of an isolated system, and the entropy change of this isolated system during a process is the sum of the entropy changes of the system and its surroundings, which is equal to the entropy generation since an isolated system involves no entropy transfer. That is,

$$S_{gen} = \Delta S_{total} = \Delta S_{sys} + \Delta S_{surr} \geq 0 \qquad (7\text{-}11)$$

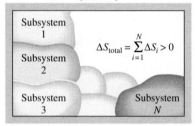

(Isolated)

$$\Delta S_{total} = \sum_{i=1}^{N} \Delta S_i > 0$$

FIGURE 7-6

The entropy change of an isolated system is the sum of the entropy changes of its components, and is never less than zero.

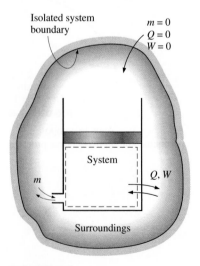

FIGURE 7-7

A system and its surroundings form an isolated system.

where the equality holds for reversible processes and the inequality for irreversible ones. Note that ΔS_{surr} refers to the change in the entropy of the surroundings as a result of the occurrence of the process under consideration.

Since no actual process is truly reversible, we can conclude that some entropy is generated during a process, and therefore the entropy of the universe, which can be considered to be an isolated system, is continuously increasing. The more irreversible a process, the larger the entropy generated during that process. No entropy is generated during reversible processes ($S_{gen} = 0$).

Entropy increase of the universe is a major concern not only to engineers but also to philosophers and theologians since entropy is viewed as a measure of the disorder (or "mixed-up-ness") in the universe.

The increase of entropy principle does not imply that the entropy of a system cannot decrease. The entropy change of a system *can* be negative during a process (Fig. 7-8), but entropy generation cannot. The increase of entropy principle can be summarized as follows:

$$S_{gen} \begin{cases} > 0 & \text{Irreversible process} \\ = 0 & \text{Reversible process} \\ < 0 & \text{Impossible process} \end{cases}$$

This relation serves as a criterion in determining whether a process is reversible, irreversible, or impossible.

Things in nature have a tendency to change until they attain a state of equilibrium. The increase of entropy principle dictates that the entropy of an isolated system will increase until the entropy of the system reaches a *maximum* value. At that point, the system is said to have reached an equilibrium state since the increase of entropy principle prohibits the system from undergoing any change of state that will result in a decrease in entropy.

Some Remarks about Entropy

In light of the preceding discussions, we can draw the following conclusions:

1. Processes can occur in a *certain* direction only, not in *any* direction. A process must proceed in the direction that complies with the increase of entropy principle, that is, $S_{gen} \geq 0$. A process that violates this principle is impossible. This principle often forces chemical reactions to come to a halt before reaching completion.

2. Entropy is a *nonconserved property,* and there is *no* such thing as the *conservation of entropy principle.* Entropy is conserved during the idealized reversible processes only and increases during *all* actual processes. Therefore, the entropy of the universe is continuously increasing.

3. The performance of engineering systems is degraded by the presence of irreversibilities, and *entropy generation* is a measure of the magnitudes of the irreversibilities present during that process. The greater the extent of irreversibilities, the greater the entropy generation. Therefore, entropy generation can be used as a quantitative measure of irreversibilities associated with a process. It is also used to establish criteria for the performance of engineering devices. This point is illustrated further in the following example.

EXAMPLE 7-2 Entropy Generation during Heat Transfer Processes

A heat source at 800 K loses 2000 kJ of heat to a sink at (a) 500 K and (b) 750 K. Determine which heat transfer process is more irreversible.

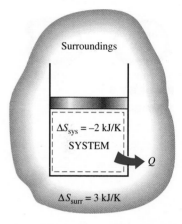

Surroundings

$\Delta S_{sys} = -2$ kJ/K

SYSTEM

Q

$\Delta S_{surr} = 3$ kJ/K

$S_{gen} = \Delta S_{total} = \Delta S_{sys} + \Delta S_{surr} = 1$ kJ/K

FIGURE 7-8

The entropy change of a system can be negative, but the entropy generation cannot.

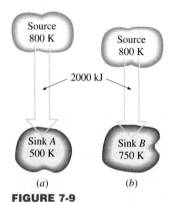

FIGURE 7-9

Schematic for Example 7-2.

Solution A sketch of the reservoirs is shown in Fig. 7-9. Both cases involve heat transfer through a finite temperature difference, and therefore both are irreversible. The magnitude of the irreversibility associated with each process can be determined by calculating the total entropy change for each case. The total entropy change for a heat transfer process involving two reservoirs (a source and a sink) is the sum of the entropy changes of each reservoir since the two reservoirs form an adiabatic system.

Or do they? The problem statement gives the impression that the two reservoirs are in direct contact during the heat transfer process. But this cannot be the case since the temperature at a point can have only one value, and thus it cannot be 800 K on one side of the point of contact and 500 K on the other side. In other words, the temperature function cannot have a jump discontinuity. Therefore, it is reasonable to assume that the two reservoirs are separated by a partition through which the temperature drops from 800 K on one side to 500 K (or 750 K) on the other. Therefore, the entropy change of the partition should also be considered when evaluating the total entropy change for this process. However, considering that entropy is a property and the values of properties depend on the state of a system, we can argue that the entropy change of the partition is zero since the partition appears to have undergone a *steady* process and thus experienced no change in its properties at any point. We base this argument on the fact that the temperature on both sides of the partition and thus throughout remained constant during this process. Therefore, we are justified to assume that $\Delta S_{\text{partition}} \cong 0$ since the entropy (as well as the energy) content of the partition essentially remained constant during this process.

The entropy change for each reservoir can be determined from Eq. 7-6 since each reservoir undergoes an internally reversible, isothermal process.

(*a*) For the heat transfer process to a sink at 500 K.

$$\Delta S_{\text{source}} = \frac{Q_{\text{source}}}{T_{\text{source}}} = \frac{-2000 \text{ kJ}}{800 \text{ K}} = -2.5 \text{ kJ/K}$$

$$\Delta S_{\text{sink}} = \frac{Q_{\text{sink}}}{T_{\text{sink}}} = \frac{+2000 \text{ kJ}}{500 \text{ K}} = +4.0 \text{ kJ/K}$$

and $\quad S_{\text{gen}} = \Delta S_{\text{total}} = \Delta S_{\text{source}} + \Delta S_{\text{sink}} = (-2.5 + 4.0) \text{ kJ/K} = \mathbf{+1.5 \text{ kJ/K}}$

Therefore, 1.5 kJ/K of entropy is generated during this process. Noting that both reservoirs have undergone internally reversible processes, the entire entropy generation took place in the partition.

(*b*) Repeating the calculations in part (*a*) for a sink temperature of 750 K, we obtain

$$\Delta S_{\text{source}} = -2.5 \text{ kJ/K}$$
$$\Delta S_{\text{sink}} = +2.7 \text{ kJ/K}$$

and $\quad S_{\text{gen}} = \Delta S_{\text{total}} = (-2.5 + 2.7) \text{ kJ/K} = \mathbf{+0.2 \text{ kJ/K}}$

The total entropy change for the process in part (*b*) is smaller, and therefore it is less irreversible. This is expected since the process in (*b*) involves a smaller temperature difference and thus a smaller irreversibility.

Discussion The irreversibilities associated with both processes could be eliminated by operating a Carnot heat engine between the source and the sink. For this case it can be easily shown that $\Delta S_{\text{total}} = 0$.

7-3 ■ ENTROPY CHANGE OF PURE SUBSTANCES

Entropy is a property, and thus the value of entropy of a system is fixed once the state of the system is specified. Specifying two intensive independent properties fixes the state of a simple compressible system, and thus the value of entropy, as well as the values of other properties at that state. Starting with its defining relation (Eq. 7-4), the entropy change of a substance can be expressed in terms of other properties (see Sec. 7-7). But in general, these relations are too complicated and are not practical to use for hand calculations. Therefore, using a suitable reference state, the entropies of substances are evaluated from measurable property data following rather involved computations, and the results are tabulated in the same manner as the other properties such as v, u, and h (Fig. 7-10).

The entropy values in the property tables are given relative to an arbitrary reference state. In steam tables the entropy of saturated liquid s_f at 0.01°C is assigned the value of zero. For refrigerant-134a, the zero value is assigned to saturated liquid at −40°C. The entropy values become negative at temperatures below the reference value.

The value of entropy at a specified state is determined just like any other property. In the compressed liquid and superheated vapor regions, it can be obtained directly from the tables at the specified state. In the saturated mixture region, it is determined from

$$s = s_f + x s_{fg} \qquad (kJ/kg \cdot K)$$

where x is the quality and s_f and s_{fg} values are listed in the saturation tables. In the absence of compressed liquid data, the entropy of the compressed liquid can be approximated by the entropy of the saturated liquid at the given temperature:

$$s_{@\,T,\,P} \cong s_{f\,@\,T} \qquad (kJ/kg \cdot K)$$

The entropy change of a specified mass m (such as a closed system) during a process is simply

$$\Delta S = m \Delta s = m(s_2 - s_1) \qquad (kJ/K) \qquad (7\text{-}12)$$

which is the difference between the entropy values at the final and initial states.

When studying the second law aspects of processes, entropy is commonly used as a coordinate on diagrams such as the T-s and h-s diagrams. The general characteristics of the T-s diagram of pure substances are shown in Fig. 7-11 using data for water. Notice from this diagram that the constant volume lines are steeper than the constant pressure lines and the constant pressure lines are parallel to the constant temperature lines in the saturated liquid–vapor mixture region. Also, the constant pressure lines almost coincide with the saturated liquid line in the compressed liquid region.

EXAMPLE 7-3 Entropy Change of a Substance in a Tank

A rigid tank contains 5 kg of refrigerant-134a initially at 20°C and 140 kPa. The refrigerant is now cooled while being stirred until its pressure drops to 100 kPa. Determine the entropy change of the refrigerant during this process.

Solution We take the refrigerant in the tank as the *system* (Fig. 7-12). This is a *closed system* since no mass crosses the system boundary during the process. We note that the change in entropy of a substance during a process is simply the difference between the entropy values at the final and initial states. The initial state of the refrigerant is completely specified.

Assumptions The volume of the tank is constant and thus $v_2 = v_1$.

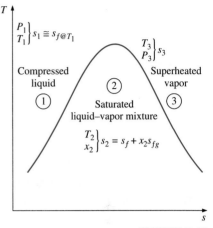

FIGURE 7-10

The entropy of a pure substance is determined from the tables, just as for any other property.

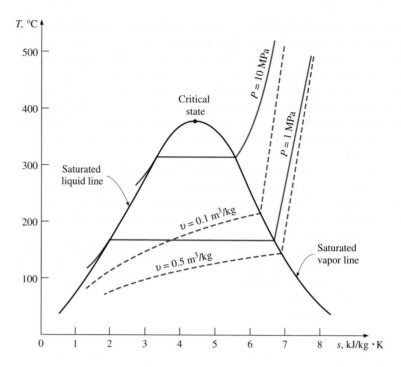

FIGURE 7-11

Schematic of the T-s diagram for water.

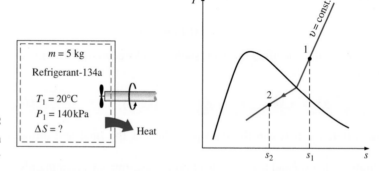

FIGURE 7-12

Schematic and T-s diagram
for Example 7-3.

Analysis Recognizing that the specific volume remains constant during this process, the properties of the refrigerant at both states are determined to be

State 1: $\left.\begin{array}{l} P_1 = 140 \text{ kPa} \\ T_1 = 20°C \end{array}\right\}$ $\begin{array}{l} s_1 = 1.0532 \text{ kJ/kg} \cdot \text{K} \\ v_1 = 0.1652 \text{ m}^3/\text{kg} \end{array}$

State 2: $\left.\begin{array}{l} P_2 = 100 \text{ kPa} \\ (v_2 = v_1) \end{array}\right\}$ $\begin{array}{l} v_f = 0.0007258 \text{ m}^3/\text{kg} \\ v_g = 0.1917 \text{ m}^3/\text{kg} \end{array}$

The refrigerant is a saturated liquid–vapor mixture at the final state since $v_f < v_2 < v_g$ at 100 kPa pressure. Therefore, we need to determine the quality first:

$$x_2 = \frac{v_2 - v_f}{v_{fg}} = \frac{0.1652 - 0.0007258}{0.1917 - 0.0007258} = 0.861$$

Thus,

$$s_2 = s_f + x_2 s_{fg} = 0.0678 + (0.861)(0.9395 - 0.0678) = 0.8183 \text{ kJ/kg} \cdot \text{K}$$

Then the entropy change of the refrigerant during this process is determined from

$$\Delta S = m(s_2 - s_1) = (5 \text{ kg})[(0.8183 - 1.0532) \text{ kJ/kg} \cdot \text{K}]$$
$$= \mathbf{-1.174 \text{ kJ/K}}$$

Discussion The negative sign indicates that the entropy of the system is decreasing during this process. This is not a violation of the second law, however, since it is the *entropy generation* S_{gen} that cannot be negative.

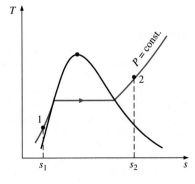

EXAMPLE 7-4 Entropy Change during a Constant Pressure Process

A piston-cylinder device initially contains 3 lbm of liquid water at 20 psia and 70°F. The water is now heated at constant pressure by the addition of 3450 Btu of heat. Determine the entropy change of the water during this process.

Solution We take the water in the cylinder as the *system* (Fig. 7-13). This is a *closed system* since no mass crosses the system boundary during the process. We note that a piston-cylinder device typically involves a moving boundary and thus boundary work W_b. Also, heat is transferred to the system and electrical work W_e is supplied to the system.

Assumptions **1** The tank is stationary and thus the kinetic and potential energy changes are zero, $\Delta KE = \Delta PE = 0$. **2** The process is quasi-equilibrium. **3** The pressure remains constant during the process and thus $P_2 = P_1$.

Analysis Water exists as a compressed liquid at the initial state since its pressure is greater than the saturation pressure of 0.3632 psia at 70°F. By approximating the compressed liquid as a saturated liquid at the given temperature, the properties at the initial state are

State 1: $\left.\begin{array}{l} P_1 = 20 \text{ psia} \\ T_1 = 70°\text{F} \end{array}\right\} \quad \begin{array}{l} s_1 \cong s_{f@\,70°\text{F}} = 0.07463 \text{ Btu/lbm} \cdot \text{R} \\ h_1 \cong h_{f@\,70°\text{F}} = 38.09 \text{ Btu/lbm} \end{array}$

At the final state, the pressure is still 20 psia, but we need one more property to fix the state. This property is determined from the energy balance applied to the system,

$$\underbrace{E_{in} - E_{out}}_{\substack{\text{Net energy transfer} \\ \text{by heat, work, and mass}}} = \underbrace{\Delta E_{system}}_{\substack{\text{Change in internal, kinetic,} \\ \text{potential, etc., energies}}}$$

$$Q_{in} - W_b = \Delta U$$

$$Q_{in} = \Delta H = m(h_2 - h_1)$$

$$3450 \text{ Btu} = (3 \text{ lbm})(h_2 - 38.09 \text{ Btu/lbm})$$

$$h_2 = 1188.1 \text{ Btu/lbm}$$

since $\Delta U + W_b = \Delta H$ for a constant pressure quasi-equilibrium process. Then,

State 2: $\left.\begin{array}{l} P_2 = 20 \text{ psia} \\ h_2 = 1188.1 \text{ Btu/lbm} \end{array}\right\} \quad \begin{array}{l} s_2 = 1.7759 \text{ Btu/lbm} \cdot \text{R} \\ \text{(Table A-6E, interpolation)} \end{array}$

FIGURE 7-13
Schematic and *T-s* diagram for Example 7-4.

Therefore, the entropy change of water during this process is

$$\Delta S = m(s_2 - s_1) = (3 \text{ lbm})(1.7759 - 0.07463) \text{ Btu/lbm} \cdot \text{R}$$
$$= \textbf{5.1038 Btu/R}$$

7-4 ■ ISENTROPIC PROCESSES

We mentioned earlier that the entropy of a fixed mass can be changed by (1) heat transfer and (2) irreversibilities. Then it follows that the entropy of a fixed mass will not change during a process that is *internally reversible* and *adiabatic* (Fig. 7-14). A process during which the entropy remains constant is called an **isentropic process.** An isentropic process is characterized by

$$Isentropic\ process: \quad \Delta s = 0 \quad \text{or} \quad s_2 = s_1 \quad (\text{kJ/kg} \cdot \text{K}) \quad (7\text{-}13)$$

That is, a substance will have the same entropy value at the end of the process as it does at the beginning if the process is carried out in an isentropic manner. We will develop some very useful isentropic relations for ideal gases later in this chapter.

Many engineering systems or devices such as pumps, turbines, nozzles, and diffusers are essentially adiabatic in their operation, and they perform best when the irreversibilities, such as the friction associated with the process, are minimized. Therefore, an isentropic process can serve as an appropriate model for actual processes. Also, isentropic processes enable us to define efficiencies for processes to compare the actual performance of these devices to the performance under idealized conditions. This should be sufficient motivation for studying the isentropic processes.

It should be recognized that a *reversible adiabatic* process is necessarily isentropic ($s_2 = s_1$), but an *isentropic* process is not necessarily a reversible adiabatic process. (The entropy increase of a substance during a process as a result of irreversibilities may be offset by a decrease in entropy as a result of heat losses, for example.) However, the term *isentropic process* is customarily used in thermodynamics to imply an *internally reversible, adiabatic process.*

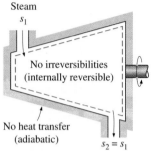

FIGURE 7-14

During an internally reversible, adiabatic (isentropic) process, the entropy of a system remains constant.

EXAMPLE 7-5   Isentropic Expansion of Steam in a Turbine
Steam enters an adiabatic turbine at 5 MPa and 450°C and leaves at a pressure of 1.4 MPa. Determine the work output of the turbine per unit mass of steam flowing through the turbine if the process is reversible and the changes in kinetic and potential energies are negligible.

Solution   We take the *turbine* as the system (Fig. 7-15). This is a *control volume* since mass crosses the system boundary during the process. We note that there is only one inlet and one exit, and thus $\dot{m}_1 = \dot{m}_2 = \dot{m}$.

Assumptions   **1** This is a steady-flow process since there is no change with time at any point and thus $\Delta m_{CV} = 0$, $\Delta E_{CV} = 0$, and $\Delta S_{CV} = 0$. **2** The process is reversible. **3** Kinetic and potential energies are negligible. **4** The turbine is adiabatic and thus there is no heat flow.

Analysis   The power output of the turbine is determined from the rate form of the energy balance,

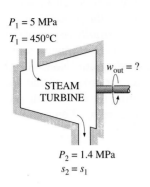

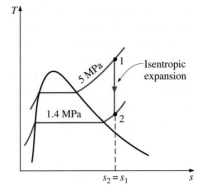

FIGURE 7-15

Schematic and T-s diagram for
Example 7-5.

$$\underbrace{\dot{E}_{in} - \dot{E}_{out}}_{\substack{\text{Rate of net energy transfer} \\ \text{by heat, work, and mass}}} = \underbrace{\Delta \dot{E}_{system}}_{\substack{\text{Rate of change in internal, kinetic,} \\ \text{potential, etc., energies}}}^{\nearrow 0 \text{ (steady)}} = 0$$

$$\dot{E}_{in} = \dot{E}_{out}$$
$$\dot{m}h_1 = \dot{W}_{out} + \dot{m}h_2 \qquad \text{(since } \dot{Q} = 0, \text{ ke} \cong \text{pe} \cong 0\text{)}$$
$$\dot{W}_{out} = \dot{m}(h_1 - h_2)$$

The inlet state is completely specified since two properties are given. But only one property (pressure) is given at the final state, and we need one more property to fix it. The second property comes from the observation that the process is reversible and adiabatic, and thus isentropic. Therefore, $s_2 = s_1$, and

State 1: $\left. \begin{array}{l} P_1 = 5 \text{ MPa} \\ T_1 = 450°C \end{array} \right\}$ $\begin{array}{l} h_1 = 3316.2 \text{ kJ/kg} \\ s_1 = 6.8186 \text{ kJ/kg} \cdot \text{K} \end{array}$

State 2: $\left. \begin{array}{l} P_2 = 1.4 \text{ MPa} \\ s_2 = s_1 \end{array} \right\}$ $h_2 = 2966.6 \text{ kJ/kg}$

Then the work output of the turbine per unit mass of the steam flowing through it becomes

$$w_{out} = h_1 - h_2 = 3316.2 - 2966.6 = \textbf{349.6 kJ/kg}$$

7-5 ■ PROPERTY DIAGRAMS INVOLVING ENTROPY

Property diagrams serve as great visual aids in the thermodynamic analysis of processes. We have used P-v and T-v diagrams extensively in previous chapters in conjunction with the first law of thermodynamics. In the second-law analysis, it is very helpful to plot the processes on diagrams for which one of the coordinates is entropy. The two diagrams commonly used in the second-law analysis are the *temperature-entropy* and the *enthalpy-entropy* diagrams.

Consider the defining equation of entropy (Eq. 7-4). It can be rearranged as

$$\delta Q_{int\,rev} = T\,dS \qquad \text{(kJ)} \qquad (7\text{-}14)$$

As shown in Fig. 7-16, $\delta Q_{rev\,int}$ corresponds to a differential area on a T-S diagram. The total heat transfer during an internally reversible process is determined by integration to be

$$Q_{int\,rev} = \int_1^2 T\,dS \qquad \text{(kJ)} \qquad (7\text{-}15)$$

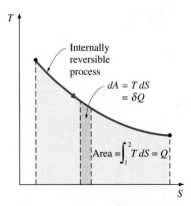

FIGURE 7-16

On a T-S diagram, the area under the process curve represents the heat transfer for internally reversible processes.

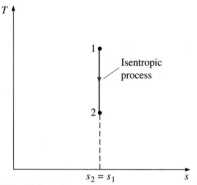

FIGURE 7-17

During an internally reversible, adiabatic (isentropic) process, the entropy of a system remains constant.

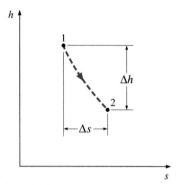

FIGURE 7-18

For adiabatic steady-flow devices, the vertical distance Δh on an h-s diagram is a measure of work, and the horizontal distance Δs is a measure of irreversibilities.

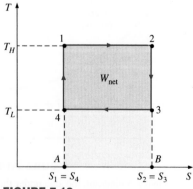

FIGURE 7-19

The T-S diagram of a Carnot cycle (Example 7-6).

which corresponds to the area under the process curve on a T-S diagram. Therefore, we conclude that *the area under the process curve on a T-S diagram represents heat transfer during an internally reversible process.* This is somewhat analogous to reversible boundary work being represented by the area under the process curve on a P-V diagram. Note that the area under the process curve represents heat transfer for processes that are internally (or totally) reversible. The area has no meaning for irreversible processes.

Equations 7-14 and 7-15 can also be expressed on a unit-mass basis as

$$\delta q_{\text{int rev}} = T \, ds \qquad \text{(kJ/kg)} \qquad (7\text{-}16)$$

and

$$q_{\text{int rev}} = \int_1^2 T \, ds \qquad \text{(kJ/kg)} \qquad (7\text{-}17)$$

To perform the integrations in Eqs. 7-15 and 7-17, one needs to know the relationship between T and s during a process. One special case for which these integrations can be performed easily is the *internally reversible isothermal process.* It yields

$$Q_{\text{int rev}} = T_0 \, \Delta S \qquad \text{(kJ)} \qquad (7\text{-}18)$$

or

$$q_{\text{int rev}} = T_0 \, \Delta s \qquad \text{(kJ/kg)} \qquad (7\text{-}19)$$

where T_0 is the constant temperature and ΔS is the entropy change of the system during the process.

In the relations above, T is the absolute temperature, which is always positive. Therefore, heat transfer during internally reversible processes is positive when entropy increases and negative when entropy decreases. An isentropic process on a T-s diagram is easily recognized as a *vertical-line segment.* This is expected since an isentropic process involves no heat transfer, and therefore the area under the process path must be zero (Fig. 7-17). The T-s diagrams serve as valuable tools for visualizing the second-law aspects of processes and cycles, and thus they are frequently used in thermodynamics. The T-s diagram of water is given in the appendix in Fig. A-9.

Another diagram commonly used in engineering is the enthalpy-entropy diagram, which is quite valuable in the analysis of steady-flow devices such as turbines, compressors, and nozzles. The coordinates of an h-s diagram represent two properties of major interest: enthalpy, which is a primary property in the first-law analysis of the steady-flow devices, and entropy, which is the property that accounts for irreversibilities during adiabatic processes. In analyzing the steady flow of steam through an adiabatic turbine, for example, the vertical distance between the inlet and the exit states (Δh) is a measure of the work output of the turbine, and the horizontal distance (Δs) is a measure of the irreversibilities associated with the process (Fig. 7-18).

The h-s diagram is also called a **Mollier diagram** after the German scientist R. Mollier (1863–1935). An h-s diagram is given in the appendix for steam in Fig. A-10.

EXAMPLE 7-6 The T-S Diagram of the Carnot Cycle

Show the Carnot cycle on a T-S diagram and indicate the areas that represent the heat supplied Q_H, heat rejected Q_L, and the net work output $W_{\text{net, out}}$ on this diagram.

Solution You will recall from Chap. 5 that the Carnot cycle is made up of two reversible isothermal (T = constant) processes and two isentropic (s = constant) processes. These four processes form a rectangle on a T-S diagram, as shown in Fig. 7-19.

On a T-S diagram, the area under the process curve represents the heat transfer for that process. Thus the area A12B represents Q_H, the area A43B represents Q_L, and the difference between these two (the area in color) represents the net work since

$$W_{net, out} = Q_H - Q_L$$

Therefore, the area enclosed by the path of a cycle (area 1234) on a T-S diagram represents the net work. Recall from Chap. 4 that the area enclosed by the path of a cycle also represents the net work on a P-V diagram.

7-6 ■ WHAT IS ENTROPY?

It is clear from the previous discussion that entropy is a useful property and serves as a valuable tool in the second-law analysis of engineering devices. But this does not mean that we know and understand entropy well. Because we do not. In fact, we cannot even give an adequate answer to the question, What is entropy? Not being able to describe entropy fully, however, does not take anything away from its usefulness. In Chap. 1, we could not define *energy* either, but it did not interfere with our understanding of energy transformations and the conservation of energy principle. Granted, entropy is not a household word like energy. But with continued use, our understanding of entropy will deepen, and our appreciation of it will grow. The discussion below will shed some light on the physical meaning of entropy by considering the microscopic nature of matter.

Entropy can be viewed as a measure of *molecular disorder,* or *molecular randomness.* As a system becomes more disordered, the positions of the molecules become less predictable and the entropy increases. Thus, it is not surprising that the entropy of a substance is lowest in the solid phase and highest in the gas phase (Fig. 7-20). In the solid phase, the molecules of a substance continually oscillate about their equilibrium positions, but they cannot move relative to each other, and their position at any instant can be predicted with good certainty. In the gas phase, however, the molecules move about at random, collide with each other, and change direction, making it extremely difficult to predict accurately the microscopic state of a system at any instant. Associated with this molecular chaos is a high value of entropy.

When viewed microscopically (from a statistical thermodynamics point of view), an isolated system that appears to be at a state of equilibrium may exhibit a high level of activity because of the continual motion of the molecules. To each state of macroscopic equilibrium there corresponds a large number of possible microscopic states or molecular configurations. The entropy of a system is related to the total number of possible microscopic states of that system, called *thermodynamic probability p*, by the **Boltzmann relation,** expressed as

$$S = k \ln p \qquad (7-20)$$

where $k = 1.3806 \times 10^{-23}$ J/K is the **Boltzmann constant.** Therefore, from a microscopic point of view, the entropy of a system increases whenever the molecular randomness or uncertainty (i.e., molecular probability) of a system increases. Thus, entropy is a measure of molecular disorder, and the molecular disorder of an isolated system increases anytime it undergoes a process.

Molecules in the gas phase possess a considerable amount of kinetic energy. But we know that no matter how large their kinetic energies are, the

Entropy, kJ/kg · K

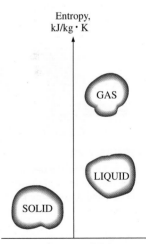

FIGURE 7-20

The level of molecular disorder (entropy) of a substance increases as it melts or evaporates.

FIGURE 7-21

Disorganized energy does not create much useful effect, no matter how large it is.

FIGURE 7-22

In the absence of friction, raising a weight by a rotating shaft does not create any disorder (entropy), and thus energy is not degraded during this process.

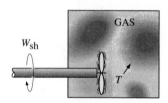

FIGURE 7-23

The paddle-wheel work done on a gas increases the level of disorder (entropy) of the gas, and thus energy is degraded during this process.

gas molecules will not rotate a paddle wheel inserted into the container and produce work. This is because the gas molecules, and the energy they possess, are disorganized. Probably the number of molecules trying to rotate the wheel in one direction at any instant is equal to the number of molecules that are trying to rotate it in the opposite direction, causing the wheel to remain motionless. Therefore, we cannot extract any useful work directly from disorganized energy (Fig. 7-21).

Now consider a rotating shaft shown in Fig. 7-22. This time the energy of the molecules is completely organized since the molecules of the shaft are rotating in the same direction together. This organized energy can readily be used to perform useful tasks such as raising a weight or generating electricity. Being an organized form of energy, work is free of disorder or randomness and thus free of entropy. *There is no entropy transfer associated with energy transfer as work.* Therefore, in the absence of any friction, the process of raising a weight by a rotating shaft (or a flywheel) will not produce any entropy. Any process that does not produce a net entropy is reversible, and thus the process described above can be reversed by lowering the weight. Therefore, energy is not degraded during this process, and no potential to do work is lost.

Instead of raising a weight, let us operate the paddle wheel in a container filled with a gas, as shown in Fig. 7-23. The paddle-wheel work in this case will be converted to the internal energy of the gas, as evidenced by a rise in gas temperature, creating a higher level of molecular chaos and disorder in the container. This process is quite different from raising a weight since the organized paddle-wheel energy is now converted to a highly disorganized form of energy, which cannot be converted back to the paddle wheel as the rotational kinetic energy. Only a portion of this energy can be converted to work by partially reorganizing it through the use of a heat engine. Therefore, energy is degraded during this process, the ability to do work is reduced, molecular disorder is produced, and associated with all this is an increase in entropy.

The *quantity* of energy is always preserved during an actual process (the first law), but the *quality* is bound to decrease (the second law). This decrease in quality is always accompanied by an increase in entropy. As an example, consider the transfer of 10 kJ of energy as heat from a hot medium to a cold one. At the end of the process, we will still have the 10 kJ of energy, but at a lower temperature and thus at a lower quality.

Heat is, in essence, a form of *disorganized energy,* and some disorganization (entropy) will flow with heat (Fig. 7-24). As a result, the entropy and the level of molecular disorder or randomness of the hot body will decrease with the entropy and the level of molecular disorder of the cold body will increase. The second law requires that the increase in entropy of the cold body be greater than the decrease in entropy of the hot body, and thus the net entropy of the combined system (the cold body and the hot body) increases. That is, the combined system is at a state of greater disorder at the final state. Thus we can conclude that processes can occur only in the direction of increased overall entropy or molecular disorder. That is, the entire universe is getting more and more chaotic every day.

From a statistical point of view, entropy is a measure of molecular randomness, that is, the uncertainty about the positions of molecules at any instant. Even in the solid phase, the molecules of a substance continually oscillate, creating an uncertainty about their position. These oscillations, however, fade as the temperature is decreased, and the molecules become completely motionless at absolute zero. This represents a state of ultimate molecular order (and minimum energy). Therefore, *the entropy of a pure*

crystalline substance at absolute zero temperature is zero* since there is no uncertainty about the state of the molecules at that instant (Fig. 7-25). This
statement is known as the **third law of thermodynamics.** The third law of
thermodynamics provides an absolute reference point for the determination of
entropy. The entropy determined relative to this point is called **absolute entropy,** and it is extremely useful in the thermodynamic analysis of chemical
reactions. Notice that the entropy of a substance that is not pure crystalline
(such as a solid solution) is not zero at absolute zero temperature. This is because more than one molecular configuration exist for such substances, which
introduces some uncertainty about the microscopic state of the substance.

The concept of entropy as a measure of disorganized energy can also be
applied to other areas. Iron molecules, for example, create a magnetic field
around themselves. In ordinary iron, molecules are randomly aligned, and
they cancel each other's magnetic effect. When iron is treated and the molecules are realigned, however, that piece of iron turns into a piece of magnet,
creating a powerful magnetic field around it.

Entropy and Entropy Generation in Daily Life

Entropy can be viewed as a measure of disorder or disorganization in a
system. Likewise, entropy generation can be viewed as a measure of disorder
or disorganization generated during a process. The concept of entropy is not
used in daily life nearly as extensively as the concept of energy, even though
entropy is readily applicable to various aspects of daily life. The extension of
the entropy concept to nontechnical fields is not a novel idea. It has been the
topic of several articles, and even some books. Below we present several ordinary events and show their relevance to the concept of entropy and entropy
generation.

Efficient people lead low-entropy (highly organized) lives. They have a
place for everything (minimum uncertainty), and it takes minimum energy for
them to locate something. Inefficient people, on the other hand, are disorganized and lead high-entropy lives. It takes them minutes (if not hours) to find
something they need, and they are likely to create a bigger disorder as they are
searching since they will probably conduct the search in a disorganized manner (Fig. 7-26). People leading high-entropy lifestyles are always on the run,
and never seem to catch up.

You probably noticed (with frustration) that some people seem to *learn*
fast and remember well what they learn. We can call this type of learning organized or low-entropy learning. These people make a conscientious effort to
file the new information properly by relating it to their existing knowledge
base and creating a solid information network in their minds. On the other
hand, people who throw the information into their minds as they study, with
no effort to secure it, may *think* they are learning. They are bound to discover
otherwise when they need to locate the information, for example during a test.
It is not easy to retrieve information from a database that is, in a sense, in the
gas phase. Students who have blackouts during tests should reexamine their
study habits.

A *library* with a good shelving and indexing system can be viewed as
a low-entropy library because of the high level of organization. Likewise, a
library with a poor shelving and indexing system can be viewed as a high-
entropy library because of the high level of disorganization. A library with no
indexing system is like no library, since a book is of no value if it cannot be
found (Fig. 6-27).

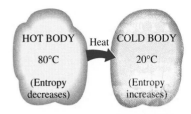

FIGURE 7-24

During a heat transfer process, the
net disorder (entropy) increases.
(The increase in the disorder of
the cold body more than offsets the
decrease in the disorder of the hot body.)

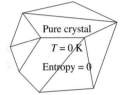

FIGURE 7-25

A pure substance at absolute
zero temperature is in perfect order,
and its entropy is zero (the third
law of thermodynamics).

FIGURE 7-26

The use of entropy
(disorganization, uncertainty)
is not limited to thermodynamics.

Consider two identical buildings, each containing one million books. In the first building, the books are *piled* on top of each other, whereas in the second building they are *highly organized, shelved, and indexed* for easy reference. There is no doubt about which building a student will prefer to go to for checking out a certain book. Yet, some may argue from the first-law point of view that these two buildings are equivalent since the mass and energy content of the two buildings are identical, despite the high level of disorganization (entropy) in the first building. This example illustrates that any realistic comparisons should involve the second-law point of view.

Two *textbooks* that seem to be identical because both cover basically the same topics and present the same information may actually be *very* different depending on *how* they cover the topics. After all, two seemingly identical cars are not so identical if one goes only half as many miles as the other one on the same amount of fuel. Likewise, two seemingly identical books are not so identical if it takes twice as long to learn a topic from one of them as it does from the other. Thus, comparisons made on the basis of the first law only may be highly misleading.

Having a disorganized (high-entropy) *army* is like having no army at all. It is no coincidence that the command centers of any armed forces are among the primary targets during a war. One army that consists of ten divisions is ten times more powerful than ten armies each consisting of a single division. Likewise, one country that consists of ten states is more powerful than ten countries, each consisting of a single state. The *United States* would not be such a powerful country if there were fifty independent countries in its place instead of a single country with fifty states. The new European common market has the potential to be a new economic superpower. The old cliché "divide and conquer" can be rephrased as "increase the entropy and conquer."

We know that mechanical friction is always accompanied by entropy generation, and thus reduced performance. We can generalize this to daily life: *friction in the work place* with fellow workers is bound to generate entropy, and thus adversely affect performance (Fig. 7-27). It will result in reduced productivity. Hopefully, someday we will be able to come up with some procedures to quantify entropy generated during nontechnical activities, and maybe even pinpoint its primary sources and magnitude.

We also know that *unrestrained expansion* (or explosion) and uncontrolled electron exchange (chemical reactions) generate entropy and are highly irreversible. Likewise, unrestrained opening of the mouth to scatter angry words is highly irreversible since this generates entropy, and it can cause considerable damage. A person who gets up in anger is bound to sit down at a loss.

FIGURE 7-27

As in mechanical systems, friction in the work place is bound to generate entropy and reduce performance.

7-7 ■ THE *T ds* RELATIONS

Earlier in this chapter, it was shown that the quantity $(\delta Q/T)_{\text{int rev}}$ corresponds to a differential change in a property, called *entropy*. The entropy change for a process, then, was evaluated by integrating $\delta Q/T$ along some imaginary internally reversible path between the actual end states (Eq. 7-5). For isothermal internally reversible processes, this integration is straightforward. But when the temperature varies during the process, we have to have a relation between δQ and T to perform this integration. Finding such relations is what we intend to do in this section.

The differential form of the conservation of energy equation for a closed stationary system (a fixed mass) containing a simple compressible substance can be expressed for an internally reversible process as

$$\delta Q_{\text{int rev}} - \delta W_{\text{int rev, out}} = dU \tag{7-21}$$

But

$$\delta Q_{\text{int rev}} = T\,dS$$

$$\delta W_{\text{int rev, out}} = P\,dV$$

Thus,

$$T\,dS = dU + P\,dV \tag{7-22}$$

or

$$T\,ds = du + P\,dv \tag{7-23}$$

per unit mass. This equation is known as the first $T\,ds$, or *Gibbs, equation*. Notice that the only type of work interaction a simple compressible system may involve as it undergoes an internally reversible process is the quasi-equilibrium boundary work.

The second $T\,ds$ equation is obtained by eliminating du from Eq. 7-23 by using the definition of enthalpy ($h = u + Pv$):

$$\left.\begin{array}{ll} h = u + Pv & \longrightarrow \quad dh = du + P\,dv + v\,dP \\ \text{(Eq. 7-23)} & \longrightarrow \quad T\,ds = du + P\,dv \end{array}\right\} \quad T\,ds = dh - v\,dP \tag{7-24}$$

Equations 7-23 and 7-24 are extremely valuable since they relate entropy changes of a system to the changes in other properties. Unlike Eq. 7-4, they are property relations and therefore are independent of the type of the processes.

The $T\,ds$ relations above are developed with an internally reversible process in mind since the entropy change between two states must be evaluated along a reversible path. But the results obtained are valid for both reversible and irreversible processes since entropy is a property and the change in a property between two states is independent of the type of process the system undergoes. Equations 7-23 and 7-24 are relations between the properties of a unit mass of a simple compressible system as it undergoes a change of state, and they are applicable whether the change occurs in a closed or an open system (Fig. 7-28).

Explicit relations for differential changes in entropy are obtained by solving for ds in Eqs. 7-23 and 7-24:

$$ds = \frac{du}{T} + \frac{P\,dv}{T} \tag{7-25}$$

and

$$ds = \frac{dh}{T} - \frac{v\,dP}{T} \tag{7-26}$$

The entropy change during a process can be determined by integrating either of these equations between the initial and the final states. To perform these integrations, however, we must know the relationship between du or dh and the temperature (such as $du = C_v\,dT$ and $dh = C_p\,dT$ for ideal gases) as well as the equation of state for the substance (such as the ideal-gas equation of state $Pv = RT$). For substances for which such relations exist, the integration of Eq. 7-25 or 7-26 is straightforward. This is done later in this chapter. For other substances, we have to rely on tabulated data.

The $T\,ds$ relations for nonsimple systems, that is, systems that involve more than one mode of quasi-equilibrium work, can be obtained in a similar manner by including all the relevant quasi-equilibrium work modes.

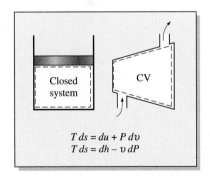

FIGURE 7-28

The $T\,ds$ relations are valid for both reversible and irreversible processes and for both closed and open systems.

7-8 ■ ENTROPY CHANGE OF LIQUIDS AND SOLIDS

We mentioned in Chap. 5 that liquids and solids can be approximated as *incompressible substances* since their specific volumes remain nearly constant during a process. Thus, $dv \cong 0$ for liquids and solids, and Eq. 7-25 for this case reduces to

$$ds = \frac{du}{T} = \frac{C\,dT}{T} \qquad (7\text{-}27)$$

since $C_p = C_v = C$ and $du = C\,dT$ for incompressible substances. Then the entropy change during a process is determined by integration to be

$$\textit{Liquids, solids:} \quad s_2 - s_1 = \int_1^2 C(T)\frac{dT}{T} \cong C_{av} \ln \frac{T_2}{T_1} \quad (\text{kJ/kg} \cdot \text{K}) \quad (7\text{-}28)$$

where C_{av} is the *average* specific heat of the substance over the given temperature range. Note that the entropy change of a truly incompressible substance depends on temperature only and is independent of pressure.

Equation 7-28 can be used to determine the entropy changes of solids and liquids with reasonable accuracy. However, for liquids that expand considerably with temperature, it may be necessary to consider the effects of volume change in calculations. This is especially the case when the temperature change is large.

A relation for isentropic processes of liquids and solids is obtained by setting the entropy change relation above equal to zero. It gives

$$\textit{Isentropic:} \quad s_2 - s_1 = C_{av} \ln \frac{T_2}{T_1} = 0 \quad \rightarrow \quad T_2 = T_1 \qquad (7\text{-}29)$$

That is, the temperature of a truly incompressible substance remains constant during an isentropic process. Therefore, the isentropic process of an incompressible substance is also isothermal. This behavior is closely approximated by liquids and solids.

EXAMPLE 7-7 Effect of Density of a Liquid on Entropy

Liquid methane is commonly used in various cryogenic applications. The critical temperature of methane is 191 K (or −82°C), and thus methane must be maintained below 191 K to keep it in liquid phase. The properties of liquid methane at various temperatures and pressures are given in Table 7-1. Determine the entropy

TABLE 7-1

Properties of liquid methane

Temp., T, K	Pressure, P, MPa	Density, ρ, kg/m³	Enthalpy, h, kJ/kg	Entropy, s, kJ/kg · K	Specific heat, C_p, kJ/kg · K
110	0.5	425.3	208.3	4.878	3.476
	1.0	425.8	209.0	4.875	3.471
	2.0	426.6	210.5	4.867	3.460
	5.0	429.1	215.0	4.844	3.432
120	0.5	410.4	243.4	5.185	3.551
	1.0	411.0	244.1	5.180	3.543
	2.0	412.0	245.4	5.171	3.528
	5.0	415.2	249.6	5.145	3.486

change of liquid methane as it undergoes a process from 110 K and 1 MPa to 120 K and 5 MPa (a) using actual data for methane and (b) approximating liquid methane as an incompressible substance. What is the error involved in the latter case?

Solution The entropy change of methane during a process is to be determined using actual data and assuming it to be incompressible.

Analysis (a) We consider a unit mass of liquid methane (Fig. 7-29). The entropies of the methane at the initial and final states are

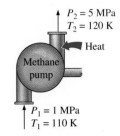

State 1: $P_1 = 1$ MPa $\}$ $s_1 = 4.875$ kJ/kg · K
             $T_1 = 110$ K $\}$ $C_{p1} = 3.471$ kJ/kg · K

State 2: $P_2 = 5$ MPa $\}$ $s_2 = 5.145$ kJ/kg · K
             $T_2 = 120$ K $\}$ $C_{p2} = 3.486$ kJ/kg · K

FIGURE 7-29

Schematic for Example 7-7.

Therefore,

$$\Delta s = s_2 - s_1 = 5.145 - 4.875 = \textbf{0.270 kJ/kg · K}$$

(b) Approximating liquid methane as an incompressible substance, its entropy change is determined to be

$$\Delta s = C_{av} \ln \frac{T_2}{T_1} = (3.4785 \text{ kJ/kg · K}) \ln \frac{120 \text{ K}}{110 \text{ K}} = \textbf{0.303 kJ/kg · K}$$

since

$$C_{p, av} = \frac{C_{p1} + C_{p2}}{2} = \frac{3.471 + 3.486}{2} = 3.4785 \text{ kJ/kg · K}$$

Therefore, the error involved in approximating liquid methane as an incompressible substance is

$$\text{Error} = \frac{|\Delta s_{actual} - \Delta s_{ideal}|}{\Delta s_{actual}} = \frac{|0.270 - 0.303|}{0.270} = \textbf{0.122 (or 12.2\%)}$$

Discussion This result is not surprising since the density of liquid methane changes during this process from 425.8 to 415.2 kg/m³ (about 3 percent), which makes us question the validity of the incompressible substance assumption. Still, this assumption enables us to obtain reasonably accurate results with less effort, which proves to be very convenient in the absence of compressed liquid data.

EXAMPLE 7-8   Economics of Replacing a Valve by a Turbine

A cryogenic manufacturing facility handles liquid methane at 115 K and 5 MPa at a rate of 0.280 m³/s . A process requires dropping the pressure of liquid methane to 1 MPa, which is done by throttling the liquid methane by passing it through a flow resistance such as a valve. A recently hired engineer proposes to replace the throttling valve by a turbine in order to produce power while dropping the pressure to 1 MPa. Using data from Table 7-1, determine the maximum amount of power that can be produced by such a turbine. Also, determine how much this turbine will save the facility from electricity usage costs per year if the turbine operates continuously (8760 h/yr) and the facility pays $0.075/kWh for electricity.

Solution We take the turbine as the system (Fig. 7-30). This is a control volume since mass crosses the system boundary during the process. We note that there is only one inlet and one exit and thus m·1 5 m·2 5 m·.

Assumptions **1** This is a steady-flow process since there is no change with time at any point and thus $\Delta m_{CV} = 0$, $\Delta E_{CV} = 0$, and $\Delta S_{CV} = 0$. **2** The turbine is

FIGURE 7-30

A 1.0-MW liquified natural gas (LNG) turbine with 95-cm turbine runner diameter being installed in a cryogenic test facility (courtesy of Ebara International Corporation, Cryodynamics Division, Sparks, Nevada).

adiabatic and thus there is no heat transfer. **3** The process is reversible. **4** Kinetic and potential energies are negligible.

Analysis The assumptions above are reasonable since a turbine is normally well-insulated and it must involve no irreversibilities for best performance and thus *maximum* power production. Therefore, the process through the turbine must be *reversible adiabatic* or *isentropic*. Then, $s_2 = s_1$ and

State 1: $\left.\begin{array}{l} P_1 = 5 \text{ MPa} \\ T_1 = 115 \text{ K} \end{array}\right\}$ $\begin{array}{l} h_1 = 232.3 \text{ kJ/kg} \\ s_1 = 4.9945 \text{ kJ/kg} \cdot \text{K} \\ \rho_1 = 422.15 \text{ kg/s} \end{array}$

State 2: $\left.\begin{array}{l} P_2 = 1 \text{ MPa} \\ s_2 = s_1 \end{array}\right\}$ $h_2 = 222.8 \text{ kJ/kg}$

Also, the mass flow rate of liquid methane is

$$\dot{m} = \rho_1 \dot{V}_1 = (422.15 \text{ kg/m}^3)(0.280 \text{ m}^3/\text{s}) = 118.2 \text{ kg/s}$$

Then the power output of the turbine is determined from the rate form of the energy balance to be

$$\underbrace{\dot{E}_{in} - \dot{E}_{out}}_{\substack{\text{Rate of net energy transfer} \\ \text{by heat, work, and mass}}} = \underbrace{\Delta\dot{E}_{system}}_{\substack{\text{Rate of change in internal, kinetic,} \\ \text{potential, etc., energies}}}\!\!\!\!\!\!\!\!\!\!\!\!\!\!{}^{\nearrow 0 \text{ (steady)}} = 0$$

$$\dot{E}_{in} = \dot{E}_{out}$$
$$\dot{m}h_1 = \dot{W}_{out} + \dot{m}h_2 \qquad (\text{since } \dot{Q} = 0, \text{ke} \cong \text{pe} \cong 0)$$
$$\dot{W}_{out} = \dot{m}(h_1 - h_2)$$
$$= (118.2 \text{ kg/s})(232.3 - 222.8) \text{ kJ/kg}$$
$$= \textbf{1123 kW}$$

For continuous operation ($365 \times 24 = 8760$ h), the amount of power produced per year will be

Annual power production $= \dot{W}_{out} \times \Delta t = (1123 \text{ kW})(8760 \text{ h/yr})$
$$= 0.9837 \times 10^7 \text{ kWh/yr}$$

At \$0.075/kWh, the amount of money this turbine will save the facility becomes

Annual power savings $=$ (Annual power production)(Unit cost of power)
$$= (0.9837 \times 10^7 \text{ kWh/yr})(\$0.075/\text{kWh})$$
$$= \textbf{\$737,800/yr}$$

That is, this turbine can save the facility \$737,800 a year by simply taking advantage of the potential that is currently being wasted by a throttling valve, and the engineer who made this observation should be rewarded.

Discussion This example shows the importance of the property entropy since it enabled us to quantify the work potential that is being wasted. In practice, the turbine will not be isentropic, and thus the power produced will be less. The analysis above gave us the upper limit. An actual turbine-generator assembly can utilize about 80 percent of the potential and produce more than 900 kW of power while saving the facility more than \$600,000 a year.

It can also be shown that the temperature of methane will drop to 113.9 K (a drop of 1.1 K) during the isentropic expansion process in the turbine instead of remaining constant at 115 K as would be the case if methane were assumed to be an incompressible substance. The temperature of methane would rise to 116.6 K (a rise of 1.6 K) during the throttling process.

An expression for the entropy change of an ideal gas can be obtained from Eq. 7-25 or 7-26 by employing the property relations for ideal gases (Fig. 7-31). By substituting $du = C_v \, dT$ and $P = RT/v$ into Eq. 7-25, the differential entropy change of an ideal gas becomes

$$ds = C_v \frac{dT}{T} + R \frac{dv}{v} \qquad (7\text{-}30)$$

The entropy change for a process is obtained by integrating this relation between the end states:

$$s_2 - s_1 = \int_1^2 C_v(T) \frac{dT}{T} + R \ln \frac{v_2}{v_1} \qquad (7\text{-}31)$$

A second relation for the entropy change of an ideal gas is obtained in a similar manner by substituting $dh = C_p \, dT$ and $v = RT/P$ into Eq. 7-26 and integrating. The result is

$$s_2 - s_1 = \int_1^2 C_p(T) \frac{dT}{T} - R \ln \frac{P_2}{P_1} \qquad (7\text{-}32)$$

The specific heats of ideal gases, with the exception of monatomic gases, depend on temperature, and the integrals in Eqs. 7-31 and 7-32 cannot be performed unless the dependence of C_v and C_p on temperature is known. Even when the $C_v(T)$ and $C_p(T)$ functions are available, performing long integrations every time entropy change is calculated is not practical. Then two reasonable choices are left: either perform these integrations by simply assuming constant specific heats or evaluate those integrals once and tabulate the results. Both approaches are presented below.

1 Constant Specific Heats (Approximate Analysis)

Assuming constant specific heats for ideal gases is a common approximation, and we used this assumption before on several occasions. It usually simplifies the analysis greatly, and the price we pay for this convenience is some loss in accuracy. The magnitude of the error introduced by this assumption depends on the situation at hand. For example, for monatomic ideal gases such as helium, the specific heats are independent of temperature, and therefore the constant-specific-heat assumption introduces no error. For ideal gases whose specific heats vary almost linearly in the temperature range of interest, the possible error is minimized by using specific-heat values evaluated at the average temperature (Fig. 7-32). The results obtained in this way usually are sufficiently accurate for most ideal gases if the temperature range is not greater than a few hundred degrees.

The entropy-change relations for ideal gases under the constant-specific-heat assumption are easily obtained by replacing $C_v(T)$ and $C_p(T)$ in Eqs. 7-31 and 7-32 by $C_{v,\,av}$ and $C_{p,\,av}$, respectively, and performing the integrations. We obtain

$$s_2 - s_1 = C_{v,\,av} \ln \frac{T_2}{T_1} + R \ln \frac{v_2}{v_1} \qquad (\text{kJ/kg} \cdot \text{K}) \qquad (7\text{-}33)$$

and $\qquad s_2 - s_1 = C_{p,\,av} \ln \frac{T_2}{T_1} - R \ln \frac{P_2}{P_1} \qquad (\text{kJ/kg} \cdot \text{K}) \qquad (7\text{-}34)$

The Entropy Change of Ideal Gases

$$Pv = RT$$
$$du = C_v \, dT$$
$$dh = C_p \, dT$$

FIGURE 7-31

A broadcast from channel IG.

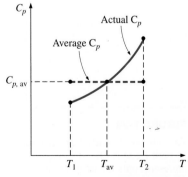

FIGURE 7-32

Under the constant-specific-heat assumption, the specific heat is assumed to be constant at some average value.

Entropy changes can also be expressed on a unit-mole basis by multiplying these relations by molar mass:

$$\bar{s}_2 - \bar{s}_1 = \bar{C}_{v,\,av} \ln \frac{T_2}{T_1} + R_u \ln \frac{v_2}{v_1} \qquad \text{(kJ/kmol} \cdot \text{K)} \qquad (7\text{-}35)$$

and

$$\bar{s}_2 - \bar{s}_1 = \bar{C}_{p,\,av} \ln \frac{T_2}{T_1} - R_u \ln \frac{P_2}{P_1} \qquad \text{(kJ/kmol} \cdot \text{K)} \qquad (7\text{-}36)$$

2 Variable Specific Heats: Exact Analysis

When the temperature change during a process is large and the specific heats of the ideal gas vary nonlinearly within the temperature range, the assumption of constant specific heats may lead to considerable errors in entropy-change calculations. For those cases, the variation of specific heats with temperature should be properly accounted for by utilizing accurate relations for the specific heats as a function of temperature. The entropy change during a process is then determined by substituting these $C_v(T)$ or $C_p(T)$ relations into Eq. 7-31 or 7-32 and performing the integrations.

Instead of performing these laborious integrals each time we have a new process, it is convenient to perform these integrals once and tabulate the results. For this purpose, we choose absolute zero as the reference temperature and define a function s° as

$$s^\circ = \int_0^T C_p(T) \frac{dT}{T} \qquad (7\text{-}37)$$

According to this definition, s° is a function of temperature alone, and its value is zero at absolute zero temperature. The values of s° are calculated at various temperatures from Eq. 7-37, and the results are tabulated in the appendix as a function of temperature for air. Given this definition, the integral Eq. 7-32 becomes

$$\int_1^2 C_p(T) \frac{dT}{T} = s_2^\circ - s_1^\circ \qquad (7\text{-}38)$$

where s_2° is the value of s° at T_2 and s_1° is the value at T_1. Thus,

$$s_2 - s_1 = s_2^\circ - s_1^\circ - R \ln \frac{P_2}{P_1} \qquad \text{(kJ/kg} \cdot \text{K)} \qquad (7\text{-}39)$$

It can also be expressed on a unit-mole basis as

$$\bar{s}_2 - \bar{s}_1 = \bar{s}_2^\circ - \bar{s}_1^\circ - R_u \ln \frac{P_2}{P_1} \qquad \text{(kJ/kmol} \cdot \text{K)} \qquad (7\text{-}40)$$

Note that unlike internal energy and enthalpy, the entropy of an ideal gas varies with specific volume or pressure as well as the temperature. Therefore, entropy cannot be tabulated as a function of temperature alone. The s° values in the tables account for the temperature dependence of entropy (Fig. 7-33). The variation of entropy with pressure is accounted for by the last term in Eq. 7-39. Another relation for entropy change can be developed based on Eq. 7-31, but this would require the definition of another function and tabulation of its values, which is not practical.

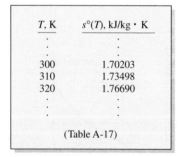

T, K	$s^\circ(T)$, kJ/kg $\cdot$ K
$\vdots$	$\vdots$
300	1.70203
310	1.73498
320	1.76690
$\vdots$	$\vdots$

(Table A-17)

FIGURE 7-33

The entropy of an ideal gas depends on both T and P. The function s° represents only the temperature-dependent part of entropy.

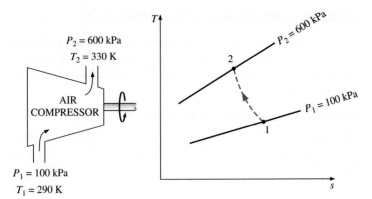

FIGURE 7-34

Schematic and *T-s* diagram for
Example 7-9.

EXAMPLE 7-9   Entropy Change of an Ideal Gas

Air is compressed from an initial state of 100 kPa and 17°C to a final state of 600 kPa and 57°C. Determine the entropy change of air during this compression process by using (a) property values from the air table and (b) average specific heats.

Solution   A sketch of the system and the *T-s* diagram for the process are given in Fig. 7-34. We note that both the initial and the final states of air are completely specified.

Assumptions   Air is an ideal gas since it is at a high temperature and low pressure relative to its critical point values. Therefore, entropy change relations developed under the ideal gas assumption are applicable.

Analysis   (a) The properties of air are given in the air table (Table A-17). Reading $s°$ values at given temperatures and substituting, we find

$$s_2 - s_1 = s_2^\circ - s_1^\circ - R \ln \frac{P_2}{P_1}$$

$$= [(1.79783 - 1.66802) \text{ kJ/kg} \cdot \text{K}] - (0.287 \text{ kJ/kg} \cdot \text{K}) \ln \frac{600 \text{ kPa}}{100 \text{ kPa}}$$

$$= -0.3844 \text{ kJ/kg} \cdot \text{K}$$

(b) The entropy change of air during this process can also be determined approximately from Eq. 7-34 by using a C_p value at the average temperature of 37°C (Table A-2b) and treating it as a constant:

$$s_2 - s_1 = C_{p,\text{av}} \ln \frac{T_2}{T_1} - R \ln \frac{P_2}{P_1}$$

$$= (1.006 \text{ kJ/kg} \cdot \text{K}) \ln \frac{330 \text{ K}}{290 \text{ K}} - (0.287 \text{ kJ/kg} \cdot \text{K}) \ln \frac{600 \text{ kPa}}{100 \text{ kPa}}$$

$$= -0.3842 \text{ kJ/kg} \cdot \text{K}$$

Discussion   The two results above are almost identical since the change in temperature during this process is relatively small (Fig. 7-35). When the temperature change is large, however, they may differ significantly. For those cases, Eq. 7-40 should be used instead of Eq. 7-34 since it accounts for the variation of specific heats with temperature.

AIR

$T_1 = 290$ K
$T_2 = 330$ K

$$s_2 - s_1 = s_2^\circ - s_1^\circ - R \ln \frac{P_2}{P_1}$$

$$= -0.3844$$

$$s_2 - s_1 = C_{p,\text{av}} \ln \frac{T_2}{T_1} - R \ln \frac{P_2}{P_1}$$

$$= -0.3842$$

FIGURE 7-35

For small temperature differences, the exact and approximate relations for entropy changes of ideal gases give almost identical results.

Isentropic Processes of Ideal Gases

Several relations for the isentropic processes of ideal gases can be obtained by setting the entropy-change relations developed above equal to zero. Again, this is done first for the case of constant specific heats and then for the case of variable specific heats.

Constant Specific Heats: Approximate Treatment

When the constant-specific-heat assumption is valid, the isentropic relations for ideal gases are obtained by setting Eqs. 7-33 and 7-34 equal to zero. From Eq. 7-33,

$$\ln \frac{T_2}{T_1} = -\frac{R}{C_v} \ln \frac{v_2}{v_1}$$

which can be rearranged as

$$\ln \frac{T_2}{T_1} = \ln \left(\frac{v_1}{v_2}\right)^{R/C_v} \tag{7-41}$$

or

$$\left(\frac{T_2}{T_1}\right)_{s=\text{const.}} = \left(\frac{v_1}{v_2}\right)^{k-1} \qquad \text{(ideal gas)} \tag{7-42}$$

since $R = C_p - C_v$, $k = C_p/C_v$, and thus $R/C_v = k - 1$.

Equation 7-42 is the *first isentropic relation* for ideal gases under the constant-specific-heat assumption. The *second isentropic relation* is obtained in a similar manner from Eq. 7-34 with the following result:

$$\left(\frac{T_2}{T_1}\right)_{s=\text{const.}} = \left(\frac{P_2}{P_1}\right)^{(k-1)/k} \qquad \text{(ideal gas)} \tag{7-43}$$

The *third isentropic relation* is obtained by substituting Eq. 7-43 into Eq. 7-42 and simplifying:

$$\left(\frac{P_2}{P_1}\right)_{s=\text{const.}} = \left(\frac{v_1}{v_2}\right)^{k} \qquad \text{(ideal gas)} \tag{7-44}$$

Equations 7-42 through 7-44 can also be expressed in a compact form as

$$Tv^{k-1} = \text{constant} \tag{7-45}$$
$$TP^{(1-k)/k} = \text{constant} \qquad \text{(ideal gas)} \tag{7-46}$$
$$Pv^{k} = \text{constant} \tag{7-47}$$

The specific heat ratio k, in general, varies with temperature, and thus an average k value for the given temperature range should be used.

Note that the ideal gas isentropic relations above, as the name implies, are strictly valid for isentropic processes only when the constant-specific-heat assumption is appropriate (Fig. 7-36).

Variable Specific Heats: Exact Treatment

When the constant-specific-heat assumption is not appropriate, the isentropic relations developed above will yield results that are not quite accurate. For

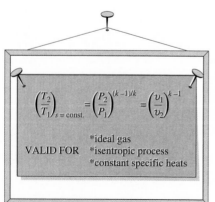

$$\left(\frac{T_2}{T_1}\right)_{s=\text{const.}} = \left(\frac{P_2}{P_1}\right)^{(k-1)/k} = \left(\frac{v_1}{v_2}\right)^{k-1}$$

VALID FOR
*ideal gas
*isentropic process
*constant specific heats

FIGURE 7-36

The isentropic relations of ideal gases are valid for the isentropic processes of ideal gases only.

such cases, we should use an isentropic relation obtained from Eq. 7-39 that accounts for the variation of specific heats with temperature. Setting this equation equal to zero gives

$$0 = s_2^\circ - s_1^\circ - R \ln \frac{P_2}{P_1}$$

or

$$s_2^\circ = s_1^\circ + R \ln \frac{P_2}{P_1} \qquad (7\text{-}48)$$

where s_2° is the s° value at the end of the isentropic process.

Relative Pressure and Relative Specific Volume

Equation 7-48 provides an accurate way of evaluating property changes of ideal gases during isentropic processes since it accounts for the variation of specific heats with temperature. However, it involves tedious iterations when the volume ratio is given instead of the pressure ratio. This is quite an inconvenience in optimization studies, which usually require numerous repetitive calculations. To remedy this deficiency, we define two new dimensionless quantities associated with isentropic processes.

The definition of the first is based on Eq. 7-48, which can be rearranged as

$$\frac{P_2}{P_1} = \exp \frac{s_2^\circ - s_1^\circ}{R}$$

or

$$\frac{P_2}{P_1} = \frac{\exp(s_2^\circ/R)}{\exp(s_1^\circ/R)}$$

The quantity $\exp(s^\circ/R)$ is defined as the **relative pressure** P_r. With this definition, the above relation becomes

$$\left(\frac{P_2}{P_1}\right)_{s=\text{const.}} = \frac{P_{r2}}{P_{r1}} \qquad (7\text{-}49)$$

Note that the relative pressure P_r is a *dimensionless* quantity that is a function of temperature only since s° depends on temperature alone. Therefore, values of P_r can be tabulated against temperature. This is done for air in Table A-17. The use of P_r data is illustrated in Fig. 7-37.

Sometimes specific volume ratios are given instead of pressure ratios. This is particularly the case when automotive engines are analyzed. In such cases, one needs to work with volume ratios. Therefore, we define another quantity related to specific volume ratios for isentropic processes. This is done by utilizing the ideal-gas relation and Eq. 7-49:

$$\frac{P_1 v_1}{T_1} = \frac{P_2 v_2}{T_2} \longrightarrow \frac{v_2}{v_1} = \frac{T_2}{T_1}\frac{P_1}{P_2} = \frac{T_2}{T_1}\frac{P_{r1}}{P_{r2}} = \frac{T_2/P_{r2}}{T_1/P_{r1}}$$

The quantity T/P_r is a function of temperature only and is defined as **relative specific volume** v_r. Thus,

$$\left(\frac{v_2}{v_1}\right)_{s=\text{const.}} = \frac{v_{r2}}{v_{r1}} \qquad (7\text{-}50)$$

Equations 7-49 and 7-50 are strictly valid for isentropic processes of ideal gases only. They account for the variation of specific heats with temperature

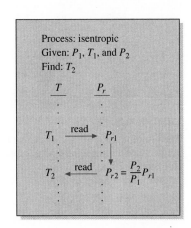

FIGURE 7-37

The use of P_r data for calculating the final temperature during an isentropic process.

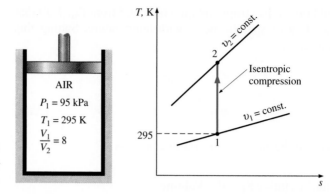

FIGURE 7-38

Schematic and *T-s* diagram for
Example 7-10.

and therefore give more accurate results than Eqs. 7-42 through 7-47. The values of P_r and v_r are listed for air in Table A-17.

EXAMPLE 7-10 Isentropic Compression of Air in a Car Engine

Air is compressed in a car engine from 22°C and 95 kPa in a reversible and adiabatic manner. If the compression ratio V_1/V_2 of this piston-cylinder device is 8, determine the final temperature of the air.

Solution A sketch of the system and the *T-s* diagram for the process are given in Fig. 7-38. We note that the process is reversible and adiabatic.

Assumptions At specified conditions, air can be treated as an ideal gas since it is at a high temperature and low pressure relative to its critical point values. Therefore, the isentropic relations developed earlier for ideal gases are applicable.

Analysis This process is easily recognized as being isentropic since it is both reversible and adiabatic. The final temperature for this isentropic process can be determined from Eq. 7-50 with the help of relative specific volume data (Table A-17), as illustrated in Fig. 7-39.

For closed systems: $$\frac{V_2}{V_1} = \frac{v_2}{v_1}$$

At $T_1 = 295$ K: $\qquad v_{r1} = 647.9$

From Eq. 7-50: $\quad v_{r2} = v_{r1}\left(\frac{v_2}{v_1}\right) = (647.9)\left(\frac{1}{8}\right) = 80.99 \quad\longrightarrow\quad T_2 = \textbf{662.7 K}$

Therefore, the temperature of air will increase by 367.7°C during this process.

Alternative solution The final temperature could also be determined from Eq. 7-42 by assuming constant specific heats for air:

$$\left(\frac{T_2}{T_1}\right)_{s=\text{const.}} = \left(\frac{V_1}{V_2}\right)^{k-1}$$

The specific heat ratio k also varies with temperature, and we need to use the value of k corresponding to the average temperature. However, the final temperature is not given, and so we cannot determine the average temperature in advance. For such cases, calculations can be started with a k value at the initial or the anticipated average temperature. This value could be refined later, if necessary, and the calculations can be repeated. We know that the temperature of the air will rise considerably during this adiabatic compression process, so we *guess* that the average temperature will be about 450 K. The k value at this anticipated average temperature is determined from Table A-2*b* to be 1.391. Then the final temperature of air becomes

$$T_2 = (295\ \text{K})(8)^{1.391-1} = 665.2\ \text{K}$$

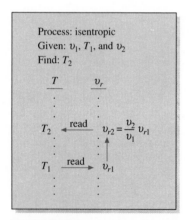

FIGURE 7-39

The use of v_r data for calculating the final temperature during an isentropic process (Example 7-10).

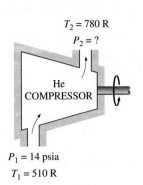

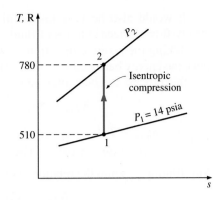

FIGURE 7-40

Schematic and *T-s* diagram for Example 7-11.

This will give an average temperature value of 480.1 K, which is sufficiently close to the assumed value of 450 K. Therefore, it is not necessary to repeat the calculations by using the *k* value at this average temperature.

The result obtained by assuming constant specific heats for this case is in error by about 0.4 percent, which is rather small. This is not surprising since the temperature change of air is relatively small (only a few hundred degrees) and the specific heats of air vary almost linearly with temperature in this temperature range.

EXAMPLE 7-11   Isentropic Compression of an Ideal Gas

Helium gas is compressed in an adiabatic compressor from an initial state of 14 psia and 50°F to a final temperature of 320°F in a reversible manner. Determine the exit pressure of helium.

Solution   A sketch of the system and the *T-s* diagram for the process are given in Fig. 7-40. We note that the process is reversible and adiabatic.

Assumptions   At specified conditions, helium can be treated as an ideal gas since it is at a high temperature relative to its critical point value of −450°F. Therefore, the isentropic relations developed earlier for ideal gases are applicable.

Analysis   The specific heat ratio *k* of helium is 1.667 and is independent of temperature in the region where it behaves as an ideal gas. Thus the final pressure of helium can be determined from Eq. 7-43:

$$P_2 = P_1\left(\frac{T_2}{T_1}\right)^{k/(k-1)} = (14 \text{ psia})\left(\frac{780 \text{ R}}{510 \text{ R}}\right)^{1.667/0.667} = \textbf{40.5 psia}$$

7-10 ■ REVERSIBLE STEADY-FLOW WORK

The work done during a process depends on the path followed as well as on the properties at the end states. In Chap. 4, we discussed reversible (quasi-equilibrium) moving boundary work associated with closed systems and expressed it in terms of the fluid properties as

$$W_b = \int_1^2 P \, dV$$

We mentioned that the quasi-equilibrium work interactions lead to the maximum work output for work-producing devices and the minimum work input for work-consuming devices.

It would also be very insightful to express the work associated with steady-flow devices in terms of fluid properties.

Taking the positive direction of work to be from the system (work output), the energy balance for a steady-flow device undergoing an internally reversible process can be expressed in differential form as

$$\delta q_{rev} - \delta w_{rev} = dh + dke + dpe$$

But
$$\begin{array}{ll} \delta q_{rev} = T\,ds & \text{(Eq. 7-16)} \\ T\,ds = dh - v\,dP & \text{(Eq. 7-24)} \end{array} \Bigg\} \quad \delta q_{rev} = dh - v\,dP$$

Substituting this into the relation above and canceling dh yield

$$-\delta w_{rev} = v\,dP + dke + dpe$$

Integrating, we find

$$w_{rev} = -\int_1^2 v\,dP - \Delta ke - \Delta pe \qquad \text{(kJ/kg)} \qquad (7\text{-}51)$$

When the changes in kinetic and potential energies are negligible, this equation reduces to

$$w_{rev} = -\int_1^2 v\,dP \qquad \text{(kJ/kg)} \qquad (7\text{-}52)$$

Equations 7-51 and 7-52 are relations for the *reversible work output* associated with an internally reversible process in a steady-flow device. They will give a negative result when work is done on the system. To avoid the negative sign, Eq. 7-51 can be written for work input to steady-flow devices such as compressors and pumps as

$$w_{rev,\,in} = \int_1^2 v\,dP + \Delta ke + \Delta pe \qquad (7\text{-}53)$$

The resemblance between the $v\,dP$ in these relations and $P\,dv$ is striking. They should not be confused with each other, however, since $P\,dv$ is associated with reversible boundary work in closed systems (Fig. 7-41).

Obviously, one needs to know v as a function of P for the given process to perform the integration in Eq. 7-51. When the working fluid is an *incompressible fluid*, the specific volume v remains constant during the process and can be taken out of the integration. Then Eq. 7-51 simplifies to

$$w_{rev} = -v(P_2 - P_1) - \Delta ke - \Delta pe \qquad \text{(kJ/kg)} \qquad (7\text{-}54)$$

For the steady flow of a liquid through a device that involves no work interactions (such as a nozzle or a pipe section), the work term is zero, and the equation above can be expressed as

$$v(P_2 - P_1) + \frac{\mathcal{V}_2^2 - \mathcal{V}_1^2}{2} + g(z_2 - z_1) = 0 \qquad (7\text{-}55)$$

which is known as the **Bernoulli equation** in fluid mechanics. It is developed for an internally reversible process and thus is applicable to incompressible

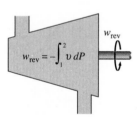

(a) Steady-flow system

$$w_{rev} = -\int_1^2 v\,dP$$

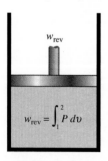

$$w_{rev} = \int_1^2 P\,dv$$

(b) Closed system

FIGURE 7-41

Reversible work relations for steady-flow and closed systems.

fluids that involve no irreversibilities such as friction or shock waves. This equation can be modified, however, to incorporate these effects.

Equation 7-52 has far-reaching implications in engineering regarding devices that produce or consume work steadily such as turbines, compressors, and pumps. It is obvious from this equation that the reversible steady-flow work is closely associated with the specific volume of the fluid flowing through the device. *The larger the specific volume, the larger the reversible work produced or consumed by the steady-flow device* (Fig. 7-42). This conclusion is equally valid for actual steady-flow devices. Therefore, every effort should be made to keep the specific volume of a fluid as small as possible during a compression process to minimize the work input and as large as possible during an expansion process to maximize the work output.

In steam or gas power plants, the pressure rise in the pump or compressor is equal to the pressure drop in the turbine if we disregard the pressure losses in various other components. In steam power plants, the pump handles liquid, which has a very small specific volume, and the turbine handles vapor, whose specific volume is many times larger. Therefore, the work output of the turbine is much larger than the work input to the pump. This is one of the reasons for the overwhelming popularity of steam power plants in electric power generation.

If we were to compress the steam exiting the turbine back to the turbine inlet pressure before cooling it first in the condenser in order to "save" the heat rejected, we would have to supply all the work produced by the turbine back to the compressor. In reality, the required work input would be even greater than the work output of the turbine because of the irreversibilities present in both processes.

In gas power plants, the working fluid (typically air) is compressed in the gas phase, and a considerable portion of the work output of the turbine is consumed by the compressor. As a result, a gas power plant delivers less net work per unit mass of the working fluid.

EXAMPLE 7-12   Compressing a Substance in the Liquid vs. Gas Phases

Determine the compressor work input required to compress steam isentropically from 100 kPa to 1 MPa, assuming that the steam exists as (*a*) saturated liquid and (*b*) saturated vapor at the initial state.

Solution   We take the turbine and then the pump as the system. Both are control volumes since mass crosses the boundary. Sketches of the pump and the turbine together with the *T-s* diagram are given in Fig. 7-43.

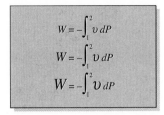

FIGURE 7-42

The larger the specific volume, the greater the work produced (or consumed) by a steady-flow device.

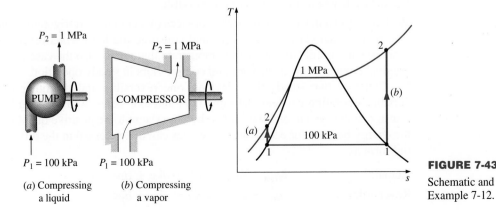

| (*a*) Compressing | (*b*) Compressing |
| a liquid | a vapor |

FIGURE 7-43

Schematic and *T-s* diagram for Example 7-12.

Assumptions **1** Steady operating conditions exist. **2** Kinetic and potential energy changes are negligible. **3** The process is given to be isentropic.

Analysis (*a*) In this case, steam is a saturated liquid initially, and its specific volume is

$$v_1 = v_{f @ 100kPa} = 0.001043 \text{ m}^3/\text{kg} \qquad \text{(Table A-5)}$$

which remains essentially constant during the process. Thus,

$$w_{rev, in} = \int_1^2 v \, dP \cong v_1(P_2 - P_1)$$

$$= (0.001043 \text{ m}^3/\text{kg})[(1000 - 100) \text{ kPa}]\left(\frac{1 \text{ kJ}}{1 \text{ kPa} \cdot \text{m}^3}\right)$$

$$= \textbf{0.94 kJ/kg}$$

(*b*) This time, steam is a saturated vapor initially and remains a vapor during the entire compression process. Since the specific volume of a gas changes considerably during a compression process, we need to know how *v* varies with *P* to perform the integration in Eq. 7-53. This relation, in general, is not readily available. But for an isentropic process, it is easily obtained from the second *T ds* relation by setting *ds* = 0:

$$\left. \begin{array}{ll} T \, ds = dh - v \, dP & \text{(Eq. 7-24)} \\ ds = 0 & \text{(isentropic process)} \end{array} \right\} \quad v \, dP = dh$$

Thus,

$$w_{rev, in} = \int_1^2 v \, dP = \int_1^2 dh = h_2 - h_1$$

This result could also be obtained from the energy balance relation for an isentropic steady-flow process. Next we determine the enthalpies:

State 1:
$$\left. \begin{array}{l} P_1 = 100 \text{ kPa} \\ \text{(sat. vapor)} \end{array} \right\} \quad \begin{array}{l} h_1 = 2675.5 \text{ kJ/kg} \\ s_1 = 7.3594 \text{ kJ/kg} \cdot \text{K} \end{array} \qquad \text{(Table A-5)}$$

State 2:
$$\left. \begin{array}{l} P_2 = 1 \text{ MPa} \\ s_2 = s_1 \end{array} \right\} \quad h_2 = 3195.5 \text{ kJ/kg} \qquad \text{(Table A-6)}$$

Thus,
$$w_{rev, in} = (3195.5 - 2675.5) \text{ kJ/kg} = \textbf{520 kJ/kg}$$

Discussion Note that compressing steam in the vapor form would require over 500 times more work than compressing it in the liquid form between the same pressure limits.

Proof that Steady-Flow Devices Deliver the Most and Consume the Least Work when the Process Is Reversible

We have shown in Chap. 6 that cyclic devices (heat engines, refrigerators, and heat pumps) deliver the most work and consume the least when reversible processes are used. Now we will demonstrate that this is also the case for individual devices such as turbines and compressors in steady operation.

Consider two steady-flow devices, one reversible and the other irreversible, operating between the same inlet and exit states. Again taking heat transfer to the system and work done by the system to be positive quantities, the energy balance for each of these devices can be expressed in the differential form as

Actual: $\qquad\qquad \delta q_{act} - \delta w_{act} = dh + d\text{ke} + d\text{pe}$

Reversible: $\qquad \delta q_{rev} - \delta w_{rev} = dh + d\text{ke} + d\text{pe}$

The right-hand sides of these two equations are identical since both devices are operating between the same end states. Thus,

$$\delta q_{act} - \delta w_{act} = \delta q_{rev} - \delta w_{rev}$$

or

$$\delta w_{rev} - \delta w_{act} = \delta q_{rev} - \delta q_{act}$$

But

$$\delta q_{rev} = T \, ds$$

Substituting this relation into the equation above and dividing each term by T, we obtain

$$\frac{\delta w_{rev} - \delta w_{act}}{T} = ds - \frac{\delta q_{act}}{T} \geq 0$$

since

$$ds \geq \frac{\delta q_{act}}{T} \qquad \text{(see Eq. 7-8)}$$

Also, T is the absolute temperature, which is always positive. Thus,

$$\delta w_{rev} \geq \delta w_{act}$$

or

$$w_{rev} \geq w_{act}$$

Thus work-producing devices such as turbines (w is positive) deliver more work, and work-consuming devices such as pumps and compressors (w is negative) require less work when they operate reversibly (Fig. 7-44).

7-11 ■ MINIMIZING THE COMPRESSOR WORK

We have shown in Sec. 7-10 that the work input to a compressor is minimized when the compression process is executed in an internally reversible manner. When the changes in kinetic and potential energies are negligible, the compressor work is given by (Eq. 7-53)

$$w_{rev,\,in} = \int_1^2 v \, dP \qquad (7\text{-}56)$$

Obviously one way of minimizing the compressor work is to approach an internally reversible process as much as possible by minimizing the irreversibilities such as friction, turbulence, and non-quasi-equilibrium compression. The extent to which this can be accomplished is limited by economic considerations. A second (and more practical) way of reducing the compressor work is to keep the specific volume of the gas as small as possible during the compression process. This is done by maintaining the temperature of the gas as low as possible during compression since the specific volume of a gas is proportional to temperature. Therefore, reducing the work input to a compressor requires that the gas be cooled as it is compressed.

To have a better understanding of the effect of cooling during the compression process, we compare the work input requirements for three kinds of processes: *an isentropic process* (involves no cooling), *a polytropic process* (involves some cooling), and *an isothermal process* (involves maximum cooling). Assuming all three processes are executed between the same pressure levels (P_1 and P_2) in an internally reversible manner and the gas behaves as an ideal gas ($Pv = RT$) with constant specific heats, we see that the compression work is determined by performing the integration in Eq. 7-56 for each case, with the following results:

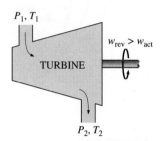

FIGURE 7-44

A reversible turbine delivers more work than an irreversible one if both operate between the same end states.

Isentropic (Pv^k = constant):

$$w_{\text{comp, in}} = \frac{kR(T_2 - T_1)}{k - 1} = \frac{kRT_1}{k - 1}\left[\left(\frac{P_2}{P_1}\right)^{(k-1)/k} - 1\right] \qquad (7\text{-}57a)$$

Polytropic (Pv^n = constant):

$$w_{\text{comp, in}} = \frac{nR(T_2 - T_1)}{n - 1} = \frac{nRT_1}{n - 1}\left[\left(\frac{P_2}{P_1}\right)^{(n-1)/n} - 1\right] \qquad (7\text{-}57b)$$

Isothermal (Pv = constant):

$$w_{\text{comp, in}} = RT\ln\frac{P_2}{P_1} \qquad (7\text{-}57c)$$

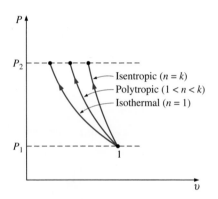

FIGURE 7-45

P-v diagrams of isentropic, polytropic, and isothermal compression processes between the same pressure limits.

The three processes are plotted on a *P-v* diagram in Fig. 7-45 for the same inlet state and exit pressure. On a *P-v* diagram, the area to the left of the process curve is the integral of $v\,dP$. Thus it is a measure of the steady-flow compression work. It is interesting to observe from this diagram that of the three internally reversible cases considered, the adiabatic compression (Pv^k = constant) requires the maximum work and the isothermal compression (T = constant or Pv = constant) requires the minimum. The work input requirement for the polytropic case (Pv^n = constant) is between these two and decreases as the polytropic exponent n is decreased, by increasing the heat rejection during the compression process. If sufficient heat is removed, the value of n approaches unity and the process becomes isothermal. One common way of cooling the gas during compression is to use cooling jackets around the casing of the compressors.

Multistage Compression with Intercooling

It is clear from the above arguments that cooling a gas as it is compressed is desirable since this reduces the required work input to the compressor. However, often it is not possible to have effective cooling through the casing of the compressor, and it becomes necessary to use other techniques to achieve effective cooling. One such technique is **multistage compression with intercooling,** where the gas is compressed in stages and cooled between each stage by passing it through a heat exchanger called an *intercooler.* Ideally, the cooling process takes place at constant pressure, and the gas is cooled to the initial temperature T_1 at each intercooler. Multistage compression with intercooling is especially attractive when a gas is to be compressed to very high pressures.

The effect of intercooling on compressor work is graphically illustrated on *P-v* and *T-s* diagrams in Fig. 7-46 for a two-stage compressor. The gas is compressed in the first stage from P_1 to an intermediate pressure P_x, cooled at constant pressure to the initial temperature T_1, and compressed in the second stage to the final pressure P_2. The compression processes, in general, can be modeled as polytropic (Pv^n = constant) where the value of n varies between k and 1. The colored area on the *P-v* diagram represents the work saved as a result of two-stage compression with intercooling. The process paths for single-stage isothermal and polytropic processes are also shown for comparison.

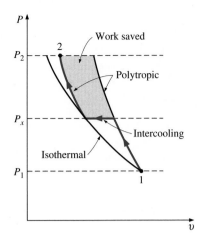
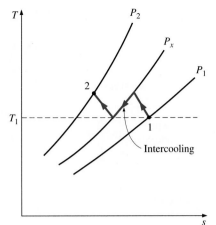

FIGURE 7-46

P-v and *T-s* diagrams for a two-stage steady-flow compression process.

The size of the colored area (the saved work input) varies with the value of the intermediate pressure P_x, and it is of practical interest to determine the conditions under which this area is maximized. The total work input for a two-stage compressor is the sum of the work inputs for each stage of compression, as determined from Eq. 7-57*b*:

$$w_{comp, in} = w_{comp\ I, in} + w_{comp\ II, in}$$

$$= \frac{nRT_1}{n-1}\left[\left(\frac{P_x}{P_1}\right)^{(n-1)/n} - 1\right] + \frac{nRT_1}{n-1}\left[\left(\frac{P_2}{P_x}\right)^{(n-1)/n} - 1\right] \quad (7\text{-}58)$$

The only variable in this equation is P_x. The P_x value that will minimize the total work is determined by differentiating this expression with respect to P_x and setting the resulting expression equal to zero. It yields

$$P_x = (P_1P_2)^{1/2} \quad \text{or} \quad \frac{P_x}{P_1} = \frac{P_2}{P_x} \quad (7\text{-}59)$$

That is, *to minimize compression work during two-stage compression, the pressure ratio across each stage of the compressor must be the same.* When this condition is satisfied, the compression work at each stage becomes identical, that is, $w_{comp\ I, in} = w_{comp\ II, in}$.

EXAMPLE 7-13 Work Input for Various Compression Processes

Air is compressed steadily by a reversible compressor from an inlet state of 100 kPa and 300 K to an exit pressure of 900 kPa. Determine the compressor work per unit mass for (*a*) isentropic compression with $k = 1.4$, (*b*) polytropic compression with $n = 1.3$, (*c*) isothermal compression, and (*d*) ideal two-stage compression with intercooling with a polytropic exponent of 1.3.

Solution We take the compressor to be the system. This is a control volume since mass crosses the boundary. A sketch of the system and the *T-s* diagram for the process are given in Fig. 7-47.

Assumptions **1** Steady operating conditions exist. **2** At specified conditions, air can be treated as an ideal gas since it is at a high temperature and low pressure relative to its critical point values. **3** Kinetic and potential energy changes are negligible.

Analysis The steady-flow compression work for all these four cases is determined by using the relations developed earlier in this section:

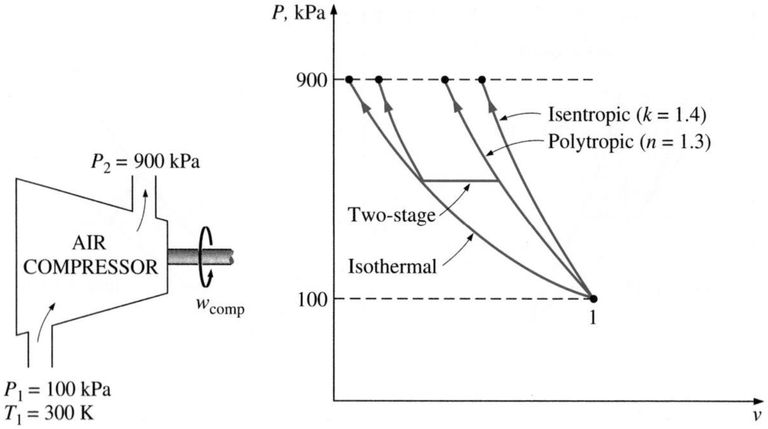

P, kPa

900

$P_2 = 900$ kPa

AIR COMPRESSOR

w_{comp}

100

Isentropic ($k = 1.4$)
Polytropic ($n = 1.3$)

Two-stage

Isothermal

1

FIGURE 7-47

Schematic and P-v diagram for Example 7-13.

$P_1 = 100$ kPa
$T_1 = 300$ K

v

(*a*) Isentropic compression with $k = 1.4$ (Eq. 7-57*a*):

$$w_{comp, in} = \frac{kRT_1}{k-1}\left[\left(\frac{P_2}{P_1}\right)^{(k-1)/k} - 1\right]$$

$$= \frac{(1.4)(0.287\ kJ/kg \cdot K)(300\ K)}{1.4 - 1}\left[\left(\frac{900\ kPa}{100\ kPa}\right)^{(1.4-1)/1.4} - 1\right]$$

$$= \textbf{263.2 kJ/kg}$$

(*b*) Polytropic compression with $n = 1.3$ (Eq. 7-57*b*):

$$w_{comp, in} = \frac{nRT_1}{n-1}\left[\left(\frac{P_2}{P_1}\right)^{(n-1)/n} - 1\right]$$

$$= \frac{(1.3)(0.287\ kJ/kg \cdot K)(300\ K)}{1.3 - 1}\left[\left(\frac{900\ kPa}{100\ kPa}\right)^{(1.3-1)/1.3} - 1\right]$$

$$= \textbf{246.4 kJ/kg}$$

(*c*) Isothermal compression (Eq. 7-57*c*):

$$w_{comp, in} = RT \ln \frac{P_2}{P_1} = (0.287\ kJ/kg \cdot K)(300\ K) \ln \frac{100\ kPa}{900\ kPa}$$

$$= \textbf{189.2 kJ/kg}$$

(*d*) Ideal two-stage compression with intercooling ($n = 1.3$): In this case, the pressure ratio across each stage is the same, and its value is determined from Eq. 7-59:

$$P_x = (P_1 P_2)^{1/2} = [(100\ kPa)(900\ kPa)]^{1/2} = 300\ kPa$$

The compressor work across each stage is also the same. Thus the total compressor work is twice the compression work for a single stage:

$$w_{comp, in} = 2w_{comp\ I, in} = 2\frac{nRT_1}{n-1}\left[\left(\frac{P_x}{P_1}\right)^{(n-1)/n} - 1\right]$$

$$= \frac{2(1.3)(0.287\ kJ/kg \cdot K)(300\ K)}{1.3 - 1}\left[\left(\frac{300\ kPa}{100\ kPa}\right)^{(1.3-1)/1.3} - 1\right]$$

$$= \textbf{215.3 kJ/kg}$$

Discussion Of all four cases considered, the isothermal compression requires the minimum work and the isentropic compression the maximum. The compressor work is decreased when two stages of polytropic compression are utilized instead of just one. As the number of compressor stages is increased, the compressor work approaches the value obtained for the isothermal case.

7-12 ■ ISENTROPIC EFFICIENCIES OF STEADY-FLOW DEVICES

We mentioned on several occasions that irreversibilities inherently accompany all actual processes and that their effect is always to downgrade the performance of devices. In engineering analysis, it would be very desirable to have some parameters that would enable us to quantify the degree of degradation of energy in these devices. In Chap. 6 we did this for cyclic devices, such as heat engines and refrigerators, by comparing the actual cycles to the idealized ones, such as the Carnot cycle. A cycle that was composed entirely of reversible processes served as the *model cycle* to which the actual cycles could be compared. This idealized model cycle enabled us to determine the theoretical limits of performance for cyclic devices under specified conditions and to examine how the performance of actual devices suffered as a result of irreversibilities.

Now we extend the analysis to discrete engineering devices working under steady-flow conditions, such as turbines, compressors, and nozzles, and we examine the degree of degradation of energy in these devices as a result of irreversibilities. But first we need to define an ideal process that will serve as a model for the actual processes.

Although some heat transfer between these devices and the surrounding medium is unavoidable, most steady-flow devices are intended to operate under adiabatic conditions. Therefore, the model process for these devices should be an adiabatic one. Furthermore, an ideal process should involve no irreversibilities since the effect of irreversibilities is always to downgrade the performance of engineering devices. Thus, the ideal process that can serve as a suitable model for most steady-flow devices is the *isentropic* process (Fig. 7-48).

The more closely the actual process approximates the idealized isentropic process, the better the device will perform. Thus, it would be desirable to have a parameter that expresses quantitatively how efficiently an actual device approximates an idealized one. This parameter is the **isentropic** or **adiabatic efficiency,** which is a measure of the deviation of actual processes from the corresponding idealized ones.

Adiabatic efficiencies are defined differently for different devices since each device is set up to perform different tasks. Below we define the adiabatic efficiencies of turbines, compressors, and nozzles by comparing the actual performance of these devices to their performance under isentropic conditions for the same inlet state and exit pressure.

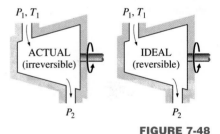

FIGURE 7-48

The isentropic process involves no irreversibilities and serves as the ideal process for adiabatic devices.

Isentropic Efficiency of Turbines

For a turbine under steady operation, the inlet state of the working fluid and the exhaust pressure are fixed. Therefore, the ideal process for an adiabatic turbine is an isentropic process between the inlet state and the exhaust pressure. The desired output of a turbine is the work produced, and the **isentropic efficiency of a turbine** is defined as *the ratio of the actual work output of the turbine to the work output that would be achieved if the process between the inlet state and the exit pressure were isentropic:*

$$\eta_T = \frac{\text{Actual turbine work}}{\text{Isentropic turbine work}} = \frac{w_a}{w_s} \qquad (7\text{-}60)$$

Usually the changes in kinetic and potential energies associated with a fluid stream flowing through a turbine are small relative to the change in enthalpy and can be neglected. Then the work output of an adiabatic turbine simply becomes the change in enthalpy, and the above relation for this case can be expressed as

$$\eta_T \cong \frac{h_1 - h_{2a}}{h_1 - h_{2s}} \qquad (7\text{-}61)$$

where h_{2a} and h_{2s} are the enthalpy values at the exit state for actual and isentropic processes, respectively. The actual and isentropic processes in a turbine are illustrated in Fig. 7-49.

The value of η_T greatly depends on the design of the individual components that make up the turbine. Well-designed, large turbines have isentropic efficiencies above 90 percent. For small turbines, however, it may drop even below 70 percent. The value of the isentropic efficiency of a turbine is determined by measuring the actual work output of the turbine and by calculating the isentropic work output for the measured inlet conditions and the exit pressure. This value may then be used conveniently in the design of power plants.

EXAMPLE 7-14   Isentropic Efficiency of a Steam Turbine

Steam enters an adiabatic turbine steadily at 3 MPa and 400°C and leaves at 50 kPa and 100°C. If the power output of the turbine is 2 MW, determine (a) the isentropic efficiency of the turbine and (b) the mass flow rate of the steam flowing through the turbine.

Solution   A sketch of the system and the *T-s* diagram of the process are given in Fig. 7-50.

Assumptions   **1** Steady operating conditions exist. **2** The changes in kinetic and potential energies are negligible. **3** The turbine is adiabatic.

Analysis   (a) The enthalpies at various states are

State 1: $\begin{array}{l} P_1 = 3 \text{ MPa} \\ T_1 = 400°C \end{array} \Big\}$ $\begin{array}{l} h_1 = 3230.9 \text{ kJ/kg} \\ s_1 = 6.9212 \text{ kJ/kg} \cdot \text{K} \end{array}$ (Table A-6)

State 2a: $\begin{array}{l} P_{2a} = 50 \text{ kPa} \\ T_{2a} = 100°C \end{array} \Big\}$ $h_{2a} = 2682.5 \text{ kJ/kg}$ (Table A-6)

The exit enthalpy of the steam for the isentropic process h_{2s} is determined from the requirement that the entropy of the steam remain constant ($s_{2s} = s_1$):

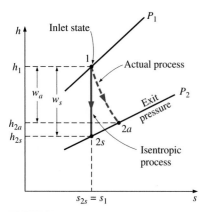

FIGURE 7-49

The *h-s* diagram for the actual and isentropic processes of an adiabatic turbine.

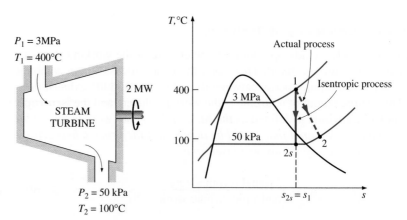

FIGURE 7-50

Schematic and *T-s* diagram for Example 7-14.

State 2s: $P_{2s} = 50$ kPa $s_f = 1.0910$ kJ/kg · K

$(s_{2s} = s_1)$ $\longrightarrow$ $s_g = 7.5939$ kJ/kg · K   (Table A-5)

Obviously, at the end of the isentropic process steam will exist as a saturated liquid–vapor mixture since $s_f < s_{2s} < s_g$. Thus we need to find the quality at state 2s first:

$$x_{2s} = \frac{s_{2s} - s_f}{s_{fg}} = \frac{6.9212 - 1.0910}{6.5029} = 0.897$$

and $h_{2s} = h_f + x_{2s}h_{fg} = 340.49 + 0.897 \, (2305.4) = 2407.4$ kJ/kg

By substituting these enthalpy values into Eq. 7-61, the isentropic efficiency of this turbine is determined to be

$$\eta_T \cong \frac{h_1 - h_{2a}}{h_1 - h_{2s}} = \frac{3230.9 - 2682.5}{3230.9 - 2407.4} = \textbf{0.666, or 66.6\%}$$

(*b*) The mass flow rate of steam through this turbine is determined from the energy balance for steady-flow systems:

$$\dot{E}_{in} = \dot{E}_{out}$$
$$\dot{m}h_1 = \dot{W}_{a,\,out} + \dot{m}h_{2a}$$
$$\dot{W}_{a,\,out} = \dot{m}(h_1 - h_{2a})$$
$$2 \text{ MW} \left(\frac{1000 \text{ KJ/s}}{1 \text{ MW}}\right) = \dot{m}(3230.9 - 2682.5) \text{ kJ/kg}$$

$$\dot{m} = \textbf{3.65 kg/s}$$

Isentropic Efficiencies of Compressors and Pumps

The **isentropic efficiency of a compressor** is defined as *the ratio of the work input required to raise the pressure of a gas to a specified value in an isentropic manner to the actual work input:*

$$\eta_C = \frac{\text{Isentropic compressor work}}{\text{Actual compressor work}} = \frac{w_s}{w_a} \qquad (7\text{-}62)$$

Notice that the isentropic compressor efficiency is defined with the *isentropic work input in the numerator* instead of in the denominator. This is because w_s is a smaller quantity than w_a, and this definition prevents η_C from becoming greater than 100 percent, which would falsely imply that the actual compressors performed better than the isentropic ones. Also notice that the inlet conditions and the exit pressure of the gas are the same for both the actual and the isentropic compressor.

When the changes in kinetic and potential energies of the gas being compressed are negligible, the work input to an adiabatic compressor becomes equal to the change in enthalpy, and Eq. 7-62 for this case becomes

$$\eta_C \cong \frac{h_{2s} - h_1}{h_{2a} - h_1} \qquad (7\text{-}63)$$

where h_{2a} and h_{2s} are the enthalpy values at the exit state for actual and isentropic compression processes, respectively, as illustrated in Fig. 7-51. Again, the value of η_C greatly depends on the design of the compressor. Well-designed compressors have isentropic efficiencies that range from 75 to 85 percent.

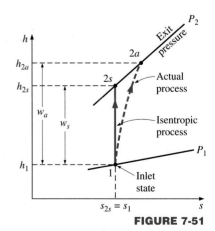

FIGURE 7-51

The *h-s* diagram of the actual and isentropic processes of an adiabatic compressor.

When the changes in potential and kinetic energies of a liquid are negligible, the isentropic efficiency of a pump is defined similarly as

$$\eta_P = \frac{w_s}{w_a} = \frac{v(P_2 - P_1)}{h_{2a} - h_1} \tag{7-64}$$

When no attempt is made to cool the gas as it is compressed, the actual compression process is nearly adiabatic and the reversible adiabatic (i.e., isentropic) process serves well as the ideal process. But sometimes *compressors are cooled intentionally* by utilizing fins or a water jacket placed around the casing to reduce the work input requirements (Fig. 7-52). In this case, the isentropic process is not suitable as the model process since the device is no longer adiabatic and the isentropic compressor efficiency defined above is meaningless. A realistic model process for compressors that are intentionally cooled during the compression process is the *reversible isothermal process*. Then we can conveniently define an **isothermal efficiency** for such cases by comparing the actual process to a reversible isothermal one:

$$\eta_C = \frac{w_t}{w_a} \tag{7-65}$$

where w_t and w_a are the required work inputs to the compressor for the reversible isothermal and actual cases, respectively.

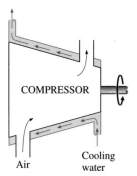

FIGURE 7-52

Compressors are sometimes intentionally cooled to minimize the work input.

EXAMPLE 7-15 Effect of Efficiency on Compressor Power Input

Air is compressed by an adiabatic compressor from 100 kPa and 12°C to a pressure of 800 kPa at a steady rate of 0.2 kg/s. If the isentropic efficiency of the compressor is 80 percent, determine (*a*) the exit temperature of air and (*b*) the required power input to the compressor.

Solution A sketch of the system and the *T-s* diagram of the process are given in Fig. 7-53.

Assumptions **1** Steady operating conditions exist. **2** Air is an ideal gas. **3** The changes in kinetic and potential energies are negligible. **4** The compressor is adiabatic.

Analysis (*a*) We know only one property (pressure) at the exit state, and we need to know one more to fix the state and thus determine the exit temperature.

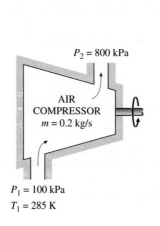

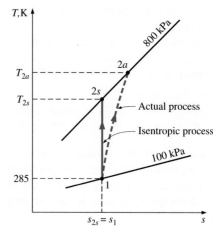

FIGURE 7-53

Schematic and *T-s* diagram for Example 7-15.

$P_1 = 100$ kPa
$T_1 = 285$ K

The property that can be determined with minimal effort in this case is h_{2a} since the adiabatic efficiency of the compressor is given.

The enthalpy of an ideal gas is a function of temperature only, and h_1 is easily determined from the air table at the inlet temperature:

$$T_1 = 285 \text{ K} \longrightarrow h_1 = 285.14 \text{ kJ/kg} \qquad \text{(Table A-17)}$$
$$(P_{r1} = 1.1584)$$

Now we need to determine h_{2s}, the enthalpy of the air at the end of the isentropic compression process. This is done by using one of the isentropic relations of ideal gases, such as Eq. 7-49:

$$P_{r2} = P_{r1}\left(\frac{P_2}{P_1}\right) = 1.1584 \left(\frac{800 \text{ kPa}}{100 \text{ kPa}}\right) = 9.2672$$

and $\qquad P_{r2} = 9.2672 \longrightarrow h_{2s} = 517.05 \text{ kJ/kg} \qquad \text{(Table A-17)}$

Substituting the known quantities into Eq. 7-63, we have

$$\eta_C \cong \frac{h_{2s} - h_1}{h_{2a} - h_1} \longrightarrow 0.80 = \frac{(517.05 - 285.14) \text{ kJ/kg}}{(h_{2a} - 285.14) \text{ kJ/kg}}$$

Thus, $\qquad h_{2a} = 575.03 \text{ kJ/kg} \longrightarrow T_{2a} = \textbf{569.5 K} \qquad \text{(Table A-17)}$

(*b*) The required power input to the compressor is determined from the energy balance for steady-flow devices,

$$\dot{E}_{in} = \dot{E}_{out}$$
$$\dot{m}h_1 + \dot{W}_{a,\,in} = \dot{m}h_{2a}$$
$$\dot{W}_{a,\,in} = \dot{m}(h_{2a} - h_1)$$
$$= (0.2 \text{ kg/s})[(575.03 - 285.14) \text{ kJ/kg}]$$
$$= \textbf{58.0 kW}$$

Discussion Notice that in determining the power input to the compressor, we used h_{2a} instead of h_{2s} since h_{2a} is the actual enthalpy of the air as it exits the compressor. The quantity h_{2s} is a hypothetical enthalpy value that the air would have if the process were isentropic.

Isentropic Efficiency of Nozzles

Nozzles are essentially adiabatic devices and are used to accelerate a fluid. Therefore, the isentropic process serves as a suitable model for nozzles. The **isentropic efficiency of a nozzle** is defined as *the ratio of the actual kinetic energy of the fluid at the nozzle exit to the kinetic energy value at the exit of an isentropic nozzle for the same inlet state and exit pressure.* That is,

$$\eta_N = \frac{\text{Actual KE at nozzle exit}}{\text{Isentropic KE at nozzle exit}} = \frac{\mathcal{V}_{2a}^2}{\mathcal{V}_{2s}^2} \qquad (7\text{-}66)$$

Note that the exit pressure is the same for both the actual and isentropic processes, but the exit state is different.

Nozzles involve no work interactions, and the fluid experiences little or no change in its potential energy as it flows through the device. If, in addition, the inlet velocity of the fluid is small relative to the exit velocity, the energy balance for this steady-flow device reduces to

$$h_1 = h_{2a} + \frac{\mathcal{V}_{2a}^2}{2}$$

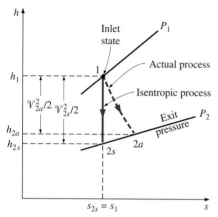

Then the adiabatic efficiency of the nozzle can be expressed in terms of enthalpies as

$$\eta_N \cong \frac{h_1 - h_{2a}}{h_1 - h_{2s}} \qquad (7-67)$$

where h_{2a} and h_{2s} are the enthalpy values at the nozzle exit for the actual and isentropic processes, respectively (Fig. 7-54). Isentropic efficiencies of nozzles are typically above 90 percent, and nozzle efficiencies above 95 percent are not uncommon.

EXAMPLE 7-16   Effect of Efficiency on Nozzle Exit Velocity

Air at 200 kPa and 950 K enters an adiabatic nozzle at low velocity and is discharged at a pressure of 80 kPa. If the isentropic efficiency of the nozzle is 92 percent, determine (*a*) the maximum possible exit velocity, (*b*) the exit temperature, and (*c*) the actual velocity of the air. Assume constant specific heats for air.

Solution   A sketch of the system and the *T-s* diagram of the process are given in Fig. 7-55.

Assumptions   **1** Steady operating conditions exist. **2** Air is an ideal gas. **3** The inlet kinetic energy is negligible. **4** The nozzle is adiabatic.

Analysis   The temperature of air will drop during this acceleration process because some of its internal energy is converted to kinetic energy. This problem can be solved accurately by using property data from the air table. But we will assume constant specific heats (thus sacrifice some accuracy) to demonstrate their use. Let us guess that the average temperature of the air will be about 800 K. Then the average values of C_p and k at this anticipated average temperature are determined from Table A-2b to be $C_p = 1.099$ kJ/kg · K and $k = 1.354$.

(*a*) The exit velocity of the air will be a maximum when the process in the nozzle involves no irreversibilities. The exit velocity in this case is determined from the steady-flow energy equation. But first we need to determine the exit temperature. For the isentropic process of an ideal gas with constant specific heats, the temperatures and pressures are related by Eq. 7-43:

$$\frac{T_{2s}}{T_1} = \left(\frac{P_{2s}}{P_1}\right)^{(k-1)/k}$$

or $T_{2s} = T_1\left(\frac{P_{2s}}{P_1}\right)^{(k-1)/k} = (950 \text{ K})\left(\frac{80 \text{ kPa}}{200 \text{ kPa}}\right)^{0.354/1.354} = 748 \text{ K}$

This will give an average temperature of 849 K, which is somewhat higher than the assumed average temperature (800 K). This result could be refined by reevaluating the k value at 749 K and repeating the calculations, but it is not warranted

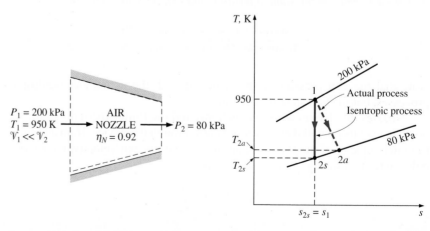

since the two average temperatures are sufficiently close (doing so would change the temperature by only 1.5 K, which is not significant).

Now we can determine the isentropic exit velocity of the air from the energy balance for this isentropic steady-flow process:

$$e_{in} = e_{out}$$

$$h_1 + \frac{\mathcal{V}_1^2}{2}^{\nearrow 0} = h_{2s} + \frac{\mathcal{V}_{2s}^2}{2}$$

or

$$\mathcal{V}_{2s} = \sqrt{2(h_{2s} - h_1)} = \sqrt{2C_{p,av}(T_1 - T_{2s})}$$

$$= \sqrt{2(1.099 \text{ kJ/kg} \cdot \text{K})[(950 - 748) \text{ K}]\left(\frac{1000 \text{ m}^2/\text{s}^2}{1 \text{ kJ/kg}}\right)}$$

$$= \textbf{666 m/s}$$

(*b*) The actual exit temperature of the air will be higher than the isentropic exit temperature evaluated above, and it is determined from Eq. 7-77. For constant specific heats,

$$\eta_N \cong \frac{h_1 - h_{2a}}{h_1 - h_{2s}} = \frac{C_{p,av}(T_1 - T_{2a})}{C_{p,av}(T_1 - T_{2s})}$$

or

$$0.92 = \frac{950 - T_{2a}}{950 - 748} \longrightarrow T_{2a} = \textbf{764 K}$$

That is, the temperature will be 16 K higher at the exit of the actual nozzle as a result of irreversibilities such as friction. It represents a loss since this rise in temperature comes at the expense of kinetic energy (Fig. 7-56).

(*c*) The actual exit velocity of air can be determined from the definition of isentropic efficiency (Eq. 7-66):

$$\eta_N = \frac{\mathcal{V}_{2a}^2}{\mathcal{V}_{2s}^2} \longrightarrow \mathcal{V}_{2a} = \sqrt{\eta_N \mathcal{V}_{2s}^2} = \textbf{639 m/s}$$

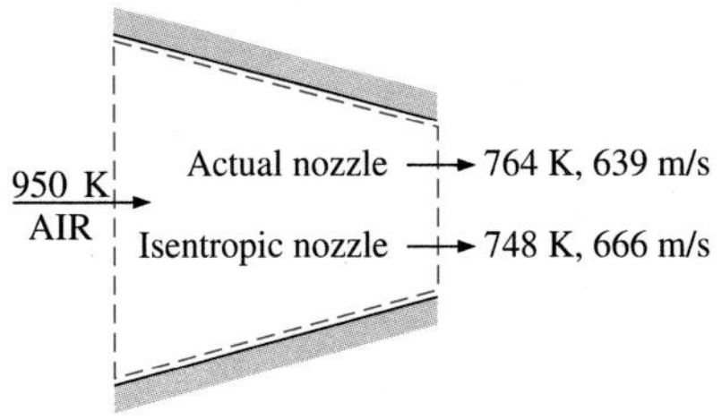

FIGURE 7-56

A substance leaves actual nozzles at a higher temperature (thus a lower velocity) as a result of friction.

7-13 ■ SUMMARY

The second law of thermodynamics leads to the definition of a new property called *entropy,* which is a quantitative measure of microscopic disorder for a system. The definition of entropy is based on the *Clausius inequality,* given by

$$\oint \frac{\delta Q}{T} \leq 0 \qquad \text{(kJ/K)}$$

where the equality holds for internally or totally reversible processes and the inequality for irreversible processes. Any quantity whose cyclic integral is zero is a property, and entropy is defined as

$$dS = \left(\frac{dQ}{T}\right)_{\text{int rev}} \qquad \text{(kJ/K)}$$

For the special case of an internally reversible, isothermal process, it gives

$$\Delta S = \frac{Q}{T_0} \qquad \text{(kJ/K)}$$

The inequality part of the Clausius inequality combined with the definition of entropy yields an inequality known as the *increase of entropy principle,* expressed as

$$S_{\text{gen}} \geq 0 \qquad \text{(kJ/K)}$$

where S_{gen} is the *entropy generated* during the process. Entropy change is caused by heat transfer, mass flow, and irreversibilities. Heat transfer to a system increases the entropy, and heat transfer from a system decreases it. The effect of irreversibilities is always to increase the entropy.

Entropy is a property, and it can be expressed in terms of more familiar properties through the *T ds* relations, expressed as

$$T\,ds = du + P\,dv$$

and

$$T\,ds = dh - v\,dP$$

These two relations have many uses in thermodynamics and serve as the starting point in developing entropy-change relations for processes. The successful use of *T ds* relations depends on the availability of property relations. Such relations do not exist for a general pure substance but are available for incompressible substances (solids, liquids) and ideal gases.

The *entropy-change* and *isentropic relations* for a process can be summarized as follows:

1. *Pure substances:*

Any process: $\qquad\qquad \Delta s = s_2 - s_1 \qquad$ (kJ/kg · K)

Isentropic process: $\qquad s_2 = s_1$

2. *Incompressible substances:*

Any process: $\qquad\qquad s_2 - s_1 = C_{\text{av}} \ln \dfrac{T_2}{T_1} \qquad$ (kJ/kg · K)

Isentropic process: $\qquad T_2 = T_1$

3. *Ideal gases:*

   a. Constant specific heats (approximate treatment):

     Any process:

$$s_2 - s_1 = C_{v,\,\text{av}} \ln \frac{T_2}{T_1} + R \ln \frac{v_2}{v_1} \qquad \text{(kJ/kg · K)}$$

and

$$s_2 - s_1 = C_{p,\,\text{av}} \ln \frac{T_2}{T_1} - R \ln \frac{P_2}{P_1} \qquad \text{(kJ/kg · K)}$$

     Or, on a unit-mole basis,

$$\bar{s}_2 - \bar{s}_1 = \bar{C}_{v,\,\text{av}} \ln \frac{T_2}{T_1} + R_u \ln \frac{v_2}{v_1} \qquad \text{(kJ/kmol · K)}$$

and

$$\bar{s}_2 - \bar{s}_1 = \bar{C}_{p,\,\text{av}} \ln \frac{T_2}{T_1} - R_u \ln \frac{P_2}{P_1} \qquad \text{(kJ/kmol · K)}$$

     Isentropic process:

$$\left(\frac{T_2}{T_1}\right)_{s=\text{const.}} = \left(\frac{v_1}{v_2}\right)^{k-1}$$

$$\left(\frac{T_2}{T_1}\right)_{s=\text{const.}} = \left(\frac{P_2}{P_1}\right)^{(k-1)/k}$$

$$\left(\frac{P_2}{P_1}\right)_{s=\text{const.}} = \left(\frac{v_1}{v_2}\right)^{k}$$

b. Variable specific heats (exact treatment):

Any process:

$$s_2 - s_1 = s_2^\circ - s_1^\circ - R \ln \frac{P_2}{P_1} \qquad \text{(kJ/kg · K)}$$

or

$$\bar{s}_2 - \bar{s}_1 = \bar{s}_2^\circ - \bar{s}_1^\circ - R_u \ln \frac{P_2}{P_1} \qquad \text{(kJ/kmol · K)}$$

Isentropic process:

$$s_2^\circ = s_1^\circ + R \ln \frac{P_2}{P_1} \qquad \text{(kJ/kg · K)}$$

$$\left(\frac{P_2}{P_1}\right)_{s=\text{const.}} = \frac{P_{r2}}{P_{r1}}$$

$$\left(\frac{v_2}{v_1}\right)_{s=\text{const.}} = \frac{v_{r2}}{v_{r1}}$$

where P_r is the *relative pressure* and v_r is the *relative specific volume*. The function s° depends on temperature only.

The *steady-flow work* for a reversible process can be expressed in terms of the fluid properties as

$$w_{\text{rev}} = -\int_1^2 v \, dP - \Delta\text{ke} - \Delta\text{pe} \qquad \text{(kJ/kg)}$$

For incompressible substances (v = constant) it simplifies to

$$w_{\text{rev}} = -v(P_2 - P_1) - \Delta\text{ke} - \Delta\text{pe} \qquad \text{(kJ/kg)}$$

The work done during a steady-flow process is proportional to the specific volume. Therefore, v should be kept as small as possible during a compression process to minimize the work input and as large as possible during an expansion process to maximize the work output.

The reversible work inputs to a compressor compressing an ideal gas from T_1, P_1 to P_2 in an isentropic (Pv^k = constant), polytropic (Pv^n = constant), or isothermal (Pv = constant) manner, are determined by integration for each case with the following results:

Isentropic:

$$w_{\text{comp, in}} = \frac{kR(T_2 - T_1)}{k - 1}$$

$$= \frac{kRT_1}{k - 1}\left[\left(\frac{P_2}{P_1}\right)^{(k-1)/k} - 1\right] \qquad \text{(kJ/kg)}$$

Polytropic:

$$w_{\text{comp, in}} = \frac{nR(T_2 - T_1)}{n - 1}$$

$$= \frac{nRT_1}{n - 1}\left[\left(\frac{P_2}{P_1}\right)^{(n-1)/n} - 1\right] \qquad \text{(kJ/kg)}$$

Isothermal:

$$w_{\text{comp, in}} = RT \ln \frac{P_2}{P_1} \qquad \text{(kJ/kg)}$$

The work input to a compressor can be reduced by using multistage compression with intercooling. For maximum savings from the work input, the pressure ratio across each stage of the compressor must be the same.

Most steady-flow devices operate under adiabatic conditions, and the ideal process for these devices is the isentropic process. The parameter that describes how efficiently a device approximates a corresponding isentropic device is called *isentropic* or *adiabatic efficiency*. It is expressed for turbines, compressors, and nozzles as follows:

$$\eta_T = \frac{\text{Actual turbine work}}{\text{Isentropic turbine work}} = \frac{w_a}{w_s} \cong \frac{h_1 - h_{2a}}{h_1 - h_{2s}}$$

$$\eta_C = \frac{\text{Isentropic compressor work}}{\text{Actual compressor work}} = \frac{w_s}{w_a} \cong \frac{h_{2s} - h_1}{h_{2a} - h_1}$$

$$\eta_N = \frac{\text{Actual KE at nozzle exit}}{\text{Isentropic KE at nozzle exit}} = \frac{V_{2a}^2}{V_{2s}^2} \cong \frac{h_1 - h_{2a}}{h_1 - h_{2s}}$$

In the relations above, h_{2a} and h_{2s} are the enthalpy values at the exit state for actual and isentropic *processes*, respectively.

REFERENCES AND SUGGESTED READING

1. A. Bejan. *Advanced Engineering Thermodynamics.* New York: John Wiley & Sons, 1988.

2. A. Bejan. *Entropy Generation through Heat and Fluid Flow.* New York: John Wiley & Sons–Interscience, 1982.

3. Y. A. Çengel and M. A. Boles. *Thermodynamics: An Engineering Approach.* 3rd ed. New York: McGraw-Hill, 1998.

4. Y. A. Çengel and H. Kimmel, "Optimization of Thermodynamic Expansion in Natural Gas Liquefaction Processes." *LNG Journal, U.K.,* May–June, 1998.

5. M. S. Moran and H. N. Shapiro. *Fundamentals of Engineering Thermodynamics.* New York: John Wiley & Sons, 1988.

6. J. Rifkin. *Entropy.* New York: The Viking Press, 1980.

PROBLEMS*

Entropy and the Increase of Entropy Principle

7-1C Does the temperature in the Clausius inequality relation have to be absolute temperature? Why?

7-2C Does a cycle for which $\oint \delta Q > 0$ violate the Clausius inequality? Why?

7-3C Is a quantity whose cyclic integral is zero necessarily a property?

7-4C Does the cyclic integral of heat have to be zero (i.e., does a system have to reject as much heat as it receives to complete a cycle)? Explain.

7-5C Does the cyclic integral of work have to be zero (i.e., does a system have to produce as much work as it consumes to complete a cycle)? Explain.

---

*Students are encouraged to answer all the concept "C" questions.

7-6C A system undergoes a process between two fixed states first in a reversible manner and then in an irreversible manner. For which case is the entropy change greater? Why?

7-7C Is the value of the integral $\int_1^2 \delta Q/T$ the same for all processes between states 1 and 2? Explain.

7-8C Is the value of the integral $\int_1^2 \delta Q/T$ the same for all reversible processes between states 1 and 2? Why?

7-9C To determine the entropy change for an irreversible process between states 1 and 2, should the integral $\int_1^2 \delta Q/T$ be performed along the actual process path or an imaginary reversible path? Explain.

7-10C Is an isothermal process necessarily internally reversible? Explain your answer with an example.

7-11C How do the values of the integral $\int_1^2 \delta Q/T$ compare for a reversible and irreversible process between the same end states?

7-12C The entropy of a hot baked potato decreases as it cools. Is this a violation of the increase of entropy principle? Explain.

7-13C Is it possible to create entropy? Is it possible to destroy it?

7-14C A piston-cylinder device contains helium gas. During a reversible, isothermal process, the entropy of the helium will (*never, sometimes, always*) increase.

7-15C A piston-cylinder device contains nitrogen gas. During a reversible, adiabatic process, the entropy of the nitrogen will (*never, sometimes, always*) increase.

7-16C A piston-cylinder device contains superheated steam. During an actual adiabatic process, the entropy of the steam will (*never, sometimes, always*) increase.

7-17C The entropy of steam will (*increase, decrease, remain the same*) as it flows through an actual adiabatic turbine.

7-18C The entropy of the working fluid of the ideal Carnot cycle (*increases, decreases, remains the same*) during the isothermal heat addition process.

7-19C The entropy of the working fluid of the ideal Carnot cycle (*increases, decreases, remains the same*) during the isothermal heat rejection process.

7-20C During a heat transfer process, the entropy of a system (*always, sometimes, never*) increases.

7-21C Is it possible for the entropy change of a closed system to be zero during an irreversible process? Explain.

7-22C What three different mechanisms can cause the entropy of a control volume to change?

7-23C Steam is accelerated as it flows through an actual adiabatic nozzle. The entropy of the steam at the nozzle exit will be (*greater than, equal to, less than*) the entropy at the nozzle inlet.

7-24C Consider a person who organizes his room, and thus decreases the entropy of the room. Does this process violate the second law of thermodynamics?

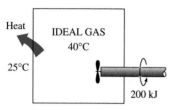

FIGURE P7-27

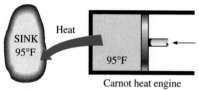

Carnot heat engine

FIGURE P7-30E

7-25C Consider a fruit tree that makes highly organized fruits out of the water and highly disorganized soil, and thus decreases the entropy of its locality. Does this process violate the second law of thermodynamics?

7-26C Consider an army unit whose soldiers are walking around at random in a field. Suddenly an order is issued and the soldiers align in a highly organized manner, decreasing the entropy. Does this process violate the second law of thermodynamics? Explain.

7-27 A rigid tank contains an ideal gas at 40°C that is being stirred by a paddle wheel. The paddle wheel does 200 kJ of work on the ideal gas. It is observed that the temperature of the ideal gas remains constant during this process as a result of heat transfer between the system and the surroundings at 25°C. Determine the entropy change of the ideal gas.

7-28 Air is compressed by a 8-kW compressor from P_1 to P_2. The air temperature is maintained constant at 25°C during this process as a result of heat transfer to the surrounding medium at 10°C. Determine the rate of entropy change of the air. State the assumptions made in solving this problem.
    Answer: -0.0268 kW/K

7-29 During the isothermal heat addition process of a Carnot cycle, 900 kJ of heat is added to the working fluid from a source at 400°C. Determine (*a*) the entropy change of the working fluid, (*b*) the entropy change of the source, and (*c*) the total entropy generation for the process.

7-30E During the isothermal heat rejection process of a Carnot cycle, the working fluid experiences an entropy change of -0.7 Btu/R. If the temperature of the energy sink is 95°F, determine (*a*) the amount of heat transfer, (*b*) the entropy change of the sink, and (*c*) the total entropy change for this process.   *Answers:* (*a*) 388.5 Btu, (*b*) 0.7 Btu/R, (*c*) 0

7-31 Refrigerant-134a enters the coils of the evaporator of a refrigeration system as a saturated liquid–vapor mixture at a pressure of 200 kPa. The refrigerant absorbs 120 kJ of heat from the cooled space, which is maintained at -5°C, and leaves as saturated vapor at the same pressure. Determine (*a*) the entropy change of the refrigerant, (*b*) the entropy change of the cooled space, and (*c*) the total entropy change for this process.

Entropy Changes of Pure Substances

7-32C Is a process that is internally reversible and adiabatic necessarily isentropic? Explain.

7-33 The radiator of a steam heating system has a volume of 20 L and is filled with superheated water vapor at 200 kPa and 200°C. At this moment both the inlet and the exit valves to the radiator are closed. After a while the temperature of the steam drops to 80°C as a result of heat transfer to the room air. Determine the entropy change of the steam during this process, in kJ/K.
    Answer: -0.0806 kJ/K

7-34 A 0.5-m³ rigid tank contains refrigerant-134a initially at 200 kPa and 40 percent quality. Heat is transferred now to the refrigerant from a source at 35°C until the pressure rises to 400 kPa. Determine (*a*) the entropy change of the refrigerant, (*b*) the entropy change of the heat source, and (*c*) the total entropy change for this process.
    Answers: (*a*) 3.883 kJ/K, (*b*) -3.441 kJ/K, (*c*) 0.422 kJ/K

7-35 A well-insulated rigid tank contains 2 kg of a saturated liquid–vapor mixture of water at 100 kPa. Initially, three-quarters of the mass is in the liquid phase. An electric resistance heater placed in the tank is now turned on and kept on until all the liquid in the tank is vaporized. Determine the entropy change of the steam during this process.   *Answer:* 8.0962 kJ/K

7-36 A rigid tank is divided into two equal parts by a partition. One part of the tank contains 1.5 kg of compressed liquid water at 300 kPa and 60°C while the other part is evacuated. The partition is now removed, and the water expands to fill the entire tank. Determine the entropy change of water during this process, if the final pressure in the tank is 15 kPa.
    Answer: −0.1134 kJ/K

7-37E A piston-cylinder device contains 3 lbm of refrigerant-134a at 120 psia and 120°F. The refrigerant is now cooled at constant pressure until it exists as a liquid at 90°F. Determine the entropy change of the refrigerant during this process.

7-38 An insulated piston-cylinder device contains 5 L of saturated liquid water at a constant pressure of l50 kPa. An electric resistance heater inside the cylinder is now turned on, and electrical work is done on the steam in the amount of 2200 kJ. Determine the entropy change of the water during this process, in kJ/K.   *Answer:* 5.72 kJ/K

7-39 An insulated piston-cylinder device contains 0.01 m³ of saturated refrigerant-134a vapor at 0.8-MPa pressure. The refrigerant is now allowed to expand in a reversible manner until the pressure drops to 0.4 MPa. Determine (*a*) the final temperature in the cylinder and (*b*) the work done by the refrigerant.

7-40 Refrigerant-134a enters an adiabatic compressor as saturated vapor at 140 kPa at a rate of 2 m³/min and is compressed to a pressure of 700 kPa. Determine the minimum power that must be supplied to the compressor.

7-41E Steam enters an adiabatic turbine at 800 psia and 900°F and leaves at a pressure of 40 psia. Determine the maximum amount of work that can be delivered by this turbine.

7-42 A heavily insulated piston-cylinder device contains 0.05 m³ of steam at 300 kPa and 150°C. Steam is now compressed in a reversible manner to a pressure of 1 MPa. Determine the work done on the steam during this process.

7-43 A piston-cylinder device contains 0.5 kg of saturated water vapor at 200°C. Heat is now transferred to steam, and steam expands reversibly and isothermally to a final pressure of 800 kPa. Determine the heat transferred and the work done during this process.

Entropy Change of Incompressible Substances

7-44C Consider two solid blocks, one hot and the other cold, brought into contact in an adiabatic container. After a while, thermal equilibrium is established in the container as a result of heat transfer. The first law requires that the amount of energy lost by the hot solid be equal to the energy gained by the cold one. Does the second law require that the decrease in entropy of the hot solid be equal to the increase in entropy of the cold one?

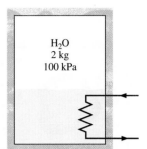

FIGURE P7-35

FIGURE P7-36

FIGURE P7-39

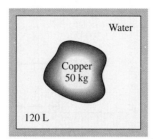

FIGURE P7-45

FIGURE P7-48

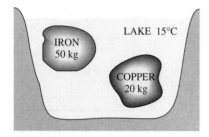

FIGURE P7-58

7-45 A 50-kg copper block initially at 80°C is dropped into an insulated tank that contains 120 L of water at 25°C. Determine the final equilibrium temperature and the total entropy change for this process.

7-46 A 5-kg iron block initially at 350°C is quenched in an insulated tank that contains 100 kg of water at 22°C. Assuming the water that vaporizes during the process condenses back in the tank, determine the amount of entropy generated during this process.

7-47 A 20-kg aluminum block initially at 200°C is brought into contact with a 20-kg block of iron at 100°C in an insulated enclosure. Determine the final equilibrium temperature and the total entropy change for this process.
Answers: 168.4°C, 0.169 kJ/K

7-48 A 50-kg iron block and a 20-kg copper block, both initially at 80°C, are dropped into a large lake at 15°C. Thermal equilibrium is established after a while as a result of heat transfer between the blocks and the lake water. Determine the total entropy generation for this process.

Entropy Change of Ideal Gases

7-49C Prove that the two relations for entropy change of ideal gases under the constant-specific-heat assumption (Eqs. 7-33 and 7-34) are equivalent.

7-50C Starting with the second $T\,ds$ relation (Eq. 7-26), obtain Eq. 7-34 for the entropy change of ideal gases under the constant-specific-heat assumption.

7-51C What does the function $s°$ in the ideal-gas tables represent?

7-52C Some properties of ideal gases such as internal energy and enthalpy vary with temperature only [that is, $u = u(T)$ and $h = h(T)$]. Is this also the case for entropy?

7-53C Starting with Eq. 7-34, obtain Eq. 7-43.

7-54C What are P_r and v_r called? Is their use limited to isentropic processes? Explain.

7-55C Can the entropy of an ideal gas change during an isothermal process?

7-56C An ideal gas undergoes a process between two specified temperatures, first at constant pressure and then at constant volume. For which case will the ideal gas experience a larger entropy change? Explain.

7-57 Oxygen gas is compressed in a piston-cylinder device from an initial state of 0.8 m³/kg and 25°C to a final state of 0.1 m³/kg and 287°C. Determine the entropy change of the oxygen during this process. Assume constant specific heats.

7-58 A 0.5-m³ insulated rigid tank contains 0.9 kg of carbon dioxide at 100 kPa. Now paddle-wheel work is done on the system until the pressure in the tank rises to 120 kPa. Determine the entropy change of carbon dioxide during this process in kJ/K. Assume constant specific heats.
Answer: 0.108 kJ/K

7-59 An insulated piston-cylinder device initially contains 300 L of air at 120 kPa and 17°C. Air is now heated for 15 min by a 200-W resistance heater placed inside the cylinder. The pressure of air is maintained constant during this process. Determine the entropy change of air, assuming (*a*) constant specific heats and (*b*) variable specific heats.

7-60 A piston-cylinder device contains 1.2 kg of nitrogen gas at 120 kPa and 27°C. The gas is now compressed slowly in a polytropic process during which $PV^{1.3}$ = constant. The process ends when the volume is reduced by one-half. Determine the entropy change of nitrogen during this process.

Answer: −0.0617 kJ/K

7-61E A mass of 8 lbm of helium undergoes a process from an initial state of 50 ft³/lbm and 80°F to a final state of 10 ft³/lbm and 200°F. Determine the entropy change of helium during this process, assuming (*a*) the process is reversible and (*b*) the process is irreversible.

7-62 Air is compressed in a piston-cylinder device from 90 kPa and 20°C to 400 kPa in a reversible isothermal process. Determine (*a*) the entropy change of air and (*b*) the work done.

7-63 Air is compressed steadily by a 5-kW compressor from 100 kPa and 17°C to 600 kPa and 167°C at a rate of 1.6 kg/min. During this process, some heat transfer takes place between the compressor and the surrounding medium at 17°C. Determine the rate of entropy change of air during this process.

Answer: −0.0025 kW/K

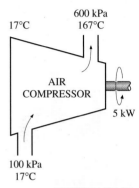

FIGURE P7-63

7-64 An insulated rigid tank is divided into two equal parts by a partition. Initially, one part contains 5 kmol of an ideal gas at 400 kPa and 50°C, and the other side is evacuated. The partition is now removed, and the gas fills the entire tank. Determine the total entropy change during this process.

Answer: 28.81 kJ/K

7-65 Air is compressed in a piston-cylinder device from 100 kPa and 17°C to 800 kPa in a reversible, adiabatic process. Determine the final temperature and the work done during this process, assuming (*a*) constant specific heats and (*b*) variable specific heats for air.

Answers: (*a*) 525.3 K, 171.1 kJ/kg; (*b*) 522.4 K, 169.3 kJ/kg

7-66 Helium gas is compressed from 100 kPa and 30°C to 500 kPa in a reversible, adiabatic process. Determine the final temperature and the work done, assuming the process takes place (*a*) in a piston-cylinder device and (*b*) in a steady-flow compressor.

7-67 An insulated, rigid tank contains 4 kg of argon gas at 450 kPa and 30°C. A valve is now opened, and argon is allowed to escape until the pressure inside drops to 150 kPa. Assuming the argon remaining inside the tank has undergone a reversible, adiabatic process, determine the final mass in the tank.

Answer: 2.07 kg

FIGURE P7-67

7-68E Air enters an adiabatic nozzle at 60 psia, 540°F, and 200 ft/s and exits at 12 psia. Assuming air to be an ideal gas with variable specific heats and disregarding any irreversibilities, determine the exit velocity of the air.

7-69 Air enters a nozzle steadily at 280 kPa and 77°C with a velocity of 50 m/s and exits at 85 kPa and 320 m/s. The heat losses from the nozzle to the surrounding medium at 20°C are estimated to be 3.2 kJ/kg. Determine (*a*) the exit temperature and (*b*) the total entropy change for this process.

Reversible Steady-Flow Work

7-70C In large compressors, the gas is frequently cooled while being compressed to reduce the power consumed by the compressor. Explain how cooling the gas during a compression process reduces the power consumption.

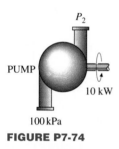

P_2

PUMP

10 kW

100 kPa

FIGURE P7-74

7-71C The turbines in steam power plants operate essentially under adiabatic conditions. A plant engineer suggests to end this practice. She proposes to run cooling water through the outer surface of the casing to cool the steam as it flows through the turbine. This way, she reasons, the entropy of the steam will decrease, the performance of the turbine will improve, and as a result the work output of the turbine will increase. How would you evaluate this proposal?

7-72C It is well known that the power consumed by a compressor can be reduced by cooling the gas during compression. Inspired by this, somebody proposes to cool the liquid as it flows through a pump, in order to reduce the power consumption of the pump. Would you support this proposal? Explain.

7-73 Water enters the pump of a steam power plant as saturated liquid at 20 kPa at a rate of 20 kg/s and exits at 6 MPa. Neglecting the changes in kinetic and potential energies and assuming the process to be reversible, determine the power input to the pump.

7-74 Liquid water enters a 10-kW pump at 100-kPa pressure at a rate of 5 kg/s. Determine the highest pressure the liquid water can have at the exit of the pump. Neglect the kinetic and potential energy changes of water, and take the specific volume of water to be 0.001 m³/kg.   *Answer:* 2100 kPa

7-75E Saturated refrigerant-134a vapor at 20 psia is compressed reversibly in an adiabatic compressor to 120 psia. Determine the work input to the compressor. What would your answer be if the refrigerant were first condensed at constant pressure before it was compressed?

7-76 Consider a steam power plant that operates between the pressure limits of 10 MPa and 20 kPa. Steam enters the pump as saturated liquid and leaves the turbine as saturated vapor. Determine the ratio of the work delivered by the turbine to the work consumed by the pump. Assume the entire cycle to be reversible and the heat losses from the pump and the turbine to be negligible.

7-77 Liquid water at 120 kPa enters a 15-kW pump where its pressure is raised to 3 MPa. If the elevation difference between the exit and the inlet levels is 10 m, determine the highest mass flow rate of liquid water this pump can handle. Neglect the kinetic energy change of water, and take the specific volume of water to be 0.001 m³/kg.

7-78E Helium gas is compressed from 14 psia and 70°F to 120 psia at a rate of 5 ft³/s. Determine the power input to the compressor, assuming the compression process to be (*a*) isentropic, (*b*) polytropic with *n* = 1.2, (*c*) isothermal, and (*d*) ideal two-stage polytropic with *n* = 1.2.

7-79 Nitrogen gas is compressed from 80 kPa and 27°C to 480 kPa by a 10-kW compressor. Determine the mass flow rate of nitrogen through the compressor, assuming the compression process to be (*a*) isentropic, (*b*) polytropic with *n* = 1.3, (*c*) isothermal, and (*d*) ideal two-stage polytropic with *n* = 1.3.
    Answers: (*a*) 0.048 kg/s, (*b*) 0.05 kg/s, (*c*) 0.063 kg/s, (*d*) 0.058 kg/s

7-80 The compression stages in the axial compressor of the industrial gas turbine are close coupled, making intercooling very impractical. To cool the air in such compressors and to reduce the compression power, it is proposed to spray water mist with drop size on the order of 5 microns into the air stream as it is compressed and to cool the air continuously as the water evaporates.

Although the collision of water droplets with turbine blades is a concern, experience with steam turbines indicates that they can cope with water droplet concentrations of up to 14 percent. Assuming air is compressed isentropically at a rate of 2 kg/s from 300 K and 100 kPa to 1200 kPa and the water is injected at a temperature of 20°C at a rate of 0.2 kg/s, determine the reduction in the exit temperature of the compressed air and the compressor power saved. Assume the water vaporizes completely before leaving the compressor, and assume an average mass flow rate of 2.1 kg/s throughout the compressor.

7-81 Reconsider Prob. 7-80. The water-injected compressor is used in a gas turbine power plant. It is claimed that the power output of a gas turbine will increase because of the increase in the mass flow rate of the gas (air + water vapor) through the turbine. Do you agree?

Isentropic Efficiencies of Steady-Flow Devices

7-82C Describe the ideal process for an (*a*) adiabatic turbine, (*b*) adiabatic compressor, and (*c*) adiabatic nozzle, and define the isentropic efficiency for each device.

7-83C Is the isentropic process a suitable model for compressors that are cooled intentionally? Explain.

7-84C On a *T-s* diagram, does the actual exit state (state 2) of an adiabatic turbine have to be on the right-hand side of the isentropic exit state (state 2*s*)? Why?

7-85 Steam enters an adiabatic turbine at 8 MPa and 500°C with a mass flow rate of 3 kg/s and leaves at 30 kPa. The isentropic efficiency of the turbine is 0.90. Neglecting the kinetic energy change of the steam, determine (*a*) the temperature at the turbine exit and (*b*) the power output of the turbine.
    Answers: (*a*) 69.1°C, (*b*) 3052 kW

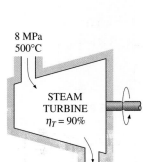

8 MPa
500°C

STEAM
TURBINE
$\eta_T = 90\%$

30 kPa

FIGURE P7-85

7-86 Steam enters an adiabatic turbine at 6 MPa, 600°C, and 80 m/s and leaves at 50 kPa, 100°C, and 140 m/s. If the power output of the turbine is 5 MW, determine (*a*) the mass flow rate of the steam flowing through the turbine and (*b*) the isentropic efficiency of the turbine.
    Answers: (*a*) 5.16 kg/s, (*b*) 83.7 percent

7-87 Argon gas enters an adiabatic turbine at 800°C and 1.5 MPa at a rate of 80 kg/min and exhausts at 200 kPa. If the power output of the turbine is 370 kW, determine the isentropic efficiency of the turbine.

7-88E Combustion gases enter an adiabatic gas turbine at 1540°F and 120 psia and leave at 60 psia with a low velocity. Treating the combustion gases as air and assuming an isentropic efficiency of 86 percent, determine the work output of the turbine.   *Answer:* 75.2 Btu/lbm

7-89 Refrigerant-134a enters an adiabatic compressor as saturated vapor at 120 kPa at a rate of 0.3 m³/min and exits at 1-MPa pressure. If the isentropic efficiency of the compressor is 80 percent, determine (*a*) the temperature of the refrigerant at the exit of the compressor and (*b*) the power input, in kW. Also, show the process on a *T-s* diagram with respect to saturation lines.
    Answers: (*a*) 57.7°C, (*b*) 1.70 kW

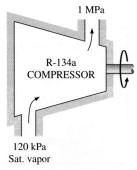

1 MPa

R-134a
COMPRESSOR

120 kPa
Sat. vapor

FIGURE P7-89

7-90 Air enters an adiabatic compressor at 100 kPa and 17°C at a rate of 1.2 m³/s, and it exits at 257°C. The compressor has an isentropic efficiency of 84 percent. Neglecting the changes in kinetic and potential energies, determine (a) the exit pressure of air and (b) the power required to drive the compressor.

7-91 Air is compressed by an adiabatic compressor from 95 kPa and 27°C to 600 kPa and 277°C. Assuming variable specific heats and neglecting the changes in kinetic and potential energies, determine (a) the isentropic efficiency of the compressor and (b) the exit temperature of air if the process were reversible.   *Answers:* (a) 81.9 percent, (b) 505.5 K

7-92E Argon gas enters an adiabatic compressor at 20 psia and 90°F with a velocity of 60 ft/s, and it exits at 200 psia and 240 ft/s. If the isentropic efficiency of the compressor is 80 percent, determine (a) the exit temperature of the argon and (b) the work input to the compressor.

7-93 Carbon dioxide enters an adiabatic compressor at 100 kPa and 300 K at a rate of 0.5 kg/s and exits at 600 kPa and 450 K. Neglecting the kinetic energy changes, determine the isentropic efficiency of the compressor.

7-94E Air enters an adiabatic nozzle at 60 psia and 1020°F with low velocity and exits at 800 ft/s. If the isentropic efficiency of the nozzle is 90 percent, determine the exit temperature and pressure of the air.

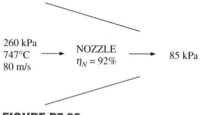

FIGURE P7-95

7-95 Hot combustion gases enter the nozzle of a turbojet engine at 260 kPa, 747°C, and 80 m/s, and they exit at a pressure of 85 kPa. Assuming an isentropic efficiency of 92 percent and treating the combustion gases as air, determine (a) the exit velocity and (b) the exit temperature.
    Answers: (a) 728.2 m/s, (b) 786.3 K

Review Problems

7-96 Show that the difference between the reversible steady-flow work and reversible moving boundary work is equal to the flow energy.

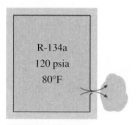

FIGURE P7-97E

7-97E A 0.8-ft³ well-insulated rigid can initially contains refrigerant-134a at 120 psia and 80°F. Now a crack develops in the can, and the refrigerant starts to leak out slowly. Assuming the refrigerant remaining in the can has undergone a reversible, adiabatic process, determine the final mass in the can when the pressure drops to 30 psia.

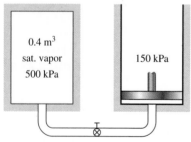

FIGURE P7-98

7-98 An insulated tank containing 0.4 m³ of saturated water vapor at 500 kPa is connected to an initially evacuated, insulated piston-cylinder device. The mass of the piston is such that a pressure of 150 kPa is required to raise it. Now the valve is opened slightly, and part of the steam flows to the cylinder, raising the piston. This process continues until the pressure in the tank drops to 150 kPa. Assuming the steam that remains in the tank to have undergone a reversible adiabatic process, determine the final temperature (a) in the rigid tank and (b) in the cylinder.

7-99 One ton (1000 kg) of liquid water at 80°C is brought into a well-insulated and well-sealed 4-m × 5-m × 6-m room initially at 22°C and 100 kPa. Assuming constant specific heats for both air and water at room temperature, determine (a) the final equilibrium temperature in the room and (b) the total entropy change during this process, in kJ/K.

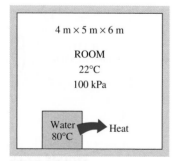

FIGURE P7-99

7-100E A piston-cylinder device initially contains 15 ft³ of helium gas at 25 psia and 70°F. Helium is now compressed in a polytropic process

(PV^n = constant) to 70 psia and 300°F. Determine (*a*) the entropy change of helium, (*b*) the entropy change of the surroundings, and (*c*) whether this process is reversible, irreversible, or impossible. Assume the surroundings are at 70°F.   *Answers:* (*a*) −0.016 Btu/R, (*b*) 0.019 Btu/R, (*c*) irreversible

7-101   Air is compressed steadily by a compressor from 100 kPa and 17°C to 700 kPa at a rate of 2 kg/min. Determine the minimum power input required if the process is (*a*) adiabatic and (*b*) isothermal. Assume air to be an ideal gas with constant specific heats, and neglect the changes in kinetic and potential energies.   *Answers:* (*a*) 7.21 kW, (*b*) 5.4 kW.

7-102   Air enters a two-stage compressor at 100 kPa and 27°C and is compressed to 900 kPa. The pressure ratio across each stage is the same, and the air is cooled to the initial temperature between the two stages. Assuming the compression process to be isentropic, determine the power input to the compressor for a mass flow rate of 0.02 kg/s. What would your answer be if only one stage of compression were used?   *Answers:* 4.44 kW, 5.26 kW

7-103   Consider a three-stage isentropic compressor with two intercoolers that cool the gas to the initial temperature between the stages. Determine the two intermediate pressures (P_x and P_y) in terms of inlet and exit pressures (P_1 and P_2) that will minimize the work input to the compressor.
      Answers: $P_x = (P_1^2 P_2)^{1/3}$, $P_y = (P_1 P_2^2)^{1/3}$

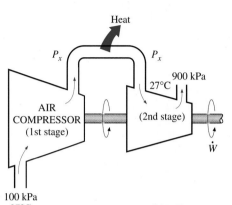

FIGURE P7-102

7-104   Steam at 7 MPa and 500°C enters a two-stage adiabatic turbine at a rate of 15 kg/s. Ten percent of the steam is extracted at the end of the first stage at a pressure of 1 MPa for other use. The remainder of the steam is further expanded in the second stage and leaves the turbine at 50 kPa. Determine the power output of the turbine, assuming (*a*) the process is reversible and (*b*) the turbine has an isentropic efficiency of 88 percent.
      Answers: (*a*) 14,928 kW, (*b*) 13,136 kW

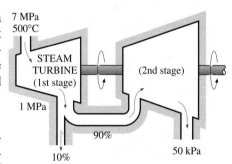

FIGURE P7-104

7-105   Steam enters a two-stage adiabatic turbine at 8 MPa and 500°C. It expands in the first stage to a pressure of 2 MPa. Then steam is reheated at constant pressure to 500°C before it is expanded in a second stage to a pressure of 100 kPa. The work output of the turbine is 40 MW. Assuming an isentropic efficiency of 84 percent for each stage of the turbine, determine the required mass flow rate of steam. Also, show the process on a *T-s* diagram with respect to saturation lines.   *Answer:* 41.0 kg/s

7-106   Refrigerant-134a at 140 kPa and −10°C is compressed by an adiabatic 0.5-kW compressor to an exit state of 700 kPa and 60°C. Neglecting the changes in kinetic and potential energies, determine (*a*) the isentropic efficiency of the compressor, (*b*) the volume flow rate of the refrigerant at the compressor inlet, in L/min, and (*c*) the maximum volume flow rate at the inlet conditions that this adiabatic 0.5-kW compressor can handle without violating the second law.

7-107E   Helium gas enters a nozzle whose isentropic efficiency is 94 percent with a low velocity, and it exits at 14 psia, 180°F, and 1000 ft/s. Determine the pressure and temperature at the nozzle inlet.

7-108   An adiabatic air compressor is to be powered by a direct-coupled adiabatic steam turbine that is also driving a generator. Steam enters the turbine at 12.5 MPa and 500°C at a rate of 25 kg/s and exits at 10 kPa and a quality of 0.92. Air enters the compressor at 98 kPa and 295 K at a rate of 10 kg/s and exits at 1 MPa and 620 K. Determine the net power delivered to the generator

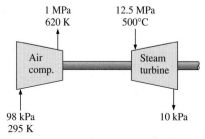

FIGURE P7-108

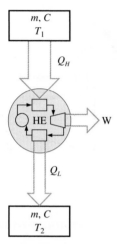

FIGURE P7-109

FIGURE P9-110

by the turbine and the rate of entropy generation within the turbine and the compressor during this process.

7-109 Consider two bodies of identical mass m and specific heat c used as thermal reservoirs (source and sink) for a heat engine. The first body is initially at an absolute temperature T_1 while the second one is at a lower absolute temperature T_2. Heat is transferred from the first body to the heat engine, which rejects the waste heat to the second body. The process continues until the final temperatures of the two bodies T_f become equal. Show that $T_f = \sqrt{T_1 T_2}$ when the heat engine produces the maximum possible work.

7-110 The explosion of a hot water tank in a school in Spencer, Oklahoma, in 1982 killed 7 people while injuring 33 others. Although the number of such explosions has decreased dramatically since the development of the ASME Pressure Vessel Code, which requires the tanks to be designed to withstand four times the normal operating pressures, they still occur as a result of the failure of the pressure relief valves and thermostats. When a tank filled with a high-pressure and high-temperature liquid ruptures, the sudden drop of the pressure of the liquid to the atmospheric level causes part of the liquid to flash into vapor, and thus to experience a huge rise in its volume. The resulting pressure wave that propagates rapidly can cause considerable damage.

Considering that the pressurized liquid in the tank eventually reaches equilibrium with its surroundings shortly after the explosion, the work that a pressurized liquid would do if allowed to expand reversibly and adiabatically to the pressure of the surroundings can be viewed as the *explosive energy* of the pressurized liquid. Because of the very short time period of the explosion and the apparent calm afterward, the explosion process can be considered to be adiabatic with no changes in kinetic and potential energies and no mixing with the air.

Consider a 100-L hot-water tank that has a working pressure of 0.5 MPa. As a result of some malfunction, the pressure in the tank rises to 2 MPa, at which point the tank explodes. Taking the atmospheric pressure to be 100 kPa and assuming the liquid in the tank to be saturated at the time of explosion, determine the total explosion energy of the tank in terms of the TNT equivalence. (The explosion energy of TNT is about 3250 kJ/kg, and 5 kg of TNT can cause total destruction of unreinforced structures within about a 7-m radius.)   *Answer:* 2.467 kg TNT

7-111 Using the arguments in the problem above, determine the total explosion energy of a 0.2-L canned drink that explodes at a pressure of 1 MPa. To how many kg of TNT is this explosion energy equivalent?

Computer, Design, and Essay Problems

7-112 Write a computer program to determine the work input to a multistage compressor for a given set of inlet and exit pressures for any number of stages. Assume that the pressure ratio across each stage is identical and the compression process is polytropic. List and plot the compressor work against the number of stages for $P_1 = 100$ kPa, $T_1 = 17°C$, $P_2 = 800$ kPa, and $n = 1.35$ for air. Based on this chart, can you justify using compressors with 3 or more stages?

7-113 It is well known that the temperature of a gas rises while it is compressed as a result of the energy input in the form of compression work. At high compression ratios, the air temperature may rise above the autoignition temperature of some hydrocarbons, including some lubricating oil. Therefore, the presence of some lubricating oil vapor in high-pressure air raises the possibility of an explosion, creating a fire hazard. The concentration of the oil within the compressor is usually too low to create a real danger. However, the oil that collects on the inner walls of exhaust piping of the compressor may cause an explosion. Such explosions have largely been eliminated by using the proper lubricating oils, carefully designing the equipment, intercooling between compressor stages, and keeping the system clean.

A compressor is to be designed for an industrial application in Los Angeles. If the compressor exit temperature is not to exceed 250°C for safety consideration, determine the maximum allowable compression ratio that is safe for all possible weather conditions for that area.

7-114 Think about all of your activities all day yesterday. List three of the activities during which you contributed considerably to the entropy increase of the universe. Explain why those activities are irreversible and how they generate entropy.

Power and Refrigeration Cycles

Two important areas of application for thermodynamics are power generation and refrigeration. Both power generation and refrigeration are usually accomplished by systems that operate on a thermodynamic cycle. Thermodynamic cycles can be divided into two general categories: *power cycles,* and *refrigeration cycles.*

The devices or systems used to produce a net power output are often called *engines,* and the thermodynamic cycles they operate on are called *power cycles.* The devices or systems used to produce refrigeration are called *refrigerators, air conditioners,* or *heat pumps,* and the cycles they operate on are called *refrigeration cycles.*

Thermodynamic cycles can also be categorized as *gas cycles* or *vapor cycles,* depending on the *phase* of the working fluid—the substance that circulates through the cyclic device. In gas cycles, the working fluid remains in the gaseous phase throughout the entire cycle, whereas in vapor cycles the working fluid exists in the vapor phase during one part of the cycle and in the liquid phase during another part.

Thermodynamic cycles can be categorized yet another way: *closed* and *open cycles.* In closed cycles, the working fluid is returned to the initial state at the end of the cycle and is recirculated. In open cycles, the working fluid is renewed at the end of each cycle instead of being recirculated. In automobile engines, for example, the combustion gases are exhausted and replaced by fresh air–fuel mixture at the end of each cycle. The engine operates on a mechanical cycle, but the working fluid in this type of device does not go through a complete thermodynamic cycle.

Heat engines are categorized as *internal combustion* or *external combustion engines,* depending on how the heat is supplied to the working fluid. In

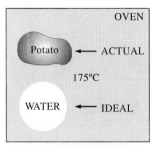

FIGURE 8-1

Modeling is a powerful engineering tool
that provides great insight and simplicity
at the expense of some loss in accuracy.

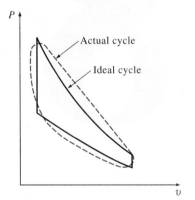

FIGURE 8-2

The analysis of many complex processes
can be reduced to a manageable level
by utilizing some idealizations.

FIGURE 8-3

Care should be exercised in the
interpretation of the results
from ideal cycles.

external combustion engines (such as steam power plants), energy is supplied
to the working fluid from an external source such as a furnace, a geothermal
well, a nuclear reactor, or even the sun. In internal combustion engines (such
as automobile engines), this is done by burning the fuel within the system
boundary. In this chapter, various gas power cycles are analyzed under some
simplifying assumptions.

Steam is the most common working fluid used in vapor power cycles
because of its many desirable characteristics, such as low cost, availability,
and high enthalpy of vaporization. Other working fluids used include sodium,
potassium, and mercury for high-temperature applications and some organic
fluids such as benzene and the freons for low-temperature applications.

Steam power plants are commonly referred to as *coal plants, nuclear
plants,* or *natural gas plants,* depending on the type of fuel used to supply heat
to the steam. But the steam goes through the same basic cycle in all of them.
Therefore, all can be analyzed in the same manner.

The most frequently used refrigeration cycle is the *vapor-compression refrigeration cycle* in which the refrigerant is vaporized and condensed alternately and is compressed in the vapor phase.

8-1 ■ BASIC CONSIDERATIONS IN THE ANALYSIS OF POWER CYCLES

Most power-producing devices operate on cycles, and the study of power
cycles is an exciting and important part of thermodynamics. The cycles encountered in actual devices are difficult to analyze because of the presence of
complicating effects, such as friction, and the absence of sufficient time for
establishment of the equilibrium conditions during the cycle. To make an
analytical study of a cycle feasible, we have to keep the complexities at a
manageable level and utilize some idealizations (Fig. 8-1). When the actual
cycle is stripped of all the internal irreversibilities and complexities, we end
up with a cycle that resembles the actual cycle closely but is made up totally
of internally reversible processes. Such a cycle is called an **ideal cycle**
(Fig. 8-2).

A simple idealized model enables engineers to study the effects of the
major parameters that dominate the cycle without getting bogged down in
the details. The cycles discussed in this chapter are somewhat idealized, but
they still retain the general characteristics of the actual cycles they represent.
The conclusions reached from the analysis of ideal cycles are also applicable
to actual cycles. The thermal efficiency of the Otto cycle, the ideal cycle for
spark-ignition automobile engines, for example, increases with the compression ratio. This is also the case for actual automobile engines. The numerical
values obtained from the analysis of an ideal cycle, however, are not necessarily representative of the actual cycles, and care should be exercised in their
interpretation (Fig. 8-3). The simplified analysis presented in this chapter for
various power cycles of practical interest may also serve as the starting point
for a more in-depth study.

Heat engines are designed for the purpose of converting other forms of
energy (usually in the form of heat) to work, and their performance is expressed in terms of the **thermal efficiency** η_{th}, which is the ratio of the net
work produced by the engine to the total heat input:

$$\eta_{th} = \frac{W_{net}}{Q_{in}} \qquad \text{or} \qquad \eta_{th} = \frac{w_{net}}{q_{in}} \qquad (8\text{-}1a, b)$$

It was pointed out in Chap. 6 that heat engines that operate on a totally reversible cycle, such as the Carnot cycle, have the highest thermal efficiency of all heat engines operating between the same temperature levels. That is, nobody can develop a cycle more efficient than the *Carnot cycle*. Then the following question arises naturally: If the Carnot cycle is the best possible cycle, why do we not use it as the model cycle for all the heat engines instead of bothering with several so-called *ideal* cycles? The answer to this question is hardware-related. Most cycles encountered in practice differ significantly from the Carnot cycle, which makes it unsuitable as a realistic model. Each ideal cycle discussed in this chapter is related to a specific work-producing device and is an *idealized* version of the actual cycle.

The ideal cycles are *internally reversible,* but, unlike the Carnot cycle, they are not necessarily externally reversible. That is, they may involve irreversibilities external to the system such as heat transfer through a finite temperature difference. Therefore, the thermal efficiency of an ideal cycle, in general, is less than that of a totally reversible cycle operating between the same temperature limits. However, it is still considerably higher than the thermal efficiency of an actual cycle because of the idealizations utilized.

The idealizations and simplifications commonly employed in the analysis of power cycles can be summarized as follows:

1.   The cycle does not involve any *friction.* Therefore, the working fluid does not experience any pressure drop as it flows in pipes or devices such as heat exchangers.

2.   All expansion and compression processes take place in a *quasi-equilibrium* manner (Fig. 8-4).

3.   The pipes connecting the various components of a system are well insulated, and *heat transfer* and *pressure drop* through them are negligible.

Neglecting the changes in *kinetic* and *potential energies* of the working fluid is another commonly utilized simplification in the analysis of power cycles. This is a reasonable assumption since in devices that involve shaft work, such as turbines, compressors, and pumps, the kinetic and potential energy terms are usually very small relative to the other terms in the energy equation. Fluid velocities encountered in devices such as condensers, boilers, and mixing chambers are typically low, and the fluid streams experience little change in their velocities, again making kinetic energy changes negligible. The only devices where the changes in kinetic energy are significant are the nozzles and diffusers, which are specifically designed to create large changes in velocity.

In the preceding chapters, *property diagrams* such as the *P-v* and *T-s* diagrams have served as valuable aids in the analysis of thermodynamic processes. On both the *P-v* and *T-s* diagrams, the area enclosed by the process curves of a cycle represents the net work produced during the cycle (Fig. 8-5), which is also equivalent to the net heat transfer for that cycle. The *T-s* diagram is particularly useful as a visual aid in the analysis of ideal power cycles. An ideal power cycle does not involve any internal irreversibilities, and so the only effect that can change the entropy of the working fluid during a process is heat transfer.

On a *T-s* diagram, a *heat-addition* process proceeds in the direction of increasing entropy, a *heat-rejection* process proceeds in the direction of decreasing entropy, and an *isentropic* (internally reversible, adiabatic) process proceeds at constant entropy. The area under the process curve on a *T-s*

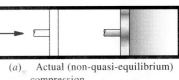

(*a*)   Actual (non-quasi-equilibrium) compression

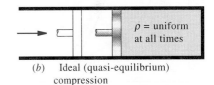

ρ = uniform at all times

(*b*)   Ideal (quasi-equilibrium) compression

FIGURE 8-4

All compression and expansion processes in ideal cycles are assumed to be quasi-equilibrium (internally reversible).

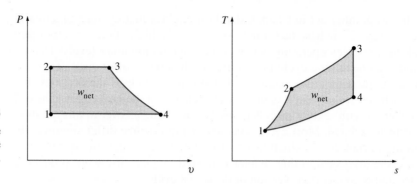

FIGURE 8-5

On both P-v and T-s diagrams, the area enclosed by the process curve represents the net work of the cycle.

diagram represents the heat transfer for that process. The area under the heat addition process on a T-s diagram is a geometric measure of the total heat supplied during the cycle q_{in}, and the area under the heat rejection process is a measure of the total heat rejected q_{out}. The difference between these two (the area enclosed by the cyclic curve) is the net heat transfer, which is also the net work produced during the cycle. Therefore, on a T-s diagram, the ratio of the area enclosed by the cyclic curve to the area under the heat-addition process curve represents the thermal efficiency of the cycle. *Any modification that will increase the ratio of these two areas will also improve the thermal efficiency of the cycle.*

Although the working fluid in an ideal power cycle operates on a closed loop, the type of individual processes that comprises the cycle depends on the individual devices used to execute the cycle. In the Rankine cycle, which is the ideal cycle for steam power plants, the working fluid flows through a series of steady-flow devices such as the turbine and condenser, whereas in the Otto cycle, which is the ideal cycle for the spark-ignition automobile engine, the working fluid is alternately expanded and compressed in a piston-cylinder device. Therefore, equations pertaining to steady-flow systems should be used in the analysis of the Rankine cycle, and equations pertaining to closed systems should be used in the analysis of the Otto cycle.

8-2 ■ CARNOT CYCLE AND ITS VALUE IN ENGINEERING

The Carnot cycle, which was introduced and discussed in Chap. 6, is composed of four totally reversible processes: isothermal heat addition, isentropic expansion, isothermal heat rejection, and isentropic compression. The P-v and T-s diagrams of a Carnot cycle are replotted in Fig. 8-6. The Carnot cycle can be executed in a closed system (a piston-cylinder device) or a steady-flow system (utilizing two turbines and two compressors, as shown in Fig. 8-7), and either a gas or a vapor can be used as the working fluid. The Carnot cycle is the most efficient cycle that can be executed between a heat source at temperature T_H and a sink at temperature T_L, and its thermal efficiency is expressed as

$$\eta_{th,\,Carnot} = 1 - \frac{T_L}{T_H} \qquad (8\text{-}2)$$

Reversible isothermal heat transfer is very difficult to achieve in reality because it would require very large heat exchangers and it would take a very long time (a power cycle in a typical engine is completed in a fraction of a second). Therefore, it is not practical to build an engine that would operate on a cycle that closely approximates the Carnot cycle.

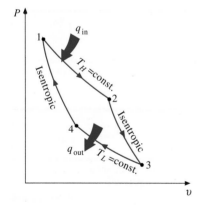

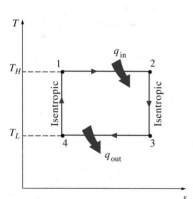

FIGURE 8-6

P-v and T-s diagrams of a Carnot cycle.

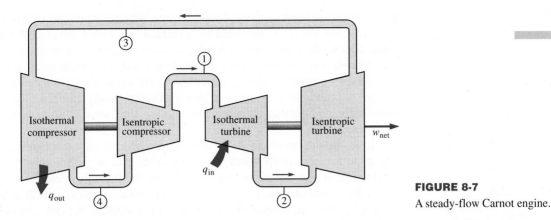

FIGURE 8-7

A steady-flow Carnot engine.

The real value of the Carnot cycle comes from its being a standard against which the actual or the ideal cycles can be compared. The thermal efficiency of the Carnot cycle is a function of the sink and source temperatures only, and the thermal efficiency relation for the Carnot cycle (Eq. 8-2) conveys an important message that is equally applicable to both ideal and actual cycles: *Thermal efficiency increases with an increase in the average temperature at which heat is supplied to the system or with a decrease in the average temperature at which heat is rejected from the system.*

The source and sink temperatures that can be used in practice are not without limits, however. The highest temperature in the cycle is limited by the maximum temperature that the components of the heat engine, such as the piston or the turbine blades, can withstand. The lowest temperature is limited by the temperature of the cooling medium utilized in the cycle such as a lake, a river, or the atmospheric air.

EXAMPLE 8-1   Derivation of the Efficiency of the Carnot Cycle

Show that the thermal efficiency of a Carnot cycle operating between the temperature limits of T_H and T_L is solely a function of these two temperatures and is given by Eq. 8-2.

Solution   The *T-s* diagram of a Carnot cycle is redrawn in Fig. 8-8. All four processes that comprise the Carnot cycle are reversible, and thus the area under each process curve represents the heat transfer for that process. Heat is transferred to the system during process 1-2 and rejected during process 3-4. Therefore, the amount of heat input and heat output for the cycle can be expressed as

$$q_{in} = T_H(s_2 - s_1) \qquad \text{and} \qquad q_{out} = T_L(s_3 - s_4) = T_L(s_2 - s_1)$$

since processes 2-3 and 4-1 are isentropic, and thus $s_2 = s_3$ and $s_4 = s_1$. Substituting these into Eq. 8-1b, we see that the thermal efficiency of a Carnot cycle is

$$\eta_{th} = \frac{w_{net}}{q_{in}} = 1 - \frac{q_{out}}{q_{in}} = 1 - \frac{T_L(s_2 - s_1)}{T_H(s_2 - s_1)} = 1 - \frac{T_L}{T_H}$$

which is the desired result. Notice that the thermal efficiency of a Carnot cycle is independent of the type of the working fluid used (an ideal gas, steam, etc.) or whether the cycle is executed in a closed or steady-flow system.

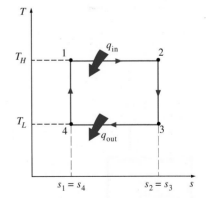

FIGURE 8-8

T-s diagram for the Carnot cycle discussed in Example 8-1.

8-3 ■ AIR-STANDARD ASSUMPTIONS

In gas power cycles, the working fluid remains a gas throughout the entire cycle. Spark-ignition automobile engines, diesel engines, and conventional gas turbines are familiar examples of devices that operate on gas cycles. In all these engines, energy is provided by burning a fuel within the system boundaries. That is, they are *internal combustion engines.* Because of this combustion process, the composition of the working fluid changes from air and fuel to combustion products during the course of the cycle. However, considering that air is predominantly nitrogen that undergoes hardly any chemical reactions in the combustion chamber, the working fluid closely resembles air at all times.

Even though internal combustion engines operate on a mechanical cycle (the piston returns to its starting position at the end of each revolution), the working fluid does not undergo a complete thermodynamic cycle. It is thrown out of the engine at some point in the cycle (as exhaust gases) instead of being returned to the initial state. Working on an open cycle is the characteristic of all internal combustion engines.

The actual gas power cycles are rather complex. To reduce the analysis to a manageable level, we utilize the following approximations, commonly known as the **air-standard assumptions:**

1. The working fluid is air, which continuously circulates in a closed loop and always behaves as an ideal gas.

2. All the processes that make up the cycle are internally reversible.

3. The combustion process is replaced by a heat-addition process from an external source (Fig. 8-9).

4. The exhaust process is replaced by a heat rejection process that restores the working fluid to its initial state.

Another assumption that is often utilized to simplify the analysis even more is that the air has constant specific heats whose values are determined at *room temperature* (25°C, or 77°F). When this assumption is utilized, the air-standard assumptions are called the **cold-air-standard assumptions.** A cycle for which the air-standard assumptions are applicable is frequently referred to as an **air-standard cycle.**

The air-standard assumptions stated above provide considerable simplification in the analysis without significantly deviating from the actual cycles. This simplified model enables us to study qualitatively the influence of major parameters on the performance of the actual engines.

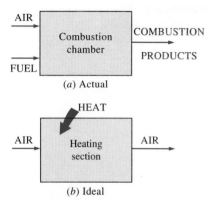

FIGURE 8-9

The combustion process is replaced by a heat-addition process in ideal cycles.

8-4 ■ AN OVERVIEW OF RECIPROCATING ENGINES

Despite its simplicity, the reciprocating engine (basically a piston-cylinder device) is one of the rare inventions that has proved to be very versatile and to have a wide range of applications. It is the powerhouse of the vast majority of automobiles, trucks, light aircraft, ships, and electric power generators, as well as many other devices.

The basic components of a reciprocating engine are shown in Fig. 8-10. The piston reciprocates in the cylinder between two fixed positions called the **top dead center** (TDC)—the position of the piston when it forms the smallest volume in the cylinder—and the **bottom dead center** (BDC)—the position of the piston when it forms the largest volume in the cylinder. The distance between the TDC and the BDC is the largest distance that the piston

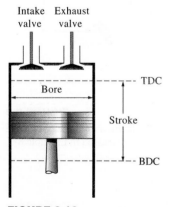

FIGURE 8-10

Nomenclature for reciprocating engines.

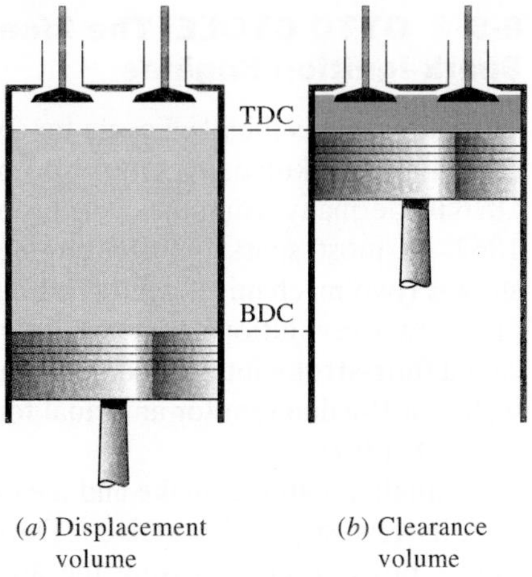

FIGURE 8-11

Displacement and clearance volumes
of a reciprocating engine.

(a) Displacement volume

(b) Clearance volume

can travel in one direction, and it is called the **stroke** of the engine. The diameter of the piston is called the **bore.** The air or air–fuel mixture is drawn into the cylinder through the **intake valve,** and the combustion products are expelled from the cylinder through the **exhaust valve.**

The minimum volume formed in the cylinder when the piston is at TDC is called the **clearance volume** (Fig. 8-11). The volume displaced by the piston as it moves between TDC and BDC is called the **displacement volume.** The ratio of the maximum volume formed in the cylinder to the minimum (clearance) volume is called the **compression ratio r** of the engine:

$$r = \frac{V_{max}}{V_{min}} = \frac{V_{BDC}}{V_{TDC}} \qquad (8\text{-}3)$$

Notice that the compression ratio is a *volume ratio* and should not be confused with the pressure ratio.

Another term frequently used in conjunction with reciprocating engines is the **mean effective pressure** (MEP). It is a fictitious pressure that, if it acted on the piston during the entire power stroke, would produce the same amount of net work as that produced during the actual cycle (Fig. 8-12). That is,

$$W_{net} = \text{MEP} \times \text{Piston area} \times \text{Stroke} = \text{MEP} \times \text{Displacement volume}$$

or

$$\text{MEP} = \frac{W_{net}}{V_{max} - V_{min}} = \frac{w_{net}}{v_{max} - v_{min}} \qquad (\text{kPa}) \qquad (8\text{-}4)$$

The mean effective pressure can be used as a parameter to compare the performances of reciprocating engines of equal size. The engine with a larger value of MEP will deliver more net work per cycle and thus will perform better.

Reciprocating engines are classified as **spark-ignition** (SI) **engines** or **compression-ignition** (CI) **engines,** depending on how the combustion process in the cylinder is initiated. In SI engines, the combustion of the air–fuel mixture is initiated by a spark plug. In CI engines, the air–fuel mixture is self-ignited as a result of compressing the mixture above its self-ignition temperature. In the next two sections, we discuss the *Otto* and *Diesel cycles,* which are the ideal cycles for the SI and CI reciprocating engines, respectively.

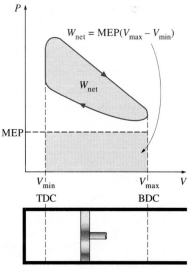

FIGURE 8-12

The net work output of a cycle is equivalent to the product of the mean effective pressure and the displacement volume.

8-5 ■ OTTO CYCLE: The Ideal Cycle for Spark-Ignition Engines

The Otto cycle is the ideal cycle for spark-ignition reciprocating engines. It is named after Nikolaus A. Otto, who built a successful four-stroke engine in 1876 in Germany using the cycle proposed by Frenchman Beau de Rochas in 1862. In most spark-ignition engines, the piston executes four complete strokes (two mechanical cycles) within the cylinder, and the crankshaft completes two revolutions for each thermodynamic cycle. These engines are called **four-stroke** internal combustion engines. A schematic of each stroke as well as a P-v diagram for an actual four-stroke spark-ignition engine is given in Fig. 8-13(a).

Initially, both the intake and the exhaust valves are closed, and the piston is at its lowest position (BDC). During the *compression stroke,* the piston moves upward, compressing the air–fuel mixture. Shortly before the piston reaches its highest position (TDC), the spark plug fires and the mixture ignites, increasing the pressure and temperature of the system. The high-pressure gases force the piston down, which in turn forces the crankshaft to rotate, producing a useful work output during the *expansion* or *power stroke.* At the end of this stroke, the piston is at its lowest position (the completion of the first mechanical cycle), and the cylinder is filled with combustion products. Now the piston moves upward one more time, purging the exhaust gases through the exhaust valve (the *exhaust stroke*), and down a second time, drawing in fresh air–fuel mixture through the intake valve (the *intake stroke*). Notice that the pressure in the cylinder is slightly above the atmospheric value during the exhaust stroke and slightly below during the intake stroke.

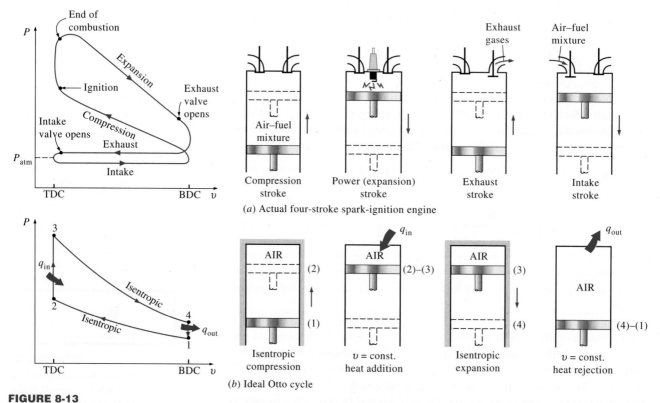

(a) Actual four-stroke spark-ignition engine

(b) Ideal Otto cycle

FIGURE 8-13

Actual and ideal cycles in spark-ignition engines and their P-v diagrams.

In **two-stroke engines,** all four functions described above are executed in just two strokes: the power stroke and the compression stroke. In these engines, the crankcase is sealed, and the outward motion of the piston is used to slightly pressurize the air–fuel mixture in the crankcase, as shown in Fig. 8-14. Also, the intake and exhaust valves are replaced by openings in the lower portion of the cylinder wall. During the latter part of the power stroke, the piston uncovers first the exhaust port, allowing the exhaust gases to be partially expelled, and then the intake port, allowing the fresh air–fuel mixture to rush in and drive most of the remaining exhaust gases out of the cylinder. This mixture is then compressed as the piston moves upward during the compression stroke and is subsequently ignited by a spark plug.

The two-stroke engines are generally less efficient than their four-stroke counterparts because of the incomplete expulsion of the exhaust gases and the partial expulsion of the fresh air–fuel mixture with the exhaust gases. However, they are relatively simple and inexpensive, and they have high power-to-weight and power-to-volume ratios, which make them suitable for applications requiring small size and weight such as for motorcycles, chain saws, and lawn mowers.

Advances in several technologies—such as direct fuel injection, stratified charge combustion, and electronic controls—brought about a renewed interest in two-stroke engines that can offer high performance and fuel economy while satisfying the future stringent emission requirements. For a given weight and displacement, a well-designed two-stroke engine can provide significantly more power than its four-stroke counterpart because two-stroke engines produce power on every engine revolution instead of every other one. In the new two-stroke engines under development, the highly atomized fuel spray that is injected into the combustion chamber toward the end of the compression stroke burns much more completely. The fuel is sprayed after the exhaust valve is closed, which prevents unburned fuel from being ejected into the atmosphere. With stratified combustion, the flame that is initiated by igniting a small amount of the rich fuel–air mixture near the spark plug propagates through the combustion chamber filled with a much leaner mixture, and this results in much cleaner combustion. Also, the advances in electronics have made it possible to ensure the optimum operation under varying engine load and speed conditions. Major car companies have research programs under way on two-stroke engines which are expected to make a comeback in the future.

The thermodynamic analysis of the actual four-stroke or two-stroke cycles described above is not a simple task. However, the analysis can be simplified significantly if the air-standard assumptions are utilized. The resulting cycle, which closely resembles the actual operating conditions, is the ideal **Otto cycle.** It consists of four internally reversible processes:

1-2   Isentropic compression

2-3   Constant volume heat addition

3-4   Isentropic expansion

4-1   Constant volume heat rejection

The execution of the Otto cycle in a piston-cylinder device together with a *P-v* diagram is illustrated in Fig. 8-13*b*. The *T-s* diagram of the Otto cycle is given in Fig. 8-15.

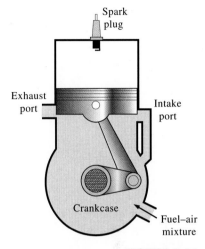

FIGURE 8-14

Schematic of a two-stroke reciprocating engine.

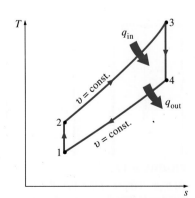

FIGURE 8-15

T-s diagram for the ideal Otto cycle.

The Otto cycle is executed in a closed system, and disregarding the changes in kinetic and potential energies, the first-law relation for any of the processes is expressed, on a unit-mass basis, as

$$(q_{in} - q_{out}) + (w_{in} - w_{out}) = \Delta u \qquad \text{(kJ/kg)} \qquad \text{(8-5)}$$

No work is involved during the two heat transfer processes since both take place at constant volume. Therefore, heat transfer to and from the working fluid can be expressed as

$$q_{in} = u_3 - u_2 = C_v(T_3 - T_2) \qquad \text{(8-6a)}$$

and
$$q_{out} = u_4 - u_1 = C_v(T_4 - T_1) \qquad \text{(8-6b)}$$

Then the thermal efficiency of the ideal Otto cycle under the cold air standard assumptions becomes

$$\eta_{th, \, Otto} = \frac{w_{net}}{q_{in}} = 1 - \frac{q_{out}}{q_{in}} = 1 - \frac{T_4 - T_1}{T_3 - T_2} = 1 - \frac{T_1(T_4/T_1 - 1)}{T_2(T_3/T_2 - 1)}$$

Processes 1-2 and 3-4 are isentropic, and $v_2 = v_3$ and $v_4 = v_1$. Thus,

$$\frac{T_1}{T_2} = \left(\frac{v_2}{v_1}\right)^{k-1} = \left(\frac{v_3}{v_4}\right)^{k-1} = \frac{T_4}{T_3} \qquad \text{(8-7)}$$

Substituting these equations into the thermal efficiency relation and simplifying give

$$\eta_{th, \, Otto} = 1 - \frac{1}{r^{k-1}} \qquad \text{(8-8)}$$

where
$$r = \frac{V_{max}}{V_{min}} = \frac{V_1}{V_2} = \frac{v_1}{v_2} \qquad \text{(8-9)}$$

is the compression ratio and k is the specific heat ratio C_p/C_v.

Equation 8-8 shows that under the cold-air-standard assumptions, the thermal efficiency of an ideal Otto cycle depends on the compression ratio of the engine and the specific heat ratio of the working fluid (if different from air). The thermal efficiency of the ideal Otto cycle increases with both the compression ratio and the specific heat ratio. This is also true for actual spark-ignition internal combustion engines. A plot of thermal efficiency versus the compression ratio is given in Fig. 8-16 for $k = 1.4$, which is the specific heat ratio value of air at room temperature. For a given compression ratio, the thermal efficiency of an actual spark-ignition engine will be less than that of an ideal Otto cycle because of the irreversibilities, such as friction, and other factors such as incomplete combustion.

We can observe from Fig. 8-16 that the thermal efficiency curve is rather steep at low compression ratios but flattens out starting with a compression ratio value of about 8. Therefore, the increase in thermal efficiency with the compression ratio is not that pronounced at high compression ratios. Also, when high compression ratios are used, the temperature of the air–fuel mixture rises above the autoignition temperature of the fuel (the temperature at which the fuel ignites without the help of a spark) during the combustion process, causing an early and rapid burn of the fuel at some point or points ahead of the flame front, followed by almost instantaneous inflammation of the end gas (Fig. 8-17). This premature ignition of the fuel, called **autoignition,** produces an audible noise, which is called **engine knock.** Autoignition in spark-ignition engines cannot be tolerated because it hurts

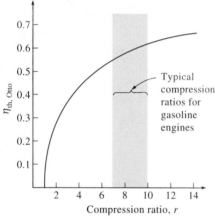

FIGURE 8-16

Thermal efficiency of the ideal Otto cycle as a function of compression ratio ($k = 1.4$).

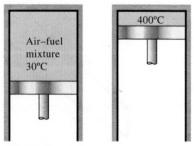

FIGURE 8-17

At high compression ratios, the air–fuel mixture temperature rises above the self-ignition temperature of the fuel during the compression process.

performance and can cause engine damage. The requirement that autoignition not be allowed places an upper limit on the compression ratios that can be used in spark-ignition internal combustion engines.

Improvement of the thermal efficiency of gasoline engines by utilizing higher compression ratios (up to about 12) without facing the autoignition problem has been made possible by using gasoline blends that have good antiknock characteristics, such as gasoline mixed with tetraethyl lead. Tetraethyl lead has been added to gasoline since the 1920s because it is the cheapest method of raising the *octane rating,* which is a measure of the engine knock resistance of a fuel. Leaded gasoline, however, has a very undesirable side effect: it forms compounds during the combustion process that are hazardous to health and pollute the environment. In an effort to combat air pollution, the government adopted a policy in the mid-1970s that resulted in the eventual phase-out of leaded gasoline. Unable to use lead, the refiners developed other, more elaborate techniques to improve the antiknock characteristics of gasoline. Most cars made since 1975 have been designed to use unleaded gasoline, and the compression ratios had to be lowered to avoid engine knock. The thermal efficiency of car engines has decreased somewhat as a result of decreased compression ratios. But, owing to the improvements in other areas (reduction in overall automobile weight, improved aerodynamic design, etc.), today's cars have better fuel economy and consequently get more miles per gallon of fuel. This is an example of how engineering decisions involve compromises, and efficiency is only one of the considerations in final design.

The second parameter affecting the thermal efficiency of an ideal Otto cycle is the specific heat ratio k. For a given compression ratio, an ideal Otto cycle using a monatomic gas (such as argon or helium, $k = 1.667$) as the working fluid will have the highest thermal efficiency. The specific heat ratio k, and thus the thermal efficiency of the ideal Otto cycle, decreases as the molecules of the working fluid get larger (Fig. 8-18). At room temperature it is 1.4 for air, 1.3 for carbon dioxide, and 1.2 for ethane. The working fluid in actual engines contains larger molecules such as carbon dioxide, and the specific heat ratio decreases with temperature, which is one of the reasons that the actual cycles have lower thermal efficiencies than the ideal Otto cycle. The thermal efficiencies of actual spark-ignition engines range from about 25 to 30 percent.

EXAMPLE 8-2 The Ideal Otto Cycle

An ideal Otto cycle has a compression ratio of 8. At the beginning of the compression process, the air is at 100 kPa and 17°C, and 800 kJ/kg of heat is

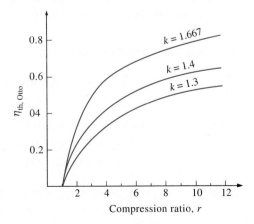

FIGURE 8-18

The thermal efficiency of the Otto cycle increases with the specific heat ratio k of the working fluid.

transferred to air during the constant-volume heat-addition process. Accounting for the variation of specific heats of air with temperature, determine (a) the maximum temperature and pressure that occur during the cycle, (b) the net work output, (c) the thermal efficiency, and (d) the mean effective pressure for the cycle.

Solution The P-v diagram of the ideal Otto cycle described is shown in Fig. 8-19. We note that the air contained in the cylinder forms a closed system.

Assumptions **1** The air-standard assumptions are applicable. **2** Kinetic and potential energy changes are negligible. **3** The variation of specific heats with temperature is to be accounted for.

Analysis (a) The maximum temperature and pressure in an Otto cycle occur at the end of the constant-volume heat-addition process (state 3). But first we need to determine the temperature and pressure of air at the end of the isentropic compression process (state 2), using data from Table A-17:

$$T_1 = 290 \text{ K} \longrightarrow u_1 = 206.91 \text{ kJ/kg}$$
$$v_{r1} = 676.1$$

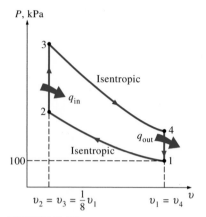

$v_2 = v_3 = \frac{1}{8} v_1$ $v_1 = v_4$

FIGURE 8-19

P-v diagram for the Otto cycle discussed in Example 8-2.

Process 1-2 (isentropic compression of an ideal gas):

$$\frac{v_{r2}}{v_{r1}} = \frac{v_2}{v_1} = \frac{1}{r} \longrightarrow v_{r2} = \frac{v_{r1}}{r} = \frac{676.1}{8} = 84.51 \longrightarrow \begin{array}{l} T_2 = 652.4 \text{ K} \\ u_2 = 475.11 \text{ kJ/kg} \end{array}$$

$$\frac{P_2 v_2}{T_2} = \frac{P_1 v_1}{T_1} \longrightarrow P_2 = P_1 \left(\frac{T_2}{T_1}\right)\left(\frac{v_1}{v_2}\right)$$
$$= (100 \text{ kPa})\left(\frac{652.4 \text{ K}}{290 \text{ K}}\right)(8) = 1799.7 \text{ kPa}$$

Process 2-3 (constant volume heat addition):

$$q_{in} = u_3 - u_2$$
$$800 \text{ kJ/kg} = u_3 - 475.11 \text{ kJ/kg}$$
$$u_3 = 1275.11 \text{ kJ/kg} \longrightarrow T_3 = \textbf{1575.1 K}$$
$$v_{r3} = 6.108$$

$$\frac{P_3 v_3}{T_3} = \frac{P_2 v_2}{T_2} \longrightarrow P_3 = P_2 \left(\frac{T_3}{T_2}\right)\left(\frac{v_2}{v_3}\right)$$
$$= (1.7997 \text{ MPa})\left(\frac{1575.1 \text{ K}}{652.4 \text{ K}}\right)(1) = \textbf{4.345 MPa}$$

(b) The net work output for the cycle is determined either by finding the boundary ($P \, dV$) work involved in each process by integration and adding them or by finding the net heat transfer that is equivalent to the net work done during the cycle. We take the latter approach. But first we need to find the internal energy of the air at state 4:

Process 3-4 (isentropic expansion of an ideal gas):

$$\frac{v_{r4}}{v_{r3}} = \frac{v_4}{v_3} = r \longrightarrow v_{r4} = r v_{r3} = (8)(6.108) = 48.864 \longrightarrow \begin{array}{l} T_4 = 795.6 \text{ K} \\ u_4 = 588.74 \text{ kJ/kg} \end{array}$$

Process 4-1 (constant volume heat rejection):

$$-q_{out} = u_1 - u_4 \longrightarrow q_{out} = u_4 - u_1$$
$$q_{out} = 588.74 - 206.91 = 381.83 \text{ kJ/kg}$$

Thus, $w_{net} = q_{net} = q_{in} - q_{out} = 800 - 381.83 = \textbf{418.17 kJ/kg}$

(c) The thermal efficiency of the cycle is determined from its definition, Eq. 8-1:

$$\eta_{th} = \frac{w_{net}}{q_{in}} = \frac{418.17 \text{ kJ/kg}}{800 \text{ kJ/kg}} = \textbf{0.523 or 52.3\%}$$

Under the cold-air-standard assumptions (constant specific heat values at room temperature), the thermal efficiency would be (Eq. 8-8)

$$\eta_{th, Otto} = 1 - \frac{1}{r^{k-1}} = 1 - r^{1-k} = 1 - (8)^{1-1.4} = 0.565 \text{ or } 56.5\%$$

which is considerably different from the value obtained above. Therefore, care should be exercised in utilizing the cold-air-standard assumptions.

(d) The mean effective pressure is determined from its definition, Eq. 8-4:

$$\text{MEP} = \frac{w_{net}}{v_1 - v_2} = \frac{w_{net}}{v_1 - v_1/r} = \frac{w_{net}}{v_1(1 - 1/r)}$$

where

$$v_1 = \frac{RT_1}{P_1} = \frac{(0.287 \text{ kPa} \cdot \text{m}^3/\text{kg} \cdot \text{K})(290 \text{ K})}{100 \text{ kPa}} = 0.832 \text{ m}^3/\text{kg}$$

Thus,

$$\text{MEP} = \frac{418.17 \text{ kJ/kg}}{(0.832 \text{ m}^3/\text{kg})(1 - \frac{1}{8})}\left(\frac{1 \text{ kPa} \cdot \text{m}^3}{1 \text{ kJ}}\right) = \textbf{574.4 kPa}$$

Therefore, a constant pressure of 574.4 kPa during the power stroke would produce the same net work output as the entire cycle.

8-6 ■ DIESEL CYCLE: The Ideal Cycle for Compression-Ignition Engines

The Diesel cycle is the ideal cycle for CI reciprocating engines. The CI engine, first proposed by Rudolph Diesel in the 1890s, is very similar to the SI engine discussed in the last section, differing mainly in the method of initiating combustion. In spark-ignition engines (also known as *gasoline engines*), the air–fuel mixture is compressed to a temperature that is below the autoignition temperature of the fuel, and the combustion process is initiated by firing a spark plug. In CI engines (also known as *diesel engines*), the air is compressed to a temperature that is above the autoignition temperature of the fuel, and combustion starts on contact as the fuel is injected into this hot air. Therefore, the spark plug and carburetor are replaced by a fuel injector in diesel engines (Fig. 8-20).

In gasoline engines, a mixture of air and fuel is compressed during the compression stroke, and the compression ratios are limited by the onset of autoignition or engine knock. In diesel engines, only air is compressed during the compression stroke, eliminating the possibility of autoignition. Therefore, diesel engines can be designed to operate at much higher compression ratios, typically between 12 and 24. Not having to deal with the problem of autoignition has another benefit: many of the stringent requirements placed on the gasoline can now be removed, and fuels that are less refined (thus less expensive) can be used in diesel engines.

The fuel injection process in diesel engines starts when the piston approaches TDC and continues during the first part of the power stroke. Therefore, the combustion process in these engines takes place over a longer interval. Because of this longer duration, the combustion process in the ideal Diesel cycle is approximated as a constant-pressure heat-addition process. In fact, this is the only process where the Otto and the Diesel cycles differ. The

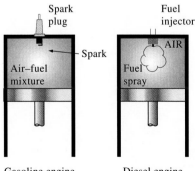

FIGURE 8-20

In diesel engines, the spark plug is replaced by a fuel injector, and only air is compressed during the compression process.

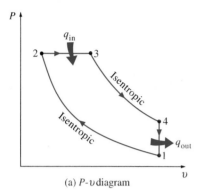

(a) P-υ diagram

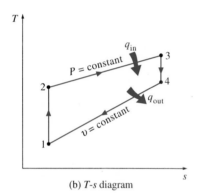

(b) T-s diagram

FIGURE 8-21

T-s and P-v diagrams for the
ideal Diesel cycle.

remaining three processes are the same for both ideal cycles. That is, process 1-2 is isentropic compression, 3-4 is isentropic expansion, and 4-1 is constant-volume heat rejection. The similarity between the two cycles is also apparent from the P-v and T-s diagrams of the Diesel cycle, shown in Fig. 8-21.

Noting that the Diesel cycle is executed in a piston-cylinder device, which forms a closed system, the amount of heat transferred to the working fluid at constant pressure and rejected from it at constant volume can be expressed as

$$q_{in} - w_{b,\,out} = u_3 - u_2 \quad \longrightarrow \quad q_{in} = P_2(v_3 - v_2) + (u_3 - u_2)$$
$$= h_3 - h_2 = C_p(T_3 - T_2) \qquad (8\text{-}10a)$$

and $\quad -q_{out} = u_1 - u_4 \quad \longrightarrow \quad q_{out} = u_4 - u_1 = C_v(T_4 - T_1) \qquad (8\text{-}10b)$

Then the thermal efficiency of the ideal Diesel cycle under the cold-air-standard assumptions becomes

$$\eta_{th,\,Diesel} = \frac{w_{net}}{q_{in}} = 1 - \frac{q_{out}}{q_{in}} = 1 - \frac{T_4 - T_1}{k(T_3 - T_2)} = 1 - \frac{T_1(T_4/T_1 - 1)}{kT_2(T_3/T_2 - 1)}$$

We now define a new quantity, the **cutoff ratio** r_c, as the ratio of the cylinder volumes after and before the combustion process:

$$r_c = \frac{V_3}{V_2} = \frac{v_3}{v_2} \qquad (8\text{-}11)$$

Utilizing this definition and the isentropic ideal-gas relations for processes 1-2 and 3-4, we see that the thermal efficiency relation reduces to

$$\eta_{th,\,Diesel} = 1 - \frac{1}{r^{k-1}} \left[\frac{r_c^k - 1}{k(r_c - 1)} \right] \qquad (8\text{-}12)$$

where r is the compression ratio defined by Eq. 8-9. Looking at Eq. 8-12 carefully, one would notice that under the cold-air-standard assumptions, the efficiency of a Diesel cycle differs from the efficiency of an Otto cycle by the quantity in the brackets. This quantity is always greater than 1. Therefore,

$$\eta_{th,\,Otto} > \eta_{th,\,Diesel} \qquad (8\text{-}13)$$

when both cycles operate on the same compression ratio. Also, as the cutoff ratio decreases, the efficiency of the Diesel cycle increases (Fig. 8-22). For the limiting case of $r_c = 1$, the quantity in the brackets becomes unity (can you prove it?), and the efficiencies of the Otto and Diesel cycles become identical. Remember, though, that diesel engines operate at much higher

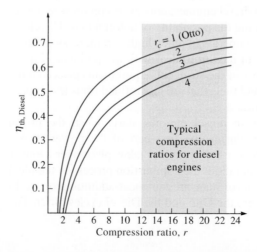

FIGURE 8-22

Thermal efficiency of the ideal Diesel cycle as a function of compression and cutoff ratios ($k = 1.4$).

compression ratios and thus are usually more efficient than the spark-ignition (gasoline) engines. The diesel engines also burn the fuel more completely since they usually operate at lower revolutions per minute than spark-ignition engines. Thermal efficiencies of large diesel engines range from about 35 to 40 percent.

The higher efficiency and lower fuel costs of diesel engines make them the clear choice in applications requiring relatively large amounts of power, such as in locomotive engines, emergency power generation units, large ships, and heavy trucks. As an example of how large a diesel engine can be, a 12-cylinder diesel engine built in 1964 by the Fiat Corporation of Italy had a normal power output of 25,200 hp (18.8 MW) at 122 rpm, a cylinder bore of 90 cm, and a stroke of 91 cm.

Approximating the combustion process in internal combustion engines as a constant-volume or a constant-pressure heat-addition process is overly simplistic and not quite realistic. Probably a better (but slightly more complex) approach would be to model the combustion process in both gasoline and diesel engines as a combination of two heat-transfer processes, one at constant volume and the other at constant pressure. The ideal cycle based on this concept is called the **dual cycle,** and a *P-v* diagram for it is given in Fig. 8-23. The relative amounts of heat transferred during each process can be adjusted to approximate the actual cycle more closely. Note that both the Otto and the Diesel cycles can be obtained as special cases of the dual cycle.

EXAMPLE 8-3   The Ideal Diesel Cycle

An ideal Diesel cycle with air as the working fluid has a compression ratio of 18 and a cutoff ratio of 2. At the beginning of the compression process, the working fluid is at 14.7 psia, 80°F, and 117 in³. Utilizing the cold-air-standard assumptions, determine (*a*) the temperature and pressure of the air at the end of each process, (*b*) the net work output and the thermal efficiency, and (*c*) the mean effective pressure.

Solution   The *P-V* diagram of the ideal Diesel cycle described is shown in Fig. 8-24. We note that the air contained in the cylinder forms a closed system.

Assumptions   **1** The cold-air-standard assumptions are applicable and thus air can be assumed to have constant specific heats at room temperature. **2** Kinetic and potential energy changes are negligible.

Analysis   The gas constant of air is $R = 0.06855$ Btu/lbm · R, and its specific heats at room temperature are $C_p = 0.240$ Btu/lbm · R and $C_v = 0.171$ Btu/lbm · R (Table A-2E*a*).

(*a*) The temperature and pressure values at the end of each process can be determined by utilizing the ideal-gas isentropic relations for processes 1-2 and 3-4. But first we determine the volumes at the end of each process from the definitions of the compression ratio and the cutoff ratio:

$$V_2 = \frac{V_1}{r} = \frac{117 \text{ in}^3}{18} = 6.5 \text{ in}^3$$

$$V_3 = r_c V_2 = (2)(6.5 \text{ in}^3) = 13 \text{ in}^3$$

$$V_4 = V_1 = 117 \text{ in}^3$$

Process 1-2 (isentropic compression of an ideal gas, constant specific heats):

$$T_2 = T_1\left(\frac{V_1}{V_2}\right)^{k-1} = (540 \text{ R})(18)^{1.4-1} = \textbf{1716 R}$$

$$P_2 = P_1\left(\frac{V_1}{V_2}\right)^{k} = (14.7 \text{ psia})(18)^{1.4} = \textbf{841 psia}$$

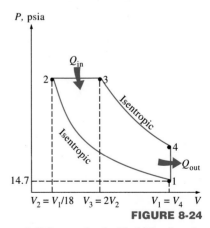

FIGURE 8-23

P-v diagram of an ideal dual cycle.

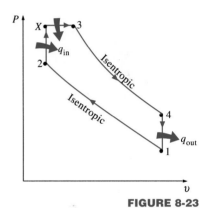

FIGURE 8-24

P-V diagram for the ideal Diesel cycle discussed in Example 8-3.

Process 2-3 (constant pressure heat addition to an ideal gas):

$$P_3 = P_2 = 841 \text{ psia}$$

$$\frac{P_2 V_2}{T_2} = \frac{P_3 V_3}{T_3} \quad \longrightarrow \quad T_3 = T_2 \left(\frac{V_3}{V_2}\right) = (1716 \text{ R})(2) = 3432 \text{ R}$$

Process 3-4 (isentropic expansion of an ideal gas, constant specific heats):

$$T_4 = T_3 \left(\frac{V_3}{V_4}\right)^{k-1} = (3432 \text{ R}) \left(\frac{13 \text{ in}^3}{117 \text{ in}^3}\right)^{1.4-1} = 1425 \text{ R}$$

$$P_4 = P_3 \left(\frac{V_3}{V_4}\right)^{k} = (841 \text{ psia}) \left(\frac{13 \text{ in}^3}{117 \text{ in}^3}\right)^{1.4} = 38.8 \text{ psia}$$

(b) The net work for a cycle is equivalent to the net heat transfer, that is, the difference between the total heat supplied and the total heat rejected. But first we find the mass of air:

$$m = \frac{P_1 V_1}{RT_1} = \frac{(14.7 \text{ psia})(117 \text{ in}^3)}{(0.3704 \text{ psia} \cdot \text{ft}^3/\text{lbm} \cdot \text{R})(540 \text{ R})} \left(\frac{1 \text{ ft}^3}{1728 \text{ in}^3}\right)$$

$$= 0.00498 \text{ lbm}$$

Process 2-3 is a constant-pressure heat-addition process, for which the boundary work and Δu terms can be combined into Δh. Thus,

$$Q_{in} = m(h_3 - h_2) = mC_p(T_3 - T_2)$$

$$= (0.00498 \text{ lbm})(0.240 \text{ Btu/lbm} \cdot \text{R})[(3432 - 1716) \text{ R}]$$

$$= 2.051 \text{ Btu}$$

Process 4-1 is a constant-volume heat-rejection process (it involves no work interactions), and the amount of heat rejected is

$$Q_{out} = m(u_4 - u_1) = mC_v(T_4 - T_1)$$

$$= (0.00498 \text{ lbm})(0.171 \text{ Btu/lbm} \cdot \text{R})[(1425 - 540) \text{ R}]$$

$$= 0.754 \text{ Btu}$$

Thus, $W_{net} = Q_{in} - Q_{out} = 2.051 - 0.754 = \mathbf{1.292 \text{ Btu}}$

Then the thermal efficiency becomes

$$\eta_{th} = \frac{W_{net}}{Q_{in}} = \frac{1.297 \text{ Btu}}{2.051 \text{ Btu}} = \mathbf{0.632 \text{ or } 63.2\%}$$

The thermal efficiency of this Diesel cycle under the cold-air-standard assumptions could also be determined from Eq. 8-12.

(c) The mean effective pressure is determined from its definition, Eq. 8-4:

$$\text{MEP} = \frac{W_{net}}{V_{max} - V_{min}} = \frac{W_{net}}{V_1 - V_2} = \frac{1.293 \text{ Btu}}{(117 - 6.5) \text{ in}^3} \left(\frac{778.17 \text{ lbf} \cdot \text{ft}}{1 \text{ Btu}}\right) \left(\frac{12 \text{ in.}}{1 \text{ ft}}\right)$$

$$= \mathbf{109.6 \text{ psia}}$$

Therefore, a constant pressure of 109.6 psia during the power stroke would produce the same net work output as the entire Diesel cycle.

8-7 ■ BRAYTON CYCLE:
The Ideal Cycle for Gas-Turbine Engines

The Brayton cycle was first proposed by George Brayton for use in the reciprocating oil-burning engine that he developed around 1870. Today, it is used for gas turbines only where both the compression and expansion processes

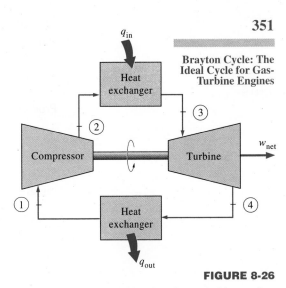

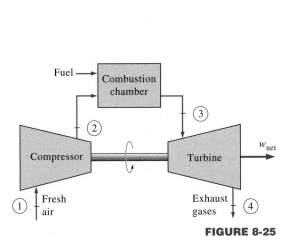

FIGURE 8-25

An open-cycle gas-turbine engine.

FIGURE 8-26

A closed-cycle gas-turbine engine.

take place in rotating machinery. Gas turbines usually operate on an *open cycle,* as shown in Fig. 8-25. Fresh air at ambient conditions is drawn into the compressor, where its temperature and pressure are raised. The high-pressure air proceeds into the combustion chamber, where the fuel is burned at constant pressure. The resulting high-temperature gases then enter the turbine, where they expand to the atmospheric pressure, thus producing power. The exhaust gases leaving the turbine are thrown out (not recirculated), causing the cycle to be classified as an open cycle.

The open gas-turbine cycle described above can be modeled as a *closed cycle,* as shown in Fig. 8-26, by utilizing the air-standard assumptions. Here the compression and expansion processes remain the same, but the combustion process is replaced by a constant-pressure heat-addition process from an external source, and the exhaust process is replaced by a constant-pressure heat-rejection process to the ambient air. The ideal cycle that the working fluid undergoes in this closed loop is the **Brayton cycle,** which is made up of four internally reversible processes:

1-2   Isentropic compression (in a compressor)

2-3   Constant pressure heat addition

3-4   Isentropic expansion (in a turbine)

4-1   Constant pressure heat rejection

The *T-s* and *P-v* diagrams of an ideal Brayton cycle are shown in Fig. 8-27. Notice that all four processes of the Brayton cycle are executed in steady-flow devices; thus, they should be analyzed as steady-flow processes. When the changes in kinetic and potential energies are neglected, the energy balance for a steady-flow process can be expressed, on a unit-mass basis, as

$$(q_{in} - q_{out}) + (w_{in} - w_{out}) = h_{exit} - h_{inlet} \qquad (8\text{-}14)$$

Therefore, heat transfers to and from the working fluid are

$$q_{in} = h_3 - h_2 = C_p(T_3 - T_2) \qquad (8\text{-}15)$$

and $\qquad\qquad q_{out} = h_4 - h_1 = C_p(T_4 - T_1) \qquad (8\text{-}16)$

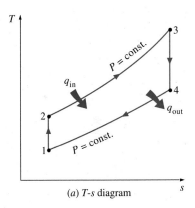

(*a*) *T-s* diagram

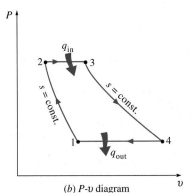

(*b*) *P-v* diagram

FIGURE 8-27

T-s and *P-v* diagrams for the ideal Brayton cycle.

Then the thermal efficiency of the ideal Brayton cycle under the cold air standard assumptions becomes

$$\eta_{th,\,Brayton} = \frac{w_{net}}{q_{in}} = 1 - \frac{q_{out}}{q_{in}} = 1 - \frac{C_p(T_4 - T_1)}{C_p(T_3 - T_2)} = 1 - \frac{T_1(T_4/T_1 - 1)}{T_2(T_3/T_2 - 1)}$$

Processes 1-2 and 3-4 are isentropic, and $P_2 = P_3$ and $P_4 = P_1$. Thus,

$$\frac{T_2}{T_1} = \left(\frac{P_2}{P_1}\right)^{(k-1)/k} = \left(\frac{P_3}{P_4}\right)^{(k-1)/k} = \frac{T_3}{T_4}$$

Substituting these equations into the thermal efficiency relation and simplifying give

$$\eta_{th,\,Brayton} = 1 - \frac{1}{r_p^{(k-1)/k}} \qquad (8\text{-}17)$$

where

$$r_p = \frac{P_2}{P_1} \qquad (8\text{-}18)$$

is the **pressure ratio** and k is the specific heat ratio. Equation 8-17 shows that under the cold-air-standard assumptions, the thermal efficiency of an ideal Brayton cycle depends on the pressure ratio of the gas turbine and the specific heat ratio of the working fluid (if different from air). The thermal efficiency increases with both of these parameters, which is also the case for actual gas turbines. A plot of thermal efficiency versus the pressure ratio is given in Fig. 8-28 for $k = 1.4$, which is the specific-heat-ratio value of air at room temperature.

The highest temperature in the cycle occurs at the end of the combustion process (state 3), and it is limited by the maximum temperature that the turbine blades can withstand. This also limits the pressure ratios that can be used in the cycle. For a fixed turbine inlet temperature T_3, the net work output per cycle increases with the pressure ratio, reaches a maximum, and then starts to decrease, as shown in Fig. 8-29. Therefore, there should be a compromise between the pressure ratio (thus the thermal efficiency) and the net work output. With less work output per cycle, a larger mass flow rate (thus a larger system) is needed to maintain the same power output which may not be economical. In most common designs, the pressure ratio of gas turbines ranges from about 11 to 16.

The air in gas turbines performs two important functions: It supplies the necessary oxidant for the combustion of the fuel, and it serves as a coolant to keep the temperature of various components within safe limits. The second function is accomplished by drawing in more air than is needed for the complete combustion of the fuel. In gas turbines, an air–fuel mass ratio of 50 or above is not uncommon. Therefore, in a cycle analysis, treating the combustion gases as air will not cause any appreciable error. Also, the mass flow rate through the turbine will be greater than that through the compressor, the difference being equal to the mass flow rate of the fuel. Thus, assuming a constant mass flow rate throughout the cycle will yield conservative results for open-loop gas-turbine engines.

The two major application areas of gas-turbine engines are *aircraft propulsion* and *electric power generation*. When it is used for aircraft propulsion, the gas turbine produces just enough power to drive the compressor and a small generator to power the auxiliary equipment. The high-velocity exhaust gases are responsible for producing the necessary thrust to propel the aircraft.

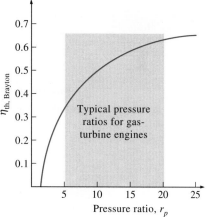

FIGURE 8-28

Thermal efficiency of the ideal Brayton cycle as a function of the pressure ratio.

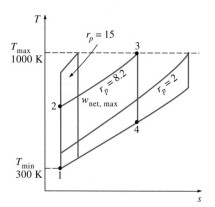

FIGURE 8-29

For fixed values of T_{min} and T_{max}, the net work of the Brayton cycle first increases with the pressure ratio, then reaches a maximum at $r_p = (T_{max}/T_{min})^{k/[2(k-1)]}$, and finally decreases.

Gas turbines are also used as stationary power plants to generate electricity as stand-alone units or in conjunction with steam power plants on the high-temperature side. In these plants, the exhaust gases of the gas turbine serve as the heat source for the steam. The gas-turbine cycle can also be executed as a closed cycle for use in nuclear power plants. This time the working fluid is not limited to air, and a gas with more desirable characteristics (such as helium) can be used.

The majority of the Western world's naval fleets already use gas-turbine engines for propulsion and electric power generation. The General Electric LM2500 gas turbines used to power ships have a simple-cycle thermal efficiency of 37 percent. The new General Electric WR-21 gas turbines equipped with intercooling and regeneration have a thermal efficiency of 43 percent and produce 21.6 MW (29,040 hp). The regeneration also reduces the exhaust temperature from 600°C (1100°F) to 350°C (650°F). Air is compressed to 3 atm before it enters the intercooler. Compared to steam-turbine and diesel-propulsion systems, the gas turbine offers greater power for a given size and weight, high reliability, long life, and more convenient operation. The engine start-up time has been reduced from 4 h required for a typical steam-propulsion system to less than 2 min for a gas turbine. Many modern marine propulsion systems use gas turbines together with diesel engines because of the high fuel consumption of simple-cycle gas-turbine engines. In combined diesel and gas-turbine systems, diesel is used to provide for efficient low-power and cruise operation, and gas turbine is used when high speeds are needed.

In gas-turbine power plants, the ratio of the compressor work to the turbine work, called the **back work ratio,** is very high (Fig. 8-30). Usually more than one-half of the turbine work output is used to drive the compressor. The situation is even worse when the adiabatic efficiencies of the compressor and the turbine are low. This is quite in contrast to steam power plants, where the back work ratio is only a few percent. This is not surprising, however, since a liquid is compressed in steam power plants instead of a gas, and the reversible steady-flow work is proportional to the specific volume of the working fluid.

A power plant with a high back work ratio requires a larger turbine to provide the additional power requirements of the compressor. Therefore, the turbines used in gas-turbine power plants are larger than those used in steam power plants of the same net power output.

Development of Gas Turbines

The gas turbine has experienced phenomenal progress and growth since its first successful development in the 1930s. The early gas turbines built in the 1940s and even 1950s had simple-cycle efficiencies of about 17 percent because of the low compressor and turbine efficiencies and low turbine inlet temperatures due to metallurgical limitations of those times. Therefore, gas turbines found only limited use despite their versatility and their ability to burn a variety of fuels. The efforts to improve the cycle efficiency concentrated in three areas:

1.   Increasing the turbine inlet (or firing) temperatures   This has been the primary approach taken to improve gas-turbine efficiency. The turbine inlet temperatures have increased steadily from about 540°C (1000°F) in the 1940s to 1425°C (2600°F) today. These increases were made possible by the development of new materials and the innovative cooling techniques for the critical components such as coating the turbine blades with ceramic layers

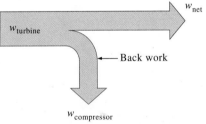

FIGURE 8-30

The fraction of the turbine work used to drive the compressor is called the back work ratio.

and cooling the blades with the discharge air from the compressor. Maintaining high turbine inlet temperatures with air-cooling technique requires the combustion temperature to be higher to compensate for the cooling effect of the cooling air. But higher combustion temperatures increase the amount of nitrogen oxides (NO_x), which are responsible for the formation of ozone at ground level and smog. Using steam as the coolant allowed an increase in the turbine inlet temperatures by 200°F without an increase in the combustion temperature. Steam is also a much more effective heat transfer medium than air.

2. Increasing the efficiencies of turbo-machinery components The performance of early turbines suffered greatly from the inefficiencies of turbines and compressors. But the advent of computers and advanced techniques for computer-aided design made it possible to design these components aerodynamically with minimal losses. The increased efficiencies of the turbines and compressors resulted in a significant increase in the cycle efficiency.

3. Adding modifications to the basic cycle The simple-cycle efficiencies of early gas turbines were practically doubled by incorporating intercooling, regeneration (or recuperation), and reheating. These improvements, of course, come at the expense of increased initial and operation costs, and they cannot be justified unless the decrease in fuel costs offsets the increase in other costs. The relatively low fuel prices, the general desire in the industry to minimize installation costs, and the tremendous increase in the simple-cycle efficiency to about 40 percent left little desire for opting for these modifications.

The first gas turbine for an electric utility was installed in 1949 in Oklahoma as part of a combined-cycle power plant. It was built by General Electric and produced 3.5 MW of power. Gas turbines installed until the mid-1970s suffered from low efficiency and poor reliability. In the past, the base-load electric power generation was dominated by large coal and nuclear power plants. However, there has been a historic shift toward natural gas–fired gas turbines because of their higher efficiencies, lower capital costs, shorter installation times, and better emission characteristics, and the abundance of natural gas supplies, and more and more electric utilities are using gas turbines for base-load power production as well as for peaking. The construction costs for gas-turbine power plants are roughly half that of comparable conventional fossil-fuel steam power plants, which were the primary base-load power plants until the early 1980s. More than half of all power plants to be installed in the foreseeable future are forecast to be gas-turbine or combined gas–steam turbine types.

A gas turbine manufactured by General Electric in the early 1990s had a pressure ratio of 13.5 and generated 135.7 MW of net power at a thermal efficiency of 33 percent in simple-cycle operation. A more recent gas turbine manufactured by General Electric uses a turbine inlet temperature of 1425°C (2600°F) and produces up to 282 MW while achieving a thermal efficiency of 39.5 percent in the simple-cycle mode. A 1.3-ton small-scale gas turbine labeled OP-16, built by the Dutch firm Opra Optimal Radial Turbine, can run on gas or liquid fuel and can replace a 16-ton diesel engine. It has a pressure ratio of 6.5 and produces up to 2 MW of power. Its efficiency is 26 percent in the simple-cycle operation, which rises to 37 percent when equipped with a regenerator.

EXAMPLE 8-4 The Simple Ideal Brayton Cycle

A stationary power plant operating on an ideal Brayton cycle has a pressure ratio of 8. The gas temperature is 300 K at the compressor inlet and 1300 K at the turbine inlet. Utilizing the air-standard assumptions, determine (a) the gas temperature at the exits of the compressor and the turbine, (b) the back work ratio, and (c) the thermal efficiency.

Solution The *T-s* diagram of the ideal Brayton cycle described is shown in Fig. 8-31. We note that the components involved in the Brayton cycle are steady-flow devices.

Assumptions **1** Steady operating conditions exist. **2** The air-standard assumptions are applicable. **3** Kinetic and potential energy changes are negligible. **4** The variation of specific heats with temperature is to be accounted for.

Analysis (a) The air temperatures at the compressor and turbine exits are determined from the isentropic relations applied to processes 1-2 and 3-4:

Process 1-2 (isentropic compression of an ideal gas);

$$T_1 = 300 \text{ K} \longrightarrow h_1 = 300.19 \text{ kJ/kg}$$
$$P_{r1} = 1.386$$

$$P_{r2} = \frac{P_2}{P_1} P_{r1} = (8)(1.386) = 11.09 \longrightarrow T_2 = \textbf{540 K} \quad \text{(at compressor exit)}$$
$$h_2 = 544.35 \text{ kJ/kg}$$

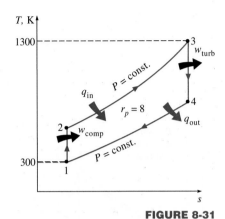

FIGURE 8-31

T-s diagram for the Brayton cycle discussed in Example 8-4.

Process 3-4 (isentropic expansion of an ideal gas):

$$T_3 = 1300 \text{ K} \longrightarrow h_3 = 1395.97 \text{ kJ/kg}$$
$$P_{r3} = 330.9$$

$$P_{r4} = \frac{P_4}{P_3} P_{r3} = \left(\frac{1}{8}\right)(330.9) = 41.36 \longrightarrow T_4 = \textbf{770 K} \quad \text{(at turbine exit)}$$
$$h_4 = 789.11 \text{ kJ/kg}$$

(b) To find the back work ratio, we need to find the work input to the compressor and the work output of the turbine:

$$w_{\text{comp, in}} = h_2 - h_1 = 544.35 - 300.19 = 244.16 \text{ kJ/kg}$$
$$w_{\text{turb, out}} = h_3 - h_4 = 1395.97 - 789.11 = 606.86 \text{ kJ/kg}$$

Thus, Back work ratio $r_{\text{bw}} = \dfrac{w_{\text{comp, in}}}{w_{\text{turb, out}}} = \dfrac{244.16 \text{ kJ/kg}}{606.86 \text{ kJ/kg}} = \textbf{0.402}$

That is, 40.2 percent of the turbine work output is used just to drive the compressor.

(c) The thermal efficiency of the cycle is the ratio of the net power output to the total heat input:

$$q_{\text{in}} = h_3 - h_2 = 1395.97 - 544.35 = 851.62 \text{ kJ/kg}$$
$$w_{\text{net}} = w_{\text{out}} - w_{\text{in}} = 606.86 - 244.16 = 362.7 \text{ kJ/kg}$$

Thus, $\eta_{\text{th}} = \dfrac{w_{\text{net}}}{q_{\text{in}}} = \dfrac{362.7 \text{ kJ/kg}}{851.62 \text{ kJ/kg}} = \textbf{0.426 or 42.6\%}$

The thermal efficiency could also be determined from

$$\eta_{\text{th}} = 1 - \frac{q_{\text{out}}}{q_{\text{in}}}$$

where $q_{\text{out}} = h_4 - h_1 = 789.11 - 300.19 = 488.92 \text{ kJ/kg}$

Discussion Under the cold-air-standard assumptions (constant specific heats, values at room temperature), the thermal efficiency would be, from Eq. 8-17,

$$\eta_{\text{th, Brayton}} = 1 - \frac{1}{r_p^{(k-1)/k}} = 1 - \frac{1}{8^{(1.4-1)/1.4}} = \mathbf{0.448}$$

which is sufficiently close to the value obtained by accounting for the variation of specific heats with temperature.

Deviation of Actual Gas-Turbine Cycles from Idealized Ones

The actual gas-turbine cycle differs from the ideal Brayton cycle on several accounts. For one thing, some pressure drop during the heat-addition and -rejection processes is inevitable. More importantly, the actual work input to the compressor will be more, and the actual work output from the turbine will be less because of irreversibilities. The deviation of actual compressor and turbine behavior from the idealized isentropic behavior can be accurately accounted for by utilizing the isentropic efficiencies of the turbine and compressor, defined as

$$\eta_C = \frac{w_s}{w_a} \cong \frac{h_{2s} - h_1}{h_{2a} - h_1} \tag{8-19}$$

and

$$\eta_T = \frac{w_a}{w_s} \cong \frac{h_3 - h_{4a}}{h_3 - h_{4s}} \tag{8-20}$$

where states 2a and 4a are the actual exit states of the compressor and the turbine, respectively, and 2s and 4s are the corresponding states for the isentropic case, as illustrated in Fig. 8-32. The effect of the turbine and compressor efficiencies on the thermal efficiency of the gas-turbine engines is illustrated below with an example.

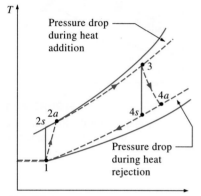

FIGURE 8-32

The deviation of an actual gas-turbine cycle from the ideal Brayton cycle as a result of irreversibilities.

EXAMPLE 8-5 An Actual Gas-Turbine Cycle

Assuming a compressor efficiency of 80 percent and a turbine efficiency of 85 percent, determine (*a*) the back work ratio, (*b*) the thermal efficiency, and (*c*) the turbine exit temperature of the gas-turbine power plant discussed in Example 8-4.

Solution (*a*) The *T-s* diagram of the cycle is shown in Fig. 8-33. The actual compressor work and turbine work are determined by using the definitions of compressor and turbine efficiencies, Eqs. 8-19 and 8-20:

Compressor: $w_{\text{comp, in}} = \dfrac{w_s}{\eta_C} = \dfrac{244.16 \text{ kJ/kg}}{0.80} = 305.20 \text{ kJ/kg}$

Turbine: $w_{\text{turb, out}} = \eta_T w_s = (0.85)(606.86 \text{ kJ/kg}) = 515.83 \text{ kJ/kg}$

Thus, $r_{\text{bw}} = \dfrac{w_{\text{comp, in}}}{w_{\text{turb, out}}} = \dfrac{305.20 \text{ kJ/kg}}{515.83 \text{ kJ/kg}} = \mathbf{0.592}$

That is, the compressor is now consuming 59.2 percent of the work produced by the turbine (up from 40.2 percent). This increase is due to the irreversibilities that occur within the compressor and the turbine.

(*b*) In this case, air will leave the compressor at a higher temperature and enthalpy, which are determined to be

$$w_{\text{comp, in}} = h_{2a} - h_1 \longrightarrow h_{2a} = h_1 + w_{\text{comp, in}}$$
$$= 300.19 + 305.20$$
$$= 605.39 \text{ kJ/kg} \quad (\text{and } T_{2a} = 598 \text{ K})$$

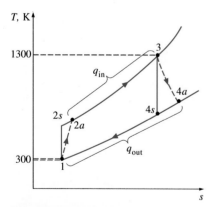

FIGURE 8-33

T-s diagram of the gas-turbine cycle discussed in Example 8-5.

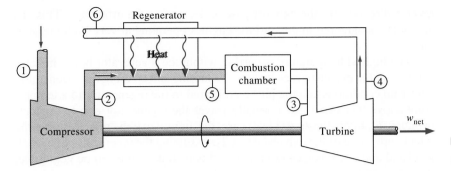

FIGURE 8-34

A gas-turbine engine with regenerator.

Thus, $q_{in} = h_3 - h_{2a} = 1395.97 - 605.39 = 790.58$ kJ/kg

$w_{net} = w_{out} - w_{in} = 515.83 - 305.20 = 210.63$ kJ/kg

and $\eta_{th} = \dfrac{w_{net}}{q_{in}} = \dfrac{210.63 \text{ kJ/kg}}{790.58 \text{ kJ/kg}} = \mathbf{0.266 \text{ or } 26.6\%}$

That is, the irreversibilities occurring within the turbine and compressor caused the thermal efficiency of the gas turbine cycle to drop from 42.6 to 26.6 percent. This example shows how sensitive the performance of a gas-turbine power plant is to the efficiencies of the compressor and the turbine. In fact, gas-turbine thermal efficiencies did not reach competitive values until significant improvements were made in the design of gas turbines and compressors.

(*c*) The air temperature at the turbine exit is determined from an energy balance on the turbine:

$w_{turb} = h_3 - h_{4a} \longrightarrow h_{4a} = h_3 - w_{turb}$

$= 1395.97 - 515.83$

$= 880.14$ kJ/kg

Then, from Table A-17,

$T_{4a} = \mathbf{853 \text{ K}}$

This value is considerably higher than the air temperature at the compressor exit ($T_{2a} = 598$ K), which suggests the use of regeneration to reduce the heat input requirements.

8-8 ■ THE BRAYTON CYCLE WITH REGENERATION

In gas-turbine engines, the temperature of the exhaust gas leaving the turbine is often considerably higher than the temperature of the air leaving the compressor. Therefore, the high-pressure air leaving the compressor can be heated by transferring heat to it from the hot exhaust gases in a counter-flow heat exchanger, which is also known as a *regenerator* or a *recuperator*. A sketch of the gas-turbine engine utilizing a regenerator and the *T-s* diagram of the new cycle are shown in Figs. 8-34 and 8-35, respectively.

The thermal efficiency of the Brayton cycle increases as a result of regeneration since the portion of energy of the exhaust gases that is normally rejected to the surroundings is now used to preheat the air entering the combustion chamber. This, in turn, decreases the heat input (thus fuel) requirements for the same net work output. Note, however, that the use of a regenerator is recommended only when the turbine exhaust temperature is higher than the compressor exit temperature. Otherwise, heat will flow in the

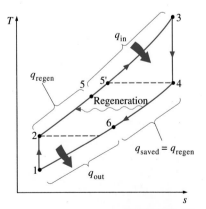

FIGURE 8-35

T-s diagram of a Brayton cycle with regeneration.

reverse direction (*to* the exhaust gases), decreasing the efficiency. This situation is encountered in gas-turbine engines operating at very high pressure ratios.

The highest temperature occurring within the regenerator is T_4, the temperature of the exhaust gases leaving the turbine and entering the regenerator. Under no conditions can the air be preheated in the regenerator to a temperature above this value. Air normally leaves the regenerator at a lower temperature, T_5. In the limiting (ideal) case, the air will exit the regenerator at the inlet temperature of the exhaust gases T_4. Assuming the regenerator to be well insulated and any changes in kinetic and potential energies to be negligible, the actual and maximum heat transfers from the exhaust gases to the air can be expressed as

$$q_{regen, act} = h_5 - h_2 \qquad (8-21)$$

and

$$q_{regen, max} = h_{5'} - h_2 = h_4 - h_2 \qquad (8-22)$$

The extent to which a regenerator approaches an ideal regenerator is called the **effectiveness** ε and is defined as

$$\varepsilon = \frac{q_{regen, act}}{q_{regen, max}} = \frac{h_5 - h_2}{h_4 - h_2} \qquad (8-23)$$

When the cold-air-standard assumptions are utilized, it reduces to

$$\varepsilon \cong \frac{T_5 - T_2}{T_4 - T_2} \qquad (8-24)$$

A regenerator with a higher effectiveness will obviously save a greater amount of fuel since it will preheat the air to a higher temperature prior to combustion. However, achieving a higher effectiveness requires the use of a larger regenerator, which carries a higher price tag and causes a larger pressure drop. Therefore, the use of a regenerator with a very high effectiveness cannot be justified economically unless the savings from the fuel costs exceed the additional expenses involved. The effectiveness of most regenerators used in practice is below 0.85.

Under the cold-air-standard assumptions, the thermal efficiency of an ideal Brayton cycle with regeneration is

$$\eta_{th, regen} = 1 - \left(\frac{T_1}{T_3}\right)(r_p)^{(k-1)/k} \qquad (8-25)$$

Therefore, the thermal efficiency of an ideal Brayton cycle with regeneration depends on the ratio of the minimum to maximum temperatures as well as the pressure ratio. The thermal efficiency is plotted in Fig. 8-36 for various pressure ratios and minimum-to-maximum temperature ratios. This figure shows that regeneration is most effective at lower pressure ratios and low minimum-to-maximum temperature ratios.

EXAMPLE 8-6 Actual Gas-Turbine Cycle with Regeneration
Determine the thermal efficiency of the gas-turbine power plant described in Example 8-5 if a regenerator having an effectiveness of 80 percent is installed.

Solution The *T-s* diagram of the cycle is shown in Fig. 8-37. We first determine the enthalpy of the air at the exit of the regenerator, using the definition of effectiveness:

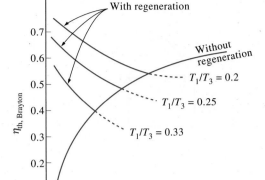

FIGURE 8-36

Thermal efficiency of the ideal Brayton cycle with and without regeneration.

$$\varepsilon = \frac{h_5 - h_{2a}}{h_{4a} - h_{2a}}$$

$$0.80 = \frac{(h_5 - 605.39)\ \text{kJ/kg}}{(880.14 - 605.39)\ \text{kJ/kg}} \longrightarrow h_5 = 825.19\ \text{kJ/kg}$$

Thus, $q_{in} = h_3 - h_5 = (1395.97 - 825.19)\ \text{kJ/kg} = 570.78\ \text{kJ/kg}$

This represents a savings of 219.8 kJ/kg from the heat input requirements. The addition of a regenerator (assumed to be frictionless) does not affect the net work output of the plant. Thus,

$$\eta_{th} = \frac{w_{net}}{q_{in}} = \frac{210.63\ \text{kJ/kg}}{570.78\ \text{kJ/kg}} = \textbf{0.369 or 36.9\%}$$

Discussion   Note that the thermal efficiency of the power plant has gone up from 26.6 to 36.9 percent as a result of installing a regenerator that helps to recuperate some of the excess energy of the exhaust gases.

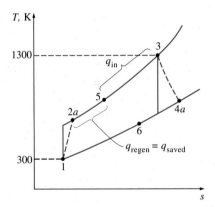

FIGURE 8-37

T-s diagram of the regenerative Brayton cycle described in Example 8-6.

8-9 ■ THE CARNOT VAPOR CYCLE

We have mentioned many times that the Carnot cycle is the most efficient cycle operating between two specified temperature levels. Thus it is natural to look at the Carnot cycle first as a prospective ideal cycle for vapor power plants. If we could, we would certainly adopt it as the ideal cycle. But as explained below, the Carnot cycle is not a suitable model for power cycles. Throughout the discussions, we assume *steam* to be the working fluid since it is the working fluid predominantly used in vapor power cycles.

Consider a steady-flow *Carnot cycle* executed within the saturation dome of a pure substance, as shown in Fig. 8-38*a*. The fluid is heated reversibly and isothermally in a boiler (process 1-2), expanded isentropically in a turbine (process 2-3), condensed reversibly and isothermally in a condenser (process 3-4), and compressed isentropically by a compressor to the initial state (process 4-1).

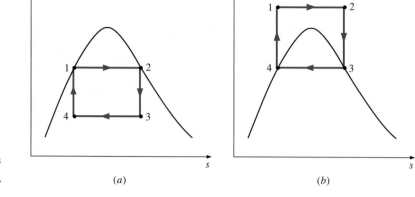

FIGURE 8-38

T-s diagram of two Carnot vapor cycles.

Several impracticalities are associated with this cycle:

1. Isothermal heat transfer to or from a two-phase system is not difficult to achieve in practice since maintaining a constant pressure in the device will automatically fix the temperature at the saturation value. Therefore, processes 1-2 and 3-4 can be approached closely in actual boilers and condensers. Limiting the heat transfer processes to two-phase systems, however, severely limits the maximum temperature that can be used in the cycle (it has to remain under the critical-point value, which is 374°C for water). Limiting the maximum temperature in the cycle also limits the thermal efficiency. Any attempt to raise the maximum temperature in the cycle will involve heat transfer to the working fluid in a single phase, which is not easy to accomplish isothermally.

2. The isentropic expansion process (process 2-3) can be approximated closely by a well-designed turbine. However, the quality of the steam decreases during this process, as shown on the *T-s* diagram in Fig. 8-38*a*. Thus the turbine will have to handle steam with low quality, that is, steam with a high moisture content. The impingement of liquid droplets on the turbine blades causes erosion and is a major source of wear. Thus steam with qualities less than about 90 percent cannot be tolerated in the operation of power plants. This problem could be eliminated by using a working fluid with a very steep saturated vapor line.

3. The isentropic compression process (process 4-1) involves the compression of a liquid–vapor mixture to a saturated liquid. There are two difficulties associated with this process. First, it is not easy to control the condensation process so precisely as to end up with the desired quality at state 4. Second, it is not practical to design a compressor that will handle two phases.

Some of these problems could be eliminated by executing the Carnot cycle in a different way, as shown in Fig. 8-38*b*. This cycle, however, presents other problems such as isentropic compression to extremely high pressures and isothermal heat transfer at variable pressures. Thus we conclude that the Carnot cycle cannot be approximated in actual devices and is not a realistic model for vapor power cycles.

8-10 ■ RANKINE CYCLE:
The Ideal Cycle for Vapor Power Cycles

Many of the impracticalities associated with the Carnot cycle can be eliminated by superheating the steam in the boiler and condensing it completely in

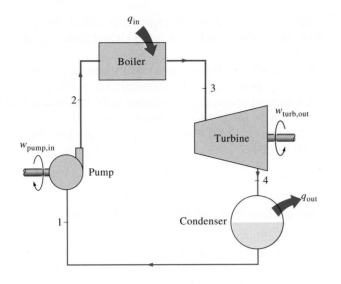

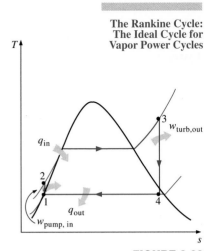

FIGURE 8-39

The simple ideal Rankine cycle.

the condenser, as shown schematically on a *T-s* diagram in Fig. 8-39. The cycle that results is the **Rankine cycle,** which is the ideal cycle for vapor power plants. The ideal Rankine cycle does not involve any internal irreversibilities and consists of the following four processes:

1-2 Isentropic compression in a pump

2-3 Constant pressure heat addition in a boiler

3-4 Isentropic expansion in a turbine

4-1 Constant pressure heat rejection in a condenser

Water enters the *pump* at state 1 as saturated liquid and is compressed isentropically to the operating pressure of the boiler. The water temperature increases somewhat during this isentropic compression process due to a slight decrease in the specific volume of the water. The vertical distance between states 1 and 2 on the *T-s* diagram is greatly exaggerated for clarity. (If water were truly incompressible, would there be a temperature change at all during this process?)

Water enters the *boiler* as a compressed liquid at state 2 and leaves as a superheated vapor at state 3. The boiler is basically a large heat exchanger where the heat originating from combustion gases, nuclear reactors, or other sources is transferred to the water essentially at constant pressure. The boiler, together with the section where the steam is superheated (the superheater), is often called the *steam generator.*

The superheated vapor at state 3 enters the *turbine,* where it expands isentropically and produces work by rotating the shaft connected to an electric generator. The pressure and the temperature of the steam drop during this process to the values at state 4, where steam enters the *condenser.* At this state, steam is usually a saturated liquid–vapor mixture with a high quality. Steam is condensed at constant pressure in the condenser, which is basically a large heat exchanger, by rejecting heat to a cooling medium such as a lake, a river, or the atmosphere. Steam leaves the condenser as saturated liquid and enters the pump, completing the cycle. In areas where water is precious, the power plants are cooled by air instead of water. This method of cooling, which is also used in car engines, is called *dry cooling*. Several power plants in the world and a few in the United States use dry cooling to conserve water.

Remembering that the area under the process curve on a *T-s* diagram represents the heat transfer for internally reversible processes, we see that the area under process curve 2-3 represents the heat transferred to the water in the boiler and the area under the process curve 4-1 represents the heat rejected in the condenser. The difference between these two (the area enclosed by the cycle) is the net work produced during the cycle.

Energy Analysis of the Ideal Rankine Cycle

All four components associated with the Rankine cycle (the pump, boiler, turbine, and condenser) are steady-flow devices, and thus all four processes that make up the Rankine cycle can be analyzed as steady-flow processes. The kinetic and potential energy changes of the steam are usually small relative to the work and heat transfer terms and are therefore usually neglected. Then the *steady-flow energy equation* per unit mass of steam reduces to

$$(q_{in} - q_{out}) + (w_{in} - w_{out}) = h_e - h_i \qquad \text{(kJ/kg)} \qquad (8\text{-}26)$$

The boiler and the condenser do not involve any work, and the pump and the turbine are assumed to be isentropic. Then the conservation of energy relation for each device can be expressed as follows:

Pump $(q = 0)$: $\qquad\qquad w_{pump,\, in} = h_2 - h_1 \qquad\qquad (8\text{-}27)$

or, $\qquad\qquad\qquad\qquad w_{pump,\, in} = v(P_2 - P_1) \qquad\qquad (8\text{-}28)$

where $\qquad h_1 = h_{f\,@\,P_1} \qquad$ and $\qquad v \cong v_1 = v_{f\,@\,P_1} \qquad (8\text{-}29)$

Boiler $(w = 0)$: $\qquad\qquad q_{in} = h_3 - h_2 \qquad\qquad (8\text{-}30)$

Turbine $(q = 0)$: $\qquad\qquad w_{turb,\, out} = h_3 - h_4 \qquad\qquad (8\text{-}31)$

Condenser $(w = 0)$: $\qquad\qquad q_{out} = h_4 - h_1 \qquad\qquad (8\text{-}32)$

The *thermal efficiency* of the Rankine cycle is determined from

$$\eta_{th} = \frac{w_{net}}{q_{in}} = 1 - \frac{q_{out}}{q_{in}} \qquad (8\text{-}33)$$

where $\qquad\qquad w_{net} = q_{in} - q_{out} = w_{turb,\, out} - w_{pump,\, in}$

The conversion efficiency of power plants in the United States is often expressed in terms of **heat rate,** which is the amount of heat supplied, in Btu, to generate 1 kWh of electricity. The smaller the heat rate, the greater the efficiency. Considering that 1 kWh = 3412 Btu, the relation between the heat rate and the thermal efficiency can be expressed as

$$\eta_{th} = \frac{3412 \text{ (Btu/kWh)}}{\text{Heat rate (Btu/kWh)}} \qquad (8\text{-}34)$$

For example, a heat rate of 11,363 Btu/kWh is equivalent to 30 percent thermal efficiency.

The thermal efficiency can also be interpreted as the ratio of the area enclosed by the cycle on a *T-s* diagram to the area under the heat-addition process. The use of these relations is illustrated in the following example.

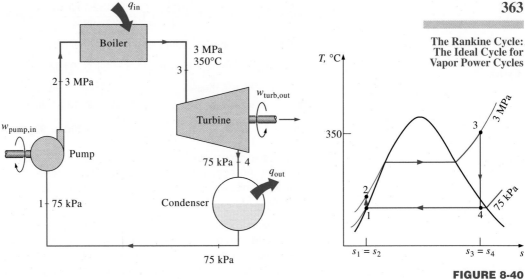

FIGURE 8-40

Schematic and *T-s* diagram for Example 8-7.

EXAMPLE 8-7   The Simple Ideal Rankine Cycle

Consider a steam power plant operating on the simple ideal Rankine cycle. The steam enters the turbine at 3 MPa and 350°C and is condensed in the condenser at a pressure of 75 kPa. Determine the thermal efficiency of this cycle.

Solution   The schematic of the power plant and the *T-s* diagram of the cycle are shown in Fig. 8-40. We note that the power plant involves steady-flow components and operates on the ideal Rankine cycle. Therefore, the pump and the turbine are isentropic, there are no pressure drops in the boiler and condenser, and steam leaves the condenser and enters the pump as saturated liquid at the condenser pressure.

Assumptions   **1** Steady operating conditions exist. **2** Kinetic and potential energy changes are negligible.

Analysis   First we determine the enthalpies at various points in the cycle, using data from steam tables (Tables A-4, A-5, and A-6):

State 1: $P_1 = 75$ kPa $\rbrace$ $h_1 = h_{f@\ 75\ kPa} = 384.39$ kJ/kg
$\quad\quad\quad$ Sat. liquid $\rbrace$ $v_1 = v_{f@\ 75\ kPa} = 0.001037$ m³/kg

State 2: $P_2 = 3$ MPa
$\quad\quad\quad$ $s_2 = s_1$

$$w_{pump,\ in} = v_1(P_2 - P_1) = (0.001037\ \text{m}^3/\text{kg})[(3000 - 75)\ \text{kPa}]\left(\frac{1\ \text{kJ}}{1\ \text{kPa} \cdot \text{m}^3}\right)$$
$$= 3.03\ \text{kJ/kg}$$

$$h_2 = h_1 + w_{pump,\ in} = (384.39 + 3.03)\ \text{kJ/kg} = 387.42\ \text{kJ/kg}$$

State 3: $P_3 = 3$ Mpa $\rbrace$ $h_3 = 3115.3$ kJ/kg
$\quad\quad\quad$ $T_3 = 350°C\rbrace$ $s_3 = 6.7428$ kJ/kg $\cdot$ K

State 4: $P_4 = 75$ kPa
$\quad\quad\quad$ $s_4 = s_3$ $\quad$ (sat. mixture)

$$x_4 = \frac{s_4 - s_f}{s_{fg}} = \frac{6.7428 - 1.213}{6.2434} = 0.886$$

$$h_4 = h_f + x_4 h_{fg} = 384.39 + 0.886(2278.6) = 2403.2\ \text{kJ/kg}$$

Thus, $q_{in} = h_3 - h_2 = (3115.3 - 387.42)\ \text{kJ/kg} = 2727.88\ \text{kJ/kg}$

$\quad\quad\quad$ $q_{out} = h_4 - h_1 = (2403.2 - 384.39)\ \text{kJ/kg} = 2018.81\ \text{kJ/kg}$

and $\qquad \eta_{th} = 1 - \dfrac{q_{out}}{q_{in}} = 1 - \dfrac{2018.81 \text{ kJ/kg}}{2727.88 \text{ kJ/kg}} = \textbf{0.260 or 26.0\%}$

The thermal efficiency could also be determined from

$$w_{turb, out} = h_3 - h_4 = (3115.3 - 2403.2) \text{ kJ/kg} = 712.1 \text{ kJ/kg}$$

$$w_{net} = w_{turb, out} - w_{pump, in} = (712.1 - 3.03) \text{ kJ/kg} = 709.07 \text{ kJ/kg}$$

or $\qquad w_{net} = q_{in} - q_{out} = (2727.88 - 2018.81) \text{ kJ/kg} = 709.07 \text{ kJ/kg}$

and $\qquad \eta_{th} = \dfrac{w_{net}}{q_{in}} = \dfrac{709.07 \text{ kJ/kg}}{2727.88 \text{ kJ/kg}} = \textbf{0.260 or 26.0\%}$

That is, this power plant converts 26 percent of the heat it receives in the boiler to net work. An actual power plant operating between the same temperature and pressure limits will have a lower efficiency because of the irreversibilities such as friction.

Discussion Notice that the back work ratio ($r_{pw} = w_{in}/w_{out}$) of this power plant is 0.004, and thus only 0.4 percent of the turbine work output is required to operate the pump. Having such low back work ratios is characteristic of vapor power cycles. This is in contrast to the gas power cycles, which typically have very high back work ratios (about 40 to 80 percent).

It is also interesting to note the thermal efficiency of a Carnot cycle operating between the same temperature limits

$$\eta_{th, Carnot} = 1 - \frac{T_{min}}{T_{max}} = 1 - \frac{(91.78 + 273) \text{ K}}{(350 + 273) \text{ K}} = 0.414$$

The difference between the two efficiencies is due to the large temperature difference between the steam and the combustion gases during the heat-addition process.

8-11 ■ DEVIATION OF ACTUAL VAPOR POWER CYCLES FROM IDEALIZED ONES

The actual vapor power cycle differs from the ideal Rankine cycle, as illustrated in Fig. 8-41a, as a result of irreversibilities in various components. Fluid friction and undesired heat loss to the surroundings are the two most common sources of irreversibilities.

Fluid friction causes pressure drops in the boiler, the condenser, and the piping between various components. As a result, steam leaves the boiler at a somewhat lower pressure. Also, the pressure at the turbine inlet is somewhat

FIGURE 8-41

(*a*) Deviation of actual vapor power cycle from the ideal Rankine cycle.
(*b*) The effect of pump and turbine irreversibilities on the ideal Rankine cycle.

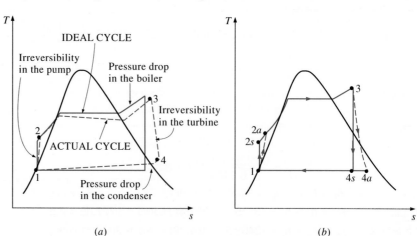

(*a*)   (*b*)

lower than that at the boiler exit due to the pressure drop in the connecting pipes. The pressure drop in the condenser is usually very small. To compensate for these pressure drops, the water must be pumped to a sufficiently higher pressure than the ideal cycle calls for. This requires a larger pump and larger work input to the pump.

The other major source of irreversibility is the *heat loss* from the steam to the surroundings as the steam flows through various components. To maintain the same level of net work output, more heat needs to be transferred to the steam in the boiler to compensate for these undesired heat losses. As a result, cycle efficiency decreases.

Of particular importance are the irreversibilities occurring within the pump and the turbine. A pump requires a greater work input, and a turbine produces a smaller work output as a result of irreversibilities. Under ideal conditions, the flow through these devices is isentropic. The deviation of actual pumps and turbines from the isentropic ones can be accurately accounted for, however, by utilizing *isentropic efficiencies,* defined as

$$\eta_P = \frac{w_s}{w_a} = \frac{h_{2s} - h_1}{h_{2a} - h_1} \tag{8-35}$$

and

$$\eta_T = \frac{w_a}{w_s} = \frac{h_3 - h_{4a}}{h_3 - h_{4s}} \tag{8-36}$$

where states $2a$ and $4a$ are the actual exit states of the pump and the turbine, respectively, and $2s$ and $4s$ are the corresponding states for the isentropic case (Fig. 8-41b).

Other factors also need to be considered in the analysis of actual vapor power cycles. In actual condensers, for example, the liquid is usually subcooled to prevent the onset of *cavitation,* the rapid vaporization and condensation of the fluid at the low-pressure side of the pump impeller, which may damage it. Additional losses occur at the bearings between the moving parts as a result of friction. Steam that leaks out during the cycle and air that leaks into the condenser represent two other sources of loss. Finally, the power consumed by the auxiliary equipment such as fans that supply air to the furnace should also be considered in evaluating the performance of actual power plants.

The effect of irreversibilities on the thermal efficiency of a steam power cycle is illustrated below with an example.

EXAMPLE 8-8   An Actual Steam Power Cycle

A steam power plant operates on the cycle shown in Fig. 8-42 If the isentropic efficiency of the turbine is 87 percent and the isentropic efficiency of the pump is 85 percent, determine (*a*) the thermal efficiency of the cycle and (*b*) the net power output of the plant for a mass flow rate of 15 kg/s.

Solution   The schematic of the power plant and the *T-s* diagram of the cycle are shown in Fig. 8-42. The temperatures and pressures of steam at various points are also indicated on the figure. We note that the power plant involves steady-flow components and operates on the Rankine cycle, but the imperfections at various components are accounted for.

Assumptions **1** Steady operating conditions exist. **2** Kinetic and potential energy changes are negligible.

Analysis (*a*) The thermal efficiency of a cycle is the ratio of the net work output to the heat input, and it is determined as follows:

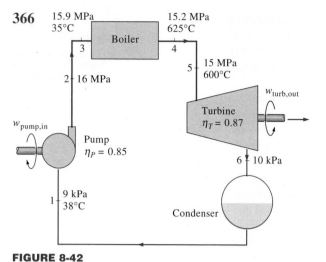

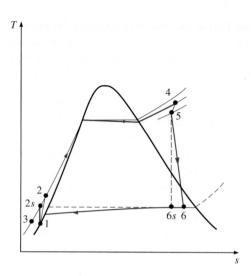

FIGURE 8-42

Schematic and *T-s* diagram for Example 8-8.

Pump work input: $w_{pump, in} = \dfrac{w_{s, pump, in}}{\eta_P} = \dfrac{v_1(P_2 - P_1)}{\eta_P}$

$$= \frac{(0.001009 \text{ m}^3/\text{kg})[(16{,}000 - 9) \text{ kPa}]}{0.85} \left(\frac{1 \text{ kJ}}{1 \text{ kPa} \cdot \text{m}^3}\right)$$

$$= 19.0 \text{ kJ/kg}$$

Turbine work output: $w_{turb, out} = \eta_T w_{s, turb, out}$

$$= \eta_T(h_5 - h_{6s}) = 0.87(3582.3 - 2114.9) \text{ kJ/kg}$$

$$= 1276.6 \text{ kJ/kg}$$

Boiler heat input: $q_{in} = h_4 - h_3 = (3647.3 - 160.1) \text{ kJ/kg} = 3487.2 \text{ kJ/kg}$

Thus, $w_{net} = w_{turb, out} - w_{pump, in} = (1276.6 - 19.0) \text{ kJ/kg} = 1257.6 \text{ kJ/kg}$

$$\eta_{th} = \frac{w_{net}}{q_{in}} = \frac{1257.6 \text{ kJ/kg}}{3487.2 \text{ kJ/kg}} = \textbf{0.361 or 36.1\%}$$

Without the irreversibilities, the thermal efficiency of this cycle would be 43.0 percent (see Example 8-9c).

(b) The power produced by this power plant is determined from

$$\dot{W}_{net} = \dot{m}(w_{net}) = (15 \text{ kg/s})(1257.6 \text{ kJ/kg}) = \textbf{18,864 kW}$$

8-12 ■ HOW CAN WE INCREASE THE EFFICIENCY OF THE RANKINE CYCLE?

Steam power plants are responsible for the production of most electric power in the world, and even small increases in thermal efficiency can mean large savings from the fuel requirements. Therefore, every effort is made to improve the efficiency of the cycle on which steam power plants operate.

The basic idea behind all the modifications to increase the thermal efficiency of a power cycle is the same: *Increase the average temperature at which heat is transferred to the working fluid in the boiler, or decrease the average temperature at which heat is rejected from the working fluid in the condenser.* That is, the average fluid temperature should be as high as possible during heat addition and as low as possible during heat rejection. Next we discuss three ways of accomplishing this for the simple ideal Rankine cycle.

1 Lowering the Condenser Pressure

(Lowers $T_{\text{low, av}}$)

Steam exists as a saturated mixture in the condenser at the saturation temperature corresponding to the pressure inside the condenser. Therefore, lowering the operating pressure of the condenser automatically lowers the temperature of the steam, and thus the temperature at which heat is rejected.

The effect of lowering the condenser pressure on the Rankine cycle efficiency is illustrated on a *T-s* diagram in Fig. 8-43. For comparison purposes, the turbine inlet state is maintained the same. The colored area on this diagram represents the increase in net work output as a result of lowering the condenser pressure from P_4 to P_4'. The heat input requirements also increase (represented by the area under curve 2'-2), but this increase is very small. Thus the overall effect of lowering the condenser pressure is an increase in the thermal efficiency of the cycle.

To take advantage of the increased efficiencies at low pressures, the condensers of steam power plants usually operate well below the atmospheric pressure. This does not present a major problem since the vapor power cycles operate in a closed loop. However, there is a lower limit on the condenser pressure that can be used. It cannot be lower than the saturation pressure corresponding to the temperature of the cooling medium. Consider, for example, a condenser that is to be cooled by a nearby river at 15°C. Allowing a temperature difference of 10°C for effective heat transfer, the steam temperature in the condenser must be above 25°C; thus the condenser pressure must be above 3.2 kPa, which is the saturation pressure at 25°C.

Lowering the condenser pressure is not without any side effects, however. For one thing, it creates the possibility of air leakage into the condenser. More importantly, it increases the moisture content of the steam at the final stages of the turbine, as can be seen from Fig. 8-43. The presence of large quantities of moisture is highly undesirable in turbines because it decreases the turbine efficiency and erodes the turbine blades. Fortunately, this problem can be corrected, as discussed later in this chapter.

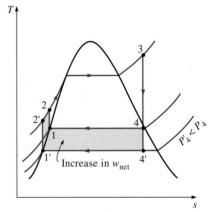

FIGURE 8-43

The effect of lowering the condenser pressure on the ideal Rankine cycle.

2 Superheating the Steam to High Temperatures

(Increases $T_{\text{high, av}}$)

The average temperature at which heat is added to the steam can be increased without increasing the boiler pressure by superheating the steam to high temperatures. The effect of superheating on the performance of vapor power cycles is illustrated on a *T-s* diagram in Fig. 8-44. The colored area on this diagram represents the increase in the net work. The total area under the process curve 3-3' represents the increase in the heat input. Thus both the net work and heat input increase as a result of superheating the steam to a higher temperature. The overall effect is an increase in thermal efficiency, however, since the average temperature at which heat is added increases.

Superheating the steam to higher temperatures has another very desirable effect: It decreases the moisture content of the steam at the turbine exit, as can be seen from the *T-s* diagram (the quality at state 4' is higher than that at state 4).

The temperature to which steam can be superheated is limited, however, by metallurgical considerations. Presently the highest steam temperature allowed at the turbine inlet is about 620°C (1150°F). Any increase in this value

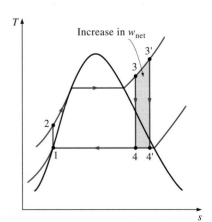

FIGURE 8-44

The effect of superheating the steam to higher temperatures on the ideal Rankine cycle.

depends on improving the present materials or finding new ones that can with-stand higher temperatures. Ceramics are very promising in this regard.

3 Increasing the Boiler Pressure (*Increases* $T_{high, av}$)

Another way of increasing the average temperature during the heat-addition process is to increase the operating pressure of the boiler, which automatically raises the temperature at which boiling takes place. This, in turn, raises the average temperature at which heat is added to the steam and thus raises the thermal efficiency of the cycle.

The effect of increasing the boiler pressure on the performance of vapor power cycles is illustrated on a *T-s* diagram in Fig. 8-45. Notice that for a fixed turbine inlet temperature, the cycle shifts to the left and the moisture content of steam at the turbine exit increases. This undesirable side effect can be corrected by reheating the steam, as discussed in the next section.

Operating pressures of boilers have gradually increased over the years from about 2.7 MPa (400 psia) in 1922 to over 30 MPa (4500 psia) today, gen-erating enough steam to produce a net power output of 1000 MW or more in a large power plant. Today many modern steam power plants operate at su-percritical pressures ($P > 22.09$ MPa) and have thermal efficiencies of about 40 percent for fossil-fuel plants and 34 percent for nuclear plants. There are about 170 supercritical-pressure steam power plants in operation in the United States. The lower efficiencies of nuclear power plants are due to the lower maximum temperatures used in those plants for safety reasons. The United States has 112 nuclear power plants, which generate about 21 percent of the nation's electricity. (In contrast, 75 percent of the electricity in France comes from nuclear plants.) The *T-s* diagram of a supercritical Rankine cycle is shown in Fig. 8-46.

The effects of lowering the condenser pressure, superheating to a higher temperature, and increasing the boiler pressure on the thermal efficiency of the Rankine cycle are illustrated below with an example.

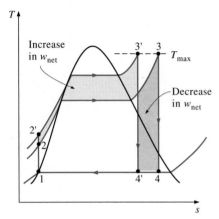

FIGURE 8-45

The effect of increasing the boiler pressure on the ideal Rankine cycle.

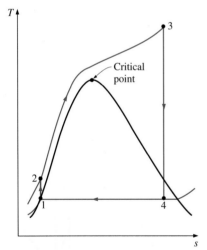

FIGURE 8-46

A supercritical Rankine cycle.

EXAMPLE 8-9 Effect of Boiler Pressure and Temperature on Efficiency

Consider a steam power plant operating on the ideal Rankine cycle. The steam enters the turbine at 3 MPa and 350°C and is condensed in the condenser at a pressure of 10 kPa. Determine (*a*) the thermal efficiency of this power plant, (*b*) the thermal efficiency if steam is superheated to 600°C instead of 350°C, and (*c*) the thermal efficiency if the boiler pressure is raised to 15 MPa while the tur-bine inlet temperature is maintained at 600°C.

Solution   The *T-s* diagrams of the cycle for all three cases are given in Fig. 8-47.

Analysis   (*a*) This is the steam power plant discussed in Example 8-7, except that the condenser pressure is lowered to 10 kPa. The thermal efficiency is deter-mined in a similar manner:

State 1: $\left. \begin{array}{l} P_1 = 10 \text{ kPa} \\ \text{Sat. liquid} \end{array} \right\}$ $\begin{array}{l} h_1 = h_{f @ 10kPa} = 191.83 \text{ kJ/kg} \\ v_1 = v_{f @ 10 \text{ kPa}} = 0.00101 \text{ m}^3\text{/kg} \end{array}$

State 2: $P_2 = 3 \text{ MPa}$

$\qquad s_2 = s_1$

$$w_{pump, in} = v_1(P_2 - P_1) = (0.00101 \text{ m}^3\text{/kg})[(3000 - 10) \text{ kPa}]\left(\frac{1 \text{ kJ}}{1 \text{ kPa} \cdot \text{m}^3}\right)$$

$$= 3.02 \text{ kJ/kg}$$

$$h_2 = h_1 + w_{pump, in} = (191.83 + 3.02) \text{ kJ/kg} = 194.85 \text{ kJ/kg}$$

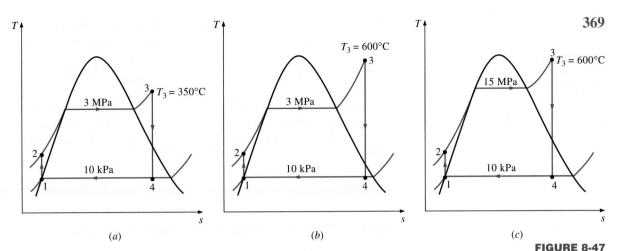

FIGURE 8-47

T-s diagrams of the three cycles discussed in Example 8-9.

State 3: $\quad P_3 = 3$ MPa $\Big\}$ $\quad h_3 = 3115.3$ kJ/kg
$\qquad\qquad T_3 = 350°C \Big\}$ $\quad s_3 = 6.7428$ kJ/kg · K

State 4: $\quad P_4 = 10$ kPa
$\qquad\qquad\qquad\qquad$ (sat. mixture)
$\qquad\quad s_4 = s_3$

$$x_4 = \frac{s_4 - s_f}{s_{fg}} = \frac{6.7428 - 0.6493}{7.5009} = 0.812$$

$\qquad\qquad h_4 = h_f + x_4 h_{fg} = 191.83 + 0.812(2392.8) = 2134.8$ kJ/kg

Thus, $\qquad q_{in} = h_3 - h_2 = (3115.3 - 194.85)$ kJ/kg $= 2920.45$ kJ/kg

$\qquad\qquad q_{out} = h_4 - h_1 = (2134.8 - 191.83)$ kJ/kg $= 1942.97$ kJ/kg

and $\qquad \eta_{th} = 1 - \dfrac{q_{out}}{q_{in}} = 1 - \dfrac{1942.97 \text{ kJ/kg}}{2920.45 \text{ kJ/kg}} = \mathbf{0.335 \text{ or } 33.5\%}$

Therefore, the thermal efficiency increases from 26.0 to 33.5 percent as a result of lowering the condenser pressure from 75 to 10 kPa. At the same time, however, the quality of the steam decreases from 0.886 to 0.812 (in other words, the moisture content increases from 11.4 to 18.8 percent).

(*b*) States 1 and 2 remain the same in this case, and the enthalpies at state 3 (3 MPa and 600°C) and state 4 (10 kPa and $s_4 = s_3$) are determined to be

$\qquad\qquad h_3 = 3682.3$ kJ/kg

$\qquad\qquad h_4 = 2378.8$ kJ/kg $\qquad (x_4 = 0.914)$

Thus, $\qquad q_{in} = h_3 - h_2 = 3682.3 - 194.85 = 3487.45$ kJ/kg

$\qquad\qquad q_{out} = h_4 - h_1 = 2378.8 - 191.83 = 2186.97$ kJ/kg

and $\qquad \eta_{th} = 1 - \dfrac{q_{out}}{q_{in}} = 1 - \dfrac{2186.97 \text{ kJ/kg}}{3487.45 \text{ kJ/kg}} = \mathbf{0.373 \text{ or } 37.3\%}$

Therefore, the thermal efficiency increases from 33.5 to 37.3 percent as a result of superheating the steam from 350 to 600°C. At the same time, the quality of the steam increases from 0.812 to 0.914 (in other words, the moisture content decreases from 18.8 to 8.6 percent).

(*c*) State 1 remains the same in this case, but the other states change. The enthalpies at state 2 (15 MPa and $s_2 = s_1$), state 3 (15 MPa and 600°C), and state 4 (10 kPa and $s_4 = s_3$) are determined in a similar manner to be

$\qquad\qquad h_2 = 206.97$ kJ/kg

$\qquad\qquad h_3 = 3582.3$ kJ/kg

$\qquad\qquad h_4 = 2115.7$ kJ/kg $\qquad (x_4 = 0.804)$

Thus, $\qquad q_{in} = h_3 - h_2 = 3582.3 - 206.97 = 3375.33$ kJ/kg

$\qquad\qquad q_{out} = h_4 - h_1 = 2115.7 - 191.83 = 1923.87$ kJ/kg

and $\qquad \eta_{th} = 1 - \dfrac{q_{out}}{q_{in}} = 1 - \dfrac{1923.87 \text{ kJ/kg}}{3375.33 \text{ kJ/kg}} = \textbf{0.430 or 43.0\%}$

Discussion   The thermal efficiency increases from 37.3 to 43.0 percent as a result of raising the boiler pressure from 3 to 15 MPa while maintaining the turbine inlet temperature at 600°C. At the same time, however, the quality of the steam decreases from 0.914 to 0.804 (in other words, the moisture content increases from 8.6 to 19.6 percent).

8-13 ■ THE IDEAL REHEAT RANKINE CYCLE

We noted in the last section that increasing the boiler pressure increases the thermal efficiency of the Rankine cycle, but it also increases the moisture content of the steam to unacceptable levels. Then it is natural to ask the following question:

> *How can we take advantage of the increased efficiencies at higher boiler pressures without facing the problem of excessive moisture at the final stages of the turbine?*

Two possibilities come to mind:

1.   Superheat the steam to very high temperatures before it enters the turbine. This would be the desirable solution since the average temperature at which heat is added would also increase, thus increasing the cycle efficiency. This is not a viable solution, however, since it will require raising the steam temperature to metallurgically unsafe levels.

2.   Expand the steam in the turbine in two stages, and reheat it in between. In other words, modify the simple ideal Rankine cycle with a **reheat** process. Reheating is a practical solution to the excessive moisture problem in turbines, and it is used frequently in modern steam power plants.

The *T-s* diagram of the ideal reheat Rankine cycle and the schematic of the power plant operating on this cycle are shown in Fig. 8-48. The ideal reheat Rankine cycle differs from the simple ideal Rankine cycle in that the expansion process takes place in two stages. In the first stage (the high-pressure turbine), steam is expanded isentropically to an intermediate pressure and sent back to the boiler where it is reheated at constant pressure, usually to the inlet temperature of the first turbine stage. Steam then expands isentropically in the second stage (low-pressure turbine) to the condenser pressure. Thus the total heat input and the total turbine work output for a reheat cycle become

$$q_{in} = q_{primary} + q_{reheat} = (h_3 - h_2) + (h_5 - h_4) \qquad (8\text{-}37)$$

and

$$w_{turb, out} = w_{turb, I} + w_{turb, II} = (h_3 - h_4) + (h_5 - h_6) \qquad (8\text{-}38)$$

The incorporation of the single reheat in a modern power plant improves the cycle efficiency by 4 to 5 percent by increasing the average temperature at which heat is added to the steam.

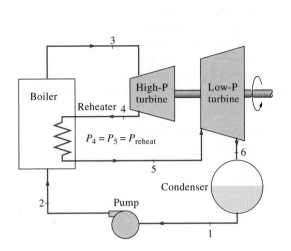

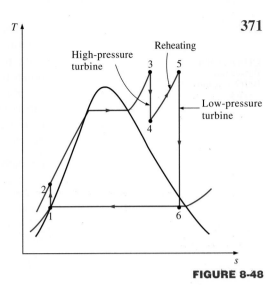

FIGURE 8-48

The ideal reheat Rankine cycle.

The average temperature during the reheat process can be increased by increasing the number of expansion and reheat stages. As the number of stages is increased, the expansion and reheat processes approach an isothermal process at the maximum temperature, as shown in Fig. 8-49. The use of more than two reheat stages, however, is not practical. The theoretical improvement in efficiency from the second reheat is about half of that which results from a single reheat. If the turbine inlet pressure is not high enough, double reheat would result in superheated exhaust. This is undesirable as it would cause the average temperature for heat rejection to increase and thus the cycle efficiency to decrease. Therefore, double reheat is used only on supercritical-pressure $(P > 22.09 \text{ MPa})$ power plants. A third reheat stage would increase the cycle efficiency by about half of the improvement attained by the second reheat. This gain is too small to justify the added cost and complexity.

The reheat cycle was introduced in the mid 1920s, but it was abandoned in the 1930s because of the operational difficulties. The steady increase in boiler pressures over the years made it necessary to reintroduce single reheat in the late 1940s and double reheat in the early 1950s.

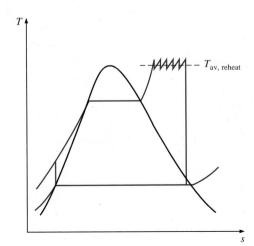

FIGURE 8-49

The average temperature at which heat is added during reheating increases as the number of reheat stages is increased.

The reheat temperatures are very close or equal to the turbine inlet temperature. The optimum reheat pressure is about one-fourth of the maximum cycle pressure. For example, the optimum reheat pressure for a cycle with a boiler pressure of 12 MPa is about 3 MPa.

Remember that the sole purpose of the reheat cycle is to reduce the moisture content of the steam at the final stages of the expansion process. If we had materials that could withstand sufficiently high temperatures, there would be no need for the reheat cycle.

EXAMPLE 8-10 The Ideal Reheat Rankine Cycle

Consider a steam power plant operating on the ideal reheat Rankine cycle. Steam enters the high-pressure turbine at 15 MPa and 600°C and is condensed in the condenser at a pressure of 10 kPa. If the moisture content of the steam at the exit of the low-pressure turbine is not to exceed 10.4 percent, determine (a) the pressure at which the steam should be reheated and (b) the thermal efficiency of the cycle. Assume the steam is reheated to the inlet temperature of the high-pressure turbine.

Solution   The schematic of the power plant and the *T-s* diagram of the cycle are shown in Fig. 8-50. We note that the power plant involves steady-flow components and operates on the ideal reheat Rankine cycle. Therefore, the pump and the turbines are isentropic, there are no pressure drops in the boiler and condenser, and steam leaves the condenser and enters the pump as saturated liquid at the condenser pressure.

Assumptions   **1** Steady operating conditions exist. **2** Kinetic and potential energy changes are negligible.

Analysis   (a) The reheat pressure is determined from the requirement that the entropies at states 5 and 6 be the same:

State 6: $P_6 = 10$ kPa

$x_6 = 0.896$ (sat. mixture)

$s_6 = s_f + x_6 s_{fg} = 0.6493 + 0.896(7.5009) = 7.370$ kJ/kg · K

Also, $h_6 = h_f + x_6 h_{fg} = 191.83 + 0.896(2392.8) = 2335.8$ kJ/kg

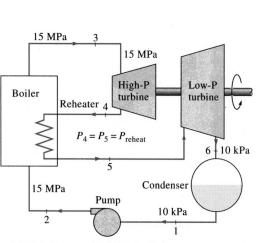

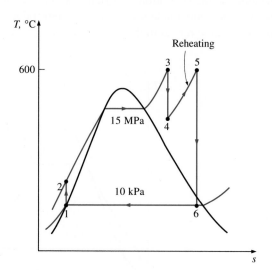

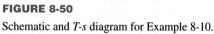

FIGURE 8-50

Schematic and *T-s* diagram for Example 8-10.

Thus,

State 5: $T_5 = 600°C$ $P_5 = \textbf{4.0 MPa}$
            $s_5 = s_6$ $h_5 = 3674.4 \text{ kJ/kg}$

Therefore, steam should be reheated at a pressure of 4 MPa or lower to prevent a moisture content above 10.4 percent.

(b) To determine the thermal efficiency, we need to know the enthalpies at all other states:

State 1: $P_1 = 10 \text{ kPa}$ $h_1 = h_{f@ 10 kPa} = 191.83 \text{ kJ/kg}$
            Sat. liquid $v_1 = v_{f@ 10 kPa} = 0.001010 \text{ m}^3/\text{kg}$

State 2: $P_2 = 15 \text{ MPa}$

            $s_2 = s_1$

$$w_{\text{pump, in}} = v_1(P_2 - P_1) = (0.001010 \text{ m}^3/\text{kg})[(15,000 - 10) \text{ kPa}]\left(\frac{1 \text{ kJ}}{1 \text{ kPa} \cdot \text{m}^3}\right)$$

$$= 15.14 \text{ kJ/kg}$$

$$h_2 = h_1 + w_{\text{pump, in}} = (191.83 + 15.14) \text{ kJ/kg} = 206.97 \text{ kJ/kg}$$

State 3: $P_3 = 15 \text{ MPa}$ $h_3 = 3582.3 \text{ kJ/kg}$
            $T_3 = 600°C$ $s_3 = 6.6776 \text{ kJ/kg} \cdot \text{K}$

State 4: $P_4 = 4 \text{ MPa}$ $h_4 = 3154.3 \text{ kJ/kg}$
            $s_4 = s_3$ $(T_4 = 375.5°C)$

Thus $q_{\text{in}} = (h_3 - h_2) + (h_5 - h_4)$

        $= (3582.3 - 206.97) \text{ kJ/kg} + (3674.4 - 3154.3) \text{ kJ/kg}$

        $= 3895.43 \text{ kJ/kg}$

        $q_{\text{out}} = h_6 - h_1 = (2335.8 - 191.83) \text{ kJ/kg}$

        $= 2143.97 \text{ kJ/kg}$

and $\eta_{\text{th}} = 1 - \dfrac{q_{\text{out}}}{q_{\text{in}}} = 1 - \dfrac{2143.97 \text{ kJ/kg}}{3895.43 \text{ kJ/kg}} = \textbf{0.450 or 45.0\%}$

Discussion   This problem was worked out in Example 8-9c for the same pressure and temperature limits but without the reheat process. A comparison of the two results reveals that reheating reduces the moisture content from 19.6 to 10.4 percent while increasing the thermal efficiency from 43.0 to 45.0 percent.

8-14 ■ REFRIGERATORS AND HEAT PUMPS

We all know from experience that heat flows in the direction of decreasing temperature, that is, from high-temperature regions to low-temperature ones. This heat-transfer process occurs in nature without requiring any devices. The reverse process, however, cannot occur by itself. The transfer of heat from a low-temperature region to a high-temperature one requires special devices called **refrigerators.**

Refrigerators are cyclic devices, and the working fluids used in the refrigeration cycles are called **refrigerants.** A refrigerator is shown schematically in Fig. 8-51a. Here Q_L is the magnitude of the heat removed from the refrigerated space at temperature T_L, Q_H is the magnitude of the heat rejected to the warm space at temperature T_H, and $W_{\text{net, in}}$ is the net work input to the refrigerator. As discussed in Chap. 6, Q_L and Q_H represent magnitudes and thus are positive quantities.

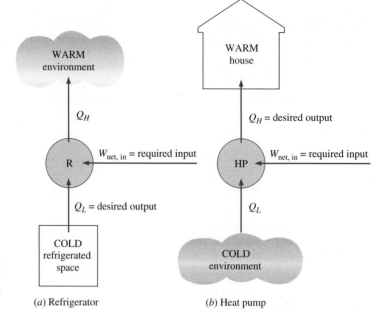

FIGURE 8-51

The objective of a refrigerator is to remove heat (Q_L) from the cold medium; the objective of a heat pump is to supply heat (Q_H) to a warm medium.

Another device that transfers heat from a low-temperature medium to a high-temperature one is the **heat pump.** Refrigerators and heat pumps are essentially the same devices; they differ in their objectives only. The objective of a refrigerator is to maintain the refrigerated space at a low temperature by removing heat from it. Discharging this heat to a higher-temperature medium is merely a necessary part of the operation, not the purpose. The objective of a heat pump, however, is to maintain a heated space at a high temperature. This is accomplished by absorbing heat from a low-temperature source, such as well water or cold outside air in winter, and supplying this heat to a warmer medium such as a house (Fig. 8-51b).

The performance of refrigerators and heat pumps is expressed in terms of the **coefficient of performance** (COP), which was defined as

$$\text{COP}_R = \frac{\text{Desired output}}{\text{Required input}} = \frac{\text{Cooling effect}}{\text{Work input}} = \frac{Q_L}{W_{\text{net, in}}} \quad (8\text{-}39)$$

$$\text{COP}_{HP} = \frac{\text{Desired output}}{\text{Required input}} = \frac{\text{Heating effect}}{\text{Work input}} = \frac{Q_H}{W_{\text{net, in}}} \quad (8\text{-}40)$$

These relations can also be expressed in the rate form by replacing the quantities Q_L, Q_H, and $W_{\text{net, in}}$ by $\dot{Q}_L$, $\dot{Q}_H$, and $\dot{W}_{\text{net, in}}$, respectively. Notice that both COP_R and COP_{HP} can be greater than 1. A comparison of Eqs. 8-39 and 8-40 reveals that

$$\text{COP}_{HP} = \text{COP}_R + 1 \quad (8\text{-}41)$$

for fixed values of Q_L and Q_H. This relation implies that $\text{COP}_{HP} > 1$ since COP_R is a positive quantity. That is, a heat pump will function, at worst, as a resistance heater, supplying as much energy to the house as it consumes. In reality, however, part of Q_H is lost to the outside air through piping and other devices, and COP_{HP} may drop below unity when the outside air temperature is too low. When this happens, the system normally switches to the fuel (natural gas, propane, oil, etc.) or resistance-heating mode.

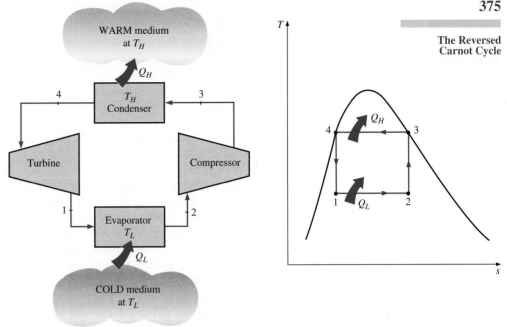

FIGURE 8-52
Schematic of a Carnot refrigerator and *T-s* diagram of the reversed Carnot cycle.

The *cooling capacity* of a refrigeration system—that is, the rate of heat removal from the refrigerated space—is often expressed in terms of **tons of refrigeration.** The capacity of a refrigeration system that can freeze 1 ton (2000 lbm) of liquid water at 0°C (32°F) into ice at 0°C in 24 h is said to be 1 ton. One ton of refrigeration is equivalent to 211 kJ/min or 200 Btu/min. The cooling load of a typical 200-m^2 residence is in the 3-ton (8-kW) range.

8-15 ■ THE REVERSED CARNOT CYCLE

You will recall from the preceding chapters that the Carnot cycle is a totally reversible cycle that consists of two reversible isothermal and two isentropic processes. It has the maximum thermal efficiency for given temperature limits, and it serves as a standard against which actual power cycles can be compared.

Since it is a reversible cycle, all four processes that comprise the Carnot cycle can be reversed. Reversing the cycle will also reverse the directions of any heat and work interactions. The result is a cycle that operates in the counterclockwise direction, which is called the **reversed Carnot cycle.** A refrigerator or heat pump that operates on the reversed Carnot cycle is called a **Carnot refrigerator** or a **Carnot heat pump.**

Consider a reversed Carnot cycle executed within the saturation dome of a refrigerant, as shown in Fig. 8-52. The refrigerant absorbs heat isothermally from a low-temperature source at T_L in the amount of Q_L (process 1-2), is compressed isentropically to state 3 (temperature rises to T_H), rejects heat isothermally to a high-temperature sink at T_H in the amount of Q_H (process 3-4), and expands isentropically to state 1 (temperature drops to T_L). The refrigerant changes from a saturated vapor state to a saturated liquid state in the condenser during process 3-4.

The coefficients of performance of Carnot refrigerators and heat pumps were determined to be

$$\text{COP}_{\text{R, Carnot}} = \frac{1}{T_H/T_L - 1} \tag{8-42}$$

and

$$\text{COP}_{\text{HP, Carnot}} = \frac{1}{1 - T_L/T_H} \tag{8-43}$$

Notice that both COPs increase as the difference between the two temperatures decreases, that is, as T_L rises or T_H falls.

The reversed Carnot cycle is the *most efficient* refrigeration cycle operating between two specific temperature levels. Therefore, it is natural to look at it first as a prospective ideal cycle for refrigerators and heat pumps. If we could, we certainly would adapt it as the ideal cycle. But as explained below, the reversed Carnot cycle is not a suitable model for refrigeration cycles.

The two isothermal heat transfer processes are not difficult to achieve in practice since maintaining a constant pressure automatically fixes the temperature of a two-phase mixture at the saturation value. Therefore, processes 1-2 and 3-4 can be approached closely in actual evaporators and condensers. However, processes 2-3 and 4-1 cannot be approximated closely in practice. This is because process 2-3 involves the compression of a liquid–vapor mixture, which requires a compressor that will handle two phases, and process 4-1 involves the expansion of high-moisture-content refrigerant.

It seems as if these problems could be eliminated by executing the reversed Carnot cycle outside the saturation region. But in this case we will have difficulty in maintaining isothermal conditions during the heat-absorption and heat-rejection processes. Therefore, we conclude that the reversed Carnot cycle cannot be approximated in actual devices and is not a realistic model for refrigeration cycles. However, the reversed Carnot cycle can serve as a standard against which actual refrigeration cycles are compared.

8-16 ■ THE IDEAL VAPOR-COMPRESSION REFRIGERATION CYCLE

Many of the impracticalities associated with the reversed Carnot cycle can be eliminated by vaporizing the refrigerant completely before it is compressed and by replacing the turbine with a throttling device, such as an expansion valve or capillary tube. The cycle that results is called the **ideal vapor-compression refrigeration cycle,** and it is shown schematically and on a *T-s* diagram in Fig. 8-53. The vapor-compression refrigeration cycle is the most widely used cycle for refrigerators, air-conditioning systems, and heat pumps. It consists of four processes:

1-2	Isentropic compression in a compressor
2-3	Constant pressure heat rejection in a condenser
3-4	Throttling in an expansion device
4-1	Constant pressure heat absorption in an evaporator

In an ideal vapor-compression refrigeration cycle, the refrigerant enters the compressor at state 1 as saturated vapor and is compressed isentropically to the condenser pressure. The temperature of the refrigerant increases during

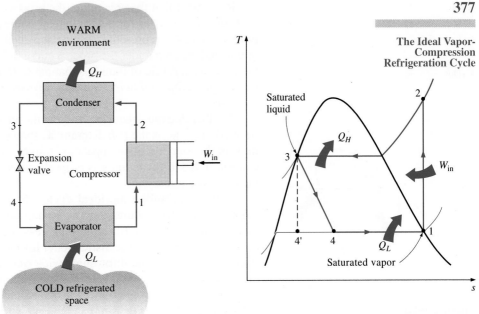

FIGURE 8-53

Schematic and T-s diagram for the ideal vapor-compression refrigeration cycle.

this isentropic compression process to well above the temperature of the surrounding medium. The refrigerant then enters the condenser as superheated vapor at state 2 and leaves as saturated liquid at state 3 as a result of heat rejection to the surroundings. The temperature of the refrigerant at this state is still above the temperature of the surroundings.

The saturated liquid refrigerant at state 3 is throttled to the evaporator pressure by passing it through an expansion valve or capillary tube. The temperature of the refrigerant drops below the temperature of the refrigerated space during this process. The refrigerant enters the evaporator at state 4 as a low-quality saturated mixture, and it completely evaporates by absorbing heat from the refrigerated space. The refrigerant leaves the evaporator as saturated vapor and reenters the compressor, completing the cycle.

In a household refrigerator, the freezer compartment where heat is absorbed by the refrigerant serves as the evaporator. The coils behind the refrigerator, where heat is dissipated to the kitchen air, serve as the condenser (Fig. 8-54).

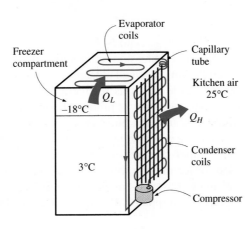

FIGURE 8-54

An ordinary household refrigerator.

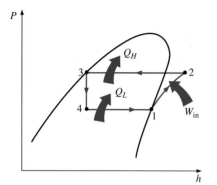

FIGURE 8-55

The *P-h* diagram of an ideal vapor-compression refrigeration cycle.

Remember that the area under the process curve on a *T-s* diagram represents the heat transfer for internally reversible processes. The area under the process curve 4-1 represents the heat absorbed by the refrigerant in the evaporator, and the area under the process curve 2-3 represents the heat rejected in the condenser. A rule of thumb is that *the COP improves by 2 to 4 percent for each °C the evaporating temperature is raised or the condensing temperature is lowered.*

Another diagram frequently used in the analysis of vapor-compression refrigeration cycles is the *P-h* diagram, as shown in Fig. 8-55. On this diagram, three of the four processes appear as straight lines, and the heat transfer in the condenser and the evaporator is proportional to the lengths of the corresponding process curves.

Notice that unlike the ideal cycles discussed before, the ideal vapor-compression refrigeration cycle is not an internally reversible cycle since it involves an irreversible (throttling) process. This process is maintained in the cycle to make it a more realistic model for the actual vapor-compression refrigeration cycle. If the throttling device were replaced by an isentropic turbine, the refrigerant would enter the evaporator at state 4′ instead of state 4. As a result, the refrigeration capacity would increase (by the area under process curve 4′-4 in Fig. 8-53) and the net work input would decrease (by the amount of work output of the turbine). Replacing the expansion valve by a turbine is not practical, however, since the added benefits cannot justify the added cost and complexity.

All four components associated with the vapor-compression refrigeration cycle are steady-flow devices, and thus all four processes that make up the cycle can be analyzed as steady-flow processes. The kinetic and potential energy changes of the refrigerant are usually small relative to the work and heat transfer terms, and therefore they can be neglected. Then the steady-flow energy equation on a unit-mass basis reduces to

$$(q_{in} - q_{out}) + (w_{in} - w_{out}) = h_e - h_i \qquad (8\text{-}44)$$

The condenser and the evaporator do not involve any work, and the compressor can be approximated as adiabatic. Then the COPs of refrigerators and heat pumps operating on the vapor-compression refrigeration cycle can be expressed as

$$\mathrm{COP_R} = \frac{q_L}{w_{net,\,in}} = \frac{h_1 - h_4}{h_2 - h_1} \qquad (8\text{-}45)$$

and

$$\mathrm{COP_{HP}} = \frac{q_H}{w_{net,\,in}} = \frac{h_2 - h_3}{h_2 - h_1} \qquad (8\text{-}46)$$

where $h_1 = h_{g\,@\,P_1}$ and $h_3 = h_{f\,@\,P_3}$ for the ideal case.

Vapor-compression refrigeration dates back to 1834 when the Englishman Jacob Perkins received a patent for a closed-cycle ice machine using ether or other volatile fluids as refrigerants. A working model of this machine was built, but it was never produced commercially. In 1850, Alexander Twining began to design and build vapor-compression ice machines using ethyl ether, which is the commercially used refrigerant in vapor-compression systems. Initially, vapor-compression refrigeration systems were large and were mainly used for ice making, brewing, and cold storage. They lacked automatic control and were steam-engine driven. In the 1890s, electric motor-driven smaller machines equipped with automatic control started to replace the older units, and refrigeration systems began to appear in butcher shops and households. By 1930, the continued improvements made it possible to have vapor-compression refrigeration systems that were relatively efficient, reliable, small, and inexpensive.

EXAMPLE 8-11   The Ideal Vapor Compression Refrigeration Cycle
A refrigerator uses refrigerant-134a as the working fluid and operates on an ideal vapor-compression refrigeration cycle between 0.14 and 0.8 MPa. If the mass flow rate of the refrigerant is 0.05 kg/s, determine (a) the rate of heat removal from the refrigerated space and the power input to the compressor, (b) the rate of heat rejection to the environment, and (c) the COP of the refrigerator.

Solution   The *T-s* diagram of the refrigeration cycle is shown in Fig. 8-56. We note that this is an ideal vapor-compression refrigeration cycle that involves steady-flow components. Therefore, the compressor is isentropic and the refrigerant leaves the condenser as a saturated liquid and enters the compressor as saturated vapor.

Assumptions   **1** Steady operating conditions exist. **2** Kinetic and potential energy changes are negligible.

Analysis   From the refrigerant-134a tables, the enthalpies of the refrigerant at all four states are determined as follows:

$$P_1 = 0.14 \text{ MPa} \longrightarrow h_1 = h_{g @ 0.14 \text{ MPa}} = 236.04 \text{ kJ/kg}$$
$$s_1 = s_{g @ 0.14 \text{ MPa}} = 0.9322 \text{ kJ/kg} \cdot \text{K}$$

$$\left. \begin{array}{l} P_2 = 0.8 \text{ MPa} \\ s_2 = s_1 \end{array} \right\} \quad h_2 = 272.05 \text{ kJ/kg}$$

$$P_3 = 0.8 \text{ MPa} \longrightarrow h_3 = h_{f @ 0.8 \text{ MPa}} = 93.42 \text{ kJ/kg}$$
$$h_4 \cong h_3 \text{ (throttling)} \longrightarrow h_4 = 93.42 \text{ kJ/kg}$$

(a) The rate of heat removal from the refrigerated space and the power input to the compressor are determined from their definitions:

$$\dot{Q}_L = \dot{m}(h_1 - h_4) = (0.05 \text{ kg/s})[(236.04 - 93.42) \text{ kJ/kg}] = \textbf{7.13 kW}$$
and $$\dot{W}_{in} = \dot{m}(h_2 - h_1) = (0.05 \text{ kg/s})[(272.05 - 236.04) \text{ kJ/kg}] = \textbf{1.80 kW}$$

(b) The rate of heat rejection from the refrigerant to the environment is determined from

$$\dot{Q}_H = \dot{m}(h_2 - h_3) = (0.05 \text{ kg/s})[(272.05 - 93.42) \text{ kJ/kg}] = \textbf{8.93 kW}$$

It could also be determined from

$$\dot{Q}_H = \dot{Q}_L + \dot{W}_{in} = 7.13 + 1.80 = 8.93 \text{ kW}$$

(c) The coefficient of performance of the refrigerator is determined from its definition:

$$\text{COP}_R = \frac{\dot{Q}_L}{\dot{W}_{in}} = \frac{7.13 \text{ kW}}{1.80 \text{ kW}} = \textbf{3.96}$$

That is, this refrigerator removes about 4 units of energy from the refrigerated space for each unit of electric energy it consumes.

Discussion   It would be interesting to see what happens if the throttling valve were replaced by an isentropic turbine. The enthalpy at state 4s (the turbine exit with $P_{4s} = 0.14$ MPa, and $s_{4s} = s_3 = 0.3459$ kJ/kg · K) in this case would be 86.92 kJ/kg, and the turbine would produce 0.34 kW of power. This would decrease the power input to the refrigerator from 1.80 to 1.46 kW and increase the

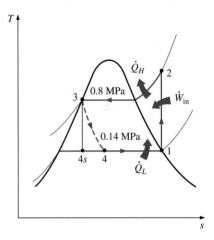

FIGURE 8-56
T-s diagram of the ideal vapor-compression refrigeration cycle described in Example 8-11.

rate of heat removal from the refrigerated space from 7.13 to 7.46 kW. As a result, the COP of the refrigerator would increase from 3.96 to 5.11, an increase of 29 percent.

8-17 ■ ACTUAL VAPOR-COMPRESSION REFRIGERATION CYCLES

An actual vapor-compression refrigeration cycle differs from the ideal one in several ways, owing mostly to the irreversibilities that occur in various components. Two common sources of irreversibilities are fluid friction (causes pressure drops) and heat transfer to or from the surroundings. The *T-s* diagram of an actual vapor-compression refrigeration cycle is shown in Fig. 8-57.

In the ideal cycle, the refrigerant leaves the evaporator and enters the compressor as *saturated vapor*. In practice, however, it may not be possible to control the state of the refrigerant so precisely. Instead, it is easier to design the system so that the refrigerant is slightly superheated at the compressor inlet. This slight overdesign ensures that the refrigerant is completely vaporized when it enters the compressor. Also, the line connecting the evaporator to the compressor is usually very long; thus the pressure drop caused by fluid friction and heat transfer from the surroundings to the refrigerant can be very significant. The result of superheating, heat gain in the connecting line, and pressure drops in the evaporator and the connecting line is an increase in the specific volume, thus an increase in the power input requirements to the compressor since steady-flow work is proportional to the specific volume.

The *compression process* in the ideal cycle is internally reversible and adiabatic, and thus isentropic. The actual compression process, however, will involve frictional effects, which increase the entropy, and heat transfer, which may increase or decrease the entropy, depending on the direction. Therefore,

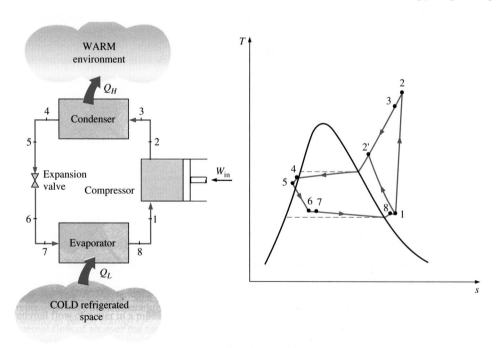

FIGURE 8-57
Schematic and *T-s* diagram for the actual vapor-compression refrigeration cycle.

the entropy of the refrigerant may increase (process 1-2) or decrease (process 1-2′) during an actual compression process, depending on which effects dominate. The compression process 1-2′ may be even more desirable than the isentropic compression process since the specific volume of the refrigerant and thus the work input requirement are smaller in this case. Therefore, the refrigerant should be cooled during the compression process whenever it is practical and economical to do so.

In the ideal case, the refrigerant is assumed to leave the condenser as *saturated liquid* at the compressor exit pressure. In actual situations, however, it is unavoidable to have some pressure drop in the condenser as well as in the lines connecting the condenser to the compressor and to the throttling valve. Also, it is not easy to execute the condensation process with such precision that the refrigerant is a saturated liquid at the end, and it is undesirable to route the refrigerant to the throttling valve before the refrigerant is completely condensed. Therefore, the refrigerant is subcooled somewhat before it enters the throttling valve. We do not mind this at all, however, since the refrigerant in this case enters the evaporator with a lower enthalpy and thus can absorb more heat from the refrigerated space. The throttling valve and the evaporator are usually located very close to each other, so the pressure drop in the connecting line is small.

EXAMPLE 8-12   The Actual Vapor-Compression Refrigeration Cycle

Refrigerant-134a enters the compressor of a refrigerator as superheated vapor at 0.14 MPa and −10°C at a rate of 0.05 kg/s and leaves at 0.8 MPa and 50°C. The refrigerant is cooled in the condenser to 26°C and 0.72 MPa and is throttled to 0.15 MPa. Disregarding any heat transfer and pressure drops in the connecting lines between the components, determine (*a*) the rate of heat removal from the refrigerated space and the power input to the compressor, (*b*) the adiabatic efficiency of the compressor, and (*c*) the coefficient of performance of the refrigerator.

Solution   The *T-s* diagram of the refrigeration cycle is shown in Fig. 8-58. We note that this is a nonideal vapor-compression refrigeration cycle that involves steady-flow components. The refrigerant leaves the condenser as a compressed liquid and enters the compressor as superheated vapor.

Assumptions   **1** Steady operating conditions exist. **2** Kinetic and potential energy changes are negligible.

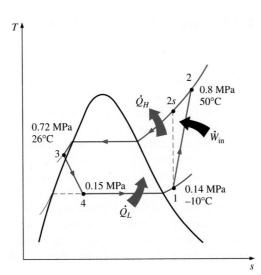

FIGURE 8-58

T-s diagram for Example 8-12.

Analysis The enthalpies of the refrigerant at various states are determined from the refrigerant tables to be

$$\left.\begin{array}{l} P_1 = 0.14\ \text{MPa} \\ T_1 = -10°\text{C} \end{array}\right\} \qquad h_1 = 243.40\ \text{kJ/kg}$$

$$\left.\begin{array}{l} P_2 = 0.8\ \text{MPa} \\ T_2 = 50°\text{C} \end{array}\right\} \qquad h_2 = 284.39\ \text{kJ/kg}$$

$$\left.\begin{array}{l} P_3 = 0.72\ \text{MPa} \\ T_3 = 26°\text{C} \end{array}\right\} \qquad h_3 \cong h_{f@\ 26°\text{C}} = 85.75\ \text{kJ/kg}$$

$$h_4 \cong h_3 (\text{throttling}) \longrightarrow h_4 = 85.75\ \text{kJ/kg}$$

(*a*) The rate of heat removal from the refrigerated space and the power input to the compressor are determined from their definitions:

$$\dot{Q}_L = \dot{m}(h_1 - h_4) = (0.05\ \text{kg/s})[(243.40 - 85.75)\ \text{kJ/kg}] = \textbf{7.88 kW}$$

and $\quad \dot{W}_{in} = \dot{m}(h_2 - h_1) = (0.05\ \text{kg/s})[(284.39 - 243.40)\ \text{kJ/kg}] = \textbf{2.05 kW}$

(*b*) The isentropic efficiency of the compressor is determined from

$$\eta_C \cong \frac{h_{2s} - h_1}{h_2 - h_1}$$

where the enthalpy at state 2s ($P_{2s} = 0.8$ MPa and $s_{2s} = s_1 = 0.9606$ kJ/kg · K) is 281.05 kJ/kg. Thus,

$$\eta_C = \frac{281.05 - 243.40}{284.39 - 243.40} = \textbf{0.919 or 91.9\%}$$

(*c*) The coefficient of performance of the refrigerator is determined from its definition:

$$\text{COP}_R = \frac{\dot{Q}_L}{\dot{W}_{in}} = \frac{7.88\ \text{kW}}{2.05\ \text{kW}} = \textbf{3.84}$$

Discussion This problem is identical to the one worked out in Example 8-11, except that the refrigerant is slightly superheated at the compressor inlet and sub-cooled at the condenser exit. Also, the compressor is not isentropic. As a result, the heat removal rate from the refrigerated space increases (by 10.5 percent), but the power input to the compressor increases even more (by 13.9 percent). Consequently, the COP of the refrigerator decreases from 3.96 to 3.84.

8-18 ■ HEAT PUMP SYSTEMS

Heat pumps are generally more expensive to purchase and install than other heating systems, but they save money in the long run in some areas because they lower the heating bills. Despite their relatively higher initial costs, the popularity of heat pumps is increasing. A large fraction of all single-family homes in the United States are heated by heat pumps.

The most common energy source for heat pumps is atmospheric air (air-to-air systems), although water and soil are also used. The major problem with air-source systems is *frosting,* which occurs in humid climates when the temperature falls below 2 to 5°C. The frost accumulation on the evaporator coils is highly undesirable since it seriously disrupts the heat transfer. The coils can be defrosted, however, by reversing the heat pump cycle (running it as an air conditioner). This results in a reduction in the efficiency of the system. Water-source systems usually use well water from depths of up to 80 m in the temperature range of 5 to 18°C, and they do not have a frosting problem. They typically have higher COPs but are more complex and require easy access to a large body of water such as underground water. Soil-source systems are also rather involved since they require long tubing placed deep in the ground where the soil temperature is relatively constant. The COP of heat pumps usually ranges between 1.5 and 4, depending on the particular system used and the temperature of the source. A new class of recently developed heat pumps that use variable-speed electric motor drives are at least twice as energy efficient as their predecessors.

Both the capacity and the efficiency of a heat pump fall significantly at low temperatures. Therefore, most air-source heat pumps require a supplementary heating system such as electric resistance heaters or an oil or gas furnace. Since water and soil temperatures do not fluctuate much, supplementary heating may not be required for water-source or soil-source systems. But the heat pump system must be large enough to meet the maximum heating load.

Heat pumps and air conditioners have the same mechanical components. Therefore, it is not economical to have two separate systems to meet the heating and cooling requirements of a building. One system can be used as a heat pump in winter and an air conditioner in summer. This is accomplished by adding a reversing valve to the cycle, as shown in Fig. 8-59. As a result of this modification, the condenser of the heat pump (located indoors) functions as the evaporator of the air conditioner in summer. Also, the evaporator of the heat pump (located outdoors) serves as the condenser of the air conditioner. This feature increases the competitiveness of the heat pump. Such dual-purpose window units are commonly used in motels.

Heat pumps are most competitive in areas that have a large cooling load during the cooling season and a relatively small heating load during the heating season, such as in the southern parts of the United States. In these areas, the heat pump can meet the entire cooling and heating needs of residential or

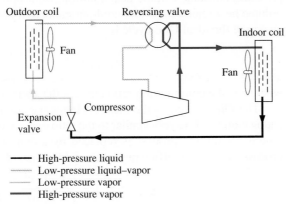

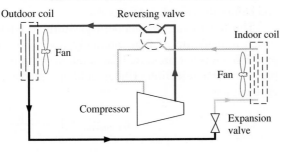

FIGURE 8-59

A heat pump can be used to heat a house in winter and to cool it in summer.

commercial buildings. The heat pump is least competitive in areas where the heating load is significant and the cooling load is small, such as in the northern parts of the United States.

8-19 ▦ SUMMARY

The most efficient cycle operating between a heat source at temperature T_H and a sink at temperature T_L is the Carnot cycle, and its thermal efficiency is given by

$$\eta_{th,\ Carnot} = 1 - \frac{T_L}{T_H}$$

The actual gas cycles are rather complex. The approximations used to simplify the analysis are known as the *air-standard assumptions*. Under these assumptions, all the processes are assumed to be internally reversible; the working fluid is assumed to be air, which behaves as an ideal gas; and the combustion and exhaust processes are replaced by heat-addition and heat-rejection processes, respectively. The air-standard assumptions are called *cold-air-standard assumptions* if, in addition, air is assumed to have constant specific heats at room temperature.

In reciprocating engines, the *compression ratio r* and the *mean effective pressure* MEP are defined as

$$r = \frac{V_{max}}{V_{min}} = \frac{V_{BDC}}{V_{TDC}}$$

$$\text{MEP} = \frac{w_{net}}{v_{max} - v_{min}} \qquad \text{(kPa)}$$

The *Otto cycle* is the ideal cycle for the spark-ignition reciprocating engines, and it consists of four internally reversible processes: isentropic compression, constant volume heat addition, isentropic expansion, and constant volume heat rejection. Under cold-air-standard assumptions, the thermal efficiency of the ideal Otto cycle is

$$\eta_{th,\ Otto} = 1 - \frac{1}{r^{k-1}}$$

where r is the compression ratio and k is the specific heat ratio C_p/C_v.

The *Diesel cycle* is the ideal cycle for the compression-ignition reciprocating engines. It is very similar to the Otto cycle, except that the constant volume heat-addition process is replaced by a constant pressure heat-addition process. Its thermal efficiency under cold-air-standard assumptions is

$$\eta_{th,\ Diesel} = 1 - \frac{1}{r^{k-1}} \left[\frac{r_c^k - 1}{k(r_c - 1)} \right]$$

where r_c is the *cutoff ratio,* defined as the ratio of the cylinder volumes after and before the combustion process.

The ideal cycle for modern gas-turbine engines is the *Brayton cycle,* which is made up of four internally reversible processes: isentropic compression, constant pressure heat addition, isentropic expansion, and constant pressure heat rejection. Under cold-air-standard assumptions, its thermal efficiency is

$$\eta_{th,\ Brayton} = 1 - \frac{1}{r_p^{(k-1)/k}}$$

where $r_p = P_{max}/P_{min}$ is the pressure ratio and k is the specific heat ratio. The thermal efficiency of the simple Brayton cycle increases with the pressure ratio.

The deviation of the actual compressor and the turbine from the idealized isentropic ones can be accurately accounted for by utilizing their isentropic efficiencies, defined as

$$\eta_C = \frac{w_s}{w_a} \cong \frac{h_{2s} - h_1}{h_{2a} - h_1}$$

and

$$\eta_T = \frac{w_a}{w_s} \cong \frac{h_3 - h_{4a}}{h_3 - h_{4s}}$$

where states 1 and 3 are the inlet states, $2a$ and $4a$ are the actual exit states, and $2s$ and $4s$ are the isentropic exit states.

In gas-turbine engines, the temperature of the exhaust gas leaving the turbine is often considerably higher than the temperature of the air leaving the compressor. Therefore, the high-pressure air leaving the compressor can be heated by transferring heat to it from the hot exhaust gases in a counter-flow heat exchanger, which is also known as a *regenerator.* The extent to which a regenerator approaches an ideal regenerator is called the *effectiveness ε* and is defined as

$$\varepsilon = \frac{q_{regen,\ act}}{q_{regen,\ max}}$$

Under cold-air-standard assumptions, the thermal efficiency of an ideal Brayton cycle with regeneration becomes

$$\eta_{th,\ regen} = 1 - \left(\frac{T_1}{T_3}\right)(r_p)^{(k-1)/k}$$

where T_1 and T_3 are the minimum and maximum temperatures, respectively, in the cycle.

The *Carnot cycle* is not a suitable model for vapor power cycles because it cannot be approximated in practice. The model cycle for vapor power cycles is the *Rankine cycle,* which is composed of four internally reversible processes: constant-pressure heat addition in a boiler, isentropic expansion in a turbine, constant-pressure heat rejection in a condenser, and isentropic compression in a pump. Steam leaves the condenser as a saturated liquid at the condenser pressure.

The thermal efficiency of the Rankine cycle can be increased by increasing the average temperature at which heat is added to the working fluid and/or by decreasing the average temperature at which heat is rejected to the cooling medium. The average temperature during heat rejection can be decreased by lowering the turbine exit pressure. Consequently, the condenser pressure of most vapor power plants is well below the atmospheric pressure. The average temperature during heat addition can be increased by raising the boiler pressure or by superheating the fluid to high temperatures. There is a limit to the degree of superheating, however, since the fluid temperature is not allowed to exceed a metallurgically safe value.

Superheating has the added advantage of decreasing the moisture content of the steam at the turbine exit. Lowering the exhaust pressure or raising the boiler pressure, however, increases the moisture content. To take advantage of the improved efficiencies at higher boiler pressures and lower condenser pressures, steam is usually *reheated* after expanding partially in the high-pressure turbine. This is done by extracting the steam after partial expansion in the high-pressure turbine, sending it back to the boiler where it is reheated at constant pressure, and returning it to the low-pressure turbine for complete expansion to the condenser pressure. The average temperature during the reheat process, and thus the thermal efficiency of the cycle, can be increased by increasing the number of expansion and reheat stages. As the number of stages is increased, the expansion and reheat processes approach an isothermal process of expansion at maximum temperature. Reheating also decreases the moisture content at the turbine exit.

The transfer of heat from lower temperature regions to higher temperature ones is called *refrigeration*. Devices that produce refrigeration are called *refrigerators*, and the cycles on which they operate are called *refrigeration cycles*. The working fluids used in refrigerators are called *refrigerants*. Refrigerators used for the purpose of heating a space by transferring heat from a cooler medium are called *heat pumps*.

The performance of refrigerators and heat pumps is expressed in terms of *coefficient of performance* (COP), defined as

$$\text{COP}_\text{R} = \frac{\text{Desired output}}{\text{Required output}} = \frac{\text{Cooling effect}}{\text{Work input}} = \frac{Q_L}{W_{\text{net, in}}}$$

$$\text{COP}_\text{HP} = \frac{\text{Desired output}}{\text{Required output}} = \frac{\text{Heating effect}}{\text{Work input}} = \frac{Q_H}{W_{\text{net, in}}}$$

The standard of comparison for refrigeration cycles is the *reversed Carnot cycle*. A refrigerator or heat pump that operates on the reversed Carnot cycle is called a *Carnot refrigerator* or a *Carnot heat pump,* and their COPs are

$$\text{COP}_\text{R, Carnot} = \frac{1}{T_H/T_L - 1}$$

$$\text{COP}_\text{HP, Carnot} = \frac{1}{1 - T_L/T_H}$$

The most widely used refrigeration cycle is the *vapor-compression refrigeration cycle*. In an ideal vapor-compression refrigeration cycle, the refrigerant enters the compressor as a saturated vapor and is cooled to the saturated liquid state in the condenser. It is then throttled to the evaporator pressure and vaporizes as it absorbs heat from the refrigerated space.

REFERENCES AND SUGGESTED READING

1. *ASHRAE, Handbook of Fundamentals.* Atlanta: American Society of Heating, Refrigerating, and Air-Conditioning Engineers, 1985.

2. R. L. Bannister and G. J. Silvestri. "The Evolution of Central Station Steam Turbines." *Mechanical Engineering,* February 1989, pp. 70–78.

3. R. L. Bannister, G. J. Silvestri, A. Hizume, and T. Fujikawa. "High Temperature Supercritical Steam Turbines." *Mechanical Engineering,* February 1987, pp. 60–65.

4. Y. A. Çengel and M. A. Boles. *Thermodynamics: An Engineering Approach.* 3rd ed. New York: McGraw-Hill, 1998.

5. M. M. El-Wakil. *Powerplant Technology.* New York: McGraw-Hill, 1984.

6. *Heat Pump Systems—A Technology Review.* OECD Report, Paris, 1982.

7. L. C. Lichty. *Combustion Engine Processes*. New York: McGraw-Hill, 1967.

8. W. Siuru. "Two-stroke Engines: Cleaner and Meaner." *Mechanical Engineering*. June 1990, pp. 66–69.

9. H. Sorensen. *Energy Conversion Systems*. New York: John Wiley & Sons, 1983.

10. *Steam, Its Generation and Use*. 39th ed. New York: Babcock and Wilcox Co., 1978.

11. W. F. Stoecker and J. W. Jones. *Refrigeration and Air Conditioning*. 2nd ed. New York: McGraw-Hill, 1982.

12. C. F. Taylor. *The Internal Combustion Engine in Theory and Practice*. Cambridge, MA: M.I.T. Press, 1968.

13. J. Weisman and R. Eckart. *Modern Power Plant Engineering*. Englewood Cliffs, NJ: Prentice Hall, 1985.

PROBLEMS*

Actual and Ideal Cycles, Carnot Cycle, Air-Standard Assumptions, Reciprocating Engines

8-1C How do gas power cycles differ from vapor power cycles?

8-2C Why is the Carnot cycle not suitable as an ideal cycle for all power-producing cyclic devices?

8-3C How does the thermal efficiency of an ideal cycle, in general, compare to that of a Carnot cycle operating between the same temperature limits?

8-4C What does the area enclosed by the cycle represent on a *P-v* diagram? How about on a *T-s* diagram?

8-5C What is the difference between air-standard assumptions and the cold-air-standard assumptions?

8-6C Do internal combustion engines operate on a closed or an open cycle? Why?

8-7C How are the combustion and exhaust processes modeled under the air-standard assumptions?

8-8C What are the air-standard assumptions?

8-9C What is the difference between the clearance volume and the displacement volume of reciprocating engines?

8-10C Define the compression ratio for reciprocating engines.

8-11C How is the mean effective pressure for reciprocating engines defined?

8-12C Can the mean effective pressure of an automobile engine in operation be less than the atmospheric pressure?

---

*Students are encouraged to answer *all* the concept "C" questions.

8-13C As a car gets older, will its compression ratio change? How about the mean effective pressure?

8-14C What is the difference between spark-ignition and compression-ignition engines?

8-15C Define the following terms related to reciprocating engines: stroke, bore, top dead center, and clearance volume.

8-16 An air-standard cycle with variable specific heats is executed in a closed system and is composed of the following four processes:
  1-2   Isentropic compression from 100 kPa and 27°C to 800 kPa
  2-3 $v = constant$ heat addition to 1800 K
  3-4   Isentropic expansion to 100 kPa
  4-1 $P = constant$ heat rejection to initial state
  (a) Show the cycle on P-v and T-s diagrams.
  (b) Calculate the net work output per unit mass.
  (c) Determine the thermal efficiency.

8-17 An air-standard cycle is executed in a closed system and is composed of the following four processes:
  1-2   Isentropic compression from 100 kPa and 27°C to 1 MPa
  2-3 $P = constant$ heat addition in amount of 2840 kJ/kg
  3-4 $v = constant$ heat rejection to 100 kPa
  4-1 $P = constant$ heat rejection to initial state
  (a) Show the cycle on P-v and T-s diagrams.
  (b) Calculate the maximum temperature in the cycle.
  (c) Determine the thermal efficiency.

Assume constant specific heats at room temperature.
Answers: (b) 3405.1 K, (c) 21.1 percent

8-18E An air-standard cycle with variable specific heats is executed in a closed system and is composed of the following four processes:
  1-2 $v = constant$ heat addition from 14.7 psia and 80°F in the amount of 300 Btu/lbm
  2-3 $P = constant$ heat addition to 3200 R
  3-4   Isentropic expansion to 14.7 psia
  4-1 $P = constant$ heat rejection to initial state
  (a) Show the cycle on P-v and T-s diagrams.
  (b) Calculate the total heat input per unit mass.
  (c) Determine the thermal efficiency.
  Answers: (b) 612.4 Btu/lbm, (c) 24.2 percent

8-19E Repeat Prob. 8-18E using constant specific heats at room temperature.

8-20 An air-standard cycle is executed in a closed system with 0.001 kg of air and consists of the following three processes:
  1-2   Isentropic compression from 100 kPa and 27°C to 1 MPa
  2-3 $P = constant$ heat addition in the amount of 1.84 kJ
  3-1 $P = c_1 v + c_2$ heat rejection to initial state (c_1 and c_2 are constants)
  (a) Show the cycle on P-v and T-s diagrams.
  (b) Calculate the heat rejected.
  (c) Determine the thermal efficiency.

Assume constant specific heats at room temperature.
Answers: (*b*) 1.422 kJ, (*c*) 22.7 percent

8-21 An air-standard cycle with variable specific heats is executed in a closed system with 0.003 kg of air and consists of the following three processes:

    1-2 $v = constant$ heat addition from 95 kPa and 17°C to 380 kPa

    2-3   Isentropic expansion to 95 kPa

    3-1 $P = constant$ heat rejection to initial state

(*a*) Show the cycle on *P-v* and *T-s* diagrams.

(*b*) Calculate the net work per cycle, in kJ.

(*c*) Determine the thermal efficiency.

8-22 Repeat Prob. 8-21 using constant specific heats at room temperature.

8-23 Consider a Carnot cycle executed in a closed system with 0.004 kg of air. The temperature limits of the cycle are 300 and 1000 K, and the minimum and maximum pressures that occur during the cycle are 20 and 1800 kPa. Assuming constant specific heats, determine the net work output per cycle.

8-24 An air-standard Carnot cycle is executed in a closed system between the temperature limits of 350 and 1200 K. The pressures before and after the isothermal compression are 150 and 300 kPa, respectively. If the net work output per cycle is 0.5 kJ, determine (*a*) the maximum pressure in the cycle, (*b*) the heat transfer to air, and (*c*) the mass of air. Assume variable specific heats for air.   *Answers:* (*a*) 30,013 kPa, (*b*) 0.706 kJ, (*c*) 0.00296 kg

8-25 Repeat Prob. 8-24 using helium as the working fluid.

Otto Cycle

8-26C What four processes make up the ideal Otto cycle?

8-27C How do the efficiencies of the ideal Otto cycle and the Carnot cycle compare for the same temperature limits? Explain.

8-28C How is the rpm (revolutions per minute) of an actual four-stroke gasoline engine related to the number of thermodynamic cycles? What would your answer be for a two-stroke engine?

8-29C Are the processes that make up the Otto cycle analyzed as closed-system or steady-flow processes? Why?

8-30C How does the thermal efficiency of an ideal Otto cycle change with the compression ratio of the engine and the specific heat ratio of the working fluid?

8-31C Why are high compression ratios not used in spark-ignition engines?

8-32C An ideal Otto cycle with a specified compression ratio is executed using (*a*) air, (*b*) argon, and (*c*) ethane as the working fluid. For which case will the thermal efficiency be the highest? Why?

8-33C What is the difference between fuel-injected gasoline engines and diesel engines?

8-34 An ideal Otto cycle has a compression ratio of 8. At the beginning of the compression process, air is at 95 kPa and 27°C, and 750 kJ/kg of heat is transferred to air during the constant-volume heat-addition process. Taking into account the variation of specific heats with temperature, determine (*a*) the pressure and temperature at the end of the heat-addition process, (*b*) the net work output, (*c*) the thermal efficiency, and (*d*) the mean effective pressure for the cycle.

Answers: (*a*) 3898 kPa, 1539 K; (*b*) 392.4 kJ/kg; (*c*) 52.3 percent; (*d*) 495 kPa

8-35 Repeat Prob. 8-34 using constant specific heats at room temperature.

8-36 The compression ratio of an air-standard Otto cycle is 9.5. Prior to the isentropic compression process, the air is at 100 kPa, 17°C, and 600 cm³. The temperature at the end of the isentropic expansion process is 800 K. Using specific heat values at room temperature, determine (*a*) the highest temperature and pressure in the cycle; (*b*) the amount of heat transferred, in kJ; (*c*) the thermal efficiency; and (*d*) the mean effective pressure.

Answers: (*a*) 1969 K, 6449 kPa; (*b*) 0.65 kJ; (*c*) 59.4 percent; (*d*) 719 kPa

8-37 Repeat Prob. 8-36, but replace the isentropic expansion process by a polytropic expansion process with the polytropic exponent $n = 1.35$.

8-38E An ideal Otto cycle with air as the working fluid has a compression ratio of 8. The minimum and maximum temperatures in the cycle are 540 and 2200 R. Accounting for the variation of specific heats with temperature, determine (*a*) the amount of heat transferred to the air during the heat-addition process, (*b*) the thermal efficiency, and (*c*) the thermal efficiency of a Carnot cycle operating between the same temperature limits.

Answers: (*a*) 198.15 Btu/lbm, (*b*) 53.4 percent, (*c*) 75.5 percent

8-39E Repeat Prob. 8-38E using argon as the working fluid.

Diesel Cycle

8-40C What is the dual cycle? How does it differ from the Otto and Diesel cycles?

8-41C How does a diesel engine differ from a gasoline engine?

8-42C How does the ideal Diesel cycle differ from the ideal Otto cycle?

8-43C For a specified compression ratio, is a diesel or gasoline engine more efficient?

8-44C Do diesel or gasoline engines operate at higher compression ratios? Why?

8-45C What is the cutoff ratio? How does it affect the thermal efficiency of a Diesel cycle?

8-46 An air-standard Diesel cycle has a compression ratio of 16 and a cutoff ratio of 2. At the beginning of the compression process, air is at 95 kPa and 27°C. Accounting for the variation of specific heats with temperature, determine (*a*) the temperature after the heat-addition process, (*b*) the thermal efficiency, and (*c*) the mean effective pressure.

Answers: (*a*) 1724.8 K, (*b*) 56.3 percent, (*c*) 675.9 kPa

8-47 Repeat Prob. 8-46 using constant specific heats at room temperature.

8-48E An air-standard Diesel cycle has a compression ratio of 18.2. Air is at 80°F and 14.7 psia at the beginning of the compression process and at 3400 R at the end of the heat-addition process. Accounting for the variation of specific heats with temperature, determine (*a*) the cutoff ratio, (*b*) the heat rejection per unit mass, and (*c*) the thermal efficiency.

    Answers: (*a*) 2.09, (*b*) 216.5 Btu/lbm, (*c*) 57.3 percent

8-49E Repeat Prob. 8-48E using constant specific heats at room temperature.

8-50 An ideal diesel engine has a compression ratio of 20 and uses air as the working fluid. The state of air at the beginning of the compression process is 95 kPa and 20°C. If the maximum temperature in the cycle is not to exceed 2200 K, determine (*a*) the thermal efficiency and (*b*) the mean effective pressure. Assume constant specific heats for air at room temperature.

    Answers: (*a*) 63.5 percent, (*b*) 933 kPa

8-51 Repeat Prob. 8-50, but replace the isentropic expansion process by the polytropic expansion process with the polytropic exponent $n = 1.35$.

8-52 A four-cylinder 4.5-L diesel engine that operates on an ideal Diesel cycle has a compression ratio of 17 and a cutoff ratio of 2.2. Air is at 27°C and 97 kPa at the beginning of the compression process. Using the cold-air-standard assumptions, determine how much power the engine will deliver at 1500 rpm.

8-53 Repeat Prob. 8-52 using nitrogen as the working fluid.

8-54 The compression ratio of an ideal dual cycle is 14. Air is at 100 kPa and 300 K at the beginning of the compression process and at 2200 K at the end of the heat-addition process. Heat transfer to air takes place partly at constant volume and partly at constant pressure, and it amounts to 1520.4 kJ/kg. Assuming variable specific heats for air, determine (*a*) the fraction of heat transferred at constant volume and (*b*) the thermal efficiency of the cycle.

8-55 Repeat Prob. 8-54 using constant specific heats at room temperature. Is the constant specific heat assumption reasonable in this case?

Ideal and Actual Gas-Turbine (Brayton) Cycles

8-56C Why are the back work ratios relatively high in gas-turbine engines?

8-57C What four processes make up the simple ideal Brayton cycle?

8-58C For fixed maximum and minimum temperatures, what is the effect of the pressure ratio on (*a*) the thermal efficiency and (*b*) the net work output of a simple ideal Brayton cycle?

8-59C Why are gas turbines operated at very high air–fuel mass ratios?

8-60C Should the processes that make up the Brayton cycle be analyzed as closed-system or steady-flow processes? Why?

8-61C What is the back work ratio? What are typical back work ratio values for gas-turbine engines?

8-62C How can the irreversibilities in the turbine and compressor of gas-turbine engines be properly accounted for?

8-63C How do the inefficiencies of the turbine and the compressor affect (*a*) the back work ratio and (*b*) the thermal efficiency of a gas-turbine engine?

8-64E A simple ideal Brayton cycle with air as the working fluid has a pressure ratio of 10. The air enters the compressor at 520 R and the turbine at 2000 R. Accounting for the variation of specific heats with temperature, determine (*a*) the air temperature at the compressor exit, (*b*) the back work ratio, and (*c*) the thermal efficiency.

8-65 A simple Brayton cycle using air as the working fluid has a pressure ratio of 8. The minimum and maximum temperatures in the cycle are 310 and 1160 K. Assuming an isentropic efficiency of 75 percent for the compressor and 82 percent for the turbine, determine (*a*) the air temperature at the turbine exit, (*b*) the net work output, and (*c*) the thermal efficiency.

8-66 Repeat Prob. 8-65 using constant specific heats at room temperature.

8-67 Air is used as the working fluid in a simple ideal Brayton cycle that has a pressure ratio of 12, a compressor inlet temperature of 300 K, and a turbine inlet temperature of 1000 K. Determine the required mass flow rate of air for a net power output of 30 MW, assuming both the compressor and the turbine have an isentropic efficiency of (*a*) 100 percent and (*b*) 80 percent. Assume constant specific heats at room temperature.
Answers: (*a*) 150.7 kg/s, (*b*) 1581 kg/s

8-68 A stationary gas-turbine power plant operates on a simple ideal Brayton cycle with air as the working fluid. The air enters the compressor at 95 kPa and 290 K and the turbine at 760 kPa and 1100 K. Heat is transferred to air at a rate of 50,000 kJ/s. Determine the power delivered by this plant (*a*) assuming constant specific heats at room temperature and (*b*) accounting for the variation of specific heats with temperature.

8-69 Air enters the compressor of a gas-turbine engine at 300 K and 100 kPa, where it is compressed to 700 kPa and 580 K. Heat is transferred to air in the amount of 950 kJ/kg before it enters the turbine. For a turbine efficiency of 86 percent, determine (*a*) the fraction of the turbine work output used to drive the compressor and (*b*) the thermal efficiency. Assume variable specific heats for air.

8-70 Repeat Prob. 8-69 using constant specific heats at room temperature.

8-71E A gas-turbine power plant operates on a simple Brayton cycle with air as the working fluid. The air enters the turbine at 120 psia and 2000 R and leaves at 15 psia and 1200 R. Heat is rejected to the surroundings at a rate of 6400 Btu/s, and air flows through the cycle at a rate of 40 lbm/s. Assuming a compressor efficiency of 80 percent, determine the net power output of the plant. Account for the variation of specific heats with temperature.
Answer: 3373 kW

8-72E For what compressor efficiency will the gas-turbine power plant in Prob. 8-71E produce zero net work?

8-73 A gas-turbine power plant operates on the simple Brayton cycle with air as the working fluid and delivers 15 MW of power. The minimum and maximum temperatures in the cycle are 310 and 900 K, and the pressure of air at the compressor exit is 8 times the value at the compressor inlet. Assuming an isentropic efficiency of 80 percent for the compressor and 86 percent for the turbine, determine the mass flow rate of air through the cycle. Account for the variation of specific heats with temperature.

8-74 Repeat Prob. 8-73 using constant specific heats at room temperature.

8-75C How does regeneration affect the efficiency of a Brayton cycle, and how does it accomplish it?

8-76C Somebody claims that at very high pressure ratios, the use of regeneration actually decreases the thermal efficiency of a gas-turbine engine. Is there any truth in this claim? Explain.

8-77C Define the effectiveness of a regenerator used in gas-turbine cycles.

8-78C In an ideal regenerator, is the air leaving the compressor heated to the temperature at (*a*) turbine inlet, (*b*) turbine exit, (*c*) slightly above turbine exit?

8-79C In 1903, Aegidius Elling of Norway designed and built an 11-hp gas turbine that used steam injection between the combustion chamber and the turbine to cool the combustion gases to a safe temperature for the materials available at the time. Currently there are several gas-turbine power plants that use steam injection to augment power and improve thermal efficiency. For example, the thermal efficiency of the General Electric LM5000 gas turbine is reported to increase from 35.8 percent in simple-cycle operation to 43 percent when steam injection is used. Explain why steam injection increases the power output and the efficiency of gas turbines. Also, explain how you would obtain the steam.

8-80E The idea of using gas turbines to power automobiles was conceived in the 1930s, and considerable research was done in the 1940s and 1950s to develop automotive gas turbines by major automobile manufacturers such as the Chrysler and Ford corporations in the United States and Rover in the United Kingdom. The world's first gas-turbine-powered automobile, the 200-hp Rover Jet 1, was built in 1950 in the United Kingdom. This was followed by the production of the Plymouth Sport Coupe by Chrysler in 1954 under the leadership of G. J. Huebner. Several hundred gas-turbine-powered Plymouth cars were built in the early 1960s for demonstration purposes and were loaned to a select group of people to gather field experience. The users had no complaints other than slow acceleration. But the cars were never mass-produced because of the high production (especially material) costs and the failure to satisfy the provisions of the 1966 Clean Air Act.

A gas-turbine-powered Plymouth car built in 1960 had a turbine inlet temperature of 1700°F, a pressure ratio of 4, and a regenerator effectiveness of 0.9. Using isentropic efficiencies of 80 percent for both the compressor and the turbine, determine the thermal efficiency of this car. Also, determine the mass flow rate of air for a net power output of 135 hp. Assume the ambient air to be at 540 R and 14.5 psia.

8-81 The 7FA gas turbine manufactured by General Electric is reported to have an efficiency of 35.9 percent in the simple-cycle mode and to produce 159 MW of net power. The pressure ratio is 14.7, the turbine inlet temperature is 1288°C, and the exhaust temperature is 589°C. The mass flow rate through the turbine is 1,536,000 kg/h. Taking the ambient conditions to be 20°C and 100 kPa, determine the isentropic efficiency of the turbine and the compressor. Also, determine the thermal efficiency of this gas turbine if a regenerator with an effectiveness of 80 percent is added.

8-82 An ideal Brayton cycle with regeneration has a pressure ratio of 10. Air enters the compressor at 300 K and the turbine at 1200 K. If the effectiveness of the regenerator is 100 percent, determine the net work output and the thermal efficiency of the cycle. Account for the variation of specific heats with temperature.

8-83   Repeat Prob. 8-82 using constant specific heats at room temperature.

8-84   A Brayton cycle with regeneration using air as the working fluid has a pressure ratio of 8. The minimum and maximum temperatures in the cycle are 310 and 1150 K. Assuming an isentropic efficiency of 75 percent for the compressor and 82 percent for the turbine and an effectiveness of 65 percent for the regenerator, determine (*a*) the air temperature at the turbine exit, (*b*) the net work output, and (*c*) the thermal efficiency.
    Answers: (*a*) 763 K, (*b*) 101.64 kJ/kg, (*c*) 21.0 percent

8-85   A stationary gas-turbine power plant operates on an ideal regenerative Brayton cycle ($\varepsilon = 100$ percent) with air as the working fluid. Air enters the compressor at 95 kPa and 290 K and the turbine at 760 kPa and 1100 K. Heat is transferred to air from an external source at a rate of 60,000 kJ/s. Determine the power delivered by this plant (*a*) assuming constant specific heats for air at room temperature and (*b*) accounting for the variation of specific heats with temperature.

8-86   Air enters the compressor of a regenerative gas-turbine engine at 300 K and 100 kPa, where it is compressed to 800 kPa and 580 K. The regenerator has an effectiveness of 72 percent, and the air enters the turbine at 1200 K. For a turbine efficiency of 86 percent, determine (*a*) the amount of heat transfer in the regenerator and (*b*) the thermal efficiency. Assume variable specific heats for air.
    Answers: (*a*) 152.5 kJ/kg, (*b*) 36.0 percent

8-87   Repeat Prob. 8-86 using constant specific heats at room temperature.

8-88   Repeat Prob. 8-86 for a regenerator effectiveness of 70 percent.

Carnot Vapor Cycle

8-89C   Why is excessive moisture in steam undesirable in steam turbines? What is the highest moisture content allowed?

8-90C   Why is the Carnot cycle not a realistic model for steam power plants?

8-91E   Water enters the boiler of a steady-flow Carnot engine as a saturated liquid at 120 psia and leaves with a quality of 0.95. Steam leaves the turbine at a pressure of 14.7 psia. Show the cycle on a *T-s* diagram relative to the saturation lines, and determine (*a*) the thermal efficiency, (*b*) the quality at the end of the isothermal heat-rejection process, and (*c*) the net work output.
    Answers: (*a*) 16.1 percent, (*b*) 0.1245, (*c*) 134.4 Btu/lbm

8-92   A steady-flow Carnot cycle uses water as the working fluid. Water changes from saturated liquid to saturated vapor as heat is transferred to it from a source at 250°C. Heat rejection takes place at a pressure of 20 kPa. Show the cycle on a *T-s* diagram relative to the saturation lines, and determine (*a*) the thermal efficiency; (*b*) the amount of heat rejected, in kJ/kg; and (*c*) the net work output.

8-93   Consider a steady-flow Carnot cycle with water as the working fluid. The maximum and minimum temperatures in the cycle are 350 and 60°C. The quality of water is 0.891 at the beginning of the heat-rejection process and

0.1 at the end. Show the cycle on a *T-s* diagram relative to the saturation lines, and determine (*a*) the thermal efficiency, (*b*) the pressure at the turbine inlet, and (*c*) the net work output.

> *Answers:* (*a*) 0.465, (*b*) 1.40 MPa, (*c*) 1624 kJ/kg

The Simple Rankine Cycle

8–94C What four processes make up the simple ideal Rankine cycle?

8–95C Consider a simple ideal Rankine cycle with fixed turbine inlet conditions. What is the effect of lowering the condenser pressure on

Pump work input:	(*a*) increases, (*b*) decreases, (*c*) remains the same
Turbine work output:	(*a*) increases, (*b*) decreases, (*c*) remains the same
Heat supplied:	(*a*) increases, (*b*) decreases, (*c*) remains the same
Heat rejected:	(*a*) increases, (*b*) decreases, (*c*) remains the same
Cycle efficiency:	(*a*) increases, (*b*) decreases, (*c*) remains the same
Moisture content at turbine exit:	(*a*) increases, (*b*) decreases, (*c*) remains the same

8–96C Consider a simple ideal Rankine cycle with fixed turbine inlet temperature and condenser pressure. What is the effect of increasing the boiler pressure on

Pump work input:	(*a*) increases, (*b*) decreases, (*c*) remains the same
Turbine work output:	(*a*) increases, (*b*) decreases, (*c*) remains the same
Heat supplied:	(*a*) increases, (*b*) decreases, (*c*) remains the same
Heat rejected:	(*a*) increases, (*b*) decreases, (*c*) remains the same
Cycle efficiency:	(*a*) increases, (*b*) decreases, (*c*) remains the same
Moisture content at turbine exit:	(*a*) increases, (*b*) decreases, (*c*) remains the same

8–97C Consider a simple ideal Rankine cycle with fixed boiler and condenser pressures. What is the effect of superheating the steam to a higher temperature on

Pump work input:	(*a*) increases, (*b*) decreases, (*c*) remains the same
Turbine work output:	(*a*) increases, (*b*) decreases, (*c*) remains the same
Heat supplied:	(*a*) increases, (*b*) decreases, (*c*) remains the same
Heat rejected:	(*a*) increases, (*b*) decreases, (*c*) remains the same
Cycle efficiency:	(*a*) increases, (*b*) decreases, (*c*) remains the same
Moisture content at turbine exit:	(*a*) increases, (*b*) decreases, (*c*) remains the same

8–98C How do actual vapor power cycles differ from idealized ones?

8–99C Compare the pressures at the inlet and the exit of the boiler for (*a*) actual and (*b*) ideal cycles.

8–100C The entropy of steam increases in actual steam turbines as a result of irreversibilities. In an effort to control entropy increase, it is proposed to cool the steam in the turbine by running cooling water around the turbine casing. It is argued that this will reduce the entropy and the enthalpy of the steam at the turbine exit and thus increase the work output. How would you evaluate this proposal?

8-101C   Is it possible to maintain a pressure of 10 kPa in a condenser that is being cooled by river water entering at 20°C?

8-102   A steam power plant operates on a simple ideal Rankine cycle between the pressure limits of 3 MPa and 50 kPa. The temperature of the steam at the turbine inlet is 400°C, and the mass flow rate of steam through the cycle is 25 kg/s. Show the cycle on a *T-s* diagram with respect to saturation lines, and determine (*a*) the thermal efficiency of the cycle and (*b*) the net power output of the power plant.

8-103   Consider a 300-MW steam power plant that operates on a simple ideal Rankine cycle. Steam enters the turbine at 10 MPa and 500°C and is cooled in the condenser at a pressure of 10 kPa. Show the cycle on a *T-s* diagram with respect to saturation lines, and determine (*a*) the quality of the steam at the turbine exit, (*b*) the thermal efficiency of the cycle, and (*c*) the mass flow rate of the steam.
   Answers: (*a*) 0.793, (*b*) 40.2 percent, (*c*) 235.4 kg/s

8-104   Repeat Prob. 8-103 assuming an isentropic efficiency of 85 percent for both the turbine and the pump.
   Answers: (*a*) 0.874, (*b*) 34.1 percent, (*c*) 277.8 kg/s

8-105E   A steam power plant operates on a simple ideal Rankine cycle between the pressure limits of 1250 and 2 psia. The mass flow rate of steam through the cycle is 75 lbm/s. The moisture content of the steam at the turbine exit is not to exceed 10 percent. Show the cycle on a *T-s* diagram with respect to saturation lines, and determine (*a*) the minimum turbine inlet temperature, (*b*) the rate of heat input in the boiler, and (*c*) the thermal efficiency of the cycle.

8-106E   Repeat Prob. 8-105E assuming an isentropic efficiency of 85 percent for both the turbine and the pump.

8-107   Consider a coal-fired steam power plant that produces 300 MW of electric power. The power plant operates on a simple ideal Rankine cycle with turbine inlet conditions of 5 MPa and 450°C and a condenser pressure of 25 kPa. The coal used has a heating value (energy released when the fuel is burned) of 29,300 kJ/kg. Assuming that 75 percent of this energy is transferred to the steam in the boiler and that the electric generator has an efficiency of 96 percent, determine (*a*) the overall plant efficiency (the ratio of net electric power output to the energy input as fuel) and (*b*) the required rate of coal supply.   *Answers:* (*a*) 24.6 percent, (*b*) 150.1 t/h

8-108   Consider a solar-pond power plant that operates on a simple ideal Rankine cycle with refrigerant-134a as the working fluid. The refrigerant enters the turbine as a saturated vapor at 1.6 MPa and leaves at 0.7 MPa. The mass flow rate of the refrigerant is 6 kg/s. Show the cycle on a *T-s* diagram with respect to saturation lines, and determine (*a*) the thermal efficiency of the cycle and (*b*) the power output of this plant.

8-109   Consider a steam power plant that operates on a simple ideal Rankine cycle and has a net power output of 30 MW. Steam enters the turbine at 7 MPa and 500°C and is cooled in the condenser at a pressure of 10 kPa by running cooling water from a lake through the tubes of the condenser at a rate of 2000 kg/s. Show the cycle on a *T-s* diagram with respect to saturation lines, and determine (*a*) the thermal efficiency of the cycle, (*b*) the mass flow rate of the steam, and (*c*) the temperature rise of the cooling water.
   Answers: (*a*) 38.9 percent, (*b*) 24.0 kg/s, (*c*) 5.63°C

8-110   Repeat Prob. 8-109 assuming an isentropic efficiency of 87 percent for both the turbine and the pump.

Answers: (*a*) 33.8 percent, (*b*) 27.65 kg/s, (*c*) 7.03°C

The Reheat Rankine Cycle

8-111C   How do the following quantities change when a simple ideal Rankine cycle is modified with reheating? Assume the mass flow rate is maintained the same.

Pump work input:	(*a*) increases, (*b*) decreases, (*c*) remains the same
Turbine work output:	(*a*) increases, (*b*) decreases, (*c*) remains the same
Heat supplied:	(*a*) increases, (*b*) decreases, (*c*) remains the same
Heat rejected:	(*a*) increases, (*b*) decreases, (*c*) remains the same
Moisture content at turbine exit:	(*a*) increases, (*b*) decreases, (*c*) remains the same

8-112C   Show the ideal Rankine cycle with three stages of reheating on a *T-s* diagram. Assume the turbine inlet temperature is the same for all stages. How does the cycle efficiency vary with the number of reheat stages?

8-113C   Consider a simple Rankine cycle and an ideal Rankine cycle with three reheat stages. Both cycles operate between the same pressure limits. The maximum temperature is 700°C in the simple cycle and 500°C in the reheat cycle. Which cycle do you think will have a higher thermal efficiency?

8-114   A steam power plant operates on the ideal reheat Rankine cycle. Steam enters the high-pressure turbine at 8 MPa and 500°C and leaves at 3 MPa. Steam is then reheated at constant pressure to 500°C before it expands to 20 kPa in the low-pressure turbine. Determine the turbine work output, in kJ/kg, and the thermal efficiency of the cycle. Also, show the cycle on a *T-s* diagram with respect to saturation lines.

8-115   Consider a steam power plant that operates on a reheat Rankine cycle and has a net power output of 150 MW. Steam enters the high-pressure turbine at 10 MPa and 500°C and the low-pressure turbine at 1 MPa and 500°C. Steam leaves the condenser as a saturated liquid at a pressure of 10 kPa. The isentropic efficiency of the turbine is 80 percent, and that of the pump is 95 percent. Show the cycle on a *T-s* diagram with respect to saturation lines, and determine (*a*) the quality (or temperature, if superheated) of the steam at the turbine exit, (*b*) the thermal efficiency of the cycle, and (*c*) the mass flow rate of the steam.   *Answers:* (*a*) 87.5°C, (*b*) 34.1 percent, (*c*) 117.5 kg/s

8-116   Repeat Prob. 8-115 assuming both the pump and the turbine are isentropic.   *Answers:* (*a*) 0.948, (*b*) 41.4 percent, (*c*) 93.8 kg/s

8-117E   Steam enters the high-pressure turbine of a steam power plant that operates on the ideal reheat Rankine cycle at 800 psia and 900°F and leaves as saturated vapor. Steam is then reheated to 800°F before it expands to a pressure of 1 psia. Heat is transferred to the steam in the boiler at a rate of 6×10^4 Btu/s. Steam is cooled in the condenser by the cooling water from a nearby river, which enters the condenser at 45°F. Show the cycle on a *T-s* diagram with respect to saturation lines, and determine (*a*) the pressure at which reheating takes place, (*b*) the net power output and thermal efficiency, and (*c*) the minimum mass flow rate of the cooling water required.

8-118 A steam power plant operates on an ideal reheat Rankine cycle between the pressure limits of 9 MPa and 10 kPa. The mass flow rate of steam through the cycle is 25 kg/s. Steam enters both stages of the turbine at 500°C. If the moisture content of the steam at the exit of the low-pressure turbine is not to exceed 10 percent, determine (a) the pressure at which reheating takes place, (b) the total rate of heat input in the boiler, and (c) the thermal efficiency of the cycle. Also, show the cycle on a T-s diagram with respect to saturation lines.

The Reversed Carnot Cycle

8-119C Why is the reversed Carnot cycle executed within the saturation dome not a realistic model for refrigeration cycles?

8-120C What is the difference between a refrigerator and a heat pump?

8-121 A steady-flow Carnot refrigeration cycle uses refrigerant-134a as the working fluid. The refrigerant changes from saturated vapor to saturated liquid at 30°C in the condenser as it rejects heat. The evaporator pressure is 120 kPa. Show the cycle on a T-s diagram relative to saturation lines, and determine (a) the coefficient of performance, (b) the amount of heat absorbed from the refrigerated space, and (c) the net work input.
Answers: (a) 4.44, *(b)* 140.4 kJ/kg, *(c)* 31.61 kJ/kg

8-122E Refrigerant-134a enters the condenser of a steady-flow Carnot refrigerator as a saturated vapor at 90 psia, and it leaves with a quality of 0.05. The heat absorption from the refrigerated space takes place at a pressure of 30 psia. Show the cycle on a T-s diagram relative to saturation lines, and determine (a) the coefficient of performance, (b) the quality at the beginning of the heat-absorption process, and (c) the net work input.

Ideal and Actual Vapor-Compression Refrigeration Cycles

8-123C Does the ideal vapor-compression refrigeration cycle involve any internal irreversibilities?

8-124C Why is the throttling valve not replaced by an isentropic turbine in the ideal vapor-compression refrigeration cycle?

8-125C It is proposed to use water instead of refrigerant-134a as the working fluid in air-conditioning applications where the minimum temperature never falls below the freezing point. Would you support this proposal? Explain.

8-126C In a refrigeration system, would you recommend condensing the refrigerant-134a at a pressure of 0.7 or 1.0 MPa if heat is to be rejected to a cooling medium at 15°C? Why?

8-127C Does the area enclosed by the cycle on a T-s diagram represent the net work input for the reversed Carnot cycle? How about for the ideal vapor-compression refrigeration cycle?

8-128C Consider two vapor-compression refrigeration cycles. The refrigerant enters the throttling valve as a saturated liquid at 30°C in one cycle and as subcooled liquid at 30°C in the other one. The evaporator pressure for both cycles is the same. Which cycle do you think will have a higher COP?

8-129C The COP of vapor-compression refrigeration cycles improves when the refrigerant is subcooled before it enters the throttling valve. Can the refrigerant be subcooled indefinitely to maximize this effect, or is there a lower limit? Explain.

8-130 A refrigerator uses refrigerant-134a as the working fluid and operates on an ideal vapor-compression refrigeration cycle between 0.12 and 0.7 MPa. The mass flow rate of the refrigerant is 0.05 kg/s. Show the cycle on a *T-s* diagram with respect to saturation lines. Determine (*a*) the rate of heat removal from the refrigerated space and the power input to the compressor, (*b*) the rate of heat rejection to the environment, and (*c*) the coefficient of performance.
    Answers: (*a*) 7.35 kW, 1.82 kW; (*b*) 9.17 kW; (*c*) 4.04

8-131 If the throttling valve in Prob. 8-130 is replaced by an isentropic turbine, determine the percentage increase in the COP and in the rate of heat removal from the refrigerated space.

8-132 Consider a 300 kJ/min refrigeration system that operates on an ideal vapor-compression refrigeration cycle with refrigerant-134a as the working fluid. The refrigerant enters the compressor as saturated vapor at 140 kPa and is compressed to 800 kPa. Show the cycle on a *T-s* diagram with respect to saturation lines, and determine (*a*) the quality of the refrigerant at the end of the throttling process, (*b*) the coefficient of performance, and (*c*) the power input to the compressor.

8-133 Repeat Prob. 8-132 assuming an isentropic efficiency of 85 percent for the compressor.

8-134 Refrigerant-134a enters the compressor of a refrigerator as superheated vapor at 0.14 MPa and −10°C at a rate of 0.04 kg/s, and it leaves at 0.7 MPa and 50°C. The refrigerant is cooled in the condenser to 24°C and 0.65 MPa, and it is throttled to 0.15 MPa. Disregarding any heat transfer and pressure drops in the connecting lines between the components, show the cycle on a *T-s* diagram with respect to saturation lines, and determine (*a*) the rate of heat removal from the refrigerated space and the power input to the compressor, (*b*) the isentropic efficiency of the compressor, and (*c*) the COP of the refrigerator.
    Answers: (*a*) 6.42 kW, 1.72 kW; (*b*) 80.7 percent; (*c*) 3.73

8-135E An ice-making machine operates on the ideal vapor-compression cycle, using refrigerant-134a. The refrigerant enters the compressor as saturated vapor at 20 psia and leaves the condenser as saturated liquid at 100 psia. Water enters the ice machine at 55°F and leaves as ice at 25°F. For an ice production rate of 20 lbm/h, determine the power input to the ice machine (169 Btu of heat needs to be removed from each lbm of water at 55°F to turn it into ice at 25°F).

8-136 Refrigerant-134a enters the compressor of a refrigerator at 140 kPa and −10°C at a rate of 0.2 m³/min and leaves at 1 MPa. The isentropic efficiency of the compressor is 78 percent. The refrigerant enters the throttling valve at 0.95 MPa and 30°C and leaves the evaporator as saturated vapor at −18.5°C. Show the cycle on a *T-s* diagram with respect to saturation lines, and determine (*a*) the power input to the compressor, (*b*) the rate of heat removal from the refrigerated space, and (*c*) the pressure drop and rate of heat gain in the line between the evaporator and the compressor.
    Answers: (*a*) 1.25 kW; (*b*) 3.31 kW; (*c*) 1.87 kPa, 0.164 kW

Heat Pump Systems

8-137C   What are the advantages and disadvantages of heat pumps? How do they compare to other heating systems?

8-138C   Do you think a heat pump system will be more cost-effective in New York or in Miami? Why?

8-139C   What is a water-source heat pump? How does the COP of a water-source heat pump system compare to that of an air-source system?

8-140E   A heat pump that operates on the ideal vapor-compression cycle with refrigerant-134a is used to heat a house and maintain it at 75°F by using underground water at 50°F as the heat source. The house is losing heat at a rate of 90,000 Btu/h. The evaporator and condenser pressures are 50 and 120 psia, respectively. Determine the power input to the heat pump and the electric power saved by using a heat pump instead of a resistance heater.
    Answers: 3.68 hp, 31.69 hp

8-141   A heat pump that operates on the ideal vapor-compression cycle with refrigerant-134a is used to heat water from 15 to 54°C at a rate of 0.18 kg/s. The condenser and evaporator pressures are 1.4 and 0.32 MPa, respectively. Determine the power input to the heat pump.

8-142   A heat pump using refrigerant-134a heats a house by using underground water at 8°C as the heat source. The house is losing heat at a rate of 60,000 kJ/h. The refrigerant enters the compressor at 280 kPa and 0°C, and it leaves at 1 MPa and 60°C. The refrigerant exits the condenser at 30°C. Determine (*a*) the power input to the heat pump, (*b*) the rate of heat absorption from the water, and (*c*) the increase in electric power input if an electric resistance heater is used instead of a heat pump.
    Answers: (*a*) 3.65 kW, (*b*) 13.02 kW, (*c*) 13.02 kW

Review Problems

8-143   A four-stroke turbocharged V-16 diesel engine built by GE Transportation Systems to power fast trains produces 4000 hp at 1050 rpm. Determine the amount of power produced per cylinder per (*a*) mechanical cycle and (*b*) thermodynamic cycle.

8-144   Consider a simple ideal Brayton cycle operating between the temperature limits of 290 K and 1500 K. Using constant specific heats at room temperature, determine the pressure ratio for which the compressor and the turbine exit temperatures of air are equal.

8-145   An air-standard cycle with variable coefficients is executed in a closed system and is composed of the following four processes:
    1-2 $v = constant$ heat addition from 100 kPa and 27°C to 300 kPa
    2-3 $P = constant$ heat addition to 1027°C
    3-4   Isentropic expansion to 100 kPa
    4-1 $P = constant$ heat rejection to initial state
    (*a*) Show the cycle on *P-v* and *T-s* diagrams.
    (*b*) Calculate the net work output per unit mass.
    (*c*) Determine the thermal efficiency.

8-146   Repeat Prob. 8-145 using constant specific heats at room temperature.

8-147 An air-standard cycle with variable specific heats is executed in a closed system with 0.002 kg of air, and it consists of the following three processes:

    1-2   Isentropic compression from 100 kPa and 27°C to 700 kPa
    2-3 $P = constant$ heat addition to initial specific volume
    3-1 $v = constant$ heat rejection to initial state
    (*a*) Show the cycle on *P-v* and *T-s* diagrams.
    (*b*) Calculate the maximum temperature in the cycle.
    (*c*) Determine the thermal efficiency.
    Answers: (*b*) 2100 K, (*c*) 15.8 percent

8-148 Repeat Prob. 8-147 using constant specific heats at room temperature.

8-149 A Carnot cycle is executed in a closed system and uses 0.002 kg of air as the working fluid. The cycle efficiency is 70 percent, and the lowest temperature in the cycle is 300 K. The pressure at the beginning of the isentropic expansion is 700 kPa, and at the end of the isentropic compression it is 1 MPa. Determine the net work output per cycle.

8-150 A four-cylinder spark-ignition engine has a compression ratio of 8, and each cylinder has a maximum volume of 0.6 L. At the beginning of the compression process, the air is at 98 kPa and 17°C, and the maximum temperature in the cycle is 1800 K. Assuming the engine to operate on the ideal Otto cycle, determine (*a*) the amount of heat supplied per cylinder, (*b*) the thermal efficiency, and (*c*) the number of revolutions per minute required for a net power output of 60 kW. Assume variable specific heats for air.

8-151 An ideal Otto cycle has a compression ratio of 9.2 and uses air as the working fluid. At the beginning of the compression process, air is at 98 kPa and 27°C. The pressure is doubled during the constant-volume heat-addition process. Accounting for the variation of specific heats with temperature, determine (*a*) the amount of heat transferred to the air, (*b*) the net work output, (*c*) the thermal efficiency, and (*d*) the mean effective pressure for the cycle.

8-152 Repeat Prob. 8-151 using constant specific heats at room temperature.

8-153 Consider an engine operating on the ideal Diesel cycle with air as the working fluid. The volume of the cylinder is 1200 cm³ at the beginning of the compression process, 75 cm³ at the end, and 150 cm³ after the heat-addition process. Air is at 17°C and 100 kPa at the beginning of the compression process. Determine (*a*) the pressure at the beginning of the heat-rejection process; (*b*) the net work per cycle, in kJ; and (*c*) the mean effective pressure.

8-154 Repeat Prob. 8-153 using argon as the working fluid.

8-155E An ideal dual cycle has a compression ratio of 12 and uses air as the working fluid. At the beginning of the compression process, air is at 14.7 psia and 90°F, and occupies a volume of 75 in³. During the heat-addition process, 0.3 Btu of heat is transferred to air at constant volume and 1.1 Btu at constant pressure. Using constant specific heats evaluated at room temperature, determine the thermal efficiency of the cycle.

8-156 Consider a simple ideal Brayton cycle with air as the working fluid. The pressure ratio of the cycle is 6, and the minimum and maximum temperatures are 300 and 1300 K, respectively. Now the pressure ratio is doubled without changing the minimum and maximum temperatures in the cycle. Determine the change in (*a*) the net work output per unit mass and (*b*) the thermal efficiency of the cycle as a result of this modification. Assume variable specific heats for air.   *Answers:* (a) 41.5 kJ/kg, (b) 10.6 percent

8-157   Repeat Prob. 8-156 using constant specific heats at room temperature.

8-158   Helium is used as the working fluid in a Brayton cycle with regeneration. The pressure ratio of the cycle is 8, the compressor inlet temperature is 300 K, and the turbine inlet temperature is 1800 K. The effectiveness of the regenerator is 75 percent. Determine the thermal efficiency and the required mass flow rate of helium for a net power output of 30 MW, assuming both the compressor and the turbine have an isentropic efficiency of (a) 100 percent and (b) 80 percent.

8-159   Consider the ideal regenerative Brayton cycle. Determine the pressure ratio that maximizes the thermal efficiency of the cycle and compare this value with the pressure ratio that maximizes the cycle net work. For the same maximum-to-minimum temperature ratios, explain why the pressure ratio for maximum efficiency is less than the pressure ratio for maximum work.

8-160   Consider an ideal gas-turbine cycle with one stage of compression and two stages of expansion and regeneration. The pressure ratio across each turbine stage is the same. The high-pressure turbine exhaust gas enters the regenerator and then enters the low-pressure turbine for expansion to the compressor inlet pressure. Determine the thermal efficiency of this cycle as a function of the compressor pressure ratio and the high-pressure turbine to compressor inlet temperature ratio. Compare your result with the efficiency of the standard regenerative cycle.

8-161   Consider a steam power plant operating on the ideal Rankine cycle with reheat between the pressure limits of 25 MPa and 10 kPa with a maximum cycle temperature of 600°C and a moisture content of 12 percent at the turbine exit. For a reheat temperature of 600°C, determine the reheat pressures of the cycle for the cases of (a) single and (b) double reheat.

8-162E   A geothermal power plant in Nevada, which started full commercial operation in 1986, is designed to operate with seven identical units. Each of these seven units consists of a pair of power cycles, labeled Level I and Level II, operating on the simple Rankine cycle using an organic fluid as the working fluid.

The heat source for the plant is geothermal water (brine) entering the vaporizer (boiler) of Level I of each unit at 325°F at a rate of 384,286 lbm/h and delivering 22.79 MBtu/h ("M" stands for "million"). The organic fluid that enters the vaporizer at 202.2°F at a rate of 157,895 lbm/h leaves it at 282.4°F and 225.8 psia as saturated vapor. This saturated vapor expands in the turbine to 95.8°F and 19.0 psia and produces 1271 kW of electric power. About 200 kW of this power is used by the pumps, the auxiliaries, and the six fans of the condenser. Subsequently, the organic working fluid is condensed in an air-cooled condenser by air that enters the condenser at 55°F at a rate of 4,195,100 lbm/h and leaves at 84.5°F. The working fluid is pumped and then preheated in a preheater to 202.4°F by absorbing 11.14 MBtu/h of heat from the geothermal water (coming from the vaporizer of Level II) entering the preheater at 211.8°F and leaving at 154.0°F.

Taking the average specific heat of the geothermal water to be 1.03 Btu/lbm · °F, determine (a) the exit temperature of the geothermal water from the vaporizer, (b) the rate of heat rejection from the working fluid to the air in the condenser, (c) the mass flow rate of the geothermal water at the pre-

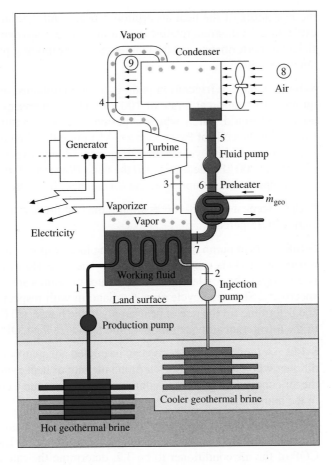

FIGURE P8-162E

Schematic of a binary geothermal power plant. (*Courtesy of ORMAT Energy Systems, Inc.*)

heater, and (*d*) the thermal efficiency of the Level I cycle of this geothermal power plant.   *Answers:* (*a*) 267.4°F, (*b*) 29.7 MBtu/h, (*c*) 187,120 lbm/h, (*d*) 10.8 percent.

8-163 Steam enters the turbine of a steam power plant that operates on a simple ideal Rankine cycle at a pressure of 6 MPa, and it leaves as a saturated vapor at 7.5 kPa. Heat is transferred to the steam in the boiler at a rate of 4×10^4 kJ/s. Steam is cooled in the condenser by the cooling water from a nearby river, which enters the condenser at 18°C. Show the cycle on a *T-s* diagram with respect to saturation lines, and determine (*a*) the turbine inlet temperature, (*b*) the net power output and thermal efficiency, and (*c*) the minimum mass flow rate of the cooling water required.

8-164 A steam power plant operates on an ideal Rankine cycle with two stages of reheat and has a net power output of 300 MW. Steam enters all three stages of the turbine at 500°C. The maximum pressure in the cycle is 15 MPa, and the minimum pressure is 5 kPa. Steam is reheated at 5 MPa the first time and at 1 MPa the second time. Show the cycle on a *T-s* diagram with respect to saturation lines, and determine (*a*) the thermal efficiency of the cycle and (*b*) the mass flow rate of the steam.
    Answers: (*a*) 45.5 percent, (*b*) 161.6 kg/s

8-165 Consider a steady-flow Carnot refrigeration cycle that uses refrigerant-134a as the working fluid. The maximum and minimum temperatures in the cycle are 20 and −20°C, respectively. The quality of the refrigerant is 0.2 at the beginning of the heat absorption process and 0.85 at the end. Show the cycle on a *T-s* diagram relative to saturation lines, and determine (*a*) the coefficient of performance, (*b*) the condenser and evaporator pressures, and (*c*) the net work input.

8-166   A large refrigeration plant is to be maintained at $-15°C$, and it requires refrigeration at a rate of 100 kW. The condenser of the plant is to be cooled by liquid water, which experiences a temperature rise of 8°C as it flows over the coils of the condenser. Assuming the plant operates on the ideal vapor-compression cycle using refrigerant-134a between the pressure limits of 120 and 700 kPa, determine (*a*) the mass flow rate of the refrigerant, (*b*) the power input to the compressor, and (*c*) the mass flow rate of the cooling water.

8-167   Repeat Prob. 8-166 assuming the compressor has an isentropic efficiency of 75 percent.

8-168   A heat pump that operates on the ideal vapor-compression cycle with refrigerant-134a is used to heat a house. The mass flow rate of the refrigerant is 0.15 kg/s. The condenser and evaporator pressures are 900 and 240 kPa, respectively. Show the cycle on a *T-s* diagram with respect to saturation lines, and determine (*a*) the rate of heat supply to the house, (*b*) the volume flow rate of the refrigerant at the compressor inlet, and (*c*) the COP of this heat pump.

8-169   A typical 200-m² house can be cooled adequately by a 3.5-ton air conditioner whose COP is 4.0. Determine the rate of heat gain of the house when the air conditioner is running continuously to maintain a constant temperature in the house.

8-170   Rooms with floor areas of up to 15-m² are cooled adequately by window air conditioners whose cooling capacity is 5000 Btu/h. Assuming the COP of the air conditioner to be 3.2, determine the rate of heat gain of the room, in Btu/h, when the air conditioner is running continuously to maintain a constant room temperature.

8-171   A heat pump water heater (HPWH) heats water by absorbing heat from the ambient air and transferring it to water. The heat pump has a COP of 2.2 and consumes 2 kW of electricity when running. Determine if this heat pump can be used to meet the cooling needs of a room most of the time for "free" by absorbing heat from the air in the room. The rate of heat gain of a room is usually less than 5000 kJ/h.

Computer, Design, and Essay Problems

8-172   Write a computer program to study the effect of variable specific heats on the thermal efficiency of the ideal Otto cycle using air as the working fluid. At the beginning of the compression process, air is at 100 kPa and 300 K. Use the equation in Table A-2*c* to account for the variation of specific heats with temperature. Determine the percentage of error involved in using constant specific-heat values at room temperature for the following combinations of compression ratios and maximum cycle temperatures: $r = 7, 8, 9, 10, 11, 12$ and $T_{max} = 1200, 1400, 1600, 1800, 2000, 2500$ K.

8-173   Write a computer program to determine the effects of pressure ratio, maximum cycle temperature, and compressor and turbine inefficiencies on the net work output per unit mass and the thermal efficiency of a simple Brayton cycle. Assume the working fluid is air that is at 100 kPa and 300 K at the compressor inlet. Also, assume constant specific heats for air at room temperature. Determine the net work output and the thermal efficiency for all combinations of the following parameters:

Pressure ratio: 5, 8, 14
Maximum cycle temperature: 800, 1200, 1600 K
Compressor adiabatic efficiency: 80, 100 percent
Turbine isentropic efficiency: 80, 100 percent

Draw conclusions from the results.

8-174   Repeat Prob. 8-173 using helium as the working fluid.

8-175   Repeat Prob. 8-173 by considering the variation of specific heats of air with temperature. Use the specific-heat expressions for air given in Table A-2(*c*).

8-176   Write a computer program to determine the effects of pressure ratio, maximum cycle temperature, regenerator effectiveness, and compressor and turbine efficiencies on the net work output per unit mass and on the thermal efficiency of a regenerative Brayton cycle. Assume the working fluid is air that is at 100 kPa and 300 K at the compressor inlet. Also, assume constant specific heats for air at room temperature. Determine the net work output and the thermal efficiency for all combinations of the following parameters:

Pressure ratio: 5, 8, 14
Maximum cycle temperature: 1000, 1400, 1600 K
Compressor isentropic efficiency: 80, 100 percent
Turbine isentropic efficiency: 80, 100 percent
Regenerator effectiveness: 70, 80 percent

8-177   Repeat Prob. 8-176 using helium as the working fluid.

8-178   Repeat Prob. 8-176 by considering the variation of specific heats of air with temperature. Use the specific-heat expressions for air given in Table A-2*c*.

8-179   Design a steam power cycle that can achieve a cycle thermal efficiency of at least 40 percent under the conditions that all turbines have isentropic efficiencies of 85 percent and all pumps have isentropic efficiencies of 60 percent. Prepare an engineering report describing your design. Your design report must include, but is not limited to, the following:

(*a*) Discussion of various cycles attempted to meet the goal as well as the positive and negative aspects of your design.

(*b*) System figures and *T-s* diagrams with labeled states and temperature, pressure, enthalpy, and entropy information for your design.

(*c*) Sample calculations.

8-180   Contact your power company and obtain information on the thermodynamic aspects of their most recently built power plant. If it is a conventional power plant, find out why it is preferred over a highly efficient combined power plant.

8-181   Several geothermal power plants are in operation in the United States and more are being built since the heat source of a geothermal plant is hot geothermal water, which is "free energy." An 8-MW geothermal power plant is being considered at a location where geothermal water at 160°C is available. Geothermal water is to serve as the heat source for a closed Rankine power cycle with refrigerant-134a as the working fluid (see Fig. P8-162E).

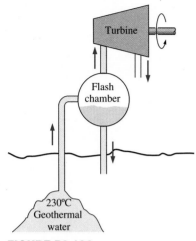

Specify suitable temperatures and pressures for the cycle, and determine the thermal efficiency of the cycle. Justify your selections.

8-182 A 10-MW geothermal power plant is being considered at a site where geothermal water at 230°C is available. Geothermal water is to be flashed into a chamber to a lower pressure where part of the water evaporates. The liquid is returned to the ground while the vapor is used to drive the steam turbine. The pressures at the turbine inlet and the turbine exit are to remain above 200 kPa and 8 kPa, respectively. High-pressure flash chambers yield a small amount of steam with high exergy whereas lower-pressure flash chambers yield considerably more steam but at a lower exergy. By trying several pressures, determine the optimum pressure of the flash chamber to maximize the power production per unit mass of geothermal water withdrawn. Also, determine the thermal efficiency for each case assuming 10 percent of the power produced is used to drive the pumps and other auxiliary equipment.

8-183 A natural gas–fired furnace in a textile plant is used to provide steam at 130°C. At times of high demand, the furnace supplies heat to the steam at a rate of 30 MJ/s. The plant also uses up to 6 MW of electrical power purchased from the local power company. The plant management is considering converting the existing process plant into a cogeneration plant to meet both their process-heat and power requirements. Your job is to come up with some designs. Designs based on a gas turbine or a steam turbine are to be considered. First decide whether a system based on a gas turbine or a steam turbine will best serve the purpose, considering the cost and the complexity. Then propose your design for the cogeneration plant complete with pressures and temperatures and the mass flow rates. Show that the proposed design meets the power and process-heat requirements of the plant.

8-184E A photographic equipment manufacturer uses a flow of 64,500 lbm/h of steam in its manufacturing process. Presently the spent steam at 3.8 psig and 224°F is exhausted to the atmosphere. Do the preliminary design of a system to use the energy in the waste steam economically. If electricity is produced, it can be generated about 8000 h/yr and its value is $0.05/kWh. If the energy is used for space heating, the value is also $0.05/kWh, but it can only be used about 3000 h/yr (only during the "heating season"). If the steam is condensed and the liquid H_2O is recycled through the process, its value is $0.50/100 gal. Make all assumptions as realistic as possible. Sketch the system you propose. Make a separate list of required components and their specifications (capacity, efficiency, etc.). The final result will be the calculated annual dollar value of the energy use plan (actually a *saving* because it will replace electricity or heat and/or water that would otherwise have to be purchased).

8-185 Design the condenser of a steam power plant that has a thermal efficiency of 40 percent and generates 10 MW of net electric power. Steam enters the condenser as saturated vapor at 10 kPa, and it is to be condensed outside horizontal tubes through which cooling water from a nearby river flows. The temperature rise of the cooling water is limited to 8°C, and the velocity of the cooling water in the pipes is limited to 6 m/s to keep the pressure drop at an acceptable level. From prior experience, the average heat flux based on the outer surface of the tubes can be taken to be 12,000 W/m². Specify the pipe diameter, total pipe length, and the arrangement of the pipes to minimize the condenser volume.

8-186   It is proposed to use a solar-powered thermoelectric system installed on the roof to cool residential buildings. The system consists of a thermoelectric refrigerator that is powered by a thermoelectric power generator whose top surface is a solar collector. Discuss the feasibility and the cost of such a system, and determine if the proposed system installed on one side of the roof can meet a significant portion of the cooling requirements of a typical house in your area.

8-187   A refrigerator using R-12 as the working fluid keeps the refrigerated space at $-15°C$ in an environment at 30°C. You are asked to redesign this refrigerator by replacing R-12 with the ozone-friendly R-134a. What changes in the pressure levels would you suggest in the new system? How do you think the COP of the new system will compare to the COP of the old system?

8-188   Solar or photovoltaic (PV) cells convert sunlight to electricity and are commonly used to power calculators, satellites, remote communication systems, and even pumps. The conversion of light to electricity is called the *photoelectric effect.* It was first discovered in 1839 by Frenchman Edmond Becquerel, and the first PV module, which consisted of several cells connected to each other, was built in 1954 by Bell Laboratories. The PV modules today have conversion efficiencies of about 12 to 15 percent. Noting that the solar energy incident on a normal surface on earth at noontime is about 1000 W/m^2 during a clear day, PV modules on a 1-m^2 surface can provide as much as 150 W of electricity. The annual average daily solar energy incident on a horizontal surface in the United States ranges from about 2 to 6 kWh/m^2.

A PV-powered pump is to be used in Arizona to pump water for wildlife from a depth of 180 m at an average rate of 400 L/day. Assuming a reasonable efficiency for the pumping system, which can be defined as the ratio of the increase in the potential energy of the water to the electrical energy consumed by the pump, and taking the conversion efficiency of the PV cells to be 0.13 to be on the conservative side, determine the size of the PV module that needs to be installed, in m^2.

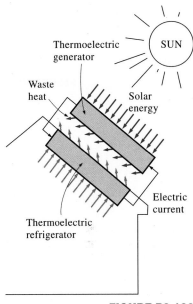

FIGURE P8-186

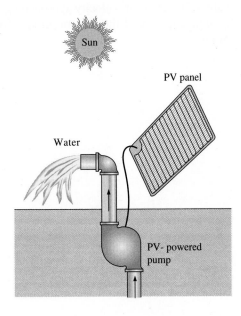

FIGURE P8-188

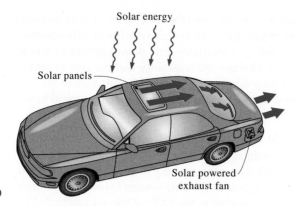

Solar energy

Solar panels

Solar powered
exhaust fan

FIGURE P8-189

8-189   The temperature in a car parked in the sun can approach 100°C when the outside air temperature is just 25°C, and it is desirable to ventilate the parked car to avoid such high temperatures. However, the ventilating fans may run down the battery if they are powered by it. To avoid that happening, it is proposed to use the PV cells discussed in Prob. 8-188 to power the fans. It is determined that the air in the car should be replaced once every minute to avoid excessive rise in the interior temperature. Determine if this can be accomplished by installing PV cells on part of the roof of the car. Also, find out if any car is currently ventilated this way.

8-190   A company owns a refrigeration system whose refrigeration capacity is 200 tons (1 ton of refrigeration = 211 kJ/min), and you are to design a forced-air cooling system for fruits whose diameters do not exceed 7 cm under the following conditions: The fruits are to be cooled from 28°C to an average temperature of 8°C. The air temperature is to remain above −2°C and below 10°C at all times, and the velocity of air approaching the fruits must remain under 2 m/s. The cooling section can be as wide as 3.5 m and as high as 2 m.

Assuming reasonable values for the average fruit density, specific heat, and porosity (the fraction of air volume in a box), recommend reasonable values for (*a*) the air velocity approaching the cooling section, (*b*) the product-cooling capacity of the system, in kg · fruit/h, and (*c*) the volume flow rate of air.

Fluid Mechanics

PART

Introduction to Fluid Mechanics

In the second part of the text we present the fundamentals of fluid mechanics. In this introductory chapter, we introduce the basic concepts commonly used in the analysis of fluid flow to avoid any misunderstandings. We start with a discussion of the numerous ways of classification of fluid flow, such as *viscous versus inviscid flow, internal versus external flow, compressible versus incompressible flow, laminar versus turbulent flow, natural versus forced flow,* and *steady versus unsteady flow.* We also discuss the no-slip condition, which is responsible for the development of boundary layers adjacent to the solid surfaces, and the no-temperature-jump condition. We continue with a brief history of the development of fluid mechanics. We then discuss the property *viscosity,* which plays a dominant role in most aspects of fluid flow. Finally, we present the property *surface tension,* and determine the *capillary rise* from static equilibrium conditions.

9-1 ■ CLASSIFICATION OF FLUID FLOWS

In Chap. 1 we defined *fluid mechanics* as the science that deals with the behavior of fluids at rest or in motion, and the interaction of fluids with solids or other fluids at the boundaries. There is a wide variety of fluid flow problems encountered in practice, and it is usually convenient to classify them on the basis of some common characteristics to make it feasible to study them in groups. There are many ways to classify the fluid flow problems, and below we present some general categories.

Viscous versus Inviscid Flow

When two fluid layers move relative to each other, a friction force develops between them and the slower layer tries to slow down the faster layer. This internal resistance to flow is called **viscosity,** which is a measure of internal stickiness of the fluid. Viscosity is caused by the cohesive forces between the molecules in liquids and by the molecular collisions in gases and is responsible for momentum transport at the molecular level. There is no fluid with zero viscosity, and thus all fluid flows involve viscous effects to some degree. The viscosity approaches zero for the so-called *superfluids* at extremely low temperatures. Flows in which the effects of viscosity are significant are called **viscous flows.** The effects of viscosity are very small in some flows, and neglecting those effects greatly simplifies the analysis without much loss in accuracy. Such idealized flows of zero-viscosity fluids are called **frictionless** or **inviscid flows.**

Internal versus External Flow

A fluid flow is classified as being either *internal* or *external,* depending on whether the fluid is forced to flow in a confined channel or over a surface. The flow of an unbounded fluid over a surface such as a plate, a wire, or a pipe is **external flow.** The flow in a pipe or duct is **internal flow** if the fluid is completely bounded by solid surfaces. Water flow in a pipe, for example, is internal flow, and air flow over an exposed pipe during a windy day is external flow. The flow of liquids in a pipe is called *open-channel flow* if the pipe is partially filled with the liquid and there is a free surface. The flows of water in rivers and irrigation ditches are examples of such flows (Fig. 9-1).

Compressible versus Incompressible Flow

A fluid is classified as being *compressible* or *incompressible,* depending on the density variation of the fluid during flow. The densities of liquids are essentially constant, and thus the flow of liquids is typically incompressible. Therefore, liquids are usually classified as *incompressible substances.* A pressure of 210 atm, for example, will cause the density of liquid water at 1 atm to change by just 1 percent. Gases, on the other hand, are highly compressible. A pressure change of just 0.01 atm, for example, will cause a change of 1 percent in the density of atmospheric air. However, gas flows can be treated as incompressible if the density changes are under 5 percent, which is usually the case when the flow velocity is less than 30 percent of the velocity of sound in that gas (i.e., the Mach number of flow is less than 0.3). The velocity of sound in air at room temperature is 346 m/s. Therefore, the compressibility effects of air can be neglected at speeds under 100 m/s. Note that a gas is a compressible

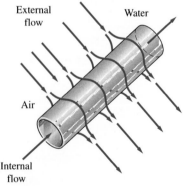

External flow

Water

Air

Internal flow

FIGURE 9-1

Internal flow of water in a pipe and the external flow of air over the same pipe.

fluid, but the flow of a gas does not necessarily need to be treated as a compressible flow.

Small density changes of liquids corresponding to large pressure changes can still have important consequences. The irritating "water hammer" in water pipes, for example, is caused by the vibrations of the pipe generated by the reflection of pressure waves following the sudden closing of valves.

Laminar versus Turbulent Flow

Some flows are smooth and orderly while others are rather chaotic. The highly ordered fluid motion characterized by smooth streamlines is called **laminar flow.** The flow of high-viscosity fluids such as oils at low velocities is typically laminar. The highly disordered fluid motion that typically occurs at high velocities is characterized by velocity fluctuations, and is called **turbulent flow.** The flow of low-viscosity fluids such as air at high velocities is typically turbulent. The flow regime greatly influences the heat transfer rates and the required power for pumping.

Natural (or unforced) versus Forced Flow

A fluid flow is said to be *natural* or *forced,* depending on how the fluid motion is initiated. In **forced flow,** a fluid is forced to flow over a surface or in a pipe by external means such as a pump or a fan. In **natural flows,** any fluid motion is due to natural means such as the buoyancy effect, which manifests itself as the rise of warmer (and thus lighter) fluid and the fall of cooler (and thus denser) fluid. This thermo-siphoning effect is commonly used to replace pumps in solar water heating systems by placing the water tank sufficiently above the solar collectors (Fig. 9-2).

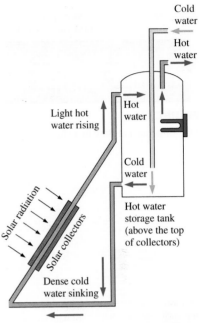

FIGURE 9-2

Natural circulation of water in a solar water heater by thermo-siphoning.

Steady versus Unsteady (transient) Flow

The terms *steady* and *uniform* are used frequently in engineering, and thus it is important to have a clear understanding of their meanings. The term **steady** implies *no change with time.* The opposite of steady is **unsteady,** or **transient.** The term *uniform,* however, implies *no change with location* over a specified region. These meanings are consistent with their everyday use (steady girlfriend, uniform distribution, etc.).

Many devices such as turbines, compressors, and nozzles operate for long periods of time under the same conditions, and they are classified as *steady-flow devices.* During steady flow, the fluid properties can change from point to point within a device, but at any fixed point they remain constant (Fig. 9-3).

One-, Two-, and Three-Dimensional Flows

A flow field is best characterized by the velocity distribution, and thus a flow is said to be *one-, two-,* or *three-dimensional* if the flow velocity $\mathcal{V}$ varies in one-, two-, or three primary dimensions, respectively. A typical fluid flow involves a three-dimensional geometry, and the velocity may vary in all three dimensions, rendering the flow three-dimensional [$\mathcal{V}(x, y, z)$ in rectangular or $\mathcal{V}(r, \theta, z)$ in cylindrical coordinates]. However, the variation of velocity in a certain direction can be small relative to the variation in other directions and can be ignored with negligible error. In such cases, the flow can be modeled conveniently as being one- or two-dimensional, which is easier to analyze.

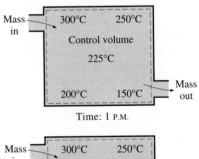

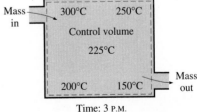

FIGURE 9-3

During steady-flow, fluid properties within the control volume may change with position, but not with time.

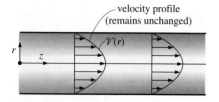

FIGURE 9-4

One-dimensional flow in a circular pipe.

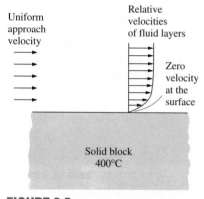

FIGURE 9-5

A fluid flowing over a stationary surface comes to a complete stop at the surface because of the no-slip condition.

When the entrance effects are disregarded, fluid flow in a circular pipe is *one-dimensional* since the velocity varies in the radial r direction but not in the angular θ or axial z directions (Fig. 9-4). That is, the velocity profile is the same at any axial z location, and it is symmetric about the axis of the pipe. This flow is more commonly called *fully developed flow.* Note that even in this simplest flow, the velocity cannot be uniform across the cross-section of the pipe because of the no-slip condition discussed below. However, for convenience in calculations, the velocity can often be modeled as constant and thus *uniform* at a cross-section. Fluid flow at inlets and exits of a pipe is usually approximated as *one-dimensional uniform flow.*

9-2 ■ BEHAVIOR OF FLUIDS PASSING SOLID WALLS

Consider the flow of a fluid in a stationary pipe or over a solid surface that is nonporous (i.e., impermeable to the fluid). All experimental observations indicate that a fluid in motion comes to a complete stop at the surface and assumes a zero velocity relative to the surface. That is, a fluid in direct contact with a solid sticks to the surface and there is no slip. This is known as the **no-slip condition,** and it is due to the viscosity of the fluid.

The no-slip condition is responsible for the development of the velocity profile. Because of friction between the fluid layers, the layer that sticks to the wall slows the adjacent fluid layer, which slows the next layer, and so on. A consequence of the no-slip condition is that all velocity profiles must have zero values at the points of contact between a fluid and a solid (Fig. 9-5). The only exception to the no-slip condition occurs in extremely rarified gases.

A similar phenomenon occurs for the temperature. When two bodies at different temperatures are brought into contact, heat transfer occurs until both bodies assume the same temperature at the point of contact. Therefore, a fluid and a solid surface will have the same temperature at the point of contact. This is known as **no-temperature-jump condition.**

9-3 ■ HISTORY OF FLUID MECHANICS

The development of fluid mechanics started in ancient times with the need to build irrigation systems and to design better ships with oars. The earliest contribution to the field was made by Archimedes (285–212 B.C.) who formulated the principles of buoyancy of submerged bodies and flotation and applied those principles to determine the gold content of the crown of King Hiero I. At about the same time, the Roman engineers built an extensive network of fresh-water supply. There was no significant development in fluid mechanics through the Middle Ages until Leonardo da Vinci (1459–1519) conducted several experiments and derived the conservation of mass equation for one-dimensional steady flow.

The development of fluid mechanics continued along two different paths: on one, the mathematicians and physicists developed the theory and applied it to "idealized" problems that did not have much practical value. On the other path, the engineers developed empirical equations that could be used in the design of fluid systems in a limited range. The lack of communication between these two groups hindered the development of fluid mechanics for a long time. The development of the laws of motion by Isaac Newton (1649–1727) and the linear law of viscosity for the so-called newtonian fluids set the stage for advances in fluid mechanics. Applying these laws to a fluid element, Leonhard Euler (1707–1783) obtained the differential equations for fluid motion in 1755. Daniel Bernoulli (1700–1782) developed the energy equation for incompressible flow in 1738. Lord Rayleigh (1849–1919) developed the powerful dimensional analysis technique. Osborn Reynolds (1849–1912) conducted extensive experiments with pipe flow and came up in 1883 with the dimensionless number that bears his name.

The general equations of fluid motion that include the effects of fluid friction, known as the Navier-Stokes equations, were developed by Claude Louis Marie Navier (1785–1836) in 1827 and independently by George Gabriel Stokes (1819–1903) in 1845. These equations were of little use at the time because they were too difficult to solve. Then in a pioneering paper in 1904, Ludwig Prandtl (1875–1953) showed that fluid flows can be divided into a layer near the walls, called the *boundary layer*, where the friction effects are significant and an outer inviscid layer where such effects are negligible, thus the Euler and Bernoulli equations are applicable. Theodore von Karman (1889–1963) and Sir Geoffrey I. Taylor (1886–1975) also contributed greatly to the development of fluid mechanics in the 20th century. The availability of high-speed computers in the last decades and the development of numerical methods have made it possible to solve a variety of real-world fluids problems and to conduct design and optimization studies through numerical simulation.

9-4 ■ VISCOSITY

Consider the flow of a fluid over a stationary plate. The fluid can be thought to consist of adjacent layers of molecules piled on top of each other. The fluid layer in contact with the plate will try to drag the plate along (with no success) via friction, exerting a **drag force** (or *friction force*) on it (Fig. 9-6). Likewise, a faster fluid layer will try to drag the adjacent slower layer and exert a drag force because of the friction between the two layers.

The friction force per unit area is called **shear stress** and is denoted by τ. Experimental studies indicate that the shear stress for most fluids is proportional to the *velocity gradient* and is expressed as

$$\text{Shear stress} \propto \text{Velocity gradient}$$

or

$$\tau = \mu \frac{d\mathcal{V}}{dy} \quad (\text{N/m}^2) \qquad (9\text{-}1)$$

where $\mathcal{V}$ is the fluid velocity and y is the direction normal to the fluid layer. The constant of proportionality μ is called the **dynamic** (or **absolute**) **viscosity** of the fluid, whose unit is kg/m · s, or equivalently, N · s/m^2 (or Pa · s, where Pa is the pressure unit pascal). A common viscosity unit is **poise**, which is equivalent to 0.1 Pa · s (or *centipoise*, which is one-hundredth of a poise).

The fluids that obey the linear relationship above in the entire range of velocities are called **newtonian fluids**, after Sir Isaac Newton, who expressed it first in 1687. Most common fluids such as water, air, gasoline, and oils are newtonian fluids. Blood and liquid plastics are examples of non-newtonian fluids. In this text, we will consider newtonian fluids only.

In fluid mechanics and heat transfer, the ratio of dynamic viscosity to density appears frequently. For convenience, this ratio is given the name **kinematic viscosity** ν and is expressed as $\nu = \mu/\rho$. Two common units of kinematic viscosity are m^2/s and **stoke** (1 stoke = 1 cm^2/s = 0.0001 m^2/s). In general, the viscosity of a fluid depends on both temperature and pressure,

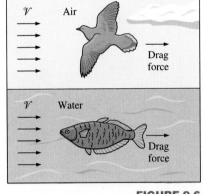

FIGURE 9-6

A moving fluid exerts a drag force on a body because of friction caused by viscosity.

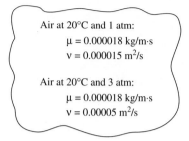

Air at 20°C and 1 atm:
$\mu = 0.000018$ kg/m·s
$\nu = 0.000015$ m²/s

Air at 20°C and 3 atm:
$\mu = 0.000018$ kg/m·s
$\nu = 0.00005$ m²/s

FIGURE 9-7

Dynamic viscosity, in general, does not depend on pressure, but kinematic viscosity does.

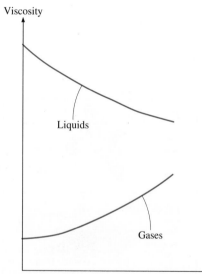

FIGURE 9-8

The viscosity of liquids decreases and the viscosity of gases increases with temperature.

although the dependence on pressure is rather weak. Also, kinematic viscosity is much less sensitive to temperature compared to dynamic viscosity. For *liquids,* both the dynamic and kinematic viscosities are practically independent of pressure, and any small variation with pressure is usually disregarded, except at extremely high pressures. For *gases,* this is also the case for dynamic viscosity (at low to moderate pressures), but not for kinematic viscosity since the density of a gas is proportional to its pressure (Fig. 9-7).

The viscosity of a fluid is a measure of its "stickiness" or "resistance to deformation." This is due to the internal frictional force that develops between fluid layers as they are forced to move relative to each other. Viscosity is caused by cohesive forces between molecules in liquids and by molecular collisions in gases, and it varies greatly with temperature. The viscosity of liquids decreases with temperature, whereas the viscosity of gases increases with temperature (Fig. 9-8). This is because, in a liquid, the molecules possess more energy at higher temperatures, and they can oppose the large cohesive intermolecular forces more strongly. As a result, the energized liquid molecules can move more freely.

In a gas, on the other hand, the intermolecular forces are negligible, and the gas molecules at high temperatures move randomly at higher velocities. This results in more molecular collisions per unit volume per unit time, and therefore in greater resistance to flow. The viscosity of a fluid is directly related to power needed to transport a fluid in a pipe.

The kinetic theory of gases predicts the viscosity of gases to be proportional to the square root of temperature. That is, $\mu_{\text{gas}} \propto \sqrt{T}$. This prediction is confirmed by practical observations, but deviations for different gases need to be accounted for by incorporating some correction factors. The viscosity of *gases* is given as a function of temperature by the **Sutherland correlation** (from *The U.S. Standard Atmosphere,* Ref. 7) as

$$\text{Gases:} \qquad \mu = \frac{aT^{1/2}}{1 + b/T} \qquad (9\text{-}2)$$

where T is absolute temperature and a and b are experimentally determined constants. Note that measuring viscosities at two different temperatures is sufficient to determine these constants. For air, the values of these constants are $a = 1.458 \times 10^{-6}$ kg/m · s · K$^{1/2}$ and $b = 110.4$ K at atmospheric conditions. The viscosity of gases is independent of pressure at low to moderate pressures (from a few percent of 1 atm to several atm). But viscosity increases at high pressures due to the increase in density.

For *liquids,* the viscosity is given as

$$\text{Liquids:} \qquad \mu = a10^{b/(T - c)} \qquad (9\text{-}3)$$

where again T is absolute temperature and a, b, and c are experimentally determined constants. For water, using the values $a = 2.414 \times 10^{-5}$ N · s/m², $b = 247.8$ K, and $c = 140$ K results in less than 2.5 percent error in viscosity in the temperature range of 0°C to 370°C (Touloukian et al., Ref. 5).

Consider a fluid layer of thickness ℓ within a small gap between two concentric cylinders, such as the thin layer of oil in a journal bearing between the rotating shaft and the stationary housing. A small section between the cylinders can be modeled as two parallel flat plates separated by the fluid, as shown in Fig. 9-9. The bottom plate is stationary, and the top plate is moving with a constant velocity $\mathcal{V}$ under the influence of a constant force F. Because of the no-slip condition, the fluid layer adjacent to the bottom plate remains stationary while the fluid layer at the top moves with the same velocity $\mathcal{V}$ as the top plate. The stationary fluid layer slows down the fluid layer just above it, and this layer slows the next fluid layer, and so on. Thus, the velocity across the fluid layer varies from zero at the bottom to $\mathcal{V}$ at the top. For newtonian fluids this variation is linear, and thus the velocity profile is a straight line. The *velocity gradient* $\mathcal{V}/\ell$ is constant in this case throughout the fluid.

For a given fluid, it is observed that the force F required to move the upper plate is proportional to the velocity $\mathcal{V}$ and the area A of the upper plate wetted by the fluid, and is inversely proportional to the thickness of the fluid layer ℓ. Again, the constant of proportionality is the dynamic viscosity μ. Then the force F can be expressed as

$$F = \mu A \frac{\mathcal{V}}{\ell} \qquad (9\text{-}4)$$

Alternately, the relation above can be used to calculate μ when the force F is measured. Therefore, the experimental setup described above can be used to measure the viscosity of fluids. Note that under identical conditions, the force F will be very different for different fluids. Also, if the shaft is not accelerating, the resistance force of the fluid due to viscosity must be equal to the applied force F.

Noting that torque is $\mathbf{T} = FR$ (force × the moment arm, which is the radius R of the inner cylinder in this case) and the tangential velocity is $\mathcal{V} = \omega R$ (angular velocity × the radius), and taking the wetted surface area of the inner cylinder to be $A = 2\pi RL$ by disregarding the shear stress acting on the two ends of the inner cylinder, Eq. 9-4 can be expressed for torque as

$$\mathbf{T} = FR = \mu \frac{2\pi R^3 \omega L}{\ell} = \mu \frac{4\pi^2 R^3 \dot{n} L}{\ell} \qquad (9\text{-}5)$$

where L is the length of the cylinder and $\dot{n}$ is the number of rotations per unit time, which is usually expressed in rpm (revolutions per minute) or rev/s. Note that the angular distance traveled during one rotation is 2π radians, and thus the relation between the angular velocity in rad/s and the rev/s is $\omega = 2\pi \dot{n}$. Equation 9-5 can be used to calculate the viscosity of a fluid by measuring the torque at a specified angular velocity. Therefore, two concentric cylinders can be used as a *viscometer,* a device that measures viscosity. If the torque is applied to the outer cylinder, then the radius of that cylinder should be used in calculations.

The viscosities of some fluids at room temperature are listed in Table 9-1. They are plotted against the temperature in Fig. 9-10. Note that the viscosities of different fluids differ by several orders of magnitude. Also note that at low velocities it is more difficult to move an object in a higher-viscosity fluid such as engine oil than it is in a lower-viscosity fluid such as water. Liquids, in general, are much more viscous than gases.

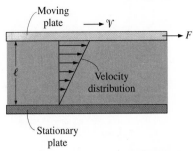

FIGURE 9-9

The motion of a fluid between two parallel plates when the upper plate is moving with a constant velocity $\mathcal{V}$.

TABLE 9-1

Dynamic (absolute) viscosities of some fluids at 1 atm and 20°C (unless otherwise stated)

Fluid	Dynamic viscosity μ, kg/m · s
Glycerin:	
−20°C	134.0
0°C	12.1
20°C	1.49
40°C	0.27
Engine oil:	
SAE 10W	0.10
SAE 10W30	0.17
SAE 30	0.29
SAE 50	0.86
Mercury	0.0015
Ethyl alcohol	0.0012
Water:	
0°C	0.0018
20°C	0.0010
100°C (liquid)	0.0003
100°C (vapor)	0.000013
Blood, 37°C	0.0004
Gasoline	0.00029
Ammonia	0.00022
Air	0.000018
Hydrogen, 0°C	0.000009

Example 9-1 Determining the Viscosity of a Fluid
The viscosity of a fluid is to be measured by a viscometer constructed of two 40-cm-long concentric cylinders (Fig. 9-11). The outer diameter of the inner cylinder is 12 cm, and the gap between the two cylinders is 0.15 cm. The inner cylinder is rotated at 300 rpm, and the torque is measured to be 1.8 N · m. Determine the viscosity of the fluid.

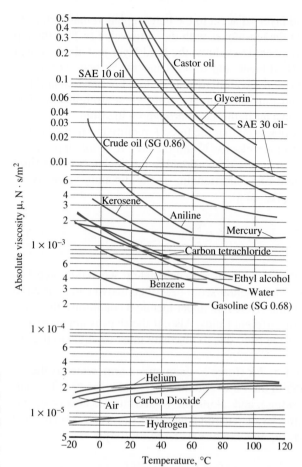

FIGURE 9-10

The variation of dynamic (absolute)
viscosities of common fluids
with temperature at 1 atm
(1 N · s/m² = 1 kg/m · s =
0.020886 lbf · s/ft²)
(from White, Ref. 8, Fig. A-1, p. 769).

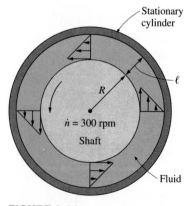

FIGURE 9-11

Schematic for Example 9-1.

Solution The torque and the rpm of a double cylinder viscometer are given. The viscosity of the fluid is to be determined.

Assumptions **1** The inner cylinder is completely submerged in the fluid. **2** The viscous effects on the two ends of the inner cylinder are negligible. **3** The fluid is newtonian.

Analysis Solving Eq. 9-5 for viscosity and substituting the given values, the viscosity of the fluid is determined to be

$$\mu = \frac{\mathbf{T}\ell}{4\pi^2 R^3 \dot{n} L} = \frac{(1.8 \text{ N} \cdot \text{m})(0.0015 \text{ m})}{4\pi^2(0.06 \text{ m})^3(300/60 \text{ s}^{-1})(0.4 \text{ m})} = \mathbf{0.158 \text{ N} \cdot \text{s/m}^2}$$

Discussion Viscosity is a strong function of temperature, and a viscosity value without a temperature is of little use. Therefore, the temperature of the fluid also should have been measured during this experiment and reported with this calculation.

9-5 ■ SURFACE TENSION AND CAPILLARY EFFECT

It is often observed that a drop of blood forms a hump on a horizontal glass, a drop of mercury forms a near-perfect sphere and can be rolled just like a steel ball over a smooth surface, water droplets from rain or dew hang from branches or leaves of trees, a liquid fuel injected into an engine forms a mist of spherical droplets, water dripping from a leaky faucet falls as spherical droplets, and a soap bubble released into the air forms a spherical shape (Fig. 9-12).

In these and other observances, liquid droplets behave like small spherical balloons filled with the liquid, and the surface of the liquid acts like a stretched elastic membrane under tension. The pulling force that causes this tension acts parallel to the surface and is due to the attractive forces between the molecules of the liquid. The magnitude of this force per unit length is called **surface tension** σ_s and is usually expressed in the unit N/m (or lbf/ft in English units). This effect is also called *surface energy* and is expressed in the equivalent unit of N · m/m^2 or J/m^2. In this case, σ_s represents the stretching work that needs to be done to increase the surface of the liquid by a unit amount.

To visualize how surface tension arises, we present a microscopic view in Fig. 9-13 by considering two liquid molecules, one at the surface and one deep within the liquid body. The attractive forces applied on the interior molecule by the surrounding molecules balance each other because of symmetry. But the attractive forces acting on the surface molecule are not symmetric, and the attractive forces applied by the gas molecules above are usually very small. Therefore, there is a net attractive force acting on the molecule at the surface of the liquid, which tends to pull the molecules on the surface toward the interior of the liquid. This force is balanced by the repulsive forces from the molecules below the surface that are being compressed. The resulting compression effect causes the liquid to minimize its surface area. This is the reason for the tendency of the liquid droplets to attain a spherical shape, which has the minimum surface area for a given volume.

You also may have observed, with amusement, that insects can land on water (even walk on water) and that small steel needles can float on water. These are again made possible by surface tension that balances the weights of these objects.

To understand the surface tension effect better, consider a liquid film (such as the film of a soap bubble) suspended on a U-shaped wire frame with a movable side (Fig. 9-14). Normally, the liquid film will tend to pull the movable wire inward in order to minimize its surface area. A force F needs to be applied on the movable wire in the opposite direction to balance this pulling effect. The thin film in the device has two surfaces (the top and bottom surfaces) exposed to air, and thus the length along which the tension acts in this case is $2b$. Then a force balance on the movable wire gives $F = 2b\sigma_s$, and thus the surface tension can be expressed as

$$\sigma_s = \frac{F}{2b} \qquad (9\text{-}6)$$

Note that for $b = 0.5$ m, the force F measured (in N) will simply be the surface tension in N/m. An apparatus of this kind with sufficient precision can be used to measure the surface tension of various fluids.

In the U-shaped wire, the force F remains constant as the movable wire is pulled to stretch the film and increase its surface area. When the movable wire is pulled a distance Δx, the surface area increases by $\Delta A = 2b \, \Delta x$, and the work done W during this stretching process is

$$W = \text{Force} \times \text{Distance} = F \, \Delta x = 2b\sigma_s \, \Delta x = \sigma_s \, \Delta A$$

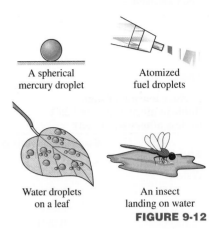

A spherical mercury droplet

Atomized fuel droplets

Water droplets on a leaf

An insect landing on water

FIGURE 9-12

Some consequences of surface tension.

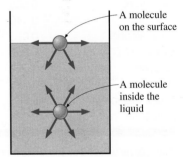

A molecule on the surface

A molecule inside the liquid

FIGURE 9-13

Attractive forces acting on a liquid molecule at the surface and deep inside the liquid.

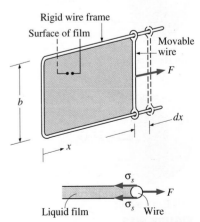

Rigid wire frame

Surface of film

Movable wire

F

b

dx

x

σ_s

F

Liquid film

σ_s

Wire

FIGURE 9-14

Stretching a liquid film with U-shaped wire, and the forces acting on the movable wire of length b.

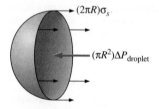

(*a*) Half a droplet

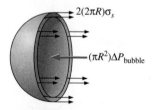

(*b*) Half a bubble

FIGURE 9-15

The free-body diagram of half a
droplet and half a bubble.

since the force remains constant in this case. This result also can be interpreted as *the surface energy of the film is increased by an amount $\sigma_s \, \Delta A$ during this stretching process,* which is consistent with the alternative interpretation of σ_s as surface energy. This is similar to a rubber band having more potential (elastic) energy after it is stretched further. In the case of liquid film, the work is used to move liquid molecules from the interior parts to the surface against the attraction forces of other molecules. Therefore, surface tension also can be defined as *the work done per unit increase in the surface area of the liquid.*

The surface tension varies greatly from substance to substance, and with temperature for a given substance, as shown in Table 9-2. At 20°C, for example, the surface tension is 0.073 N/m for water and 0.440 N/m for mercury surrounded by atmospheric air. The very high surface tension also explains why mercury droplets form spherical balls that can be rolled on a surface like a solid ball without wetting the surface. The surface tension of a liquid, in general, decreases with temperature and becomes zero at the critical point (and thus there is no distinct liquid–vapor interface at temperatures above the critical point). The effect of pressure on tension is usually negligible.

The surface tension of a substance can be changed considerably by *impurities.* Therefore, certain chemicals, called *surfactants,* can be added to a liquid to decrease its surface tension. For example, soaps and detergents lower the surface tension of water and enable it to penetrate through the small openings between fibers for more effective washing. But this also means that devices whose operation depends on surface tension (such as heat pipes) can be destroyed by the presence of impurities due to poor workmanship.

We speak of surface tension for liquids only at liquid–liquid or liquid–gas interfaces. Therefore, it is important to specify the adjacent liquid or gas when specifying surface tension. Also, surface tension determines the size of the liquid droplets that form. A droplet that keeps growing by the addition of more mass will break down when the surface tension can no longer hold it together. This is like a balloon that will burst while being inflated when the pressure inside rises above the strength of the balloon material.

A curved interface indicates a pressure difference (or "pressure jump") across the interface with pressure being higher on the concave side. The excess pressure ΔP inside a droplet or bubble above the atmospheric pressure, for example, can be determined by considering the free-body diagram of half a droplet or bubble (Fig. 9-15). Noting that surface tension acts along the circumference and the pressure acts on the area, horizontal force balances for the droplet and the bubble give

$$Droplet: \quad (2\pi R)\sigma_s = (\pi R^2)\Delta P_{droplet} \quad \rightarrow \quad \Delta P_{droplet} = P_i - P_o = \frac{2\sigma_s}{R} \quad (9\text{-}7)$$

$$Bubble: \quad 2(2\pi R)\sigma_s = (\pi R^2)\Delta P_{bubble} \quad \rightarrow \quad \Delta P_{bubble} = P_i - P_o = \frac{4\sigma_s}{R} \quad (9\text{-}8)$$

where P_i and P_o are the pressures inside and outside the droplet or bubble, respectively. When the droplet or bubble is in the atmosphere, P_o is simply atmospheric pressure. The factor 2 in the force balance for the bubble is due to the bubble consisting of a film with *two* surfaces (inner and outer surfaces) and thus two circumferences in the cross section.

The excess pressure in a droplet (or bubble) also can be determined by considering a differential increase in the radius of the droplet due to the addition of a differential amount of mass and interpreting the surface tension as the increase in the surface energy per unit area. Then the increase in the surface energy of the droplet during this differential expansion process becomes

$$\delta W_{\text{surface}} = \sigma_s \, dA = \sigma_s \, d(4\pi R^2) = 8\pi R \sigma_s \, dR$$

The expansion work done during this differential process is determined by multiplying the force by distance to obtain

$$\delta W_{\text{expansion}} = \text{Force} \times \text{Distance} = F \, dR = (\Delta PA) \, dR = 4\pi R^2 \, \Delta P \, dR$$

Equating the two expressions above gives $\Delta P_{\text{droplet}} = 2\sigma_s/R$, which is the same relation obtained before and given in Eq. 9-7. Note that the excess pressure in a droplet or bubble is inversely proportional to the radius.

Capillary Effect

Another interesting consequence of surface tension is the **capillary effect,** which is the rise or fall of a liquid in a small-diameter tube inserted into the liquid. Such narrow tubes or confined flow channels are called **capillaries.** The rise of kerosene through a cotton wick inserted into the reservoir of a kerosene lamp is due to this effect. The capillary effect is also partially responsible for the rise of water to the top of tall trees.

It is commonly observed that water in a glass container curves up slightly at the edges where it touches the glass surface; but the opposite occurs for mercury: it curves down at the edges (Fig. 9-16). This effect is usually expressed by saying that water *wets* the glass (by sticking to it) while mercury does not. The strength of the capillary effect is quantified by the **contact** (or *wetting*) **angle** ϕ, defined as *the angle that the tangent to the liquid surface makes with the solid surface at the point of contact.* The surface tension force acts along this tangent line toward the solid surface. A liquid is said to wet the surface when $\phi < 90°$ and not to wet the surface when $\phi > 90°$. In atmospheric air, the contact angle of water (and most other organic liquids) with glass is nearly zero, $\phi \approx 0°$. Therefore, the surface tension force acts upwards on water in a glass tube along the circumference, tending to pull the water up. As a result, water rises in the tube until the weight of the liquid in the tube above the liquid level of the reservoir balances the surface tension force. The contact angle is 130° for mercury–glass and 26° for kerosene–glass in air. Note that the contact angle, in general, will be different in different environments (such as another gas or liquid in place of air).

The phenomenon of capillary effect can be explained microscopically by considering *cohesive forces* (the forces between like molecules, like water) and *adhesive forces* (the forces between dislike molecules, like water and glass). The liquid molecules at the solid–liquid interface are subjected to both cohesive forces by other liquid molecules and adhesive forces by the molecules of the solid. The relative magnitudes of these forces determine whether a liquid wets a solid surface or not. Obviously, the water molecules are more strongly attracted to the glass molecules than they are to other water molecules, and thus water tends to rise along the glass surface. The opposite occurs for mercury, which causes the liquid surface near the glass wall to be suppressed (Fig. 9-17).

The magnitude of the capillary rise in a circular tube can be determined from a force balance on the cylindrical liquid column of height h in the tube (Fig. 9-18). The bottom of the liquid column is at the same level as the free surface of the reservoir, and thus the pressure there must be atmospheric pressure. This will balance the atmospheric pressure acting at the top surface,

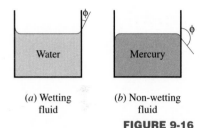

(a) Wetting fluid   (b) Non-wetting fluid

FIGURE 9-16

The contact angle for wetting and nonwetting fluids.

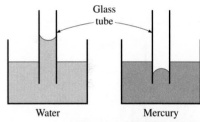

FIGURE 9-17

The capillary rise of water and the capillary fall of mercury in a small-diameter glass.

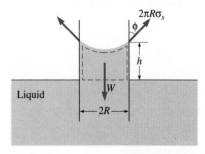

FIGURE 9-18

The forces acting on a liquid column that has risen in a tube due to the capillary effect.

and thus these two effects will cancel each other. The weight of the liquid column is

$$W = mg = \rho V g = \rho g (\pi R^2 h)$$

Equating the vertical component of the surface tension force to the weight gives

$$W = F_{surface} \quad \rightarrow \quad \rho g (\pi R^2 h) = 2\pi R \sigma_s \cos \phi$$

Solving for h gives the capillary rise to be

Capillary rise: $\qquad h = \dfrac{2\sigma_s}{\rho g R} \cos \phi \qquad (R = \text{constant}) \qquad (9\text{-}9)$

The relation above is also valid for nonwetting liquids (such as mercury in glass) and gives the capillary drop. In this case $\phi > 90°$ and thus $\cos \phi < 0$, which makes h negative. Therefore, a negative value of capillary rise corresponds to a capillary drop.

Note that the capillary rise is inversely proportional to the radius of the tube. Therefore, the thinner the tube is, the greater the rise (or fall) of the liquid will be in the tube. In practice, the capillary effect is usually negligible in tubes whose diameter is greater than 1 cm. When pressure measurements are made using manometers and barometers, it is important to use sufficiently large tubes to minimize the capillary effect. The capillary rise is also inversely proportional to the density of the liquid, as expected. Therefore, lighter liquids will experience greater capillary rises. Finally, it should be kept in mind that Eq. 9-9 is derived for constant diameter tubes, and should not be used for tubes of variable cross section.

Example 9-2 The Capillary Rise of Water in a Tube

A 0.6-mm-diameter glass tube is inserted into water at 20°C in a cup. Determine the capillary rise of water in the tube (Fig. 9-19).

Solution The rise of water in a slender tube as a result of the capillary effect is to be determined.

Assumptions **1** There are no impurities in the water and no contamination on the surfaces of the glass tube. **2** The experiment is conducted in atmospheric air.

Properties The surface tension of water at 20°C is 0.073 N/m (Table 9-2). The contact angle of water with glass is 0° (from above text). We take the density of liquid water to be 1000 kg/m³.

Analysis The capillary rise is determined directly from Eq. 9-9 by substituting the given values to be

$$h = \frac{2\sigma_s}{\rho g R} \cos \phi = \frac{2(0.073 \text{ N/m})}{(1000 \text{ kg/m}^3)(9.81 \text{ m/s}^2)(0.3 \times 10^{-3}\text{ m})}\left(\frac{1 \text{ kg} \cdot \text{m/s}^2}{1 \text{ N}}\right)\cos 0°$$

$$= 0.050 \text{ m} = \mathbf{5.0 \text{ cm}}$$

Therefore, water will rise in the tube 5 cm above the liquid level in the cup.

Discussion Note that if the tube diameter were 1 cm, the capillary rise would be 0.3 mm, which is hardly noticeable to the eye. Actually, the capillary rise in a large-diameter tube occurs only at the rim. The center does not rise at all. Therefore, the capillary effect can be ignored for large-diameter tubes.

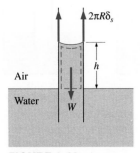

FIGURE 9-19

Schematic for Example 9-2.

In this chapter, the basic concepts of fluid mechanics are introduced and discussed. *Fluid mechanics* is the science that deals with the behavior of fluids at rest or in motion and the interaction of fluids with solids or other fluids at the boundaries. The flow of an unbounded fluid over a surface is *external flow,* and the flow in a pipe or duct is *internal flow* if the fluid is completely bounded by solid surfaces. A fluid flow is classified as being *compressible* or *incompressible,* depending on the density variation of the fluid during flow. The densities of liquids are essentially constant, and thus the flow of liquids is typically incompressible. The term *steady* implies *no change with time.* The opposite of steady is *unsteady,* or *transient.* The term *uniform* implies *no change with location* over a specified region. A flow is said to be *one-dimensional* when the velocity changes in one dimension only. A fluid in direct contact with a solid surface sticks to the surface and there is no slip. This is known as the *no-slip condition,* and it is due to the viscosity of the fluid.

The *viscosity* of a fluid is a measure of its "stickiness" or "resistance to deformation." The drag force per unit area is called *shear stress* and is expressed as

$$\tau = \mu \frac{dV}{dy}$$

where μ is the *dynamic* (or *absolute*) *viscosity* of the fluid. Fluids that obey this linear relationship are called *newtonian fluids.* The ratio of dynamic viscosity to density is called the *kinematic viscosity* ν. For liquids, both the dynamic and kinematic viscosities are essentially independent of pressure. For gases, this is also the case for dynamic viscosity, but not for kinematic viscosity since the density of a gas is proportional to its pressure.

The pulling effect on liquid molecules at an interface caused by the attractive forces of molecules per unit length is called *surface tension* σ_s. Surface tension also represents the stretching work that needs to be done to increase the surface area of a liquid by a unit amount and is called *surface energy.* A curved interface indicates a pressure difference across the interface, with pressure being higher on the concave side. The excess pressure ΔP inside a droplet or bubble is given by

$$\Delta P_{\text{droplet}} = P_i - P_o = \frac{2\sigma_s}{R} \qquad \text{and} \qquad \Delta P_{\text{bubble}} = P_i - P_o = \frac{4\sigma_s}{R}$$

where P_i and P_o are the pressures inside and outside the droplet or bubble. The rise or fall of a liquid in a small-diameter tube inserted into the liquid due to surface tension is called the *capillary effect,* and such narrow tubes or confined flow channels are called *capillaries.* The strength of the capillary effect is quantified by the *contact angle* ϕ, defined as the angle that the tangent to the liquid surface makes with the solid surface at the point of contact. A liquid is said to wet the surface when $\phi < 90°$ and not to wet the surface when $\phi > 90°$. The capillary rise or drop is given by

$$h = \frac{2\sigma_s}{\rho g R} \cos \phi$$

The capillary rise is inversely proportional to the radius of the tube and is negligible for tubes whose diameter is larger than about 1 cm.

REFERENCES AND SUGGESTED READING

1. R. W. Fox and A. T. McDonald. *Introduction to Fluid Mechanics.* 5th ed. New York: John Wiley & Sons, 1999.

2. D. C. Giancoli. *Physics.* 3rd ed. Upper Saddle River, NJ: Prentice Hall, 1991.

3. M. C. Potter and D. C. Wiggert. *Mechanics of Fluids.* 2nd ed. Upper Saddle River, NJ: Prentice Hall, 1997.

4. J. A. Roberson and C. L. Grove. *Engineering Fluid Mechanics.* 6th ed. New York: John Wiley & Sons, 1997.

5. Y. S. Touloukian, S. C. Saxena, and P. Hestermans. *Thermophysical Properties of Matter: The TPRC Data Series.* Vol. 11, *Viscosity.* New York: Plenum Publishing Corp., 1975.

6. L. Trefethen. "Surface Tension in Fluid Mechanics." In *Illustrated Experiments in Fluid Mechanics.* Cambridge, MA: The MIT Press, 1972.

7. *The U.S. Standard Atmosphere.* Washington, D.C.: U.S. Government Printing Office, 1976.

8. F. M. White. *Fluid Mechanics.* 4th ed. New York: McGraw-Hill, 1999.

9. C. L. Yaws. *Handbook of Viscosity.* 3 vols. Houston, TX: Gulf Publishing, 1994.

10. C. L. Yaws, X. Lin, and L. Bu. "Calculate Viscosities for 355 Compounds. An Equation Can Be Used to Calculate Liquid Viscosity as a Function of Temperature." *Chemical Engineering* 101, no. 4 (April 1994), pp. 119–28.

PROBLEMS*

Classification of Fluid Flows

9-1C Define internal, external, and open-channel flows.

9-2C Define incompressible flow and incompressible fluid. Must the flow of a compressible fluid necessarily be treated as compressible?

9-3C What is the no-slip condition? What causes it?

Viscosity

9-4C What is viscosity? What is the cause of it in liquids and in gases? Do liquids or gases have higher dynamic viscosities?

9-5C What is newtonian fluid? Is water a newtonian fluid?

9-6C Consider two identical small glass balls dropped into two identical containers, one filled with water and the other with oil. Which ball will reach the bottom of the container first? Why?

9-7C How does the dynamic viscosity of (*a*) liquids and (*b*) gases vary with temperature?

---

*Students are encouraged to answer *all* the concept "C" questions.

9-8C How does the kinematic viscosity of (*a*) liquids and (*b*) gases vary with pressure?

9-9 The viscosity of a fluid is to be measured by a viscometer constructed of two 75-cm-long concentric cylinders. The outer diameter of the inner cylinder is 15 cm, and the gap between the two cylinders is 0.12 cm. The inner cylinder is rotated at 200 rpm, and the torque is measured to be 0.8 N · m. Determine the viscosity of the fluid.

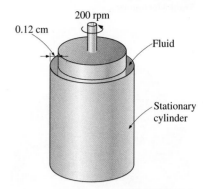

FIGURE P9-9

9-10E The viscosity of a fluid is to be measured by a viscometer constructed of two 3-ft-long concentric cylinders. The inner diameter of the outer cylinder is 6 in., and the gap between the two cylinders is 0.05 in. The outer cylinder is rotated at 250 rpm, and the torque is measured to be 1.2 lbf · ft. Determine the viscosity of the fluid.   *Answer:* 0.000648 lbs · s/ft²

9-11 In regions far from the entrance, fluid flow through a circular pipe is one-dimensional, and the velocity profile for laminar flow is given by

$$\mathcal{V}(r) = \mathcal{V}_{max}\left(1 - \frac{r^2}{R^2}\right)$$

where R is the radius of the pipe, r is the radial distance from the center of the pipe, and $\mathcal{V}_{max}$ is the maximum flow velocity, which occurs at the center. Obtain (*a*) a relation for the drag force applied by the fluid on a section of the pipe of length L and (*b*) the value of the drag force for water flow at 20°C with $R = 0.08$ m, $L = 15$ m, $\mathcal{V}_{max} = 3$ m/s, and $\mu = 0.0010$ kg/m · s.

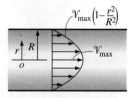

FIGURE P9-11

9-12 Repeat Prob. 9-11 for $\mathcal{V}_{max} = 5$ m/s.   *Answer:* (*b*) 0.942 N

Surface Tension and Capillary Effect

9-13C What is surface tension? What is it caused by? Why is the surface tension also called surface energy?

9-14C Consider a soap bubble. Is the pressure inside the bubble higher or lower than the pressure outside?

9-15C What is the capillary effect? What is it caused by? How is it affected by the contact angle?

9-16C A small-diameter tube is inserted into a liquid whose contact angle is 110°. Will the level of liquid in the tube rise or drop? Explain.

9-17C Is the capillary rise greater in small- or large-diameter tubes?

9-18E A 0.03-in.-diameter glass tube is inserted into kerosene at 68°F. Determine the capillary rise of kerosene in the tube. The contact angle of kerosene with a glass surface is 26°.   *Answer:* 0.65 in

9-19 A 1.9-mm-diameter tube is inserted into an unknown liquid whose density is 960 kg/m³, and it is observed that the liquid rises 5 mm in the tube, making a contact angle of 15°. Determine the surface tension of the liquid.

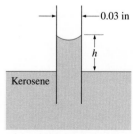

FIGURE P9-18E

9-20 Determine the gage pressure inside a soap bubble of diameter (*a*) 0.2 cm and (*b*) 5 cm at 20°C.

9-21 Nutrients dissolved in water are carried to upper parts of plants by tiny tubes partly because of the capillary effect. Determine how high the water

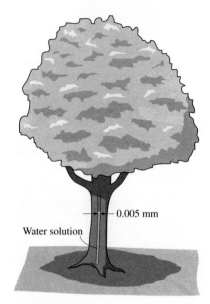

Water solution — 0.005 mm

solution will rise in a tree in a 0.005-mm-diameter tube as a result of the capillary effect. Treat the solution as water at 20°C with a contact angle of 15°.
Answer: 5.75 m

9-22 The surface tension of a liquid is to be measured using a liquid film suspended on a U-shaped wire frame with an 8-cm-long movable side. If the force needed to move the wire is 0.012 N, determine the surface tension of this liquid in air.

Review Problems

9-23E A 0.9-in.-diameter glass tube is inserted into mercury, which makes a contact angle of 140° with glass. Determine the capillary drop of mercury in the tube at 68°F.   *Answer:* 0.0175 m

9-24 Derive a relation for the capillary rise of a liquid between two large parallel plates a distance t apart inserted into the liquid vertically. Take the contact angle to be ϕ.

9-25 Consider a 30-cm-long journal bearing that is lubricated with oil whose viscosity is 0.1 kg/m · s at 20°C at the beginning of operation and 0.008 kg/m · s at the anticipated steady operating temperature of 80°C. The diameter of the shaft is 8 cm, and the average gap between the shaft and the journal is 0.08 cm. Determine the torque needed to overcome the bearing friction initially and during steady operation when the shaft is rotated at 500 rpm.

Computer, Design, and Essay Problems

9-26 Write an essay on the rise of the fluid to the top of the trees by the capillary and other effects.

9-27 Write an essay on the oils used in car engines in different seasons and their viscosities.

Fluid Statics

This chapter deals with forces applied by fluids at rest in a gravitational field. The fluid property responsible for those forces is *pressure,* which is the force exerted by a fluid per unit area. We start this chapter with a discussion of the *hydrostatic forces* applied on submerged bodies with plane or curved surfaces and determine the pressure center. We then consider the *buoyant force* applied by fluids on submerged or floating bodies. Finally, we discuss the *stability* of such bodies. This chapter makes extensive use of force balances for bodies in static equilibrium, and it will be helpful if the students first review the relevant topics from Statics.

10-1 ■ INTRODUCTION

Fluid statics deals with problems associated with fluids at rest. The fluid can be either gaseous or liquid. Fluid statics is generally referred to as *hydrostatics* when the fluid is a liquid and as *aerostatics* when the fluid is a gas. In fluid statics, there is no relative motion between adjacent fluid layers, and thus there are no shear (tangential) stresses in the fluid trying to deform it. The only stress we deal with in fluid statics is the *normal stress,* which is the pressure, and the variation of pressure is due only to the weight of the fluid. Therefore, the topic of fluid statics has significance only in gravity fields, and the force relations developed naturally involve the gravitational acceleration *g*. The force exerted by a fluid at rest to a surface is normal to the surface at the point of contact since there is no relative motion between the fluid and the solid surface, and thus no shear forces acting parallel to the surface.

Fluid statics is used to determine the forces acting on floating or submerged bodies and the forces developed by devices like hydraulic presses and car jacks. The design of many engineering systems such as water dams and liquid storage tanks requires the determination of the forces acting on the surfaces using fluid statics. The complete description of the resultant hydrostatic force acting on a submerged surface requires the determination of the magnitude, the direction, and the line of action of the force. In the next two sections, we will consider the forces acting on both plane and curved surfaces of submerged bodies due to pressure.

10-2 ■ HYDROSTATIC FORCES ON SUBMERGED PLANE SURFACES

A flat plate submerged in a liquid, such as a gate valve in a dam, the wall of a liquid storage tank, or the surface of a ship, is subjected to fluid pressure distributed over its surface. On a *plane* surface, the hydrostatic forces form a system of parallel forces, and we often need to determine the *magnitude* of the force and its *point of application,* which is called the **center of pressure.** In most cases, the other side of the plate is open to the atmosphere (such as the dry side of a gate), and thus atmospheric pressure acts on both sides of the plate, yielding a zero resultant. In such cases, it is convenient to disregard the atmospheric pressure and work with the gage pressure $P_{\text{gage}} = \rho g h$ only (Fig. 10-1).

Consider the top surface of a flat plate of arbitrary shape completely submerged in a liquid, as shown in Fig. 10-2 together with its top view. The plane

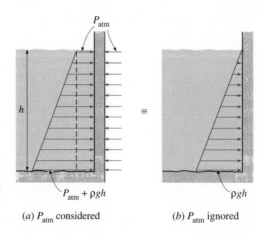

FIGURE 10-1

When analyzing hydrostatic forces on submerged surfaces, the atmospheric pressure can be ignored for simplicity when it acts on both sides of the body.

(a) P_{atm} considered

(b) P_{atm} ignored

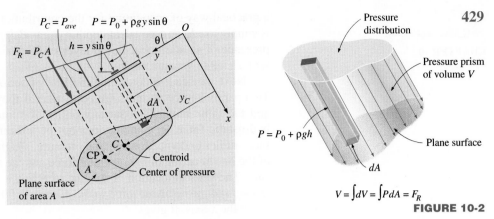

429

$$V = \int dV = \int P dA = F_R$$

FIGURE 10-2

Hydrostatic force on an inclined plane surface completely submerged in a liquid.

of this surface (normal to the paper) intersects the horizontal free surface with an angle θ, and we take the line of intersection to be the x axis. The absolute pressure above the liquid is P_0, which is the local atmospheric pressure P_{atm} if the liquid is open to the atmosphere (but P_0 may be different than P_{atm} if the space above the liquid is evacuated or pressurized). Then the absolute pressure at any point on the plate is

$$P = P_0 + \rho gh = P_0 + \rho gy \sin \theta \qquad (10\text{-}1)$$

where h is the vertical distance of the point from the free surface and y is the distance of the point from the x axis (from point O in Fig. 10-2). The resultant hydrostatic force F_R acting on the surface is determined by integrating the force $P\, dA$ acting on a differential area dA over the entire surface area,

$$F_R = \int_A P\, dA = \int_A (P_0 + \rho gy \sin \theta)\, dA = P_0 A + \rho g \sin \theta \int_A y\, dA \qquad (10\text{-}2)$$

But the *first moment of area* $\int_A y\, dA$ is related to the y coordinate of the centroid (or center) of the surface by

$$y_C = \frac{1}{A} \int_A y\, dA \qquad (10\text{-}3)$$

Substituting,

$$F_R = (P_0 + \rho gy_C \sin \theta)A = (P_0 + \rho gh_C)A = P_C A = P_{ave} A \qquad (10\text{-}4)$$

where $P_C = P_0 + \rho gh_C$ is the pressure at the centroid of the surface, which is equivalent to the *average* pressure on the surface, and $h_C = y_C \sin \theta$ is the *vertical distance* of the centroid from the free surface of the liquid (Fig. 10-3). Thus we conclude that:

The magnitude of the resultant force acting on a plane surface of a completely submerged plate in a homogeneous (constant density) fluid is equal to the product of the pressure P_C at the centroid of the surface and the area A of the surface (Fig. 10-4).

The pressure P_0 is usually the atmospheric pressure, which can be ignored in most cases since it acts on both sides of the plate. When this is not the case,

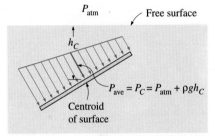

FIGURE 10-3

The pressure at the centroid of a surface is equivalent to the *average* pressure on the surface.

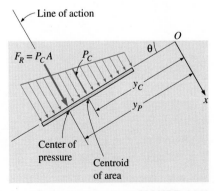

FIGURE 10-4

The resultant force acting on a plane surface is equal to the product of the pressure at the centroid of the surface and the surface area, and its line of action passes through the center of pressure.

a practical way of accounting for the contribution of P_0 to the resultant force is simply to add an equivalent depth $h_{equiv} = P_0/\rho g$ to h_C; that is, to assume the presence of an additional liquid layer of thickness h_{equiv} on top of the liquid with absolute vacuum above.

Next we need to determine the line of action of the resultant force F_R. Two parallel force systems are equivalent if they have the same magnitude and the same moment about any point. The line of action of the resultant hydrostatic force, in general, does not pass through the centroid of the surface—it lies underneath where the pressure is higher. The point of intersection of the line of action of the resultant force and the surface is the *center of pressure*. The vertical location of the line of action is determined by equating the moment of the resultant force to the moment of the distributed pressure force about the x axis. It gives

$$y_P F_R = \int_A yP \, dA = \int_A y(P_0 + \rho gy \sin \theta) \, dA = P_0 \int_A y \, dA + \rho g \sin \theta \int_A y^2 \, dA$$

or

$$y_P F_R = P_0 y_C A + \rho g \sin \theta \, I_{xx, O} \tag{10-5}$$

where y_P is the distance of the center of pressure from the x axis (point O in the figure) and $I_{xx, O} = \int_A y^2 \, dA$ is the *second moment of area* (also called the *area moment of inertia*) about the x axis. The second moments of area are widely available for common shapes in engineering handbooks, but they are usually given about the axes passing through the centroid of the area. Fortunately, the second moments of area about two parallel axes are related to each other by the *parallel axis theorem,* which in this case is expressed as

$$I_{xx, O} = I_{xx, C} + y_C^2 A \tag{10-6}$$

where $I_{xx, C}$ is the second moment of area about the x axis passing through the centroid of the area and y_C (the y coordinate of the centroid) is the distance between the two parallel axes. Substituting the F_R relation from Eq. 10-4 and the $I_{xx, O}$ relation from Eq. 10-6 into Eq. 10-5 and solving for y_P gives

$$y_P = y_C + \frac{I_{xx, C}}{[y_C + P_0/(\rho g \sin \theta)]A} \tag{10-7a}$$

For $P_0 = 0$, which is usually the case when the atmospheric pressure is ignored, it simplifies to

$$y_P = y_C + \frac{I_{xx, C}}{y_C A} \tag{10-7b}$$

Knowing y_P, the vertical distance of the center of pressure from the free surface is determined from $h_P = y_P \sin \theta$.

The $I_{xx, C}$ for some common areas are given in Fig. 10-5. For these and other areas that possess symmetry about the y axis, the center of pressure lies on the y axis directly below the centroid. The location of the center of pressure in such cases is simply the point on the surface of the vertical plane of symmetry at a distance h_P from the free surface.

Pressure acts normal to the surface, and the hydrostatic forces acting on a flat plate of any shape form a volume whose base is the plate area and whose

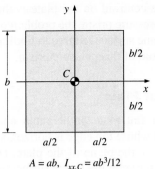

$A = ab, \ I_{xx,C} = ab^3/12$

(a) Rectangle

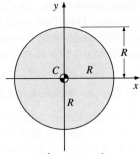

$A = \pi R^2, \ I_{xx,C} = \pi R^4/4$

(b) Circle

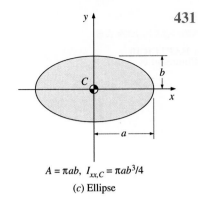

$A = \pi ab, \ I_{xx,C} = \pi ab^3/4$

(c) Ellipse

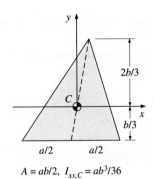

$A = ab/2, \ I_{xx,C} = ab^3/36$

(d) Triangle

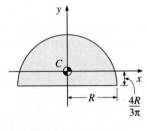

$A = \pi R^2/2, \ I_{xx,C} = 0.109757R^4$

(e) Semicircle

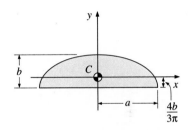

$A = \pi ab/2, \ I_{xx,C} = 0.109757ab^3$

(f) Semiellipse

FIGURE 10-5

The centroid and the centroidal moments of inertia for some common geometries.

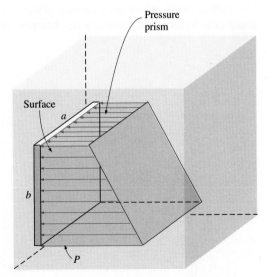

FIGURE 10-6

The hydrostatic forces acting on a plane surface form a volume whose base is the surface and whose height is the pressure.

height is the linearly varying pressure, as shown in Fig. 10-6. This virtual **pressure prism** has an interesting physical interpretation: its *volume* is equal to the *magnitude* of the resultant hydrostatic force acting on the plate since $V = \int P \, dA$, and the line of action of this force passes through the *centroid* of this homogeneous prism. The projection of the centroid on the plate is the *pressure center.* Therefore, with the concept of pressure prism, the problem of describing the resultant hydrostatic force on a plane surface reduces to finding the volume and the two coordinates of the centroid of this pressure prism.

Special Case: Submerged Rectangular Plate

Consider a completely submerged rectangular flat plate of height b and width a tilted an angle θ from the horizontal and whose top edge is horizontal and is at a distance s from the free surface along the plane of the plate, as shown in Fig. 10-7a. The resultant hydrostatic force on the upper surface is equal to the average pressure, which is the pressure at the midpoint of the surface, times the surface area A. That is,

Tilted rectangular plate: $\quad F_R = P_C A = [P_0 + \rho g(s + b/2) \sin \theta]ab \quad$ (10-8)

The force acts at a vertical distance of $h_P = y_P \sin \theta$ from the free surface directly beneath the centroid of the plate where, from Eq. 10-7a,

$$
\begin{aligned}
y_P &= s + \frac{b}{2} + \frac{ab^3/12}{[s + b/2 + P_0/(\rho g \sin \theta)]ab} \\
&= s + \frac{b}{2} + \frac{b^2}{12[s + b/2 + P_0/(\rho g \sin \theta)]}
\end{aligned} \quad (10\text{-}9)
$$

When the upper edge of the plate is at the free surface and thus $s = 0$, Eq. 10-8 reduces to

Tilted rectangular plate (s = 0): $\quad F_R = [P_0 + \rho g(b \sin \theta)/2] ab \quad$ (10-10)

For a completely submerged vertical plate ($\theta = 90°$) whose top edge is horizontal, the hydrostatic force can be obtained by setting $\sin \theta = 1$ (Fig. 10-7b)

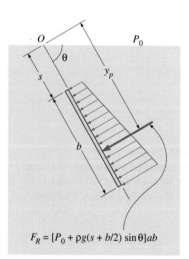

$F_R = [P_0 + \rho g(s + b/2) \sin \theta]ab$

(a) Tilted plate

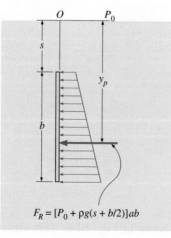

$F_R = [P_0 + \rho g(s + b/2)]ab$

(b) Vertical plate

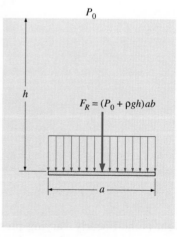

$F_R = (P_0 + \rho gh)ab$

(c) Horizontal plate

FIGURE 10-7

Hydrostatic force acting on the top surface of a submerged rectangular plate for tilted, vertical, and horizontal cases.

| Vertical rectangular plate: | $F_R = [P_0 + \rho g(s + b/2)]ab$ | (10-11) |

| Vertical rectangular plate ($s = 0$): | $F_R = (P_0 + \rho gb/2)ab$ | (10-12) |

Therefore, for $P_0 = 0$, the hydrostatic force on a vertical rectangular sur-face of height b whose top edge is horizontal and at the free surface is $F_R = \rho gab^2/2$ acting at a distance of $2b/3$ from the free surface directly beneath the centroid of the plate.

Pressure distribution on a submerged horizontal surface is uniform, and its magnitude is $P = P_0 + \rho gh$, where h is the distance of the surface from the free surface. Therefore, the hydrostatic force acting on a horizontal rectangu-lar surface is

| Horizontal rectangular plate: | $F_R = (P_0 + \rho gh)ab$ | (10-13) |

and it acts through the midpoint of the plate (Fig. 10-7c).

Example 10-1   Hydrostatic Force Acting on the Door of a Submerged Car
A car plunges into a lake during an accident and lands at the bottom of the lake on its wheels (Fig. 10-8). The door is 1.2-m high and 1-m wide, and the top edge of the door is 8 m below the free surface of the water. Determine the hydrostatic force on the door and the location of the pressure center, and discuss if the driver can open the door.

Solution   A car is submerged in water. The hydrostatic force on the door is to be determined, and the likelihood of the driver opening the door is to be assessed.

Assumptions   **1** The bottom surface of the lake is horizontal. **2** The car is well-sealed so that no water leaks inside. **3** The door can be approximated as a ver-tical rectangular plate. **4** The pressure in the car remains at atmospheric value since there is no water leaking in, and thus no compression of the air inside. Therefore, we can ignore the atmospheric pressure in calculations since it acts on both sides of the door.

Properties   We take the density of lake water to be 1000 kg/m^3 throughout.

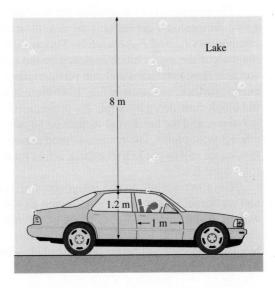

FIGURE 10-8

Schematic for Example 10-1.

Analysis The average pressure on the door is the pressure value at the centroid (midpoint) of the door and is determined to be

$$P_{ave} = P_C = \rho g h_C = \rho g(s + b/2)$$

$$= (1000 \text{ kg/m}^3)(9.81 \text{ m/s}^2)(8 + 1.2/2 \text{ m})\left(\frac{1 \text{ kN}}{1000 \text{ kg} \cdot \text{m/s}^2}\right)$$

$$= 84.4 \text{ kN/m}^2$$

Then the resultant hydrostatic force on the door becomes

$$F_R = P_{ave}A = (84.4 \text{ kN/m}^2)(1 \text{ m} \times 1.2 \text{ m}) = 101.3 \text{ kN}$$

The pressure center is directly under the midpoint of the plate, and its distance from the surface of the lake is determined from Eq. 10-9 by setting $P_0 = 0$ to be

$$y_P = s + \frac{b}{2} + \frac{b^2}{12(s + b/2)} = 8 + \frac{1.2}{2} + \frac{1.2^2}{12(8 + 1.2/2)} = 8.61 \text{ m}$$

Discussion A strong person can lift 100 kg, whose weight is 981 N or about 1 kN. Also, the person can apply the force at a point furthest from the hinges (1 m further) for maximum effect and generate a moment of 1 kNm. The resultant hydrostatic force acts under the midpoint of the door, and thus a distance of 0.5 m from the hinges. This generates a moment of 50.6 kNm, which is about 50 times the moment the driver can possibly generate. Therefore, it is impossible for the driver to open the door of the car. The driver's best bet is to let some water in (by rolling the window down a little, for example) and to keep his head close to the ceiling. The driver should be able to open the door shortly before the car is filled with water since at that point the pressures on both sides of the door will nearly be the same and opening the door in water will be almost as easy as opening it in air.

10-3 ■ HYDROSTATIC FORCES ON SUBMERGED CURVED SURFACES

For a submerged curved surface, the determination of the resultant hydrostatic force is more involved since it typically requires the integration of the pressure forces that change direction along the curved surface. The concept of the pressure prism in this case is not much help either because of the complicated shapes involved.

The easiest way to determine the resultant hydrostatic force F_R acting on a two-dimensional curved surface is to determine the horizontal and vertical components F_H and F_V separately. This is done by considering the free-body diagram of the liquid block enclosed by the curved surface and the two plane surfaces (one horizontal and one vertical) passing through the two ends of the curved surface, as shown in Fig. 10-9. Note that the vertical surface of the liquid block considered is simply the projection of the curved surface on a *vertical plane,* and the horizontal surface is the projection of the curved surface on a *horizontal plane.* The resultant force acting on the curved solid surface is then equal and opposite to the force acting on the curved liquid surface (Newton's third law).

The force acting on the imaginary horizontal or vertical plane surface and its line of action can be determined as discussed in the last section. The weight of the enclosed liquid block of volume V is simply $W = \rho g V$, and it acts downward through the centroid of this volume. Noting that the fluid block is in static equilibrium, the force balances in the horizontal and vertical directions give

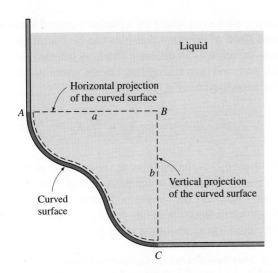

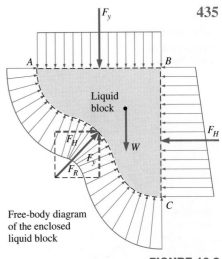

FIGURE 10-9

Determination of the hydrostatic force acting on a submerged curved surface.

Horizontal force component on curved surface: $\qquad F_H = F_x \qquad$ (10-14)

Vertical force component on curved surface: $\qquad F_V = F_y + W \qquad$ (10-15)

where the summation $F_y + W$ is a vector addition (i.e., add magnitudes if both act in the same direction and subtract if they act in opposite directions). Thus, we conclude the following:

> *1 The horizontal component of the hydrostatic force acting on a curved surface is equal (in both magnitude and the line of action) to the hydrostatic force acting on the vertical projection of the curved surface.*
>
> *2 The vertical component of the hydrostatic force acting on a curved surface is equal to the hydrostatic force acting on the horizontal projection of the curved surface, plus (minus, if acting in the opposite direction) the weight of the fluid block.*

The magnitude of the resultant hydrostatic force acting on the curved surface is $F_R = \sqrt{F_H^2 + F_V^2}$ and the tangent of the angle it makes with the horizontal is $\tan \alpha = F_V/F_H$. The exact location of the line of action of the resultant force (e.g., its distance from one of the end points of the curved surface) can be determined by taking a moment about an appropriate point. The discussions above are valid for all curved surfaces regardless of whether they are above or below the liquid. Note that in the case of a *curved surface above a liquid,* the weight of the liquid is subtracted from the vertical component of the hydrostatic force since they act in opposite directions (Fig. 10-10).

When the curved surface is a *circular arc* (full circle or any part of it), the resultant hydrostatic force acting on the surface always passes through the center of the circle. This is because the pressure forces are normal to the surface, and all lines normal to the surface of a circle pass through the center of the circle. Thus, the pressure forces form a concurrent force system at the center, which can be reduced to a single equivalent force at that point (Fig. 10-11).

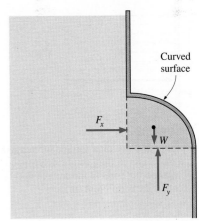

FIGURE 10-10

When a curved surface is above the liquid, the weight of the liquid and the vertical component of the hydrostatic force act in the opposite directions.

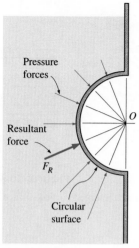

FIGURE 10-11

The hydrostatic force acting on a circular surface always passes through the center of the circle since the pressure forces are normal to the surface and they all pass through the center.

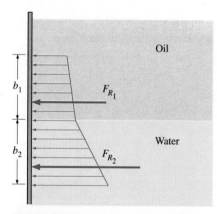

FIGURE 10-12

Hydrostatic force on a surface submerged in a multilayered fluid can be determined by considering parts of the surface in different fluids as different surfaces.

Finally, hydrostatic forces acting on a plane or curved surface submerged in a **multilayered fluid** of different densities can be determined by considering different parts of surfaces in different fluids as different surfaces, finding the force on each part, and then adding them using vector addition. For a plane surface, it can be expressed as (Fig. 10-12)

$$\text{Plane surface in a multilayered fluid:} \quad F_R = \sum F_{R,i} = \sum P_{C,i} A_i \quad (10\text{-}16)$$

where $P_{C,i} = P_0 + \rho_i g h_{C,i}$ is the pressure at the centroid of the portion of the surface in fluid i and A_i is the area of the plate in that fluid. The line of action of this equivalent force can be determined from the requirement that the moment of the equivalent force about any point is equal to the sum of the moments of the individual forces about the same point.

Example 10-2 A Gravity-Controlled Cylindrical Gate

A long solid cylinder of radius 0.8 m hinged at point A is used as an automatic gate, as shown in Fig. 10-13. When the water level reaches 5 m, the gate opens by turning about the hinge at point A. Determine (a) the hydrostatic force acting on the cylinder and its line of action when the gate opens and (b) the weight of the cylinder per m length of the cylinder.

Solution The height of a water reservoir is controlled by a cylindrical gate hinged to the reservoir. The hydrostatic force on the cylinder and the weight of the cylinder per m length are to be determined.

Assumptions **1** The hinge is frictionless. **2** Atmospheric pressure acts on both sides of the gate, and thus it can be ignored.

Properties We take the density of water to be 1000 kg/m³ throughout.

Analysis (a) We consider the free-body diagram of the liquid block enclosed by the circular surface of the cylinder and its vertical and horizontal projections. The hydrostatic forces acting on the vertical and horizontal plane surfaces as well as the weight of the liquid block are determined as follows:

Horizontal force on vertical surface:

$$F_H = F_x = P_{ave} A = \rho g h_C A = \rho g(s + R/2)A$$
$$= (1000 \text{ kg/m}^3)(9.81 \text{ m/s}^2)(4.2 + 0.8/2 \text{ m})(0.8 \text{ m} \times 1 \text{ m})\left(\frac{1 \text{ kN}}{1000 \text{ kg} \cdot \text{m/s}^2}\right)$$
$$= 36.1 \text{ kN}$$

Vertical force on horizontal surface (upward):

$$F_y = P_{ave} A = \rho g h_C A = \rho g h_{bottom} A$$
$$= (1000 \text{ kg/m}^3)(9.81 \text{ m/s}^2)(5 \text{ m})(0.8 \text{ m} \times 1 \text{ m})\left(\frac{1 \text{ kN}}{1000 \text{ kg} \cdot \text{m/s}^2}\right)$$
$$= 39.2 \text{ kN}$$

Weight of fluid block per m length (downward):

$$W = mg = \rho g V = \rho g(R^2 - \pi R^2/4)(1 \text{ m})$$
$$= (1000 \text{ kg/m}^3)(9.81 \text{ m/s}^2)(0.8 \text{ m})^2(1 - \pi/4)(1 \text{ m})\left(\frac{1 \text{ kN}}{1000 \text{ kg} \cdot \text{m/s}^2}\right)$$
$$= 1.3 \text{ kN}$$

Therefore, the net upward vertical force is

$$F_V = F_y - W = 39.2 - 1.3 = 37.9 \text{ kN}$$

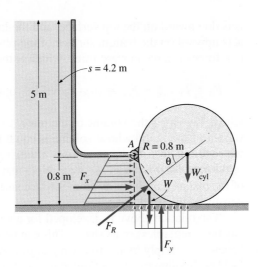

FIGURE 10-13

Schematic for Example 10-2.

Then the magnitude and direction of the hydrostatic force acting on the cylindrical surface become

$$F_R = \sqrt{F_H^2 + F_V^2} = \sqrt{36.1^2 + 37.9^2} = \textbf{52.3 kN}$$

$$\tan\theta = F_V/F_H = 37.9/36.1 = 1.05 \quad \rightarrow \quad \theta = 46.4°$$

Therefore, the magnitude of the hydrostatic force acting on the cylinder is 52.3 kN per m length of the cylinder, and its line of action passes through the center of the cylinder making an angle 46.4° with the horizontal.

(*b*) When the water level is 5-m high, the gate is about to open and thus the reaction force at the bottom of the cylinder is zero. Then the forces other than those at the hinge acting on the cylinder are its weight, acting through the center, and the hydrostatic force exerted by water. Taking a moment about the point *A* where the hinge is and equating it to zero gives

$$F_R R \sin\theta - W_{cyl} R = 0 \quad \rightarrow \quad W_{cyl} = F_R \sin\theta = (52.3\text{ kN}) \sin 46.4° = \textbf{37.9 kN}$$

Discussion The weight of the cylinder per m length is determined to be 37.9 kN. It can be shown that this corresponds to a mass of 3863 kg per m length and to a density of 1921 kg/m³ for the material of the cylinder.

10-4 ■ BUOYANCY AND STABILITY

It is a common experience that an object feels lighter and weighs less in a liquid than it does in air. This can be demonstrated easily by weighing a heavy object in water by a waterproof spring scale. Also, objects made of wood or other light materials float on water. These and other observations suggest that a fluid exerts an upward force on a body immersed in it. This force that tends to lift the body is called the **buoyant force** and is denoted by F_B.

The buoyant force is caused by the increase of pressure in a fluid with depth. Consider, for example, a flat plate of thickness *h* submerged in a liquid whose density is ρ_f parallel to the free surface, as shown in Fig. 10-14. The area of the top (and also bottom) surface of the plate is *A,* and its distance to the free surface is *s*. The pressures at the top and bottom surfaces of the plate are $\rho_f gs$ and $\rho_f g(s + h)$, respectively. Then the hydrostatic force $F_{top} = \rho_f gsA$

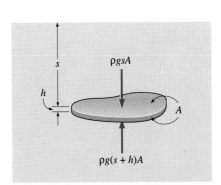

FIGURE 10-14

A flat plate of uniform thickness *h* submerged in a liquid parallel to the free surface

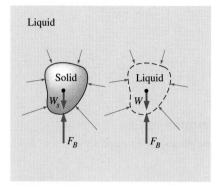

FIGURE 10-15

The buoyant forces acting on a solid body submerged in a fluid and a fluid body of the same shape at the same depth are identical.

acts downward on the top surface, and the larger force $F_{\text{bottom}} = \rho_f g(s + h)A$ acts upward on the bottom surface of the plate. The difference between these two forces is a net upward force, which is the *buoyant force*,

$$F_B = F_{\text{bottom}} - F_{\text{top}} = \rho_f g(s + h)A - \rho_f ghA = \rho_f ghA = \rho_f gV \quad (10\text{-}17)$$

where $V = hA$ is the volume of the plate. But the relation $\rho_f gV$ is simply the weight of the liquid whose volume is equal to the volume of the plate. Thus, we conclude that *the buoyant force acting on the plate is equal to the weight of the liquid displaced by the plate.* Note that the buoyant force is independent of the distance of the body from the free surface. It is also independent of the density of the solid body.

The relation above is developed for a simple geometry, but it is valid for any body regardless of its shape. This can be shown mathematically by a force balance, or simply by the following argument: Consider an arbitrarily shaped solid body submerged in a fluid at rest and compare it to a body of fluid of the same shape indicated by dotted lines at the same distance from the free surface (Fig. 10-15). The buoyant forces acting on these two bodies are the same since the pressure distributions, which depend on depth only, are the same at the boundaries of both. The imaginary fluid body is in static equilibrium, and thus the net force acting on it is zero. Therefore, the upward buoyant force must be equal to the weight of the imaginary fluid body whose volume is equal to the volume of the solid body. Further, the weight and the buoyant force must have the same line of action to have a zero moment. This is known as the **Archimedes' principle** and is expressed as follows:

The buoyant force acting on a body immersed in a fluid is equal to the weight of the fluid displaced by the body, and it acts upward through the centroid of the displaced volume.

For *floating* bodies, the weight of the entire body must be equal to the buoyant force, which is the weight of the fluid whose volume is equal to the volume of the submerged portion of the floating body. That is,

$$F_B = W \quad \rightarrow \quad \rho_f g V_{\text{sub}} = \rho_{\text{ave, body}} g V_{\text{total}} \quad \rightarrow \quad \frac{V_{\text{sub}}}{V_{\text{total}}} = \frac{\rho_{\text{ave, body}}}{\rho_f} \quad (10\text{-}18)$$

Therefore, the submerged volume fraction of a floating body is equal to the ratio of the average density of the body to the density of the fluid. Note that when the density ratio equals one, the floating body becomes completely submerged.

It follows from the discussions above that a body immersed in a fluid (1) will remain at rest at any point in the fluid when its density is equal to the density of the fluid, (2) will sink to the bottom when its density is greater than the density of the fluid, and (3) will rise to the surface of the fluid and float when the density of the body is less than the density of the fluid (Fig. 10-16).

The buoyant force is proportional to the density of the fluid, and thus we might think that the buoyant force exerted by gases such as air is negligible. This is certainly the case in general, but there are significant exceptions. For example, the volume of a person is about 0.1 m^3, and taking the density of air to be 1.2 kg/m^3, the buoyant force exerted by air on the person is

$$F_B = \rho_f gV = (1.2 \text{ kg/m}^3)(9.81 \text{ m/s}^2)(0.1 \text{ m}^3) \cong 1.2 \text{ N}$$

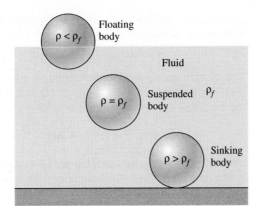

FIGURE 10-16

A solid body dropped into a fluid will sink, float, or remain at rest at any point in the fluid, depending on its density relative to the density of the fluid.

The weight of an 80-kg person is $80 \times 9.81 = 788$ N. Therefore, ignoring the buoyancy in this case will result in an error in weight of just 0.15 percent, which is negligible. But the buoyancy effects in gases dominate some important natural phenomena such as the rise of warm air in a cooler environment and thus the onset of natural convection currents, the rise of hot air or helium balloons, the rise of water vapor to high elevations, and air movements in the atmosphere. A helium balloon, for example, will rise as a result of the buoyancy effect until it reaches an altitude where the density of air (which decreases with altitude) equals the density of helium in the balloon—assuming the balloon does not burst by then.

Archimedes' principle is also used in modern geology by considering the continents to be floating on a sea of magma.

Example 10-3 Measuring Specific Gravity by a Hydrometer

If you have a seawater aquarium, you have probably used a small cylindrical glass tube with some lead-weight at its bottom to measure the salinity of the water by simply watching how deep the tube sinks. Such a device that floats in a vertical position and is used to measure the specific gravity of a liquid is called a *hydrometer* (Fig. 10-17). The top part of the hydrometer extends above the liquid surface, and the divisions on it allow one to read the specific gravity directly. The hydrometer is calibrated such that in pure water it reads exactly 1.0 at the air–water interface. (*a*) Obtain a relation for the specific gravity of a liquid as a function of distance Δz from the mark corresponding to pure water and (*b*) determine the mass of lead that must be poured into a 1-cm-diameter, 20-cm-long hydrometer if it is to float halfway (the 10-cm mark) in pure water.

Solution The specific gravity of a liquid is to be measured by a hydrometer. A relation between specific gravity and the vertical distance from the reference level is to be obtained, and the amount of lead that needs to be added into the tube for a certain hydrometer is to be determined.

Assumptions The weight of the glass tube is negligible relative to the weight of the lead added.

Properties We take the density of pure water to be 1000 kg/m^3.

Analysis (*a*) Noting that the hydrometer is in static equilibrium, the buoyant force F_B exerted by the liquid must always be equal to the weight W of the hydrometer. In pure water, let the vertical distance between the bottom of the hydrometer and the free surface of water be z_0. Setting $F_B = W$ in this case gives

$$W_{\text{hydro}} = F_{B,\,w} = \rho_w g V_{\text{sub}} = \rho_w g A z_0 \tag{1}$$

where A is the cross-sectional area of the tube.

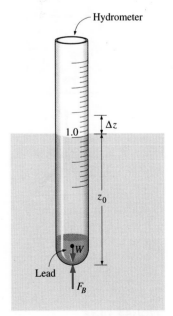

FIGURE 10-17

Schematic for Example 10-3.

In a fluid lighter than water, the hydrometer will sink deeper, and the liquid level will be a distance of Δz above z_0. Again setting $F_B = W$ gives

$$W_{hydro} = F_{B, f} = \rho_f g V_{sub} = \rho_f g A(z_0 + \Delta z) \tag{2}$$

This relation is also valid for fluids heavier than water by taking the Δz below z_0 to be a negative quantity. Setting Eqs. (1) and (2) above equal to each other since the weight of the hydrometer is constant and rearranging give

$$\rho_w g A z_0 = \rho_f g A(z_0 + \Delta z) \quad \rightarrow \quad \rho_s = \frac{\rho_f}{\rho_w} = \frac{z_0}{z_0 + \Delta z}$$

which is the relation between the specific gravity of the fluid and Δz. Note that z_0 is constant for a given hydrometer and Δz is negative for fluids heavier than pure water.

(*b*) Disregarding the weight of the glass tube, the amount of lead that needs to be added to the tube is determined from the requirement that the weight of the lead be equal to the buoyant force. When the hydrometer is floating with half of it submerged in water, the buoyant force acting on it is

$$F_B = \rho_w g V_{sub}$$

Equating F_B to the weight of lead gives

$$W = mg = \rho_w g V_{sub}$$

Solving for m and substituting, the mass of lead is determined to be

$$m = \rho_w V_{sub} = \rho_w[\pi R^2 h_{sub}] = (1000 \text{ kg/m}^3)[\pi(0.005 \text{ m})^2(0.1 \text{ m})] = \textbf{0.00785 kg}$$

Discussion Note that if the hydrometer was required to sink only 5 cm in water, the required mass of lead would be one-half of this amount. Also, the assumption that the weight of the glass tube is negligible needs to be verified since the mass of lead is only 7.85 g.

Example 10-4 Weight Loss of an Object in Seawater
A crane is used to lower weights into the sea (density = 1025 kg/m³) for an underwater construction project (Fig. 10-18). Determine the tension in the rope of the crane due to a rectangular 0.4 m × 0.4 m × 3 m concrete block (density = 2300 kg/m³) when it is (*a*) suspended in the air and (*b*) completely immersed in water.

Solution A concrete block is lowered into the sea. The tension in the rope is to be determined before and after the block is in water.

Assumptions **1** The buoyancy of air is negligible. **2** The weight of the ropes is negligible.

Analysis (*a*) Consider the free-body diagram of the concrete block. The forces acting on the concrete block in air are its weight and the upward pull action (tension) by the rope. These two forces must balance each other, and thus the tension in the rope must be equal to the weight of the block:

$$V = (0.4 \text{ m})(0.4 \text{ m})(3 \text{ m}) = 1.28 \text{ m}^3$$

$$F_{T, air} = W = \rho_{concrete} g V$$

$$= (2300 \text{ kg/m}^3)(9.81 \text{ m/s}^2)(1.28 \text{ m}^3)\left(\frac{1 \text{ kN}}{1000 \text{ kg} \cdot \text{m/s}^2}\right) = \textbf{28.9 kN}$$

(*b*) When the block is immersed in water, there is the additional force of buoyancy acting upward. The force balance in this case gives

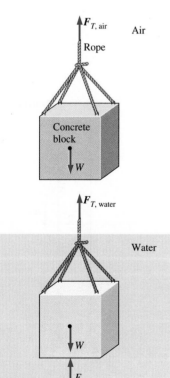

FIGURE 10-18
Schematic for Example 10-4.

$$F_B = \rho_f g V = (1025 \text{ kg/m}^3)(9.81 \text{ m/s}^2)(1.28 \text{ m}^3)\left(\frac{1 \text{ kN}}{1000 \text{ kg} \cdot \text{m/s}^2}\right) = 12.9 \text{ kN}$$

$$F_{T,\,water} = W - F_B = 28.9 - 12.9 = \textbf{16.0 kN}$$

Discussion Note that the weight of the concrete block, and thus the tension of the rope, decreases by $(28.9 - 16.0)/28.9 = 44.6$ percent in water.

Stability of Immersed and Floating Bodies

An important application of the buoyancy concept is the assessment of the stability of immersed and floating bodies with no external attachments. This topic is of great importance in the design of ships and submarines. Below we provide some general qualitative discussions on vertical and rotational stability.

For an immersed or floating body in static equilibrium, the weight and the buoyant force acting on the body balance each other, and such bodies are inherently stable in the *vertical direction.* If an immersed body is raised or lowered to a different depth, the body will remain in equilibrium at that location. If a floating body is raised or lowered somewhat by a vertical force, the body will return to its original position as soon as the external effect is removed. Therefore, immersed and floating bodies possess vertical stability. In fact, the immersed body is neutrally stable since it does not return to its original position after a disturbance.

The *rotational stability* of an *immersed body* depends on the relative locations of the *center of gravity G* of the body and the *center of buoyancy B,* which is the centroid of the displaced volume. An immersed body is *stable* if the body is bottom-heavy and thus the point G is below the point B (Fig. 10-19). A rotational disturbance of the body in such cases produces a *restoring moment* to return the body to its original stable position. Thus, a stable design for a submarine calls for the engines and the cabins for the crew to be located at the lower half in order to shift the weight to the bottom as much as possible. Hot-air or helium balloons (which can be viewed as being immersed in air) are also stable since the cage that carries the load is at the bottom. An immersed body whose center of gravity G is above point B is *unstable,* and any disturbance will cause this body to turn upside down. A body for which G and B coincide is *neutrally stable.* This is the case for bodies whose density is constant throughout. For such bodies, there is no tendency to overturn.

The rotational stability criteria are similar for *floating bodies.* Again, if the floating body is bottom-heavy and thus the center of gravity G is below the center of buoyancy $B,$ the body is always stable. But unlike immersed bodies, a floating body may still be stable when G is above B (Fig. 10-20). This is because the centroid of the displaced volume shifts to the side to a

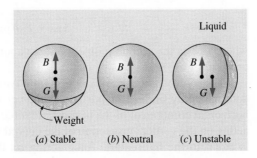

(*a*) Stable   (*b*) Neutral   (*c*) Unstable

FIGURE 10-19

An immersed body is stable if the body is bottom-heavy and thus the center of gravity G is below the centroid B of the body.

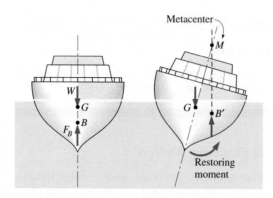

FIGURE 10-20

A floating body is stable if the metacenter M is above the center of gravity G, and thus GM is positive, and unstable if the M is below G, and thus GM is negative.

point B' during a rotational disturbance while the center of gravity G of the body remains unchanged. If the point B' is sufficiently far, these two forces create a restoring moment and return the body to the original position. A measure of stability for floating bodies is the **metacentric height** GM, which is the distance between the center of gravity G and the metacenter M—the intersection point of the lines of action of the buoyant force before and after rotation. The metacenter may be considered to be a fixed point for most hull shapes for small rolling angles up to about 20°. A floating body is stable if the point M is above the point G, and thus GM is positive, and unstable if the point M is below the point G, and thus GM is negative. In the latter case, the weight and the buoyant force acting on the tilted body generate an overturning moment instead of a restoring moment, causing the body to capsize. The length of the metacentric height GM above G is a measure of the stability: the larger it is, the more stable the floating body will be.

10-5 ■ SUMMARY

Fluid statics deals with problems associated with fluids at rest, and it is called *hydrostatics* when the fluid is a liquid. The magnitude of the resultant force acting on a plane surface of a completely submerged plate in a homogeneous fluid is equal to the product of the pressure P_C at the centroid of the surface and the area A of the surface and is expressed as

$$F_R = (P_0 + \rho g h_C)A = P_C A = P_{\text{ave}} A$$

where $h_C = y_C \sin \theta$ is the *vertical distance* of the centroid from the free surface of the liquid. The pressure P_0 is usually the atmospheric pressure, which can be ignored in most cases since it acts on both sides of the plate. The point of intersection of the line of action of the resultant force and the surface is the *center of pressure*. The vertical location of the line of action of the resultant force is given by

$$y_P = y_C + \frac{I_{xx,\,C}}{[y_C + P_0/(\rho g \sin \theta)]A}$$

where $I_{xx,\,C}$ is the second moment of area about the x axis passing through the centroid of the area.

The hydrostatic force F_R acting on a two-dimensional curved surface is determined by finding the horizontal and vertical components F_H and F_V. The horizontal component of the hydrostatic force is equal to the hydrostatic force acting on the vertical projection of the curved surface. The vertical component is equal to the hydrostatic force acting on the horizontal projection of the curved surface, plus (minus, if acting in the opposite direction) the weight of the fluid block. When the curved surface is a *circular arc,* the resultant hydrostatic force acting on the surface always passes through the center of the circle.

A fluid exerts an upward force on a body immersed in it. This force is called the *buoyant force* and is expressed as

$$F_B = \rho_f g V$$

where V is the volume of the body. This is known as the *Archimedes' principle* and is expressed as: the buoyant force acting on a body immersed in a fluid is equal to the weight of the fluid displaced by the body; it acts upward through the centroid of the displaced volume. The buoyant force is independent of the distance of the body from the free surface. For *floating* bodies, the submerged volume fraction of the body is equal to the ratio of the average density of the body to the density of the fluid.

REFERENCES AND SUGGESTED READING

1. R. W. Fox and A. T. McDonald. *Introduction to Fluid Mechanics.* 5th ed. New York: John Wiley & Sons, 1999.

2. D. C. Giancoli. *Physics.* 3rd ed. Upper Saddle River, NJ: Prentice Hall, 1991.

3. J. L. Meriam. *Statics.* 2nd ed. New York: John Wiley & Sons, 1971.

4. M. C. Potter and D. C. Wiggert. *Mechanics of Fluids.* 2nd ed. Upper Saddle River, NJ: Prentice Hall, 1997.

5. J. A. Roberson and C. L. Grove. *Engineering Fluid Mechanics.* 6th ed. New York: John Wiley & Sons, 1997.

6. F. M. White. *Fluid Mechanics.* 4th ed. New York: McGraw-Hill, 1999.

PROBLEMS*

Fluid Statics: Hydrostatic Forces on Plane and Curved Surfaces

10-1C Define the resultant hydrostatic force acting on a submerged surface, and the center of pressure.

10-2C Someone claims that she can determine the magnitude of the hydrostatic force acting on a plane surface submerged in water regardless of its shape and orientation if she knew the vertical distance of the centroid of the surface from the free surface and the area of the surface. Is this a valid claim? Explain.

10-3C A submerged horizontal flat plate is suspended in water by a string attached at the centroid of its upper surface. Now the plate is rotated 45° about an axis that passes through its centroid. Discuss the change on the hydrostatic force acting on the top surface of this plate as a result of this rotation. Assume the plate remains submerged at all times.

*Students are encouraged to answer *all* concept "C" questions.

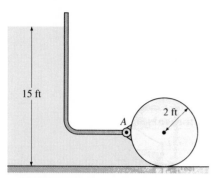

FIGURE P10-8E

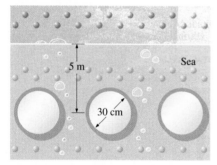

FIGURE P10-11

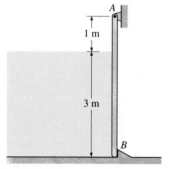

FIGURE P10-13

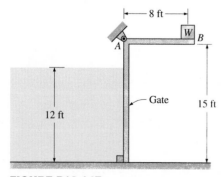

FIGURE P10-14E

10-4C Consider a submerged curved surface. Explain how you would determine the horizontal component of the hydrostatic force acting on this surface.

10-5C Consider a submerged curved surface. Explain how you would determine the vertical component of the hydrostatic force acting on this surface.

10-6C Consider a circular surface subjected to hydrostatic forces by a constant density liquid. If the magnitudes of the horizontal and vertical components of the resultant hydrostatic force are determined, explain how you would determine the line of action of this force.

10-7 Consider a car submerged in water in a lake with a flat bottom. The driver's side door of the car is 1.1-m high and 0.9-m wide, and the top edge of the door is 6 m below the water surface. Determine the net force acting on the door (normal to its surface) and the location of the pressure center if (*a*) the car is well-sealed and it contains air at atmospheric pressure and (*b*) the car is filled with water through cracks.

10-8E A long solid cylinder of radius 2 ft hinged at point *A* is used as an automatic gate, as shown in the figure. When the water level reaches 15 ft, the cylindrical gate opens by turning about the hinge at point *A*. Determine (*a*) the hydrostatic force acting on the cylinder and its line of action when the gate opens and (*b*) the weight of the cylinder per ft length of the cylinder.

10-9 Consider a 4-m-long, 4-m-wide, and 1.5-m-high above-the-ground swimming pool that is filled with water to the rim. (*a*) Determine the hydrostatic force on each wall and the distance of the line of action of this force from the ground. (*b*) If the height of the walls of the pool is doubled, will the hydrostatic force on each wall double or quadruple? Why?   *Answer:* (*a*) 44.1 kN

10-10E Consider a 200-ft-high, 800-ft-wide dam filled to capacity. Determine (*a*) the hydrostatic force on the dam and (*b*) the force per unit area of the dam near the top and near the bottom. Can you explain why the lower sections of the dams are usually much thicker?

10-11 A room in the lower level of a cruise ship has a 30-cm-diameter circular window. If the midpoint of the window is 5 m below the water surface, determine the hydrostatic force acting on the window, and the pressure center. Take the specific gravity of seawater to be 1.025.
    Answers: 3554 N, 5.001 m

10-12 The wet surface of the side wall of a 100-m-long dam is a quarter circle with a radius of 10 m. Determine the hydrostatic force on the dam and its line of action when the dam is filled to the rim.

10-13 A 4-m-high, 5-m-wide rectangular plate blocks the end of a 3-m-deep fresh-water channel, as shown in the figure. The plate is hinged about a horizontal axis along its upper edge through a point *A* and is restrained from opening by a fixed ridge at point *B*. Determine the force exerted on the plate by the ridge.

10-14E The flow of water from a reservoir is controlled by a 5-ft-wide L-shaped gate hinged at point *A,* as shown in the figure. If it is desired that the gate open when the water height is 12 ft, determine the mass of the required weight *W.*   *Answer:* 30,900 lbm

10-15E Repeat Prob. 10-14E for a water height of 8 ft.

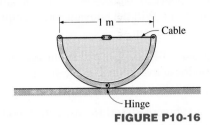

FIGURE P10-16

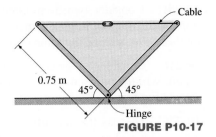

FIGURE P10-17

10-16   A water trough of semi-circular cross section of radius 0.5 m consists of two symmetric parts hinged to each other at the bottom, as shown in the figure. The two parts are held together by a cable and turnbuckle placed every 3 m along the length of the trough. Calculate the tension in each cable when the trough is filled to the rim.

10-17   The two sides of a V-shaped water trough are hinged to each other at the bottom where they meet, as shown in the figure, making an angle of 45° with the ground from both sides. Each side is 0.75 m wide, and the two parts are held together by a cable and turnbuckle placed every 3 m along the length of the trough. Calculate the tension in each cable when the trough is filled to the rim.   *Answer:* 2755 N

10-18   Repeat Prob. 10-17 for the case of a partially filled trough with a water height of 0.4 m.

10-19   A retaining wall against mud slide is to be constructed by placing 0.8-m-high and 0.2-m-wide rectangular concrete blocks ($\rho = 2700$ kg/m³) side by side, as shown in the figure. The friction coefficient between the ground and the concrete blocks is $f = 0.3$, and the density of the mud is about 1800 kg/m³. There is concern that the concrete blocks may slide or tip over the lower left edge as the mud level rises. Determine the mud height at which (*a*) the blocks will overcome friction and start sliding and (*b*) the blocks will tip over.

10-20   Repeat Prob. 10-19 for 0.3-m-wide concrete blocks.

10-21   A 4-m-long quarter-circular gate of radius 3 m and of negligible weight is hinged about its upper edge *A*. The gate controls the flow of water over the ledge at *B,* where the gate is pressed by a spring. Determine the minimum spring force required to keep the gate closed when the water level rises to *A* at the upper edge of the gate.

10-22   Repeat Prob. 10-21 for a radius of 4 m for the gate.
    Answer: 314 kN

Buoyancy

10-23C   What is buoyant force? What causes it? What is the magnitude of the buoyant force acting on a submerged body whose volume is *V?* What are the direction and the line of action of the buoyant force?

10-24C   Consider two identical spherical balls submerged in water at different depths. Will the buoyant forces acting on these two balls be the same or different? Explain.

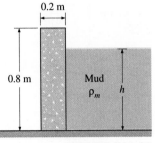

FIGURE P10-19

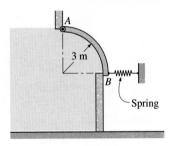

FIGURE P10-21

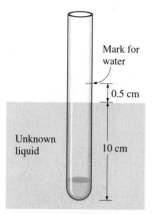

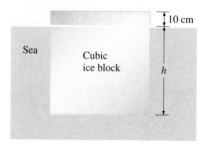

FIGURE P10-28

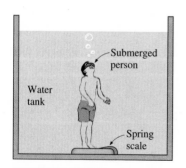

FIGURE P10-31

FIGURE P10-34E

10-25C Consider two 5-cm-diameter spherical balls—one made of aluminum, the other of iron—submerged in water. Will the buoyant forces acting on these two balls be the same or different? Explain.

10-26C Consider a 3-kg copper cube and a 3-kg copper ball submerged in a liquid. Will the buoyant forces acting on these two bodies be the same or different? Explain.

10-27C Discuss the stability of (a) a submerged and (b) a floating body whose center of gravity is above the center of buoyancy.

10-28 The density of a liquid is to be determined by an old 1-cm-diameter cylindrical hydrometer whose division marks are completely wiped out. The hydrometer is first dropped in water, and the water level is marked. The hydrometer is then dropped into the other liquid, and it is observed that the mark for water has risen 0.5 cm above the liquid–air interface. If the height of the water mark is 10 cm, determine the density of the liquid.

10-29E A crane is used to lower weights into a lake for an underwater construction project. Determine the tension in the rope of the crane due to a 3-ft-diameter spherical steel block (density = 494 lbm/ft^3) when it is (a) suspended in the air and (b) completely immersed in water.

10-30 The volume and the average density of an irregularly shaped body are to be determined by using a spring scale. The body weighs 6800 N in air and 4790 N in water. Determine the volume and the density of the body. State your assumptions.

10-31 Consider a large cubic ice block floating in a sea. The specific gravities of ice and seawater are 0.92 and 1.025, respectively. If a 10-cm-high portion of the ice block extends above the surface of the water, determine the height of the ice block below the surface.   *Answer:* 87.6 cm

10-32 A 110-kg granite rock (ρ = 2700 kg/m^3) is dropped into a lake. A man dives in and tries to lift the rock. Determine how much force the man needs to apply to lift it from the bottom of the lake. Do you think he can do it?

10-33 It is said that Archimedes discovered his principle during a bath while thinking about how he could determine if King Hiero's crown was actually made of pure gold. While in the bathtub, he conceived the idea that he could determine the average density of an irregularly shaped object by weighing it in air and also in water. If the crown weighed 3.20 kgf (= 31.4 N) in air and 2.95 kgf (= 28.9 N) in water, determine if the crown is made of pure gold. The density of gold is 19,300 kg/m^3. Discuss how you can solve this problem without weighing the crown in water but by using an ordinary bucket with no calibration for volume. You may weigh anything in air.

10-34E One of the common procedures in fitness programs is to determine the fat-to-muscle ratio of the body. This is based on the principle that the muscle tissue is denser than the fat tissue, and, thus, the higher the average density of the body, the higher is the fraction of muscle tissue. The average density of the body can be determined by weighing the person in air and also while submerged in water in a tank. Treating all tissues and bones (other than fat) as muscle with an equivalent density of ρ_{muscle}, obtain a relation for the volume fraction of body fat x_{fat}.   *Answer:* $x_{fat} = (\rho_{muscle} - \rho_{ave})/(\rho_{muscle} - \rho_{fat})$.

10-35   The hull of a boat has a volume of 150 m³, and the total mass of the boat when empty is 8560 kg. Determine how much load this boat can carry without sinking (*a*) in a lake and (*b*) in a sea with a specific gravity of 1.03.

Review Problems

10-36   The density of a floating body can be determined by tying weights to the body until both the body and the weights are completely submerged, and then weighing them separately in air. Consider a wood log that weighs 1540 N in air. If it takes 34 kg of lead ($\rho = 11,300$ kg/m³) to completely sink the log and the lead in water, determine the average density of the log.
Answer: 835 kg/m³

10-37   A 200-kg, 5-m-wide rectangular gate shown in the figure is hinged at *B* and leans against the floor at *A* making an angle of 45° with the horizontal. The gate is to be opened from its lower edge by applying a normal force at its center. Determine the minimum force *F* required to open the water gate.
Answer: 520 kN

10-38   Repeat Prob. 10-37 for a water height of 2 m above the hinge at *B*.

10-39   A 3-m-high, 6-m-wide rectangular gate is hinged at the top edge at *A* and is restrained by a fixed ridge at *B*. Determine the hydrostatic force exerted on the gate by the 5-m-high water and the location of the pressure center.

10-40   Repeat Prob. 10-39 for a total water height of 2 m.

10-41E   A semicircular 30-ft-diameter tunnel is to be built under a 150-ft-deep, 2000-ft-long lake, as shown in the figure. Determine the total hydrostatic force acting on the roof of the tunnel.

10-42   A 500-kg, 6-m-diameter hemispherical dome on a level surface is filled with water, as shown in the figure. Someone claims that he can lift this dome by making use of Pascal's principle by attaching a long tube to the top and filling it with water. Determine the required height of water in the tube to lift the dome. Disregard the weight of the tube and the water in it.
Answer: 2.02 m

10-43   The water in a 25-m-deep reservoir is kept inside by a 150-m-wide wall whose cross section is an equilateral triangle, as shown in the figure.

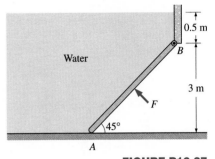

FIGURE P10-37

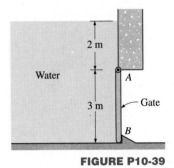

FIGURE P10-39

FIGURE P10-41E

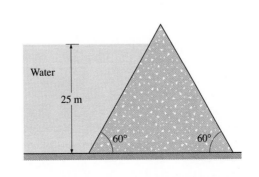

FIGURE P10-42   **FIGURE P10-43**

Determine (*a*) the total force (hydrostatic + atmospheric) acting on the inner surface of the wall and its line of action and (*b*) the magnitude of the horizontal component of this force. Take $P_{atm} = 100$ kPa.

Computer, Design, and Essay Problems

10-44 Shoes are to be designed to enable people of up to 80 kg to walk on fresh or seawater. The shoes are to be made of blown plastic in the shape of a sphere, (American) football, or French bread. Determine the equivalent diameter of each shoe and comment on the proposed shapes from the stability point of view. What is your assessment of the marketability of these shoes?

10-45 The volume of a rock is to be determined without using any volume measurement devices. Explain how you would do this with a waterproof spring scale.

Bernoulli, Energy, and Momentum Equations

This chapter deals with three fundamental equations of fluid mechanics: the Bernoulli equation, the energy equation, and the linear momentum equation. The *Bernoulli equation* is concerned with the conservation of kinetic, potential, and pressure energies of a fluid stream, and their conversion to each other during idealized frictionless flow, and is applicable under some restrictive conditions. The *energy equation* is a statement of the conservation of energy principle and is applicable under all conditions. In fluid mechanics, it is found to be convenient to separate *mechanical energy* from *thermal energy* and to consider the conversion of mechanical energy to thermal energy as a result of frictional effects as *mechanical energy loss*. Then the energy equation is usually expressed as the *conservation of mechanical energy*. The *momentum equation* is a statement of Newton's second law for fluid streams and is used to determine the forces caused by fluid flow.

We start this chapter with a discussion of various forms of mechanical energy and the efficiency of mechanical work devices such as pumps and turbines. Then we derive the momentum equation by applying Newton's second law to a fluid element along a streamline and demonstrate its use in a variety of applications. We continue with the development of the energy equation in a form suitable for use in fluid mechanics and introduce the concept of *head loss*. Next we derive the general *Reynolds transport theorem* and obtain the *momentum equation* from it as a special case. Finally, we apply the momentum equation, which is a *vector* equation, to several practical problems.

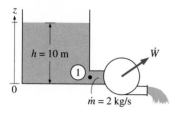

$$\dot{W}_{max} = \dot{m}\,\frac{P_1}{\rho} = \dot{m}\,\frac{\rho gh}{\rho} = \dot{m}gh$$
$$= (2\ \text{kg/s})(9.81\ \text{m/s}^2)(10\ \text{m})$$
$$= 196\ \text{W}$$

FIGURE 11-1

Pressure energy P/ρ can be converted to mechanical work completely by a reversible turbine, and thus it is a form of mechanical energy.

11-1 ■ MECHANICAL ENERGY AND PUMP EFFICIENCY

Most fluid systems are designed to transport a fluid from one location to another at a specified rate, velocity, and elevation, and a system may generate mechanical work in a turbine or it may consume mechanical work in a pump or fan during this process. These systems do not involve the conversion of nuclear, chemical, or thermal energy to mechanical energy. Also, they do not involve any heat transfer in any significant amount, and they operate essentially at constant temperature. Such systems can be analyzed conveniently by considering the *mechanical forms of energy* only and the frictional effects that cause the mechanical energy to be lost (i.e., to be converted to thermal energy that usually cannot be used for any useful purpose).

The **mechanical energy** can be defined as *the form of energy that can be converted to mechanical work completely and directly by a mechanical device such as a turbine.* The kinetic and potential energies are the familiar forms of mechanical energy. Thermal energy is not mechanical energy, however, since it cannot be converted to work directly and completely (the second law of thermodynamics). Another form of mechanical energy is the *pressure energy* discussed below (Fig. 11-1). Then the mechanical energy can be expressed as

Mechanical energy = Kinetic energy + Potential energy + Pressure energy

Consider a container of height h filled with water, as shown in Fig. 11-2, with reference level selected at the bottom surface. The gage pressure and the potential energy per unit mass are $P_A = 0$ and $pe_A = gh$ at point A at the free surface, and $P_B = \rho gh$ and $pe_B = 0$ at point B at the bottom of the container. A perfect hydraulic turbine will produce the same work $w_{turbine} = gh$ whether it receives water (or any other fluid with constant density) from the top or the bottom of the container. Therefore, the mechanical energy of water at the bottom surface is equivalent to that at the top surface, which is the potential energy. The mechanical energy that is due to the pressure of the fluid is called the **pressure energy** and is expressed as

$$\text{Pressure energy} = Pv = \frac{P}{\rho} \qquad \text{(kJ/kg)} \qquad (11\text{-}1)$$

The pressure energy of the liquid at a point in a reservoir is gh, where h is the vertical distance of the point from the free surface. Note that the *pressure energy* is equivalent to the *flow energy* for a flowing fluid, except that usually *gage pressure* is used in pressure energy instead of absolute pressure. Also, the pressure unit Pa is equivalent to Pa = N/m^2 = N · m/m^3 = J/m^3, which leads us to view pressure as *energy per unit volume,* and the product Pv (whose unit is J/kg) as *energy per unit mass.* For a stationary fluid with

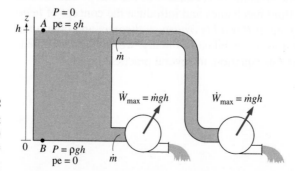

FIGURE 11-2

The mechanical energy of water at the bottom of a lake (pressure energy) is equal to the mechanical energy at the free surface of the lake (potential energy).

constant density, the sum of the pressure energy and the potential energy is constant.

The transfer of mechanical energy is usually accomplished by a rotating shaft, and thus mechanical work is often referred to as *shaft work*. A pump or a fan receives shaft work (usually from an electric motor) and transfers it to the fluid as mechanical energy (less frictional losses). A turbine, on the other hand, converts the mechanical energy of a fluid to shaft work.

The transfer of mechanical energy is usually accomplished by a rotating shaft, and thus mechanical work is often referred to as *shaft work*. A pump or a fan receives shaft work (usually from an electric motor) and transfers it to the fluid as mechanical energy (less frictional losses). A turbine, on the other hand, converts the mechanical energy of a fluid to shaft work.

Mechanical energy is entropy free, and 100 percent of it can be converted from one mechanical form to another in the absence of any irreversibilities such as friction. Therefore, the **mechanical efficiency** of a device or process can be defined as (Fig. 11-3)

$$\eta_{mech} = \frac{\text{Mechanical energy output}}{\text{Mechanical energy input}} = \frac{E_{mech, out}}{E_{mech, in}} = 1 - \frac{E_{mech, loss}}{E_{mech, in}} \qquad (11\text{-}2)$$

A conversion efficiency of less than 100 percent indicates that conversion is less than perfect and some losses have occurred during conversion. A mechanical efficiency of 97 percent indicates that 3 percent of the mechanical energy input is converted to heat as a result of frictional heating, and this will manifest itself as a slight rise in the temperature of the fluid.

In fluid systems, we are usually interested in increasing the pressure, velocity, and/or elevation of a fluid. This is done by *supplying mechanical energy to the fluid by a pump, a fan, or a compressor* (we will refer to all of them as pumps). Or we are interested in the reverse process of *extracting mechanical energy from a fluid by a turbine*, and producing mechanical power in the form of a rotating shaft that can drive a generator or any other rotary device. The degree of success of the conversion process between the mechanical work supplied or extracted and the mechanical energy of the fluid is expressed by the **pump efficiency** and **turbine efficiency,** defined as

$$\eta_{pump} = \frac{\text{Mechanical energy supplied to the fluid}}{\text{Mechanical power input}} = \frac{\Delta\dot{E}_{mech, fluid}}{\dot{W}_{shaft, in}} = \frac{\dot{W}_{pump, u}}{\dot{W}_{shaft, in}} \qquad (11\text{-}3)$$

where $\Delta\dot{E}_{mech, fluid} = \dot{E}_{mech, out} - \dot{E}_{mech, in}$ is the rate of increase in the mechanical energy of the fluid, which is equivalent to the useful pumping power $\dot{W}_{pump, u}$ supplied to the fluid, and

$$\eta_{turbine} = \frac{\text{Mechanical power output}}{\text{Mechanical energy extracted from the fluid}} = \frac{\dot{W}_{shaft, out}}{|\Delta\dot{E}_{mech, fluid}|} \qquad (11\text{-}4)$$

where $|\Delta\dot{E}_{mech, fluid}| = \dot{E}_{mech, in} - \dot{E}_{mech, out}$ is the rate of decrease in the mechanical energy of the fluid, and we used the absolute value sign to avoid negative values for efficiencies. A pump or turbine efficiency of 100 percent indicates perfect conversion between the shaft work and the mechanical energy of the fluid, and this value can be approached as the frictional effects are minimized.

The mechanical efficiency should not be confused with the **motor efficiency** and the **generator efficiency,** which are defined as

$$\textit{Motor:} \qquad \eta_{motor} = \frac{\text{Mechanical power output}}{\text{Electrical power input}} = \frac{\dot{W}_{shaft, out}}{\dot{W}_{elect, in}} \qquad (11\text{-}5)$$

451

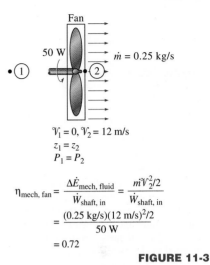

Mechanical Energy and Pump Efficiency

$$\mathcal{V}_1 = 0, \mathcal{V}_2 = 12 \text{ m/s}$$
$$z_1 = z_2$$
$$P_1 = P_2$$

$$\eta_{mech, fan} = \frac{\Delta\dot{E}_{mech, fluid}}{\dot{W}_{shaft, in}} = \frac{\dot{m}\mathcal{V}_2^2/2}{\dot{W}_{shaft, in}}$$
$$= \frac{(0.25 \text{ kg/s})(12 \text{ m/s})^2/2}{50 \text{ W}}$$
$$= 0.72$$

FIGURE 11-3

The mechanical efficiency of a fan is the ratio of the kinetic energy of air at the fan exit to the mechanical power input.

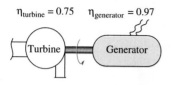

$\eta_{turbine} = 0.75 \quad \eta_{generator} = 0.97$

$\eta_{turbine-gen} = \eta_{turbine}\, \eta_{generator}$
$= 0.75 \times 0.97$
$= 0.73$

FIGURE 11-4

The overall efficiency of a hydraulic turbine-generator is the product of the efficiency of the turbine and the efficiency of the generator and represents the fraction of the mechanical energy of the water converted to electrical energy.

and

$$Generator: \qquad \eta_{generator} = \frac{\text{Electrical power output}}{\text{Mechanical power input}} = \frac{\dot{W}_{elect,\,out}}{\dot{W}_{shaft,\,in}} \qquad (11\text{-}6)$$

A pump is usually packaged together with its motor, and a hydraulic turbine with its generator. Therefore, we are usually interested in the **combined** or **overall efficiency** of pump/motor and turbine/generator combinations (Fig. 11-4), which are defined as

$$\eta_{pump\text{-}motor} = \eta_{pump}\,\eta_{motor} = \frac{\dot{E}_{mech,\,out} - \dot{E}_{mech,\,in}}{\dot{W}_{elect,\,in}} = \frac{\Delta\dot{E}_{mech,\,fluid}}{\dot{W}_{elect,\,in}} \qquad (11\text{-}7)$$

and

$$\eta_{turbine\text{-}gen} = \eta_{turbine}\,\eta_{generator} = \frac{\dot{W}_{elect,\,out}}{\dot{E}_{mech,\,in} - \dot{E}_{mech,\,out}} = \frac{\dot{W}_{elect,\,out}}{|\Delta\dot{E}_{mech,\,fluid}|} \qquad (11\text{-}8)$$

All the efficiencies defined above range between 0 and 100 percent. The lower limit of 0 percent corresponds to the conversion of the entire mechanical or electrical energy input to heat, and the device in this case functions like a resistance heater. The upper limit of 100 percent corresponds to the case of perfect conversion with no friction or other irreversibilities, and thus no conversion of mechanical or electrical energy to heat.

EXAMPLE 11-1   Performance of a Hydraulic Turbine-Generator

The water in a large lake is to be used to generate electricity by installing a hydraulic turbine-generator at a location where the depth of the water is 50 m (Fig. 11-5). Water is to be supplied at a rate of 5000 kg/s. If the electric power generated is measured to be 1862 kW and the generator efficiency is 95 percent, determine (*a*) the overall efficiency of the turbine-generator, (*b*) the mechanical efficiency of the turbine, and (*c*) the shaft power supplied by the turbine to the generator.

Solution   A hydraulic turbine-generator is to generate electricity from the water of a deep lake. The overall efficiency, the turbine efficiency, and the shaft power are to be determined.

Assumptions   **1** The elevation of the lake remains constant. **2** The mechanical energy of water at the turbine exit is negligible.

Properties   The density of water can be taken to be ρ = 1000 kg/m³.

Analysis   (*a*) We take the bottom of the lake as the reference level for convenience. Then kinetic and potential energies of water are zero, and the mechanical energy of water consists of pressure energy only, which is

$$e_{mech,\,in} - e_{mech,\,out} = \frac{P}{\rho} - 0 = gh = (9.81 \text{ m/s}^2)(50 \text{ m})\left(\frac{1 \text{ kJ/kg}}{1000 \text{ m}^2/\text{s}^2}\right) = 0.491 \text{ kJ/kg}$$

Then the rate at which mechanical energy is supplied to the turbine by the fluid and the overall efficiency become

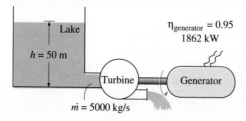

FIGURE 11-5

Schematic for Example 11-1.

$$|\Delta \dot{E}_{\text{mech, fluid}}| = \dot{m}(e_{\text{mech, in}} - e_{\text{mech, out}}) = (5000 \text{ kg/s})(0.491 \text{ kJ/kg}) = 2455 \text{ kW}$$

$$\eta_{\text{overall}} = \eta_{\text{turbine-gen}} = \frac{\dot{W}_{\text{elect, out}}}{|\Delta \dot{E}_{\text{mech, fluid}}|} = \frac{1862 \text{ kW}}{2455 \text{ kW}} = \mathbf{0.76}$$

(*b*) Knowing the overall and generator efficiencies, the mechanical efficiency of the turbine is determined from

$$\eta_{\text{turbine-gen}} = \eta_{\text{turbine}} \, \eta_{\text{generator}} \qquad \rightarrow \qquad \eta_{\text{turbine}} = \frac{\eta_{\text{turbine-gen}}}{\eta_{\text{generator}}} = \frac{0.76}{0.95} = \mathbf{0.80}$$

(*c*) The shaft power output is determined from the definition of mechanical efficiency,

$$\dot{W}_{\text{shaft, out}} = \eta_{\text{turbine}} \, |\Delta \dot{E}_{\text{mech, fluid}}| = (0.80)(2455 \text{ kW}) = \mathbf{1964 \text{ kW}}$$

Discussion   Note that the lake supplies 2455 kW of mechanical energy to the turbine, which converts 1964 kW of it to shaft work that drives the generator, which generates 1862 kW of electric power.

EXAMPLE 11-2   Conservation of Energy for an Oscillating Steel Ball

The motion of a steel ball in a hemispherical bowl of radius *h* shown in Fig. 11-6 is to be analyzed. The ball is initially held at the highest location at point *A*, and then it is released. Obtain relations for the conservation of energy of the ball for the cases of frictionless and actual motions.

Solution   A steel ball is released in a bowl. A relation for the energy balance is to be obtained.

Assumptions   The motion is frictionless, and thus friction between the ball, the bowl, and the air is negligible.

Analysis   When the ball is released, it will accelerate under the influence of gravity, reach a maximum velocity (and minimum elevation) at point *B* at the bottom of the bowl, and move up toward point *C* on the opposite side. In the ideal case of frictionless motion, the ball will oscillate between points *A* and *C*. The actual motion involves the conversion of the kinetic and potential energies of the ball to each other, together with overcoming resistance to motion due to friction (doing frictional work). The general energy balance for any system undergoing any process is

$$\underbrace{E_{\text{in}} - E_{\text{out}}}_{\substack{\text{Net energy transfer} \\ \text{by heat, work, and mass}}} = \underbrace{\Delta E_{\text{system}}}_{\substack{\text{Change in internal, kinetic,} \\ \text{potential, etc., energies}}}$$

Then the energy balance for the ball for a process from point 1 to point 2 becomes

$$-w_{\text{friction}} = (\text{ke}_2 + \text{pe}_2) - (\text{ke}_1 + \text{pe}_1)$$

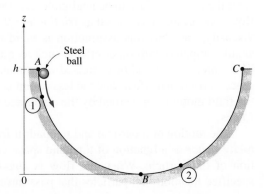

FIGURE 11-6

Schematic for Example 11-2.

or

$$\frac{\mathcal{V}_1^2}{2} + gz_1 = \frac{\mathcal{V}_2^2}{2} + gz_2 + w_{\text{friction}}$$

since there is no energy transfer by heat or mass, and no change in the internal energy of the ball (the heat generated by frictional heating is dissipated to the surrounding air). The frictional work term w_{friction} is often expressed as e_{loss} to represent the loss (conversion) of mechanical energy into the thermal (heat) energy.

For the idealized case of frictionless motion, the relation above reduces to

$$\frac{\mathcal{V}_1^2}{2} + gz_1 = \frac{\mathcal{V}_2^2}{2} + gz_2 \qquad \text{or} \qquad \frac{\mathcal{V}^2}{2} + gz = C = \text{constant}$$

where the value of the constant is $C = gh$. That is, *when the frictional effects are negligible, the sum of the kinetic and potential energies of the ball remains constant.*

Discussion   This is certainly a more intuitive and convenient form of the conservation of energy equation for this and other similar processes such as the swinging motion of the pendulum of a wall clock. The relation obtained is analogous to the Bernoulli equation derived in the next section.

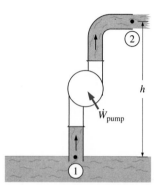

Steady flow
$\mathcal{V}_1 = \mathcal{V}_2$
$z_2 = z_1 + h$
$P_1 \cong P_2 \cong P_{\text{atm}}$

$\dot{E}_{\text{mech, in}} = \dot{E}_{\text{mech, out}} + \dot{E}_{\text{mech, loss}}$

$\dot{W}_{\text{pump}} + \dot{m}gz_1 = \dot{m}gz_2 + \dot{E}_{\text{mech, loss}}$

$\dot{W}_{\text{pump}} = \dot{m}gh + \dot{E}_{\text{mech, loss}}$

FIGURE 11-7

Most fluid flow problems involve mechanical forms of energy only, and such problems are conveniently solved by using a *mechanical energy* balance.

Most processes encountered in practice involve only certain forms of energy, and in such cases it is more convenient to work with the simplified versions of the energy balance. For systems that involve only *mechanical forms of energy* and its transfer as *shaft work*, the conservation of energy principle can be expressed conveniently as

$$E_{\text{mech, in}} - E_{\text{mech, out}} = \Delta E_{\text{mech, system}} + E_{\text{mech, loss}} \tag{11-9}$$

where $E_{\text{mech, loss}}$ represents the conversion of mechanical energy to heat due to irreversibilities such as friction. For a system in steady operation, the mechanical energy balance becomes $E_{\text{mech, in}} = E_{\text{mech, out}} + E_{\text{mech, loss}}$ (Fig. 11-7).

11-2 ▪ THE BERNOULLI EQUATION

The Bernoulli equation is *a relation between pressure, velocity, and elevation in steady, incompressible, frictionless flow.* Despite its simplicity, it has proven to be a very powerful tool in fluid mechanics. In this section, we derive the Bernoulli equation by applying the *conservation of linear momentum principle* and demonstrate its use.

A key assumption in the derivation of the Bernoulli equation is to consider the *viscous effects* to be negligible, and thus the fluid to be *inviscid.* Such flows are usually designated as *frictionless flow.* There is no fluid with zero viscosity, and thus this assumption is valid only when viscous effects are small compared with other effects such as gravity and pressure (Fig. 11-8). Therefore, care should be exercised when making this assumption. In the absence of frictional effects and the less common effects such as surface tension, the fluid motion is governed by the combined effects of pressure and gravity forces.

The motion of a particle and the path it follows are described by the *velocity vector* as a function of time and space coordinates and the initial position of the particle. When the flow is *steady* (no change with time at a specified location), all particles that pass through the same point will follow

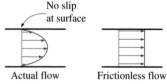

No slip
at surface

Actual flow   Frictionless flow

FIGURE 11-8

There is no fluid with zero viscosity, and thus the frictionless flow idealization is appropriate only when the viscous effects are relatively small.

the same path (which is the *streamline*), and the velocity vectors remain tangent to the path at every point.

Acceleration of a Fluid Particle

Often it is convenient to describe the motion of a particle in terms of its distance s from the origin together with the radius of curvature along the streamline. The velocity of the particle is related to the distance by $\mathcal{V} = ds/dt$, which may vary along the streamline. In two-dimensional flow, the acceleration can be decomposed into two components: *streamwise acceleration* a_s along the streamline and *normal acceleration* a_n (also called the *centrifugal acceleration*) in the direction normal to the streamline, which is given as $a_n = \mathcal{V}^2/R$. Note that streamwise acceleration is due to a change in speed along a streamline, and normal acceleration is due to a change in direction. For particles that move along a *straight path,* $a_n = 0$ since the radius of curvature is infinity and thus there is no change in direction. The Bernoulli equation results from a force balance along a streamline.

One may be tempted to think that acceleration is zero in steady flow since acceleration is the rate of change of velocity with time, and in steady flow there is no change with time. Well, a garden hose nozzle will tell us that this understanding is not correct. Even in steady flow and thus constant mass flow rate, water will accelerate through the nozzle (Fig. 11-9). *Steady* simply means no change with time at a specified location, but the value of a quantity may change from one location to another. In the case of a nozzle, the velocity of water remains constant at a specified point, but it changes from the inlet to the exit (water accelerates along the nozzle).

Mathematically, this can be expressed as follows: We take the velocity $\mathcal{V}$ to be a function of s and t. Taking the total differential of $\mathcal{V}(s, t)$ and dividing both sides by dt give

$$d\mathcal{V} = \frac{\partial \mathcal{V}}{\partial s} \, ds + \frac{\partial \mathcal{V}}{\partial t} \, dt \qquad \text{and} \qquad \frac{d\mathcal{V}}{dt} = \frac{\partial \mathcal{V}}{\partial s} \frac{ds}{dt} + \frac{\partial \mathcal{V}}{\partial t} \qquad \text{(11-10a, b)}$$

In steady flow $\partial\mathcal{V}/\partial t = 0$ and thus $\mathcal{V} = \mathcal{V}(s)$, and the acceleration in the s direction becomes

$$a_s = \frac{d\mathcal{V}}{dt} = \frac{\partial \mathcal{V}}{\partial s} \frac{ds}{dt} = \frac{\partial \mathcal{V}}{\partial s} \mathcal{V} = \mathcal{V} \frac{d\mathcal{V}}{ds} \qquad \text{(11-11)}$$

Therefore, acceleration in steady flow is due to the change of velocity with position.

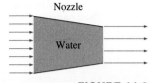

FIGURE 11-9

During steady flow, a fluid may not accelerate in time at a fixed point, but it may accelerate in space.

Derivation of the Bernoulli Equation

Consider the motion of a fluid particle in a flow field in steady flow. Applying Newton's second law (which is referred to as the *conservation of linear momentum* relation in fluid mechanics) in the s direction on a particle moving along a streamline gives

$$\sum F_s = ma_s \qquad \text{(11-12)}$$

When the friction forces are disregarded, the significant forces acting in the s direction are the pressure (acting on both sides) and the component of the weight of the particle in the s direction (Fig. 11-10). Therefore,

$$P \, dA - (P + dP) \, dA - W \sin \theta = m\mathcal{V} \frac{d\mathcal{V}}{ds} \qquad \text{(11-13)}$$

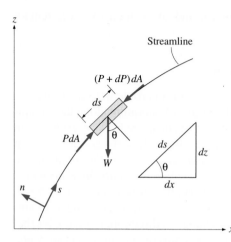

FIGURE 11-10

The forces acting on a
fluid particle along a streamline.

(Steady flow)

General:

$$\int \frac{dP}{\rho} + \frac{V^2}{2} + gz = \text{constant}$$

Incompressible flow (ρ = constant):

$$\frac{P}{\rho} + \frac{V^2}{2} + gz = \text{constant}$$

FIGURE 11-11

The Bernoulli equation is derived
assuming incompressible flow, and
thus it should not be used for flows with
significant compressibility effects.

where θ is the angle between the normal of the streamline and the vertical z axis at that point, $m = \rho V = \rho \, dA \, ds$ is the mass, $W = mg = \rho g \, dA \, ds$ is the weight of the fluid particle, and $\sin \theta = dz/ds$. Substituting,

$$-dP \, dA - \rho g \, dA \, ds \frac{dz}{ds} = \rho \, dA \, ds \, V \frac{dV}{ds} \qquad (11\text{-}14)$$

Canceling dA from each term and simplifying,

$$-dP - \rho g \, dz = \rho V \, dV \qquad (11\text{-}15)$$

Noting that $V \, dV = \frac{1}{2} d(V^2)$ and dividing each term by ρ gives

$$\frac{dP}{\rho} + \frac{1}{2} d(V^2) + g \, dz = 0 \qquad (11\text{-}16)$$

Integrating (Fig. 11-11),

Steady flow: $\qquad \int \frac{dP}{\rho} + \frac{V^2}{2} + gz = \text{constant}$ (along a streamline) $\qquad (11\text{-}17)$

since the last two terms are exact differentials. In the case of incompressible flow (ρ = constant), the first term also becomes an exact differential, and its integration gives

Steady, incompressible: $\qquad \dfrac{P}{\rho} + \dfrac{V^2}{2} + gz = \text{constant} \qquad$ (kJ/kg) $\qquad (11\text{-}18)$

This is the famous **Bernoulli equation,** which is commonly used in fluid mechanics for steady, frictionless, incompressible flow. The value of the constant can be evaluated at any point on the streamline where the pressure, density, velocity, and elevation are known. The Bernoulli equation can also be written between any two points on the same streamline as

Steady, incompressible: $\qquad \dfrac{P_1}{\rho} + \dfrac{V_1^2}{2} + gz_1 = \dfrac{P_2}{\rho} + \dfrac{V_2^2}{2} + gz_2 \qquad (11\text{-}19)$

The Bernoulli equation is obtained from the conservation of momentum for a fluid particle during steady, incompressible, and inviscid flow along a streamline. It can also be obtained from the *first law of thermodynamics* applied to a steady-flow system, as shown later in this chapter.

The Bernoulli equation was first stated in words in a textbook by Daniel Bernoulli in 1738 and was derived later by Leonhard Euler in 1755. We recognize $\mathcal{V}^2/2$ as *kinetic energy*, gz as *potential energy*, and P/ρ as *pressure energy* per unit mass. Therefore, the Bernoulli equation is an expression of *mechanical energy balance* and can be stated as follows (Fig. 11-12):

> *The sum of the kinetic, potential, and pressure energies of a fluid particle is constant along a streamline during steady flow when the compressibility and frictional effects are negligible.*

The kinetic, potential, and pressure energies are the mechanical forms of energy, as discussed earlier, and the Bernoulli equation can be viewed as the "conservation of mechanical energy principle." This is equivalent to the general conservation of energy principle for systems that do not involve any conversion of mechanical energy and thermal energy to each other, and thus the mechanical energy and thermal energy are conserved separately. The Bernoulli equation states that during steady, inviscid, incompressible flow, the various forms of mechanical energy are converted to each other, but their sum remains constant. In other words, there is no dissipation of mechanical energy during such flows since there is no friction that converts mechanical energy to sensible thermal (internal) energy.

Recall that energy is transferred to a system as work when a force is applied to a system through a distance. In the light of Newton's second law of motion, the Bernoulli equation can also be viewed as *the work done by the pressure and gravity forces on the fluid particle is equal to the increase in the kinetic energy of the particle.*

Despite the highly restrictive assumptions used in its derivation (steady, frictionless, incompressible flow), the Bernoulli equation is commonly used in practice since a variety of fluid flow problems can be analyzed with it with reasonable accuracy. This is because many flows are very nearly steady, and the compressibility and frictional effects are relatively small.

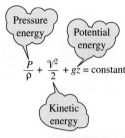

FIGURE 11-12

The Bernoulli equation states that the sum of the kinetic, potential, and pressure energies of a fluid particle is constant along a streamline during steady flow.

Unsteady Compressible Flow

Similarly, using both terms in the acceleration expression, it can be shown that the Bernoulli equation for *unsteady, compressible flow* is

Unsteady, compressible: $\quad \displaystyle\int \frac{dP}{\rho} + \int \frac{\partial \mathcal{V}}{\partial t}\, ds + \frac{\mathcal{V}^2}{2} + gz = \text{constant} \quad$ (11-20)

Force Balance across Streamlines

It is left as an exercise to show that a force balance in the direction n normal to the streamline yields the following relation applicable *across* the streamlines for steady, incompressible, frictionless flow:

$$\frac{P}{\rho} + \int \frac{\mathcal{V}^2}{R}\, dn + gz = \text{constant} \qquad \text{(across streamlines)} \quad (11\text{-}21)$$

For flow along a straight line, $R \to \infty$ and thus the relation above reduces to $P/\rho + gz = \text{constant}$ or $P = -\rho gz + \text{constant}$, which is an expression for the variation of hydrostatic pressure with vertical distance for a stationary fluid body. Therefore, the variation of pressure with elevation in steady,

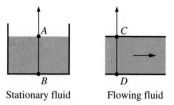

$$P_A = P_C$$
$$P_B = P_D$$

FIGURE 11-13

The variation of pressure with elevation in steady, incompressible, frictionless flow along a straight line is the same as that in the stationary fluid (but this is not the case for a curved flow section).

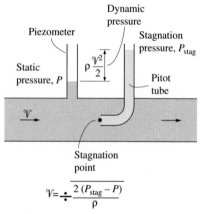

$$V = \pm \sqrt{\frac{2\,(P_{stag} - P)}{\rho}}$$

FIGURE 11-14

The static, dynamic, and stagnation pressures.

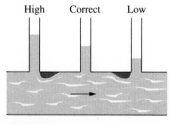

FIGURE 11-15

Careless placement of the pressure measurement device may result in erroneous reading of the static pressure.

incompressible, frictionless flow along a straight line is the same as that in the stationary fluid (Fig. 11-13).

Static, Dynamic, and Stagnation Pressures

The Bernoulli equation states that the sum of the pressure, kinetic, and potential energies of a fluid particle along a streamline is constant. Therefore, the kinetic and potential energies of the fluid can be converted to pressure energy (and vice versa) during flow, causing the pressure to change. This phenomenon can be made more visible by multiplying the Bernoulli equation by the density ρ,

$$P + \rho \frac{V^2}{2} + \rho g z = \text{constant} \quad \text{(kPa)} \qquad (11\text{-}22)$$

Each term in the equation above has pressure units, and thus each term represents some kind of pressure:

- P is the **static pressure** (it does not incorporate any dynamic effects); it represents the actual pressure of the fluid. This is the same as the pressure used in thermodynamics and property tables.
- $\rho V^2/2$ is the **dynamic pressure;** it represents the pressure rise when the fluid in motion is brought to a stop.
- $\rho g z$ is the **hydrostatic pressure,** which is not pressure in a real sense since its value depends on the reference level selected; it accounts for the elevation effects, i.e., of fluid weight on pressure.

The sum of the static, dynamic, and hydrostatic pressures is called the **total pressure.** Therefore, the Bernoulli equation states that *the total pressure along a streamline is constant.*

The sum of the static and dynamic pressures is called the **stagnation pressure,** and it is expressed as

$$P_{stag} = P + \rho \frac{V^2}{2} \quad \text{(kPa)} \qquad (11\text{-}23)$$

The stagnation pressure represents the pressure at a point where the fluid is brought to a complete stop in a frictionless manner. The static, dynamic, and stagnation pressures are shown in Fig. 11-14. When static and stagnation pressures are measured at a specified location, the fluid velocity at that location can be calculated from

$$V = \sqrt{\frac{2(P_{stag} - P)}{\rho}} \qquad (11\text{-}24)$$

Measurement of flow velocities by a *pitot tube* is based on this principle. When the static pressure is measured by drilling a hole in the tube wall, care must be exercised to ensure that the opening of the hole is flush with the wall surface, with no extrusions before or after the hole (Fig. 11-15). Otherwise the reading will incorporate some dynamic effects, and thus it will be in error.

When a stationary body is immersed in a fluid stream, some fluid will flow over and some under the body. The line (actually, the surface) that separates the upper and lower flows terminates at the stagnation point and is called the **stagnation streamline.**

Limitations on the Use of the Bernoulli Equation

The Bernoulli equation is one of the most frequently used and misused equations in fluid mechanics. Its versatility, simplicity, and ease of use make it a very valuable tool for use in analysis, but the same attributes also make it very tempting to misuse. Therefore, it is important to understand the restrictions on its applicability and observe the limitations on its use, as explained below:

1. Steady flow The first limitation on the Bernoulli equation is that it is applicable to *steady flow*. Therefore, it should not be used during the transient start-up and shut-down periods, or during periods of change in the flow conditions.

2. Frictionless flow Every flow involves some friction, no matter how small, and the *frictional effects* may or may not be negligible. The situation is complicated even more by the amount of error that can be tolerated. In general, the frictional effects are negligible for short flow sections with large cross sections, especially at low flow velocities. The frictional effects are usually significant for long and narrow flow passages and for *diverging flow sections* such as diffusers because of the increased possibility of the fluid separating from the walls in such geometries. The frictional effects are also significant near solid surfaces, and thus the Bernoulli equation is more accurate in the core region of the flow (Fig. 11-16).

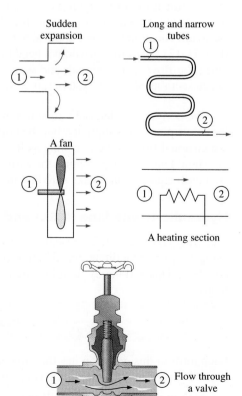

FIGURE 11-16

Frictional effects and components that disturb the streamlined structure of flow in a flow section can make the Bernoulli equation invalid.

A component that disturbs the streamlined structure of flow and thus causes considerable mixing and back flow such as a sharp entrance of a tube or a partially closed valve in a flow section can make the Bernoulli equation inapplicable.

3. No shaft work The Bernoulli equation was derived from a force balance on a particle moving along a streamline. Therefore, the Bernoulli equation is not applicable in a flow section that involves a pump, turbine, fan, or any other machine or impeller since such devices destroy the streamlines and carry out energy interactions with the fluid particles. When the flow section considered involves any of these devices, the energy equation should be used instead to account for the shaft work input or output. However, the Bernoulli equation can still be applied to a flow section prior to or past a machine (assuming, of course, that the other restrictions on its use are satisfied).

4. Incompressible flow One of the assumptions used in the derivation of the Bernoulli equation is that ρ = constant and thus the flow is incompressible. This condition is readily satisfied by liquids and also by gases at Mach numbers less that about 0.3 since compressibility effects and thus density variations of gases are negligible at such relatively low velocities.

5. No heat transfer The density of a gas is inversely proportional to temperature, and thus the Bernoulli equation should not be used for flow sections that involve significant temperature change such as heating or cooling sections.

6. Flow along a streamline Strictly speaking, the Bernoulli equation $P/\rho + \mathcal{V}^2/2 + gz = C$ is applicable along a streamline, and the value of the constant C, in general, is different for different streamlines. But when the flow is irrotational (and thus there is no swirling in the flow field), the value of the constant C remains the same for all streamlines, and, therefore, the Bernoulli equation becomes applicable between any two points along the flow (Fig. 11-17). Therefore, we do not need to be concerned about the streamlines when the flow is irrotational, and we can apply the Bernoulli equation between sections of a flow system rather than between points.

We derived the Bernoulli equation by considering two-dimensional flow in the x-z plane for simplicity, but the equation is valid for the general three-dimensional flow as well, as long as it is applied along the same streamline. We should always keep in mind the assumptions used in the derivation of the Bernoulli equation and make sure that they are not violated.

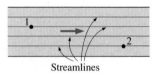

$$\frac{P_1}{\rho} + \frac{\mathcal{V}_1^2}{2} + gz_1 = \frac{P_2}{\rho} + \frac{\mathcal{V}_2^2}{2} + gz_2$$

FIGURE 11-17

When the flow is irrotational, the Bernoulli equation becomes applicable between any two points along the flow (not necessarily on the same streamline).

Hydraulic Grade Line (HGL) and Energy Line (EL)

It is often convenient to represent the level of mechanical energy graphically using *heights* to facilitate visualization of the various terms of the Bernoulli equation. This is done by dividing each term of the Bernoulli equation by g to give

$$\frac{P}{\rho g} + \frac{\mathcal{V}^2}{2g} + z = H = \text{constant} \qquad \text{(m)} \qquad (11\text{-}25)$$

Each term in this equation has the dimension of length and represents some kind of "head" of a flowing fluid as follows:

- $P/\rho g$ is the **pressure head;** it represents the height of a fluid column that produces the static pressure P.

- $V^2/2g$ is the **velocity head;** it represents the elevation needed for a fluid to reach the velocity V during frictionless free fall.
- z is the **elevation head;** it represents the potential energy of the fluid.

Also, H is the **total head** for the flow. Therefore, the Bernoulli equation can be expressed in terms of heads as: *the sum of the pressure, velocity, and elevation heads along a streamline is constant during steady flow when the compressibility and frictional effects are negligible* (Fig. 11-18).

If a piezometer (measures static pressure) is tapped into a pipe, as shown in Fig. 11-19, the liquid would rise to a height of $P/\rho g$ above the pipe center. The *hydraulic grade line* (HGL) is obtained by doing this at several locations along the pipe and drawing a line through the liquid levels in piezometers. The vertical distance above the pipe center is a measure of pressure within the pipe. Similarly, if a pitot tube (measures static + dynamic pressure) is tapped into a pipe, the liquid would rise to a height of $P/\rho g + V^2/2g$ above the pipe center, or a distance of $V^2/2g$ above the HGL. The *energy line* (EL), also called the *energy grade line* (EGL), is obtained by doing this at several locations along the pipe and drawing a line through the liquid levels in pitot tubes.

Noting that the fluid also has elevation head z (unless the reference level is taken to be the centerline of the pipe), the HGL and EL can be defined as follows: The line that represents the sum of the static pressure and the elevation heads $P/\rho g + z$, is called the **hydraulic grade line.** The line that represents the total head of the fluid, $P/\rho g + V^2/2g + z$, is called the **energy line.** The difference between the heights of EL and HGL is equal to the dynamic head, $V^2/2g$. We note the following about the HGL and EL:

- For *stationary bodies* such as reservoirs or lakes, the EL and HGL coincide with the free surface of the liquid. The elevation of the free surface z in such cases represents both the EL and the HGL since the velocity is zero and the static pressure (gage) is zero.
- The EL is always a distance $V^2/2g$ above the HGL. These two lines approach each other as the velocity decreases, and they diverge as the velocity increases. The height of the HGL decreases as the velocity increases, and vice versa.
- In *frictionless Bernoulli-type flow,* EL is horizontal and its height remains constant. This is also the case for HGL when the flow velocity is constant (Fig. 11-20).
- For *open channel flow,* the HGL coincides with the free surface of the liquid, and the EL is a distance $V^2/2g$ above the free surface.
- At a *pipe exit,* the pressure head is zero (atmospheric pressure) and thus the HGL coincides with the pipe exit.
- The *mechanical energy loss* due to frictional effects (conversion to thermal energy) will cause the EL and HGL to slope downward in the direction of flow. The slope is a measure of the pipe loss (discussed in detail in Chap. 12). A component that generates significant frictional effects such as a valve will cause a sudden drop in EL and HGL at that location.
- A *steep jump* occurs in EL and HGL whenever mechanical energy is added to the fluid (by a pump, for example). Likewise, a *steep drop* occurs in EL and HGL whenever mechanical energy is removed from the fluid (by a turbine, for example), as shown in Fig. 11-21.
- The pressure (gage) of a fluid is zero at locations where the HGL *intersects* the fluid. The pressure in a flow section that lies above the HGL is negative, and the pressure in a section that lies below the HGL is

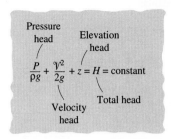

FIGURE 11-18

An alternative form of the Bernoulli equation is expressed in terms of heads as *the sum of the pressure, velocity, and elevation heads is constant.*

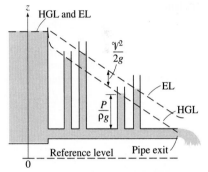

FIGURE 11-19

The hydraulic grade line (HGL) and the energy line (EL) for free discharge from a reservoir through a horizontal pipe (both HGL and EL decline due to pipe friction).

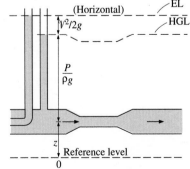

FIGURE 11-20

In frictionless Bernoulli-type flow, EL is horizontal and its height remains constant. But this is not the case for HGL when the flow velocity varies along the flow.

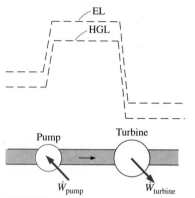

FIGURE 11-21

A *steep jump* occurs in EL and HGL
whenever mechanical energy is added to
the fluid by a pump, and a *steep drop*
occurs whenever mechanical energy is
removed from the fluid by a turbine.

positive (Fig. 11-22). Therefore, an accurate drawing of a piping system and the HGL can be used to determine the regions where the pressure in the pipe is negative (below the atmospheric pressure).

The last remark indicates that an accurate drawing of a piping system and the HGL can be used to identify potential locations where the pressure in the pipe can be below the atmospheric pressure. This is important to avoid situations in which the pressure drops below the vapor pressure of the liquid (which causes the alternate evaporation and condensation of the fluid, called *cavitation*). Proper consideration is necessary in the placement of a liquid pump to assure that the suction side pressure does not fall too low, especially at elevated temperatures where vapor pressure is greater.

11-3 ■ APPLICATIONS OF THE BERNOULLI EQUATION

In the previous section, we discussed the fundamental aspects of the Bernoulli equation. In this section, we will demonstrate its use in a wide range of applications through examples.

EXAMPLE 11-3 Spraying Water into the Air

Water is flowing from a hose attached to a water main at 400 kPa gage (Fig. 11-23). A child places his thumb to cover most of the hose outlet, increasing the pressure at the outlet and causing a thin jet of high-speed water to emerge. If the hose is held upward, what is the maximum height that the jet could achieve?

Solution Water from a hose attached to the water main is sprayed into the air. The maximum height the water jet can rise is to be determined.

Assumptions   **1** The flow is steady, frictionless, incompressible, and irrotational (so that the Bernoulli equation is applicable). **2** The water pressure in the hose near the outlet is equal to the water main pressure. **3** The surface tension effects are negligible. **4** The friction between the water and air is negligible. **5** The irreversibilities that may occur at the outlet of the hose due to abrupt expansion are negligible.

Properties   We take the density of water to be 1000 kg/m³.

Analysis   This problem involves the conversion of pressure, kinetic, and potential energies to each other without involving any pumps, turbines, and wasteful components with large frictional losses, and thus it is suitable for the use of the Bernoulli equation. The water height will be maximum under the stated assumptions. The velocity inside the hose is relatively low ($\mathcal{V}_1 \cong 0$) and we take the hose outlet as the reference level ($z_1 = 0$). At the top of the water trajectory $\mathcal{V}_2 = 0$, and atmospheric pressure pertains. Then the Bernoulli equation simplifies to

$$\frac{P_1}{\rho g} + \cancelto{0}{\frac{\mathcal{V}_1^2}{2g}} + \cancelto{0}{z_1} = \frac{P_2}{\rho g} + \cancelto{0}{\frac{\mathcal{V}_2^2}{2g}} + z_2 \quad \rightarrow \quad \frac{P_1}{\rho g} = \frac{P_{atm}}{\rho g} + z_2$$

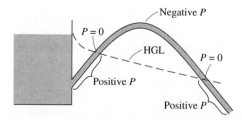

FIGURE 11-22

The pressure (gage) of a fluid
is zero at locations where the HGL
intersects the fluid, and the pressure is
negative (vacuum) in a flow section
that lies above the HGL.

Solving for z_2 and substituting,

$$z_2 = \frac{P_1 - P_{atm}}{\rho g} = \frac{P_{1,\,gage}}{\rho g} = \frac{400 \text{ kPa}}{(1000 \text{ kg/m}^3)(9.81 \text{ m/s}^2)}\left(\frac{1000 \text{ N/m}^2}{1 \text{ kPa}}\right)\left(\frac{1 \text{ kg} \cdot \text{m/s}^2}{1 \text{ N}}\right)$$

$$= \textbf{40.8 m}$$

Therefore, the water jet can rise as high as 40.8 m into the sky in this case.

Discussion   The result obtained by the Bernoulli equation represents the upper limit and should be interpreted accordingly. It tells us that the water cannot possibly rise more than 40.8 m, and, in all likelihood, the rise will be much less than 40.8 m.

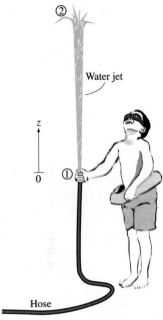

Water jet

FIGURE 11-23

Schematic for Example 11-3.

EXAMPLE 11-4   Water Discharge from a Large Tank

A large tank open to the atmosphere is filled with water to a height of 5 m from the bottom (Fig. 11-24). A tap near the bottom of the tank is now opened, and water flows out from the smooth and rounded outlet. Determine the water velocity at the outlet.

Solution   A tap near the bottom of a tank is opened. The exit velocity of water from the tank is to be determined.

Assumptions   The flow is steady, frictionless, incompressible, and irrotational (so that the Bernoulli equation is applicable).

Analysis   This problem involves the conversion of pressure, kinetic, and potential energies to each other without involving any pumps, turbines, and wasteful components with large frictional losses, and thus it is suitable for the use of the Bernoulli equation. We take point 1 to be at the free surface of water so that $P_1 = P_{atm}$ (open to the atmosphere), $\mathcal{V}_1 \cong 0$ (the tank is large relative to the outlet), and $z_1 = 5$ m and $z_2 = 0$ (we take the reference level at the center of the outlet). Also, $P_2 = P_{atm}$ (water discharges into the atmosphere). Then the Bernoulli equation simplifies to

$$\frac{P_1}{\rho g} + \cancel{\frac{\mathcal{V}_1^2}{2g}}^{0} + z_1 = \frac{P_2}{\rho g} + \frac{\mathcal{V}_2^2}{2g} + \cancel{z_2}^{0} \quad \rightarrow \quad z_1 = \frac{\mathcal{V}_2^2}{2g}$$

Solving for $\mathcal{V}_2$ and substituting,

$$\mathcal{V}_2 = \sqrt{2gz_1} = \sqrt{2(9.81 \text{ m/s}^2)(5 \text{ m})} = \textbf{9.9 m/s}$$

The relation $\mathcal{V} = \sqrt{2gz}$ is called the **Toricelli equation.**
    Therefore, the water will leave the tank with a velocity of 9.9 m/s. This is the same velocity that would manifest if a solid were dropped a distance of 5 m in the absence of air friction drag. (What would the velocity be if the tap were at the bottom of the tank instead of on the side?)

Discussion   Note that the Bernoulli equation applies along a streamline, and streamlines generally do not make abrupt turns. Therefore, if the orifice is sharp-edged instead of rounded, then the flow will be disturbed, and the velocity will be less than 9.9 m/s, especially near the edges. Care must be exercised when attempting to apply the Bernoulli equation to situations where abrupt expansions or contractions occur since the friction and flow disturbance in such cases may not be negligible.

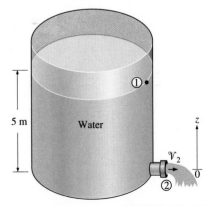

5 m   Water

FIGURE 11-24

Schematic for Example 11-4.

EXAMPLE 11-5   Siphoning out Gasoline from a Fuel Tank

During a trip to the beach ($P_{atm} = 1$ atm $= 101.3$ kPa), a car runs out of gasoline, and it becomes necessary to siphon gas out of the car of a good Samaritan (Fig. 11-25). The siphon is a small-diameter hose, and to start the siphon it is

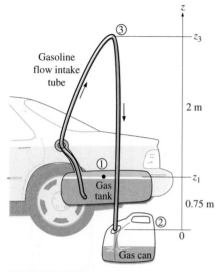

FIGURE 11-25
Schematic for Example 11-5.

necessary to insert one siphon end in the full gas tank, fill the hose with gasoline via suction, and then place the other end in a gas can below the level of the gas tank. The difference in pressure between point 1 (at the free surface of the gasoline in the tank) and point 2 (at the outlet of the tube) will cause the liquid to flow from the higher to the lower elevation. Point 2 is located 0.75 m below point 1 in this case, and point 3 is located 2 m above. The siphon diameter is 4 mm, and the friction loss of the liquid inside the siphon is to be disregarded. Determine (a) the minimum time to withdraw 4 L of gasoline from the tank to the can and (b) the pressure at point 3. The density of gasoline is 750 kg/m³.

Solution Gasoline is to be siphoned from a tank. The time it takes to withdraw 4 L of gasoline and the pressure at the highest point in the system are to be determined.

Assumptions **1** The flow is steady, frictionless, incompressible, and irrotational (so that the Bernoulli equation is applicable). **2** The gasoline level in the tank remains constant. **3** Frictional losses in the tube are negligible.

Properties The density of gasoline is given to be 750 kg/m³.

Analysis We take point 1 to be at the free surface of gasoline in the tank so that $P_1 = P_{atm}$ (open to the atmosphere), $V_1 \cong 0$ (the tank is large relative to the tube diameter), and $z_2 = 0$ (point 2 is taken as the reference level). Also, $P_2 = P_{atm}$ (gasoline discharges into the atmosphere). Then the Bernoulli equation simplifies to

$$\frac{\cancel{P_1}}{\rho g} + \frac{\cancel{V_1^2}^{\,0}}{2g} + z_1 = \frac{\cancel{P_2}}{\rho g} + \frac{V_2^2}{2g} + \cancel{z_2}^{\,0} \quad \rightarrow \quad z_1 = \frac{V_2^2}{2g}$$

Solving for V_2 and substituting,

$$V_2 = \sqrt{2gz_1} = \sqrt{2(9.81 \text{ m/s}^2)(0.75 \text{ m})} = 3.84 \text{ m/s}$$

The cross-sectional area of the tube and the flow rate of gasoline are

$$A = \pi D^2/4 = \pi(5 \times 10^{-3} \text{ m})^2/4 = 1.96 \times 10^{-5} \text{ m}^2$$
$$\dot{V} = V_2 A = (3.84 \text{ m/s})(1.96 \times 10^{-5} \text{ m}^2) = 7.53 \times 10^{-5} \text{ m}^3/\text{s} = 0.0753 \text{ L/s}$$

Then the time needed to siphon 4 L of gasoline becomes

$$\Delta t = \frac{V}{\dot{V}} = \frac{4 \text{ L}}{0.0753 \text{ L/s}} = \mathbf{53.1 \text{ s}}$$

(b) The pressure at point 3 can be determined by writing the Bernoulli equation between points 2 and 3. Noting that $V_2 = V_3$ (conservation of mass), $z_2 = 0$, and $P_2 = P_{atm}$,

$$\frac{P_2}{\rho g} + \frac{V_2^2}{2g} + \cancel{z_2}^{\,0} = \frac{P_3}{\rho g} + \frac{V_3^2}{2g} + z_3 \quad \rightarrow \quad \frac{P_{atm}}{\rho g} = \frac{P_3}{\rho g} + z_3$$

Solving for P_3 and substituting,

$$P_3 = P_{atm} - \rho g z_3$$

$$= 101.3 \text{ kPa} - (750 \text{ kg/m}^3)(9.81 \text{ m/s}^2)(2.75 \text{ m})\left(\frac{1 \text{ N}}{1 \text{ kg} \cdot \text{m/s}^2}\right)\left(\frac{1 \text{ kPa}}{1000 \text{ N/m}^2}\right)$$

$$= \mathbf{81.1 \text{ kPa}}$$

Discussion The siphoning time is determined assuming frictionless flow, and thus this is the *minimum time* required. In reality, the time will be longer than 53.1 s because of the friction between the gasoline and the tube surface. Also, the pressure at point 3 is below the atmospheric pressure. If the elevation difference between points 1 and 3 is too high, the pressure at point 3 may drop below the vapor pressure of gasoline at the gasoline temperature, and some gasoline may evaporate. The vapor then may form a pocket at the top and halt the flow of gasoline.

EXAMPLE 11-6   Velocity Measurement by a Pitot Tube

A piezometer and a pitot tube are tapped into a horizontal water pipe, as shown in Fig. 11-26, to measure static and stagnation (static + dynamic) pressures. For the indicated water column heights, determine the velocity at the center of the pipe.

Solution   The static and stagnation pressures in a horizontal pipe are measured. The velocity at the center of the pipe is to be determined.

Assumptions   The flow is steady, frictionless, incompressible, and irrotational (so that the Bernoulli equation is applicable).

Analysis   We take points 1 and 2 along the centerline of the pipe, with point 1 directly under the piezometer and point 2 at the entrance of the pitot tube. This is a steady flow with straight and parallel streamlines, and thus the static pressure at any point is equal to the hydrostatic pressure at that point. Therefore, the gage pressures at points 1 and 2 can be expressed as

$$P_1 = \rho g(h_1 + h_2)$$
$$P_2 = \rho g(h_1 + h_2 + h_3)$$

Noting that point 2 is a stagnation point and thus $\mathcal{V}_2 = 0$ and $z_1 = z_2$, the application of the Bernoulli equation between points 1 and 2 gives

$$\frac{P_1}{\rho g} + \frac{\mathcal{V}_1^2}{2g} + z_1 = \frac{P_2}{\rho g} + \cancelto{0}{\frac{\mathcal{V}_2^2}{2g}} + z_2 \quad \rightarrow \quad \frac{\mathcal{V}_1^2}{2g} = \frac{P_2 - P_1}{\rho g}$$

Substituting the P_1 and P_2 expressions gives

$$\frac{\mathcal{V}_1^2}{2g} = \frac{P_2 - P_1}{\rho g} = \frac{\rho g(h_1 + h_2 + h_3) - \rho g(h_1 + h_2)}{\rho g} = h_3$$

Solving for $\mathcal{V}_1$ and substituting,

$$\mathcal{V}_1 = \sqrt{2gh_3} = \sqrt{2(9.81 \text{ m/s}^2)(0.12 \text{ m})} = \textbf{1.53 m/s}$$

Discussion   Note that to determine the flow velocity, all we need is to measure the height of the excess fluid column in the pitot tube.

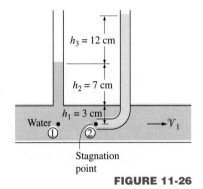

FIGURE 11-26

Schematic for Example 11-6.

EXAMPLE 11-7   The Rise of the Ocean Due to Hurricanes

A hurricane is a tropical storm formed over the ocean by low atmospheric pressures. As a hurricane approaches land, inordinate ocean swells (very high tides) accompany the hurricane. A Class-5 hurricane features winds in excess of 155 mph, although the wind velocity at the center "eye" is very low.

    Figure 11-27 depicts a hurricane hovering over the ocean swell below. The atmospheric pressure 200 miles from the eye is 30.0 inHg (at point 1, generally normal for the ocean) and the winds are calm. The hurricane atmospheric pressure at the eye of the storm is 22.0 inHg. Estimate the ocean swell at (*a*) the eye of the hurricane at point 3 and (*b*) Point 2, where the wind velocity is 155 mph. Take the density of seawater and mercury to be 64 lbm/ft³ and 848 lbm/ft³, respectively, and the density of air at normal sea level temperature and pressure to be 0.076 lbm/ft³.

Solution   A hurricane is moving over the ocean. The amount of ocean swell at the eye and active regions of the hurricane are to be determined.

Assumptions   **1** The air flow within the hurricane is steady, frictionless, incompressible, and irrotational (so that the Bernoulli equation is applicable). (This is certainly a very questionable assumption for a highly turbulent flow, but it is justified in the solution.) **2** The effect of water drifted into the air is negligible.

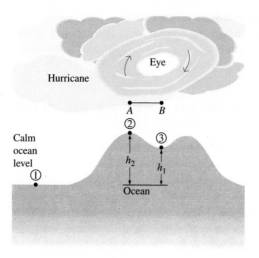

FIGURE 11-27
Schematic for Example 11-7.

Properties   The densities of air at normal conditions, seawater, and mercury are given to be 0.076 lbm/ft³, 64 lbm/ft³, and 848 lbm/ft³, respectively.

Analysis   (*a*) Reduced atmospheric pressure over the water causes the water to rise. Thus, decreased pressure at point 2 relative to point 1 will cause the ocean water to rise at point 2. The same is true at point 3, where the storm air velocity is negligible. The pressure difference given in terms of the mercury column height can be expressed in terms of the seawater column height by

$$\Delta P = (\rho g h)_{\text{Hg}} = (\rho g h)_{\text{sw}} \quad \rightarrow \quad h_{\text{sw}} = \frac{\rho_{\text{Hg}}}{\rho_{\text{sw}}} h_{\text{Hg}} \quad \text{or} \quad \Delta P_{\text{sw}} = \frac{\rho_{\text{Hg}}}{\rho_{\text{sw}}} \Delta P_{\text{Hg}}$$

Then the pressure difference between points 1 and 2 in terms of the seawater column height becomes

$$h_1 = \Delta P_{\text{sw}} = \frac{\rho_{\text{Hg}}}{\rho_{\text{sw}}} \Delta P_{\text{Hg}} = \left(\frac{848 \ \text{lbm/ft}^3}{64 \ \text{lbm/ft}^3}\right)[(30 - 22) \ \text{inHg}]\left(\frac{1 \ \text{ft}}{12 \ \text{in.}}\right) = \mathbf{8.83 \ ft}$$

which is equivalent to the storm surge at the *eye of the hurricane* since the wind velocity there is negligible and there are no dynamic effects.

(*b*) To determine the additional rise of ocean water at point 2 due to the high winds at that point, we write the Bernoulli equation between points *A* and *B*, which are on top of the points 2 and 3, respectively. Noting that $\mathcal{V}_B \cong 0$ (the eye region of the hurricane is relatively calm) and $z_A = z_B$ (both points are on the same horizontal line), the Bernoulli equation simplifies to

$$\frac{P_A}{\rho g} + \frac{\mathcal{V}_A^2}{2g} + \cancel{z_A} = \frac{P_B}{\rho g} + \cancel{\frac{\mathcal{V}_B^2}{2g}}^{\,0} + \cancel{z_B} \quad \rightarrow \quad \frac{P_B - P_A}{\rho g} = \frac{\mathcal{V}_A^2}{2g}$$

Substituting,

$$\frac{P_B - P_A}{\rho g} = \frac{\mathcal{V}_A^2}{2g} = \frac{(155 \ \text{mph})^2}{2(32.2 \ \text{ft/s}^2)}\left(\frac{1.4667 \ \text{ft/s}}{1 \ \text{mph}}\right)^2 = 803 \ \text{ft}$$

where ρ is the density of air in the hurricane. Noting that the density of an ideal gas at constant temperature is proportional to absolute pressure and the density of air at the normal atmospheric pressure of 14.7 psia $\cong$ 30 inHg is 0.076 lbm/ft³, the density of air in the hurricane is

$$\rho_{\text{air}} = \frac{P_{\text{air}}}{P_{\text{atm air}}} \rho_{\text{atm air}} = \left(\frac{22 \ \text{inHg}}{30 \ \text{inHg}}\right)(0.076 \ \text{lbm/ft}^3) = 0.056 \ \text{lbm/ft}^3$$

Using the relation developed above in part (*a*), the equivalent seawater column height of 803 ft of air column height is determined to be

$$h_{\text{dynamic}} = \Delta P_{\text{sw, dynamic}} = \frac{\rho_{\text{air}}}{\rho_{\text{sw}}} \Delta P_{\text{air}} = \left(\frac{0.056 \text{ lbm/ft}^3}{64 \text{ lbm/ft}^3}\right)(803 \text{ ft}) = 0.70 \text{ ft}$$

Therefore, the pressure at point 2 is 0.70 ft seawater column lower than the pressure at point 3 due to the high wind velocities, causing the ocean to rise an additional 0.70 ft. Then the total storm surge at point 2 becomes

$$h_2 = h_1 + h_{\text{dynamic}} = 8.83 + 0.70 = \textbf{9.53 ft}$$

Discussion This problem involves highly turbulent flow and the intense breakdown of the streamlines, and thus the applicability of the Bernoulli equation in part (*b*) is questionable. However, less than 10 percent of the result is due to the use of the Bernoulli equation, and thus the results obtained are still highly reliable. The Bernoulli analysis shows that the rise of seawater due to high-velocity winds cannot be more than 0.70 ft.

The wind power of hurricanes is not the only cause of damage to coastal areas. Ocean flooding and erosion from excessive tides is just as serious, so are high waves generated by the storm turbulence and energy.

Example 11-8 Bernoulli Equation for Compressible Flow
Derive the Bernoulli equation when the compressibility effects are not negligible for an ideal gas undergoing (*a*) an isothermal process and (*b*) an isentropic process.

Solution The Bernoulli equation for compressible flow is to be obtained for an ideal gas for isothermal and isentropic processes.

Assumptions **1** The flow is steady, frictionless, and irrotational. **2** The fluid is an ideal gas, so the relation $P = \rho RT$ is applicable. **3** The specific heats are constant so that $P/\rho^k = $ *constant* during an isentropic process.

Analysis (*a*) When the compressibility effects are significant and the flow cannot be assumed to be compressible, the Bernoulli equation is given by Eq. 11-17 as

$$\int \frac{dP}{\rho} + \frac{V^2}{2} + gz = \text{constant} \tag{1}$$

The compressibility effects can be properly accounted for by expressing ρ in terms of pressure, and then performing the integration $\int dP/\rho$ in Eq. (1). But this requires a relation between P and ρ for the process. For the *isothermal* expansion or compression of an ideal gas, the integral in Eq. (1) can be performed easily by noting that $T = $ constant and substituting $\rho = P/RT$. It gives

$$\int \frac{dP}{\rho} = \int \frac{dP}{P/RT} = RT \ln P$$

Substituting into Eq. (1) gives the desired relation,

Isothermal process: $\qquad\qquad RT \ln P + \frac{V^2}{2} + gz = \text{constant} \tag{2}$

(*b*) A more practical case of compressible flow is the *isentropic flow of ideal gases* through equipment that involves high-speed fluid flow such as nozzles, diffusers, and the passages between turbine blades. Isentropic (i.e., reversible and adiabatic) flow is closely approximated by these devices, and it is characterized by the relation $P/\rho^k = C = $ *constant* where k is the specific heat ratio of the gas, as discussed in Chap. 7. Solving for ρ from $P/\rho^k = C$ gives $\rho = C^{-1/k} P^{1/k}$. Performing the integration,

$$\int \frac{dP}{\rho} = \int C^{1/k} P^{-1/k} \, dP = C^{1/k} \frac{P^{-1/k+1}}{-1/k + 1} = P^{1/k} \frac{P^{-1/k+1}}{-1/k + 1} = \left(\frac{k}{k-1}\right) \frac{P}{\rho} \tag{3}$$

Substituting, the Bernoulli equation for steady, isentropic, compressible flow of an ideal gas becomes

Isentropic flow:

$$\left(\frac{k}{k-1}\right)\frac{P}{\rho} + \frac{\mathcal{V}^2}{2} + gz = \text{constant} \qquad (4a)$$

or

$$\left(\frac{k}{k-1}\right)\frac{P_1}{\rho_1} + \frac{\mathcal{V}_1^2}{2} + gz_1 = \left(\frac{k}{k-1}\right)\frac{P_2}{\rho_2} + \frac{\mathcal{V}_2^2}{2} + gz_2 \qquad (4b)$$

A common practical situation involves the acceleration of a gas from rest (stagnation conditions at state 1) with negligible change in elevation. In that case we have $z_1 = z_2$, $\mathcal{V}_1 = 0$. Noting that $\rho = P/RT$ for ideal gases, $P/\rho^k = constant$ for isentropic flow, and the Mach number is defined as $M = \mathcal{V}/C$ where $C = \sqrt{kRT}$ is the local speed of sound for ideal gases, the relation above simplifies to

$$\frac{P_1}{P_2} = \left[1 + \left(\frac{k-1}{2}\right)M_2^2\right]^{k/(k-1)} \qquad (4c)$$

where state 1 is the stagnation state and state 2 is any state along the flow.

Discussion It can be shown that the results obtained using the compressible and incompressible equations deviate no more than 2 percent when the Mach number is less than 0.3. Therefore, the flow of an ideal gas can be considered to be incompressible when $M \lesssim 0.3$. For atmospheric air at normal conditions, this corresponds to a flow speed of about 100 m/s or 360 km/h, which covers our range of interest. In this text we will not deal with compressible flow.

11-4 ■ ENERGY EQUATION FOR FLOW SYSTEMS

In Chap. 5 the energy balance for any system undergoing any process was expressed as $E_{in} - E_{out} = \Delta E_{system}$, which states that *the change in the energy content of a system during a process is equal to the difference between the energy input and the energy output.* During a *steady-flow process,* the total energy content of a control volume remains constant (and thus $\Delta E_{system} = 0$) and the amount of energy entering a control volume in all forms (by heat, work, and mass) must be equal to the amount of energy leaving it. Then the rate form of the general energy equation reduces for a steady-flow process to $\dot{E}_{in} - \dot{E}_{out} = \Delta \dot{E}_{system} = 0$ or

$$\underbrace{\dot{E}_{in}}_{\substack{\text{Rate of net energy transfer in} \\ \text{by heat, work, and mass}}} = \underbrace{\dot{E}_{out}}_{\substack{\text{Rate of net energy transfer out} \\ \text{by heat, work, and mass}}} \qquad (11\text{-}26)$$

Noting that energy can be transferred by heat, work, and mass only, the energy balance above for a control volume that consists of a *flow section* with one inlet and one exit (such as a pipe) can be written explicitly as (Fig. 11-28)

$$\dot{Q}_{in} + \dot{W}_{in} + \dot{m}\theta_1 = \dot{Q}_{out} + \dot{W}_{out} + \dot{m}\theta_2 \qquad (11\text{-}27)$$

FIGURE 11-28

The energy transfers by heat, work, and mass associated with a steady-flow system with one inlet and one exit.

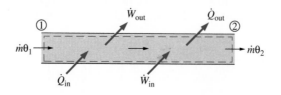

where $\dot{m}$ is the mass flow rate of the fluid and

$$\theta = h + \text{ke} + \text{pe} = h + \mathcal{V}^2/2 + gz = u + P/\rho + \mathcal{V}^2/2 + gz \quad (11\text{-}28)$$

is the energy of a flowing fluid per unit mass. Here h is the *enthalpy,* u is the *internal energy,* P/ρ is the *flow energy,* $\mathcal{V}^2/2$ is the *kinetic energy,* and gz is the *potential energy* of the fluid per unit mass. Substituting this into Eq. 11-27 after dividing it by $\dot{m}$ gives the general steady-flow energy balance for a flow section as

$$q_{\text{in}} + w_{\text{in}} + \left(u_1 + \frac{P_1}{\rho_1} + \frac{\mathcal{V}_1^2}{2} + gz_1 \right) = q_{\text{out}} + w_{\text{out}} + \left(u_2 + \frac{P_2}{\rho_2} + \frac{\mathcal{V}_2^2}{2} + gz_2 \right)$$
$$(11\text{-}29)$$

The relation above is valid for compressible or incompressible flow and adiabatic or nonadiabatic processes.

Most fluid flow systems encountered in practice (those intended to transport fluids from one location to another) do not involve heat transfer ($q_{\text{in}} = q_{\text{out}} = 0$) and work interactions other than the mechanical work ($w = w_{\text{shaft}}$), and the compressibility effects are negligible ($\rho = \text{constant}$). Also, the internal energy u in that case depends on temperature only [$u = u(T)$ instead of $u = u(T, v)$] and thus $\Delta u = u_2 - u_1 = C_v(T_2 - T_1)$. The internal energy increases as a result of friction and other imperfections that occur during the conversion of mechanical energy to thermal energy. The conservation of energy principle requires this *increase in internal energy* to be equal to the *decrease in the mechanical energy.* This represents a loss of mechanical energy indicated by (Fig. 11-29)

$$e_{\text{mech, loss}} = u_2 - u_1 = C_v(T_2 - T_1) \quad \text{(adiabatic flow)} \quad (11\text{-}30a)$$

or

$$\dot{E}_{\text{mech, loss}} = \dot{m}(u_2 - u_1) = \dot{m}C_v(T_2 - T_1) \quad \text{(adiabatic flow)} \quad (11\text{-}30b)$$

Then the energy balance relation above reduces to

$$\frac{P_1}{\rho} + \frac{\mathcal{V}_1^2}{2} + gz_1 + w_{\text{pump, u}} = \frac{P_2}{\rho} + \frac{\mathcal{V}_2^2}{2} + gz_2 + w_{\text{turbine}} + e_{\text{mech, loss}} \quad (11\text{-}31)$$

where we used $w_{\text{pump, u}}$ for the useful work input and w_{turbine} for the work output. Either absolute or gage pressure can be used for P since P_{atm}/ρ would appear on both sides if it does and would cancel out.

Multiplying Eq. 11-31 by the mass flow rate $\dot{m}$ gives

$$\dot{m}\left(\frac{P_1}{\rho} + \frac{\mathcal{V}_1^2}{2} + gz_1 \right) + \dot{W}_{\text{pump, u}} = \dot{m}\left(\frac{P_2}{\rho} + \frac{\mathcal{V}_2^2}{2} + gz_2 \right) + \dot{W}_{\text{turbine}} + \dot{E}_{\text{mech, loss}}$$
$$(11\text{-}32)$$

where $\dot{W}_{\text{pump, u}} = \dot{m}w_{\text{pump, u}}$ is the useful pumping power input and $\dot{W}_{\text{turbine}} = \dot{m}w_{\text{turbine}}$ is the mechanical (shaft) turbine power output. Equations 11-31 and 11-32 are statements for the conservation of mechanical energy, which is valid for systems that do not involve any forms or interactions of energy other than various forms of mechanical energy and its dissipation into thermal energy. It can be expressed compactly as

$$e_{\text{mech, in}} = e_{\text{mech, out}} + e_{\text{mech, loss}} \quad (11\text{-}33)$$

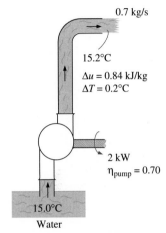

0.7 kg/s

15.2°C

$\Delta u = 0.84$ kJ/kg
$\Delta T = 0.2$°C

2 kW
$\eta_{\text{pump}} = 0.70$

15.0°C

Water

FIGURE 11-29

The lost mechanical energy in a fluid flow system results in an increase in the internal energy of the fluid, and thus in a rise of fluid temperature.

Energy equation:

$$\frac{P_1}{\rho g} + \frac{\mathcal{V}_1^2}{2g} + z_1 + h_{\text{pump, u}} \overset{0}{\nearrow} = \frac{P_2}{\rho g} + \frac{\mathcal{V}_2^2}{2g} + z_2 + h_{\text{turbine}} \overset{0}{\nearrow} + h_L \overset{0}{\nearrow}$$

Bernoulli equation:

$$\frac{P_1}{\rho g} + \frac{\mathcal{V}_1^2}{2g} + z_1 = \frac{P_2}{\rho g} + \frac{\mathcal{V}_2^2}{2g} + z_2$$

FIGURE 11-30

When the frictional losses are negligible and there are no work devices involved, the steady-flow energy equation reduces to the Bernoulli equation.

Dividing each term of Eq. 11-31 by g gives the energy equation in terms of heads as

$$\frac{P_1}{\rho g} + \frac{\mathcal{V}_1^2}{2g} + z_1 + h_{\text{pump, u}} = \frac{P_2}{\rho g} + \frac{\mathcal{V}_2^2}{2g} + z_2 + h_{\text{turbine}} + h_L \quad (11\text{-}34)$$

where

$$h_{\text{pump, u}} = \frac{w_{\text{pump, u}}}{g} = \frac{\dot{W}_{\text{pump, u}}}{\dot{m}g} = \frac{\Delta\dot{E}_{\text{mech, fluid}}}{\dot{m}g} \text{ is the } \textit{useful pump head input}$$

$$h_{\text{turbine}} = \frac{w_{\text{turbine}}}{g} = \frac{\dot{W}_{\text{turbine}}}{\dot{m}g} = \frac{|\Delta\dot{E}_{\text{mech, fluid}}|}{\dot{m}g} \text{ is the } \textit{turbine head output}$$

$$h_L = \frac{e_{\text{mech, loss}}}{g} = \frac{\dot{E}_{\text{mech, loss}}}{\dot{m}g} \text{ is the } \textit{head loss} \text{ between points 1 and 2}$$

The *pump head* is zero if the piping system does not involve a pump, a fan, or a compressor; the *turbine head* is zero if the system does not involve a turbine; and both are zero if the flow section analyzed does not involve any mechanical work-producing or consuming devices. Also, the *head loss* is zero when the system is "perfect" so that it does not involve any mechanical losses.

Special Case: Frictionless Flow with No Work Devices

When the frictional effects are negligible, there is no dissipation of mechanical energy into thermal energy, and thus $h_L = e_{\text{mech, loss}}/g = 0$, as shown in Example 11-8. Also, $h_{\text{pump, u}} = h_{\text{turbine}} = 0$ when there are no mechanical work devices such as fans, pumps, or turbines. Then Eq. 11-33 reduces to (Fig. 11-30)

$$\frac{P_1}{\rho g} + \frac{\mathcal{V}_1^2}{2g} + z_1 = \frac{P_2}{\rho g} + \frac{\mathcal{V}_2^2}{2g} + z_2 \qquad \text{or} \qquad \frac{P}{\rho g} + \frac{\mathcal{V}^2}{2g} + z = \text{constant}$$

which is the **Bernoulli equation** derived earlier using Newton's second law of motion.

Kinetic Energy Correction Factor, α

The average flow velocity $\mathcal{V}$ was defined such that the relation $\rho\mathcal{V}A$ gives the actual mass flow rate. Therefore, there is no such thing as a correction factor for mass flow rate. However, the kinetic energy of a fluid stream obtained from $\mathcal{V}^2/2$ is not the same as the actual kinetic energy of the fluid stream since the square of a sum is not equal to the sum of the squares of its components (Fig. 11-31). This error can be corrected by replacing the kinetic energy terms $\mathcal{V}^2/2$ in the energy equation by $\alpha\mathcal{V}^2/2$ where α is the **kinetic energy correction factor**. By using equations for the variation of velocity with the radial

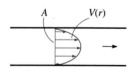

$$\dot{m} = \rho\mathcal{V}_{\text{ave}}A, \ \rho = \text{constant}$$

$$\dot{\text{KE}}_{\text{act}} = \int\!\text{ke}\,\delta\dot{m} = \int_A \frac{1}{2}\mathcal{V}^2(r)[\rho\mathcal{V}(r)\,dA]$$

$$= \frac{1}{2}\,\rho\!\int_A\mathcal{V}^3(r)\,dA$$

$$\dot{\text{KE}}_{\text{ave}} = \frac{1}{2}\,\dot{m}\mathcal{V}_{\text{ave}}^2 = \frac{1}{2}\,\rho A\mathcal{V}_{\text{ave}}^3$$

$$\alpha = \frac{\dot{\text{KE}}_{\text{act}}}{\dot{\text{KE}}_{\text{ave}}} = \frac{1}{A}\int_A\left(\frac{\mathcal{V}(r)}{\mathcal{V}_{\text{ave}}}\right)^3 dA$$

FIGURE 11-31

The determination of the *kinetic energy correction factor* using actual velocity distribution $\mathcal{V}(r)$ and the average velocity $\mathcal{V}_{\text{ave}}$ at a cross section.

distance, it can be shown that the correction factor is 2.0 for the fully developed laminar pipe flow, and it ranges between 1.04 and 1.11 for turbulent flow in a circular pipe.

The kinetic energy correction factors are usually disregarded in an elementary analysis since (1) most flows encountered in practice are turbulent, for which the correction factor is near unity, and (2) the kinetic energy terms are usually small relative to the other terms in the energy equation, and multiplying them by a factor less than 2.0 does not make much difference. Besides, when the velocity and thus the kinetic energy are high, the flow turns turbulent. Therefore, we will not consider the kinetic energy correction factor in the analysis. However, the reader should keep in mind that he or she may encounter situations for which these factors are significant, especially when the flow is laminar.

EXAMPLE 11-9 Effect of Friction on Fluid Temperature and Head Loss
Show that during steady and incompressible flow of a fluid in an adiabatic flow section (a) the temperature remains constant and there is no head loss when the flow is frictionless and (b) the temperature increases and some head loss occurs when there are frictional effects. Discuss if it is possible for the fluid temperature to decrease during such flow (Fig. 11-32).

FIGURE 11-32

Schematic for Example 11-9.

Solution Steady and incompressible flow through an adiabatic section is considered. The effects of friction on the temperature and the heat loss are to be determined.

Assumptions **1** The flow is steady and incompressible. **2** The flow section is adiabatic and thus there is no heat transfer.

Analysis The density of a fluid remains constant during incompressible flow, and the entropy change of an incompressible system was given in Chap. 7 as

$$\Delta s = C_v \ln \frac{T_2}{T_1}$$

This relation represents the entropy change of the fluid per unit mass as it flows through the flow section from state 1 at the inlet to state 2 at the exit. It was mentioned that entropy change is caused by two effects: (1) heat transfer and (2) irreversibilities. Therefore, in the absence of heat transfer, entropy change is due to irreversibilities only whose effect is always to increase entropy.

(a) The entropy change of the fluid is zero when the process does not involve any irreversibilities such as skin friction and swirling, and thus for *frictionless flow* we have

Temperature change: $\qquad \Delta s = C_v \ln \frac{T_2}{T_1} = 0 \qquad \rightarrow \qquad T_2 = T_1$

Mechanical energy loss: $\qquad e_{mech,\,loss} = u_2 - u_1 = C_v(T_2 - T_1) = 0$

Head loss: $\qquad\qquad\qquad h_L = e_{mech,\,loss}/g = 0$

Thus we conclude that during steady, adiabatic, *reversible* flow, (1) the temperature of the fluid remains constant, (2) no mechanical energy is converted to thermal energy, and (3) there is no head loss.

(b) When there are irreversibilities such as friction, the entropy change is positive and thus we have:

Temperature change: $\qquad \Delta s = C_v \ln \frac{T_2}{T_1} > 0 \qquad \rightarrow \qquad T_2 > T_1$

Mechanical energy loss: $\qquad e_{mech,\,loss} = u_2 - u_1 = C_v(T_2 - T_1) > 0$

Head loss: $\qquad\qquad\qquad h_L = e_{mech,\,loss}/g > 0$

Thus we conclude that during steady, adiabatic, *irreversible* flow, (1) the temperature of the fluid increases, (2) some mechanical energy is converted to thermal energy, and (3) some head loss occurs.

Discussion It is impossible for the fluid temperature to decrease during steady, incompressible, adiabatic flow since this would require the entropy of an adiabatic system to decrease, which would be a violation of the second law of thermodynamics.

EXAMPLE 11-10   Pumping Power and Frictional Heating in a Pump

The pump of a water distribution system is powered by a 15-kW electric motor whose efficiency is 90 percent (Fig. 11-33). The water flow rate through the pump is 50 L/s. The diameters of the inlet and exit pipes are the same, and the elevation difference across the pump is negligible. If the pressures at the inlet and outlet of the pump are measured to be 100 kPa and 300 kPa (absolute), respectively, determine (*a*) the mechanical efficiency of the pump and (*b*) the temperature rise of water as it flows through the pump due to the mechanical inefficiency.

Solution   The pressures across a pump are measured. The mechanical efficiency of the pump and the temperature rise of water are to be determined.

Assumptions   **1** The flow is steady, one-dimensional, adiabatic, and incompressible. **2** The pump is driven by an external motor so that the heat generated by the motor is dissipated to the atmosphere. **3** The elevation difference between the inlet and outlet of the pump is negligible, $z_1 \cong z_2$. **4** The inlet and outlet diameters are the same and thus the inlet and exit velocities are equal, $\mathcal{V}_1 = \mathcal{V}_2$.

Properties   We take the density of water to be 1 kg/L = 1000 kg/m³ and its specific heat to be 4.18 kJ/kg · °C (Table A-3).

Analysis   (*a*) The mass flow rate of water through the pump is

$$\dot{m} = \rho \dot{V} = (1 \text{ kg/L})(50 \text{ L/s}) = 50 \text{ kg/s}$$

The motor draws 15 kW of power and is 90 percent efficient. Then the mechanical (shaft) power it delivers to the pump is

$$\dot{W}_{\text{pump, shaft}} = \eta_{\text{motor}} \dot{W}_{\text{electric}} = (0.90)(15 \text{ kW}) = 13.5 \text{ kW}$$

To determine the mechanical efficiency of the pump, we need to know the increase in the mechanical energy of the fluid as it flows through the pump, which is

$$\Delta\dot{E}_{\text{mech, fluid}} = \dot{E}_{\text{mech, out}} - \dot{E}_{\text{mech, in}} = \dot{m}\left(\frac{P_2}{\rho} + \frac{\mathcal{V}_2^2}{2} + gz_2\right) - \dot{m}\left(\frac{P_1}{\rho} + \frac{\mathcal{V}_1^2}{2} + gz_1\right)$$

Simplifying it for this case and substituting the given values,

$$\Delta\dot{E}_{\text{mech, fluid}} = \dot{m}\left(\frac{P_2 - P_1}{\rho}\right) = (50 \text{ kg/s})\left(\frac{(300 - 100) \text{ kPa}}{1000 \text{ kg/m}^3}\right)\left(\frac{1 \text{ kJ}}{1 \text{ kPa} \cdot \text{m}^3}\right) = 10 \text{ kW}$$

Then the mechanical efficiency of the pump becomes

$$\eta_{\text{pump}} = \frac{\Delta\dot{E}_{\text{mech, fluid}}}{\dot{W}_{\text{pump, shaft}}} = \frac{\dot{W}_{\text{pump, u}}}{\dot{W}_{\text{pump, shaft}}} = \frac{10 \text{ kW}}{13.5 \text{ kW}} = 0.741 \quad \text{or} \quad \textbf{74.1\%}$$

(*b*) Of the 13.5-kW mechanical power supplied by the pump, only 10 kW is imparted to the fluid as mechanical energy. The remaining 3.5 kW is converted to thermal energy due to frictional effects, and this "lost" mechanical energy manifests itself as a heating effect in the fluid,

$$\dot{E}_{\text{mech, loss}} = \dot{W}_{\text{pump, shaft}} - \Delta\dot{E}_{\text{mech, fluid}} = 13.5 - 10 = 3.5 \text{ kW}$$

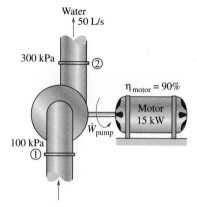

Water
50 L/s

300 kPa
②

$\eta_{\text{motor}} = 90\%$

Motor
15 kW

$\dot{W}_{\text{pump}}$

100 kPa
①

FIGURE 11-33

Schematic for Example 11-10.

The temperature rise of water due to this mechanical inefficiency is determined from the thermal energy balance, $\dot{E}_{mech,\,loss} = \dot{m}(u_2 - u_1) = \dot{m}C\Delta T$. Solving for ΔT,

$$\Delta T = \frac{\dot{E}_{mech,\,loss}}{\dot{m}C} = \frac{3.5\text{ kW}}{(50\text{ kg/s})(4.18\text{ kJ/kg}\cdot\text{°C})} = \mathbf{0.017°C}$$

Therefore, the water will experience a temperature rise of 0.017°C, which is very small, as it flows through the pump.

Discussion In an actual application, the temperature rise of water will probably be less since part of the heat generated will be transferred to the casing of the pump, and from the casing to the surrounding air. If the entire pump motor were submerged in water, then the 1.5 kW dissipated to the air due to motor inefficiency would also be transferred to the surrounding water as heat. This would cause the incoming water temperature to rise somewhat.

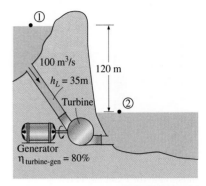

FIGURE 11-34
Schematic for Example 11-11.

EXAMPLE 11-11   Hydroelectric Power Generation from a Dam

In a hydroelectric power plant, 100 m³/s of water flows from an elevation of 120 m to a turbine, where electric power is generated (Fig. 11-34). The total head loss in the system from point 1 to point 2 (excluding the turbine unit) is determined to be 35 m. If the overall efficiency of the turbine-generator is 80 percent, estimate the electrical power output.

Solution The available head, flow rate, head loss, and efficiency of a hydroelectric turbine are given. The electrical power output is to be determined.

Assumptions **1** The flow is steady, one-dimensional, adiabatic, and incompressible. **2** The water levels at the reservoir and the discharge site remain constant.

Properties We take the density of water to be 1000 kg/m³.

Analysis The mass flow rate of water through the turbine is

$$\dot{m} = \rho\dot{V} = (1000\text{ kg/m}^3)(100\text{ m}^3/\text{s}) = 10^5\text{ kg/s}$$

We take point 2 as the reference level, and thus $z_2 = 0$. Also, both points 1 and 2 are open to the atmosphere ($P_1 = P_2 = P_{atm}$) and the flow velocities are negligible at both points ($\mathcal{V}_1 = \mathcal{V}_2 = 0$). Then the energy equation for steady incompressible flow reduces to

$$\frac{P_1}{\rho g} + \frac{\mathcal{V}_1^2}{2g} + z_1 + \cancel{h_{pump,\,u}}^{0} = \cancel{\frac{P_2}{\rho g}} + \cancel{\frac{\mathcal{V}_2^2}{2g}} + \cancel{z_2}^{0} + h_{turbine} + h_L \quad \rightarrow \quad h_{turbine} = z_1 - h_L$$

Substituting, the turbine head and the turbine power output are determined to be

$$h_{turbine} = z_1 - h_L = 120 - 35 = 85\text{ m}$$

$$\dot{W}_{turbine} = \dot{m}gh_{turbine} = (10^5\text{ kg/s})(9.81\text{ m/s}^2)(85\text{ m})\left(\frac{1\text{ kJ/kg}}{1000\text{ m}^2/\text{s}^2}\right) = 83{,}400\text{ kW}$$

Therefore, a perfect turbine-generator would generate 83,400 kW of electricity from this resource. The electric power generated by the actual unit is

$$\dot{W}_{electric} = \eta_{turbine-gen}\dot{W}_{turbine} = (0.80)(83.4\text{ MW}) = \mathbf{66.7\text{ MW}}$$

Discussion Note that the power generation will increase by almost 1 MW for each percentage point improvement in the efficiency of the turbine-generator unit.

EXAMPLE 11-12   Fan Selection for Air Cooling of a Computer

A fan is to be selected to cool a computer case whose dimensions are 12 cm × 40 cm × 40 cm (Fig. 11-35). Half of the volume in the case is expected to be filled

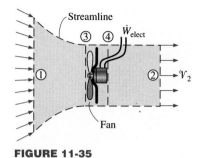

FIGURE 11-35

Schematic for Example 11-12.

with components and the other half to be air space. A 6-cm-diameter hole is available at the front of the case for the installment of the fan that is to replace the air in the void spaces of the case once every second. Small low-power fan-motor-combined units are available in the market and their efficiency is estimated to be 30 percent. Determine (a) the wattage of the fan-motor unit to be purchased and (b) the pressure difference across the fan. Take the air density to be 1.20 kg/m³.

Solution   A fan is to air cool a computer case by replacing the air inside completely once every second. The power of the fan and the pressure difference across it are to be determined.

Assumptions   **1** The flow is steady, one-dimensional, adiabatic, and incompressible. **2** Losses other than those due to the inefficiency of the fan-motor unit are negligible ($h_L = 0$).

Properties   The density of air is given to be 1.20 kg/m³.

Analysis   (a) Noting that half of the volume of the case is occupied by the components, the air volume in the computer case is

$$V = \text{(Void fraction)(Total case volume)}$$
$$= 0.5(12 \text{ cm} \times 40 \text{ cm} \times 40 \text{ cm}) = 9600 \text{ cm}^3$$

Therefore, the volume and mass flow rates of air through the case are

$$\dot{V} = \frac{V}{\Delta t} = \frac{9600 \text{ cm}^3}{1 \text{ s}} = 9600 \text{ cm}^3/\text{s} = 9.6 \times 10^{-3} \text{ m}^3/\text{s}$$
$$\dot{m} = \rho\dot{V} = (1.20 \text{ kg/m}^3)(9.6 \times 10^{-3} \text{ m}^3/\text{s}) = 0.0115 \text{ kg/s}$$

The cross-sectional area of the opening in the case and the average air velocity are

$$A = \frac{\pi D^2}{4} = \frac{\pi (0.06 \text{ m})^2}{14} = 2.83 \times 10^{-5} \text{ m}^2$$
$$\mathcal{V} = \frac{\dot{V}}{A} = \frac{9.6 \times 10^{-3} \text{ m}^3/\text{s}}{2.83 \times 10^{-5} \text{ m}^2} = 3.39 \text{ m/s}$$

We draw the control volume around the fan such that both the inlet and the exit are at the atmospheric pressure ($P_1 = P_2 = P_{atm}$), as shown in the figure, and the inlet section 1 is large and far from the fan so that the flow velocity at the inlet section is negligible ($\mathcal{V}_1 \cong 0$). Noting that $z_1 = z_2$ and the head losses due to flow are negligible, the energy equation simplifies to

$$\dot{m}\left(\frac{P_1}{\rho} + \frac{\mathcal{V}_1^2}{2}^{\,0} + gz_1^{\,0}\right) + \dot{W}_{\text{fan, u}} = \dot{m}\left(\frac{P_2}{\rho} + \frac{\mathcal{V}_2^2}{2} + gz_2^{\,0}\right) + \dot{W}_{\text{turbine}}^{\,0} + \dot{E}_{\text{mech, loss}}^{\,0}$$

Solving for $\dot{W}_{\text{fan, u}}$ and substituting,

$$\dot{W}_{\text{fan, u}} = \dot{m}\frac{\mathcal{V}_2^2}{2} = (0.0115 \text{ kg/s})\frac{(3.39 \text{ m/s})^2}{2}\left(\frac{1 \text{ N}}{1 \text{ kg} \cdot \text{m/s}^2}\right) = 0.066 \text{ W}$$

Then the required electrical power input to the fan is determined to be

$$\dot{W}_{\text{elect}} = \frac{\dot{W}_{\text{fan, u}}}{\eta_{\text{fan–motor}}} = \frac{0.066 \text{ W}}{0.3} = \textbf{0.22 W}$$

Therefore, a fan–motor rated at 0.22 W is adequate for this job.

(b) To determine the pressure difference across the fan unit, we take points 3 and 4 to be on the two sides of the fan on a horizontal line. This time again $z_3 = z_4$ and $\mathcal{V}_3 = \mathcal{V}_4$ since the fan is a narrow cross section. Assuming the frictional losses to be negligible, the energy equation reduces to

$$\dot{m}\frac{P_3}{\rho} + \dot{W}_{\text{fan, u}} = \dot{m}\frac{P_4}{\rho} + \dot{E}_{\text{mech, loss}}^{\,0} \quad \rightarrow \quad \dot{W}_{\text{fan, u}} = \dot{m}\frac{P_4 - P_3}{\rho}$$

Solving for $P_4 - P_3$ and substituting,

$$P_4 - P_3 = \frac{\rho \dot{W}_{\text{fan, u}}}{\dot{m}} = \frac{(1.2 \text{ kg/m}^3)(0.066 \text{ W})}{0.0115 \text{ kg/s}} \left(\frac{1 \text{ Pa} \cdot \text{m}^3}{1 \text{ Ws}} \right) = \textbf{6.9 Pa}$$

Therefore, the fan will raise the pressure of air 6.9 Pa before discharging it.

Discussion   The efficiency of the fan–motor unit is given to be 30 percent, which means 30 percent of the electric power $\dot{W}_{\text{electric}}$ consumed by the unit will be converted to useful mechanical energy while the rest (70 percent) will be "lost" and converted to thermal energy.

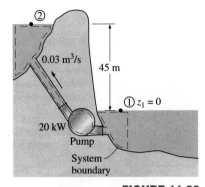

FIGURE 11-36

Schematic for Example 11-13.

EXAMPLE 11-13   Head and Power Loss during Water Pumping

Water is pumped from a lower reservoir to a higher reservoir by a pump that supplies 20 kW of useful mechanical power to water (Fig. 11-36). The free surface of the upper reservoir is 45 m higher than the surface of the lower reservoir. If the flow rate of water is measured to be 0.03 m³/s, determine the head loss of the system and the lost mechanical power during this process.

Solution   Water is pumped from a lower reservoir to a higher one. The head and power loss associated with this process are to be determined.

Assumptions   **1** The flow is steady, one-dimensional, adiabatic, and incompressible. **2** The elevation difference between the reservoirs is constant.

Properties   We take the density of water to be 1000 kg/m³.

Analysis   The mass flow rate of water through the system is

$$\dot{m} = \rho \dot{V} = (1000 \text{ kg/m}^3)(0.03 \text{ m}^3/\text{s}) = 30 \text{ kg/s}$$

We choose points 1 and 2 at the free surfaces of the lower and upper reservoirs, respectively, and take the surface of the lower reservoir as the reference level ($z_1 = 0$). Both points are open to the atmosphere ($P_1 = P_2 = P_{\text{atm}}$) and the velocities at both locations are negligible ($\mathcal{V}_1 = \mathcal{V}_2 = 0$). Then the energy equation for steady incompressible flow for a control volume between 1 and 2 reduces to

$$\dot{m}\left(\frac{P_1^{\nearrow 0}}{\rho} + \frac{\mathcal{V}_1^{2\,\nearrow 0}}{2} + gz_1^{\nearrow 0} \right) + \dot{W}_{\text{pump, u}} = \dot{m}\left(\frac{P_2^{\nearrow 0}}{\rho} + \frac{\mathcal{V}_2^{2\,\nearrow 0}}{2} + gz_2 \right) + \dot{W}_{\text{turbine}}^{\nearrow 0} + \dot{E}_{\text{mech, loss}}$$

$$\dot{W}_{\text{pump, u}} = \dot{m}gz_2 + \dot{E}_{\text{mech, loss}} \quad \rightarrow \quad \dot{E}_{\text{mech, loss}} = \dot{W}_{\text{pump, u}} - \dot{m}gz_2$$

Substituting, the lost mechanical power and head loss are determined to be

$$\dot{E}_{\text{mech, loss}} = 20 \text{ kW} - (30 \text{ kg/s})(9.81 \text{ m/s}^2)(45 \text{ m})\left(\frac{1 \text{ N}}{1 \text{ kg} \cdot \text{m/s}^2} \right)\left(\frac{1 \text{ kW}}{1000 \text{ N} \cdot \text{m/s}} \right)$$

$$= \textbf{6.76 kW}$$

$$h_L = \frac{\dot{E}_{\text{mech, loss}}}{\dot{m}g} = \frac{6.76 \text{ kW}}{(30 \text{ kg/s})(9.81 \text{ m/s}^2)} \left(\frac{1 \text{ kg} \cdot \text{m/s}^2}{1 \text{ N}} \right)\left(\frac{1000 \text{ N} \cdot \text{m/s}}{1 \text{ kW}} \right) = \textbf{23.0 m}$$

Discussion   The 6.76 kW of power is used to overcome the friction in the piping system. Note that the pump could raise the water an additional 23 m if there were no losses in the system. Also in that case, the pump would function as a turbine, drawing 20 kW of power from the water when the water is allowed to flow from the upper reservoir to the lower reservoir.

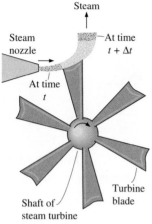

FIGURE 11-37

A system (a fixed quantity of fluid) can change in shape and size, but no mass crosses its boundary.

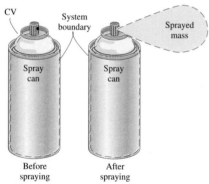

FIGURE 11-38

A spray can can be analyzed by taking either the contents of a can (a closed system) or the interior volume of the can (a control volume) as the system.

11-5 ■ THE REYNOLDS TRANSPORT THEOREM

In Chap. 1 we defined the system as a *quantity of matter* (called a *closed system* or just *system*) *or a region in space* (called an *open system* or a *control volume*) *chosen for study.* The size and shape of a closed system may change during a process, but no mass crosses its boundaries. A control volume, on the other hand, allows mass to flow in or out during a process across its boundaries, called the *control surface.* Also, a control volume may move and deform during a process, but most real-world applications involve fixed, nondeformable control volumes. In solid mechanics, we normally work with a system, which is a collection of specified mass. In fluid mechanics, however, we mostly work with control volumes.

Historically, there have been two approaches to the selection of a system in fluid mechanics: Eulerian and Lagrangian. Each approach has advantages in certain situations. The **Lagrangian approach** identifies a fixed quantity of fluid and follows it through as it moves. The Lagrangian boundary can change in shape and size, but no mass crosses its boundary (Fig. 11-37). An observer moves with the system and notes how external forces manifest their changes on the system. In the **Eulerian approach,** a fixed volume and its boundary surface are identified, and the stationary observer analyzes movement of fluid entering and leaving this volume. The Eulerian control surface can often take the shape of a container, such as a nozzle or pipe, so that surface effects can be included readily in the analysis. The mass and other properties within the chosen volume can change during the process.

As an example, consider the contents of a *hair spray can* (Fig. 11-38). When analyzing the spraying process, a natural choice for the system is either the contents of the can (a closed system) or the entire cavity bounded by the inner surfaces of the can (a control volume). These two choices are identical before the hair spray is used. When some contents of the can are discharged, the system approach considers the discharged mass as part of the system and tracks it (a difficult job indeed), and thus the mass of the system remains constant. Conceptually this is equivalent to attaching a flat balloon to the nozzle of the can and letting the spray inflate the balloon. The inner surface of the balloon now becomes part of the boundary of the system. The control volume approach, however, is not concerned at all with the mass that has escaped the can (other than its properties at the exit), and thus the mass of the control volume decreases during this process while its volume remains constant. Therefore, the *system approach* treats the spraying process as an expansion of the system, whereas the *control volume approach* considers it as a fluid discharge.

Most principles of fluid mechanics are adopted from solid mechanics, where the physical laws dealing with the time rates of change of extensive properties are expressed for systems. In fluid mechanics, it is usually more convenient to work with control volumes, and thus there is a need to relate the changes in a control volume to the changes in a system. The relationship between the time rates of change of an extensive property for a system and for a control volume is expressed by the **Reynolds transport theorem,** which provides the link between the system and the control volume concepts.

The general form of the Reynolds transport theorem can be derived by considering a general system with an arbitrary shape and arbitrary interactions, but the process is rather involved and difficult to follow. Instead, we will derive it in a straightforward manner using a simple geometry and then generalize the results.

Consider one-dimensional flow through a diverging flow section shown in Fig. 11-39. We choose the control volume to be between the sections (1) and (2) of the flow section. Both (1) and (2) are normal to the direction of flow. At some initial time t, the system coincides with the control volume, and thus the system and control volume are identical. During a very short time interval Δt, the system moves slightly in the flow direction with a uniform velocity $\mathcal{V}_1$ at section (1) and $\mathcal{V}_2$ at section (2). The region uncovered by the system during this motion is designated as section I (which is still part of the CV), and the new region covered by the system is designated as section II. Therefore, at time $t + \Delta t$, the system consists of the same fluid, but it occupies the region $CV - I + II$. The control volume is fixed in space, and thus it remains as CV at all times.

Let B represent any extensive property (such as mass, energy, and momentum) and let $b = B/m$ represent the corresponding intensive property. Noting that extensive properties are additive, the extensive property B of the system at times t and $t + \Delta t$ can be expressed as

$$B_{\text{sys},\,t} = B_{\text{CV},\,t} \qquad \text{(the system and the CV coincide at time } t\text{)} \qquad (11\text{-}35a)$$

$$B_{\text{sys},\,t+\Delta t} = B_{\text{CV},\,t+\Delta t} - B_{\text{I},\,t+\Delta t} + B_{\text{II},\,t+\Delta t} \qquad (11\text{-}35b)$$

Subtracting the first equation above from the second one and dividing by Δt gives

$$\frac{B_{\text{sys},\,t+\Delta t} - B_{\text{sys},\,t}}{\Delta t} = \frac{B_{\text{CV},\,t+\Delta t} - B_{\text{CV},\,t}}{\Delta t} - \frac{B_{\text{I},\,t+\Delta t}}{\Delta t} + \frac{B_{\text{II},\,t+\Delta t}}{\Delta t} \qquad (11\text{-}36)$$

Taking the limit as $\Delta t \to 0$ and using the definition of derivative gives

$$\frac{dB_{\text{sys}}}{dt} = \frac{dB_{\text{CV}}}{dt} - \dot{B}_{\text{in}} + \dot{B}_{\text{out}} \qquad (11\text{-}37)$$

or

$$\frac{dB_{\text{sys}}}{dt} = \frac{dB_{\text{CV}}}{dt} - b_1 \rho_1 \mathcal{V}_1 A_1 + b_2 \rho_2 \mathcal{V}_2 A_2 \qquad (11\text{-}38)$$

since

$$B_{\text{I},\,t+\Delta t} = b_1 m_{\text{I},\,t+\Delta t} = b_1 \rho_1 V_{\text{I},\,t+\Delta t} = b_1 \rho_1 \mathcal{V}_1 \Delta t\, A_1 \qquad (11\text{-}39a)$$

$$B_{\text{II},\,t+\Delta t} = b_2 m_{\text{II},\,t+\Delta t} = b_2 \rho_2 V_{\text{II},\,t+\Delta t} = b_2 \rho_2 \mathcal{V}_2 \Delta t\, A_2 \qquad (11\text{-}39b)$$

and

$$\dot{B}_{\text{in}} = \dot{B}_{\text{I}} = \lim_{\Delta t \to 0} \frac{B_{\text{I},\,t+\Delta t}}{\Delta t} = \lim_{\Delta t \to 0} \frac{b_1 \rho_1 \mathcal{V}_1 \Delta t\, A_1}{\Delta t} = b_1 \rho_1 \mathcal{V}_1 A_1 \qquad (11\text{-}40a)$$

$$\dot{B}_{\text{out}} = \dot{B}_{\text{II}} = \lim_{\Delta t \to 0} \frac{B_{\text{II},\,t+\Delta t}}{\Delta t} = \lim_{\Delta t \to 0} \frac{b_2 \rho_2 \mathcal{V}_2 \Delta t\, A_2}{\Delta t} = b_2 \rho_2 \mathcal{V}_2 A_2 \qquad (11\text{-}40b)$$

Equation 11-37 states that *the time rate of change of the property B of the system is equal to the time rate of change of B of the control volume and the net flux of B by mass across the control surface.* This is the desired relation since it relates the change of a property of a system to the change of that property for a control volume.

The influx $\dot{B}_{\text{in}}$ and outflux $\dot{B}_{\text{out}}$ of the property B in this case were easy to determine since there were one inlet and one exit and the velocities were normal to the surfaces at sections (1) and (2). In general, however, we may have several inlet and exit ports and the velocity may not be normal to the control

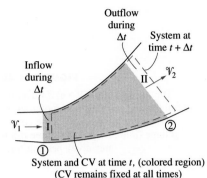

System and CV at time t, (colored region)
(CV remains fixed at all times)

At time t: Sys = CV
At time $t + \Delta t$: Sys = CV – I + II

FIGURE 11-39

A moving system and a fixed control volume in a flow section at times t and $t + \Delta t$.

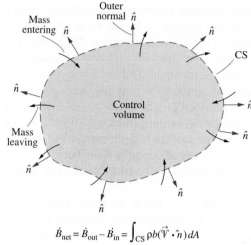

FIGURE 11-40

The integral of $b\rho(\vec{V} \cdot \hat{n})\, dA$ over the control surface gives the net amount of the property b flowing out of the control volume (into the control volume if it is negative), per unit time.

$$\dot{B}_{net} = \dot{B}_{out} - \dot{B}_{in} = \int_{CS} \rho b(\vec{V} \cdot \hat{n})\, dA$$

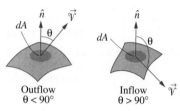

$\vec{V}$: Velocity vector
$\hat{n}$: Outer normal vector

$\vec{V} \cdot \hat{n} = |\vec{V}||\hat{n}|\cos\theta = \mathcal{V}\cos\theta$
If $\theta < 90°$, then $\cos\theta > 0$ (outflow).
If $\theta > 90°$, then $\cos\theta < 0$ (inflow).
If $\theta = 90°$, then $\cos\theta = 0$ (no flow).

FIGURE 11-41

Inflow and outflow of mass across the differential area of a control surface.

surface at the point of entry. Also, the velocity may not be uniform. To generalize the process, we consider a differential surface area dA on the control surface and denote its outer normal by $\hat{n}$. The flow rate of the property b through dA is $b\rho(\vec{V} \cdot \hat{n})\, dA$ since the dot product $\vec{V} \cdot \hat{n}$ gives the normal component of the velocity. Then the net rate of flow through the entire control surface is determined by integration to be (Fig. 11-40)

$$\dot{B}_{net} = \dot{B}_{out} - \dot{B}_{in} = \int_{CS} \rho b(\vec{V} \cdot \hat{n})\, dA \qquad \text{(inflow if negative)} \qquad (11\text{-}41)$$

An important aspect of this relation is that it automatically subtracts the inflow from the outflow, as explained next. The dot product of the velocity vector at a point on the control surface and the outer normal at that point is $\vec{V} \cdot \hat{n} = |\vec{V}||\hat{n}|\cos\theta = \mathcal{V}\cos\theta$ where θ is the angle between the velocity vector and the outer normal, as shown in Fig. 11-41. For $\theta < 90°$, we have $\cos\theta > 0$ and thus $\vec{V} \cdot \hat{n} > 0$ for the outflow of mass from the control volume, and for $\theta > 90°$, we have $\cos\theta < 0$ and thus $\vec{V} \cdot \hat{n} < 0$ for inflow of mass into the control volume. Therefore, the differential quantity $b\rho(\vec{V} \cdot \hat{n})\, dA$ is positive for mass flowing out of the control volume and negative for mass flowing into the control volume, and its integral over the entire control surface gives the rate of net outflow of the property B by mass.

The properties of the control volume, in general, may vary with position. Then the property B content of the control volume can be determined by integration from

$$B_{CV} = \int_{CV} \rho b\, dV \qquad (11\text{-}42)$$

Substituting Eqs. 11-41 and 11-42 into Eq. 11-37 gives

General: $\qquad \dfrac{dB_{sys}}{dt} = \dfrac{d}{dt}\int_{CV} \rho b\, dV + \int_{CS} \rho b(\vec{V} \cdot \hat{n})\, dA \qquad (11\text{-}43)$

This is the general form of the **Reynolds transport theorem,** also known as the *system-to-control-volume transformation.* For a fixed control volume, the time derivative on the right-hand side can be moved under the integral since the domain of integration does not change with time. But the time derivative in that case should be expressed as a partial derivative since density and the quantity b may depend on the position within the control volume. For a control volume that deforms with time, the time derivative must be applied after integration.

The term $\dfrac{dB_{CV}}{dt} = \dfrac{d}{dt}\displaystyle\int_{CV} \rho b\, dV$ represents the time rate of change of the property B content of the control volume. A positive value for dB_{CV}/dt indicates an increase in the B content, and a negative value indicates a decrease.

Equation 11-43 is also valid for control volumes moving at a constant velocity $\vec{V}_c$ provided that the fluid velocity $\vec{V}$ in the last term be replaced by the relative velocity $\vec{V}_r = \vec{V} - \vec{V}_c$ (a vector subtraction), which is the velocity at which a person moving with the control volume would observe the fluid crossing the control surface (Fig. 11-42). In other words, the fluid velocity must be expressed relative to a coordinate system fixed on the control volume. No other change is required. Many practical systems such as turbine and propeller blades involve moving control volumes.

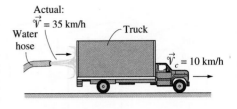

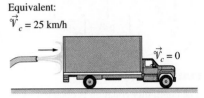

FIGURE 11-42

Reynolds transport theorem can also be applied to control volumes moving at constant velocity by using the relative fluid velocity.

Special Case 1: Steady Flow

During steady flow, the amount of the property B within the control volume remains constant, and thus the time derivative in Eq. 11-43 becomes zero. Then the Reynolds transport theorem reduces to

Steady flow:
$$\frac{dB_{sys}}{dt} = \int_{CS} \rho b(\vec{V} \cdot \hat{n})\, dA \qquad (11\text{-}44)$$

Note that unlike the control volume, the property B content of the system may still change with time during a steady process, but in this case the change must be equal to the net property transported by mass across the control surface.

Special Case 2: One-Dimensional Flow

In many practical applications, the fluid crosses the boundaries of the control volume at a given number of inlets and exits with nearly *uniform properties* over the cross sections where the fluid enters or leaves the control volume. In such cases, the last term in the Reynolds transport theorem can be expressed as the difference between the properties of outgoing and incoming streams, like Eq. 11-38 for the one-inlet and one-exit case. The Reynolds transport theorem in this case reduces to

One-dimensional flow:
$$\frac{dB_{sys}}{dt} = \frac{d}{dt}\int_{CV} \rho b\, dV + \sum_{out} \underbrace{\rho_e b_e \mathcal{V}_e A_e}_{\text{for each exit}} - \sum_{in} \underbrace{\rho_i b_i \mathcal{V}_i A_i}_{\text{for each inlet}} \qquad (11\text{-}45)$$

or

$$\frac{dB_{sys}}{dt} = \frac{d}{dt}\int_{CV} \rho b\, dV + \sum_{out} \dot{m}_e b_e - \sum_{in} \dot{m}_i b_i \qquad (11\text{-}46)$$

where the subscripts e and i stand for *exit* and *inlet,* respectively. Note that the one-dimensional flow assumption simplifies the analysis greatly.

$$\frac{dB_{sys}}{dt} = \frac{d}{dt}\int_{CV}\rho b\, dV + \int_{CS}\rho b(\vec{V}\cdot\hat{n})\, dA$$

$B_{sys} = m_{sys}$ $b = 1$ $b = 1$

$$\frac{\overset{0}{\overbrace{dm_{sys}}}}{dt} = \frac{d}{dt}\int_{CV}\rho\, dV + \int_{CS}\rho(\vec{V}\cdot\hat{n})\, dA$$

FIGURE 11-43

The continuity equation is obtained by replacing B in the Reynolds transport theorem by the mass m and b by 1 (m per unit mass = m/m = 1).

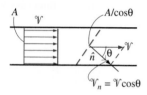

$\dot{m} = \rho(\mathcal{V}\cos\theta)(A/\cos\theta) = \rho\mathcal{V}A$

(a) Control surface *at an angle* to flow

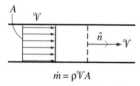

$\dot{m} = \rho\mathcal{V}A$

(b) Control surface *normal* to flow

FIGURE 11-44

A control surface should always be selected *normal to flow* at all locations it crosses fluid flow to avoid complications.

An Application: The Continuity Equation

We can demonstrate the use of the Reynolds transport theorem by taking the property B to be the mass, m. Then we have $b = 1$ since dividing the mass by mass to get the property per unit mass gives unity. Also, the mass of a system is constant, and thus its time derivative is zero. Then the Reynolds transport equation in this case reduces to (Fig. 11-43)

Continuity equation:	$0 = \dfrac{d}{dt}\displaystyle\int_{CV}\rho\, dV + \displaystyle\int_{CS}\rho(\vec{V}\cdot\hat{n})\, dA$	(11-47)

It states that *the time rate of change of the mass within the control volume plus the net mass flow rate through the control surface is equal to zero.* The continuity equation is also valid for moving control volumes provided that velocities relative to the control volume be used for the fluid streams. After all, a moving control volume can be viewed as a stationary one if everything is expressed relative to a reference frame attached to the control volume.

Some practical problems (such as the injection of medication through the needle of a syringe by the forced motion of the plunger) involve deforming control volumes. The continuity equation can still be used for such deforming control volumes provided that the velocity of the fluid crossing a deforming part of the control surface be expressed relative to the control surface (that is, the fluid velocity should be expressed relative to a reference frame attached to the deforming part of the control surface). Problems involving moving control volumes with deforming surfaces often challenge the power of imagination.

There is considerable flexibility in the selection of a control volume when solving a problem. Several control volume choices may be correct, but some are more convenient to work with. A control volume should not introduce any unnecessary complications. The proper choice of a control volume can make the solution of a seemingly complicated problem rather easy. A simple rule in selecting a control volume is to make the control surface *normal to flow* at all locations it crosses fluid flow. This way the dot product $\vec{V}\cdot\hat{n}$ simply becomes the magnitude of the velocity and, in one-dimensional flow, the integral $\int\rho(\vec{V}\cdot\hat{n})\, dA$ is simply $\rho\mathcal{V}A$ (Fig. 11-44).

The term $\dfrac{d}{dt}\displaystyle\int_{CV}\rho\, dV$ becomes zero when the flow is *steady* or when the control volume is filled with an incompressible fluid at all times. The continuity in that case reduces to (Fig. 11-45)

Steady flow:	$\displaystyle\sum_{out}\dot{m}_e = \sum_{in}\dot{m}_i$	(11-48)

For *steady one-dimensional incompressible flow* with *one inlet* and *one exit*, the continuity equation simplifies to (see Chap. 1 for details and examples)

Single stream, ρ = constant:	$\rho\mathcal{V}_1 A_1 = \rho\mathcal{V}_2 A_2$ or $\mathcal{V}_1 A_1 = \mathcal{V}_2 A_2$	(11-49)

where $\mathcal{V}$ is the normal velocity and A is the cross-sectional area of the inlet or exit. Therefore, the Reynolds transport theorem is a very powerful tool, and we can use it with confidence. In the next section, we will apply the Reynolds transport theorem to obtain the linear momentum equation for flow systems.

Newton's second law for a system subjected to a net force F is expressed as

$$\sum \vec{F} = m\vec{a} = m\frac{d\vec{V}}{dt} = \frac{d}{dt}(m\vec{V}) \tag{11-50}$$

where $m\vec{V}$ is the **linear momentum** of the system. Noting that both the density and the velocity may change from point to point within the system, Newton's second law can be expressed more generally as

$$\sum \vec{F} = \frac{d}{dt}\int_{\text{sys}} \vec{V}\rho \, dV \tag{11-51}$$

where $\delta m = \rho \, dV$ is the mass of a differential volume element dV and $\vec{V}\rho \, dV$ is its momentum. Therefore, Newton's second law can be stated as *the sum of all external forces acting on a system is equal to the time rate of change of the linear momentum of the system.* This statement is valid for a fixed coordinate system or a coordinate system that moves with a constant velocity on a straight path, called an *inertial reference frame.* Note that this is a vector relation, and thus the quantities $\vec{F}$, $\vec{a}$, and $\vec{V}$ have direction as well as magnitude.

The relation above is for a given mass of a solid or fluid and is of limited use in fluid mechanics since most flow systems are analyzed using control volumes. The *Reynolds transport theorem* developed in the previous section provides the necessary tools to shift from the system formulation to the control volume formulation. Setting $b = \vec{V}$ and thus $B = m\vec{V}$, the Reynolds transport theorem (Eq. 11-44) can be expressed for momentum as (Fig. 11-46)

$$\frac{d(m\vec{V})_{\text{sys}}}{dt} = \frac{d}{dt}\int_{\text{CV}} \rho\vec{V} \, dV + \int_{\text{CS}} \rho\vec{V}(\vec{V} \cdot \hat{n}) \, dA \tag{11-52}$$

But the left-hand side of this equation is, from Eq. 11-50, equal to $\sum \vec{F}$. Substituting, Newton's second law is obtained to be

General: $\qquad \sum \vec{F} = \dfrac{d}{dt}\displaystyle\int_{\text{CV}} \rho\vec{V} \, dV + \int_{\text{CS}} \rho\vec{V}(\vec{V} \cdot \hat{n}) \, dA \tag{11-53}$

which can be stated as

$$\begin{pmatrix} \text{The sum of} \\ \text{all external} \\ \text{forces acting} \\ \text{on a CV} \end{pmatrix} = \begin{pmatrix} \text{The time rate of change} \\ \text{of the linear} \\ \text{momentum of the} \\ \text{contents of the CV} \end{pmatrix} + \begin{pmatrix} \text{The net flow rate} \\ \text{of linear momentum} \\ \text{through the control} \\ \text{surface by mass} \end{pmatrix}$$

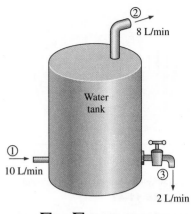

The Linear Momentum Equation

8 L/min

Water tank

① 10 L/min

③ 2 L/min

$$\sum_{\text{out}} \dot{V}_e = \sum_{\text{in}} \dot{V}_i \rightarrow \dot{V}_2 + \dot{V}_3 = \dot{V}_1$$

FIGURE 11-45

During steady incompressible flow, the total (volume) flow rate into the control volume is equal to the total flow rate out of the control volume.

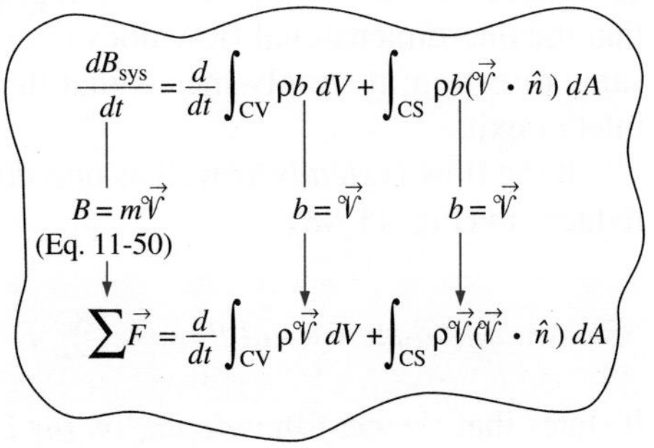

$$\frac{dB_{\text{sys}}}{dt} = \frac{d}{dt}\int_{\text{CV}} \rho b \, dV + \int_{\text{CS}} \rho b(\vec{V} \cdot \hat{n}) \, dA$$

$B = m\vec{V}$
(Eq. 11-50) $\qquad b = \vec{V} \qquad b = \vec{V}$

$$\sum \vec{F} = \frac{d}{dt}\int_{\text{CV}} \rho\vec{V} \, dV + \int_{\text{CS}} \rho\vec{V}(\vec{V} \cdot \hat{n}) \, dA$$

FIGURE 11-46

The linear momentum equation is obtained by replacing B in the Reynolds transport theorem by the momentum $m\vec{V}$ and b by the velocity $\vec{V}$.

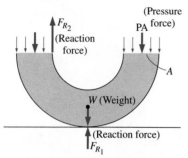

F_{R_2}
(Reaction
force)

PA (Pressure force)

A

W (Weight)

(Reaction force)
F_{R_1}

An 180° elbow supported by the ground

FIGURE 11-47

In most cases, the force $\vec{F}$ consists of
weights, hydrostatic pressure forces,
and reaction forces.

In fluid mechanics, Newton's second law is usually referred to as the **linear momentum equation** (or just the *momentum equation*). Note that the momentum equation is a *vector equation,* and thus each term should be treated as a vector. Also, the components of this equation can be resolved along orthogonal coordinates (such as x, y, and z in the rectangular coordinate system) for convenience.

The forces acting on the control volume consist of *body forces* that act throughout the entire body of the control volume (such as gravity, electric, and magnetic forces) and *surface forces* that act on the control surface (such as the pressure forces and reaction forces at points of contact). The surface forces appear as the control volume is isolated from its surroundings for analysis, and the effect of the detached body is replaced by a force at that location. This is similar to drawing the free-body diagrams of bodies in statics. A well-chosen control volume exposes only the forces that are to be determined (such as the reaction forces to support a component) and the minimum number of other forces. In most cases, the force $\vec{F}$ consist of weights, hydrostatic pressure forces, and reaction forces (Fig. 11-47). The momentum equation is commonly used to calculate the forces (usually on the anchors) induced by the flow.

Special Cases

During *steady flow,* the amount of momentum within the control volume remains constant, and thus the time rate of change of linear momentum of the contents of the control volume (the derivative in Eq. 11-53) is zero. It gives

Steady flow:
$$\sum \vec{F} = \int_{CS} \rho \vec{V}(\vec{V} \cdot \hat{n})\, dA \tag{11-54}$$

Most momentum problems considered in this text will be steady.

In many practical applications, the fluid crosses the boundaries of the control volume at a certain number of inlets and exits with nearly *uniform properties* over the cross-sectional areas where the fluid enters or leaves the control volume. In such cases, the last term of the Reynolds transport theorem can be expressed as the difference between the properties of outgoing and incoming streams, as given in Eq. 11-38 for the one-inlet and one-exit case. The Reynolds transport equation for this *one-dimensional* flow case reduces to

One-dimensional flow:
$$\sum \vec{F} = \frac{d}{dt}\int_{CV} \rho\vec{V}\, dV + \sum_{out} \dot{m}_e \vec{V}_e - \sum_{in} \dot{m}_i \vec{V}_i \tag{11-55}$$

Note that the one-dimensional flow assumption simplifies the analysis greatly, and therefore it is commonly used in practice. But it should be kept in mind that the one-dimensional flow does not mean that all fluid streams are in the same direction; it simply means that the flow properties are uniform at any inlet or exit.

If the flow is *steady* as well as *one-dimensional,* the relation above further reduces to (Fig. 11-48)

Steady, one-dimensional flow:
$$\sum \vec{F} = \sum_{out} \dot{m}_e \vec{V}_e - \sum_{in} \dot{m}_i \vec{V}_i \tag{11-56}$$

It states that *the net force acting on the control volume during steady flow is equal to the difference between the outgoing and the incoming momentum fluxes by mass.* This statement can also be expressed for any direction.

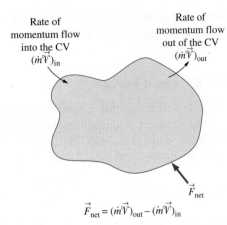

$$\vec{F}_{net} = (\dot{m}\vec{V})_{out} - (\dot{m}\vec{V})_{in}$$

FIGURE 11–48

The net force acting on the control
volume during steady flow is equal to
the difference between the outgoing and
the incoming momentum fluxes by mass.

Many practical problems involve stationary control volumes with just one inlet and one exit. The mass flow rate for such single-stream systems remains constant, and the relation above reduces to

Steady, one-dimensional flow
(one-inlet, one-exit): $\qquad \sum \vec{F} = \dot{m}(\vec{V}_2 - \vec{V}_1)$ $\qquad$ (11–57)

where $\vec{V}_1$ and $\vec{V}_2$ are the inlet and the exit velocities of the fluid stream relative to the control surface, respectively. We emphasize again that all the relations above are *vector* equations, and thus all the additions and subtractions are *vector* additions and subtractions. Recall that subtracting a vector is equivalent to adding it after reversing its direction (Fig. 11–49). Also, when writing the momentum equation along a specified coordinate (such as the *x* axis), we use the projections of the vectors on that axis. For example, the last equation written along the *x* coordinate is

Along x coordinate: $\qquad \sum \vec{F}_x = \dot{m}(\vec{V}_{2,x} - \vec{V}_{1,x})$ $\qquad$ (11–58)

where $\sum \vec{F}_x$ is the vector sum of the *x* components of the forces and $\vec{V}_{2,x}$ and $\vec{V}_{1,x}$ are the *x* components of the exit and inlet velocities of the fluid stream, respectively. The force or velocity components in the positive *x* direction are positive quantities, and those in the negative *x* direction are negative quantities. Also, it is a good practice to take the direction of unknown forces in the positive direction (unless the problem is very straightforward). This way, the assumed direction is correct if we obtain a positive value for the unknown force. A negative value obtained indicates that the assumed direction is wrong, and it should be reversed.

A common simplification in the application of the momentum equation is to ignore the *atmospheric pressure* and work with gage pressures. This is because the atmospheric pressure acts in all directions and its effect cancels out in every direction (Fig. 11–50). This means we also can ignore the pressure forces at exit sections where the fluid is discharged to the atmosphere since the discharge pressures in such cases will be very nearly atmospheric pressure at subsonic velocities.

The internal forces (such as the pressure force between a fluid and the inner surfaces of the flow section) are not considered in the momentum analysis unless they are exposed by passing the control surface through that area. Only external forces are considered in the analysis. We should choose the control

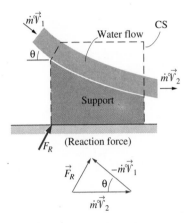

Note: $\vec{V}_2 \neq \vec{V}_1$ even if $|\vec{V}_2| = |\vec{V}_1|$ (which will be the case when $z_2 = z_1$) unless $\theta = 0$.

FIGURE 11–49

The determination of the reaction force on the support caused by a change of direction of water by vector addition. (The viscous, friction, and gravity effects are ignored for simplicity. Note that the reaction force will be zero only if $\theta = 0$.)

Water discharge from the nozzle of a hose

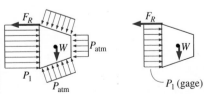

With atmospheric $\qquad$ With atmospheric
pressure considered $\qquad$ pressure cancelled out

FIGURE 11–50

The atmospheric pressure acts in all directions, and thus it can be ignored when performing force balances since its effect cancels out in every direction.

volume such that the forces in which we are not interested remain internal, and thus they do not complicate the analysis.

No External Forces

An interesting situation arises when there are no net external forces such as gravity, pressure, and reaction forces acting on the body—a common situation for space vehicles and satellites. For a control volume with a uniform velocity and one-dimensional flow, Eq. 11-55 reduces in this case to

$$\textit{No external forces:} \qquad \frac{d(m\vec{V})_{CV}}{dt} = \sum_{\text{in}} \dot{m}_i \vec{V}_i - \sum_{\text{out}} \dot{m}_e \vec{V}_e \qquad (11\text{-}59)$$

This is an expression of the conservation of momentum principle, which can be stated as *in the absence of external forces, the rate of change of the momentum of a control volume is equal to the difference between the incoming and the outgoing momentum fluxes by mass.*

When the mass m of the control volume remains nearly constant, the left-hand side of the equation above simply becomes mass times acceleration since

$$m_{CV} = \textit{constant:} \qquad \frac{d(m\vec{V})_{CV}}{dt} = m_{CV} \frac{d\vec{V}_{CV}}{dt} = (m\vec{a})_{CV} \qquad (11\text{-}60)$$

Therefore, the control volume in this case can be treated as a solid body, with a net force $\vec{F} = m\vec{a}$ (due to a change of momentum) acting on it. This approach can be used to determine the linear acceleration of space vehicles when a rocket is fired.

Momentum-Flux Correction Factor, β

Under the one-dimensional flow assumption, the momentum flux at the inlets and the exits of a control volume is evaluated using the *average* flow velocity. However, this is not equivalent to the integral that involves the square of the actual velocity distribution. This error can be corrected by replacing the momentum-flux terms $\dot{m}\vec{V}$ in the momentum equation by $\beta\dot{m}\vec{V}$ where β is the *momentum-flux correction factor.* By using equations for the variation of velocity with the radial distance, it can be shown that this correction factor is 1.33 for the fully developed laminar flow, and it ranges between 1.01 and 1.04 for turbulent flow in a circular pipe (note that these values are much smaller than the kinetic energy correction factors discussed in Sec. 11-4). The momentum-flux correction factors are normally disregarded since their values are close to unity and their effect on the final result is usually insignificant.

EXAMPLE 11-14   The Force to Hold a Deflector Elbow in Place

A reducing elbow is used to deflect water flow at a rate of 14 kg/s in a horizontal pipe upwards 30° while accelerating it (Fig. 11-51). The elbow discharges water into the atmosphere. The cross-sectional area of the elbow is 113 cm² at the inlet and 7 cm² at the exit. The elevation difference between the centers of the exit and the inlet is 30 cm. The weight of the elbow and the water in it is considered to be negligible. Determine (a) the gage pressure at the center of the inlet of the elbow and (b) the anchoring force needed to hold the elbow in place.

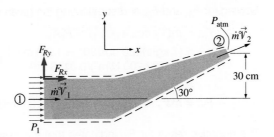

FIGURE 11-51
Schematic for Example 11-14.

Solution A reducing elbow deflects water upwards and discharges it to the atmosphere. The pressure at the inlet of the elbow and the force needed to hold the elbow in place are to be determined.

Assumptions **1** The flow is steady, frictionless, one-dimensional, and irrotational (so the Bernoulli equation is applicable). **2** The weight of the elbow and the water in it is negligible. **3** The water is discharged to the atmosphere, and thus the gage pressure at the exit is zero.

Properties We take the density of water to be 1000 kg/m³.

Analysis (a) We take the elbow as the control volume and designate the entrance by 1 and the exit by 2. We also take the x and y coordinates as shown. The continuity equation for this one-inlet, one-exit, steady-flow system is $\dot{m}_1 = \dot{m}_2 = \dot{m} = 14$ kg/s. Noting that $\dot{m} = \rho A \mathscr{V}$, the inlet and the exit velocities of water are

$$\mathscr{V}_1 = \frac{\dot{m}}{\rho A_1} = \frac{14 \text{ kg/s}}{(1000 \text{ kg/m}^3)(0.0113 \text{ m}^2)} = 1.24 \text{ m/s}$$

$$\mathscr{V}_2 = \frac{\dot{m}}{\rho A_2} = \frac{14 \text{ kg/s}}{(1000 \text{ kg/m}^3)(7 \times 10^{-4} \text{ m}^2)} = 20.0 \text{ m/s}$$

Taking the center of the inlet cross section as the reference level ($y_1 = 0$) and noting that $P_2 = P_{atm}$, the Bernoulli equation for a streamline going through the center of the elbow is expressed as

$$\frac{P_1}{\rho g} + \frac{\mathscr{V}_1^2}{2g} + y_1 = \frac{P_2}{\rho g} + \frac{\mathscr{V}_2^2}{2g} + y_2$$

$$P_1 - P_2 = \rho g \left(\frac{\mathscr{V}_2^2 - \mathscr{V}_1^2}{2g} + y_2 - y_1 \right)$$

$$P_1 - P_{atm} = (1000 \text{ kg/m}^3)(9.81 \text{ m/s}^2)$$
$$\times \left(\frac{(20 \text{ m/s})^2 - (1.24 \text{ m/s})^2}{2(9.81 \text{ m/s}^2)} + 0.3 - 0 \right) \left(\frac{1 \text{ kN}}{1000 \text{ kg} \cdot \text{m/s}^2} \right)$$

$$P_{1, \text{ gage}} = 202.2 \text{ kN/m}^2 = \textbf{202.2 kPa}$$

(b) The momentum equation for steady one-dimensional flow is

$$\sum \vec{F} = \sum_{\text{out}} \dot{m}_e \vec{\mathscr{V}}_e - \sum_{\text{in}} \dot{m}_i \vec{\mathscr{V}}_i$$

We let the x and y components of the anchoring force of the elbow be F_{Rx} and F_{Ry} and assume them to be in the positive direction. We also use gage pressure since the atmospheric pressure acts on the entire control surface. Then the momentum equations along the x and y axes become

$$F_{Rx} + P_1 A_1 = \dot{m} \mathscr{V}_2 \cos \theta - \dot{m} \mathscr{V}_1$$
$$F_{Ry} = \dot{m} \mathscr{V}_2 \sin \theta$$

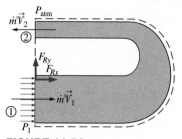

FIGURE 11-52
Schematic for Example 11-15.

Solving for F_{Rx} and F_{Ry} and substituting the given values,

$$F_{Rx} = \dot{m}(\mathcal{V}_2 \cos\theta - \mathcal{V}_1) - P_1 A_1$$
$$= (7 \text{ kg/s})[(20 \cos 30° - 1.24) \text{ m/s}]\left(\frac{1 \text{ N}}{1 \text{ kg} \cdot \text{m/s}^2}\right)$$
$$- (202{,}200 \text{ N/m}^2)(0.0113 \text{ m}^2)$$
$$= 113 - 2285 = \mathbf{-2172\ N}$$
$$F_{Ry} = \dot{m}\mathcal{V}_2 \sin\theta = (7 \text{ kg/s})(20 \sin 30° \text{ m/s})\left(\frac{1 \text{ N}}{1 \text{ kg} \cdot \text{m/s}^2}\right) = \mathbf{70\ N}$$

The negative result for F_{Rx} indicates that the assumed direction is wrong and it should be reversed. Therefore, F_{Rx} acts in the negative x direction.

EXAMPLE 11-15   The Force to Hold a Reversing Elbow in Place

The deflector elbow in the previous example is replaced by a reversing elbow such that the fluid makes a 180° U-turn before it is discharged, as shown in Fig. 11-52. The distance between the centers of the inlet and the exit sections is still 0.3 m. Determine the anchoring force needed to hold the elbow in place.

Solution   The inlet and the exit velocities and the pressure at the inlet of the elbow remain the same, but the vertical component of the anchoring force at the connection of the elbow to the pipe is zero in this case ($F_{Ry} = 0$) since there is no other force or momentum flux in the vertical direction. The horizontal component of the anchoring force is determined from the momentum equation written in the x direction. Noting that the exit velocity is negative since it is in the negative x direction, we have

$$F_{Rx} + P_1 A_1 = \dot{m}(-\mathcal{V}_2) - \dot{m}\mathcal{V}_1$$

Solving for F_{Rx} and substituting the known values,

$$F_{Rx} = -\dot{m}(\mathcal{V}_2 + \mathcal{V}_1) - P_1 A_1$$
$$= -(7 \text{ kg/s})[(20 + 1.24) \text{ m/s}]\left(\frac{1 \text{ N}}{1 \text{ kg} \cdot \text{m/s}^2}\right) - (202{,}200 \text{ N/m}^2)(0.0113 \text{ m}^2)$$
$$= -149 - 2285 = \mathbf{-2434\ N}$$

Therefore, the horizontal force on the flange is 2434 N acting in the negative x direction (trying to separate the elbow from the pipe). This force is equivalent to the weight of about 250 kg mass, and thus the connectors (such as bolts) used must be strong enough to withstand this force.

Discussion   If the reversing elbow is replaced by a straight nozzle (like one used by fire fighters) such that water is discharged in the positive x direction, the momentum equation in the x direction becomes

$$F_{Rx} + P_1 A_1 = \dot{m}\mathcal{V}_2 - \dot{m}\mathcal{V}_1 \quad \rightarrow \quad F_{Rx} = \dot{m}(\mathcal{V}_2 - \mathcal{V}_1) - P_1 A_1$$

since both $\mathcal{V}_1$ and $\mathcal{V}_2$ are in the positive x direction. This shows the importance of using the correct sign (positive if in the positive direction and negative if in the opposite direction) for velocities and forces.

EXAMPLE 11-16   Water Jet Striking a Stationary Plate

Water accelerated by a nozzle strikes a stationary vertical plate at a rate of 10 kg/s with a normal velocity of 20 m/s (Fig. 11-53). After the strike, the water stream splatters off in all directions in the plane of the plate. Determine the force needed to prevent the plate from moving horizontally due to the water stream.

Solution   A water jet strikes a vertical stationary plate normally. The force needed to hold the plate in place is to be determined.

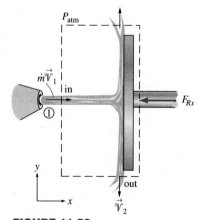

FIGURE 11-53
Schematic for Example 11-16.

Assumptions **1** The flow is steady and one-dimensional. **2** The water splatters in directions normal to the approach direction of the water jet. **3** The water jet is exposed to the atmosphere, and thus the pressure of the water jet and the splattered water is the atmospheric pressure, which is disregarded since it acts on the entire system. **4** The vertical forces and momentum fluxes are not considered since they have no effect on the horizontal reaction force.

Analysis We draw the control volume for this problem such that it contains the entire plate and cuts through the water jet and the support bar normally. The momentum equation for steady one-dimensional flow is given as

$$\sum \vec{F} = \sum_{out} \dot{m}_e \vec{\mathcal{V}}_e - \sum_{in} \dot{m}_i \vec{\mathcal{V}}_i$$

Writing it for this problem along the x direction (without forgetting the negative sign for forces and velocities in the negative x direction) and noting that $\mathcal{V}_{1,x} = \mathcal{V}_1$ and $\mathcal{V}_{2,x} = 0$ give

$$F_R = -\dot{m}\mathcal{V}_1$$

Substituting the given values,

$$F_R = -\dot{m}\mathcal{V}_1 = -(10 \text{ kg/s})(+20 \text{ m/s})\left(\frac{1 \text{ N}}{1 \text{ kg} \cdot \text{m/s}^2}\right) = -200 \text{ N}$$

Therefore, the support must apply a 200-N horizontal force (equivalent to the weight of about a 20-kg mass) in the negative x direction (the opposite direction of the water jet) to hold the plate in place.

EXAMPLE 11-17   Power Generation and Wind Loading of a Wind Turbine

A wind generator with a 30-foot-diameter blade span has a cut-in wind speed (minimum speed for power generation) of 7 mph, at which velocity the turbine generates 0.4 kW of electric power (Fig. 11-54). Determine (*a*) the efficiency of the wind turbine–generator set and (*b*) the horizontal force exerted by the wind on the supporting mast of the wind turbine. What is the effect of doubling the wind velocity to 14 mph on power generation and the force exerted? Assume the efficiency remains the same, and take the density of air to be 0.076 lbm/ft³.

Solution The power generation and loading of a wind turbine are to be analyzed. The efficiency and the force exerted on the mast are to be determined, and the effects of doubling the wind velocity are to be investigated.

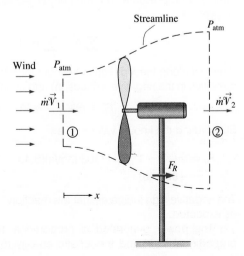

FIGURE 11-54

Schematic for Example 11-17.

Assumptions **1** The wind flow is steady, one-dimensional, and incompressible. **2** The efficiency of the turbine-generator is independent of wind speed. **3** The frictional effects are negligible, and thus none of the incoming kinetic energy is converted to thermal energy. **4** The average velocity of air through the wind turbine is the same as the wind velocity (actually, it is considerably less—see the discussion that follows the example).

Properties The density of air is given to be 0.076 lbm/ft³.

Analysis Kinetic energy is a mechanical form of energy, and thus it can be converted to work entirely. Therefore, the power potential of the wind is its kinetic energy, which is $\mathcal{V}^2/2$ per unit mass and $\dot{m}\mathcal{V}^2/2$ for a given mass flow rate:

$$\mathcal{V}_1 = (7 \text{ mph})\left(\frac{1.4667 \text{ ft/s}}{1 \text{ mph}}\right) = 10.27 \text{ ft/s}$$

$$\dot{m} = \rho_1 \mathcal{V}_1 A_1 = \rho_1 \mathcal{V}_1 \frac{\pi D^2}{4} = (0.076 \text{ lbm/ft}^3)(10.27 \text{ ft/s})\frac{\pi(30 \text{ ft})^2}{4} = 1103 \text{ lbm/s}$$

$$\dot{W}_{max} = \dot{m}\text{ke}_1 = \dot{m}\frac{\mathcal{V}_1^2}{2}$$

$$= (1103 \text{ lbm/s})\frac{(10.27 \text{ ft/s})^2}{2}\left(\frac{1 \text{ lbf}}{32.2 \text{ lbm} \cdot \text{ft/s}^2}\right)\left(\frac{1 \text{ kW}}{737.56 \text{ lbf} \cdot \text{ft/s}}\right) = 1.225 \text{ kW}$$

Therefore, the available power to the wind turbine is 1.225 kW at the wind velocity of 7 mph. Then the turbine-generator efficiency becomes

$$\eta_{\text{wind turbine}} = \frac{\dot{W}_{act}}{\dot{W}_{max}} = \frac{0.4 \text{ kW}}{1.225 \text{ kW}} = \textbf{0.327} \quad (\text{or } \textbf{32.7\%})$$

(*b*) The frictional effects are assumed to be negligible, and thus the portion of incoming kinetic energy not converted to electric power leaves the wind turbine as outgoing kinetic energy. Noting that the mass flow rate remains constant, the exit velocity is determined to be

$$\dot{m}\text{ke}_2 = \dot{m}\text{ke}_1(1 - \eta_{\text{wind turbine}}) \quad \rightarrow \quad \dot{m}\frac{\mathcal{V}_2^2}{2} = \dot{m}\frac{\mathcal{V}_1^2}{2}(1 - \eta_{\text{wind turbine}})$$

or

$$\mathcal{V}_2 = \mathcal{V}_1 \sqrt{1 - \eta_{\text{wind turbine}}} = (10.27 \text{ ft/s})\sqrt{1 - 0.327} = 8.43 \text{ ft/s}$$

We draw a control volume around the wind turbine such that the wind is normal to the control surface at the inlet and the exit and the entire control surface is at the atmospheric pressure. The momentum equation for steady one-dimensional flow is given as

$$\sum \vec{F} = \sum_{\text{out}} \dot{m}_e \vec{\mathcal{V}}_e - \sum_{\text{in}} \dot{m}_i \vec{\mathcal{V}}_i$$

Writing it along the *x* direction (without forgetting the negative sign for forces and velocities in the negative *x* direction) and noting that $\mathcal{V}_{1,x} = \mathcal{V}_1$ and $\mathcal{V}_{2,x} = \mathcal{V}_2$ give

$$F_R = \dot{m}\mathcal{V}_2 - \dot{m}\mathcal{V}_1 = \dot{m}(\mathcal{V}_2 - \mathcal{V}_1)$$

Substituting the known values gives

$$F_R = \dot{m}(\mathcal{V}_2 - \mathcal{V}_1) = (1103 \text{ lbm/s})(8.43 - 10.27 \text{ ft/s})\left(\frac{1 \text{ lbf}}{32.2 \text{ lbm} \cdot \text{ft/s}^2}\right)$$
$$= \textbf{-63.1 lbf}$$

The negative sign indicates that the reaction force acts in the negative *x* direction, as expected.

The power generated is proportional to $\mathcal{V}^3$ since the mass flow rate is proportional to $\mathcal{V}$ and the kinetic energy to $\mathcal{V}^2$. Therefore, doubling the wind velocity to 14 mph will increase the power generation by a factor of $2^3 = 8$ to $0.4 \times 8 = 3.2$ kW. The force exerted by the wind on the support mast is proportional to $\mathcal{V}^3$. Therefore, doubling the wind velocity to 14 mph will increase the wind force by a factor of $2^2 = 4$ to $63.1 \times 4 = 252.4$ lbf.

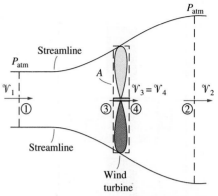

Discussion   To gain more insight into the operation of devices with propellers such as helicopters, wind turbines, hydraulic turbines, and turbofan engines, we reconsider the wind turbine and draw the streamlines, as shown in Fig. 11-55. (In the case of power-consuming devices such as a fan and a helicopter, the streamlines converge rather than diverge since the exit velocity will be higher and thus the exit area will be lower.) The upper and lower streamlines can be considered to form an "imaginary duct" for the flow of air through the propeller. Sections 1 and 2 are sufficiently far from the propeller so that $P_1 = P_2 = P_{atm}$. The momentum equation for this large control volume between sections 1 and 2 was obtained to be

$$F_R = \dot{m}(\mathcal{V}_2 - \mathcal{V}_1) \qquad (1)$$

The smaller control volume between sections 3 and 4 encloses the propeller, and $A_3 = A_4 = A$ and $\mathcal{V}_3 = \mathcal{V}_4$ since it is so slim. The propeller is a device that causes a pressure change, and thus the pressures P_3 and P_4 are different. The momentum equation applied to the smaller control volume gives

$$F_R + P_3 A - P_4 A = 0 \qquad \rightarrow \qquad F_R = (P_4 - P_3)A \qquad (2)$$

FIGURE 11-55

The large and small control volumes for the analysis of a propeller bounded by the streamlines.

The Bernoulli equation is not applicable between sections 1 and 2 since the path crosses a propeller, but it is applicable separately between sections 1 and 3 and sections 4 and 2:

$$\frac{P_1}{\rho g} + \frac{\mathcal{V}_1^2}{2g} + z_1 = \frac{P_3}{\rho g} + \frac{\mathcal{V}_3^2}{2g} + z_3 \qquad \text{and} \qquad \frac{P_4}{\rho g} + \frac{\mathcal{V}_4^2}{2g} + z_4 = \frac{P_2}{\rho g} + \frac{\mathcal{V}_2^2}{2g} + z_2$$

Adding these two equations and noting that $z_1 = z_2 = z_3 = z_4$, $\mathcal{V}_3 = \mathcal{V}_4$, and $P_1 = P_2 = P_{atm}$ gives

$$\frac{\mathcal{V}_2^2 - \mathcal{V}_1^2}{2} = \frac{P_4 - P_3}{\rho} \qquad (3)$$

Substituting $\dot{m} = \rho A \mathcal{V}_3$ into (1) and then combining it with (2) and (3) gives

$$\mathcal{V}_3 = \frac{\mathcal{V}_1 + \mathcal{V}_2}{2} \qquad (4)$$

Thus we conclude that *the average velocity of a fluid through a propeller is the arithmetic average of the upstream and downstream velocities.* Of course, the validity of this result is limited by the applicability of the Bernoulli equation.

Now back to the wind turbine. The velocity through the propeller can be expressed as $\mathcal{V}_3 = \mathcal{V}_1(1 - a)$, where $a < 1$ since $\mathcal{V}_3 < \mathcal{V}_1$. Combining this expression with (4) gives $\mathcal{V}_2 = \mathcal{V}_1(1 - 2a)$. Also, the mass flow rate through the propeller becomes $\dot{m} = \rho A \mathcal{V}_3 = \rho A \mathcal{V}_1(1 - a)$. When the frictional effects and losses are neglected, the power generated by a wind turbine is simply the difference between the incoming and the outgoing kinetic energies:

$$\dot{W} = \dot{m}(\text{ke}_1 - \text{ke}_2) = \frac{\dot{m}(\mathcal{V}_1^2 - \mathcal{V}_2^2)}{2} = \frac{\rho A \mathcal{V}_1(1 - a)[\mathcal{V}_1^2 - \mathcal{V}_1^2(1 - 2a)^2]}{2}$$
$$= 2\rho A \mathcal{V}_1^3 a(1 - a)^2$$

Dividing this by the available power of the wind gives the efficiency of the wind turbine in terms of a,

$$\eta_{\text{wind turbine}} = \frac{\dot{W}}{\dot{W}_{max}} = \frac{2\rho A \mathcal{V}_1^3 a(1 - a)^2}{\dot{m}_{max} \mathcal{V}_1^2/2} = \frac{2\rho A \mathcal{V}_1^3 a(1 - a)^2}{(\rho A \mathcal{V}_1) \mathcal{V}_1^2/2} = 4a(1 - a)^2$$

The value of a that maximizes the efficiency is determined by setting the derivative of $\eta_{\text{wind turbine}}$ with respect to a equal to zero and solving for a. It gives

$a = ⅓$. Substituting this value into the efficiency relation above gives $\eta_{\text{wind turbine}} =$ $^{16}\!/_{27} = 0.593$, which is the upper limit for the efficiency of wind turbines and other propellers. This is known as the **Betz limit.** The efficiency of actual wind turbines is about half of this ideal value.

Example 11-18   Repositioning of a Satellite

An orbiting satellite system has a mass of $m_{\text{sat}} = 5000$ kg and is traveling at a constant velocity of $\mathcal{V}_0$. To alter its orbit, an attached rocket discharges $m_f = 100$ kg of solid fuel at a velocity $\mathcal{V}_f = 3000$ m/s relative to $\mathcal{V}_0$ in a direction opposite to $\mathcal{V}_0$ (Fig. 11-56). The fuel discharge rate is constant for two seconds. Determine (a) the acceleration of the system during this two-second period, (b) the change of velocity of the satellite system during this time period, and (c) the thrust exerted on the system.

Solution   The rocket of a satellite is fired in the opposite direction to motion. The acceleration, the velocity change, and the thrust are to be determined.

Assumptions   **1** The flow of combustion gases is steady and one-dimensional during the firing period. **2** There are no external forces acting on the satellite, and the effect of the pressure force at the nozzle exit is negligible. **3** The mass of discharged fuel is negligible relative to the mass of the satellite, and thus the satellite may be treated as a solid body with a constant mass.

Analysis   (a) A body moving at constant velocity can be considered to be stationary for convenience. Then the velocities of fluid streams become simply their velocities relative to the moving body. We take the direction of motion of the satellite as the positive direction along the x axis. There are no external forces acting on the satellite and its mass is nearly constant. Therefore, the satellite can be treated as a solid body with constant mass, and the momentum equation in this case is simply Eq. 11-60,

$$\frac{d(m\vec{\mathcal{V}})_{\text{CV}}}{dt} = \sum_{\text{in}} \dot{m}_i\vec{\mathcal{V}}_i - \sum_{\text{out}} \dot{m}_e\vec{\mathcal{V}}_e \quad \rightarrow \quad m_{\text{sat}}\frac{d\vec{\mathcal{V}}_{\text{sat}}}{dt} = -\dot{m}_f\vec{\mathcal{V}}_f$$

Noting that the motion is on a straight line and the discharged gases move in the negative x direction, we can write the momentum equation using magnitudes as

$$m_{\text{sat}}\frac{d\mathcal{V}_{\text{sat}}}{dt} = \dot{m}_f\mathcal{V}_f \quad \rightarrow \quad \frac{d\mathcal{V}_{\text{sat}}}{dt} = \frac{\dot{m}_f}{m_{\text{sat}}}\mathcal{V}_f = \frac{m_f/\Delta t}{m_{\text{sat}}}\mathcal{V}_f$$

Substituting, the acceleration of the satellite during the first two seconds is determined to be

$$a_{\text{sat}} = \frac{d\mathcal{V}_{\text{sat}}}{dt} = \frac{m_f/\Delta t}{m_{\text{sat}}}\mathcal{V}_f = \frac{(100\ \text{kg})/(2\ \text{s})}{5000\ \text{kg}}(3000\ \text{m/s}) = \textbf{30 m/s}^2$$

(b) Knowing acceleration, which is constant, the velocity change of the satellite during the first two seconds is determined from the definition of acceleration $a_{\text{sat}} = d\mathcal{V}_{\text{sat}}/dt$ to be

$$d\mathcal{V}_{\text{sat}} = a_{\text{sat}}\, dt \quad \rightarrow \quad \Delta\mathcal{V}_{\text{sat}} = a_{\text{sat}}\Delta t = (30\ \text{m/s}^2)(2\ \text{s}) = \textbf{60 m/s}$$

(c) The thrust exerted on the system is simply the momentum flux of the combustion gases in the reverse direction:

$$\text{Thrust} = F_R = -\dot{m}_f\mathcal{V}_f = -(100/2\ \text{kg/s})(-3000\ \text{m/s})\left(\frac{1\ \text{kN}}{1000\ \text{kg} \cdot \text{m/s}^2}\right) = \textbf{150 kN}$$

Discussion   Note that if this satellite were attached somewhere, it would exert a force of 150 kN (equivalent to the weight of 15 tons of mass) to its support. This can be verified by taking the satellite as the system, and applying the momentum equation.

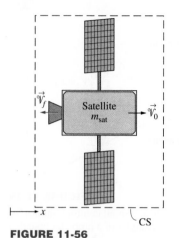

FIGURE 11-56

Schematic for Example 11-18.

This chapter deals with the Bernoulli, the energy, and the momentum equations and their applications. The *mechanical energy* is the form of energy such as kinetic, potential, and pressure energies that can be converted to mechanical work completely and directly by a mechanical device. The efficiencies of various devices are defined as

$$\eta_{pump} = \frac{\text{Mechanical energy supplied to the fluid}}{\text{Mechanical power input}} = \frac{\Delta \dot{E}_{mech, fluid}}{\dot{W}_{shaft, in}} = \frac{\dot{W}_{pump, u}}{\dot{W}_{shaft, in}}$$

$$\eta_{turbine} = \frac{\text{Mechanical power output}}{\text{Mechanical energy extracted from the fluid}} = \frac{\dot{W}_{shaft, out}}{\left| \Delta \dot{E}_{mech, fluid} \right|}$$

$$\eta_{motor} = \frac{\text{Mechanical power output}}{\text{Electrical power input}} = \frac{\dot{W}_{shaft, out}}{\dot{W}_{elect, in}}$$

$$\eta_{generator} = \frac{\text{Electrical power output}}{\text{Mechanical power input}} = \frac{\dot{W}_{elect, out}}{\dot{W}_{shaft, in}}$$

$$\eta_{pump\text{-}motor} = \eta_{pump} \, \eta_{motor} = \frac{\dot{E}_{mech, out} - \dot{E}_{mech, in}}{\dot{W}_{elect, in}} = \frac{\Delta \dot{E}_{mech, fluid}}{\dot{W}_{elect, in}} = \frac{\dot{W}_{pump, in}}{\dot{W}_{elect, in}}$$

$$\eta_{turbine\text{-}gen} = \eta_{turbine} \, \eta_{generator} = \frac{\dot{W}_{elect, out}}{\dot{E}_{mech, in} - \dot{E}_{mech, out}} = \frac{\dot{W}_{elect, out}}{\left| \Delta \dot{E}_{mech, fluid} \right|}$$

The *Bernoulli equation* is a relation between pressure, velocity, and elevation in steady, incompressible, frictionless flow and is expressed as

$$\frac{P}{\rho} + \frac{V^2}{2} + gz = \text{constant}$$

or, between any two points on a streamline, as

$$\frac{P_1}{\rho} + \frac{V_1^2}{2} + gz_1 = \frac{P_2}{\rho} + \frac{V_2^2}{2} + gz_2$$

The Bernoulli equation is an expression of mechanical energy balance and can be stated *as the sum of the kinetic, potential, and pressure energies of a fluid particle is constant along a streamline during steady flow when the compressibility and frictional effects are negligible.* Multiplying the Bernoulli equation by density gives

$$P + \rho \frac{V^2}{2} + \rho gz = \text{constant}$$

where P is the *static pressure,* which represents the actual pressure of the fluid; $\rho V^2/2$ is the *dynamic pressure,* which represents the pressure rise when the fluid in motion is brought to a stop; and ρgz is the *hydrostatic pressure,* which accounts for the effects of fluid weight on pressure. The sum of the static, dynamic, and hydrostatic pressures is called the *total pressure.* The Bernoulli equation states that *the total pressure along a streamline is constant.* The sum of the static and dynamic pressures is called the *stagnation pressure,* which represents the pressure at a point where the fluid is brought to a complete stop in a frictionless manner. The Bernoulli equation can also be represented in terms of "heads" by dividing each term by g,

$$\frac{P}{\rho g} + \frac{V^2}{2g} + z = H = \text{constant}$$

where $P/\rho g$ is the *pressure head,* which represents the height of a fluid column that produces the static pressure P; $\mathcal{V}^2/2$ is the *velocity head,* which represents the elevation needed for a fluid to reach the velocity $\mathcal{V}$ during frictionless free fall; and z is the *elevation head,* which represents the potential energy of the fluid. Also, H is the *total head* for the flow. The line that represents the sum of the static pressure and the elevation heads, $P/\rho g + z$, is called the *hydraulic grade line* (HGL), and the line that represents the total head of the fluid, $P/\rho g + \mathcal{V}^2/2g + z$, is called the *energy line* (EL).

In the absence of any heat transfer and other thermal effects, the *energy equation* for a fluid stream can be expressed as

$$\frac{P_1}{\rho} + \frac{\mathcal{V}_1^2}{2} + gz_1 + w_{\text{pump, u}} = \frac{P_2}{\rho} + \frac{\mathcal{V}_2^2}{2} + gz_2 + w_{\text{turbine}} + e_{\text{mech, loss}}$$

$$\dot{m}\left(\frac{P_1}{\rho} + \frac{\mathcal{V}_1^2}{2} + gz_1\right) + \dot{W}_{\text{pump, u}} = \dot{m}\left(\frac{P_2}{\rho} + \frac{\mathcal{V}_2^2}{2} + gz_2\right) + \dot{W}_{\text{turbine}} + \dot{E}_{\text{mech, loss}}$$

$$\frac{P_1}{\rho g} + \frac{\mathcal{V}_1^2}{2g} + z_1 + h_{\text{pump, u}} = \frac{P_2}{\rho g} + \frac{\mathcal{V}_2^2}{2g} + z_2 + h_{\text{turbine}} + h_L$$

where

$$h_{\text{pump}} = \frac{w_{\text{pump, u}}}{g} = \frac{\dot{W}_{\text{pump, u}}}{\dot{m}g}$$

$$h_{\text{turbine}} = \frac{w_{\text{turbine}}}{g} = \frac{\dot{W}_{\text{turbine}}}{\dot{m}g}$$

$$h_L = \frac{e_{\text{mech, loss}}}{g} = \frac{\dot{E}_{\text{mech, loss}}}{\dot{m}g}$$

$$\dot{E}_{\text{mech, loss}} = \dot{m}(u_2 - u_1) = \dot{m}C_v(T_2 - T_1).$$

The changes in the properties of a system and of a control volume are related to each other by the *Reynolds transport theorem* (also known as the *system-to-control-volume transformation*), expressed as

$$\frac{dB_{\text{sys}}}{dt} = \frac{d}{dt}\int_{\text{CV}} \rho b \, dV + \int_{\text{CS}} \rho b(\vec{\mathcal{V}} \cdot \hat{n}) \, dA$$

Setting $b = \vec{\mathcal{V}}$ and thus $B = m\vec{\mathcal{V}}$ in the Reynolds transport theorem and utilizing Newton's second law gives the *linear momentum equation,* expressed as

General:
$$\sum \vec{F} = \frac{d}{dt}\int_{\text{CV}} \rho\vec{\mathcal{V}} \, dV + \int_{\text{CS}} \rho\vec{\mathcal{V}}(\vec{\mathcal{V}} \cdot \hat{n}) \, dA$$

It reduces to the following special cases:

One-dimensional flow:
$$\sum \vec{F} = \frac{d}{dt}\int_{\text{CV}} \rho\vec{\mathcal{V}} \, dV + \sum_{\text{out}} \dot{m}_e\vec{\mathcal{V}}_e - \sum_{\text{in}} \dot{m}_i\vec{\mathcal{V}}_i$$

Steady, one-dimensional flow:
$$\sum \vec{F} = \sum_{\text{out}} \dot{m}_e\vec{\mathcal{V}}_e - \sum_{\text{in}} \dot{m}_i\vec{\mathcal{V}}_i$$

Steady, one-dimensional flow (one inlet, one exit):
$$\sum \vec{F} = \dot{m}(\vec{\mathcal{V}}_2 - \vec{\mathcal{V}}_1)$$

No external forces:
$$\frac{d(m\vec{\mathcal{V}})_{\text{CV}}}{dt} = \sum_{\text{in}} \dot{m}_i\vec{\mathcal{V}}_i - \sum_{\text{out}} \dot{m}_e\vec{\mathcal{V}}_e$$

The last relation is an expression for the conservation of momentum, which can be stated as *in the absence of external forces, the rate of change of the momentum of a control volume is equal to the difference between the incoming and the outgoing momentum fluxes by mass.*

REFERENCES AND SUGGESTED READING

1. Y. A. Çengel and M. A. Boles. *Thermodynamics: An Engineering Approach.* 3rd ed. New York: McGraw-Hill, 1998.

2. R. C. Dorf, ed. in chief. *The Engineering Handbook.* Boca Raton, FL: CRC Press, 1995.

3. B. R. Munson, D. F. Young, and T. Okiishi. *Fundamentals of Fluid Mechanics.* 3rd ed. New York: John Wiley & Sons, 1998.

4. M. C. Potter and D. C. Wiggert. *Mechanics of Fluids.* 2d ed. Upper Saddle River, NJ: Prentice Hall, 1997.

5. J. A. Roberson and C. L. Grove. *Engineering Fluid Mechanics.* 6th ed. New York: John Wiley & Sons, 1997.

PROBLEMS*

Mechanical Energy and Pump Efficiency

11-1C What is mechanical energy? How does it differ from thermal energy? What are the forms of mechanical energy of a fluid stream?

11-2C What is mechanical efficiency? What does a mechanical efficiency of 100 percent mean for a hydraulic turbine?

11-3C How is the combined pump-motor efficiency of a pump and motor system defined? Can the combined pump-motor efficiency be greater than either of the pump or the motor efficiency?

11-4C Define turbine efficiency, generator efficiency, and combined turbine-generator efficiency.

11-5 Consider a river flowing toward a lake at an average velocity of 3 m/s at a rate of 500 m³/s at a location 180 m above the lake surface. Determine the total mechanical energy of the river water per unit mass and the power generation potential of the entire river at that location. *Answer:* 885 MW

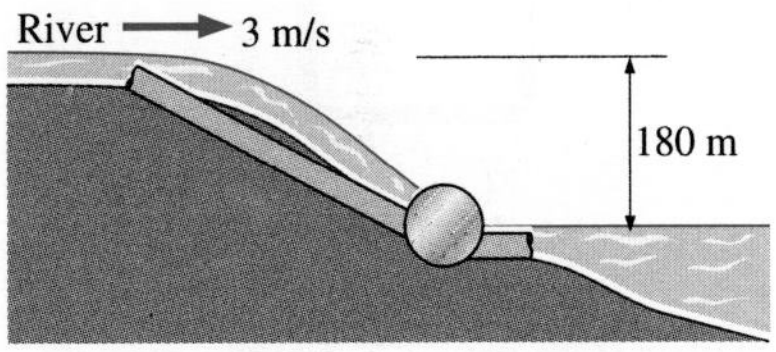

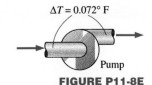

River → 3 m/s

180 m

FIGURE P11-5

11-6 Electrical power is to be generated by installing a hydraulic turbine-generator at a site 70 m below the free surface of a large water reservoir that can supply water at a rate of 1500 kg/s steadily. If the mechanical power output of the turbine is 800 kW and the electrical power generation is 750 kW, determine the turbine efficiency and the combined turbine-generator efficiency of this plant. Neglect losses in the pipes.

11-7 At a certain location, wind is blowing steadily at 25 m/s. Determine the mechanical energy of air per unit mass and the power generation potential of a wind turbine with 50-m-diameter blades at that location. Also determine the actual electric power generation assuming an overall efficiency of 30 percent. Take the air density to be 1.25 kg/m³.

11-8E A differential thermocouple with sensors at the inlet and exit of a pump indicate that the temperature of water rises 0.072°F as it flows through the pump at a rate of 1.5 ft³/s. If the shaft power input to the pump is 27 hp, determine the mechanical efficiency of the pump. *Answer:* 64.7%

$\Delta T = 0.072°$ F

Pump

FIGURE P11-8E

---

*Students are encouraged to answer *all* the concept "C" questions.

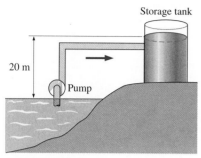

FIGURE P11-9

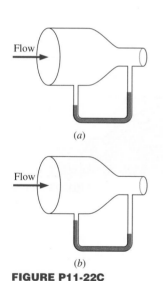

(a)

(b)
FIGURE P11-22C

FIGURE P11-23C

11-9 Water is pumped from a lake to a storage tank 20 m above at a rate of 70 L/s while consuming 20.4 kW of electric power. Disregarding any frictional losses in the pipes and any changes in kinetic energy, determine (a) the overall efficiency of the pump-motor unit and (b) the pressure difference between the inlet and the exit of the pump.

Bernoulli Equation

11-10C What is streamwise acceleration? How does it differ from normal acceleration? Can a fluid particle accelerate in steady flow?

11-11C Express the Bernoulli equation in three different ways using (a) energies, (b) pressures, and (c) heads.

11-12C What are the three major assumptions used in the derivation of the Bernoulli equation?

11-13C Define static, dynamic, and hydrostatic pressure. Under what conditions is their sum constant for a flow stream?

11-14C What is stagnation pressure? Explain how it can be measured.

11-15C Define pressure head, velocity head, and elevation head for a fluid stream and express them for a fluid stream whose pressure is P, velocity is $\mathcal{V}$, and elevation is z.

11-16C What is the hydraulic grade line? How does it differ from the energy line? Under what conditions do both lines coincide with the free surface of a liquid?

11-17C How is the location of the hydraulic grade line determined for open channel flow? How is it determined at the exit of a pipe discharging to the atmosphere?

11-18C The water level of a tank on a building roof is 20 meters above the ground. A hose leads from the tank bottom to the ground. The end of the hose has a nozzle, which is pointed straight up. What is the maximum height to which the water could rise? What factors would reduce this height?

11-19C In a certain application, a siphon must go over a high wall. Can water or oil with a specific gravity of 0.8 go over a higher wall? Why?

11-20C Explain how and why a siphon works. Someone proposes siphoning cold water over a 7-m-high wall. Is this feasible? Explain.

11-21C A student siphons water over a 8.5-m wall at sea level. He then climbs to the summit of Mount Shasta (elevation 4390 m, $P_{atm} = 58.5$ kPa) and attempts the same experiment. Comment on his prospects for success.

11-22C A glass manometer with oil as the working fluid is connected to an air duct as shown in the figure. Will the oil in the manometer move as in figure (a) or in figure (b)? Explain. What would your response be if the flow direction is reversed?

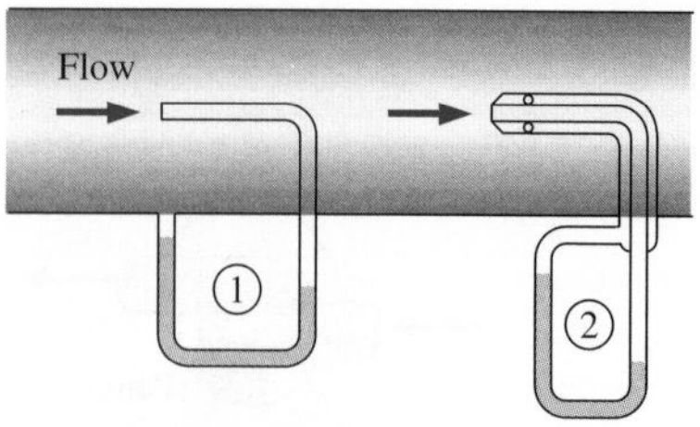

11-23C The velocity of a fluid flowing in a pipe is to be measured by two different pitot-type mercury manometers shown in the figure. Would you expect both manometers to predict the same velocity for flowing water? If not, which would be more accurate? Explain. What would your response be if air were flowing in the pipe instead of water?

11-24 In cold climates, the water pipes may freeze and burst if proper precautions are not taken. In such an occurrence, the exposed part of a pipe on the ground ruptures, and water shoots up to 22 m. Estimate the gage pressure of water in the pipe. State your assumptions and discuss if the actual pressure is more or less than the value you predicted.

11-25 A pitot tube is used to measure the velocity of an aircraft flying at 3000 m. If the differential pressure reading is 3 kPa, determine the velocity of the aircraft.

11-26 While traveling on a dirt road, the bottom of a car hits a sharp rock and a small hole develops at the bottom of its gas tank. If the height of the gasoline in the tank is 30 cm, determine the initial velocity of the gasoline at the hole. Discuss how the velocity will change with time and how the flow will be affected if the lid of the tank is closed tightly. *Answer:* 2.43 m/s

11-27E The drinking water needs of an office are met by large water bottles. One end of a 0.25-in.-diameter plastic hose is inserted into the bottle placed on a high stand, while the other end with an on/off valve is maintained 2 ft below the bottom of the bottle. If the water level in the bottle is 1.5 ft when it is full, determine how long it will take at the minimum to fill an 8-oz glass ($= 0.00835$ ft^3) (*a*) when the bottle is first opened and (*b*) when the bottle is almost empty.

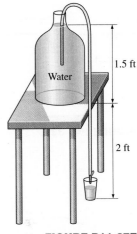

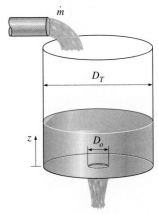

FIGURE P11-27E

11-28 A piezometer and a pitot tube are tapped into a 3-cm-diameter horizontal water pipe, and the height of the water columns are measured to be 15 cm in the piezometer and 35 cm in the pitot tube (both measured from the top surface of the pipe). Determine the velocity at the center of the pipe.

11-29 The diameter of a cylindrical water tank is D_o and its height is H. The tank is filled with water, which is open to the atmosphere. An orifice of diameter D_o with a smooth entrance (i.e., no losses) is open at the bottom. Develop a relation for the time required for the tank (*a*) to empty halfway and (*b*) to empty completely.

11-30 A pressurized tank of water has a 10-cm-diameter orifice at the bottom, where water discharges to the atmosphere. The water level is 3 meters above the outlet. The tank air pressure above the water level is 300 kPa (absolute) while the atmospheric pressure is 100 kPa. Assuming frictionless flow, determine the discharge rate of water from the tank. *Answer:* 0.168 m^3/s

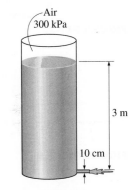

FIGURE P11-30

11-31E A siphon pumps water from a large reservoir to a lower tank that is initially empty. The tank also has a rounded orifice 20 ft below the reservoir surface where the water leaves the tank. Both the siphon and the orifice diameters are 2 in. Ignoring frictional losses, determine to what height the water will rise in the tank at equilibrium.

11-32 Water enters a tank of diameter D_T steadily at a mass flow rate of $\dot{m}_{in}$. An orifice at the bottom with diameter D_o allows water to escape. The orifice has a rounded entrance, so the frictional losses are negligible. If the tank is initially empty, (*a*) determine the maximum height that the water will reach in the tank and (*b*) obtain a relation for water height z as a function of time.

11-33E Water flows through a horizontal pipe at a rate of 1 gal/s. The pipe consists of two sections of diameters 4 in. and 2 in. with a smooth reducing section. The pressure difference between the two pipe sections is measured by

FIGURE P11-32

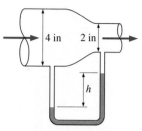

FIGURE P11-33E

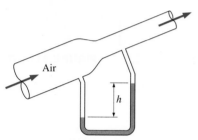

FIGURE P11-38

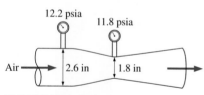

FIGURE P11-39E

a mercury manometer. Assuming frictionless flow, determine the differential height of mercury between the two pipe sections. *Answer:* 0.52 in

11-34 An airplane is flying at an altitude of 12,000 m. Determine the gage pressure at the stagnation point on the nose of the plane if the speed of the plane is 200 km/h. How would you solve this problem if the speed were 900 km/h? Explain.

11-35 The air velocity in the duct of a heating system is to be measured by a pitot tube inserted into the duct parallel to flow. If the differential height between the water columns connected to the two outlets of the pitot tube is 1.2 cm, determine (*a*) the flow velocity and (*b*) the pressure rise at the tip of the pitot tube. The air temperature and pressure in the duct are 45°C and 98 kPa.

11-36 The water in a 10-m-diameter, 2-m-high above-the-ground swimming pool is to be emptied by unplugging a 3-cm-diameter, 25-m-long horizontal pipe attached to the bottom of the pool. Determine the maximum discharge rate of water through the pipe. Also, explain why the actual flow rate will be less.

11-37 Reconsider Prob. 11-36. Determine how long it will take to empty the swimming pool completely. *Answer:* 19.7 h

11-38 Air at 110 kPa and 50°C flows upward through a 6-cm-diameter inclined duct at a rate of 30 L/s. The duct diameter is then reduced to 4 cm through a reducer. The pressure change across the reducer is measured by a water manometer. The elevation difference between the two points on the pipe where the two arms of the manometer are attached is 0.20 m. Determine the differential height between fluid levels of the two arms of the manometer.

11-39E Air is flowing through a venturi meter whose diameter is 2.6 in. at the entrance part (location 1) and 1.8 in. at the throat (location 2). The gage pressure is measured to be 12.2 psia at the entrance and 11.8 psia at the throat. Assuming frictionless flow, show that the volume flow rate can be expressed as

$$\dot{V} = A_2 \sqrt{\frac{2(P_1 - P_2)}{\rho(1 - A_2^2/A_1^2)}}$$

and determine the flow rate of air. Take the air density to be 0.075 lbm/ft³.

11-40 The water pressure in the mains of a city at a particular location is 400 kPa gage. Determine if this main can serve water to neighborhoods that are 50 m above this location.

11-41 A hand-held bicycle pump can be used as an atomizer to generate a fine mist of paint or pesticide by forcing air at a high velocity through a small hole and placing a short tube between the liquid reservoir and the high-speed air jet whose low pressure drives the liquid up through the tube. In such an

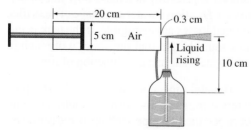

FIGURE P11-41

atomizer, the hole diameter is 0.3 cm, the vertical distance between the liquid level in the tube and the hole is 10 cm, and the bore (diameter) and the stroke of the air pump are 5 cm and 20 cm, respectively. If the atmospheric conditions are 20°C and 95 kPa, determine the minimum speed that the piston must be moved in the cylinder during pumping to initiate the atomizing effect. The liquid reservoir is open to the atmosphere.

11-42   The water level in a tank is 20 meters above the ground. A hose is connected to the bottom of the tank, and the nozzle at the end of the hose is pointed straight up. The tank cover is airtight, and the air pressure above the water surface is 2 atm gage. The system is at sea level. Determine the maximum height to which the water stream could rise.   *Answer:* 40.7 m

11-43   A pitot tube connected to a water manometer is used to measure the velocity of air. If the deflection (the vertical distance between the fluid levels in the two arms) is 5.2 cm, determine the air velocity. Take the density of air to be 1.25 kg/m³.

11-44E   The air velocity in a duct is measured by a pitot tube connected to a differential pressure gage. If the air is at 13.4 psia absolute and 70°F and the reading of the differential pressure gage is 0.15 psi, determine the air velocity.
    Answer: 142.7 ft/s

11-45   In a hydroelectric power plant, water enters the turbine nozzles at 500 kPa absolute with a low velocity. If the exit pressure is the atmospheric pressure of 100 kPa, determine the maximum velocity to which water can be accelerated by the nozzles before striking the turbine blades.

Energy Equation

11-46C   Consider the steady adiabatic flow of an incompressible fluid. Can the temperature of the fluid decrease during flow? Explain.

11-47C   Consider the steady adiabatic flow of an incompressible fluid. If the temperature of the fluid remains constant during flow, is it accurate to say that the flow is frictionless?

11-48C   What is head loss? How is it related to the mechanical energy loss?

11-49C   What is pump head? How is it related to useful power input to the pump?

11-50C   What is the kinetic energy correction factor? Is it significant?

11-51C   The water level in a tank is 20 meters above the ground. A hose is connected to the bottom of the tank, and the nozzle at the end of the hose is pointed straight up. The water stream from the nozzle is observed to rise 25 meters above the ground. Explain what may cause the water from the hose to rise above the tank level.

11-52   Underground water is to be pumped by a 70 percent efficient 3-kW submerged pump to a pool whose free surface is 30 m above the underground water level. The diameter of the pipe is 5 cm on the intake side and 7 cm on the supply side. Determine (*a*) the maximum flow rate of water and (*b*) the pressure difference across the pump. Assume the elevation difference between the pump inlet and the outlet to be negligible.

11-53   Reconsider Prob. 11-52. Determine the flow rate of water and the pressure difference across the pump if the head loss of the piping system is 5 m.

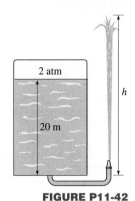
FIGURE P11-42

FIGURE P11-43

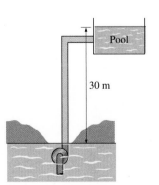
FIGURE P11-52

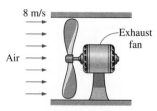

FIGURE P11-56

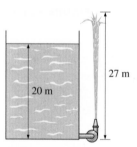

FIGURE P11-60

11-54E In a hydroelectric power plant, water flows from an elevation of 240 ft to a turbine, where electric power is generated. For an overall turbine-generator efficiency of 83 percent, determine the minimum flow rate required to generate 100 kW of electricity.   *Answer:* 370 lbm/s

11-55E Reconsider Prob. 11-54E. Determine the flow rate of water if the head loss of the piping system between the free surfaces of the source and the sink is 28 ft.

11-56 A fan is to be selected to ventilate a bathroom whose dimensions are 2 m × 3 m × 3 m. The air velocity is not to exceed 8 m/s to minimize vibration and noise. The combined efficiency of the fan–motor unit to be used can be taken to be 50 percent. If the fan is to replace the entire volume of air in 10 min, determine (*a*) the wattage of the fan–motor unit to be purchased, (*b*) the diameter of the fan casing, and (*c*) the pressure difference across the fan. Take the air density to be 1.25 kg/m³.

11-57 Water is being pumped from a large lake to a reservoir 25 m above at a rate of 25 L/s by a 10-kW (shaft) pump. If the head loss of the piping system is 7 m, determine the mechanical efficiency of the pump.   *Answer:* 78.5%

11-58 A 7-hp (shaft) pump is used to raise water to a 15-m higher elevation. If the mechanical efficiency of the pump is 75 percent, determine the maximum volume flow rate of water.

11-59 Water flows at a rate of 0.035 m³/s in a horizontal pipe whose diameter is reduced from 15 cm to 8 cm by a reducer. If the pressure at the centerline is measured to be 470 kPa and 440 kPa before and after the reducer, respectively, determine the head loss in the reducer.   *Answer:* 0.79 m

11-60 The water level in a tank is 20 m above the ground. A hose is connected to the bottom of the tank, and the nozzle at the end of the hose is pointed straight up. The tank is at sea level, and the water surface is open to the atmosphere. In the line leading from the tank to the nozzle is a pump, which increases the pressure of water. If the water jet rises to a height of 27 m from the ground, determine the minimum pressure rise supplied by the pump to the water line.

11-61 A hydraulic turbine has 60 m of head available at a flow rate of 0.25 m³/s, and its overall turbine-generator efficiency is 78 percent. Determine the electric power output of this turbine.

11-62 The demand for electric power is usually much higher during the day than it is at night, and utility companies often sell power at night at much lower prices to encourage consumers to use the available power generation

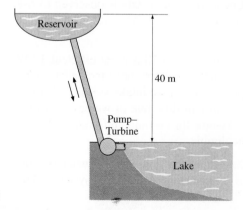

FIGURE P11-62

capacity and to avoid building new expensive power plants that will be used only a short time during peak periods. Utilities are also willing to purchase power produced during the day from private parties at a high price.

Suppose a utility company is selling electric power for $0.03/kWh at night and is willing to pay $0.08/kWh for power produced during the day. To take advantage of this opportunity, an entrepreneur is considering building a large reservoir 40 m above the lake level, pumping water from the lake to the reservoir at night using cheap power, and letting the water flow from the reservoir back to the lake during the day, producing power as the pump-motor operates as a turbine-generator during reverse flow. Preliminary analysis shows that a water flow rate of 2 m³/s can be used in either direction, and the head loss of the piping system is 4 m. The combined pump-motor and turbine-generator efficiencies are expected to be 75 percent each. Assuming the system operates for 10 h each in the pump and turbine modes during a typical day, determine the potential revenue this pump-turbine system can generate per year.

11-63 Water flows at a rate of 20 L/s through a horizontal pipe whose diameter is constant at 3 cm. The pressure drop across a valve in the pipe is measured to be 2 kPa. Determine the head loss of the valve, and the useful pumping power needed to overcome the resulting pressure drop.
Answers: 0.204 m, 40 W

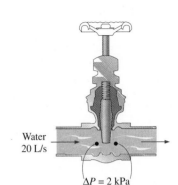

Water
20 L/s

$\Delta P = 2$ kPa
FIGURE P11-63

11-64E The water level in a tank is 66 ft above the ground. A hose is connected to the bottom of the tank, and the nozzle at the end of the hose is pointed straight up. The tank cover is airtight, but the pressure over the water surface is unknown. Determine the minimum tank air pressure (gage) that will cause a water stream from the nozzle to rise 90 ft from the ground.

11-65 A large tank is initially filled with water 2 m above the center of a sharp-edged 10-cm-diameter orifice. The tank water surface is open to the atmosphere, and the orifice drains to the atmosphere. If the total head loss in the system is 0.3 m, determine the initial discharge velocity of water from the tank.

11-66 Water enters a hydraulic turbine through a 30-cm-diameter pipe at a rate of 0.8 m³/s and exits through a 25-cm-diameter pipe. The pressure drop in the turbine is measured by a mercury manometer to be 1.2 m. For a combined turbine-generator efficiency of 83 percent, determine the net electric power output.

11-67 The velocity profile for turbulent flow in a circular pipe is usually expressed as $\mathcal{V}(r) = \mathcal{V}_{max} (1 - r/R)^{1/n}$ where $n = 7$. Determine the kinetic energy correction factor for this flow. *Answer:* 1.06

11-68 An oil pump is drawing 35 kW of electric power while pumping oil with $\rho = 860$ kg/m³ at a rate of 0.1 m³/s. The inlet and outlet diameters of the pipe are 8 cm and 12 cm, respectively. If the pressure rise of oil in the pump is measured to be 400 kPa and the motor efficiency is 90 percent, determine the mechanical efficiency of the pump.

11-69E A 73-percent efficient 8-hp pump is pumping water from a lake to a nearby pool at a rate of 1.2 ft³/s through a constant-diameter pipe. The free surface of the pool is 35 ft above that of the lake. Determine the head loss of the piping system, in ft, and the mechanical power used to overcome it.

11-70 A fireboat is to fight fires at coastal areas by drawing seawater with a density of 1030 kg/m³ through a 20-cm-diameter pipe at a rate of 0.1 m³/s and discharging it through a hose nozzle with an exit diameter of 5 cm. The total

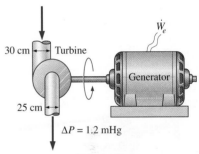

30 cm — Turbine
$\dot{W}_e$
Generator
25 cm
$\Delta P = 1.2$ mHg
FIGURE P11-66

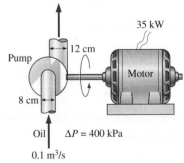

35 kW
Pump — 12 cm
Motor
8 cm
Oil $\Delta P = 400$ kPa
0.1 m³/s
FIGURE P11-68

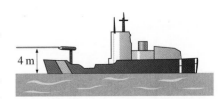

4 m
FIGURE P11-70

head loss of the system is 3 m, and the position of the nozzle is 4 m above the sea level. For a pump efficiency of 70 percent, determine the required shaft power input to the pump and the water discharge velocity.
Answers: 200 kW, 50.9 m/s

Momentum Equation

11-71C Explain the difference between the Lagrangian and Eulerian approaches to fluid mechanics.

11-72C A test engineer observes an operating jet engine bolted to a slab and draws a control volume. Is this control volume Eulerian or Lagrangian?

11-73C Is momentum a vector? If so, in what direction does it point?

11-74C What is the importance of the Reynolds transport theorem in fluid mechanics? How is the linear momentum equation obtained from it?

11-75C Write the momentum equation for steady one-dimensional flow for the case of no external forces and explain the physical significance of its terms.

11-76C In the application of the momentum equation, explain why we can ignore the atmospheric pressure and work with gage pressures only.

11-77C Two firemen are fighting a fire with identical water hoses and nozzles, except that one is holding the hose straight so that the water leaves the nozzle in the same direction it comes, while the other holds it backwards so that the water makes a U-turn before being discharged. Which fireman will experience a greater reaction force?

11-78C A rocket in space (no friction or resistance to motion) can expel gases relative to itself at some high velocity $\mathcal{V}$. Is $\mathcal{V}$ the upper limit to the rocket's ultimate velocity?

11-79C Describe in terms of momentum and air flow why a helicopter hovers.

11-80C Does it take more, equal, or less power for a helicopter to hover at the top of a high mountain than it does at sea level? Explain.

11-81C In a given location, would a helicopter require more energy to obtain the same performance in summer or winter? Explain.

11-82C A horizontal water jet from a nozzle of constant exit cross section impinges normally on a stationary vertical flat plate. A certain force F is required to hold the plate against the water stream. If the water velocity is doubled, will the necessary holding force also be doubled? Explain.

11-83C A constant velocity horizontal water jet from a stationary nozzle impinges normally on a vertical flat plate that is held in a frictionless track. As the water jet hits the plate, it begins to move due to the water force. Will the acceleration of the plate remain constant or change? Explain.

11-84C A horizontal water jet of constant velocity $\mathcal{V}$ from a stationary nozzle impinges normally on a vertical flat plate that is held in a frictionless track. As the water jet hits the plate, it begins to move due to the water force. What is the highest velocity the plate can attain? Explain.

11-85 Show that the force exerted by a liquid jet on a stationary nozzle as it leaves with a velocity $\mathcal{V}$ is proportional to $\mathcal{V}^2$ or, alternatively, to $\dot{m}^2$.

FIGURE P11-79C

FIGURE P11-83C

11-86 A horizontal water jet of constant velocity V impinges normally on a vertical flat plate and splashes off the sides in the vertical plane. The plate is moving toward the oncoming water jet with velocity $\frac{1}{2}V$. If a force F is required to maintain the plate stationary, how much force is required to move the plate toward the water jet?

11-87 A 90° elbow is used to direct water flow at a rate of 25 kg/s in a horizontal pipe upwards. The diameter of the entire elbow is 10 cm. The elbow discharges water into the atmosphere, and thus the pressure at the exit is the local atmospheric pressure. The elevation difference between the centers of the exit and the inlet of the elbow is 35 cm. The weight of the elbow and the water in it is considered to be negligible. Determine (a) the gage pressure at the center of the inlet of the elbow and (b) the anchoring force needed to hold the elbow in place.

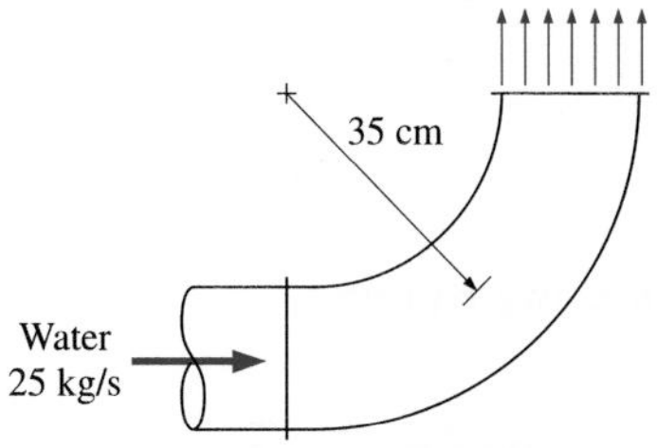

11-88 Repeat Prob. 11-87 for the case of another (identical) elbow being attached to the existing elbow so that the fluid makes a U-turn.
    Answers: (a) 6.87 kPa, (b) 213 N

11-89E A horizontal water jet impinges against a vertical flat plate at 30 ft/s, and splashes off the sides in the vertical plane. If a horizontal force of 200 lbf is required to hold the plate against the water stream, determine the volume flow rate of the water.

11-90 A reducing elbow is used to deflect water flow at a rate of 30 kg/s in a horizontal pipe upwards by an angle $\theta = 45°$ from the flow direction while accelerating it. The elbow discharges water into the atmosphere. The cross-sectional area of the elbow is 150 cm² at the inlet and 25 cm² at the exit. The elevation difference between the centers of the exit and the inlet is 40 cm. The mass of the elbow and the water in it is 50 kg. Determine the anchoring force needed to hold the elbow in place.

11-91 Repeat Prob. 11-90 for the case of $\theta = 135°$.

11-92 Water accelerated by a nozzle to 15 m/s strikes the vertical back surface of a cart moving horizontally at a constant velocity of 5 m/s in the flow direction. The mass flow rate of water is 25 kg/s. After the strike, the water stream splatters off in all directions in the plane of the back surface. (a) Determine the force that needs to be applied on the brakes of the cart to prevent it from accelerating. (b) If this force were used to generate power instead of wasting it on the brakes, determine the maximum amount of power that can be generated.   *Answers:* (a) 250 N, (b) 1.25 kW

11-93 Reconsider Prob. 11-92. If the mass of the cart is 300 kg and the brakes fail, determine the acceleration of the cart when the water first strikes it. Assume the mass of water that wets the back surface is negligible.

11-94E A 100-ft³/s water jet is moving in the positive x direction at 20 ft/s. The stream hits a stationary splitter, such that half of the flow is diverted upward at 45° and the other half is directed downward, and both streams have a final speed of 20 ft/s. Disregarding gravitational effects, determine the x and y components of the force required to hold the splitter in place against the water force.

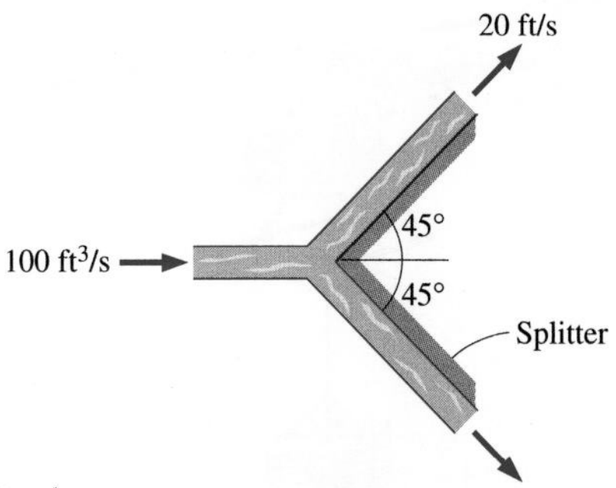

11-95 A horizontal 5-cm-diameter water jet with a velocity of 25 m/s impinges normally upon a vertical plate of mass 1000 kg. The plate is held in a frictionless track and is initially stationary. When the jet hits the plate, the plate begins to move in the direction of the jet. The water always splatters in the plane of the retreating plate. Determine (a) the acceleration of the plate

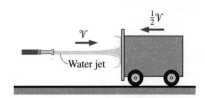
FIGURE P11-86

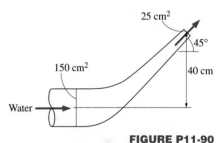
FIGURE P11-87

FIGURE P11-90

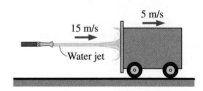
FIGURE P11-92

FIGURE P11-94E

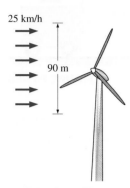
25 km/h

90 m

FIGURE P11-97

Water jet
100 ft/s
3 in
100 ft/s

FIGURE P11-98E

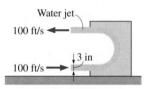
5 m³/min

FIGURE P11-100

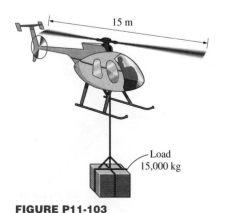
15 m

Load
15,000 kg

FIGURE P11-103

when the jet first hits it (time = 0), (b) the time it will take for the plate to reach a velocity of 12.5 m/s, and (c) the plate velocity 20 s after the jet first hits the plate. Assume the velocity of the jet relative to the plate remains constant.

11-96   Water flowing in a horizontal 30-cm-diameter pipe at 5 m/s and 300 kPa gage enters a 90° bend reducing section, which connects to a 15-cm-diameter vertical pipe. The inlet of the bend is 50 cm above the exit. Neglecting any frictional and gravitational effects, determine the net resultant force exerted on the reducer by the water.

11-97   Commercially available large wind turbines have blade span diameters as large as 100 m and generate over 3 MW of electric power at peak design conditions. Consider a wind turbine with a 90-m blade span subjected to 25 km/h steady winds. If the combined turbine-generator efficiency of the wind turbine is 32 percent, determine (a) the power generated by the turbine and (b) the horizontal force exerted by the wind on the supporting mast of the turbine. Take the density of air to be 1.25 kg/m³, and disregard frictional effects.

11-98E   A 3-in.-diameter horizontal water jet having a velocity of 100 ft/s strikes a curved plate, which deflects the water back in its original direction. How much force is required to hold the plate against the water stream?

11-99E   A 3-in.-diameter horizontal jet of water, with velocity 100 ft/s, strikes a bent plate, which deflects the water by 135° from its original direction. How much force is required to hold the plate against the water stream and what is its direction? Disregard frictional and gravitational effects.

11-100   Firemen are holding a nozzle at the end of a hose while trying to extinguish a fire. If the nozzle exit diameter is 6 cm and the water flow rate is 5 m³/min, determine (a) the average water exit velocity and (b) the horizontal resistance force required of the firemen to hold the nozzle.
   Answers: (a) 29.5 m/s, (b) 2457 N

11-101   An 8-cm-diameter horizontal jet of water with a velocity of 30 m/s strikes a flat plate that is moving in the same direction as the jet at a velocity of 10 m/s. The water splatters in all directions in the plane of the plate. How much force does the water stream exert against the plate?

11-102E   A fan with 24-in.-diameter blades moves 2000 cfm (cubic feet per minute) of air at 70°F at sea level. Determine (a) the force required to hold the fan and (b) the minimum power input required for the fan. Choose the control volume sufficiently large to contain the fan, and the gage pressure and the air velocity on the inlet side to be zero. Assume air approaches the fan through a large area with negligible velocity, and air exits the fan with a uniform velocity at atmospheric pressure through an imaginary cylinder whose diameter is the fan blade diameter.
   Answers: (a) 0.82 lbf, (b) 5.91 W

11-103   An unloaded helicopter of mass 10,000 kg hovers at sea level while it is being loaded. In the unloaded hover mode, the blades rotate at 400 rpm. The horizontal blades above the helicopter cause a 15-m-diameter air mass to move downwards at an average velocity proportional to the overhead blade rotational velocity (rpm). A load of 15,000 kg is loaded onto the helicopter, and the helicopter slowly rises. Determine (a) the volumetric air flow rate downdraft that the helicopter generates during unloaded hover and the

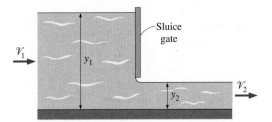

FIGURE P11-105

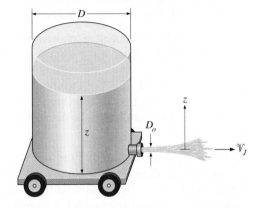

FIGURE P11-106

required power input and (*b*) the rpm of the helicopter blades to hover with the 15,000-kg load and the required power input. Take the density of atmospheric air to be 1.18 kg/m³. Assume air approaches the blades from the top through a large area with negligible velocity, and air is forced by the blades to move down with a uniform velocity through an imaginary cylinder whose base is the blade span area.

11-104 Reconsider the helicopter in Prob. 11-103, except that it is hovering on top of a 3000-m-high mountain where the air density is 0.79 kg/m³. Noting that the unloaded helicopter blades must rotate at 400 rpm to hover at sea level, determine the blade rotational velocity to hover at the higher altitude. Also determine the percent increase in the required power input to hover at 3000-m altitude relative to that at sea level.   *Answers:* 489 rpm, 22%

11-105 A sluice gate, which controls flow rate in a channel by simply raising or lowering a vertical plate, is commonly used in irrigation systems. A force is exerted on the gate due to the difference between the water heights y_1 and y_2 and the flow velocities V_1 and V_2 upstream and downstream from the gate, respectively. Disregarding the wall shear forces at the channel surfaces, develop relations for V_1, V_2, and the force acting on a sluice gate of width w during steady and uniform flow.
    Answer: $F_R = \dot{m}(V_1 - V_2) + \frac{w}{2}\rho g(y_1^2 - y_2^2)$

Review Problems

11-106 A stationary water tank of diameter D is mounted on wheels and is placed on a frictionless level surface. A smooth hole of diameter D_o near the bottom of the tank allows water to jet horizontally and rearward, and the water jet force propels the system forward. The water in the tank is much heavier than the tank-and-wheel assembly, so only the mass of water remaining in the tank needs to be considered in this problem. Considering the decrease in

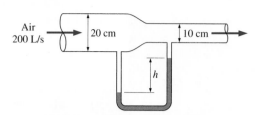

Air
200 L/s

20 cm

10 cm

h

FIGURE P11-108

the mass of water with time, develop relations for (a) the acceleration, (b) the velocity, and (c) the distance traveled by the system as a function of time.

11-107 A pressurized 2-m diameter tank of water has a 10-cm-diameter orifice at the bottom, where water discharges to the atmosphere. The water level initially is 3 meters above the outlet. The tank air pressure above the water level is maintained at 300 kPa absolute and the atmospheric pressure is 100 kPa. Assuming frictionless flow, determine (a) how long it will take for half of the water in the tank to be discharged and (b) the water level in the tank after 10 seconds.

11-108 Air flows through a pipe at a rate of 200 L/s. The pipe consists of two sections of diameters 20 cm and 10 cm with a smooth reducing section that connects them. The pressure difference between the two pipe sections is measured by a water manometer. Assuming frictionless flow, determine the differential height of water between the two pipe sections. Take the air density to be 1.20 kg/m³. *Answer:* 3.7 cm

11-109 Air at 100 kPa and 25°C flows in a horizontal duct of variable cross section. The water column in the manometer that measures the difference between two sections has a vertical displacement of 5.0 cm. If the velocity in the first section is low and the friction is negligible, determine the velocity at the second section. Also, if the manometer reading has a possible error of ±2 mm, conduct an error analysis to estimate the range of validity for the velocity found.

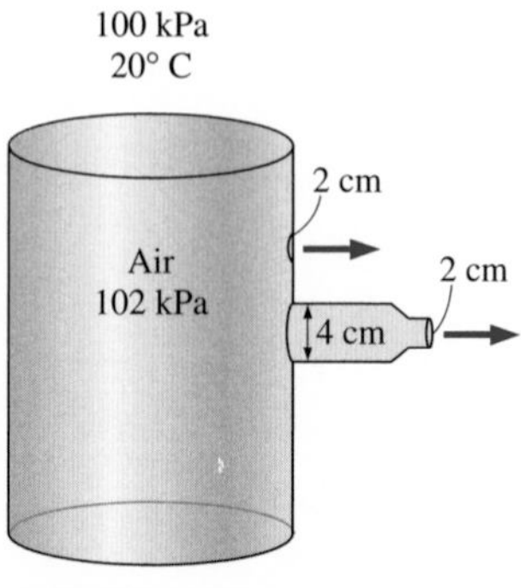

100 kPa
20° C

2 cm

Air
102 kPa

2 cm

4 cm

FIGURE P11-110

11-110 A very large tank contains air at 102 kPa at a location where the atmospheric air is at 100 kPa and 20°C. Now a 2-cm diameter tap is opened. Determine the maximum flow rate of air through the hole. What would your response be if air is discharged through a 2-m-long, 4-cm-diameter tube with a 2-cm-diameter nozzle? Would you solve the problem the same way if the pressure in the storage tank were 300 kPa?

11-111 Water is flowing through a venturi meter whose diameter is 7 cm at the entrance part and 4 cm at the throat. The pressure is measured to be 430 kPa at the entrance and 120 kPa at the throat. Assuming frictionless flow, determine the flow rate of water. *Answer:* 0.538 m³/s

11-112E The water level in a tank is 60 ft above the ground. A hose is connected to the bottom of the tank, and the nozzle at the end of the hose is pointed straight up. The tank is at sea level, and the water surface is open to the atmosphere. In the line leading from the tank to the nozzle is a pump, which increases the water pressure by 10 psia. Determine the maximum height to which the water stream could rise.

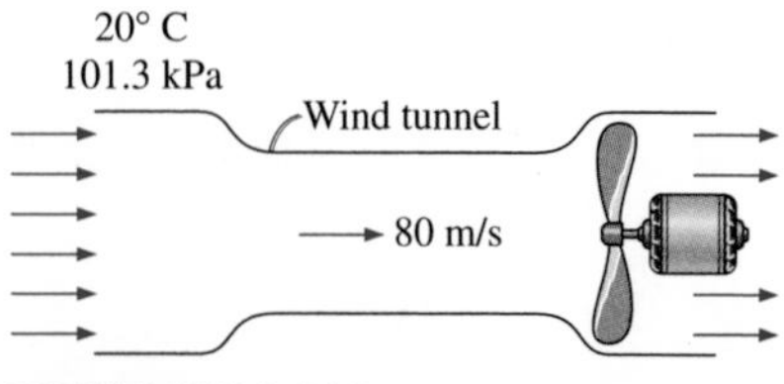

20° C
101.3 kPa

Wind tunnel

80 m/s

FIGURE P11-113

11-113 A wind tunnel draws atmospheric air at 20°C and 101.3 kPa by a large fan located near the exit of the tunnel. If the air velocity in the tunnel is 80 m/s, determine the pressure in the tunnel.

11-114 Water flows at a rate of 0.025 m³/s in a horizontal pipe whose diameter increases from 6 cm to 11 cm by an enlargement section. If the head loss across the enlargement section is 0.8 m, determine the pressure change.

11-115 A 2-m-high large tank is initially filled with water. The tank water surface is open to the atmosphere, and a sharp-edged 10-cm-diameter orifice at the bottom drains to the atmosphere through a horizontal 100-m-long pipe. If the total head loss of the system is determined to be 1.5 m, determine the initial velocity of the water from the tank. *Answer:* 3.13 m/s

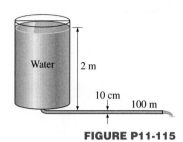

FIGURE P11-115

11-116 Reconsider Prob. 11-115. In order to drain the tank faster, a pump is installed near the tank exit. Determine the pump head input necessary to establish an average water velocity of 4 m/s when the tank is full.

11-117 Water is flowing into and discharging from a pipe U-section. The geometry, with diameters and flow rates, is indicated in the figure. At flange (1), the total absolute pressure is 200 kPa, and 30 kg/s flows into the pipe. At flange (2), the total pressure is 150 kPa. At location (3), 8 kg/s of water discharges to the atmosphere, which is at 100 kPa. Determine the total x and y forces at the two flanges connecting the pipe. Discuss the significance of gravity forces for this problem.

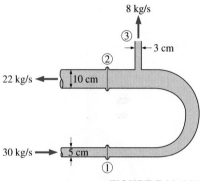

FIGURE P11-117

11-118 A tripod holding a nozzle, which directs a 5-cm-diameter stream of water from a hose, is shown in the figure. The nozzle mass is 10 kg when filled with water. The tripod is rated to provide 1800 N of holding force. A fireman was standing 60 cm behind the nozzle and was hit by the nozzle when the tripod suddenly failed and released the nozzle. You have been hired as an accident reconstructionist and, after testing the tripod, have determined that as water flow rate increased, it did collapse at 1800 N. In your final report you must state the water velocity and the flow rate consistent with the failure and the nozzle velocity when it hit the fireman.
    Answers: 30.2 m/s, 0.0593 m³/s, 14.7 m/s

11-119 Consider an airplane with a jet engine attached to the tail section that expels combustion gases at a rate of 10 kg/s with a velocity of $\mathcal{V} = 250$ m/s relative to the plane. During landing, a thrust reverser (which serves as a brake for the aircraft and facilitates landing on a short runway) is lowered in the path of the exhaust jet, which deflects the exhaust from rearward to 160°. Determine (*a*) the thrust (forward force) that the engine produces prior to the insertion of the thrust reverser and (*b*) the braking force produced after the thrust reverser is deployed.

FIGURE P11-118

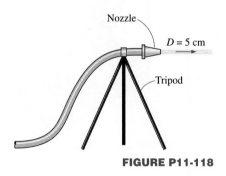

FIGURE P11-119

11-120E A spacecraft cruising in space at a constant velocity of 1500 ft/s has a mass of 18,000 lbm. To slow down the spacecraft, a solid fuel rocket is fired, and the combustion gases leave the rocket at a constant rate of 150 lbm/s at a velocity of 5000 ft/s in the same direction as the spacecraft for a period of 5 seconds. Assuming the mass of the spacecraft remains constant, determine (*a*) the deceleration of the spacecraft during this 5-second period, (*b*) the change of velocity of the spacecraft during this time period, and (*c*) the thrust exerted on the spacecraft.

11-121 An 8-cm-diameter horizontal water jet having a velocity of 30 m/s strikes a stationary flat plate. The water splatters in all directions in the plane of the plate. How much force is required to hold the plate against the water stream?

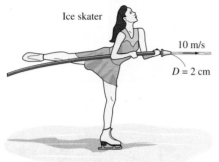

Ice skater

10 m/s

$D = 2$ cm

FIGURE P11-123

$D = 5$ cm

15 m/s

FIGURE P11-124

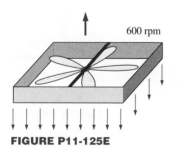

600 rpm

FIGURE P11-125E

11-122 An 8-cm-diameter horizontal jet of water, with velocity 30 m/s, strikes a cone, which deflects the water by 45° from its original direction. How much force is required to hold the cone against the water stream?

11-123 A 60-kg ice skater is standing on ice with ice skates (no friction). She is holding a flexible hose (essentially weightless) that directs a 2-cm-diameter stream of water horizontally parallel to her skates. The hose exit water velocity is 10 m/s. If she is initially standing still, determine (a) the velocity of the skater and the distance she has traveled in 5 seconds and (b) how long it will take to move 5 meters and the velocity at that moment.
Answers: (a) 2.62 m/s, 6.54 m, (b) 4.4 s, 2.3 m/s

11-124 The apocryphal Indiana Jones needs to ascend a 10-m-high building. There is a large hose filled with pressurized water hanging down from the building top. He builds a square platform and mounts four 5-cm-diameter nozzles pointing down at each corner. By connecting hose branches, a water jet with 15 m/s velocity can be produced from each nozzle. Jones, the platform, and the nozzles have a combined mass of 150 kg. Determine (a) the minimum water jet velocity needed to raise the system, (b) how long it will take for the system to rise 10 m when the water jet velocity is 15 m/s and the velocity of the system at that moment, and (c) how much higher the momentum will raise Jones if he shuts off the water at the moment the platform reaches 10 m above the ground. How much time does he have to jump from the platform to the roof? *Answers:* (a) 13.7 m/s, (b) 3.2 s, (c) 2.1 m, 1.3 s

11-125E An engineering student is considering using a fan as a levitation demonstration. He plans to face the box-enclosed fan so the air blast is directed face down through a 3-ft-diameter blade span area. The system weights 5 lbf, and he will secure the system from rotating. By increasing the power to the fan, he plans to increase the blade rpm and air exit velocity until the exhaust provides sufficient upward force to cause the box fan to hover in the air. Determine (a) the air exit velocity to produce 5 lbf, (b) the volumetric flow rate needed, and (c) the minimum mechanical power that must be supplied to the air stream. Take the air density to be 0.070 lbm/ft^3.

11-126 A soldier jumps from a plane and opens his parachute when his velocity reaches the terminal velocity $\mathcal{V}_T$. The parachute slows him down to his landing velocity of $\mathcal{V}_F$. After the parachute is deployed, the air resistance is proportional to the velocity squared (i.e., $F = k\mathcal{V}^2$). The soldier, his parachute, and his gear have a total mass of m. Show that $k = \dfrac{mg}{\mathcal{V}_F^2}$ and develop a relation for the soldier's velocity after he opens the parachute at time $t = 0$.

$$Answer: \mathcal{V} = \mathcal{V}_F \frac{\mathcal{V}_T + \mathcal{V}_F + (\mathcal{V}_T - \mathcal{V}_F)e^{-2gt/\mathcal{V}_F}}{\mathcal{V}_T + \mathcal{V}_F - (\mathcal{V}_T - \mathcal{V}_F)e^{-2gt/\mathcal{V}_F}}$$

11-127 A horizontal water jet with a flow rate of $\dot{V}$ and cross-sectional area of A will drive a covered cart of mass m_c along a level and frictionless path. The jet enters a hole at the rear of the cart, and all water that enters the cart is retained, increasing the system mass. The relative velocity between the jet of constant velocity V_J and the cart of variable velocity V is $V_J - V$. If the cart is initially empty and stationary when the jet action is initiated, develop a relation (integral form is acceptable) for cart velocity versus time.

11-128 Frictionless vertical guide rails maintain a plate of mass m_p in a horizontal position, such that it can slide freely in the vertical direction. A nozzle can direct a water stream of area A against the plate underside. The water jet splatters in the plate plane, applying an upward force against the plate. The water flow rate $\dot{m}$ (kg/s) can be controlled. Assume that times are short, so the velocity of the rising jet can be considered constant with height. (*a*) Determine the minimum mass flow rate $\dot{m}_{min}$ necessary to just levitate the plate and obtain a relation for the steady-state velocity of the upward moving plate for $\dot{m} > \dot{m}_{min}$. (*b*) At time $t = 0$, the plate is at rest, and the water jet with $\dot{m} > \dot{m}_{min}$ is suddenly turned on. Apply a force balance to the plate and obtain the integral that relates velocity to time (do not solve).

Computer, Design, and Essay Problems

11-129 Your company is setting up an experiment that involves the measurement of air flow rate in a duct, and you are to come up with proper instrumentation. Research the available techniques and devices for air flow rate measurement, discuss the advantages and disadvantages of each technique, and make a recommendation.

11-130 Computer-aided designs, the use of better materials, and better manufacturing techniques have resulted in a tremendous increase in the efficiency of pumps, turbines, and electric motors. Contact several pump, turbine, and motor manufacturers and obtain information about the efficiency of their products. In general, how does efficiency vary with rated power of these devices?

11-131 Using a hand-held bicycle pump to generate an air jet, a soda can as the water reservoir, and a straw as the tube, design and build an atomizer. Study the effects of various parameters such as the tube length, the diameter of the exit hole, and the pumping speed on performance.

11-132 Using a flexible drinking straw and a ruler, explain how you would measure the water flow velocity in a river.

FIGURE P11-126

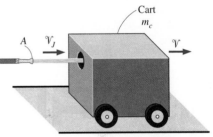

FIGURE P11-127

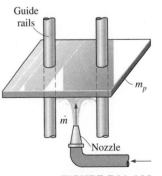

FIGURE P11-128

Flow in Pipes

Fluid flow in circular and noncircular pipes is commonly encountered in practice. The hot and cold water that we use in our homes is pumped through pipes. Water in a city is distributed by extensive piping networks. Oil and natural gas are transported hundreds of miles by large pipelines. Blood is carried throughout our bodies by veins. The cooling water in an engine is transported by hoses to the tubes in the radiator where it is cooled as it flows. Thermal energy in a hydronic space heating system is transferred to the circulating water in the boiler, and then it is transported to the desired locations in pipes.

Fluid flow is classified as *external* and *internal,* depending on whether the fluid is forced to flow over a surface or in a pipe. Internal and external flows exhibit very different characteristics. In this chapter we consider *internal flow* where the pipe is completely filled with the fluid and flow is driven primarily by a pressure difference. This should not be confused with the *open-channel flow* where the pipe is partially filled by the fluid and thus the flow is partially bounded by solid surfaces, as in an irrigation ditch, and flow is driven by gravity alone.

We start this chapter with a general physical description of internal flow and the *velocity boundary layer.* We continue with the discussion of the dimensionless *Reynolds number* and its physical significance. We then discuss the characteristics of flow inside pipes and introduce the *pressure drop* correlations associated with it for both laminar and turbulent flows. Finally, we present the minor losses and determine the pressure drop and pumping power requirements for real-world piping systems.

Circular pipe

Water
50 atm

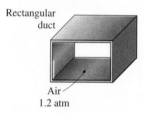

Rectangular
duct

Air
1.2 atm

FIGURE 12-1

Circular pipes can withstand large pressure differences between the inside and the outside without undergoing any distortion, but the noncircular pipes cannot.

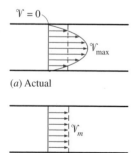

$\mathcal{V} = 0$

$\mathcal{V}_{max}$

(a) Actual

$\mathcal{V}_m$

(b) Idealized

FIGURE 12-2

Actual and idealized velocity profiles for flow in a tube (the mass flow rate of the fluid is the same for both cases).

12-1 ■ INTRODUCTION

Liquid or gas flow through *pipes* or *ducts* is commonly used in heating and cooling applications and fluid distribution networks. The fluid in such applications is forced to flow by a fan or pump through a flow section. We will pay particular attention to *friction*, which is directly related to the *pressure drop* and *head loss* during flow through pipes and ducts. The pressure drop is then used to determine the pumping power requirement. A typical piping system involves pipes of different diameters connected to each other by various fittings or elbows, valves to control the flow rate, and pumps to pressurize the fluid.

The terms *pipe, duct, tube,* and *conduit* are usually used interchangeably for flow sections. In general, flow sections of circular cross section are referred to as *pipes* (especially when the fluid is a liquid), and the flow sections of noncircular cross section as *ducts* (especially when the fluid is a gas). Small-diameter pipes are usually referred to as *tubes*. Given this uncertainty, we will use more descriptive phrases (such as *a circular pipe* or *a rectangular duct*) whenever necessary to avoid any misunderstandings.

You have probably noticed that most fluids, especially liquids, are transported in *circular pipes*. This is because pipes with a circular cross section can withstand large pressure differences between the inside and the outside without undergoing any distortion. *Noncircular pipes* are usually used in applications such as the heating and cooling systems of buildings where the pressure difference is relatively small, the manufacturing and installation costs are lower, and the available space is limited for duct work (Fig. 12-1).

Although the theory of fluid flow is reasonably well understood, theoretical solutions are obtained only for a few simple cases such as fully developed laminar flow in a circular pipe. Therefore, we must rely on the experimental results and empirical relations obtained for most fluid flow problems rather than closed-form analytical solutions. Noting that experimental results are obtained under carefully controlled laboratory conditions, and that no two piping systems are exactly alike, we must not be so naive as to view the results obtained as "exact." An error of 10 percent (or more) in the friction coefficient calculated using the relations in this chapter is the "norm" rather than the "exception."

The fluid velocity in a pipe changes from *zero* at the surface because of the *no-slip condition* to a maximum at the pipe center. In fluid flow, it is convenient to work with an *average* or *mean* velocity $\mathcal{V}_m$, which remains constant in incompressible flow when the cross-sectional area of the pipe is constant. The mean velocity in actual heating and cooling applications may change somewhat because of changes in density with temperature. But, in practice, we evaluate the fluid properties at some average temperature and treat them as constants. The convenience in working with constant properties usually more than justifies the slight loss in accuracy.

Also, the friction between the fluid layers in a pipe does cause a slight rise in fluid temperature as a result of the mechanical energy being converted to sensible heat energy. But this temperature rise due to *fictional heating* is usually too small to warrant any consideration in calculations and thus is disregarded. For example, in the absence of any heat transfer, no noticeable difference will be detected between the inlet and exit temperatures of water flowing in a pipe. The primary consequence of friction in fluid flow is pressure drop, and thus any significant temperature change in the fluid is due to heat transfer.

The value of the mean velocity $\mathcal{V}_m$ is determined from the requirement that the *conservation of mass* principle be satisfied (Fig. 12-2). That is, the mass flow rate through the pipe is evaluated using the mean velocity $\mathcal{V}_m$ from

$$\dot{m} = \rho \mathcal{V}_m A_c \qquad \text{(kg/s)} \qquad (12\text{-}1)$$

where ρ is the density of the fluid and A_c is the cross-sectional area, which is equal to $A_c = \pi D^2/4$ for a circular pipe.

12-2 ■ LAMINAR AND TURBULENT FLOW

If you have been around smokers, you probably have noticed that the cigarette smoke rises in a smooth plume for the first few centimeters and then starts fluctuating randomly in all directions as it continues its journey toward the lungs of others (Fig. 12-3). Likewise, a careful inspection of flow in a pipe reveals that the fluid flow is streamlined at low velocities but turns chaotic as the velocity is increased beyond a critical value. The flow regime in the first case is said to be **laminar,** characterized by *smooth streamlines* and *highly ordered motion* due to the fluid flowing in laminae or layers, and **turbulent** in the second case, where it is characterized by *velocity fluctuations* and *highly disordered motion.* The **transition** from laminar to turbulent flow does not occur suddenly; rather, it occurs over some range of velocity where the flow hesitates between laminar and turbulent flows before it becomes fully turbulent. Most flows encountered in practice are turbulent. Laminar flow is encountered when highly viscous fluids such as oils flow in small tubes or narrow passages.

We can verify the existence of these laminar, transition, and turbulent flow regimes by injecting a dye streak into the flow in a glass tube, as the British scientist Osborn Reynolds (1842–1912) did for the first time. We will observe that the dye streak will form a *straight and smooth line* at low flow rates when the flow is laminar (we may see some blurring because of molecular diffusion), will have *bursts of fluctuations* in the transition regime, and will *zigzag rapidly and randomly* when the flow is turbulent (Fig. 12-4). These zigzags and the dispersion of the dye are indicative of the fluctuations in the main flow and the rapid mixing of fluid chunks from adjacent layers.

Reynolds Number

The transition from laminar to turbulent flow depends on the *geometry, surface roughness, mean fluid velocity, surface temperature,* and *type of fluid,* among other things. After exhaustive experiments, Osborn Reynolds discovered in 1883 that the flow regime depends mainly on the ratio of the *inertia forces* to *viscous forces* in the fluid. This ratio is called the **Reynolds number** and is expressed for internal flow in circular pipes as (Fig. 12-5)

$$\text{Re} = \frac{\text{Inertia forces}}{\text{Viscous forces}} = \frac{\mathcal{V}_m D}{\nu} \qquad (12\text{-}2)$$

where

$\mathcal{V}_m$ = mean fluid velocity, m/s
D = characteristic length of the geometry (diameter in this case), m
$\nu = \mu/\rho$ = kinematic viscosity of the fluid, m²/s

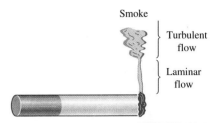

FIGURE 12-3

Laminar and turbulent flow regimes of cigarette smoke.

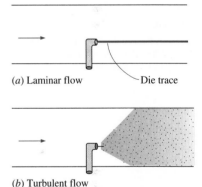

(a) Laminar flow   Die trace

(b) Turbulent flow

FIGURE 12-4

The behavior of colored fluid injected into the flow in laminar and turbulent flows in a tube.

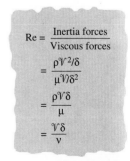

$$\text{Re} = \frac{\text{Inertia forces}}{\text{Viscous forces}}$$
$$= \frac{\rho \mathcal{V}^2/\delta}{\mu \mathcal{V}/\delta^2}$$
$$= \frac{\rho \mathcal{V} \delta}{\mu}$$
$$= \frac{\mathcal{V} \delta}{\nu}$$

FIGURE 12-5

The Reynolds number can be viewed as the ratio of the inertia forces to viscous forces acting on a fluid volume element.

Note that the Reynolds number is a *dimensionless* quantity. Also note that *kinematic viscosity* ν differs from dynamic viscosity μ by the factor ρ. Kinematic viscosity has the unit m²/s, and can be viewed as *viscous diffusivity.*

At *large* Reynolds numbers, the inertia forces, which are proportional to the density and the velocity of the fluid, are large relative to the viscous forces, and thus the viscous forces cannot prevent the random and rapid fluctuations of the fluid. At *small* Reynolds numbers, however, the viscous forces are large enough to overcome the inertia forces and to keep the fluid "in line." Thus, the flow is *turbulent* in the first case and *laminar* in the second.

It certainly is desirable to have precise values of Reynolds numbers for laminar, transitional, and turbulent flows, but this is not the case in practice. This is because the transition from laminar to turbulent flow also depends on the degree of disturbance of the flow by *surface roughness, pipe vibrations,* and the *fluctuations in the flow.* Under most practical conditions, the flow in a pipe is laminar at Re < 2300, turbulent at Re > 4000, and transitional in between. That is,

Laminar flow:	Re < 2300
Transitional flow:	2300 ≤ Re ≤ 4000
Turbulent flow:	Re > 4000

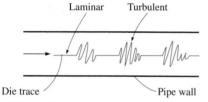

Laminar   Turbulent

Die trace   Pipe wall

FIGURE 12-6

In the transitional flow region of 2300 ≤ Re ≤ 4000, the flow switches between laminar and turbulent randomly.

In transitional flow, the flow switches between laminar and turbulent randomly (Fig. 12-6). It should be kept in mind that laminar flow can be maintained at much higher Reynolds numbers in very smooth pipes by avoiding flow disturbances and pipe vibrations. In such carefully controlled experiments, laminar flow has been maintained at Reynolds numbers of up to 100,000. For inviscid flow, the Reynolds number is "infinity" since the viscosity is assumed to be zero.

Entrance Region and the Entry Length

Consider a fluid entering a circular pipe with uniform velocity. The fluid particles in the layer in contact with the surface of the pipe will come to a complete stop because of the no-slip condition. This motionless layer will *slow down* the particles of the neighboring fluid layer as a result of friction between the particles of these two adjoining fluid layers at different velocities. This fluid layer will then slow down the molecules of the next layer, and so on. To make up for this velocity reduction, the velocity of the fluid at the midsection of the pipe will have to increase to keep the mass flow rate through the pipe constant. As a result, a velocity gradient develops along the pipe. The region of the flow in which the effects of the viscous shearing forces caused by fluid viscosity are felt is called the **velocity boundary layer** or just the **boundary layer.** The hypothetical boundary surface divides the flow in a cross section of the pipe into two regions: the **boundary layer region,** in which the viscous effects and the velocity changes are significant, and the **inviscid flow region,** in which the frictional effects are negligible and the velocity remains essentially constant.

The thickness of the boundary layer increases in the flow direction until the boundary layer reaches the pipe center and thus fills the entire pipe, as shown in Fig. 12-7. The region from the pipe inlet to the point at which the boundary layer merges at the centerline is called the **hydrodynamic entry region,** and the length of this region is called the **hydrodynamic entry length** L_h, also called the **entrance length.** The region beyond the hydrodynamic

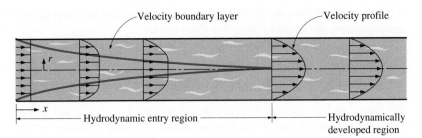

Velocity boundary layer   Velocity profile

Hydrodynamic entry region —|— Hydrodynamically developed region

FIGURE 12-7

The development of the velocity boundary layer in a tube. (The developed mean velocity profile will be parabolic in laminar flow, as shown, but somewhat blunt in turbulent flow.)

entry region in which the velocity profile is fully developed and remains unchanged is called the **hydrodynamically developed region.** The flow is said to be **fully developed** when the normalized temperature profile also remains constant. Hydrodynamically developed flow is equivalent to fully developed flow when the fluid in the pipe is not heated or cooled since the fluid temperature in this case remains constant throughout. The mean velocity profile in the hydrodynamically developed region is *parabolic* in laminar flow and somewhat *flatter* in turbulent flow.

The hydrodynamic entry length in *laminar flow* in a pipe is given approximately as

$$L_{h,\,\text{laminar}} \approx 0.06\,\text{Re}\,D \qquad (12\text{-}3)$$

For Re = 20, the hydrodynamic entry length is about the size of the diameter, but increases linearly with the velocity. In the limiting case of Re ≈ 2300, the hydrodynamic entry length is 138D.

In *turbulent flow,* the intense mixing during random fluctuations usually overshadows the effects of momentum diffusion. For smooth pipes, the hydrodynamic entry length is expressed as

$$L_{h,\,\text{turbulent}} \approx 4.4D(\text{Re})^{1/6} \qquad (12\text{-}4)$$

The entry length is much shorter in turbulent flow, as expected. It is 18D at Re = 4000 and increases to 44D at Re = 1,000,000. The entry length in turbulent flow is usually less than 50 diameters, and its dependence on the Reynolds number is weaker.

The friction factor (defined later by Eq. 12-19) is related to the shear stress at the surface, which is related to the slope of the velocity profile at the surface. Noting that the mean velocity profile remains unchanged in the hydrodynamically developed region, the friction factor also remains constant in that region. Thus, we conclude that *the friction factor in the developed flow region remains constant* (Fig. 12-8).

Consider a fluid flowing through a pipe. The friction factor is *highest* at the pipe inlet where the thickness of the boundary layer is zero and decreases gradually to the fully developed value, as shown in Fig. 12-9. Therefore, the pressure drop is *higher* in the entry regions of a pipe, and the effect of the entry region is always to *enhance* the average friction factor for the entire pipe. This enhancement can be significant for short pipes but negligible for long ones.

Correlations for the friction factor for the entry regions are available in the literature. However, the pipes used in practice are usually many times the length of entry regions and thus the flow through the pipes is usually assumed to be fully developed for the entire length of the pipe. This approach, which

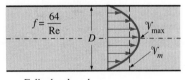

$$f = \frac{64}{\text{Re}}$$

Fully developed laminar flow

FIGURE 12-8

In a laminar flow in a pipe, the friction factor remains constant in the developed region.

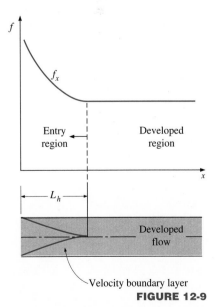

FIGURE 12-9

The variation of the friction factor in the flow direction for flow in a pipe.

we will also use for simplicity, gives *reasonable* values for long pipes, but it underpredicts the friction factor for short ones.

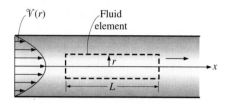

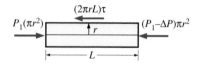

FIGURE 12-10

Free-body diagram of a cylindrical fluid element of radius r and length L oriented coaxially with a horizontal pipe in fully developed steady flow.

12-3 ■ FULLY DEVELOPED LAMINAR FLOW IN PIPES

We mentioned earlier that flow in pipes is laminar for Re $\lesssim 2300$. Further, the flow is fully developed if the pipe is sufficiently long (relative to the entry length) and its diameter is constant. The velocity profile at a cross section in that case remains unchanged in the flow direction (this is also true for fully developed turbulent flow). Below we will obtain a relation for the velocity distribution. Then we will use it to obtain relations for the pressure drop and head loss in the pipe. Another important aspect of the analysis below is that it is one of the few available for viscous flow.

Consider incompressible, steady, fully developed flow in a straight horizontal circular pipe. Each fluid particle moves at a constant velocity along a streamline. There is no acceleration in this case since the flow is steady and the velocity profile remains unchanged in fully developed flow.

Now consider a *cylindrical fluid element* of radius r and length L oriented coaxially with the pipe, as shown in Fig. 12-10. Noting that the pressure force acting on a submerged plane surface is the product of the pressure at the centroid of the surface and the surface area, the pressure force acting on either end surface of the element will be equal to the pressure at the center of the element (which is on the centerline of the pipe) times the area of the face of the element. Pressure decreases in the flow direction because of viscous effects, and we let the pressure drop across the fluid element be ΔP (a positive quantity). If the pressure on the left surface of the element is P_1, then the pressure on the right surface becomes $P_2 = P_1 - \Delta P$.

There is also the shear stress τ acting on the lateral surface of the cylindrical element in the opposite direction to flow. Noting that $\Sigma F_x = ma_x = 0$ since there is no acceleration, a force balance on the fluid element in the flow (or x) direction gives

$$P_1 \pi r^2 - (P_1 - \Delta P)\pi r^2 - 2\pi r L \tau = 0 \qquad (12\text{-}5)$$

which indicates that in fully developed pipe flow, the viscous and pressure forces balance each other. Simplifying gives

$$\frac{\Delta P}{L} = \frac{2\tau}{r} \qquad \text{or} \qquad \tau = \frac{\Delta P}{2L} r \qquad (12\text{-}6\text{a,b})$$

Thus we conclude that *the shear stress is proportional to the radial distance* since both ΔP and L are independent of r. Therefore, the shear stress varies linearly from zero at the centerline ($\tau = 0$ at $r = 0$) to a maximum of τ_w at the pipe surface ($r = R = D/2$). Expressing Eq. 12-6b at $r = D/2$ gives

$$\frac{\Delta P}{L} = \frac{2\tau_w}{D/2} \qquad \rightarrow \qquad \Delta P = \frac{4L}{D} \tau_w \qquad (12\text{-}7)$$

where τ_w is the wall shear stress. Note that *the pressure drop along the pipe is proportional to the wall shear stress and the pipe length, and is inversely proportional to the pipe diameter* (Fig. 12-11).

The analysis so far is valid for both laminar and turbulent flow. For the laminar flow of a Newtonian fluid, we can carry the analysis further by utilizing the proportionality of the shear stress to the velocity gradient expressed as

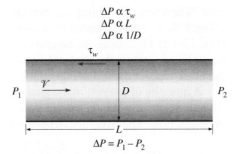

$$\Delta P \, \alpha \, \tau_w$$
$$\Delta P \, \alpha \, L$$
$$\Delta P \, \alpha \, 1/D$$

$$\Delta P = P_1 - P_2$$

FIGURE 12-11

Pressure drop in pipe flow is proportional to wall shear stress and pipe length and is inversely proportional to pipe diameter.

$$\tau = -\mu \frac{d\mathcal{V}}{dr} \tag{12-8}$$

The quantity $d\mathcal{V}/dr$ is negative in pipe flow, and the negative sign is included to obtain positive values for τ. Combining Eqs. 12-6a and 12-8 gives

$$\frac{d\mathcal{V}}{dr} = -\left(\frac{\Delta P}{2\mu L}\right)r \tag{12-9}$$

Integrating both sides with respect to r gives

$$\mathcal{V}(r) = -\left(\frac{\Delta P}{4\mu L}\right)r^2 + C \tag{12-10}$$

where C is the integration constant. Applying the no-slip condition at the wall ($\mathcal{V} = 0$ at $r = R$) gives $C = [\Delta P/(4\pi L)]R^2$. Substituting, the *velocity profile* is determined to be

$$\mathcal{V}(r) = \frac{\Delta P R^2}{4\mu L}\left(1 - \frac{r^2}{R^2}\right) = \mathcal{V}_{max}\left(1 - \frac{r^2}{R^2}\right) \tag{12-11}$$

where

$$\mathcal{V}_{max} = \mathcal{V}_{center} = \frac{\Delta P R^2}{4\mu L} \qquad \text{(Laminar flow)} \tag{12-12}$$

is the **centerline velocity.** Therefore, the velocity profile in fully developed laminar flow in a pipe is *parabolic* with a maximum at the centerline and a minimum at the wall. The **average velocity** is determined from

$$\mathcal{V}_m = \frac{\int_{A_c} \mathcal{V}(r)\, dA}{A_c} = \frac{\int_0^R \frac{\Delta P R^2}{4\mu L}\left(1 - \frac{r^2}{R_2}\right)2\pi r\, dr}{\pi R^2} = \frac{\Delta P R^2}{8\mu L} = \frac{\mathcal{V}_{max}}{2} \tag{12-13}$$

Therefore, *the average velocity is one-half of the maximum velocity.* Then the **volume flow rate** becomes

$$\dot{\mathcal{V}} = \mathcal{V}_m A_c = \frac{\Delta P R^2}{8\mu L}\pi R^2 = \frac{\pi R^4 \, \Delta P}{8\mu L} = \frac{\pi D^4 \, \Delta P}{128\mu L} \tag{12-14}$$

This equation is known as the **Poiseuille's Law,** and this flow is called the **Hagen-Poiseuille flow** in honor of the works of G. Hagen (1797–1839) and J. Poiseuille (1799–1869) on the subject. Note from Eq. 12-14 that *for a specified flow rate, the pressure drop and thus the required pumping power are proportional to the length of the pipe and the viscosity of the fluid but are inversely proportional to the fourth power of the radius (or diameter) of the*

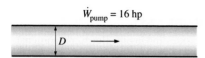

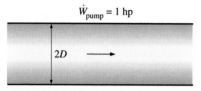

FIGURE 12-12

The pumping power requirement for a laminar flow piping system can be reduced by a factor of 16 by doubling the pipe diameter.

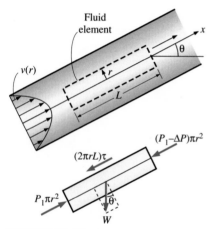

FIGURE 12-13

Free-body diagram of a cylindrical fluid element of radius r and length L oriented coaxially with an inclined pipe in fully developed steady flow.

Horizontal pipe: $\dot{V} = \dfrac{\Delta P \pi D^4}{32\mu L}$

Inclined pipe: $\dot{V} = \dfrac{(\Delta P - \rho gL \sin\theta)\pi D^4}{128\mu L}$

Uphill flow: $\theta > 0$ and $\sin\theta > 0$
Downhill flow: $\theta < 0$ and $\sin\theta < 0$

FIGURE 12-14

The relations developed for horizontal pipes can also be used for inclined pipes by replacing ΔP with $\Delta P - \rho gL \sin\theta$.

pipe. Therefore, the pumping power requirement for a piping system in laminar flow can be reduced by a factor of 16 by doubling the pipe diameter (Fig. 12-12). Of course, the benefits of the reduction in the energy costs must be weighed against the increased cost of construction due to using a larger diameter pipe.

A quantity of interest in the analysis of pipe flow is the *pressure drop* ΔP since it is directly related to the power requirements of the fan or pump to maintain flow. The **pressure drop** can be expressed from Eq. 12-13 as

$$\Delta P = \frac{8\mu L \mathcal{V}_m}{R^2} = \frac{32\mu L \mathcal{V}_m}{D^2} \qquad \text{(Laminar flow)} \qquad (12\text{-}15)$$

In most practical problems, we are not concerned with the variation of velocity with radial distance, and the mean velocity is simply indicated by $\mathcal{V}$ without the subscript m.

Inclined Pipes

Relations for inclined pipes are obtained in a similar manner from a force balance in the x direction (along the flow), as shown in Fig. 12-13. The only additional force in this case is the component of the fluid weight in the flow direction, whose magnitude is

$$W_x = W \sin\theta = \rho g(\pi r^2 L)\sin\theta \qquad (12\text{-}16)$$

where θ is the angle between the flow direction and the horizontal. Note that $\theta > 0$ and thus $\sin\theta > 0$ for uphill flow, and $\theta < 0$ and thus $\sin\theta < 0$ for downhill flow. A force balance in this case gives, after simplification,

$$\frac{\Delta P - \rho gL \sin\theta}{L} = \frac{2\tau}{r} \qquad (12\text{-}17)$$

which is identical to Eq. 12-7, except that ΔP is replaced by $\Delta P - \rho gL \sin\theta$. Therefore, *all the results obtained above for horizontal pipes can also be used for inclined pipes provided that ΔP is replaced by $\Delta P - \rho gL \sin\theta$* (Fig. 12-14). For example, the *average velocity* and the *volume flow rate* relations for inclined pipes are

$$\mathcal{V}_m = \frac{(\Delta P - \rho gL \sin\theta)D^2}{32\mu L} \quad \text{and} \quad \dot{V} = \frac{(\Delta P - \rho gL \sin\theta)\pi D^4}{128\mu L} \qquad (12\text{-}18\text{a,b})$$

In inclined pipes, the combined effect of pressure difference and gravity drives the flow. Gravity helps downhill flow but opposes uphill flow. Therefore, much greater pressure differences need to be applied to maintain a specified flow rate in uphill flow although this becomes important only for liquids, because the density of gases is generally low. In the special case of *no flow* ($\mathcal{V} = 0$), we have $\Delta P = \rho gL \sin\theta$, which is what we would obtain from fluid statics.

Pressure Drop and Head Loss

In practice, it is found convenient to express the *pressure drop* due to frictional effects for internal flows (laminar or turbulent flows, circular or noncircular pipes, smooth or rough surfaces) as (Fig. 12-15)

$$\Delta P = f \frac{L}{D} \frac{\rho \mathcal{V}_m^2}{2} \qquad (12\text{-}19)$$

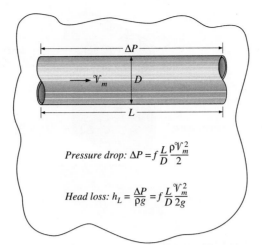

Pressure drop: $\Delta P = f \dfrac{L}{D}\dfrac{\rho \mathcal{V}_m^2}{2}$

Head loss: $h_L = \dfrac{\Delta P}{\rho g} = f \dfrac{L}{D}\dfrac{\mathcal{V}_m^2}{2g}$

FIGURE 12-15

The relation for pressure drop (and head loss) is one of the most general relations in fluid mechanics, and it is valid for laminar or turbulent flows, circular or noncircular pipes, and smooth or rough surfaces.

where $\rho \mathcal{V}_m^2/2$ is the *dynamic pressure* and the dimensionless quantity f is the **friction factor** (also called the **Darcy friction factor** after French engineer Henry Darcy, 1803–1858, who first studied experimentally the effects of roughness on pipe resistance). It should not be confused with the *Fanning friction factor,* which is defined as $f/4$.

Equation 12-19 gives the pressure drop for a flow section of length L provided that (1) the flow section is horizontal so that there are no hydrostatic or gravity effects, (2) the flow section does not involve any work devices such as a pump or a turbine since they change the fluid pressure, and (3) the cross-sectional area of the flow section is constant and thus the mean flow velocity is constant. If one or more of these conditions are not satisfied, the pressure drop should be determined from the energy equation.

Setting Eqs. 12-15 and 12-19 equal to each other and solving for f gives the friction factor for the *fully developed laminar flow in a circular pipe* to be

$$f = \frac{64\mu}{\rho D \mathcal{V}_m} = \frac{64}{\mathrm{Re}} \qquad \text{(circular pipe)} \qquad (12\text{-}20)$$

Alternately, the friction factor can also be expressed in terms of the wall shear stress (Eq. 12-7) as $f = 8\tau_w/(\rho \mathcal{V}_m^2)$. Equation (12-20) shows that *in laminar flow, the friction factor is a function of the Reynolds number only and is independent of the surface roughness of the pipe surface.* Once the pressure drop is available, the required pumping power is determined from

$$\dot{W}_{\text{pump}} = \dot{V}\Delta P \qquad (12\text{-}21)$$

where $\dot{V}$ is the volume flow rate of flow.

In the analysis of piping systems, pressure losses are commonly expressed in terms of the *equivalent fluid column height,* called the **head loss** h_L. Noting that $\Delta P = \rho g h$ in fluid statics, the *pipe head loss* is obtained by dividing ΔP by ρg to give

$$h_L = \frac{\Delta P}{\rho g} = f \frac{L}{D}\frac{\mathcal{V}_m^2}{2g} \qquad (12\text{-}22)$$

The *head loss* h_L represents *the additional height that the fluid needs to be raised by a pump in order to overcome the frictional losses in the pipe.* The

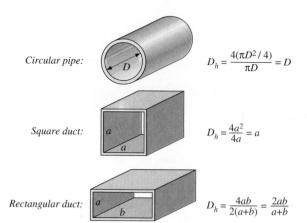

Circular pipe: $D_h = \dfrac{4(\pi D^2/4)}{\pi D} = D$

Square duct: $D_h = \dfrac{4a^2}{4a} = a$

Rectangular duct: $D_h = \dfrac{4ab}{2(a+b)} = \dfrac{2ab}{a+b}$

FIGURE 12-16

The hydraulic diameter
$D_h = 4A_c/p$ is defined such that it reduces
to ordinary diameter for circular pipes.

TABLE 12-1

Friction factor for fully developed laminar flow in noncircular pipes ($D_h = 4A_c/p$ and $\mathrm{Re} = \mathcal{V}_m D_h / \nu$)

Cross section of pipe	a/b or θ°	Friction factor f
Rectangle	a/b	
	1	56.92/Re
	2	62.20/Re
	3	68.36/Re
	4	72.92/Re
	6	78.80/Re
	8	82.32/Re
	∞	96.00/Re
Ellipse	a/b	
	1	64.00/Re
	2	67.28/Re
	4	72.96/Re
	8	76.60/Re
	16	78.16/Re
Triangle	θ	
	10°	50.80/Re
	30°	52.28/Re
	60°	53.32/Re
	90°	52.60/Re
	120°	50.96/Re

head loss is caused by viscosity, and it is directly related to the wall shear stress. For the ideal inviscid flow, the head loss is zero since there is no viscous dissipation. Again, Eqs. 12-21 and 12-22 are valid for both laminar and turbulent flows in circular and noncircular horizontal or inclined pipes.

The friction factor f is given in Table 12-1 for *fully developed laminar flow* in pipes of various cross sections. The Reynolds number for flow in these pipes is based on the **hydraulic diameter** D_h defined as (Fig. 12-16)

$$D_h = \frac{4A_c}{p} \qquad (12\text{-}23)$$

where A_c is the cross-sectional area of the pipe and p is its perimeter. The hydraulic diameter is defined such that it reduces to ordinary diameter D for circular pipes since $A_c = \pi D^2/4$ and $p = \pi D$. In laminar flow, the effect of *surface roughness* on the friction factor is negligible.

EXAMPLE 12-1 Flow Rates in Horizontal and Inclined Pipes

Oil at 20°C ($\rho = 888$ kg/m³ and $\mu = 0.8$ kg/m · s) is flowing in a 5-cm-diameter, 40-m-long pipe steadily (Fig. 12-17). During the flow, the pressures at the pipe inlet and exit are measured to be 745 kPa and 97 kPa, respectively. Determine the flow rate of oil through the pipe assuming the pipe is (*a*) horizontal, (*b*) inclined 15° upwards, and (*c*) inclined 15° downwards. Also verify that the flow through the pipe is laminar.

Solution The pressure readings across a pipe are given. The flow rates are to be determined for three different orientations, and the flow is to be shown to be laminar.

Assumptions **1** The flow is steady and incompressible. **2** The entrance effects are negligible, and thus the flow is fully developed. **3** The flow is laminar (to be verified). **4** The pipe involves no components such as bends, valves, and connectors. **5** The piping section involves no work devices such as a pump or a turbine.

Properties The density and dynamic viscosity of oil are given to be $\rho = 888$ kg/m³ and $\mu = 0.8$ kg/m · s.

Analysis The pressure drop across the pipe and the cross-sectional area are

$$\Delta P = P_1 - P_2 = 745 - 97 = 648 \text{ kPa}$$
$$A_c = \pi D^2/4 = \pi(0.05 \text{ m})^2/4 = 0.001963 \text{ m}^2$$

(*a*) The flow rate for all three cases can be determined from Eq. 12-18b,

$$\dot V = \frac{(\Delta P - \rho g L \sin \theta)\pi D^4}{128\mu L}$$

where θ is the angle the pipe makes with the horizontal. For the horizontal case, $\theta = 0$ and thus $\sin \theta = 0$. Therefore,

$$\dot V_{horiz} = \frac{\Delta P \pi D^4}{128\mu L} = \frac{(648 \text{ kPa})\pi(0.05 \text{ m})^4}{128(0.8 \text{ kg/m} \cdot \text{s})(40 \text{ m})}\left(\frac{1000 \text{ N/m}^2}{1 \text{ kPa}}\right)\left(\frac{1 \text{ kg} \cdot \text{m/s}^2}{1 \text{ N}}\right)$$

$$= 0.00311 \text{ m}^3\text{/s}$$

(b) For uphill flow with an inclination of 15°, we have $\theta = +15°$ and

$$\dot V_{uphill} = \frac{(\Delta P - \rho g L \sin \theta)\pi D^4}{128\mu L}$$

$$= \frac{[648,000 \text{ Pa} - (888 \text{ kg/m}^3)(9.81 \text{ m/s}^2)(40 \text{ m}) \sin 15°]\pi(0.05 \text{ m})^4}{128(0.8 \text{ kg/m} \cdot \text{s})(40 \text{ m})}$$

$$= 0.00267 \text{ m}^3\text{/s}$$

since 1 Pa = 1 N/m² = 1 kg/m · s².

(c) For downhill flow with an inclination of 15°, we have $\theta = -15°$ and

$$\dot V_{downhill} = \frac{(\Delta P - \rho g L \sin \theta)\pi D^4}{128\mu L}$$

$$= \frac{[648,000 \text{ Pa} - (888 \text{ kg/m}^3)(9.81 \text{ m/s}^2)(40 \text{ m}) \sin(-15°)]\pi(0.05 \text{ m})^4}{128(0.8 \text{ kg/m} \cdot \text{s})(40 \text{ m})}$$

$$= 0.00354 \text{ m}^3\text{/s}$$

since 1 Pa = 1 N/m² = 1 kg/m · s².

The flow rate is the highest for the downhill flow case, as expected. The average fluid velocity and the Reynolds number in this case are

$$\mathcal{V} = \frac{\dot V}{A_c} = \frac{0.00354 \text{ m}^3\text{/s}}{0.001963 \text{ m}^2} = 1.80 \text{ m/s}$$

$$\text{Re} = \frac{\rho \mathcal{V} D}{\mu} = \frac{(888 \text{ kg/m}^3)(1.80 \text{ m/s})(0.05 \text{ m})}{0.8 \text{ kg/m} \cdot \text{s}} = 100$$

which is less than 2300. Therefore, the flow is **laminar** for all three cases and the analysis above is valid.

Discussion   Note that the flow is driven by the combined effect of the pressure difference and gravity. As can be seen from the calculated rates above, gravity opposes uphill flow but helps downhill flow. Gravity has no effect on the flow rate in the horizontal case. Downhill flow can occur even in the absence of an applied pressure difference. For the case of $P_1 = P_2 = 97$ kPa (i.e., no applied pressure difference), the pressure throughout the entire pipe would remain constant at 97 Pa, and the fluid would flow through the pipe at a rate of 0.00043 m³/s under the influence of gravity. The flow rate will increase as the tilt angle of the pipe from the horizontal is increased, and would reach the maximum value when the pipe is vertical.

Horizontal

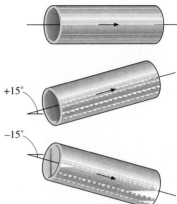

FIGURE 12-17
Schematic for Example 12-1.

EXAMPLE 12-2   Pressure Drop and Head Loss in a Pipe
Water at 40°F ($\rho = 62.42$ lbm/ft³ and $\mu = 3.74$ lbm/ft · h) is flowing in a 0.15-in.-diameter, 30-ft-long horizontal pipe steadily at an average velocity of 3 ft/s (Fig. 12-18). Determine (a) the pressure drop, (b) the head loss, and (c) the pumping power requirement to overcome this pressure drop.

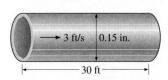

FIGURE 12-18
Schematic for Example 12-2.

Solution The average flow velocity in a pipe is given. The pressure drop, the head loss, and the pumping power are to be determined.

Assumptions **1** The flow is steady and incompressible. **2** The entrance effects are negligible, and thus the flow is fully developed. **3** The flow is laminar (to be verified). **4** The pipe involves no components such as bends, valves, and connectors. **5** The piping section involves no work devices such as a pump or a turbine.

Properties The density and dynamic viscosity of water are given to be $\rho = 62.42$ lbm/ft³ and $\mu = 3.74$ lbm/ft · h = 0.00104 lbm/ft · s, respectively.

Analysis (*a*) First we need to determine the flow regime. The Reynolds number of the flow is

$$\text{Re} = \frac{\rho \mathcal{V} D}{\mu} = \frac{(62.42 \text{ lbm/ft}^3)(3 \text{ ft/s})(0.12/12 \text{ ft})}{3.74 \text{ lbm/ft} \cdot \text{h}} \left(\frac{3600 \text{ s}}{1 \text{ h}}\right) = 1803$$

which is less than 2300. Therefore, the flow is laminar. Then the friction factor and the pressure drop become

$$f = \frac{64}{\text{Re}} = \frac{64}{1803} = 0.0355$$

$$\Delta P = f \frac{L}{D} \frac{\rho \mathcal{V}^2}{2} = 0.0355 \frac{30 \text{ ft}}{0.12/12 \text{ ft}} \frac{(62.42 \text{ lbm/ft}^3)(3 \text{ ft/s})^2}{2} \left(\frac{1 \text{ lbf}}{32.2 \text{ lbm} \cdot \text{ft/s}^2}\right)$$
$$= 930 \text{ lbf/ft}^2 = \textbf{6.46 psi}$$

(*b*) The head loss in the pipe is determined from

$$h_L = \frac{\Delta P}{\rho g} = f \frac{L}{D} \frac{\mathcal{V}^2}{2g} = 0.0355 \frac{30 \text{ ft}}{0.12/12 \text{ ft}} \frac{(3 \text{ ft/s})^2}{2(32.2 \text{ ft/s}^2)} = \textbf{29.8 ft}$$

(*c*) The volume flow rate and the pumping power requirements are

$$\dot{V} = \mathcal{V} A_c = \mathcal{V}(\pi D^2/4) = (3 \text{ ft/s})[\pi(0.12/12 \text{ ft})^2/4] = 0.000236 \text{ ft}^3/\text{s}$$

$$\dot{W}_{\text{pump}} = \dot{V}\Delta P = (0.000236 \text{ ft}^3/\text{s})(930 \text{ lbf/ft}^2)\left(\frac{1 \text{ W}}{0.737 \text{ lbf} \cdot \text{ft/s}}\right) = \textbf{0.30 W}$$

Therefore, power input in the amount of 0.30 W is needed to overcome the frictional losses in the flow due to viscosity.

12-4 ■ FULLY DEVELOPED TURBULENT FLOW IN PIPES

Turbulent flow occurs more frequently in practice than laminar flow, and thus we wish to obtain similar relations for the friction factor. However, turbulent flow is a complex mechanism dominated by random fluctuations, and despite tremendous amount of work in this area by the researchers, the theory of turbulent flow remains mostly undeveloped. Therefore, we must rely on experiments and the empirical relations developed for friction factor correlations. The chaotic fluctuations of fluid particles play a dominant role in pressure drop, and these random motions must be considered in analysis together with the mean velocity.

Perhaps the first thought that comes to mind is to determine the shear stress in an analogous manner to laminar flow from $\tau = -\mu \, d\mathcal{V}/dr$ where $\mathcal{V}(r)$ is the mean velocity profile for turbulent flow. But the experimental studies show that this is not the case, and the shear stress is much larger due to the turbulent fluctuations. Therefore, it is convenient to think of the turbulent shear stress as consisting of two parts: the *laminar component*, which accounts for the friction between layers in the flow direction (expressed as $\tau_{lam} = -\mu\, d\mathcal{V}/dr$), and the *turbulent component,* which accounts for the friction between the fluctuating fluid chunks and the fluid body (expressed as τ_{turb} and is related to the fluctuation components of velocity). Then the *total shear stress* in turbulent flow can be expressed as

$$\tau_{total} = \tau_{lam} + \tau_{turb} \tag{12-24}$$

The typical mean velocity profile and relative magnitudes of laminar and turbulent components of shear stress for turbulent flow in a pipe are given in Fig. 12-19. Note that although the velocity profile is approximately parabolic in laminar flow, it becomes flatter in turbulent flow, with a sharp drop near the side wall. The flatness increases with the Reynolds number, and the velocity profile appears to be nearly uniform, lending support to the commonly utilized uniform velocity profile assumption.

The turbulent boundary layer can be considered to consist of three layers. The very thin layer next to the wall where the viscous effects are dominant is the **laminar sublayer.** The velocity profile in this layer is nearly linear and the flow is streamlined. This region is dominated by the shear stress τ_{lam}. Next to the laminar sublayer is the **buffer** (or **overlap**) **layer,** in which the turbulent effects are significant but not dominant. The magnitudes of laminar and turbulent shear stresses are comparable in this layer. Next to it is the **turbulent** (or **outer**) **layer,** in which the turbulent effects and thus the turbulent component of shear stress τ_{turb} dominate.

Despite the small thickness of the laminar sublayer (usually much less than 1 percent of the pipe diameter), the characteristics of the flow in this layer are very important since they set the stage for flow in the rest of the pipe. Any irregularity or roughness on the surface disturbs this layer and affects the flow. Therefore, unlike laminar flow, the friction factor in turbulent flow is a strong function of surface roughness.

It should be kept in mind that roughness is a relative concept, and it has significance when its height ε is comparable to the thickness of the laminar sublayer (which is a function of the Reynolds number). All materials appear "rough" under a microscope with sufficient magnification. In fluid mechanics, a surface is characterized as being rough when the hills of roughness protrude out of the laminar sublayer. A surface is said to be smooth when the sublayer submerges the roughness elements. Glass and plastic surfaces are considered to be hydrodynamically smooth.

The Moody Chart

The friction factor in fully developed turbulent flow depends on the Reynolds number and the **relative roughness** ε/D, which is the ratio of the mean height of roughness of the pipe to the pipe diameter. The functional form of this dependence cannot be obtained from a theoretical analysis, and all available results are obtained from painstaking experiments using artificially roughened surfaces (usually by gluing sand grains of a known size on the inner surfaces of the pipes). Most such experiments were conducted by Prandtl's student J. Nikuradse in 1933, followed by the works of others. The friction factor was calculated from the measurements of the flow rate and the pressure drop.

The experimental results obtained are presented in tabular, graphical, and functional forms obtained by curve-fitting experimental data. In 1939, C. F. Colebrook combined all the data for transition and turbulent flow in smooth

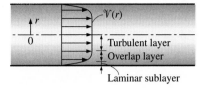

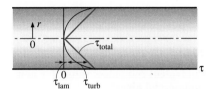

FIGURE 12-19

The velocity profile and the variation of shear stress with radial distance for turbulent flow in a pipe.

as well as rough pipes into the following implicit relation known as the **Cole-brook equation:**

$$\frac{1}{\sqrt{f}} = -2.0 \log\left(\frac{\varepsilon/D}{3.7} + \frac{2.51}{\mathrm{Re}\ \sqrt{f}}\right) \qquad \text{(turbulent flow)} \qquad (12\text{-}25)$$

In 1944, L. F. Moody plotted this formula into the now famous **Moody chart,** given in the appendix in Fig. A-27. It presents the friction factors for pipe flow as a function of the Reynolds number and ε/D over a wide range. It is probably one of the most widely accepted and used charts in engineering. Although it is developed for circular pipes, it can also be used for noncircular pipes by replacing the diameter by the hydraulic diameter.

Commercially available pipes differ from those used in the experiments in that the roughness of pipes in the market is not uniform, and it is difficult to give a precise description of it. Equivalent roughness values for some commercial pipes are given in Table 12-2 as well as on the Moody chart. But it should be kept in mind that these values are for new pipes, and the relative roughness of pipes may increase with use as a result of corrosion, scale buildup, and precipitation. As a result, the friction factor may increase by a factor of 5 to 10. Actual operating conditions must be considered in the design of piping systems. Also, the Moody chart and its equivalent Colebrook equation involve several uncertainties (the roughness size, experimental error, curve fitting of data, etc.), and thus the results obtained should not be treated as "exact." It is usually considered to be accurate to ± 15 percent over the entire range in the figure.

The Colebrook equation is implicit in f, and thus the determination of the friction factor requires tedious iteration unless an equation solver such as EES is used. An approximate explicit relation for f is given by S. E. Haaland in 1983 as

$$\frac{1}{\sqrt{f}} \approx -1.8 \log\left[\frac{6.9}{\mathrm{Re}} + \left(\frac{\varepsilon/D}{3.7}\right)^{1.11}\right] \qquad (12\text{-}26)$$

The results obtained from this relation are within 2 percent of those obtained from the Colebrook equation.

We make the following observations from the Moody chart:

- For laminar flow, the friction factor decreases with increasing Reynolds number, and it is independent of surface roughness.
- The friction factor is a minimum for a smooth pipe (but still not zero because of the no-slip condition) and increases with roughness. The Colebrook equation in this case ($\varepsilon = 0$) reduces to $1/\sqrt{f} = 2.0 \log (\mathrm{Re}\ \sqrt{f}) - 0.8$ (Fig. 12-20).
- The transition region from laminar to turbulent regime ($2300 < \mathrm{Re} < 4000$) is indicated by the shaded area in the chart. The flow in this region alternates between laminar and turbulent, and thus the friction factor may also alternate between the values for laminar and turbulent flow. The data in this range are the least reliable. At small relative roughnesses, the friction factor increases in the transition region and approaches the value for smooth pipes.
- At very large Reynolds numbers (to the right of the dashed line on the chart) the friction factor curves corresponding to specified relative roughness curves are nearly horizontal, and thus the friction factors are independent of the Reynolds number (Fig. 12-21). The flow in that region is called *fully rough flow,* or *completely* (or *fully*) *turbulent flow.* This is because the thickness of the laminar sublayer decreases with

TABLE 12-2

Equivalent roughness values for new commercial pipes*

Material	Roughness, ε	
	ft	mm
Glass, plastic	0 (smooth)	
Concrete	0.003–0.03	0.9–9
Wood stave	0.0016	0.5
Rubber, smoothed	0.000033	0.01
Copper or brass tubing	0.000005	0.0015
Cast iron	0.00085	0.26
Galvanized iron	0.0005	0.15
Wrought iron	0.00015	0.046
Stainless steel	0.000007	0.002
Commercial steel	0.00015	0.045

*The uncertainty in these values can be as much as ± 60 percent.

Relative roughness, ε/L	Friction factor, f
0.0*	0.0119
0.00001	0.0119
0.0001	0.0134
0.0005	0.0172
0.001	0.0199
0.005	0.0305
0.01	0.0380
0.05	0.0716

*Smooth surface. All values are for $\mathrm{Re} = 10^6$, and are calculated from Eq. 12-25.

FIGURE 12-20

The friction factor is minimum for a smooth pipe and increases with roughness.

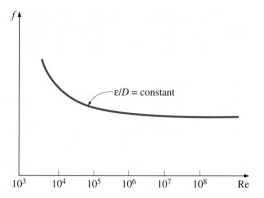

FIGURE 12-21

At very large Reynolds numbers, the friction factor curves are nearly horizontal, and thus the friction factors are independent of the Reynolds number.

increasing Reynolds number, and it becomes so thin that the surface roughness protrudes into the flow. The viscous effects in this case are produced in the main flow primarily by the protruding roughness elements, and the contribution of the laminar sublayer is negligible. The Colebrook equation in the completely turbulent zone (Re $\rightarrow \infty$) reduces to $1/\sqrt{f} = -2.0 \log[(\varepsilon/D)/3.7]$, which is explicit in f.

In calculations, we should make sure that we use the internal diameter of the pipe, which may be different than the nominal diameter. For example, the internal diameter of a steel pipe whose nominal diameter is 1 in. is 1.049 in. (Table 12-3).

Types of Fluid Flow Problems

In the design and analysis of piping systems that involve the use of the Moody chart (or the Colebrook equation), we usually encounter three types of problems (the fluid and the roughness of the pipe are assumed to be specified in all cases) (Fig. 12-22):

1. Determining the **pressure drop** (or head loss) when the pipe length and diameter are given for a specified flow rate (or velocity).

2. Determining the **flow rate** when the pipe length and diameter are given for a specified pressure drop.

3. Determining the **pipe diameter** when the pipe length and flow rate are given for a specified pressure drop (or head loss).

The *first type* of problems is straightforward and can be solved directly by using the Moody chart. The *second and third types* of problems are commonly encountered in engineering design (in the selection of pipe diameter, for example, that minimizes the sum of the construction and pumping costs), but the use of the Moody chart with such problems requires a trial-and-error approach unless an equation solver is used.

In the *third type* of problems, the diameter is not known and thus the Reynolds number and the relative roughness cannot be calculated. Therefore, we start calculations by assuming a pipe diameter. The pressure drop calculated for the assumed diameter is then compared to the specified pressure drop, and calculations are repeated with another pipe diameter if needed. In the second kind of problems, the diameter is given but the flow rate is unknown. A good guess for the friction factor in that case is obtained from the completely turbulent flow region for the given roughness. This is true for large Reynolds numbers, which is usually the case in practice.

TABLE 12-3

Standard sizes for Schedule 40 steel pipes

Nominal size, in.	Actual inside diameter, in.
⅛	0.269
¼	0.364
⅜	0.493
½	0.622
¾	0.824
1	1.049
1½	1.610
2	2.067
2½	2.469
3	3.068
5	5.047
10	10.02

Given	Find
1. $L, D, \dot{V}$	ΔP (or h_L)
2. $L, D, \Delta P$	$\dot{V}$
3. $L, \Delta P, \dot{V}$	D

FIGURE 12-22

The three types of problems encountered in pipe flow.

To avoid tedious iterations in head loss, flow rate, and diameter calculations, Swamee and Jain (Ref. 10) proposed the following explicit relations in 1976 that are accurate to within 2 percent of the Moody chart:

$$h_L = 1.07 \frac{\dot{V}^2 L}{gD^5}\left\{ \ln\left[\frac{\varepsilon}{3.7D} + 4.62\left(\frac{\nu D}{\dot{V}}\right)^{0.9}\right]\right\}^{-2} \quad \begin{array}{l} 10^{-6} < \varepsilon/D < 10^{-2} \\ 3000 < Re < 3 \times 10^8 \end{array} \quad (12\text{-}27)$$

$$\dot{V} = -0.965\left(\frac{gD^5 h_L}{L}\right)^{0.5} \ln\left[\frac{\varepsilon}{3.7D} + \left(\frac{3.17\nu^2 L}{gD^3 h_L}\right)^{0.5}\right] \quad Re > 2000 \quad (12\text{-}28)$$

$$D = 0.66\left[\varepsilon^{1.25}\left(\frac{L\dot{V}^2}{gh_L}\right)^{4.75} + \nu\dot{V}^{9.4}\left(\frac{L}{gh_L}\right)^{5.2}\right]^{0.04} \quad \begin{array}{l} 10^{-6} < \varepsilon/D < 10^{-2} \\ 5000 < Re < 3 \times 10^8 \end{array} \quad (12\text{-}29)$$

Note that all quantities are dimensional and the units simplify to the desired unit (for example, to m or ft in the last relation) when consistent units are used. Noting that the Moody chart is accurate to within 5 percent of experimental data, we should have no reservation in using the approximate relations above in the design of piping systems.

EXAMPLE 12-3 Determining the Head Loss in a Water Pipe

Water at 60°F (ρ = 62.36 lbm/ft³ and μ = 2.713 lbm/ft · h) is flowing in a 2-in.-diameter horizontal pipe made of stainless steel steadily at a rate of 0.2 ft³/s (Fig. 12-23). Determine the pressure drop, the head loss, and the required pumping power input for flow over a 200-ft-long section of the pipe.

Solution The flow rate through a specified water pipe is given. The pressure drop, the head loss, and the pumping power requirements are to be determined.

Assumptions **1** The flow is steady and incompressible. **2** The entrance effects are negligible, and thus the flow is fully developed. **3** The pipe involves no components such as bends, valves, and connectors. **4** The piping section involves no work devices such as a pump or a turbine.

Properties The density and dynamic viscosity of water are given to be ρ = 62.36 lbm/ft³ and μ = 2.713 lbm/ft · h = 0.0007536 lbm/ft · s, respectively.

Analysis First we calculate the mean velocity and the Reynolds number to determine the flow regime:

$$\mathcal{V} = \frac{\dot{V}}{A_c} = \frac{\dot{V}}{\pi D^2/4} = \frac{0.2 \text{ ft}^3/\text{s}}{\pi(2/12 \text{ ft})^2/4} = 9.17 \text{ ft/s}$$

$$Re = \frac{\rho \mathcal{V} D}{\mu} = \frac{(62.36 \text{ lbm/ft}^3)(9.17 \text{ ft/s})(2/12 \text{ ft})}{2.713 \text{ lbm/ft} \cdot \text{h}}\left(\frac{3600 \text{ s}}{1 \text{ h}}\right) = 126,400$$

which is greater than 4000. Therefore, the flow is turbulent. The relative roughness of the pipe is

$$\varepsilon/D = \frac{0.000007 \text{ ft}}{2/12 \text{ ft}} = 0.000042$$

The friction factor corresponding to this relative roughness and the Reynolds number can simply be determined from the Moody chart. To avoid the reading error, we determine it from the Colebrook equation:

$$\frac{1}{\sqrt{f}} = -2.0 \log\left(\frac{\varepsilon/D}{3.7} + \frac{2.51}{Re\sqrt{f}}\right) \quad \rightarrow \quad \frac{1}{\sqrt{f}} = -2.0 \log\left(\frac{0.000042}{3.7} + \frac{2.51}{126,400\sqrt{f}}\right)$$

Using an equation solver or an iterative scheme, the friction factor is determined to be f = 0.0174. Then the pressure drop, head loss, and the required power input become

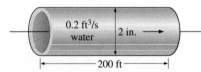

FIGURE 12-23
Schematic for Example 12-3.

$$\Delta P = f \frac{L}{D} \frac{\rho \mathcal{V}^2}{2} = 0.0174 \frac{200 \text{ ft}}{2/12 \text{ ft}} \frac{(62.36 \text{ lbm/ft}^3)(9.17 \text{ ft/s})^2}{2} \left(\frac{1 \text{ lbf}}{32.2 \text{ lbm} \cdot \text{ft/s}} \right)$$

$$= 1700 \text{ lbf/ft}^2 = 11.8 \text{ psi}$$

$$h_L = \frac{\Delta P}{\rho g} = f \frac{L}{D} \frac{\mathcal{V}^2}{2g} = 0.0174 \frac{200 \text{ ft}}{2/12 \text{ ft}} \frac{(9.17 \text{ ft/s})^2}{2(32.2 \text{ ft/s}^2)} = 27.3 \text{ ft}$$

$$\dot{W}_{pump} = \dot{V}\Delta P = (0.2 \text{ ft}^3/\text{s})(1700 \text{ lbf/ft}^2) \left(\frac{1 \text{ W}}{0.737 \text{ lbf} \cdot \text{ft/s}} \right) = 461 \text{ W}$$

Therefore, power input in the amount of 461 W is needed to overcome the frictional losses in the pipe.

Discussion The friction factor also could be determined easily from the explicit Haaland relation. It would give $f = 0.0172$, which is sufficiently close to 0.0174. Also, the friction factor corresponding to $\varepsilon = 0$ in this case is 0.0171, which indicates that stainless steel pipes can be assumed to be smooth with negligible error.

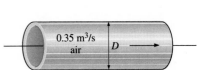

EXAMPLE 12-4   Determining the Diameter of an Air Duct
Heated air at 1 atm and 37°C is to be transported in a 150-m-long circular plastic duct at a rate of 0.35 m³/s (Fig. 12-24). If the head loss in the pipe is not to exceed 20 m, determine the minimum diameter of the duct.

FIGURE 12-24

Schematic for Example 12-4.

Solution   The flow rate and the head loss in an air duct are given. The diameter of the duct is to be determined.

Assumptions   **1** The flow is steady and incompressible. **2** The entrance effects are negligible, and thus the flow is fully developed. **3** The duct involves no components such as bends, valves, and connectors. **4** Air is an ideal gas. **5** The duct is smooth since it is made of plastic. **6** The flow is turbulent (to be verified).

Properties   The density, dynamic viscosity, and kinematic viscosity of air at 37°C = 310 K are $\rho = 1.143 \text{ kg/m}^3$, $\mu = 1.90 \times 10^{-5} \text{ kg/m} \cdot \text{s}$, and $\nu = 1.67 \times 10^{-5} \text{ m}^2/\text{s}$ (Table A-18).

Analysis   This is a third-type problem since it involves the determination of diameter for specified flow rate and head loss. We can solve this problem by three different approaches: (1) a trial-and-error approach by assuming a pipe diameter, calculating the head loss, comparing the result to the specified head loss, and repeating calculations until the calculated head loss matches the specified value; (2) writing all the relevant equations (leaving the diameter as an unknown) and solving them simultaneously using an equation solver; and (3) using the third Swamee-Jain formula. Below we will demonstrate the use of the last two approaches.

The average velocity, the Reynolds number, the friction factor, and the head loss relations can be expressed as (D is in m, and $\mathcal{V}$ is in m/s, Re and f are dimensionless)

$$\mathcal{V} = \frac{\dot{V}}{A_c} = \frac{\dot{V}}{\pi D^2/4} = \frac{0.35 \text{ m}^3/\text{s}}{\pi D^2/4}$$

$$\text{Re} = \frac{\mathcal{V}D}{\nu} = \frac{\mathcal{V}D}{1.67 \times 10^{-5} \text{ m}^2/\text{s}}$$

$$\frac{1}{\sqrt{f}} = -2.0 \log\left(\frac{\varepsilon/D}{3.7} + \frac{2.51}{\text{Re} \sqrt{f}} \right) = -2.0 \log\left(\frac{2.51}{\text{Re} \sqrt{f}} \right)$$

$$h_L = f \frac{L}{D} \frac{\mathcal{V}^2}{2g} \quad \rightarrow \quad 20 = f \frac{150 \text{ m}}{D} \frac{\mathcal{V}^2}{2(9.81 \text{ m/s}^2)}$$

This is a set of four equations in four unknowns, and solving them with the equation solver EES gives

$$D = \mathbf{0.267\ m}, \qquad f = 0.0180, \qquad \mathcal{V} = 6.23\ \text{m/s}, \qquad \text{and} \qquad \text{Re} = \mathbf{99{,}800}$$

Therefore, the diameter of the duct should be more than 26.7 cm if the head loss is not to exceed 20 m. Note that Re > 4000, and thus the turbulent flow assumption is verified.

The diameter can also be determined directly from the third Swamee-Jain formula to be

$$D = 0.66\left[\varepsilon^{1.25}\left(\frac{L\dot{V}^2}{gh_L}\right)^{4.75} + \nu\dot{V}^{9.4}\left(\frac{L}{gh_L}\right)^{5.2}\right]^{0.04}$$

$$= 0.66\left[0 + (1.67 \times 10^{-5}\ \text{m}^2/\text{s})(0.35\ \text{m}^3/\text{s})^{9.4}\left(\frac{150\ \text{m}}{(9.81\ \text{m/s}^2)(20\ \text{m})}\right)^{5.2}\right]^{0.04}$$

$$= 0.271\ \text{m}$$

Discussion Note that the difference between the two results is less than 2 percent. Therefore, the simple Swamee-Jain relation can be used with confidence.

12-5 ■ MINOR LOSSES

The fluid in a typical piping system passes through various fittings, valves, bends, elbows, tees, inlets, exits, enlargements, and contractions in addition to the pipes. These components interrupt the smooth flow of the fluid and cause additional losses because of the flow separation and mixing they induce. In a typical system with long pipes, these losses are minor compared to the total head loss in the pipes (the *major losses*) and are called **minor losses.** Although this is generally true, in some cases the minor losses may be greater than the major losses. This will be the case in systems with several turns and valves in a short distance. The head loss introduced by a completely open valve, for example, may be negligible. But a partially closed valve may cause the largest head loss in the system, as evidenced by the drop in the flow rate. Flow through valves and fittings is very complex, and a theoretical analysis is not available. Therefore, minor losses are determined experimentally, usually by the manufacturers of the components.

Minor losses are usually expressed in terms of the **loss coefficient** K_L, defined as (Fig. 12-25)

Loss coefficient:	$K_L = \dfrac{h_L}{\mathcal{V}^2/(2g)}$	(12-30)

When the inlet diameter equals outlet diameter, the loss coefficient of a component can also be determined by measuring the pressure drop across the component and dividing it by the dynamic pressure, $K_L = \Delta P/(0.5\ \rho\mathcal{V}^2)$. When the loss coefficient for a component is available, the head loss for that component is determined from

Minor loss:	$h_L = K_L\dfrac{\mathcal{V}^2}{2g}$	(12-31)

The loss coefficient, in general depends on the geometry of the component and the Reynolds number, just like the friction factor. However, it is usually assumed to be independent of the Reynolds number. This is a reasonable approximation since most flows in practice have large Reynolds numbers and

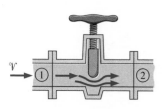

$$\Delta P = P_1 - P_2$$

$$K_L = \frac{\Delta P}{\frac{1}{2}\rho\mathcal{V}^2}$$

FIGURE 12-25

The loss coefficient of a component (such as the gate valve shown) is determined by measuring the pressure drop it causes and dividing it by the dynamic pressure in the pipe.

the loss coefficients (including the friction factor) are independent of the Reynolds number at large Reynolds numbers.

Minor losses are also expressed in terms of the **equivalent length** L_{equiv}, defined as (Fig. 12-26)

Equivalent length: $\quad h_L = K_L \dfrac{\mathcal{V}^2}{2g} = f \dfrac{L_{equiv}}{D} \dfrac{\mathcal{V}^2}{2g} \quad \rightarrow \quad L_{equiv} = \dfrac{D}{f} K_L \quad$ (12-32)

where f is the friction factor and D is the diameter of the pipe that contains the component. The head loss caused by the component is equivalent to the head loss caused by a section of the pipe whose length is L_{equiv}. Therefore, the contribution of a component to the head loss can be accounted for by simply adding L_{equiv} to the total pipe length.

Both approaches are used in practice, but the use of loss coefficients is more common. Therefore, we will also use that approach in this book. Once all the loss coefficients are available, the total head loss in a piping system can be determined from

Total head loss (general): $\quad h_{L,\,total} = h_{L,\,major} + h_{L,\,minor}$

$$= \sum f_i \frac{L_i}{D_i} \frac{\mathcal{V}_i^2}{2g} + \sum K_{L,j} \frac{\mathcal{V}_j^2}{2g} \quad (12\text{-}33)$$

where i represents each pipe section with constant diameter and j represents each component that causes a minor loss. If the entire piping system analyzed has a constant diameter, the relation above reduces to

Total head loss (D = constant): $\quad h_{L,\,total} = \left(f \dfrac{L}{D} + \sum K_L \right) \dfrac{\mathcal{V}^2}{2g} \quad$ (12-34)

where $\mathcal{V}$ is the average flow velocity through the entire system (note that $\mathcal{V}$ = constant since D = constant).

Representative loss coefficients K_L are given in Table 12-4 for inlets, exits, bends, sudden and gradual area changes, and valves. There is considerable uncertainty in these values since the loss coefficients, in general, vary with the pipe diameter, the surface roughness, the Reynolds number, and the details of the design. The loss coefficients of two seemingly identical valves by two different manufacturers, for example, can differ by a factor of 2 or more. Therefore, the particular manufacturer's data should be consulted in the final design of piping systems rather than relying on the representative values in handbooks.

The head loss at the inlet of a pipe is a strong function of geometry. It is almost negligible for well-rounded inlets ($K_L = 0.03$ for $r/D > 0.2$), but increases to about 0.50 for sharp-edged inlets (Fig. 12-27). That is, a sharp-edged inlet causes half of the velocity head to be lost as the fluid enters the pipe. This is because the fluid cannot make sharp 90° turns easily, especially at high velocities. As a result, the flow separates at the corners, and the flow is constricted into the *vena contracta* region formed in the mid section of the pipe (Fig. 12-28). Therefore, a sharp-edged inlet acts like a flow constriction. The velocity increases in the vena contracta region (and the pressure decreases) because of the reduced effective flow area and then decreases as the flow fills the entire cross section of the pipe. There would be negligible loss if the pressure were increased in accordance with Bernoulli's equation (the velocity head would simply be converted into pressure head). However, this

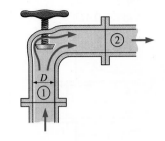

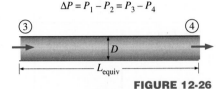

$\Delta P = P_1 - P_2 = P_3 - P_4$

FIGURE 12-26

The head loss caused by the component (such as the angle valve shown) is equivalent to the head loss caused by a section of the pipe whose length is the equivalent length.

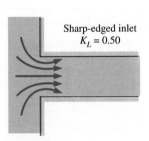

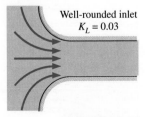

FIGURE 12-27

The head loss at the inlet of a pipe is almost negligible for well-rounded inlets ($K_L = 0.03$ for $r/D > 0.2$) but increases to about 0.50 for sharp-edged inlets.

TABLE 12-4

Loss coefficients K_L of various pipe components for turbulent flow
(for use in the relation $h_L = K_L V^2/(2g)$ where V is the mean velocity in the pipe that contains the component)*

Pipe Entrance		
Reentrant: $K_L = 0.80$ ($t \ll D$ and $l \sim 0.1D$)	*Sharp-edged:* $K_L = 0.50$	*Well-rounded* ($r/D > 0.2$): $K_L = 0.03$ *Slightly rounded* ($r/D = 0.1$): $K_L =$ 0.12 (see Fig. 12-29)

Pipe Exit		
Reentrant: $K_L = 1.0$	*Sharp-edged:* $K_L = 1.0$	*Rounded:* $K_L = 1.0$

Sudden Expansion and Contraction (based on the velocity in the smaller-diameter pipe)

Sudden expansion: $K_L = \left(1 - \dfrac{d^2}{D^2}\right)^2$

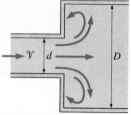

Sudden contraction: See chart.

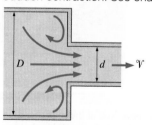

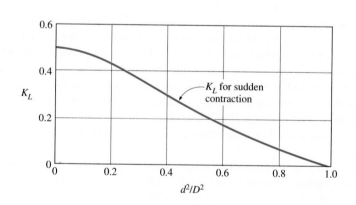

Gradual Expansion and Contraction (based on the velocity in the smaller-diameter pipe)

Expansion:
$K_L = 0.02$ for $\theta = 20°$
$K_L = 0.04$ for $\theta = 45°$
$K_L = 0.07$ for $\theta = 60°$

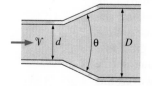

Contraction (for $\theta = 20°$):
$K_L = 0.30$ for $d/D = 0.2$
$K_L = 0.25$ for $d/D = 0.4$
$K_L = 0.15$ for $d/D = 0.6$
$K_L = 0.10$ for $d/D = 0.8$

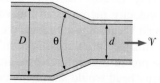

TABLE 12-4 *(Concluded)*

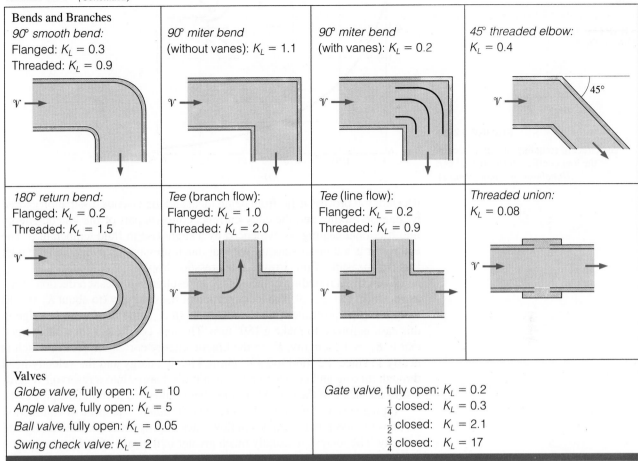

Bends and Branches			
90° smooth bend: Flanged: $K_L = 0.3$ Threaded: $K_L = 0.9$	**90° miter bend** (without vanes): $K_L = 1.1$	**90° miter bend** (with vanes): $K_L = 0.2$	**45° threaded elbow:** $K_L = 0.4$
180° return bend: Flanged: $K_L = 0.2$ Threaded: $K_L = 1.5$	**Tee (branch flow):** Flanged: $K_L = 1.0$ Threaded: $K_L = 2.0$	**Tee (line flow):** Flanged: $K_L = 0.2$ Threaded: $K_L = 0.9$	**Threaded union:** $K_L = 0.08$

Valves	
Globe valve, fully open: $K_L = 10$	*Gate valve*, fully open: $K_L = 0.2$
Angle valve, fully open: $K_L = 5$	$\frac{1}{4}$ closed: $K_L = 0.3$
Ball valve, fully open: $K_L = 0.05$	$\frac{1}{2}$ closed: $K_L = 2.1$
Swing check valve: $K_L = 2$	$\frac{3}{4}$ closed: $K_L = 17$

*These are representative values for loss coefficients. Actual values strongly depend on the design and manufacture of the components and may differ from the given values considerably (especially for valves). Actual manufacturer's data should be used in the final design.

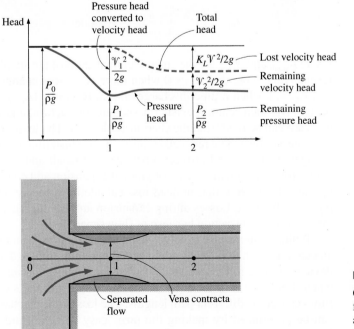

FIGURE 12-28

Graphical representation of flow contraction and the associated heat loss at a sharp-edged pipe inlet.

529

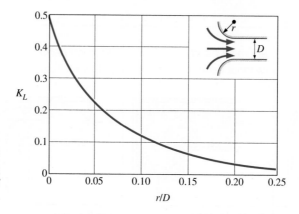

FIGURE 12-29

The effect of rounding of a pipe inlet on the loss coefficient (from *ASHRAE Handbook of Fundamentals*).

deceleration process is far from being ideal and the viscous dissipation caused by intense mixing and the eddy currents converts part of the kinetic energy into frictional heating, as evidenced by a slight rise in fluid temperature. The end result is a drop in velocity without much pressure recovery, and the inlet loss is a measure of this irreversible pressure drop.

Even slight rounding of the edges can result in significant reduction of K_L, as shown in Fig. 12-29. The loss coefficient rises sharply (to about $K_L = 0.8$) when the pipe protrudes into the reservoir since some fluid near the edge in this case is forced to make a 180° turn. The loss coefficient for a submerged exit is $K_L = 1$ (actually, $K_L =$ the kinetic energy correction factor, which is nearly 1) since the fluid loses its entire kinetic energy and the velocity head through mixing and comes to rest when it discharges into a reservoir regardless of the shape of the exit. Therefore, there is no need to round the pipe exits.

Piping systems often involve *sudden* or *gradual* expansion or contraction sections to incorporate changes in flow rates or properties such as density and velocity. The losses are usually much greater in the case of *sudden* expansion and contraction (or wide-angle expansion) because of flow separation. By combining the conservation of mass, momentum, and energy equations, the loss coefficient for the case of **sudden expansion** is determined to be

$$K_L = \left(1 - \frac{A_{\text{small}}}{A_{\text{large}}}\right)^2 \quad \text{(sudden expansion)} \qquad (12\text{-}35)$$

where A_{small} and A_{large} are the cross-sectional areas of the small and large pipes, respectively. Note that $K_L = 0$ when there is no area change ($A_{\text{small}} = A_{\text{large}}$) and $K_L = 1$ when a pipe discharges into a reservoir ($A_{\text{large}} \gg A_{\text{small}}$), as expected. No such relation exists for sudden contraction, and the K_L values in that case can be read from the chart in Table 12-4. The losses due to expansion and contraction can be reduced significantly by installing conical gradual area changers (nozzles and diffusers) between the small and large pipes. The K_L values for representative cases of gradual expansion and contraction are given in Table 12-4. Note that in head loss calculations, the velocity in the *small pipe* is to be used. Losses during expansion are usually much higher than the losses during contraction because of flow separation.

Piping systems also involve changes in direction without a change in diameter, and such flow sections are called *bends* or *elbows*. The losses in these devices are due to flow separation (just like a car being thrown off the road when it enters a turn too fast) on the inner side and the swirling secondary flows caused by different path lengths. The losses during changes of direction can be minimized by making the turn "easy" on the fluid by using circular arcs (like the 90° elbow) instead of sharp turns (like the miter bends) (Fig. 12–30). But the use of sharp turns (and thus suffering a penalty in lost coefficient) may be necessary when the turning space is limited. In such cases, the losses can be minimized by utilizing properly placed guide vanes to help the flow turn in an orderly manner without being thrown off the course. The loss coefficients for some elbows and miter bends as well as tees are given in Table 12–4. These coefficients do not involve the frictional losses along the pipe bend. Such losses should be calculated as in straight pipes (using the length of the centerline as the pipe length) and added to other losses.

Valves are commonly used in piping systems to control the flow rates by simply altering the head loss until the desired flow rate is achieved. For valves it is desirable to have a very low loss coefficient when they are fully open so that they cause minimal head loss during full load operation. Several different valve designs, each with its own advantages and disadvantages, are in common use today. The *gate valve* slides up and down like a gate, the *globe valve* closes a hole placed in the valve, the *angle valve* is a globe valve with a 90° turn, and the *check valve* allows the fluid to flow only in one way like a diode in an electric circuit. Table 12–4 lists the representative loss coefficients of the popular designs. Note that the loss coefficient increases drastically as a valve is closed (Fig. 12–31). Also, the deviation in the loss coefficients for different manufacturers is greatest for valves.

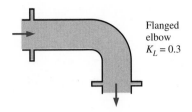

Flanged elbow
$K_L = 0.3$

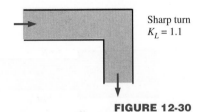

Sharp turn
$K_L = 1.1$

FIGURE 12–30

The losses during changes of direction can be minimized by making the turn "easy" on the fluid by using circular arcs instead of sharp turns.

EXAMPLE 12-5   Head Loss and Pressure Rise During Gradual Expansion

A 6-cm-diameter horizontal water pipe expands gradually making 30° from the horizontal to a 9-cm-diameter pipe (Fig. 12–32). The mean velocity and pressure of water before the expansion section are 7 m/s and 150 kPa, respectively. Determine the head loss in the expansion section, and the pressure in the larger-diameter pipe.

Solution   A horizontal water pipe expands gradually into a larger-diameter pipe. The head loss and pressure after the expansion are to be determined.

Assumptions   The flow is steady and incompressible.

Properties   We take the density of water to be $\rho = 1000$ kg/m³. The loss coefficient for gradual expansion at 30° from the horizontal is $K_L = 0.07$ (Table 12–4).

Analysis   Noting that the density of water remains constant, the downstream velocity of water is determined from conservation of mass to be

$$\dot{m}_1 = \dot{m}_2 \quad \rightarrow \quad \rho \mathcal{V}_1 A_1 = \rho \mathcal{V}_2 A_2 \quad \rightarrow \quad \mathcal{V}_2 = \frac{A_1}{A_2} \mathcal{V}_1 = \frac{D_1^2}{D_2^2} \mathcal{V}_1$$

$$\mathcal{V}_2 = \frac{(0.06 \text{ m})^2}{(0.09 \text{ m})^2} (7 \text{ m/s}) = 3.11 \text{ m/s}$$

Then the head loss in the expansion section becomes

$$h_L = K_L \frac{\mathcal{V}_1^2}{2g} = (0.07) \frac{(7 \text{ m/s})^2}{2(9.81 \text{ m/s}^2)} = \textbf{0.050 m}$$

Noting that $z_1 = z_2$ and there are no pumps or turbines involved, the energy equation for the expansion section can be expressed in terms of heads as

$$\frac{P_1}{\rho g} + \frac{\mathcal{V}_1^2}{2g} + z_1 + h_{pump,\,u}^{\,0} = \frac{P_2}{\rho g} + \frac{\mathcal{V}_2^2}{2g} + z_2 + h_{turbine}^{\,0} + h_L \quad \rightarrow \quad \frac{P_1}{\rho g} + \frac{\mathcal{V}_1^2}{2g} = \frac{P_2}{\rho g} + \frac{\mathcal{V}_2^2}{2g} + h_L$$

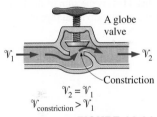

A globe valve

$\mathcal{V}_1$ $\mathcal{V}_2$

Constriction

$\mathcal{V}_2 = \mathcal{V}_1$
$\mathcal{V}_{constriction} > \mathcal{V}_1$

FIGURE 12–31

The large head loss in a partially closed valve is due to irreversible deceleration and mixing of high-velocity fluid coming from the narrow valve passage.

①   6 cm   9 cm   ②

Water
7 m/s
150 kPa

FIGURE 12–32

Schematic for Example 12–5.

Solving for P_2 and substituting,

$$P_2 = P_1 + \rho\left\{\frac{\mathcal{V}_1^2 - \mathcal{V}_2^2}{2} - gh_L\right\} = (150 \text{ kPa}) + (1000 \text{ kg/m}^3)$$

$$\times \left\{\frac{(7 \text{ m/s})^2 - (3.11 \text{ m/s})^2}{2} - (9.81 \text{ m/s}^2)(0.05 \text{ m})\right\}\left(\frac{1 \text{ kN}}{1000 \text{ kg} \cdot \text{m/s}}\right)\left(\frac{1 \text{ kPa}}{1 \text{ kN/m}^2}\right)$$

$$= \mathbf{169.2 \text{ kPa}}$$

Therefore, despite the head (and pressure) loss, the pressure increases from 150 kPa to 169.2 kPa after the expansion. This is due to the conversion of dynamic pressure to static pressure when the mean flow velocity is decreased in the larger pipe.

Discussion It is common knowledge that higher pressure upstream is necessary to cause flow, and it may come as a surprise to you that the downstream pressure has *increased* after the expansion, despite the loss. This is because the flow is driven by the sum of the three heads that comprise the total head (namely, the pressure head, velocity head, and elevation head). During flow expansion, the higher velocity head upstream is converted to pressure head downstream, and this increase outweighs the nonrecoverable head loss. Also, you may be tempted to solve this problem using the Bernoulli equation. Such a solution would ignore the head (and the associated pressure) loss, and result in a higher pressure for the fluid downstream.

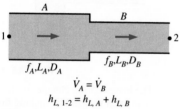

FIGURE 12-33

For pipes *in series,* the flow rate is the same in each pipe, and the total head loss is the sum of the head losses in individual pipes.

12-6 ■ PIPING NETWORKS AND PUMP SELECTION

Most piping systems encountered in practice such as the water distribution systems in cities or commercial or residential establishments involve numerous parallel and series connections as well as several sources (discharge of fluid into the system) and loads (withdrawal of fluid from the system). A piping project may involve the design of a new system or the expansion of an existing system. The engineering objective in such projects is to design a piping system that will deliver the specified flow rates at specified pressures reliably at minimum total (initial plus operating and maintenance) cost. Once the layout of the system is prepared, the determination of the pipe diameters and the pressures throughout the system, while remaining within the budget constraints, typically requires solving the system repeatedly until the optimal solution is reached. Computer modeling and analysis of such systems with colorful real-life graphics make this tedious task a simple chore.

Piping systems typically involve several pipes connected to each other in series or in parallel, as shown in Figs. 12-33 and 12-34. When the pipes are

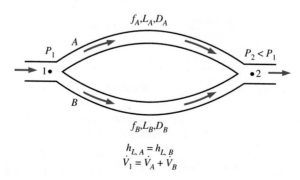

FIGURE 12-34

For pipes *in parallel,* the head loss is the same in each pipe, and the total flow rate is the sum of the flow rates in individual pipes.

connected **in series,** the flow rate through the entire system remains constant regardless of the diameters of the individual pipes in the system. This is a natural consequence of the conservation of mass principle for steady incompressible flow. The total head loss in this case is equal to the sum of the head losses in individual pipes in the system, including the minor losses. The expansion or contraction losses at connections are considered to belong to the smaller-diameter pipe since the expansion and contraction loss coefficients are defined on the basis of the average velocity in the smaller-diameter pipe.

For a pipe that branches out into two (or more) **parallel pipes** and then rejoins at a junction downstream, the total flow rate is the sum of the flow rates in the individual pipes. The pressure drop (or head loss) in each individual pipe connected in parallel must be the same since $\Delta P = P_A - P_B$ and the junction pressures P_A and P_B are the same for all of the individual pipes. For a system of two parallel pipes 1 and 2 between junctions A and B, this can be expressed as

$$h_{L,1} = h_{L,2} \qquad \rightarrow \qquad f_1 \frac{L_1}{D_1} \frac{V_1^2}{2g} = f_2 \frac{L_2}{D_2} \frac{V_2^2}{2g}$$

Then the ratio of the mean velocities and the flow rates in the two parallel pipes become

$$\frac{V_1}{V_2} = \left(\frac{f_2}{f_1} \frac{L_2}{L_1} \frac{D_1}{D_2} \right)^{0.5} \qquad \text{and} \qquad \frac{\dot{V}_1}{\dot{V}_2} = \frac{A_{c,1} V_1}{A_{c,2} V_2} = \frac{D_1^2}{D_2^2} \left(\frac{f_2}{f_1} \frac{L_2}{L_1} \frac{D_1}{D_2} \right)^{0.5}$$

Therefore, the relative flow rates in parallel pipes are established from the requirement that the head loss in each pipe be the same. This result can be extended to any number of pipes connected in parallel. The result is also valid for pipes for which the minor losses are significant if the equivalent lengths for components that contribute to minor losses are added to the pipe length. Note that the flow rate in one of the parallel branches is proportional to the 2.5th power of the diameter and is inversely proportional to the square root of its length and friction factor.

The analysis of piping networks, no matter how complex they are, is based on two simple principles:

1. *Conservation of mass throughout the system must be satisfied.* This is done by requiring the total flow into a junction to be equal to the total flow out of the junction for all junctions in the system. Also, the flow rate must remain constant in pipes connected in series regardless of the changes in diameters.

2. *Pressure drop (and thus head loss) between two junctions must be the same for all paths between the two junctions.* This is because pressure is a point function and it cannot have two values at a specified point. In practice this rule is used by requiring the algebraic sum of head losses in a loop (for all loops) be equal to zero. (A head loss is taken to be positive for flow in the clockwise direction and negative for flow in the counterclockwise direction.)

Therefore, the analysis of piping networks is very similar to the analysis of electric circuits, with flow rate corresponding to electric current and pressure corresponding to electric potential. However, the situation is much more complex here since, unlike the electric resistance, the "flow resistance" is a highly nonlinear function. Therefore, the analysis of piping networks requires the solution of a system of nonlinear equations simultaneously. The analysis of such systems is beyond the scope of this introductory text.

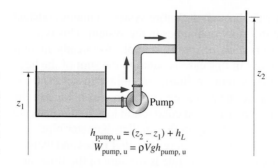

FIGURE 12-35

When a pump moves a fluid from one reservoir to another, the useful pump head requirement is equal to the elevation difference between the two reservoirs plus the head loss.

$$h_{\text{pump, u}} = (z_2 - z_1) + h_L$$
$$\dot{W}_{\text{pump, u}} = \rho \dot{V} g h_{\text{pump, u}}$$

Energy Equation Revisited

When a piping system involves a pump and/or turbine, the steady flow energy equation on a unit mass basis can be expressed as (see Sec. 11-4)

$$\frac{P_1}{\rho} + \frac{V_1^2}{2} + g z_1 + w_{\text{pump, u}} = \frac{P_2}{\rho} + \frac{V_2^2}{2} + g z_2 + w_{\text{turbine}} + g h_L \quad (12\text{-}36)$$

Dividing each term by g gives the energy equation in terms of head losses as

$$\frac{P_1}{\rho g} + \frac{V_1^2}{2g} + z_1 + h_{\text{pump, u}} = \frac{P_2}{\rho g} + \frac{V_2^2}{2g} + z_2 + h_{\text{turbine}} + h_L \quad (12\text{-}37)$$

where $h_{\text{pump, u}} = w_{\text{pump, u}}/g$ is the useful pump (or fan) head, $h_{\text{turbine}} = w_{\text{turbine}}/g$ is the turbine head, and h_L is the head loss (including the minor losses if they are significant) between points 1 and 2. The pump head is zero if the piping system does not involve a pump or a fan, the turbine head is zero if the system does not involve a turbine, and both are zero if the system does not involve any mechanical work-producing or work-consuming devices.

Many practical piping systems involve a pump to move a fluid from one reservoir to another. Taking points 1 and 2 to be at the *free surfaces* of the reservoirs, the energy equation in this case reduces for the useful pump head required to (Fig. 12-35)

$$h_{\text{pump, u}} = (z_2 - z_1) + h_L \quad (12\text{-}38)$$

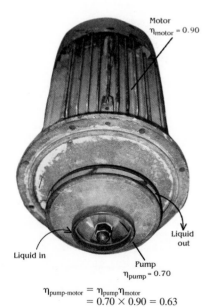

Motor
$\eta_{\text{motor}} = 0.90$

Liquid in

Liquid out

Pump
$\eta_{\text{pump}} = 0.70$

$\eta_{\text{pump-motor}} = \eta_{\text{pump}}\eta_{\text{motor}}$
$= 0.70 \times 0.90 = 0.63$

FIGURE 12-36

The efficiency of the pump-motor combination is the product of the pump and the motor efficiencies.

since the velocities at free surfaces are negligible and the pressures are at-mospheric pressure. Therefore, the pump head is equal to the elevation differ-ence between the two reservoirs plus the head loss. In the case of zero head loss (ideal frictionless flow), the pump head is simply equal to the elevation difference between the two reservoirs. In the case of $z_1 > z_2$ (the first reservoir being at a higher elevation than the second one) with no pump, the flow is dri-ven by gravity at a flow rate that causes a head loss equal to the elevation dif-ference. A similar argument can be given for the turbine head for a hydroelectric power plant by replacing $h_{\text{pump, u}}$ in Eq. 12-38 by $-h_{\text{turbine}}$.

Once the useful pump head is known, the *mechanical power that needs to be delivered by the pump to the fluid* and the *electric power consumed by the motor of the pump* for a specified flow rate are determined from

$$\dot{W}_{\text{pump, shaft}} = \frac{\dot{W}_{\text{pump, u}}}{\eta_{\text{pump}}} = \frac{\rho \dot{V} g h_{\text{pump, u}}}{\eta_{\text{pump}}} \quad \text{and} \quad \dot{W}_{\text{elect}} = \frac{\dot{W}_{\text{pump, u}}}{\eta_{\text{pump-motor}}} \quad (12\text{-}39)$$

where $\eta_{\text{pump-motor}}$ is the *efficiency of the pump-motor combination*, which is the product of the pump and the motor efficiencies (Fig. 12-36). The pump-motor

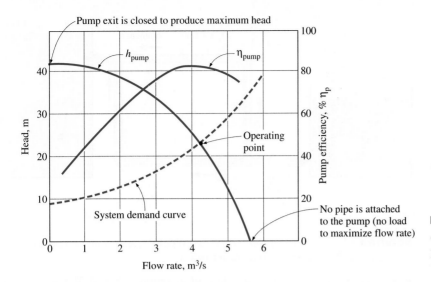

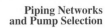

FIGURE 12-37

Characteristic pump curves for centrifugal pumps, the system curve for a piping system, and the operating point.

efficiency is defined as the ratio of the net mechanical energy delivered to the fluid by the pump to the electric energy consumed by the motor of the pump, and it usually ranges between 50 and 80 percent.

The head loss of a piping system increases (usually quadratically) with the flow rate. The plot of the head loss versus the flow rate is called the **system** (or **demand**) **curve.** The head produced by a pump is not a constant either. Both the pump head and the pump efficiency vary with the flow rate, and pump manufacturers supply this variation in graphical form, as shown in Fig. 12-37. These experimentally determined h_{pump} and η_{pump} versus $\dot{V}$ curves are called **characteristic** (or **supply**) **curves.** Note that the flow rate of a pump increases as the required head decreases. The intersection point of the pump head curve with the vertical axis represents the *maximum head* the pump can provide, while the intersection point with the horizontal axis indicates the *maximum flow rate* the pump can supply.

The *efficiency* of a pump is sufficiently high for a certain range of head and flow rate combination. Therefore, a pump that can supply the required head and flow rate is not necessarily a good choice for a piping system unless the efficiency of the pump at those conditions is sufficiently high. The pump installed in a piping system will operate at the point where the *system curve* and the *characteristic curve* intersect. This point of intersection is called the **operating point,** as shown in Fig. 12-37. The head produced by the pump at this point matches the head requirements of the system at that flow rate. Also, the efficiency of the pump during operation is the value corresponding to that flow rate.

EXAMPLE 12-6   Pumping Water through Two Parallel Pipes

Water at 20°C is to be pumped from a reservoir ($z_A = 5$ m) to another reservoir at a higher elevation ($z_B = 13$ m) through two 36-m-long pipes connected in parallel, as shown in Fig. 12-38. The pipes are made of commercial steel, and the diameters of the two pipes are 4 cm and 8 cm. Water is to be pumped by a 70 percent efficient motor-pump combination that draws 8 kW of electric power during operation. The minor losses and the head loss in pipes that connect the parallel pipes to the two reservoirs are considered to be negligible. Determine the total flow rate between the reservoirs and the flow rates through each of the parallel pipes.

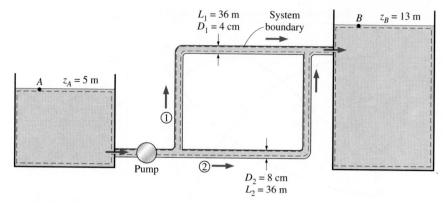

FIGURE 12-38

The piping system discussed in Example 12-6.

Solution The pumping power input to a piping system with two parallel pipes is given. The flow rates are to be determined.

Assumptions **1** The flow is steady and incompressible. **2** The entrance effects are negligible, and thus the flow is fully developed. **3** The elevations of the reservoirs remain constant. **4** The minor losses and the head loss in pipes other than the parallel pipes are said to be negligible. **5** The flows through both pipes are turbulent (to be verified).

Properties The density and dynamic viscosity of water at 20°C are $\rho = 998$ kg/m³ and $\mu = 1.002 \times 10^{-3}$ kg/m · s (Table A-15). The roughness of commercial steel pipe is $\varepsilon = 0.000045$ m (Fig. A-27).

Analysis This problem cannot be solved directly since the velocities (or flow rates) in the pipes are not known. Therefore, we would normally use a trial-and-error approach here. However, nowadays equation solvers such as EES are widely available, and thus below we will simply set up the equations to be solved by an equation solver. The head supplied by the pump to the fluid is determined from

$$\dot{W}_{elect} = \frac{\rho \dot{V} g h_{pump,\,u}}{\eta_{pump\text{-}motor}} \quad \rightarrow \quad 8000 \text{ W} = \frac{(998 \text{ kg/m}^3)\dot{V}(9.81 \text{ m/s}^2)h_{pump,\,u}}{0.70} \quad (1)$$

We choose points A and B at the free surfaces of the two reservoirs. Noting that the fluid at both points is open to the atmosphere (and thus $P_A = P_B = P_{atm}$) and that the fluid velocities at both points are zero ($\mathcal{V}_A = \mathcal{V}_B = 0$), the energy equation between these two points simplifies to

$$\frac{\cancel{P_A}}{\rho g} + \frac{\cancel{\mathcal{V}_A^2}^{\,0}}{2g} + z_A + h_{pump,\,u} = \frac{\cancel{P_B}}{\rho g} + \frac{\cancel{\mathcal{V}_B^2}^{\,0}}{2g} + z_B + h_L \quad \rightarrow \quad h_{pump,\,u} = (z_B - z_A) + h_L$$

or

$$h_{pump,\,u} = (13 - 5) + h_L \quad (2)$$

where

$$h_L = h_{L,\,1} = h_{L,\,2} \quad (3)(4)$$

We designate the 4-cm-diameter pipe by 1 and the 8-cm-diameter pipe by 2. The average velocity, the Reynolds number, the friction factor, and the head loss in each pipe are expressed as

$$\mathcal{V}_1 = \frac{\dot{V}_1}{A_{c,\,1}} = \frac{\dot{V}_1}{\pi D_1^2/4} \quad \rightarrow \quad \mathcal{V}_1 = \frac{\dot{V}_1}{\pi (0.04 \text{ m})^2/4} \quad (5)$$

$$\mathcal{V}_2 = \frac{\dot{V}_2}{A_{c,\,2}} = \frac{\dot{V}_2}{\pi D_2^2/4} \quad \rightarrow \quad \mathcal{V}_2 = \frac{\dot{V}_2}{\pi (0.08 \text{ m})^2/4} \quad (6)$$

$$Re_1 = \frac{\rho \mathcal{V}_1 D_1}{\mu} \quad \rightarrow \quad Re_1 = \frac{(998 \text{ kg/m}^3)\mathcal{V}_1(0.04 \text{ m})}{1.002 \times 10^{-3} \text{ kg/m} \cdot \text{s}} \tag{7}$$

$$Re_2 = \frac{\rho \mathcal{V}_2 D_2}{\mu} \quad \rightarrow \quad Re_2 = \frac{(998 \text{ kg/m}^3)\mathcal{V}_2(0.08 \text{ m})}{1.002 \times 10^{-3} \text{ kg/m} \cdot \text{s}} \tag{8}$$

$$\frac{1}{\sqrt{f_1}} = -2.0 \log\left(\frac{\varepsilon/D_1}{3.7} + \frac{2.51}{Re_1 \sqrt{f_1}}\right) \quad \rightarrow \quad \frac{1}{\sqrt{f_1}} = -2.0 \log\left(\frac{0.000045}{3.7 \times 0.04} + \frac{2.51}{Re_1 \sqrt{f_1}}\right) \tag{9}$$

$$\frac{1}{\sqrt{f_2}} = -2.0 \log\left(\frac{\varepsilon/D_2}{3.7} + \frac{2.51}{Re_2 \sqrt{f_2}}\right) \quad \rightarrow \quad \frac{1}{\sqrt{f_2}} = -2.0 \log\left(\frac{0.000045}{3.7 \times 0.08} + \frac{2.51}{Re_2 \sqrt{f_2}}\right) \tag{10}$$

$$h_{L,1} = f_1 \frac{L_1}{D_1}\frac{\mathcal{V}_1^2}{2g} \quad \rightarrow \quad h_{L,1} = f_1 \frac{36 \text{ m}}{0.04 \text{ m}}\frac{\mathcal{V}_1^2}{2(9.81 \text{ m/s}^2)} \tag{11}$$

$$h_{L,2} = f_2 \frac{L_2}{D_2}\frac{\mathcal{V}_2^2}{2g} \quad \rightarrow \quad h_{L,2} = f_2 \frac{36 \text{ m}}{0.08 \text{ m}}\frac{\mathcal{V}_2^2}{2(9.81 \text{ m/s}^2)} \tag{12}$$

$$\dot{V} = \dot{V}_1 + \dot{V}_2 \tag{13}$$

This is a system of 13 equations in 13 unknowns, and their simultaneous solution by an equation solver gives

$$\dot{V} = 0.0300 \text{ m}^3/\text{s}, \quad \dot{V}_1 = 0.00415 \text{ m}^3/\text{s}, \quad \dot{V}_2 = 0.0259 \text{ m}^3/\text{s}$$

$$\mathcal{V}_1 = 3.30 \text{ m/s}, \quad \mathcal{V}_2 = 5.15 \text{ m/s}, \quad h_L = h_{L,1} = h_{L,2} = 11.1 \text{ m}, \quad h_{\text{pump, u}} = 19.1 \text{ m}$$

$$Re_1 = 131{,}600, \quad Re_2 = 410{,}000, \quad f_1 = 0.0221, \quad f_2 = 0.0182$$

Note that Re > 4000 for both pipes, and thus the assumption of turbulent flow is verified.

Discussion The two parallel pipes are identical, except the diameter of the first pipe is half the diameter of the second one. But only 14 percent of the water flows through the first pipe. This shows the strong dependence of the flow rate (and the head loss) to diameter. Also, it can be shown that if the free surfaces of the two reservoirs were at the same elevation (and thus $z_A = z_B$), the flow rate would increase by 20 percent from 0.0300 to 0.0361 m³/s. Alternately, if the reservoirs were as given but the flows were frictionless (and thus zero head loss), the flow rate would become 0.0715 m³/s (an increase of 138 percent).

EXAMPLE 12-7 Gravity-Driven Water Flow in a Pipe

Water at 10°C flows from a large reservoir to a smaller one through a 5-cm-diameter cast iron piping system, shown in Fig. 12-39. Determine the elevation z_1 for a flow rate of 6 L/s.

Solution The flow rate through a piping system connecting two reservoirs is given. The elevation of the source is to be determined.

Assumptions **1** The flow is steady and incompressible. **2** The entrance effects are negligible, and thus the flow is fully developed. **3** The elevations of the reservoirs remain constant. **4** There are no pumps or turbines in the line.

Properties The density and dynamic viscosity of water at 10°C are $\rho = 999.7$ kg/m³ and $\mu = 1.307 \times 10^{-3}$ kg/m · s (Table A-15). The roughness of cast iron pipe is $\varepsilon = 0.00026$ m (Fig. A-27).

Analysis The piping system involves 89 m of piping, a sharp-edged entrance ($K_L = 0.5$), 2 standard flanged elbows ($K_L = 0.3$ each), a fully open gate valve

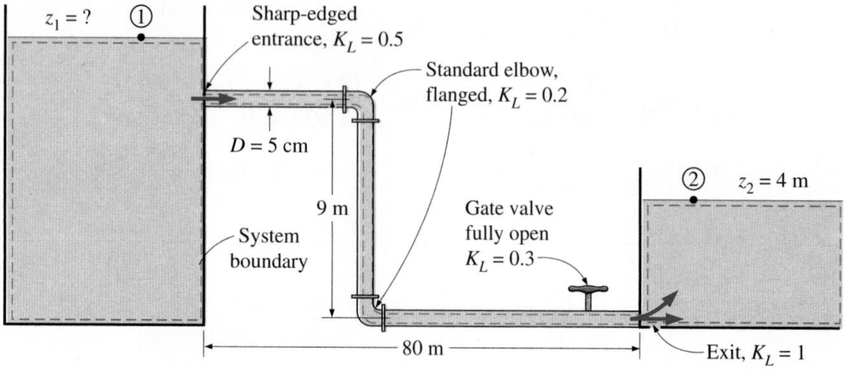

FIGURE 12-39

The piping system discussed in Example 12-7.

($K_L = 0.2$), and an exit ($K_L = 1.0$). We choose points 1 and 2 at the free surfaces of the two reservoirs. Noting that the fluid at both points is open to the atmosphere (and thus $P_1 = P_2 = P_{atm}$) and that the fluid velocities at both points are zero ($V_1 = V_2 = 0$), the energy equation for a control volume between these two points simplifies to

$$\frac{P_1}{\rho g} + \frac{V_1^2}{2g} \cancelto{0}{} + z_1 = \frac{P_2}{\rho g} + \frac{V_2^2}{2g} \cancelto{0}{} + z_2 + h_L \quad \rightarrow \quad z_1 = z_2 + h_L$$

where

$$h_L = h_{L,\,total} = h_{L,\,major} + h_{L,\,minor} = \left(f\frac{L}{D} + \sum K_L\right)\frac{V^2}{2g}$$

since the diameter of the piping system is constant. The average velocity in the pipe and the Reynolds number are

$$V = \frac{\dot{V}}{A_c} = \frac{\dot{V}}{\pi D^2/4} = \frac{0.006 \text{ m}^3/\text{s}}{\pi(0.05 \text{ m})^2/4} = 3.06 \text{ m/s}$$

$$Re = \frac{\rho V D}{\mu} = \frac{(999.7 \text{ kg/m}^3)(3.06 \text{ m/s})(0.05 \text{ m})}{1.307 \times 10^{-3}\text{ kg/m} \cdot \text{s}} = 117{,}000$$

The flow is turbulent since Re > 4000. Noting that $\varepsilon/D = 0.00026/0.05 = 0.052$, the friction factor can be determined from the Colebrook equation (or the Moody chart),

$$\frac{1}{\sqrt{f}} = -2.0 \log\left(\frac{\varepsilon/D}{3.7} + \frac{2.51}{Re \sqrt{f}}\right) \quad \rightarrow \quad \frac{1}{\sqrt{f}} = -2.0 \log\left(\frac{0.0052}{3.7} + \frac{2.51}{117{,}000 \sqrt{f}}\right)$$

It gives $f = 0.0315$. The sum of the loss coefficients is

$$\sum K_L = K_{L,\,entrance} + 2K_{L,\,elbow} + K_{L,\,valve} + K_{L,\,exit} = 0.5 + 2 \times 0.3 + 0.2 + 1.0 = 2.3$$

Then the total head loss and the elevation of the source become

$$h_L = \left(f\frac{L}{D} + \sum K_L\right)\frac{V^2}{2g} = \left(0.0315\frac{89 \text{ m}}{0.05 \text{ m}} + 2.3\right)\frac{(3.06 \text{ m/s})^2}{2(9.81 \text{ m/s}^2)} = 27.9 \text{ m}$$

$$z_1 = z_2 + h_L = 4 + 27.9 = \textbf{31.9 m}$$

Therefore, the free surface of the first reservoir must be 31.9 m above the ground level to ensure water flow between the two reservoirs at the specified rate.

Discussion Note that $fL/D = 56.1$ in this case, which is about 24 times the total minor loss coefficient. Therefore, ignoring the sources of minor losses in this case would result in about 4 percent error.

It can be shown that the total head loss would be 35.9 m (instead of 27.9 m) if the valve were three-fourths closed, and it would drop to 24.8 m if the pipe between the two reservoirs were straight at the ground level (thus eliminating the elbows and the vertical section of the pipe). The head loss could be reduced further (from 24.8 to 24.6 m) by rounding the entrance. The head loss can be reduced from 27.9 to 16.0 m by replacing the cast iron pipes by smooth pipes such as those made of plastic.

12-7 ■ SUMMARY

In *internal flow,* a pipe is completely filled with fluid. *Laminar flow* is characterized by smooth streamlines and highly ordered motion, and *turbulent flow* is characterized by velocity fluctuations and highly disordered motion. The *Reynolds number* is defined as

$$\text{Re} = \frac{\text{Inertia forces}}{\text{Viscous forces}} = \frac{\mathcal{V}_m D}{\nu}$$

Under most practical conditions, the flow in a pipe is laminar at $\text{Re} < 2300$, turbulent at $\text{Re} > 4000$, and transitional in between.

The region of the flow in which the effects of the viscous shearing forces are felt is called the *velocity boundary layer.* The region from the pipe inlet to the point at which the boundary layer merges at the centerline is called the *hydrodynamic entry region,* and the length of this region is called the *hydrodynamic entry length L_h.* It is given by

$$L_{h,\,\text{laminar}} \approx 0.06\,\text{Re}\,D \qquad \text{and} \qquad L_{h,\,\text{turbulent}} \approx 4.4D(\text{Re})^{1/6}$$

The friction coefficient in the developed flow region remains constant. The *maximum* and *mean* velocities in fully developed laminar flow in a circular pipe are

$$\mathcal{V}_{\text{max}} = \mathcal{V}_{\text{center}} = \frac{\Delta P R^2}{4\mu L} \qquad \text{and} \qquad \mathcal{V}_m = \frac{\Delta P R^2}{8\mu L} = \frac{\mathcal{V}_{\text{max}}}{2}$$

The *volume flow rate* and the *pressure drop* for laminar flow in a horizontal pipe are

$$\dot{V} = \mathcal{V}_{\text{ave}} A_c = \frac{\pi D^4 \Delta P}{32\mu L} \qquad \text{and} \qquad \Delta P = \frac{8\mu L \mathcal{V}_m}{R^2} = \frac{32\mu L \mathcal{V}_m}{D^2}$$

The results obtained above for horizontal pipes can also be used for inclined pipes provided that ΔP is replaced by $\Delta P - \rho g L \sin\theta$,

$$\mathcal{V}_m = \frac{(\Delta P - \rho g L \sin\theta)D^2}{32\mu L} \qquad \text{and} \qquad \dot{V} = \frac{(\Delta P - \rho g L \sin\theta)\pi D^4}{128\mu L}$$

The *pressure drop* and *head loss* for all types of internal flows (laminar or turbulent, in circular or noncircular pipes, smooth or rough surfaces) are

$$\Delta P = f \frac{L}{D} \frac{\rho \mathcal{V}_m^2}{2} \qquad \text{and} \qquad h_L = \frac{\Delta P}{\rho g} = f \frac{L}{D} \frac{\mathcal{V}_m^2}{2g}$$

where $\rho \mathcal{V}_m^2/2$ is the *dynamic pressure* and the dimensionless quantity f is the *friction factor.* For fully developed laminar flow in a circular pipe, the friction factor is $f = 64/\text{Re}$. When the pressure drop is available, the required pumping power is

$$\dot{W}_{\text{pump}} = \dot{V} \Delta P$$

For noncircular pipes, the diameter in the above relations is replaced by the *hydraulic diameter D_h,* defined as $D_h = 4A_c/p$, where A_c is the cross-sectional area of the pipe and p is its perimeter.

In fully developed turbulent flow, the friction coefficient depends on the Reynolds number and the *relative roughness* ε/D. The friction coefficient in turbulent flow is given by the *Colebrook equation,* expressed as

$$\frac{1}{\sqrt{f}} = -2.0 \log\left(\frac{\varepsilon/D}{3.7} + \frac{2.51}{\text{Re}\,\sqrt{f}}\right)$$

The plot of this formula is known as the Moody chart. The design and analysis of piping systems involve the determination of the head loss, the flow rate, or the pipe diameter. Tedious iterations in these calculations can be avoided by the approximate Swamee and Jain formulas expressed as

$$h_L = 1.07\frac{\dot{V}^2 L}{gD^5}\left\{\ln\left[\frac{\varepsilon}{3.7D} + 4.62\left(\frac{vD}{\dot{V}}\right)^{0.9}\right]\right\}^{-2} \qquad \begin{array}{l} 10^{-6} < \varepsilon/D < 10^{-2} \\ 3000 < \text{Re} < 3\times10^8 \end{array}$$

$$\dot{V} = -0.965\left(\frac{gD^5 h_L}{L}\right)^{0.5}\ln\left[\frac{\varepsilon}{3.7D} + \left(\frac{3.17v^2 L}{gD^3 h_L}\right)^{0.5}\right] \qquad \text{Re} > 2000$$

$$D = 0.66\left[\varepsilon^{1.25}\left(\frac{L\dot{V}^2}{gh_L}\right)^{4.75} + v\dot{V}^{9.4}\left(\frac{L}{gh_L}\right)^{5.2}\right]^{0.04} \qquad \begin{array}{l} 10^{-6} < \varepsilon/D < 10^{-2} \\ 5000 < \text{Re} < 3\times10^8 \end{array}$$

The losses that occur in the piping components such as the fittings, valves, bends, elbows, tees, inlets, exits, enlargements, and contractions are called *minor losses.* The minor losses are usually expressed in terms of the *loss coefficient* K_L. The head loss for a component is determined from

Minor loss: $$h_L = K_L\frac{V^2}{2g}$$

When all the loss coefficients are available, the total head loss in a piping system is determined from

$$h_{L,\text{total}} = h_{L,\text{major}} + h_{L,\text{minor}} = \sum f_i\frac{L_i}{D_i}\frac{V_i^2}{2g} + \sum K_{L,j}\frac{V_j^2}{2g}$$

If the entire piping system has a constant diameter, it reduces to

$$h_{L,\text{total}} = \left(f\frac{L}{D} + \sum K_L\right)\frac{V^2}{2g}$$

The analysis of a piping system is based on two simple principles: (1) The conservation of mass throughout the system must be satisfied and (2) the pressure drop (and thus head loss) between two points must be the same for all paths between the two points. When the pipes are connected *in series,* the flow rate through the entire system remains constant regardless of the diameters of the individual pipes. For a pipe that branches out into two (or more) *parallel pipes* and then rejoins at a junction downstream, the total flow rate is the sum of the flow rates in the individual pipes.

When a piping system involves a pump and/or turbine, the steady flow energy equation can be expressed as

$$\frac{P_1}{\rho g} + \frac{V_1^2}{2g} + z_1 + h_{\text{pump, u}} = \frac{P_2}{\rho g} + \frac{V_2^2}{2g} + z_2 + h_{\text{turbine}} + h_L$$

When the useful pump head $h_{\text{pump, u}}$ is known, the mechanical power that needs to be supplied by the pump to the fluid and the electric power consumed by the motor of the pump for a specified flow rate are determined from

$$\dot{W}_{\text{pump, shaft}} = \frac{\dot{W}_{\text{pump, u}}}{\eta_{\text{pump}}} = \frac{\rho\dot{V}gh_{\text{pump, u}}}{\eta_{\text{pump}}} \qquad \text{and} \qquad \dot{W}_{\text{elect}} = \frac{\dot{W}_{\text{pump, u}}}{\eta_{\text{pump-motor}}}$$

where $\eta_{\text{pump-motor}}$ is the *efficiency of the pump-motor combination,* which is the product of the pump and the motor efficiencies.

The plot of the head loss versus the flow rate $\dot{V}$ is called the *system curve.* The head produced by a pump is not a constant either. The h_{pump} and η_{pump} versus $\dot{V}$ curves of pumps are called the *characteristic curves.* The pump installed in a piping system will operate at the *operating point,* which is the point of intersection of the system curve and the characteristic curve.

REFERENCES AND SUGGESTED READING

1. C. F. Colebrook. "Turbulent Flow in Pipes, with Particular Reference to the Transition between the Smooth and Rough Pipe Laws." *Journal of the Institute of Civil Engineers London.* 11 (1939), pp. 133–56.

2. R. W. Fox and A. T. McDonald. *Introduction to Fluid Mechanics.* 5th ed. New York: John Wiley & Sons, 1999.

3. S. E. Haaland. "Simple and Explicit Formulas for the Friction Factor in Turbulent Pipe Flow." *Journal of Fluids Engineering,* March 1983, pp. 89–90.

4. L. F. Moody. "Friction Factors for Pipe Flows." *Transactions of the ASME* 66 (1944), pp. 671–84.

5. B. R. Munson, D. F. Young, and T. Okiishi. *Fundamentals of Fluid Mechanics.* 3rd ed. New York: John Wiley & Sons, 1998.

6. M. C. Potter and D. C. Wiggert. *Mechanics of Fluids.* 2nd ed. Upper Saddle River, NJ: Prentice Hall, 1997.

7. O. Reynolds. "On the Experimental Investigation of the Circumstances Which Determine Whether the Motion of Water Shall Be Direct or Sinuous, and the Law of Resistance in Parallel Channels." *Philosophical Transactions of the Royal Society of London* 174 (1883), pp. 935–82.

8. J. A. Roberson and C. L. Grove. *Engineering Fluid Mechanics.* 6th ed. New York: John Wiley & Sons, 1997.

9. H. Schlichting. *Boundary Layer Theory,* 7th ed. New York: McGraw-Hill, 1979.

10. P. K. Swamee and A. K. Jain. "Explicit Equations for Pipe-Flow Problems." *Journal of the Hydraulics Division,* ASCE 102, no. HY5 (May 1976), pp. 657–64.

11. F. M. White. *Fluid Mechanics.* 4th ed. New York: McGraw-Hill, 1999.

PROBLEMS*

Laminar and Turbulent Flow

12-1C Why are liquids usually transported in circular pipes?

12-2C What is the physical significance of the Reynolds number? How is it defined for (*a*) flow in a circular tube of inner diameter D and (*b*) flow in a rectangular duct of cross section $a \times b$?

FIGURE P12-2C

*Students are encouraged to answer *all* the concept "C" questions.

12-3C Consider a person walking first in air and then in the water at the same speed. For which motion will the Reynolds number be higher?

12-4C Show that the Reynolds number for flow in a circular tube of diameter D can be expressed as Re $= 4\dot{m}/\pi D\mu$.

12-5C Which fluid at room temperature requires a larger pump to move at a specified velocity in a specified tube: water or engine oil? Why?

12-6C What is the generally accepted value of the Reynolds number above which the flow in smooth pipes is turbulent?

12-7C In the fully developed region of flow in a circular tube, will the velocity profile change in the flow direction?

12-8C How is the hydrodynamic entry length defined for flow in a tube? Is the entry length longer in laminar or turbulent flow?

12-9C Consider laminar flow in a circular tube. Will the friction factor be higher near the inlet of the tube or near the exit? Why? What would your response be if the flow were turbulent?

12-10C How does surface roughness affect the pressure drop in a tube if the flow is turbulent? What would your response be if the flow were laminar?

12-11C How does the friction factor f vary along the flow direction in the fully developed region in (*a*) laminar flow and (*b*) turbulent flow?

Fully Developed Flow in Pipes

12-12C What fluid property is responsible for the development of the velocity boundary layer? For what kinds of fluids will there be no velocity boundary layer in a pipe?

12-13C How is the friction factor for flow in a tube related to the pressure drop? How is the pressure drop related to the pumping power requirement for a given mass flow rate?

12-14C Someone claims that the shear stress at the center of a circular pipe during fully developed laminar flow is zero. Do you agree with this claim? Explain.

12-15C Consider fully developed flow in a circular pipe with negligible entrance effects. If the length of the pipe is doubled, the pressure drop will (*a*) double, (*b*) more than double, (*c*) less than double, (*d*) reduce by half, or (*e*) remain constant.

12-16C Someone claims that the volume flow rate in a circular pipe with laminar flow can be determined by measuring the velocity at the centerline in the fully developed region, multiplying it by the cross-sectional area, and dividing the result by 2. Do you agree? Explain.

12-17C Someone claims that the average velocity in a circular pipe in fully developed laminar flow can be determined by simply measuring the velocity at $R/2$ (midway between the wall surface and the centerline). Do you agree? Explain.

12-18C Consider fully developed laminar flow in a circular pipe. If the diameter of the pipe is reduced by half while the flow rate and the pipe length are held constant, the pressure drop will (*a*) double, (*b*) triple, (*c*) quadruple, (*d*) increase by a factor of 8, (*e*) increase by a factor of 16.

12-19C Consider fully developed laminar flow in a circular pipe. If the viscosity of the fluid is reduced by half by heating while the flow rate is held constant, how will the pressure drop change?

12-20C How is head loss related to pressure drop? For a given fluid, explain how you would convert head loss to pressure drop.

12-21C What is the physical mechanism that causes the friction factor to be higher in turbulent flow?

12-22C Explain why the friction factor is independent of the Reynolds number at very large Reynolds numbers.

12-23E Oil at 80°F ($\rho = 56.8$ lbm/ft^3 and $\mu = 0.0278$ lbm/ft · s) is flowing steadily in a 0.5-in.-diameter, 120-ft-long pipe. During the flow, the pressure at the pipe inlet and exit is measured to be 120 psi and 14 psi, respectively. Determine the flow rate of oil through the pipe assuming the pipe is (*a*) horizontal, (*b*) inclined 20° upwards, and (*c*) inclined 20° downwards.

12-24 Oil with a density of 850 kg/m^3 and kinematic viscosity of 0.00062 m^2/s is being discharged by a 5-mm-diameter, 40-m-long horizontal pipe from a storage tank open to the atmosphere. The height of the liquid level above the center of the pipe is 2 m. Disregarding the minor losses, determine the flow rate of oil through the pipe.

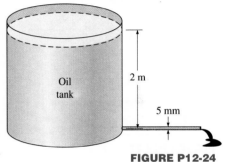

FIGURE P12-24

12-25 Water at 10°C ($\rho = 999.7$ kg/m^3 and $\mu = 1.307 \times 10^{-3}$ kg/m · s) is flowing steadily in a 0.20-cm-diameter, 15-m-long pipe at an average velocity of 1.2 m/s. Determine (*a*) the pressure drop, (*b*) the head loss, and (*c*) the pumping power requirement to overcome this pressure drop.
   Answers: (*a*) 188 kPa, (*b*) 19.2 m, (*c*) 0.71 W

12-26 Water at 15°C ($\rho = 999.1$ kg/m^3 and $\mu = 1.138 \times 10^{-3}$ kg/m · s) is flowing steadily in a 30-m-long and 4-cm-diameter horizontal pipe made of stainless steel at a rate of 5 L/s. Determine (*a*) the pressure drop, (*b*) the head loss, and (*c*) the pumping power requirement to overcome this pressure drop.

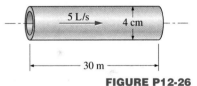

FIGURE P12-26

12-27E Heated air at 1 atm and 100°F is to be transported in a 400-ft-long circular plastic duct at a rate of 12 ft^3/s. If the head loss in the pipe is not to exceed 50 ft, determine the minimum diameter of the duct.

12-28 In fully developed laminar flow in a circular pipe, the velocity at $R/2$ (midway between the wall surface and the centerline) is measured to be 6 m/s. Determine the velocity at the center of the pipe.   *Answer:* 8 m/s

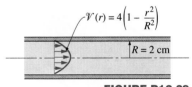

FIGURE P12-29

12-29 The velocity profile in fully developed laminar flow in a circular pipe of inner radius $R = 2$ cm, in m/s, is given by $\mathcal{V}(r) = 4(1 - r^2/R^2)$. Determine the mean and maximum velocities in the pipe and the volume flow rate.

12-30 Repeat Prob. 12-29 for a pipe of inner radius 5 cm.

12-31 Consider an air solar collector that is 1 m wide and 5 m long and has a constant spacing of 3 cm between the glass cover and the collector plate. Air flows at an average temperature of 47°C at a rate of 0.15 m^3/s through the 1-m-wide edge of the collector along the 5-m-long passageway. Disregarding the entrance and roughness effects, determine the pressure drop in the collector.   *Answer:* 29 Pa

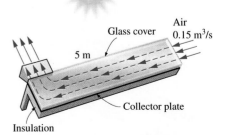

FIGURE P12-31

12-32 Consider the flow of oil with $\rho = 894$ kg/m^3 and $\mu = 2.33$ kg/m · s in a 40-cm-diameter pipeline at an average velocity of 0.5 m/s. A 300-m-long section of the pipeline passes through the icy waters of a lake. Disregarding the entrance effects, determine the pumping power required to overcome the pressure losses and to maintain the flow of oil in the pipe.

12-33 Consider laminar flow of a fluid through a square channel with smooth surfaces. Now the mean velocity of the fluid is doubled. Determine the change in the pressure drop of the fluid. Assume the flow regime remains unchanged.

12-34 Repeat Prob. 12-33 for turbulent flow in smooth pipes for which the friction factor is given as $f = 0.184 \, Re^{-0.2}$. What would your answer be for fully turbulent flow in a rough pipe?

12-35 Air enters a 7-m-long section of a rectangular duct of cross section 15 cm × 20 cm made of commercial steel at 1 atm and 37°C at an average velocity of 7 m/s. Disregarding the entrance effects, determine the fan power needed to overcome the pressure losses in this section of the duct.
Answer: 4.9 W

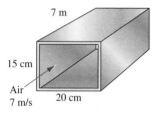

7 m

15 cm

Air
7 m/s   20 cm

FIGURE P12-35

12-36E Water at 60°F passes through 0.75-in.-internal-diameter copper tubes at a rate of 0.7 lbm/s. Determine the pumping power per ft of pipe length required to maintain this flow at the specified rate.

12-37 Oil with $\rho = 876$ kg/m³ and $\mu = 0.24$ kg/m · s is flowing through a 1.5-cm-diameter pipe that discharges into the atmosphere at 88 kPa. The absolute pressure 15 m before the exit is measured to be 135 kPa. Determine the flow rate of oil through the pipe if the pipe is (*a*) horizontal, (*b*) inclined 8° upwards from the horizontal, and (*c*) inclined 8° downwards from the horizontal.

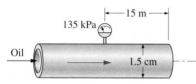

135 kPa   ⊢—15 m—⊣

Oil

1.5 cm

FIGURE P12-37

12-38 Glycerin at 40°C with $\rho = 1252$ kg/m³ and $\mu = 0.27$ kg/m· s is flowing through a 2-cm-diameter, 25-m-long pipe that discharges into the atmosphere at 100 kPa. The flow rate through the pipe is 0.035 L/s. (*a*) Determine the absolute pressure 25 m before the pipe exit. (*b*) At what angle θ must the pipe be inclined downwards from the horizontal for the pressure in the entire pipe to be atmospheric pressure and the flow rate to be maintained the same?

12-39 In an air heating system, heated air at 43°C and 105 kPa absolute is distributed through a 0.2 m × 0.3 m rectangular duct made of commercial steel duct at a rate of 0.5 m³/s. Determine the pressure drop and head loss through a 40-m-long section of the duct. *Answers:* 128 Pa, 93.8 m

12-40 Glycerin at 40°C with $\rho = 1252$ kg/m³ and $\mu = 0.27$ kg/m · s is flowing through a 5-cm-diameter horizontal smooth pipe with a mean velocity of 5 m/s. Determine the pressure drop per 10 m of the pipe.

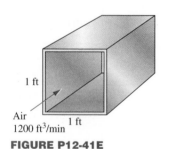

1 ft

Air
1200 ft³/min   1 ft

FIGURE P12-41E

12-41E Air at 1 atm and 60°F is flowing through a 1 ft × 1 ft square duct made of commercial steel at a rate of 1200 cfm. Determine the pressure drop and head loss per ft of the duct.

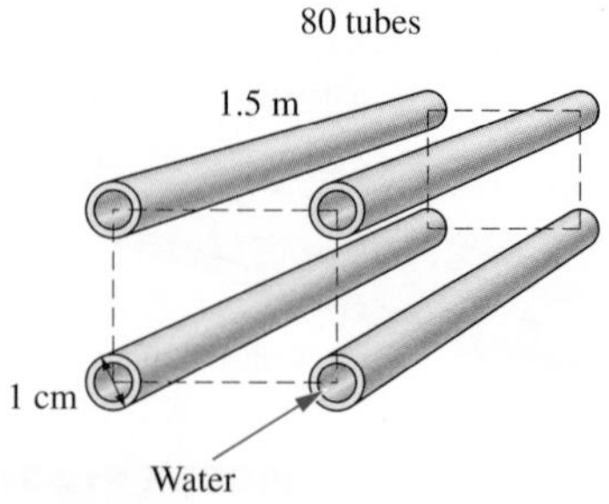

12-42 Liquid ammonia at −20°C is flowing through a 30-m-long section of a 5-mm-diameter copper tube at a rate of 0.15 kg/s. Determine the pressure drop, the head loss, and the pumping power required to overcome the frictional losses in the tube. *Answers:* 4827 kPa, 738 m, 1.09 kW

12-43 Shell-and-tube heat exchangers with hundreds of tubes housed in a shell are commonly used in practice for heat transfer between two fluids. Such a heat exchanger used in an active solar hot water system transfers heat from a water-antifreeze solution flowing through the shell and the solar collector to fresh water flowing through the tubes at an average temperature of 60°C at a rate of 15 L/s. The heat exchanger contains 80 brass tubes 1 cm in diameter

80 tubes

1.5 m

1 cm

Water

FIGURE P12-43

and 1.5 m in length. Disregarding inlet, exit, and header losses, determine the pressure drop across a single tube and the pumping power required by the tube-side fluid of the heat exchanger.

After operating a long time, 1-mm-thick scale builds up on the inner surfaces with an equivalent roughness of 0.4 mm. For the same pumping power input, determine the percent reduction in the flow rate of water through the tubes.

Minor Losses

12-44C What is minor loss in pipe flow? How is the minor loss coefficient K_L defined?

12-45C Define equivalent length for minor loss in pipe flow. How is it related to the minor loss coefficient?

12-46C The effect of rounding of a pipe inlet on the loss coefficient is (a) negligible, (b) somewhat significant, (c) very significant.

12-47C The effect of rounding of a pipe exit on the loss coefficient is (a) negligible, (b) somewhat significant, (c) very significant.

12-48C Which has a greater minor loss coefficient during pipe flow: gradual expansion or gradual contraction? Why?

12-49C A piping system involves sharp turns, and thus large minor head losses. One way of reducing the head loss is to replace the sharp turns by circular elbows. What is another way?

12-50C During a retrofitting project of a fluid flow system to reduce the pumping power, it is proposed to install vanes into the miter elbows or to replace the sharp turns in 90° miter elbows by smooth curved bends. Which approach will result in a greater reduction in pumping power requirements?

12-51 Water is to be withdrawn from a 3-m-high water reservoir by drilling a 3-cm-diameter hole at the bottom surface. Determine the flow rate of water through the hole if (a) the entrance of the hole is well-rounded and (b) the entrance is sharp-edged.

12-52 Consider flow from a water reservoir through a circular hole of diameter D at the side wall at a vertical distance H from the free surface. The flow rate through an actual hole with a sharp-edged entrance ($K_L = 0.5$) will be considerably less than the flow rate calculated assuming "frictionless" flow and thus zero loss for the hole. Obtain a relation for the "equivalent diameter" of the sharp-edged hole for use in frictionless flow relations.

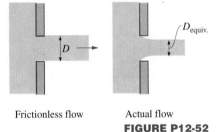

Frictionless flow　　Actual flow

FIGURE P12-52

12-53 Repeat Prob. 12-52 for a slightly rounded entrance ($K_L = 0.12$).

12-54 A horizontal pipe has an abrupt expansion from $D_1 = 10$ cm to $D_2 = 20$ cm. The water velocity in the smaller section is 10 m/s, and the flow is turbulent. The pressure in the smaller section is $P_1 = 300$ kPa. Determine the downstream pressure P_2, and estimate the error that would have occurred if Bernoulli's equation had been used.   *Answers:* 319 kPa, 28 kPa

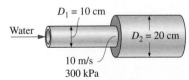

FIGURE P12-54

Piping Systems and Pump Selection

12-55C A piping system involves two pipes of different diameters (but of identical length, material, and roughness) connected in series. How would you compare the (a) flow rates and (b) pressure drops in these two pipes?

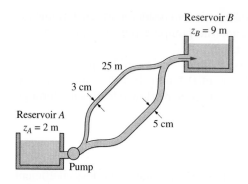

FIGURE P12-61

12-56C A piping system involves two pipes of different diameters (but of identical length, material, and roughness) connected in parallel. How would you compare the (*a*) flow rates and (*b*) pressure drops in these two pipes?

12-57C A piping system involves two pipes of identical diameters but of different lengths connected in parallel. How would you compare the pressure drops in these two pipes?

12-58C Water is pumped from a large lower reservoir to a higher reservoir. Someone claims that if the head loss is negligible, the required pump head is equal to the elevation difference between the free surfaces of the two reservoirs. Do you agree?

12-59C A piping system equipped with a pump is operating steadily. Explain how the operating point (the flow rate and the head loss) is established.

12-60C For a piping system, define the system curve, the characteristic curve, and the operating point on a head versus flow rate chart.

12-61 Water at 20°C is to be pumped from a reservoir ($z_A = 2$ m) to another reservoir at a higher elevation ($z_B = 9$ m) through two 25-m-long plastic pipes connected in parallel. The diameters of the two pipes are 3 cm and 5 cm. Water is to be pumped by a 68 percent efficient motor/pump unit that draws 7 kW of electric power during operation. The minor losses and the head loss in pipes that connect the parallel pipes to the two reservoirs are considered to be negligible. Determine the total flow rate between the reservoirs and the flow rates through each of the parallel pipes.

12-62E Water at 70°F flows by gravity from a large reservoir at a high elevation to a smaller one through a 120-ft-long, 2-in.-diameter cast iron piping system that involves four standard flanged elbows, a well-rounded entrance, a sharp-edged exit, and a fully open gate valve. Taking the free surface of the lower reservoir as the reference level, determine the elevation z_1 of the higher reservoir for a flow rate of 10 ft³/min.   *Answer:* 23.1 ft

12-63 A 10-m-diameter tank is initially filled with water 2 m above the center of a sharp-edged 10-cm-diameter orifice. The tank water surface is open to the atmosphere, and the orifice drains to the atmosphere. Calculate (*a*) the initial velocity from the tank and (*b*) the time required to empty the tank. Does the loss coefficient of the orifice cause a significant increase in the draining time of the tank?

12-64 A 10-m-diameter tank is initially filled with water 2 m above a sharp-edged 10-cm-diameter orifice. The tank water surface is open to the atmosphere, and the orifice drains to the atmosphere through a 100-m-long pipe.

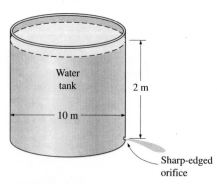

FIGURE P12-63

The friction coefficient of the pipe can be taken to be 0.015. Determine (*a*) the initial velocity from the tank and (*b*) the time required to empty the tank.

12-65   Reconsider Prob. 12-64. In order to drain the tank faster, a pump is installed near the tank exit. Determine how much pump power input is necessary to establish an average water velocity of 4 m/s when the tank is full at $z = 2$ m. Also, assuming the discharge velocity to remain constant, estimate the time required to drain the tank.

Someone suggested that it makes no difference whether the pump is located at the beginning or at the end of the pipe, and that the performance will be the same in either case, but another person argued that placing the pump near the end of the pipe may cause cavitation. The water temperature is 30°C, so the water vapor pressure is $P_v = 4.246$ kPa $= 0.43$ m-H_2O, and the system is located at sea level. Investigate if there is the possibility of cavitation and if we should be concerned about the location of the pump.

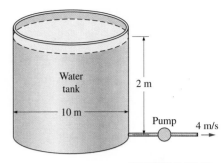

FIGURE P12-65

12-66   Oil at 20°C is flowing through a vertical glass funnel that consists of a 15-cm-high cylindrical reservoir and a 1-cm-diameter, 25-cm-high pipe. The funnel is always maintained full by the addition of oil from a tank. Assuming the entrance effects to be negligible, determine the flow rate of oil through the funnel and calculate the "funnel effectiveness," which can be defined as the ratio of the actual flow rate through the funnel to the maximum flow rate for the "frictionless" case.   *Answers:* 4.23×10^{-6} m³/s, 1.95 percent

12-67   Repeat Prob. 12-66 assuming (*a*) the diameter of the pipe is doubled and (*b*) the length of the pipe is doubled.

FIGURE P12-66

12-68   Water at 15°C is drained from a large reservoir using two horizontal plastic pipes connected in series. The first pipe is 20 m long and has a 10-cm diameter while the second pipe is 35 m long and has a 4-cm diameter. The water level in the reservoir is 30 m above the centerline of the pipe. The pipe entrance is sharp-edged, and the contraction between the two pipes is sudden. Determine the discharge rate of water from the reservoir.

12-69E   A farmer is to pump water at 70°F from a river to a water storage tank nearby using a total of 125 ft-long, 5-in.-diameter plastic pipes with three flanged 90° smooth bends. The water velocity near the river surface is 6 ft/s, and the pipe inlet is placed in the river normal to the flow direction of water to take advantage of the dynamic pressure. The elevation difference between the free surface of the tank and the river is 12 ft. For a flow rate of 1.5 ft³/s and an overall pump efficiency of 70 percent, determine the required electric power input to the pump.

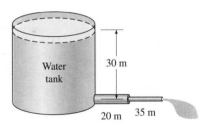

FIGURE P12-68

12-70   A water tank filled with solar-heated water is to be used for showers in a field using gravity-driven flow. The system involves 20 m of 1.5-cm-diameter galvanized iron piping with four miter bends (90°) without vanes and a wide-open globe valve. If water is to flow at a rate of 0.4 L/s through the shower head, determine how high the water level in the tank must be from the exit level of the shower. Disregard the losses at the entrance and at the shower head, and take the water temperature to be 40°C.

12-71   Two water reservoirs *A* and *B* are connected to each other through a 40-m-long, 2-cm-diameter cast iron pipe with a sharp-edged entrance. The pipe also involves a swing check valve and a fully open gate valve. The water level in both reservoirs is the same, but reservoir *A* is pressurized by compressed air while reservoir *B* is open to the atmosphere at 88 kPa. If the initial

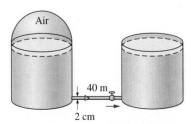

FIGURE P12-71

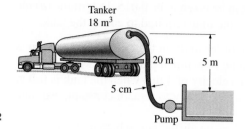

FIGURE P12-72

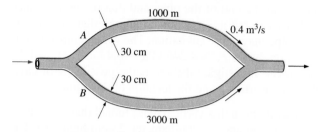

FIGURE P12-74

flow rate through the pipe is 1.2 L/s, determine the absolute air pressure on top of reservoir A. Take the water temperature to be 10°C. *Answer:* 733 kPa

12-72   A tanker is to be filled with fuel oil with $\rho = 920$ kg/m³ and $\mu = 0.045$ kg/m · s from an underground reservoir using a 20-m-long, 5-cm-diameter plastic hose with a slightly rounded entrance and two 90° smooth bends. The elevation difference between the oil level in the reservoir and the top of the tanker where the hose is discharged is 5 m. The capacity of the tanker is 18 m³, and the filling time is 30 minutes. Assuming an overall pump efficiency of 75 percent, determine the required power input to the pump.

12-73   Two pipes of identical length and material are connected in parallel. The diameter of pipe A is twice the diameter of pipe B. Assuming the friction factor to be the same in both cases and disregarding minor losses, determine the ratio of the flow rates in the two pipes.

12-74   A certain part of cast iron piping of a water distribution system involves a parallel section. Both parallel pipes have a diameter of 30 cm, and the flow is fully turbulent. One of the branches (pipe A) is 1000 m long while the other branch (pipe B) is 3000 m long. If the flow rate through pipe A is 0.4 m³/s, determine the flow rate through pipe B. Disregard minor losses and assume the water temperature to be 15°C. Show that the flow is fully turbulent, and thus the friction factor is independent of Reynolds number.
Answer: 0.231 m³/s

12-75   Repeat Prob. 12-74 assuming pipe A has a halfway-closed gate valve $(K_L = 2.1)$ while pipe B has a fully open globe valve $(K_L = 10)$, and the other minor losses are negligible. Assume the flow to be fully turbulent.

12-76   A geothermal district heating system involves the transport of geothermal water at 110°C from a geothermal well to a city at about the same elevation for a distance of 12 km at a rate of 1.5 m³/s in 60-cm-diameter stainless steel pipes. The fluid pressures at the wellhead and the arrival point in the city are to be the same. The minor losses are negligible because of the large length-to-diameter ratio and the relatively small number of components that cause minor losses. (*a*) Assuming the pump-motor efficiency to be 65 percent, determine the electric power consumption of the system for pumping. Would you recommend the use of a single large pump or several smaller pumps of the same total pumping power scattered along the pipeline? Explain.

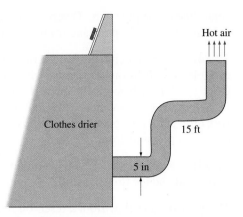

FIGURE P12-78E

(*b*) Determine the daily cost of power consumption of the system if the unit cost of electricity is $0.06/kWh. (*c*) The temperature of geothermal water is estimated to drop 0.5°C during this long flow. Determine if the frictional heating during flow can make up for this drop in temperature.

12-77 Repeat Prob. 12-76 for cast iron pipes of the same diameter.

12-78E A clothes drier discharges air at 1 atm and 120°F at a rate of 1.2 ft³/s when its 5-in.-diameter, well-rounded vent with negligible loss is not connected to any duct. Determine the flow rate when the vent is connected to a 15-ft-long, 5-in.-diameter duct made of galvanized iron, with three 90° flanged smooth bends. Take the density of air to be 0.068 lbm/ft³ and the friction factor of the duct to be 0.028, and assume the fan power input to remain constant.

12-79 In large buildings, hot water in a water tank is circulated through a loop so that the user doesn't have to wait for all the water in long piping to drain before hot water starts coming out. A certain recirculating loop involves 40-m-long, 1.2-cm-diameter cast iron pipes with six 90° threaded smooth bends and two fully open gate valves. If the mean flow velocity through the loop is 2.5 m/s, determine the required power input for the recirculating pump. Take the average water temperature to be 60°C and the efficiency of the pump to be 70 percent.   *Answer:* 2.18 kW

12-80 Repeat Prob. 12-79 for plastic pipes.

Review Problems

12-81 The compressed air requirements of a manufacturing facility are met by a 150-hp compressor that draws in air from the outside through an 8-m-long, 20-cm-diameter duct made of thin galvanized iron sheets. The compressor takes in air at a rate of 0.27 m³/s at the outdoor conditions of 17°C and 95 kPa. Disregarding any minor losses, determine the useful power used by the compressor to overcome the frictional losses in this duct.
     Answer: 9.16 kW

12-82 A house built on a riverside is to be cooled in summer by utilizing the cool water of the river. A 15-m-long section of a circular stainless steel duct of 20-cm diameter passes through the water. Air flows through the underwater section of the duct at 3 m/s at an average temperature of 17°C. For an overall fan efficiency of 55%, determine the fan power needed to overcome the flow resistance in this section of the duct.

12-83 The velocity profile in fully developed laminar flow in a circular pipe, in m/s, is given by $V(r) = 6(1 - 100r^2)$ where r is the radial distance from the

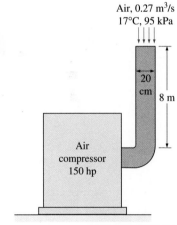

FIGURE P12-81

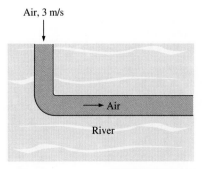

FIGURE P12-82

centerline of the pipe in m. Determine (a) the radius of the pipe, (b) the mean velocity through the pipe, and (c) the maximum velocity in the pipe.

12-84E The velocity profile in fully developed laminar flow of water at 40°F in a 80-ft-long horizontal circular pipe, in ft/s, is given by $\mathcal{V}(r) = 0.8(1 - 625r^2)$ where r is the radial distance from the centerline of the pipe in ft. Determine (a) the volume flow rate of water through the pipe, (b) the pressure drop across the pipe, and (c) the useful pumping power required to overcome this pressure drop.

12-85E Repeat Prob. 12-84E assuming the pipe is inclined 12° from the horizontal and the flow is uphill.

12-86 Consider flow from a reservoir through a horizontal pipe of length L and diameter D that penetrates into the side wall at a vertical distance H from the free surface. The flow rate through an actual pipe with a reentrant section ($K_L = 0.8$) will be considerably less than the flow rate through the hole calculated assuming "frictionless" flow and thus zero loss. Obtain a relation for the "equivalent diameter" of the reentrant pipe for use in relations for frictionless flow through a hole and determine its value for a pipe friction factor, length, and diameter of 0.023, 10 m, and 0.04 m, respectively. Assume the friction factor of the pipe to remain constant.

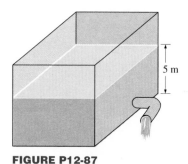

12-87 Water is to be withdrawn from a 5-m-high water reservoir by drilling a well-rounded 3-cm-diameter hole with negligible loss at the bottom surface and attaching a horizontal 90° bend of negligible length. Determine the flow rate of water through the bend if (a) the bend is a flanged smooth bend and (b) the bend is a miter bend without vanes.
Answers: (a) 0.00614 m³/s, (b) 0.00483 m³/s

FIGURE P12-87

12-88 In a geothermal district heating system, 10,000 kg/s of hot water must be delivered a distance of 10 km in a horizontal pipe. The minor losses are negligible, and the only significant energy loss will arise from pipe friction. The friction factor can be taken to be 0.015. Specifying a larger diameter pipe would reduce water velocity, velocity head, pipe friction, and thus power consumption. But a larger pipe also would cost more money initially to purchase and install. Otherwise stated, there is an optimum pipe diameter that will minimize the sum of pipe cost and future electric power cost.

Assume the system will run 24 hours/day, every day, for 30 years. During this time the cost of electricity will remain constant at $0.06/kWh. Assume system performance stays constant over the decades (this may not be true, especially if highly mineralized water is passed through the pipeline—scale may form). The pump has an overall efficiency of 80 percent. The cost to purchase, install, and insulate a 10-km pipe depends upon the diameter D and is given by $Cost = \$10^6 D^2$ where D is in meters. Assuming zero inflation and interest rate for simplicity and zero salvage value and zero maintenance cost, determine the optimum pipe diameter.

12-89 Water at 15°C is to be discharged from a reservoir at a rate of 10 L/s using two horizontal cast iron pipes connected in series and a pump between them. The first pipe is 20 m long and has a 6-cm diameter, while the second pipe is 35 m long and has a 4-cm diameter. The water level in the reservoir is 30 m above the centerline of the pipe. The pipe entrance is sharp-edged, and losses associated with the connection of the pump are negligible. Determine the required pumping head and the minimum pumping power to maintain the indicated flow rate.

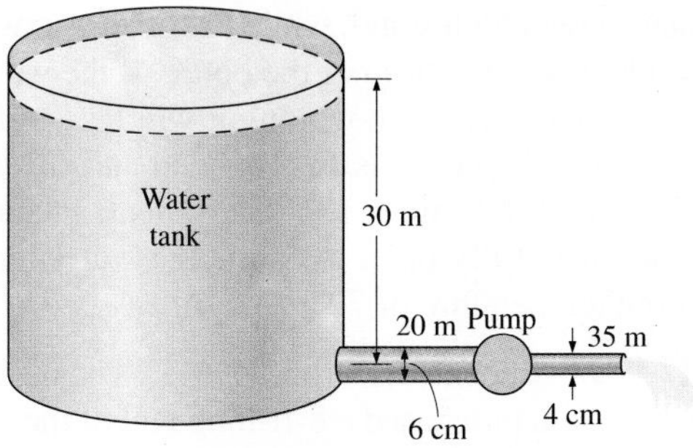

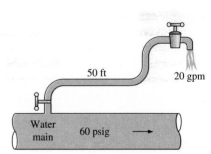

FIGURE P12-89

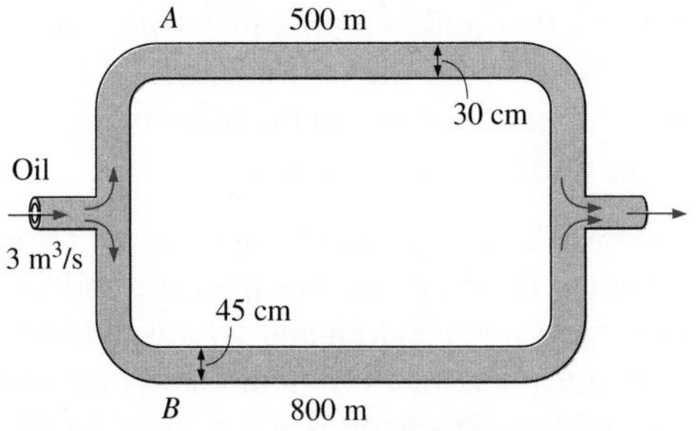

FIGURE P12-91

12-90 Two pipes of identical diameter and material are connected in parallel. The length of pipe A is twice the length of pipe B. Assuming the flow is fully turbulent in both pipes and thus the friction factor is independent of the Reynolds number and disregarding minor losses, determine the ratio of the flow rates in the two pipes. *Answer:* 0.707

12-91 A pipeline that transports oil at 40°C at a rate of 3 m³/s branches out into two parallel pipes made of commercial steel that reconnect downstream. Pipe A is 500 m long and has a diameter of 30 cm while pipe B is 800 m long and has a diameter of 45 cm. The minor losses are considered to be negligible. Determine the flow rate through each of the parallel pipes.

12-92 Repeat Prob. 12-91 for hot water flow of a district heating system at 80°C.

12-93E A water fountain is to be installed at a remote location by attaching a cast iron pipe directly to a water main through which water is flowing at a pressure of 60 psig. The entrance to the pipe is sharp-edged, and the 50-ft-long piping system involves three 90° miter bends without vanes, a fully open gate valve, and an angle valve with a loss coefficient of 5 when fully open. If the system is to provide water at a rate of 20 gal/min, determine the minimum diameter of the piping system. *Answer:* 0.78 in

FIGURE P12-93E

12-94E Repeat Prob. 12-93E for plastic pipes.

12-95 In a hydroelectric power plant, water at 20°C is supplied to the turbine at a rate of 0.8 m³/s through a 200-m-long, 0.35-m-diameter cast iron pipe. The elevation difference between the free surface of the reservoir and the turbine discharge is 70 m, and the combined turbine-generator efficiency is 75 percent. Disregarding the minor losses because of the large length-to-diameter ratio, determine the electric power output of this plant.

12-96 In Prob. 12-95, the pipe diameter is doubled in order to reduce the pipe losses. Determine the percent increase in the net power output as a result of this modification.

12-97E The drinking water needs of an office are met by large water bottles. One end of a 0.35-in.-diameter, 6-ft-long plastic hose is inserted into the bottle

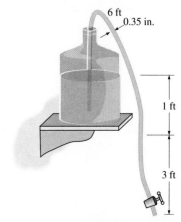

FIGURE P12-97E

placed on a high stand, while the other end with an on/off valve is maintained 3 ft below the bottom of the bottle. If the water level in the bottle is 1 ft when it is full, determine how long it will take to fill an 8-oz glass (= 0.00835 ft³) (*a*) when the bottle is first opened and (*b*) when the bottle is almost empty. Take the total minor loss coefficient, including the on/off valve, to be 2.8 when it is fully open. Assume the water temperature to be the same as the room temperature of 70°F.   *Answers:* (*a*) 2.6 s, (*b*) 3.0 s

12-98E Reconsider Prob. 12-97E. The office worker who set up the siphoning system purchased a 8-ft-long reel of the plastic tube and wanted to use the whole thing to avoid cutting it in pieces, thinking that it is the elevation difference that makes siphoning work, and the length of the tube is not important. So he used the entire 8-ft-long tube. Assuming there are no additional turns or constrictions in the tube (being very optimistic), determine the time it takes to fill a glass of water.

12-99 A circular water pipe has an abrupt expansion from diameter $D_1 = 10$ cm to $D_2 = 20$ cm. The pressure and the mean water velocity in the smaller pipe are $P_1 = 120$ kPa and 10 m/s, and the flow is turbulent. By applying the continuity, momentum, and energy equations, show that the loss coefficient for sudden expansion is $K_L = (1 - D_1^2/D_2^2)^2$, and calculate K_L and P_2 for the given case.

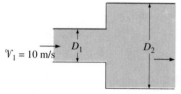

FIGURE P12-99

12-100 The water at 20°C in a 10-m-diameter, 2-m-high above-the-ground swimming pool is to be emptied by unplugging a 3-cm-diameter, 25-m-long horizontal plastic pipe attached to the bottom of the pool. Determine the initial rate of discharge of water through the pipe and the time it will take to empty the swimming pool completely assuming the entrance to the pipe is well-rounded with negligible loss. Take the friction factor of the pipe to be 0.022. Using the initial discharge velocity, check if this is a reasonable value for the friction factor.

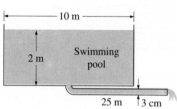

FIGURE P12-100

12-101 Repeat Prob. 12-100 for a sharp-edged entrance to the pipe with $K_L = 0.5$.

Computer, Design, and Essay Problems

12-102 Electronic boxes such as computers are commonly cooled by a fan. Write an essay on forced air cooling of electronic boxes and on the selection of the fan for electronic devices.

12-103 Design an experiment to measure the viscosity of liquids using a vertical funnel with a cylindrical reservoir of height *h* and a narrow flow section of diameter *D* and length *L*. Making appropriate assumptions, obtain a relation for viscosity in terms of easily measurable quantities such as density and volume flow rate. Is there a need for the use of a correction factor?

12-104 A pump is to be selected for a waterfall in a garden. The water collects in a pond at the bottom, and the elevation difference between the free surface of the pond and the location where the water is discharged is 3 m. The flow rate of water is to be at least 8 L/s. Select an appropriate motor-pump unit for this job and identify three manufacturers with product model numbers and prices. Make a selection and explain why you selected that particular product. Also estimate the cost of annual power consumption of this unit assuming continuous operation.

Flow over Bodies: Drag and Lift

In the preceding chapter, we considered the flow of fluids inside pipes, with emphasis on pressure drop and head losses and their relations to flow rate. In this chapter, we consider the flow of fluids over bodies that are immersed in a fluid, with emphasis on the resulting lift and drag forces.

When a fluid moves over a solid body, it exerts pressure forces normal to the surface and shear forces parallel to the surface along the outer surface of the body. We are usually interested in the *resultant* of the pressure and shear forces acting on the body rather than the details of the distributions of these forces along the entire surface of the body. The component of the resultant pressure and shear forces that acts in the flow direction is called the *drag force,* and the component that acts normal to the flow direction is called the *lift.*

We start this chapter with a discussion of drag and lift and explore the concepts of pressure drag, friction drag, and flow separation. We continue with the drag coefficients of various two- and three-dimensional geometries encountered in practice and determine the drag force using experimentally determined drag coefficients. We then examine the development of the velocity boundary layer over a flat surface and develop relations for the skin friction coefficient for flow over flat plates, cylinders, and spheres. Finally, we discuss the lift developed by airfoils and the factors that affect the lift characteristics of bodies.

13-1 ■ INTRODUCTION

Fluid flow over solid bodies frequently occurs in practice, and it is responsible for numerous physical phenomena such as the drag force acting on automobiles, power lines, trees, and underwater pipelines; the lift developed by airplane wings; the upward draft of rain, snow, hail, and dust particles in high winds; the transportation of red blood cells by blood flow; the entrainment and disbursement of liquid droplets by sprays; the vibration and noise generated by bodies moving in a fluid; and the power generated by wind turbines (Fig. 13-1). Therefore, developing a good understanding of external flow is important in the design of many engineering systems such as aircraft, automobiles, buildings, ships, submarines, and all kinds of turbines. Late-model cars, for example, have been designed with particular emphasis on aerodynamics. This has resulted in significant reductions in fuel consumption and noise and considerable improvement in handling.

Sometimes a fluid moves over a stationary body (such as the wind blowing over a building), and other times a body moves through a quiescent fluid (such as a car moving through air). These two seemingly different processes are equivalent to each other; what matters is the relative motion between the fluid and the body. Such motions are conveniently analyzed by fixing the coordinate system on the body and are referred to as **flow over bodies** or **external flow.** The aerodynamic aspects of different airplane wing designs, for example, are studied conveniently in a lab by placing the wings in a wind tunnel and blowing air over them by large fans.

The flow fields and geometries for most external flow problems are too complicated to be solved analytically, and thus we have to rely on correlations based on experimental data. The availability of high-speed computers has made it possible to conduct series of "numerical experimentations" quickly by solving the governing equations numerically and to resort to the expensive and time-consuming testing and experimentation only in the final stages of design. Such testing is done in wind tunnels. H. F. Phillips (1845–1912) built the first wind tunnel in 1894 and measured lift and drag. In this chapter, we will rely mostly on relations developed experimentally.

The velocity of the fluid relative to the immersed solid body sufficiently far from the body (outside the boundary layer) is called the **free-stream velocity** and is denoted by $\mathcal{V}_\infty$. It is usually taken to be equal to the **upstream velocity** $\mathcal{V}$ (also called the **approach velocity**), which is the velocity of the approaching fluid far ahead of the body. This idealization is nearly exact for very thin bodies, such as a flat plate parallel to flow, but approximate for thick

FIGURE 13-1

Flow past bodies is commonly encountered in practice.

bodies such as a circular cylinder. The fluid velocity ranges from zero at the surface (the no-slip condition) to the free-stream value away from the surface, and the subscript "infinity" serves as a reminder that this is the value at a distance where the presence of the body is not felt. The upstream velocity, in general, may vary with location and time (e.g., the wind blowing past a building). But in the design and analysis, the upstream velocity is usually assumed to be *uniform* and *steady* for convenience, and this is what we will do in this chapter.

The shape of a body has a profound influence on the flow over the body and the velocity field. The flow over a body is said to be **two-dimensional** when the body is very long and of constant cross section and the flow is normal to the body. The wind blowing over a long pipe perpendicular to its axis is an example of two-dimensional flow. Note that the velocity component in the axial direction is zero in this case, and thus the velocity is two-dimensional. The two-dimensional idealization is appropriate when the body is sufficiently long so that the end effects are negligible and the approach flow is uniform. Another simplification occurs when the body possesses symmetry along an axis in the flow direction. The flow in this case is also two-dimensional and is said to be **axisymmetric.** A bullet piercing through air is an example of axisymmetric flow. The velocity in this case varies with the axial distance x and the radial distance r. Flow over a body that cannot be modeled as two-dimensional or axisymmetric such as flow over a car is **three-dimensional** (Fig. 13-2).

Flow over bodies can also be classified as **incompressible flows** (e.g., flows over automobiles, submarines, and buildings) and **compressible flows** (e.g., flows over high-speed aircraft, rockets, and missiles). Compressibility effects are negligible at velocities below about 100 m/s (or 360 km/h), and such flows can be treated as incompressible. Compressible flow and flows that involve partially immersed bodies with a free surface (such as a ship cruising in water) are beyond the scope of this introductory text.

Bodies subjected to fluid flow are classified as being streamlined or blunt, depending on their overall shape. A body is said to be **streamlined** if a conscious effort is made to align its shape with the anticipated streamlines in the flow. Streamlined bodies such as race cars and airplanes appear to be contoured and sleek. Otherwise, a body (such as a building) tends to block the flow and is said to be **bluff** or **blunt.** Usually it is much easier to force a streamlined body through a fluid, and thus streamlining has been of great importance in the design of vehicles and airplanes (Fig. 13-3).

Variations in velocity during internal or external flow, in general, are accompanied by changes in pressure in accordance with the Bernoulli equation (when the viscous effects are negligible). For liquid flow, the pressure at some points may drop below the vapor pressure of the liquid, causing the liquid to

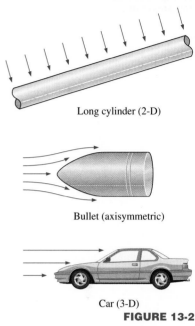

Long cylinder (2-D)

Bullet (axisymmetric)

Car (3-D)

FIGURE 13-2

Two-dimensional, axisymmetric, and three-dimensional flows.

60 mph ←   70 hp

60 mph ←   50 hp

FIGURE 13-3

Usually it is much easier to force a streamlined body through a fluid than it is a blunt body

vaporize or "boil" at those locations and to form small vapor bubbles. The bubbles collapse implosively when they are swept into a region of higher pressure, producing pressures over 800 MPa (115,000 psi). The phenomenon of forming vapor cavities in regions of low pressure in liquid flow is called **cavitation.** At 15°C, for example, the vapor pressure of water is 1.7 kPa. Therefore, water may cavitate at locations where the pressure drops below this value (due to high velocities) such as in the constrictions in a valve or the tips of impeller blades.

Cavitation must be avoided (or at least minimized) in flow systems since it reduces performance, generates annoying vibrations and noise, and causes damage to equipment. The pressure spikes resulting from the large number of bubbles collapsing near a solid surface over a long period of time may cause erosion, surface pitting, fatigue failure, and the eventual destruction of the components or machinery. The presence of cavitation in a flow system can be sensed by its characteristic tumbling sound.

13-2 ■ DRAG AND LIFT

It is a common experience that a body meets some resistance when it is forced to move through a fluid, especially a liquid. As you may have noticed, it is very difficult to walk in water because of the much greater resistance it offers to motion compared to air. Also, you may have seen high winds knocking down trees, power lines, and even trailers and felt the strong "push" the wind exerts on your body (Fig. 13-4). You experience the same feeling when you extend your arm out of the window of a moving car. A fluid may exert forces and moments on a body in and about various directions. The force a flowing fluid exerts on a body in the flow direction is called **drag.** The drag force can be measured directly by attaching the body subjected to fluid flow to a calibrated spring and measuring the displacement in the flow direction (just like measuring weight with a spring scale).

Drag is usually an undesirable effect, like friction, and we do our best to minimize it. Reduction of drag is closely associated with the reduction of fuel consumption in automobiles, submarines, and aircraft; improved safety and durability of structures subjected to high winds; and reduction of noise and

FIGURE 13-4

High winds knock down trees,
power lines, and even people
as a result of the drag force.

vibration. But in some cases drag produces a very beneficial effect and we try to maximize it. Friction, for example, is a "life saver" in the brakes of automobiles. Likewise, it is the drag that makes it possible for people to parachute, for pollens to fly to distant locations, and for us all to enjoy the waves of the oceans and the relaxing movements of the leaves of trees.

A stationary fluid exerts only normal pressure forces on the surface of a body immersed in it. A moving fluid, however, also exerts tangential shear forces on the surface because of the no-slip condition caused by viscous effects. Both of these forces, in general, have components in the direction of flow, and thus the drag force is due to the combined effects of pressure and wall shear forces in the flow direction. The components of the pressure and wall shear forces in the direction normal to flow tend to move the body in that direction and are called **lift.**

For two-dimensional flows, the resultant of the pressure and shear forces can be resolved into two components: one in the direction of flow, which is the drag force, and another in the direction normal to flow, which is the lift, as shown in Fig. 13-5. For three-dimensional flows, there is also a side force component in the direction normal to the paper that will tend to move the body in that direction. The fluid forces also may generate moments and cause the body to rotate. The moment about the flow direction is called the *rolling moment,* the moment about the lift direction is called the *yawing moment,* and the moment about the side force direction is called the *pitching moment.* For bodies that possess symmetry about the lift-drag plane such as cars, airplanes, and ships, the side force, the yawing moment, and the rolling moment are zero when the wind and wave forces are aligned with the body. What remain for such bodies are the drag and lift forces and the pitching moment. For axisymmetric bodies such as a bullet, the only effect fluid exerts on the body is the drag force.

The pressure and shear forces acting on a differential area dA on the surface are $P\,dA$ and $\tau_w\,dA$, respectively. The differential drag force F_D and the lift force F_L acting on dA in two-dimensional flow are (Fig. 13-5)

Drag force:
$$dF_D = -P\,dA\cos\theta + \tau_w\,dA\sin\theta \qquad (13\text{-}1)$$

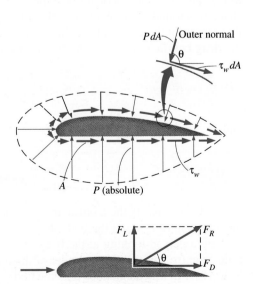

FIGURE 13-5

The pressure and viscous forces acting on a two-dimensional body and the resultant lift and drag forces.

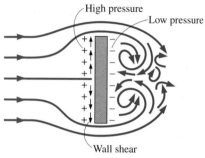

FIGURE 13-6

Drag force acting on a flat plate normal to
flow depends on the pressure only and is
independent of the wall shear, which acts
normal to flow.

and

Lift force: $$dF_L = -P \, dA \sin \theta - \tau_w \, dA \cos \theta \qquad (13\text{-}2)$$

where θ is the angle the outer normal of dA makes with the positive flow direction. The total drag and lift acting on the body can be determined by integrating the relations above over the entire surface of the body. This is not practical, however, since the detailed distributions of pressure and shear forces are difficult to obtain by analysis or measurements. Fortunately, this information is often not needed in practice. Usually all we need to know is the resultant drag force and lift acting on the entire body, which can be measured directly and easily in a wind tunnel.

Equations 13-1 and 13-2 show that both the skin friction (wall shear) and pressure, in general, contribute to the drag and the lift. In the special case of a thin *flat plate* aligned parallel to the flow direction, the drag force depends on the wall shear only and is independent of pressure since $\theta = 90°$. When the flat plate is placed normal to the flow direction, however, the drag force depends on the pressure only and is independent of the wall shear since the shear stress in this case acts in the direction normal to flow and $\theta = 0°$ (Fig. 13-6). If the flat plate is tilted at an angle θ relative to the flow direction, then the drag force will depend on both the pressure and the shear stress.

The wings of airplanes are shaped specifically to generate lift. This is done by making the top surface of the wing curved and the bottom surface nearly flat. The fluid flowing over the wing travels a longer distance to reach the end of the wing (compared to the fluid flowing under the wing), and thus it must have a larger average velocity. The Bernoulli effect will then cause the pressure over the top surface of the wing to be lower. The pressure difference between the top and bottom surfaces of the wing generates an upward force that tends to lift the wing and thus the airplane to which it is connected. For slender bodies such as wings, the shear force acts nearly parallel to the flow direction, and thus its contribution to the lift is small. The drag force for such slender bodies is mostly due to shear forces (the skin friction).

The drag and lift forces depend on the density ρ of the fluid, the upstream velocity $\mathcal{V}$, and the size, shape, and orientation of the body, among other things, and it is not practical to list these forces for a variety of situations. Instead, it is found convenient to work with appropriate dimensionless numbers that represent the drag and lift characteristics of the body. These numbers are the **drag coefficient** C_D and the **lift coefficient** C_L, and they are defined as

Drag coefficient: $$C_D = \frac{F_D}{\frac{1}{2}\rho \mathcal{V}^2 A} \qquad (13\text{-}3)$$

and

Lift coefficient: $$C_L = \frac{F_L}{\frac{1}{2}\rho \mathcal{V}^2 A} \qquad (13\text{-}4)$$

where A is ordinarily the **frontal area** (the area projected on a plane normal to the direction of flow) of the body. In other words, A is the area that would be seen by a person looking at the body from the direction of the approaching fluid. The frontal area of a cylinder of diameter D and length L, for example, is $A = LD$. In lift calculations of some bodies, such as airfoils, A is taken to be the **planform area,** which is the area that would be seen by a person looking at the body from above in a direction normal to the body. The drag and lift coefficients are primarily functions of the shape of the body, but in some cases they also depend on the Reynolds number and the surface roughness. The term $\frac{1}{2}\rho\mathcal{V}^2$ is the **dynamic pressure.**

When a body is dropped into the atmosphere or a lake, it first accelerates under the influence of its weight. The motion of the body is resisted by the drag force, which acts in the direction opposite to the motion. As the velocity of the body increases, so does the drag force. This continues until all the forces balance each other and the net force acting on the body (and thus its acceleration) is zero. Then the velocity of the body remains constant during the rest of its fall if the properties of the fluid in the path of the body remain essentially constant. This is the maximum velocity a falling body can attain and is called the **terminal velocity** (Fig. 13-7). The forces acting on a falling body are usually the drag force, the buoyant force, and the weight of the body.

EXAMPLE 13-1 Measuring the Drag Coefficient of a Car

The drag coefficient of a car at the design conditions of 1 atm, 70°F, and 60 mph is to be determined experimentally in a large wind tunnel in a full-scale testing. The height and width of the car are 4.2 ft and 5.3 ft, respectively (Fig. 13-8). If the force acting on the car in the flow direction is measured to be 68 lbf, determine the drag coefficient of this car.

Solution   The drag force acting on a car is measured in a wind tunnel. The drag coefficient of the car at the test conditions is to be determined.

Assumptions   **1** The flow of air is steady and incompressible. **2** The cross section of the tunnel is large enough to simulate free flow over the car. **3** The bottom of the tunnel is also moving at the speed of air to approximate actual driving conditions or this effect is negligible.

Properties   The density of air at 1 atm and 70°F is $\rho = 0.0755$ lbm/ft³ (Table A-18E).

Analysis   The drag force acting on a body and the drag coefficient are given by

$$F_D = C_D A \frac{\rho\mathcal{V}^2}{2} \quad \text{and} \quad C_D = \frac{2F_D}{\rho A \mathcal{V}^2}$$

where A is the frontal area. Substituting and noting that 1 mph = 1.467 ft/s, the drag coefficient of the car is determined to be

$$C_D = \frac{2 \times (68 \text{ lbf})}{(0.0755 \text{ lbm/ft}^3)(4.2 \times 5.3 \text{ ft}^2)(60 \times 1.467 \text{ ft/s})^2}\left(\frac{32.2 \text{ lbm} \cdot \text{ft/s}^2}{1 \text{ lbf}}\right) = \mathbf{0.34}$$

Discussion   Note that the drag coefficient depends on the design conditions, and its value will be different at different conditions. Therefore, the published drag coefficients of different vehicles can be compared meaningfully only if they are determined under identical conditions. This shows the importance of developing standard testing procedures in industry.

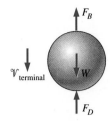

$$F_D = W - F_B$$
(No acceleration)
FIGURE 13-7

During a free fall, a body reaches its constant *terminal velocity* when the drag force equals the weight of the body less the buoyant force.

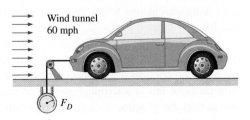

FIGURE 13-8

Schematic for Example 13-1.

13-3 ■ FRICTION AND PRESSURE DRAG

As mentioned earlier, the drag force is the net force exerted by a fluid on a body in the direction of flow due to the combined effects of wall shear and pressure forces. It is often instructive to separate the two effects and study them separately.

The part of drag that is due directly to wall shear stress τ_w is called the **skin friction drag** (or just **friction drag**) $F_{D,\text{friction}}$ since it is caused by frictional effects, and the part that is due directly to pressure P is called the **pressure drag** $F_{D,\text{pressure}}$ (also called the **form drag** because of its strong dependence on the form or shape of the body). The friction and pressure drag coefficients are defined as

$$C_{D,\text{friction}} = \frac{F_{D,\text{friction}}}{\frac{1}{2}\rho V^2 A} \qquad \text{and} \qquad C_{D,\text{pressure}} = \frac{F_{D,\text{pressure}}}{\frac{1}{2}\rho V^2 A} \qquad (13\text{-}5)$$

When the friction and pressure drag coefficients or forces are available, the total drag coefficient or force can be determined by simply adding them:

$$C_D = C_{D,\text{friction}} + C_{D,\text{pressure}} \qquad \text{and} \qquad F_D = F_{D,\text{friction}} + F_{D,\text{pressure}} \qquad (13\text{-}6)$$

The *friction drag* is the component of the wall shear force in the direction of flow and thus depends on the orientation of the body as well as the magnitude of the wall shear stress τ_w. The friction drag is *zero* for a surface normal to flow and *maximum* for a surface parallel to flow since the friction drag in this case equals the total shear force on the surface.

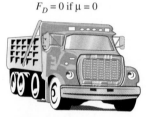

$F_D = 0$ if $\mu = 0$

FIGURE 13-9

For the flow of an "idealized" fluid with zero viscosity past a body, both the friction drag and pressure drag are zero regardless of the shape of the body.

Friction drag is a strong function of viscosity, and an "idealized" fluid with zero viscosity would produce zero friction drag since the wall shear would be zero (Fig. 13-9). The pressure drag also would be zero in this case during steady flow regardless of the shape of the body since there would be no pressure losses. This can be shown using Bernoulli's equation. For flow in the horizontal direction, for example, the pressure along a horizontal line will be constant (just like stationary fluids) since the upstream velocity is constant, and thus there will be no net pressure force acting on the body in the horizontal direction. During flow past a blunt body such as a cylinder, the pressure changes along curved streamlines even in inviscid flow, but the net force acting on the body is still zero. Therefore, the total drag is zero for the case of ideal inviscid fluid flow.

The Reynolds number is inversely proportional to the viscosity of the fluid. Therefore, the contribution of friction drag to total drag for blunt bodies is less at higher Reynolds numbers and may be negligible at very high Reynolds numbers. The drag in such cases is mostly due to pressure drag. At low Reynolds numbers, most drag is due to friction drag. This is especially the case for highly streamlined bodies such as airfoils. The friction drag is also proportional to the surface area. Therefore, bodies with a larger surface area will experience a larger friction drag. Large commercial airplanes, for example, to save fuel reduce their total surface area and thus drag by retracting their wing extensions when they reach the cruising altitudes where they need less lift.

The *friction drag coefficient* is analogous to the *friction factor* in pipe flow discussed in the preceding chapter, and its value depends on the flow regime. The friction drag coefficient is independent of **surface roughness** in laminar flow but is a strong function of surface roughness in turbulent flow due to surface roughness elements protruding further into the highly viscous laminar sublayer.

The pressure drag is proportional to the *difference* between the pressures acting on the front and back sides of the immersed body and the frontal area. Therefore, the pressure drag is usually dominant for "blunt" bodies, often negligible for streamlined bodies such as airfoils, and zero for thin flat plates parallel to the flow. The pressure drag becomes most significant when the velocity of the fluid is too high for the fluid to be able to follow the curvature of the body, and thus the fluid *separates* from the body at some point and creates a very low pressure region in the back. The pressure drag in this case is due to the large pressure difference between the front and back sides of the body.

Reducing Drag by Streamlining

The first thought that comes to mind to reduce drag is to streamline a body. Even car salesmen are quick to point out the low drag coefficients of their cars, owing to streamlining. But streamlining has opposite effects on pressure and friction drags. It decreases pressure drag by delaying boundary layer separation and thus reducing the pressure difference between the front and back of the body and increases the friction drag by increasing the surface area. The end result depends on which effect dominates. Therefore, any optimization study to reduce the drag of a body must consider both effects and must attempt to minimize the *sum* of the two, as shown in Fig. 13-10. The minimum total drag occurs at $D/L = 0.25$ in this case. For the case of a circular cylinder with the same thickness, the drag coefficient would be about five times as much. Therefore, it is possible to reduce the drag of a cylindrical component to one-fifth by the use of proper fairings.

The effect of streamlining on the drag coefficient can be described best by considering long elliptical cylinders with different aspect (or length-to-width) ratios L/D, where L is the length in the flow direction and D is the thickness, as shown in Fig. 13-11. Note that the drag coefficient decreases drastically as the ellipse becomes slimmer. For the special case of $L/D = 1$ (a circular cylinder), the drag coefficient is $C_D \approx 1$ at this Reynolds number. As the aspect ratio is decreased and the cylinder resembles a flat plate, the drag coefficient increases to 1.9, the value for a flat plate normal to flow. Note that the curve becomes nearly flat for aspect ratios greater than about 4. Therefore, for a given diameter D, elliptical shapes with an aspect ratio of about $L/D \approx 4$ usually offer good compromise between the total drag coefficient and length L. The reduction in the drag coefficient at high aspect ratios is primarily due to

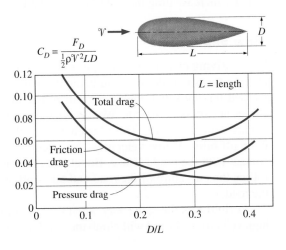

FIGURE 13-10

The variation of friction, pressure, and total drag coefficients of a streamlined strut with thickness-to-chord-length ratio for Re = 4 × 10⁴ (from Abbott and von Doenhoff, Ref. 2).

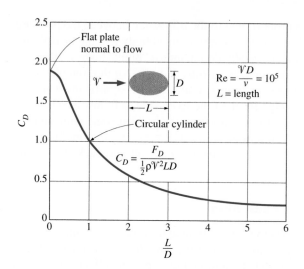

FIGURE 13-11

The variation of the drag coefficient of
a long elliptical cylinder with aspect
ratio based on the frontal area bD
(from Blevins, Ref. 5).

the boundary layer staying attached to the surface longer and the resulting pressure recovery. The friction drag on an elliptical cylinder with an aspect ratio of 4 is negligible (less than 2 percent of total drag).

As the aspect ratio of an elliptical cylinder is increased by flattening it (i.e., decreasing D while holding L constant), the drag starts increasing and tends to infinity as $L/D \to \infty$ (i.e., as the ellipse resembles a flat plate parallel to flow). This is due to the frontal area, which appears in the denominator in the definition of C_D, approaching zero. It does not mean that the drag force increases drastically (actually, the drag force decreases) as the body becomes flat. This shows that the frontal area is inappropriate for use in the drag force relations for slim bodies such as thin airfoils and flat plates. In such cases, the drag coefficient is defined on the basis of the *planform area,* which is simply the surface area for a flat plate. This is quite appropriate since for slim bodies the drag is almost entirely due to friction drag, which is proportional to the surface area.

Streamlining has the added benefit of *reducing vibration and noise.* Streamlining should be considered only for blunt bodies that are subjected to high-velocity fluid flow (and thus high Reynolds numbers) for which flow separation is a real possibility. It is not necessary for bodies that typically involve low Reynolds number flows (Re < 1) since the drag in those cases is almost entirely due to friction drag, and streamlining will only increase the surface area and thus the total drag. Therefore, careless streamlining may actually increase drag instead of decreasing it.

Flow Separation

When driving on country roads, it is a common safety measure to slow down at sharp turns in order to avoid being thrown off the road. Many drivers have learned the hard way that a car will refuse to comply when forced to turn curves at excessive speeds. We can view this phenomenon as "the separation of cars" from the roads. This phenomenon is also observed when fast vehicles jump off hills. At low velocities, the wheels of the vehicle always remain in contact with the road surface. But at high velocities, the vehicle is too fast to follow the curvature of the road and takes off at the hill, losing contact with the road.

A fluid acts the same way when forced to flow over a curved surface at high velocities. A fluid will climb the uphill portion of the curved surface with

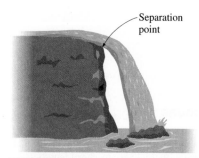

Separation
point

FIGURE 13-12

Flow separation in a waterfall.

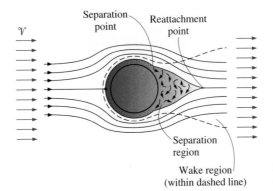

Separation point

Reattachment point

Separation region

Wake region (within dashed line)

FIGURE 13-13

Separation and reattachment during flow over a cylinder and the wake region.

no problem, but it will have difficulty remaining attached to the surface on the downhill side. At sufficiently high velocities, the fluid stream will detach itself from the surface of the body. This is called **separation** (Fig. 13-12). The location of the separation point depends on several factors such as the Reynolds number (and thus velocity), the surface roughness, and the level of fluctuations in the free stream, and it is usually difficult to predict exactly where separation will occur, unless there are sharp corners or abrupt changes in the shape of the body.

When a fluid separates from the body, it forms a *separated region* between the body and the fluid stream. This low pressure region behind the body where recirculating and back flows occur is called the **separated region.** The larger the separation area is, the larger the pressure drag will be. The effects of flow separation are felt far downstream in the form of reduced velocity (relative to the upstream velocity). The region of flow trailing the body where the effects of the body on velocity is felt is called the **wake** (Fig. 13-13). The separated region comes to an end when the two flow streams reattach. Therefore, the separated region is an enclosed volume whereas the wake keeps growing behind the body until the fluid in the wake region regains its velocity and the velocity profile becomes nearly flat again. Viscous effects are the most significant in the boundary layer, the separated region, and the wake. The flow outside these regions can be considered to be inviscid.

The occurrence of separation is not limited to blunt bodies. Separation may also occur on a streamlined body such as an airplane wing at a sufficiently large **angle of attack** (larger than about 16° for most airfoils), which is the angle the incoming fluid stream makes with the **chord** (the line that connects the nose and the end) of the body. Flow separation on the top surface of a wing reduces lift drastically and may cause the airplane to **stall.** Stalling has been blamed for many airplane accidents and loss of efficiencies in turbomachinery (Fig. 13-14).

Note that drag and lift are strongly dependent on the shape of the body, and any effect that causes the shape to change will have a profound effect on the drag and lift. For example, snow accumulation and ice formation on airplane wings may change the shape of the wings sufficiently to cause significant loss in lift. This phenomenon has caused many airplanes to loose altitude and crash and many others to abort takeoff. Therefore, it has become a routine safety measure to check for ice or snow build-up on critical components of airplanes before takeoff in bad weather. This is especially important for airplanes that have waited a long time on the runway before takeoff because of heavy traffic.

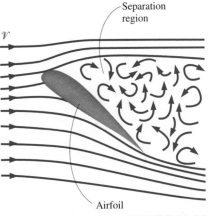

Separation region

Airfoil

FIGURE 13-14

At large angles of attack, flow may separate completely from the top surface of an airfoil, reducing lift drastically and causing the airfoil to stall.

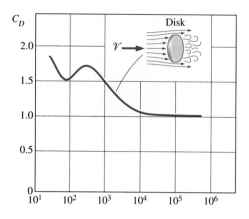

An important consequence of flow separation is the formation and shedding of circulating fluid chunks, called **vortices,** in the wake region. The continual generation of these vortices downstream is referred to as **vortex shedding.** This phenomenon usually occurs during normal flow over long cylinders or spheres for Re $\geq$ 90. The vibrations generated by vortices near the body may cause the body to resonate to dangerous levels if the frequency of the vortices is close to the natural frequency of the body—a situation that must be avoided in the design of equipment that is subjected to high-velocity fluid flow such as suspended bridges subjected to steady high winds and the wings of airplanes.

13-4 ■ DRAG COEFFICIENTS OF COMMON GEOMETRIES

The concept of drag has important consequences in daily life, and the drag behavior of various natural and manmade bodies is characterized by their drag coefficients measured under typical operating conditions. Although drag is caused by two different effects (friction and pressure), it is usually difficult to determine them separately. Besides, in most cases, we are interested in the *total* drag rather than the individual drag components, and thus usually the *total* drag coefficient is reported. The determination of drag coefficients has been the topic of numerous studies (mostly experimental), and there is a huge amount of drag coefficient data in the literature for just about any geometry of practical interest.

The drag coefficient, in general, depends on the *Reynolds number,* especially for Reynolds numbers below about 10^4. At higher Reynolds numbers, the drag coefficients for most geometries remain essentially constant (Fig. 13-15). This is due to the flow at high Reynolds numbers becoming fully turbulent. The reported drag coefficients are for such high Reynolds numbers. However, this is not the case for rounded bodies such as circular cylinders and spheres, as we will discuss later.

The drag coefficient exhibits different behavior in the low (creeping), moderate (laminar), and high (turbulent) regions of Reynolds number. The inertia effects are negligible in low Reynolds number flows (Re < 1), called the **creeping flow,** and the fluid wraps around the body smoothly. The drag coefficient in this case is inversely proportional to the Reynolds number, and for a sphere it is determined to be

$$\textit{Sphere:} \qquad\qquad C_D = \frac{24}{\text{Re}} \qquad (\text{for Re} \leqslant 1) \qquad\qquad (13\text{-}7)$$

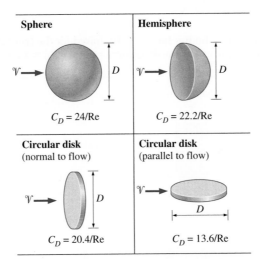

FIGURE 13-16

Drag coefficients C_D at low velocities
(Re $\leqslant 1$ where Re $= \mathcal{V}D/\nu$ and
$A = \pi D^2/4$).

Then the drag force acting on a spherical object at low Reynolds numbers becomes

$$F_D = C_D A \frac{\rho \mathcal{V}^2}{2} = \frac{24}{\text{Re}} A \frac{\rho \mathcal{V}^2}{2} = \frac{24}{\rho \mathcal{V}D/\mu} \frac{\pi D^2}{4} \frac{\rho \mathcal{V}^2}{2} = 3\pi\mu\mathcal{V}D \quad (13\text{-}8)$$

which is known as the **Stokes law,** after British mathematician and physicist G. G. Stokes (1819–1903). This relation shows that at very low Reynolds numbers, the drag force acting on spherical objects is proportional to the diameter, the velocity, and the viscosity of the fluid. This relation is often applicable to dust particles in the air and suspended solid particles in water.

The drag coefficients for low Reynolds number flows past some other geometries are given in Fig. 13-16. Note that at low Reynolds numbers, the shape of the body does not have a major influence on the drag coefficient.

The drag coefficients for various two- and three-dimensional bodies are given in Tables 13-1 and 13-2 for large Reynolds numbers. We can make several observations from these tables about the drag coefficient at high Reynolds numbers. First of all, the *orientation* of the body relative to the direction of flow has a major influence on the drag coefficient. For example, the drag coefficient for flow over a hemisphere is 0.4 when the spherical side faces the flow, but it increases threefold to 1.2 when the flat side faces the flow (Fig. 13-17). This shows that the rounded nose of a *bullet* serves another purpose in addition to piercing: reducing drag and thus increasing the range of the gun. Also, for blunt bodies with sharp corners, such as flow over a rectangular block or a flat plate normal to flow, separation occurs at the edges of the front and back surfaces, with no significant change in the character of flow. Therefore, the drag coefficient of such bodies is nearly independent of the Reynolds number. Note that the drag coefficient of a long rectangular rod can be reduced almost by half from 2.2 to 1.2 by rounding the corners.

A hemisphere at two different orientations
for Re $> 10^4$

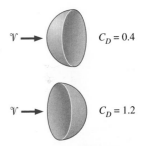

FIGURE 13-17

The drag coefficient of a body may change
drastically by changing its orientation
(and thus shape) relative to the
direction of flow.

Biological Systems and Drag

The concept of drag also has important consequences for biological systems. For example, the bodies of *fish,* especially the ones that swim fast for long distances (such as dolphins), are highly streamlined to minimize drag (the drag coefficient of dolphins based on the wetted skin area is about 0.0035, comparable to the value for a flat plate in turbulent flow). So it is no surprise that we build submarines that mimic large fish. The tropical fish with fascinating

TABLE 13-1

Drag coefficients C_D of various two-dimensional bodies for $Re > 10^4$ based on the frontal area $A = bD$, where b is the length normal to the direction of the paper (for use in the drag force relation $F_D = C_D A \rho \mathcal{V}^2/2$ where $\mathcal{V}$ is the free-stream velocity away from the body)

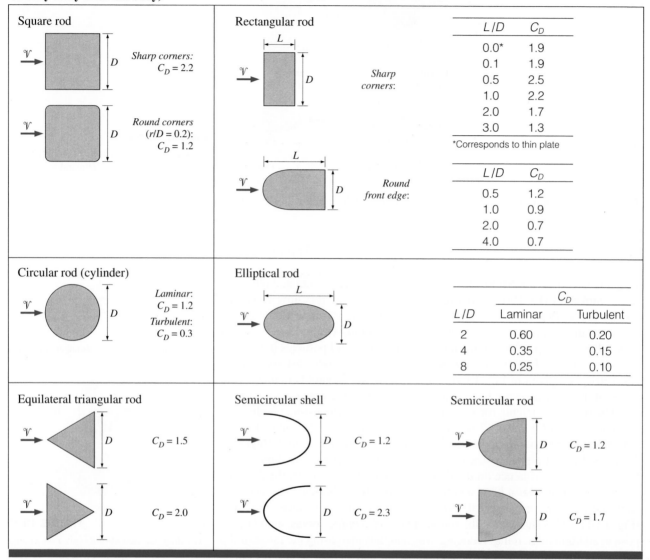

L/D	C_D
0.0*	1.9
0.1	1.9
0.5	2.5
1.0	2.2
2.0	1.7
3.0	1.3

*Corresponds to thin plate

L/D	C_D
0.5	1.2
1.0	0.9
2.0	0.7
4.0	0.7

	C_D	
L/D	Laminar	Turbulent
2	0.60	0.20
4	0.35	0.15
8	0.25	0.10

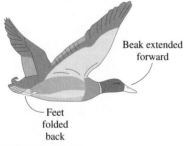

Beak extended forward

Feet folded back

FIGURE 13-18

Birds teach us a lesson on drag reduction by extending their beak forward and folding their feet backward during flight.

beauty and elegance, on the other hand, swim gracefully short distances only. Obviously grace, not high speed and drag, was the primary consideration in their design. Birds teach us a lesson on drag reduction by extending their beak forward and folding their feet backward during flight (Fig. 13-18). Airplanes, which look like big birds, retract their wheels after takeoff in order to reduce drag and thus fuel consumption.

The flexible structure of plants enables them to reduce drag at high winds by changing their shapes. Large flat leaves, for example, curl into a low-drag conical shape at high wind speeds, while tree branches cluster to reduce drag. Flexible trunks bend under the influence of the wind to reduce drag and the bending moment by reducing frontal area.

TABLE 13-2   567

Representative drag coefficients C_D for various three-dimensional bodies for Re > 10^4 based on the frontal area
(for use in the drag force relation $F_D = C_D A \rho \mathcal{V}^2/2$ where $\mathcal{V}$ is the free-stream velocity away from the body)

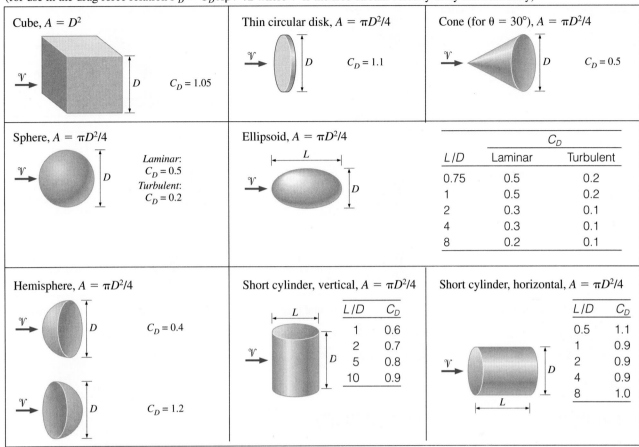

Cube, $A = D^2$ $C_D = 1.05$

Thin circular disk, $A = \pi D^2/4$ $C_D = 1.1$

Cone (for $\theta = 30°$), $A = \pi D^2/4$ $C_D = 0.5$

Sphere, $A = \pi D^2/4$
Laminar: $C_D = 0.5$
Turbulent: $C_D = 0.2$

Ellipsoid, $A = \pi D^2/4$

	C_D	
L/D	Laminar	Turbulent
0.75	0.5	0.2
1	0.5	0.2
2	0.3	0.1
4	0.3	0.1
8	0.2	0.1

Hemisphere, $A = \pi D^2/4$ $C_D = 0.4$ $C_D = 1.2$

Short cylinder, vertical, $A = \pi D^2/4$

L/D	C_D
1	0.6
2	0.7
5	0.8
10	0.9

Short cylinder, horizontal, $A = \pi D^2/4$

L/D	C_D
0.5	1.1
1	0.9
2	0.9
4	0.9
8	1.0

If you watch the Olympic games, you have probably observed many instances of conscious effort by the competitors to reduce drag. Some examples: During 100-m running, the runners hold their fingers together and straight and move their hands parallel to the direction of motion to reduce the hand drag. Swimmers with long hair cover their head with a tight and smooth cover to reduce head drag. They also wear well-fitting one-piece swimming suits. Horse and bicycle riders lean forward as much as they can to reduce drag (by reducing both the drag coefficient and frontal area). Speed skiers do the same thing. Fairings are commonly used in motorcycles to reduce drag.

Drag Coefficients of Vehicles

The term *drag coefficient* is commonly used in various areas of daily life. Car manufacturers try to attract consumers by pointing out the *low drag coefficients* of their cars (Fig. 13-19). The drag coefficients of vehicles range from about 1.0 for large semitrucks to 0.4 for minivans, and to 0.3 for passenger cars. In general, the more blunt the vehicle, the higher the drag coefficient. Installing a fairing reduces the drag coefficient of tractor-trailer rigs by about 25 percent by making the frontal surface more streamlined. As a rule of

FIGURE 13-19
Modern vehicles are shaped so as to minimize the drag coefficient and thus maximize the fuel efficiency.

TABLE 13-2 *(Concluded)*

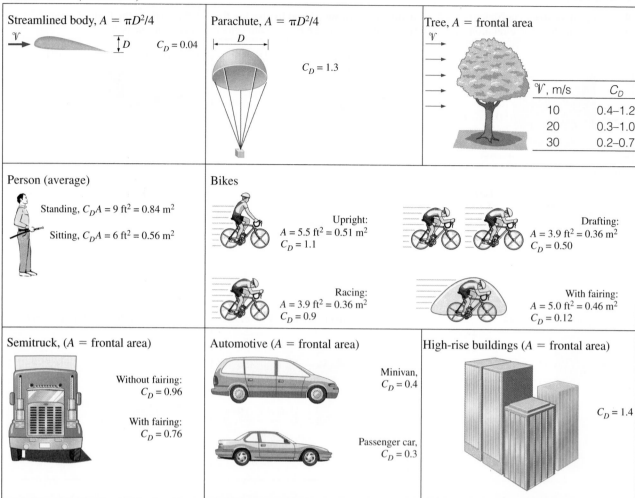

Streamlined body, $A = \pi D^2/4$

$C_D = 0.04$

Parachute, $A = \pi D^2/4$

$C_D = 1.3$

Tree, $A = $ frontal area

$\mathcal{V}$, m/s	C_D
10	0.4–1.2
20	0.3–1.0
30	0.2–0.7

Person (average)

Standing, $C_D A = 9$ ft$^2 = 0.84$ m^2

Sitting, $C_D A = 6$ ft$^2 = 0.56$ m^2

Bikes

Upright:
$A = 5.5$ ft$^2 = 0.51$ m^2
$C_D = 1.1$

Racing:
$A = 3.9$ ft$^2 = 0.36$ m^2
$C_D = 0.9$

Drafting:
$A = 3.9$ ft$^2 = 0.36$ m^2
$C_D = 0.50$

With fairing:
$A = 5.0$ ft$^2 = 0.46$ m^2
$C_D = 0.12$

Semitruck, ($A = $ frontal area)

Without fairing:
$C_D = 0.96$

With fairing:
$C_D = 0.76$

Automotive ($A = $ frontal area)

Minivan,
$C_D = 0.4$

Passenger car,
$C_D = 0.3$

High-rise buildings ($A = $ frontal area)

$C_D = 1.4$

thumb, the percentage of fuel savings due to reduced drag is about half the percentage of drag reduction.

From the drag point of view, the ideal shape of a *vehicle* is the basic *teardrop,* with a drag coefficient of about 0.1 for the turbulent flow case. But this shape needs to be modified to accommodate several necessary external components such as wheels, mirrors, axles, antennas, and so on. Also, the vehicle must be high enough for comfort and there must be a minimum clearance from the road. Further, a vehicle cannot be too long to fit in garages and parking spaces. Controlling the material and manufacturing costs requires minimizing or eliminating any "dead" volume that cannot be utilized. The result is a shape that resembles more of a "box" than a "teardrop," and this was the shape of early cars with a drag coefficient of 0.8 in the 1920s. This wasn't a problem in those days since the velocities were low and drag was not a major design consideration.

The average drag coefficients of cars dropped to about 0.70 in the 1940s, to 0.55 in the 1970s, to 0.45 in the 1980s, and to 0.30 in the 1990s as a result of improved manufacturing techniques for metal forming and paying more attention to the shape of the car and streamlining. The drag coefficient for well-built racing cars is about 0.2, but this is achieved after making the comfort of

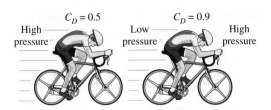

$C_D = 0.5$ High pressure Low pressure $C_D = 0.9$ High pressure

FIGURE 13-20

The drag coefficients of bodies following other moving bodies closely can be reduced considerably due to drafting (i.e., falling into the vacuum created by the body in front).

drivers a secondary consideration. Noting that the theoretical lower limit of C_D is about 0.1 and the value for racing cars is 0.2, it appears that there is only little room for further improvement in the drag coefficient of passenger cars from the current value of 0.3. For trucks and buses, the drag coefficient can be reduced further by optimizing the front and rear contours (by rounding, for example) to the extent it is practical while keeping the overall length of the vehicle the same.

When traveling as a group, a sneaky way of reducing drag is **drafting,** a phenomenon well-known by bicycle riders and car racers. It involves approaching a moving body from behind and *being drafted* into the low pressure region in the rear of the body. The drag coefficient of a racing bicyclist, for example, can be reduced from 0.9 to 0.5 by drafting, as shown in Table 13-2 (Fig. 13-20).

We also can help reduce the overall drag of a vehicle and thus fuel consumption by being more conscious drivers. For example, drag force is proportional to the square of velocity. Therefore, driving over the speed limit on the highways not only increases the chances of getting speeding tickets, but it also increases the amount of fuel consumption per mile. Therefore, driving at moderate speeds is safe and economical. Also, anything that extends from the car, even an arm, increases the drag coefficient. Driving with the windows rolled down also increases the drag and fuel consumption. At highway speeds, a driver can save gas in hot weather by running the air conditioner instead of driving with the windows rolled down. Usually the turbulence and additional drag generated by open windows consume more gasoline than does the air conditioner.

Superposition

The shapes of many bodies encountered in practice are not simple. But such bodies can be treated conveniently in drag force calculations by considering them to be composed of two or more simple bodies. A satellite dish mounted on a roof with a cylindrical bar, for example, can be considered to be a combination of a hemispherical body and a cylinder. Then the drag coefficient of the body can be determined approximately by using **superposition.** Such a simplistic approach does not account for the effects of components on each other, and thus the results obtained should be interpreted accordingly.

EXAMPLE 13-2 Effect of Mirror Design on the Fuel Consumption of a Car

As part of the continuing efforts to reduce the drag coefficient and thus to improve the fuel efficiency of cars, the design of side rear-view mirrors has changed drastically from a simple circular plate to a streamlined shape. Determine the amount of fuel and money saved per year as a result of replacing a 13-cm-diameter flat mirror by one with a hemispherical back (Fig. 13-21). Assume the car is driven 24,000 km a year at an average speed of 95 km/h. Take the density and price of gasoline to be 0.8 kg/L and $0.60/L, respectively; the heating value of gasoline to be 44,000 kJ/kg; and the overall efficiency of the engine to be 30 percent.

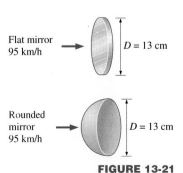

Flat mirror 95 km/h $D = 13$ cm

Rounded mirror 95 km/h $D = 13$ cm

FIGURE 13-21

Schematic for Example 13-2.

Solution The flat mirror of a car is replaced by one with a hemispherical back. The amount of fuel and money saved per year as a result are to be determined.

Assumptions **1** The car is driven 24,000 km a year at an average speed of 95 km/h. **2** The effect of the car body on the flow around the mirror is negligible (no interference). **3** The average density of air is 1.20 kg/m³.

Properties The densities of air and gasoline are taken to be 1.20 kg/m³ and 800 kg/m³, respectively. The heating value of gasoline is given to be 44,000 kJ/kg. The drag coefficients C_D are 1.1 for a circular disk and 0.40 for a hemispherical body (Table 13-2).

Analysis The drag force acting on a body is determined from

$$F_D = C_D \, A \, \frac{\rho \mathcal{V}^2}{2}$$

where A is the frontal area of the body, which is $A = \pi D^2/4$ for both the flat and rounded mirrors. The drag force acting on the flat mirror is

$$F_D = 1.1 \, \frac{\pi(0.13 \text{ m})^2}{4} \, \frac{(1.20 \text{ kg/m}^3)(95 \text{ km/h})^2}{2} \left(\frac{1 \text{ m/s}}{3.6 \text{ km/h}}\right)^2 \left(\frac{1 \text{ N}}{1 \text{ kg} \cdot \text{m/s}^2}\right) = 6.10 \text{ N}$$

Noting that work is force times distance, the amount of work done to overcome this drag force and the required energy input for a distance of 24,000 km are

$$W_{\text{drag}} = F_D \times L = (6.10 \text{ N})(24,000 \text{ km/year}) = 146,400 \text{ kJ/year}$$

$$E_{\text{in}} = \frac{W_{\text{drag}}}{\eta_{\text{car}}} = \frac{146,400 \text{ kJ/year}}{0.3} = 488,000 \text{ kJ/year}$$

Then the amount and costs of the fuel that supplies this much energy are

$$\text{Amount of fuel} = \frac{m_{\text{fuel}}}{\rho_{\text{fuel}}} = \frac{E_{\text{in}}/HV}{\rho_{\text{fuel}}} = \frac{(488,000 \text{ kJ/year})/(44,000 \text{ kJ/kg})}{0.8 \text{ kg/L}} = 13.9 \text{ L/year}$$

$$\text{Cost} = (\text{Amount of fuel})(\text{Unit cost}) = (13.9 \text{ L/year})(\$0.60/\text{L}) = \$8.32/\text{year}$$

That is, the car uses 13.9 L of gasoline at a cost of $8.32 per year to overcome the drag generated by a flat mirror extending out from the side of a car.

The drag force and the work done to overcome it are directly proportional to the drag coefficient. Then the percent reduction in the fuel consumption due to replacing the mirror is equal to the percent reduction in the drag coefficient:

$$\text{Reduction ratio} = \frac{C_{D,\text{ flat}} - C_{D,\text{ hemisp}}}{C_{D,\text{ flat}}} = \frac{1.1 - 0.4}{1.1} = 0.636$$

$$\text{Fuel reduction} = (\text{Reduction ratio})(\text{Amount of fuel})$$
$$= 0.636(13.9 \text{ L/year}) = \textbf{8.84 L/year}$$

$$\text{Cost reduction} = (\text{Reduction ratio})(\text{Cost}) = 0.636(\$8.32/\text{year}) = \textbf{\$5.29/year}$$

Therefore, replacing a flat mirror by a hemispherical one reduces the fuel consumption due to mirror drag by 63.6 percent.

Discussion Note from this example that significant reductions in drag and fuel consumption can be achieved by streamlining the shape of various components and the entire car. So it is no surprise that the sharp corners are replaced in late model cars by rounded contours. This also explains why large airplanes retract their wheels after takeoff and small airplanes use contoured fairings around their wheels.

The example above is indicative of the tremendous amount of effort put in recent years into redesigning various parts of the cars such as the window

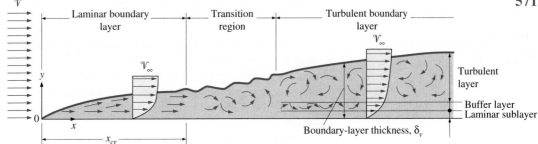

FIGURE 13-22

The development of the boundary layer for flow over a flat plate, and the different flow regimes.

moldings, the door handles, the windshield, and the front and rear ends in order to reduce aerodynamic drag. For a car moving on a level road at constant speed, the power developed by the engine is used to overcome rolling resistance, friction between moving components, aerodynamic drag, and driving the auxiliary equipment. The drag is negligible at low speeds, but becomes significant at speeds above 30 mph. Reduction of the frontal area of the cars (to the dislike of tall drivers) has also contributed greatly to the reduction of drag and fuel consumption.

13-5 ■ FLOW OVER FLAT PLATES

Consider the flow of a fluid over a *flat plate,* as shown in Fig. 13-22. Surfaces that are slightly contoured such as turbine blades can also be approximated as flat plates with reasonable accuracy. The x coordinate is measured along the plate surface from the *leading edge* of the plate in the direction of the flow, and y is measured from the surface in the normal direction. The fluid approaches the plate in the x direction with a uniform upstream velocity $\mathcal{V}$. For the sake of discussion, we can consider the fluid to consist of *adjacent layers* piled on top of each other.

The velocity of fluid particles in the first fluid layer adjacent to the plate is *zero* because of the no-slip condition. This motionless layer *slows down* the particles of the neighboring fluid layer as a result of friction between the particles of these two adjoining fluid layers at different velocities. This fluid layer then slows down the molecules of the next layer, and so on. Thus, the presence of the plate is felt up to some distance δ_v from the plate beyond which the free-stream velocity $\mathcal{V}_\infty$, which is very nearly equal to the approach velocity $\mathcal{V}$, remains essentially unchanged. As a result, the fluid velocity at any x location will vary from 0 at $y = 0$ to nearly $\mathcal{V}_\infty$ at $y = \delta_v$ (Fig. 13-23).

The region of the flow above the plate bounded by δ_v in which the effects of the viscous shearing forces caused by fluid viscosity are felt is called the **velocity boundary layer** or just the **boundary layer.** The *thickness* of the boundary layer, δ_v, is arbitrarily defined as the distance from the surface at which the velocity is $0.99\mathcal{V}_\infty$.

The hypothetical line of velocity $0.99\mathcal{V}_\infty$ divides the flow over a plate into two regions: the **boundary layer region,** in which the viscous effects and the velocity changes are significant, and the **inviscid flow region,** in which the frictional effects are negligible and the velocity remains essentially constant.

For parallel flow over a flat plate, the pressure drag is zero, and thus the drag coefficient is equal to the *friction drag coefficient,* or simply the *friction coefficient* (Fig. 13-24). That is,

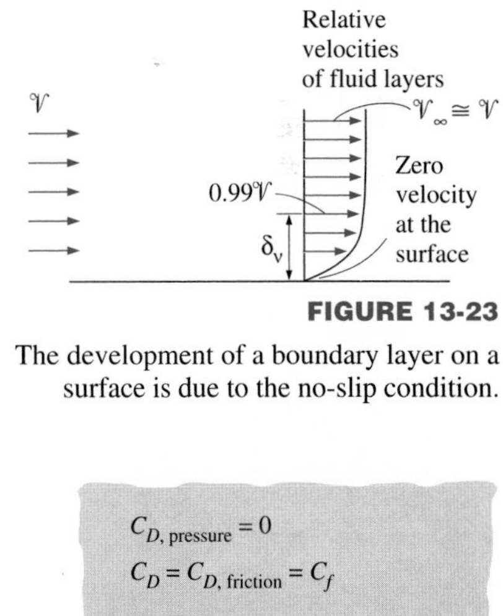

FIGURE 13-23

The development of a boundary layer on a surface is due to the no-slip condition.

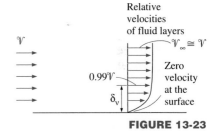

$$C_{D,\text{pressure}} = 0$$
$$C_D = C_{D,\text{friction}} = C_f$$

$$F_{D,\text{pressure}} = 0$$
$$F_D = F_{D,\text{friction}} = F_f = C_f A \frac{\rho \mathcal{V}^2}{2}$$

FIGURE 13-24

For a flat plate, the pressure drag is zero, and thus the drag coefficient is equal to the friction coefficient and the drag force is equal to the friction force.

Flat plate: $$C_D = C_{D,\text{friction}} = C_f \tag{13-9}$$

Once the average friction coefficient C_f is available, the drag (or friction) force over the surface can be determined from Eq. 13-3 where A is the surface area of the plate exposed to fluid flow. When both sides of a thin plate are subjected to flow, A becomes the total area of the top and bottom surfaces. Note that the friction coefficient, in general, will vary with location along the surface.

Typical mean velocity profiles in laminar and turbulent flow are also given in Fig. 13-22. Note that the velocity profile is approximately parabolic in laminar flow and becomes flatter in turbulent flow, with a sharp drop near the surface. The turbulent boundary layer can be considered to consist of three layers. The very thin layer next to the wall where the viscous effects are dominant is the **laminar sublayer.** The velocity profile in this layer is nearly linear and the flow is streamlined. Next to the laminar sublayer is the **buffer layer,** in which the turbulent effects are significant but not dominant, and next to it is the **turbulent layer,** in which the turbulent effects dominate.

The transition from laminar to turbulent flow depends on the *surface geometry, surface roughness, upstream velocity, surface temperature,* and *type of fluid,* among other things, and is best characterized by the Reynolds number. For external flow, the Reynolds number is expressed as

$$\text{Re}_L = \frac{\rho \mathcal{V} L}{\mu} = \frac{\mathcal{V} L}{\nu} \tag{13-10}$$

where $\mathcal{V}$ is the upstream velocity and L is the characteristic length of the geometry, which, for a flat plate, is the length of the plate in the flow direction. Note that unlike pipe flow, the Reynolds number varies for a flat plate along the flow. For any point on a flat plate, the characteristic length is the distance x of the point from the leading edge in the flow direction.

For flow over a *flat plate,* transition from laminar to turbulent begins at about $\text{Re} \approx 1 \times 10^5$, but does not become fully turbulent before the Reynolds number reaches much higher values. A generally accepted value for the critical Reynolds number is

$$\text{Re}_{L,\text{critical}} \approx 5 \times 10^5$$

This generally accepted value of the critical Reynolds number for a flat plate may vary somewhat depending on the surface roughness, the turbulence level, and the variation of pressure along the surface.

The friction coefficient for a flat plate can be determined theoretically by solving the conservation of mass and momentum equations approximately or numerically. It can also be determined experimentally and expressed by empirical correlations.

The local friction coefficient *varies* along the surface of the flat plate as a result of the changes in the velocity boundary layers in the flow direction. We are usually interested in the drag force on the *entire* surface, which can be determined using the *average* friction coefficient. But sometimes we are also interested in the drag force at a certain location, and in such cases, we need to know the *local* value of the friction coefficient. With this in mind, below we present correlations for both local (identified with the subscript x) and average friction coefficients over a flat plate for *laminar, turbulent,* and *combined laminar and turbulent* flow conditions. Once the local values are available, the *average* friction coefficient for the entire plate can be determined by integration from

$$C_f = \frac{1}{L} \int_0^L C_{f,x} \, dx \qquad (13\text{-}11)$$

Based on analysis and experimental studies, the boundary layer thickness and the *local* friction coefficients at location x for laminar and turbulent flows over a flat plate are given by

Laminar: $\quad \delta_{v,x} = \dfrac{5x}{\text{Re}_x^{1/2}} \quad$ and $\quad C_{f,x} = \dfrac{0.664}{\text{Re}_x^{1/2}} \quad \text{Re}_x < 5 \times 10^5$

$$(13\text{-}12)$$

Turbulent: $\quad \delta_{v,x} = \dfrac{0.382x}{\text{Re}_x^{1/5}} \quad$ and $\quad C_{f,x} = \dfrac{0.0592}{\text{Re}_x^{1/5}} \quad 5 \times 10^5 \le \text{Re}_x \le 10^7$

$$(13\text{-}13)$$

where x is the distance from the leading edge of the plate and $\text{Re}_x = \mathcal{V}x/v$ is the Reynolds number at location x. Note that $C_{f,x}$ is proportional to $1/\text{Re}_x^{1/2}$ and thus to $x^{-1/2}$ for laminar flow. Therefore, $C_{f,x}$ is supposedly *infinite* at the leading edge ($x = 0$) and decreases by a factor of $x^{-1/2}$ in the flow direction. The variation of the boundary layer thickness δ and the friction coefficient C_f along a flat plate is shown in Fig. 13-25. The local friction coefficient is higher in turbulent flow than it is in laminar flow because of the intense mixing that occurs in the turbulent boundary layer. Note that $C_{f,x}$ reaches its highest values when the flow becomes fully turbulent and then decreases by a factor of $x^{-1/5}$ in the flow direction, as shown in the figure.

The *average* friction coefficient over the entire plate is determined by substituting the relations above into Eq. 13-11 and performing the simple integrations (Fig. 13-26). We get

Laminar flow:	$C_f = \dfrac{1.328}{\text{Re}_L^{1/2}}$	$\text{Re}_L < 5 \times 10^5$	(13-14)
Turbulent flow:	$C_f = \dfrac{0.074}{\text{Re}_L^{1/5}}$	$5 \times 10^5 \le \text{Re}_L \le 10^7$	(13-15)

The first relation above gives the average friction coefficient for the entire plate when the flow is *laminar* over the *entire* plate. The second relation gives the average friction coefficient for the entire plate only when the flow is *turbulent* over the *entire* plate, or when the laminar flow region of the plate is too small relative to the turbulent flow region (that is, $x_{cr} \ll L$ where the length of the plate x_{cr} over which the flow is laminar can be determined from $\text{Re}_{cr} = 5 \times 10^5 = \mathcal{V}x_{cr}/v$.)

Combined Laminar and Turbulent Flow

In some cases, a flat plate is sufficiently long for the flow to become turbulent, but not long enough to disregard the laminar flow region. In such cases, the *average* friction coefficient over the entire plate is determined by performing the integration in Eq. 13-11 over two parts: the laminar region $0 \le x \le x_{cr}$ and the turbulent region $x_{cr} < x < L$ as

$$C_f = \frac{1}{L} \left(\int_0^{x_{cr}} C_{f,x,\,\text{laminar}} \, dx + \int_{x_{cr}}^L C_{f,x,\,\text{turbulent}} \, dx \right) \qquad (13\text{-}16)$$

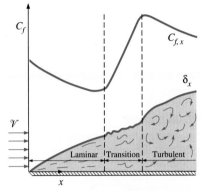

FIGURE 13-25

The variation of the local friction coefficient for flow over a flat plate.

$$C_f = \frac{1}{L} \int_0^L C_{f,x} \, dx$$

$$= \frac{1}{L} \int_0^L \frac{0.664}{\text{Re}_x^{1/2}} \, dx$$

$$= \frac{0.664}{L} \int_0^L \left(\frac{\mathcal{V}x}{v} \right)^{-1/2} dx$$

$$= \frac{0.664}{L} \left(\frac{\mathcal{V}}{v} \right)^{-1/2} \frac{x^{1/2}}{\frac{1}{2}} \Big|_0^L$$

$$= \frac{2 \times 0.664}{L} \left(\frac{\mathcal{V}L}{v} \right)^{-1/2}$$

$$= \frac{1.328}{\text{Re}_L^{1/2}}$$

FIGURE 13-26

The average friction coefficient over a surface is determined by integrating the local friction coefficient over the entire surface.

Relative roughness, ε/L	Friction coefficient C_f
0.0*	0.0029
1×10^{-5}	0.0032
1×10^{-4}	0.0049
1×10^{-3}	0.0084

*Smooth surface for Re = 10^7. Others calculated from Eq. 13-18.

FIGURE 13-27

For turbulent flow, surface roughness may cause the friction coefficient to increase severalfold.

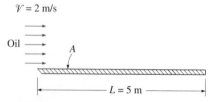

$\mathcal{V}$ = 2 m/s

Oil

A

L = 5 m

FIGURE 13-28

Schematic for Example 13-3.

Note that we included the transition region with the turbulent region. Again taking the critical Reynolds number to be $Re_{cr} = 5 \times 10^5$ and performing the integrations above after substituting the indicated expressions, the *average* friction coefficient over the *entire* plate is determined to be

$$C_f = \frac{0.074}{Re_L^{1/5}} - \frac{1742}{Re_L} \qquad 5 \times 10^5 \leq Re_L \leq 10^7 \qquad (13\text{-}17)$$

The constants in the relation above will be different for different critical Reynolds numbers. Also, the surfaces are assumed to be *smooth* and the free stream to be *turbulent free*. For laminar flow, the friction coefficient depends only on the Reynolds number; the surface roughness has no effect. For turbulent flow, however, surface roughness causes the friction coefficient to increase severalfold, to the point that in a fully turbulent regime, the friction coefficient is a function of surface roughness alone and independent of the Reynolds number (Fig. 13-27). This is analogous to flow in pipes. A curve fit of experimental data for the average friction coefficient in this regime is given by Schlichting (Ref. 15) as

$$\textit{Rough surface:} \qquad C_f = \left(1.89 - 1.62 \log \frac{\varepsilon}{L}\right)^{-2.5} \qquad (13\text{-}18)$$

where ε is the surface roughness and L is the length of the plate in the flow direction. In the absence of a better relation, the relation above can be used for turbulent flow on rough surfaces for Re $> 10^6$, especially when $\varepsilon/L > 10^{-4}$.

EXAMPLE 13-3   Flow of Hot Oil over a Flat Plate
Engine oil at 40°C flows over a 5-m-long flat plate with a free-stream velocity of 2 m/s (Fig. 13-28). Determine the drag force acting on the plate per unit width.

Solution   Engine oil flows over a flat plate. The drag force per unit width of the plate is to be determined.

Assumptions   1 The flow is steady and incompressible. 2 The critical Reynolds number is $Re_{cr} = 5 \times 10^5$.

Properties   The density and kinematic viscosity of engine oil at 40°C are $\rho = 876$ kg/m³ and $\nu = 242 \times 10^{-6}$ m²/s (Table A-16).

Analysis   Noting that $L = 5$ m, the Reynolds number at the end of the plate is

$$Re_L = \frac{\mathcal{V}L}{\nu} = \frac{(2 \text{ m/s})(5 \text{ m})}{242 \times 10^{-6} \text{ m}^2/\text{s}} = 4.13 \times 10^3$$

which is less than the critical Reynolds number. Thus we have *laminar flow* over the entire plate, and the average friction coefficient is determined from

$$C_f = 1.328 \, Re_L^{-0.5} = 1.328 \times (4.13 \times 10^3)^{-0.5} = 0.0207$$

Noting that the pressure drag is zero and thus $C_D = C_f$ for a flat plate, the drag force acting on the plate per unit width becomes

$$F_D = C_f A \frac{\rho \mathcal{V}^2}{2} = 0.0207 \times (5 \times 1 \text{ m}^2) \frac{(876 \text{ kg/m}^3)(2 \text{ m/s})^2}{2} \left(\frac{1 \text{ N}}{1 \text{ kg} \cdot \text{m/s}^2}\right)$$

$$= \textbf{181 N}$$

The total drag force acting on the entire plate can be determined by multiplying the value obtained above by the width of the plate.

Discussion   The force per unit width corresponds to the weight of a mass of about 18 kg. Therefore, a person who applies an equal and opposite force to the plate to keep it from moving will feel like he or she is using as much force as is necessary to hold a 18-kg mass from dropping.

13-6 ■ FLOW ACROSS CYLINDERS AND SPHERES

Flow across cylinders and spheres is frequently encountered in practice. For example, the tubes in a shell-and-tube heat exchanger involve both *internal flow* through the tubes and *external flow* over the tubes, and both flows must be considered in the analysis of the heat exchanger. Also, many sports such as soccer, tennis, and golf involve flow over spherical balls.

The characteristic length for a circular cylinder or sphere is taken to be the *external diameter D*. Thus, the Reynolds number is defined as $Re = \mathcal{V}D/\nu$ where $\mathcal{V}$ is the uniform velocity of the fluid as it approaches the cylinder or sphere. The critical Reynolds number for flow across a circular cylinder or sphere is about $Re_{cr} \approx 2 \times 10^5$. That is, the boundary layer remains laminar for about $Re \lesssim 2 \times 10^5$ and becomes turbulent for $Re \gtrsim 2 \times 10^5$.

Cross flow over a cylinder exhibits complex flow patterns, as shown in Fig. 13-29. The fluid approaching the cylinder branches out and encircles the cylinder, forming a boundary layer that wraps around the cylinder. The fluid particles on the midplane strike the cylinder at the stagnation point, bringing the fluid to a complete stop and thus raising the pressure at that point. The pressure decreases in the flow direction while the fluid velocity increases.

At very low upstream velocities ($Re \lesssim 1$), the fluid completely wraps around the cylinder and the two arms of the fluid meet on the rear side of the cylinder in an orderly manner. Thus, the fluid follows the curvature of the cylinder. At higher velocities, the fluid still hugs the cylinder on the frontal side, but it is too fast to remain attached to the surface as it approaches the top of the cylinder. As a result, the boundary layer detaches from the surface, forming a separation region behind the cylinder. Flow in the wake region is characterized by random vortex formation and pressures much lower than the stagnation point pressure.

The nature of the flow across a cylinder or sphere strongly affects the total drag coefficient C_D. Both the *friction drag* and the *pressure drag* can be significant. The high pressure in the vicinity of the stagnation point and the low pressure on the opposite side in the wake produce a net force on the body in the direction of flow. The drag force is primarily due to friction drag at low Reynolds numbers ($Re < 10$) and to pressure drag at high Reynolds numbers ($Re > 5000$). Both effects are significant at intermediate Reynolds numbers.

The average drag coefficients C_D for cross flow over a smooth single circular cylinder and a sphere are given in Fig. 13-30. The curves exhibit different behaviors in different ranges of Reynolds numbers:

- For $Re \lesssim 1$, we have creeping flow, and the drag coefficient decreases with increasing Reynolds number. For a sphere, it is $C_D = 24/Re$. There is no flow separation in this regime.
- At about $Re = 10$, separation starts occurring on the rear of the body with vortex shedding starting at about $Re \approx 90$. The region of separation increases with increasing Reynolds number up to about $Re = 10^3$. At this point, the drag is mostly (about 95 percent) due to pressure drag. The drag coefficient continues to decrease with increasing Reynolds number in this range of $10 < Re < 10^3$. (A decrease in the drag coefficient does not necessarily indicate a decrease in drag. The drag force is proportional to the square of the velocity, and the increase in velocity at

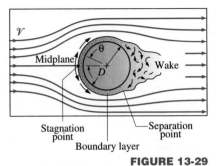

FIGURE 13-29

Typical flow patterns in cross flow over a cylinder.

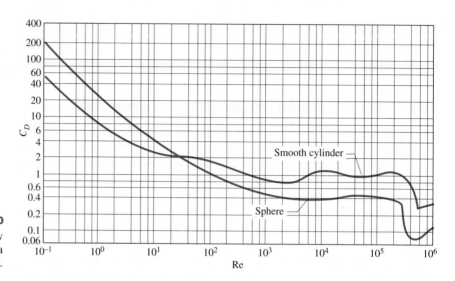

FIGURE 13-30

Average drag coefficient for cross flow over a smooth circular cylinder and a smooth sphere (from Schlichting, Ref. 15).

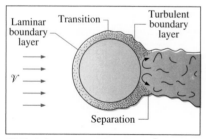

(a) Laminar flow (Re < 2×10^5)

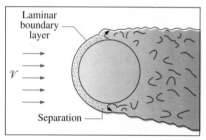

(b) Turbulence occurs (Re > 2×10^5)

FIGURE 13-31

Turbulence delays flow separation.

higher Reynolds numbers usually more than offsets the decrease in the drag coefficient.)

- In the moderate range of $10^3 <$ Re $< 10^5$, the drag coefficient remains relatively constant. This behavior is characteristic of blunt bodies. The flow in the boundary layer is laminar in this range, but the flow in the separated region past the cylinder or sphere is highly turbulent with a wide turbulent wake.

- There is a sudden drop in the drag coefficient somewhere in the range of $10^5 <$ Re $< 10^6$ (usually, at about 2×10^5). This large reduction in C_D is due to the flow in the boundary layer becoming *turbulent,* which moves the separation point further on the rear of the body, reducing the size of the wake and thus the magnitude of the pressure drag. This is in contrast to streamlined bodies, which experience an increase in the drag coefficient (mostly due to friction drag) when the boundary layer becomes turbulent.

Flow separation occurs at about $\theta \approx 80°$ (measured from the stagnation point) when the boundary layer is *laminar* and at about $\theta \approx 140°$ when it is *turbulent* (Fig. 13-31). The delay of separation in turbulent flow is caused by the rapid fluctuations of the fluid in the transverse direction, which enables the turbulent boundary layer to travel further along the surface before separation occurs, resulting in a narrower wake and a smaller pressure drag. In the range of Reynolds numbers where the flow changes from laminar to turbulent, even the drag force F_D decreases as the velocity (and thus Reynolds number) increases. This results in a sudden decrease in drag of a flying body and instabilities in flight.

Effect of Surface Roughness

We mentioned earlier that *surface roughness,* in general, increases the drag coefficient in turbulent flow. This is especially the case for streamlined bodies. For blunt bodies such as a circular cylinder or sphere, however, an increase in the surface roughness may actually *decrease* the drag coefficient, as shown in Fig. 13-32 for a sphere. This is done by tripping the flow into turbulence at a lower Reynolds number, and thus causing the fluid to close in behind the body, narrowing the wake and reducing pressure drag considerably. This

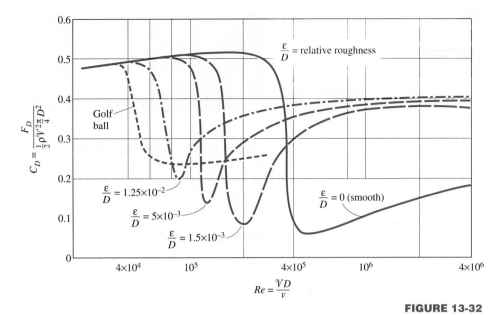

$$C_D = \frac{F_D}{\frac{1}{2}\rho V^2 \frac{\pi}{4}D^2}$$

$$Re = \frac{V D}{v}$$

FIGURE 13-32

The effect of surface roughness on the drag coefficient of a sphere (from Blevins, Ref. 5).

results in a much smaller drag coefficient and thus drag force for a rough-sur-faced cylinder or sphere in a certain range of Reynolds number compared to a smooth one of identical size at the same velocity. At $Re = 10^5$, for example, $C_D = 0.1$ for a rough sphere with $\varepsilon/D = 0.0015$, whereas $C_D = 0.5$ for a smooth one. Therefore, the drag coefficient in this case is reduced by a fac-tor of 5 by simply roughening the surface. Note, however, that at $Re = 10^6$, $C_D = 0.4$ for the rough sphere while $C_D = 0.1$ for the smooth one. Obviously, roughening the sphere in this case will increase the drag by a factor of 4 (Fig. 13-33).

The discussion above shows that roughening the surface can be used to great advantage in reducing drag, but it can also backfire on us if we are not careful—specifically, if we do not operate in the right range of Reynolds num-ber. With this consideration, golf balls are intentionally roughened to induce *turbulence* at a lower Reynolds number to take advantage of the sharp *drop* in the drag coefficient at the onset of turbulence in the boundary layer (the typi-cal velocity range of golf balls is 15 to 150 m/s, and the Reynolds number is less than 4×10^5). The critical Reynolds number of dimpled golf balls is about 4×10^4. The occurrence of turbulent flow at this Reynolds number reduces the drag coefficient of a golf ball by half, as shown in Fig. 13-32. For a given hit, this means a longer distance for the ball. Experienced golfers also give the ball a spin during the hit, which helps the rough ball develop a lift and thus travel higher and further. A similar argument can be given for a tennis ball. For a table tennis ball, however, the distances are very short, and the balls never reach the speeds in the turbulent range. Therefore, the surfaces of table tennis balls are made smooth.

Once the drag coefficient is available, the drag force acting on a body in cross flow can be determined from Eq. 13-1 where A is the *frontal area* ($A = LD$ for a cylinder of length L and $A = \pi D^2/4$ for a sphere). It should be kept in mind that the free-stream turbulence and disturbances by other bodies in flow (such as flow over tube bundles) may affect the drag coefficients significantly.

	C_D	
Re	**Smooth surface**	**Rough surface, $\varepsilon/L = 0.0015$**
10^5	0.5	0.1
10^6	0.1	0.4

FIGURE 13-33

Surface roughness may increase or decrease the drag coefficient of a spherical object, depending on the value of the Reynolds number.

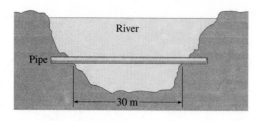

FIGURE 13-34

Schematic for Example 13-4.

EXAMPLE 13-4 Drag Force Acting on a Pipe in a River

A 2.2-cm-outer-diameter pipe is to cross a river at a 30-m-wide section while being completely immersed in water (Fig. 13-34). The average flow velocity of water is 4 m/s and the water temperature is 15°C. Determine the drag force exerted on the pipe by the river.

Solution A pipe is crossing a river. The drag force that acts on the pipe is to be determined.

Assumptions **1** The outer surface of the pipe is smooth so that Fig. 13-30 can be used to determine the drag coefficient. **2** Water flow in the river is steady. **3** The direction of water flow is normal to the pipe. **4** Turbulence in river flow is not considered.

Properties The density and dynamic viscosity of water at 15°C are $\rho = 999.1$ kg/m^3 and $\mu = 1.138 \times 10^{-3}$ kg/m · s (Table A-15).

Analysis Noting that $D = 0.022$ m, the Reynolds number for flow over the pipe is

$$\text{Re} = \frac{\mathcal{V}D}{\nu} = \frac{\rho \mathcal{V} D}{\mu} = \frac{(999.1 \text{ kg/m}^3)(4 \text{ m/s})(0.022 \text{ m})}{1.138 \times 10^{-3} \text{ kg/m} \cdot \text{s}} = 7.73 \times 10^4$$

The drag coefficient corresponding to this value is, from Fig 13-30, $C_D = 1.0$. Also, the frontal area for flow past a cylinder is $A = LD$. Then the drag force acting on the pipe becomes

$$F_D = C_D A \frac{\rho \mathcal{V}^2}{2} = 1.0(30 \times 0.022 \text{ m}^2) \frac{(999.1 \text{ kg/m}^3)(4 \text{ m/s})^2}{2} \left(\frac{1 \text{ N}}{1 \text{ kg} \cdot \text{m/s}^2}\right)$$

$$= \textbf{5275 N}$$

Discussion Note that this force is equivalent to the weight of a mass over 500 kg. Therefore, the drag force the river exerts on the pipe is equivalent to hanging a total of over 500 kg in mass on the pipe supported at its ends 30 m apart. The necessary precautions should be taken if the pipe cannot support this force.

13-7 ■ LIFT

Lift was defined earlier as the component of the net force (due to viscous and pressure forces) that is perpendicular to the flow direction, and the lift coefficient was defined as

$$C_L = \frac{F_L}{\frac{1}{2}\rho \mathcal{V}^2 A} \qquad \text{(Eq. 13-4)}$$

where A in this case is normally the *planform area,* which is the area that would be seen by a person looking at the body from above in a direction normal to the body, and $\mathcal{V}$ is the upstream velocity of the fluid (or, equivalently, the velocity of a flying body in a quiescent fluid). For an airfoil of width (or span) b and chord length c (the length between the leading and trailing edges),

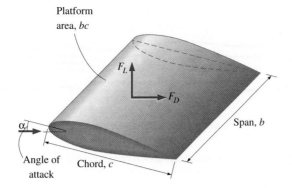

Platform area, bc

F_L

F_D

Span, b

α

Angle of attack

Chord, c

FIGURE 13-35

Definition of various terms associated with an airfoil.

the planform area is $A = bc$. The distance between the two ends of a wing or airfoil is called the **wingspan** or just **span.** For an aircraft, the wingspan is taken to be the total distance between the tips of the two wings, which includes the width of the fuselage between the wings (Fig. 13-35). The average lift per unit planform area F_L/A is called the **wing loading,** which is simply the ratio of the weight of the aircraft to the planform area of the wings (since lift equals the weight during flying at constant altitude).

The flying of airplanes is based on lift, and thus developing a better understanding of lift as well as improving the lift characteristics of bodies have been the focus of numerous studies. Our emphasis in this section will be on devices such as *airfoils* that are specifically designed to generate lift while keeping the drag at a minimum. But it should be kept in mind that some devices such as the *spoilers* and *inverted airfoils* on racing cars are designed for the opposite purpose of avoiding lift or even generating negative lift to improve traction and control (some early cars actually "took off" at high speeds as a result of the lift produced, which alerted the engineers to come up with ways to reduce lift in their design).

For devices that are intended to generate lift such as airfoils, the contribution of *viscous effects* to lift is usually negligible since wall shear is parallel to the surfaces of such devices and thus nearly normal to the direction of lift (Fig. 13-36). Therefore, lift in practice can be taken to be due entirely to the pressure distribution on the surfaces of the body, and thus the shape of the body has the primary effect on lift. Then the primary consideration in the design of airfoils is minimizing the average pressure at the upper surface while maximizing it at the lower surface. The Bernoulli equation can be used as a guide in identifying the high- and low-pressure regions: *Pressure is low at locations where the flow velocity is high, and pressure is high at locations where the flow velocity is low.* Also, lift is practically independent of the surface roughness since roughness affects the wall shear, not the pressure. The contribution of shear to lift is usually significant for very small (lightweight) bodies that can fly at low velocities (and thus very low Reynolds numbers).

Noting that the contribution of viscous effects to lift is negligible, we should be able to determine the lift acting on an airfoil by simply integrating the pressure distribution around the airfoil. The pressure changes in the flow direction along the surface, but it essentially remains constant in a direction normal to the surface. Therefore, it seems reasonable to ignore the very thin boundary layer on the airfoil and calculate the pressure distribution around the airfoil from the relatively simple potential flow theory (zero vorticity, irrotational flow) for which net viscous forces are zero for flow past an airfoil.

The flow fields obtained from such calculations are sketched in Fig. 13-37 for both symmetrical and nonsymmetrical airfoils by ignoring the

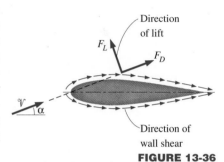

Direction of lift

F_L

F_D

$\mathcal{V}$

α

Direction of wall shear

FIGURE 13-36

For airfoils, the contribution of viscous effects to lift is usually negligible since wall shear is parallel to the surfaces and thus nearly normal to the direction of lift.

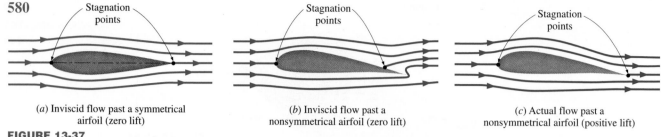

(a) Inviscid flow past a symmetrical airfoil (zero lift)

(b) Inviscid flow past a nonsymmetrical airfoil (zero lift)

(c) Actual flow past a nonsymmetrical airfoil (positive lift)

FIGURE 13-37

Inviscid (ideal) and actual flow past symmetrical and nonsymmetrical airfoils at zero angle of attack.

thin boundary layer. At zero angle of attack, the lift produced by the symmetrical airfoil is zero, as expected because of symmetry, and the stagnation points are at the leading and trailing edges. For the nonsymmetrical airfoil, the front stagnation point has moved down below the leading edge, and the rear stagnation point has moved up to the upper surface close to the trailing edge. To our surprise, the lift produced is calculated again to be zero—a clear contradiction of experimental observations and measurements. Obviously, the theory needs to be modified to bring it in line with the observed phenomenon.

The source of inconsistency is the rear stagnation point being at the upper surface instead of the trailing edge. This requires the lower side fluid to make a nearly U-turn and flow around the trailing edge toward the stagnation point while remaining attached to the surface, which is a physical impossibility since the observed phenomenon is the separation of flow at sharp turns (imagine a car making this turn at high speed). Therefore, if separation is to occur at a point (instead of over a region, which may occur at large angles of attack), it must occur at the trailing edge, and the stagnation point at the upper surface must move to the trailing edge. This way the two flow streams from the top and the bottom sides of the airfoil meet at the trailing edge, yielding a smooth flow downstream parallel to the chord line. The lift is generated from the requirement that the flow velocity at the top side must be higher (since the top surface has a larger curvature and thus a longer flow path), and thus the pressure on that side must be lower due to the Bernoulli effect.

The potential flow theory and the observed phenomenon can be reconciled as follows: Flow starts out as predicted by theory, with no lift, but the lower fluid stream separates at the trailing edge when the velocity reaches a certain value. This forces the separated upper fluid stream to close in at the trailing edge, initiating a clockwise swirl or vortex. This swirling motion increases the velocity of the upper stream while decreasing the velocity of the lower stream. The starting vortex is then shed downstream, and a smooth streamlined flow is established and lift is fully developed. When the potential theory is modified by the addition of a sufficient amount of swirling to move the stagnation point down to the trailing edge, excellent agreement is obtained between theory and experiments for the flow field and the lift.

It is desirable for airfoils to generate the most lift while producing the least drag. Therefore, a measure of performance for airfoils is the **lift-to-drag ratio,** which is equivalent to the ratio of the lift to drag coefficients C_L/C_D. This information is provided by either plotting C_L versus C_D for different values of the angle of attack (a lift-drag polar) or by plotting the ratio C_L/C_D versus the angle of attack. The latter is done for a particular airfoil design in Fig. 13-38. Note that the C_L/C_D ratio increases with the angle of attack until the airfoil stalls, and the value of the lift-to-drag can be in the order of 100.

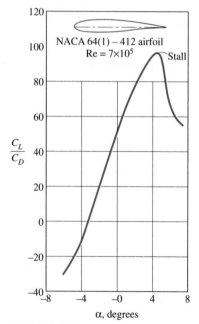

FIGURE 13-38

The variation of the lift-to-drag ratio with the angle of attack for an airfoil (from Abbott, von Doenhoff, and Stivers, Ref. 3).

(*a*) Flaps extended (takeoff)

(*b*) Flaps retracted (cruising)

FIGURE 13-39

The lift and drag characteristics of an airfoil during takeoff and landing can be changed by changing the shape of the airfoil by the use of movable flaps.

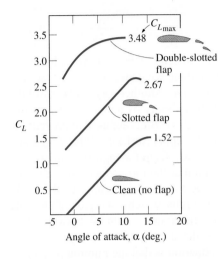

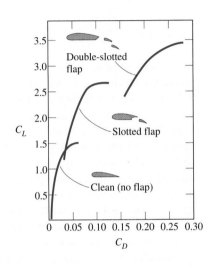

FIGURE 13-40

Effect of flaps on the lift and drag coefficients of an airfoil (from Abbott and von Doenhoff, Ref. 2, for NACA 23012).

Obviously one way of changing the lift and drag characteristics of an airfoil is to change the angle of attack. But usually this is not practical. A better approach is to change the shape of the airfoil by the use of movable *leading edge* and *trailing edge flaps,* as is commonly done in modern large aircraft (Fig. 13-39). The flaps are used to alter the shape of the wings during takeoff and landing to maximize lift and to enable the aircraft to land or take off at low speeds. The increase in drag during this takeoff and landing is not much of a concern because of the relatively short time periods involved. Once at cruising altitude, the flaps are retracted, and the wing is returned to its "normal" shape with minimal drag coefficient and adequate lift coefficient to minimize fuel consumption while cruising at a constant altitude. Note that even a small lift coefficient can generate a large lift force during normal operation because of the large cruising velocities of aircraft and the proportionality of lift to the square of flow velocity.

The effects of flaps on the lift and drag coefficients are shown in Fig. 13-40 for an airfoil. Note that the maximum lift coefficient increases from about 1.5 for the airfoil with no flaps to 3.5 for the double-slotted flap case. But also note that the drag coefficient increases from about 0.06 for the airfoil with no flaps to about 0.3 for the double-slotted flap case. This is a fivefold increase in the drag coefficient, and the engines must work much harder to provide the necessary thrust to overcome this drag. The angle of attack of the flaps can be increased to maximize the lift coefficient. Also, the leading and trailing edges extend the chord length, and thus enlarge the wing area A. The Boeing 727 uses a triple-slotted flap at the trailing edge and a slot at the leading edge.

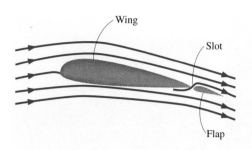

FIGURE 13-41

A flapped airfoil with a slot
to increase the lift coefficient.

The minimum flight velocity can be determined from the requirement that the total weight W of the aircraft be equal to lift and $C_L = C_{L,\,max}$. That is,

$$W = F_L = \tfrac{1}{2} C_{L,\,max}\, \rho \mathcal{V}_{min}^2\, A \qquad \rightarrow \qquad \mathcal{V}_{min} = \sqrt{\frac{2W}{\rho C_{L,\,max}\, A}} \qquad (13\text{-}19)$$

For a given weight, the landing or takeoff speed can be minimized by maximizing the product of the lift coefficient and the wing area, $C_{L,\,max}\, A$. One way of doing that is to use flaps, as discussed above. Another way is to control the boundary layer, which can be accomplished simply by leaving flow sections (slots) between the flaps, as shown in Fig. 13-41. Slots are used to prevent the separation of the boundary layer from the upper surface of the wings and the flaps. This is done by allowing air to move from the high-pressure region under the wing into the low-pressure region at the top surface. Note that the lift coefficient reaches its maximum value $C_L = C_{L,\,max}$, and thus the flight velocity reaches its minimum, at stall conditions, which is a region of unstable operation and must be avoided. The Federal Aviation Administration (FAA) does not allow operation below 1.2 times the stall speed for safety.

Another thing we notice from this equation is that the minimum velocity for takeoff or landing is inversely proportional to the square root of density. Noting that air density decreases with altitude (by about 15 percent at 1500 m), longer runways are required at airports at higher altitudes such as Denver to accommodate higher minimum takeoff and landing velocities. The situation becomes even more critical on hot summer days since the density of air is inversely proportional to temperature.

End Effects of Wing Tips

For airplane wings and other airfoils of finite size, the end effects at the tips become important because of the fluid leakage between the lower and upper surfaces. The pressure difference between the lower surface (high-pressure region) and the upper surface (low-pressure region) drives the fluid at the tips upwards while the fluid is swept toward the back because of the relative motion between the fluid and the wing. This results in a swirling motion that spirals along the flow, called the **tip vortex,** at the tips of both wings. Vortices are also formed along the airfoil between the tips of the wings. These distributed vortices collect toward the edges after being shed from the trailing edges of the wings and combine with the tip vortices to form two streaks of powerful **trailing vortices** along the tips of the wings (Fig. 13-42). Trailing vortices generated by large aircraft continue to exist for a long time for long distances (over 10 km) before they gradually disappear due to viscous dissipation. Such vortices and the accompanying downdraft are strong enough to cause a small aircraft to lose control and flip over. Therefore, following a large aircraft closely (within 10 km) poses a real danger for smaller aircraft. In nature, this effect is used to advantage by birds that migrate in V-formation by utilizing

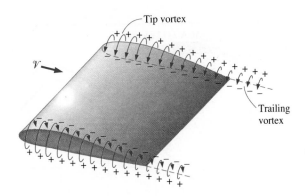

FIGURE 13-42

The tip and trailing vortices that form during flow past a wing due to leakage around wing tips.

the updraft generated by the bird in front. It is determined that the birds in a typical flock can fly to their destination in V-formation with one-third less energy. Military jets also occasionally fly in V-formation for the same reason.

Tip vortices that interact with the free stream impose forces on the wing tips in all directions, including the flow direction. The component of the force in the flow direction adds to drag and is called **induced drag.** The total drag of a wing is then the sum of the induced drag (3-D effects) and the drag of the airfoil section.

The ratio of the square of the average span of an airfoil to the planform area is call the **aspect ratio.** For an airfoil with a rectangular planform of chord c and span b, it is expressed as

$$AR = \frac{b^2}{A} = \frac{b^2}{bc} = \frac{b}{c} \qquad (13\text{-}20)$$

Therefore, the aspect ratio is a measure of how narrow an airfoil is in the flow direction. The lift coefficient of wings, in general, increases while the drag coefficient decreases with increasing aspect ratio. This is because a long narrow wing (large aspect ratio) has a shorter tip length and thus smaller tip losses and smaller induced drag than a short and wide wing of the same planform area. Therefore, bodies with large aspect ratios fly more efficiently, but they are less maneuverable because of their larger moment of inertia (owing to the greater distance from the center). Bodies with smaller aspect ratios maneuver better since the wings are closer to the central part. So it is no surprise that *fighter planes* (and fighter birds like falcons) have short and wide wings while *large commercial planes* (and soaring birds like albatrosses) have long and narrow wings.

The end effects can be minimized by attaching **endplates** at the tips of the wings perpendicular to the top surface. The endplates function by blocking the leakage around the wing tips, which results in considerable reduction in the strength of the tip vortices and the induced drag. Wing tip feathers on birds fan out for the same purpose.

The development of efficient (low-drag) airfoils was the subject of intense experimental investigations in the 1930s. These airfoils were standardized by the National Advisory Committee for Aeronautics (NACA, which is now NASA), and extensive lists of data on lift coefficients were reported. The variation of the lift coefficient C_L with the angle of attack for two airfoils (NACA 0012 and NACA 2412) is given in Fig. 13-43. We make the following observations from this figure:

> The lift coefficient increases almost linearly with the angle of attack α, reaches a maximum at about $\alpha = 16°$, and then starts to decrease sharply. The decrease of lift with an increase in the angle of attack is

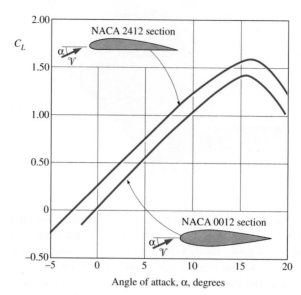

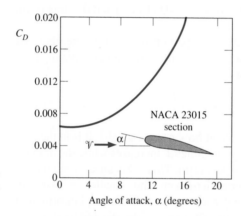

called *stall,* and it is caused by flow separation and the formation of a wide wake region over the top surface of the airfoil. Stall is highly undesirable since it also increases drag.

- At zero angle of attack ($\alpha = 0°$), the lift coefficient is zero for symmetrical airfoils but nonzero for nonsymmetrical ones with greater curvature at the top surface. Therefore, planes with symmetrical wing sections must fly with their wings at an angle of attack.
- The lift coefficient can be increased by severalfold by adjusting the angle of attack (from 0.25 at $\alpha = 0°$ for the nonsymmetrical airfoil to 1.25 at $\alpha = 10°$).
- The drag coefficient also increases with the angle of attack, often exponentially (Fig. 13-44). Therefore, large angles of attack should be used sparingly for short periods of time for fuel efficiency.

Lift Generated by Spinning

You have probably heard about giving a spin to a tennis ball or making a drop shot on a tennis or ping pong ball by giving a fore spin in order to alter the lift characteristics and cause the ball to produce a more desirable trajectory and bounce of the shot. Golf, soccer, and baseball players also utilize spin in their

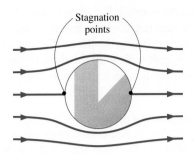

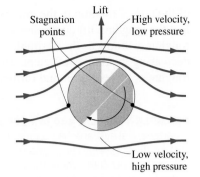

(a) Uniform flow over a stationary cylinder   (b) Uniform flow over a rotating cylinder

FIGURE 13-45

Generation of lift in uniform flow through rotation during "idealized" potential flow (the actual flow involves flow separation in the waste region).

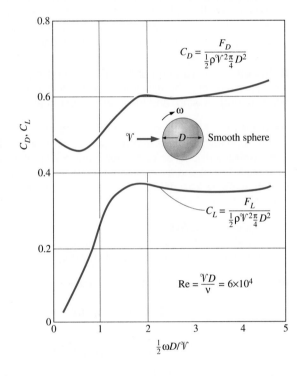

FIGURE 13-46

The variation of lift and drag coefficients of a smooth sphere with the rate of rotation for $\text{Re} = \mathcal{V}D/\nu = 6 \times 10^4$ (from Goldstein, Ref. 8).

games. The phenomenon of producing lift by the rotation of a solid body is called the **Magnus effect** after the German scientist Heinrich Magnus (1802–1870), who was the first to study the lift of rotating bodies, which is illustrated in Fig. 13-45. When there is uniform flow with no rotation, the lift is zero because of symmetry. But when the cylinder is rotated about its axis, the cylinder will drag some fluid around because of the no-slip condition and the flow field will reflect the superposition of the uniform and rotational flows. The stagnation points will shift down, and the flow will no longer be symmetric about the horizontal plane that passes through the center of the cylinder. The average pressure at the upper half will be less than the average pressure at the lower half because of the Bernoulli effect, and thus there will be a *net upward force* (lift) acting on the cylinder. A similar argument can be given for the lift generated on a spinning ball.

The effect of the rate of rotation on the lift and drag coefficients of a smooth sphere is shown in Fig. 13-46. Note that the lift coefficient strongly depends on the rate of rotation, especially at low angular velocities. The effect

70,000 kg

150 m², double-flapped

558 km/h   12,000 m

FIGURE 13-47

Schematic for Example 13-5.

of the rate of rotation in the drag coefficient is small. Roughness will also affect the drag and lift coefficients. In a certain range of the Reynolds number, roughness produces the desirable effect of increasing the lift coefficient while decreasing the drag coefficient. Therefore, golf balls with the right amount of roughness travel higher and further than smooth balls for the same hit.

EXAMPLE 13-5   Lift and Drag of a Commercial Airplane

A commercial airplane has a total mass of 70,000 kg and a wing planform area of 150 m² (Fig. 13-47). The plane has a cruising speed of 558 km/h and a cruising altitude of 12,000 m, where the air density is 0.312 kg/m³. The plane has double-slotted flaps for use during takeoff and landing, but it cruises with all flaps retracted. Assuming the lift and the drag characteristics of the wings can be approximated by NACA 23012 (Fig. 13-40), determine (a) the minimum safe speed for takeoff and landing with and without extending the flaps, (b) the angle of attack to cruise steadily at the cruising altitude, and (c) the power that needs to be supplied to provide enough thrust to overcome wing drag.

Solution   The cruising conditions of a passenger plane and its wing characteristics are given. The minimum safe landing and takeoff speeds, the angle of attack during cruising, and the power required are to be determined.

Assumptions   **1** The drag and lift produced by parts of the plane other than the wings, such as the fuselage drag, are not considered. **2** The wings are assumed to be two-dimensional airfoil sections, and the tip effects of the wings are not considered. **3** The lift and the drag characteristics of the wings can be approximated by NACA 23012 so that Fig. 13-40 is applicable. **4** The average density of air on the ground is 1.20 kg/m³.

Properties   The densities of air are 1.20 kg/m³ on the ground and 0.312 kg/m³ at cruising altitude. The maximum lift coefficients $C_{L, max}$ of the wings are 3.48 and 1.52 with and without flaps, respectively (Fig. 13-40).

Analysis   (a) The weight and cruising speed of the airplane are

$$W = mg = (70{,}000 \text{ kg})(9.81 \text{ m/s}^2)\left(\frac{1 \text{ N}}{1 \text{ kg} \cdot \text{m/s}^2}\right) = 686{,}700 \text{ N}$$

$$\mathcal{V} = (558 \text{ km/h})\left(\frac{1 \text{ m/s}}{3.6 \text{ km/h}}\right) = 155 \text{ m/s}$$

The minimum velocity corresponding to the stall conditions without and with flaps are

$$\mathcal{V}_{min\,1} = \sqrt{\frac{2W}{\rho C_{L,\,max\,1}\, A}} = \sqrt{\frac{2(686{,}700 \text{ N})}{(1.2 \text{ kg/m}^3)(1.52)(150 \text{ m}^3)}\left(\frac{1 \text{ kg} \cdot \text{m/s}^2}{1 \text{ N}}\right)} = 70.9 \text{ m/s}$$

$$\mathcal{V}_{min\,2} = \sqrt{\frac{2W}{\rho C_{L,\,max\,2}\, A}} = \sqrt{\frac{2(686{,}700 \text{ N})}{(1.2 \text{ kg/m}^3)(3.48)(150 \text{ m}^2)}\left(\frac{1 \text{ kg} \cdot \text{m/s}^2}{1 \text{ N}}\right)} = 46.8 \text{ m/s}$$

Then the "safe" minimum velocities to avoid the stall region are obtained by multiplying the values above by 1.2:

Without flaps: $\mathcal{V}_{min\,1,\,safe} = 1.2\mathcal{V}_{min\,1} = 1.2 \times (70.9 \text{ m/s}) = 85.1 \text{ m/s} = $ **306 km/h**

With flaps: $\mathcal{V}_{min\,2,\,safe} = 1.2\mathcal{V}_{min\,2} = 1.2 \times (46.8 \text{ m/s}) = 56.2 \text{ m/s} = $ **202 km/h**

since 1 m/s = 3.6 km/h. Note that the use of flaps allows the plane to take off and land at considerably lower velocities, and thus on a shorter runway.

(b) When an aircraft is cruising steadily at a constant altitude, the lift must be equal to the weight of the aircraft, $F_L = W$. Then the lift coefficient is determined to be

$$C_L = \frac{F_L}{\frac{1}{2}\rho \mathcal{V}^2 A} = \frac{686{,}700 \text{ N}}{\frac{1}{2}(0.312 \text{ kg/m}^3)(155 \text{ m/s})^2 (150 \text{ m}^2)} \left(\frac{1 \text{ kg} \cdot \text{m/s}^2}{1 \text{ N}}\right) = 1.22$$

For the case of no flaps, the angle of attack corresponding to this value of C_L is determined from Fig. 13-40 to be $\alpha \approx$ **10°**.

(c) When the aircraft is cruising steadily at a constant altitude, the net force acting on the aircraft is zero, and thus thrust provided by the engines must be equal to the drag force. The drag coefficient corresponding to the cruising lift coefficient of 1.22 is determined from Fig. 13-40 to be $C_D \approx 0.03$. Then the drag force acting on the wings becomes

$$F_D = C_D A \frac{\rho \mathcal{V}^2}{2} = (0.03)(150 \text{ m}^2)\frac{(0.312 \text{ kg/m}^3)(155 \text{ m/s})^2}{2}\left(\frac{1 \text{ kN}}{1000 \text{ kg} \cdot \text{m/s}^2}\right) = 16.9 \text{ kN}$$

Noting that power is force times velocity (distance per unit time), the power required to overcome this drag is equal to the thrust times the cruising velocity:

$$\text{Power} = \text{Thrust} \times \text{Velocity} = F_D \mathcal{V} = (16.9 \text{ kN})(155 \text{ m/s})\left(\frac{1 \text{ kW}}{1 \text{ kN} \cdot \text{m/s}}\right) = \textbf{2620 kW}$$

Therefore, the engines must supply 2620 kW of power to overcome the drag during cruising. For a propulsion efficiency of 30 percent (i.e., 30 percent of the energy of the fuel is utilized to propel the aircraft), the plane requires energy input at a rate of 8733 kJ/s.

Discussion The power determined above is the power to overcome the drag that acts on the wings only and does not include the drag that acts on the remaining parts of the aircraft (the fuselage, the tail, etc). Therefore, the total power required during cruising will be much greater. Also, it does not consider induced drag which can be dominant during takeoff when the angle of attack is high (Fig. 13-40 is for a 2-D airfoil, and does not include 3-D effects).

EXAMPLE 13-6 Effect of Spin on a Tennis Ball
A tennis ball with a mass of 0.125 lbm and a diameter of 2.52 in. is hit at 45 mph with a backspin of 4800 rpm (Fig. 13-48). Determine if the ball will fall or rise under the combined effect of gravity and lift due to spinning shortly after being hit in air at 1 atm and 80°F.

Solution A tennis ball is hit with a backspin. It is to be determined whether the ball will fall or rise after being hit.

Assumptions **1** The surfaces of the ball are smooth enough for Fig. 13-46 to be applicable. **2** The ball is hit horizontally so that it starts its motion horizontally.

Properties The density and kinematic viscosity of air at 1 atm and 80°F are $\rho = 0.074$ lbm/ft³ and $\nu = 0.170 \times 10^{-3}$ ft²/s (Table A-18).

Analysis The ball is hit horizontally, and thus it would normally fall under the effect of gravity without the spin. The backspin will generate a lift, and the ball will rise if the lift is greater than the weight of the ball. The lift can be determined from

$$F_L = C_L A \frac{\rho \mathcal{V}^2}{2}$$

where A is the frontal area of the ball, which is $A = \pi D^2/4$. The regular and angular velocities of the ball are

$$\mathcal{V} = (45 \text{ mi/h})\left(\frac{5280 \text{ ft}}{1 \text{ mi}}\right)\left(\frac{1 \text{ h}}{3600 \text{ s}}\right) = 66 \text{ ft/s}$$

$$\omega = (4800 \text{ rev/min})\left(\frac{2\pi \text{ rad}}{1 \text{ rev}}\right)\left(\frac{1 \text{ min}}{60 \text{ s}}\right) = 502 \text{ rad/s}$$

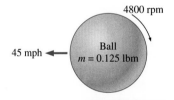

FIGURE 13-48
Schematic for Example 13-6.

Then,

$$\frac{\omega D}{2V} = \frac{(502 \text{ rad/s})(2.52/12 \text{ ft})}{2(66 \text{ ft/s})} = 0.80 \text{ rad}$$

From Fig. 13-46, the lift coefficient corresponding to this value is $C_L = 0.21$. Then the lift acting on the ball is

$$F_L = (0.21)\frac{\pi(2.52/12 \text{ ft})^2}{4}\frac{(0.074 \text{ lbm/ft}^3)(66 \text{ ft/s})^2}{2}\left(\frac{1 \text{ lbf}}{32.2 \text{ lbm} \cdot \text{ft/s}^2}\right) = 0.036 \text{ lbf}$$

The weight of the ball is

$$W = mg = (0.125 \text{ lbm})(32.2 \text{ ft/s}^2)\left(\frac{1 \text{ lbf}}{32.2 \text{ lbm} \cdot \text{ft/s}^2}\right) = 0.125 \text{ lbf}$$

which is more than the lift. Therefore, the ball will drop under the combined effect of gravity and lift due to spinning after hitting with a net force of 0.125 − 0.036 = 0.089 lbf.

Discussion   This example shows that a ball can be thrown much further by giving it a backspin. Note that a topspin will have the opposite effect (negative lift) and will speed up the drop of the ball to the ground. Also, the Reynolds number for this problem is 8×10^4, which is sufficiently close to the 6×10^4 for which Fig. 13-46 is prepared.

13-8 ■ SUMMARY

In this chapter, we studied flow of fluids over immersed bodies with emphasis on the resulting lift and drag forces. A fluid may exert forces and moments on a body in and about various directions. The force a flowing fluid exerts on a body in the flow direction is called *drag*. The components of the pressure and wall shear forces in the normal direction to flow tend to move the body in that direction and are called *lift*. The part of drag that is due directly to wall shear stress τ_w is called the *skin friction drag* since it is caused by frictional effects, and the part that is due directly to pressure P is called the *pressure drag* or *form drag* because of its strong dependence on the form or shape of the body.

The *drag coefficient* C_D and the *lift coefficient* C_L are dimensionless numbers that represent the drag and the lift characteristics of a body and are defined as

$$C_D = \frac{F_D}{\frac{1}{2}\rho V^2 A} \quad \text{and} \quad C_L = \frac{F_L}{\frac{1}{2}\rho V^2 A}$$

where A is usually the *frontal area* (the area projected on a plane normal to the direction of flow) of the body. For plates and airfoils, A is taken to be the *planform area*, which is the area that would be seen by a person looking at the body from above in a direction normal to the body. The drag coefficient, in general, depends on the *Reynolds number*, especially for Reynolds numbers below 10^4. At higher Reynolds numbers, the drag coefficients for most geometries remain essentially constant.

A body is said to be *streamlined* if a conscious effort is made to align its shape with the anticipated streamlines in the flow in order to reduce drag. Otherwise, a body (such as a building) tends to block the flow and is said to be *blunt* or *bluff*. Streamlining has the added benefit of reducing vibration and noise. At sufficiently high velocities, the fluid stream will detach itself from the surface of the body. This is called *separation*. When a fluid stream separates from the body, it forms a *separated region* between the body and the fluid stream. Separation also may occur on a streamlined body such as an airplane wing at a sufficiently large *angle of attack,* which is the angle the incoming fluid stream makes with the *chord* (the line that connects the nose and the end) of the body. Flow separation on the top surface of a wing reduces lift drastically and may cause the airplane to *stall*.

The region of the flow above the plate bounded by δ_v in which the effects of the viscous shearing forces caused by fluid viscosity are felt is called the *velocity boundary layer* or just the *boundary layer*. The *thickness* of the boundary layer, δ_v, is defined as the distance from the surface at which the velocity is $0.99\mathcal{V}_\infty$. The hypothetical line of velocity $0.99\mathcal{V}_\infty$ divides the flow over a plate into two regions: the *boundary layer region,* in which the viscous effects and the velocity changes are significant, and the *inviscid flow region,* in which the frictional effects are negligible and the velocity remains essentially constant.

For external flow, the Reynolds number is expressed as

$$\mathrm{Re}_L = \frac{\rho \mathcal{V} L}{\mu} = \frac{\mathcal{V} L}{\nu}$$

where $\mathcal{V}$ is the upstream velocity and L is the characteristic length of the geometry, which is the length of the plate in the flow direction for a flat plate and the diameter D for a cylinder or sphere. The *average* friction coefficients over the entire plate are

Laminar flow: $\qquad C_f = \dfrac{1.328}{\mathrm{Re}_L^{1/2}} \qquad \mathrm{Re}_L < 5 \times 10^5$

Turbulent flow: $\qquad C_f = \dfrac{0.074}{\mathrm{Re}_L^{1/5}} \qquad 5 \times 10^5 \leq \mathrm{Re}_L \leq 10^7$

The first relation above gives the average friction coefficient for the entire plate when the flow is laminar over the entire plate. For a critical Reynolds number of $\mathrm{Re}_{cr} = 5 \times 10^5$, the average friction coefficient over the *entire* plate is

$$C_f = \frac{0.074}{\mathrm{Re}_L^{1/5}} - \frac{1742}{\mathrm{Re}_L} \qquad 5 \times 10^5 \leq \mathrm{Re}_L \leq 10^7$$

A curve fit of experimental data for the average friction coefficient in this regime is

Rough surface: $\qquad C_f = \left(1.89 - 1.62 \log \dfrac{\varepsilon}{L}\right)^{-2.5}$

where ε is the surface roughness and L is the length of the plate in the flow direction. In the absence of a better relation, the relation above can be used for turbulent flow on rough surfaces for $\mathrm{Re} > 10^6$, especially when $\varepsilon/L > 10^{-4}$.

Surface roughness, in general, increases the drag coefficient in turbulent flow. For blunt bodies such as a circular cylinder or sphere, however, an increase in the surface roughness may *decrease* the drag coefficient. This is done by tripping the flow into turbulence at a lower Reynolds number, and thus causing the fluid to close in behind the body, narrowing the wake and reducing pressure drag considerably. It is desirable for airfoils to generate the most lift while producing the least drag. Therefore, a measure of performance for airfoils is the *lift-to-drag ratio, C_L/C_D*.

The minimum flight velocity of an aircraft can be determined from

$$\mathcal{V}_{min} = \sqrt{\frac{2W}{\rho C_{L,\,max}\, A}}$$

For a given weight, the landing or takeoff speed can be minimized by maximizing the product of the lift coefficient and the wing area, $C_{L,\,max}\,A$. For airplane wings and other airfoils of finite size, the pressure difference between the lower and the upper surfaces drives the fluid at the tips upwards. This results in a swirling motion that spirals along the flow, called the *tip vortex*. Tip vortices that interact with the free stream impose forces on the wing tips in all directions, including the flow direction. The component of the force in the flow direction adds to drag and is called *induced drag*. The total drag of a wing is then the sum of the induced drag (3-D effects) and the drag of the airfoil section. It is observed that lift develops when a cylinder or sphere in flow is rotated at a sufficiently high rate. The phenomenon of producing lift by the rotation of a solid body is called the *Magnus effect*.

REFERENCES AND SUGGESTED READING

1. I. H. Abbott. "The Drag of Two Streamline Bodies as Affected by Protuberances and Appendages." *NACA Report* 451 (1932).

2. I. H. Abbott and A. E. von Doenhoff. *Theory of Wing Sections, Including a Summary of Airfoil Data.* New York: Dover, 1959.

3. I. H. Abbott, A. E. von Doenhoff, and L. S. Stivers. "Summary of Airfoil Data." *NACA Report* 824, Langley Field, VA, 1945.

4. J. D. Anderson. *Fundamentals of Aerodynamics.* 2nd ed. New York: McGraw-Hill, 1991.

5. R. D. Blevins. *Applied Fluid Dynamics Handbook.* New York: Van Nostrand Reinhold, 1984.

6. S. W. Churchill and M. Bernstein. "A Correlating Equation for Forced Convection from Gases and Liquids to a Circular Cylinder in Cross Flow." *Journal of Heat Transfer* 99 (1977), pp. 300–6.

7. R. W. Fox and A. T. McDonald. *Introduction to Fluid Mechanics.* 5th ed. New York: John Wiley & Sons, 1999.

8. S. Goldstein. *Modern Developments in Fluid Dynamics.* London: Oxford Press, 1938.

9. J. Happel. *Low Reynolds Number Hydrocarbons.* Englewood Cliffs, NJ: Prentice Hall, 1965.

10. S. F. Hoerner. *Fluid-Dynamic Drag.* [Published by the author.] Library of Congress No. 64, 1966, 1965.

11. W. H. Hucho. *Aerodynamics of Road Vehicles.* London: Butterworth-Heinemann, 1987.

12. B. R. Munson, D. F. Young, and T. Okiishi. *Fundamentals of Fluid Mechanics.* 3rd ed. New York: John Wiley & Sons, 1998.

13. M. C. Potter and D. C. Wiggert. *Mechanics of Fluids.* 2nd ed. Upper Saddle River, NJ: Prentice Hall, 1997.

14. J. A. Roberson and C. L. Grove. *Engineering Fluid Mechanics.* 6th ed. New York: John Wiley & Sons, 1997.

15. H. Schlichting. *Boundary Layer Theory.* 7th ed. New York: McGraw-Hill, 1979.

16. J. Vogel. *Life in Moving Fluids.* 2nd ed. Boston: Willard Grand Press, 1994.

17. F. M. White. *Fluid Mechanics.* 4th ed. New York: McGraw-Hill, 1999.

PROBLEMS*

Drag, Lift, and Drag Coefficients of Common Geometries

13-1C What is the difference between internal and external flows?

13-2C Explain when an external flow is two-dimensional, three-dimensional, and axisymmetric. What type of flow is the flow of air over a car?

13-3C What is the difference between the upstream velocity and the free-stream velocity? For what types of flow are these two velocities equal to each other?

13-4C What is the difference between streamlined and blunt bodies? Is a tennis ball a streamlined or blunt body?

13-5C What is cavitation? Under what conditions does it occur? Why do we try to avoid cavitation?

13-6C What is drag? What causes it? Why do we usually try to minimize it?

13-7C What is lift? What causes it? Does wall shear contribute to the lift?

13-8C During flow over a given body, the drag force, the upstream velocity, and the fluid density are measured. Explain how you would determine the drag coefficient. What area would you use in calculations?

13-9C During flow over a given slender body such as a wing, the lift force, the upstream velocity, and the fluid density are measured. Explain how you would determine the lift coefficient. What area would you use in calculations?

13-10C Define frontal area of a body subjected to external flow. When is it appropriate to use the frontal area in drag and lift calculations?

13-11C Define planform area of a body subjected to external flow. When is it appropriate to use the planform area in drag and lift calculations?

13-12C What is terminal velocity? How is it determined?

13-13C What is the difference between skin friction drag and pressure drag? Which is usually more significant for slender bodies such as airfoils?

---

*Students are encouraged to answer *all* concept "C" questions.

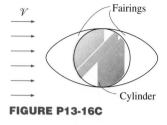

FIGURE P13-16C

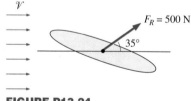

FIGURE P13-24

13-14C What is the effect of surface roughness on the friction drag coefficient in laminar and turbulent flows?

13-15C In general, how does the drag coefficient vary with the Reynolds number at (*a*) low and moderate Reynolds numbers and (*b*) at high Reynolds numbers (Re > 10^4)?

13-16C Fairings are attached to the front and back of a cylindrical body to make it look like an airfoil. What is the effect of this modification on the (*a*) friction drag, (*b*) pressure drag, and (*c*) total drag? Assume the Reynolds number is high enough so that the flow is turbulent for both cases.

13-17C What is the effect of streamlining on (*a*) friction drag and (*b*) pressure drag? Does the total drag acting on a body necessarily decrease as a result of streamlining? Explain.

13-18C What is flow separation? What causes it? What is the effect of flow separation on the drag coefficient?

13-19C What is drafting? How does it affect the drag coefficient of the drafted body?

13-20C Which car is more likely to be more fuel-efficient: the one with sharp corners or the one that is contoured to resemble an ellipse? Why?

13-21C Which bicyclist is more likely to go faster: the one who keeps his head and his body in the most upright position or the one who leans down and brings his body closer to his knees? Why?

13-22 The drag coefficient of a car at the design conditions of 1 atm, 25°C, and 90 km/h is to be determined experimentally in a large wind tunnel in a full-scale testing. The height and width of the car are 1.40 m and 1.65 m, respectively. If the horizontal force acting on the car is measured to be 350 N, determine the total drag coefficient of this car.   *Answer:* 0.41

13-23 A car is moving at a constant velocity of 80 km/h. Determine the upstream velocity to be used in fluid flow analysis if (*a*) the air is calm, (*b*) wind is blowing against the direction of motion of the car at 30 km/h, and (*c*) wind is blowing in the same direction of motion of the car at 50 km/h.

13-24 The resultant of the pressure and wall shear forces acting on a body is measured to be 500 N, making 35° with the direction of flow. Determine the drag and the lift forces acting on the body.

13-25 During a high Reynolds number experiment, the total drag force acting on a spherical body of diameter $D = 12$ cm subjected to air flow at 1 atm and 280 K is measured to be 5.2 N. The pressure drag acting on the body is calculated by integrating the pressure distribution (measured by the use of pressure sensors throughout the surface) to be 4.9 N. Determine the friction drag coefficient of the sphere.   *Answer:* 0.0115

13-26E To reduce the drag coefficient and thus to improve the fuel efficiency, the frontal area of a car is to be reduced. Determine the amount of fuel and money saved per year as a result of reducing the frontal area from 18 ft^2 to 15 ft^2. Assume the car is driven 12,000 miles a year at an average speed of 55 mph. Take the density and price of gasoline to be 50 lbm/ft^3 and \$2.20/gal, respectively; the density of air to be 0.075 lbm/ft^3, the heating value of gasoline to be 20,000 Btu/lbm; and the overall efficiency of the engine to be 32 percent.

13-27   A circular stop sign has a diameter of 50 cm and is subjected to winds up to 150 km/h at 10°C and 100 kPa. Determine the drag force acting on the sign. Also determine the bending moment at the bottom of its pole whose height from the ground to the bottom of the sign is 1.5 m. Disregard the drag on the pole.

13-28E   Wind loading is a primary consideration in the design of the supporting mechanisms of billboards, as evidenced by many billboards being knocked down during high winds. Determine the wind force acting on an 8-ft-high 20-ft-wide billboard due to 90-mph winds in the normal direction when the atmospheric conditions are 14.3 psia and 40°F.   *Answer: 6684 lbf*

13-29   Advertisement signs are commonly carried by taxicabs for additional income, but they also increase the fuel cost. Consider a sign that consists of a 0.30-m-high, 1.5-m-wide, and 1.5-m-long rectangular block mounted on top of a taxicab such that the sign has a frontal area of 0.3 m by 1.5 m from all four sides. Determine the increase in the annual fuel cost of this taxicab due to this sign. Assume the taxicab is driven 60,000 km a year at an average speed of 50 km/h and the overall efficiency of the engine is 28 percent. Take the density, unit price, and heating value of gasoline to be 0.75 kg/L, $0.50/L, and 42,000 kJ/kg, respectively, and the density of air to be 1.25 kg/m³.

13-30   It is proposed to meet the water needs of a recreational vehicle (RV) by installing a 2-m-long, 0.5-m-diameter cylindrical tank on top of the vehicle. Determine the additional power requirement of the RV at a speed of 95 km/h when the tank is installed such that its circular surfaces face (*a*) the front and back and (*b*) the sides of the RV. Assume atmospheric conditions are 87 kPa and 22°C.   *Answers: (a) 1.67 kW, (b) 7.55 kW*

13-31E   At highway speeds, the power generated by car engines is mostly used to overcome aerodynamic drag, and thus the fuel consumption is nearly proportional to the drag force on a level road. Determine the percentage increase in fuel consumption of a car per unit time when a person who normally drives at 55 mph now starts driving at 70 mph.

13-32   A 4-mm-diameter plastic sphere whose density is 1150 kg/m³ is dropped into water at 20°C. Determine the terminal velocity of the sphere in water.

13-33   During major windstorms, high vehicles such as RVs and semi trucks may be thrown off the road and boxcars off their tracks, especially when they are empty and in open areas. Consider a 5000-kg semi truck that is 8 m long, 2 m high, and 2 m wide. The distance between the bottom of the truck and the road is 0.75 m. Now the truck is exposed to winds from its side surface. Determine the wind velocity that will tip the truck over to its side. Take the air density to be 120 kg/m³.

13-34   An 80-kg bicyclist is riding his 15-kg bicycle downhill on a road with a slope of 12° without pedaling or breaking. The bicyclist has a frontal area of 0.45 m² and a drag coefficient of 1.1 in the upright position, and a frontal area of 0.4 m² and a drag coefficient of 0.9 in the racing position. Disregarding the rolling resistance and friction at the bearings, determine the terminal velocity of the bicyclist for both positions. Take the air density to be 1.25 kg/m³.
    Answers: 90 km/h, 106 km/h

13-35   A wind turbine with two or four hollow hemispherical cups connected to a pivot is commonly used to measure wind speed. Consider a wind turbine

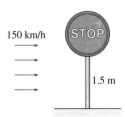

150 km/h

1.5 m

FIGURE P13-27

FIGURE P13-29

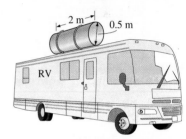

2 m   0.5 m

RV

FIGURE P13-30

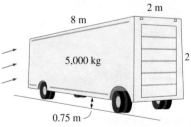

2 m

8 m

5,000 kg

2

0.75 m

FIGURE P13-33

25 cm

FIGURE P13-35

with two 5-cm-diameter cups with a center-to-center distance of 25 cm, as shown in the figure. The pivot is stuck as a result of some malfunction, and the cups stopped rotating. For a wind speed of 15 m/s and air density of 1.25 kg/m^3, determine the maximum torque this turbine applies on the pivot.

13-36E   A 5-ft-diameter spherical tank completely submerged in fresh water is being towed by a ship at 12 ft/s. Assuming turbulent flow, determine the required towing power.

13-37   During steady motion of a vehicle on a level road, the power delivered to the wheels is used to overcome aerodynamic drag and rolling resistance (the product of the rolling resistance coefficient and the weight of the vehicle), assuming the friction at the bearings of the wheels is negligible. Consider a car that has a total mass of 950 kg, a drag coefficient of 0.32, a frontal area of 1.8 m^2, and a rolling resistance coefficient of 0.04. The maximum power the engine can deliver to the wheels is 80 kW. Determine (*a*) the velocity at which the rolling resistance is equal to the aerodynamic drag force and (*b*) the maximum velocity of this car. Take the air density to be 1.20 kg/m^3.

13-38   A submarine can be treated as an ellipsoid with a diameter of 5 m and a length of 10 m. Determine the power required for this submarine to cruise horizontally and steadily at 40 km/h in seawater whose density is 1025 kg/m^3. Also determine the power required to tow this submarine in air whose density is 1.30 kg/m^3. Assume the flow is turbulent in both cases.

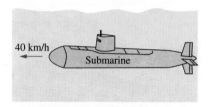

FIGURE P13-38

13-39   An 0.80-m-diameter, 1.2-m-high garbage can is found in the morning tipped over due to high winds during the night. Assuming the average density of the garbage inside to be 150 kg/m^3 and and taking the air density to be 1.25 kg/m^3, estimate the wind velocity during the night when the can was tipped over. Take the drag coefficient of the can to be 0.7.   *Answer:* 186 km/h

13-40E   The drag coefficient of a vehicle increases when its windows are rolled down or its sunroof is opened. A sports car has a frontal area of 22 ft^2 and a drag coefficient of 0.32 when the windows and sunroof are closed. The drag coefficient increases to 0.41 when the sunroof is open. Determine the additional power consumption of the car at (*a*) 35 mph and (*b*) 70 mph when the sunroof is opened. Take the density of air to be 0.075 lbm/ft^3.

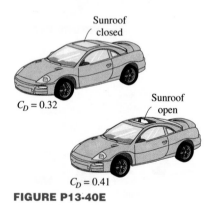

Sunroof closed

$C_D = 0.32$

Sunroof open

$C_D = 0.41$

FIGURE P13-40E

Flow over Flat Plates

13-41C   What fluid property is responsible for the development of the velocity boundary layer? For what kind of fluids will there be no velocity boundary layer on a flat plate?

13-42C   What does the friction coefficient represent in flow over a flat plate? How is it related to the drag force acting on the plate?

13-43C   Consider laminar flow over a flat plate. Will the friction coefficient change with position?

13-44C   How is the average friction coefficient determined in flow over a flat plate?

13-45E   Light oil at 80°F flows over a 15-ft-long flat plate with a free-stream velocity of 6 ft/s. Determine the total drag force per unit width of the plate.

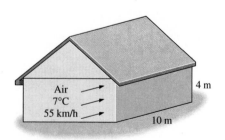

Air
7°C
55 km/h

4 m

10 m

FIGURE P13-47

13-46   The local atmospheric pressure in Denver, Colorado (elevation 1610 m), is 83.4 kPa. Air at this pressure and at 27°C flows with a velocity of

6 m/s over a 2.5-m × 8-m flat plate. Determine the drag force acting on the top surface of the plate if the air flows parallel to the (a) 8-m-long side and (b) the 2.5-m-long side.

13-47 During a winter day, wind at 55 km/h and 7°C is blowing parallel to a 4-m-high and 10-m-long wall of a house. Assuming the wall surfaces to be smooth, determine the drag force acting on the wall. What would your answer be if the wind velocity has doubled? *Answers:* 16 N, 58 N

13-48E Air at 65°F flows over a 10-ft-long flat plate at 7 ft/s. Determine the local friction coefficient at intervals of 1 ft and plot the results against the distance from the leading edge.

13-49 The forming section of a plastics plant puts out a continuous sheet of plastic that is 1.2 m wide and 2 mm thick at a rate of 15 m/min. The sheet is subjected to air flow at a velocity of 3 m/s on both sides along its surfaces normal to the direction of motion of the sheet. The width of the air cooling section is such that a fixed point on the plastic sheet passes through that section in 2 s. Using properties of air at 1 atm and 330 K, determine the drag force the air exerts on the plastic sheet in the direction of air flow.

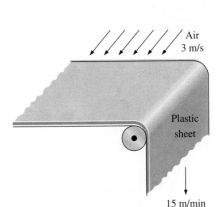

FIGURE P13-49

13-50 The top surface of the passenger car of a train moving at a velocity of 70 km/h is 2.8 m wide and 8 m long. If the outdoors air is at 1 atm and 25°C, determine the drag force acting on the top surface of the car.

FIGURE P13-50

13-51 The weight of a thin flat plate 50 cm × 50 cm in size is balanced by a counterweight that has a mass of 2 kg, as shown in the figure. Now a fan is turned on, and air at 1 atm and 25°C flows downward over both surfaces of the plate with a free-stream velocity of 10 m/s. Determine the mass of the counterweight that needs to be added in order to balance the plate in this case.

13-52 Consider laminar flow of a fluid over a flat plate. Now the free-stream velocity of the fluid is doubled. Determine the change in the drag force on the plate. Assume the flow to remain laminar. *Answer:* A 2.83-fold increase

13-53E Consider a refrigeration truck traveling at 55 mph at a location where the air temperature is at 1 atm and 80°F. The refrigerated compartment of the truck can be considered to be a 9-ft-wide, 8-ft-high, and 20-ft-long rectangular box. Assuming the air flow over the entire outer surface to be turbulent, determine the drag force acting on the top and side surfaces and the power required to overcome this drag.

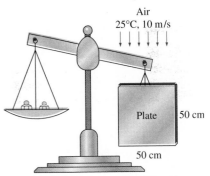

FIGURE P13-51

13-54 Air at 25°C and 1 atm is flowing over a long flat plate with a velocity of 8 m/s. Determine the distance from the leading edge of the plate where the flow becomes turbulent, and the thickness of the boundary layer at that location.

13-55 Repeat Prob. 13-54 for water.

Flow across Cylinders and Spheres

13-56C In flow over cylinders, why does the drag coefficient suddenly drop when the flow becomes turbulent? Isn't turbulence supposed to increase the drag coefficient instead of decreasing it?

13-57C In flow over blunt bodies such as a cylinder, how does the pressure drag differ from the friction drag?

13-58C Why is flow separation in flow over cylinders delayed in turbulent flow?

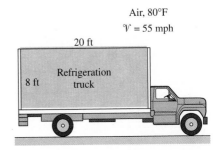

FIGURE P13-53E

Air
60°F, 20 mph

FIGURE P13-61E

13-59E A 1.2-in.-outer-diameter pipe is to cross a river at a 70-ft-wide section while being completely immersed in water. The average flow velocity of water is 10 ft/s, and the water temperature is 70°F. Determine the drag force exerted on the pipe by the river. *Answer:* 880 lbf

13-60 A long 8-cm-diameter steam pipe passes through some area open to the winds. Determine the drag force acting on the pipe per unit of its length when the air is at 1 atm and 7°C and the wind is blowing across the pipe at a velocity of 50 km/h.

13-61E A person extends his uncovered arms into the windy air outside at 1 atm and 60°F and 20 mph in order to feel nature closely. Treating the arm as a 2-ft-long and 3-in.-diameter cylinder, determine the drag force on both arms. *Answer:* 1.02 lbf

13-62 A 6-mm-diameter electrical transmission line is exposed to windy air. Determine the drag force exerted on a 50-m-long section of the wire during a windy day when the air is at 1 atm and 17°C and the wind is blowing across the transmission line at 40 km/h.

13-63 Consider 0.8-cm-diameter hail that is falling freely in atmospheric air at 1 atm and 7°C. Determine the terminal velocity of the hail. Take the density of hail to be 910 kg/m³.

13-64 A 0.1-mm-diameter dust particle whose density is 2.1 g/cm³ is observed to be suspended in the air at 1 atm and 25°C at a fixed point. Estimate the updraft velocity of air motion at that location. Assume the Stokes law to be applicable. Is this a valid assumption? *Answer:* 0.62 m/s

13-65 Dust particles of diameter 0.05 mm and density 1.8 g/cm³ are unsettled during high winds and rise to a height of 500 m by the time things calm down. Estimate how long it will take for the dust particles to fall back to the ground in air at 1 atm and 17°C, and their velocity. Disregard the initial transient period during which the dust particles accelerate to their terminal velocity, and assume Stokes law to be applicable.

13-66 A 2-m-long, 0.2-m-diameter cylindrical pine log (density = 513 kg/m³) is suspended by a crane in the horizontal position. The log is subjected to normal winds of 40 km/h at 7°C and 88 kPa. Disregarding the weight of the cable and its drag, determine the angle θ the cable will make with the horizontal and the tension on the cable.

13-67 One of the popular demonstrations in science museums involves the suspension of a ping-pong ball by an upward air jet. Children are amused by the ball always coming back to the center when it is pushed by a finger to the

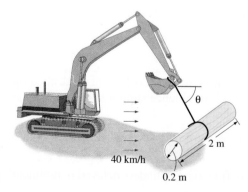

θ

2 m

40 km/h

0.2 m

FIGURE P13-66

side of the jet. Explain this phenomenon using the Bernoulli equation. Also determine the velocity of air if the ball has a mass of 2.6 g and a diameter of 3.8 cm. Assume air is at 1 atm and 25°C.

Lift

13-68C Why is the contribution of viscous effects to lift usually negligible for airfoils?

13-69C Air is flowing past a symmetrical airfoil at zero angle of attack. Will the (*a*) lift and (*b*) drag acting on the airfoil be zero or nonzero?

13-70C Air is flowing past a nonsymmetrical airfoil at zero angle of attack. Will the (*a*) lift and (*b*) drag acting on the airfoil be zero or nonzero?

13-71C Air is flowing past a symmetrical airfoil at an angle of attack of 5°. Will the (*a*) lift and (*b*) drag acting on the airfoil be zero or nonzero?

13-72C What is stall? What causes an airfoil to stall? Why are commercial aircraft not allowed to fly at conditions near stall?

13-73C Both the lift and the drag of an airfoil increase with an increase in the angle of attack. In general, which increases at a much higher rate, the lift or the drag?

13-74C Why are flaps used at the leading and trailing edges of the wings of large aircraft during takeoff and landing? Can't an aircraft take off or land without them?

13-75C How do the flaps affect the lift and the drag of the wings?

13-76C What is the effect of wing tip vortices (the air circulation from the lower part of the wings to the upper part) on the drag and the lift?

13-77C What is induced drag on wings? Can induced drag be minimized by using long and narrow wings or short and wide wings?

13-78C Air is flowing past a spherical ball. Will the lift exerted on the ball be zero or nonzero? Answer the same question if the ball is spinning.

13-79 A tennis ball with a mass of 57 g and a diameter of 6.4 cm is hit with an initial velocity of 75 km/h and a backspin of 4200 rpm. Determine if the ball will fall or rise under the combined effect of gravity and lift due to spinning shortly after hitting. Assume air is at 1 atm and 25°C.

13-80 Consider an aircraft, which takes off at 190 km/h when it is fully loaded. If the weight of the aircraft is increased by 20 percent as a result of overloading, determine the speed at which the overloaded aircraft will take off. *Answer:* 208 km/h

13-81 Consider an airplane whose takeoff speed is 220 km/h and that takes 15 s to take off at sea level. For an airport at an elevation of 1600 m (such as Denver), determine (*a*) the takeoff speed, (*b*) the takeoff time, and (*c*) the additional runway length required for this airplane. Assume constant acceleration for both cases.

13-82E An airplane is consuming fuel at a rate of 2 gal/min when cruising at a constant altitude of 10,000 ft at constant speed. Assuming the drag coefficient to remain the same, determine the rate of fuel consumption at an altitude of 30,000 ft at the same speed.

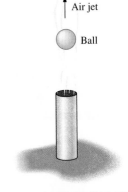

Air jet

Ball

FIGURE P13-67

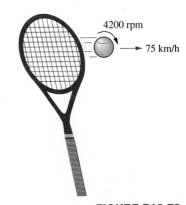

4200 rpm

75 km/h

FIGURE P13-79

220 km/h

FIGURE P13-81

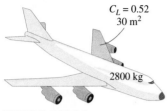

$C_L = 0.52$
30 m²

2800 kg

FIGURE P13-84

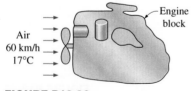

Engine
block

Air
60 km/h
17°C

FIGURE P13-90

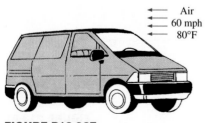

Air
60 mph
80°F

FIGURE P13-92E

13-83 A jumbo jet airplane has a mass of about 400,000 kg when fully loaded with over 400 passengers and takes off at a speed of 250 km/h. Determine the takeoff speed when the airplane has 100 empty seats. Assume each passenger with luggage is 140 kg and the wing and flap settings are maintained the same.   *Answer:* 246 km/h

13-84 A small aircraft has a wing area of 30 m², a lift coefficient of 0.52 at takeoff settings, and a total mass of 2800 kg. Determine (*a*) the takeoff speed of this aircraft at sea level at standard atmospheric conditions, (*b*) the wing loading, and (*c*) the required power to maintain a constant cruising speed of 300 km/h for a cruising drag coefficient of 0.035.

13-85 A small airplane has a total mass of 1800 kg and a wing area of 42 m². Determine the lift and drag coefficients of this airplane while cruising at an altitude of 4000 m at a constant speed of 280 km/h and generating 190 kW of power.

13-86 The NACA 64(1)-412 airfoil has a lift-to-drag ratio of 50 at 0° angle of attack, as shown in Fig. 13-38. At what angle of attack will this ratio increase to 80?

13-87 Consider a light plane that has a total weight of 15,000 N and a wing area of 34 m² and whose wings resemble the NACA 23012 airfoil with no flaps. Using data from Fig. 13-40, determine the takeoff speed at an angle of attack of 5° at sea level. Also determine the stall speed.   *Answer:* 94 km/h

13-88 An airplane has a mass of 50,000 kg, a wing area of 300 m², a maximum lift coefficient of 3.2, and a cruising drag coefficient of 0.03 at an altitude of 12,000 m. Determine (*a*) the takeoff speed at sea level, assuming it is 20 percent over the stall speed, and (*b*) the thrust that the engines must deliver for a cruising speed of 700 km/h.

13-89E A 2.4-in-diameter smooth ball rotating at 500 rpm is dropped in a water stream at 60°F flowing at 4 ft/s. Determine the lift and the drag force acting on the ball when it is first dropped in water.

Review Problems

13-90 An automotive engine can be approximated as a 0.4-m-high, 0.60-m-wide, and 0.7-m-long rectangular block. The ambient air is at 1 atm and 17°C. Determine the drag force acting on the bottom surface of the engine block as the car travels at a velocity of 60 km/h. Assume the flow to be turbulent over the entire surface because of the constant agitation of the engine block.
      Answer: 0.35 N

13-91 Calculate the thickness of the boundary layer during flow over a 2.5-m-long flat plate at intervals of 25 cm and plot the boundary layer over the plate for the flow of (*a*) air, (*b*) water, and (*c*) engine oil at 1 atm and 20°C at an upstream velocity of 3 m/s.

13-92E The passenger compartment of a minivan traveling at 60 mph in ambient air at 1 atm and 80°F can be modeled as a 3.2-ft-high, 6-ft-wide, and 11-ft-long rectangular box. The air flow over the exterior surfaces can be assumed to be turbulent because of the intense vibrations involved. Determine the drag force acting on the top and the two side surfaces of the van and the power required to overcome it.

13-93 A 2-m-external-diameter spherical tank is located outdoors at 1 atm and 27°C and is subjected to winds at 25 km/h. Determine the drag force exerted on it by the wind. *Answer:* 12 N

13-94 A 2-m-high, 4-m-wide rectangular advertisement panel is attached to a 4-m-wide, 0.15-m-high rectangular concrete block (density = 2300 kg/m³) by two 5-cm-diameter, 4-m-high (exposed part) poles, as shown in the figure. If the sign is to withstand 150 km/h winds from any direction, determine (a) the maximum drag force on the panel, (b) the drag force acting on the poles, and (c) the minimum length L of the concrete block for the panel to resist the winds. Take the density of air to be 1.30 kg/m³.

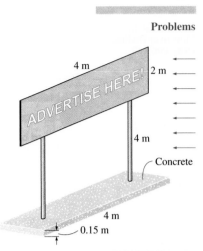

FIGURE P13-94

13-95 A plastic boat whose bottom surface can be approximated as a 1.5-m-wide, 2-m-long flat surface is to move through water at 15°C at speeds up to 30 km/h. Determine the friction drag exerted on the boat by water and the power needed to overcome it.

FIGURE P13-95

13-96E A commercial airplane has a total mass of 150,000 lbm and a wing planform area of 1800 ft². The plane has a cruising speed of 550 mph and a cruising altitude of 38,000 ft where the air density is 0.0208 lbm/ft³. The plane has double-slotted flaps for use during takeoff and landing, but it cruises with all flaps retracted. Assuming the lift and drag characteristics of the wings can be approximated by NACA 23012, determine (a) the minimum safe speed for takeoff and landing with and without extending the flaps, (b) the angle of attack to cruise steadily at the cruising altitude, and (c) the power that needs to be supplied to provide enough thrust to overcome drag. Take the air density on the ground to be 0.075 lbm/ft³.

13-97 A 9.5-cm smooth ball has a velocity of 36 km/h during a typical hit. Determine the percent increase in the drag coefficient if the ball is given a spin of 3500 rpm in air at 1 atm and 25°C.

13-98 A paratrooper and his 8-m-diameter parachute weigh 950 N. Taking the average air density to be 1.2 kg/m³, determine the terminal velocity of the paratrooper. *Answer:* 4.9 m/s

FIGURE P13-98

13-99 A 17,000-kg tractor-trailer rig has a frontal area of 9.2 m², a drag coefficient of 0.96, a rolling resistance coefficient of 0.05 (multiplying the weight of a vehicle by the rolling resistance coefficient gives the rolling resistance), a bearing friction resistance of 350 N, and a maximum speed of 110 km/h on a level road during steady cruising in calm weather with an air density of 1.25 kg/m³. Now a fairing is installed to the front of the rig to suppress separation and to streamline the flow to the top surface, and the drag coefficient is reduced to 0.76. Determine the maximum speed of the rig with the fairing. *Answer:* 133 km/h

13-100 Stokes law can be used to determine the viscosity of a fluid by dropping a spherical object in it and measuring the terminal velocity of the object in that fluid. This can be done by plotting the distance traveled against time and observing when the curve becomes linear. During such an experiment a 5-mm-diameter glass ball (ρ = 2500 kg/m³) is dropped into a fluid whose density is 875 kg/m³, and the terminal velocity is measured to be 0.12 m/s. Determine the viscosity of the fluid.

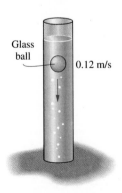

FIGURE P13-100

Computer, Design, and Essay Problems

13-101   Engine oil at 40°C is flowing over a long flat plate with a velocity of 4 m/s. Determine the distance x_{cr} from the leading edge of the plate where the flow becomes turbulent and calculate and plot the thickness of the boundary layer over a length of $2x_{cr}$.

13-102   Write a report on the history of the reduction of the drag coefficients of cars and obtain the drag coefficient data for some recent car models from the catalogs or car manufacturers.

13-103   Write a report on the flaps used at the leading and trailing edges of the wings of large commercial aircraft. Discuss how the flaps affect the drag and lift coefficients during takeoff and landing.

13-104   Large commercial airplanes cruise at high altitudes (up to about 40,000 ft) to save fuel. Discuss how flying at high altitudes reduces drag and saves fuel. Also discuss why small planes fly at relatively low altitudes.

13-105   Many drivers turn off their air conditioners and roll down the car windows in hopes of saving fuel. But it is claimed that this apparent "free cooling" actually increases the fuel consumption of the car. Investigate this matter and write a report on which practice will save gasoline under what conditions.

Heat Transfer

PART **III**

Mechanisms of Heat Transfer

14

The science of thermodynamics deals with the *amount* of heat transfer as a system undergoes a process from one equilibrium state to another, and makes no reference to *how long* the process will take. But in engineering, we are often interested in the *rate* of heat transfer, which is the topic of the science of *heat transfer.*

In this chapter we present an overview of the three basic mechanisms of heat transfer, which are conduction, convection, and radiation, and discuss thermal conductivity. *Conduction* is the transfer of energy from the more energetic particles of a substance to the adjacent, less energetic ones as a result of interactions between the particles. *Convection* is the mode of heat transfer between a solid surface and the adjacent liquid or gas that is in motion, and it involves the combined effects of conduction and fluid motion. *Radiation* is the energy emitted by matter in the form of electromagnetic waves (or photons) as a result of the changes in the electronic configurations of the atoms or molecules. We close this chapter with a discussion of simultaneous heat transfer. The topics introduced in this chapter are treated in more detail in the following chapters.

14-1 ■ INTRODUCTION

In Chapter 2 we discussed various forms of energy. In heat transfer, we are primarily interested in **heat,** which is *the form of energy that can be transferred from one system to another as a result of temperature difference.* A thermodynamic analysis is concerned with the *amount* of heat transfer as a system undergoes a process from one equilibrium state to another. The science that deals with the determination of the *rates* of such energy transfers is the **heat transfer.** The transfer of energy as heat is always from the higher-temperature medium to the lower-temperature one, and heat transfer stops when the two mediums reach the same temperature.

Heat can be transferred in three different ways: *conduction, convection,* and *radiation.* All modes of heat transfer require the existence of a temperature difference, and all modes of heat transfer are from the high-temperature medium to a lower-temperature one. Below we give a brief description of each mode. A detailed study of these heat transfer modes is given in later chapters of this text.

14-2 ■ CONDUCTION

Conduction is the transfer of energy from the more energetic particles of a substance to the adjacent less energetic ones as a result of interactions between the particles. Conduction can take place in solids, liquids, or gases. In *gases* and *liquids,* conduction is due to the *collisions* and *diffusion* of the molecules during their random motion. In *solids,* it is due to the combination of *vibrations* of the molecules in a lattice and the energy transport by *free electrons.* A cold canned drink in a warm room, for example, eventually warms up to the room temperature as a result of heat transfer from the room to the drink through the aluminum can by conduction.

The *rate* of heat conduction through a medium depends on the *geometry* of the medium, its *thickness,* and the *material* of the medium, as well as the *temperature difference* across the medium. We know that wrapping a hot water tank with glass wool (an insulating material) reduces the rate of heat loss from the tank. The thicker the insulation, the smaller the heat loss. We also know that a hot water tank will lose heat at a higher rate when the temperature of the room housing the tank is lowered. Further, the larger the tank, the larger the surface area and thus the rate of heat loss.

Consider steady heat conduction through a large plane wall of thickness $\Delta x = L$ and area A, as shown in Fig. 14-1. The temperature difference across the wall is $\Delta T = T_2 - T_1$. Experiments have shown that the rate of heat transfer $\dot{Q}$ through the wall is *doubled* when the temperature difference ΔT across the wall or the area A normal to the direction of heat transfer is doubled, but is *halved* when the wall thickness L is doubled. Thus we conclude that *the rate of heat conduction through a plane layer is proportional to the temperature difference across the layer and the heat transfer area, but is inversely proportional to the thickness of the layer.* That is,

$$\text{Rate of heat conduction} \propto \frac{(\text{Area})(\text{Temperature difference})}{\text{Thickness}}$$

or,

$$\dot{Q}_{\text{cond}} = kA \frac{\Delta T}{\Delta x} \qquad (\text{W}) \qquad (14\text{-}1)$$

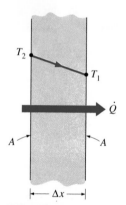

FIGURE 14-1

Heat conduction through a large plane wall of thickness Δx and area A.

where the constant of proportionality k is the **thermal conductivity** of the material, which is a *measure of the ability of a material to conduct heat* (Fig. 14-2). In the limiting case of $\Delta x \to 0$, the equation above reduces to the differential form

$$\dot{Q}_{cond} = -kA\frac{dT}{dx} \qquad \text{(W)} \qquad (14\text{-}2)$$

which is called **Fourier's law of heat conduction** after J. Fourier, who expressed it first in his heat transfer text in 1822. Here dT/dx is the **temperature gradient,** which is the slope of the temperature curve on a T-x diagram (the rate of change of T with x), at location x. The relation above indicates that the rate of heat conduction in a direction is proportional to the temperature gradient in that direction. Heat is conducted in the direction of decreasing temperature, and the temperature gradient becomes negative when temperature decreases with increasing x. Therefore, a *negative sign* is added to Eq. 14-2 to make heat transfer in the positive x direction a positive quantity.

The heat transfer area A is always *normal* to the direction of heat transfer. For heat loss through a 5-m-long, 3-m-high, and 25-cm-thick wall, for example, the heat transfer area is $A = 15$ m². Note that the thickness of the wall has no effect on A (Fig. 14-3).

EXAMPLE 14-1 The Cost of Heat Loss through the Roof
The roof of an electrically heated home is 6 m long, 8 m wide, and 0.25 m thick, and is made of a flat layer of concrete whose thermal conductivity is $k = 0.8$ W/m · °C (Fig. 14-4). On a certain winter night, the temperatures of the inner and the outer surfaces of the roof are measured to be 15°C and 4°C, respectively, for a period of 10 hours. Determine (a) the rate of heat loss through the roof that night and (b) the cost of that heat loss to the home owner if the cost of electricity is $0.08/kWh.

Solution The inner and outer surfaces of the flat concrete roof of an electrically heated home are maintained at 15°C and 4°C, respectively, during a night. The heat loss through the roof and its cost that night are to be determined.

Assumptions **1** Steady operating conditions exist during the entire night since the surface temperatures of the roof remain constant at the specified values. **2** Constant properties can be used for the roof.

Properties The thermal conductivity of the roof is given to be $k = 0.8$ W/m · °C.

Analysis (a) Noting that heat transfer through the roof is by conduction and the area of the roof is $A = 6$ m $\times$ 8 m $= 48$ m², the steady rate of heat transfer through the roof is determined to be

$$\dot{Q} = kA\frac{\Delta T}{L} = (0.8 \text{ W/m} \cdot °\text{C})(48 \text{ m}^2)\frac{(15-4)°\text{C}}{0.25 \text{ m}} = \textbf{1690 W} = \textbf{1.690 kW}$$

(b) The amount of heat lost through the roof during a 10-hour period and its cost are determined from

$$Q = \dot{Q}\,\Delta t = (1.690 \text{ kW})(10 \text{ h}) = 16.90 \text{ kWh}$$

$$\text{Cost} = (\text{Amount of energy})(\text{Unit cost of energy})$$

$$= (16.90 \text{ kWh})(\$0.08/\text{kWh}) = \textbf{\$1.35}$$

Discussion The cost to the home owner of the heat loss through the roof that night was $1.35. The total heating bill of the house will be much larger since the heat losses through the walls are not considered in the above calculations.

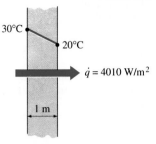

(a) Copper ($k = 401$ W/m·°C)

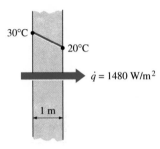

(b) Silicon ($k = 148$ W/m·°C)

FIGURE 14-2

The rate of heat conduction through a solid is directly proportional to its thermal conductivity.

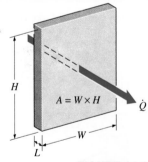

FIGURE 14-3

In heat conduction analysis, A represents the area *normal* to the direction of heat transfer.

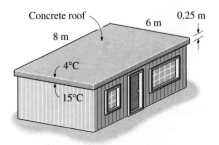

FIGURE 14-4

Schematic for Example 14-1.

TABLE 14-1

The thermal conductivities of some materials at room temperature

Material	k, W/m · °C*
Diamond	2300
Silver	429
Copper	401
Gold	317
Aluminum	237
Iron	80.2
Mercury (l)	8.54
Glass	0.78
Brick	0.72
Water (l)	0.613
Human skin	0.37
Wood (oak)	0.17
Helium (g)	0.152
Soft rubber	0.13
Glass fiber	0.043
Air (g)	0.026
Urethane, rigid foam	0.026

*Multiply by 0.5778 to convert to Btu/h · ft · °F.

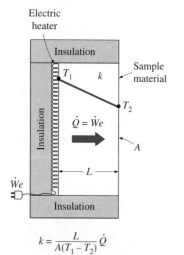

$$k = \frac{L}{A(T_1 - T_2)} \dot{Q}$$

FIGURE 14-5

A simple experimental setup to determine the thermal conductivity of a material.

14-3 ■ THERMAL CONDUCTIVITY

We have seen that different materials store heat differently, and we have defined the property specific heat C_p as a measure of a material's ability to store thermal energy. For example, $C_p = 4.18$ kJ/kg · °C for water and $C_p = 0.45$ kJ/kg · °C for iron at room temperature, which indicates that water can store almost 10 times the energy that iron can per unit mass. Likewise, the thermal conductivity k is a measure of a material's ability to conduct heat. For example, $k = 0.608$ W/m · °C for water and $k = 80.2$ W/m · °C for iron at room temperature, which indicates that iron conducts heat more than 100 times faster than water can. Thus we say that water is a poor heat conductor relative to iron, although water is an excellent medium to store thermal energy.

Equation 14-1 for the rate of conduction heat transfer under steady conditions can also be viewed as the defining equation for thermal conductivity. Thus the **thermal conductivity** of a material can be defined as *the rate of heat transfer through a unit thickness of the material per unit area per unit temperature difference.* The thermal conductivity of a material is a measure of how fast heat will flow in that material. A high value for thermal conductivity indicates that the material is a good heat conductor, and a low value indicates that the material is a poor heat conductor or *insulator.* The thermal conductivities of some common materials at room temperature are given in Table 14-1. The thermal conductivity of pure copper at room temperature is $k = 401$ W/m · °C, which indicates that a 1-m-thick copper wall will conduct heat at a rate of 401 W per m² area per °C temperature difference across the wall. Note that materials such as copper and silver that are good electric conductors are also good heat conductors, and have high values of thermal conductivity. Materials such as rubber, wood, and styrofoam are poor conductors of heat and have low conductivity values.

A layer of material of known thickness and area can be heated from one side by an electric resistance heater of known output. If the outer surfaces of the heater are well insulated, all the heat generated by the resistance heater will be transferred through the material whose conductivity is to be determined. Then measuring the two surface temperatures of the material when steady heat transfer is reached and substituting them into Eq. 14-1 together with other known quantities give the thermal conductivity (Fig. 14-5).

The thermal conductivities of materials vary over a wide range, as shown in Fig. 14-6. The thermal conductivities of gases such as air vary by a factor of 10^4 from those of pure metals such as copper. Note that pure crystals and metals have the highest thermal conductivities, and gases and insulating materials the lowest.

Temperature is a measure of the kinetic energies of the particles such as the molecules or atoms of a substance. In a liquid or gas, the kinetic energy of the molecules is due to their random translational motion as well as their vibrational and rotational motions. When two molecules possessing different kinetic energies collide, part of the kinetic energy of the more energetic (higher-temperature) molecule is transferred to the less energetic (lower-temperature) molecule, much the same as when two elastic balls of the same mass at different velocities collide, part of the kinetic energy of the faster ball is transferred to the slower one. The higher the temperature, the faster the molecules move and the higher the number of such collisions, and the better the heat transfer.

The *kinetic theory* of gases predicts and the experiments confirm that the thermal conductivity of gases is proportional to the *square root of the abso-*

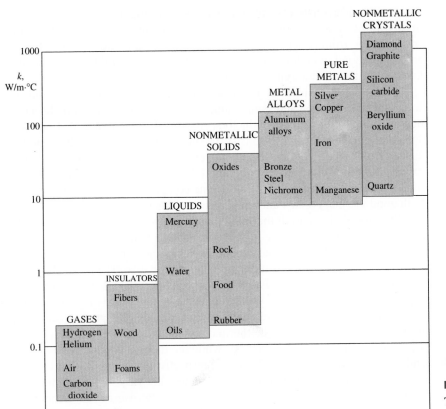

FIGURE 14-6

The range of thermal conductivity of various materials at room temperature.

lute temperature *T,* and inversely proportional to the *square root of the molar mass M.* Therefore, the thermal conductivity of a gas increases with increasing temperature and decreasing molar mass. So it is not surprising that the thermal conductivity of helium ($M = 4$) is much higher than those of air ($M = 29$) and argon ($M = 40$).

The thermal conductivities of *gases* at 1 atm pressure are listed in Table A-18. However, they can also be used at pressures other than 1 atm, since the thermal conductivity of gases is *independent of pressure* in a wide range of pressures encountered in practice.

The mechanism of heat conduction in a *liquid* is complicated by the fact that the molecules are more closely spaced, and they exert a stronger intermolecular force field. The thermal conductivities of liquids usually lie between those of solids and gases. The thermal conductivity of a substance is normally highest in the solid phase and lowest in the gas phase. Unlike gases, the thermal conductivities of most liquids decrease with increasing temperature, with water being a notable exception. Like gases, the conductivity of liquids decreases with increasing molar mass. Liquid metals such as mercury and sodium have high thermal conductivities and are very suitable for use in applications where a high heat transfer rate to a liquid is desired, as in nuclear power plants.

In *solids,* heat conduction is due to two effects: the *lattice vibrational waves* induced by the vibrational motions of the molecules positioned at relatively fixed positions in a periodic manner called a lattice, and the energy transported via the *free flow of electrons* in the solid (Fig. 14-7). The thermal conductivity of a solid is obtained by adding the lattice and electronic

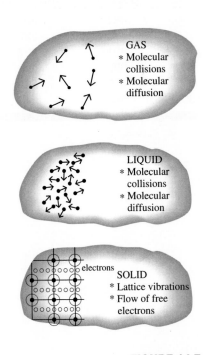

FIGURE 14-7

The mechanisms of heat conduction in different phases of a substance.

TABLE 14-2

The thermal conductivity of an alloy is usually much lower than the thermal conductivity of either metal of which it is composed

Pure metal or alloy	k, W/m · °C, at 300 K
Copper	401
Nickel	91
Constantan (55% Cu, 45% Ni)	23
Copper	401
Aluminum	237
Commercial bronze (90% Cu, 10% Al)	52

TABLE 14-3

Thermal conductivities of materials vary with temperature

T, K	Copper	Aluminum
100	482	302
200	413	237
300	401	237
400	393	240
600	379	231
800	366	218

components. The relatively high thermal conductivities of pure metals are primarily due to the electronic component. The lattice component of thermal conductivity strongly depends on the way the molecules are arranged. For example, diamond, which is a highly ordered crystalline solid, has the highest known thermal conductivity at room temperature.

Unlike metals, which are good electrical and heat conductors, *crystalline solids* such as diamond and semiconductors such as silicon are good heat conductors but poor electrical conductors. As a result, such materials find widespread use in the electronics industry. Despite their higher price, diamond heat sinks are used in the cooling of sensitive electronic components because of the excellent thermal conductivity of diamond. Silicon oils and gaskets are commonly used in the packaging of electronic components because they provide both good thermal contact and good electrical insulation.

Pure metals have high thermal conductivities, and one would think that *metal alloys* should also have high conductivities. One would expect an alloy made of two metals of thermal conductivities k_1 and k_2 to have a conductivity k between k_1 and k_2. But this turns out not to be the case. The thermal conductivity of an alloy of two metals is usually much lower than that of either metal, as shown in Table 14-2. Even small amounts in a pure metal of "foreign" molecules that are good conductors themselves seriously disrupt the flow of heat in that metal. For example, the thermal conductivity of steel containing just 1 percent of chrome is 62 W/m · °C, while the thermal conductivities of iron and chromium are 83 and 95 W/m · °C, respectively.

The thermal conductivities of materials vary with temperature (Table 14-3). The variation of thermal conductivity over certain temperature ranges is negligible for some materials, but significant for others, as shown in Fig. 14-8. The thermal conductivities of certain solids exhibit dramatic

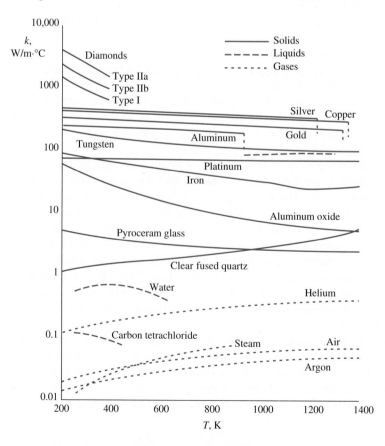

FIGURE 14-8

The variation of the thermal conductivity of various solids, liquids, and gases with temperature.

increases at temperatures near absolute zero, when these solids become *super-conductors.* For example, the conductivity of copper reaches a maximum value of about 20,000 W/m · °C at 20 K, which is about 50 times the conductivity at room temperature. The thermal conductivities and other thermal properties of various materials are given in Tables A-19 to A-23.

The temperature dependence of thermal conductivity causes considerable complexity in conduction analysis. Therefore, it is common practice to evaluate the thermal conductivity k at the *average temperature* and treat it as a *constant* in calculations.

In heat transfer analysis, a material is normally assumed to be *isotropic;* that is, to have uniform properties in all directions. This assumption is realistic for most materials, except those that exhibit different structural characteristics in different directions, such as laminated composite materials and wood. The thermal conductivity of wood across the grain, for example, is different than that parallel to the grain.

Thermal Diffusivity

The product ρC_p, which is frequently encountered in heat transfer analysis, is called the **heat capacity** of a material. Both the specific heat C_p and the heat capacity ρC_p represent the heat storage capability of a material. But C_p expresses it *per unit mass* whereas ρC_p expresses it *per unit volume,* as can be noticed from their units J/kg · °C and J/m³ · °C, respectively.

Another material property that appears in the transient heat conduction analysis is the **thermal diffusivity,** which represents how fast heat diffuses through a material and is defined as

$$\alpha = \frac{\text{Heat conducted}}{\text{Heat stored}} = \frac{k}{\rho C_p} \quad \text{(m}^2\text{/s)} \quad (14\text{-}3)$$

Note that the thermal conductivity k represents how well a material conducts heat, and the heat capacity ρC_p represents how much energy a material stores per unit volume. Therefore, the thermal diffusivity of a material can be viewed as the ratio of the *heat conducted* through the material to the *heat stored* per unit volume. A material that has a high thermal conductivity or a low heat capacity will obviously have a large thermal diffusivity. The larger the thermal diffusivity, the faster the propagation of heat into the medium. A small value of thermal diffusivity means that heat is mostly absorbed by the material and a small amount of heat will be conducted further.

The thermal diffusivities of some common materials at 20°C are given in Table 14-4. Note that the thermal diffusivity ranges from $\alpha = 0.14 \times 10^{-6}$ m²/s for water to 174×10^{-6} m²/s for silver, which is a difference of more than a thousand times. Also note that the thermal diffusivities of beef and water are the same. This is not surprising, since meat as well as fresh vegetables and fruits are mostly water, and thus they possess the thermal properties of water.

EXAMPLE 14-2   Measuring the Thermal Conductivity of a Material
A common way of measuring the thermal conductivity of a material is to sandwich an electric thermofoil heater between two identical samples of the material, as shown in Fig. 14-9. The thickness of the resistance heater, including its cover, which is made of thin silicon rubber, is usually less than 0.5 mm. A circulating fluid such as tap water keeps the exposed ends of the samples at constant temperature. The lateral surfaces of the samples are well insulated to ensure that heat transfer through the samples is one-dimensional. Two thermocouples are embedded into each sample some distance L apart, and a differential

TABLE 14-4

The thermal diffusivities of some materials at room temperature

Material	α, m²/s*
Silver	149×10^{-6}
Gold	127×10^{-6}
Copper	113×10^{-6}
Aluminum	97.5×10^{-6}
Iron	22.8×10^{-6}
Mercury (l)	4.7×10^{-6}
Marble	1.2×10^{-6}
Ice	1.2×10^{-6}
Concrete	0.75×10^{-6}
Brick	0.52×10^{-6}
Heavy soil (dry)	0.52×10^{-6}
Glass	0.34×10^{-6}
Glass wool	0.23×10^{-6}
Water (l)	0.14×10^{-6}
Beef	0.14×10^{-6}
Wood (oak)	0.13×10^{-6}

*Multiply by 10.76 to convert to ft²/s.

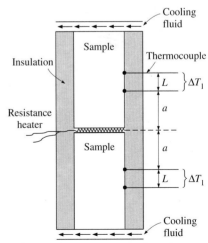

FIGURE 14-9

Apparatus to measure the thermal conductivity of a material using two identical samples and a thin resistance heater (Example 14-2).

thermometer reads the temperature drop ΔT across this distance along each sample. When steady operating conditions are reached, the total rate of heat transfer through both samples becomes equal to the electric power drawn by the heater, which is determined by multiplying the electric current by the voltage.

In a certain experiment, cylindrical samples of diameter 5 cm and length 10 cm are used. The two thermocouples in each sample are placed 3 cm apart. After initial transients, the electric heater is observed to draw 0.4 A at 110 V, and both differential thermometers read a temperature difference of 15°C. Determine the thermal conductivity of the sample.

Solution   The thermal conductivity of a material is to be determined by ensuring one-dimensional heat conduction, and by measuring temperatures when steady operating conditions are reached.

Assumptions   **1** Steady operating conditions exist since the temperature readings do not change with time. **2** Heat losses through the lateral surfaces of the apparatus are negligible since those surfaces are well insulated, and thus the entire heat generated by the heater is conducted through the samples. **3** The apparatus possesses thermal symmetry.

Analysis   The electrical power consumed by the resistance heater and converted to heat is

$$\dot{W}_e = VI = (110 \text{ V})(0.4 \text{ A}) = 44 \text{ W}$$

The rate of heat flow through each sample is

$$\dot{Q} = \tfrac{1}{2}\,\dot{W}_e = \tfrac{1}{2} \times (44 \text{ W}) = 22 \text{ W}$$

since only half of the heat generated will flow through each sample because of symmetry. Reading the same temperature difference across the same distance in each sample also confirms that the apparatus possesses thermal symmetry. The heat transfer area is the area normal to the direction of heat flow, which is the cross-sectional area of the cylinder in this case:

$$A = \tfrac{1}{4}\,\pi D^2 = \tfrac{1}{4}\,\pi (0.05 \text{ m})^2 = 0.00196 \text{ m}^2$$

Noting that the temperature drops by 15°C within 3 cm in the direction of heat flow, the thermal conductivity of the sample is determined to be

$$\dot{Q} = kA\frac{\Delta T}{L} \quad \rightarrow \quad k = \frac{\dot{Q}L}{A\,\Delta T} = \frac{(22 \text{ W})(0.03 \text{ m})}{(0.00196 \text{ m}^2)(15°C)} = \textbf{22.4 W/m} \cdot \textbf{°C}$$

Discussion   Perhaps you are wondering if we really need to use two samples in the apparatus, since the measurements on the second sample do not give any additional information. It seems like we can replace the second sample by insulation. Indeed, we do not need the second sample; however, it enables us to verify the temperature measurements on the first sample and provides thermal symmetry, which reduces experimental error.

EXAMPLE 14-3   Conversion between SI and English Units

An engineer who is working on the heat transfer analysis of a brick building in English units needs the thermal conductivity of brick. But the only value he can find from his handbooks is 0.72 W/m · °C, which is in SI units. To make matters worse, the engineer does not have a direct conversion factor between the two unit systems for thermal conductivity (he should have kept his heat transfer textbook instead of selling it back to the bookstore). Can you help him out?

Solution   The situation this engineer is facing is not unique, and most engineers often find themselves in a similar position. A person must be very careful during unit conversion in order not to fall into some common pitfalls and to avoid some costly mistakes. Although unit conversion is a simple process, it requires utmost care and careful reasoning.

The conversion factors for W and m are straightforward and are given in conversion tables to be

$$1 \text{ W} = 3.41214 \text{ Btu/h}$$

$$1 \text{ m} = 3.2808 \text{ ft}$$

But the conversion of °C into °F is not so simple, and it can be a source of error if one is not careful. Perhaps the first thought that comes to mind is to replace °C by (°F − 32)/1.8 since $T(°C) = [T(°F) − 32]/1.8$. But this will be wrong since the °C in the unit W/m · °C represents *per °C change in temperature*. Noting that 1°C change in temperature corresponds to 1.8°F, the proper conversion factor to be used is

$$1°C = 1.8°F$$

Substituting, we get

$$1 \text{ W/m} \cdot °C = \frac{3.41214 \text{ Btu/h}}{(3.2808 \text{ ft})(1.8°F)} = 0.5778 \text{ Btu/h} \cdot \text{ft} \cdot °F$$

which is the desired conversion factor. Therefore, the thermal conductivity of the brick in English units is

$$k_{\text{brick}} = 0.72 \text{ W/m} \cdot °C$$
$$= 0.72 \times (0.5778 \text{ Btu/h} \cdot \text{ft} \cdot °F)$$
$$= \mathbf{0.42 \text{ Btu/h} \cdot \text{ft} \cdot °F}$$

Discussion Note that the thermal conductivity value of a material in English units is about half that in SI units (Fig. 14-10). Also note that we rounded the result to two significant digits (the same number in the original value) since expressing the result in more significant digits (such as 0.4160 instead of 0.42) would falsely imply a more accurate value than the original one.

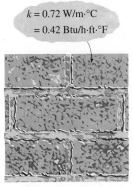

FIGURE 14-10

The thermal conductivity value in English units is obtained by multiplying the value in SI units by 0.5778.

14-4 ■ CONVECTION

Convection is the mode of energy transfer between a solid surface and the adjacent liquid or gas that is in motion, and it involves the combined effects of *conduction* and *fluid motion*. The faster the fluid motion, the greater the convection heat transfer. In the absence of any bulk fluid motion, heat transfer between a solid surface and the adjacent fluid is by pure conduction. The presence of bulk motion of the fluid enhances the heat transfer between the solid surface and the fluid, but it also complicates the determination of heat transfer rates.

Consider the cooling of a hot block by blowing cool air over its top surface (Fig. 14-11). Energy is first transferred to the air layer adjacent to the block by conduction. This energy is then carried away from the surface by convection, that is, by the combined effects of conduction within the air that is due to random motion of air molecules and the bulk or macroscopic motion of the air that removes the heated air near the surface and replaces it by the cooler air.

Convection is called **forced convection** if the fluid is forced to flow over the surface by external means such as a fan, pump, or the wind. In contrast, convection is called **natural** (or **free**) **convection** if the fluid motion is caused by buoyancy forces that are induced by density differences due to the variation of temperature in the fluid (Fig. 14-12). For example, in the absence of a

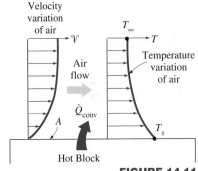

FIGURE 14-11

Heat transfer from a hot surface to air by convection.

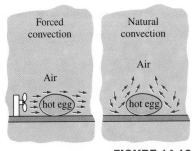

FIGURE 14-12

The cooling of a boiled egg by forced and natural convection.

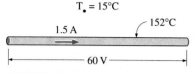

$T_\bullet = 15°C$

1.5 A

152°C

60 V

FIGURE 14-13
Schematic for Example 14-4.

TABLE 14-5

Typical values of convection heat transfer coefficient

Type of convection	h, W/m² · °C*
Free convection of gases	2–25
Free convection of liquids	10–1000
Forced convection of gases	25–250
Forced convection of liquids	50–20,000
Boiling and condensation	2500–100,000

*Multiply by 0.176 to convert to Btu/h · ft² · °F.

fan, heat transfer from the surface of the hot block in Fig. 14-13 will be by natural convection since any motion in the air in this case will be due to the rise of the warmer (and thus lighter) air near the surface and the fall of the cooler (and thus heavier) air to fill its place. Heat transfer between the block and the surrounding air will be by conduction if the temperature difference between the air and the block is not large enough to overcome the resistance of air to movement and thus to initiate natural convection currents.

Heat transfer processes that involve *change of phase* of a fluid are also considered to be convection because of the fluid motion induced during the process, such as the rise of the vapor bubbles during boiling or the fall of the liquid droplets during condensation.

Despite the complexity of convection, the rate of *convection heat transfer* is observed to be proportional to the temperature difference, and is conveniently expressed by **Newton's law of cooling** as

$$\dot{Q}_{conv} = hA\,(T_s - T_\infty) \qquad \text{(W)} \qquad (14\text{-}4)$$

where h is the *convection heat transfer coefficient* in W/m² · °C or Btu/h · ft² · °F, A is the surface area through which convection heat transfer takes place, T_s is the surface temperature, and T_∞ is the temperature of the fluid sufficiently far from the surface. Note that at the surface, the fluid temperature equals the surface temperature of the solid.

The convection heat transfer coefficient h is not a property of the fluid. It is an experimentally determined parameter whose value depends on all the variables influencing convection such as the surface geometry, the nature of fluid motion, the properties of the fluid, and the bulk fluid velocity. Typical values of h are given in Table 14-5.

Some people do not consider convection to be a fundamental mechanism of heat transfer since it is essentially heat conduction in the presence of fluid motion. But we still need to give this combined phenomenon a name, unless we are willing to keep referring to it as "conduction with fluid motion." Thus, it is practical to recognize convection as a separate heat transfer mechanism despite the valid arguments to the contrary.

EXAMPLE 14-4   Measuring Convection Heat Transfer Coefficient
A 2-m-long, 0.3-cm-diameter electrical wire extends across a room at 15°C, as shown in Fig. 14-13. Heat is generated in the wire as a result of resistance heating, and the surface temperature of the wire is measured to be 152°C in steady operation. Also, the voltage drop and electric current through the wire are measured to be 60 V and 1.5 A, respectively. Disregarding any heat transfer by radiation, determine the convection heat transfer coefficient for heat transfer between the outer surface of the wire and the air in the room.

Solution   The convection heat transfer coefficient for heat transfer from an electrically heated wire to air is to be determined by measuring temperatures when steady operating conditions are reached and the electric power consumed.

Assumptions   **1** Steady operating conditions exist since the temperature readings do not change with time. **2** Radiation heat transfer is negligible.

Analysis   When steady operating conditions are reached, the rate of heat loss from the wire will equal the rate of heat generation in the wire as a result of resistance heating. That is,

$$\dot{Q} = \dot{E}_{generated} = VI = (60\text{ V})(1.5\text{ A}) = 90\text{ W}$$

The surface area of the wire is

$$A = \pi D L = \pi(0.003 \text{ m})(2 \text{ m}) = 0.01885 \text{ m}^2$$

Newton's law of cooling for convection heat transfer is expressed as

$$\dot{Q}_{\text{conv}} = hA\,(T_s - T_\infty)$$

Disregarding any heat transfer by radiation and thus assuming all the heat loss from the wire to occur by convection, the convection heat transfer coefficient is determined to be

$$h = \frac{\dot{Q}_{\text{conv}}}{A(T_s - T_\infty)} = \frac{90 \text{ W}}{(0.01885 \text{ m}^2)(152 - 15)^\circ\text{C}} = \mathbf{34.9 \text{ W/m}^2 \cdot {}^\circ\text{C}}$$

Discussion Note that the simple setup described above can be used to determine the average heat transfer coefficients from a variety of surfaces in air. Also, heat transfer by radiation can be eliminated by keeping the surrounding surfaces at the temperature of the wire.

14-5 ■ RADIATION

Radiation is the energy emitted by matter in the form of *electromagnetic waves* (or *photons*) as a result of the changes in the electronic configurations of the atoms or molecules. Unlike conduction and convection, the transfer of energy by radiation does not require the presence of an *intervening medium.* In fact, energy transfer by radiation is fastest (at the speed of light) and it suffers no attenuation in a vacuum. This is how the energy of the sun reaches the earth.

In heat transfer studies we are interested in *thermal radiation,* which is the form of radiation emitted by bodies because of their temperature. It differs from other forms of electromagnetic radiation such as x-rays, gamma rays, microwaves, radio waves, and television waves that are not related to temperature. All bodies at a temperature above absolute zero emit thermal radiation.

Radiation is a *volumetric phenomenon,* and all solids, liquids, and gases emit, absorb, or transmit radiation to varying degrees. However, radiation is usually considered to be a *surface phenomenon* for solids that are opaque to thermal radiation such as metals, wood, and rocks since the radiation emitted by the interior regions of such material can never reach the surface, and the radiation incident on such bodies is usually absorbed within a few microns from the surface.

The maximum rate of radiation that can be emitted from a surface at an absolute temperature T_s (in K or R) is given by the **Stefan-Boltzmann law** as

$$\dot{Q}_{\text{emit, max}} = \sigma A T_s^4 \qquad \text{(W)} \qquad (14\text{-}5)$$

where $\sigma = 5.67 \times 10^{-8} \text{ W/m}^2 \cdot \text{K}^4$ or $0.1714 \times 10^{-8} \text{ Btu/h} \cdot \text{ft}^2 \cdot \text{R}^4$ is the *Stefan-Boltzmann constant.* The idealized surface that emits radiation at this maximum rate is called a **blackbody,** and the radiation emitted by a blackbody is called **blackbody radiation** (Fig. 14-14). The radiation emitted by all real surfaces is less than the radiation emitted by a blackbody at the same temperature, and is expressed as

$$\dot{Q}_{\text{emit}} = \varepsilon \sigma A T_s^4 \qquad \text{(W)} \qquad (14\text{-}6)$$

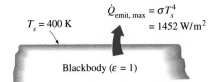

$T_s = 400 \text{ K}$

$\dot{Q}_{\text{emit, max}} = \sigma T_s^4$
$= 1452 \text{ W/m}^2$

Blackbody ($\varepsilon = 1$)

FIGURE 14-14

Blackbody radiation represents the *maximum amount of radiation that can be emitted from a surface at a specified temperature.*

TABLE 14-6

Emissivities of some materials at 300 K

Material	Emissivity
Aluminum foil	0.07
Anodized aluminum	0.82
Polished copper	0.03
Polished gold	0.03
Polished silver	0.02
Polished stainless steel	0.17
Black paint	0.98
White paint	0.90
White paper	0.92–0.97
Asphalt pavement	0.85–0.93
Red brick	0.93–0.96
Human skin	0.95
Wood	0.82–0.92
Soil	0.93–0.96
Water	0.96
Vegetation	0.92–0.96

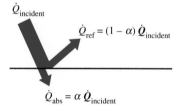

FIGURE 14-15

The absorption of radiation incident on an opaque surface of absorptivity α.

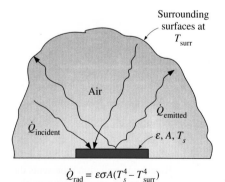

$$\dot{Q}_{rad} = \varepsilon\sigma A(T_s^4 - T_{surr}^4)$$

FIGURE 14-16

Radiation heat transfer between a surface and the surfaces surrounding it.

where ε is the **emissivity** of the surface. The property emissivity, whose value is in the range $0 \leq \varepsilon \leq 1$, is a measure of how closely a surface approximates a blackbody for which $\varepsilon = 1$. The emissivities of some surfaces are given in Table 14-6.

Another important radiation property of a surface is its **absorptivity** α, which is the fraction of the radiation energy incident on a surface that is absorbed by the surface. Like emissivity, its value is in the range $0 \leq \alpha \leq 1$. A blackbody absorbs the entire radiation incident on it. That is, a blackbody is a perfect absorber ($\alpha = 1$) as it is a perfect emitter.

In general, both ε and α of a surface depend on the temperature and the wavelength of the radiation. **Kirchhoff's law** of radiation states that the emissivity and the absorptivity of a surface are equal at the same temperature and wavelength. In most practical applications, the dependence of ε and α on the temperature and wavelength is ignored, and the average absorptivity of a surface is taken to be equal to its average emissivity. The rate at which a surface absorbs radiation is determined from (Fig. 14-15)

$$\dot{Q}_{absorbed} = \alpha\dot{Q}_{incident} \qquad \text{(W)} \qquad (14\text{-}7)$$

where $\dot{Q}_{incident}$ is the rate at which radiation is incident on the surface and α is the absorptivity of the surface. For opaque (nontransparent) surfaces, the portion of incident radiation not absorbed by the surface is reflected back.

The difference between the rates of radiation emitted by the surface and the radiation absorbed is the *net* radiation heat transfer. If the rate of radiation absorption is greater than the rate of radiation emission, the surface is said to be *gaining* energy by radiation. Otherwise, the surface is said to be *losing* energy by radiation. In general, the determination of the net rate of heat transfer by radiation between two surfaces is a complicated matter since it depends on the properties of the surfaces, their orientation relative to each other, and the interaction of the medium between the surfaces with radiation.

When a surface of emissivity ε and surface area A at an *absolute temperature* T_s is *completely enclosed* by a much larger (or black) surface at absolute temperature T_{surr} separated by a gas (such as air) that does not intervene with radiation, the net rate of radiation heat transfer between these two surfaces is given by (Fig. 14-16)

$$\dot{Q}_{rad} = \varepsilon\sigma A\,(T_s^4 - T_{surr}^4) \qquad \text{(W)} \qquad (14\text{-}8)$$

In this special case, the emissivity and the surface area of the surrounding surface do not have any effect on the net radiation heat transfer.

Radiation heat transfer to or from a surface surrounded by a gas such as air occurs *parallel* to conduction (or convection, if there is bulk gas motion) between the surface and the gas. Thus the total heat transfer is determined by *adding* the contributions of both heat transfer mechanisms. For simplicity and convenience, this is often done by defining a **combined heat transfer coefficient** $h_{combined}$ that includes the effects of both convection and radiation. Then the *total* heat transfer rate to or from a surface by convection and radiation is expressed as

$$\dot{Q}_{total} = h_{combined}A(T_s - T_\infty) \qquad \text{(W)} \qquad (14\text{-}9)$$

Note that the combined heat transfer coefficient is essentially a convection heat transfer coefficient modified to include the effects of radiation.

Radiation is usually significant relative to conduction or natural convection, but negligible relative to forced convection. Thus radiation in forced convection applications is normally disregarded, especially when the surfaces involved have low emissivities and low to moderate temperatures.

EXAMPLE 14-5 Radiation Effect on Thermal Comfort

It is a common experience to feel "chilly" in winter and "warm" in summer in our homes even when the thermostat setting is kept the same. This is due to the so called "radiation effect" resulting from radiation heat exchange between our bodies and the surrounding surfaces of the walls and the ceiling.

Consider a person standing in a room maintained at 22°C at all times. The inner surfaces of the walls, floors, and the ceiling of the house are observed to be at an average temperature of 10°C in winter and 25°C in summer. Determine the rate of radiation heat transfer between this person and the surrounding surfaces if the exposed surface area and the average outer surface temperature of the person are 1.4 m² and 30°C, respectively (Fig. 14-17).

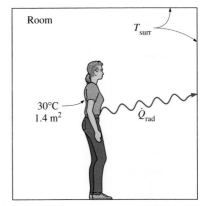

FIGURE 14-17

Schematic for Example 14-5.

Solution   The rates of radiation heat transfer between a person and the surrounding surfaces at specified temperatures are to be determined in summer and winter.

Assumptions   **1** Steady operating conditions exist. **2** Heat transfer by convection is not considered. **3** The person is completely surrounded by the interior surfaces of the room. **4** The surrounding surfaces are at a uniform temperature.

Properties   The emissivity of a person is $\varepsilon = 0.95$ (Table 14-6).

Analysis   The net rates of radiation heat transfer from the body to the surrounding walls, ceiling, and floor in winter and summer are

$$
\begin{aligned}
\dot{Q}_{rad, \, winter} &= \varepsilon\sigma A(T_s^4 - T_{surr, \, winter}^4) \\
&= (0.95)(5.67 \times 10^{-8} \, \text{W/m}^2 \cdot \text{K}^4)(1.4 \, \text{m}^2) \\
&\quad \times [(30 + 273)^4 - (10 + 273)^4] \, \text{K}^4 \\
&= \mathbf{152 \ W}
\end{aligned}
$$

and

$$
\begin{aligned}
\dot{Q}_{rad, \, summer} &= \varepsilon\sigma A(T_s^4 - T_{surr, \, summer}^4) \\
&= (0.95)(5.67 \times 10^{-8} \, \text{W/m}^2 \cdot \text{K}^4)(1.4 \, \text{m}^2) \\
&\quad \times [(30 + 273)^4 - (25 + 273)^4] \, \text{K}^4 \\
&= \mathbf{40.9 \ W}
\end{aligned}
$$

Discussion   Note that we must use *absolute temperatures* in radiation calculations. Also note that the rate of heat loss from the person by radiation is almost four times as large in winter than it is in summer, which explains the "chill" we feel in winter even if the thermostat setting is kept the same.

14-6 ■ SIMULTANEOUS HEAT TRANSFER MECHANISMS

We mentioned that there are three mechanisms of heat transfer, but not all three can exist simultaneously in a medium. For example, heat transfer is only by conduction in *opaque solids,* but by conduction and radiation in *semitransparent solids.* Thus, a solid may involve conduction and radiation but not convection. However, a solid may involve heat transfer by convection and/or radiation on its surfaces exposed to a fluid or other surfaces. For example, the outer surfaces of a cold piece of rock will warm up in a warmer environment as a result of heat gain by convection (from the air) and radiation

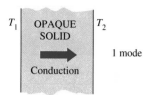

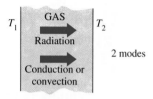

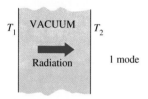

FIGURE 14-18

Although there are three mechanisms of
heat transfer, a medium may involve only
two of them simultaneously.

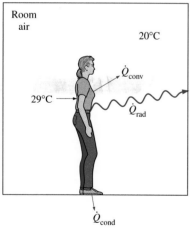

FIGURE 14-19

Heat transfer from the person described in
Example 14-6.

(from the sun or the warmer surrounding surfaces). But the inner parts of the rock will warm up as this heat is transferred to the inner region of the rock by conduction.

Heat transfer is by conduction and possibly by radiation in a *still fluid* (no bulk fluid motion) and by convection and radiation in a *flowing fluid*. In the absence of radiation, heat transfer through a fluid is either by conduction or convection, depending on the presence of any bulk fluid motion. Convection can be viewed as combined conduction and fluid motion, and conduction in a fluid can be viewed as a special case of convection in the absence of any fluid motion (Fig. 14-18).

Thus, when we deal with heat transfer through a *fluid,* we have either *conduction* or *convection,* but not both. Also, gases are practically transparent to radiation, except that some gases are known to absorb radiation strongly at certain wavelengths. Ozone, for example, strongly absorbs ultraviolet radiation. But in most cases, a gas between two solid surfaces does not interfere with radiation and acts effectively as a vacuum. Liquids, on the other hand, are usually strong absorbers of radiation.

Finally, heat transfer through a *vacuum* is by radiation only since conduction or convection requires the presence of a material medium.

EXAMPLE 14-6 Heat Loss from a Person

Consider a person standing in a breezy room at 20°C. Determine the total rate of heat transfer from this person if the exposed surface area and the average outer surface temperature of the person are 1.6 m² and 29°C, respectively, and the convection heat transfer coefficient is 6 W/m² · °C (Fig. 14-19).

Solution The total rate of heat transfer from a person by both convection and radiation to the surrounding air and surfaces at specified temperatures is to be determined.

Assumptions **1** Steady operating conditions exist. **2** The person is completely surrounded by the interior surfaces of the room. **3** The surrounding surfaces are at the same temperature as the air in the room. **4** Heat conduction to the floor through the feet is negligible.

Properties The emissivity of a person is $\varepsilon = 0.95$ (Table 14-6).

Analysis The heat transfer between the person and the air in the room will be by convection (instead of conduction) since it is conceivable that the air in the vicinity of the skin or clothing will warm up and rise as a result of heat transfer from the body, initiating natural convection currents. It appears that the experimentally determined value for the rate of convection heat transfer in this case is 6 W per unit surface area (m²) per unit temperature difference (in K or °C) between the person and the air away from the person. Thus, the rate of convection heat transfer from the person to the air in the room is

$$\dot{Q}_{conv} = hA(T_s - T_\infty)$$
$$= (6 \text{ W/m}^2 \cdot °C)(1.6 \text{ m}^2)(29 - 20)°C$$
$$= 86.4 \text{ W}$$

The person will also lose heat by radiation to the surrounding wall surfaces. We take the temperature of the surfaces of the walls, ceiling, and floor to be equal to the air temperature in this case for simplicity, but we recognize that this does not need to be the case. These surfaces may be at a higher or lower temperature than the average temperature of the room air, depending on the outdoor conditions and the structure of the walls. Considering that air does not intervene with radiation and the person is completely enclosed by the surrounding surfaces, the net rate of radiation heat transfer from the person to the surrounding walls, ceiling, and floor is

$$\dot{Q}_{rad} = \varepsilon \sigma A(T_s^4 - T_{surr}^4)$$
$$= (0.95)(5.67 \times 10^{-8} \text{ W/m}^2 \cdot \text{K}^4)(1.6 \text{ m}^2)$$
$$\times [(29 + 273)^4 - (20 + 273)^4] \text{ K}^4$$
$$= 81.7 \text{ W}$$

Note that we must use *absolute* temperatures in radiation calculations. Also note that we used the emissivity value for the skin and clothing at room temperature since the emissivity is not expected to change significantly at a slightly higher temperature.

Then the rate of total heat transfer from the body is determined by adding these two quantities:

$$\dot{Q}_{total} = \dot{Q}_{conv} + \dot{Q}_{rad} = (86.4 + 81.7) \text{ W} = \textbf{168.1 W}$$

Discussion The heat transfer would be much higher if the person were not dressed since the exposed surface temperature would be higher. Thus, an important function of the clothes is to serve as a barrier against heat transfer.

In the above calculations, heat transfer through the feet to the floor by conduction, which is usually very small, is neglected. Heat transfer from the skin by perspiration, which is the dominant mode of heat transfer in hot environments, is not considered here.

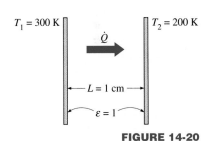

$T_1 = 300 \text{ K}$ $T_2 = 200 \text{ K}$

$L = 1 \text{ cm}$

$\varepsilon = 1$

FIGURE 14-20
Schematic for Example 14-7.

EXAMPLE 14-7   Heat Transfer between Two Isothermal Plates

Consider steady heat transfer between two large parallel plates at constant temperatures of $T_1 = 300$ K and $T_2 = 200$ K that are $L = 1$ cm apart, as shown in Fig. 14-20. Assuming the surfaces to be black (emissivity $\varepsilon = 1$), determine the rate of heat transfer between the plates per unit surface area assuming the gap between the plates is (*a*) filled with atmospheric air, (*b*) evacuated, (*c*) filled with urethane insulation, and (*d*) filled with superinsulation that has an apparent thermal conductivity of 0.00002 W/m · °C.

Solution The total rate of heat transfer between two large parallel plates at specified temperatures is to be determined for four different cases.

Assumptions **1** Steady operating conditions exist. **2** There are no natural convection currents in the air between the plates. **3** The surfaces are black and thus $\varepsilon = 1$.

Properties The thermal conductivity at the average temperature of 250 K is $k = 0.0223$ W/m · °C for air (Table A-18), 0.026 W/m · °C for urethane insulation (Table A-22), and 0.00002 W/m · °C for the superinsulation.

Analysis (*a*) The rates of conduction and radiation heat transfer between the plates through the air layer are

$$\dot{Q}_{cond} = kA \frac{T_1 - T_2}{L} = (0.0223 \text{ W/m} \cdot {}^\circ\text{C})(1 \text{ m}^2)\frac{(300 - 200){}^\circ\text{C}}{0.01 \text{ m}} = 223 \text{ W}$$

and

$$\dot{Q}_{rad} = \varepsilon \sigma A(T_1^4 - T_2^4)$$
$$= (1)(5.67 \times 10^{-8} \text{ W/m}^2 \cdot \text{K}^4)(1 \text{ m}^2)[(300 \text{ K})^4 - (200 \text{ K})^4] = 368 \text{ W}$$

Therefore,

$$\dot{Q}_{total} = \dot{Q}_{cond} + \dot{Q}_{rad} = 223 + 368 = \textbf{591 W}$$

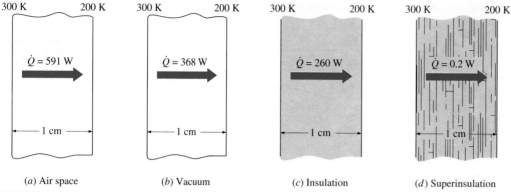

FIGURE 14-21

Different ways of reducing heat transfer between two isothermal plates, and their effectiveness.

The heat transfer rate in reality will be higher because of the natural convection currents that are likely to occur in the air space between the plates.

(*b*) When the air space between the plates is evacuated, there will be no conduction or convection, and the only heat transfer between the plates will be by radiation. Therefore,

$$\dot{Q}_{total} = \dot{Q}_{rad} = \textbf{368 W}$$

(*c*) An opaque solid material placed between two plates blocks direct radiation heat transfer between the plates. Also, the thermal conductivity of an insulating material accounts for the radiation heat transfer that may be occurring through the voids in the insulating material. The rate of heat transfer through the urethane insulation is

$$\dot{Q}_{total} = \dot{Q}_{cond} = kA\frac{T_1 - T_2}{L} = (0.026 \text{ W/m} \cdot °\text{C})(1 \text{ m}^2)\frac{(300 - 200)°\text{C}}{0.01 \text{ m}} = \textbf{260 W}$$

Note that heat transfer through the urethane material is less than the heat transfer through the air determined in (*a*), although the thermal conductivity of the insulation is higher than that of air. This is because the insulation blocks the radiation whereas air transmits it.

(*d*) The layers of the superinsulation prevents any direct radiation heat transfer between the plates. However, radiation heat transfer between the sheets of superinsulation does occur, and the apparent thermal conductivity of the superinsulation accounts for this effect. Therefore,

$$\dot{Q}_{total} = kA\frac{T_1 - T_2}{L} = (0.00002 \text{ W/m} \cdot °\text{C})(1 \text{ m}^2)\frac{(300 - 200)°\text{C}}{0.01 \text{ m}} = \textbf{0.2 W}$$

which is $\frac{1}{1840}$ of the heat transfer through the vacuum. The results of this example are summarized in Fig. 14-21 to put them into perspective.

Discussion This example demonstrates the effectiveness of superinsulations, which are discussed in the next chapter, and explains why they are the insulation of choice in critical applications despite their high cost.

FIGURE 14-22

A chicken being cooked in a microwave oven (Example 14-8).

EXAMPLE 14-8   Heat Transfer in Conventional and Microwave Ovens

The fast and efficient cooking of microwave ovens made them one of the essential appliances in modern kitchens (Fig. 14-22). Discuss the heat transfer mechanisms associated with the cooking of a chicken in microwave and conventional ovens, and explain why cooking in a microwave oven is more efficient.

Solution Food is cooked in a microwave oven by absorbing the electromagnetic radiation energy generated by the microwave tube, called the magnetron. The radiation emitted by the magnetron is not thermal radiation, since its emission is not due to the temperature of the magnetron; rather, it is due to the conversion of electrical energy into electromagnetic radiation at a specified wavelength. The wavelength of the microwave radiation is such that it is *reflected* by metal surfaces; *transmitted* by the cookware made of glass, ceramic, or plastic; and *absorbed* and converted to internal energy by food (especially the water, sugar, and fat) molecules.

In a microwave oven, the *radiation* that strikes the chicken is absorbed by the skin of the chicken and the outer parts. As a result, the temperature of the chicken at and near the skin rises. Heat is then *conducted* toward the inner parts of the chicken from its outer parts. Of course, some of the heat absorbed by the outer surface of the chicken is lost to the air in the oven by *convection*.

In a conventional oven, the air in the oven is first heated to the desired temperature by the electric or gas heating element. This preheating may take several minutes. The heat is then transferred from the air to the skin of the chicken by *natural convection* in most ovens or by *forced convection* in the newer convection ovens that utilize a fan. The air motion in convection ovens increases the convection heat transfer coefficient and thus decreases the cooking time. Heat is then *conducted* toward the inner parts of the chicken from its outer parts as in microwave ovens.

Microwave ovens replace the slow convection heat transfer process in conventional ovens by the instantaneous radiation heat transfer. As a result, microwave ovens transfer energy to the food at full capacity the moment they are turned on, and thus they cook faster while consuming less energy.

EXAMPLE 14-9   Heating of a Plate by Solar Energy

A thin metal plate is insulated on the back and exposed to solar radiation at the front surface (Fig. 14-23). The exposed surface of the plate has an absorptivity of 0.6 for solar radiation. If solar radiation is incident on the plate at a rate of 700 W/m^2 and the surrounding air temperature is 25°C, determine the surface temperature of the plate when the heat loss by convection and radiation equals the solar energy absorbed by the plate. Assume the combined convection and radiation heat transfer coefficient to be 50 W/m$^2 \cdot$ °C.

Solution The back side of the thin metal plate is insulated and the front side is exposed to solar radiation. The surface temperature of the plate is to be determined when it stabilizes.

Assumptions **1** Steady operating conditions exist. **2** Heat transfer through the insulated side of the plate is negligible. **3** The heat transfer coefficient remains constant.

Properties The solar absorptivity of the plate is given to be $\alpha = 0.6$.

Analysis The absorptivity of the plate is 0.6, and thus 60 percent of the solar radiation incident on the plate will be absorbed continuously. As a result, the temperature of the plate will rise, and the temperature difference between the plate and the surroundings will increase. This increasing temperature difference will cause the rate of heat loss from the plate to the surroundings to increase. At some point, the rate of heat loss from the plate will equal the rate of solar energy absorbed, and the temperature of the plate will no longer change. The temperature of the plate when steady operation is established is determined from

$$\dot{E}_{gained} = \dot{E}_{lost} \qquad \text{or} \qquad \alpha A \, \dot{q}_{incident, \, solar} = h_{combined} A \, (T_s - T_\infty)$$

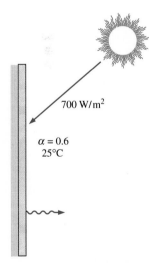

FIGURE 14-23
Schematic for Example 14-9.

700 W/m^2

$\alpha = 0.6$
25°C

Solving for T_s and substituting, the plate surface temperature is determined to be

$$T_s = T_\infty + \alpha \frac{\dot{q}_{\text{incident, solar}}}{h_{\text{combined}}} = 25°C + \frac{0.6 \times (700 \text{ W/m}^2)}{50 \text{ W/m}^2 \cdot °C} = \textbf{33.4°C}$$

Discussion   Note that the heat losses will prevent the plate temperature from rising above 33.4°C. Also, the combined heat transfer coefficient accounts for the effects of both convection and radiation, and thus it is very convenient to use in heat transfer calculations when its value is known with reasonable accuracy.

14-7 ■ SUMMARY

Heat can be transferred in three different ways: Conduction, convection, and radiation. *Conduction* is the transfer of energy from the more energetic particles of a substance to the adjacent less energetic ones as a result of interactions between the particles, and is expressed by *Fourier's law of heat conduction* as

$$\dot{Q}_{\text{cond}} = -kA\frac{dT}{dx} \qquad \text{(W)}$$

where k is the *thermal conductivity* of the material, A is the *area* normal to the direction of heat transfer, and dT/dx is the *temperature gradient*. The rate of heat conduction across a plane layer of thickness L is given by

$$\dot{Q}_{\text{cond}} = kA\frac{\Delta T}{L} \qquad \text{(W)}$$

where ΔT is the temperature difference across the layer.

Convection is the mode of heat transfer between a solid surface and the adjacent liquid or gas that is in motion, and involves the combined effects of conduction and fluid motion. The rate of convection heat transfer is expressed by *Newton's law of cooling* as

$$\dot{Q}_{\text{conv}} = hA(T_s - T_\infty) \qquad \text{(W)}$$

where h is the *convection heat transfer coefficient* in W/m² · °C or Btu/h · ft² · °F, A is the surface area through which convection heat transfer takes place, T_s is the surface temperature, and T_∞ is the temperature of the fluid sufficiently far from the surface.

Radiation is the energy emitted by matter in the form of electromagnetic waves (or photons) as a result of the changes in the electronic configurations of the atoms or molecules. The maximum rate of radiation that can be emitted from a surface at an absolute temperature T_s is given by the *Stefan-Boltzmann law* as $\dot{Q}_{\text{emit, max}} = \sigma A T_s^4$, where $\sigma = 5.67 \times 10^{-8}$ W/m² · K⁴ or 0.1714×10^{-8} Btu/h · ft² · R⁴ is the *Stefan-Boltzmann constant*.

When a surface of emissivity ε and surface area A at an absolute temperature T_s is completely enclosed by a much larger (or black) surface at absolute temperature T_{surr} separated by a gas (such as air) that does not intervene with radiation, the net rate of radiation heat transfer between these two surfaces is given by

$$\dot{Q}_{\text{rad}} = \varepsilon\sigma A(T_s^4 - T_{\text{surr}}^4) \qquad \text{(W)}$$

In this special case, the emissivity and the surface area of the surrounding surface do not have any effect on the net radiation heat transfer.

The rate at which a surface absorbs radiation is determined from $\dot{Q}_{\text{absorbed}} = \alpha\dot{Q}_{\text{incident}}$ where $\dot{Q}_{\text{incident}}$ is the rate at which radiation is incident on the surface and α is the absorptivity of the surface.

REFERENCES AND SUGGESTED READING

1. Y. A. Çengel. *Heat Transfer: A Practical Approach.* New York: McGraw-Hill, 1998.

2. J. P. Holman. *Heat Transfer.* 8th ed. New York: McGraw-Hill, 1997.

3. F. P. Incropera and D. P. DeWitt. *Fundamentals of Heat and Mass Transfer.* 4th ed. New York: Wiley, 1996.

4. F. Kreith and M. S. Bohn. *Principles of Heat Transfer.* 5th ed. St. Paul, MN: West Publishing, 1993.

5. A. F. Mills. *Basic Heat and Mass Transfer.* 2nd ed. Prentice Hall, Upper Saddle River, 1999.

6. M. N. Ozisik. *Heat Transfer—A Basic Approach.* New York: McGraw-Hill, 1985.

PROBLEMS*

Heat Transfer Mechanisms

14-1C Define thermal conductivity and explain its significance in heat transfer.

14-2C What are the mechanisms of heat transfer? How are they distinguished from each other?

14-3C What is the physical mechanism of heat conduction in a solid, a liquid, and a gas?

14-4C Consider heat transfer through a windowless wall of a house in a winter day. Discuss the parameters that affect the rate of heat conduction through the wall.

14-5C Write down the expressions for the physical laws that govern each mode of heat transfer, and identify the variables involved in each relation.

14-6C How does heat conduction differ from convection?

14-7C Does any of the energy of the sun reach the earth by conduction or convection?

14-8C How does forced convection differ from natural convection?

14-9C Define emissivity and absorptivity. What is Kirchhoff's law of radiation?

14-10C What is a blackbody? How do real bodies differ from blackbodies?

14-11C Judging from its unit W/m · °C, can we define thermal conductivity of a material as the rate of heat transfer through the material per unit thickness per unit temperature difference? Explain.

14-12C Consider heat loss through the two walls of a house on a winter night. The walls are identical, except that one of them has a tightly fit glass window. Through which wall will the house lose more heat? Explain.

---

*Students are encouraged to answer *all* the concept "C" questions.

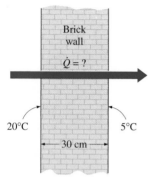

FIGURE P14-19

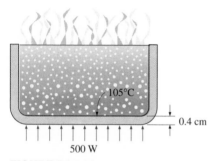

FIGURE P14-21

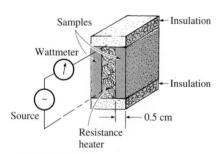

FIGURE P14-24

14-13C   Which is a better heat conductor, diamond or silver?

14-14C   Consider two walls of a house that are identical except that one is made of 10-cm-thick wood, while the other is made of 25-cm-thick brick. Through which wall will the house lose more heat in winter?

14-15C   How do the thermal conductivity of gases and liquids vary with temperature?

14-16C   Why is the thermal conductivity of superinsulation orders of magnitude lower than the thermal conductivity of ordinary insulation?

14-17C   Why do we characterize the heat conduction ability of insulators in terms of their apparent thermal conductivity instead of the ordinary thermal conductivity?

14-18C   Consider an alloy of two metals whose thermal conductivities are k_1 and k_2. Will the thermal conductivity of the alloy be less than k_1, greater than k_2, or between k_1 and k_2?

14-19   The inner and outer surfaces of a 5-m $\times$ 6-m brick wall of thickness 30 cm and thermal conductivity 0.69 W/m $\cdot$ °C are maintained at temperatures of 20°C and 5°C, respectively. Determine the rate of heat transfer through the wall, in W.   *Answer:* 1035 W

14-20   The inner and outer surfaces of a 0.5-cm-thick 2-m $\times$ 2-m window glass in winter are 10°C and 3°C, respectively. If the thermal conductivity of the glass is 0.78 W/m $\cdot$ °C, determine the amount of heat loss, in kJ, through the glass over a period of 5 hours. What would your answer be if the glass were 1 cm thick?   *Answers:* 78,624 kJ,   39,312 kJ

14-21   An aluminum pan whose thermal conductivity is 237 W/m $\cdot$ °C has a flat bottom with diameter 20 cm and thickness 0.4 cm. Heat is transferred steadily to boiling water in the pan through its bottom at a rate of 500 W. If the inner surface of the bottom of the pan is at 105°C, determine the temperature of the outer surface of the bottom of the pan.

14-22E   The north wall of an electrically heated home is 20 ft long, 10 ft high, and 1 ft thick, and is made of brick whose thermal conductivity is $k = 0.42$ Btu/h $\cdot$ ft $\cdot$ °F. On a certain winter night, the temperatures of the inner and the outer surfaces of the wall are measured to be at about 62°F and 25°F, respectively, for a period of 8 hours. Determine (*a*) the rate of heat loss through the wall that night and (*b*) the cost of that heat loss to the home owner if the cost of electricity is $0.07/kWh.

14-23   In a certain experiment, cylindrical samples of diameter 4 cm and length 7 cm are used (see Fig. 1-29). The two thermocouples in each sample are placed 3 cm apart. After initial transients, the electric heater is observed to draw 0.6 A at 110 V, and both differential thermometers read a temperature difference of 10°C. Determine the thermal conductivity of the sample.
    Answer: 78.8 W/m $\cdot$ °C

14-24   One way of measuring the thermal conductivity of a material is to sandwich an electric thermofoil heater between two identical rectangular samples of the material and to heavily insulate the four outer edges, as shown in the figure. Thermocouples attached to the inner and outer surfaces of the samples record the temperatures.

During an experiment, two 0.5-cm-thick samples 10 cm × 10 cm in size are used. When steady operation is reached, the heater is observed to draw 35 W of electric power, and the temperature of each sample is observed to drop from 82°C at the inner surface to 74°C at the outer surface. Determine the thermal conductivity of the material at the average temperature.

14-25   Repeat Prob. 14-24 for an electric power consumption of 20 W.

14-26   A heat flux meter attached to the inner surface of a 3-cm-thick refrigerator door indicates a heat flux of 25 W/m² through the door. Also, the temperatures of the inner and the outer surfaces of the door are measured to be 7°C and 15°C, respectively. Determine the average thermal conductivity of the refrigerator door.   *Answer:* 0.0938 W/m · °C

14-27   Consider a person standing in a room maintained at 20°C at all times. The inner surfaces of the walls, floors, and ceiling of the house are observed to be at an average temperature of 12°C in winter and 23°C in summer. Determine the rates of radiation heat transfer between this person and the surrounding surfaces in both summer and winter if the exposed surface area, emissivity, and the average outer surface temperature of the person are 1.6 m², 0.95, and 32°C, respectively.

14-28   For heat transfer purposes, a standing man can be modeled as a 30-cm-diameter, 170-cm-long vertical cylinder with both the top and bottom surfaces insulated and with the side surface at an average temperature of 34°C. For a convection heat transfer coefficient of 15 W/m² · °C, determine the rate of heat loss from this man by convection in an environment at 20°C.
    Answer: 336 W

14-29   Hot air at 80°C is blown over a 2-m × 4-m flat surface at 30°C. If the average convection heat transfer coefficient is 55 W/m² · °C, determine the rate of heat transfer from the air to the plate, in kW.   *Answer:* 22 kW

14-30   The heat generated in the circuitry on the surface of a silicon chip (k = 130 W/m · °C) is conducted to the ceramic substrate to which it is attached. The chip is 6 mm × 6 mm in size and 0.5 mm thick and dissipates 3 W of power. Disregarding any heat transfer through the 0.5-mm-high side surfaces, determine the temperature difference between the front and back surfaces of the chip in steady operation.

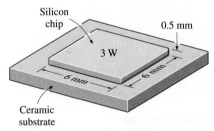

FIGURE P14-30

14-31   A 50-cm-long, 800-W electric resistance heating element with diameter 0.5 cm and surface temperature 120°C is immersed in 40 kg of water initially at 20°C. Determine how long it will take for this heater to raise the water temperature to 80°C. Also, determine the convection heat transfer coefficients at the beginning and at the end of the heating process.

14-32   A 5-cm-external-diameter, 10-m-long hot water pipe at 80°C is losing heat to the surrounding air at 5°C by natural convection with a heat transfer coefficient of 25 W/m² · °C. Determine the rate of heat loss from the pipe by natural convection, in W.   *Answer:* 2945 W

14-33   A hollow spherical iron container with outer diameter 20 cm and thickness 0.4 cm is filled with iced water at 0°C. If the outer surface temperature is 5°C, determine the approximate rate of heat loss from the sphere, in kW, and the rate at which ice melts in the container. The heat from fusion of water is 333.7 kJ/kg.

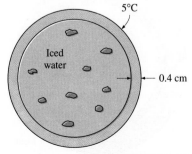

FIGURE P14-33

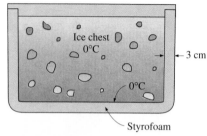

FIGURE P14-37

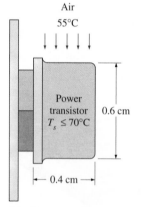

FIGURE P14-38

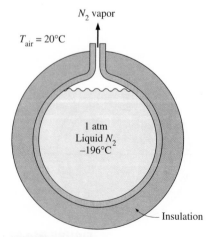

FIGURE P14-40

14-34E The inner and outer glasses of a 6-ft × 6-ft double-pane window are at 60°F and 42°F, respectively. If the 0.5-in. space between the two glasses is filled with still air, determine the rate of heat transfer through the window, in Btu/h. *Answer:* 224 Btu/h

14-35 Two surfaces of a 2-cm-thick plate are maintained at 0°C and 100°C, respectively. If it is determined that heat is transferred through the plate at a rate of 500 W/m², determine its thermal conductivity.

14-36 Four power transistors, each dissipating 15 W, are mounted on a thin vertical aluminum plate 22 cm × 22 cm in size. The heat generated by the transistors is to be dissipated by both surfaces of the plate to the surrounding air at 25°C, which is blown over the plate by a fan. The entire plate can be assumed to be nearly isothermal, and the exposed surface area of the transistor can be taken to be equal to its base area. If the average convection heat transfer coefficient is 25 W/m² · °C, determine the temperature of the aluminum plate. Disregard any radiation effects.

14-37 An ice chest whose outer dimensions are 30 cm × 40 cm × 40 cm is made of 3-cm-thick Styrofoam ($k = 0.033$ W/m · °C). Initially, the chest is filled with 40 kg of ice at 0°C, and the inner surface temperature of the ice chest can be taken to be 0°C at all times. The heat of fusion of ice at 0°C is 333.7 kJ/kg, and the surrounding ambient air is at 30°C. Disregarding any heat transfer from the 40-cm × 40-cm base of the ice chest, determine how long it will take for the ice in the chest to melt completely if the outer surfaces of the ice chest are at 8°C. *Answer:* 32.7 days

14-38 A transistor with a height of 0.4 cm and a diameter of 0.6 cm is mounted on a circuit board. The transistor is cooled by air flowing over it with an average heat transfer coefficient of 30 W/m² · °C. If the air temperature is 55°C and the transistor case temperature is not to exceed 70°C, determine the amount of power this transistor can dissipate safely. Disregard any heat transfer from the transistor base.

14-39E A 200-ft-long section of a steam pipe whose outer diameter is 4 inches passes through an open space at 50°F. The average temperature of the outer surface of the pipe is measured to be 280°F, and the average heat transfer coefficient on that surface is determined to be 6 Btu/h · ft² · °F. Determine (*a*) the rate of heat loss from the steam pipe and (*b*) the annual cost of this energy loss if steam is generated in a natural gas furnace having an efficiency of 86 percent, and the price of natural gas is $0.58/therm (1 therm = 100,000 Btu). *Answers:* (*a*) 289,000 Btu/h, (*b*) $17,074/yr

14-40 The boiling temperature of nitrogen at atmospheric pressure at sea level (1 atm) is −196°C. Therefore, nitrogen is commonly used in low temperature scientific studies since the temperature of liquid nitrogen in a tank open to the atmosphere will remain constant at −196°C until the liquid nitrogen in the tank is depleted. Any heat transfer to the tank will result in the evaporation of some liquid nitrogen, which has a heat of vaporization of 198 kJ/kg and a density of 810 kg/m³ at 1 atm.

Consider a 4-m-diameter spherical tank initially filled with liquid nitrogen at 1 atm and −196°C. The tank is exposed to 20°C ambient air with a heat transfer coefficient of 25 W/m² · °C. The temperature of the thin-shelled spherical tank is observed to be almost the same as the temperature of the nitrogen inside. Disregarding any radiation heat exchange, determine the rate of evaporation of the liquid nitrogen in the tank as a result of the heat transfer from the ambient air.

14-41 Repeat Prob. 14-40 for liquid oxygen, which has a boiling temperature of $-183°C$, a heat of vaporization of 213 kJ/kg, and a density of 1140 kg/m^3 at 1 atm pressure.

14-42 Consider a person whose exposed surface area is 1.7 m^2, emissivity is 0.7, and surface temperature is 32°C. Determine the rate of heat loss from that person by radiation in a large room having walls at a temperature of (*a*) 300 K and (*b*) 280 K. *Answers:* (*a*) 37.4 W, (*b*) 169.2 W

14-43 A 0.3-cm-thick, 12-cm-high, and 18-cm-long circuit board houses 80 closely spaced logic chips on one side, each dissipating 0.04 W. The board is impregnated with copper fillings and has an effective thermal conductivity of 16 W/m · °C. All the heat generated in the chips is conducted across the circuit board and is dissipated from the back side of the board to the ambient air. Determine the temperature difference between the two sides of the circuit board. *Answer:* 0.028°C

14-44 Consider a sealed 20-cm-high electronic box whose base dimensions are 40 cm × 40 cm placed in a vacuum chamber. The emissivity of the outer surface of the box is 0.95. If the electronic components in the box dissipate a total of 100 W of power and the outer surface temperature of the box is not to exceed 55°C, determine the temperature at which the surrounding surfaces must be kept if this box is to be cooled by radiation alone. Assume the heat transfer from the bottom surface of the box to the stand to be negligible.

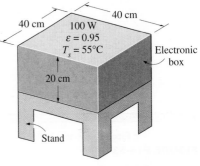

FIGURE P14-44

14-45 Using the conversion factors between W and Btu/h, m and ft, and K and R, express the Stefan-Boltzmann constant $\sigma = 5.67 \times 10^{-8}$ W/m^2 · K^4 in the English unit Btu/h · ft^2 · R^4.

14-46 An engineer who is working on the heat transfer analysis of a house in English units needs the convection heat transfer coefficient on the outer surface of the house. But the only value he can find from his handbooks is 20 W/m^2 · °C, which is in SI units. The engineer does not have a direct conversion factor between the two unit systems for the convection heat transfer coefficient. Using the conversion factors between W and Btu/h, m and ft, and °C and °F, express the given convection heat transfer coefficient in Btu/h · ft^2 · °F. *Answer:* 3.52 Btu/h · ft^2 · °F

Simultaneous Heat Transfer Mechanisms

14-47C Can all three modes of heat transfer occur simultaneously (in parallel) in a medium?

14-48C Can a medium involve (*a*) conduction and convection, (*b*) conduction and radiation, or (*c*) convection and radiation simultaneously? Give examples for the "yes" answers.

14-49C The deep human body temperature of a healthy person remains constant at 37°C while the temperature and the humidity of the environment change with time. Discuss the heat transfer mechanisms between the human body and the environment both in summer and winter, and explain how a person can keep cooler in summer and warmer in winter.

14-50C We often turn the fan on in summer to help us cool. Explain how a fan makes us feel cooler in the summer. Also explain why some people use ceiling fans also in winter.

14-51 Consider a person standing in a room at 23°C. Determine the total rate of heat transfer from this person if the exposed surface area and the skin temperature of the person are 1.7 m² and 32°C, respectively, and the convection heat transfer coefficient is 5 W/m² · °C. Take the emissivity of the skin and the clothes to be 0.9, and assume the temperature of the inner surfaces of the room to be the same as the air temperature. *Answer:* 161.3 W

14-52 Consider steady heat transfer between two large parallel plates at constant temperatures of $T_1 = 290$ K and $T_2 = 150$ K that are $L = 2$ cm apart. Assuming the surfaces to be black (emissivity $\varepsilon = 1$), determine the rate of heat transfer between the plates per unit surface area assuming the gap between the plates is (*a*) filled with atmospheric air, (*b*) evacuated, (*c*) filled with fiberglass insulation, and (*d*) filled with superinsulation having an apparent thermal conductivity of 0.00015 W/m · °C.

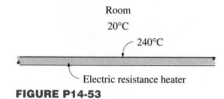

Room
20°C

240°C

Electric resistance heater

FIGURE P14-53

14-53 A 1.4-m-long, 0.2-cm-diameter electrical wire extends across a room that is maintained at 20°C. Heat is generated in the wire as a result of resistance heating, and the surface temperature of the wire is measured to be 240°C in steady operation. Also, the voltage drop and electric current through the wire are measured to be 110 V and 3 A, respectively. Disregarding any heat transfer by radiation, determine the convection heat transfer coefficient for heat transfer between the outer surface of the wire and the air in the room. *Answer:* 170.5 W/m² · °C

14-54E A 2-in-diameter spherical ball whose surface is maintained at a temperature of 170°F is suspended in the middle of a room at 70°F. If the convection heat transfer coefficient is 3 Btu/h · ft² · °F and the emissivity of the surface is 0.8, determine the total rate of heat transfer from the ball.

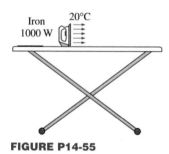

Iron   20°C
1000 W

FIGURE P14-55

14-55 A 1000-W iron is left on the iron board with its base exposed to the air at 20°C. The convection heat transfer coefficient between the base surface and the surrounding air is 35 W/m² · °C. If the base has an emissivity of 0.6 and a surface area of 0.02 m², determine the temperature of the base of the iron. *Answer:* 674°C

14-56 The outer surface of a spacecraft in space has an emissivity of 0.8 and an absorptivity of 0.3 for solar radiation. If solar radiation is incident on the spacecraft at a rate of 1000 W/m², determine the surface temperature of the spacecraft when the radiation emitted equals the solar energy absorbed.

14-57 A 3-m-internal-diameter spherical tank made of 1-cm-thick stainless steel is used to store iced water at 0°C. The tank is located outdoors at 25°C. Assuming the entire steel tank to be at 0°C and thus the thermal resistance of the tank to be negligible, determine (*a*) the rate of heat transfer to the iced water in the tank and (*b*) the amount of ice at 0°C that melts during a 24-hour period. The heat of fusion of water at atmospheric pressure is $h_{if} = 333.7$ kJ/kg. The emissivity of the outer surface of the tank is 0.6, and the convection heat transfer coefficient on the outer surface can be taken to be 30 W/m² · °C. Assume the average surrounding surface temperature for radiation exchange to be 15°C. *Answer:* 5898 kg

14-58 The roof of a house consists of a 15-cm-thick concrete slab ($k = 2$ W/m · °C) that is 15 m wide and 20 m long. The emissivity of the outer surface of the roof is 0.9, and the convection heat transfer coefficient on that surface is estimated to be 15 W/m$^2 \cdot$ °C. The inner surface of the roof is maintained at 15°C. On a clear winter night, the ambient air is reported to be at 10°C while the night sky temperature for radiation heat transfer is 255 K. Considering both radiation and convection heat transfer, determine the outer surface temperature and the rate of heat transfer through the roof.

If the house is heated by a furnace burning natural gas with an efficiency of 85 percent, and the unit cost of natural gas is $0.60/therm (1 therm = 105,500 kJ of energy content), determine the money lost through the roof that night during a 14-hour period.

14-59E Consider a flat plate solar collector placed horizontally on the flat roof of a house. The collector is 5 ft wide and 15 ft long, and the average temperature of the exposed surface of the collector is 100°F. The emissivity of the exposed surface of the collector is 0.9. Determine the rate of heat loss from the collector by convection and radiation during a calm day when the ambient air temperature is 70°F and the effective sky temperature for radiation exchange is 50°F. Take the convection heat transfer coefficient on the exposed surface to be 2.5 Btu/h $\cdot$ ft$^2 \cdot$ °F.

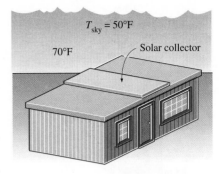

FIGURE P14-59E

Review Problems

14-60 It is well known that wind makes the cold air feel much colder as a result of the *windchill* effect that is due to the increase in the convection heat transfer coefficient with increasing air velocity. The windchill effect is usually expressed in terms of the *windchill factor*, which is the difference between the actual air temperature and the equivalent calm-air temperature. For example, a windchill factor of 20°C for an actual air temperature of 5°C means that the windy air at 5°C feels as cold as the still air at −15°C. In other words, a person will lose as much heat to air at 5°C with a windchill factor of 20°C as he or she would in calm air at −15°C.

For heat transfer purposes, a standing man can be modeled as a 30-cm-diameter, 170-cm-long vertical cylinder with both the top and bottom surfaces insulated and with the side surface at an average temperature of 34°C. For a convection heat transfer coefficient of 15 W/m$^2 \cdot$ °C, determine the rate of heat loss from this man by convection in still air at 20°C. What would your answer be if the convection heat transfer coefficient is increased to 50 W/m$^2 \cdot$ °C as a result of winds? What is the windchill factor in this case? *Answers:* 336 W, 1120 W, 32.7°C

14-61 A thin metal plate is insulated on the back and exposed to solar radiation on the front surface. The exposed surface of the plate has an absorptivity of 0.7 for solar radiation. If solar radiation is incident on the plate at a rate of 800 W/m^2 and the surrounding air temperature is 10°C, determine the surface temperature of the plate when the heat loss by convection equals the solar energy absorbed by the plate. Take the convection heat transfer coefficient to be 30 W/m$^2 \cdot$ °C, and disregard any heat loss by radiation.

14-62 Consider a 3-m × 3-m × 3-m cubical furnace whose top and side surfaces closely approximate black surfaces at a temperature of 1200 K. The base surface has an emissivity of $\varepsilon = 0.7$, and is maintained at 800 K. Determine the net rate of radiation heat transfer to the base surface from the top and side surfaces. *Answer:* 594,400 W

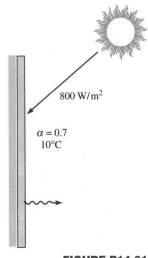

FIGURE P14-61

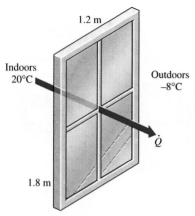

FIGURE P14-63

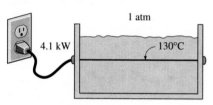

FIGURE P14-65

14-63 The rate of heat loss through a unit surface area of a window per unit temperature between the indoors and the outdoors is called the U-factor. The value of the U-factor ranges from about 1.25 W/m² · °C (or 0.22 Btu/h · ft² · °F) for low-e coated, argon-filled, quadruple-pane windows to 6.25 W/m² · °C (or 1.1 Btu/h · ft² · °F) for a single-pane window with aluminum frames. Determine the range for the rate of heat loss through a 1.2-m × 1.8-m window of a house that is maintained at 20°C when the outdoor air temperature is −8°C.

14-64 Consider a house in Atlanta, Georgia, that is maintained at 22°C and has a total of 20 m² of window area. The windows are double-door type with wood frames and metal spacers and have a U-factor of 2.5 W/m² · °C (see the previous problem for the definition of U-factor). The winter average temperature of Atlanta is 11.3°C. Determine the average rate of heat loss through the windows in winter.

14-65 A 50-cm-long, 2-mm-diameter electric resistance wire submerged in water is used to determine the boiling heat transfer coefficient in water at 1 atm experimentally. The wire temperature is measured to be 130°C when a wattmeter indicates the electric power consumed to be 4.1 kW. Using Newton's law of cooling, determine the boiling heat transfer coefficient.

Computer, Design, and Essay Problems

14-66 Write an essay on how microwave ovens work, and explain how they cook much faster than conventional ovens. Discuss whether conventional electric or microwave ovens consume more electricity for the same task.

14-67 Conduct the following experiment to determine the heat transfer coefficient between an incandescent light bulb and the surrounding air using a 60-W light bulb. You will need an indoor–outdoor thermometer, which can be purchased for about $10 in a hardware store, and a metal glue. You will also need a piece of string and a ruler to calculate the surface area of the light bulb. First, measure the air temperature in the room, and then glue the tip of the thermocouple wire of the thermometer to the glass of the light bulb. Turn the light on and wait until the temperature reading stabilizes. The temperature reading will give the surface temperature of the light bulb. Assuming 10 percent of the rated power of the bulb is converted to light, calculate the heat transfer coefficient from Newton's law of cooling.

Steady Heat Conduction

15

In heat transfer analysis, we are often interested in the rate of heat transfer through a medium under steady conditions and surface temperatures. Such problems can be solved easily without involving any differential equations by the introduction of *thermal resistance concepts* in an analogous manner to electrical circuit problems. In this case, the thermal resistance corresponds to electrical resistance, temperature difference corresponds to voltage, and the heat transfer rate corresponds to electric current.

We start this chapter with *one-dimensional steady heat conduction* in a plane wall, a cylinder, and a sphere, and develop relations for *thermal resistances* in these geometries. We also develop thermal resistance relations for convection and radiation conditions at the boundaries. We apply this concept to heat conduction problems in *multilayer* plane walls, cylinders, and spheres and generalize it to systems that involve heat transfer in two or three dimensions. We also discuss the *thermal contact resistance* and the *overall heat transfer coefficient* and develop relations for the critical radius of insulation for a cylinder and a sphere. We then present extensive discussions on *thermal insulations* because of their importance and widespread use, including the optimum thickness of insulation. Finally, we discuss steady heat transfer from *finned surfaces*.

15-1 ■ STEADY HEAT CONDUCTION IN PLANE WALLS

Consider steady heat conduction through the walls of a house during a winter day. We know that heat is continuously lost to the outdoors through the wall. We intuitively feel that heat transfer through the wall is in the *normal direction* to the wall surface, and no significant heat transfer takes place in the wall in other directions (Fig. 15-1).

Recall that heat transfer in a certain direction is driven by the *temperature gradient* in that direction. There will be no heat transfer in a direction in which there is no change in temperature. Temperature measurements at several locations on the inner or outer wall surface will confirm that a wall surface is nearly *isothermal*. That is, the temperatures at the top and bottom of a wall surface as well as at the right or left ends are almost the same. Therefore, there will be no heat transfer through the wall from the top to the bottom, or from left to right, but there will be considerable temperature difference between the inner and the outer surfaces of the wall, and thus significant heat transfer in the direction from the inner surface to the outer one.

The small thickness of the wall causes the temperature gradient in that direction to be large. Further, if the air temperatures in and outside the house remain constant, then heat transfer through the wall of a house can be modeled as *steady* and *one-dimensional*. The temperature of the wall in this case will depend on one direction only (say the *x*-direction) and can be expressed as $T(x)$.

Noting that heat transfer is the only energy interaction involved in this case and there is no heat generation, the *energy balance* for the wall can be expressed as

$$\begin{pmatrix} \text{Rate of} \\ \text{heat transfer} \\ \text{into the wall} \end{pmatrix} - \begin{pmatrix} \text{Rate of} \\ \text{heat transfer} \\ \text{out of the wall} \end{pmatrix} = \begin{pmatrix} \text{Rate of change} \\ \text{of the energy} \\ \text{of the wall} \end{pmatrix}$$

or

$$\dot{Q}_{\text{in}} - \dot{Q}_{\text{out}} = \frac{dE_{\text{wall}}}{dt} \qquad (15\text{-}1)$$

But $dE_{\text{wall}}/dt = 0$ for *steady* operation, since there is no change in the temperature of the wall with time at any point. Therefore, the rate of heat transfer into the wall must be equal to the rate of heat transfer out of it. In other words, *the rate of heat transfer through the wall must be constant*, $\dot{Q}_{\text{cond, wall}}$ = constant.

Consider a plane wall of thickness L and average thermal conductivity k. The two surfaces of the wall are maintained at constant temperatures of T_1 and T_2. For one-dimensional steady heat conduction through the wall, we have $T(x)$. Then Fourier's law of heat conduction for the wall can be expressed as

$$\dot{Q}_{\text{cond, wall}} = -kA\frac{dT}{dx} \qquad (\text{W}) \qquad (15\text{-}2)$$

where the rate of conduction heat transfer $\dot{Q}_{\text{cond wall}}$ and the surface area A are constant. Thus we have dT/dx = constant, which means that *the temperature through the wall varies linearly with x*. That is, the temperature distribution in the wall under steady conditions is a *straight line* (Fig. 15-2).

Separating the variables in the above equation and integrating from $x = 0$, where $T(0) = T_1$, to $x = L$, where $T(L) = T_2$, we get

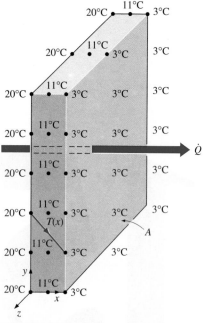

FIGURE 15-1

Heat flow through a wall is one-dimensional when the temperature of the wall varies in one direction only.

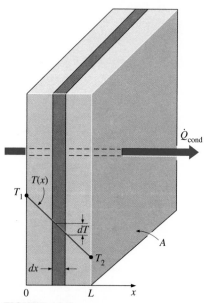

FIGURE 15-2

Under steady conditions, the temperature distribution in a plane wall is a straight line.

$$\int_{x=0}^{L} \dot{Q}_{\text{cond, wall}} \, dx = -\int_{T=T_1}^{T_2} kA \, dT$$

Performing the integrations and rearranging gives

$$\dot{Q}_{\text{cond, wall}} = kA \frac{T_1 - T_2}{L} \qquad \text{(W)} \qquad (15\text{-}3)$$

which is identical to Eq. 14-1. Again, *the rate of heat conduction through a plane wall is proportional to the average thermal conductivity, the wall area, and the temperature difference, but is inversely proportional to the wall thickness.* Also, once the rate of heat conduction is available, the temperature $T(x)$ at any location x can be determined by replacing T_2 in Eq. 15-3 by T, and L by x.

The Thermal Resistance Concept

Equation 15-3 for heat conduction through a plane wall can be rearranged as

$$\dot{Q}_{\text{cond, wall}} = \frac{T_1 - T_2}{R_{\text{wall}}} \qquad \text{(W)} \qquad (15\text{-}4)$$

where

$$R_{\text{wall}} = \frac{L}{kA} \qquad \text{(°C/W)} \qquad (15\text{-}5)$$

is the *thermal resistance* of the wall against heat conduction or simply the **conduction resistance** of the wall. Note that the thermal resistance of a medium depends on the *geometry* and the *thermal properties* of the medium.

The equation above for heat flow is analogous to the relation for *electric current flow I*, expressed as

$$I = \frac{\mathbf{V}_1 - \mathbf{V}_2}{R_e} \qquad (15\text{-}6)$$

where $R_e = L/\sigma_e A$ is the *electric resistance* and $\mathbf{V}_1 - \mathbf{V}_2$ is the *voltage difference* across the resistance (σ_e is the electrical conductivity). Thus, the *rate of heat transfer* through a layer corresponds to the *electric current*, the *thermal resistance* corresponds to *electrical resistance*, and the *temperature difference* corresponds to *voltage difference* across the layer (Fig. 15-3).

Consider convection heat transfer from a solid surface of area A and temperature T_s to a fluid whose temperature sufficiently far from the surface is T_∞, with a convection heat transfer coefficient h. Newton's law of cooling for convection heat transfer rate $\dot{Q}_{\text{conv}} = hA(T_s - T_\infty)$ can be rearranged as

$$\dot{Q}_{\text{conv}} = \frac{T_s - T_\infty}{R_{\text{conv}}} \qquad \text{(W)} \qquad (15\text{-}7)$$

where

$$R_{\text{conv}} = \frac{1}{hA} \qquad \text{(°C/W)} \qquad (15\text{-}8)$$

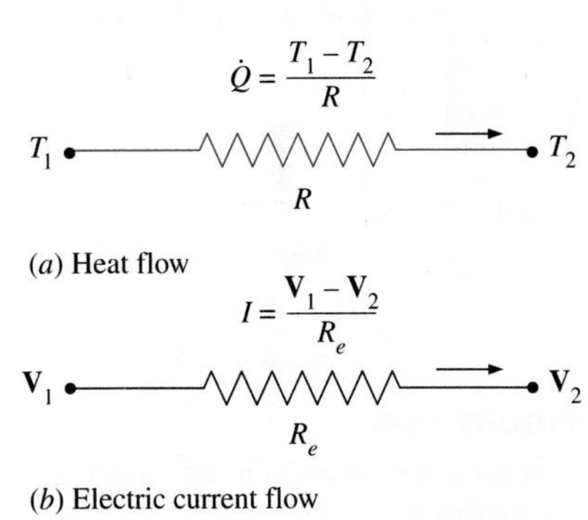

$$\dot{Q} = \frac{T_1 - T_2}{R}$$

$T_1 \bullet\!\!-\!\!\!\wedge\!\!\wedge\!\!\wedge\!\!\wedge\!\!\!-\!\!\longrightarrow\!\bullet T_2$

R

(a) Heat flow

$$I = \frac{\mathbf{V}_1 - \mathbf{V}_2}{R_e}$$

$\mathbf{V}_1 \bullet\!\!-\!\!\!\wedge\!\!\wedge\!\!\wedge\!\!\wedge\!\!\!-\!\!\longrightarrow\!\bullet \mathbf{V}_2$

R_e

(b) Electric current flow

FIGURE 15-3

Analogy between thermal
and electrical resistance concepts.

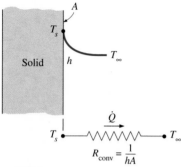

FIGURE 15-4

Schematic for convection
resistance at a surface.

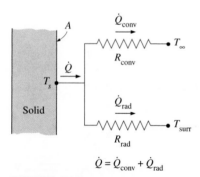

FIGURE 15-5

Schematic for convection and radiation
resistances at a surface.

is the *thermal resistance* of the surface against heat convection, or simply the **convection resistance** of the surface (Fig. 15-4). Note that when the convection heat transfer coefficient is very large ($h \rightarrow \infty$), the convection resistance becomes *zero* and $T_s \approx T_\infty$. That is, the surface offers *no resistance to convection,* and thus it does not slow down the heat transfer process. This situation is approached in practice at surfaces where boiling and condensation occur. Also note that the surface does not have to be a plane surface. Equation 15-8 for convection resistance is valid for surfaces of any shape, provided that the assumption of h = constant and uniform is reasonable.

When the wall is surrounded by a gas, the *radiation effects,* which we have ignored so far, can be significant and may need to be considered. The rate of radiation heat transfer between a surface of emissivity ε and area A at temperature T_s and the surrounding surfaces at some average temperature T_{surr} can be expressed as

$$\dot{Q}_{rad} = \varepsilon\sigma A(T_s^4 - T_{surr}^4) = h_{rad} A(T_s - T_{surr}) = \frac{T_s - T_{surr}}{R_{rad}} \quad \text{(W)} \quad (15\text{-}9)$$

where

$$R_{rad} = \frac{1}{h_{rad} A} \quad \text{(K/W)} \quad (15\text{-}10)$$

is the *thermal resistance* of a surface against radiation, or the *radiation resistance,* and

$$h_{rad} = \frac{\dot{Q}_{rad}}{A(T_s - T_{surr})} = \varepsilon\sigma(T_s^2 + T_{surr}^2)(T_s + T_{surr}) \quad \text{(W/m}^2 \cdot \text{K)} \quad (15\text{-}11)$$

is the **radiation heat transfer coefficient.** Note that both T_s and T_{surr} *must* be in K in the evaluation of h_{rad}. The definition of the radiation heat transfer coefficient enables us to express radiation conveniently in an analogous manner to convection in terms of a temperature difference. But h_{rad} depends strongly on temperature while h_{conv} usually does not.

A surface exposed to the surrounding air involves convection and radiation simultaneously, and the total heat transfer at the surface is determined by adding (or subtracting, if in the opposite direction) the radiation and convection components. The convection and radiation resistances are parallel to each other, as shown in Fig. 15-5, and may cause some complication in the thermal resistance network. When $T_{surr} \approx T_\infty$, the radiation effect can properly be accounted for by replacing h in the convection resistance relation by

$$h_{combined} = h_{conv} + h_{rad} \quad \text{(W/m}^2 \cdot \text{K)} \quad (15\text{-}12)$$

where $h_{combined}$ is the **combined heat transfer coefficient.** This way all the complications associated with radiation are avoided.

Thermal Resistance Network

Now consider steady one-dimensional heat flow through a plane wall of thickness L and thermal conductivity k that is exposed to convection on both sides to fluids at temperatures $T_{\infty 1}$ and $T_{\infty 2}$ with heat transfer coefficients h_1 and h_2, respectively, as shown in Fig. 15-6. Assuming $T_{\infty 2} < T_{\infty 1}$, the variation of temperature will be as shown in the figure. Note that the temperature varies

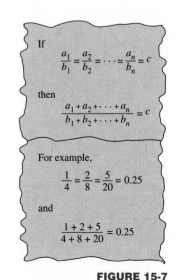

$$\dot{Q} = \frac{T_{\infty 1} - T_{\infty 2}}{R_{conv,\,1} + R_{wall} + R_{conv,\,2}}$$

$$I = \frac{\mathcal{V}_1 - \mathcal{V}_2}{R_{e,\,1} + R_{e,\,2} + R_{e,\,3}}$$

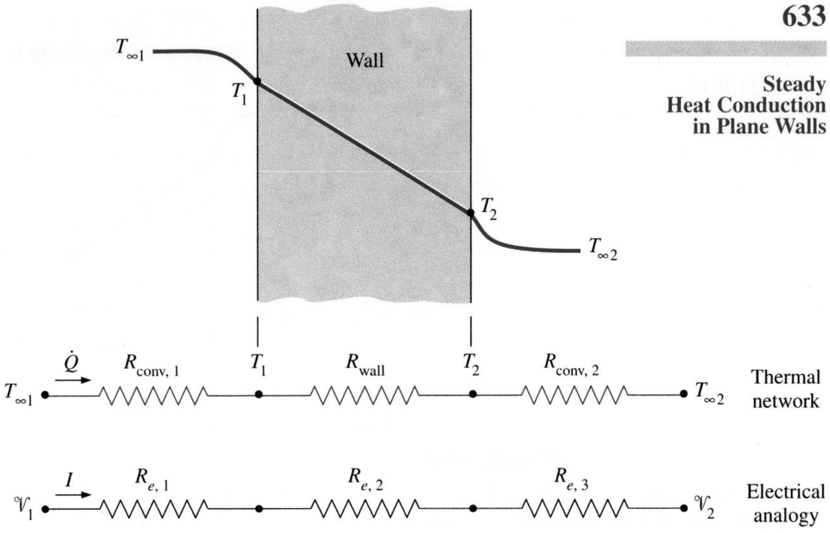

FIGURE 15-6

The thermal resistance network for heat transfer through a plane wall subjected to convection on both sides, and the electrical analogy.

linearly in the wall, and asymptotically approaches $T_{\infty 1}$ and $T_{\infty 2}$ in the fluids as we move away from the wall.

Under steady conditions we have

$$\begin{pmatrix} \text{Rate of} \\ \textit{heat convection} \\ \text{into the wall} \end{pmatrix} = \begin{pmatrix} \text{Rate of} \\ \textit{heat conduction} \\ \text{through the wall} \end{pmatrix} = \begin{pmatrix} \text{Rate of} \\ \textit{heat convection} \\ \text{from the wall} \end{pmatrix}$$

or

$$\dot{Q} = h_1 A(T_{\infty 1} - T_1) = kA\frac{T_1 - T_2}{L} = h_2 A(T_2 - T_{\infty 2}) \tag{15-13}$$

which can be rearranged as

$$\begin{aligned} \dot{Q} &= \frac{T_{\infty 1} - T_1}{1/h_1 A} = \frac{T_1 - T_2}{L/kA} = \frac{T_2 - T_{\infty 2}}{1/h_2 A} \\ &= \frac{T_{\infty 1} - T_1}{R_{conv,\,1}} = \frac{T_1 - T_2}{R_{wall}} = \frac{T_2 - T_{\infty 2}}{R_{conv,\,2}} \end{aligned} \tag{15-14}$$

Adding the numerators and denominators yields (Fig. 15-7)

$$\dot{Q} = \frac{T_{\infty} - T_{\infty 2}}{R_{total}} \quad \text{(W)} \tag{15-15}$$

where

$$R_{total} = R_{conv,\,1} + R_{wall} + R_{conv,\,2} = \frac{1}{h_1 A} + \frac{L}{kA} + \frac{1}{h_2 A} \quad (^\circ\text{C/W}) \tag{15-16}$$

Note that the heat transfer area A is constant for a plane wall, and the rate of heat transfer through a wall separating two mediums is equal to the temperature difference divided by the total thermal resistance between the mediums. Also note that the thermal resistances are in *series,* and the equivalent thermal resistance is determined by simply *adding* the individual resistances, just like

If

$$\frac{a_1}{b_1} = \frac{a_2}{b_2} = \cdots = \frac{a_n}{b_n} = c$$

then

$$\frac{a_1 + a_2 + \cdots + a_n}{b_1 + b_2 + \cdots + b_n} = c$$

For example,

$$\frac{1}{4} = \frac{2}{8} = \frac{5}{20} = 0.25$$

and

$$\frac{1 + 2 + 5}{4 + 8 + 20} = 0.25$$

FIGURE 15-7

A useful mathematical identity.

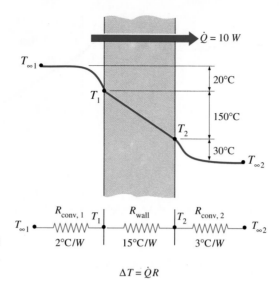

FIGURE 15-8

The temperature drop across a layer is proportional to its thermal resistance.

the electrical resistances connected in series. Thus, the electrical analogy still applies. We summarize this as *the rate of steady heat transfer between two surfaces is equal to the temperature difference divided by the total thermal resistance between those two surfaces.*

Another observation that can be made from Eq. 15-15 is that the ratio of the temperature drop to the thermal resistance across any layer is constant, and thus the temperature drop across any layer is proportional to the thermal resistance of the layer. The larger the resistance, the larger the temperature drop. In fact, the equation $\dot{Q} = \Delta T/R$ can be rearranged as

$$\Delta T = \dot{Q} R \qquad (^\circ C) \qquad (15\text{-}17)$$

which indicates that the *temperature drop* across any layer is equal to the *rate of heat transfer* times the *thermal resistance* across that layer (Fig. 15-8). You may recall that this is also true for voltage drop across an electrical resistance when the electric current is constant.

It is sometimes convenient to express heat transfer through a medium in an analogous manner to Newton's law of cooling as

$$\dot{Q} = UA\,\Delta T \qquad (W) \qquad (15\text{-}18)$$

where U is the **overall heat transfer coefficient.** A comparison of Eqs. 15-15 and 15-18 reveals that

$$UA = \frac{1}{R_{\text{total}}} \qquad (15\text{-}19)$$

Therefore, for a unit area, the overall heat transfer coefficient is equal to the inverse of the total thermal resistance.

Note that we do not need to know the surface temperatures of the wall in order to evaluate the rate of steady heat transfer through it. All we need to know is the convection heat transfer coefficients and the fluid temperatures on both sides of the wall. The *surface temperature* of the wall can be determined as described above using the thermal resistance concept, but by taking the

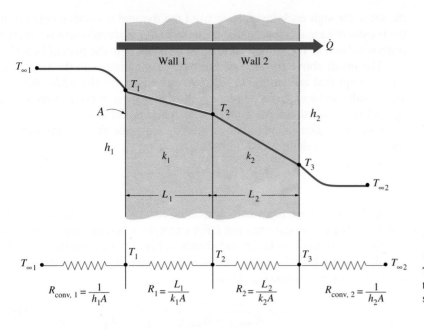

FIGURE 15-9
The thermal resistance network for heat transfer through a two-layer plane wall subjected to convection on both sides.

surface at which the temperature is to be determined as one of the terminal surfaces. For example, once $\dot{Q}$ is evaluated, the surface temperature T_1 can be determined from

$$\dot{Q} = \frac{T_{\infty 1} - T_1}{R_{\text{conv, 1}}} = \frac{T_{\infty 1} - T_1}{1/h_1 A} \tag{15-20}$$

Multilayer Plane Walls

In practice we often encounter plane walls that consist of several layers of different materials. The thermal resistance concept can still be used to determine the rate of steady heat transfer through such *composite* walls. As you may have already guessed, this is done by simply noting that the conduction resistance of each wall is *L/kA* connected in series, and using the electrical analogy, that is, by dividing the *temperature difference* between two surfaces at known temperatures by the *total thermal resistance* between them.

Consider a plane wall that consists of two layers (such as a brick wall with a layer of insulation). The rate of steady heat transfer through this two-layer composite wall can be expressed as (Fig. 15-9)

$$\dot{Q} = \frac{T_{\infty 1} - T_{\infty 2}}{R_{\text{total}}} \tag{15-21}$$

where R_{total} is the *total thermal resistance,* expressed as

$$R_{\text{total}} = R_{\text{conv, 1}} + R_{\text{wall, 1}} + R_{\text{wall, 2}} + R_{\text{conv, 2}}$$
$$= \frac{1}{h_1 A} + \frac{L_1}{k_1 A} + \frac{L_2}{k_2 A} + \frac{1}{h_2 A} \tag{15-22}$$

The subscripts 1 and 2 in the R_{wall} relations above indicate the first and the second layers, respectively. We could also obtain this result by following the approach used above for the single-layer case by noting that the rate of steady heat transfer $\dot{Q}$ through a multilayer medium is constant, and thus it must be

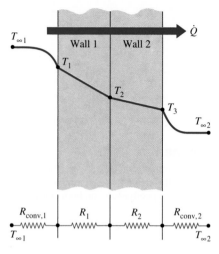

$$\text{To find } T_1: \quad \dot{Q} = \frac{T_{\infty1} - T_1}{R_{conv,1}}$$

$$\text{To find } T_2: \quad \dot{Q} = \frac{T_{\infty1} - T_2}{R_{conv,1} + R_1}$$

$$\text{To find } T_3: \quad \dot{Q} = \frac{T_3 - T_{\infty2}}{R_{conv,2}}$$

FIGURE 15-10

The evaluation of the surface and interface temperatures when $T_{\infty1}$ and $T_{\infty2}$ are given and $\dot{Q}$ is calculated.

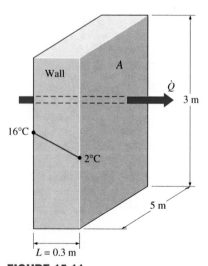

FIGURE 15-11

Schematic for Example 15-1.

the same through each layer. Note from the thermal resistance network that the resistances are *in series,* and thus the *total thermal resistance* is simply the *arithmetic sum* of the individual thermal resistances in the path of heat flow.

The result above for the *two-layer* case is analogous to the *single-layer* case, except that an *additional resistance* is added for the *additional layer.* This result can be extended to plane walls that consist of *three* or *more* layers by adding an *additional resistance* for each *additional layer.*

Once $\dot{Q}$ is *known,* an unknown surface temperature T_j at any surface or interface j can be determined from

$$\dot{Q} = \frac{T_i - T_j}{R_{\text{total},\, i-j}} \tag{15-23}$$

where T_i is a *known* temperature at location i and $R_{\text{total},\, i-j}$ is the total thermal resistance between locations i and j. For example, when the fluid temperatures $T_{\infty1}$ and $T_{\infty2}$ for the two-layer case shown in Fig. 15-9 are available and $\dot{Q}$ is calculated from Eq. 15-23, the interface temperature T_2 between the two walls can be determined from (Fig. 15-10)

$$\dot{Q} = \frac{T_{\infty1} - T_2}{R_{conv,\,1} + R_{wall,\,1}} = \frac{T_{\infty1} - T_2}{\dfrac{1}{h_1 A} + \dfrac{L_1}{k_1 A}} \tag{15-24}$$

The temperature drop across a layer is easily determined from Eq. 15-17 by multiplying $\dot{Q}$ by the thermal resistance of that layer.

The thermal resistance concept is widely used in practice because it is intuitively easy to understand and it has proven to be a powerful tool in the solution of a wide range of heat transfer problems. But its use is limited to systems through which the rate of heat transfer $\dot{Q}$ remains *constant*; that is, to systems involving *steady* heat transfer with *no heat generation* (such as resistance heating or chemical reactions) within the medium.

EXAMPLE 15-1   Heat Loss through a Wall

Consider a 3-m-high, 5-m-wide, and 0.3-m-thick wall whose thermal conductivity is $k = 0.9$ W/m · °C (Fig. 15-11). On a certain day, the temperatures of the inner and the outer surfaces of the wall are measured to be 16°C and 2°C, respectively. Determine the rate of heat loss through the wall on that day.

Solution   The two surfaces of a wall are maintained at specified temperatures. The rate of heat loss through the wall is to be determined.

Assumptions   **1** Heat transfer through the wall is steady since the surface temperatures remain constant at the specified values. **2** Heat transfer is one-dimensional since any significant temperature gradients will exist in the direction from the indoors to the outdoors. **3** Thermal conductivity is constant.

Properties   The thermal conductivity is given to be $k = 0.9$ W/m · °C.

Analysis   Noting that the heat transfer through the wall is by conduction and the surface area of the wall is $A = 3$ m $\times$ 5 m $= 15$ m², the steady rate of heat transfer through the wall can be determined from Eq. 15-3 to be

$$\dot{Q} = kA\frac{T_1 - T_2}{L} = (0.9 \text{ W/m} \cdot {}^\circ\text{C})(15 \text{ m}^2)\frac{(16 - 2){}^\circ\text{C}}{0.3 \text{ m}} = \textbf{630 W}$$

We could also determine the steady rate of heat transfer through the wall by making use of the thermal resistance concept from

$$\dot{Q} = \frac{\Delta T_{wall}}{R_{wall}}$$

where

$$R_{wall} = \frac{L}{kA} = \frac{0.3 \text{ m}}{(0.9 \text{ W/m} \cdot °C)(15 \text{ m}^2)} = 0.02222°C/W$$

Substituting, we get

$$\dot{Q} = \frac{(16 - 2)°C}{0.02222°C/W} = 630 \text{ W}$$

Discussion This is the same result obtained earlier. Note that heat conduction through a plane wall with specified surface temperatures can be determined directly and easily without utilizing the thermal resistance concept. However, the thermal resistance concept serves as a valuable tool in more complex heat transfer problems, as you will see in the following examples.

EXAMPLE 15-2 Heat Loss through a Single-Pane Window

Consider a 0.8-m-high and 1.5-m-wide glass window with a thickness of 8 mm and a thermal conductivity of $k = 0.78$ W/m · °C. Determine the steady rate of heat transfer through this glass window and the temperature of its inner surface for a day during which the room is maintained at 20°C while the temperature of the outdoors is −10°C. Take the heat transfer coefficients on the inner and outer surfaces of the window to be $h_1 = 10$ W/m² · °C and $h_2 = 40$ W/m² · °C, which includes the effects of radiation.

Solution Heat loss through a window glass is considered. The rate of heat transfer through the window and the inner surface temperature are to be determined.

Assumptions **1** Heat transfer through the window is steady since the surface temperatures remain constant at the specified values. **2** Heat transfer through the wall is one-dimensional since any significant temperature gradients will exist in the direction from the indoors to the outdoors. **3** Thermal conductivity is constant. **4** Heat transfer by radiation is negligible.

Properties The thermal conductivity is given to be $k = 0.78$ W/m · °C.

Analysis This problem involves conduction through the glass window and convection at its surfaces, and can best be handled by making use of the thermal resistance concept and drawing the thermal resistance network, as shown in Fig. 15-12. Noting that the surface area of the window is $A = 0.8 \text{ m} \times 1.5 \text{ m} = 1.2 \text{ m}^2$, the individual resistances are evaluated from their definitions to be

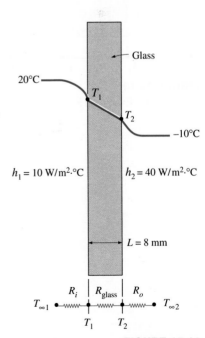

$$R_i = R_{conv, 1} = \frac{1}{h_1 A} = \frac{1}{(10 \text{ W/m}^2 \cdot °C)(1.2 \text{ m}^2)} = 0.08333°C/W$$

$$R_{glass} = \frac{L}{kA} = \frac{0.008 \text{ m}}{(0.78 \text{ W/m} \cdot °C)(1.2 \text{ m}^2)} = 0.00855°C/W$$

$$R_o = R_{conv, 2} = \frac{1}{h_2 A} = \frac{1}{(40 \text{ W/m}^2 \cdot °C)(1.2 \text{ m}^2)} = 0.02083°C/W$$

Noting that all three resistances are in series, the total resistance is determined to be

$$R_{total} = R_{conv, 1} + R_{glass} + R_{conv, 2} = 0.08333 + 0.00855 + 0.02083$$
$$= 0.1127°C/W$$

Then the steady rate of heat transfer through the window becomes

$$\dot{Q} = \frac{T_{\infty 1} - T_{\infty 2}}{R_{total}} = \frac{[20 - (-10)]°C}{0.1127°C/W} = \mathbf{266 \text{ W}}$$

FIGURE 15-12
Schematic for Example 15-2.

Knowing the rate of heat transfer, the inner surface temperature of the window glass can be determined from

$$\dot{Q} = \frac{T_{\infty 1} - T_1}{R_{conv, 1}} \quad \longrightarrow \quad T_1 = T_{\infty 1} - \dot{Q}R_{conv, 1}$$
$$= 20°C - (266 \text{ W})(0.08333°C/W)$$
$$= -2.2°C$$

Discussion   Note that the inner surface temperature of the window glass will be $-2.2°C$ even though the temperature of the air in the room is maintained at 20°C. Such low surface temperatures are highly undesirable since they cause the formation of fog or even frost on the inner surfaces of the glass when the humidity in the room is high.

EXAMPLE 15-3   Heat Loss through Double-Pane Windows

Consider a 0.8-m-high and 1.5-m-wide double-pane window consisting of two 4-mm-thick layers of glass ($k = 0.78$ W/m · °C) separated by a 10-mm-wide stagnant air space ($k = 0.026$ W/m · °C). Determine the steady rate of heat transfer through this double-pane window and the temperature of its inner surface for a day during which the room is maintained at 20°C while the temperature of the outdoors is $-10°C$. Take the convection heat transfer coefficients on the inner and outer surfaces of the window to be $h_1 = 10$ W/m^2 · °C and $h_2 = 40$ W/m^2 · °C, which includes the effects of radiation.

Solution   A double-pane window is considered. The rate of heat transfer through the window and the inner surface temperature are to be determined.

Analysis   This example problem is identical to the previous one except that the single 8-mm-thick window glass is replaced by two 4-mm-thick glasses that enclose a 10-mm-wide stagnant air space. Therefore, the thermal resistance network of this problem will involve two additional conduction resistances corresponding to the two additional layers, as shown in Fig. 15-13. Noting that the surface area of the window is again $A = 0.8$ m $\times$ 1.5 m = 1.2 m^2, the individual resistances are evaluated from their definitions to be

$$R_i = R_{conv, 1} = \frac{1}{h_1 A} = \frac{1}{(10 \text{ W/m}^2 \cdot °C)(1.2 \text{ m}^2)} = 0.08333°C/W$$

$$R_1 = R_3 = R_{glass} = \frac{L_1}{k_1 A} = \frac{0.004 \text{ m}}{(0.78 \text{ W/m} \cdot °C)(1.2 \text{ m}^2)} = 0.00427°C/W$$

$$R_2 = R_{air} = \frac{L_2}{k_2 A} = \frac{0.01 \text{ m}}{(0.026 \text{ W/m} \cdot °C)(1.2 \text{ m}^2)} = 0.3205°C/W$$

$$R_o = R_{conv, 2} = \frac{1}{h_2 A} = \frac{1}{(40 \text{ W/m}^2 \cdot °C)(1.2 \text{ m}^2)} = 0.02083°C/W$$

Noting that all three resistances are in series, the total resistance is

$$R_{total} = R_{conv, 1} + R_{glass, 1} + R_{air} + R_{glass, 2} + R_{conv, 2}$$
$$= 0.08333 + 0.00427 + 0.3205 + 0.00427 + 0.02083$$
$$= 0.4332°C/W$$

Then the steady rate of heat transfer through the window becomes

$$\dot{Q} = \frac{T_{\infty 1} - T_{\infty 2}}{R_{total}} = \frac{[20 - (-10)]°C}{0.4332°C/W} = \textbf{69.2 W}$$

which is about one-fourth of the result obtained in the previous example. This explains the popularity of the double- and even triple-pane windows in cold climates. The drastic reduction in the heat transfer rate in this case is due to

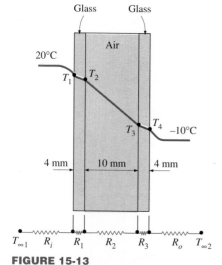

FIGURE 15-13
Schematic for Example 15-3.

the large thermal resistance of the air layer between the glasses. In reality, the thermal resistance of the air layer will be somewhat lower because of the natural convection currents that are likely to occur in the air space.

The inner surface temperature of the window in this case will be

$$T_1 = T_{\infty 1} - \dot{Q}R_{\text{conv}, 1} = 20°C - (69.2 \text{ W})(0.08333°C/W) = \textbf{14.2°C}$$

which is considerably higher than the $-2.2°C$ obtained in the previous example. Therefore, a double-pane window will rarely get fogged. A double-pane window will also reduce the heat gain in summer, and thus reduce the air-conditioning costs.

15-2 ■ THERMAL CONTACT RESISTANCE

In the analysis of heat conduction through multilayer solids, we assumed "perfect contact" at the interface of two layers, and thus no temperature drop at the interface. This would be the case when the surfaces are perfectly smooth and they produce a perfect contact at each point. In reality, however, even flat surfaces that appear smooth to the eye turn out to be rather rough when examined under a microscope, as shown in Fig. 15-14, with numerous peaks and valleys. That is, a surface is *microscopically rough* no matter how smooth it appears to be.

When two such surfaces are pressed against each other, the peaks will form good material contact but the valleys will form voids filled with air. As a result, an interface will contain numerous *air gaps* of varying sizes that act as *insulation* because of the low thermal conductivity of air. Thus, an interface offers some resistance to heat transfer, and this resistance per unit interface area is called the **thermal contact resistance,** R_c. The value of R_c is determined experimentally using a setup like the one shown in Fig. 15-15, and as expected, there is considerable scatter of data because of the difficulty in characterizing the surfaces.

Consider heat transfer through two metal rods of cross-sectional area A that are pressed against each other. Heat transfer through the interface of these

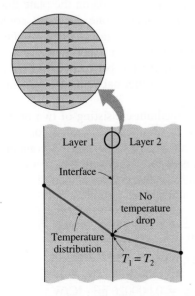

(a) Ideal (perfect) thermal contact

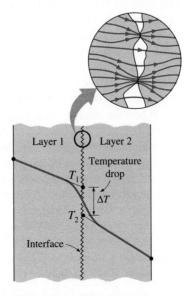

(b) Actual (imperfect) thermal contact

FIGURE 15-14

Temperature distribution and heat flow lines along two solid plates pressed against each other for the case of perfect and imperfect contact.

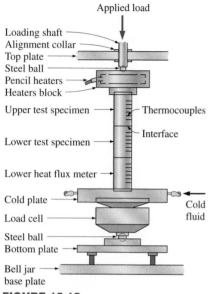

Applied load

Loading shaft
Alignment collar
Top plate
Steel ball
Pencil heaters
Heaters block

Upper test specimen — Thermocouples

— Interface

Lower test specimen

Lower heat flux meter

Cold plate
Cold
fluid
Load cell

Steel ball
Bottom plate

Bell jar
base plate

FIGURE 15-15

A typical experimental setup for
the determination of thermal contact
resistance (from Song et al., Ref. 20).

two rods is the sum of the heat transfers through the *solid contact spots* and
the *gaps* in the noncontact areas and can be expressed as

$$\dot{Q} = \dot{Q}_{\text{contact}} + \dot{Q}_{\text{gap}} \qquad (15\text{-}25)$$

It can also be expressed in an analogous manner to Newton's law of cooling as

$$\dot{Q} = h_c A \, \Delta T_{\text{interface}} \qquad (15\text{-}26)$$

where A is the apparent interface area (which is the same as the cross-sectional
area of the rods) and $\Delta T_{\text{interface}}$ is the effective temperature difference at the
interface. The quantity h_c, which corresponds to the convection heat transfer
coefficient, is called the **thermal contact conductance** and is expressed as

$$h_c = \frac{\dot{Q}/A}{\Delta T_{\text{interface}}} \qquad (\text{W/m}^2 \cdot {}^\circ\text{C}) \qquad (15\text{-}27)$$

It is related to thermal contact resistance by

$$R_c = \frac{1}{h_c} = \frac{\Delta T_{\text{interface}}}{\dot{Q}/A} \qquad (\text{m}^2 \cdot {}^\circ\text{C/W}) \qquad (15\text{-}28)$$

That is, thermal contact resistance is the inverse of thermal contact conductance. Usually, thermal contact conductance is reported in the literature, but
the concept of thermal contact resistance serves as a better vehicle for explaining the effect of interface on heat transfer. Note that R_c represents thermal contact resistance *per unit area*. The thermal resistance for the entire
interface is obtained by dividing R_c by the apparent interface area A.

The thermal contact resistance can be determined from Eq. 15-28 by
measuring the temperature drop at the interface and dividing it by the heat
flux under steady conditions. The value of thermal contact resistance depends
on the *surface roughness* and the *material properties* as well as the *temperature* and *pressure* at the interface and the *type of fluid* trapped at the
interface. The situation becomes more complex when plates are fastened by
bolts, screws, or rivets since the interface pressure in this case is nonuniform.
The thermal contact resistance in that case also depends on the plate thickness, the bolt radius, and the size of the contact zone. Thermal contact
resistance is observed to *decrease* with *decreasing surface roughness*
and *increasing interface pressure,* as expected. Most experimentally determined values of the thermal contact resistance fall between 0.000005 and
0.0005 m² · °C/W (the corresponding range of thermal contact conductance
is 2000 to 200,000 W/m² · °C).

When we analyze heat transfer in a medium consisting of two or more
layers, the first thing we need to know is whether the thermal contact resistance is *significant* or not. We can answer this question by comparing
the magnitudes of the thermal resistances of the layers with typical values of
thermal contact resistance. For example, the thermal resistance of a 1-cm-
thick layer of an insulating material per unit surface area is

$$R_{c,\,\text{insulation}} = \frac{L}{k} = \frac{0.01 \text{ m}}{0.04 \text{ W/m} \cdot {}^\circ\text{C}} = 0.25 \text{ m}^2 \cdot {}^\circ\text{C/W}$$

whereas for a 1-cm-thick layer of copper, it is

$$R_{c,\,\text{copper}} = \frac{L}{k} = \frac{0.01 \text{ m}}{386 \text{ W/m} \cdot {}^\circ\text{C}} = 0.000026 \text{ m}^2 \cdot {}^\circ\text{C/W}$$

Comparing the values above with typical values of thermal contact resistance, we conclude that thermal contact resistance is significant and can even dominate the heat transfer for good heat conductors such as metals, but can be disregarded for poor heat conductors such as insulations. This is not surprising since insulating materials consist mostly of air space just like the interface itself.

The thermal contact resistance can be minimized by applying a thermally conducting liquid called a *thermal grease* such as silicon oil on the surfaces before they are pressed against each other. This is commonly done when attaching electronic components such as power transistors to heat sinks. The thermal contact resistance can also be reduced by replacing the air at the interface by a *better conducting gas* such as helium or hydrogen, as shown in Table 15-1.

Another way to minimize the contact resistance is to insert a *soft metallic foil* such as tin, silver, copper, nickel, or aluminum between the two surfaces. Experimental studies show that the thermal contact resistance can be reduced by a factor of up to 7 by a metallic foil at the interface. For maximum effectiveness, the foils must be very thin. The effect of metallic coatings on thermal contact conductance is shown in Fig. 15-16 for various metal surfaces.

There is considerable uncertainty in the contact conductance data reported in the literature, and care should be exercised when using them. In Table 15-2 some experimental results are given for the contact conductance between similar and dissimilar metal surfaces for use in preliminary design calculations. Note that the *thermal contact conductance* is *highest* (and thus the contact resistance is lowest) for *soft metals* with *smooth surfaces* at *high pressure*.

EXAMPLE 15-4   Equivalent Thickness for Contact Resistance

The thermal contact conductance at the interface of two 1-cm-thick aluminum plates is measured to be 11,000 W/m² · °C. Determine the thickness of the aluminum plate whose thermal resistance is equal to the thermal resistance of the interface between the plates (Fig. 15-17).

Solution   The thickness of aluminum plate whose thermal resistance is equal to the thermal contact resistance is to be determined.

Properties   The thermal conductivity of aluminum is $k = 237$ W/m · °C (Table A-19).

Analysis   Noting that thermal contact resistance is the inverse of thermal contact conductance, the thermal contact resistance is determined to be

$$R_c = \frac{1}{h_c} = \frac{1}{11{,}000 \text{ W/m}^2 \cdot {}^{\circ}\text{C}} = 0.909 \times 10^{-4} \text{ m}^2 \cdot {}^{\circ}\text{C/W}$$

For a unit surface area, the thermal resistance of a flat plate is defined as

$$R = \frac{L}{k}$$

where L is the thickness of the plate and k is the thermal conductivity. Setting $R = R_c$, the equivalent thickness is determined from the relation above to be

$$L = kR_c = (237 \text{ W/m} \cdot {}^{\circ}\text{C})(0.909 \times 10^{-4} \text{ m}^2 \cdot {}^{\circ}\text{C/W}) = 0.0215 \text{ m} = \textbf{2.15 cm}$$

Discussion   Note that the interface between the two plates offers as much resistance to heat transfer as a 2.15-cm-thick aluminum plate. It is interesting that the thermal contact resistance in this case is greater than the sum of the thermal resistances of both plates.

TABLE 15-1

Thermal contact conductance for aluminum plates with different fluids at the interface for a surface roughness of 10 μm and interface pressure of 1 atm (from Fried, Ref. 8)

Fluid at the interface	Contact conductance, h_c, W/m² · °C
Air	3640
Helium	9520
Hydrogen	13,900
Silicone oil	19,000
Glycerin	37,700

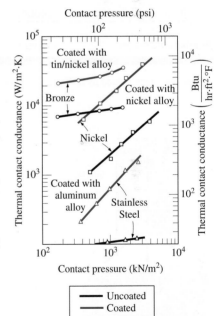

FIGURE 15-16

Effect of metallic coatings on thermal contact conductance (from Peterson, Ref. 18).

TABLE 15-2

Thermal contact conductance of some metal surfaces in air (from Holman, Ref. 10, and Kreith and Bohn, Ref. 14)

Material	Surface condition	Rough- ness, μm	Tempera- ture, °C	Pressure, MPa	h_c* W/m² · °C
Identical Metal Pairs					
416 Stainless steel	Ground	2.54	90–200	0.3–2.5	3800
304 Stainless steel	Ground	1.14	20	4–7	1900
Aluminum	Ground	2.54	150	1.2–2.5	11,400
Copper	Ground	1.27	20	1.2–20	143,000
Copper	Milled	3.81	20	1–5	55,500
Copper (vacuum)	Milled	0.25	30	0.7–7	11,400
Dissimilar Metal Pairs					
Stainless steel–				10	2900
Aluminum		20–30	20	20	3600
Stainless steel–				10	16,400
Aluminum		1.0–2.0	20	20	20,800
Steel Ct-30–				10	50,000
Aluminum	Ground	1.4–2.0	20	15–35	59,000
Steel Ct-30–				10	4800
Aluminum	Milled	4.5–7.2	20	30	8300
Aluminum-Copper	Ground	1.3–1.4	20	5	42,000
				15	56,000
Aluminum-Copper	Milled	4.4–4.5	20	10	12,000
				20–35	22,000

*Divide the given values by 5.678 to convert to Btu/h · ft² · °F.

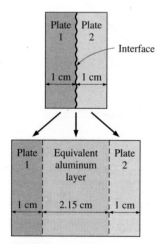

FIGURE 15-17
Schematic for Example 15-4.

EXAMPLE 15-5 Contact Resistance of Transistors

Four identical power transistors with aluminum casing are attached on one side of a 1-cm-thick 20-cm × 20-cm square copper plate (k = 386 W/m · °C) by screws that exert an average pressure of 6 MPa (Fig. 15-18). The base area of each transistor is 8 cm², and each transistor is placed at the center of a 10-cm × 10-cm quarter section of the plate. The interface roughness is estimated to be about 1.5 μm. All transistors are covered by a thick Plexiglas layer, which is a poor conductor of heat, and thus all the heat generated at the junction of the transistor must be dissipated to the ambient at 20°C through the back surface of the copper plate. The combined convection/radiation heat transfer coefficient at the back surface can be taken to be 25 W/m² · °C. If the case temperature of the transistor is not to exceed 70°C, determine the maximum power each transistor can dissipate safely, and the temperature jump at the case-plate interface.

Solution Four identical power transistors are attached on a copper plate. For a maximum case temperature of 70°C, the maximum power dissipation and the temperature jump at the interface are to be determined.

Assumptions **1** Steady operating conditions exist. **2** Heat transfer can be approximated as being one-dimensional, although it is recognized that heat conduction in some parts of the plate will be two-dimensional since the plate area is much larger than the base area of the transistor. But the large thermal conductivity of copper will minimize this effect. **3** All the heat generated at the junction is dissipated through the back surface of the plate since the transistors are covered by a thick Plexiglas layer. **4** Thermal conductivities are constant.

Properties The thermal conductivity of copper is given to be $k = 386$ W/m · °C. The contact conductance is obtained from Table 15-2 to be $h_c = 42{,}000$ W/m² · °C, which corresponds to copper-aluminum interface for the case of 1.3–1.4 μm roughness and 5 MPa pressure, which is sufficiently close to what we have.

Analysis The contact area between the case and the plate is given to be 8 cm², and the plate area for each transistor is 100 cm². The thermal resistance network of this problem consists of three resistances in series (interface, plate, and convection), which are determined to be

$$R_{interface} = \frac{1}{h_c A_c} = \frac{1}{(42{,}000 \text{ W/m}^2 \cdot °\text{C})(8 \times 10^{-4} \text{ m}^2)} = 0.030°\text{C/W}$$

$$R_{plate} = \frac{L}{kA} = \frac{0.01 \text{ m}}{(386 \text{ W/m} \cdot °\text{C})(0.01 \text{ m}^2)} = 0.0026°\text{C/W}$$

$$R_{conv} = \frac{1}{h_o A} = \frac{1}{(25 \text{ W/m}^2 \cdot °\text{C})(0.01 \text{ m}^2)} = 4.0°\text{C/W}$$

The total thermal resistance is then

$$R_{total} = R_{interface} + R_{plate} + R_{ambient} = 0.030 + 0.0026 + 4.0 = 4.0326°\text{C/W}$$

Note that the thermal resistance of a copper plate is very small and can be ignored altogether. Then the rate of heat transfer is determined to be

$$\dot{Q} = \frac{\Delta T}{R_{total}} = \frac{(70 - 20)°\text{C}}{4.0326°\text{C/W}} = \mathbf{12.4 \text{ W}}$$

Therefore, the power transistor should not be operated at power levels greater than 12.4 W if the case temperature is not to exceed 70°C.

The temperature jump at the interface is determined from

$$\Delta T_{interface} = \dot{Q} R_{interface} = (12.4 \text{ W})(0.030°\text{C/W}) = \mathbf{0.37°C}$$

which is not very large. Therefore, even if we eliminate the thermal contact resistance at the interface completely, we will lower the operating temperature of the transistor in this case by less than 0.4°C.

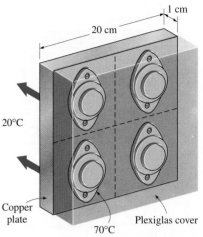

FIGURE 15-18

Schematic for Example 15-5.

15-3 ■ GENERALIZED THERMAL RESISTANCE NETWORKS

The *thermal resistance* concept or the *electrical analogy* can also be used to solve steady heat transfer problems that involve parallel layers or combined series-parallel arrangements. Although such problems are often two- or even three-dimensional, approximate solutions can be obtained by assuming one-dimensional heat transfer and using the thermal resistance network.

Consider the composite wall shown in Fig. 15-19, which consists of two parallel layers. The thermal resistance network, which consists of two parallel resistances, can be represented as shown in the figure. Noting that the total heat transfer is the sum of the heat transfers through each layer, we have

$$\dot{Q} = \dot{Q}_1 + \dot{Q}_2 = \frac{T_1 - T_2}{R_1} + \frac{T_1 - T_2}{R_2} = (T_1 - T_2)\left(\frac{1}{R_1} + \frac{1}{R_2}\right) \quad (15\text{-}29)$$

Utilizing electrical analogy, we get

$$\dot{Q} = \frac{T_1 - T_2}{R_{total}} \quad (15\text{-}30)$$

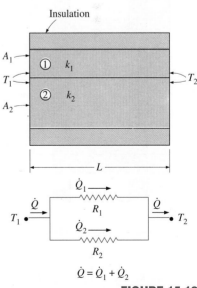

$$\dot{Q} = \dot{Q}_1 + \dot{Q}_2$$

FIGURE 15-19

Thermal resistance network for two parallel layers.

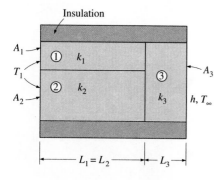

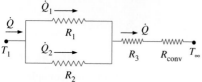

FIGURE 15-20

Thermal resistance network for combined series-parallel arrangement.

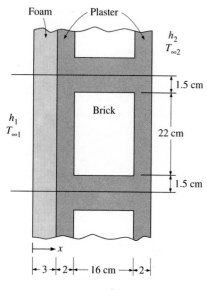

FIGURE 15-21

Schematic for Example 15-6.

where

$$\frac{1}{R_{total}} + \frac{1}{R_1} + \frac{1}{R_2} \longrightarrow R_{total} = \frac{R_1 R_2}{R_1 + R_2} \qquad (15\text{-}31)$$

since the resistances are in parallel.

Now consider the combined series-parallel arrangement shown in Fig. 15-20. The total rate of heat transfer through this composite system can again be expressed as

$$\dot{Q} = \frac{T_1 - T_\infty}{R_{total}} \qquad (15\text{-}32)$$

where

$$R_{total} = R_{12} + R_3 + R_{conv} = \frac{R_1 R_2}{R_1 + R_2} + R_3 + R_{conv} \qquad (15\text{-}33)$$

and

$$R_1 = \frac{L_1}{k_1 A_1}, \qquad R_2 = \frac{L_2}{k_2 A_2}, \qquad R_3 = \frac{L_3}{k_3 A_3}, \qquad R_{conv} = \frac{1}{h A_3} \qquad (15\text{-}34)$$

Once the individual thermal resistances are evaluated, the total resistance and the total rate of heat transfer can easily be determined from the relations above.

The result obtained will be somewhat approximate, since the surfaces of the third layer will probably not be isothermal, and heat transfer between the first two layers is likely to occur.

Two assumptions commonly used in solving complex multidimensional heat transfer problems by treating them as one-dimensional (say, in the x-direction) using the thermal resistance network are (1) any plane wall normal to the x-axis is *isothermal* (i.e., to assume the temperature to vary in the x-direction only) and (2) any plane parallel to the x-axis is *adiabatic* (i.e., to assume heat transfer to occur in the x-direction only). These two assumptions result in different resistance networks, and thus different (but usually close) values for the total thermal resistance and thus heat transfer. The actual result lies between these two values. In geometries in which heat transfer occurs predominantly in one direction, either approach gives satisfactory results.

EXAMPLE 15-6 Heat Loss through a Composite Wall

A 3-m-high and 5-m-wide wall consists of long 16-cm × 22-cm cross-section horizontal bricks ($k = 0.72$ W/m · °C) separated by 3-cm-thick plaster layers ($k = 0.22$ W/m · °C). There are also 2-cm-thick plaster layers on each side of the brick and a 3-cm-thick rigid foam ($k = 0.026$ W/m · °C) on the inner side of the wall, as shown in Fig. 15-21. The indoor and the outdoor temperatures are 20°C and −10°C, and the convection heat transfer coefficients on the inner and the outer sides are $h_1 = 10$ W/m² · °C and $h_2 = 25$ W/m² · °C, respectively. Assuming one-dimensional heat transfer and disregarding radiation, determine the rate of heat transfer through the wall.

Solution The composition of a composite wall is given. The rate of heat transfer through the wall is to be determined.

Assumptions **1** Heat transfer is steady since there is no indication of change with time. **2** Heat transfer can be approximated as being one-dimensional since it is predominantly in the x-direction. **3** Thermal conductivities are constant. **4** Heat transfer by radiation is negligible.

Properties The thermal conductivities are given to be $k = 0.72$ W/m · °C for bricks, $k = 0.22$ W/m · °C for plaster layers, and $k = 0.026$ W/m · °C for the rigid foam.

Analysis There is a pattern in the construction of this wall that repeats itself every 25-cm distance in the vertical direction. There is no variation in the horizontal direction. Therefore, we consider a 1-m-deep and 0.25-m-high portion of the wall, since it is representative of the entire wall.

Assuming any cross-section of the wall normal to the *x*-direction to be *isothermal*, the thermal resistance network for the representative section of the wall becomes as shown in Fig. 15-21. The individual resistances are evaluated as follows:

$$R_i = R_{conv,\, 1} = \frac{1}{h_1 A} = \frac{1}{(10 \text{ W/m}^2 \cdot \text{°C})(0.25 \times 1 \text{ m}^2)} = 0.4\text{°C/W}$$

$$R_1 = R_{foam} = \frac{L}{kA} = \frac{0.03 \text{ m}}{(0.026 \text{ W/m} \cdot \text{°C})(0.25 \times 1 \text{ m}^2)} = 4.6\text{°C/W}$$

$$R_2 = R_6 = R_{plaster,\, side} = \frac{L}{kA} = \frac{0.02 \text{ m}}{(0.22 \text{ W/m} \cdot \text{°C})(0.25 \times 1 \text{ m}^2)}$$
$$= 0.36\text{°C/W}$$

$$R_3 = R_5 = R_{plaster,\, center} = \frac{L}{kA} = \frac{0.16 \text{ m}}{(0.22 \text{ W/m} \cdot \text{°C})(0.015 \times 1 \text{ m}^2)}$$
$$= 48.48\text{°C/W}$$

$$R_4 = R_{brick} = \frac{L}{kA} = \frac{0.16 \text{ m}}{(0.72 \text{ W/m} \cdot \text{°C})(0.22 \times 1 \text{ m}^2)} = 1.01\text{°C/W}$$

$$R_o = R_{conv,\, 2} = \frac{1}{h_2 A} = \frac{1}{(25 \text{ W/m}^2 \cdot \text{°C})(0.25 \times 1 \text{ m}^2)} = 0.16\text{°C/W}$$

The three resistances R_3, R_4, and R_5 in the middle are parallel, and their equivalent resistance is determined from

$$\frac{1}{R_{mid}} = \frac{1}{R_3} + \frac{1}{R_4} + \frac{1}{R_5} = \frac{1}{48.48} + \frac{1}{1.01} + \frac{1}{48.48} = 1.03 \text{ W/°C}$$

which gives

$$R_{mid} = 0.97\text{°C/W}$$

Now all the resistances are in series, and the total resistance is determined to be

$$R_{total} = R_i + R_1 + R_2 + R_{mid} + R_6 + R_o$$
$$= 0.4 + 4.6 + 0.36 + 0.97 + 0.36 + 0.16$$
$$= 6.85\text{°C/W}$$

Then the steady rate of heat transfer through the wall becomes

$$\dot{Q} = \frac{T_{\infty 1} - T_{\infty 2}}{R_{total}} = \frac{[20 - (-10)]\text{°C}}{6.85\text{°C/W}} = 4.38 \text{ W} \quad \text{(per 0.25 m}^2 \text{ surface area)}$$

or $4.38/0.25 = 17.5$ W per m^2 surface area. The total surface area of the wall is $A = 3 \text{ m} \times 5 \text{ m} = 15 \text{ m}^2$. Then the rate of heat transfer through the entire wall becomes

$$\dot{Q}_{total} = (17.5 \text{ W/m}^2)(15 \text{ m}^2) = \textbf{262.5 W}$$

Of course, this result is approximate, since we assumed the temperature within the wall to vary in one direction only and ignored any temperature change (and thus heat transfer) in the other two directions.

Discussion In the above solution, we assumed the temperature at any cross-section of the wall normal to the *x*-direction to be *isothermal*. We could also solve

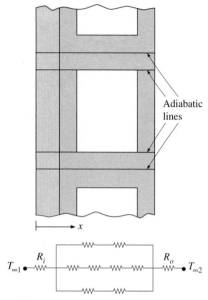

Adiabatic
lines

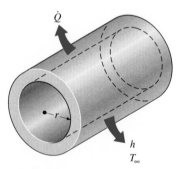

FIGURE 15-22

Alternative thermal resistance network for
Example 15-6 for the case of surfaces
parallel to the primary direction of heat
transfer being adiabatic.

FIGURE 15-23

Heat is lost from a hot water pipe to the air
outside in the radial direction, and thus
heat transfer from a long pipe is very
nearly one-dimensional.

this problem by going to the other extreme and assuming the surfaces parallel to
the x-direction to be *adiabatic*. The thermal resistance network in this case will be
as shown in Fig. 15-22. By following the approach outlined above, the total thermal resistance in this case is determined to be R_{total} = 6.97°C/W, which is very
close to the value 6.85°C/W obtained before. Thus either approach would give
roughly the same result in this case. This example demonstrates that either approach can be used in practice to obtain satisfactory results.

15-4 ■ HEAT CONDUCTION IN CYLINDERS AND SPHERES

Consider steady heat conduction through a hot water pipe. Heat is continuously lost to the outdoors through the wall of the pipe, and we intuitively feel
that heat transfer through the pipe is in the normal direction to the wall surface
and no significant heat transfer takes place in the pipe in other directions
(Fig. 15-23). The wall of the pipe, whose thickness is rather small, separates
two fluids at different temperatures, and thus the temperature gradient in the
radial direction will be relatively large. Further, if the fluid temperatures inside and outside the pipe remain constant, then heat transfer through the pipe
can be modeled as *steady*. Thus heat transfer through the pipe can be modeled
as *steady* and *one-dimensional*. The temperature of the pipe in this case will
depend on one direction only (the radial r-direction) and can be expressed as
$T = T(r)$. The temperature is independent of the azimuthal angle or the axial
distance. This situation is approximated in practice in long cylindrical pipes
and spherical containers.

In *steady* operation, there is no change in the temperature of the pipe with
time at any point. Therefore, the rate of heat transfer into the pipe must be
equal to the rate of heat transfer out of it. In other words, heat transfer through
the pipe must be constant, $\dot{Q}_{cond, cyl}$ = constant.

Consider a long cylindrical layer (such as a circular pipe) of inner radius
r_1, outer radius r_2, length L, and average thermal conductivity k (Fig. 15-24).
The two surfaces of the cylindrical layer are maintained at constant temperatures T_1 and T_2. There is no heat generation in the layer and the thermal
conductivity is constant. For one-dimensional heat conduction through the
cylindrical layer, we have $T(r)$. Then Fourier's law of heat conduction for heat
transfer through the cylindrical layer can be expressed as

$$\dot{Q}_{cond, cyl} = -kA\frac{dT}{dr} \quad \text{(W)} \quad (15\text{-}35)$$

where $A = 2\pi rL$ is the heat transfer surface area at location r. Note that A depends on r, and thus it *varies* in the direction of heat transfer. Separating the
variables in the above equation and integrating from $r = r_1$, where $T(r_1) = T_1$,
to $r = r_2$, where $T(r_2) = T_2$, gives

$$\int_{r=r_1}^{r_2} \frac{\dot{Q}_{cond, cyl}}{A} dr = -\int_{T=T_1}^{T_2} k \, dT \quad (15\text{-}36)$$

Substituting $A = 2\pi rL$ and performing the integrations give

$$\dot{Q}_{cond, cyl} = 2\pi Lk\frac{T_1 - T_2}{\ln(r_2/r_1)} \quad \text{(W)} \quad (15\text{-}37)$$

since $\dot{Q}_{cond, cyl}$ = constant. This equation can be rearranged as

$$\dot{Q}_{\text{cond, cyl}} = \frac{T_1 - T_2}{R_{\text{cyl}}} \qquad \text{(W)} \qquad \text{(15-38)}$$

where

$$R_{\text{cyl}} = \frac{\ln(r_2/r_1)}{2\pi L k} = \frac{\ln(\text{Outer radius/Inner radius})}{2\pi \times (\text{Length}) \times (\text{Thermal conductivity})} \qquad \text{(15-39)}$$

is the *thermal resistance* of the cylindrical layer against heat conduction, or simply the **conduction resistance** of the cylinder layer.

We can repeat the analysis above for a *spherical layer* by taking $A = 4\pi r^2$ and performing the integrations in Eq. 15-36. The result can be expressed as

$$\dot{Q}_{\text{cond, sph}} = \frac{T_1 - T_2}{R_{\text{sph}}} \qquad \text{(15-40)}$$

where

$$R_{\text{sph}} = \frac{r_2 - r_1}{4\pi r_1 r_2 k}$$
$$= \frac{\text{Outer radius} - \text{Inner radius}}{4\pi(\text{Outer radius})(\text{Inner radius})(\text{Thermal conductivity})} \qquad \text{(15-41)}$$

is the *thermal resistance* of the spherical layer against heat conduction, or simply the **conduction resistance** of the spherical layer.

Now consider steady one-dimensional heat flow through a cylindrical or spherical layer that is exposed to convection on both sides to fluids at temperatures $T_{\infty 1}$ and $T_{\infty 2}$ with heat transfer coefficients h_1 and h_2, respectively, as shown in Fig. 15-25. The thermal resistance network in this case consists of one conduction and two convection resistances in series, just like the one for the plane wall, and the rate of heat transfer under steady conditions can be expressed as

$$\dot{Q} = \frac{T_{\infty 1} - T_{\infty 2}}{R_{\text{total}}} \qquad \text{(15-42)}$$

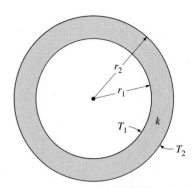

FIGURE 15-24

A long cylindrical pipe (or spherical shell)
with specified inner and outer surface
temperatures T_1 and T_2.

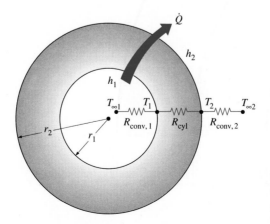

$$R_{\text{total}} = R_{\text{conv,1}} + R_{\text{cyl}} + R_{\text{conv,2}}$$

FIGURE 15-25

The thermal resistance network for a
cylindrical (or spherical) shell subjected
to convection from both the inner and the
outer sides.

where

$$R_{total} = R_{conv,\,1} + R_{cyl} + R_{conv,\,2}$$
$$= \frac{1}{(2\pi r_1 L)h_1} + \frac{\ln(r_2/r_1)}{2\pi L k} + \frac{1}{(2\pi r_2 L)h_2} \tag{15-43}$$

for a *cylindrical* layer, and

$$R_{total} = R_{conv,\,1} + R_{sph} + R_{conv,\,2}$$
$$= \frac{1}{(4\pi r_1^2)h_1} + \frac{r_2 - r_1}{4\pi r_1 r_2 k} + \frac{1}{(4\pi r_2^2)h_2} \tag{15-44}$$

for a *spherical* layer. Note that A in the convection resistance relation $R_{conv} = 1/hA$ is the *surface area at which convection occurs*. It is equal to $A = 2\pi r L$ for a cylindrical surface and $A = 4\pi r^2$ for a spherical surface of radius r. Also note that the thermal resistances are in series, and thus the total thermal resistance is determined by simply adding the individual resistances, just like the electrical resistances connected in series.

Multilayered Cylinders and Spheres

Steady heat transfer through multilayered cylindrical or spherical shells can be handled just like multilayered plane walls discussed earlier by simply adding an *additional resistance* in series for each *additional layer*. For example, the steady heat transfer rate through the three-layered composite cylinder of length L shown in Fig. 15-26 with convection on both sides can be expressed as

$$\dot{Q} = \frac{T_{\infty 1} - T_{\infty 2}}{R_{total}} \tag{15-45}$$

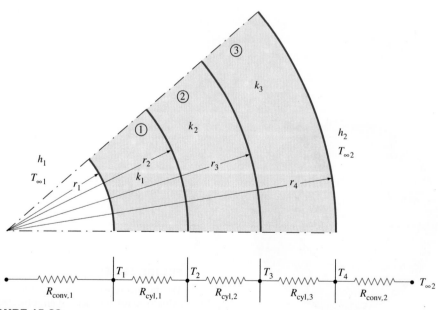

FIGURE 15-26

The thermal resistance network for heat transfer through a three-layered composite cylinder subjected to convection on both sides.

where R_{total} is the *total thermal resistance*, expressed as

$$R_{total} = R_{conv,\,1} + R_{cyl,\,1} + R_{cyl,\,2} + R_{cyl,\,3} + R_{conv,\,2}$$
$$= \frac{1}{h_1 A_1} + \frac{\ln(r_2/r_1)}{2\pi L k_1} + \frac{\ln(r_3/r_2)}{2\pi L k_2} + \frac{\ln(r_4/r_3)}{2\pi L k_3} + \frac{1}{h_2 A_4} \qquad (15\text{-}46)$$

where $A_1 = 2\pi r_1 L$ and $A_4 = 2\pi r_4 L$. Equation 15-46 can also be used for a three-layered spherical shell by replacing the thermal resistances of cylindrical layers by the corresponding spherical ones. Again, note from the thermal resistance network that the resistances are in series, and thus the total thermal resistance is simply the *arithmetic sum* of the individual thermal resistances in the path of heat flow.

Once $\dot{Q}$ is known, we can determine any intermediate temperature T_j by applying the relation $\dot{Q} = (T_i - T_j)/R_{total,\,i-j}$ across any layer or layers such that T_i is a *known* temperature at location i and $R_{total,\,i-j}$ is the total thermal resistance between locations i and j (Fig. 15-27). For example, once $\dot{Q}$ has been calculated, the interface temperature T_2 between the first and second cylindrical layers can be determined from

$$\dot{Q} = \frac{T_{\infty 1} - T_2}{R_{conv,\,1} + R_{cyl,\,1}} = \frac{T_{\infty 1} - T_2}{\dfrac{1}{h_1(2\pi r_1 L)} + \dfrac{\ln(r_2/r_1)}{2\pi L k_1}} \qquad (15\text{-}47)$$

We could also calculate T_2 from

$$\dot{Q} = \frac{T_2 - T_{\infty 2}}{R_2 + R_3 + R_{conv,\,2}} = \frac{T_2 - T_{\infty 2}}{\dfrac{\ln(r_3/r_2)}{2\pi L k_2} + \dfrac{\ln(r_4/r_3)}{2\pi L k_3} + \dfrac{1}{h_o(2\pi r_4 L)}} \qquad (15\text{-}48)$$

Although both relations will give the same result, we prefer the first one since it involves fewer terms and thus less work.

The thermal resistance concept can also be used for *other geometries*, provided that the proper conduction resistances and the proper surface areas in convection resistances are used.

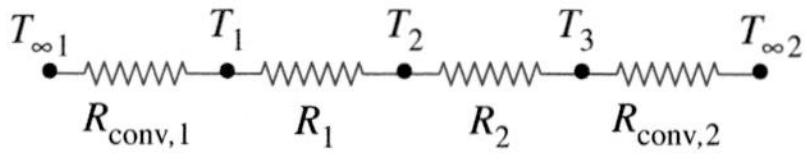

$$\dot{Q} = \frac{T_{\infty 1} - T_1}{R_{conv,\,1}}$$
$$= \frac{T_{\infty 1} - T_2}{R_{conv,\,1} + R_1}$$
$$= \frac{T_1 - T_3}{R_1 + R_2}$$
$$= \frac{T_2 - T_3}{R_2}$$
$$= \frac{T_2 - T_{\infty 2}}{R_2 + R_{conv,\,2}}$$
$$= \cdots$$

FIGURE 15-27

The ratio $\Delta T/R$ across any layer is equal to $\dot{Q}$, which remains constant in one-dimensional steady conduction.

EXAMPLE 15-7 Heat Transfer to a Spherical Container

A 3-m internal diameter spherical tank made of 2-cm-thick stainless steel ($k = 15$ W/m · °C) is used to store iced water at $T_{\infty 1} = 0$°C. The tank is located in a room whose temperature is $T_{\infty 2} = 22$°C. The walls of the room are also at 22°C. The outer surface of the tank is black and heat transfer between the outer surface of the tank and the surroundings is by natural convection and radiation. The convection heat transfer coefficients at the inner and the outer surfaces of the tank are $h_1 = 80$ W/m² · °C and $h_2 = 10$ W/m² · °C, respectively. Determine (a) the rate of heat transfer to the iced water in the tank and (b) the amount of ice at 0°C that melts during a 24-h period.

Solution A spherical container filled with iced water is subjected to convection and radiation heat transfer at its outer surface. The rate of heat transfer and the amount of ice that melts per day are to be determined.

Assumptions **1** Heat transfer is steady since the specified thermal conditions at the boundaries do not change with time. **2** Heat transfer is one-dimensional since there is thermal symmetry about the midpoint. **3** Thermal conductivity is constant.

Properties The thermal conductivity of steel is given to be $k = 15$ W/m · °C. The heat of fusion of water at atmospheric pressure is $h_{if} = 333.7$ kJ/kg. The outer surface of the tank is black and thus its emissivity is $\varepsilon = 1$.

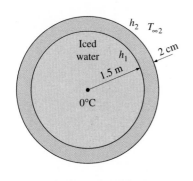

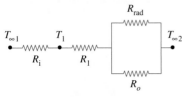

FIGURE 15-28

Schematic for Example 15-7.

Analysis (*a*) The thermal resistance network for this problem is given in Fig. 15-28. Noting that the inner diameter of the tank is $D_1 = 3$ m and the outer diameter is $D_2 = 3.04$ m, the inner and the outer surface areas of the tank are

$$A_1 = \pi D_1^2 = \pi(3 \text{ m})^2 = 28.3 \text{ m}^2$$
$$A_2 = \pi D_2^2 = \pi(3.04 \text{ m})^2 = 29.0 \text{ m}^2$$

Also, the radiation heat transfer coefficient is given by

$$h_{rad} = \varepsilon\sigma(T_2^2 + T_{\infty 2}^2)(T_2 + T_{\infty 2})$$

But we do not know the outer surface temperature T_2 of the tank, and thus we cannot calculate h_{rad}. Therefore, we need to assume a T_2 value now and check the accuracy of this assumption later. We will repeat the calculations if necessary using a revised value for T_2.

We note that T_2 must be between 0°C and 22°C, but it must be closer to 0°C, since the heat transfer coefficient inside the tank is much larger. Taking $T_2 = 5°C = 278$ K, the radiation heat transfer coefficient is determined to be

$$h_{rad} = (1)(5.67 \times 10^{-8} \text{ W/m}^2 \cdot \text{K}^4)[(295 \text{ K})^2 + (278 \text{ K})^2][(295 + 278) \text{ K}]$$
$$= 5.34 \text{ W/m}^2 \cdot \text{K} = 5.34 \text{ W/m}^2 \cdot °C$$

Then the individual thermal resistances become

$$R_i = R_{conv, 1} = \frac{1}{h_1 A_1} = \frac{1}{(80 \text{ W/m}^2 \cdot °C)(28.3 \text{ m}^2)} = 0.000442°C/W$$

$$R_1 = R_{sphere} = \frac{r_2 - r_1}{4\pi k r_1 r_2} = \frac{(1.52 - 1.50) \text{ m}}{4\pi (15 \text{ W/m} \cdot °C)(1.52 \text{ m})(1.50 \text{ m})}$$
$$= 0.000047°C/W$$

$$R_o = R_{conv, 2} = \frac{1}{h_2 A_2} = \frac{1}{(10 \text{ W/m}^2 \cdot °C)(29.0 \text{ m}^2)} = 0.00345°C/W$$

$$R_{rad} = \frac{1}{h_{rad} A_2} = \frac{1}{(5.34 \text{ W/m}^2 \cdot °C)(29.0 \text{ m}^2)} = 0.00646°C/W$$

The two parallel resistances R_o and R_{rad} can be replaced by an equivalent resistance R_{equiv} determined from

$$\frac{1}{R_{equiv}} = \frac{1}{R_o} + \frac{1}{R_{rad}} = \frac{1}{0.00345} + \frac{1}{0.00646} = 444.7 \text{ W/°C}$$

which gives

$$R_{equiv} = 0.00225°C/W$$

Now all the resistances are in series, and the total resistance is determined to be

$$R_{total} = R_i + R_1 + R_{equiv} = 0.000442 + 0.000047 + 0.00225 = 0.00274°C/W$$

Then the steady rate of heat transfer to the iced water becomes

$$\dot{Q} = \frac{T_{\infty 2} - T_{\infty 1}}{R_{total}} = \frac{(22 - 0)°C}{0.00274°C/W} = \textbf{8029 W} \qquad (\text{or } \dot{Q} = 8.027 \text{ kJ/s})$$

To check the validity of our original assumption, we now determine the outer surface temperature from

$$\dot{Q} = \frac{T_{\infty 2} - T_2}{R_{equiv}} \longrightarrow T_2 = T_{\infty 2} - \dot{Q}R_{equiv}$$
$$= 22°C - (8029 \text{ W})(0.00225°C/W) = 4°C$$

which is sufficiently close to the 5°C assumed in the determination of the radiation heat transfer coefficient. Therefore, there is no need to repeat the calculations using 4°C for T_2.

(b) The total amount of heat transfer during a 24-h period is

$$Q = \dot{Q} \, \Delta t = (8.029 \text{ kJ/s})(24 \times 3600 \text{ s}) = 673{,}700 \text{ kJ}$$

Noting that it takes 333.7 kJ of energy to melt 1 kg of ice at 0°C, the amount of ice which will melt during a 24-h period is

$$m_{ice} = \frac{Q}{h_{if}} = \frac{673{,}700 \text{ kJ}}{333.7 \text{ kJ/kg}} = \mathbf{2079 \text{ kg}}$$

Therefore, about 2 metric tons of ice will melt in the tank every day.

Discussion An easier way to deal with combined convection and radiation at a surface when the surrounding medium and surfaces are at the same temperature is to add the radiation and convection heat transfer coefficients and to treat the result as the convection heat transfer coefficient. That is, to take $h = 10 + 5.34 = 15.34 \text{ W/m}^2 \cdot °C$ in this case. This way, we can ignore radiation since its contribution is accounted for in the convection heat transfer coefficient. The convection resistance of the outer surface in this case would be

$$R_{combined} = \frac{1}{h_{combined} A_2} = \frac{1}{(15.34 \text{ W/m}^2 \cdot °C)(29.0 \text{ m}^2)} = 0.00225°C/W$$

which is identical to the value obtained for the equivalent resistance for the parallel convection and the radiation resistances.

EXAMPLE 15-8 Heat Loss through an Insulated Steam Pipe

Steam at $T_{\infty 1} = 320°C$ flows in a cast iron pipe ($k = 80 \text{ W/m} \cdot °C$) whose inner and outer diameters are $D_1 = 5$ cm and $D_2 = 5.5$ cm, respectively. The pipe is covered with 3-cm-thick glass wool insulation with $k = 0.05 \text{ W/m} \cdot °C$. Heat is lost to the surroundings at $T_{\infty 2} = 5°C$ by natural convection and radiation, with a combined heat transfer coefficient of $h_2 = 18 \text{ W/m}^2 \cdot °C$. Taking the heat transfer coefficient inside the pipe to be $h_1 = 60 \text{ W/m}^2 \cdot °C$, determine the rate of heat loss from the steam per unit length of the pipe. Also determine the temperature drops across the pipe shell and the insulation.

Solution A steam pipe covered with glass wool insulation is subjected to convection on its surfaces. The rate of heat transfer per unit length and the temperature drops across the pipe and the insulation are to be determined.

Assumptions **1** Heat transfer is steady since there is no indication of any change with time. **2** Heat transfer is one-dimensional since there is thermal symmetry about the centerline and no variation in the axial direction. **3** Thermal conductivities are constant. **4** The thermal contact resistance at the interface is negligible.

Properties The thermal conductivities are given to be $k = 80 \text{ W/m} \cdot °C$ for cast iron and $k = 0.05 \text{ W/m} \cdot °C$ for glass wool insulation.

Analysis The thermal resistance network for this problem involves four resistances in series and is given in Fig. 15-29. Taking $L = 1$ m, the areas of the surfaces exposed to convection are determined to be

$$A_1 = 2\pi r_1 L = 2\pi(0.025 \text{ m})(1 \text{ m}) = 0.157 \text{ m}^2$$
$$A_3 = 2\pi r_3 L = 2\pi(0.0575 \text{ m})(1 \text{ m}) = 0.361 \text{ m}^2$$

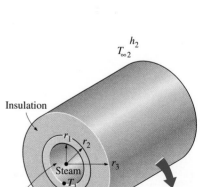

FIGURE 15-29

Schematic for Example 15-8.

Then the individual thermal resistances become

$$R_i = R_{\text{conv, 1}} = \frac{1}{h_1 A} = \frac{1}{(60 \text{ W/m}^2 \cdot {}^\circ\text{C})(0.157 \text{ m}^2)} = 0.106{}^\circ\text{C/W}$$

$$R_1 = R_{\text{pipe}} = \frac{\ln(r_2/r_1)}{2\pi k_1 L} = \frac{\ln(2.75/2.5)}{2\pi(80 \text{ W/m} \cdot {}^\circ\text{C})(1 \text{ m})} = 0.0002{}^\circ\text{C/W}$$

$$R_2 = R_{\text{insulation}} = \frac{\ln(r_3/r_2)}{2\pi k_2 L} = \frac{\ln(5.75/2.75)}{2\pi(0.05 \text{ W/m} \cdot {}^\circ\text{C})(1 \text{ m})} = 2.35{}^\circ\text{C/W}$$

$$R_o = R_{\text{conv, 2}} = \frac{1}{h_2 A_3} = \frac{1}{(18 \text{ W/m}^2 \cdot {}^\circ\text{C})(0.361 \text{ m}^2)} = 0.154{}^\circ\text{C/W}$$

Noting that all resistances are in series, the total resistance is determined to be

$$R_{\text{total}} = R_i + R_1 + R_2 + R_o = 0.106 + 0.0002 + 2.35 + 0.154 = 2.61{}^\circ\text{C/W}$$

Then the steady rate of heat loss from the steam becomes

$$\dot{Q} = \frac{T_{\infty 1} - T_{\infty 2}}{R_{\text{total}}} = \frac{(320 - 5){}^\circ\text{C}}{2.61{}^\circ\text{C/W}} = \textbf{120.7 W} \qquad \text{(per m pipe length)}$$

The heat loss for a given pipe length can be determined by multiplying the above quantity by the pipe length L.

The temperature drops across the pipe and the insulation are determined from Eq. 15-17 to be

$$\Delta T_{\text{pipe}} = \dot{Q} R_{\text{pipe}} = (120.7 \text{ W})(0.0002{}^\circ\text{C/W}) = \textbf{0.02}{}^\circ\textbf{C}$$

$$\Delta T_{\text{insulation}} = \dot{Q} R_{\text{insulation}} = (120.7 \text{ W})(2.35{}^\circ\text{C/W}) = \textbf{284}{}^\circ\textbf{C}$$

That is, the temperatures between the inner and the outer surfaces of the pipe differ by 0.02°C, whereas the temperatures between the inner and the outer surfaces of the insulation differ by 284°C.

Discussion Note that the thermal resistance of the pipe is too small relative to the other resistances and can be neglected without causing any significant error. Also note that the temperature drop across the pipe is practically zero, and thus the pipe can be assumed to be isothermal. The resistance to heat flow in insulated pipes is primarily due to the insulation.

15-5 ■ CRITICAL RADIUS OF INSULATION

We know that adding more insulation to a wall or to the attic always decreases heat transfer. The thicker the insulation, the lower the heat transfer rate. This is expected, since the heat transfer area A is constant, and adding insulation always increases the thermal resistance of the wall without affecting the convection resistance.

Adding insulation to a cylindrical pipe or a spherical shell, however, is a different matter. The additional insulation increases the conduction resistance of the insulation layer but decreases the convection resistance of the surface because of the increase in the outer surface area for convection. The heat transfer from the pipe may increase or decrease, depending on which effect dominates.

Consider a cylindrical pipe of outer radius r_1 whose outer surface temperature T_1 is maintained constant (Fig. 15-30). The pipe is now insulated with a material whose thermal conductivity is k and outer radius is r_2. Heat is lost from the pipe to the surrounding medium at temperature T_∞, with a convection

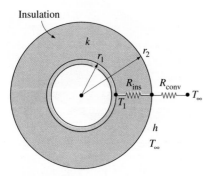

FIGURE 15-30
An insulated cylindrical pipe exposed to convection from the outer surface and the thermal resistance network associated with it.

heat transfer coefficient h. The rate of heat transfer from the insulated pipe to the surrounding air can be expressed as (Fig. 15-31)

$$\dot{Q} = \frac{T_1 - T_\infty}{R_{ins} + R_{conv}} = \frac{T_1 - T_\infty}{\dfrac{\ln(r_2/r_1)}{2\pi Lk} + \dfrac{1}{h(2\pi r_2 L)}} \qquad (15\text{-}49)$$

The variation of $\dot{Q}$ with the outer radius of the insulation r_2 is plotted in Fig. 15-31. The value of r_2 at which $\dot{Q}$ reaches a maximum is determined from the requirement that $d\dot{Q}/dr_2 = 0$ (zero slope). Performing the differentiation and solving for r_2 yields the **critical radius of insulation** for a cylindrical body to be

$$r_{cr, \text{cylinder}} = \frac{k}{h} \qquad \text{(m)} \qquad (15\text{-}50)$$

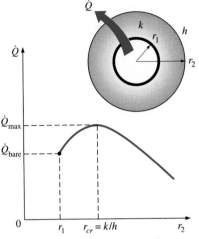

FIGURE 15-31

Note that the critical radius of insulation depends on the thermal conductivity of the insulation k and the external convection heat transfer coefficient h. The rate of heat transfer from the cylinder increases with the addition of insulation for $r_2 < r_{cr}$, reaches a maximum when $r_2 = r_{cr}$, and starts to decrease for $r_2 > r_{cr}$. Thus, insulating the pipe may actually increase the rate of heat transfer from the pipe instead of decreasing it when $r_2 < r_{cr}$.

The important question to answer at this point is whether we need to be concerned about the critical radius of insulation when insulating hot water pipes or even hot water tanks. Should we always check and make sure that the outer radius of insulation exceeds the critical radius before we install any insulation? Probably not, as explained below.

The value of the critical radius r_{cr} will be the largest when k is large and h is small. Noting that the lowest value of h encountered in practice is about 5 W/m$^2 \cdot$ °C for the case of natural convection of gases, and that the thermal conductivity of common insulating materials is about 0.05 W/m$^2 \cdot$ °C, the largest value of the critical radius we are likely to encounter is

$$r_{cr, \text{max}} = \frac{k_{\text{max, insulation}}}{h_{\text{min}}} \approx \frac{0.05 \text{ W/m} \cdot \text{°C}}{5 \text{ W/m}^2 \cdot \text{°C}} = 0.01 \text{ m} = 1 \text{ cm}$$

This value would be even smaller when the radiation effects are considered. The critical radius would be much less in forced convection, often less than 1 mm, because of much larger h values associated with forced convection. Therefore, we can insulate hot water or steam pipes freely without worrying about the possibility of increasing the heat transfer by insulating the pipes.

The radius of electric wires may be smaller than the critical radius. Therefore, the plastic electrical insulation may actually *enhance* the heat transfer from electric wires and thus keep their steady operating temperatures at lower and thus safer levels.

The discussions above can be repeated for a sphere, and it can be shown in a similar manner that the critical radius of insulation for a spherical shell is

$$r_{cr, \text{sphere}} = \frac{2k}{h} \qquad (15\text{-}51)$$

where k is the thermal conductivity of the insulation and h is the convection heat transfer coefficient on the outer surface.

EXAMPLE 15-9 Heat Loss from an Insulated Electric Wire

A 3-mm-diameter and 5-m-long electric wire is tightly wrapped with a 2-mm-thick plastic cover whose thermal conductivity is $k = 0.15$ W/m · °C. Electrical measurements indicate that a current of 10 A passes through the wire and there is a voltage drop of 8 V along the wire. If the insulated wire is exposed to a medium at $T_\infty = 30$°C with a heat transfer coefficient of $h = 12$ W/m² · °C, determine the temperature at the interface of the wire and the plastic cover in steady operation. Also determine whether doubling the thickness of the plastic cover will increase or decrease this interface temperature.

Solution An electric wire is tightly wrapped with a plastic cover. The interface temperature and the effect of doubling the thickness of the plastic cover on the interface temperature are to be determined.

Assumptions **1** Heat transfer is steady since there is no indication of any change with time. **2** Heat transfer is one-dimensional since there is thermal symmetry about the centerline and no variation in the axial direction. **3** Thermal conductivities are constant. **4** The thermal contact resistance at the interface is negligible. **5** Heat transfer coefficient incorporates the radiation effects, if any.

Properties The thermal conductivity of plastic is given to be $k = 0.15$ W/m · °C.

Analysis Heat is generated in the wire and its temperature rises as a result of resistance heating. We assume heat is generated uniformly throughout the wire and is transferred to the surrounding medium in the radial direction. In steady operation, the rate of heat transfer becomes equal to the heat generated within the wire, which is determined to be

$$\dot{Q} = \dot{W}_e = VI = (8 \text{ V})(10 \text{ A}) = 80 \text{ W}$$

The thermal resistance network for this problem involves a conduction resistance for the plastic cover and a convection resistance for the outer surface in series, as shown in Fig. 15-32. The values of these two resistances are determined to be

$$A_2 = (2\pi r_2)L = 2\pi(0.0035 \text{ m})(5 \text{ m}) = 0.110 \text{ m}^2$$

$$R_{conv} = \frac{1}{hA_2} = \frac{1}{(12 \text{ W/m}^2 \cdot \text{°C})(0.110 \text{ m}^2)} = 0.76 \text{°C/W}$$

$$R_{plastic} = \frac{\ln(r_2/r_1)}{2\pi kL} = \frac{\ln(3.5/1.5)}{2\pi(0.15 \text{ W/m} \cdot \text{°C})(5 \text{ m})} = 0.18 \text{°C/W}$$

and therefore

$$R_{total} = R_{plastic} + R_{conv} = 0.76 + 0.18 = 0.94 \text{°C/W}$$

Then the interface temperature can be determined from

$$\dot{Q} = \frac{T_1 - T_\infty}{R_{total}} \longrightarrow T_1 = T_\infty + \dot{Q}R_{total}$$

$$= 30\text{°C} + (80 \text{ W})(0.94\text{°C/W}) = \textbf{105°C}$$

Note that we did not involve the electrical wire directly in the thermal resistance network, since the wire involves heat generation.

To answer the second part of the question, we need to know the critical radius of insulation of the plastic cover. It is determined from Eq. 15-50 to be

$$r_{cr} = \frac{k}{h} = \frac{0.15 \text{ W/m} \cdot \text{°C}}{12 \text{ W/m}^2 \cdot \text{°C}} = 0.0125 \text{ m} = 12.5 \text{ mm}$$

which is larger than the radius of the plastic cover. Therefore, increasing the thickness of the plastic cover will *enhance* heat transfer until the outer radius of the cover reaches 12.5 mm. As a result, the rate of heat transfer $\dot{Q}$ will *increase* when the interface temperature T_1 is held constant, or T_1 will *decrease* when $\dot{Q}$ is held constant, which is the case here.

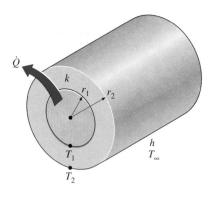

FIGURE 15-32
Schematic for Example 15-9.

Discussion It can be shown by repeating the calculations above for a 4-mm-thick plastic cover that the interface temperature drops to 90.6°C when the thickness of the plastic cover is doubled. It can also be shown in a similar manner that the interface reaches a minimum temperature of 83°C when the outer radius of the plastic cover equals the critical radius.

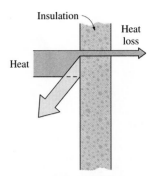

FIGURE 15-33

Thermal insulation retards heat transfer by acting as a barrier in the path of heat flow.

15-6 ▓ THERMAL INSULATION

Thermal insulations are *materials or combinations of materials that are used primarily to provide resistance to heat flow* (Fig. 15-33). You are probably familiar with several kinds of insulation available in the market. Most insulations are heterogeneous materials made of low thermal conductivity materials, and they involve air pockets. This is not surprising since air has one of the lowest thermal conductivities and is readily available. The *Styrofoam* commonly used as a packaging material for TVs, VCRs, computers, and just about anything because of its light weight is also an excellent insulator.

Temperature difference is the driving force for heat flow, and the greater the temperature difference, the larger the rate of heat transfer. We can slow down the heat flow between two mediums at different temperatures by putting "barriers" on the path of heat flow. Thermal insulations serve as such barriers, and they play a major role in the design and manufacture of all energy-efficient devices or systems, and they are usually the cornerstone of all energy conservation projects. A 1991 Drexel University study of the energy-intensive U.S. industries revealed that insulation saves the U.S. industry nearly 2 billion barrels of oil per year, valued at $60 billion a year in energy costs, and more can be saved by practicing better insulation techniques and retrofitting the older industrial facilities.

Heat is generated in *furnaces* or *heaters* by burning a fuel such as coal, oil, or natural gas or by passing electric current through a *resistance heater.* Electricity is rarely used for heating purposes since its unit cost is much higher. The heat generated is absorbed by the medium in the furnace and its surfaces, causing a temperature rise above the ambient temperature. This temperature difference drives heat transfer from the hot medium to the ambient, and insulation reduces the amount of heat loss and thus saves fuel and money. Therefore, insulation *pays for itself* from the energy it saves. Insulating properly requires a one-time capital investment, but its effects are dramatic and long term. The payback period of insulation is usually under two years. That is, the money insulation saves during the first two years is usually greater than its initial material and installation costs. On a broader perspective, insulation also helps the environment and fights air pollution and the greenhouse effect by reducing the amount of fuel burned and thus the amount of CO_2 and other gases released into the atmosphere (Fig. 15-34).

Saving energy with insulation is not limited to hot surfaces. We can also save energy and money by insulating *cold surfaces* (surfaces whose temperature is below the ambient temperature) such as chilled water lines, cryogenic storage tanks, refrigerated trucks, and air-conditioning ducts. The source of "coldness" is *refrigeration,* which requires energy input, usually electricity. In this case, heat is transferred from the surroundings to the cold surfaces, and the refrigeration unit must now work harder and longer to make up for this heat gain and thus it must consume more electrical energy. A cold canned drink can be kept cold much longer by wrapping it in a blanket. A refrigerator with well-insulated walls will consume much less electricity than a similar refrigerator with little or no insulation. Insulating a house will result in

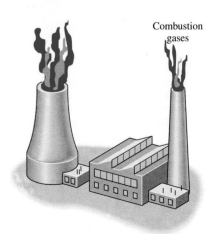

FIGURE 15-34

Insulation also helps the environment by reducing the amount of fuel burned and the air pollutants released.

FIGURE 15-35

In cold weather, we minimize heat loss from our bodies by putting on thick layers of insulation (coats or furs).

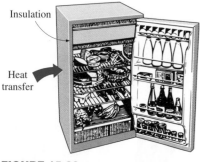

FIGURE 15-36

The insulation layers in the walls of a refrigerator reduce the amount of heat flow into the refrigerator and thus the running time of the refrigerator, saving electricity.

reduced cooling load, and thus reduced electricity consumption for air-conditioning.

Whether we realize it or not, we have an *intuitive* understanding and appreciation of thermal insulation. As babies we feel much better in our blankies, and as children we know we should wear a sweater or coat when going outside in cold weather (Fig. 15-35). When getting out of a pool after swimming on a windy day, we quickly wrap in a towel to stop shivering. Similarly, early man used animal furs to keep warm and built shelters using mud bricks and wood. Cork was used as a roof covering for centuries. The need for effective thermal insulation became evident with the development of mechanical refrigeration later in the 19th century, and a great deal of work was done at universities and government and private laboratories in the 1910s and 1920s to identify and characterize thermal insulation (Powell, Ref. 19).

Thermal insulation in the form of *mud, clay, straw, rags,* and *wood strips* was first used in the 18th century on steam engines to keep workmen from being burned by hot surfaces. As a result, boiler room temperatures dropped and it was noticed that fuel consumption was also reduced. The realization of improved engine efficiency and energy savings prompted the search for materials with improved thermal efficiency. One of the first such materials was *mineral wool* insulation, which, like many materials, was discovered by accident. About 1840, an iron producer in Wales aimed a stream of high-pressure steam at the slag flowing from a blast furnace, and manufactured mineral wool was born. In the early 1860s, this slag wool was a by-product of manufacturing cannons for the Civil War and quickly found its way into many industrial uses. By 1880, builders began installing mineral wool in houses, with one of the most notable applications being General Grant's house. The insulation of this house was described in an article: "it keeps the house cool in summer and warm in winter; it prevents the spread of fire; and it deadens the sound between floors" (Edmunds, Ref. 6). An article published in 1887 in *Scientific American* detailing the benefits of insulating the entire house gave a major boost to the use of insulation in residential buildings.

The energy crisis of the 1970s had a tremendous impact on the public awareness of energy and limited energy reserves and brought an emphasis on *energy conservation.* We have also seen the development of new and more effective insulation materials since then, and a considerable increase in the use of insulation. Thermal insulation is used in more places than you may be aware of. The walls of your house are probably filled with some kind of insulation, and the roof is likely to have a thick layer of insulation. The "thickness" of the walls of your refrigerator is due to the insulation layer sandwiched between two layers of sheet metal (Fig. 15-36). The walls of your range are also insulated to conserve energy, and your hot water tank contains less water than you think because of the 2- to 4-cm-thick insulation in the walls of the tank. Also, your hot water pipe may look much thicker than the cold water pipe because of insulation.

Reasons for Insulating

If you examine the engine compartment of your car, you will notice that the firewall between the engine and the passenger compartment as well as the inner surface of the hood are insulated. The reason for insulating the hood is not to conserve the waste heat from the engine but to protect people from burning themselves by touching the hood surface, which will be too hot if not insulated. As this example shows, the use of insulation is not limited to energy conservation. Various reasons for using insulation can be summarized as follows:

- **Energy Conservation** Conserving energy by reducing the rate of heat flow is the primary reason for insulating surfaces. Insulation materials that will perform satisfactorily in the temperature range of $-268°C$ to $1000°C$ ($-450°F$ to $1800°F$) are widely available.

- **Personnel Protection and Comfort** A surface that is too hot poses a danger to people who are working in that area of accidentally touching the hot surface and burning themselves (Fig. 15-37). To prevent this danger and to comply with the OSHA (Occupational Safety and Health Administration) standards, the temperatures of hot surfaces should be reduced to below $60°C$ ($140°F$) by insulating them. Also, the excessive heat coming off the hot surfaces creates an unpleasant environment in which to work, which adversely affects the performance or productivity of the workers, especially in summer months.

- **Maintaining Process Temperature** Some processes in the chemical industry are temperature-sensitive, and it may become necessary to insulate the process tanks and flow sections heavily to maintain the same temperature throughout.

- **Reducing Temperature Variation and Fluctuations** The temperature in an enclosure may vary greatly between the midsection and the edges if the enclosure is not insulated. For example, the temperature near the walls of a poorly insulated house is much lower than the temperature at the midsections. Also, the temperature in an uninsulated enclosure will follow the temperature changes in the environment closely and fluctuate. Insulation minimizes temperature nonuniformity in an enclosure and slows down fluctuations.

- **Condensation and Corrosion Prevention** Water vapor in the air condenses on surfaces whose temperature is below the dew point, and the outer surfaces of the tanks or pipes that contain a cold fluid frequently fall below the dew-point temperature unless they have adequate insulation. The liquid water on exposed surfaces of the metal tanks or pipes may promote corrosion as well as algae growth.

- **Fire Protection** Damage during a fire may be minimized by keeping valuable combustibles in a safety box that is well insulated. Insulation may lower the rate of heat flow to such levels that the temperature in the box never rises to unsafe levels during fire.

- **Freezing Protection** Prolonged exposure to subfreezing temperatures may cause water in pipes or storage vessels to freeze and burst as a result of heat transfer from the water to the cold ambient. The bursting of pipes as a result of freezing can cause considerable damage. Adequate insulation will slow down the heat loss from the water and prevent freezing during limited exposure to subfreezing temperatures. For example, covering vegetables during a cold night will protect them from freezing, and burying water pipes in the ground at a sufficient depth will keep them from freezing during the entire winter. Wearing thick gloves will protect the fingers from possible frostbite. Also, a molten metal or plastic in a container will solidify on the inner surface if the container is not properly insulated.

- **Reducing Noise and Vibration** An added benefit of thermal insulation is its ability to dampen noise and vibrations (Fig. 15-38). The insulation materials differ in their ability to reduce noise and vibration,

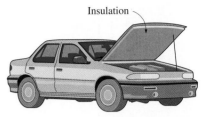

FIGURE 15-37

The hood of the engine compartment of a car is insulated to reduce its temperature and to protect people from burning themselves.

FIGURE 15-38

Insulation materials absorb vibration and sound waves, and are used to minimize sound transmission.

and the proper kind can be selected if noise reduction is an important consideration.

There are a wide variety of insulation materials available in the market, but most are primarily made of fiberglass, mineral wool, polyethylene, foam, or calcium silicate. They come in various trade names such as Ethafoam Polyethylene Foam Sheeting, Solimide Polimide Foam Sheets, FPC Fiberglass Reinforced Silicone Foam Sheeting, Silicone Sponge Rubber Sheets, fiberglass/mineral wool insulation blankets, wire-reinforced mineral wool insulation, Reflect-All Insulation, granulated bulk mineral wool insulation, cork insulation sheets, foil-faced fiberglass insulation, blended sponge rubber sheeting, and numerous others. The unit costs of various insulations in 1995 dollars for 2.54-cm (1-in)-thick blankets are about $9/m^2 for foil-faced fiberglass, $12/m^2 for wire-reinforced mineral wool, and $13/m^2 for cork.

Today various forms of **fiberglass insulation** are widely used in process industries and heating and air-conditioning applications because of their low cost, light weight, resiliency, and versatility. But they are not suitable for some applications because of their low resistance to moisture and fire and their limited maximum service temperature. Fiberglass insulations come in various forms such as unfaced fiberglass insulation, vinyl-faced fiberglass insulation, foil-faced fiberglass insulation, and fiberglass insulation sheets. The reflective foil-faced fiberglass insulation resists vapor penetration and retards radiation because of the aluminum foil on it and is suitable for use on pipes, ducts, and other surfaces.

Mineral wool is resilient, lightweight, fibrous, wool-like, thermally efficient, fire resistant up to 1100°C (2000°F), and forms a sound barrier. Mineral wool insulation comes in the form of blankets, rolls, or blocks. It is also available in composite forms such as fiberglass/mineral wool insulation blankets, wire-reinforced mineral wool insulation blankets, high-temperature mineral wool insulation, and granulated bulk mineral wool insulation. **Calcium silicate** is a solid material that is suitable for use at high temperatures, but it is more expensive. Also, it needs to be cut with a saw during installation, and thus it takes longer to install and there is more waste.

Superinsulators

You may be tempted to think that the most effective way to reduce heat transfer is to use insulating materials that are known to have very low thermal conductivities such as urethane or rigid foam ($k = 0.026$ W/m · °C) or fiberglass ($k = 0.035$ W/m · °C). After all, they are widely available, inexpensive, and easy to install. Looking at the thermal conductivities of materials, you may also notice that the thermal conductivity of air at room temperature is 0.026 W/m · °C, which is lower than the conductivities of practically all of the ordinary insulating materials. Thus you may think that a layer of enclosed air space is as effective as any of the common insulating materials of the same thickness. Of course, heat transfer through the air will probably be higher than what a pure conduction analysis alone would indicate because of the natural convection currents that are likely to occur in the air layer. Besides, air is transparent to radiation, and thus heat will also be lost from the surface by radiation. The thermal conductivity of air is practically independent of pressure unless the pressure is extremely high or extremely low. Therefore, we can reduce the thermal conductivity of air and thus the conduction heat transfer through the air by evacuating the air space. In the limiting case of absolute vacuum, the thermal conductivity will be zero since there will be no particles in this case to "conduct" heat from one surface to the other, and thus the conduction heat transfer will be zero. Noting that the thermal conductivity cannot be negative, an absolute vacuum must be the ultimate insulator, right? Well, not quite.

The purpose of insulation is to reduce "total" heat transfer from a surface, not just conduction. A vacuum totally eliminates conduction but offers zero resistance to radiation, whose magnitude can be comparable to conduction or natural convection in gases (Fig. 15–39). Thus, a vacuum is no more effective in reducing heat transfer than sealing off one of the lanes of a two-lane road is in reducing the flow of traffic on a one-way road.

Insulation against radiation heat transfer between two surfaces is achieved by placing "barriers" between the two surfaces, which are highly reflective thin metal sheets. Radiation heat transfer between two surfaces is inversely proportional to the number of such sheets placed between the surfaces. Very effective insulations are obtained by using closely packed layers of highly reflective thin metal sheets such as aluminum foil (usually 25 sheets per cm) separated by fibers made of insulating material such as glass fiber (Fig. 15–40). Further, the space between the layers is evacuated to form a vacuum under 0.000001 atm pressure to minimize conduction or convection heat transfer through the air space between the layers. The result is an insulating material whose apparent thermal conductivity is below 2×10^{-5} W/m · °C, which is one thousand times less than the conductivity of air or any common insulating material. These specially built insulators are called **superinsulators,** and they are commonly used in space applications and cryogenics, which is the branch of heat transfer dealing with temperatures below 100 K (−173°C) such as those encountered in the liquefaction, storage, and transportation of gases, with helium, hydrogen, nitrogen, and oxygen being the most common ones.

The *R*-value of Insulation

The effectiveness of insulation materials is given by some manufacturers in terms of their **R-value,** which is the *thermal resistance* of the material *per unit surface area.* For *flat insulation* the R-value is obtained by simply dividing the thickness of the insulation by its thermal conductivity. That is,

$$R\text{-value} = \frac{L}{k} \qquad \text{(flat insulation)} \qquad (15\text{-}52)$$

where L is the thickness and k is the thermal conductivity of the material. Note that doubling the thickness L doubles the R-value of flat insulation. For *pipe insulation,* the R-value is determined using the thermal resistance relation from

$$R\text{-value} = \frac{r_2}{k} \ln \frac{r_2}{r_1} \qquad \text{(pipe insulation)} \qquad (15\text{-}53)$$

where r_1 is the inside radius of insulation and r_2 is the outside radius of insulation. Once the R-value is available, the rate of heat transfer through the insulation can be determined from

$$\dot{Q} = \frac{\Delta T}{R\text{-value}} \times \text{Area} \qquad (15\text{-}54)$$

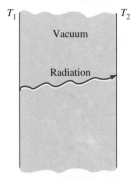

FIGURE 15-39

Evacuating the space between two surfaces completely eliminates heat transfer by conduction or convection but leaves the door wide open for radiation.

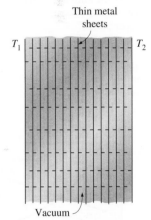

FIGURE 15-40

Superinsulators are built by closely packing layers of highly reflective thin metal sheets and evacuating the space between them.

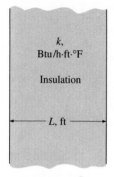

$$R\text{-value} = \frac{L}{k}$$

FIGURE 15-41

The R-value of an insulating material is simply the ratio of the thickness of the material to its thermal conductivity in proper units.

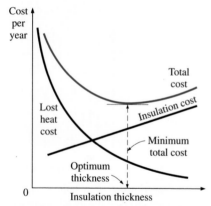

FIGURE 15-42

Determination of the optimum thickness of insulation on the basis of minimum total cost.

where ΔT is the temperature difference across the insulation and Area is the outer surface area for a cylinder.

In the United States, the R-values of insulation are expressed without any units, such as R-19 and R-30. These R-values are obtained by dividing the thickness of the material in *feet* by its thermal conductivity in the unit Btu/h · ft · °F so that the R-values actually have the unit h · ft^2 · °F/Btu. For example, the R-value of 6-in.-thick glass fiber insulation whose thermal conductivity is 0.025 Btu/h · ft · °F is (Fig. 15-41)

$$R\text{-value} = \frac{L}{k} = \frac{0.5 \text{ ft}}{0.025 \text{ Btu/h } \cdot \text{ ft } \cdot \text{ °F}} = 20 \text{ h} \cdot \text{ ft}^2 \cdot \text{°F/Btu}$$

Thus, this 6-in.-thick glass fiber insulation would be referred to as R-20 insulation by the builders. The unit of R-value is m^2 · °C/W in SI units, with the conversion relation 1 m^2 · °C/W = 5.678 h · ft^2 · °F/Btu. Therefore, a small R-value in SI corresponds to a large R-value in English units.

Optimum Thickness of Insulation

It should be realized that insulation does not eliminate heat transfer; it merely reduces it. The thicker the insulation, the lower the rate of heat transfer but also the higher the cost of insulation. Therefore, there should be an *optimum* thickness of insulation that corresponds to a minimum combined cost of insulation and heat lost. The determination of the optimum thickness of insulation is illustrated in Fig. 15-42. Notice that the cost of insulation increases roughly linearly with thickness while the cost of heat loss decreases exponentially. The total cost, which is the sum of the insulation cost and the lost heat cost, decreases first, reaches a minimum, and then increases. The thickness corresponding to the minimum total cost is the optimum thickness of insulation, and this is the recommended thickness of insulation to be installed.

If you are mathematically inclined, you can determine the *optimum thickness* by obtaining an expression for the *total cost*, which is the sum of the expressions for the lost heat cost and insulation cost as a function of thickness; *differentiating* the total cost expression with respect to the thickness; and *setting* it equal to zero. The thickness value satisfying the resulting equation is the optimum thickness. The cost values can be determined from an annualized lifetime analysis or simply from the requirement that the insulation pay for itself within two or three years. Note that the optimum thickness of insulation depends on the fuel cost, and the higher the fuel cost, the larger the optimum thickness of insulation. Considering that insulation will be in service for many years and the fuel prices are likely to escalate, a reasonable increase in fuel prices must be assumed in calculations. Otherwise, what is optimum insulation today will be inadequate insulation in the years to come, and we may have to face the possibility of costly retrofitting projects. This is what happened in the 1970s and 1980s to insulations installed in the 1960s.

The discussion above on optimum thickness is valid when the type and manufacturer of insulation are already selected, and the only thing to be determined is the most economical thickness. But often there are several suitable insulations for a job, and the selection process can be rather confusing since each insulation can have a different thermal conductivity, different installation cost, and different service life. In such cases, a selection can be made by preparing an annualized cost versus thickness chart like

Fig. 15-43 for each insulation, and determining the one having the *lowest* minimum cost. The insulation with the lowest annual cost is obviously the most economical insulation, and the insulation thickness corresponding to the *minimum total cost* is the *optimum thickness*. When the optimum thickness falls between two commercially available thicknesses, it is a good practice to be conservative and choose the thicker insulation. The extra thickness will provide a little safety cushion for any possible decline in performance over time and will help the environment by reducing the production of greenhouse gases such as CO_2.

The determination of the optimum thickness of insulation requires a heat transfer and economic analysis, which can be tedious and time-consuming. But a selection can be made in a few minutes using the tables and charts prepared by TIMA (Thermal Insulation Manufacturers Association) and member companies. The primary inputs required for using these tables or charts are the operating and ambient temperatures, pipe diameter (in the case of pipe insulation), and the unit fuel cost. Recommended insulation thicknesses for hot surfaces at specified temperatures are given in Table 15-3. Recommended thicknesses of *pipe insulations* as a function of service temperatures are 0.5 to 1 in. for 150°F, 1 to 2 in. for 250°F, 1.5 to 3 in. for 350°F, 2 to 4.5 in. for 450°F, 2.5 to 5.5 in. for 550°F, and 3 to 6 in. for 650°F for nominal pipe diameters of 0.5 to 36 in. The lower recommended insulation thicknesses are for pipes with small diameters, and the larger ones are for pipes with large diameters.

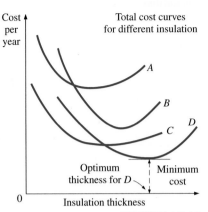

FIGURE 15-43

Determination of the most economical type of insulation and its optimum thickness.

TABLE 15-3

Recommended insulation thicknesses for flat hot surfaces as a function of surface temperature (from TIMA *Energy Savings Guide*)

Surface temperature	Insulation thickness
150°F (66°C)	2″ (5.1 cm)
250°F (121°C)	3″ (7.6 cm)
350°F (177°C)	4″ (10.2 cm)
550°F (288°C)	6″ (15.2 cm)
750°F (400°C)	9″ (22.9 cm)
950°F (510°C)	10″ (25.44 cm)

EXAMPLE 15-10 Effect of Insulation on Surface Temperature

Hot water at T_i = 120°C flows in a stainless steel pipe (k = 15 W/m · °C) whose inner diameter is 1.6 cm and thickness is 0.2 cm. The pipe is to be covered with adequate insulation so that the temperature of the outer surface of the insulation does not exceed 40°C when the ambient temperature is T_o = 25°C. Taking the heat transfer coefficients inside and outside the pipe to be h_i = 70 W/m² · °C and h_o = 20 W/m² · °C, respectively, determine the thickness of fiberglass insulation (k = 0.038 W/m · °C) that needs to be installed on the pipe.

Solution A steam pipe is to be covered with enough insulation to reduce the exposed surface temperature to 40°C. The thickness of insulation that needs to be installed is to be determined.

Assumptions **1** Heat transfer is steady since there is no indication of any change with time. **2** Heat transfer is one-dimensional since there is thermal symmetry about the centerline and no variation in the axial direction. **3** Thermal conductivities are constant. **4** The thermal contact resistance at the interface is negligible.

Properties The thermal conductivities are given to be k = 15 W/m · °C for the steel pipe and k = 0.038 W/m · °C for fiberglass insulation.

Analysis The thermal resistance network for this problem involves four resistances in series and is given in Fig. 15-44. The inner radius of the pipe is r_1 = 0.8 cm and the outer radius of the pipe and thus the inner radius of the insulation is r_2 = 1.0 cm. Letting r_3 represent the outer radius of the insulation, the areas of the surfaces exposed to convection for an L = 1-m-long section of the pipe become

$$A_1 = 2\pi r_1 L = 2\pi(0.008 \text{ m})(1 \text{ m}) = 0.0503 \text{ m}^2$$

$$A_3 = 2\pi r_3 L = 2\pi r_3 (1 \text{ m}) = 6.28 r_3 \text{ m}^2$$

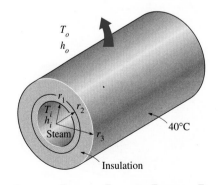

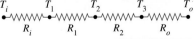

FIGURE 15-44

Schematic for Example 15-10.

Then the individual thermal resistances are determined to be

$$R_i = R_{conv, 1} = \frac{1}{h_i A_1} = \frac{1}{(70 \text{ W/m}^2 \cdot {}^\circ\text{C})(0.0503 \text{ m}^2)} = 0.284{}^\circ\text{C/W}$$

$$R_1 = R_{pipe} = \frac{\ln(r_2/r_1)}{2\pi k_1 L} = \frac{\ln(0.01/0.008)}{2\pi(15 \text{ W/m} \cdot {}^\circ\text{C})(1 \text{ m})} = 0.0024{}^\circ\text{C/W}$$

$$R_2 = R_{insulation} = \frac{\ln(r_3/r_2)}{2\pi k_2 L} = \frac{\ln(r_3/0.01)}{2\pi(0.038 \text{ W/m} \cdot {}^\circ\text{C})(1 \text{ m})}$$

$$= 4.188 \ln(r_3/0.01){}^\circ\text{C/W}$$

$$R_o = R_{conv, 2} = \frac{1}{h_o A_3} = \frac{1}{(20 \text{ W/m}^2 \cdot {}^\circ\text{C})(6.28r_3 \text{ m}^2)} = \frac{1}{125.6r_3}{}^\circ\text{C/W}$$

Noting that all resistances are in series, the total resistance is determined to be

$$R_{total} = R_i + R_1 + R_2 + R_o$$
$$= [0.284 + 0.0024 + 4.188 \ln(r_3/0.01) + 1/125.6r_3]{}^\circ\text{C/W}$$

Then the steady rate of heat loss from the steam becomes

$$\dot{Q} = \frac{T_i - T_o}{R_{total}} = \frac{(120 - 125){}^\circ\text{C}}{[0.284 + 0.0024 + 4.188 \ln(r_3/0.01) + 1/125.6r_3]{}^\circ\text{C/W}}$$

Noting that the outer surface temperature of insulation is specified to be 40°C, the rate of heat loss can also be expressed as

$$\dot{Q} = \frac{T_3 - T_o}{R_o} = \frac{(40 - 25){}^\circ\text{C}}{(1/125.6r_3){}^\circ\text{C/W}} = 1884r_3$$

Setting the two relations above equal to each other and solving for r_3 gives $r_3 = 0.0170$ m. Then the minimum thickness of fiberglass insulation required is

$$t = r_3 - r_2 = 0.0170 - 0.0100 = 0.0070 \text{ m} = \textbf{0.70 cm}$$

Discussion   Insulating the pipe with at least 0.70-cm-thick fiberglass insulation will ensure that the outer surface temperature of the pipe will be at 40°C or below.

EXAMPLE 15-11   Optimum Thickness of Insulation

During a plant visit, you notice that the outer surface of a cylindrical curing oven is very hot, and your measurements indicate that the average temperature of the exposed surface of the oven is 180°F when the surrounding air temperature is 75°F. You suggest to the plant manager that the oven should be insulated, but the manager does not think it is worth the expense. Then you propose to the manager to pay for the insulation yourself if he lets you keep the savings from the fuel bill for one year. That is, if the fuel bill is $5000/yr before insulation and drops to $2000/yr after insulation, you will get paid $3000. The manager agrees since he has nothing to lose, and a lot to gain. Is this a smart bet on your part?

The oven is 12 ft long and 8 ft in diameter, as shown in Fig. 15-45. The plant operates 16 h a day 365 days a year, and thus 5840 h/yr. The insulation to be used is fiberglass ($k_{ins} = 0.024$ Btu/h · ft · °F), whose cost is $0.70/ft² per inch of thickness for materials, plus $2.00/ft² for labor regardless of thickness. The combined heat transfer coefficient on the outer surface is estimated to be $h_o = 3.5$ Btu/h · ft² · °F. The oven uses natural gas, whose unit cost is $0.75/therm input (1 therm = 100,000 Btu), and the efficiency of the oven is 80 percent. Disregarding any inflation or interest, determine how much money you will make out of this venture, if any, and the thickness of insulation (in whole inches) that will maximize your earnings.

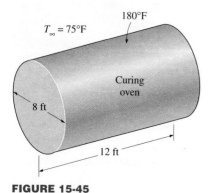

FIGURE 15-45

Schematic for Example 15-11.

Solution A cylindrical oven is to be insulated to reduce heat losses. The optimum thickness of insulation and the potential earnings are to be determined.

Assumptions **1** Steady operating conditions exist. **2** Heat transfer through the insulation is one-dimensional. **3** Thermal conductivities are constant. **4** The thermal contact resistance at the interface is negligible. **5** The surfaces of the cylindrical oven can be treated as plain surfaces since its diameter is greater than 3 ft.

Properties The thermal conductivity of insulation is given to be $k = 0.024$ Btu/h · ft · °F.

Analysis The exposed surface area of the oven is

$$A = 2A_{base} + A_{side} = 2\pi r^2 + 2\pi rL = 2\pi(4 \text{ ft})^2 + 2\pi(4 \text{ ft})(12 \text{ ft}) = 402 \text{ ft}^2$$

The rate of heat loss from the oven before the insulation is installed is determined from

$$\dot{Q} = h_o A(T_s - T_\infty) = (3.5 \text{ Btu/h} \cdot \text{ft}^2 \cdot °F)(402 \text{ ft}^2)(180 - 75)°F = 147{,}700 \text{ Btu/h}$$

Noting that the plant operates 5840 h/yr, the total amount of heat loss from the oven per year is

$$Q = \dot{Q}\Delta t = (147{,}700 \text{ Btu/h})(5840 \text{ h/yr}) = 0.863 \times 10^9 \text{ Btu/yr}$$

The efficiency of the oven is given to be 80 percent. Therefore, to generate this much heat, the oven must consume energy (in the form of natural gas) at a rate of

$$Q_{in} = Q/\eta_{oven} = (0.863 \times 10^9 \text{ Btu/yr})/0.80 = 1.079 \times 10^9 \text{ Btu/yr}$$
$$= 10{,}790 \text{ therms}$$

since 1 therm = 100,000 Btu. Then the annual fuel cost of this oven before insulation becomes

$$\text{Annual cost} = Q_{in} \times \text{Unit cost}$$
$$= (10{,}790 \text{ therm/yr})(\$0.75/\text{therm}) = \$8093/\text{yr}$$

That is, the heat losses from the exposed surfaces of the oven are currently costing the plant just over \$8000/yr.

When insulation is installed, the rate of heat transfer from the oven can be determined from

$$\dot{Q}_{ins} = \frac{T_s - T_\infty}{R_{total}} = \frac{T_s - T_\infty}{R_{ins} + R_{conv}} = A\frac{T_s - T_\infty}{\dfrac{t_{ins}}{k_{ins}} + \dfrac{1}{h_o}}$$

We expect the surface temperature of the oven to increase and the heat transfer coefficient to decrease somewhat when insulation is installed. We assume these two effects to counteract each other. Then the relation above for 1-in.-thick insulation gives the rate of heat loss to be

$$\dot{Q}_{ins} = \frac{A(T_s - T_\infty)}{\dfrac{t_{ins}}{k_{ins}} + \dfrac{1}{h_o}} = \frac{(402 \text{ ft}^2)(180 - 75)°F}{\dfrac{1/12 \text{ ft}}{0.024 \text{ Btu/h} \cdot \text{ft} \cdot °F} + \dfrac{1}{3.5 \text{ Btu/h} \cdot \text{ft}^2 \cdot °F}}$$
$$= 11{,}230 \text{ Btu/h}$$

Also, the total amount of heat loss from the oven per year and the amount and cost of energy consumption of the oven become

$$Q_{ins} = \dot{Q}_{ins}\Delta t = (11{,}230 \text{ Btu/h})(5840 \text{ h/yr}) = 0.6558 \times 10^8 \text{ Btu/yr}$$

$$Q_{in, ins} = Q_{ins}/\eta_{oven} = (0.6558 \times 10^8 \text{ Btu/yr})/0.80 = 0.820 \times 10^8 \text{ Btu/yr}$$
$$= 820 \text{ therms}$$

$$\text{Annual cost} = Q_{in, ins} \times \text{Unit cost}$$
$$= (820 \text{ therm/yr})(\$0.75/\text{therm}) = \$615/\text{yr}$$

Therefore, insulating the oven by 1-in.-thick fiberglass insulation will reduce the fuel bill by $8093 − $615 = $7362 per year. The unit cost of insulation is given to be $2.70/ft². Then the installation cost of insulation becomes

Insulation cost = (Unit cost)(Surface area) = ($2.70/ft²)(402 ft²) = $1085

The sum of the insulation and heat loss costs is

Total cost = Insulation cost + Heat loss cost = $1085 + $615 = $1700

Then the net earnings will be

Earnings = Income − Expenses = $8093 − $1700 = $6393

To determine the thickness of insulation that maximizes your earnings, we repeat the calculations above for 2-, 3-, 4-, and 5-in.-thick insulations, and list the results in Table 15-4. Note that the total cost of insulation decreases first with increasing insulation thickness, reaches a minimum, and then starts to increase.

TABLE 15-4

The variation of total insulation cost with insulation thickness

Insulation thickness	Heat loss, Btu/h	Lost fuel, therms/yr	Lost fuel cost, $/yr	Insulation cost, $	Total cost, $
1 in.	11,230	820	615	1085	1700
2 in.	5838	426	320	1367	1687
3 in.	3944	288	216	1648	1864
4 in.	2978	217	163	1930	2093
5 in.	2392	175	131	2211	2342

We observe that the total insulation cost is a minimum at $1687 for the case of **2-in-thick** insulation. The earnings in this case are

Maximum earnings = Income − Minimum expenses

= $8093 − $1687 = **$6406**

which is not bad for a day's worth of work. The plant manager is also a big winner in this venture since the heat losses will cost him only $320/yr during the second and consequent years instead of $8093/yr. A thicker insulation could probably be justified in this case if the cost of insulation is annualized over the lifetime of insulation, say 20 years. Several energy conservation measures are being marketed as explained above by several power companies and private firms.

15-7 ■ HEAT TRANSFER FROM FINNED SURFACES

The rate of heat transfer from a surface at a temperature T_s to the surrounding medium at T_∞ is given by Newton's law of cooling as

$$\dot{Q}_{conv} = hA(T_s - T_\infty)$$

where A is the heat transfer surface area and h is the convection heat transfer coefficient. When the temperatures T_s and T_∞ are fixed by design considerations, as is often the case, there are *two ways* to increase the rate of heat transfer: to increase the *convection heat transfer coefficient* h or to increase the *surface area* A. Increasing h may require the installation of a pump or fan, or replacing the existing one with a larger one, but this approach may or may not be practical. Besides, it may not be adequate. The alternative is to increase the surface area by attaching to the surface *extended surfaces* called *fins* made of

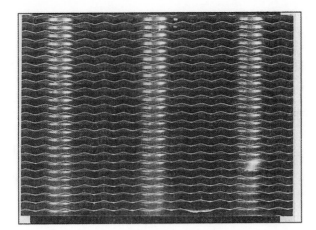

FIGURE 15-46

The thin plate fins of a car radiator greatly increase the rate of heat transfer to the air (courtesy of James Kleiser).

highly conductive materials such as aluminum. Finned surfaces are manufactured by extruding, welding, or wrapping a thin metal sheet on a surface. Fins enhance heat transfer from a surface by exposing a larger surface area to convection and radiation.

Finned surfaces are commonly used in practice to enhance heat transfer, and they often increase the rate of heat transfer from a surface severalfold. The *car radiator* shown in Fig. 15-46 is an example of a finned surface. The closely packed thin metal sheets attached to the hot water tubes increase the surface area for convection and thus the rate of convection heat transfer from the tubes to the air many times. There are a variety of innovative fin designs available in the market, and they seem to be limited only by imagination (Fig. 15-47).

In the analysis of the fins, we consider *steady* operation with *no heat generation* in the fin, and we assume the thermal conductivity k of the material to remain constant. We also assume the convection heat transfer coefficient h to be *constant* and *uniform* over the entire surface of the fin for convenience in the analysis. We recognize that the convection heat transfer coefficient h, in general, varies along the fin as well as its circumference, and its value at a point is a strong function of the *fluid motion* at that point. The value of h is usually much lower at the *fin base* than it is at the *fin tip* because the fluid is surrounded by solid surfaces near the base, which seriously disrupt its motion to the point of "suffocating" it, while the fluid near the fin tip has little contact with a solid surface and thus encounters little resistance to flow. Therefore, adding too many fins on a surface may actually decrease the overall heat transfer when the decrease in h offsets any gain resulting from the increase in the surface area.

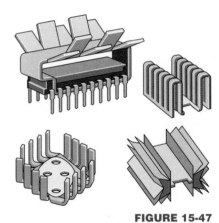

FIGURE 15-47

Some innovative fin designs.

Fin Equation

Consider a volume element of a fin at location x having a length of Δx, cross-sectional area of A_c, and a perimeter of p, as shown in Fig. 15-48. Under steady conditions, the energy balance on this volume element can be expressed as

$$\begin{pmatrix} \text{Rate of } heat \\ conduction \text{ into} \\ \text{the element at } x \end{pmatrix} = \begin{pmatrix} \text{Rate of } heat \\ conduction \text{ from the} \\ \text{element at } x + \Delta x \end{pmatrix} + \begin{pmatrix} \text{Rate of } heat \\ convection \text{ from} \\ \text{the element} \end{pmatrix}$$

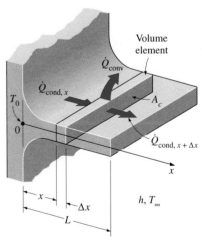

FIGURE 15-48

Volume element of a fin at location x having a length of Δx, cross-sectional area of A_c, and perimeter of p.

or

$$\dot{Q}_{\text{cond}, x} = \dot{Q}_{\text{cond}, x + \Delta x} + \dot{Q}_{\text{conv}}$$

where

$$\dot{Q}_{\text{conv}} = h(p\,\Delta x)(T - T_\infty)$$

Substituting and dividing by Δx, we obtain

$$\frac{\dot{Q}_{\text{cond}, x + \Delta x} - \dot{Q}_{\text{cond}, x}}{\Delta x} + hp(T - T_\infty) = 0$$

Taking the limit as $\Delta x \to 0$ and using the definition of the derivative gives

$$\frac{d\dot{Q}_{\text{cond}}}{dx} + hp(T - T_\infty) = 0 \qquad (15\text{-}55)$$

From Fourier's law of heat conduction we have

$$\dot{Q}_{\text{cond}} = -kA_c \frac{dT}{dx}$$

where A_c is the cross-sectional area of the fin at location x. Substitution of this relation into Eq. 15-55 gives the differential equation governing heat transfer in fins,

$$\frac{d}{dx}\left(kA_c \frac{dT}{dx}\right) - hp(T - T_\infty) = 0 \qquad (15\text{-}56)$$

In general, the cross-sectional area A_c and the perimeter p of a fin vary with x, which makes this differential equation difficult to solve. In the special case of *constant cross-section* and *constant thermal conductivity,* the differential equation above reduces to

$$\frac{d^2\theta}{dx^2} - a^2\theta = 0 \qquad (15\text{-}57)$$

where

$$a^2 = \frac{hp}{kA_c}$$

and $\theta = T - T_\infty$ is the *temperature excess.* At the fin base we have $\theta_b = T_b - T_\infty$.

Equation 15-57 is a linear, homogeneous, second-order differential equation with constant coefficients. A fundamental theory of differential equations states that such an equation has two linearly independent solution functions, and its general solution is the linear combination of those two solution functions. A careful examination of the differential equation reveals that subtracting a constant multiple of the solution function θ from its second derivative yields zero. Thus we conclude that the function θ and its second derivative must be *constant multiples* of each other. The only functions whose derivatives are constant multiples of the functions themselves are the *exponential functions* (or a linear combination of exponential functions such as sine and cosine hyperbolic functions). Therefore, the solution functions of the differential equation above are the exponential functions e^{-ax} or e^{ax} or constant multiples of them. This can be verified by direct substitution. For example, the second derivative of e^{-ax} is a^2e^{-ax}, and its substitution into Eq. 15-57 yields zero. Therefore, the general solution of the differential equation Eq. 15-57 is

$$\theta(x) = C_1 e^{ax} + C_2 e^{-ax} \qquad (15\text{-}58)$$

where C_1 and C_2 are arbitrary constants whose values are to be determined from the boundary conditions at the base and at the tip of the fin. Note that we need only two conditions to determine C_1 and C_2 uniquely.

The temperature of the plate to which the fins are attached is normally known in advance. Therefore, at the fin base we have a *specified temperature* boundary condition, expressed as

Boundary condition at fin base: $\qquad \theta(0) = \theta_b = T_b - T_\infty \qquad (15\text{-}59)$

At the fin tip we have several possibilities, including specified temperature, negligible heat loss (idealized as an insulated tip), convection, and combined convection and radiation (Fig. 15-49). Below we consider each case separately.

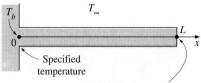

(a) Specified temperature
(b) Negligible heat loss
(c) Convection
(d) Convection and radiation

FIGURE 15-49

Boundary conditions at the fin base and the fin tip.

1   Infinitely Long Fin ($T_{\text{fin tip}} = T_\infty$)

For a sufficiently long fin of *uniform* cross-section ($A_c = $ constant), the temperature of the fin at the fin tip will approach the environment temperature T_∞ and thus θ will approach zero. That is,

Boundary condition at fin tip: $\quad \theta(L) = T(L) - T_\infty = 0 \quad$ as $\quad L \rightarrow \infty$

This condition will be satisfied by the function e^{-ax}, but not by the other prospective solution function e^{ax} since it tends to infinity as x gets larger. Therefore, the general solution in this case will consist of a constant multiple of e^{-ax}. The value of the constant multiple is determined from the requirement that at the fin base where $x = 0$ the value of θ will be θ_b. Noting that $e^{-ax} = e^0 = 1$, the proper value of the constant is θ_b, and the solution function we are looking for is $\theta(x) = \theta_b e^{-ax}$. This function satisfies the differential equation as well as the requirements that the solution reduce to θ_b at the fin base and approach zero at the fin tip for large x. Noting that $\theta = T - T_\infty$ and $a = \sqrt{hp/kA_c}$, the variation of temperature along the fin in this case can be expressed as

Very long fin: $\qquad \dfrac{T(x) - T_\infty}{T_b - T_\infty} = e^{-ax} = e^{-x\sqrt{hp/kA_c}} \qquad (15\text{-}60)$

Note that the temperature along the fin in this case decreases *exponentially* from T_b to T_∞, as shown in Fig. 15-50. The steady rate of *heat transfer* from the entire fin can be determined from Fourier's law of heat conduction

Very long fin: $\qquad \dot{Q}_{\text{log fin}} = -kA_c \left. \dfrac{dT}{dx} \right|_{x=0} = \sqrt{hpkA_c}\,(T_b - T_\infty) \qquad (15\text{-}61)$

where p is the perimeter, A_c is the cross-sectional area of the fin, and x is the distance from the fin base. Alternately, the rate of heat transfer from the fin could also be determined by considering heat transfer from a differential volume element of the fin and integrating it over the entire surface of the fin. That is,

$$\dot{Q}_{\text{fin}} = \int_{A_{\text{fin}}} h[T(x) - T_\infty]\, dA_{\text{fin}} = \int_{A_{\text{fin}}} h\theta(x)\, dA_{\text{fin}} \qquad (15\text{-}62)$$

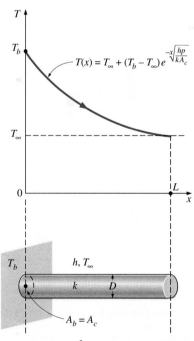

($p = \pi D$, $A_c = \pi D^2/4$ for a cylindrical fin)

FIGURE 15-50

A long circular fin of uniform cross-section and the variation of temperature along it.

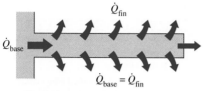

FIGURE 15-51

Under steady conditions, heat transfer from the exposed surfaces of the fin is equal to heat conduction to the fin at the base.

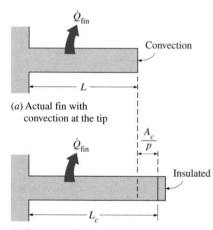

(a) Actual fin with convection at the tip

(b) Equivalent fin with insulated tip

FIGURE 15-52

Corrected fin length L_c is defined such that heat transfer from a fin of length L_c with insulated tip is equal to heat transfer from the actual fin of length L with convection at the fin tip.

The two approaches described above are equivalent and give the same result since, under steady conditions, the heat transfer from the exposed surfaces of the fin is equal to the heat transfer to the fin at the base (Fig. 15-51).

2   Negligible Heat Loss from the Fin Tip (Insulated fin tip, $\dot{Q}_{\text{fin tip}} = 0$)

Fins are not likely to be so long that their temperature approaches the surrounding temperature at the tip. A more realistic situation is for heat transfer from the fin tip to be negligible since the heat transfer from the fin is proportional to its surface area, and the surface area of the fin tip is usually a negligible fraction of the total fin area. Then the fin tip can be assumed to be insulated, and the condition at the fin tip can be expressed as

$$\text{Boundary condition at fin tip:} \qquad \frac{d\theta}{dx}\bigg|_{x=L} = 0 \qquad (15\text{-}63)$$

The condition at the fin base remains the same as expressed in Eq. 15-59. The application of these two conditions on the general solution (Eq. 15-58) yields, after some manipulations, the following relation for the temperature distribution:

$$\text{Adiabatic fin tip:} \qquad \frac{T(x) - T_\infty}{T_b - T_\infty} = \frac{\cosh a(L - x)}{\cosh aL} \qquad (15\text{-}64)$$

The rate of heat transfer from the fin can be determined again from Fourier's law of heat conduction:

$$\text{Adiabatic fin tip:} \qquad \dot{Q}_{\text{insulated tip}} = -kA_c \frac{dT}{dx}\bigg|_{x=0} \qquad (15\text{-}65)$$
$$= \sqrt{hpkA_c}\,(T_b - T_\infty)\tanh aL$$

Note that the heat transfer relations for the very long fin and the fin with negligible heat loss at the tip differ by the factor $\tanh aL$, which approaches 1 as L becomes very large.

3   Convection (or Combined Convection and Radiation) from Fin Tip

The fin tips, in practice, are exposed to the surroundings, and thus the proper boundary condition for the fin tip is convection that also includes the effects of radiation. The fin equation can still be solved in this case using the convection at the fin tip as the second boundary condition, but the analysis becomes more involved, and it results in rather lengthy expressions for the temperature distribution and the heat transfer. Yet, in general, the fin tip area is a small fraction of the total fin surface area, and thus the complexities involved can hardly justify the improvement in accuracy.

A practical way of accounting for the heat loss from the fin tip is to replace the *fin length L* in the relation for the *insulated tip* case by a **corrected length** defined as (Fig. 15-52)

$$\text{Corrected fin length:} \qquad L_c = L + \frac{A_c}{p} \qquad (15\text{-}66)$$

where A_c is the cross-sectional area and p is the perimeter of the fin at the tip. Multiplying the relation above by the perimeter gives $A_{\text{corrected}} = A_{\text{fin (lateral)}} + A_{\text{tip}}$, which indicates that the fin area determined using the corrected length is equivalent to the sum of the lateral fin area plus the fin tip area.

The corrected length approximation gives very good results when the variation of temperature near the fin tip is small (which is the case when aL ≥ 1) and the heat transfer coefficient at the fin tip is about the same as that at the lateral surface of the fin. Therefore, *fins subjected to convection at their tips can be treated as fins with insulated tips by replacing the actual fin length by the corrected length in Eqs. 15-64 and 15-65.*

Using the proper relations for A_c and p, the corrected lengths for rectangular and cylindrical fins are easily determined to be

$$L_{c, \text{rectangular fin}} = L + \frac{t}{2} \quad \text{and} \quad L_{c, \text{cylindrical fin}} = L + \frac{D}{4}$$

where t is the thickness of the rectangular fins and D is the diameter of the cylindrical fins.

Fin Efficiency

Consider the surface of a *plane wall* at temperature T_b exposed to a medium at temperature T_∞. Heat is lost from the surface to the surrounding medium by convection with a heat transfer coefficient of h. Disregarding radiation or accounting for its contribution in the convection coefficient h, heat transfer from a surface area A is expressed as $\dot{Q} = hA(T_s - T_\infty)$.

Now let us consider a fin of constant cross-sectional area $A_c = A_b$ and length L that is attached to the surface with a perfect contact (Fig. 15-53). This time heat will flow from the surface to the fin *by conduction* and from the fin to the surrounding medium *by convection* with the same heat transfer coefficient h. The temperature of the fin will be T_b at the fin base and gradually decrease toward the fin tip. Convection from the fin surface causes the temperature at any cross-section to drop somewhat from the midsection toward the outer surfaces. However, the cross-sectional area of the fins is usually very small, and thus the temperature at any cross-section can be considered to be uniform. Also, the fin tip can be assumed for convenience and simplicity to be insulated by using the corrected length for the fin instead of the actual length.

In the limiting case of *zero thermal resistance* or *infinite thermal conductivity* ($k \rightarrow \infty$), the temperature of the fin will be uniform at the base value of T_b. The heat transfer from the fin will be *maximum* in this case and can be expressed as

$$\dot{Q}_{\text{fin, max}} = hA_{\text{fin}}(T_b - T_\infty) \tag{15-67}$$

In reality, however, the temperature of the fin will drop along the fin, and thus the heat transfer from the fin will be less because of the decreasing temperature difference $T(x) - T_\infty$ toward the fin tip, as shown in Fig. 15-54. To account for the effect of this decrease in temperature on heat transfer, we define a **fin efficiency** as

$$\eta_{\text{fin}} = \frac{\dot{Q}_{\text{fin}}}{\dot{Q}_{\text{fin, max}}} = \frac{\text{Actual heat transfer rate from the fin}}{\begin{array}{c}\text{Ideal heat transfer rate from the fin}\\ \text{if the entire fin were at base temperature}\end{array}} \tag{15-68}$$

or

$$\dot{Q}_{\text{fin}} = \eta_{\text{fin}} \dot{Q}_{\text{fin, max}} = \eta_{\text{fin}} hA_{\text{fin}}(T_b - T_\infty) \tag{15-69}$$

where A_{fin} is the total surface area of the fin. This relation enables us to determine the heat transfer from a fin when its efficiency is known. For the cases

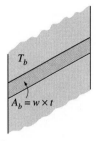

(a) Surface without fins

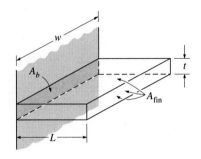

(b) Surface with a fin

$$A_{\text{fin}} = 2 \times w \times L + w \times t$$
$$\cong 2 \times w \times L$$

FIGURE 15-53

Fins enhance heat transfer from a surface by enhancing surface area.

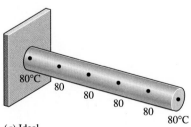

(a) Ideal

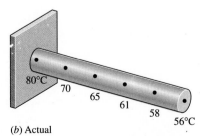

(b) Actual

FIGURE 15-54

Ideal and actual temperature distribution in a fin.

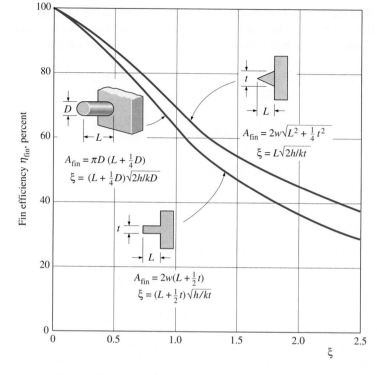

FIGURE 15-55

Efficiency of circular, rectangular, and triangular fins on a plain surface of width w (from Gardner, Ref. 9).

of constant cross-section of *very long fins* and *fins with insulated tips,* the fin efficiency can be expressed as

$$\eta_{\text{long fin}} = \frac{\dot{Q}_{\text{fin}}}{\dot{Q}_{\text{fin, max}}} = \frac{\sqrt{hpkA_c}\,(T_b - T_\infty)}{hA_{\text{fin}}\,(T_b - T_\infty)} = \frac{1}{L}\sqrt{\frac{kA_c}{hp}} = \frac{1}{aL} \qquad (15\text{-}70)$$

and

$$\eta_{\text{insulated tip}} = \frac{\dot{Q}_{\text{fin}}}{\dot{Q}_{\text{fin, max}}} = \frac{\sqrt{hpkA_c}\,(T_b - T_\infty)\tanh aL}{hA_{\text{fin}}\,(T_b - T_\infty)}$$
$$= \frac{\tanh aL}{L}\sqrt{\frac{kp}{hA_c}} = \frac{\tanh aL}{aL} \qquad (15\text{-}71)$$

since $A_{\text{fin}} = pL$ for fins with constant cross-section. Equation 15-71 can also be used for fins subjected to convection provided that the fin length L is replaced by the corrected length L_c.

Fin efficiency relations are developed for fins of various profiles and are plotted in Fig. 15-55 for fins on a *plain surface* and in Fig. 15-56 for *circular fins* of constant thickness. The fin surface area associated with each profile is also given on each figure. For most fins of constant thickness encountered in practice, the fin thickness t is too small relative to the fin length L, and thus the fin tip area is negligible.

Note that fins with triangular and parabolic profiles contain less material and are more efficient than the ones with rectangular profiles, and thus are more suitable for applications requiring minimum weight such as space applications.

An important consideration in the design of finned surfaces is the selection of the proper *fin length L.* Normally the *longer* the fin, the *larger* the heat

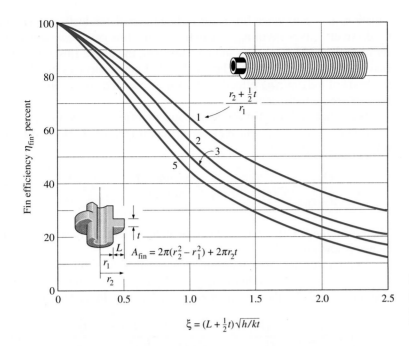

FIGURE 15-56

Efficiency of circular fins of length L and constant thickness t (from Gardner, Ref. 9).

transfer area and thus the *higher* the rate of heat transfer from the fin. But also the larger the fin, the bigger the mass, the higher the price, and the larger the fluid friction. Therefore, increasing the length of the fin beyond a certain value cannot be justified unless the added benefits outweigh the added cost. Also, the fin efficiency decreases with increasing fin length because of the decrease in fin temperature with length. Fin lengths that cause the fin efficiency to drop below 60 percent usually cannot be justified economically and should be avoided. The efficiency of most fins used in practice is above 90 percent.

Fin Effectiveness

Fins are used to *enhance* heat transfer, and the use of fins on a surface cannot be recommended unless the enhancement in heat transfer justifies the added cost and complexity associated with the fins. In fact, there is no assurance that adding fins on a surface will *enhance* heat transfer. The performance of the fins is judged on the basis of the enhancement in heat transfer relative to the no-fin case. The performance of fins expressed in terms of the *fin effectiveness* ε_{fin} is defined as (Fig. 15-57)

$$\varepsilon_{fin} = \frac{\dot{Q}_{fin}}{\dot{Q}_{no\ fin}} = \frac{\dot{Q}_{fin}}{hA_b(T_b - T_\infty)} = \frac{\text{Heat transfer rate from the fin of } base\ area\ A_b}{\text{Heat transfer rate from the surface of } area\ A_b} \quad (15\text{-}72)$$

Here, A_b is the cross-sectional area of the fin at the base and $\dot{Q}_{no\ fin}$ represents the rate of heat transfer from this area if no fins are attached to the surface. An effectiveness of $\varepsilon_{fin} = 1$ indicates that the addition of fins to the surface does not affect heat transfer at all. That is, heat conducted to the fin through the base area A_b is equal to the heat transferred from the same area A_b to the surrounding medium. An effectiveness of $\varepsilon_{fin} < 1$ indicates that the fin actually acts as *insulation*, slowing down the heat transfer from the surface. This

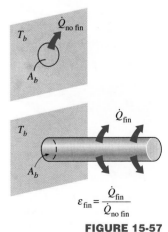

$$\varepsilon_{fin} = \frac{\dot{Q}_{fin}}{\dot{Q}_{no\ fin}}$$

FIGURE 15-57

The effectiveness of a fin.

situation can occur when fins made of low thermal conductivity materials are used. An effectiveness of $\varepsilon_{\text{fin}} > 1$ indicates that fins are *enhancing* heat transfer from the surface, as they should. However, the use of fins cannot be justified unless ε_{fin} is sufficiently larger than 1. Finned surfaces are designed on the basis of *maximizing* effectiveness for a specified cost or *minimizing* cost for a desired effectiveness.

Note that both the fin efficiency and fin effectiveness are related to the performance of the fin, but they are different quantities. However, they are related to each other by

$$\varepsilon_{\text{fin}} = \frac{\dot{Q}_{\text{fin}}}{\dot{Q}_{\text{no fin}}} = \frac{\dot{Q}_{\text{fin}}}{hA_b\,(T_b - T_\infty)} = \frac{\eta_{\text{fin}} hA_{\text{fin}}\,(T_b - T_\infty)}{hA_b\,(T_b - T_\infty)} = \frac{A_{\text{fin}}}{A_b}\,\eta_{\text{fin}} \quad (15\text{-}73)$$

Therefore, the fin effectiveness can be determined easily when the fin efficiency is known, or vice versa.

The rate of heat transfer from a sufficiently *long* fin of *uniform* cross-section under steady conditions is given by Eq. 15-61. Substituting this relation into Eq. 15-72, the effectiveness of such a long fin is determined to be

$$\varepsilon_{\text{long fin}} = \frac{\dot{Q}_{\text{fin}}}{\dot{Q}_{\text{no fin}}} = \frac{\sqrt{hpkA_c}\,(T_b - T_\infty)}{hA_b\,(T_b - T_\infty)} = \sqrt{\frac{kp}{hA_c}} \quad (15\text{-}74)$$

since $A_c = A_b$ in this case. We can draw several important conclusions from the fin effectiveness relation above for consideration in the design and selection of the fins:

- The *thermal conductivity k* of the fin material should be as high as possible. Thus it is no coincidence that fins are made from metals, with copper, aluminum, and iron being the most common ones. Perhaps the most widely used fins are made of aluminum because of its low cost and weight and its resistance to corrosion.
- The ratio of the *perimeter* to the *cross-sectional area* of the fin p/A_c should be as high as possible. This criterion is satisfied by *thin* plate fins or *slender* pin fins.
- The use of fins is *most effective* in applications involving a *low convection heat transfer coefficient*. Thus, the use of fins is more easily justified when the medium is a *gas* instead of a *liquid* and the heat transfer is by *natural convection* instead of by forced convection. Therefore, it is no coincidence that in liquid-to-gas heat exchangers such as the car radiator, fins are placed on the *gas* side.

When determining the rate of heat transfer from a finned surface, we must consider the *unfinned portion* of the surface as well as the *fins*. Therefore, the rate of heat transfer for a surface containing *n* fins can be expressed as

$$\begin{aligned}
\dot{Q}_{\text{total, fin}} &= \dot{Q}_{\text{unfin}} + \dot{Q}_{\text{fin}} \\
&= hA_{\text{unfin}}\,(T_b - T_\infty) + \eta_{\text{fin}} hA_{\text{fin}}\,(T_b - T_\infty) \quad (15\text{-}75) \\
&= h(A_{\text{unfin}} + \eta_{\text{fin}} A_{\text{fin}})(T_b - T_\infty)
\end{aligned}$$

We can also define an **overall effectiveness** for a finned surface as the ratio of the total heat transfer from the finned surface to the heat transfer from the same surface if there were no fins,

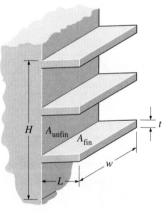

$A_{\text{no fin}} = w \times H$
$A_{\text{unfin}} = w \times H - 3 \times (t \times w)$
$A_{\text{fin}} = 2 \times L \times w + t \times w$ (one fin)
$\quad\quad \approx 2 \times L \times w$

FIGURE 15-58

Various surface areas associated with a rectangular surface with three fins.

$$\varepsilon_{\text{fin, overall}} = \frac{\dot{Q}_{\text{total, fin}}}{\dot{Q}_{\text{total, no fin}}} = \frac{h(A_{\text{unfin}} + \eta_{\text{fin}} A_{\text{fin}})(T_b - T_\infty)}{hA_{\text{no fin}}(T_b - T_\infty)} \quad (15\text{-}76)$$

where $A_{\text{no fin}}$ is the area of the surface when there are no fins, A_{fin} is the total surface area of all the fins on the surface, and A_{unfin} is the area of the unfinned portion of the surface (Fig. 15-58). Note that the overall fin effectiveness depends on the fin density (number of fins per unit length) as well as the effectiveness of the individual fins. The overall effectiveness is a better measure of the performance of a finned surface than the effectiveness of the individual fins.

Proper Length of a Fin

An important step in the design of a fin is the determination of the appropriate length of the fin once the fin material and the fin cross-section are specified. You may be tempted to think that the longer the fin, the larger the surface area and thus the higher the rate of heat transfer. Therefore, for maximum heat transfer, the fin should be infinitely long. However, the temperature drops along the fin exponentially and reaches the environment temperature at some length. The part of the fin beyond this length does not contribute to heat transfer since it is at the temperature of the environment, as shown in Fig. 15-59. Therefore, designing such an "extra long" fin is out of the question since it results in material waste, excessive weight, and increased size and thus increased cost with no benefit in return (in fact, such a long fin will hurt performance since it will suppress fluid motion and thus reduce the convection heat transfer coefficient). Fins that are so long that the temperature approaches the environment temperature cannot be recommended either since the little increase in heat transfer at the tip region cannot justify the large increase in the weight and cost.

To get a sense of the proper length of a fin, we compare heat transfer from a fin of finite length to heat transfer from an infinitely long fin under the same conditions. The ratio of these two heat transfers is

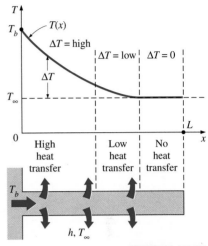

FIGURE 15-59

Because of the gradual temperature drop along the fin, the region near the fin tip makes little or no contribution to heat transfer.

$$\text{Heat transfer} \atop \text{ratio:} \quad \frac{\dot{Q}_{\text{fin}}}{\dot{Q}_{\text{long fin}}} = \frac{\sqrt{hpkA_c}\,(T_b - T_\infty)\tanh aL}{\sqrt{hpkA_c}\,(T_b - T_\infty)} = \tanh aL \quad (15\text{-}77)$$

Using a hand calculator, the values of $\tanh aL$ are evaluated for some values of aL and the results are given in Table 15-5. We observe from the table that heat transfer from a fin increases with aL almost linearly at first, but the curve reaches a plateau later and reaches a value for the infinitely long fin at about $aL = 5$. Therefore, a fin whose length is $L = \frac{1}{5}a$ can be considered to be an infinitely long fin. We also observe that reducing the fin length by half in that case (from $aL = 5$ to $aL = 2.5$) causes a drop of just 1 percent in heat transfer. We certainly would not hesitate sacrificing 1 percent in heat transfer performance in return for 50 percent reduction in the size and possibly the cost of the fin. In practice, a fin length that corresponds to about $aL = 1$ will transfer 76.2 percent of the heat that can be transferred by an infinitely long fin, and thus it should offer a good compromise between heat transfer performance and the fin size.

TABLE 15-5

The variation of heat transfer from a fin relative to that from an infinitely long fin

aL	$\dfrac{\dot{Q}_{\text{fin}}}{\dot{Q}_{\text{long fin}}} = \tanh aL$
0.1	0.100
0.2	0.197
0.5	0.462
1.0	0.762
1.5	0.905
2.0	0.964
2.5	0.987
3.0	0.995
4.0	0.999
5.0	1.000

A common approximation used in the analysis of fins is to assume the fin temperature varies in one direction only (along the fin length) and the temperature variation along other directions is negligible. Perhaps you are wondering if this one-dimensional approximation is a reasonable one. This is certainly the case for fins made of thin metal sheets such as the fins on a car radiator, but we wouldn't be so sure for fins made of thick materials. Studies have shown that the error involved in one-dimensional fin analysis is negligible (less than about 1 percent) when

$$\frac{h\delta}{k} < 0.2$$

where δ is the characteristic thickness of the fin, which is taken to be the plate thickness t for rectangular fins and the diameter D for cylindrical ones.

Specially designed finned surfaces called *heat sinks,* which are commonly used in the cooling of electronic equipment, involve one-of-a-kind complex geometries, as shown in Table 15-6. The heat transfer performance of heat sinks is usually expressed in terms of their *thermal resistances R* in °C/W, which is defined as

$$\dot{Q}_{\text{fin}} = \frac{T_b - T_\infty}{R} = hA_{\text{fin}}\,\eta_{\text{fin}}\,(T_b - T_\infty) \qquad (15\text{-}78)$$

A small value of thermal resistance indicates a small temperature drop across the heat sink, and thus a high fin efficiency.

EXAMPLE 15-12   Maximum Power Dissipation of a Transistor

Power transistors that are commonly used in electronic devices consume large amounts of electric power. The failure rate of electronic components increases almost exponentially with operating temperature. As a rule of thumb, the failure rate of electronic components is halved for each 10°C reduction in the junction operating temperature. Therefore, the operating temperature of electronic components is kept below a safe level to minimize the risk of failure.

The sensitive electronic circuitry of a power transistor at the junction is protected by its case, which is a rigid metal enclosure. Heat transfer characteristics of a power transistor are usually specified by the manufacturer in terms of the case-to-ambient thermal resistance, which accounts for both the natural convection and radiation heat transfers.

The case-to-ambient thermal resistance of a power transistor that has a maximum power rating of 10 W is given to be 20°C/W. If the case temperature of the transistor is not to exceed 85°C, determine the power at which this transistor can be operated safely in an environment at 25°C.

Solution   The maximum power rating of a transistor whose case temperature is not to exceed 85°C is to be determined.

Assumptions   **1** Steady operating conditions exist. **2** The transistor case is isothermal at 85°C.

Properties   The case-to-ambient thermal resistance is given to be 20°C/W.

Analysis   The power transistor and the thermal resistance network associated with it are shown in Fig. 15-60. We notice from the thermal resistance network that there is a single resistance of 20°C/W between the case at $T_c = 85$°C and the ambient at $T_\infty = 25$°C, and thus the rate of heat transfer is

$$\dot{Q} = \left(\frac{\Delta T}{R}\right)_{\text{case-ambient}} = \frac{T_c - T_\infty}{R_{\text{case-ambient}}} = \frac{(85 - 25)\text{°C}}{20\text{°C/W}} = \textbf{3 W}$$

Therefore, this power transistor should not be operated at power levels above 3 W if its case temperature is not to exceed 85°C.

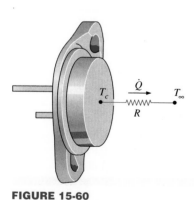

FIGURE 15-60

Schematic for Example 15-12.

TABLE 15-6

Combined natural convection and radiation thermal resistance of various
heat sinks used in the cooling of electronic devices between the heat sink and the
surroundings. All fins are made of aluminum 6063T-5, are black anodized, and
are 76 mm (3 in.) long (courtesy of Vemaline Products, Inc.).

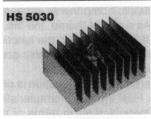
HS 5030

$R = 0.9°C/W$ (vertical)
$R = 1.2°C/W$ (horizontal)

Dimensions: 76 mm $\times$ 105 mm $\times$ 44 mm
Surface area: 677 cm^2

HS 6065

$R = 5°C/W$

Dimensions: 76 mm $\times$ 38 mm $\times$ 24 mm
Surface area: 387 cm^2

HS 6071

$R = 1.4°C/W$ (vertical)
$R = 1.8°C/W$ (horizontal)

Dimensions: 76 mm $\times$ 92 mm $\times$ 26 mm
Surface area: 968 cm^2

HS 6105

$R = 1.8°C/W$ (vertical)
$R = 2.1°C/W$ (horizontal)

Dimensions: 76 mm $\times$ 127 mm $\times$ 91 mm
Surface area: 677 cm^2

HS 6115

$R = 1.1°C/W$ (vertical)
$R = 1.3°C/W$ (horizontal)

Dimensions: 76 mm $\times$ 102 mm $\times$ 25 mm
Surface area: 929 cm^2

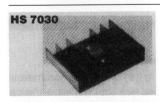
HS 7030

$R = 2.9°C/W$ (vertical)
$R = 3.1°C/W$ (horizontal)

Dimensions: 76 mm $\times$ 97 mm $\times$ 19 mm
Surface area: 290 cm^2

Discussion This transistor can be used at higher power levels by attaching it to
a heat sink (which lowers the thermal resistance by increasing the heat transfer
surface area, as discussed in the next example) or by using a fan (which lowers
the thermal resistance by increasing the convection heat transfer coefficient).

EXAMPLE 15-13 Selecting a Heat Sink for a Transistor

A 60-W power transistor is to be cooled by attaching it to one of the commercially available heat sinks shown in Table 15-6. Select a heat sink that will allow the case temperature of the transistor not to exceed 90°C in the ambient air at 30°C.

Solution A commercially available heat sink from Table 15-6 is to be selected to keep the case temperature of a transistor below 90°C.

Assumptions **1** Steady operating conditions exist. **2** The transistor case is isothermal at 90°C. **3** The contact resistance between the transistor and the heat sink is negligible.

Analysis The rate of heat transfer from a 60-W transistor at full power is $\dot{Q} = 60$ W. The thermal resistance between the transistor attached to the heat sink and the ambient air for the specified temperature difference is determined to be

$$\dot{Q} = \frac{\Delta T}{R} \longrightarrow R = \frac{\Delta T}{\dot{Q}} = \frac{(90 - 30)°C}{60 \text{ W}} = 1.0°C/W$$

Therefore, the thermal resistance of the heat sink should be below 1.0°C/W. An examination of Table 15-6 reveals that the HS 5030, whose thermal resistance is 0.9°C/W in the vertical position, is the only heat sink that will meet this requirement.

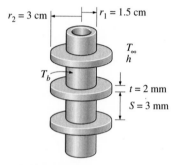

FIGURE 15-61

Schematic for Example 15-14.

EXAMPLE 15-14 Effect of Fins on Heat Transfer from Steam Pipes

Steam in a heating system flows through tubes whose outer diameter is $D_1 = 3$ cm and whose walls are maintained at a temperature of 120°C. Circular aluminum fins ($k = 180$ W/m · °C) of outer diameter $D_2 = 6$ cm and constant thickness $t = 2$ mm are attached to the tube, as shown in Fig. 15-61. The space between the fins is 3 mm, and thus there are 200 fins per meter length of the tube. Heat is transferred to the surrounding air at $T_\infty = 25°C$, with a combined heat transfer coefficient of $h = 60$ W/m² · °C. Determine the increase in heat transfer from the tube per meter of its length as a result of adding fins.

Solution Circular aluminum fins are to be attached to the tubes of a heating system. The increase in heat transfer from the tubes per unit length as a result of adding fins is to be determined.

Assumptions **1** Steady operating conditions exist. **2** The heat transfer coefficient is uniform over the entire fin surfaces. **3** Thermal conductivity is constant. **4** Heat transfer by radiation is negligible.

Properties The thermal conductivity of the fins is given to be $k = 180$ W/m · °C.

Analysis In the case of no fins, heat transfer from the tube per meter of its length is determined from Newton's law of cooling to be

$$A_{\text{no fin}} = \pi D_1 L = \pi(0.03 \text{ m})(1 \text{ m}) = 0.0942 \text{ m}^2$$
$$\dot{Q}_{\text{no fin}} = hA_{\text{no fin}}(T_b - T_\infty)$$
$$= (60 \text{ W/m}^2 \cdot °C)(0.0942 \text{ m}^2)(120 - 25)°C$$
$$= 537 \text{ W}$$

The efficiency of the circular fins attached to a circular tube is plotted in Fig. 15-56. Noting that $L = \frac{1}{2}(D_2 - D_1) = \frac{1}{2}(0.06 - 0.03) = 0.015$ m in this case, we have

$$\left. \begin{array}{l} \dfrac{r_2 + \frac{1}{2}t}{r_1} = \dfrac{(0.03 + \frac{1}{2} \times 0.002) \text{ m}}{0.015 \text{ m}} = 2.07 \\[3mm] (L + \frac{1}{2}t) \sqrt{\dfrac{h}{kt}} = (0.015 + \frac{1}{2} \times 0.002) \text{ m} \times \sqrt{\dfrac{60 \text{ W/m}^2 \cdot °C}{(180 \text{ W/m} \cdot °C)(0.002 \text{ m})}} = 0.207 \end{array} \right\} \eta_{\text{fin}} = 0.95$$

In the figure (Fig. 15-61):

$r_2 = 3$ cm $r_1 = 1.5$ cm

T_∞, h

T_b

$t = 2$ mm

$S = 3$ mm

$$A_{fin} = 2\pi(r_2^2 - r_1^2) + 2\pi r_2 t$$
$$= 2\pi[(0.03 \text{ m})^2 - (0.015 \text{ m})^2] + 2\pi(0.03 \text{ m})(0.002 \text{ m})$$
$$= 0.00462 \text{ m}^2$$

$$\dot{Q}_{fin} = \eta_{fin}\dot{Q}_{fin, max} = \eta_{fin}hA_{fin}(T_b - T_\infty)$$
$$= 0.95(60 \text{ W/m}^2 \cdot °C)(0.00462 \text{ m}^2)(120 - 25)°C$$
$$= 25.0 \text{ W}$$

Noting that the space between the two fins is 3 mm, heat transfer from the unfinned portion of the tube is

$$A_{unfin} = \pi D_1 S = \pi(0.03 \text{ m})(0.003 \text{ m}) = 0.000283 \text{ m}^2$$
$$\dot{Q}_{unfin} = hA_{unfin}(T_b - T_\infty)$$
$$= (60 \text{ W/m}^2 \cdot °C)(0.000283 \text{ m}^2)(120 - 25)°C$$
$$= 1.60 \text{ W}$$

Noting that there are 200 fins and thus 200 interfin spacings per meter length of the tube, the total heat transfer from the finned tube becomes

$$\dot{Q}_{total, fin} = n(\dot{Q}_{fin} + \dot{Q}_{unfin}) = 200(25.0 + 1.6) \text{ W} = 5320 \text{ W}$$

Therefore, the increase in heat transfer from the tube per meter of its length as a result of the addition of fins is

$$\dot{Q}_{increase} = \dot{Q}_{total, fin} - \dot{Q}_{no fin} = 5320 - 537 = \textbf{4783 W} \qquad \text{(per m tube length)}$$

Discussion   The overall effectiveness of the finned tube is

$$\varepsilon_{fin, overall} = \frac{\dot{Q}_{total, fin}}{\dot{Q}_{total, no fin}} = \frac{5320 \text{ W}}{537 \text{ W}} = 9.9$$

That is, the rate of heat transfer from the steam tube increases by a factor of almost 10 as a result of adding fins. This explains the widespread use of the finned surfaces.

EXAMPLE 15-15   Cost of Heat Loss through Walls in Winter

Consider an electrically heated house whose walls are 9 ft high and have an *R*-value of insulation of 13 (i.e., a thickness-to-thermal conductivity ratio of $L/k = 13 \text{ h} \cdot \text{ft}^2 \cdot °F/\text{Btu}$). Two of the walls of the house are 40 ft long and the others are 30 ft long. The house is maintained at 75°F at all times, while the temperature of the outdoors varies. Determine the amount of heat lost through the walls of the house on a certain day during which the average temperature of the outdoors is 45°F. Also, determine the cost of this heat loss to the homeowner if the unit cost of electricity is $0.075/kWh. For combined convection and radiation heat transfer coefficients, use the ASHRAE (American Society of Heating, Refrigeration, and Air Conditioning Engineers) recommended values of $h_i = 1.46 \text{ Btu/h} \cdot \text{ft}^2 \cdot °F$ for the inner surface of the walls and $h_o = 4.0 \text{ Btu/h} \cdot \text{ft}^2 \cdot °F$ for the outer surface of the walls under 15 mph wind conditions in winter.

Solution   An electrically heated house with R-13 insulation is considered. The amount of heat lost through the walls and its cost are to be determined.

Assumptions   **1** The indoor and outdoor air temperatures have remained at the given values for the entire day so that heat transfer through the walls is steady. **2** Heat transfer through the walls is one-dimensional since any significant temperature gradients in this case will exist in the direction from the indoors to the outdoors. **3** The radiation effects are accounted for in the heat transfer coefficients.

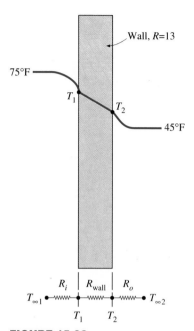

FIGURE 15-62
Schematic for Example 15-15.

Analysis This problem involves conduction through the wall and convection at its surfaces and can best be handled by making use of the thermal resistance concept and drawing the thermal resistance network, as shown in Fig. 15-62. The heat transfer surface area of the walls is

$$A = \text{Circumference} \times \text{Height} = (2 \times 30 \text{ ft} + 2 \times 40 \text{ ft})(9 \text{ ft}) = 1260 \text{ ft}^2$$

Then the individual resistances are evaluated from their definitions to be

$$R_i = R_{\text{conv}, i} = \frac{1}{h_i A} = \frac{1}{(1.46 \text{ Btu/h} \cdot \text{ft}^2 \cdot \text{°F})(1260 \text{ ft}^2)} = 0.00054 \text{ h} \cdot \text{°F/Btu}$$

$$R_{\text{wall}} = \frac{L}{kA} = \frac{R\text{-value}}{A} = \frac{13 \text{ h} \cdot \text{ft}^2 \cdot \text{°F/Btu}}{1260 \text{ ft}^2} = 0.01032 \text{ h} \cdot \text{°F/Btu}$$

$$R_o = R_{\text{conv}, o} = \frac{1}{h_c A} = \frac{1}{(4.0 \text{ Btu/h} \cdot \text{ft}^2 \cdot \text{°F})(1260 \text{ ft}^2)} = 0.00020 \text{ h} \cdot \text{°F/Btu}$$

Noting that all three resistances are in series, the total resistance is

$$R_{\text{total}} = R_i + R_{\text{wall}} + R_o = 0.00054 + 0.01032 + 0.00020 = 0.01106 \text{ h} \cdot \text{°F/Btu}$$

Then the steady rate of heat transfer through the walls of the house becomes

$$\dot{Q} = \frac{T_{\infty 1} - T_{\infty 2}}{R_{\text{total}}} = \frac{(75 - 45)\text{°F}}{0.01106 \text{ h} \cdot \text{°F/Btu}} = 2712 \text{ Btu/h}$$

Finally, the total amount of heat lost through the walls during a 24-h period and its cost to the home owner are

$$Q = \dot{Q} \, \Delta t = (2712 \text{ Btu/h})(24\text{-h/day}) = \textbf{65,099 Btu/day} = \textbf{19.1 kWh/day}$$

since 1 kWh = 3412 Btu, and

$$\text{Heating cost} = (\text{Energy lost})(\text{Cost of energy}) = (19.1 \text{ kWh/day})(\$0.075/\text{kWh})$$
$$= \textbf{\$1.43/day}$$

Discussion The heat losses through the walls of the house will cost the home owner that day $1.43 worth of electricity.

15-8 ■ SUMMARY

One-dimensional heat transfer through a simple or composite body exposed to convection from both sides to mediums at temperatures $T_{\infty 1}$ and $T_{\infty 2}$ can be expressed as

$$\dot{Q} = \frac{T_{\infty 1} - T_{\infty 2}}{R_{\text{total}}} \qquad \text{(W)}$$

where R_{total} is the total thermal resistance between the two mediums. For a plane wall exposed to convection on both sides, the total resistance is expressed as

$$R_{\text{total}} = R_{\text{conv}, 1} + R_{\text{wall}} + R_{\text{conv}, 2} = \frac{1}{h_1 A} + \frac{L}{kA} + \frac{1}{h_2 A} \qquad \text{(°C/W)}$$

This relation can be extended to plane walls that consist of two or more layers by adding an additional resistance for each additional layer. The elementary thermal resistance relations can be expressed as follows:

Conduction resistance (plane wall): $R_{wall} = \dfrac{L}{kA}$

Conduction resistance (cylinder): $R_{cyl} = \dfrac{\ln(r_2/r_1)}{2\pi Lk}$

Conduction resistance (sphere): $R_{sph} = \dfrac{r_2 - r_1}{4\pi r_1 r_2 k}$

Convection resistance: $R_{conv} = \dfrac{1}{hA}$

Interface resistance: $R_{interface} = \dfrac{1}{h_c A} = \dfrac{R_c}{A}$

Radiation resistance: $R_{rad} = \dfrac{1}{h_{rad} A}$

where h_c is the thermal contact conductance, R_c is the thermal contact resistance, and the radiation heat transfer coefficient is defined as

$$h_{rad} = \varepsilon\sigma(T_s^2 + T_{surr}^2)(T_s + T_{surr}) \qquad (\text{W/m}^2 \cdot \text{K})$$

Once the rate of heat transfer is available, the *temperature drop* across any layer can be determined from

$$\Delta T = \dot{Q} R \qquad (°\text{C})$$

The thermal resistance concept can also be used to solve steady heat transfer problems involving parallel layers or combined series-parallel arrangements.

Adding insulation to a cylindrical pipe or a spherical shell will increase the rate of heat transfer if the outer radius of the insulation is less than the *critical radius of insulation,* defined as

$$r_{cr,\, cylinder} = \frac{k_{ins}}{h}$$

$$r_{cr,\, sphere} = \frac{2k_{ins}}{h}$$

Thermal insulations are materials or a combination of materials that are used primarily to provide resistance to heat flow. Thermal insulations are used for various reasons, such as energy conservation, personnel protection and comfort, maintenance of process temperature, reduction of temperature variation and fluctuations, condensation and corrosion prevention, fire protection, freezing protection, and reduction of noise and vibration.

The effectiveness of an insulation is often given in terms of its *R-value,* the thermal resistance of the material per unit surface area, expressed as

$$R\text{-value} = \frac{L}{k} \qquad (\text{flat insulation})$$

where L is the thickness and k is the thermal conductivity of the material.

Optimum thickness of insulation is usually determined on the basis of minimum combined cost of insulation and heat lost.

Finned surfaces are commonly used in practice to enhance heat transfer. Fins enhance heat transfer from a surface by exposing a larger surface area to convection. The temperature distribution along the fin for very long fins and for fins with negligible heat transfer at the fin are given by

Very long fin:
$$\frac{T(x) - T_\infty}{T_b - T_\infty} = e^{-x\sqrt{hp/kA_c}}$$

Adiabatic fin tip:
$$\frac{T(x) - T_\infty}{T_b - T_\infty} = \frac{\cosh a(L - x)}{\cosh aL}$$

where $a = \sqrt{hp/kA_c}$, p is the perimeter, and A_c is the cross-sectional area of the fin. The rates of heat transfer for both cases are given to be

Very long fin:
$$\dot{Q}_{\text{long fin}} = -kA_c \frac{dT}{dx}\bigg|_{x=0} = \sqrt{hpkA_c}\,(T_b - T_\infty)$$

Adiabatic fin tip:
$$\dot{Q}_{\text{insulated tip}} = -kA_c \frac{dT}{dx}\bigg|_{x=0} = \sqrt{hpkA_c}\,(T_b - T_\infty)\tanh aL$$

Fins exposed to convection at their tips can be treated as fins with insulated tips by using the corrected length $L_c = L + A_c/p$ instead of the actual fin length.

The temperature of a fin drops along the fin, and thus the heat transfer from the fin will be less because of the decreasing temperature difference toward the fin tip. To account for the effect of this decrease in temperature on heat transfer, we define *fin efficiency* as

$$\eta_{\text{fin}} = \frac{\dot{Q}_{\text{fin}}}{\dot{Q}_{\text{fin, max}}} = \frac{\text{Actual heat transfer rate from the fin}}{\begin{array}{c}\text{Ideal heat transfer rate from the fin if}\\ \text{the entire fin were at base temperature}\end{array}}$$

When the fin efficiency is available, the rate of heat transfer from a fin can be determined from

$$\dot{Q}_{\text{fin}} = \eta_{\text{fin}}\dot{Q}_{\text{fin, max}} = \eta_{\text{fin}}hA_{\text{fin}}\,(T_b - T_\infty)$$

The performance of the fins is judged on the basis of the enhancement in heat transfer relative to the no-fin case and is expressed in terms of the *fin effectiveness* ε_{fin}, defined as

$$\varepsilon_{\text{fin}} = \frac{\dot{Q}_{\text{fin}}}{\dot{Q}_{\text{no fin}}} = \frac{\dot{Q}_{\text{fin}}}{hA_b\,(T_b - T_\infty)} = \frac{\begin{array}{c}\text{Heat transfer rate from}\\ \text{the fin of } base\ area\ A_b\end{array}}{\begin{array}{c}\text{Heat transfer rate from}\\ \text{the surface of } area\ A_b\end{array}}$$

Here, A_b is the cross-sectional area of the fin at the base and $\dot{Q}_{\text{no fin}}$ represents the rate of heat transfer from this area if no fins are attached to the surface. The *overall effectiveness* for a finned surface is defined as the ratio of the total heat transfer from the finned surface to the heat transfer from the same surface if there were no fins,

$$\varepsilon_{\text{fin, overall}} = \frac{\dot{Q}_{\text{total, fin}}}{\dot{Q}_{\text{total, no fin}}} = \frac{h(A_{\text{unfin}} + \eta_{\text{fin}}A_{\text{fin}})(T_b - T_\infty)}{hA_{\text{no fin}}\,(T_b - T_\infty)}$$

Fin efficiency and fin effectiveness are related to each other by

$$\varepsilon_{\text{fin}} = \frac{A_{\text{fin}}}{A_b}\,\eta_{\text{fin}}$$

1. American Society of Heating, Refrigeration, and Air Conditioning Engineers. *Handbook of Fundamentals.* Atlanta: ASHRAE, 1993.

2. R. V. Andrews. "Solving Conductive Heat Transfer Problems with Electrical-Analogue Shape Factors." *Chemical Engineering Progress* 5 (1955), p. 67.

3. R. Barron. *Cryogenic Systems.* New York: McGraw-Hill, 1967.

4. Y. A. Çengel. *Heat Transfer: A Practical Approach.* New York: McGraw-Hill, 1998.

5. C. Danish. "Factors to Consider in Planning Reinsulation Projects." *Plant Engineering,* November 1985, pp. 81–83.

6. W. M. Edmunds. "Residential Insulation." *ASTM Standardization News,* January 1989, pp. 36–39.

7. L. S. Fletcher. "Recent Developments in Contact Conductance Heat Transfer." *Journal of Heat Transfer* 110, no. 4B (1988), pp. 1059–79.

8. E. Fried. "Thermal Conduction Contribution to Heat Transfer at Contacts." *Thermal Conductivity,* vol. 2, ed. R. P. Tye. London: Academic Press, 1969.

9. K. A. Gardner. "Efficiency of Extended Surfaces." *Trans. ASME* 67 (1945), pp. 621–31.

10. J. P. Holman. *Heat Transfer.* 7th ed. New York: McGraw-Hill, 1990.

11. F. P. Incropera and D. P. DeWitt. *Introduction to Heat Transfer.* 2nd ed. New York: John Wiley & Sons, 1990.

12. W. I. Irwin. "Insulate Intelligently." *Chemical Engineering Progress,* May 1991, pp. 51–55.

13. D. Q. Kern and A. D. Kraus. *Extended Surface Heat Transfer.* New York: McGraw-Hill, 1972.

14. F. Kreith and M. S. Bohn. *Principles of Heat Transfer.* 5th ed. St. Paul, MN: West Publishing, 1993.

15. V. M. Liss. "Selecting Thermal Insulation." *Chemical Engineering,* May 26, 1986, pp. 103–5.

16. W. A. Lotz. "Facts About Thermal Insulation." *ASHRAE Journal,* June 1969, pp. 83–84.

17. M. N. Özişik. *Heat Transfer—A Basic Approach.* New York: McGraw-Hill, 1985.

18. G. P. Peterson. "Thermal Contact Resistance in Waste Heat Recovery Systems." *Proceedings of the 18th ASME/ETCE Hydrocarbon Processing Symposium.* Dallas, TX, 1987, pp. 45–51.

19. F. J. Powell. "Thermal Insulation—Still Number One." *ASTM Standardization News,* January 1989, pp. 32–35.

20. S. Song, M. M. Yovanovich, and F. O. Goodman. "Thermal Gap Conductance of Conforming Surfaces in Contact." *Journal of Heat Transfer* 115 (1993), p. 533.

21. N. V. Suryanarayana. *Engineering Heat Transfer.* St. Paul, MN: West Publishing, 1995.

22. L. C. Thomas. *Heat Transfer.* Englewood Cliffs, NJ: Prentice Hall, 1992.

PROBLEMS*

Steady Heat Conduction in Plane Walls

15-1C Consider one-dimensional heat conduction through a cylindrical rod of diameter D and length L. What is the heat transfer area of the rod if (*a*) the lateral surfaces of the rod are insulated and (*b*) the top and bottom surfaces of the rod are insulated?

15-2C Consider a 1.5-m $\times$ 2-m glass window whose thickness is 0.01 m. What is the heat transfer area of the window?

15-3C Consider heat conduction through a plane wall. Does the energy content of the wall change during steady heat conduction? How about during transient conduction? Explain.

15-4C Consider heat conduction through a wall of thickness L and area A. Under what conditions will the temperature distributions in the wall be a straight line?

15-5C What does the thermal resistance of a medium represent?

15-6C How is the combined heat transfer coefficient defined? What convenience does it offer in heat transfer calculations?

15-7C Can we define the convection resistance per unit surface area as the inverse of the convection heat transfer coefficient?

15-8C Why are the convection and the radiation resistances at a surface in parallel instead of being in series?

15-9C Consider a surface of area A at which the convection and radiation heat transfer coefficients are h_{conv} and h_{rad}, respectively. Explain how you would determine (*a*) the single equivalent heat transfer coefficient, and (*b*) the equivalent thermal resistance. Assume the medium and the surrounding surfaces are at the same temperature.

15-10C How does the thermal resistance network associated with a single-layer plane wall differ from the one associated with a five-layer composite wall?

15-11C Consider steady one-dimensional heat transfer through a multilayer medium. If the rate of heat transfer $\dot{Q}$ is known, explain how you would determine the temperature drop across each layer.

15-12C Consider steady one-dimensional heat transfer through a plane wall exposed to convection from both sides to environments at known temperatures $T_{\infty 1}$ and $T_{\infty 2}$ with known heat transfer coefficients h_1 and h_2. Once the rate of heat transfer $\dot{Q}$ has been evaluated, explain how you would determine the temperature of each surface.

15-13C Someone comments that a microwave oven can be viewed as a conventional oven with zero convection resistance at the surface of the food. Is this an accurate statement?

---

*Students are encouraged to answer *all* the concept "C" questions.

15-14C Consider a window glass consisting of two 4-mm-thick glass sheets pressed tightly against each other. Compare the heat transfer rate through this window with that of one consisting of a single 8-mm-thick glass sheet under identical conditions.

15-15C Consider steady heat transfer through the wall of a room in winter. The convection heat transfer coefficient at the outer surface of the wall is three times that of the inner surface as a result of the winds. On which surface of the wall do you think the temperature will be closer to the surrounding air temperature? Explain.

15-16C The bottom of a pan is made of a 4-mm-thick aluminum layer. In order to increase the rate of heat transfer through the bottom of the pan, someone proposes a design for the bottom that consists of a 3-mm-thick copper layer sandwiched between two 2-mm-thick aluminum layers. Will the new design conduct heat better? Explain. Assume perfect contact between the layers.

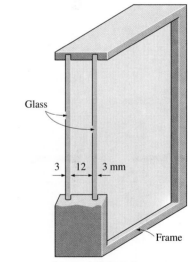

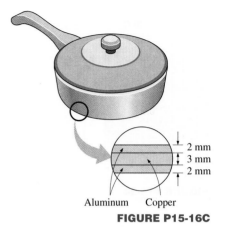

FIGURE P15-16C

15-17C Consider two cold canned drinks, one wrapped in a blanket and the other placed on a table in the same room. Which drink will warm up faster?

15-18 Consider a 4-m-high, 6-m-wide, and 0.3-m-thick brick wall whose thermal conductivity is $k = 0.8$ W/m · °C . On a certain day, the temperatures of the inner and the outer surfaces of the wall are measured to be 14°C and 6°C, respectively. Determine the rate of heat loss through the wall on that day.

15-19 Consider a 1.2-m-high and 2-m-wide glass window whose thickness is 6 mm and thermal conductivity is $k = 0.78$ W/m · °C. Determine the steady rate of heat transfer through this glass window and the temperature of its inner surface for a day during which the room is maintained at 24°C while the temperature of the outdoors is −5°C. Take the convection heat transfer coefficients on the inner and outer surfaces of the window to be $h_1 = 10$ W/m² · °C and $h_2 = 25$ W/m² · °C, and disregard any heat transfer by radiation.

15-20 Consider a 1.2-m-high and 2-m-wide double-pane window consisting of two 3-mm-thick layers of glass ($k = 0.78$ W/m · °C) separated by a 12-mm-wide stagnant air space ($k = 0.026$ W/m · °C). Determine the steady rate of heat transfer through this double-pane window and the temperature of its inner surface for a day during which the room is maintained at 24°C while the temperature of the outdoors is −5°C. Take the convection heat transfer coefficients on the inner and outer surfaces of the window to be $h_1 = 10$ W/m² · °C and $h_2 = 25$ W/m² · °C, and disregard any heat transfer by radiation.
Answers: 113 W, 19.3°C

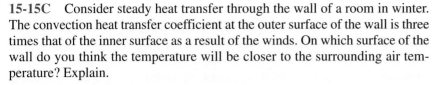

FIGURE P15-20

15-21 Repeat Prob. 15-20, assuming the space between the two glass layers is evacuated.

15-22E Consider an electrically heated brick house ($k = 0.40$ Btu/h · ft · °F) whose walls are 9 ft high and 1 ft thick. Two of the walls of the house are 40 ft long and the others are 30 ft long. The house is maintained at 70°F at all times while the temperature of the outdoors varies. On a certain day, the temperature of the inner surface of the walls is measured to be at 55°F while the average temperature of the outer surface is observed to remain at 45°F during the day for 10 h and at 35°F at night for 14 h. Determine the amount of heat lost from the house that day. Also determine the cost of that heat loss to the homeowner for an electricity price of $0.075/kWh.

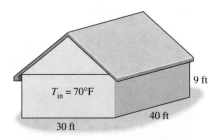

FIGURE P15-22E

15-23 A cylindrical resistor element on a circuit board dissipates 0.15 W of power in an environment at 40°C. The resistor is 1.2 cm long, and has a

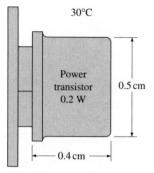

30°C

Power
transistor
0.2 W

0.5 cm

0.4 cm

FIGURE P15-24

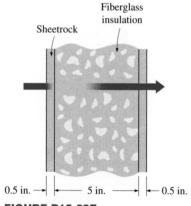

Fiberglass
insulation

Sheetrock

0.5 in. 5 in. 0.5 in.

FIGURE P15-28E

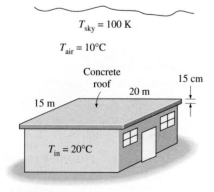

$T_{sky} = 100$ K

$T_{air} = 10$°C

Concrete
roof

15 cm

20 m

15 m

$T_{in} = 20$°C

FIGURE P15-29

diameter of 0.3 cm. Assuming heat to be transferred uniformly from all surfaces, determine (*a*) the amount of heat this resistor dissipates during a 24-h period, (*b*) the heat flux on the surface of the resistor, in W/m^2, and (*c*) the surface temperature of the resistor for a combined convection and radiation heat transfer coefficient of 9 W/m$^2 \cdot$ °C.

15-24 Consider a power transistor that dissipates 0.2 W of power in an environment at 30°C. The transistor is 0.4 cm long and has a diameter of 0.5 cm. Assuming heat to be transferred uniformly from all surfaces, determine (*a*) the amount of heat this resistor dissipates during a 24-h period, in kWh; (*b*) the heat flux on the surface of the transistor, in W/m^2; and (*c*) the surface temperature of the resistor for a combined convection and radiation heat transfer coefficient of 12 W/m$^2 \cdot$ °C.

15-25 A 12-cm × 18-cm circuit board houses on its surface 100 closely spaced logic chips, each dissipating 0.07 W. The heat transfer from the back surface of the board is negligible. If the heat transfer coefficient on the surface of the board is 10 W/m$^2 \cdot$ °C, determine (*a*) the heat flux on the surface of the circuit board, in W/m^2; (*b*) the surface temperature of the chips; and (*c*) the thermal resistance between the surface of the circuit board and the cooling medium, in °C/W.

15-26 Consider a person standing in a room at 20°C with an exposed surface area of 1.7 m^2. The deep body temperature of the human body is 37°C, and the thermal conductivity of the human tissue near the skin is about 0.3 W/m $\cdot$ °C. The body is losing heat at a rate of 150 W by natural convection and radiation to the surroundings. Taking the body temperature 0.5 cm beneath the skin to be 37°C, determine the skin temperature of the person.
Answer: 35.5° C

15-27 Water is boiling in a 25-cm-diameter aluminum pan ($k = 237$ W/m $\cdot$ °C) at 95°C. Heat is transferred steadily to the boiling water in the pan through its 0.5-cm-thick flat bottom at a rate of 600 W. If the inner surface temperature of the bottom of the pan is 108°C, determine (*a*) the boiling heat transfer coefficient on the inner surface of the pan, and (*b*) the outer surface temperature of the bottom of the pan.

15-28E A wall is constructed of two layers of 0.5-in-thick sheetrock ($k = 0.10$ Btu/h $\cdot$ ft $\cdot$ °F), which is a plasterboard made of two layers of heavy paper separated by a layer of gypsum, placed 5 in. apart. The space between the sheetrocks is filled with fiberglass insulation ($k = 0.020$ Btu/h $\cdot$ ft $\cdot$ °F). Determine (*a*) the thermal resistance of the wall, and (*b*) its R-value of insulation in English units.

15-29 The roof of a house consists of a 15-cm-thick concrete slab ($k = 2$ W/m $\cdot$ °C) that is 15 m wide and 20 m long. The convection heat transfer coefficients on the inner and outer surfaces of the roof are 5 and 12 W/m$^2 \cdot$ °C, respectively. On a clear winter night, the ambient air is reported to be at 10°C, while the night sky temperature is 100 K. The house and the interior surfaces of the wall are maintained at a constant temperature of 20°C. The emissivity of both surfaces of the concrete roof is 0.9. Considering both radiation and convection heat transfers, determine the rate of heat transfer through the roof, and the inner surface temperature of the roof.

If the house is heated by a furnace burning natural gas with an efficiency of 80 percent, and the price of natural gas is $0.60/therm (1 therm =

105,500 kJ of energy content), determine the money lost through the roof that night during a 14-h period.

15-30 A 2-m × 1.5-m section of wall of an industrial furnace burning natural gas is not insulated, and the temperature at the outer surface of this section is measured to be 80°C. The temperature of the furnace room is 30°C, and the combined convection and radiation heat transfer coefficient at the surface of the outer furnace is 10 W/m² · °C. It is proposed to insulate this section of the furnace wall with glass wool insulation ($k = 0.038$ W/m · °C) in order to reduce the heat loss by 90 percent. Assuming the outer surface temperature of the metal section still remains at about 80°C, determine the thickness of the insulation that needs to be used.

The furnace operates continuously and has an efficiency of 78 percent. The price of the natural gas is $0.55/therm (1 therm = 105,500 kJ of energy content). If the installation of the insulation will cost $250 for materials and labor, determine how long it will take for the insulation to pay for itself from the energy it saves.

15-31 Repeat Prob. 15-30 for expanded perlite insulation assuming conductivity is $k = 0.052$ W/m · °C.

15-32E Consider a house whose walls are 12 ft high and 40 ft long. Two of the walls of the house have no windows, while each of the other two walls has four windows made of 0.25-in.-thick glass ($k = 0.45$ Btu/h · ft · °F), 3 ft × 5 ft in size. The walls are certified to have an R-value of 19 (i.e., an L/k value of 19 h · ft² · °F/Btu). Disregarding any direct radiation gain or loss through the windows and taking the heat transfer coefficients at the inner and outer surfaces of the house to be 2 and 4 Btu/h · ft² · °F, respectively, determine the ratio of the heat transfer through the walls with and without windows.

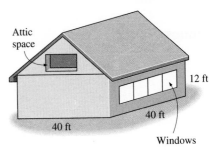

FIGURE P15-32E

15-33 Consider a house that has a 10-m × 20-m base and a 4-m-high wall. All four walls of the house have an R-value of 2.31 m² · °C/W. The two 10-m × 4-m walls have no windows. The third wall has five windows made of 0.5-cm-thick glass ($k = 0.78$ W/m · °C), 1.2 m × 1.8 m in size. The fourth wall has the same size and number of windows, but they are double-paned with a 1.5-cm-thick stagnant air space ($k = 0.026$ W/m · °C) enclosed between two 0.5-cm-thick glass layers. The thermostat in the house is set at 22°C and the average temperature outside at that location is 5°C during the seven-month-long heating season. Disregarding any direct radiation gain or loss through the windows and taking the heat transfer coefficients at the inner and outer surfaces of the house to be 7 and 15 W/m² · °C, respectively, determine the average rate of heat transfer through each wall.

If the house is electrically heated and the price of electricity is $0.09/kWh, determine the amount of money this household will save per heating season by converting the single-pane windows to double-pane windows.

15-34 The wall of a refrigerator is constructed of fiberglass insulation ($k = 0.035$ W/m · °C) sandwiched between two layers of 1-mm-thick sheet metal ($k = 15.1$ W/m · °C). The refrigerated space is maintained at 3°C, and the average heat transfer coefficients at the inner and outer surfaces of the wall are 4 W/m² · °C and 9 W/m² · °C, respectively. The kitchen temperature averages 25°C. It is observed that condensation occurs on the outer surfaces of the refrigerator when the temperature of the outer surface drops to 20°C. Determine the minimum thickness of fiberglass insulation that needs to be used in the wall in order to avoid condensation on the outer surfaces.

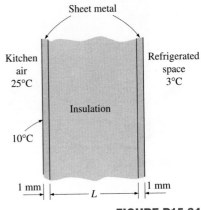

FIGURE P15-34

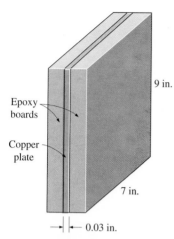

Epoxy
boards

Copper
plate

9 in.

7 in.

0.03 in.

FIGURE P15-36E

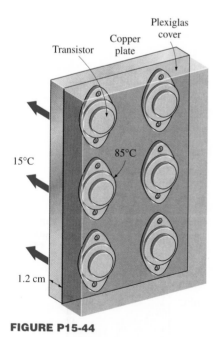

Plexiglas
cover

Copper
plate

Transistor

15°C

85°C

1.2 cm

FIGURE P15-44

15-35 Heat is to be conducted along a circuit board that has a copper layer on one side. The circuit board is 15 cm long and 15 cm wide, and the thicknesses of the copper and epoxy layers are 0.1 mm and 1.2 mm, respectively. Disregarding heat transfer from side surfaces, determine the percentages of heat conduction along the copper ($k = 386$ W/m · °C) and epoxy ($k = 0.26$ W/m · °C) layers. Also determine the effective thermal conductivity of the board. *Answers:* 0.8 percent, 99.2 percent, and 29.9 W/m · °C

15-36E A 0.03-in-thick copper plate ($k = 223$ Btu/h · ft · °F) is sandwiched between two 0.1-in.-thick epoxy boards ($k = 0.15$ Btu/h · ft · °F) that are 7 in. × 9 in. in size. Determine the effective thermal conductivity of the board along its 9-in.-long side. What fraction of the heat conducted along that side is conducted through copper?

Thermal Contact Resistance

15-37C What is thermal contact resistance? How is it related to thermal contact conductance?

15-38C Will the thermal contact resistance be greater for smooth or rough plain surfaces?

15-39C A wall consists of two layers of insulation pressed against each other. Do we need to be concerned about the thermal contact resistance at the interface in a heat transfer analysis or can we just ignore it?

15-40C A plate consists of two thin metal layers pressed against each other. Do we need to be concerned about the thermal contact resistance at the interface in a heat transfer analysis or can we just ignore it?

15-41C Consider two surfaces pressed against each other. Now the air at the interface is evacuated. Will the thermal contact resistance at the interface increase or decrease as a result?

15-42C Explain how the thermal contact resistance can be minimized.

15-43 The thermal contact conductance at the interface of two 1-cm-thick copper plates is measured to be 18,000 W/m² · °C. Determine the thickness of the copper plate whose thermal resistance is equal to the thermal resistance of the interface between the plates.

15-44 Six identical power transistors with aluminum casing are attached on one side of a 1.2-cm-thick 20-cm × 30-cm copper plate ($k = 386$ W/m · °C) by screws that exert an average pressure of 10 MPa. The base area of each transistor is 9 cm², and each transistor is placed at the center of a 10-cm × 10-cm section of the plate. The interface roughness is estimated to be about 1.4 μm. All transistors are covered by a thick Plexiglas layer, which is a poor conductor of heat, and thus all the heat generated at the junction of the transistor must be dissipated to the ambient at 15°C through the back surface of the copper plate. The combined convection/radiation heat transfer coefficient at the back surface can be taken to be 30 W/m² · °C. If the case temperature of the transistor is not to exceed 85°C, determine the maximum power each transistor can dissipate safely, and the temperature jump at the case-plate interface.

15-45 Two 5-cm-diameter, 15-cm-long aluminum bars ($k = 176$ W/m · °C) with ground surfaces are pressed against each other with a pressure of 20 atm.

The bars are enclosed in an insulation sleeve and, thus, heat transfer from the lateral surfaces is negligible. If the top and bottom surfaces of the two-bar system are maintained at temperatures of 150°C and 20°C, respectively, determine (a) the rate of heat transfer along the cylinders under steady conditions and (b) the temperature drop at the interface.
Answers: (a) 142.4 W, (b) 5.1°C

15-46 A 1-mm-thick copper plate (k = 386 W/m · °C) is sandwiched between two 5-mm-thick epoxy boards (k = 0.26 W/m · °C) that are 15 cm × 20 cm in size. If the thermal contact conductance on both sides of the copper plate is estimated to be 4000 W/m · °C, determine the error involved in the total thermal resistance of the plate if the thermal contact conductances are ignored.

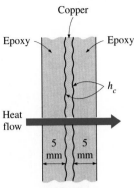

FIGURE P15-46

Generalized Thermal Resistance Networks

15-47C When plotting the thermal resistance network associated with a heat transfer problem, explain when two resistances are in series and when they are in parallel.

15-48C The thermal resistance networks can also be used approximately for multidimensional problems. For what kind of multidimensional problems will the thermal resistance approach give adequate results?

15-49C What are the two approaches used in the development of the thermal resistance network for two-dimensional problems?

15-50 A 4-m-high and 6-m-wide wall consists of a long 18-cm × 30-cm cross-section of horizontal bricks (k = 0.72 W/m · °C) separated by 3-cm-thick plaster layers (k = 0.22 W/m · °C). There are also 2-cm-thick plaster layers on each side of the wall, and a 2-cm-thick rigid foam (k = 0.026 W/m · °C) on the inner side of the wall. The indoor and the outdoor temperatures are 22°C and −4°C, and the convection heat transfer coefficients on the inner and the outer sides are h_1 = 10 W/m² · °C and h_2 = 20 W/m² · °C, respectively. Assuming one-dimensional heat transfer and disregarding radiation, determine the rate of heat transfer through the wall.

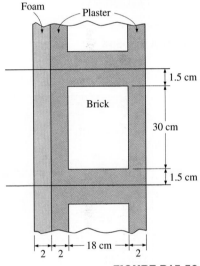

FIGURE P15-50

15-51 A 10-cm-thick wall is to be constructed with 2.5-m-long wood studs (k = 0.11 W/m · °C) that have a cross-section of 10 cm × 10 cm. At some point the builder ran out of those studs and started using pairs of 2.5-m-long wood studs that have a cross-section of 5 cm × 10 cm nailed to each other instead. The manganese steel nails (k = 50 W/m · °C) are 10 cm long and have a diameter of 0.4 cm. A total of 50 nails are used to connect the two studs, which are mounted to the wall such that the nails cross the wall. The temperature difference between the inner and outer surfaces of the wall is 15°C. Assuming the thermal contact resistance between the two layers to be negligible, determine the rate of heat transfer (a) through a solid stud and (b) through a stud pair of equal length and width nailed to each other. (c) Also determine the effective conductivity of the nailed stud pair.

15-52 A 12-m-long and 5-m-high wall is constructed of two layers of 1-cm-thick sheetrock (k = 0.17 W/m · °C) spaced 12 cm by wood studs (k = 0.11 W/m · °C) whose cross-section is 12 cm × 5 cm. The studs are placed vertically 60 cm apart, and the space between them is filled with fiberglass insulation (k = 0.034 W/m · °C). The house is maintained at 20°C and the ambient temperature outside is −5°C. Taking the heat transfer coefficients

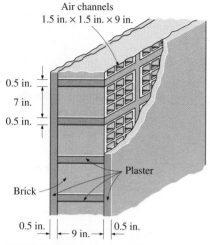

Air channels
1.5 in. × 1.5 in. × 9 in.

0.5 in.

7 in.

0.5 in.

Plaster

Brick

0.5 in. | 9 in. | 0.5 in.

FIGURE P15-53E

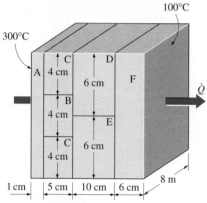

100°C

300°C

A | C 4 cm | D
| 6 cm | F
B 4 cm |
| E
C 6 cm |
4 cm |

1 cm | 5 cm | 10 cm | 6 cm | 8 m

$\dot{Q}$

FIGURE P15-54

Multilayered
ski jacket

FIGURE P15-56

at the inner and outer surfaces of the house to be 8.3 and 34 W/m² · °C, respectively, determine (*a*) the thermal resistance of the wall considering a representative section of it and (*b*) the rate of heat transfer through the wall.

15-53E A 10-in.-thick, 30-ft-long, and 10-ft-high wall is to be constructed using 9-in.-long solid bricks (k = 0.40 Btu/h · ft · °F) of cross-section 7 in. × 7 in., or identical size bricks with nine square air holes (k = 0.015 Btu/h · ft · °F) that are 9 in. long and have a cross-section of 1.5 in. × 1.5 in. There is a 0.5-in.-thick plaster layer (k = 0.10 Btu/h · ft · °F) between two adjacent bricks on all four sides and on both sides of the wall. The house is maintained at 75°F and the ambient temperature outside is 35°F. Taking the heat transfer coefficients at the inner and outer surfaces of the wall to be 1.5 and 4 Btu/h · ft² · °F, respectively, determine the rate of heat transfer through the wall constructed of (*a*) solid bricks and (*b*) bricks with air holes.

15-54 Consider a 5-m-high, 8-m-long, and 0.22-m-thick wall whose representative cross-section is as given in Fig. P15-54. The thermal conductivities of various materials used, in W/m · °C, are $k_A = k_F = 2$, $k_B = 8$, $k_C = 20$, $k_D = 15$, and $k_E = 35$. The left and right surfaces of the wall are maintained at uniform temperatures of 300°C and 100°C, respectively. Assuming heat transfer through the wall to be one-dimensional, determine (*a*) the rate of heat transfer through the wall; (*b*) the temperature at the point where the sections *B*, *D*, and *E* meet; and (*c*) the temperature drop across the section *F*. Disregard any contact resistances at the interfaces.

15-55 Repeat Prob. 15-54 assuming that the thermal contact resistance at the interfaces D-F and E-F is 0.00012 m² · °C/W.

15-56 Clothing made of several thin layers of fabric with trapped air in between, often called ski clothing, is commonly used in cold climates because it is light, fashionable, and a very effective thermal insulator. So it is no surprise that such clothing has largely replaced thick and heavy old-fashioned coats.

Consider a jacket made of five layers of 0.1-mm-thick synthetic fabric (k = 0.13 W/m · °C) with 1.5-mm-thick air space (k = 0.026 W/m · °C) between the layers. Assuming the inner surface temperature of the jacket to be 28°C and the surface area to be 1.1 m², determine the rate of heat loss through the jacket when the temperature of the outdoors is −5°C and the heat transfer coefficient at the outer surface is 25 W/m² · °C.

What would your response be if the jacket is made of a single layer of 0.5-mm-thick synthetic fabric? What should be the thickness of a wool fabric (k = 0.035 W/m · °C) if the person is to achieve the same level of thermal comfort wearing a thick wool coat instead of a ski jacket?

15-57 Repeat Prob. 15-56 assuming the layers of the jacket are made of cotton fabric (k = 0.06 W/m · °C).

15-58 A 5-m-wide, 4-m-high, and 40-m-long kiln used to cure concrete pipes is made of 20-cm-thick concrete walls and ceiling (k = 0.9 W/m · °C). The kiln is maintained at 40°C by injecting hot steam into it. The two ends of the kiln, 4 m × 5 m in size, are made of a 3-mm-thick sheet metal covered with 2-cm-thick Styrofoam (k = 0.033 W/m · °C). The convection heat transfer coefficients on the inner and the outer surfaces of the kiln are 3000 W/m² · °C and 25 W/m² · °C, respectively. Disregarding any heat loss through the floor, determine the rate of heat loss from the kiln when the ambient air is at −4°C.

15-59E Consider a 6-in. × 8-in. epoxy glass laminate ($k = 0.10$ Btu/h · ft · °F) whose thickness is 0.05 in. In order to reduce the thermal resistance across its thickness, cylindrical copper fillings ($k = 223$ Btu/h · ft · °F) of 0.02 in. diameter are to be planted throughout the board, with a center-to-center distance of 0.06 in. Determine the new value of the thermal resistance of the epoxy board for heat conduction across its thickness as a result of this modification.

Answer: 0.0099 h · °F/Btu

Heat Conduction in Cylinders and Spheres

15-60C Consider one-dimensional heat conduction through a plane wall, a long cylinder, and a sphere. For which of these geometries is the heat transfer area constant, and for which of them is it variable? Explain.

15-61C What is an infinitely long cylinder? When is it proper to treat an actual cylinder as being infinitely long, and when is it not?

15-62C Consider a short cylinder whose top and bottom surfaces are insulated. The cylinder is initially at a uniform temperature T_i and is subjected to convection from its side surface to a medium at temperature T_∞, with a heat transfer coefficient of h. Is the heat transfer in this short cylinder one- or two-dimensional? Explain.

15-63C Can the thermal resistance concept be used for a solid cylinder or sphere in steady operation? Explain.

15-64 A 5-m-internal-diameter spherical tank made of 1.5-cm-thick stainless steel ($k = 15$ W/m · °C) is used to store iced water at 0°C. The tank is located in a room whose temperature is 20°C. The walls of the room are also at 20°C. The outer surface of the tank is black (emissivity $\varepsilon = 1$), and heat transfer between the outer surface of the tank and the surroundings is by natural convection and radiation. The convection heat transfer coefficients at the inner and the outer surfaces of the tank are 80 W/m² · °C and 10 W/m² · °C, respectively. Determine (*a*) the rate of heat transfer to the iced water in the tank and (*b*) the amount of ice at 0°C that melts during a 24-h period. The heat of fusion of water at atmospheric pressure is $h_{if} = 333.7$ kJ/kg.

15-65 Steam at 320°C flows in a stainless steel pipe ($k = 15$ W/m · °C) whose inner and outer diameters are 5 cm and 5.5 cm, respectively. The pipe is covered with 3-cm-thick glass wool insulation ($k = 0.038$ W/m · °C). Heat is lost to the surroundings at 5°C by natural convection and radiation, with a combined natural convection and radiation heat transfer coefficient of 15 W/m² · °C. Taking the heat transfer coefficient inside the pipe to be 80 W/m² · °C, determine the rate of heat loss from the steam per unit length of the pipe. Also determine the temperature drops across the pipe shell and the insulation.

15-66 A 50-m-long section of a steam pipe whose outer diameter is 10 cm passes through an open space at 15°C. The average temperature of the outer surface of the pipe is measured to be 150°C. If the combined heat transfer coefficient on the outer surface of the pipe is 20 W/m² · °C, determine (*a*) the rate of heat loss from the steam pipe, (*b*) the annual cost of this energy lost if steam is generated in a natural gas furnace that has an efficiency of 75 percent and the price of natural gas is $0.52/therm (1 therm = 105,500 kJ), and (*c*) the thickness of fiberglass insulation ($k = 0.035$ W/m · °C) needed in order to

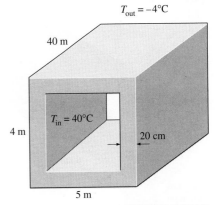

$T_{out} = -4°C$

40 m

$T_{in} = 40°C$

4 m

20 cm

5 m

FIGURE P15-58

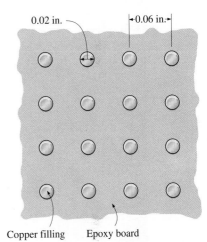

0.02 in.

0.06 in.

Copper filling   Epoxy board

FIGURE P15-59E

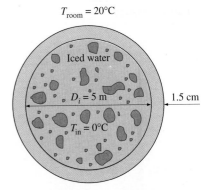

$T_{room} = 20°C$

Iced water

$D_i = 5$ m

1.5 cm

$T_{in} = 0°C$

FIGURE P15-64

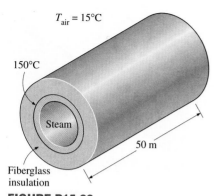

$T_{air} = 15°C$

150°C

Steam

50 m

Fiberglass
insulation

FIGURE P15-66

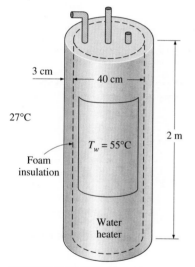

3 cm

40 cm

27°C

$T_w = 55°C$

2 m

Foam
insulation

Water
heater

FIGURE P15-67

3°C

12.5 cm

$T_{air} = 25°C$

Soda

6 cm

FIGURE P15-68

save 90 percent of the heat lost. Assume the pipe temperature to remain constant at 150°C.

15-67 Consider a 2-m-high electric hot water heater that has a diameter of 40 cm and maintains the hot water at 55°C. The tank is located in a small room whose average temperature is 27°C, and the heat transfer coefficients on the inner and outer surfaces of the heater are 50 and 12 W/m² · °C, respectively. The tank is placed in another 46-cm-diameter sheet metal tank of negligible thickness, and the space between the two tanks is filled with foam insulation ($k = 0.03$ W/m · °C). The thermal resistances of the water tank and the outer thin sheet metal shell are very small and can be neglected. The price of electricity is $0.08/kWh, and the home owner pays $280 a year for water heating. Determine the fraction of the hot water energy cost of this household that is due to the heat loss from the tank.

Hot water tank insulation kits consisting of 3-cm-thick fiberglass insulation ($k = 0.035$ W/m · °C) large enough to wrap the entire tank are available in the market for about $30. If such an insulation is installed on this water tank by the home owner himself, how long will it take for this additional insulation to pay for itself?   *Answers:* 17.5 percent, 1.5 years

15-68 Consider a cold aluminum canned drink that is initially at a uniform temperature of 3°C. The can is 12.5 cm high and has a diameter of 6 cm. If the combined convection/radiation heat transfer coefficient between the can and the surrounding air at 25°C is 10 W/m² · °C, determine how long it will take for the average temperature of the drink to rise to 10°C.

In an effort to slow down the warming of the cold drink, a person puts the can in a perfectly fitting 1-cm-thick cylindrical rubber insulation ($k = 0.13$ W/m · °C). Now how long will it take for the average temperature of the drink to rise to 10°C? Assume the top of the can is not covered.

15-69 Repeat Prob. 15-68, assuming a thermal contact resistance of 0.00008 m² · °C/W between the can and the insulation.

15-70E Steam at 600°F is flowing through a steel pipe ($k = 8.7$ Btu/h · ft · °F) whose inner and outer diameters are 3.5 in. and 4.0 in., respectively, in an environment at 60°F. The pipe is insulated with 2-in.-thick fiberglass insulation ($k = 0.020$ Btu/h · ft · °F). If the heat transfer coefficients on the inside and the outside of the pipe are 30 and 5 Btu/h · ft² · °F, respectively, determine the rate of heat loss from the steam per foot length of the pipe. What is the error involved in neglecting the thermal resistance of the steel pipe in calculations?

15-71 Hot water at an average temperature of 90°C is flowing through a 15-m section of a cast iron pipe ($k = 52$ W/m · °C) whose inner and outer diameters are 4 cm and 4.6 cm, respectively. The outer surface of the pipe, whose emissivity is 0.7, is exposed to the cold air at 10°C in the basement, with a heat transfer coefficient of 15 W/m² · °C. The heat transfer coefficient at the inner surface of the pipe is 120 W/m² · °C. Taking the walls of the basement to be at 10°C also, determine the rate of heat loss from the hot water. Also, determine the average velocity of the water in the pipe if the temperature of the water drops by 3°C as it passes through the basement.

15-72 Repeat Prob. 15-71 for a pipe made of copper ($k = 386$ W/m · °C) instead of cast iron.

15-73E Steam exiting the turbine of a steam power plant at 100°F is to be condensed in a large condenser by cooling water flowing through copper pipes ($k = 223$ Btu/h · ft · °F) of inner diameter 0.4 in. and outer diameter 0.6 in. at an average temperature of 70°F. The heat of vaporization of water at 100°F is 1037 Btu/lbm. The heat transfer coefficients are 1500 Btu/h · ft² · °F on the steam side and 35 Btu/h · ft² · °F on the water side. Determine the length of the tube required to condense steam at a rate of 400 lbm/h.

Answer: 3830 ft

15-74E Repeat Prob. 15-73E, assuming that a 0.01-in.-thick layer of mineral deposit ($k = 0.5$ Btu/h · ft · °F) has formed on the inner surface of the pipe.

15-75 The boiling temperature of nitrogen at atmospheric pressure at sea level (1 atm pressure) is −196°C. Therefore, nitrogen is commonly used in low-temperature scientific studies since the temperature of liquid nitrogen in a tank open to the atmosphere will remain constant at −196°C until it is depleted. Any heat transfer to the tank will result in the evaporation of some liquid nitrogen, which has a heat of vaporization of 198 kJ/kg and a density of 810 kg/m³ at 1 atm.

Consider a 3-m-diameter spherical tank that is initially filled with liquid nitrogen at 1 atm and −196°C. The tank is exposed to ambient air at 15°C, with a combined convection and radiation heat transfer coefficient of 35 W/m² · °C. The temperature of the thin-shelled spherical tank is observed to be almost the same as the temperature of the nitrogen inside. Determine the rate of evaporation of the liquid nitrogen in the tank as a result of the heat transfer from the ambient air if the tank is (*a*) not insulated, (*b*) insulated with 5-cm-thick fiberglass insulation ($k = 0.035$ W/m · °C), and (*c*) insulated with 2-cm-thick superinsulation which has an effective thermal conductivity of 0.00005 W/m · °C.

15-76 Repeat Prob. 15-75 for liquid oxygen, which has a boiling temperature of −183°C, a heat of vaporization of 213 kJ/kg, and a density of 1140 kg/m³ at 1 atm pressure.

Critical Radius of Insulation

15-77C What is the critical radius of insulation? How is it defined for a cylindrical layer?

15-78C A pipe is insulated such that the outer radius of the insulation is less than the critical radius. Now the insulation is taken off. Will the rate of heat transfer from the pipe increase or decrease for the same pipe surface temperature?

15-79C A pipe is insulated to reduce the heat loss from it. However, measurements indicate that the rate of heat loss has increased instead of decreasing. Can the measurements be right?

15-80C Consider a pipe at a constant temperature whose radius is greater than the critical radius of insulation. Someone claims that the rate of heat loss from the pipe has increased when some insulation is added to the pipe. Is this claim valid?

15-81C Consider an insulated pipe exposed to the atmosphere. Will the critical radius of insulation be greater on calm days or on windy days? Why?

15-82 A 2-mm-diameter and 10-m-long electric wire is tightly wrapped with a 1-mm-thick plastic cover whose thermal conductivity is $k = 0.15$ W/m · °C.

FIGURE P15-70E

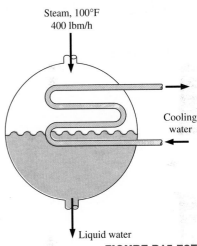

FIGURE P15-73E

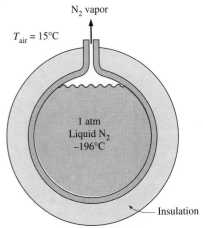

FIGURE P15-75

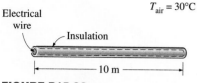

Electrical
wire

Insulation

$T_{air} = 30°C$

10 m

FIGURE P15-82

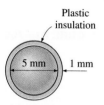

Plastic
insulation

5 mm   1 mm

FIGURE P15-85

Electrical measurements indicate that a current of 10 A passes through the wire and there is a voltage drop of 8 V along the wire. If the insulated wire is exposed to a medium at $T_\infty = 30°C$ with a heat transfer coefficient of $h = 18$ W/m$^2 \cdot$ °C, determine the temperature at the interface of the wire and the plastic cover in steady operation. Also determine if doubling the thickness of the plastic cover will increase or decrease this interface temperature.

15-83E A 0.083-in.-diameter electrical wire at 115°F is covered by 0.02-in.-thick plastic insulation ($k = 0.075$ Btu/h $\cdot$ ft $\cdot$ °F). The wire is exposed to a medium at 50°F, with a combined convection and radiation heat transfer coefficient of 2.5 Btu/h $\cdot$ ft$^2 \cdot$ °F. Determine if the plastic insulation on the wire will increase or decrease heat transfer from the wire.
  Answer: It helps

15-84E Repeat Prob. 15-83E, assuming a thermal contact resistance of 0.001 h $\cdot$ ft$^2 \cdot$ °F/Btu at the interface of the wire and the insulation.

15-85 A 5-mm-diameter spherical ball at 50°C is covered by a 1-mm-thick plastic insulation ($k = 0.13$ W/m $\cdot$ °C). The ball is exposed to a medium at 15°C, with a combined convection and radiation heat transfer coefficient of 20 W/m$^2 \cdot$ °C. Determine if the plastic insulation on the ball will help or hurt heat transfer from the ball.

Thermal Insulation

15-86C What is thermal insulation? How does a thermal insulator differ in purpose from an electrical insulator and from a sound insulator?

15-87C Does insulating cold surfaces save energy? Explain.

15-88C What is the R-value of insulation? How is it determined? Will doubling the thickness of flat insulation double its R-value?

15-89C How does the R-value of an insulation differ from its thermal resistance?

15-90C Why is the thermal conductivity of superinsulation orders of magnitude lower than the thermal conductivities of ordinary insulations?

15-91C Someone suggests that one function of hair is to insulate the head. Do you agree with this suggestion?

15-92C Name five different reasons for using insulation in industrial facilities.

15-93C What is optimum thickness of insulation? How is it determined?

15-94 What is the thickness of flat R-8 (in SI units) insulation whose thermal conductivity is 0.04 W/m $\cdot$ °C?

15-95E What is the thickness of flat R-20 (in English units) insulation whose thermal conductivity is 0.02 Btu/h $\cdot$ ft $\cdot$ °F?

15-96 Hot water at 110°C flows in a cast iron pipe ($k = 52$ W/m $\cdot$ °C) whose inner radius is 2.0 cm and thickness is 0.3 cm. The pipe is to be covered with adequate insulation so that the temperature of the outer surface of the insulation does not exceed 30°C when the ambient temperature is 22°C. Taking the

heat transfer coefficients inside and outside the pipe to be $h_i = 80$ W/m$^2 \cdot$ °C and $h_o = 22$ W/m$^2 \cdot$ °C, respectively, determine the thickness of fiber glass insulation ($k = 0.038$ W/m $\cdot$ °C) that needs to be installed on the pipe.
    Answer: 0.84 cm

15-97   Consider a furnace whose average outer surface temperature is measured to be 90°C when the average surrounding air temperature is 27°C. The furnace is 6 m long and 3 m in diameter. The plant operates 80 hours per week for 52 weeks per year. You are to insulate the furnace using fiberglass insulation ($k_{ins} = 0.038$ W/m $\cdot$ °C) whose cost is $10/m^2 per cm of thickness for materials, plus $30/m^2 for labor regardless of thickness. The combined heat transfer coefficient on the outer surface is estimated to be $h_o = 30$ W/m$^2 \cdot$ °C. The furnace uses natural gas whose unit cost is $0.50/therm input (1 therm = 105,500 kJ), and the efficiency of the furnace is 78 percent. The management is willing to authorize the installation of the thickest insulation (in whole cm) that will pay for itself (materials and labor) in one year. That is, the total cost of insulation should be roughly equal to the drop in the fuel cost of the furnace for one year. Determine the thickness of insulation to be used and the money saved per year. Assume the surface temperature of the furnace and the heat transfer coefficient are to remain constant.   *Answer:* 14 cm

15-98   Repeat Prob. 15-97 for an outer surface temperature of 75°C for the furnace.

15-99E   Steam at 400°F is flowing through a steel pipe ($k = 8.7$ Btu/h $\cdot$ ft $\cdot$ °F) whose inner and outer diameters are 3.5 in. and 4.0 in., respectively, in an environment at 60°F. The pipe is insulated with 1-in.-thick fiberglass insulation ($k = 0.020$ Btu/h $\cdot$ ft $\cdot$ °F), and the heat transfer coefficients on the inside and the outside of the pipe are 30 Btu/h $\cdot$ ft$^2 \cdot$ °F and 5 Btu/h $\cdot$ ft$^2 \cdot$ °F, respectively. It is proposed to add another 1-in.-thick layer of fiberglass insulation on top of the existing one to reduce the heat losses further and to save energy and money. The total cost of new insulation is $7 per ft length of the pipe, and the net fuel cost of energy in the steam is $0.01 per 1000 Btu (therefore, each 1000 Btu reduction in the heat loss will save the plant $0.01). The policy of the plant is to implement energy conservation measures that pay for themselves within two years. Assuming continuous operation (8760 h/year), determine if the proposed additional insulation is justified.

15-100   The plumbing system of a plant involves some section of a plastic pipe ($k = 0.16$ W/m $\cdot$ °C) of inner diameter 6 cm and outer diameter 6.6 cm exposed to the ambient air. You are to insulate the pipe with adequate weatherjacketed fiberglass insulation ($k = 0.035$ W/m $\cdot$ °C) to prevent freezing of water in the pipe. The plant is closed for the weekends for a period of 60 h, and the water in the pipe remains still during that period. The ambient temperature in the area gets as low as -10°C in winter, and the high winds can cause heat transfer coefficients as high as 30 W/m$^2 \cdot$ °C. Also, the water temperature in the pipe can be as cold as 15°C, and water starts freezing when its temperature drops to 0°C. Disregarding the convection resistance inside the pipe, determine the thickness of insulation that will protect the water from freezing under worst conditions.

15-101   Repeat Prob. 15-100 assuming 20 percent of the water in the pipe is allowed to freeze without jeopardizing safety.   *Answer:* 27.9 cm

Heat Transfer from Finned Surfaces

15-102C What is the reason for the widespread use of fins on surfaces?

15-103C What is the difference between the fin effectiveness and the fin efficiency?

15-104C The fins attached to a surface are determined to have an effectiveness of 0.9. Do you think the rate of heat transfer from the surface has increased or decreased as a result of the addition of these fins?

15-105C Explain how the fins enhance heat transfer from a surface. Also, explain how the addition of fins may actually decrease heat transfer from a surface.

15-106C How does the overall effectiveness of a finned surface differ from the effectiveness of a single fin?

15-107C Hot water is to be cooled as it flows through the tubes exposed to atmospheric air. Fins are to be attached in order to enhance heat transfer. Would you recommend attaching the fins inside or outside the tubes? Why?

15-108C Hot air is to be cooled as it is forced to flow through the tubes exposed to atmospheric air. Fins are to be added in order to enhance heat transfer. Would you recommend attaching the fins inside or outside the tubes? Why? When would you recommend attaching fins both inside and outside the tubes?

15-109C Consider two finned surfaces that are identical except that the fins on the first surface are formed by casting or extrusion, whereas they are attached to the second surface afterwards by welding or tight fitting. For which case do you think the fins will provide greater enhancement in heat transfer? Explain.

15-110C The heat transfer surface area of a fin is equal to the sum of all surfaces of the fin exposed to the surrounding medium, including the surface area of the fin tip. Under what conditions can we neglect heat transfer from the fin tip?

15-111C Does the (a) efficiency and (b) effectiveness of a fin increase or decrease as the fin length is increased?

15-112C Two pin fins are identical, except that the diameter of one of them is twice the diameter of the other. For which fin will the (a) fin effectiveness and (b) fin efficiency be higher? Explain.

15-113C Two plate fins of constant rectangular cross-section are identical, except that the thickness of one of them is twice the thickness of the other. For which fin will the (a) fin effectiveness and (b) fin efficiency be higher? Explain.

15-114C Two finned surfaces are identical, except that the convection heat transfer coefficient of one of them is twice that of the other. For which finned surface will the (a) fin effectiveness and (b) fin efficiency be higher? Explain.

15-115 Obtain a relation for the fin efficiency for a fin of constant cross-sectional area A_c, perimeter p, length L, and thermal conductivity k exposed to convection to a medium at T_∞ with a heat transfer coefficient h. Assume the fins are sufficiently long so that the temperature of the fin at the tip is nearly T_∞. Take the temperature of the fin at the base to be T_b and neglect heat trans-

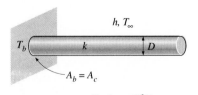

$p = \pi D, \ A_c = \pi D^2/4$

FIGURE P15-115

fer from the fin tips. Simplify the relation for (a) a circular fin of diameter D and (b) rectangular fins of thickness t.

15-116 The case-to-ambient thermal resistance of a power transistor that has a maximum power rating of 15 W is given to be 25°C/W. If the case temperature of the transistor is not to exceed 80°C, determine the power at which this transistor can be operated safely in an environment at 30°C.

15-117 A 40-W power transistor is to be cooled by attaching it to one of the commercially available heat sinks shown in Table 15-6. Select a heat sink that will allow the case temperature of the transistor not to exceed 90° in the ambient air at 20°.

$T_{air} = 20°C$

90°C

40 W

FIGURE P15-117

15-118 A 30-W power transistor is to be cooled by attaching it to one of the commercially available heat sinks shown in Table 15-6. Select a heat sink that will allow the case temperature of the transistor not to exceed 80°C in the ambient air at 35°C.

15-119 Steam in a heating system flows through tubes whose outer diameter is 5 cm and whose walls are maintained at a temperature of 180°C. Circular aluminum alloy 2024-T6 fins ($k = 186$ W/m · °C) of outer diameter 6 cm and constant thickness 1 mm are attached to the tube. The space between the fins is 3 mm, and thus there are 250 fins per meter length of the tube. Heat is transferred to the surrounding air at $T_\infty = 25°C$, with a heat transfer coefficient of 40 W/m² · °C. Determine the increase in heat transfer from the tube per meter of its length as a result of adding fins. *Answer:* 2274 W

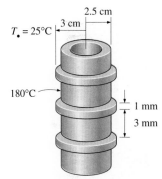

2.5 cm

3 cm

$T_\cdot = 25°C$

180°C

1 mm

3 mm

FIGURE P15-119

15-120E Consider a stainless steel spoon ($k = 8.7$ Btu/h · ft · °F) partially immersed in boiling water at 200°F in a kitchen at 75°F. The handle of the spoon has a cross-section of 0.08 in. × 0.5 in., and extends 7 in. in the air from the free surface of the water. If the heat transfer coefficient at the exposed surfaces of the spoon handle is 3 Btu/h · ft² · °F, determine the temperature difference across the exposed surface of the spoon handle. State your assumptions. *Answer:* 124.6°F

15-121E Repeat Prob. 15-120 for a silver spoon ($k = 247$ Btu/h · ft · °F).

15-122 A 0.3-cm-thick, 12-cm-high, and 18-cm-long circuit board houses 80 closely spaced logic chips on one side, each dissipating 0.04 W. The board is impregnated with copper fillings and has an effective thermal conductivity of 20 W/m · °C. All the heat generated in the chips is conducted across the circuit board and is dissipated from the back side of the board to a medium at 40°C, with a heat transfer coefficient of 50 W/m² · °C. (a) Determine the temperatures on the two sides of the circuit board. (b) Now a 0.2-cm-thick, 12-cm-high, and 18-cm-long aluminum plate ($k = 237$ W/m · °C) with 864 2-cm-long aluminum pin fins of diameter 0.25 cm is attached to the back side of the circuit board with a 0.02-cm-thick epoxy adhesive ($k = 1.8$ W/m · °C). Determine the new temperatures on the two sides of the circuit board.

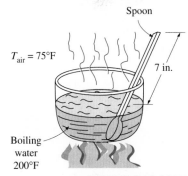

Spoon

$T_{air} = 75°F$

7 in.

Boiling water 200°F

FIGURE P15-120E

15-123 Repeat Prob. 15-122 using a copper plate with copper fins ($k = 386$ W/m · °C) instead of aluminum ones.

15-124 A hot surface at 100°C is to be cooled by attaching 15-cm-long, 0.25-cm-diameter aluminum pin fins ($k = 237$ W/m · °C) to it, with a center-to-center distance of 0.6 cm. The temperature of the surrounding medium is 30°C, and the heat transfer coefficient on the surfaces is 35 W/m² · °C. Determine the rate of heat transfer from the surface for a 1-m × 1-m section of the plate. Also determine the overall effectiveness of the fins.

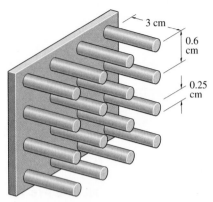

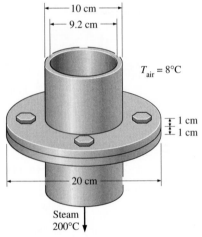

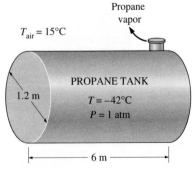

15-125   Repeat Prob. 15-124 using copper fins ($k = 386$ W/m · °C) instead of aluminum ones.

15-126   Two 3-m-long and 0.4-cm-thick cast iron ($k = 52$ W/m · °C) steam pipes of outer diameter 10 cm are connected to each other through two 1-cm-thick flanges of outer diameter 20 cm. The steam flows inside the pipe at an average temperature of 200°C with a heat transfer coefficient of 180 W/m² · °C. The outer surface of the pipe is exposed to an ambient at 8°C, with a heat transfer coefficient of 25 W/m² · °C. (*a*) Disregarding the flanges, determine the average outer surface temperature of the pipe. (*b*) Using this temperature for the base of the flange and treating the flanges as the fins, determine the fin efficiency and the rate of heat transfer from the flanges. (*c*) What length of pipe is the flange section equivalent to for heat transfer purposes?

Review Problems

15-127E   Steam is produced in the copper tubes ($k = 223$ Btu/h · ft · °F) of a heat exchanger at a temperature of 250°F by another fluid condensing on the outside surfaces of the tubes at 350°F. The inner and outer diameters of the tube are 1 in. and 1.3 in., respectively. When the heat exchanger was new, the rate of heat transfer per foot length of the tube was 2×10^4 Btu/h. Determine the rate of heat transfer per foot length of the tube when a 0.01-in.-thick layer of limestone ($k = 1.7$ Btu/h · ft · °F) has formed on the inner surface of the tube after extended use.

15-128E   Repeat Prob. 15-127E, assuming that a 0.01-in.-thick limestone layer has formed on both the inner and outer surfaces of the tube.

15-129   A 1.2-m-diameter and 6-m-long cylindrical propane tank is initially filled with liquid propane whose density is 581 kg/m³. The tank is exposed to the ambient air at 15°C, with a heat transfer coefficient of 20 W/m² · °C. Now a crack develops at the top of the tank and the pressure inside drops to 1 atm while the temperature drops to −42°C, which is the boiling temperature of propane at 1 atm. The heat of vaporization of propane at 1 atm is 425 kJ/kg. The propane is slowly vaporized as a result of the heat transfer from the ambient air into the tank, and the propane vapor escapes the tank at −42°C through the crack. Assuming the propane tank to be at about the same temperature as the propane inside at all times, determine how long it will take for the propane tank to empty if the tank is (*a*) not insulated and (*b*) insulated with 7.5-cm-thick glass wool insulation ($k = 0.038$ W/m · °C).

15-130   Hot water is flowing at an average velocity of 1.5 m/s through a cast iron pipe ($k = 52$ W/m · °C) whose inner and outer diameters are 3 cm and 3.5 cm, respectively. The pipe passes through a 15-m-long section of a basement whose temperature is 15°C. If the temperature of the water drops from 70°C to 67°C as it passes through the basement and the heat transfer coefficient on the inner surface of the pipe is 400 W/m² · °C, determine the combined convection and radiation heat transfer coefficient at the outer surface of the pipe.   *Answer:* 272.5 W/m² · °C

15-131   Newly formed concrete pipes are usually cured first overnight by steam in a curing kiln maintained at a temperature of 45°C before the pipes are cured for several days outside. The heat and moisture to the kiln is provided by steam flowing in a pipe whose outer diameter is 12 cm. During a plant inspection, it was noticed that the pipe passes through a 10-m section that is completely exposed to the ambient air before it reaches the kiln. The temperature measurements indicate that the average temperature of the outer surface of the steam pipe is 82°C when the ambient temperature is 5°C. The combined convection and radiation heat transfer coefficient at the outer surface of the pipe is estimated to be 25 W/m² · °C. Determine the amount of heat lost from the steam during a 10-h curing process that night.

Steam is supplied by a gas-fired steam generator that has an efficiency of 80 percent, and the plant pays $0.60/therm of natural gas (1 therm = 105,500 kJ). If the pipe is insulated and 90 percent of the heat loss is saved as a result, determine the amount of money this facility will save a year as a result of insulating the steam pipes. Assume that the concrete pipes are cured 110 nights a year. State your assumptions.

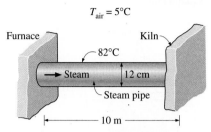

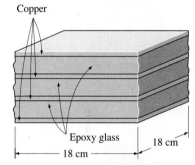

FIGURE P15-131

15-132 Consider an 18-cm × 18-cm multilayer circuit board dissipating 27 W of heat. The board consists of four layers of 0.2-mm-thick copper ($k = 386$ W/m · °C) and three layers of 1.5-mm-thick epoxy glass ($k = 0.26$ W/m · °C) sandwiched together, as shown in the figure. The circuit board is attached to a heat sink from both ends, and the temperature of the board at those ends is 35°C. Heat is considered to be uniformly generated in the epoxy layers of the board at a rate of 0.5 W per 1-cm × 18-cm epoxy laminate strip (or 1.5 W per 1-cm × 18-cm strip of the board). Considering only a portion of the board because of symmetry, determine the magnitude and location of the maximum temperature that occurs in the board. Assume heat transfer from the top and bottom faces of the board to be negligible.

FIGURE P15-132

15-133 The plumbing system of a house involves a 0.5-m section of a plastic pipe ($k = 0.16$ W/m · °C) of inner diameter 2 cm and outer diameter 2.4 cm exposed to the ambient air. During a cold and windy night, the ambient air temperature remains at about −5°C for a period of 14 h. The combined convection and radiation heat transfer coefficient on the outer surface of the pipe is estimated to be 40 W/m² · °C, and the heat of fusion of water is 333.7 kJ/kg. Assuming the pipe to contain stationary water initially at 0°C, determine if the water in that section of the pipe will completely freeze that night.

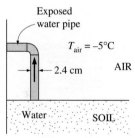

FIGURE P15-133

15-134 Repeat Prob. 15-133 for the case of a heat transfer coefficient of 10 W/m² · °C on the outer surface as a result of putting a fence around the pipe that blocks the wind.

15-135E The surface temperature of a 3-in.-diameter baked potato is observed to drop from 300°F to 200°F in 5 minutes in an environment at 70°F. Determine the average heat transfer coefficient between the potato and its surroundings. Using this heat transfer coefficient and the same surface temperature, determine how long it will take for the potato to experience the same temperature drop if it is wrapped completely in a 0.12-in.-thick towel ($k = 0.035$ Btu/h · ft · °F). You may use the properties of water for potato.

15-136E Repeat Prob. 15-135E assuming there is a 0.02-in.-thick air space ($k = 0.015$ Btu/h · ft · °F) between the potato and the towel.

15-137 An ice chest whose outer dimensions are 30 cm × 40 cm × 50 cm is made of 3-cm-thick Styrofoam ($k = 0.033$ W/m · °C). Initially, the chest is filled with 45 kg of ice at 0°C, and the inner surface temperature of the ice chest can be taken to be 0°C at all times. The heat of fusion of ice at 0°C is 333.7 kJ/kg, and the heat transfer coefficient between the outer surface of the ice chest and surrounding air at 30°C is 20 W/m² · °C. Disregarding any heat

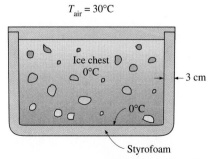

FIGURE P15-137

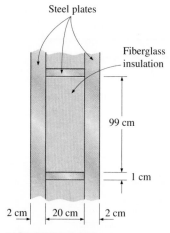

Steel plates

Fiberglass insulation

99 cm

1 cm

2 cm | 20 cm | 2 cm

FIGURE P15-138

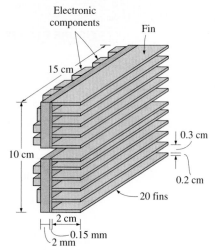

Electronic components

Fin

15 cm

10 cm

0.3 cm

0.2 cm

20 fins

2 cm

0.15 mm

2 mm

FIGURE P15-139

transfer from the 40-cm × 50-cm base of the ice chest, determine how long it will take for the ice in the chest to melt completely.

15-138 A 4-m-high and 6-m-long wall is constructed of two large 2-cm-thick steel plates ($k = 15$ W/m · °C) separated by 1-cm-thick and 20-cm-wide steel bars placed 99 cm apart. The remaining space between the steel plates is filled with fiberglass insulation ($k = 0.035$ W/m · °C). If the temperature difference between the inner and the outer surfaces of the walls is 15°C, determine the rate of heat transfer through the wall. Can we ignore the steel bars between the plates in heat transfer analysis since they occupy only 1 percent of the heat transfer surface area?

15-139 A 0.2-cm-thick, 10-cm-high, and 15-cm-long circuit board houses electronic components on one side that dissipate a total of 15 W of heat uniformly. The board is impregnated with conducting metal fillings and has an effective thermal conductivity of 12 W/m · °C. All the heat generated in the components is conducted across the circuit board and is dissipated from the back side of the board to a medium at 37°C, with a heat transfer coefficient of 45 W/m² · °C. (*a*) Determine the surface temperatures on the two sides of the circuit board. (*b*) Now a 0.1-cm-thick, 10-cm-high, and 15-cm-long aluminum plate ($k = 237$ W/m · °C) with 20 0.2-cm-thick, 2-cm-long, and 15-cm-wide aluminum fins of rectangular profile are attached to the back side of the circuit board with a 0.015-cm-thick epoxy adhesive ($k = 1.8$ W/m · °C). Determine the new temperatures on the two sides of the circuit board.

15-140 Repeat Prob. 15-139 using a copper plate with copper fins ($k = 386$ W/m · °C) instead of aluminum ones.

Computer, Design, and Essay Problems

15-141 The temperature in deep space is close to absolute zero, which presents thermal challenges for the astronauts who do space walks. Propose a design for the clothing of the astronauts that will be most suitable for the thermal environment in space. Defend the selections in your design.

15-142 In the design of electronic components, it is very desirable to attach the electronic circuitry to a substrate material that is a very good thermal conductor but also a very effective electrical insulator. If the high cost is not a major concern, what material would you propose for the substrate?

15-143 Using cylindrical samples of the same material, devise an experiment to determine the thermal contact resistance. Cylindrical samples are available at any length, and the thermal conductivity of the material is known.

15-144 What are the considerations in determining the proper length and right number of fins attached to a surface?

15-145 Find out about the wall construction of the cabins of large commercial airplanes, the range of ambient conditions under which they operate, typical heat transfer coefficients on the inner and outer surfaces of the wall, and the heat generation rates inside. Determine the size of the heating and air-conditioning system that will be able to maintain the cabin at 20°C at all times for an airplane capable of carrying 400 people.

Transient Heat Conduction

CHAPTER

16

The temperature of a body, in general, varies with time as well as position. In rectangular coordinates, this variation is expressed as $T(x, y, z, t)$, where (x, y, z) indicates variation in the x, y, and z directions, respectively, and t indicates variation with time. In the preceding chapter, we considered heat conduction under *steady* conditions, for which the temperature of a body at any point does not change with time. This certainly simplified the analysis, especially when the temperature varied in one direction only, and we were able to obtain analytical solutions. In this chapter, we consider the variation of temperature with *time* as well as *position* in one- and multidimensional systems.

We start this chapter with the analysis of *lumped systems* in which the temperature of a solid varies with time but remains uniform throughout the solid at any time. Then we consider the variation of temperature with time as well as position for one-dimensional heat conduction problems such as those associated with a large plane wall, a long cylinder, a sphere, and a semi-infinite medium using *transient temperature charts* and analytical solutions. Finally, we consider transient heat conduction in multidimensional systems by utilizing the *product solution*.

70°C
70°C   70°C
70°C   70°C

(a) Copper ball

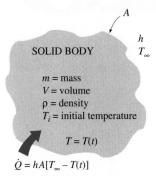

110°C
90°C
40°C

(b) Roast beef
FIGURE 16-1

A small copper ball can be modeled as a lumped system, but a roast beef cannot.

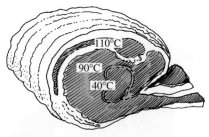

A

h
T_∞

SOLID BODY

m = mass
V = volume
ρ = density
T_i = initial temperature

$T = T(t)$

$\dot{Q} = hA[T_\infty - T(t)]$

FIGURE 16-2

The geometry and parameters involved in the lumped system analysis.

16-1 ■ LUMPED SYSTEM ANALYSIS

In heat transfer analysis, some bodies are observed to behave like a "lump" whose interior temperature remains essentially uniform at all times during a heat transfer process. The temperature of such bodies can be taken to be a function of time only, $T(t)$. Heat transfer analysis that utilizes this idealization is known as **lumped system analysis,** which provides great simplification in certain classes of heat transfer problems without much sacrifice from accuracy.

Consider a small hot copper ball coming out of an oven (Fig. 16-1). Measurements indicate that the temperature of the copper ball changes with time, but it does not change much with position at any given time. Thus the temperature of the ball remains uniform at all times, and we can talk about the temperature of the ball with no reference to a specific location.

Now let us go to the other extreme and consider a large roast in an oven. If you have done any roasting, you must have noticed that the temperature distribution within the roast is not even close to being uniform. You can easily verify this by taking the roast out before it is completely done and cutting it in half. You will see that the outer parts of the roast are well done while the center part is barely warm. Thus, lumped system analysis is not applicable in this case. Before presenting a criterion about applicability of lumped system analysis, we develop the formulation associated with it.

Consider a body of arbitrary shape of mass m, volume V, surface area A, density ρ, and specific heat C_p initially at a uniform temperature T_i (Fig. 16-2). At time $t = 0$, the body is placed into a medium at temperature T_∞, and heat transfer takes place between the body and its environment, with a heat transfer coefficient h. For the sake of discussion, we will assume that $T_\infty > T_i$, but the analysis is equally valid for the opposite case. We assume lumped system analysis to be applicable, so that the temperature remains uniform within the body at all times and changes with time only, $T = T(t)$.

During a differential time interval dt, the temperature of the body rises by a differential amount dT. An energy balance of the solid for the time interval dt can be expressed as

$$\begin{pmatrix}\text{Heat transfer into the body} \\ \text{during } dt\end{pmatrix} = \begin{pmatrix}\text{The increase in the} \\ \text{energy of the body} \\ \text{during } dt\end{pmatrix}$$

or

$$hA(T_\infty - T)\, dt = mC_p\, dT \qquad (16\text{-}1)$$

Noting that $m = \rho V$ and $dT = d(T - T_\infty)$ since T_∞ = constant, Eq. 16-1 can be rearranged as

$$\frac{d(T - T_\infty)}{T - T_\infty} = -\frac{hA}{\rho V C_p}\, dt \qquad (16\text{-}2)$$

Integrating from $t = 0$, at which $T = T_i$, to any time t, at which $T = T(t)$, gives

$$\ln \frac{T(t) - T_\infty}{T_i - T_\infty} = -\frac{hA}{\rho V C_p}\, t \qquad (16\text{-}3)$$

Taking the exponential of both sides and rearranging, we obtain

$$\frac{T(t) - T_\infty}{T_i - T_\infty} = e^{-bt} \qquad (16\text{-}4)$$

where

$$b = \frac{hA}{\rho V C_p} \quad \text{(1/s)} \quad \text{(16-5)}$$

is a positive quantity whose dimension is $(\text{time})^{-1}$. Equation 16-4 is plotted in Fig. 16-3 for different values of b. There are two observations that can be made from this figure and the relation above:

1. Equation 16-4 enables us to determine the temperature $T(t)$ of a body at time t, or alternatively, the time t required for the temperature to reach a specified value $T(t)$.

2. The temperature of a body approaches the ambient temperature T_∞ exponentially. The temperature of the body changes rapidly at the beginning, but rather slowly later on. A large value of b indicates that the body will approach the environment temperature in a short time. The larger the value of the exponent b, the higher the rate of decay in temperature. Note that b is proportional to the surface area, but inversely proportional to the mass and the specific heat of the body. This is not surprising since it takes longer to heat or cool a larger mass, especially when it has a large specific heat.

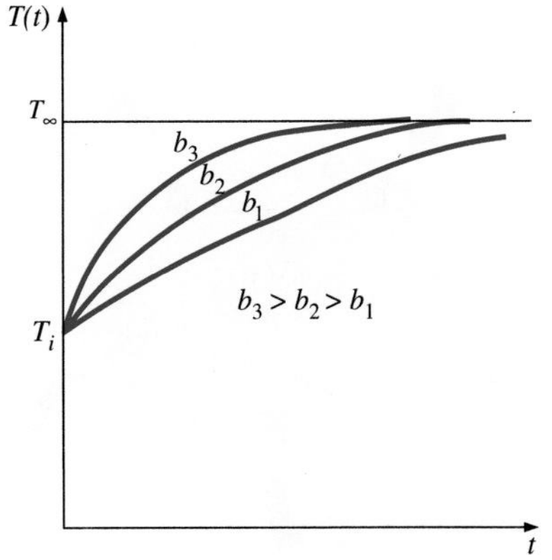

Once the temperature $T(t)$ at time t is available from Eq. 16-4, the *rate* of convection heat transfer between the body and its environment at that time can be determined from Newton's law of cooling as

$$\dot{Q}(t) = hA[T(t) - T_\infty] \quad \text{(W)} \quad \text{(16-6)}$$

The *total amount* of heat transfer between the body and the surrounding medium over the time interval $t = 0$ to t is simply the change in the energy content of the body:

$$Q = mC_p[T(t) - T_i] \quad \text{(kJ)} \quad \text{(16-7)}$$

The amount of heat transfer reaches its *upper limit* when the body reaches the surrounding temperature T_∞. Therefore, the *maximum* heat transfer between the body and its surroundings is (Fig. 16-4)

$$Q_{\max} = mC_p(T_\infty - T_i) \quad \text{(kJ)} \quad \text{(16-8)}$$

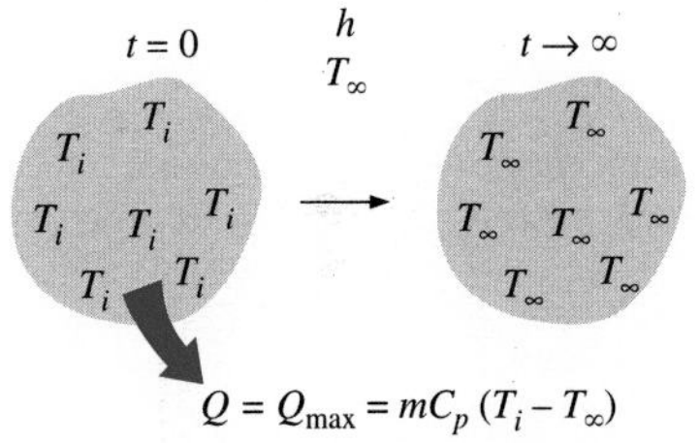

We could also obtain this equation by substituting the $T(t)$ relation from Eq. 16-4 into the $\dot{Q}(t)$ relation in Eq. 16-6 and integrating it from $t = 0$ to $t \to \infty$.

FIGURE 16-3

The temperature of a lumped system approaches the environment temperature as time gets larger.

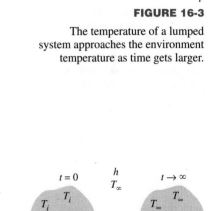

$$Q = Q_{\max} = mC_p\,(T_i - T_\infty)$$

FIGURE 16-4

Heat transfer to or from a body reaches its maximum value when the body reaches the environment temperature.

Criteria for Lumped System Analysis

The lumped system analysis certainly provides great convenience in heat transfer analysis, and naturally we would like to know when it is appropriate to use it. The first step in establishing a criterion for the applicability of the lumped system analysis is to define a **characteristic length** as

$$L_c = \frac{V}{A}$$

and a **Biot number** Bi as

$$\text{Bi} = \frac{hL_c}{k} \tag{16-9}$$

It can also be expressed as (Fig. 16-5)

$$\text{Bi} = \frac{h}{k/L_c}\frac{\Delta T}{\Delta T} = \frac{\text{Convection at the surface of the body}}{\text{Conduction within the body}}$$

or

$$\text{Bi} = \frac{L_c/k}{1/h} = \frac{\text{Conduction resistance within the body}}{\text{Convection resistance at the surface of the body}}$$

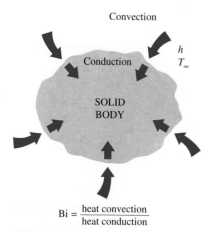

Convection

Conduction

h
T_∞

SOLID
BODY

$$\text{Bi} = \frac{\text{heat convection}}{\text{heat conduction}}$$

FIGURE 16-5

The Biot number can be viewed as the ratio of the convection at the surface to conduction within the body.

When a solid body is being heated by the hotter fluid surrounding it (such as a potato being baked in an oven), heat is first *convected* to the body and subsequently *conducted* within the body. The Biot number is the *ratio* of the internal resistance of a body to *heat conduction* to its external resistance to *heat convection.* Therefore, a small Biot number represents small resistance to heat conduction, and thus small temperature gradients within the body.

Lumped system analysis assumes a *uniform* temperature distribution throughout the body, which will be the case only when the thermal resistance of the body to heat conduction (the *conduction resistance*) is zero. Thus, lumped system analysis is *exact* when Bi = 0 and *approximate* when Bi > 0. Of course, the smaller the Bi number, the more accurate the lumped system analysis. Then the question we must answer is, How much accuracy are we willing to sacrifice for the convenience of the lumped system analysis?

Before answering this question, we should mention that a 20 percent uncertainty in the convection heat transfer coefficient h in most cases is considered "normal" and "expected." Assuming h to be *constant* and *uniform* is also an approximation of questionable validity, especially for irregular geometries. Therefore, in the absence of sufficient experimental data for the specific geometry under consideration, we cannot claim our results to be better than ±20 percent, even when Bi = 0. This being the case, introducing another source of uncertainty in the problem will hardly have any effect on the overall uncertainty, provided that it is minor. It is generally accepted that lumped system analysis is *applicable* if

$$\text{Bi} \leq 0.1$$

When this criterion is satisfied, the temperatures within the body relative to the surroundings (i.e., $T - T_\infty$) remain within 5 percent of each other even for well-rounded geometries such as a spherical ball. Thus, when Bi < 0.1, the variation of temperature with location within the body will be slight and can reasonably be approximated as being uniform.

The first step in the application of lumped system analysis is the calculation of the *Biot number,* and the assessment of the applicability of this approach. One may still wish to use lumped system analysis even when the criterion Bi < 0.1 is not satisfied, if high accuracy is not a major concern.

Note that the Biot number is the ratio of the *convection* at the surface to *conduction* within the body, and this number should be as small as possible for lumped system analysis to be applicable. Therefore, *small bodies* with *high thermal conductivity* are good candidates for lumped system analysis, especially when they are in a medium that is a poor conductor of heat (such as air or another gas) and motionless. Thus, the hot small copper ball placed in quiescent air, discussed earlier, is most likely to satisfy the criterion for lumped system analysis (Fig. 16-6).

Some Remarks on Heat Transfer in Lumped Systems

To understand the heat transfer mechanism during the heating or cooling of a solid by the fluid surrounding it, and the criterion for lumped system analysis, consider the following analogy (Fig. 16-7). People from the mainland are to go *by boat* to an island whose entire shore is a harbor, and from the harbor to their destinations on the island *by bus*. The overcrowding of people at the harbor depends on the boat traffic to the island and the ground transportation system on the island. If there is an excellent ground transportation system with plenty of buses, there will be no overcrowding at the harbor, especially when the boat traffic is light. But when the opposite is true, there will be a huge overcrowding at the harbor, creating a large difference between the populations at the harbor and inland. The chance of overcrowding is much lower in a small island with plenty of fast buses.

In heat transfer, a poor ground transportation system corresponds to poor heat conduction in a body, and overcrowding at the harbor to the accumulation of heat and the subsequent rise in temperature near the surface of the body relative to its inner parts. Lumped system analysis is obviously not applicable when there is overcrowding at the surface. Of course, we have disregarded radiation in this analogy and thus the air traffic to the island. Like passengers at the harbor, heat changes *vehicles* at the surface from *convection* to *conduction*. Noting that a surface has zero thickness and thus cannot store any energy, heat reaching the surface of a body by convection must continue its journey within the body by conduction.

Consider heat transfer from a hot body to its cooler surroundings. Heat will be transferred from the body to the surrounding fluid as a result of a temperature difference. But this energy will come from the region near the surface, and thus the temperature of the body near the surface will drop. This creates a *temperature gradient* between the inner and outer regions of the body and initiates heat flow by conduction from the interior of the body toward the outer surface.

When the convection heat transfer coefficient h and thus convection heat transfer from the body are high, the temperature of the body near the surface will drop quickly (Fig. 16-8). This will create a larger temperature difference between the inner and outer regions unless the body is able to transfer heat from the inner to the outer regions just as fast. Thus, the magnitude of the maximum temperature difference within the body depends strongly on the ability of a body to conduct heat toward its surface relative to the ability of the surrounding medium to convect this heat away from the surface. The Biot number is a measure of the relative magnitudes of these two competing effects.

Recall that heat conduction in a specified direction n per unit surface area is expressed as $\dot{q} = -k\, \partial T/\partial n$, where $\partial T/\partial n$ is the temperature gradient and k is the thermal conductivity of the solid. Thus, the temperature distribution in the body will be *uniform* only when its thermal conductivity is *infinite,* and no such material is known to exist. Therefore, temperature gradients and thus temperature differences must exist within the body, no matter how small, in order for heat conduction to take place. Of course, the temperature gradient and the thermal conductivity are inversely proportional for a given heat flux. Therefore, the larger the thermal conductivity, the smaller the temperature gradient.

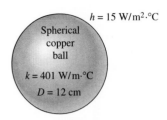

$$L_c = \frac{V}{A} = \frac{\frac{1}{6}\pi D^3}{\pi D^2} = \frac{1}{6}D = 0.02 \text{ m}$$

$$\text{Bi} = \frac{hL_c}{k} = \frac{15 \times 0.02}{401} = 0.00075 < 0.1$$

FIGURE 16-6

Small bodies with high thermal conductivities and low convection coefficients are most likely to satisfy the criterion for lumped system analysis.

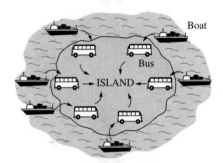

FIGURE 16-7

Analogy between heat transfer to a solid and passenger traffic to an island.

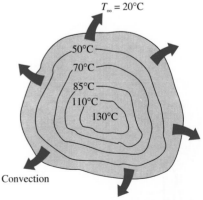

FIGURE 16-8

When the convection coefficient h is high and k is low, large temperature differences occur between the inner and outer regions of a large solid.

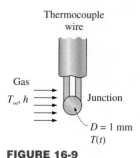

FIGURE 16-9

Schematic for Example 16-1.

EXAMPLE 16-1   Temperature Measurement by Thermocouples

The temperature of a gas stream is to be measured by a thermocouple whose junction can be approximated as a 1-mm-diameter sphere, as shown in Fig. 16-9. The properties of the junction are $k = 35$ W/m · °C, $\rho = 8500$ kg/m³, and $C_p = 320$ J/kg · °C, and the convection heat transfer coefficient between the junction and the gas is $h = 210$ W/m² · °C. Determine how long it will take for the thermocouple to read 99 percent of the initial temperature difference.

Solution   The temperature of a gas stream is to be measured by a thermocouple. The time it takes to register 99 percent of the initial ΔT is to be determined.

Assumptions   **1** The junction is spherical in shape with a diameter of $D = 0.001$ m. **2** The thermal properties of the junction and the heat transfer coefficient are constant. **3** Radiation effects are negligible.

Properties   The properties of the junction are given in the problem statement.

Analysis   The characteristic length of the junction is

$$L_c = \frac{V}{A} = \frac{\frac{1}{6}\pi D^3}{\pi D^2} = \frac{1}{6} D = \frac{1}{6}(0.001 \text{ m}) = 1.67 \times 10^{-4} \text{ m}$$

Then the Biot number becomes

$$\text{Bi} = \frac{hL_c}{k} = \frac{(210 \text{ W/m}^2 \cdot °\text{C})(1.67 \times 10^{-4} \text{ m})}{35 \text{ W/m} \cdot °\text{C}} = 0.001 < 0.1$$

Therefore, lumped system analysis is applicable, and the error involved in this approximation is negligible.

In order to read 99 percent of the initial temperature difference $T_i - T_\infty$ between the junction and the gas, we must have

$$\frac{T(t) - T_\infty}{T_i - T_\infty} = 0.01$$

For example, when $T_i = 0$°C and $T_\infty = 100$°C, a thermocouple is considered to have read 99 percent of this applied temperature difference when its reading indicates $T(t) = 99$°C.

The value of the exponent b is

$$b = \frac{hA_s}{\rho C_p V} = \frac{h}{\rho C_p L_c} = \frac{210 \text{ W/m}^2 \cdot °\text{C}}{(8500 \text{ kg/m}^3)(320 \text{ J/kg} \cdot °\text{C})(1.67 \times 10^{-4} \text{ m})} = 0.462 \text{ s}^{-1}$$

We now substitute these values into Eq. 16-4 and obtain

$$\frac{T(t) - T_\infty}{T_i - T_\infty} = e^{-bt} \longrightarrow 0.01 = e^{-(0.462 \text{ s}^{-1})t}$$

which yields

$$t = \mathbf{10 \ s}$$

Therefore, we must wait at least 10 s for the temperature of the thermocouple junction to approach within 1 percent of the initial junction-gas temperature difference.

FIGURE 16-10

Schematic for Example 16-2.

EXAMPLE 16-2   Predicting the Time of Death

A person is found dead at 5 PM in a room whose temperature is 20°C. The temperature of the body is measured to be 25°C when found, and the heat transfer coefficient is estimated to be $h = 8$ W/m² · °C. Modeling the body as a 30-cm-diameter, 1.70-m-long cylinder, estimate the time of death of that person (Fig. 16-10).

Solution   A body is found while still warm. The time of death is to be estimated.

705

**Transient Heat
Conduction in Large
Plane Walls, Long
Cylinders, and
Spheres**

Assumptions **1** The body can be modeled as a 30-cm-diameter, 1.70-m-long cylinder. **2** The thermal properties of the body and the heat transfer coefficient are constant. **3** The radiation effects are negligible. **4** The person was healthy(!) when he or she died with a body temperature of 37°C.

Properties The average human body is 72 percent water by mass, and thus we can assume the body to have the properties of water at the average temperature of $(37 + 25)/2 = 31°C$; $k = 0.617$ W/m · °C, $\rho = 996$ kg/m³, and $C_p = 4178$ J/kg · °C (Table A-15).

Analysis The characteristic length of the body is

$$L_c = \frac{V}{A} = \frac{\pi r_o^2 L}{2\pi r_o L + 2\pi r_o^2} = \frac{\pi (0.15 \text{ m})^2 (1.7 \text{ m})}{2\pi (0.15 \text{ m})(1.7 \text{ m}) + 2\pi (0.15 \text{ m})^2} = 0.0689 \text{ m}$$

Then the Biot number becomes

$$\text{Bi} = \frac{hL_c}{k} = \frac{(8 \text{ W/m}^2 \cdot °C)(0.0689 \text{ m})}{0.617 \text{ W/m} \cdot °C} = 0.89 > 0.1$$

Therefore, lumped system analysis is *not* applicable. However, we can still use it to get a "rough" estimate of the time of death. The exponent b in this case is

$$b = \frac{hA}{\rho C_p V} = \frac{h}{\rho C_p L_c} = \frac{8 \text{ W/m}^2 \cdot °C}{(996 \text{ kg/m}^3)(4178 \text{ J/kg} \cdot °C)(0.0689 \text{ m})}$$

$$= 2.79 \times 10^{-5} \text{ s}^{-1}$$

We now substitute these values into Eq. 16-4,

$$\frac{T(t) - T_\infty}{T_i - T_\infty} = e^{-bt} \longrightarrow \frac{25 - 20}{37 - 20} = e^{-(2.79 \times 10^{-5} \text{ s}^{-1})t}$$

which yields

$$t = 43,860 \text{ s} = \textbf{12.2 h}$$

Therefore, as a rough estimate, the person died about 12 h before the body was found, and thus the time of death is 5 AM. This example demonstrates how to obtain "ball park" values using a simple analysis.

16-2 ▪ TRANSIENT HEAT CONDUCTION IN LARGE PLANE WALLS, LONG CYLINDERS, AND SPHERES

In the preceding section, we considered bodies in which the variation of temperature within the body was negligible; that is, bodies that remain nearly *isothermal* during a process. Relatively *small* bodies of *highly conductive* materials approximate this behavior. In general, however, the temperature within a body will change from point to point as well as with time. In this section, we consider the variation of temperature with *time* and *position* in one-dimensional problems such as those associated with a large plane wall, a long cylinder, and a sphere.

Consider a plane wall of thickness $2L$, a long cylinder of radius r_o, and a sphere of radius r_o initially at a *uniform temperature* T_i, as shown in Fig. 16-11. At time $t = 0$, each geometry is placed in a large medium that is at a constant temperature T_∞ and kept in that medium for $t > 0$. Heat transfer takes place between these bodies and their environments by convection with a *uniform* and *constant* heat transfer coefficient h. Note that all three cases possess geometric and thermal symmetry: the plane wall is symmetric about its *center plane* ($x = 0$), the cylinder is symmetric about its *centerline* ($r = 0$), and the

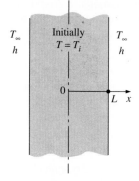

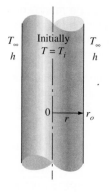

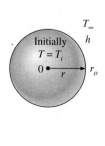

FIGURE 16-11

Schematic of the simple geometries in which heat transfer is one-dimensional.

(a) A large plane wall

(b) A long cylinder

(c) A sphere

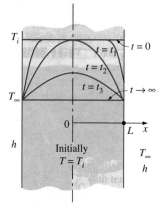

FIGURE 16-12

Transient temperature profiles in a plane wall exposed to convection from its surfaces for $T_i > T_\infty$.

sphere is symmetric about its *center point* ($r = 0$). We neglect *radiation* heat transfer between these bodies and their surrounding surfaces, or incorporate the radiation effect into the convection heat transfer coefficient h.

The variation of the temperature profile with *time* in the plane wall is illustrated in Fig. 16-12. When the wall is first exposed to the surrounding medium at $T_\infty < T_i$ at $t = 0$, the entire wall is at its initial temperature T_i. But the wall temperature at and near the surfaces starts to drop as a result of heat transfer from the wall to the surrounding medium. This creates a *temperature gradient* in the wall and initiates heat conduction from the inner parts of the wall toward its outer surfaces. Note that the temperature at the center of the wall remains at T_i until $t = t_2$, and that the temperature profile within the wall remains symmetric at all times about the center plane. The temperature profile gets flatter and flatter as time passes as a result of heat transfer, and eventually becomes uniform at $T = T_\infty$. That is, the wall reaches *thermal equilibrium* with its surroundings. At that point, the heat transfer stops since there is no longer a temperature difference. Similar discussions can be given for the long cylinder or sphere.

The formulation of the problems for the determination of the one-dimensional transient temperature distribution $T(x, t)$ in a wall results in a partial differential equation, which can be solved using advanced mathematical techniques. The solution, however, normally involves infinite series, which are inconvenient and time-consuming to evaluate. Therefore, there is clear motivation to present the solution in *tabular* or *graphical* form. However, the solution involves the parameters x, L, t, k, α, h, T_i, and T_∞, which are too many to make any graphical presentation of the results practical. In order to reduce the number of parameters, we nondimensionalize the problem by defining the following dimensionless quantities:

Dimensionless temperature: $\quad \theta(x, t) = \dfrac{T(x, t) - T_\infty}{T_i - T_\infty}$

Dimensionless distance from the center: $\quad X = \dfrac{x}{L}$

Dimensionless heat transfer coefficient: $\quad \text{Bi} = \dfrac{hL}{k} \quad$ **(Biot number)**

Dimensionless time: $\quad \tau = \dfrac{\alpha t}{L^2} \quad$ **(Fourier number)**

The nondimensionalization enables us to present the temperature in terms of three parameters only: X, Bi, and τ. This makes it practical to present the

707

**Transient Heat
Conduction in Large
Plane Walls, Long
Cylinders, and
Spheres**

solution in graphical form. The dimensionless quantities defined above for a plane wall can also be used for a *cylinder* or *sphere* by replacing the space variable x by r and the half-thickness L by the outer radius r_o. Note that the characteristic length in the definition of the Biot number is taken to be the *half-thickness L* for the plane wall, and the *radius r_o* for the long cylinder and sphere instead of V/A used in lumped system analysis.

The one-dimensional transient heat conduction problem described above can be solved exactly for any of the three geometries, but the solution involves infinite series, which are difficult to deal with. However, the terms in the solutions converge rapidly with increasing time, and for $\tau > 0.2$, keeping the first term and neglecting all the remaining terms in the series results in an error under 2 percent. We are usually interested in the solution for times with $\tau > 0.2$, and thus it is very convenient to express the solution using this **one-term approximation,** given as

Plane wall:	$\theta(x, t)_{\text{wall}} = \dfrac{T(x, t) - T_\infty}{T_i - T_\infty} = A_1 e^{-\lambda_1^2 \tau} \cos(\lambda_1 x/L),$	$\tau > 0.2$ (16-10)
Cylinder:	$\theta(r, t)_{\text{cyl}} = \dfrac{T(r, t) - T_\infty}{T_i - T_\infty} = A_1 e^{-\lambda_1^2 \tau} J_0(\lambda_1 r/r_o),$	$\tau > 0.2$ (16-11)
Sphere:	$\theta(r, t)_{\text{sph}} = \dfrac{T(r, t) - T_\infty}{T_i - T_\infty} = A_1 e^{-\lambda_1^2 \tau} \dfrac{\sin(\lambda_1 r/r_o)}{\lambda_1 r/r_o},$	$\tau > 0.2$ (16-12)

where the constants A_1 and λ_1 are functions of the Bi number only, and their values are listed in Table 16-1 against the Bi number for all three geometries. The function J_0 is the zeroth-order Bessel function of the first kind, whose value can be determined from Table 16-2. Noting that $\cos(0) = J_0(0) = 1$ and the limit of $(\sin x)/x$ is also 1, the above relations simplify to the following at the center of a plane wall, cylinder, or sphere:

Center of plane wall $(x = 0)$:	$\theta_{0, \text{wall}} = \dfrac{T_o - T_\infty}{T_i - T_\infty} = A_1 e^{-\lambda_1^2 \tau}$	(16-13)
Center of cylinder $(r = 0)$:	$\theta_{0, \text{cyl}} = \dfrac{T_o - T_\infty}{T_i - T_\infty} = A_1 e^{-\lambda_1^2 \tau}$	(16-14)
Center of sphere $(r = 0)$:	$\theta_{0, \text{sph}} = \dfrac{T_o - T_\infty}{T_i - T_\infty} = A_1 e^{-\lambda_1^2 \tau}$	(16-15)

Once the Bi number is known, the above relations can be used to determine the temperature anywhere in the medium. The determination of the constants A_1 and λ_1 usually requires interpolation. For those who prefer reading charts to interpolating, the relations above are plotted and the one-term approximation solutions are presented in graphical form, known as the *transient temperature charts*. Note that the charts are sometimes difficult to read, and they are subject to reading errors. Therefore, the relations above should be preferred to the charts.

The transient temperature charts in Figs. 16-13, 16-14, and 16-15 for a large plane wall, long cylinder, and sphere were presented by M. P. Heisler in 1947 and are called **Heisler charts.** They were supplemented in 1961 with transient heat transfer charts by H. Gröber. There are *three* charts associated with each geometry: the first chart is to determine the temperature T_o at the *center* of the geometry at a given time t. The second chart is to determine the temperature at *other locations* at the same time in terms of T_o. The third chart

TABLE 16-1

Coefficients used in the one-term approximate solution of transient one-dimensional heat conduction in plane walls, cylinders, and spheres ($Bi = hL/k$ for a plane wall of thickness $2L$, and $Bi = hr_o/k$ for a cylinder or sphere of radius r_o)

Bi	Plane wall λ_1	Plane wall A_1	Cylinder λ_1	Cylinder A_1	Sphere λ_1	Sphere A_1
0.01	0.0998	1.0017	0.1412	1.0025	0.1730	1.0030
0.02	0.1410	1.0033	0.1995	1.0050	0.2445	1.0060
0.04	0.1987	1.0066	0.2814	1.0099	0.3450	1.0120
0.06	0.2425	1.0098	0.3438	1.0148	0.4217	1.0179
0.08	0.2791	1.0130	0.3960	1.0197	0.4860	1.0239
0.1	0.3111	1.0161	0.4417	1.0246	0.5423	1.0298
0.2	0.4328	1.0311	0.6170	1.0483	0.7593	1.0592
0.3	0.5218	1.0450	0.7465	1.0712	0.9208	1.0880
0.4	0.5932	1.0580	0.8516	1.0931	1.0528	1.1164
0.5	0.6533	1.0701	0.9408	1.1143	1.1656	1.1441
0.6	0.7051	1.0814	1.0184	1.1345	1.2644	1.1713
0.7	0.7506	1.0918	1.0873	1.1539	1.3525	1.1978
0.8	0.7910	1.1016	1.1490	1.1724	1.4320	1.2236
0.9	0.8274	1.1107	1.2048	1.1902	1.5044	1.2488
1.0	0.8603	1.1191	1.2558	1.2071	1.5708	1.2732
2.0	1.0769	1.1785	1.5995	1.3384	2.0288	1.4793
3.0	1.1925	1.2102	1.7887	1.4191	2.2889	1.6227
4.0	1.2646	1.2287	1.9081	1.4698	2.4556	1.7202
5.0	1.3138	1.2403	1.9898	1.5029	2.5704	1.7870
6.0	1.3496	1.2479	2.0490	1.5253	2.6537	1.8338
7.0	1.3766	1.2532	2.0937	1.5411	2.7165	1.8673
8.0	1.3978	1.2570	2.1286	1.5526	2.7654	1.8920
9.0	1.4149	1.2598	2.1566	1.5611	2.8044	1.9106
10.0	1.4289	1.2620	2.1795	1.5677	2.8363	1.9249
20.0	1.4961	1.2699	2.2880	1.5919	2.9857	1.9781
30.0	1.5202	1.2717	2.3261	1.5973	3.0372	1.9898
40.0	1.5325	1.2723	2.3455	1.5993	3.0632	1.9942
50.0	1.5400	1.2727	2.3572	1.6002	3.0788	1.9962
100.0	1.5552	1.2731	2.3809	1.6015	3.1102	1.9990
∞	1.5708	1.2732	2.4048	1.6021	3.1416	2.0000

TABLE 16-2

The zeroth- and first-order Bessel functions of the first kind

ξ	$J_0(\xi)$	$J_1(\xi)$
0.0	1.0000	0.0000
0.1	0.9975	0.0499
0.2	0.9900	0.0995
0.3	0.9776	0.1483
0.4	0.9604	0.1960
0.5	0.9385	0.2423
0.6	0.9120	0.2867
0.7	0.8812	0.3290
0.8	0.8463	0.3688
0.9	0.8075	0.4059
1.0	0.7652	0.4400
1.1	0.7196	0.4709
1.2	0.6711	0.4983
1.3	0.6201	0.5220
1.4	0.5669	0.5419
1.5	0.5118	0.5579
1.6	0.4554	0.5699
1.7	0.3980	0.5778
1.8	0.3400	0.5815
1.9	0.2818	0.5812
2.0	0.2239	0.5767
2.1	0.1666	0.5683
2.2	0.1104	0.5560
2.3	0.0555	0.5399
2.4	0.0025	0.5202
2.6	−0.0968	−0.4708
2.8	−0.1850	−0.4097
3.0	−0.2601	−0.3391
3.2	−0.3202	−0.2613

is to determine the total amount of *heat transfer* up to the time t. These plots are valid for $\tau > 0.2$.

Note that the case $1/Bi = k/hL = 0$ corresponds to $h \to \infty$, which corresponds to the case of *specified surface temperature* T_∞. That is, the case in which the surfaces of the body are suddenly brought to the temperature T_∞ at $t = 0$ and kept at T_∞ at all times can be handled by setting h to infinity (Fig. 16-16).

The temperature of the body changes from the initial temperature T_i to the temperature of the surroundings T_∞ at the end of the transient heat conduction process. Thus, the *maximum* amount of heat that a body can gain (or lose if $T_i > T_\infty$) is simply the *change* in the *energy content* of the body. That is,

$$Q_{max} = mC_p(T_\infty - T_i) = \rho V C_p(T_\infty - T_i) \quad \text{(kJ)} \quad (16\text{-}16)$$

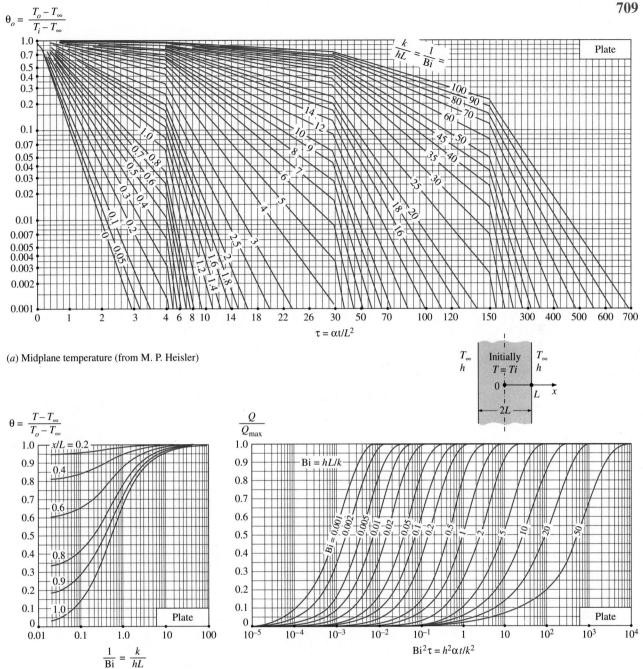

$\theta_o = \dfrac{T_o - T_\infty}{T_i - T_\infty}$

$\tau = \alpha t/L^2$

(a) Midplane temperature (from M. P. Heisler)

$\theta = \dfrac{T - T_\infty}{T_o - T_\infty}$

$\dfrac{Q}{Q_{max}}$

$\dfrac{1}{Bi} = \dfrac{k}{hL}$

$Bi^2\tau = h^2\alpha t/k^2$

(b) Temperature distribution (from M. P. Heisler)

(c) Heat transfer (from H. Gröber et al.)

FIGURE 16-13

Transient temperature and heat transfer charts for a plane wall of thickness $2L$ initially at a uniform temperature T_i subjected to convection from both sides to an environment at temperature T_∞ with a convection coefficient of h.

where m is the mass, V is the volume, ρ is the density, and C_p is the specific heat of the body. Thus, Q_{max} represents the amount of heat transfer for $t \to \infty$. The amount of heat transfer Q at a finite time t will obviously be less than this maximum. The ratio Q/Q_{max} is plotted in Figures 16-13c, 16-14c, and 16-15c against the variables Bi and $h^2\alpha t/k^2$ for the large plane wall, long cylinder, and

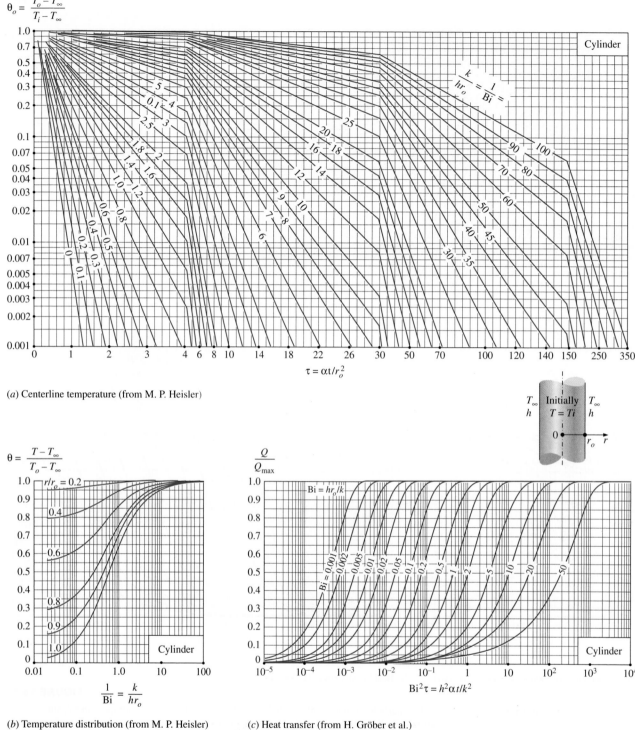

$\theta_o = \dfrac{T_o - T_\infty}{T_i - T_\infty}$

(a) Centerline temperature (from M. P. Heisler)

$\theta = \dfrac{T - T_\infty}{T_o - T_\infty}$

$\dfrac{Q}{Q_{max}}$

(b) Temperature distribution (from M. P. Heisler)   (c) Heat transfer (from H. Gröber et al.)

FIGURE 16-14

Transient temperature and heat transfer charts for a long cylinder of radius r_o initially at a uniform temperature T_i subjected to convection from all sides to an environment at temperature T_∞ with a convection coefficient of h.

sphere, respectively. Note that once the *fraction* of heat transfer Q/Q_{max} has been determined from these charts for the given t, the actual amount of heat

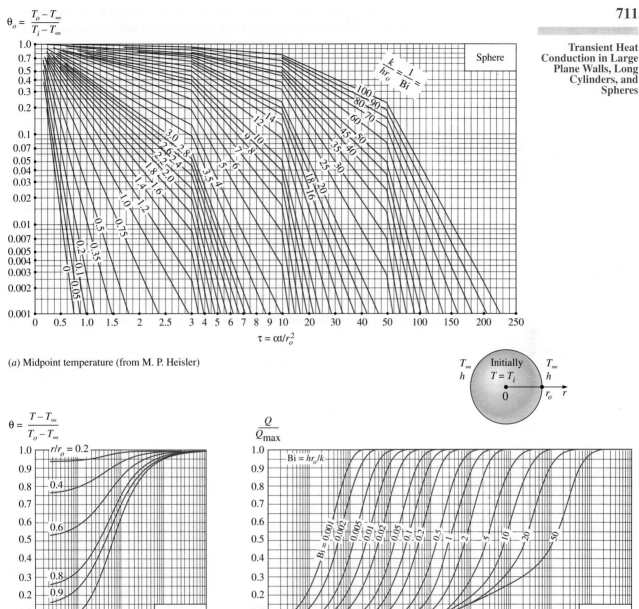

$$\theta_o = \frac{T_o - T_\infty}{T_i - T_\infty}$$

(a) Midpoint temperature (from M. P. Heisler)

$$\theta = \frac{T - T_\infty}{T_o - T_\infty}$$

$$\frac{Q}{Q_{\text{max}}}$$

(b) Temperature distribution (from M. P. Heisler)

(c) Heat transfer (from H. Gröber et al.)

FIGURE 16-15

Transient temperature and heat transfer charts for a sphere of radius r_o initially at a uniform temperature T_i subjected to convection from all sides to an environment at temperature T_∞ with a convection coefficient of h.

transfer by that time can be evaluated by multiplying this fraction by Q_{max}. A *negative* sign for Q_{max} indicates that heat is *leaving* the body (Fig. 16-17).

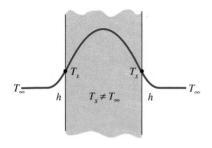

(a) Finite convection coefficient

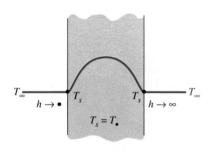

(b) Infinite convection coefficient

FIGURE 16-16

The specified surface temperature corresponds to the case of convection to an environment at T_∞ with a convection coefficient h that is *infinite*.

The fraction of heat transfer can also be determined from the following relations, which are based on the one-term approximations discussed above:

Plane wall: $\left(\dfrac{Q}{Q_{max}}\right)_{wall} = 1 - \theta_{0,\,wall}\dfrac{\sin\lambda_1}{\lambda_1}$ (16-17)

Cylinder: $\left(\dfrac{Q}{Q_{max}}\right)_{cyl} = 1 - 2\theta_{0,\,cyl}\dfrac{J_1(\lambda_1)}{\lambda_1}$ (16-18)

Sphere: $\left(\dfrac{Q}{Q_{max}}\right)_{sph} = 1 - 3\theta_{0,\,sph}\dfrac{\sin\lambda_1 - \lambda_1\cos\lambda_1}{\lambda_1^3}$ (16-19)

The use of the Heisler/Gröber charts and the one-term solutions discussed above is limited to the conditions specified at the beginning of this section: the body is initially at a *uniform* temperature, the temperature of the medium surrounding the body and the convection heat transfer coefficient are *constant* and *uniform,* and there is no *energy generation* in the body.

We discussed the physical significance of the *Biot number* earlier and indicated that it is a measure of the relative magnitudes of the two heat transfer mechanisms: *convection* at the surface and *conduction* through the solid. A *small* value of Bi indicates that the inner resistance of the body to heat conduction is *small* relative to the resistance to convection between the surface and the fluid. As a result, the temperature distribution within the solid becomes fairly uniform, and lumped system analysis becomes applicable. Recall that when Bi < 0.1, the error in assuming the temperature within the body to be *uniform* is negligible.

To understand the physical significance of the *Fourier number* τ, we express it as (Fig. 16-18)

$$\tau = \frac{\alpha t}{L^2} = \frac{kL^2\,(1/L)}{\rho C_p L^3/t}\frac{\Delta T}{\Delta T} = \frac{\begin{array}{c}\text{The rate at which heat is }conducted\\ \text{across }L\text{ of a body of volume }L^3\end{array}}{\begin{array}{c}\text{The rate at which heat is }stored\\ \text{in a body of volume }L^3\end{array}} \quad (16\text{-}20)$$

Therefore, the Fourier number is a measure of *heat conducted* through a body relative to *heat stored*. Thus, a large value of the Fourier number indicates faster propagation of heat through a body.

Perhaps you are wondering about what constitutes an infinitely large plate or an infinitely long cylinder. After all, nothing in this world is infinite. A plate whose thickness is small relative to the other dimensions can be modeled as an infinitely large plate, except very near the outer edges. But the edge effects on large bodies are usually negligible, and thus a large plane wall such as the wall of a house can be modeled as an infinitely large wall for heat transfer purposes. Similarly, a long cylinder whose diameter is small relative to its length can be analyzed as an infinitely long cylinder. The use of the transient temperature charts and the one-term solutions is illustrated in the following examples.

EXAMPLE 16-3   Boiling Eggs

An ordinary egg can be approximated as a 5-cm-diameter sphere (Fig. 16-19). The egg is initially at a uniform temperature of 5°C and is dropped into boiling water at 95°C. Taking the convection heat transfer coefficient to be $h = 1200$ W/m² · °C, determine how long it will take for the center of the egg to reach 70°C.

Solution   An egg is cooked in boiling water. The cooking time of the egg is to be determined.

Assumptions **1** The egg is spherical in shape with a radius of $r_0 = 2.5$ cm. **2** Heat conduction in the egg is one-dimensional because of thermal symmetry about the midpoint. **3** The thermal properties of the egg and the heat transfer coefficient are constant. **4** The Fourier number is $\tau > 0.2$ so that the one-term approximate solutions are applicable.

Properties The water content of eggs is about 74 percent, and thus the thermal conductivity and diffusivity of eggs can be approximated by those of water at the average temperature of (5 1 70)/2 5 37.5°C; k 5 0.627 W/m · °C and a 5 k/rCp 5 0.151 3 1026 m2/s (Table A-15).

Analysis The temperature within the egg varies with radial distance as well as time, and the temperature at a specified location at a given time can be determined from the Heisler charts or the one-term solutions. Here we will use the latter to demonstrate their use. The Biot number for this problem is

$$\text{Bi} = \frac{hr_0}{k} = \frac{(1200 \text{ W/m}^2 \cdot °\text{C})(0.025 \text{ m})}{0.627 \text{ W/m} \cdot °\text{C}} = 47.8$$

which is much greater than 0.1, and thus the lumped system analysis is not applicable. The coefficients λ_1 and A_1 for a sphere corresponding to this Bi are, from Table 16-1,

$$\lambda_1 = 3.0753, \qquad A_1 = 1.9958$$

Substituting these and other values into Eq. 16-15 and solving for τ gives

$$\frac{T_o - T_\infty}{T_i - T_\infty} = A_1 e^{-\lambda_1^2 \tau} \longrightarrow \frac{70 - 95}{5 - 95} = 1.9958 e^{-(3.0753)^2 \tau} \longrightarrow \tau = 0.209$$

which is greater than 0.2, and thus the one-term solution is applicable with an error of less than 2 percent. Then the cooking time is determined from the definition of the Fourier number to be

$$t = \frac{\tau r_o^2}{\alpha} = \frac{(0.209)(0.025 \text{ m})^2}{0.151 \times 10^{-6} \text{ m}^2/\text{s}} = 865 \text{ s} \approx \textbf{14.4 min}$$

Therefore, it will take about 15 min for the center of the egg to be heated from 5°C to 70°C.

Discussion Note that the Biot number in lumped system analysis was defined differently as Bi = hL_c/k = $h(r/3)/k$. However, either definition can be used in determining the applicability of the lumped system analysis unless Bi ≈ 0.1.

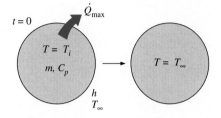

(a) Maximum heat transfer ($t \to \bullet$)

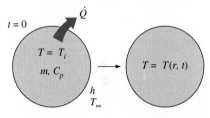

(b) Actual heat transfer for time t

FIGURE 16-17

The fraction of total heat transfer Q/Q_{max} up to a specified time t is determined using the Gröber charts.

EXAMPLE 16-4 Heating of Large Brass Plates in an Oven

In a production facility, large brass plates of 4 cm thickness that are initially at a uniform temperature of 20°C are heated by passing them through an oven that is maintained at 500°C (Fig. 16-20). The plates remain in the oven for a period of 7 min. Taking the combined convection and radiation heat transfer coefficient to be $h = 120$ W/m² · °C, determine the surface temperature of the plates when they come out of the oven.

Solution Large brass plates are heated in an oven. The surface temperature of the plates leaving the oven is to be determined.

Assumptions **1** Heat conduction in the plate is one-dimensional since the plate is large relative to its thickness and there is thermal symmetry about the center plane. **2** The thermal properties of the plate and the heat transfer coefficient are constant. **3** The Fourier number is $\tau > 0.2$ so that the one-term approximate solutions are applicable.

Properties The properties of brass at room temperature are $k = 110$ W/m · °C, $\rho = 8530$ kg/m³, $C_p = 380$ J/kg · °C, and $\alpha = 33.9 \times 10^{-6}$ m²/s (Table A-19). More accurate results are obtained by using properties at average temperature.

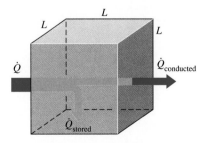

Fourier number: $\tau = \dfrac{\alpha t}{L^2} = \dfrac{\dot{Q}_{\text{conducted}}}{\dot{Q}_{\text{stored}}}$

FIGURE 16-18

Fourier number at time t can be viewed as the ratio of the rate of heat conducted to the rate of heat stored at that time.

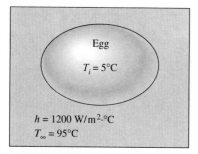

FIGURE 16-19
Schematic for Example 16-3.

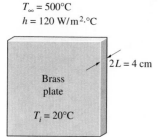

FIGURE 16-20
Schematic for Example 16-4.

$T_\infty = 200°C$
$h = 80 \text{ W/m}^2 \cdot °C$

Stainless steel
shaft

$T_i = 600°C$ $D = 20$ cm

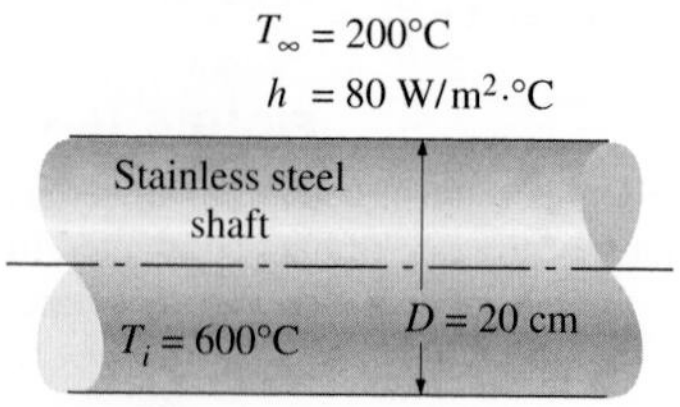

FIGURE 16-21
Schematic for Example 16-5.

Analysis The temperature at a specified location at a given time can be determined from the Heisler charts or one-term solutions. Here we will use the charts to demonstrate their use. Noting that the half-thickness of the plate is $L = 0.02$ m, from Fig. 16-13 we have

$$\frac{1}{\text{Bi}} = \frac{k}{hL} = \frac{100 \text{ W/m} \cdot °C}{(120 \text{ W/m}^2 \cdot °C)(0.02 \text{ m})} = 45.8$$

$$\tau = \frac{\alpha t}{L^2} = \frac{(33.9 \times 10^{-6} \text{ m}^2/\text{s})(7 \times 60 \text{ s})}{(0.02 \text{ m})^2} = 35.6$$

$$\left.\right\} \frac{T_o - T_\infty}{T_i - T_\infty} = 0.46$$

Also,

$$\frac{1}{\text{Bi}} = \frac{k}{hL} = 45.8$$

$$\frac{x}{L} = \frac{L}{L} = 1$$

$$\left.\right\} \frac{T - T_\infty}{T_o - T_\infty} = 0.99$$

Therefore,

$$\frac{T - T_\infty}{T_i - T_\infty} = \frac{T - T_\infty}{T_o - T_\infty}\frac{T_o - T_\infty}{T_i - T_\infty} = 0.46 \times 0.99 = 0.455$$

and

$$T = T_\infty + 0.455(T_i - T_\infty) = 500 + 0.455(20 - 500) = \textbf{282°C}$$

Therefore, the surface temperature of the plates will be 282°C when they leave the oven.

Discussion We notice that the Biot number in this case is Bi = 1/45.8 = 0.022, which is much less than 0.1. Therefore, we expect the lumped system analysis to be applicable. This is also evident from $(T - T_\infty)/(T_o - T_\infty) = 0.99$, which indicates that the temperatures at the center and the surface of the plate relative to the surrounding temperature are within 1 percent of each other. Noting that the error involved in reading the Heisler charts is typically at least a few percent, the lumped system analysis in this case may yield just as accurate results with less effort.

The heat transfer surface area of the plate is 2A, where A is the face area of the plate (the plate transfers heat through both of its surfaces), and the volume of the plate is $V = (2L)A$, where L is the half-thickness of the plate. The exponent b used in the lumped system analysis is determined to be

$$b = \frac{hA}{\rho C_p V} = \frac{h(2A)}{\rho C_p(2LA)} = \frac{h}{\rho C_p L}$$

$$= \frac{120 \text{ W/m}^2 \cdot °C}{(8530 \text{ kg/m}^3)(380 \text{ J/kg} \cdot °C)(0.02 \text{ m})} = 0.00185 \text{ s}^{-1}$$

Then the temperature of the plate at $t = 7$ min = 420 s is determined from

$$\frac{T(t) - T_\infty}{T_i - T_\infty} = e^{-bt} \longrightarrow \frac{T(t) - 500}{20 - 500} = e^{-(0.00185 \text{ s}^{-1})(420 \text{ s})}$$

It yields

$$T(t) = 279°C$$

which is practically identical to the result obtained above using the Heisler charts. Therefore, we can use lumped system analysis with confidence when the Biot number is sufficiently small.

EXAMPLE 16-5   Cooling of a Long Stainless Steel Cylindrical Shaft
A long 20-cm-diameter cylindrical shaft made of stainless steel 304 comes out of an oven at a uniform temperature of 600°C (Fig. 16-21). The shaft is then allowed

to cool slowly in an environment chamber at 200°C with an average heat transfer coefficient of $h = 80$ W/m² · °C. Determine the temperature at the center of the shaft 45 min after the start of the cooling process. Also, determine the heat transfer per unit length of the shaft during this time period.

Solution A long cylindrical shaft at 600°C is allowed to cool slowly. The center temperature and the heat transfer per unit length are to be determined.

Assumptions **1** Heat conduction in the shaft is one-dimensional since it is long and it has thermal symmetry about the centerline. **2** The thermal properties of the shaft and the heat transfer coefficient are constant. **3** The Fourier number is $\tau > 0.2$ so that the one-term approximate solutions are applicable.

Properties The properties of stainless steel 304 at room temperature are $k = 14.9$ W/m · °C, $\rho = 7900$ kg/m³, $C_p = 477$ J/kg · °C, and $\alpha = 3.95 \times 10^{-6}$ m²/s (Table A-19). More accurate results can be obtained by using properties at average temperature.

Analysis The temperature within the shaft may vary with the radial distance r as well as time, and the temperature at a specified location at a given time can be determined from the Heisler charts. Noting that the radius of the shaft is $r_o = 0.1$ m, from Fig. 16-14 we have

$$\left. \begin{array}{l} \dfrac{1}{Bi} = \dfrac{k}{hr_o} = \dfrac{14.9\ \text{W/m} \cdot \text{°C}}{(80\ \text{W/m}^2 \cdot \text{°C})(0.1\ \text{m})} = 1.86 \\[2mm] \tau = \dfrac{\alpha t}{r_o^2} = \dfrac{(3.95 \times 10^{-6}\ \text{m}^2/\text{s})(45 \times 60\ \text{s})}{(0.1\ \text{m})^2} = 1.07 \end{array} \right\} \dfrac{T_o - T_\infty}{T_i - T_\infty} = 0.40$$

and

$$T_o = T_\infty + 0.4(T_i - T_\infty) = 200 + 0.4(600 - 200) = \mathbf{360°C}$$

Therefore, the center temperature of the shaft will drop from 600°C to 360°C in 45 min.

To determine the actual heat transfer, we first need to calculate the maximum heat that can be transferred from the cylinder, which is the sensible energy of the cylinder relative to its environment. Taking $L = 1$ m,

$$m = \rho V = \rho \pi r_o^2 L = (7900\ \text{kg/m}^3)\pi(0.1\ \text{m})^2(1\ \text{m}) = 248.2\ \text{kg}$$

$$Q_{max} = mC_p(T_\infty - T_i) = (248.2\ \text{kg})(0.477\ \text{kJ/kg} \cdot \text{°C})(600 - 200)\text{°C}$$
$$= 47{,}354\ \text{kJ}$$

The dimensionless heat transfer ratio is determined from Fig. 16-14c for a long cylinder to be

$$\left. \begin{array}{l} Bi = \dfrac{1}{1/Bi} = \dfrac{1}{1.86} = 0.537 \\[2mm] \dfrac{h^2 \alpha t}{k^2} = Bi^2\tau = (0.537)^2(1.07) = 0.309 \end{array} \right\} \dfrac{Q}{Q_{max}} = 0.62$$

Therefore,

$$Q = 0.62 Q_{max} = 0.62 \times (47{,}354\ \text{kJ}) = \mathbf{29{,}360\ kJ}$$

which is the total heat transfer from the shaft during the first 45 min of the cooling.

Alternative solution We could also solve this problem using the one-term solution relation instead of the transient charts. First we find the Biot number

$$Bi = \dfrac{hr_o}{k} = \dfrac{(80\ \text{W/m}^2 \cdot \text{°C})(0.1\ \text{m})}{14.9\ \text{W/m} \cdot \text{°C}} = 0.537$$

The coefficients λ_1 and A_1 for a cylinder corresponding to this Bi are determined from Table 16-1 to be

$$\lambda_1 = 0.970, \qquad A_1 = 1.122$$

Substituting these values into Eq. 16-14 gives

$$\theta_0 = \frac{T_o - T_\infty}{T_i - T_\infty} = A_1 e^{-\lambda_1^2 \tau} = 1.122 e^{-(0.970)^2(1.07)} = 0.41$$

and thus

$$T_o = T_\infty + 0.41(T_i - T_\infty) = 200 + 0.41(600 - 200) = \textbf{364°C}$$

The value of $J_1(\lambda_1)$ for $\lambda_1 = 0.970$ is determined from Table 16-2 to be 0.430. Then the fractional heat transfer is determined from Eq. 16-18 to be

$$\frac{Q}{Q_{max}} = 1 - 2\theta_0 \frac{J_1(\lambda_1)}{\lambda_1} = 1 - 2 \times 0.41 \frac{0.430}{0.970} = 0.636$$

and thus

$$Q = 0.636 Q_{max} = 0.636 \times (47,354 \text{ kJ}) = \textbf{30,120 kJ}$$

Discussion   The slight difference between the two results is due to the reading error of the charts.

16-3 ■ TRANSIENT HEAT CONDUCTION IN SEMI-INFINITE SOLIDS

A semi-infinite solid is an idealized body that has a *single plane surface* and extends to infinity in all directions, as shown in Fig. 16-22. This idealized body is used to indicate that the temperature change in the part of the body in which we are interested (the region close to the surface) is due to the thermal conditions on a single surface. The earth, for example, can be considered to be a semi-infinite medium in determining the variation of temperature near its surface. Also, a thick wall can be modeled as a semi-infinite medium if all we are interested in is the variation of temperature in the region near one of the surfaces, and the other surface is too far to have any impact on the region of interest during the time of observation.

Consider a semi-infinite solid that is at a uniform temperature T_i. At time $t = 0$, the surface of the solid at $x = 0$ is exposed to convection by a fluid at a constant temperature T_∞, with a heat transfer coefficient h. This problem can be formulated as a partial differential equation, which can be solved analytically for the transient temperature distribution $T(x, t)$. The solution obtained is presented in Fig. 16-23 graphically for the *nondimensionalized temperature* defined as

$$1 - \theta(x, t) = 1 - \frac{T(x, t) - T_\infty}{T_i - T_\infty} = \frac{T(x, t) - T_i}{T_\infty - T_i} \qquad (16\text{-}21)$$

against the dimensionless variable $x/(2\sqrt{\alpha t})$ for various values of the parameter $h\sqrt{\alpha t}/k$.

Note that the values on the vertical axis correspond to $x = 0$, and thus represent the surface temperature. The curve $h\sqrt{\alpha t}/k = \infty$ corresponds to $h \to \infty$, which corresponds to the case of *specified temperature* T_∞ at the surface at $x = 0$. That is, the case in which the surface of the semi-infinite body is suddenly brought to temperature T_∞ at $t = 0$ and kept at T_∞ at all times can be handled by setting h to infinity. The specified surface temperature case is closely

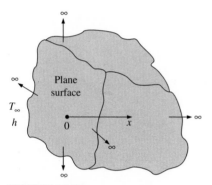

FIGURE 16-22

Schematic of a semi-infinite body.

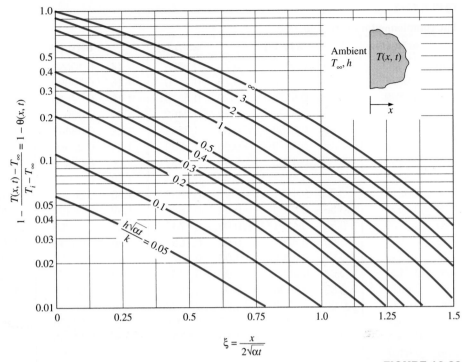

FIGURE 16-23

Variation of temperature with position and time in a semi-infinite solid initially at T_i subjected to convection to an environment at T_∞ with a convection heat transfer coefficient of h (from P. J. Schneider, Ref. 11).

approximated in practice when condensation or boiling takes place on the surface. For a *finite* heat transfer coefficient h, the surface temperature approaches the fluid temperature T_∞ as the time t approaches infinity.

The exact solution of the transient one-dimensional heat conduction problem in a semi-infinite medium that is initially at a uniform temperature of T_i and is suddenly subjected to convection at time $t = 0$ has been obtained, and is expressed as

$$\frac{T(x, t) - T_i}{T_\infty - T_i} = \operatorname{erfc}\left(\frac{x}{2\sqrt{\alpha t}}\right) - \exp\left(\frac{hx}{k} + \frac{h^2\alpha t}{k^2}\right)\left[\operatorname{erfc}\left(\frac{x}{2\sqrt{\alpha t}} + \frac{h\sqrt{\alpha t}}{k}\right)\right]$$

$$(16\text{-}22)$$

where the quantity erfc (ξ) is the **complementary error function,** defined as

$$\operatorname{erfc}(\xi) = 1 - \frac{2}{\sqrt{\pi}} \int_0^\xi e^{-u^2}\, du \qquad (16\text{-}23)$$

Despite its simple appearance, the integral that appears in the above relation cannot be performed analytically. Therefore, it is evaluated numerically for different values of ξ, and the results are listed in Table 16-3. For the special case of $h \to \infty$, the surface temperature T_s becomes equal to the fluid temperature T_∞, and Eq. 16-22 reduces to

$$\frac{T(x, t) - T_i}{T_s - T_i} = \operatorname{erfc}\left(\frac{x}{2\sqrt{\alpha t}}\right) \qquad (16\text{-}24)$$

This solution corresponds to the case when the temperature of the exposed surface of the medium is suddenly raised (or lowered) to T_s at $t = 0$ and is

TABLE 16-3

The complementary error function

ξ	erfc (ξ)	ξ	erfc (ξ)	ξ	erfc (ξ)	ξ	erfc (ξ)	ξ	erfc (ξ)	ξ	erfc (ξ)
0.00	1.00000	0.38	0.5910	0.76	0.2825	1.14	0.1069	1.52	0.03159	1.90	0.00721
0.02	0.9774	0.40	0.5716	0.78	0.2700	1.16	0.10090	1.54	0.02941	1.92	0.00662
0.04	0.9549	0.42	0.5525	0.80	0.2579	1.18	0.09516	1.56	0.02737	1.94	0.00608
0.06	0.9324	0.44	0.5338	0.82	0.2462	1.20	0.08969	1.58	0.02545	1.96	0.00557
0.08	0.9099	0.46	0.5153	0.84	0.2349	1.22	0.08447	1.60	0.02365	1.98	0.00511
0.10	0.8875	0.48	0.4973	0.86	0.2239	1.24	0.07950	1.62	0.02196	2.00	0.00468
0.12	0.8652	0.50	0.4795	0.88	0.2133	1.26	0.07476	1.64	0.02038	2.10	0.00298
0.14	0.8431	0.52	0.4621	0.90	0.2031	1.28	0.07027	1.66	0.01890	2.20	0.00186
0.16	0.8210	0.54	0.4451	0.92	0.1932	1.30	0.06599	1.68	0.01751	2.30	0.00114
0.18	0.7991	0.56	0.4284	0.94	0.1837	1.32	0.06194	1.70	0.01612	2.40	0.00069
0.20	0.7773	0.58	0.4121	0.96	0.1746	1.34	0.05809	1.72	0.01500	2.50	0.00041
0.22	0.7557	0.60	0.3961	0.98	0.1658	1.36	0.05444	1.74	0.01387	2.60	0.00024
0.24	0.7343	0.62	0.3806	1.00	0.1573	1.38	0.05098	1.76	0.01281	2.70	0.00013
0.26	0.7131	0.64	0.3654	1.02	0.1492	1.40	0.04772	1.78	0.01183	2.80	0.00008
0.28	0.6921	0.66	0.3506	1.04	0.1413	1.42	0.04462	1.80	0.01091	2.90	0.00004
0.30	0.6714	0.68	0.3362	1.06	0.1339	1.44	0.04170	1.82	0.01006	3.00	0.00002
0.32	0.6509	0.70	0.3222	1.08	0.1267	1.46	0.03895	1.84	0.00926	3.20	0.00001
0.34	0.6306	0.72	0.3086	1.10	0.1198	1.48	0.03635	1.86	0.00853	3.40	0.00000
0.36	0.6107	0.74	0.2953	1.12	0.1132	1.50	0.03390	1.88	0.00784	3.60	0.00000

maintained at that value at all times. Although the graphical solution given in Fig. 16-23 is simply a plot of the exact analytical solution given by Eq. 16-23, it is subject to reading errors, and thus is of limited accuracy.

EXAMPLE 16-6   Minimum Burial Depth of Water Pipes to Avoid Freezing
In areas where the air temperature remains below 0°C for prolonged periods of time, the freezing of water in underground pipes is a major concern. Fortunately, the soil remains relatively warm during those periods, and it takes weeks for the subfreezing temperatures to reach the water mains in the ground. Thus, the soil effectively serves as an insulation to protect the water from subfreezing temperatures in winter.

The ground at a particular location is covered with snow pack at $-10°C$ for a continuous period of three months, and the average soil properties at that location are $k = 0.4$ W/m $\cdot$ °C and $\alpha = 0.15 \times 10^{-6}$ m²/s (Fig. 16-24). Assuming an initial uniform temperature of 15°C for the ground, determine the minimum burial depth to prevent the water pipes from freezing.

Solution   The water pipes are buried in the ground to prevent freezing. The minimum burial depth at a particular location is to be determined.

Assumptions   **1** The temperature in the soil is affected by the thermal conditions at one surface only, and thus the soil can be considered to be a semi-infinite medium with a specified surface temperature of $-10°C$. **2** The thermal properties of the soil are constant.

Properties   The properties of the soil are as given in the problem statement.

Analysis   The temperature of the soil surrounding the pipes will be 0°C after three months in the case of minimum burial depth. Therefore, from Fig. 16-23, we have

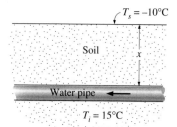

FIGURE 16-24
Schematic for Example 16-6.

$$\left.\begin{array}{l} \dfrac{h\sqrt{\alpha t}}{k} = \infty \qquad \text{(since } h \to \infty) \\[2ex] 1 - \dfrac{T(x,\,t) - T_\infty}{T_i - T_\infty} = 1 - \dfrac{0 - (-10)}{15 - (-10)} = 0.6 \end{array}\right\} \quad \xi = \dfrac{x}{2\sqrt{\alpha t}} = 0.36$$

We note that

$$t = (90 \text{ days})(24 \text{ h/day})(3600 \text{ s/h}) = 7.78 \times 10^6 \text{ s}$$

and thus

$$x = 2\xi\sqrt{\alpha t} = 2 \times 0.36\sqrt{(0.15 \times 10^{-6}\text{ m}^2/\text{s})(7.78 \times 10^6\text{ s})} = \textbf{0.77 m}$$

Therefore, the water pipes must be buried to a depth of at least 77 cm to avoid freezing under the specified harsh winter conditions.

Alternative solution   The solution of this problem could also be determined from Eq. 16-24:

$$\frac{T(x,\,t) - T_i}{T_s - T_i} = \text{erfc}\left(\frac{x}{2\sqrt{\alpha t}}\right) \quad \longrightarrow \quad \frac{0 - 15}{-10 - 15} = \text{erfc}\left(\frac{x}{2\sqrt{\alpha t}}\right) = 0.60$$

The argument that corresponds to this value of the complementary error function is determined from Table 16-3 to be $\xi = 0.37$. Therefore,

$$x = 2\xi\sqrt{\alpha t} = 2 \times 0.37\sqrt{(0.15 \times 10^{-6}\text{ m}^2/\text{s})(7.78 \times 10^6\text{ s})} = \textbf{0.80 m}$$

Again, the slight difference is due to the reading error of the chart.

16-4 ■ TRANSIENT HEAT CONDUCTION IN MULTIDIMENSIONAL SYSTEMS

The transient temperature charts presented earlier can be used to determine the temperature distribution and heat transfer in *one-dimensional* heat conduction problems associated with a large plane wall, a long cylinder, a sphere, and a semi-infinite medium. Using a clever superposition principle called the **product solution,** these charts can also be used to construct solutions for the *two-dimensional* transient heat conduction problems encountered in geometries such as a short cylinder, a long rectangular bar, or a semi-infinite cylinder or plate, and even *three-dimensional* problems associated with geometries such as a rectangular prism or a semi-infinite rectangular bar, provided that *all* surfaces of the solid are subjected to convection to the *same* fluid at temperature T_∞, with the *same* heat transfer coefficient h, and the body involves no heat generation (Fig. 16-25). The solution in such multidimensional geometries can be expressed as the *product* of the solutions for the one-dimensional geometries whose intersection is the multidimensional geometry.

Consider a *short cylinder* of height a and radius r_o initially at a uniform temperature T_i. There is no heat generation in the cylinder. At time $t = 0$, the cylinder is subjected to convection from all surfaces to a medium at temperature T_∞ with a heat transfer coefficient h. The temperature within the cylinder will change with x as well as r and time t since heat transfer will occur from the top and bottom of the cylinder as well as its side surfaces. That is, $T = T(r, x, t)$ and thus this is a two-dimensional transient heat conduction

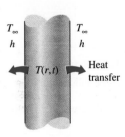

(a) Long cylinder

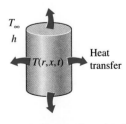

(b) Short cylinder (two-dimensional)

FIGURE 16-25

The temperature in a short cylinder exposed to convection from all surfaces varies in both the radial and axial directions, and thus heat is transferred in both directions.

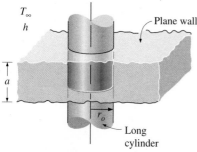

FIGURE 16-26

A short cylinder of radius r_o and height a is the *intersection* of a long cylinder of radius r_o and a plane wall of thickness a.

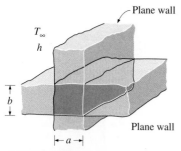

FIGURE 16-27

A long solid bar of rectangular profile $a \times b$ is the *intersection* of two plane walls of thicknesses a and b.

problem. When the properties are assumed to be constant, it can be shown that the solution of this two-dimensional problem can be expressed as

$$\left(\frac{T(r, x, t) - T_\infty}{T_i - T_\infty}\right)_{\substack{\text{short} \\ \text{cylinder}}} = \left(\frac{T(x, t) - T_\infty}{T_i - T_\infty}\right)_{\substack{\text{plane} \\ \text{wall}}} \left(\frac{T(r, t) - T_\infty}{T_i - T_\infty}\right)_{\substack{\text{infinite} \\ \text{cylinder}}} \quad (16\text{-}25)$$

That is, the solution for the two-dimensional short cylinder of height a and radius r_o is equal to the *product* of the nondimensionalized solutions for the one-dimensional plane wall of thickness a and the long cylinder of radius r_o, which are the two geometries whose intersection is the short cylinder, as shown in Fig. 16-26. We generalize this as follows: *the solution for a multidimensional geometry is the product of the solutions of the one-dimensional geometries whose intersection is the multidimensional body.*

For convenience, the one-dimensional solutions are denoted by

$$\theta_{\text{wall}}(x, t) = \left(\frac{T(x, t) - T_\infty}{T_i - T_\infty}\right)_{\substack{\text{plane} \\ \text{wall}}}$$

$$\theta_{\text{cyl}}(r, t) = \left(\frac{T(r, t) - T_\infty}{T_i - T_\infty}\right)_{\substack{\text{infinite} \\ \text{cylinder}}} \quad (16\text{-}26)$$

$$\theta_{\text{semi-inf}}(x, t) = \left(\frac{T(x, t) - T_\infty}{T_i - T_\infty}\right)_{\substack{\text{semi-infinite} \\ \text{solid}}}$$

For example, the solution for a long solid bar whose cross section is an $a \times b$ rectangle is the intersection of the two infinite plane walls of thicknesses a and b, as shown in Fig. 16-27, and thus the transient temperature distribution for this rectangular bar can be expressed as

$$\left(\frac{T(x, y, t) - T_\infty}{T_i - T_\infty}\right)_{\substack{\text{rectangular} \\ \text{bar}}} = \theta_{\text{wall}}(x, t)\theta_{\text{wall}}(y, t) \quad (16\text{-}27)$$

The proper forms of the product solutions for some other geometries are given in Table 16-4. It is important to note that the x-coordinate is measured from the *surface* in a semi-infinite solid, and from the *midplane* in a plane wall. The radial distance r is always measured from the centerline.

Note that the solution of a *two-dimensional* problem involves the product of *two* one-dimensional solutions, whereas the solution of a *three-dimensional* problem involves the product of *three* one-dimensional solutions.

A modified form of the product solution can also be used to determine the total transient heat transfer to or from a multidimensional geometry by using the one-dimensional values, as shown by L. S. Langston in 1982. The transient heat transfer for a two-dimensional geometry formed by the intersection of two one-dimensional geometries 1 and 2 is

$$\left(\frac{Q}{Q_{\max}}\right)_{\text{total, 2D}} = \left(\frac{Q}{Q_{\max}}\right)_1 + \left(\frac{Q}{Q_{\max}}\right)_2 \left[1 - \left(\frac{Q}{Q_{\max}}\right)_1\right] \quad (16\text{-}28)$$

Transient heat transfer for a three-dimensional body formed by the intersection of three one-dimensional bodies 1, 2, and 3 is given by

TABLE 16-4

Multidimensional solutions expressed as products of one-dimensional solutions for bodies that are initially at a uniform temperature T_i and exposed to convection from all surfaces to a medium at T_∞

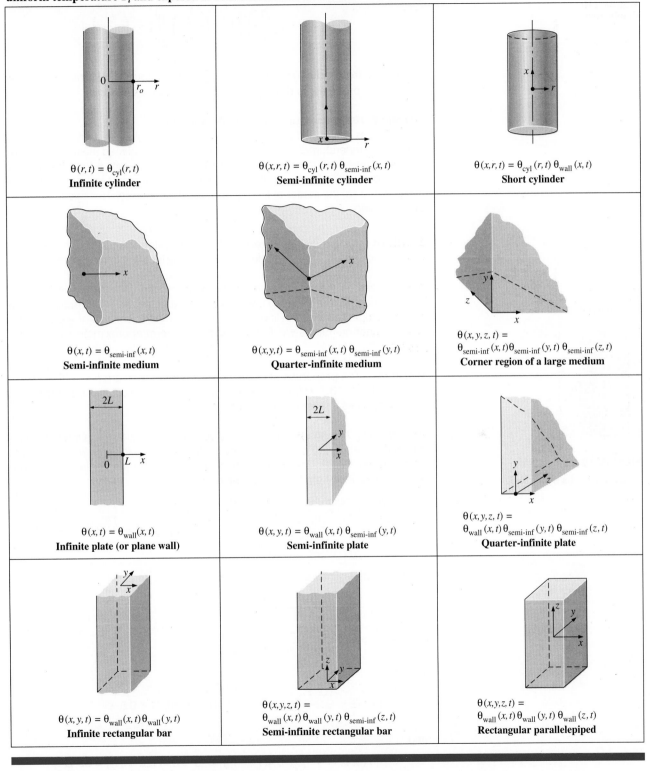

$\theta(r, t) = \theta_{cyl}(r, t)$
Infinite cylinder

$\theta(x, r, t) = \theta_{cyl}(r, t)\, \theta_{semi\text{-}inf}(x, t)$
Semi-infinite cylinder

$\theta(x, r, t) = \theta_{cyl}(r, t)\, \theta_{wall}(x, t)$
Short cylinder

$\theta(x, t) = \theta_{semi\text{-}inf}(x, t)$
Semi-infinite medium

$\theta(x, y, t) = \theta_{semi\text{-}inf}(x, t)\, \theta_{semi\text{-}inf}(y, t)$
Quarter-infinite medium

$\theta(x, y, z, t) =$
$\theta_{semi\text{-}inf}(x, t)\theta_{semi\text{-}inf}(y, t)\,\theta_{semi\text{-}inf}(z, t)$
Corner region of a large medium

$\theta(x, t) = \theta_{wall}(x, t)$
Infinite plate (or plane wall)

$\theta(x, y, t) = \theta_{wall}(x, t)\, \theta_{semi\text{-}inf}(y, t)$
Semi-infinite plate

$\theta(x, y, z, t) =$
$\theta_{wall}(x, t)\theta_{semi\text{-}inf}(y, t)\,\theta_{semi\text{-}inf}(z, t)$
Quarter-infinite plate

$\theta(x, y, t) = \theta_{wall}(x, t)\theta_{wall}(y, t)$
Infinite rectangular bar

$\theta(x, y, z, t) =$
$\theta_{wall}(x, t)\, \theta_{wall}(y, t)\, \theta_{semi\text{-}inf}(z, t)$
Semi-infinite rectangular bar

$\theta(x, y, z, t) =$
$\theta_{wall}(x, t)\, \theta_{wall}(y, t)\, \theta_{wall}(z, t)$
Rectangular parallelepiped

$$\left(\frac{Q}{Q_{max}}\right)_{total, 3D} = \left(\frac{Q}{Q_{max}}\right)_1 + \left(\frac{Q}{Q_{max}}\right)_2 \left[1 - \left(\frac{Q}{Q_{max}}\right)_1\right]$$

$$+ \left(\frac{Q}{Q_{max}}\right)_3 \left[1 - \left(\frac{Q}{Q_{max}}\right)_1\right]\left[1 - \left(\frac{Q}{Q_{max}}\right)_2\right]$$

(16-29)

The use of the product solution in transient two- and three-dimensional heat conduction problems is illustrated in the following examples.

EXAMPLE 16-7 Cooling of a Short Brass Cylinder

A short brass cylinder of diameter $D = 10$ cm and height $H = 12$ cm is initially at a uniform temperature $T_i = 120°C$. The cylinder is now placed in atmospheric air at 25°C, where heat transfer takes place by convection, with a heat transfer co-efficient of $h = 60$ W/m² · °C. Calculate the temperature at (a) the center of the cylinder and (b) the center of the top surface of the cylinder 15 min after the start of the cooling.

Solution A short cylinder is allowed to cool in atmospheric air. The tempera-tures at the centers of the cylinder and the top surface are to be determined.

Assumptions **1** Heat conduction in the short cylinder is two-dimensional, and thus the temperature varies in both the axial x- and the radial r-directions. **2** The thermal properties of the cylinder and the heat transfer coefficient are constant. **3** The Fourier number is $\tau > 0.2$ so that the one-term approximate solutions are applicable.

Properties The properties of brass at room temperature are $k = 110$ W/m · °C and $\alpha = 33.9 \times 10^{-6}$ m²/s (Table A-19). More accurate results can be obtained by using properties at average temperature.

Analysis (a) This short cylinder can physically be formed by the intersection of a long cylinder of radius $r_o = 5$ cm and a plane wall of thickness $2L = 12$ cm, as shown in Fig. 16-28. The dimensionless temperature at the center of the plane wall is determined from Figure 16-13a to be

$$\left.\begin{array}{l} \tau = \dfrac{\alpha t}{L^2} = \dfrac{(3.39 \times 10^{-5} \text{ m}^2/\text{s})(900 \text{ s})}{(0.06 \text{ m})^2} = 8.48 \\[3mm] \dfrac{1}{Bi} = \dfrac{k}{hL} = \dfrac{110 \text{ W/m} \cdot °C}{(60 \text{ W/m}^2 \cdot °C)(0.06 \text{ m})} = 30.6 \end{array}\right\} \quad \theta_{wall}(0, t) = \dfrac{T(0, t) - T_\infty}{T_i - T_\infty} = 0.8$$

Similarly, at the center of the cylinder, we have

$$\left.\begin{array}{l} \tau = \dfrac{\alpha t}{r_o^2} = \dfrac{(3.39 \times 10^{-5} \text{ m}^2/\text{s})(900 \text{ s})}{(0.05 \text{ m})^2} = 12.2 \\[3mm] \dfrac{1}{Bi} = \dfrac{k}{hr_o} = \dfrac{110 \text{ W/m} \cdot °C}{(60 \text{ W/m}^2 \cdot °C)(0.05 \text{ m})} = 36.7 \end{array}\right\} \quad \theta_{cyl}(0, t) = \dfrac{T(0, t) - T_\infty}{T_i - T_\infty} = 0.5$$

Therefore,

$$\left(\frac{T(0, 0, t) - T_\infty}{T_i - T_\infty}\right)_{\substack{short \\ cylinder}} = \theta_{wall}(0, t) \times \theta_{cyl}(0, t) = 0.8 \times 0.5 = 0.4$$

and

$$T(0, 0, t) = T_\infty + 0.4(T_i - T_\infty) = 25 + 0.4(120 - 25) = \mathbf{63°C}$$

This is the temperature at the center of the short cylinder, which is also the center of both the long cylinder and the plate.

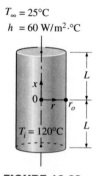

$T_\infty = 25°C$
$h = 60$ W/m²·°C

FIGURE 16-28

Schematic for Example 16-7.

(b) The center of the top surface of the cylinder is still at the center of the long cylinder ($r = 0$), but at the outer surface of the plane wall ($x = L$). Therefore, we first need to find the surface temperature of the wall. Noting that $x = L = 0.06$ m,

$$\left.\begin{array}{l} \dfrac{x}{L} = \dfrac{0.06 \text{ m}}{0.06 \text{ m}} = 1 \\[2mm] \dfrac{1}{\text{Bi}} = \dfrac{k}{hL} = \dfrac{110 \text{ W/m} \cdot \text{°C}}{(60 \text{ W/m}^2 \cdot \text{°C})(0.06 \text{ m})} = 30.6 \end{array}\right\} \quad \dfrac{T(L, t) - T_\infty}{T_o - T_\infty} = 0.98$$

Then

$$\theta_{\text{wall}}(L, t) = \frac{T(L, t) - T_\infty}{T_i - T_\infty} = \left(\frac{T(L, t) - T_\infty}{T_o - T_\infty}\right)\left(\frac{T_o - T_\infty}{T_i - T_\infty}\right) = 0.98 \times 0.8 = 0.784$$

Therefore,

$$\left(\frac{T(L, 0, t) - T_\infty}{T_i - T_\infty}\right)_{\substack{\text{short} \\ \text{cylinder}}} = \theta_{\text{wall}}(L, t)\theta_{\text{cyl}}(0, t) = 0.784 \times 0.5 = 0.392$$

and

$$T(L, 0, t) = T_\infty + 0.392(T_i - T_\infty) = 25 + 0.392(120 - 25) = \textbf{62.2°C}$$

which is the temperature at the center of the top surface of the cylinder.

EXAMPLE 16-8   Heat Transfer from a Short Cylinder

Determine the total heat transfer from the short brass cylinder ($\rho = 8530$ kg/m³, $C_p = 0.380$ kJ/kg $\cdot$ °C) discussed in Example 16-7.

Solution   We first determine the maximum heat that can be transferred from the cylinder, which is the sensible energy content of the cylinder relative to its environment:

$$m = \rho V = \rho \pi r_o^2 L = (8530 \text{ kg/m}^3)\pi(0.05 \text{ m})^2(0.06 \text{ m}) = 4.02 \text{ kg}$$
$$Q_{\max} = mC_p(T_i - T_\infty) = (4.02 \text{ kg})(0.380 \text{ kJ/kg} \cdot \text{°C})(120 - 25)\text{°C} = 145.1 \text{ kJ}$$

Then we determine the dimensionless heat transfer ratios for both geometries. For the plane wall, it is determined from Fig. 16-13c to be

$$\left.\begin{array}{l} \text{Bi} = \dfrac{1}{1/\text{Bi}} = \dfrac{1}{30.6} = 0.0327 \\[2mm] \dfrac{h^2\alpha t}{k^2} = \text{Bi}^2\tau = (0.0327)^2(8.48) = 0.0091 \end{array}\right\} \quad \left(\frac{Q}{Q_{\max}}\right)_{\substack{\text{plane} \\ \text{wall}}} = 0.23$$

Similarly, for the cylinder, we have

$$\left.\begin{array}{l} \text{Bi} = \dfrac{1}{1/\text{Bi}} = \dfrac{1}{36.7} = 0.0272 \\[2mm] \dfrac{h^2\alpha t}{k^2} = \text{Bi}^2\tau = (0.0272)^2(12.2) = 0.0090 \end{array}\right\} \quad \left(\frac{Q}{Q_{\max}}\right)_{\substack{\text{infinite} \\ \text{cylinder}}} = 0.47$$

Then the heat transfer ratio for the short cylinder is, from Eq. 16-28,

$$\left(\frac{Q}{Q_{\max}}\right)_{\text{short cyl}} = \left(\frac{Q}{Q_{\max}}\right)_1 + \left(\frac{Q}{Q_{\max}}\right)_2\left[1 - \left(\frac{Q}{Q_{\max}}\right)_1\right]$$
$$= 0.23 + 0.47(1 - 0.23) = 0.592$$

Therefore, the total heat transfer from the cylinder during the first 15 min of cooling is

$$Q = 0.592Q_{\max} = 0.592 \times (145.1 \text{ kJ}) = \textbf{85.9 kJ}$$

EXAMPLE 16-9 Cooling of a Long Cylinder by Water

A semi-infinite aluminum cylinder of diameter $D = 20$ cm is initially at a uniform temperature $T_i = 200°C$. The cylinder is now placed in water at 15°C where heat transfer takes place by convection, with a heat transfer coefficient of $h = 120$ W/m² · °C. Determine the temperature at the center of the cylinder 15 cm from the end surface 5 min after the start of the cooling.

Solution A semi-infinite aluminum cylinder is cooled by water. The temperature at the center of the cylinder 15 cm from the end surface is to be determined.

Assumptions **1** Heat conduction in the semi-infinite cylinder is two-dimensional, and thus the temperature varies in both the axial x- and the radial r-directions. **2** The thermal properties of the cylinder and the heat transfer coefficient are constant. **3** The Fourier number is $\tau > 0.2$ so that the one-term approximate solutions are applicable.

Properties The properties of aluminum at room temperature are $k = 237$ W/m · °C and $\alpha = 9.71 \times 10^{-6}$ m²/s (Table A-19). More accurate results can be obtained by using properties at average temperature.

Analysis This semi-infinite cylinder can physically be formed by the intersection of an infinite cylinder of radius $r_o = 10$ cm and a semi-infinite medium, as shown in Fig. 16-29.

We will solve this problem using the one-term solution relation for the cylinder and the analytic solution for the semi-infinite medium. First we consider the infinitely long cylinder and evaluate the Biot number:

$$Bi = \frac{hr_o}{k} = \frac{(120 \text{ W/m}^2 \cdot °C)(0.1 \text{ m})}{237 \text{ W/m} \cdot °C} = 0.05$$

The coefficients λ_1 and A_1 for a cylinder corresponding to this Bi are determined from Table 16-1 to be $\lambda_1 = 0.3126$ and $A_1 = 1.0124$. The Fourier number in this case is

$$\tau = \frac{\alpha t}{r_o^2} = \frac{(9.71 \times 10^{-5} \text{ m}^2/\text{s})(5 \times 60 \text{ s})}{(0.1 \text{ m})^2} = 2.91 > 0.2$$

and thus the one-term approximation is applicable. Substituting these values into Eq. 16-14 gives

$$\theta_0 = \theta_{cyl}(0, t) = A_1 e^{-\lambda_1^2 \tau} = 1.0124 e^{-(0.3126)^2(2.91)} = 0.762$$

The solution for the semi-infinite solid can be determined from

$$1 - \theta_{semi\text{-}inf}(x, t) = \text{erfc}\left(\frac{x}{2\sqrt{\alpha t}}\right) - \exp\left(\frac{hx}{k} + \frac{h^2 \alpha t}{k^2}\right)\left[\text{erfc}\left(\frac{x}{2\sqrt{\alpha t}} + \frac{h\sqrt{\alpha t}}{k}\right)\right]$$

First we determine the various quantities in parentheses:

$$\xi = \frac{x}{2\sqrt{\alpha t}} = \frac{0.15 \text{ m}}{2\sqrt{(9.71 \times 10^{-5} \text{ m}^2/\text{s})(5 \times 60 \text{ s})}} = 0.44$$

$$\frac{h\sqrt{\alpha t}}{k} = \frac{(120 \text{ W/m}^2 \cdot °C)\sqrt{(9.71 \times 10^{-5} \text{ m}^2/\text{s})(300 \text{ s})}}{237 \text{ W/m} \cdot °C} = 0.086$$

$$\frac{hx}{k} = \frac{(120 \text{ W/m}^2 \cdot °C)(0.15 \text{ m})}{237 \text{ W/m} \cdot °C} = 0.0759$$

$$\frac{h^2 \alpha t}{k^2} = \left(\frac{h\sqrt{\alpha t}}{k}\right)^2 = (0.086)^2 = 0.0074$$

Substituting and evaluating the complementary error functions from Table 16-3,

$$\theta_{semi\text{-}inf}(x, t) = 1 - \text{erfc}(0.44) + \exp(0.0759 + 0.0074)\, \text{erfc}(0.44 + 0.086)$$
$$= 1 - 0.5338 + \exp(0.0833) \times 0.457$$
$$= 0.963$$

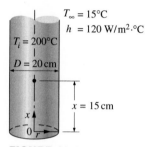

$T_\infty = 15°C$
$h = 120$ W/m²·°C
$T_i = 200°C$
$D = 20$ cm
$x = 15$ cm
x
0

FIGURE 16-29

Schematic for Example 16-9.

Now we apply the product solution to get

$$\left(\frac{T(x, 0, t) - T_\infty}{T_i - T_\infty}\right)_{\substack{\text{semi-infinite} \\ \text{cylinder}}} = \theta_{\text{semi-inf}}(x, t)\theta_{\text{cyl}}(0, t) = 0.963 \times 0.762 = 0.734$$

and

$$T(x, 0, t) = T_\infty + 0.734(T_i - T_\infty) = 15 + 0.734(200 - 15) = \textbf{151°C}$$

which is the temperature at the center of the cylinder 15 cm from the exposed bottom surface.

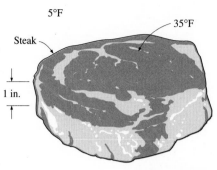

FIGURE 16-30
Schematic for Example 16-10.

EXAMPLE 16-10   Refrigerating Steaks while Avoiding Frostbite

In a meat processing plant, 1-in.-thick steaks initially at 75°F are to be cooled in the racks of a large refrigerator that is maintained at 5°F (Fig. 16-30). The steaks are placed close to each other, so that heat transfer from the 1-in.-thick edges is negligible. The entire steak is to be cooled below 45°F, but its temperature is not to drop below 35°F at any point during refrigeration to avoid "frostbite." The convection heat transfer coefficient and thus the rate of heat transfer from the steak can be controlled by varying the speed of a circulating fan inside. Determine the heat transfer coefficient h that will enable us to meet both temperature constraints while keeping the refrigeration time to a minimum. The steak can be treated as a homogeneous layer having the properties $\rho = 74.9$ lbm/ft³, $C_p = 0.98$ Btu/lbm · °F, $k = 0.26$ Btu/h · ft · °F, and $\alpha = 0.0035$ ft²/h.

Solution   Steaks are to be cooled in a refrigerator maintained at 5°F. The heat transfer coefficient that will allow cooling the steaks below 45°F while avoiding frostbite is to be determined.

Assumptions   **1** Heat conduction through the steaks is one-dimensional since the steaks form a large layer relative to their thickness and there is thermal symmetry about the center plane. **2** The thermal properties of the steaks and the heat transfer coefficient are constant. **3** The Fourier number is $\tau > 0.2$ so that the one-term approximate solutions are applicable.

Properties   The properties of the steaks are as given in the problem statement.

Analysis   The lowest temperature in the steak will occur at the surfaces and the highest temperature at the center at a given time, since the inner part will be the last place to be cooled. In the limiting case, the surface temperature at $x = L = 0.5$ in. from the center will be 35°F, while the midplane temperature is 45°F in an environment at 5°F. Then, from Fig. 16-13b, we obtain

$$\left.\begin{array}{l} \dfrac{x}{L} = \dfrac{0.5 \text{ in.}}{0.5 \text{ in.}} = 1 \\[2mm] \dfrac{T(L, t) - T_\infty}{T_o - T_\infty} = \dfrac{35 - 5}{45 - 5} = 0.75 \end{array}\right\} \quad \dfrac{1}{\text{Bi}} = \dfrac{k}{hL} = 1.5$$

which gives

$$h = \frac{1}{1.5}\frac{k}{L} = \frac{0.26 \text{ Btu/h} \cdot \text{ft} \cdot \text{°F}}{1.5(0.5/12 \text{ ft})} = 4.16 \text{ Btu/h} \cdot \text{ft}^2 \cdot \text{°F}$$

Discussion   The convection heat transfer coefficient should be kept below this value to satisfy the constraints on the temperature of the steak during refrigeration. We can also meet the constraints by using a lower heat transfer coefficient, but doing so would extend the refrigeration time unnecessarily.

16-5 ■ SUMMARY

In this chapter we considered the variation of temperature with time as well as position in one- or multidimensional systems. We first considered the *lumped systems* in which the temperature varies with time but remains uniform throughout the system at any time. The temperature of a lumped body of arbitrary shape of mass m, volume V, surface area A_s, density ρ, and specific heat C_p initially at a uniform temperature T_i that is exposed to convection at time $t = 0$ in a medium at temperature T_∞ with a heat transfer coefficient h is expressed as

$$\frac{T(t) - T_\infty}{T_i - T_\infty} = e^{-bt}$$

where

$$b = \frac{hA}{\rho C_p V} = \frac{h}{\rho C_p L_c} \quad \text{(1/s)}$$

is a positive quantity whose dimension is $(\text{time})^{-1}$. This relation can be used to determine the temperature $T(t)$ of a body at time t or, alternately, the time t required for the temperature to reach a specified value $T(t)$. Once the temperature $T(t)$ at time t is available, the *rate* of convection heat transfer between the body and its environment at that time can be determined from Newton's law of cooling as

$$\dot{Q}(t) = hA[T(t) - T_\infty] \quad \text{(W)}$$

The *total amount* of heat transfer between the body and the surrounding medium over the time interval $t = 0$ to t is simply the change in the energy content of the body,

$$Q = mC_p[T(t) - T_i] \quad \text{(kJ)}$$

The amount of heat transfer reaches its upper limit when the body reaches the surrounding temperature T_∞. Therefore, the *maximum* heat transfer between the body and its surroundings is

$$Q_{max} = mC_p (T_\infty - T_i) \quad \text{(kJ)}$$

The error involved in lumped system analysis is negligible when

$$\text{Bi} = \frac{hL_c}{k} < 0.1$$

where Bi is the *Biot number* and $L_c = V/A$ is the *characteristic length.*

When the lumped system analysis is not applicable, the variation of temperature with position as well as time can be determined using the *transient temperature charts* given in Figs. 16-13, 16-14, 16-15, and 16-23 for a large plane wall, a long cylinder, a sphere, and a semi-infinite medium, respectively. These charts are applicable for one-dimensional heat transfer in those geometries. Therefore, their use is limited to situations in which the body is initially at a uniform temperature, all surfaces are subjected to the same thermal conditions, and the body does not involve any heat generation. These charts can also be used to determine the total heat transfer from the body up to a specified time t.

Using a *one-term approximation,* the solutions of one-dimensional transient heat conduction problems are expressed analytically as

Plane wall: $\qquad \theta(x, t)_{\text{wall}} = \dfrac{T(x, t) - T_\infty}{T_i - T_\infty} = A_1 e^{-\lambda_1^2 \tau} \cos(\lambda_1 x/L), \quad \tau > 0.2$

Cylinder: $\qquad \theta(r, t)_{\text{cyl}} = \dfrac{T(r, t) - T_\infty}{T_i - T_\infty} = A_1 e^{-\lambda_1^2 \tau} J_0(\lambda_1 r/r_o), \quad \tau > 0.2$

Sphere: $\qquad \theta(r, t)_{\text{sph}} = \dfrac{T(r, t) - T_\infty}{T_i - T_\infty} = A_1 e^{-\lambda_1^2 \tau} \dfrac{\sin(\lambda_1 r/r_o)}{\lambda_1 r/r_o}, \quad \tau > 0.2$

where the constants A_1 and λ_1 are functions of the Bi number only, and their values are listed in Table 16-1 against the Bi number for all three geometries. The error involved in one-term solutions is less than 2 percent when $\tau > 0.2$.

Using the one-term solutions, the fractional heat transfers in different geometries are expressed as

Plane wall: $\qquad \left(\dfrac{Q}{Q_{\text{max}}}\right)_{\text{wall}} = 1 - \theta_{0,\text{wall}} \dfrac{\sin \lambda_1}{\lambda_1}$

Cylinder: $\qquad \left(\dfrac{Q}{Q_{\text{max}}}\right)_{\text{cyl}} = 1 - 2\theta_{0,\text{cyl}} \dfrac{J_1(\lambda_1)}{\lambda_1}$

Sphere: $\qquad \left(\dfrac{Q}{Q_{\text{max}}}\right)_{\text{sph}} = 1 - 3\theta_{0,\text{sph}} \dfrac{\sin \lambda_1 - \lambda_1 \cos \lambda_1}{\lambda_1^3}$

The analytic solution for one-dimensional transient heat conduction in a semi-infinite solid subjected to convection is given by

$$\dfrac{T(x, t) - T_i}{T_\infty - T_i} = \text{erfc}\left(\dfrac{x}{2\sqrt{\alpha t}}\right) - \exp\left(\dfrac{hx}{k} + \dfrac{h^2 \alpha t}{k^2}\right)\left[\text{erfc}\left(\dfrac{x}{2\sqrt{\alpha t}} + \dfrac{h\sqrt{\alpha t}}{k}\right)\right]$$

where the quantity erfc (ξ) is the *complementary error function*. For the special case of $h \to \infty$, the surface temperature T_s becomes equal to the fluid temperature T_∞, and the above equation reduces to

$$\dfrac{T(x, t) - T_i}{T_s - T_i} = \text{erfc}\left(\dfrac{x}{2\sqrt{\alpha t}}\right) \qquad (T_s = \text{constant})$$

Using a clever superposition principle called the *product solution* these charts can also be used to construct solutions for the *two-dimensional* transient heat conduction problems encountered in geometries such as a short cylinder, a long rectangular bar, or a semi-infinite cylinder or plate, and even *three-dimensional* problems associated with geometries such as a rectangular prism or a semi-infinite rectangular bar, provided that all surfaces of the solid are subjected to convection to the same fluid at temperature T_∞, with the same convection heat transfer coefficient h, and the body involves no heat generation. The solution in such multidimensional geometries can be expressed as the product of the solutions for the one-dimensional geometries whose intersection is the multidimensional geometry.

The total heat transfer to or from a multidimensional geometry can also be determined by using the one-dimensional values. The transient heat transfer for a two-dimensional geometry formed by the intersection of two one-dimensional geometries 1 and 2 is

$$\left(\dfrac{Q}{Q_{\text{max}}}\right)_{\text{total, 2D}} = \left(\dfrac{Q}{Q_{\text{max}}}\right)_1 + \left(\dfrac{Q}{Q_{\text{max}}}\right)_2 \left[1 - \left(\dfrac{Q}{Q_{\text{max}}}\right)_1\right]$$

Transient heat transfer for a three-dimensional body formed by the intersection of three one-dimensional bodies 1, 2, and 3 is given by

$$\left(\frac{Q}{Q_{max}}\right)_{total,\,3D} = \left(\frac{Q}{Q_{max}}\right)_1 + \left(\frac{Q}{Q_{max}}\right)_2 \left[1 - \left(\frac{Q}{Q_{max}}\right)_1\right]$$

$$+ \left(\frac{Q}{Q_{max}}\right)_3 \left[1 - \left(\frac{Q}{Q_{max}}\right)_1\right]\left[1 - \left(\frac{Q}{Q_{max}}\right)_2\right]$$

REFERENCES AND SUGGESTED READING

1. Y. Bayazitoglu and M. N. Özişik. *Elements of Heat Transfer.* New York: McGraw-Hill, 1988.

2. H. S. Carslaw and J. C. Jaeger. *Conduction of Heat in Solids.* 2nd ed. London: Oxford University Press, 1959.

3. H. Gröber, S. Erk, and U. Grigull. *Fundamentals of Heat Transfer.* New York: McGraw-Hill, 1961.

4. M. P. Heisler. "Temperature Charts for Induction and Constant Temperature Heating." *ASME Transactions* 69 (1947), pp. 227–36.

5. J. P. Holman. *Heat Transfer.* 7th ed. New York: McGraw-Hill, 1990.

6. F. P. Incropera and D. P. DeWitt. *Introduction to Heat Transfer.* 2nd ed. New York: John Wiley & Sons, 1990.

7. M. Jakob. *Heat Transfer.* Vol. 1. New York: John Wiley & Sons, 1949.

8. F. Kreith and M. S. Bohn. *Principles of Heat Transfer.* 5th ed. St. Paul, MN: West Publishing, 1993.

9. L. S. Langston. "Heat Transfer from Multidimensional Objects Using One-Dimensional Solutions for Heat Loss." *International Journal of Heat and Mass Transfer* 25 (1982), pp. 149–50.

10. M. N. Özişik, *Heat Transfer—A Basic Approach.* New York: McGraw-Hill, 1985.

11. P. J. Schneider. *Conduction Heat Transfer.* Reading, MA: Addison-Wesley, 1955.

12. L. C. Thomas. *Heat Transfer.* Englewood Cliffs, NJ: Prentice Hall, 1992.

13. F. M. White. *Heat and Mass Transfer.* Reading, MA: Addison-Wesley, 1988.

PROBLEMS*

Lumped System Analysis

16-1C How does transient heat conduction differ from steady conduction? How do two-dimensional heat transfer problems differ from one-dimensional ones?

---

*Students are encouraged to answer *all* the concept "C" questions.

16-2C   What is lumped system analysis? When is it applicable?

16-3C   Consider heat transfer between two identical hot solid bodies and the air surrounding them. The first solid is being cooled by a fan while the second one is allowed to cool naturally. For which solid is the lumped system analysis more likely to be applicable? Why?

16-4C   Consider heat transfer between two identical hot solid bodies and their environments. The first solid is dropped in a large container filled with water, while the second one is allowed to cool naturally in the air. For which solid is the lumped system analysis more likely to be applicable? Why?

16-5C   Consider a hot baked potato on a plate. The temperature of the potato is observed to drop by 4°C during the first minute. Will the temperature drop during the second minute be less than, equal to, or more than 4°C? Why?

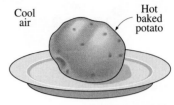

Cool
air

Hot
baked
potato

FIGURE P16-5C

16-6C   Consider a potato being baked in an oven that is maintained at a constant temperature. The temperature of the potato is observed to rise by 5°C during the first minute. Will the temperature rise during the second minute be less than, equal to, or more than 5°C? Why?

16-7C   What is the physical significance of the Biot number? Is the Biot number more likely to be larger for highly conducting solids or poorly conducting ones?

16-8C   Consider two identical 16-kg pieces of roast beef. The first piece is baked as a whole, while the second is baked after being cut into two equal pieces in the same oven. Will there be any difference between the cooking times of the whole and cut roasts? Why?

16-9C   Consider a sphere and a cylinder of equal volume made of copper. Both the sphere and the cylinder are initially at the same temperature and are exposed to convection in the same environment. Which do you think will cool faster, the cylinder or the sphere? Why?

16-10C   In what medium is the lumped system analysis more likely to be applicable: in water or in air? Why?

16-11C   For which solid is the lumped system analysis more likely to be applicable: an actual apple or a golden apple of the same size? Why?

16-12C   For which kind of bodies made of the same material is the lumped system analysis more likely to be applicable: slender ones or well-rounded ones of the same volume? Why?

16-13   Obtain relations for the characteristic lengths of a large plane wall of thickness $2L$, a very long cylinder of radius r_o, and a sphere of radius r_o.

16-14   Obtain a relation for the time required for a lumped system to reach the average temperature $\frac{1}{2}(T_i + T_\infty)$, where T_i is the initial temperature and T_∞ is the temperature of the environment.

16-15   The temperature of a gas stream is to be measured by a thermocouple whose junction can be approximated as a 1.2-mm-diameter sphere. The properties of the junction are $k = 35$ W/m · °C, $\rho = 8500$ kg/m³, and $C_p = 320$ J/kg · °C, and the heat transfer coefficient between the junction and the gas is $h = 65$ W/m² · °C. Determine how long it will take for the thermocouple to read 99 percent of the initial temperature difference.
   Answer: 38.5 s

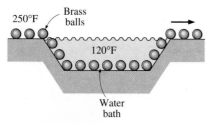

250°F Brass balls
120°F
Water bath

FIGURE P16-16E

FIGURE P16-20E

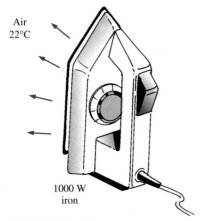

Air 22°C

1000 W iron

FIGURE P16-21

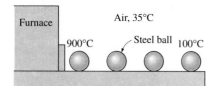

Furnace   Air, 35°C
900°C   Steel ball   100°C

FIGURE P16-23

16-16E In a manufacturing facility, 2-in.-diameter brass balls ($k = 64.1$ Btu/h · ft · °F, $\rho = 532$ lbm/ft^3, and $C_p = 0.092$ Btu/lbm · °F) initially at 250°F are quenched in a water bath at 120°F for a period of 2 min at a rate of 100 balls per minute. If the convection heat transfer coefficient is 42 Btu/h · ft^2 · °F, determine (a) the temperature of the balls after quenching and (b) the rate at which heat needs to be removed from the water in order to keep its temperature constant at 120°F.

16-17E Repeat Prob. 16-16E for aluminum balls.

16-18 To warm up some milk for a baby, a mother pours milk into a thin-walled glass whose diameter is 6 cm. The height of the milk in the glass is 7 cm. She then places the glass into a large pan filled with hot water at 60°C. The milk is stirred constantly, so that its temperature is uniform at all times. If the heat transfer coefficient between the water and the glass is 120 W/m^2 · °C, determine how long it will take for the milk to warm up from 3°C to 38°C. Take the properties of the milk to be the same as those of water. Can the milk in this case be treated as a lumped system? Why? *Answer:* 5.9 min

16-19 Repeat Prob. 16-18 for the case of water also being stirred, so that the heat transfer coefficient is doubled to 240 W/m^2 · °C.

16-20E During a picnic on a hot summer day, all the cold drinks disappeared quickly, and the only available drinks were those at the ambient temperature of 75°F. In an effort to cool a 12-fluid-oz drink in a can, which is 5 in. high and has a diameter of 2.5 in., a person grabs the can and starts shaking it in the iced water of the chest at 32°F. The temperature of the drink can be assumed to be uniform at all times, and the heat transfer coefficient between the iced water and the aluminum can is 30 Btu/h · ft^2 · °F. Using the properties of water for the drink, estimate how long it will take for the canned drink to cool to 45°F.

16-21 Consider a 1000-W iron whose base plate is made of 0.5-cm-thick aluminum alloy 2024-T6 ($\rho = 2770$ kg/m^3, $C_p = 875$ J/kg · °C, $\alpha = 7.3 \times 10^{-5}$ m^2/s). The base plate has a surface area of 0.03 m^2. Initially, the iron is in thermal equilibrium with the ambient air at 22°C. Taking the heat transfer coefficient at the surface of the base plate to be 12 W/m^2 · °C and assuming 85 percent of the heat generated in the resistance wires is transferred to the plate, determine how long it will take for the plate temperature to reach 140°C. Is it realistic to assume the plate temperature to be uniform at all times?

16-22 Stainless steel ball bearings ($\rho = 8085$ kg/m^3, $k = 15.1$ W/m · °C, $C_p = 0.480$ kJ/kg · °C, and $\alpha = 3.91 \times 10^{-6}$ m^2/s) having a diameter of 1.2 cm are to be quenched in water. The balls leave the oven at a uniform temperature of 900°C and are exposed to air at 30°C for a while before they are dropped into the water. If the temperature of the balls is not to fall below 850°C prior to quenching and the heat transfer coefficient in the air is 125 W/m^2 · °C, determine how long they can stand in the air before being dropped into the water. *Answer:* 3.7 s

16-23 Carbon steel balls ($\rho = 7833$ kg/m^3, $k = 54$ W/m · °C, $C_p = 0.465$ kJ/kg · °C, and $\alpha = 1.474 \times 10^{-6}$ m^2/s) 8 mm in diameter are annealed by heating them first to 900°C in a furnace and then allowing them to cool slowly to 100°C in ambient air at 35°C. If the average heat transfer coefficient is 75 W/m^2 · °C, determine how long the annealing process will take. If 2500 balls are to be annealed per hour, determine the total rate of heat transfer from the balls to the ambient air.

16-24 An electronic device dissipating 30 W has a mass of 20 g, a specific heat of 850 J/kg · °C, and a surface area of 5 cm². The device is lightly used, and it is on for 5 min and then off for several hours, during which it cools to the ambient temperature of 25°C. Taking the heat transfer coefficient to be 12 W/m² · °C, determine the temperature of the device at the end of the 5-min operating period. What would your answer be if the device were attached to an aluminum heat sink having a mass of 200 g and a surface area of 50 cm²? Assume the device and the heat sink to be nearly isothermal.

Transient Heat Conduction in Large Plane Walls, Long Cylinders, and Spheres

16-25C What is an infinitely long cylinder? When is it proper to treat an actual cylinder as being infinitely long, and when is it not? For example, is it proper to use this model when finding the temperatures near the bottom or top surfaces of a cylinder? Explain.

16-26C Can the transient temperature charts in Fig. 16-13 for a plane wall exposed to convection on both sides be used for a plane wall with one side exposed to convection while the other side is insulated? Explain.

16-27C Why are the transient temperature charts prepared using non-dimensionalized quantities such as the Biot and Fourier numbers instead of the actual variables such as thermal conductivity and time?

16-28C What is the physical significance of the Fourier number? Will the Fourier number for a specified heat transfer problem double when the time is doubled?

16-29C How can we use the transient temperature charts when the surface temperature of the geometry is specified instead of the temperature of the surrounding medium and the convection heat transfer coefficient?

16-30C A body at an initial temperature of T_i is brought into a medium at a constant temperature of T_∞. How can you determine the maximum possible amount of heat transfer between the body and the surrounding medium?

16-31C The Biot number during a heat transfer process between a sphere and its surroundings is determined to be 0.02. Would you use lumped system analysis or the transient temperature charts when determining the midpoint temperature of the sphere? Why?

16-32 A student calculates that the total heat transfer from a spherical copper ball of diameter 15 cm initially at 200°C and its environment at a constant temperature of 25°C during the first 20 min of cooling is 4200 kJ. Is this result reasonable? Why?

16-33 An ordinary egg can be approximated as a 5.5-cm-diameter sphere whose properties are roughly those of water at room temperature ($k = 0.6$ W/m · °C and $\alpha = 0.14 \times 10^{-6}$ m²/s). The egg is initially at a uniform temperature of 8°C and is dropped into boiling water at 97°C. Taking the convection heat transfer coefficient to be $h = 1400$ W/m² · °C, determine how long it will take for the center of the egg to reach 70°C.

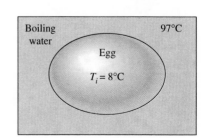

Boiling water

97°C

Egg

$T_i = 8$°C

FIGURE P16-33

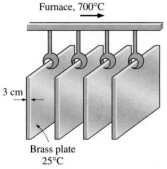

Furnace, 700°C

3 cm

Brass plate
25°C

FIGURE P16-34

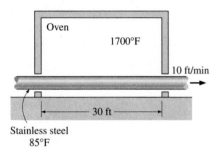

Oven
1700°F

10 ft/min

30 ft

Stainless steel
85°F

FIGURE P16-36E

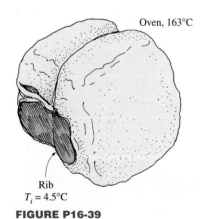

Oven, 163°C

Rib
$T_i = 4.5°C$

FIGURE P16-39

16-34 In a production facility, 3-cm-thick large brass plates ($k = 110$ W/m · °C, $\rho = 8530$ kg/m^3, $C_p = 380$ J/kg · °C, and $\alpha = 33.9 \times 10^{-6}$ m^2/s) that are initially at a uniform temperature of 25°C are heated by passing them through an oven maintained at 700°C. The plates remain in the oven for a period of 10 min. Taking the convection heat transfer coefficient to be $h = 80$ W/m^2 · °C, determine the surface temperature of the plates when they come out of the oven.

16-35 A long 35-cm-diameter cylindrical shaft made of stainless steel 304 ($k = 14.9$ W/m · °C, $\rho = 7900$ kg/m^3, $C_p = 477$ J/kg · °C, and $\alpha = 3.95 \times 10^{-6}$ m^2/s) comes out of an oven at a uniform temperature of 400°C. The shaft is then allowed to cool slowly in a chamber at 150°C with an average convection heat transfer coefficient of $h = 60$ W/m^2 · °C. Determine the temperature at the center of the shaft 20 min after the start of the cooling process. Also, determine the heat transfer per unit length of the shaft during this time period.
Answers: 390°C, 15,680 kJ

16-36E Long cylindrical AISI stainless steel rods ($k = 7.74$ Btu/h · ft · °F and $\alpha = 0.135$ ft^2/h) of 4-in. diameter are heat-treated by drawing them at a velocity of 10 ft/min through a 30-ft-long oven maintained at 1700°F. The heat transfer coefficient in the oven is 20 Btu/h · ft^2 · °F. If the rods enter the oven at 85°F, determine their centerline temperature when they leave.

16-37 In a meat processing plant, 2-cm-thick steaks ($k = 0.45$ W/m · °C and $\alpha = 0.91 \times 10^{-7}$ m^2/s) that are initially at 25°C are to be cooled by passing them through a refrigeration room at -10°C. The heat transfer coefficient on both sides of the steaks is 9 W/m^2 · °C. If both surfaces of the steaks are to be cooled to 3°C, determine how long the steaks should be kept in the refrigeration room.

16-38 A long cylindrical wood log ($k = 0.17$ W/m · °C and $\alpha = 1.28 \times 10^{-7}$ m^2/s) is 10 cm in diameter and is initially at a uniform temperature of 10°C. It is exposed to hot gases at 500°C in a fireplace with a heat transfer coefficient of 13.6 W/m^2 · °C on the surface. If the ignition temperature of the wood is 420°C, determine how long it will be before the log ignites.

16-39 In *Betty Crocker's Cookbook,* it is stated that it takes 2 h 45 min to roast a 3.2-kg rib initially at 4.5°C "rare" in an oven maintained at 163°C. It is recommended that a meat thermometer be used to monitor the cooking, and the rib is considered rare done when the thermometer inserted into the center of the thickest part of the meat registers 60°C. The rib can be treated as a homogeneous spherical object with the properties $\rho = 1200$ kg/m^3, $C_p = 4.1$ kJ/kg · °C, $k = 0.45$ W/m · °C, and $\alpha = 0.91 \times 10^{-7}$ m^2/s. Determine (a) the heat transfer coefficient at the surface of the rib, (b) the temperature of the outer surface of the rib when it is done, and (c) the amount of heat transferred to the rib. (d) Using the values obtained, predict how long it will take to roast this rib to "medium" level, which occurs when the innermost temperature of the rib reaches 71°C. Compare your result to the listed value of 3 h 20 min.

If the roast rib is to be set on the counter for about 15 min before it is sliced, it is recommended that the rib be taken out of the oven when the thermometer registers about 4°C below the indicated value because the rib will continue cooking even after it is taken out of the oven. Do you agree with this recommendation?
Answers: (a) 156.9 W/m^2 · °C, (b) 159.5°C, (c) 1629 kJ, (d) 3.0 h

16-40 Repeat Prob. 16-39 for a roast rib that is to be "well-done" instead of "rare." A rib is considered to be well-done when its center temperature reaches 77°C, and the roasting in this case takes about 4 h 15 min.

16-41 For heat transfer purposes, an egg can be considered to be a 5.5-cm-diameter sphere having the properties of water. An egg that is initially at 8°C is dropped into the boiling water at 100°C. The heat transfer coefficient at the surface of the egg is estimated to be 500 W/m² · °C. If the egg is considered cooked when its center temperature reaches 60°C, determine how long the egg should be kept in the boiling water.

16-42 Repeat Prob. 16-41 for a location at 1610-m elevation such as Denver, Colorado, where the boiling temperature of water is 94.4°C.

16-43 The author and his 6-year-old son have conducted the following experiment to determine the thermal conductivity of a hot dog. They first boiled water in a large pan and measured the temperature of the boiling water to be 94°C, which is not surprising, since they live at an elevation of about 1650 m in Reno, Nevada. They then took a hot dog that is 12.5 cm long and 2.2 cm in diameter and inserted a thermocouple into the midpoint of the hot dog and another thermocouple just under the skin. They waited until both thermocouples read 20°C, which is the ambient temperature. They then dropped the hot dog into boiling water and observed the changes in both temperatures. Exactly 2 min after the hot dog was dropped into the boiling water, they recorded the center and the surface temperatures to be 59°C and 88°C, respectively. The density of the hot dog can be taken to be 980 kg/m³, which is slightly less than the density of water, since the hot dog was observed to be floating in water while being almost completely immersed. The specific heat of a hot dog can be taken to be 3900 J/kg · °C, which is slightly less than that of water, since a hot dog is mostly water. Using transient temperature charts, determine (a) the thermal diffusivity of the hot dog, (b) the thermal conductivity of the hot dog, and (c) the convection heat transfer coefficient.

Answers: (a) 2×10^{-7} m²/s, (b) 0.76 W/m · °C, (c) 461 W/m² · °C.

16-44 Using the data and the answers given in Prob. 16-43, determine the center and the surface temperatures of the hot dog 4 min after the start of the cooking. Also determine the amount of heat transferred to the hot dog.

16-45E In a chicken processing plant, whole chickens averaging 5 lb each and initially at 72°F are to be cooled in the racks of a large refrigerator that is maintained at 5°F. The entire chicken is to be cooled below 45°F, but the temperature of the chicken is not to drop below 35°F at any point during refrigeration. The convection heat transfer coefficient and thus the rate of heat transfer from the chicken can be controlled by varying the speed of a circulating fan inside. Determine the heat transfer coefficient that will enable us to meet both temperature constraints while keeping the refrigeration time to a minimum. The chicken can be treated as a homogeneous spherical object having the properties $\rho = 74.9$ lbm/ft³, $C_p = 0.98$ Btu/lbm · °F, $k = 0.26$ Btu/h · ft · °F, and $\alpha = 0.0035$ ft²/h.

16-46 A person puts a few apples into the freezer at −15°C to cool them quickly for guests who are about to arrive. Initially, the apples are at a uniform temperature of 20°C, and the heat transfer coefficient on the surfaces is 8 W/m² · °C. Treating the apples as 9-cm-diameter spheres and taking their properties to be $\rho = 840$ kg/m³, $C_p = 3.81$ kJ/kg · °C, $k = 0.418$ W/m · °C,

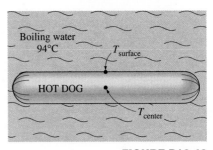

Boiling water
94°C

$T_{surface}$

HOT DOG

T_{center}

FIGURE P16-43

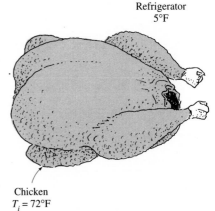

Refrigerator
5°F

Chicken
$T_i = 72°F$

FIGURE P16-45E

Ambient air
−15°C

Orange
$T_i = 15°C$

FIGURE P16-47

and $\alpha = 1.3 \times 10^{-7}$ m²/s, determine the center and surface temperatures of the apples in 1 h. Also, determine the amount of heat transfer from each apple.

16-47   Citrus fruits are very susceptible to cold weather, and extended exposure to subfreezing temperatures can destroy them. Consider an 8-cm-diameter orange that is initially at 15°C. A cold front moves in one night, and the ambient temperature suddenly drops to −6°C, with a heat transfer coefficient of 15 W/m² · °C. Using the properties of water for the orange and assuming the ambient conditions to remain constant for 4 h before the cold front moves out, determine if any part of the orange will freeze that night.

16-48   An 8-cm-diameter potato ($\rho = 1100$ kg/m³, $C_p = 3900$ J/kg · °C, $k = 0.6$ W/m · °C, and $\alpha = 1.4 \times 10^{-7}$ m²/s) that is initially at a uniform temperature of 25°C is baked in an oven at 170°C until a temperature sensor inserted to the center of the potato indicates a reading of 70°C. The potato is then taken out of the oven and wrapped in thick towels so that almost no heat is lost from the baked potato. Assuming the heat transfer coefficient in the oven to be 25 W/m² · °C, determine (a) how long the potato is baked in the oven and (b) the final equilibrium temperature of the potato after it is wrapped.

Transient Heat Conduction in Semi-Infinite Solids

16-49C   What is a semi-infinite medium? Give examples of solid bodies that can be treated as semi-infinite mediums for heat transfer purposes.

16-50C   Under what conditions can a plane wall be treated as a semi-infinite medium?

16-51C   Consider a hot semi-infinite solid at an initial temperature of T_i that is exposed to convection to a cooler medium at a constant temperature of T_∞, with a heat transfer coefficient of h. Explain how you can determine the total amount of heat transfer from the solid up to a specified time t_o.

16-52   In areas where the air temperature remains below 0°C for prolonged periods of time, the freezing of water in underground pipes is a major concern. Fortunately, the soil remains relatively warm during those periods, and it takes weeks for the subfreezing temperatures to reach the water mains in the ground. Thus, the soil effectively serves as an insulation to protect the water from the freezing atmospheric temperatures in winter.

The ground at a particular location is covered with snow pack at −8°C for a continuous period of 60 days, and the average soil properties at that location are $k = 0.4$ W/m · °C and $\alpha = 0.15 \times 10^{-6}$ m²/s. Assuming an initial uniform temperature of 10°C for the ground, determine the minimum burial depth to prevent the water pipes from freezing.

16-53   The soil temperature in the upper layers of the earth varies with the variations in the atmospheric conditions. Before a cold front moves in, the earth at a location is initially at a uniform temperature of 10°C. Then the area is subjected to a temperature of −10°C and high winds that resulted in a convection heat transfer coefficient of 40 W/m² · °C on the earth's surface for a period of 10 h. Taking the properties of the soil at that location to be $k = 0.9$ W/m · °C and $\alpha = 1.6 \times 10^{-5}$ m²/s, determine the soil temperature at distances 0, 10, 20, and 50 cm from the earth's surface at the end of this 10-h period.

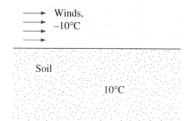

Winds,
−10°C

Soil

10°C

FIGURE P16-53

16-54E The walls of a furnace are made of 1.5-ft-thick concrete ($k = 0.64$ Btu/h · ft · °F and $\alpha = 0.023$ ft²/h). Initially, the furnace and the surrounding air are in thermal equilibrium at 70°F. The furnace is then fired, and the inner surfaces of the furnace are subjected to hot gases at 1800°F with a very large heat transfer coefficient. Determine how long it will take for the temperature of the outer surface of the furnace walls to rise to 70.1°F.
Answer: 181 min

16-55 A thick wood slab ($k = 0.17$ W/m · °C and $\alpha = 1.28 \times 10^{-7}$ m²/s) that is initially at a uniform temperature of 25°C is exposed to hot gases at 550°C for a period of 5 minutes. The heat transfer coefficient between the gases and the wood slab is 35 W/m² · °C. If the ignition temperature of the wood is 420°C, determine if the wood will ignite.

16-56 A large cast iron container ($k = 52$ W/m · °C and $\alpha = 1.70 \times 10^{-5}$ m²/s) with 5-cm-thick walls is initially at a uniform temperature of 0°C and is filled with ice at 0°C. Now the outer surfaces of the container are exposed to hot water at 60°C with a very large heat transfer coefficient. Determine how long it will be before the ice inside the container starts melting. Also, taking the heat transfer coefficient on the inner surface of the container to be 250 W/m² · °C, determine the rate of heat transfer to the ice through a 1.2-m-wide and 2-m-high section of the wall when steady operating conditions are reached. Assume the ice starts melting when its inner surface temperature rises to 0.1°C.

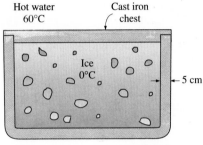

FIGURE P16-56

Transient Heat Conduction in Multidimensional Systems

16-57C What is the product solution method? How is it used to determine the transient temperature distribution in a two-dimensional system?

16-58C How is the product solution used to determine the variation of temperature with time and position in three-dimensional systems?

16-59C A short cylinder initially at a uniform temperature T_i is subjected to convection from all of its surfaces to a medium at temperature T_∞. Explain how you can determine the temperature of the midpoint of the cylinder at a specified time t.

16-60C Consider a short cylinder whose top and bottom surfaces are insulated. The cylinder is initially at a uniform temperature T_i and is subjected to convection from its side surface to a medium at temperature T_∞ with a heat transfer coefficient of h. Is the heat transfer in this short cylinder one- or two-dimensional? Explain.

16-61 A short brass cylinder ($\rho = 8530$ kg/m³, $C_p = 0.389$ kJ/kg · °C, $k = 110$ W/m · °C, and $\alpha = 3.39 \times 10^{-5}$ m²/s) of diameter $D = 8$ cm and height $H = 15$ cm is initially at a uniform temperature of $T_i = 150$°C. The cylinder is now placed in atmospheric air at 20°C, where heat transfer takes place by convection with a heat transfer coefficient of $h = 40$ W/m² · °C. Calculate (*a*) the center temperature of the cylinder, (*b*) the center temperature of the top surface of the cylinder, and (*c*) the total heat transfer from the cylinder 15 min after the start of the cooling.

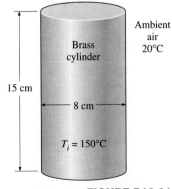

FIGURE P16-61

16-62 A semi-infinite aluminum cylinder ($k = 237$ W/m · °C, $\alpha = 9.71 \times 10^{-5}$ m²/s) of diameter $D = 15$ cm is initially at a uniform temperature of $T_i = 150$°C. The cylinder is now placed in water at 10°C, where heat transfer takes

Room
air
18°C

Ice
block
−20°C

FIGURE P16-65

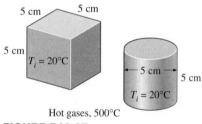

5 cm   5 cm

5 cm $T_i = 20°C$

← 5 cm →   5 cm

$T_i = 20°C$

Hot gases, 500°C

FIGURE P16-67

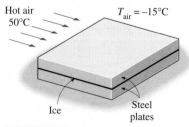

Hot air
50°C

$T_{air} = -15°C$

Ice   Steel
plates

FIGURE P16-71

place by convection with a heat transfer coefficient of $h = 140$ W/m² · °C. Determine the temperature at the center of the cylinder 10 cm from the end surface 8 min after the start of cooling.

16-63E   A hot dog can be considered to be a cylinder 5 in. long and 0.8 in. in diameter whose properties are $\rho = 61.2$ lbm/ft³, $C_p = 0.93$ Btu/lbm · °F, $k = 0.44$ Btu/h · ft · °F, and $\alpha = 0.0077$ ft²/h. A hot dog initially at 40°F is dropped into boiling water at 212°F. If the heat transfer coefficient at the surface of the hot dog is estimated to be 120 Btu/h · ft² · °F, determine the center temperature of the hot dog after 5, 10, and 15 min by treating the hot dog as (a) a finite cylinder and (b) an infinitely long cylinder.

16-64E   Repeat Prob. 16-63E for a location at 5300 ft elevation such as Denver, Colorado, where the boiling temperature of water is 202°F.

16-65   A 5-cm-high rectangular ice block ($k = 2.22$ W/m · °C and $\alpha = 0.124 \times 10^{-7}$ m²/s) initially at −20°C is placed on a table on its square base 4 cm × 4 cm in size in a room at 18°C. The heat transfer coefficient on the exposed surfaces of the ice block is 12 W/m² · °C. Disregarding any heat transfer from the base to the table, determine how long it will be before the ice block starts melting. Where on the ice block will the first liquid droplets appear?

16-66   A 2-cm-high cylindrical ice block ($k = 2.22$ W/m · °C and $\alpha = 0.124 \times 10^{-7}$ m²/s) is placed on a table on its base of diameter 2 cm in a room at 20°C. The heat transfer coefficient on the exposed surfaces of the ice block is 13 W/m² · °C, and heat transfer from the base of the ice block to the table is negligible. If the ice block is not to start melting at any point for at least 2 h, determine what the initial temperature of the ice block should be.

16-67   Consider a cubic block whose sides are 5 cm long and a cylindrical block whose height and diameter are also 5 cm. Both blocks are initially at 20°C and are made of granite ($k = 2.5$ W/m · °C and $\alpha = 1.15 \times 10^{-6}$ m²/s). Now both blocks are exposed to hot gases at 500°C in a furnace on all of their surfaces with a heat transfer coefficient of 40 W/m² · °C. Determine the center temperature of each geometry after 10, 20, and 60 min.

16-68   Repeat Prob. 16-67 with the heat transfer coefficient at the top and the bottom surfaces of each block being doubled to 80 W/m² · °C.

16-69   A 20-cm-long cylindrical aluminum block ($\rho = 2702$ kg/m³, $C_p = 0.896$ kJ/kg · °C, $k = 236$ W/m · °C, and $\alpha = 9.75 \times 10^{-5}$ m²/s), 15 cm in diameter, is initially at a uniform temperature of 20°C. The block is to be heated in a furnace at 1200°C until its center temperature rises to 300°C. If the heat transfer coefficient on all surfaces of the block is 50 W/m² · °C, determine how long the block should be kept in the furnace. Also, determine the amount of heat transfer from the aluminum block if it is allowed to cool in the room until its temperature drops to 20°C throughout.

16-70   Repeat Prob. 16-69 for the case where the aluminum block is inserted into the furnace on a low-conductivity material so that the heat transfer to or from the bottom surface of the block is negligible.

Review Problems

16-71   Consider two 2-cm-thick large steel plates ($k = 43$ W/m · °C and $\alpha = 1.17 \times 10^{-5}$ m²/s) that were put on top of each other while wet and left outside during a cold winter night at $-15°C$. The next day, a worker needs one of the plates, but the plates are stuck together because the freezing of the water between the two plates has bonded them together. In an effort to melt the ice between the plates and separate them, the worker takes a large hairdryer and blows hot air at 50°C all over the exposed surface of the plate on the top. The convection heat transfer coefficient at the top surface is estimated to be 40 W/m² · °C. Determine how long the worker must keep blowing hot air before the two plates separate.   *Answer:* 507 s

16-72   Consider a curing kiln whose walls are made of 30-cm-thick concrete whose properties are $k = 0.9$ W/m · °C and $\alpha = 0.23 \times 10^{-5}$ m²/s. Initially, the kiln and its walls are in equilibrium with the surroundings at 5°C. Then all the doors are closed and the kiln is heated by steam so that the temperature of the inner surface of the walls is raised to 45°C and is maintained at that level for 3 h. The curing kiln is then opened and exposed to the atmospheric air after the stream flow is turned off. If the outer surfaces of the walls of the kiln were insulated, would it save any energy that day during the period the kiln was used for curing for 3 h only, or would it make no difference? Base your answer on calculations.

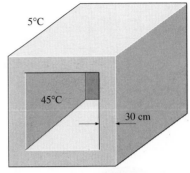

FIGURE P16-72

16-73   The water main in the cities must be placed at sufficient depth below the earth's surface to avoid freezing during extended periods of subfreezing temperatures. Determine the minimum depth at which the water main must be placed at a location where the soil is initially at 15°C and the earth's surface temperature under the worst conditions is expected to remain at $-10°C$ for a period of 75 days. Take the properties of soil at that location to be $k = 0.7$ W/m · °C and $\alpha = 1.4 \times 10^{-5}$ m²/s.   *Answer:* 7.05 m

16-74   A hot dog can be considered to be a 12-cm-long cylinder whose diameter is 2 cm and whose properties are $\rho = 980$ kg/m³, $C_p = 3.9$ kJ/kg · °C, $k = 0.76$ W/m · °C, and $\alpha = 2 \times 10^{-7}$ m²/s. A hot dog initially at 5°C is dropped into boiling water at 100°C. The heat transfer coefficient at the surface of the hot dog is estimated to be 600 W/m² · °C. If the hot dog is considered cooked when its center temperature reaches 80°C, determine how long it will take to cook it in the boiling water.

FIGURE P16-74

16-75   A long roll of 2-m-wide and 0.5-cm-thick 1-Mn manganese steel plate coming off a furnace at 820°C is to be quenched in an oil bath ($C_p = 2.0$ kJ/kg · °C) at 45°C. The metal sheet is moving at a steady velocity of 10 m/min, and the oil bath is 8 m long. Taking the convection heat transfer coefficient on both sides of the plate to be 860 W/m² · °C, determine the temperature of the sheet metal when it leaves the oil bath. Also, determine the required rate of heat removal from the oil to keep its temperature constant at 45°C.

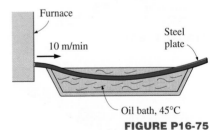

FIGURE P16-75

16-76E   In *Betty Crocker's Cookbook,* it is stated that it takes 5 h to roast a 14-lb stuffed turkey initially at 40°F in an oven maintained at 325°F. It is recommended that a meat thermometer be used to monitor the cooking, and the turkey is considered done when the thermometer inserted deep into the thickest part of the breast or thigh without touching the bone registers 185°F. The turkey can be treated as a homogeneous spherical object with the properties $\rho = 75$ lbm/ft³, $C_p = 0.98$ Btu/lbm · °F, $k = 0.26$ Btu/h · ft · °F, and $\alpha = 0.0035$ ft²/h. Assuming the tip of the thermometer is at one-third radial distance from the center of the turkey, determine (*a*) the average heat transfer coefficient at the surface of the turkey, (*b*) the temperature of the skin of

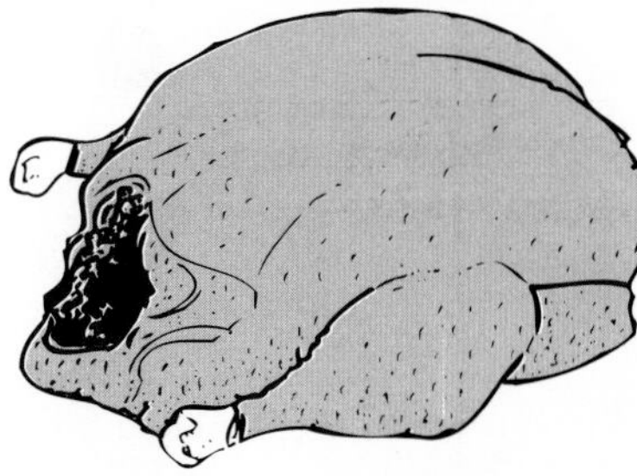

FIGURE P16-76E

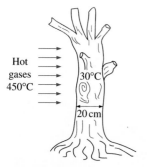

FIGURE P16-77

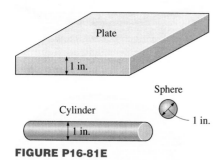

Watermelon, 25°C
FIGURE P16-78

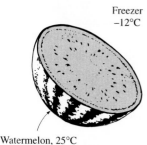

FIGURE P16-79

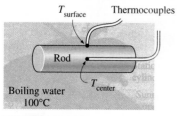

FIGURE P16-81E

the turkey when it is done, and (c) the total amount of heat transferred to the turkey in the oven. Will the reading of the thermometer be more or less than 185°F 5 min after the turkey is taken out of the oven?

16-77 During a fire, the trunks of some dry oak trees ($k = 0.17$ W/m · °C and $\alpha = 1.28 \times 10^{-7}$ m²/s) that are initially at a uniform temperature of 30°C are exposed to hot gases at 450°C for a period of 4 h, with a heat transfer coefficient of 65 W/m² · °C on the surface. The ignition temperature of the trees is 410°C. Treating the trunks of the trees as long cylindrical rods of diameter 20 cm, determine if these dry trees will ignite as the fire sweeps through them.

16-78 We often cut a watermelon in half and put it into the freezer to cool it quickly. But usually we forget to check on it and end up having a watermelon with a frozen layer on the top. To avoid this potential problem a person wants to set the timer such that it will go off when the temperature of the exposed surface of the watermelon drops to 3°C.

Consider a 30-cm-diameter spherical watermelon that is cut into two equal parts and put into a freezer at −12°C. Initially, the entire watermelon is at a uniform temperature of 25°C, and the heat transfer coefficient on the surfaces is 30 W/m² · °C. Assuming the watermelon to have the properties of water, determine how long it will take for the center of the exposed cut surfaces of the watermelon to drop to 3°C.

16-79 The thermal conductivity of a solid whose density and specific heat are known can be determined from the relation $k = \alpha/\rho C_p$ after evaluating the thermal diffusivity α.

Consider a 2-cm-diameter cylindrical rod made of a sample material whose density and specific heat are 3700 kg/m³ and 920 J/kg · °C, respectively. The sample is initially at a uniform temperature of 25°C. In order to measure the temperatures of the sample at its surface and its center, a thermocouple is inserted to the center of the sample along the centerline, and another thermocouple is welded into a small hole drilled on the surface. The sample is dropped into boiling water at 100°C. After 3 min, the surface and the center temperatures are recorded to be 93°C and 75°C, respectively. Determine the thermal diffusivity and the thermal conductivity of the material.

16-80 In desert climates, rainfall is not a common occurrence since the rain droplets formed in the upper layer of the atmosphere often evaporate before they reach the ground. Consider a raindrop that is initially at a temperature of 5°C and has a diameter of 5 mm. Determine how long it will take for the diameter of the raindrop to reduce to 3 mm as it falls through ambient air at 25°C with a heat transfer coefficient of 400 W/m² · °C. The water temperature can be assumed to remain constant and uniform at 5°C at all times, and the heat of vaporization of water at 5°C is 2490 kJ/kg.

16-81E Consider a plate of thickness 1 in., a long cylinder of diameter 1 in., and a sphere of diameter 1 in., all initially at 400°F and all made of bronze ($k = 15.0$ Btu/h · ft · °F and $\alpha = 0.333$ ft²/h). Now all three of these geometries are exposed to cool air at 75°F on all of their surfaces, with a heat transfer coefficient of 7 Btu/h · ft² · °F. Determine the center temperature of each geometry after 5, 10, and 30 min. Explain why the center temperature of the sphere is always the lowest.

16-82E Repeat Prob. 16-81E for cast iron geometries ($k = 29$ Btu/h · ft · °F and $\alpha = 0.61$ ft²/h).

16-83 Long aluminum wires of diameter 3 mm ($\rho = 2702$ kg/m^3, $C_p = 0.896$ kJ/kg · °C, $k = 236$ W/m · °C, and $\alpha = 9.75 \times 10^{-5}$ m^2/s) are extruded at a temperature of 350°C and exposed to atmospheric air at 30°C with a heat transfer coefficient of 35 W/m^2 · °C. (*a*) Determine how long it will take for the wire temperature to drop to 50°C. (*b*) If the wire is extruded at a velocity of 10 m/min, determine how far the wire travels after extrusion by the time its temperature drops to 50°C. What change in the cooling process would you propose to shorten this distance? (*c*) Assuming the aluminum wire leaves the extrusion room at 50°C, determine the rate of heat transfer from the wire to the extrusion room. *Answers: (a)* 144 s, *(b)* 24 m, *(c)* 855 W

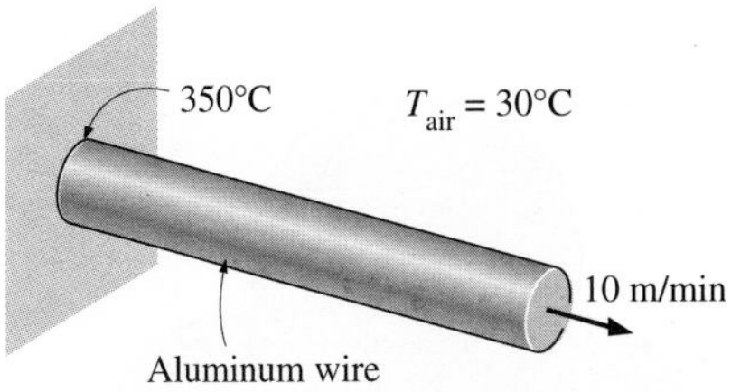

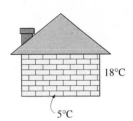

FIGURE P16-83

16-84 Repeat Prob. 16-83 for a copper wire ($\rho = 8950$ kg/m^3, $C_p = 0.383$ kJ/kg · °C, $k = 386$ W/m · °C, and $\alpha = 1.13 \times 10^{-4}$ m^2/s).

16-85 Consider a brick house ($k = 0.72$ W/m · °C and $\alpha = 0.45 \times 10^{-6}$ m^2/s) whose walls are 10 m long, 3 m high, and 0.3 m thick. The heater of the house broke down one night, and the entire house, including its walls, was observed to be 5°C throughout in the morning. The outdoors warmed up as the day progressed, but no change was felt in the house, which was tightly sealed. Assuming the outer surface temperature of the house to remain constant at 18°C, determine how long it would take for the temperature of the inner surfaces of the walls to rise to 5.1°C.

FIGURE P16-85

Computer, Design, and Essay Problems

16-86 Conduct the following experiment at home to determine the combined convection and radiation heat transfer coefficient at the surface of an apple exposed to the room air. You will need two thermometers and a clock.

First, weigh the apple and measure its diameter. You may measure its volume by placing it in a large measuring cup halfway filled with water, and measuring the change in volume when it is completely immersed in the water. Refrigerate the apple overnight so that it is at a uniform temperature in the morning and measure the air temperature in the kitchen. Then take the apple out and stick one of the thermometers to its middle and the other just under the skin. Record both temperatures every 5 min for an hour. Using these two temperatures, calculate the heat transfer coefficient for each interval and take their average. The result is the combined convection and radiation heat transfer coefficient for this heat transfer process. Using your experimental data, also calculate the thermal conductivity and thermal diffusivity of the apple and compare them to the values given above.

16-87 Repeat Prob. 16-86 using a banana instead of an apple. The thermal properties of bananas are practically the same as those of apples.

16-88 Conduct the following experiment to determine the lumped exponent *b* in Eq. 16-5 for a can of soda and then predict the temperature of the soda at different times. Leave the soda in the refrigerator overnight. Measure the air temperature in the kitchen and the temperature of the soda while it is still in the refrigerator by taping the sensor of the thermometer to the outer surface of the can. Then take the soda out and measure its temperature again in 5 min. Using these values, calculate the exponent *b*. Using this *b*-value, predict the temperatures of the soda in 10, 15, 20, 30, and 60 min and compare the results with the actual temperature measurements. Do you think the lumped system analysis is valid in this case?

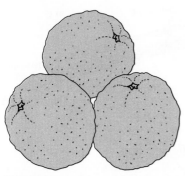

FIGURE P16-90

16-89 Whole ready-to-cook turkeys range from about 2 to 11 kg. Roasted turkeys are considered done when a thermometer inserted deep into the turkey registers 85°C. From your favorite cookbook, obtain the instructions to bake a large stuffed turkey and evaluate the average heat transfer coefficient during baking. Using this heat transfer coefficient, estimate the baking time for a turkey that is only half as large.

16-90 Citrus trees are very susceptible to cold weather, and extended exposure to subfreezing temperatures can destroy the crop. In order to protect the trees from occasional cold fronts with subfreezing temperatures, tree growers in Florida usually install water sprinklers on the trees. When the temperature drops below a certain level, the sprinklers spray water on the trees and their fruits to protect them against the damage the subfreezing temperatures can cause. Explain the basic mechanism behind this protection measure and write an essay on how the system works in practice.

Forced Convection

So far, we have considered *conduction,* which is the mechanism of heat transfer through a solid or fluid in the absence of any fluid motion. We now consider *convection,* which is the mechanism of heat transfer through a fluid in the presence of bulk fluid motion.

Convection is classified as *natural* (or *free*) or *forced convection,* depending on how the fluid motion is initiated. In forced convection, the fluid is forced to flow over a surface or in a pipe by external means such as a pump or a fan. In natural convection, any fluid motion is caused by natural means such as the buoyancy effect, which manifests itself as the rise of warmer fluid and the fall of the cooler fluid. Convection is also classified as *external* and *internal,* depending on whether the fluid is forced to flow over a surface or in a channel.

We start this chapter with a general physical description of the *convection* mechanism and the *thermal boundary layer.* We continue with the discussion of the dimensionless *Prandtl* and *Nusselt numbers,* and their physical significance. We then present empirical relations for the *heat transfer coefficients* for flow over various geometries such as a flat plate, cylinder, and sphere, for both laminar and turbulent flow conditions. Finally, we discuss the characteristics of flow inside tubes and present the heat transfer correlations associated with it. The relevant concepts from Chaps. 12 and 13 should be reviewed before this chapter is studied.

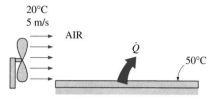

20°C
5 m/s

AIR

$\dot{Q}$

50°C

(a) Forced convection

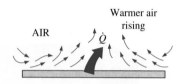

Warmer air
rising

AIR

$\dot{Q}$

(b) Free convection

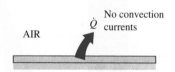

No convection
currents

AIR

$\dot{Q}$

(c) Conduction

FIGURE 17-1

Heat transfer from a hot surface
to the surrounding fluid by
convection and conduction.

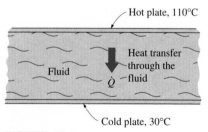

Hot plate, 110°C

Fluid

Heat transfer
through the
fluid
$\dot{Q}$

Cold plate, 30°C

FIGURE 17-2

Heat transfer through a fluid sandwiched
between two parallel plates.

17-1 ■ PHYSICAL MECHANISM OF FORCED CONVECTION

We mentioned earlier that there are three basic mechanisms of heat transfer: conduction, convection, and radiation. Conduction and convection are similar in that both mechanisms require the presence of a material medium. But they are different in that convection requires the presence of *fluid motion.*

Heat transfer through a *solid* is always by *conduction,* since the molecules of a solid remain at relatively fixed positions. Heat transfer through a *liquid* or *gas,* however, can be by *conduction* or *convection,* depending on the presence of any bulk fluid motion. Heat transfer through a fluid is by *convection* in the presence of bulk fluid motion and by *conduction* in the absence of it. Therefore, conduction in a fluid can be viewed as the *limiting case* of convection, corresponding to the case of quiescent fluid (Fig. 17-l).

Convection heat transfer is complicated by the fact that it involves *fluid motion* as well as *heat conduction.* The fluid motion *enhances* heat transfer, since it brings hotter and cooler chunks of fluid into contact, initiating higher rates of conduction at a greater number of sites in a fluid. Therefore, the rate of heat transfer through a fluid is much higher by convection than it is by conduction. In fact, the higher the *fluid velocity,* the higher the rate of *heat transfer.*

To clarify this point further, consider steady heat transfer through a fluid contained between two *parallel plates* maintained at different temperatures, as shown in Fig. 17-2. The temperatures of the fluid and the plate will be the same at the points of contact because of the *continuity of temperature.* Assuming no fluid motion, the energy of the hotter fluid molecules near the hot plate will be transferred to the adjacent cooler fluid molecules. This energy will then be transferred to the next layer of the cooler fluid molecules. This energy will then be transferred to the next layer of the cooler fluid, and so on, until it is finally transferred to the other plate. This is what happens during *conduction* through a fluid. Now let us use a syringe to draw some fluid near the hot plate and inject it near the cold plate repeatedly. You can imagine that this will speed up the heat transfer process considerably, since some energy is *carried* to the other side as a result of fluid motion.

Consider the cooling of a *hot iron block* with a fan blowing air over its top surface, as shown in Fig. 17-3. We know that heat will be transferred from the hot block to the surrounding cooler air, and the block will eventually cool. We also know that the block will cool faster if the fan is switched to a higher speed. Replacing air by water will enhance the convection heat transfer even more.

Experience shows that convection heat transfer strongly depends on the fluid properties *dynamic viscosity* μ, *thermal conductivity* k, *density* ρ, and *specific heat* C_p, as well as the *fluid velocity* $\mathcal{V}$. It also depends on the *geometry* and *roughness* of the solid surface, in addition to the *type of fluid flow* (such as being laminar or turbulent). Thus, we expect the convection heat transfer relations to be rather complex because of the dependence of convection on so many variables. This is not surprising, since convection is the most complex mechanism of heat transfer.

Despite the complexity of convection, the rate of *convection heat transfer* is observed to be proportional to the temperature difference and is conveniently expressed by **Newton's law of cooling** as

$$\dot{q}_{conv} = h(T_s - T_\infty) \qquad (W/m^2) \qquad (17\text{-}1)$$

or

$$\dot{Q}_{conv} = hA(T_s - T_\infty) \qquad \text{(W)} \qquad (17\text{-}2)$$

where

h = convection heat transfer coefficient, W/m² · °C
A = heat transfer surface area, m²
T_s = temperature of the surface, °C
T_∞ = temperature of the fluid sufficiently far from the surface, °C

Judging from its units, the **convection heat transfer coefficient** can be defined as *the rate of heat transfer between a solid surface and a fluid per unit surface area per unit temperature difference.*

You should not be deceived by the simple appearance of this relation, because the convection heat transfer coefficient h depends on the several variables mentioned above, and thus is difficult to determine.

An implication of the no-slip condition is that heat transfer from the solid surface to the fluid layer adjacent to the surface is by *pure conduction,* since the fluid layer is motionless, and can be expressed as

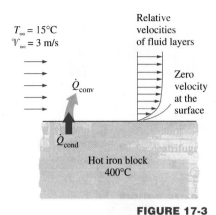

$$\dot{q}_{conv} = \dot{q}_{cond} = -k_{fluid} \left. \frac{\partial T}{\partial y} \right|_{y=0} \qquad \text{(W/m}^2\text{)} \qquad (17\text{-}3)$$

FIGURE 17-3

The cooling of a hot block by forced convection.

where T represents the temperature distribution in the fluid and $(\partial T/\partial y)_{y=0}$ is the *temperature gradient* at the surface. This heat is then *convected away* from the surface as a result of fluid motion. Note that convection heat transfer from a solid surface to a fluid is merely the conduction heat transfer from the solid surface to the fluid layer adjacent to the surface. Therefore, we can equate the expressions 17-1 and 17-3 for the heat flux to obtain

$$h = \frac{-k_{fluid}(\partial T/\partial y)_{y=0}}{T_s - T_\infty} \qquad \text{(W/m}^2 \cdot \text{°C)} \qquad (17\text{-}4)$$

for the determination of the *convection heat transfer coefficient* when the temperature distribution within the fluid is known.

The convection heat transfer coefficient, in general, varies along the flow (or x-) direction. The *average* or *mean* convection heat transfer coefficient for a surface in such cases is determined by properly averaging the *local* convection heat transfer coefficients over the entire surface.

In convection studies, it is common practice to nondimensionalize the governing equations and combine the variables, which group together into *dimensionless numbers* in order to reduce the number of total variables. It is also common practice to *nondimensionalize* the heat transfer coefficient h with the **Nusselt number,** defined as

$$\text{Nu} = \frac{h\delta}{k} \qquad (17\text{-}5)$$

where k is the thermal conductivity of the fluid and δ is the *characteristic length.* The Nusselt number is named after Wilhelm Nusselt, who made significant contributions to convective heat transfer in the first half of the 20th century, and it is viewed as the *dimensionless convection heat transfer coefficient.*

To understand the physical significance of the Nusselt number, consider a fluid layer of thickness δ and temperature difference $\Delta T = T_2 - T_1$, as shown

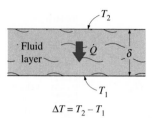

FIGURE 17-4

Heat transfer through a fluid layer of thickness δ and temperature difference ΔT.

FIGURE 17-5

We resort to forced convection whenever we need to increase the rate of heat transfer.

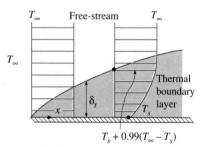

FIGURE 17-6

Thermal boundary layer on a flat plate (the fluid is hotter than the plate surface).

in Fig. 17-4. Heat transfer through the fluid layer will be by *convection* when the fluid involves some motion and by *conduction* when the fluid layer is motionless. Heat flux (the rate of heat transfer per unit time per unit surface area) in either case will be

$$\dot{q}_{conv} = h\,\Delta T \tag{17-6}$$

and

$$\dot{q}_{cond} = k\,\frac{\Delta T}{\delta} \tag{17-7}$$

Taking their ratio gives

$$\frac{\dot{q}_{conv}}{\dot{q}_{cond}} = \frac{h\,\Delta T}{k\,\Delta T/\delta} = \frac{h\delta}{k} = \text{Nu} \tag{17-8}$$

which is the **Nusselt number.** Therefore, the Nusselt number represents the enhancement of heat transfer through a fluid layer as a result of convection relative to conduction across the same fluid layer. The larger the Nusselt number, the more effective the convection. A Nusselt number of Nu = 1 for a fluid layer represents heat transfer by pure conduction.

We use forced convection in daily life more often than you might think (Fig. 17-5). We resort to forced convection whenever we want to increase the rate of heat transfer from a hot object. For example, we turn on the *fan* on hot summer days to help our body cool more effectively. The higher the fan speed, the better we feel. We *stir* our soup and *blow* on a hot slice of pizza to make them cool faster. The air on *windy* winter days feels much colder than it actually is. The simplest solution to heating problems in electronics packaging is to use a large enough fan.

17-2 ■ THERMAL BOUNDARY LAYER

We have seen in Chap. 13 that a velocity boundary layer develops when a fluid flows over a surface as a result of the fluid layer adjacent to the surface assuming the surface velocity (i.e., zero velocity relative to the surface). Also, we defined the velocity boundary layer as the region in which the fluid velocity varies from zero to $0.99\mathcal{V}_\infty$. Likewise, a *thermal boundary layer* develops when a fluid at a specified temperature flows over a surface that is at a different temperature, as shown in Fig. 17-6.

Consider the flow of a fluid at a uniform temperature of T_∞ over an isothermal flat plate at a temperature T_s. The fluid particles in the layer adjacent to the surface will reach thermal equilibrium with the plate and assume the surface temperature T_s. These fluid particles will then exchange energy with the particles in the adjoining fluid layer, and so on. As a result, a temperature profile will develop in the flow field that ranges from T_s at the surface to T_∞ sufficiently far from the surface. The flow region over the surface in which the temperature variation in the direction normal to the surface is significant is the **thermal boundary layer.** The *thickness* of the thermal boundary layer δ_t at any location along the surface is defined as *the distance from the surface at which the temperature difference $T - T_s$ equals $0.99(T_\infty - T_s)$.* Note that for the special case of $T_s = 0$, we have $T = 0.99T_\infty$ at the outer edge of the thermal boundary layer, which is analogous to $\mathcal{V} = 0.99\mathcal{V}_\infty$ for the velocity boundary layer.

The thickness of the thermal boundary layer increases in the flow direction, since the effects of heat transfer are felt at greater distances from the surface further down stream.

Convection heat transfer along a surface is directly related to the *temperature gradient* at that location. Therefore, the shape of the temperature profile in the thermal boundary layer dictates the convection heat transfer between a solid surface and the fluid flowing over it. In flow over a heated (or cooled) surface, both velocity and thermal boundary layers will develop simultaneously. Noting that the fluid velocity will have a strong influence on the temperature profile, the development of the velocity boundary layer relative to the thermal boundary layer will have a strong effect on convection heat transfer.

The relative thickness of the velocity and the thermal boundary layers is best described by the *dimensionless* parameter **Prandtl number,** defined as

$$\text{Pr} = \frac{\text{Molecular diffusivity of momentum}}{\text{Molecular diffusivity of heat}} = \frac{\nu}{\alpha} = \frac{\mu C_p}{k} \quad (17\text{-}9)$$

It is named after Ludwig Prandtl, who introduced the concept of boundary layer in 1904 and made significant contributions to boundary layer theory. The Prandtl numbers of fluids range from less than 0.01 for liquid metals to more than 100,000 for heavy oils (Table 17-1). Note that the Prandtl number is in the order of 10 for water.

The Prandtl numbers of gases are about 1, which indicates that both momentum and heat diffuse through the fluid at about the same rate. Heat diffuses very quickly in liquid metals ($\text{Pr} \ll 1$) and very slowly in oils ($\text{Pr} \gg 1$) relative to momentum. Consequently the thermal boundary layer is much thicker for liquid metals and much thinner for oils relative to the velocity boundary layer (Fig. 17-7).

17-3 ■ FLOW OVER FLAT PLATES

So far we have discussed the physical aspects of forced convection over surfaces. In this section, we will discuss the determination of the *heat transfer rate* to or from a flat plate for both laminar and turbulent flow cases. The Reynolds number for external flow was expressed in Chap. 13 as

$$\text{Re}_L = \frac{\rho \mathcal{V} L}{\mu} = \frac{\mathcal{V} L}{\nu} \quad (17\text{-}10)$$

where $\mathcal{V}$ is the upstream velocity and L is the characteristic length of the geometry, which, for a flat plate, is the length of the plate in the flow direction. For any point on a flat plate, the characteristic length is the distance x of the point from the leading edge in the flow direction. For flow over a flat plate, the transition from laminar to turbulent occurs at the generally accepted critical Reynolds number of $\text{Re}_{\text{cr}} \approx 5 \times 10^5$.

The heat transfer coefficient for a flat plate can be determined theoretically by solving the conservation of mass, momentum, and energy equations approximately or numerically. It can also be determined experimentally and expressed by empirical correlations. In either approach, it is found that the *average* Nusselt number can be expressed in terms of the Reynolds and Prandtl numbers in the form

$$\text{Nu} = \frac{hL}{k} = C \, \text{Re}_L^m \, \text{Pr}^n \quad (17\text{-}11)$$

TABLE 17-1

Typical ranges of Prandtl numbers for common fluids

Fluid	Pr
Liquid metals	0.004–0.030
Gases	0.7–1.0
Water	1.7–13.7
Light organic fluids	5–50
Oils	50–100,000
Glycerin	2000–100,000

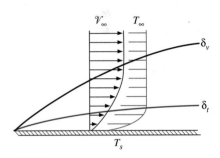

(a) Oils

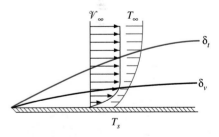

(b) Liquid metals (like mercury)

FIGURE 17-7

The relative thicknesses of the velocity and thermal boundary layers for liquid metals and oils.

where C, m, and n are constants and L is the *length* of the plate in the flow direction. The *local* Nusselt number at any point on the plate will depend on the distance of that point from the leading edge.

The fluid temperature in the thermal boundary layer varies from T_s at the surface to about T_∞ at the outer edge of the boundary. The fluid properties also vary with temperature, and thus with position across the boundary layer. In order to account for the variation of the properties with temperature properly, the fluid properties are usually evaluated at the so-called **film temperature,** defined as

$$T_f = \frac{T_s + T_\infty}{2} \tag{17-12}$$

which is the *arithmetic average* of the surface and the free-stream temperatures. The fluid properties are then assumed to remain constant at those values during the entire flow.

The local heat transfer coefficient *varies* along the surface of the flat plate as a result of the changes in the velocity and thermal boundary layers in the flow direction. We are usually interested in the heat transfer on the *entire* surface, which can be determined using the *average* heat transfer coefficient. But sometimes we are also interested in the local heat flux, and we need to know the *local* value of the heat transfer coefficient in such cases. With this in mind, below we present correlations for both local and average heat transfer coefficients. The *local* quantities are identified with the subscript x.

The *average* heat transfer coefficient for the entire plate can be determined from the corresponding *local* values by integration from

$$h = \frac{1}{L} \int_0^L h_x \, dx \tag{17-13}$$

Once h is available, the heat transfer rate can be determined from Eq. 17-2. Next we discuss the local and average heat transfer coefficients over a flat plate for *laminar, turbulent,* and *combined laminar and turbulent* flow conditions.

Based on analysis and experimental studies, the *local* Nusselt number at location x for laminar and turbulent flows over a flat plate are given by

Laminar: $\quad \mathrm{Nu}_x = \dfrac{h_x x}{k} = 0.332 \, \mathrm{Re}_x^{0.5} \, \mathrm{Pr}^{1/3} \quad \mathrm{Pr} \geq 0.6, \mathrm{Re} < 5 \times 10^5 \tag{17-14}$

Turbulent: $\quad \mathrm{Nu}_x = \dfrac{h_x x}{k} = 0.0296 \, \mathrm{Re}_x^{0.8} \, \mathrm{Pr}^{1/3} \quad \begin{matrix} 0.6 \leq \mathrm{Pr} \leq 60 \\ 5 \times 10^5 \leq \mathrm{Re}_x \leq 10^7 \end{matrix} \tag{17-15}$

where x is the distance from the leading edge of the plate and $\mathrm{Re}_x = \mathcal{V}x/\nu$ is the Reynolds number at location x. Note that h_x is proportional to $1/\mathrm{Re}_x^{0.5}$ and thus to $x^{-0.5}$ for laminar flow. Therefore, h_x is supposedly *infinite* at the leading edge ($x = 0$) and decreases by a factor of $x^{-0.5}$ in the flow direction. The variation of the boundary layer thickness δ and the friction and heat transfer coefficients along an isothermal flat plate are shown in Fig. 17-8. The local heat transfer coefficients are higher in turbulent flow than they are in laminar flow because of the intense mixing that occurs in the turbulent boundary layer. Note that h_x reaches its highest values when the flow becomes fully turbulent, and then decreases by a factor of $x^{-0.2}$ in the flow direction, as shown in the figure.

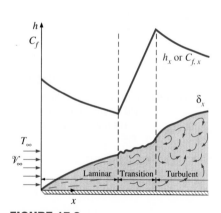

FIGURE 17-8

The variation of the local friction and heat transfer coefficients for flow over a flat plate.

The *average* Nusselt number over the entire plate is determined by substituting the relations above into Eq. 17-13 and performing the simple integrations. We get

Laminar flow: $\quad \mathrm{Nu} = \dfrac{hL}{k} = 0.664\ \mathrm{Re}_L^{0.5}\ \mathrm{Pr}^{1/3} \quad \mathrm{Re}_L < 5 \times 10^5 \quad$ (17-16)

Turbulent flow: $\quad \mathrm{Nu} = \dfrac{hL}{k} = 0.037\ \mathrm{Re}_L^{0.8}\ \mathrm{Pr}^{1/3} \quad \begin{array}{l} 0.6 \leq \mathrm{Pr} \leq 60 \\ 5 \times 10^5 \leq \mathrm{Re}_x \leq 10^7 \end{array}$

$$(17\text{-}17)$$

The first relation above gives the average heat transfer coefficient for the entire plate when the flow is *laminar* over the *entire* plate. The second relation gives the average heat transfer coefficient for the entire plate only when the flow is *turbulent* over the *entire* plate, or when the laminar flow region of the plate is too small relative to the turbulent flow region (that is, $x_{cr} \ll L$ where the length of the plate x_{cr} over which the flow is laminar can be determined from $\mathrm{Re}_{cr} = 5 \times 10^5 = \mathcal{V} x_{cr}/\nu$.)

Combined Laminar and Turbulent Flow

In some cases, a flat plate is sufficiently long for the flow to become turbulent, but not long enough to disregard the laminar flow region. In such cases, the *average* heat transfer coefficient over the entire plate is determined by performing the integration in Eq. 17-13 over two parts: the laminar region $0 \leq x \leq x_{cr}$ and the turbulent region $x_{cr} < x \leq L$ as

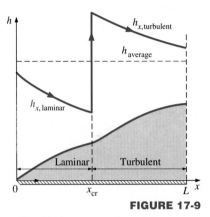

FIGURE 17-9

Graphical representation of the average heat transfer coefficient for a flat plate with combined laminar and turbulent flow.

$$h = \frac{1}{L}\left(\int_0^{x_{cr}} h_{x,\ \text{laminar}}\ dx + \int_{x_{cr}}^{L} h_{x,\ \text{turbulent}}\ dx \right) \qquad (17\text{-}18)$$

Note that we included the transition region with the turbulent region. Again taking the critical Reynolds number to be $\mathrm{Re}_{cr} = 5 \times 10^5$ and performing the integrations above after substituting the indicated expressions, the *average* Nusselt number over the *entire* plate is determined to be (Fig. 17-9)

$$\mathrm{Nu} = \frac{hL}{k} = (0.037\ \mathrm{Re}_L^{0.8} - 871)\mathrm{Pr}^{1/3} \quad \begin{array}{l} 0.6 \leq \mathrm{Pr} \leq 60 \\ 5 \times 10^5 \leq \mathrm{Re}_x \leq 10^7 \end{array} \qquad (17\text{-}19)$$

The constants in the relation above will be different for different critical Reynolds numbers.

The relations above have been obtained for the case of *isothermal* surfaces but could also be used approximately for the case of nonisothermal surfaces by assuming the surface temperature to be constant at some average value. Also, the surfaces are assumed to be *smooth,* and the free stream to be *turbulence free.*

When a flat plate is subjected to *uniform heat flux* instead of uniform temperature, the local Nusselt number is given by

Laminar flow: $\qquad \mathrm{Nu}_x = 0.453\ \mathrm{Re}_x^{0.5}\ \mathrm{Pr}^{1/3}$ $\qquad\qquad\qquad$ (17-20)

Turbulent flow: $\qquad \mathrm{Nu}_x = 0.0308\ \mathrm{Re}_x^{0.8}\ \mathrm{Pr}^{1/3}$ $\qquad\qquad\quad$ (17-21)

The relations above give values that are 36 percent higher for laminar flow and 4 percent higher for turbulent flow relative to the isothermal plate case.

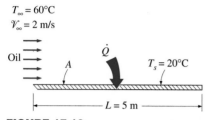

$T_\infty = 60°C$
$\mathcal{V}_\infty = 2$ m/s

Oil

$\dot{Q}$

A

$T_s = 20°C$

$L = 5$ m

FIGURE 17-10

Schematic for Example 17-1.

EXAMPLE 17-1   Flow of Hot Oil over a Flat Plate

Engine oil at 60°C flows over a 5-m-long flat plate whose temperature is 20°C with a velocity of 2 m/s (Fig. 17-10). Determine the total drag force and the rate of heat transfer per unit width of the entire plate.

Solution   Engine oil flows over a flat plate. The total rate of heat transfer per unit width of the plate is to be determined.

Assumptions   1 The flow is steady and incompressible. **2** The critical Reynolds number is $Re_{cr} = 5 \times 10^5$.

Properties   The properties of engine oil at the film temperature of $T_f = (T_s + T_\infty)/2 = (20 + 60)/2 = 40°C$ are (Table A-16).

$$\rho = 876 \text{ kg/m}^3 \qquad\qquad Pr = 2870$$
$$k = 0.144 \text{ W/m} \cdot °C \qquad\qquad \nu = 242 \times 10^{-6} \text{ m}^2/\text{s}$$

Analysis   Noting that $L = 5$ m, the Reynolds number at the end of the plate is

$$Re_L = \frac{\mathcal{V}L}{\nu} = \frac{(2 \text{ m/s})(5 \text{ m})}{0.242 \times 10^{-5} \text{ m}^2/\text{s}} = 4.13 \times 10^4$$

which is less than the critical Reynolds number, Thus we have *laminar flow* over the entire plate, and the average Nusselt number is determined from

$$Nu = \frac{hL}{k} = 0.664 \ Re_L^{0.5} \ Pr^{1/3} = 0.664 \times (4.13 \times 10^4)^{0.5} \times 2870^{1/3} = 1918$$

Then,

$$h = \frac{k}{L} Nu = \frac{0.144 \text{ W/m} \cdot °C}{5 \text{ m}} (1918) = 55.2 \text{ W/m}^2 \cdot °C$$

and

$$\dot{Q} = hA(T_\infty - T_s) = (55.2 \text{ W/m}^2 \cdot °C)(5 \times 1 \text{ m}^2)(60 - 20)°C = \mathbf{11{,}040 \text{ W}}$$

Discussion   Note that heat transfer is always from the higher-temperature medium to the lower-temperature one. In this case, it is from the oil to the plate. The heat transfer rate is per m width of the plate. The heat transfer for the entire plate can be obtained by multiplying the value obtained by the actual width of the plate.

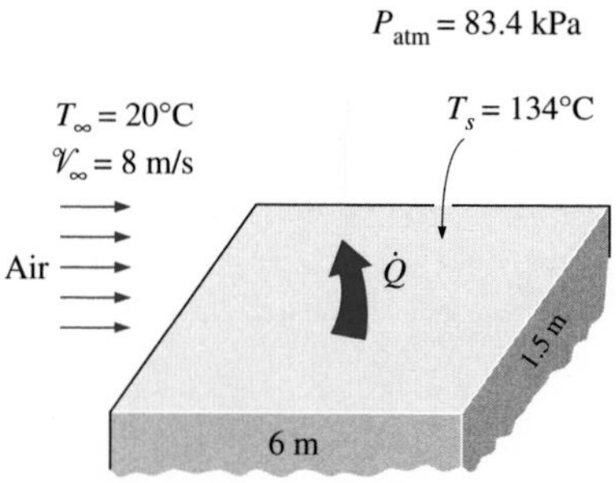

$P_{atm} = 83.4$ kPa

$T_\infty = 20°C$
$\mathcal{V}_\infty = 8$ m/s

$T_s = 134°C$

Air

$\dot{Q}$

1.5 m

6 m

FIGURE 17-11

Schematic for Example 17-2.

EXAMPLE 17-2   Cooling of a Hot Block by Forced Air at High Elevation

The local atmospheric pressure in Denver, Colorado (elevation 1610 m), is 83.4 kPa. Air at this pressure and 20°C flows with a velocity of 8 m/s over a 1.5 m × 6 m flat plate whose temperature is 134°C (Fig. 17-11). Determine the rate of heat transfer from the plate if the air flows parallel to the (a) 6-m-long side and (b) the 1.5-m side.

Solution   The top surface of a hot block is to be cooled by forced air. The rate of heat transfer is to be determined for two cases.

Assumptions   1 Steady operating conditions exist. **2** The critical Reynolds number is $Re_{cr} = 5 \times 10^5$. **3** Radiation effects are negligible. **4** Air is an ideal gas.

Properties   The properties k, μ, C_p, and Pr of ideal gases are independent of pressure, while the properties ν and α are inversely proportional to density and thus pressure. The properties of air at the film temperature of $T_f = (T_s + T_\infty)/2 = (134 + 20)/2 = 77°C = 350$ K and 1 atm pressure are (Table A-18)

$$k = 0.0297 \text{ W/m} \cdot °\text{C} \qquad \text{Pr} = 0.706$$
$$\nu_{@ 1 \text{ atm}} = 2.06 \times 10^{-5} \text{ m}^2/\text{s}$$

The atmospheric pressure in Denver is P = (83.4 kPa)/(101.325 kPa/atm) = 0.823 atm. Then the kinematic viscosity of air in Denver becomes

$$\nu = \nu_{@ 1 \text{ atm}}/P = (2.06 \times 10^{-5} \text{ m}^2/\text{s})/0.823 = 2.50 \times 10^{-5} \text{ m}^2/\text{s}$$

Analysis (*a*) When air flow is parallel to the long side, we have $L = 6$ m, and the Reynolds number at the end of the plate becomes

$$\text{Re}_L = \frac{\mathcal{V}_\infty L}{\nu} = \frac{(8 \text{ m/s})(6 \text{ m})}{2.50 \times 10^{-5} \text{ m}^2/\text{s}} = 1.92 \times 10^6$$

which is greater than the critical Reynolds number. Thus, we have combined laminar and turbulent flow, and the average Nusselt number for the entire plate is determined from

$$\text{Nu} = \frac{hL}{k} = (0.037 \, \text{Re}_L^{0.8} - 871)\text{Pr}^{1/3}$$
$$= [0.037(1.92 \times 10^6)^{0.8} - 871]0.706^{1/3}$$
$$= 2727$$

Then

$$h = \frac{k}{L}\text{Nu} = \frac{0.0297 \text{ W/m} \cdot °\text{C}}{6 \text{ m}}(2727) = 13.5 \text{ W/m}^2 \cdot °\text{C}$$
$$A = wL = (1.5 \text{ m})(6 \text{ m}) = 9 \text{ m}^2$$

and

$$\dot{Q} = hA(T_s - T_\infty) = (13.5 \text{ W/m}^2 \cdot °\text{C})(9 \text{ m}^2)(134 - 20)°\text{C} = \textbf{13,850 W}$$

Note that if we disregarded the laminar region and assumed turbulent flow over the entire plate, we would get Nu = 3520 from Eq. 17-17, which is 28 percent higher than the value calculated above. Therefore, that assumption would cause a 28 percent error in the heat transfer calculation in this case.

(*b*) When air flow is parallel to the short side, we have $L = 1.5$ m, and the Reynolds number at the end of the plate becomes

$$\text{Re}_L = \frac{\mathcal{V}_\infty L}{\nu} = \frac{(8 \text{ m/s})(1.5 \text{ m})}{2.50 \times 10^{-5} \text{ m}^2/\text{s}} = 4.80 \times 10^5$$

which is less than the critical Reynolds number. Thus we have laminar flow over the entire plate, and the average Nusselt number is determined from

$$\text{Nu} = \frac{hL}{k} = 0.664 \, \text{Re}_L^{0.5} \, \text{Pr}^{1/3} = 0.664 \times (4.8 \times 10^5)^{0.5} \times 0.706^{1/3} = 410$$

Then

$$h = \frac{k}{L}\text{Nu} = \frac{0.0297 \text{ W/m} \cdot °\text{C}}{1.5 \text{ m}}(410) = 8.12 \text{ W/m}^2 \cdot °\text{C}$$

and

$$\dot{Q} = hA(T_s - T_\infty) = (8.12 \text{ W/m}^2 \cdot °\text{C})(9 \text{ m}^2)(134 - 20)°\text{C} = \textbf{8330 W}$$

which is considerably less than the heat transfer rate determined in case (*a*).

Discussion Note that the *direction* of fluid flow can have a significant effect on convection heat transfer to or from a surface (Fig. 17-12). In this case, we can increase the heat transfer rate by 67 percent by simply blowing the air parallel to the long side of the rectangular plate instead of the short side.

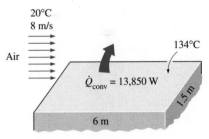

(*a*) Flow along the long side

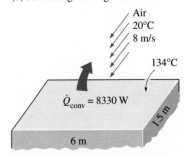

(*b*) Flow along the short side

FIGURE 17-12

The direction of fluid flow can have a significant effect on convection heat transfer.

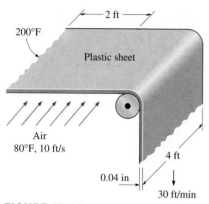

FIGURE 17-13

Schematic for Example 17-3.

EXAMPLE 17-3 Cooling of Plastic Sheets by Forced Air

The forming section of a plastics plant puts out a continuous sheet of plastic that is 4 ft wide and 0.04 in. thick at a velocity of 30 ft/min. The temperature of the plastic sheet is 200°F when it is exposed to the surrounding air, and a 2-ft-long section of the plastic sheet is subjected to air flow at 80°F at a velocity of 10 ft/s on both sides along its surfaces normal to the direction of motion of the sheet, as shown in Fig. 17-13. Determine (a) the rate of heat transfer from the plastic sheet to air by forced convection and radiation and (b) the temperature of the plastic sheet at the end of the cooling section. Take the density, specific heat, and emissivity of the plastic sheet to be $\rho = 75$ lbm/ft³, $C_p = 0.4$ Btu/lbm · °F, and $\varepsilon = 0.9$.

Solution Plastic sheets are cooled as they leave the forming section of a plastics plant. The rate of heat loss from the plastic sheet by convection and radiation and the exit temperature of the plastic sheet are to be determined.

Assumptions **1** Steady operating conditions exist. **2** The critical Reynolds number is $Re_{cr} = 5 \times 10^5$. **3** Air is an ideal gas. **4** The local atmospheric pressure is 1 atm.

Properties The properties of the plastic sheet are given in the problem statement. The properties of air at the film temperature of $T_f = (T_s + T_\infty)/2 = (200 + 80)/2 = 140°F$ and 1 atm pressure are (Table A-18E)

$$k = 0.0162 \text{ Btu/h} \cdot \text{ft} \cdot °F \qquad Pr = 0.72$$
$$\nu = 0.204 \times 10^{-3} \text{ ft}^2/\text{s}$$

Analysis (a) We expect the temperature of the plastic sheet to drop somewhat as it flows through the 2-ft-long cooling section, but at this point we do not know the magnitude of that drop. Therefore, we assume the plastic sheet to be isothermal at 200°F to get started. We will repeat the calculations if necessary to account for the temperature drop of the plastic sheet.

Noting that $L = 4$ ft, the Reynolds number at the end of the air flow across the plastic sheet is

$$Re_L = \frac{V_\infty L}{\nu} = \frac{(10 \text{ ft/s})(4 \text{ ft})}{0.204 \times 10^{-3} \text{ ft}^2/\text{s}} = 1.96 \times 10^5$$

which is less than the critical Reynolds number. Thus, we have *laminar flow* over the entire sheet, and the Nusselt number is determined from the laminar flow relations for a flat plate to be

$$Nu = \frac{hL}{k} = 0.664 \, Re_L^{0.5} \, Pr^{1/3} = 0.664 \times (1.96 \times 10^5)^{0.5} \times 0.72^{1/3} = 263.5$$

Then,

$$h = \frac{k}{L} Nu = \frac{0.0162 \text{ Btu/h} \cdot \text{ft} \cdot °F}{4 \text{ ft}}(263.5) = 1.07 \text{ Btu/h} \cdot \text{ft}^2 \cdot °F$$
$$A = (2 \text{ ft})(4 \text{ ft})(2 \text{ sides}) = 16 \text{ ft}^2$$

and

$$\dot{Q}_{conv} = hA(T_s - T_\infty)$$
$$= (1.07 \text{ Btu/h} \cdot \text{ft}^2 \cdot °F)(16 \text{ ft}^2)(200 - 80)°F$$
$$= 2054 \text{ Btu/h}$$
$$\dot{Q}_{rad} = \varepsilon \sigma A(T_s^4 - T_{surr}^4)$$
$$= (0.9)(0.1714 \text{ Btu/h} \cdot \text{ft}^2 \cdot R^4)(16 \text{ ft}^2)[(660 \text{ R})^4 - (540 \text{ R})^4]$$
$$= 2584 \text{ Btu/h}$$

Therefore, the rate of cooling of the plastic sheet by combined convection and radiation is

$$\dot{Q}_{\text{total}} = \dot{Q}_{\text{conv}} + \dot{Q}_{\text{rad}} = 2054 + 2584 = \textbf{4638 Btu/h}$$

(b) To find the temperature of the plastic sheet at the end of the cooling section, we need to know the mass of the plastic rolling out per unit time (or the mass flow rate), which is determined from

$$\dot{m} = \rho A_c \mathscr{V} = (75 \text{ lbm/ft}^3)\left(\frac{4 \times 0.04}{12} \text{ ft}^3\right)\left(\frac{30}{60} \text{ ft /s}\right) = 0.5 \text{ lbm/s}$$

Then, an energy balance on the cooled section of the plastic sheet yields

$$\dot{Q} = \dot{m} C_p (T_2 - T_1) \quad \rightarrow \quad T_2 = T_1 + \frac{\dot{Q}}{\dot{m} C_p}$$

Noting that $\dot{Q}$ is a negative quantity (heat loss) for the plastic sheet and substituting, the temperature of the plastic sheet as it leaves the cooling section is determined to be

$$T_2 = 200°F + \frac{-4638 \text{ Btu/h}}{(0.5 \text{ lbm/s})(0.4 \text{ Btu/lbm} \cdot °F)}\left(\frac{1 \text{ h}}{3600 \text{ s}}\right) = \textbf{193.6°F}$$

Discussion The average temperature of the plastic sheet drops by about 6.4°F as it passes through the cooling section. The calculations now can be repeated by taking the average temperature of the plastic sheet to be 196.8°F instead of 200°F for better accuracy, but the change in the results will be insignificant because of the small change in temperature.

17-4 ■ FLOW ACROSS CYLINDERS AND SPHERES

Flows across cylinders and spheres, in general, involve *flow separation,* which is difficult to handle analytically. Therefore, such flows must be studied experimentally or numerically. Indeed, flow across cylinders and spheres has been studied experimentally by numerous investigators, and several empirical correlations are developed for the heat transfer coefficient.

The complicated flow pattern across a cylinder greatly influences heat transfer. The variation of the local Nusselt number Nu_θ around the periphery of a cylinder subjected to cross flow of air is given in Fig. 17-14. Note that, for all cases, the value of Nu_θ starts out relatively high at the stagnation point ($\theta = 0°$) but decreases with increasing θ as a result of the thickening of the laminar boundary layer. On the two curves at the bottom corresponding to Re = 70,800 and 101,300, Nu_θ reaches a minimum at $\theta \approx 80°$, which is the separation point in laminar flow. Then Nu_θ increases with increasing θ as a result of the intense mixing in the separated flow region (the wake). The curves at the top corresponding to Re = 140,000 to 219,000 differ from the first two curves in that they have *two* minima for Nu_θ. The sharp increase in Nu_θ at about $\theta \approx 90°$ is due to the transition from laminar to turbulent flow. The later decrease in Nu_θ is again due to the thickening of the boundary layer. Nu_θ reaches its second minimum at about $\theta \approx 140°$, which is the flow separation point in turbulent flow, and increases with θ as a result of the intense mixing in the turbulent wake region.

The discussions above on the local heat transfer coefficients are insightful; however, they are of little value in heat transfer calculations since the calculation of heat transfer requires the *average* heat transfer coefficient over the entire surface. Of the several such relations available in the literature for the average Nusselt number for cross flow over a cylinder, we present the one proposed by Churchill and Bernstein:

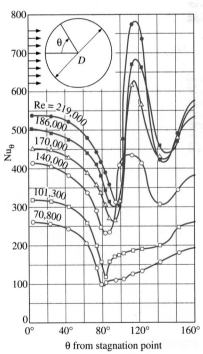

FIGURE 17-14

Variation of the local heat transfer coefficient along the circumference of a circular cylinder in cross flow of air (from Giedt, Ref. 5).

$$\text{Nu}_{\text{cyl}} = \frac{hD}{k} = 0.3 + \frac{0.62\,\text{Re}^{1/2}\,\text{Pr}^{1/3}}{[1 + (0.4/\text{Pr})^{2/3}]^{1/4}}\left[1 + \left(\frac{\text{Re}}{282{,}000}\right)^{5/8}\right]^{4/5} \quad (17\text{-}22)$$

This relation is quite comprehensive in that it correlates all available data well for Re Pr > 0.2. The fluid properties are evaluated at the *film temperature* $T_f = \frac{1}{2}(T_\infty + T_s)$, which is the average of the free-stream and surface temperatures.

For flow over a *sphere*, Whitaker recommends the following comprehensive correlation:

$$\text{Nu}_{\text{sph}} = \frac{hD}{k} = 2 + [0.4\,\text{Re}^{1/2} + 0.06\,\text{Re}^{2/3}]\,\text{Pr}^{0.4}\left(\frac{\mu_\infty}{\mu_s}\right)^{1/4} \quad (17\text{-}23)$$

which is valid for $3.5 \le \text{Re} \le 80{,}000$ and $0.7 \le \text{Pr} \le 380$. The fluid properties in this case are evaluated at the free-stream temperature T_∞, except for μ_s, which is evaluated at the surface temperature T_s. Although the two relations above are considered to be quite accurate, the results obtained from them can be off by as much as 30 percent.

The average Nusselt number for flow across cylinders can be expressed compactly as

$$\text{Nu}_{\text{cyl}} = \frac{hD}{k} = C\,\text{Re}^m\,\text{Pr}^n \quad (17\text{-}24)$$

where $n = \frac{1}{3}$ and the experimentally determined constants C and m are given in Table 17-2 for circular as well as various noncircular cylinders. The characteristic length D for use in the calculation of the Reynolds and the Nusselt numbers for different geometries is as indicated on the figure. All fluid properties are evaluated at the *film temperature* $T_f = \frac{1}{2}(T_\infty + T_s)$.

The relations for cylinders above are for *single* cylinders or cylinders oriented such that the flow over them is not affected by the presence of others. Also, they are applicable to *smooth* surfaces. *Surface roughness* and the *free-stream turbulence* may affect the drag and heat transfer coefficients significantly. Drag and heat transfer coefficients for flow over *tube bundles* can be obtained from some of the standard heat transfer texts listed at the end of this chapter. Eq. 17-24 provides a simpler alternative to Eq. 17-22 for flow over cylinders. However, Eq. 17-22 is more accurate, and thus should be preferred in calculations whenever possible.

EXAMPLE 17-4   Heat Loss from a Steam Pipe in Windy Air

A long 10-cm-diameter steam pipe whose external surface temperature is 110°C passes through some open area that is not protected against the winds (Fig. 17-15). Determine the rate of heat loss from the pipe per unit of its length when the air is at 1 atm pressure and 4°C and the wind is blowing across the pipe at a velocity of 8 m/s.

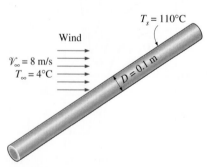

$T_s = 110°C$

Wind

$V_\infty = 8$ m/s
$T_\infty = 4°C$

$D = 0.1$ m

FIGURE 17-15
Schematic for Example 17-4.

Solution   A steam pipe is exposed to windy air. The rate of heat loss from the steam is to be determined.

Assumptions   **1** Steady operating conditions exist. **2** Radiation effects are negligible. **3** Air is an ideal gas.

Properties   The properties of air at the film temperature of $T_f = (T_s + T_\infty)/2 = (110 + 4)/2 = 57°C = 330$ K and 1 atm pressure are (Table A-18)

TABLE 17-2

Empirical correlations for the average Nusselt number for forced convection over circular and noncircular cylinders in cross flow (from Zhukauskas, Ref. 18, and Jakob, Ref. 8)

Cross-section of the cylinder	Fluid	Range of Re	Nusselt number
Circle	Gas or liquid	0.4–4 4–40 40–4000 4000–40,000 40,000–400,000	$Nu = 0.989 Re^{0.330} Pr^{1/3}$ $Nu = 0.911 Re^{0.385} Pr^{1/3}$ $Nu = 0.683 Re^{0.466} Pr^{1/3}$ $Nu = 0.193 Re^{0.618} Pr^{1/3}$ $Nu = 0.027 Re^{0.805} Pr^{1/3}$
Square	Gas	5000–100,000	$Nu = 0.102 Re^{0.675} Pr^{1/3}$
Square (tilted 45°)	Gas	5000–100,000	$Nu = 0.246 Re^{0.588} Pr^{1/3}$
Hexagon	Gas	5000–100,000	$Nu = 0.153 Re^{0.638} Pr^{1/3}$
Hexagon (tilted 45°)	Gas	5000–19,500 19,500–100,000	$Nu = 0.160 Re^{0.638} Pr^{1/3}$ $Nu = 0.0385 Re^{0.782} Pr^{1/3}$
Vertical plate	Gas	4000–15,000	$Nu = 0.228 Re^{0.731} Pr^{1/3}$
Ellipse	Gas	2500–15,000	$Nu = 0.248 Re^{0.612} Pr^{1/3}$

$$k = 0.0283 \text{ W/m} \cdot °\text{C} \qquad Pr = 0.708$$
$$\nu = 1.86 \times 10^{-5} \text{ m}^2/\text{s}$$

Analysis This is an *external flow* problem, since we are interested in the heat transfer from the pipe to the air that is flowing outside the pipe. The Reynolds number is

$$Re = \frac{\mathcal{V}_\infty D}{\nu} = \frac{(8 \text{ m/s})(0.1 \text{ m})}{1.86 \times 10^{-5} \text{ m}^2/\text{s}} = 43{,}011$$

Then the Nusselt number in this case can be determined from

$$Nu = \frac{hD}{k} = 0.3 + \frac{0.62\,Re^{1/2}\,Pr^{1/3}}{[1 + (0.4/Pr)^{2/3}]^{1/4}} \left[1 + \left(\frac{Re}{282,000}\right)^{5/8}\right]^{4/5}$$

$$= 0.3 + \frac{0.62(43,011)^{1/2}\,(0.708)^{1/3}}{[1 + (0.4/0.708)^{2/3}]^{1/4}} \left[1 + \left(\frac{43,001}{282,000}\right)^{5/8}\right]^{4/5}$$

$$= 125.1$$

and

$$h = \frac{k}{D}\,Nu = \frac{0.0283\ \text{W/m} \cdot \text{°C}}{0.1\ \text{m}}(125.1) = 35.4\ \text{W/m}^2 \cdot \text{°C}$$

Then the rate of heat transfer from the pipe per unit of its length becomes

$$A = pL = \pi DL = \pi(0.1\ \text{m})(1\ \text{m}) = 0.314\ \text{m}^2$$
$$\dot{Q} = hA(T_s - T_\infty) = (35.4\ \text{W/m}^2 \cdot \text{C})(0.314\ \text{m}^2)(110 - 4)\text{°C} = \textbf{1178 W}$$

The rate of heat loss from the entire pipe can be obtained by multiplying the value above by the length of the pipe in m.

Discussion The simpler Nusselt number relation in Table 17-2 in this case would give Nu = 129, which is 3 percent higher than the value obtained above using Eq. 17-22.

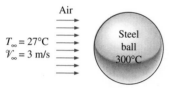

$T_\infty = 27\text{°C}$
$\mathcal{V}_\infty = 3$ m/s

Air

Steel ball 300°C

FIGURE 17-16

Schematic for Example 17-5.

EXAMPLE 17-5   Cooling of a Steel Ball by Forced Air

A 25-cm-diameter stainless steel ball (ρ = 8055 kg/m³, C_p = 480 J/kg · °C) is removed from the oven at a uniform temperature of 300°C (Fig. 17-16). The ball is then subjected to the flow of air at 1 atm pressure and 27°C with a velocity of 3 m/s. The surface temperature of the ball eventually drops to 200°C. Determine the average convection heat transfer coefficient during this cooling process and estimate how long this cooling process will take.

Solution   A hot stainless steel ball is cooled by forced air. The average convection heat transfer coefficient and the cooling time are to be determined.

Assumptions   **1** Steady operating conditions exist. **2** Radiation effects are negligible. **3** Air is an ideal gas. **4** The outer surface temperature of the ball is uniform at all times. **5** The surface temperature of the ball during cooling is changing. Therefore, the convection heat transfer coefficient between the ball and the air will also change. To avoid this complexity, we take the surface temperature of the ball to be constant at the average temperature of (300 + 200)/2 = 250°C in the evaluation of the heat transfer coefficient and use the value obtained for the entire cooling process.

Properties   The dynamic viscosity of air at the surface temperature is $\mu_s = \mu_{@\,250°C} = 2.76 \times 10^{-5}$ kg/m · s. The properties of air at the free-stream temperature of 27°C and 1 atm are (Table A-18)

$$k = 0.0261\ \text{W/m} \cdot \text{°C} \qquad \nu = 1.57 \times 10^{-5}\ \text{m}^2/\text{s}$$
$$\mu = 1.85 \times 10^{-5}\ \text{kg/m} \cdot \text{s} \qquad Pr = 0.712$$

Analysis   This is an *external flow* problem since the air flows outside the ball. The Reynolds number of the flow is determined from

$$Re = \frac{\mathcal{V}_\infty D}{\nu} = \frac{(3\ \text{m/s})(0.25\ \text{m})}{1.57 \times 10^{-5}\ \text{m}^2/\text{s}} = 47,800$$

The Nusselt number is

$$Nu = \frac{hD}{k} = 2 + [0.4\ Re^{1/2} + 0.06\ Re^{2/3}]\ Pr^{0.4} \left(\frac{\mu_\infty}{\mu_s}\right)^{1/4}$$

$$= 2 + [0.4(47{,}800)^{1/2} + 0.06(47{,}800)^{2/3}](0.712)^{0.4} \left(\frac{1.85 \times 10^{-5}}{2.76 \times 10^{-5}}\right)^{1/4}$$

$$= 139$$

Then the average convection heat transfer coefficient becomes

$$h = \frac{k}{D}\ Nu = \frac{0.0261\ W/m \cdot °C}{0.25\ m}(139) = \textbf{14.5 W/m}^2 \cdot \textbf{°C}$$

In order to estimate the time of cooling of the ball from 300°C to 200°C, we determine the *average* rate of heat transfer from Newton's law of cooling by using the *average* surface temperature. That is,

$$A = \pi D^2 = \pi(0.25\ m)^2 = 0.196\ m^2$$

$$\dot{Q}_{ave} = hA(T_{s,\,ave} - T_\infty) = (14.5\ W/m^2 \cdot °C)(0.196\ m^2)(250 - 27)°C = 634\ W$$

Next we determine the *total* heat transferred from the ball, which is simply the change in the energy of the ball as it cools from 300°C to 200°C:

$$m = \rho V = \rho \tfrac{1}{6}\pi D^3 = (8055\ kg/m^3)\ \tfrac{1}{6}\pi(0.25\ m)^3 = 65.9\ kg$$

$$Q_{total} = mC_p(T_2 - T_1) = (65.9\ kg)(480\ J/kg \cdot °C)(300 - 200)°C = 3{,}163{,}000\ J$$

In the above calculation, we assumed that the *entire ball* is at 200°C, which is not necessarily true. The inner region of the ball will probably be at a higher temperature than its surface. With this assumption, the time of cooling is determined to be

$$\Delta t \approx \frac{Q}{\dot{Q}_{ave}} = \frac{3{,}163{,}000\ J}{594\ J/s} = 5325\ s = \textbf{1 h 29 min}$$

Discussion   The time of cooling could also be determined more accurately using the transient temperature charts or relations introduced in Chap. 16. But the simplifying assumptions we made above can be justified if all we need is a ballpark value. It will be naive to expect the time of cooling to be exactly 1 h 29 min, but, using our engineering judgment, it is realistic to expect the time of cooling to be somewhere between one and two hours.

17-5 ■ GENERAL CONSIDERATIONS FOR FLOW IN PIPES

Liquid or gas flow through *pipes* or *ducts* is commonly used in practice in heating and cooling applications. The fluid in such applications is forced to flow by a fan or pump through a tube that is sufficiently long to accomplish the desired heat transfer.

When a fluid is heated or cooled as it flows through a tube, the temperature of a fluid at any cross-section changes from T_s at the surface of the wall at that cross-section to some maximum (or minimum in the case of heating) at the tube center. In fluid flow it is convenient to work with an *average* or *mean* temperature T_m that remains constant at a cross-section. The mean temperature T_m will change in the flow direction, however, whenever the fluid is heated or cooled.

The value of the mean temperature T_m is determined from the requirement that the *conservation of energy* principle be satisfied. That is, the energy transported by the fluid through a cross-section in actual flow will be equal to the energy that would be transported through the same cross-section if the fluid

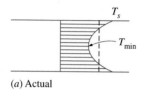

(a) Actual

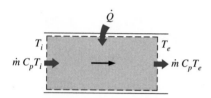

(b) Idealized

FIGURE 17-17

Actual and idealized temperature profiles for flow in a tube (the rate at which energy is transported with the fluid is the same for both cases).

Energy balance:
$$\dot{Q} = \dot{m} C_p (T_e - T_i)$$

FIGURE 17-18

The heat transfer to a fluid flowing in a tube is equal to the increase in the energy of the fluid.

were at a constant temperature T_m. This can be expressed mathematically as (Fig. 17-17)

$$\dot{E}_{\text{fluid}} = \dot{m} C_p T_m = \int_{\dot{m}} C_p T \, \delta\dot{m} = \int_{A_c} C_p T \rho \mathcal{V} \, dA_c \qquad \text{(kJ/s)} \qquad (17\text{-}25)$$

where C_p is the specific heat of the fluid and $\dot{m}$ is the mass flow rate. Note that the product $\dot{m} C_p T_m$ at any cross-section along the tube represents the *energy flow* with the fluid at that cross-section. You will recall that in the absence of any work interactions (such as electric resistance heating), the conservation of energy equation for the steady flow of a fluid in a tube can be expressed as (Fig. 17-18)

$$\dot{Q} = \dot{m} C_p (T_e - T_i) \qquad \text{(kJ/s)} \qquad (17\text{-}26)$$

where T_i and T_e are the mean fluid temperatures at the inlet and exit of the tube, respectively, and $\dot{Q}$ is the rate of heat transfer to or from the fluid. Note that the temperature of a fluid flowing in a tube remains constant in the absence of any energy interactions through the wall of the tube.

The thermal conditions at the surface of a tube can usually be approximated with reasonable accuracy to be *constant surface temperature* (T_s = constant) or *constant surface heat flux* ($\dot{q}_s$ = constant). For example, the constant surface temperature condition is realized when a phase change process such as boiling or condensation occurs at the outer surface of a tube. The constant surface heat flux condition is realized when the tube is subjected to radiation or electric resistance heating uniformly from all directions.

The convection heat flux at any location on the tube can be expressed as

$$\dot{q} = h(T_s - T_m) \qquad \text{(W/m}^2\text{)} \qquad (17\text{-}27)$$

where h is the *local* heat transfer coefficient and T_s and T_m are the surface and the mean fluid temperatures at that location. Note that the mean fluid temperature T_m of a fluid flowing in a tube must change during heating or cooling. Therefore, when h = constant, the surface temperature T_s must change when $\dot{q}_s$ = constant, and the surface heat flux $\dot{q}_s$ must change when T_s = constant. Thus we may have either T_s = constant or $\dot{q}_s$ = constant at the surface of a tube, but not both. Below we consider convection heat transfer for these two common cases.

Constant Surface Heat Flux ($\dot{q}_s$ = constant)

In the case of $\dot{q}_s$ = constant, the rate of heat transfer can also be expressed as

$$\dot{Q} = \dot{q}_s A = \dot{m} C_p (T_e - T_i) \qquad \text{(W)} \qquad (17\text{-}28)$$

Then the mean fluid temperature at the tube exit becomes

$$T_e = T_i + \frac{\dot{q}_s A}{\dot{m} C_p} \qquad (17\text{-}29)$$

Note that the mean fluid temperature increases *linearly* in the flow direction in the case of constant surface heat flux, since the surface area increases linearly in the flow direction (A is equal to the perimeter, which is constant, times the tube length).

The surface temperature in this case can be determined from $\dot{q} = h(T_s - T_m)$. Note that when h is constant, $T_s - T_m = $ constant, and thus the surface temperature will also increase *linearly* in the flow direction (Fig. 17-19). Of course, this is true when the variation of the specific heat C_p with T is disregarded and C_p is assumed to remain constant.

Constant Surface Temperature (T_s = constant)

From Newton's law of cooling, the rate of heat transfer to or from a fluid flowing in a tube can be expressed as

$$\dot{Q} = hA\,\Delta T_{ave} = hA(T_s - T_m)_{ave} \tag{17-30}$$

where h is the average convection heat transfer coefficient, A is the heat transfer surface area (it is equal to πDL for a circular pipe of length L), and ΔT_{ave} is some appropriate *average* temperature difference between the fluid and the surface. Below we discuss two suitable ways of expressing ΔT_{ave}.

In the constant surface temperature (T_s = constant) case, ΔT_{ave} can be expressed *approximately* by the **arithmetic mean temperature difference** ΔT_{am} as

$$\Delta T_{ave} \approx \Delta T_{am} = \frac{\Delta T_i + \Delta T_e}{2} = \frac{(T_s - T_i) + (T_s - T_e)}{2}$$
$$= T_s - \frac{T_i + T_e}{2} = T_s - T_b \tag{17-31}$$

where $T_b = \frac{1}{2}(T_i + T_e)$ is the *bulk mean fluid temperature,* which is the *arithmetic average* of the mean fluid temperatures at the inlet and the exit of the tube.

Note that the *arithmetic mean temperature difference* ΔT_{am} is simply the *average* of the *temperature differences* between the surface and the fluid at the inlet and the exit of the tube. Inherent in this definition is the assumption that the mean fluid temperature varies linearly along the tube, which is hardly ever the case when T_s = constant. This simple approximation often gives acceptable results, but not always. Therefore, we need a better way to evaluate ΔT_{ave}.

Consider the heating of a fluid in a tube of constant cross-section whose inner surface is maintained at a constant temperature of T_s. We know that the mean temperature of the fluid T_m will increase in the flow direction as a result of heat transfer. The energy balance on a differential control volume shown in Fig. 17-20 gives

$$\dot{m}C_p\,dT_m = h(T_s - T_m)\,dA \tag{17-32}$$

That is, the increase in the energy of the fluid (represented by an increase in its mean temperature by dT_m) is equal to the heat transferred to the fluid from the tube surface by convection. Noting that the differential surface area is $dA = p\,dx$ where p is the perimeter of the tube, and that $dT_m = -d(T_s - T_m)$, since T_s is constant, the relation above can be rearranged as

$$\frac{d(T_s - T_m)}{T_s - T_m} = -\frac{hp}{\dot{m}C_p}\,dx \tag{17-33}$$

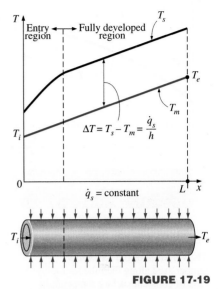

FIGURE 17-19

Variation of the *tube surface* and the *mean fluid* temperatures along the tube for the case of constant surface heat flux.

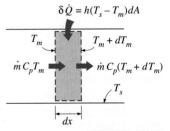

FIGURE 17-20

Energy interactions for a differential control volume in a tube.

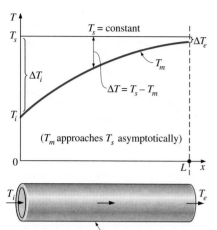

FIGURE 17-21

The variation of the *mean fluid* temperature along the tube for the case of constant surface temperature.

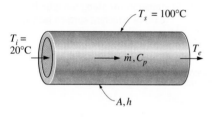

NTU $= hA / \dot{m}C_p$	T_e, °C
0.01	20.8
0.05	23.9
0.10	27.6
0.50	51.5
1.00	70.6
5.00	99.5
10.00	100.0

FIGURE 17-22

An NTU greater than 5 indicates that the fluid flowing in a tube will reach the surface temperature at the exit regardless of the inlet temperature.

Integrating from $x = 0$ (tube inlet where $T_m = T_i$) to $x = L$ (tube exit where $T_m = T_e$) gives

$$\ln \frac{T_s - T_e}{T_s - T_i} = -\frac{hA}{\dot{m}C_p} \qquad (17\text{-}34)$$

where $A = pL$ is the surface area of the tube and h is the constant *average* convection heat transfer coefficient. Taking the exponential of both sides and solving for T_e gives the following very useful relation for the determination of the *mean fluid temperature at the tube exit*:

$$T_e = T_s - (T_s - T_i)e^{-hA/\dot{m}C_p} \qquad (17\text{-}35)$$

This relation can also be used to determine the mean fluid temperature $T_m(x)$ at any x by replacing $A = pL$ by px.

Note that the temperature difference between the fluid and the surface *decays exponentially* in the flow direction, and the rate of decay depends on the magnitude of the exponent $hA/\dot{m}C_p$, as shown in Fig. 17-21. This dimensionless parameter is called the *number of transfer units,* denoted by NTU, and is a measure of the effectiveness of the heat transfer systems. For NTU > 5, the exit temperature of the fluid becomes almost equal to the surface temperature, $T_e \approx T_s$ (Fig. 17-22). Noting that the fluid temperature can approach the surface temperature but cannot cross it, an NTU of about 5 indicates that the limit is reached for heat transfer, and the heat transfer will not increase no matter how much we extend the length of the tube. A small value of NTU, on the other hand, indicates more opportunities for heat transfer, and the heat transfer will continue increasing as the tube length is increased. A large NTU and thus a large heat transfer surface area (which means a large tube) may be desirable from a heat transfer point of view, but it may be unacceptable from an economic point of view. The selection of heat transfer equipment usually reflects a compromise between heat transfer performance and cost.

Solving Eq. 17-34 for $\dot{m}C_p$ gives

$$\dot{m}C_p = \frac{hA}{\ln \dfrac{T_s - T_e}{T_s - T_i}} \qquad (17\text{-}36)$$

Substituting this into Eq. 17-28, we obtain

$$\dot{Q} = hA \, \Delta T_{\ln} \qquad (17\text{-}37)$$

where

$$\Delta T_{\ln} = \frac{T_e - T_i}{\ln \dfrac{T_s - T_e}{T_s - T_i}} = \frac{\Delta T_e - \Delta T_i}{\ln (\Delta T_e / \Delta T_i)} \qquad (17\text{-}38)$$

is the **logarithmic mean temperature difference.** Note that $\Delta T_i = T_s - T_i$ and $\Delta T_e = T_s - T_e$ are the temperature differences between the surface and the fluid at the inlet and the exit of the tube, respectively. The $\Delta T_{\ln}$ relation above appears to be prone to misuse, but it is practically failsafe, since using T_i in place of T_e and vice versa in the numerator and/or the denominator will, at most, affect the sign, not the magnitude. Also, it can be used for both heating ($T_s > T_i$ and T_e) and cooling ($T_s < T_i$ and T_e) of a fluid in a tube.

The logarithmic mean temperature difference ΔT_{ln} is obtained by tracing the actual temperature profile of the fluid along the tube, and is an *exact* representation of the *average temperature difference* between the fluid and the surface. It truly reflects the exponential decay of the local temperature difference. When ΔT_e differs from ΔT_i by no more than 40 percent, the error in using the arithmetic mean temperature difference is less than 1 percent. But the error increases to undesirable levels when ΔT_e differs from ΔT_i by greater amounts. Therefore, we should always use the logarithmic mean temperature difference when determining the convection heat transfer in a tube whose surface is maintained at a constant temperature T_s.

17-6 ■ THERMAL ENTRY LENGTH

The Reynolds number for flow inside pipes of diameter D was defined in Chap. 12 as

$$\mathrm{Re} = \frac{\mathscr{V}_m D}{\nu} \qquad (17\text{-}39)$$

where $\mathscr{V}_m$ is mean fluid velocity and $\nu = \mu/\rho$ is the kinematic viscosity of the fluid. Under most practical conditions, the flow in a pipe was said to be laminar for $\mathrm{Re} < 2300$, turbulent for $\mathrm{Re} > 4000$, and transitional in between.

Relations for hydrodynamic entry length were developed in Chap. 12 by considering a fluid entering a circular pipe at a uniform velocity. We now consider a fluid at a uniform temperature entering a circular tube that is at a different temperature. This time, the fluid particles in the layer in contact with the surface of the tube will assume the surface temperature. This will initiate convection heat transfer in the tube and the development of a **thermal boundary layer** along the tube. The thickness of this boundary layer also increases in the flow direction until the boundary layer reaches the tube center and thus fills the entire tube, as shown in Fig. 17-23. The region of flow over which the thermal boundary layer develops and reaches the tube center is called the **thermal entry region,** and the length of this region is called the **thermal entry length** L_t. The region beyond the thermal entry region in which the dimensionless temperature profile expressed as $(T - T_s)/(T_m - T_s)$ remains unchanged is called the **thermally developed region.** The region in which the flow is both hydrodynamically and thermally developed is called the **fully developed flow.**

Note that the *temperature profile* in the thermally developed region may *vary* with x in the flow direction. That is, unlike the velocity profile, the temperature profile can be different at different cross-sections of the tube in the developed region, and it usually is. However, it can be shown that the

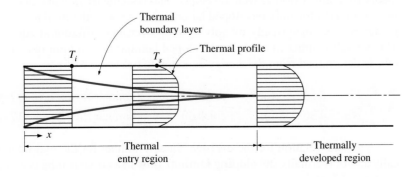

FIGURE 17-23

The development of the thermal boundary layer in a tube. (The fluid in the tube is being cooled.)

dimensionless temperature profile defined above remains unchanged in the thermally developed region when the temperature or heat flux at the tube surface remains constant.

In laminar flow in a tube, the magnitude of the dimensionless Prandtl number Pr is a measure of the relative growth of the velocity and thermal boundary layers. For fluids with Pr $\approx$ 1, such as gases, the two boundary layers essentially coincide with each other. For fluids with Pr $\gg$ 1, such as oils, the velocity boundary layer outgrows the thermal boundary layer. As a result, the hydrodynamic entry length is smaller than the thermal entry length. The opposite is true for fluids with Pr $\ll$ 1 such as liquid metals.

The hydrodynamic and thermal entry lengths in laminar and turbulent flows are given approximately as

$$L_{h,\,\text{laminar}} \approx 0.06 \,\text{Re}\, D \tag{17-40}$$

$$L_{t,\,\text{laminar}} \approx 0.06 \,\text{Re}\,\text{Pr}\, D = \text{Pr}\, L_{h,\,\text{laminar}} \tag{17-41}$$

$$L_{h,\,\text{turbulent}} \approx L_{t,\,\text{turbulent}} \approx 4.4D\,(\text{Re})^{1/6} \tag{17-42}$$

In *turbulent flow,* the intense mixing during random fluctuations usually overshadows the effects of momentum and heat diffusion, and therefore the hydrodynamic and thermal entry lengths are of about the same size. Also, *the friction factor and the heat transfer coefficient remain constant in fully developed laminar or turbulent flow* since the velocity and normalized temperature profiles do not vary in the flow direction.

Consider a fluid that is being heated (or cooled) in a pipe as it flows through it. The friction factor and the heat transfer coefficient are *highest* at the pipe inlet where the thickness of the boundary layers is zero, and decrease gradually to the fully developed values, as shown in Fig. 17-24. Therefore, the pressure drop and heat flux are *higher* in the entry regions of a tube, and the effect of the entry region is always to *enhance* the average friction and heat transfer coefficients for the entire tube. This enhancement can be significant for short tubes but negligible for long ones.

Precise correlations for the heat transfer coefficient for the entry regions are available in the literature. However, the tubes used in practice in forced convection are usually many times the length of either entry region, and thus the flow through the tubes is assumed to be fully developed for the entire length of the tube. This approach, which we will also use for simplicity, gives *reasonable* results for long tubes and *conservative* results for short ones.

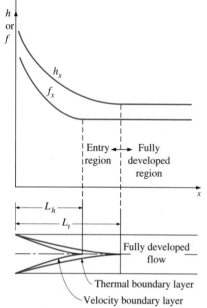

FIGURE 17-24

Variation of the friction factor and the convection heat transfer coefficient in the flow direction for flow in a tube (Pr > 1).

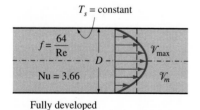

Fully developed
laminar flow

FIGURE 17-25

In laminar flow in a tube with constant surface temperature, both the *friction factor* and the *heat transfer coefficient* remain constant in the fully developed region.

17-7 ◾ FORCED CONVECTION IN PIPES

We mentioned earlier that flow in smooth tubes is laminar for Re < 2300. The theory for laminar flow is well developed, and both the friction and heat transfer coefficients for fully developed laminar flow in smooth circular tubes can be determined analytically by solving the governing differential equations. The Nusselt number in the fully developed laminar flow region in a circular tube is determined from the energy equation to be (Fig. 17-25)

Nu = 3.66	for T_s = constant	(laminar flow)	(17-43)
Nu = 4.36	for $\dot{q}_s$ = constant	(laminar flow)	(17-44)

A general relation for the *average* Nusselt number for the hydrodynamically and/or thermally **developing laminar flow** in a *circular* tube is given by Sieder and Tate as

$$\text{Nu} = 1.86\left(\frac{\text{Re Pr } D}{L}\right)^{1/3}\left(\frac{\mu_b}{\mu_s}\right)^{0.14} \qquad \text{Pr} > 0.5 \qquad (17\text{-}45)$$

All properties are evaluated at the bulk mean fluid temperature, except for μ_s, which is evaluated at the surface temperature.

The Nusselt number relations are given in Table 17-3 for *fully developed laminar flow* in tubes of various cross-sections. The Reynolds and Nusselt numbers for flow in these tubes are based on the **hydraulic diameter D_h** defined as

$$D_h = \frac{4A_c}{p} \qquad (17\text{-}46)$$

where A_c is the cross-sectional area of the tube and p is its perimeter. Once the Nusselt number is available, the convection heat transfer coefficient is determined from $h = k\,\text{Nu}/D_h$. It turns out that for a fixed surface area, the *circular*

TABLE 17-3

Nusselt number for fully developed laminar flow in tubes of various cross-sections ($D_h = 4A_c/p$, $\text{Re} = \mathcal{V}_m D_h/\nu$, and $\text{Nu} = hD_h/k$)

Cross-section of tube	a/b or $\theta°$	Nusselt number	
		T_s = const.	q_s = const.
Circle	—	3.66	4.36
Rectangle	a/b		
	1	2.98	3.61
	2	3.39	4.12
	3	3.96	4.79
	4	4.44	5.33
	6	5.14	6.05
	8	5.60	6.49
	∞	7.54	8.24
Ellipse	a/b		
	1	3.66	4.36
	2	3.74	4.56
	4	3.79	4.88
	8	3.72	5.09
	16	3.65	5.18
Triangle	θ		
	10°	1.61	2.45
	30°	2.26	2.91
	60°	2.47	3.11
	90°	2.34	2.98
	120°	2.00	2.68

tube gives the most heat transfer for the least pressure drop, which explains the overwhelming popularity of circular tubes in heat transfer equipment.

The effect of *surface roughness* on the friction factor and the heat transfer coefficient in laminar flow is negligible.

Turbulent Flow

We mentioned earlier that flow in smooth tubes is turbulent at Re > 4000. Turbulent flow is commonly utilized in practice because of the higher heat transfer coefficients associated with it. Most correlations for the friction and heat transfer coefficients in turbulent flow are based on experimental studies because of the difficulty in dealing with turbulent flow theoretically.

The friction factor for fully developed turbulent flow in tubes with *smooth* as well as *rough surfaces* over a wide range of Reynolds numbers can be determined from the implicit *Colebrook equation* given in Chap. 12 or the *Moody chart* (which is a plot of this equation) given in Fig. A-27. For *smooth* tubes, the friction factor in turbulent flow can be determined from the explicit **first Petukhov equation** given as

$$f = (0.790 \ln \text{Re} - 164)^{-2} \qquad \text{(smooth tubes)} \qquad (17\text{-}47)$$

The agreement between the Petukhov and Colebrook equations is very good, as shown in Table 17-4. Therefore, the Petukhov equation can be used with confidence.

The Nusselt number in turbulent flow is related to the friction factor through the famous **Chilton–Colburn analogy** expressed as

$$\text{Nu} = 0.125f \, \text{Re} \, \text{Pr}^{1/3} \qquad \text{(turbulent flow)} \qquad (17\text{-}48)$$

Once the friction factor is available, this equation can be used conveniently to evaluate the Nusselt number for both smooth and rough tubes. For fully developed turbulent flow in *smooth tubes,* a simple relation for the Nusselt number can be obtained by substituting the simplified friction factor relation $f = 0.184 \, \text{Re}^{-0.2}$ into Eq. 17-48. It gives

$$\text{Nu} = 0.023 \, \text{Re}^{0.8} \, \text{Pr}^{1/3} \qquad \begin{matrix} 0.7 \leq \text{Pr} \leq 160 \\ \text{Re} > 10{,}000 \end{matrix} \qquad (17\text{-}49)$$

which is known as the **Colburn equation.** The accuracy of this equation can be improved by modifying it as

$$\text{Nu} = 0.023 \, \text{Re}^{0.8} \, \text{Pr}^n \qquad (17\text{-}50)$$

where $n = 0.4$ for *heating* and 0.3 for *cooling* of the fluid flowing through the tube. This equation is known as the **Dittus–Boulter equation,** and it is preferred to the Colburn equation. The fluid properties are evaluated at the *bulk mean fluid temperature* $T_b = \frac{1}{2}(T_i + T_e)$, which is the arithmetic average of the mean fluid temperatures at the inlet and the exit of the tube. When the temperature difference between the fluid and the wall is very large, it may be necessary to use a correction factor to account for the different viscosities near the wall and at the tube center.

The Nusselt number relations above are fairly simple, but they may give errors as large as 25 percent. This error can be reduced considerably to less

TABLE 17-4

Comparison of the Petukhov and Colebrook equations for the determination of friction factor for turbulent flow in smooth tubes

Reynolds number	Friction factor, f	
	Petukhov	**Colebook**
5×10^3	0.0386	0.0374
1×10^4	0.0315	0.0309
5×10^4	0.0210	0.0209
1×10^5	0.0180	0.0180
5×10^5	0.0131	0.0132
1×10^6	0.0116	0.0117
5×10^6	0.0090	0.0090
1×10^7	0.0081	0.0081

than 10 percent by using more complex but accurate relations such as the **second Petukhov equation** expressed as

$$\text{Nu} = \frac{\text{Re Pr}(f/8)}{1.07 + 12.7(f/8)^{0.5}(\text{Pr}^{2/3} - 1)} \qquad \begin{array}{l} 0.5 \le \text{Pr} \le 2000 \\ 10^4 < \text{Re} < 5 \times 10^6 \end{array} \qquad (17\text{-}51)$$

The relations above are not very sensitive to the *thermal conditions* at the tube surfaces and can be used for both T_s = constant and $\dot{q}_s$ = constant cases. Despite their simplicity, the correlations above give sufficiently accurate results for most engineering purposes. They can also be used to obtain rough estimates of the friction factor and the heat transfer coefficients in the transition region $2300 \le \text{Re} \le 4000$, especially when the Reynolds number is closer to 4000 than it is to 2300.

Tubes with rough surfaces have much higher heat transfer coefficients than tubes with smooth surfaces. Therefore, tube surfaces are often intentionally *roughened, corrugated,* or *finned* in order to *enhance* the convection heat transfer coefficient and thus the convection heat transfer rate (Fig. 17-26). Heat transfer in turbulent flow in a tube has been increased by as much as 400 percent by roughening the surface. Roughening the surface, of course, also increases the friction factor and thus the power requirement for the pump or the fan.

The turbulent flow relations above can also be used for *noncircular tubes* with reasonable accuracy by replacing the diameter D in the evaluation of the Reynolds number by the hydraulic diameter $D_h = 4A_c/p$.

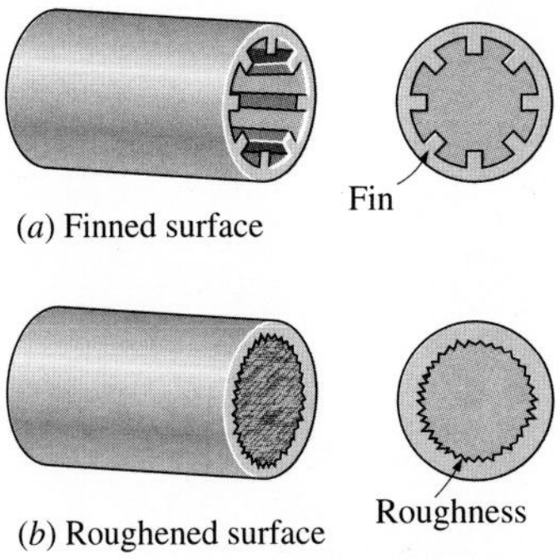

(a) Finned surface — Fin

(b) Roughened surface — Roughness

FIGURE 17-26

Tube surfaces are often *roughened, corrugated,* or *finned* in order to *enhance* convection heat transfer.

EXAMPLE 17-6 Heating of Water by Resistance Heaters in a Tube

Water is to be heated from 15°C to 65°C as it flows through a 3-cm-internal-diameter 5-m-long tube (Fig. 17-27). The tube is equipped with an electric resistance heater that provides uniform heating throughout the surface of the tube. The outer surface of the heater is well insulated, so that in steady operation all the heat generated in the heater is transferred to the water in the tube. If the system is to provide hot water at a rate of 10 L/min, determine the power rating of the resistance heater. Also, estimate the inner surface temperature of the pipe at the exit.

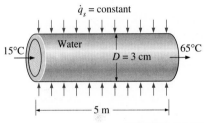

FIGURE 17-27

Schematic for Example 17-6.

Solution Water is to be heated in a tube equipped with an electric resistance heater on its surface. The power rating of the heater and the inner surface temperature are to be determined.

Assumptions **1** Steady flow conditions exist. **2** The surface heat flux is uniform. **3** The inner surfaces of the tube are smooth.

Properties The properties of water at the bulk mean temperature of $T_b = (T_i + T_e)/2 = (15 + 65)/2 = 40°C$ are (Table A-15).

$$\rho = 992.1 \text{ kg/m}^3 \qquad\qquad C_p = 4179 \text{ J/kg} \cdot °C$$
$$k = 0.631 \text{ W/m} \cdot °C \qquad\qquad \text{Pr} = 4.32$$
$$\nu = \mu/\rho = 0.658 \times 10^{-6} \text{ m}^2/\text{s}$$

Analysis This is an *internal flow* problem since the water is flowing in a pipe. The cross-sectional and heat transfer surface areas are

$$A_c = \tfrac{1}{4}\pi D^2 = \tfrac{1}{4}\pi(0.03 \text{ m})^2 = 7.069 \times 10^{-4} \text{ m}^2$$
$$A = pL = \pi DL = \pi(0.03 \text{ m})(5 \text{ m}) = 0.471 \text{ m}^2$$

The volume flow rate of water is given as $\dot{V} = 10 \text{ L/min} = 0.01 \text{ m}^3/\text{min}$. Then the mass flow rate of water becomes

$$\dot{m} = \rho\dot{V} = (992.1 \text{ kg/m}^3)(0.01 \text{ m}^3/\text{min}) = 9.921 \text{ kg/min} = 0.1654 \text{ kg/s}$$

To heat the water at this mass flow rate from 15°C to 65°C, heat must be supplied to the water at a rate of

$$\dot{Q} = \dot{m} C_p (T_e - T_i)$$
$$= (0.1654 \text{ kg/s})(4.179 \text{ kJ/kg} \cdot °\text{C})(65 - 15)°\text{C}$$
$$= 34.6 \text{ kJ/s} = 34.6 \text{ kW}$$

All of this energy must come from the resistance heater. Therefore, the power rating of the heater must be **34.6 kW**.

The surface temperature T_s of the tube at any location can be determined from

$$\dot{q}_s = h(T_s - T_m) \quad \rightarrow \quad T_s = T_m + \frac{\dot{q}_s}{h}$$

where h is the heat transfer coefficient and T_m is the mean temperature of the fluid at that location. The surface heat flux is constant in this case, and its value can be determined from

$$\dot{q}_s = \frac{\dot{Q}}{A} = \frac{34.6 \text{ kW}}{0.471 \text{ m}^2} = 73.46 \text{ kW/m}^2$$

To determine the heat transfer coefficient, we first need to find the mean velocity of water and the Reynolds number:

$$\mathcal{V}_m = \frac{\dot{V}}{A_c} = \frac{0.010 \text{ m}^3/\text{min}}{7.069 \times 10^{-4} \text{ m}^2} = 14.15 \text{ m/min} = 0.236 \text{ m/s}$$

$$\text{Re} = \frac{\mathcal{V}_m D}{\nu} = \frac{(0.236 \text{ m/s})(0.03 \text{ m})}{0.658 \times 10^{-6} \text{ m}^2/\text{s}} = 10,760$$

which is greater than 4000. Therefore, the flow is turbulent in this case and the entry lengths are roughly

$$L_h \approx L_t \approx 4.4 D \, \text{Re}^{1/6} = 4.4 \, (0.03 \text{ m})(10,760)^{1/6} = 0.62 \text{ m}$$

which is much shorter than the total length of the pipe. Therefore, we can assume fully developed turbulent flow in the entire pipe and determine the Nusselt number from

$$\text{Nu} = \frac{hD}{k} = 0.023 \, \text{Re}^{0.8} \, \text{Pr}^{0.4} = 0.023(10,760)^{0.8} \, (4.34)^{0.4} = 69.5$$

Then,

$$h = \frac{k}{D} \text{Nu} = \frac{0.631 \text{ W/m} \cdot °\text{C}}{0.03 \text{ m}} (69.5) = 1462 \text{ W/m}^2 \cdot °\text{C}$$

and the surface temperature of the pipe at the exit becomes

$$T_s = T_m + \frac{\dot{q}_s}{h} = 65°\text{C} + \frac{73,460 \text{ W/m}^2}{1462 \text{ W/m}^2 \cdot °\text{C}} = \mathbf{115°C}$$

Discussion Note that the inner surface temperature of the pipe will be 50°C higher than the mean water temperature at the pipe exit. This temperature difference of 50°C between the water and the surface will remain constant throughout the fully developed flow region.

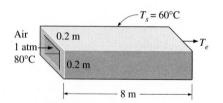

FIGURE 17-28
Schematic for Example 17-7.

EXAMPLE 17-7   Heat Loss from the Ducts of a Heating System in the Attic
Hot air at atmospheric pressure and 80°C enters an 8-m-long uninsulated square duct of cross-section 0.2 m × 0.2 m that passes through the attic of a house at a rate of 0.15 m³/s (Fig. 17-28). The duct is observed to be nearly isothermal at

60°C. Determine the exit temperature of the air and the rate of heat loss from the duct to the attic space.

Solution Heat loss from uninsulated square ducts of a heating system in the attic is considered. The exit temperature and the rate of heat loss are to be determined.

Assumptions **1** Steady operating conditions exist. **2** The inner surfaces of the duct are smooth. **3** Air is an ideal gas.

Properties We do not know the exit temperature of the air in the duct, and thus we cannot determine the bulk mean temperature of air, which is the temperature at which the properties are to be determined. The mean temperature of air at the inlet is 80°C or 353 K, and we expect this temperature to drop somewhat as a result of heat loss through the duct whose surface is at a lower temperature. Thus it is reasonable to assume a bulk mean temperature of 350 K for air (we will check this assumption later) for the purpose of evaluating the properties of air. At this temperature and 1 atm we read (Table A-18)

$$\rho = 1.009 \text{ kg/m}^3 \qquad C_p = 1008 \text{ J/kg} \cdot °C$$
$$k = 0.0297 \text{ W/m} \cdot °C \qquad Pr = 0.706$$
$$\nu = 2.06 \times 10^{-5} \text{ m}^2/\text{s}$$

Analysis This is an *internal flow* problem since the air is flowing in a duct. The characteristic length (which is the hydraulic diameter), the mean velocity, and the Reynolds number in this case are

$$D_h = \frac{4A_c}{p} = \frac{4a^2}{4a} = a = 0.2 \text{ m}$$

$$\mathcal{V}_m = \frac{\dot{V}}{A_c} = \frac{0.15 \text{ m}^3/\text{s}}{(0.2 \text{ m})^2} = 3.75 \text{ m/s}$$

$$\text{Re} = \frac{\mathcal{V}_m D_h}{\nu} = \frac{(3.75 \text{ m/s})(0.2 \text{ m})}{2.06 \times 10^{-5} \text{ m}^2/\text{s}} = 36{,}408$$

which is greater than 4000. Therefore, the flow is turbulent and the entry lengths in this case are roughly

$$L_h \approx L_t \approx 4.4D \, \text{Re}^{1/6} = 4.4 \, (0.2 \text{ m})(36{,}408)^{1/6} = 5.1 \text{ m}$$

which is much shorter than the total length of the duct. Therefore, we can assume fully developed turbulent flow in the entire duct and determine the Nusselt number from

$$\text{Nu} = \frac{hD_h}{k} = 0.023 \, \text{Re}^{0.8} \, \text{Pr}^{0.3} = 0.023(36{,}408)^{0.8}(0.706)^{0.3} = 92.3$$

Then,

$$h = \frac{k}{D_h} \text{Nu} = \frac{0.0297 \text{ W/m} \cdot °C}{0.2 \text{ m}} (92.3) = 13.7 \text{ W/m}^2 \cdot °C$$
$$A = pL = 4aL = 4 \times (0.2 \text{ m})(8 \text{ m}) = 6.4 \text{ m}^2$$
$$\dot{m} = \rho\dot{V} = (1.009 \text{ kg/m}^3)(0.15 \text{ m}^3/\text{s}) = 0.151 \text{ kg/s}$$

Next, we determine the exit temperature of air from

$$T_e = T_s - (T_s - T_i)e^{-hA/\dot{m}C_p}$$

$$= 60°C - [(60 - 80)°C] \exp\left[-\frac{(13.7 \text{ W/m}^2 \cdot °C)(6.4 \text{ m}^2)}{(0.151 \text{ kg/s})(1008 \text{ J/kg} \cdot °C)}\right]$$

$$= \mathbf{71.2°C}$$

Then the logarithmic mean temperature difference and the rate of heat loss from the air become

$$\Delta T_{\ln} = \frac{T_e - T_i}{\ln \dfrac{T_s - T_e}{T_s - T_i}} = \frac{71.2 - 80}{\ln \dfrac{60 - 71.2}{60 - 80}} = 15.2°C$$

$$\dot{Q} = hA\,\Delta T_{\ln} = (13.7\ \text{W/m}^2 \cdot °C)(6.4\ \text{m}^2)(15.2°C) = \textbf{1368 W}$$

Therefore, air will lose heat at a rate of 1368 W as it flows through the duct in the attic.

Discussion Having calculated the exit temperature of the air, we can now determine the actual bulk mean fluid temperature from

$$T_b = \frac{T_i + T_e}{2} = \frac{80 + 71.2}{2} = 75.6°C = 348.6\ \text{K}$$

which is sufficiently close to the assumed value of 350 K at which we evaluated the properties of air. Therefore, it is not necessary to re-evaluate the properties at this T_b and to repeat the calculations.

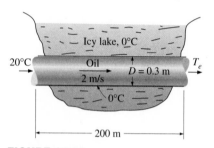

FIGURE 17-29

Schematic for Example 17-8.

EXAMPLE 17-8   Flow of Oil in a Pipeline through the Icy Waters of a Lake
Consider the flow of oil at 20°C in a 30-cm-diameter pipeline at an average velocity of 2 m/s (Fig. 17-29). A 200-m-long section of the pipeline passes through icy waters of a lake at 0°C. Measurements indicate that the surface temperature of the pipe is very nearly 0°C. Disregarding the thermal resistance of the pipe material, determine (*a*) the temperature of the oil when the pipe leaves the lake, (*b*) the rate of heat transfer from the oil, and (*c*) the pumping power required to overcome the pressure losses and to maintain the flow of the oil in the pipe.

Solution   Oil flows in a pipeline that passes through icy waters of a lake at 0°C. The exit temperature of the oil, the rate of heat loss, and the pumping power needed to overcome pressure losses are to be determined.

Assumptions   **1** Steady operating conditions exist. **2** The surface temperature of the pipe is very nearly 0°C. **3** The thermal resistance of the pipe is negligible. **4** The inner surfaces of the pipeline are smooth. **5** The flow is hydrodynamically developed when the pipeline reaches the lake.

Properties   We do not know the exit temperature of the oil, and thus we cannot determine the bulk mean temperature, which is the temperature at which the properties of oil are to be evaluated. The mean temperature of the oil at the inlet is 20°C, and we expect this temperature to drop somewhat as a result of heat loss to the icy waters of the lake. We evaluate the properties of the oil at the inlet temperature, but we will repeat the calculations, if necessary, using properties at the evaluated bulk mean temperature. At 20°C we read (Table A-16)

$$\rho = 888\ \text{kg/m}^3 \qquad \nu = 901 \times 10^{-6}\ \text{m}^2/\text{s}$$
$$k = 0.145\ \text{W/m} \cdot °C \qquad C_p = 1880\ \text{J/kg} \cdot °C$$
$$\mu = 0.800\ \text{kg/m} \cdot \text{s} \qquad \text{Pr} = 10{,}400$$

Analysis   (*a*) This is an *internal flow* problem since the oil is flowing in a pipe. The Reynolds number in this case is

$$\text{Re} = \frac{\mathcal{V}_m D_h}{\nu} = \frac{(2\ \text{m/s})(0.3\ \text{m})}{901 \times 10^{-6}\ \text{m}^2/\text{s}} = 666$$

which is less than the critical Reynolds number of 2300. Therefore, the flow is laminar, and the thermal entry length in this case is roughly

$$L_t \approx 0.05 \, \text{Re Pr} \, D = 0.05 \times 666 \times 10{,}400 \times (0.3 \, \text{m}) \approx 104{,}000 \, \text{m}$$

which is much greater than the total length of the pipe. This is typical of fluids with high Prandtl numbers. Therefore, we assume thermally developing flow and determine the Nusselt number from

$$\text{Nu} = \frac{hD}{k} = 1.86\left(\frac{\text{Re Pr } D}{L}\right)^{1/3}\left(\frac{\mu_b}{\mu_s}\right)^{0.14}$$

$$= 1.86\left(\frac{666 \times 10{,}400 \times 0.3 \, \text{m}}{200 \, \text{m}}\right)^{1/3}\left(\frac{0.8}{3.85}\right)^{0.14} = 32.6$$

where the dynamic viscosity μ_s is determined at the surface temperature of 0°C. Note that this Nusselt number is considerably higher than the fully developed value of 3.66. Then,

$$h = \frac{k}{D}\text{Nu} = \frac{0.145 \, \text{W/m} \cdot {}^\circ\text{C}}{0.3 \, \text{m}}(32.6) = 15.8 \, \text{W/m}^2 \cdot {}^\circ\text{C}$$

Also,

$$A = pL = \pi DL = \pi(0.3 \, \text{m})(200 \, \text{m}) = 188.5 \, \text{m}^2$$
$$\dot{m} = \rho A_c \mathcal{V}_m = (888 \, \text{kg/m}^3)[\tfrac{1}{4}\pi(0.3 \, \text{m})^2](2 \, \text{m/s}) = 125.5 \, \text{kg/s}$$

Next we determine the exit temperature of oil from

$$T_e = T_s - (T_s - T_i)e^{-hA/\dot{m}C_p}$$

$$= 0°C - [(0 - 20)°C] \exp\left[-\frac{(15.8 \, \text{W/m}^2 \cdot {}^\circ\text{C})(188.5 \, \text{m}^2)}{(125.5 \, \text{kg/s})(1880 \, \text{J/kg} \cdot {}^\circ\text{C})}\right]$$

$$= \mathbf{19.75°C}$$

Thus, the mean temperature of oil drops by a mere 0.25°C as it crosses the lake. This makes the bulk mean oil temperature 19.875°C, which is practically identical to the inlet mean temperature of 20°C. Therefore, we do not need to re-evaluate the properties at this bulk temperature and repeat the calculations.

(b) The logarithmic mean temperature difference and the rate of heat loss from the oil are

$$\Delta T_{\text{ln}} = \frac{T_e - T_i}{\ln\dfrac{T_s - T_e}{T_s - T_i}} = \frac{19.75 - 20}{\ln\dfrac{0 - 19.75}{0 - 20}} = 19.875°C$$

$$\dot{Q} = hA \, \Delta T_{\text{ln}} = (15.8 \, \text{W/m}^2 \cdot {}^\circ\text{C})(188.5 \, \text{m}^2)(19.875°C) = \mathbf{59{,}190 \, W}$$

Therefore, the oil will lose heat at a rate of 59,190 W as it flows through the pipe in the icy waters of the lake. Note that ΔT_{ln} is identical to the arithmetic mean temperature in this case, since $\Delta T_i \approx \Delta T_e$.

(c) The laminar flow of oil is hydrodynamically developed. Therefore, the friction factor can be determined from (see Chap. 12)

$$f = \frac{64}{\text{Re}} = \frac{64}{666} = 0.0961$$

Then the pressure drop in the pipe and the required pumping power become

$$\Delta P = f\frac{L}{D}\frac{\rho \mathcal{V}_m^2}{2} = 0.0961\frac{200 \, \text{m}}{0.3 \, \text{m}}\frac{(888 \, \text{kg/m}^3)(2 \, \text{m/s})^2}{2} = 113{,}780 \, \text{N/m}^2$$

$$\dot{W}_{\text{pump}} = \frac{\dot{m}\Delta P}{\rho} = \frac{(125.5 \, \text{kg/s})(113{,}780 \, \text{N/m}^2)}{888 \, \text{kg/m}^3} = \mathbf{16.1 \, kW}$$

Discussion We will need a 16.1-kW pump just to overcome the friction in the pipe as the oil flows in the 200-m-long pipe through the lake.

17-8 ■ SUMMARY

Convection is the mode of heat transfer that involves conduction as well as bulk fluid motion. The rate of convection heat transfer in external flow is expressed by *Newton's law of cooling* as

$$\dot{Q} = hA(T_s - T_\infty)$$

where T_s is the surface temperature and T_∞ is the free-stream temperature. The heat transfer coefficient h is usually expressed in the dimensionless form as the *Nusselt number* as $Nu = h\delta/k$ where δ is the *characteristic length*. The characteristic length for noncircular tubes is the *hydraulic diameter* D_h defined as $D_h = 4A_c/p$ where A_c is the cross-sectional area of the tube and p is its perimeter. The value of the critical Reynolds number is about 5×10^5 for flow over a flat plate, 2×10^5 for flow over cylinders and spheres, and 2300 for flow inside tubes.

For flow over a flat plate, the average Nusselt number is determined from

Laminar flow: $Nu = \dfrac{hL}{k} = 0.664 \, Re_L^{0.5} \, Pr^{1/3}$ $\qquad\qquad$ $Re_L < 5 \times 10^5$

Turbulent flow: $Nu = \dfrac{hL}{k} = 0.037 \, Re_L^{0.8} \, Pr^{1/3}$ $\qquad$ $\begin{array}{l} 0.6 \le Pr \le 60 \\ 5 \times 10^5 \le Re_x \le 10^7 \end{array}$

Combined: $Nu = \dfrac{hL}{k} = (0.037 \, Re_L^{0.8} - 871)Pr^{1/3}$ $\begin{array}{l} 0.6 \le Pr \le 60 \\ 5 \times 10^5 \le Re_x \le 10^7 \end{array}$

In order to properly account for the variation of the properties with temperature, the fluid properties are usually evaluated at the *film temperature,* defined as $T_f = (T_s + T_\infty)/2$, which is the *arithmetic average* of the surface and the free-stream temperatures.

The average Nusselt numbers for cross flow over a *cylinder* and *sphere* can be determined from

$$Nu_{cyl} = \frac{hD}{k} = 0.3 + \frac{0.62 \, Re^{1/2} \, Pr^{1/3}}{[1 + (0.4/Pr)^{2/3}]^{1/4}} \left[1 + \left(\frac{Re}{282{,}000}\right)^{5/8}\right]^{4/5}$$

which is valid for $Re \, Pr > 0.2$, and

$$Nu_{sph} = \frac{hD}{k} = 2 + [0.4 \, Re^{1/2} + 0.06 \, Re^{2/3}] \, Pr^{0.4} \left(\frac{\mu_\infty}{\mu_s}\right)^{1/4}$$

which is valid for $3.5 \le Re \le 80{,}000$ and $0.7 \le Pr \le 380$. The fluid properties are evaluated at the film temperature $T_f = \frac{1}{2}(T_\infty + T_s)$ in the case of a cylinder, and at the free-stream temperature T_∞ (except for μ_s, which is evaluated at the surface temperature T_s) in the case of a sphere.

The heat transfer to a fluid during steady flow in a tube can be expressed as

$$\dot{Q} = \dot{m}C_p(T_e - T_i)$$

where T_i and T_e are the mean fluid temperatures at the inlet and exit of the tube. The conditions at the surface of a tube can usually be approximated with reasonable accuracy to be *constant surface temperature* (T_s = constant) or *constant surface heat flux* ($\dot{q}_s$ = constant). In the case of $\dot{q}_s$ = constant, the rate of heat transfer can be expressed as

$$\dot{Q} = \dot{q}_s A = \dot{m}C_p(T_e - T_i)$$

Then mean fluid temperature at the tube exit becomes

$$T_e = T_i + \frac{\dot{q}_s A}{\dot{m}C_p}$$

In the case of T_s = constant, the rate of heat transfer is expressed as

$$\dot{Q} = hA\,\Delta T_{\ln} \quad \text{where} \quad \Delta T_{\ln} = \frac{T_e - T_i}{\ln\dfrac{T_s - T_e}{T_s - T_i}} = \frac{\Delta T_e - \Delta T_i}{\ln(\Delta T_e/\Delta T_i)}$$

is the *logarithmic mean temperature difference*. Note that $\Delta T_i = T_s - T_i$ and $\Delta T_e = T_s - T_e$ are the temperature differences between the surface and the fluid at the inlet and the exit of the tube, respectively. Then the mean fluid temperature at the tube exit in this case can be determined from

$$T_e = T_s - (T_s - T_i)e^{-hA/\dot{m}C_p}$$

The Nusselt number for *fully developed laminar flow* in a circular tube is given by

$$\text{Nu} = 3.66 \qquad \text{for } T_s = \text{constant} \qquad \text{(laminar flow)}$$
$$\text{Nu} = 4.36 \qquad \text{for } \dot{q}_s = \text{constant} \qquad \text{(laminar flow)}$$

For the hydrodynamically and/or thermally *developing laminar flow* in a *circular* tube the Nusselt number is given by Sieder and Tate as

$$\text{Nu} = 1.86\left(\frac{\text{Re Pr } D}{L}\right)^{1/3}\left(\frac{\mu_b}{\mu_s}\right)^{0.14} \qquad \text{Pr} > 0.5$$

The Nusselt number in *turbulent flow* is related to the friction factor by the *Chilton–Colburn analogy* expressed as

$$\text{Nu} = 0.125f\,\text{Re Pr}^{1/3} \qquad \text{(turbulent flow)}$$

For fully developed turbulent flow in a *smooth tube,* the Nusselt number is given by the *Dittus–Boulter equation* expressed as

$$\text{Nu} = 0.023\,\text{Re}^{0.8}\,\text{Pr}^n \qquad \text{(smooth tubes)}$$

where $n = 0.4$ for *heating* and 0.3 for *cooling* of the fluid flowing through the tube. A more complex but more accurate relation for turbulent flow is the **second Petukhov equation** expressed as

$$\text{Nu} = \frac{\text{Re Pr}(f/8)}{1.07 + 12.7(f/8)^{0.5}\,(\text{Pr}^{2/3} - 1)} \qquad \begin{array}{l} 0.5 \le \text{Pr} \le 2000 \\ 10^4 < \text{Re} < 5 \times 10^6 \end{array}$$

The relations above are not very sensitive to the *thermal conditions* at the tube surfaces and can be used for both T_s = constant and $\dot{q}_s$ = constant cases. The fluid properties are evaluated at the *bulk mean fluid temperature* $T_b = (T_i + T_e)/2$, which is the arithmetic average of the mean fluid temperatures at the inlet and the exit of the tube.

REFERENCES AND SUGGESTED READING

1. Y. Bayazitoglu and M. N. Özişik. *Elements of Heat Transfer.* New York: McGraw-Hill, 1988.

2. S. W. Churchill and M. Bernstein. "A Correlating Equation for Forced Convection from Gases and Liquids to a Circular Cylinder in Cross Flow." *Journal of Heat Transfer* 99 (1977), pp. 300–6.

3. A. P. Colburn. *Transactions of the AIChE* 26 (1933), p. 174.

4. F. W. Dittus and L. M. K. Boelter. *University of California Publications on Engineering* 2 (1930), p. 433.

5.   W. H. Giedt. "Investigation of Variation of Point Unit-Heat Transfer Co-efficient around a Cylinder Normal to an Air Stream." *Transactions of the ASME* 71 (1949), pp. 375–81.

6.   J. P. Holman. *Heat Transfer.* 8th ed. New York: McGraw-Hill, 1997.

7.   F. P. Incropera and D. P. DeWitt. *Introduction to Heat Transfer.* 3rd ed. New York: John Wiley & Sons, 1996.

8.   M. Jakob. *Heat Transfer.* Vol 1. NewYork: John Wiley & Sons, 1949.

9.   F. Kreith and M. S. Bohn. *Principles of Heat Transfer.* 5th ed. St. Paul, MN: West Publishing, 1993.

10.   L. F. Moody. "Friction Factor for Pipe Flow." *Transactions of the ASME* 66 (1944), pp. 671–84.

11.   B. S. Petukhov. "Heat Transfer and Friction in Turbulent Pipe Flow with Variable Physical Properties." In *Advances in Heat Transfer,* eds. J. P. Hart-nett and T. F. Irvine, Jr. Vol. 6. New York: Academic Press, 1970.

12.   O. Reynolds. "On the Experimental Investigation of the Circumstances Which Determine Whether the Motion of Water Shall Be Direct or Sinuous, and the Law of Resistance in Parallel Channels." *Philosophical Transactions of the Royal Society of London* 174 (1883), pp. 935–82.

13.   H. Schlichting. *Boundary Layer Theory.* 7th ed. New York: McGraw-Hill, 1979.

14.   E. N. Sieder and G. E. Tate. "Heat Transfer and Pressure Drop of Liquids in Tubes." *Industrial Engineering Chemistry* 28 (1936), pp. 1429–35.

15.   L. C. Thomas. *Heat Transfer.* Englewood Cliffs, NJ: Prentice Hall, 1992.

16.   S. Whitaker. "Forced Convection Heat Transfer Correlations for Flow in Pipe, Past Flat Plates, Single Cylinders, and for Flow in Packed Beds and Tube Bundles." *AIChE Journal* 18 (1972), pp. 361–71.

17.   F. M. White. *Heat and Mass Transfer.* Reading, MA: Addison-Wesley, 1988.

18.   A. Zhukauskas. "Heat Transfer from Tubes in Cross Flow." In *Advances in Heat Transfer,* eds. J. P. Hartnett and T. F. Irvine, Jr. Vol. 8. New York: Academic Press, 1972.

PROBLEMS*

Physical Mechanism of Forced Convection

17-1C   What is forced convection? How does it differ from natural convection? Is convection caused by winds forced or natural convection?

17-2C   What is external forced convection? How does it differ from internal forced convection? Can a heat transfer system involve both internal and external convection at the same time? Give an example.

17-3C   In which mode of heat transfer is the convection heat transfer coefficient usually higher, natural convection or forced convection? Why?

*Students are encouraged to answer *all* the concept "C" questions.

17-4C Consider a hot baked potato. Will the potato cool faster or slower when we blow the warm air coming from our lungs on it instead of letting it cool naturally in the cooler air in the room? Explain.

17-5C What is the physical significance of the Prandtl number? Does the value of the Prandtl number depend on the type of flow or the flow geometry? Does the Prandtl number of air change with pressure? Does it change with temperature?

17-6C What is the physical significance of the Nusselt number? How is it defined for (a) flow over a flat plate of length L, (b) flow over a cylinder of outer diameter D_o, (c) flow in a circular tube of inner diameter D_i, and (d) flow in a rectangular tube of cross-section $a \times b$?

17-7C When is heat transfer through a fluid conduction and when is it convection? For what case is the rate of heat transfer higher? How does the convection heat transfer coefficient differ from the thermal conductivity of a fluid?

17-8C How does turbulent flow differ from laminar flow? For which flow is the heat transfer coefficient higher?

17-9C Will a thermal boundary layer develop in flow over a surface even if both the fluid and the surface are at the same temperature?

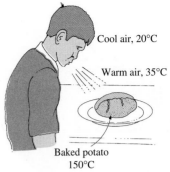

FIGURE P17-4C

Flow over Flat Plates

17-10C Consider laminar forced convection from a horizontal flat plate. Will the heat flux be higher at the leading edge or at the tail of the plate? Why?

17-11C For flow over a flat plate, how does the flow in the thermal boundary layer differ from the flow outside the thermal boundary layer?

17-12 Engine oil at 80°C flows over a 6-m-long flat plate whose temperature is 30°C with a velocity of 3 m/s. Determine the rate of heat transfer over the entire plate per unit width.

17-13 The local atmospheric pressure in Denver, Colorado (elevation 1610 m), is 83.4 kPa. Air at this pressure and at 30°C flows with a velocity of 6 m/s over a 2.5-m × 8-m flat plate whose temperature is 120°C. Determine the rate of heat transfer from the plate if the air flows parallel to the (a) 8-m-long side and (b) the 2.5-m side.

17-14 During a cold winter day, wind at 55 km/h is blowing parallel to a 4-m-high and 10-m-long wall of a house. If the air outside is at 5°C and the surface temperature of the wall is 12°C, determine the rate of heat loss from that wall by convection. What would your answer be if the wind velocity has doubled?   *Answers: 9184 W, 16,400 W*

17-15E Air at 65°F flows over a 10-ft-long flat plate at 7 ft/s. Determine the local heat transfer coefficients at intervals of 1 ft, and plot the results against the distance from the leading edge.

17-16 Consider a hot automotive engine, which can be approximated as a 0.5-m-high, 0.40-m-wide, and 0.8-m-long rectangular block. The bottom surface of the block is at a temperature of 80°C and has an emissivity of 0.95. The ambient air is at 30°C, and the road surface is at 25°C. Determine the

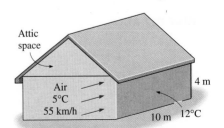

FIGURE P17-14

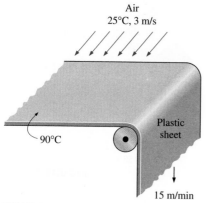

FIGURE P17-17

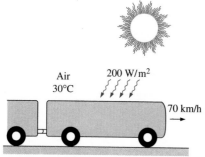

FIGURE P17-18

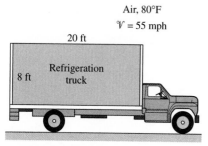

FIGURE P17-21E

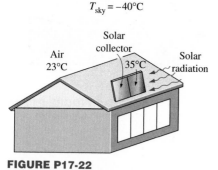

FIGURE P17-22

rate of heat transfer from the bottom surface of the engine block by convection and radiation as the car travels at a velocity of 80 km/h. Assume the flow to be turbulent over the entire surface because of the constant agitation of the engine block.

17-17 The forming section of a plastics plant puts out a continuous sheet of plastic that is 1.2 m wide and 2 mm thick at a rate of 15 m/min. The temperature of the plastic sheet is 90°C when it is exposed to the surrounding air, and the sheet is subjected to air flow at 25°C at a velocity of 3 m/s on both sides along its surfaces normal to the direction of motion of the sheet. The width of the air cooling section is such that a fixed point on the plastic sheet passes through that section in 2 s. Determine the rate of heat transfer from the plastic sheet to the air.

17-18 The top surface of the passenger car of a train moving at a velocity of 70 km/h is 2.8 m wide and 8 m long. The top surface is absorbing solar radiation at a rate of 200 W/m², and the temperature of the ambient air is 30°C. Assuming the roof of the car to be perfectly insulated and the radiation heat exchange with the surroundings to be small relative to convection, determine the equilibrium temperature of the top surface of the car.
Answer: 35°C

17-19 A 15-cm × 15-cm circuit board dissipating 15 W of power uniformly is cooled by air, which approaches the circuit board at 50°C with a velocity of 5 m/s. Disregarding any heat transfer from the back surface of the board, determine the surface temperature of the electronic components (*a*) at the leading edge and (*b*) at the end of the board. Assume the flow to be turbulent since the electronic components are expected to act as turbulators.

17-20 Consider laminar flow of a fluid over a flat plate maintained at a constant temperature. Now the free-stream velocity of the fluid is doubled. Determine the change in the rate of heat transfer between the fluid and the plate. Assume the flow to remain laminar.

17-21E Consider a refrigeration truck traveling at 55 mph at a location where the air temperature is 80°F. The refrigerated compartment of the truck can be considered to be a 9-ft-wide, 8-ft-high, and 20-ft-long rectangular box. The refrigeration system of the truck can provide 3 tons of refrigeration (i.e., it can remove heat at a rate of 600 Btu/min). The outer surface of the truck is coated with a low-emissivity material, and thus radiation heat transfer is very small. Determine the average temperature of the outer surface of the refrigeration compartment of the truck if the refrigeration system is observed to be operating at half the capacity. Assume the air flow over the entire outer surface to be turbulent and the heat transfer coefficient at the front and rear surfaces to be equal to that on side surfaces.

17-22 Solar radiation is incident on the glass cover of a solar collector at a rate of 700 W/m². The glass transmits 88 percent of the incident radiation and has an emissivity of 0.90. The entire hot water needs of a family in summer can be met by two collectors 1.2 m high and 1 m wide. The two collectors are attached to each other on one side so that they appear like a single collector 1.2 m × 2 m in size. The temperature of the glass cover is measured to be 35°C on a day when the surrounding air temperature is 23°C and the wind is blowing at 30 km/h. The effective sky temperature for radiation exchange between the glass cover and the open sky is −40°C. Water enters the tubes attached to the absorber plate at a rate of 1 kg/min. Assuming the back surface of the absorber plate to be heavily insulated and the only heat loss to occur through the glass cover, determine (a) the total rate of heat loss from the collector, (b) the collector efficiency, which is the ratio of the amount of heat transferred to the water to the solar energy incident on the collector, and (c) the temperature rise of water as it flows through the collector.

Answers: (a) 1262 W, (b) 0.15, (c) 3.1°C

17-23 A transformer that is 10 cm long, 6.2 cm wide, and 5 cm high is to be cooled by attaching a 10 cm × 6.2 cm wide polished aluminum heat sink (emissivity = 0.03) to its top surface. The heat sink has seven fins, which are 5 mm high, 2 mm thick, and 10 cm long. A fan blows air at 25°C parallel to the passages between the fins. The heat sink is to dissipate 20 W of heat and the base temperature of the heat sink is not to exceed 60°C. Assuming the fins and the base plate to be nearly isothermal and the radiation heat transfer to be negligible, determine the minimum free-stream velocity the fan needs to supply to avoid overheating.

17-24 Repeat Prob. 17-23 assuming the heat sink to be black-anodized and thus to have an effective emissivity of 0.90. Note that in radiation calculations the base area (10 cm × 6.2 cm) is to be used, not the total surface area.

17-25 An array of power transistors, dissipating 3 W of power each, are to be cooled by mounting them on a 25-cm × 25-cm square aluminum plate and blowing air at 35°C over the plate with a fan at a velocity of 4 m/s. The average temperature of the plate is not to exceed 65°C. Assuming the heat transfer from the back side of the plate to be negligible and disregarding radiation, determine the number of transistors that can be placed on this plate.

17-26 Repeat Prob. 17-25 for a location at an elevation of 1610 m where the atmospheric pressure is 83.4 kPa. *Answer:* 8

Flow across Cylinders and Spheres

17-27C Consider laminar flow of air across a hot circular cylinder. At what point on the cylinder will the heat transfer be highest? What would your answer be if the flow were turbulent?

17-28 A long 8-cm-diameter steam pipe whose external surface temperature is 90°C passes through some open area that is not protected against the winds. Determine the rate of heat loss from the pipe per unit of its length when the air is at 1 atm pressure and 7°C and the wind is blowing across the pipe at a velocity of 50 km/h.

17-29 A stainless steel ball (ρ = 8055 kg/m^3, C_p = 480 J/kg · °C) of diameter D = 15 cm is removed from the oven at a uniform temperature of 350°C. The ball is then subjected to the flow of air at 1 atm pressure and 30°C with a velocity of 6 m/s. The surface temperature of the ball eventually drops to 250°C. Determine the average convection heat transfer coefficient during this cooling process and estimate how long this process has taken.

17-30E A person extends his uncovered arms into the windy air outside at 40°F and 20 mph in order to feel nature closely. Initially, the skin temperature of the arm is 86°F. Treating the arm as a 2-ft-long and 3-in.-diameter cylinder, determine the rate of heat loss from the arm.

17-31 An average person generates heat at a rate of 84 W while resting. Assuming one-quarter of this heat is lost from the head and disregarding

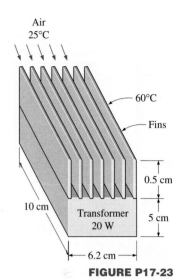

FIGURE P17-23

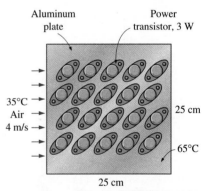

FIGURE P17-25

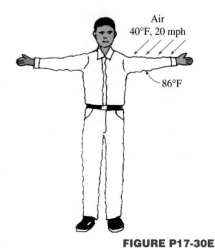

FIGURE P17-30E

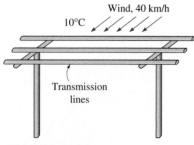

FIGURE P17-33

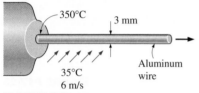

FIGURE P17-35

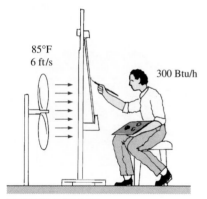

FIGURE P17-36E

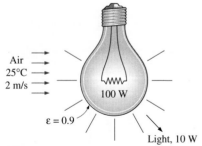

FIGURE P17-37

radiation, determine the average surface temperature of the head when it is not covered and is subjected to winds at 10°C and 35 km/h. The head can be approximated as a 30-cm-diameter sphere.   *Answer:* 12.7°C

17-32   Consider the flow of a fluid across a cylinder maintained at a constant temperature. Now the free-stream velocity of the fluid is doubled. Determine the change in the rate of heat transfer between the fluid and the cylinder.

17-33   A 6-mm-diameter electrical transmission line carries an electric current of 50 A and has a resistance of 0.002 ohm per meter length. Determine the surface temperature of the wire during a windy day when the air temperature is 10°C and the wind is blowing across the transmission line at 40 km/h.

17-34   A heating system is to be designed to keep the wings of an aircraft cruising at a velocity of 900 km/h above freezing temperatures during flight at 12,200-m altitude where the standard atmospheric conditions are −55.4°C and 18.8 kPa. Approximating the wing as a cylinder of elliptical cross-section whose minor axis is 30 cm and disregarding radiation, determine the average convection heat transfer coefficient on the wing surface and the average rate of heat transfer per unit surface area.

17-35   A long aluminum wire of diameter 3 mm is extruded at a temperature of 350°C. The wire is subjected to cross air flow at 35°C at a velocity of 6 m/s. Determine the rate of heat transfer from the wire to the air per meter length when it is first exposed to the air.

17-36E   Consider a person who is trying to keep cool on a hot summer day by turning a fan on and exposing his entire body to air flow. The air temperature is 85°F and the fan is blowing air at a velocity of 6 ft/s. If the person is doing light work and generating sensible heat at a rate of 300 Btu/h, determine the average temperature of the outer surface (skin or clothing) of the person. The average human body can be treated as a 1-ft-diameter cylinder with an exposed surface area of 18 ft². Disregard any heat transfer by radiation. What would your answer be if the air velocity were doubled?
        Answers: 95.0°F, 91.5°F

17-37   An incandescent light bulb is an inexpensive but highly inefficient device that converts electrical energy into light. It converts about 10 percent of the electrical energy it consumes into light while converting the remaining 90 percent into heat. (A fluorescent light bulb will give the same amount of light while consuming only one-fourth of the electrical energy, and it will last 10 times longer than an incandescent light bulb.) The glass bulb of the lamp heats up very quickly as a result of absorbing all that heat and dissipating it to the surroundings by convection and radiation.

Consider a 10-cm-diameter 100-W light bulb cooled by a fan that blows air at 25°C to the bulb at a velocity of 2 m/s. The surrounding surfaces are also at 25°C, and the emissivity of the glass is 0.9. Assuming 10 percent of the energy passes through the glass bulb as light with negligible absorption and the rest of the energy is absorbed and dissipated by the bulb itself, determine the equilibrium temperature of the glass bulb.

17-38   During a plant visit, it was noticed that a 12-m-long section of a 10-cm-diameter steam pipe is completely exposed to the ambient air. The temperature measurements indicate that the average temperature of the outer surface of the steam pipe is 82°C when the ambient temperature is 5°C. There are also light winds in the area at 10 km/h. The emissivity of the outer surface of the pipe is 0.8, and the average temperature of the surfaces surrounding the pipe, including the sky, is estimated to be 0°C. Determine the amount of heat lost from the steam during a 10-h-long work day.

Steam is supplied by a gas-fired steam generator that has an efficiency of 80 percent, and the plant pays $0.54/therm of natural gas (1 therm = 105,500 kJ). If the pipe is insulated and 90 percent of the heat loss is saved, determine the amount of money this facility will save a year as a result of insulating the steam pipes. Assume the plant operates every day of the year for 10 h. State your assumptions.

17-39 Reconsider Prob. 17-38. There seems to be some uncertainty about the average temperature of the surfaces surrounding the pipe used in radiation calculations, and you are asked to determine if it makes any significant difference in overall heat transfer. Repeat the calculations in Prob. 17-38 for average surrounding and surface temperatures of −20°C and 25°C, respectively, and determine the change in the values obtained.

17-40E A 12-ft-long, 1.5-kW electrical resistance wire is made of 0.1-in.-diameter stainless steel ($k = 8.7$ Btu/h · ft · °F). The resistance wire operates in an environment at 85°F. Determine the surface temperature of the wire if it is cooled by a fan blowing air at a velocity of 20 ft/s.

17-41 The components of an electronic system are located in a 1.5-m-long horizontal duct whose cross-section is 20 cm × 20 cm. The components in the duct are not allowed to come into direct contact with cooling air, and thus are cooled by air at 30°C flowing over the duct with a velocity of 200 m/min. If the surface temperature of the duct is not to exceed 65°C, determine the total power rating of the electronic devices that can be mounted into the duct.
Answer: 643 W

17-42 Repeat Prob. 17-41 for a location at 4000-m altitude where the atmospheric pressure is 61.66 kPa.

17-43 A 0.4-W cylindrical electronic component with diameter 0.3 cm and length 1.8 cm and mounted on a circuit board is cooled by air flowing across it at a velocity of 150 m/min. If the air temperature is 45°C, determine the surface temperature of the component.

Flow in Tubes

17-44C In the fully developed region of flow in a circular tube, will the velocity profile change in the flow direction? How about the temperature profile?

17-45C Consider the flow of oil in a tube. How will the hydrodynamic and thermal entry lengths compare if the flow is laminar? How would they compare if the flow were turbulent?

17-46C Consider the flow of mercury (a liquid metal) in a tube. How will the hydrodynamic and thermal entry lengths compare if the flow is laminar? How would they compare if the flow were turbulent?

17-47C What do the mean velocity $\mathcal{V}_m$ and the mean temperature T_m represent in flow through circular tubes of constant diameter?

17-48C Consider fluid flow in a tube whose surface temperature remains constant. What is the appropriate temperature difference for use in Newton's law of cooling with an average heat transfer coefficient?

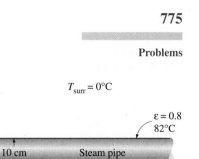
FIGURE P17-38

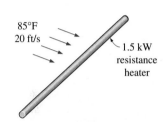
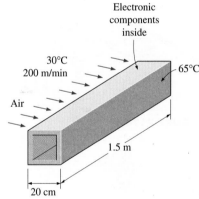
FIGURE P17-40E

FIGURE P17-41

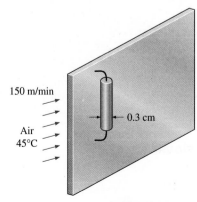
FIGURE P17-43

17-49C What is the physical significance of the number of transfer units NTU = $hA/\dot{m}C_p$? What do a small and a large NTU tell the heat transfer engineer about a heat transfer system?

17-50C What does the logarithmic mean temperature difference represent for flow in a tube whose surface temperature is constant? Why do we use the logarithmic mean temperature instead of the arithmetic mean temperature?

17-51C How is the thermal entry length defined for flow in a tube? In what region is the flow in a tube fully developed?

17-52C Consider laminar forced convection in a circular tube. Will the heat flux be higher near the inlet of the tube or near the exit? Why?

17-53C Consider turbulent forced convection in a circular tube. Will the heat flux be higher near the inlet of the tube or near the exit? Why?

17-54C How does surface roughness affect the heat transfer in a tube if the fluid flow is turbulent? What would your response be if the flow in the tube were laminar?

17-55 Water is to be heated from 12°C to 70°C as it flows through a 2-cm-internal-diameter, 7-m-long tube. The tube is equipped with an electric resistance heater, which provides uniform heating throughout the surface of the tube. The outer surface of the heater is well insulated, so that in steady operation all the heat generated in the heater is transferred to the water in the tube. If the system is to provide hot water at a rate of 8 L/min, determine the power rating of the resistance heater. Also, estimate the inner surface temperature of the pipe at the exit.

17-56 Hot air at atmospheric pressure and 85°C enters a 10-m-long uninsulated square duct of cross-section 0.15 m × 0.15 m that passes through the attic of a house at a rate of 0.10 m³/s. The duct is observed to be nearly isothermal at 70°C. Determine the exit temperature of the air and the rate of heat loss from the duct to the air space in the attic.
  Answers: 75.6°C, 946 W

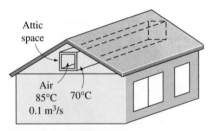

Attic
space

Air
85°C   70°C
0.1 m³/s

FIGURE P17-56

17-57 Consider an air solar collector that is 1 m wide and 5 m long and has a constant spacing of 3 cm between the glass cover and the collector plate. Air enters the collector at 30°C at a rate of 0.15 m³/s through the 1-m-wide edge and flows along the 5-m-long passage way. If the average temperatures of the glass cover and the collector plate are 20°C and 60°C, respectively, determine (*a*) the net rate of heat transfer to the air in the collector and (*b*) the temperature rise of air as it flows through the collector.

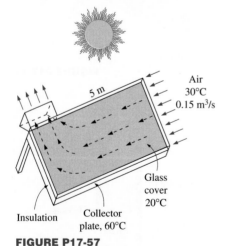

5 m

Air
30°C
0.15 m³/s

Glass
cover
20°C

Insulation   Collector
plate, 60°C

FIGURE P17-57

17-58 Consider the flow of oil at 10°C in a 40-cm-diameter pipeline at an average velocity of 0.5 m/s. A 300-m-long section of the pipeline passes through icy waters of a lake at 0°C. Measurements indicate that the surface temperature of the pipe is very nearly 0°C. Disregarding the thermal resistance of the pipe material, determine (*a*) the temperature of the oil when the pipe leaves the lake and (*b*) the rate of heat transfer from the oil.

17-59 Consider laminar flow of a fluid through a square channel maintained at a constant temperature. Now the mean velocity of the fluid is doubled. Determine the change in the rate of heat transfer between the fluid and the walls of the channel. Assume the flow regime remains unchanged.
  Answer: 1.26

17-60 Repeat Prob. 17-59 for turbulent flow.

17-61E The hot water needs of a household are to be met by heating water at 55°F to 200°F by a parabolic solar collector at a rate of 4 lbm/s. Water flows through a 1.25-in.-diameter thin aluminum tube whose outer surface is black-anodized in order to maximize its solar absorption ability. The centerline of the tube coincides with the focal line of the collector, and a glass sleeve is placed outside the tube to minimize the heat losses. If solar energy is transferred to water at a net rate of 350 Btu/h per ft length of the tube, determine the required length of the parabolic collector to meet the hot water requirements of this house. Also, determine the surface temperature of the tube at the exit.

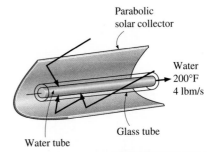

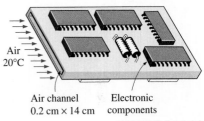

FIGURE P17-61E

17-62 A 15-cm × 20-cm printed circuit board whose components are not allowed to come into direct contact with air for reliability reasons is to be cooled by passing cool air through a 20-cm-long channel of rectangular cross-section 0.2 cm × 14 cm drilled into the board. The heat generated by the electronic components is conducted across the thin layer of the board to the channel, where it is removed by air that enters the channel at 20°C. The heat flux at the top surface of the channel can be considered to be uniform, and heat transfer through other surfaces is negligible. If the velocity of the air at the inlet of the channel is not to exceed 4 m/s and the surface temperature of the channel is to remain under 50°C, determine the maximum total power of the electronic components that can safely be mounted on this circuit board.

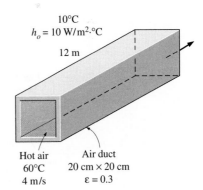

FIGURE P17-62

17-63 Repeat Prob. 17-62 by replacing air with helium, which has six times the thermal conductivity of air.

17-64 Air enters a 7-m-long section of a rectangular duct of cross-section 15 cm × 20 cm at 50°C at an average velocity of 7 m/s. If the walls of the duct are maintained at 10°C, determine (a) the outlet temperature of the air and (b) the rate of heat transfer from the air. *Answers:* (a) 32.8°C, (b) 3674 W

17-65 Hot air at 60°C leaving the furnace of a house enters a 12-m-long section of a sheet metal duct of rectangular cross-section 20 cm × 20 cm at an average velocity of 4 m/s. The thermal resistance of the duct is negligible, and the outer surface of the duct, whose emissivity is 0.3, is exposed to the cold air at 10°C in the basement, with a convection heat transfer coefficient of 10 W/m² · °C. Taking the walls of the basement to be at 10°C also, determine (a) the temperature at which the hot air will leave the basement and (b) the rate of heat loss from the hot air in the duct to the basement.

17-66 The components of an electronic system dissipating 90 W are located in a 1-m-long horizontal duct whose cross-section is 16 cm × 16 cm. The components in the duct are cooled by forced air, which enters at 32°C at a rate of 0.65 m³/min. Assuming 85 percent of the heat generated inside is transferred to air flowing through the duct and the remaining 15 percent is lost through the outer surfaces of the duct, determine (a) the exit temperature of air and (b) the highest component surface temperature in the duct.

17-67 Repeat Prob. 17-66 for a circular horizontal duct of 15-cm diameter.

17-68 Consider a hollow-core printed circuit board 12 cm high and 18 cm long, dissipating a total of 20 W. The width of the air gap in the middle of the PCB is 0.25 cm. The cooling air enters the 12-cm-wide core at 32°C at a rate

FIGURE P17-65

of 0.8 L/s. Assuming the heat generated to be uniformly distributed over the two side surfaces of the PCB, determine (*a*) the temperature at which the air leaves the hollow core and (*b*) the highest temperature on the inner surface of the core.   *Answers:* (*a*) 53.7°C, (*b*) 64.0°C

17-69   Repeat Prob. 17-68 for a hollow-core PCB dissipating 35 W.

17-70E   Water at 54°F is heated by passing it through 0.75-in.-internal-diameter thin-walled copper tubes. Heat is supplied to the water by steam that condenses outside the copper tubes at 250°F. If water is to be heated to 140°F at a rate of 0.7 lbm/s, determine the length of the copper tube that needs to be used. Assume the entire copper tube to be at the steam temperature of 250°F.

17-71   A computer cooled by a fan contains eight PCBs, each dissipating 10 W of power. The height of the PCBs is 12 cm and the length is 18 cm. The clearance between the tips of the components on the PCB and the back surface of the adjacent PCB is 0.3 cm. The cooling air is supplied by a 25-W fan mounted at the inlet. If the temperature rise of air as it flows through the case of the computer is not to exceed 10°C, determine (*a*) the flow rate of the air that the fan needs to deliver, (*b*) the fraction of the temperature rise of air that is due to the heat generated by the fan and its motor, and (*c*) the highest allowable inlet air temperature if the surface temperature of the components is not to exceed 70°C anywhere in the system. Use air properties at 300 K.

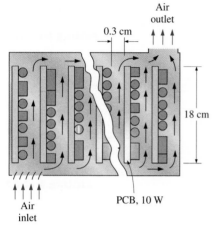

FIGURE P17-71

Review Problems

17-72   Consider a house that is maintained at 22°C at all times. The walls of the house have R-3.38 insulation in SI units (i.e., an L/k value or a thermal resistance of 3.38 m² · °C/W). During a cold winter night, the outside air temperature is 4°C and wind at 50 km/h is blowing parallel to a 3-m-high and 8-m-long wall of the house. If the heat transfer coefficient on the interior surface of the wall is 8 W/m² · °C, determine the rate of heat loss from that wall of the house. Draw the thermal resistance network and disregard radiation heat transfer.   *Answer:* 122 W

17-73   An automotive engine can be approximated as a 0.4-m-high, 0.60-m-wide, and 0.7-m-long rectangular block. The bottom surface of the block is at a temperature of 75°C and has an emissivity of 0.92. The ambient air is at 20°C, and the road surface is at 10°C. Determine the rate of heat transfer from the bottom surface of the engine block by convection and radiation as the car travels at a velocity of 60 km/h. Assume the flow to be turbulent over the entire surface because of the constant agitation of the engine block. How will the heat transfer be affected when a 2-mm-thick gunk ($k = 3$ W/m · °C) has formed at the bottom surface as a result of the dirt and oil collected at that surface over time? Assume the metal temperature under the gunk still to be 75°C.

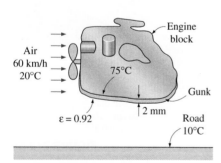

FIGURE P17-73

17-74E   The passenger compartment of a minivan traveling at 60 mph can be modeled as 3.2-ft-high, 6-ft-wide, and 11-ft-long rectangular box whose walls have an insulating value of R-3 (i.e., a wall thickness–to–thermal conductivity ratio of 3 h · ft² · °F/Btu). The interior of a minivan is maintained at an average temperature of 70°F during a trip at night while the outside air temperature is 90°F. The average heat transfer coefficient on the interior surfaces of the van is 1.2 Btu/h · ft² · °F. The air flow over the exterior surfaces can be assumed to be turbulent because of the intense vibrations involved, and

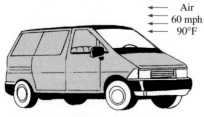

FIGURE P17-74E

the heat transfer coefficient on the front and back surfaces can be taken to be equal to that on the top surface. Disregarding any heat gain or loss by radiation, determine the rate of heat transfer from the ambient air to the van.

17-75 Consider a house that is maintained at a constant temperature of 22°C. One of the walls of the house has three single-pane glass windows that are 1.5 m high and 1.2 m long. The glass ($k = 0.78$ W/m · °C) is 0.5 cm thick, and the heat transfer coefficient on the inner surface of the glass is 8 W/m² · C. Now winds at 60 km/h start to blow parallel to the surface of this wall. If the air temperature outside is -2°C, determine the rate of heat loss through the windows of this wall. Assume radiation heat transfer to be negligible.

17-76 The compressed air requirements of a manufacturing facility are met by a 150-hp compressor located in a room that is maintained at 25°C. In order to minimize the compressor work, the intake port of the compressor is connected to the outside through an 8-m-long, 20-cm-diameter duct made of thin aluminum sheet. The compressor takes in air at a rate of 0.27 m³/s at the outdoor conditions of 10°C and 95 kPa. Disregarding the thermal resistance of the duct and taking the heat transfer coefficient on the outer surface of the duct to be 10 W/m² · °C, determine (*a*) the power used by the compressor to overcome the pressure drop in this duct, (*b*) the rate of heat transfer to the incoming cooler air, and (*c*) the temperature rise of air as it flows through the duct.

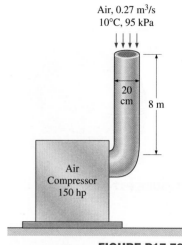

Air, 0.27 m³/s
10°C, 95 kPa

20 cm

8 m

Air Compressor 150 hp

FIGURE P17-76

17-77 Consider a person who is trying to keep cool on a hot summer day by turning a fan on and exposing his body to air flow. The air temperature is 32°C, and the fan is blowing air at a velocity of 5 m/s. The surrounding surfaces are at 40°C, and the emissivity of the person can be taken to be 0.9. If the person is doing light work and generating sensible heat at a rate of 90 W, determine the average temperature of the outer surface (skin or clothing) of the person. The average human body can be treated as a 30-cm-diameter cylinder with an exposed surface area of 1.7 m². *Answer:* 36.3°C

17-78 A house built on a riverside is to be cooled in summer by utilizing the cool water of the river, which flows at an average temperature of 15°C. A 15-m-long section of a circular duct of 20-cm diameter passes through the water. Air enters the underwater section of the duct at 25°C at a velocity of 3 m/s. Assuming the surface of the duct to be at the temperature of the water, determine the outlet temperature of air as it leaves the underwater portion of the duct. Also, determine the fan power needed to overcome the flow resistance in this section of the duct.

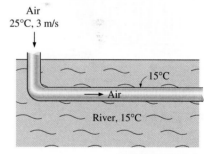

Air 25°C, 3 m/s

15°C

Air

River, 15°C

FIGURE P17-78

17-79 Repeat Prob. 17-78 assuming that a 0.15-mm-thick layer of mineral deposit ($k = 3$ W/m · °C) formed on the inner surface of the pipe.

17-80E The exhaust gases of an automotive engine leave the combustion chamber and enter a 8-ft-long and 3.5-in.-diameter thin-walled steel exhaust pipe at 940°F and 16.1 psia at a rate of 0.2 lbm/s. The surrounding ambient air is at a temperature of 75°F, and the heat transfer coefficient on the outer surface of the exhaust pipe is 3 Btu/h · ft² · °F. Assuming the exhaust gases to have the properties of air, determine (*a*) the velocity of the exhaust gases at the inlet of the exhaust pipe and (*b*) the temperature at which the exhaust gases will leave the pipe and enter the air.

17-81 Hot water at 90°C enters a 15-m section of a cast iron pipe ($k = 52$ W/m · °C) whose inner and outer diameters are 4 and 4.6 cm, respectively, at an average velocity of 0.8 m/s. The outer surface of the pipe,

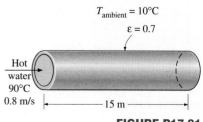

$T_{ambient} = 10$°C

$\varepsilon = 0.7$

Hot water 90°C 0.8 m/s

15 m

FIGURE P17-81

whose emissivity is 0.7, is exposed to the cold air at 10°C in a basement, with a convection heat transfer coefficient of 15 W/m² · °C. Taking the walls of the basement to be at 10°C also, determine (a) the rate of heat loss from the water and (b) the temperature at which the water leaves the basement.

17-82 Repeat Prob. 17-81 for a pipe made of copper ($k = 386$ W/m · °C) instead of cast iron.

17-83 Four power transistors, each dissipating 15 W, are mounted on a thin vertical aluminum plate ($k = 237$ W/m · °C) 22 cm × 22 cm in size. The heat generated by the transistors is to be dissipated by both surfaces of the plate to the surrounding air at 25°C, which is blown over the plate by a fan at a velocity of 250 m/min. The entire plate can be assumed to be nearly isothermal, and the exposed surface area of the transistor can be taken to be equal to its base area. Determine the temperature of the aluminum plate.

17-84 A 3-m-internal-diameter spherical tank made of 1-cm-thick stainless steel ($k = 15$ W/m · °C) is used to store iced water at 0°C. The tank is located outdoors at 30°C and is subjected to winds at 25 km/h. Assuming the entire steel tank to be at 0°C and thus its thermal resistance to be negligible, determine (a) the rate of heat transfer to the iced water in the tank and (b) the amount of ice at 0°C that melts during a 24-h period. The heat of fusion of water at atmospheric pressure is $h_{if} = 333.7$ kJ/kg. Disregard any heat transfer by radiation.

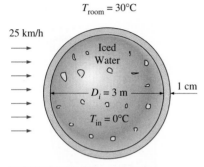

$T_{room} = 30°C$

25 km/h

Iced Water

$D_i = 3$ m

$T_{in} = 0°C$

1 cm

FIGURE P17-84

17-85 Repeat Prob. 17-84, assuming the inner surface of the tank to be at 0°C but by taking the thermal resistance of the tank and heat transfer by radiation into consideration. Assume the average surrounding surface temperature for radiation exchange to be 15°C and the outer surface of the tank to have an emissivity of 0.9.   *Answers: (a)* 13,630 W, *(b)* 3529 kg

17-86 D. B. Tuckerman and R. F. Pease of Stanford University demonstrated in the early 1980s that integrated circuits can be cooled very effectively by fabricating a series of microscopic channels 0.3 mm high and 0.05 mm wide in the back of the substrate and covering them with a plate to confine the fluid flow within the channels. They were able to dissipate 790 W of power generated in a 1-cm² silicon chip at a junction-to-ambient temperature difference of 71°C using water as the coolant flowing at a rate of 0.01 L/s through 100 such channels under a 1-cm × 1-cm silicon chip. Heat is transferred primarily through the base area of the channel, and it was found that the increased surface area and thus the fin effect are of lesser importance. Disregarding the entrance effects and ignoring any heat transfer from the side and cover surfaces, determine (a) the temperature rise of water as it flows through the microchannels and (b) the average surface temperature of the base of the microchannels for a power dissipation of 50 W. Assume the water enters the channels at 20°C.

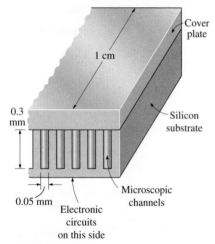

Cover plate

1 cm

0.3 mm

Silicon substrate

0.05 mm

Electronic circuits on this side

Microscopic channels

FIGURE P17-86

17-87E A transistor with a height of 0.25 in. and a diameter of 0.22 in. is mounted on a circuit board. The transistor is cooled by air flowing over it at a velocity of 500 ft/min. If the air temperature is 130°F and the transistor case temperature is not to exceed 165°F, determine the amount of power this transistor can dissipate safely.

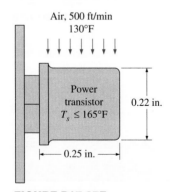

Air, 500 ft/min
130°F

Power transistor
$T_s \leq 165°F$

0.22 in.

0.25 in.

FIGURE P17-87E

17-88 Liquid-cooled systems have high heat transfer coefficients associated with them, but they have the inherent disadvantage that they present potential leakage problems. Therefore, air is proposed to be used as the microchannel coolant. Repeat Prob. 17-86 using air as the cooling fluid instead of water, entering at a rate of 0.5 L/s.

17-89 A desktop computer is to be cooled by a fan. The electronic components of the computer consume 45 W of power under full-load conditions. The computer is to operate in environments at temperatures up to 50°C and at elevations up to 3000 m where the atmospheric pressure is 70.12 kPa. The exit temperature of air is not to exceed 60°C to meet the reliability requirements. Also, the average velocity of air is not to exceed 120 m/min at the exit of the computer case, where the fan is installed to keep the noise level down. Determine the flow rate of the fan that needs to be installed and the diameter of the casing of the fan.

17-90 The roof of a house consists of a 15-cm-thick concrete slab ($k = 2$ W/m · °C) 15 m wide and 20 m long. The convection heat transfer coefficient on the inner surface of the roof is 5 W/m² · °C. On a clear winter night, the ambient air is reported to be at 10°C, while the night sky temperature is 100 K. The house and the interior surfaces of the wall are maintained at a constant temperature of 20°C. The emissivity of both surfaces of the concrete roof is 0.9. Considering both radiation and convection heat transfer, determine the rate of heat transfer through the roof when wind at 60 km/h is blowing over the roof.

If the house is heated by a furnace burning natural gas with an efficiency of 85 percent, and the price of natural gas is $0.60/therm (1 therm = 105,500 kJ of energy content), determine the money lost through the roof that night during a 14-h period. *Answers:* 28 kW, $9.44

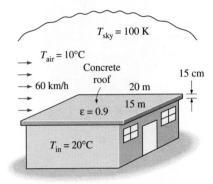

FIGURE P17-90

17-91 Steam at 250°C flows in a stainless steel pipe ($k = 15$ W/m · °C) whose inner and outer diameters are 4 cm and 4.6 cm, respectively. The pipe is covered with 3.5-cm-thick glass wool insulation ($k = 0.038$ W/m · °C) whose outer surface has an emissivity of 0.3. Heat is lost to the surrounding air and surfaces at 3°C by convection and radiation. Taking the heat transfer coefficient inside the pipe to be 80 W/m² · °C, determine the rate of heat loss from the steam per unit length of the pipe when air is flowing across the pipe at 4 m/s.

17-92 The boiling temperature of nitrogen at atmospheric pressure at sea level (1 atm pressure) is −196°C. Therefore, nitrogen is commonly used in low-temperature scientific studies, since the temperature of liquid nitrogen in a tank open to the atmosphere will remain constant at −196°C until it is depleted. Any heat transfer to the tank will result in the evaporation of some liquid nitrogen, which has a heat of vaporization of 198 kJ/kg and a density of 810 kg/m³ at 1 atm.

Consider a 4-m-diameter spherical tank that is initially filled with liquid nitrogen at 1 atm and −196°C. The tank is exposed to 20°C ambient air and 40 km/h winds. The temperature of the thin-shelled spherical tank is observed to be almost the same as the temperature of the nitrogen inside. Disregarding any radiation heat exchange, determine the rate of evaporation of the liquid nitrogen in the tank as a result of heat transfer from the ambient air if the tank is (*a*) not insulated, (*b*) insulated with 5-cm-thick fiberglass insulation ($k = 0.035$ W/m · °C), and (*c*) insulated with 2-cm-thick superinsulation that has an effective thermal conductivity of 0.00005 W/m · °C.

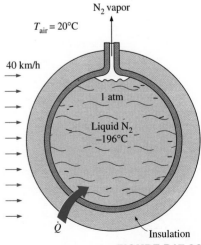

FIGURE P17-92

Winds
40°F
35 mph

It feels
like 11°F

FIGURE P17-95E

17-93 Repeat Prob. 17-92 for liquid oxygen, which has a boiling temperature of −183°C, a heat of vaporization of 213 kJ/kg, and a density of 1140 kg/m³ at 1 atm pressure.

17-94 A 0.3-cm-thick, 12-cm-high, and 18-cm-long circuit board houses 80 closely spaced logic chips on one side, each dissipating 0.04 W. The board is impregnated with copper fillings and has an effective thermal conductivity of 16 W/m · °C. All the heat generated in the chips is conducted across the circuit board and is dissipated from the back side of the board to the ambient air at 40°C, which is forced to flow over the surface by a fan at a free-stream velocity of 400 m/min. Determine the temperatures on the two sides of the circuit board.   *Answers:* 46.28°C, 46.31°C

17-95E It is well known that cold air feels much colder in windy weather than what the thermometer reading indicates because of the "chilling effect" of the wind. This effect is due to the increase in the convection heat transfer coefficient with increasing air velocities. The *equivalent windchill temperature* in °F is given by (1993 *ASHRAE Handbook of Fundamentals,* Atlanta, GA, p. 8.15)

$$T_{equiv} = 91.4 - (91.4 - T_{ambient})(0.475 - 0.0203\mathcal{V} + 0.304\sqrt{\mathcal{V}})$$

where $\mathcal{V}$ is the wind velocity in mph and $T_{ambient}$ is the ambient air temperature in °F in calm air, which is taken to be air with light winds at speeds up to 4 mph. The constant 91.4°F in the above equation is the mean skin temperature of a resting person in a comfortable environment. Windy air at a temperature $T_{ambient}$ and velocity $\mathcal{V}$ will feel as cold as calm air at a temperature T_{equiv}. The equation above is valid for winds up to 43 mph. Winds at higher velocities produce little additional chilling effect. Determine the equivalent wind chill temperature of an environment at 10°F at wind speeds of 10, 20, 30, and 40 mph. Exposed flesh can freeze within one minute at a temperature below −25°F in calm weather. Does a person need to be concerned about this possibility in any of the cases above?

Computer, Design, and Essay Problems

17-96 Obtain information on frostbite and the conditions under which it occurs. Using the relation in Prob.0 17-95E, prepare a table that shows how long people can stay in cold and windy weather for specified temperatures and wind speeds before the exposed flesh is in danger of experiencing frostbite.

17-97 Write an article on forced convection cooling with air, helium, water, and a dielectric liquid. Discuss the advantages and disadvantages of each fluid in heat transfer. Explain the circumstances under which a certain fluid will be most suitable for the cooling job.

Natural Convection

In the preceding chapter, we considered heat transfer by *forced convection* where a fluid was *forced* to move over a surface or in a tube by external means such as a pump or a fan. In this chapter, we consider *natural convection,* where any fluid motion occurs by natural means such as buoyancy. The fluid motion in forced convection is quite *noticeable,* since a fan or a pump can transfer enough momentum to the fluid to move it in a certain direction. The fluid motion in natural convection, however, is often not noticeable because of the low velocities involved.

Convection heat transfer coefficient is a strong function of *velocity:* the higher the velocity, the higher the convection heat transfer coefficient. The fluid velocities associated with natural convection are low, typically under 1 m/s. Therefore, the *heat transfer coefficients* encountered in natural convection are usually *much lower* than those encountered in forced convection. Yet several types of heat transfer equipment are designed to operate under natural convection conditions instead of forced convection, because natural convection does not require the use of a fluid mover.

We start this chapter with a discussion of the physical mechanism of *natural convection* and the *Grashof number.* We then present the correlations to evaluate heat transfer by natural convection for various geometries, including enclosures. Finally, we discuss simultaneous forced and natural convection.

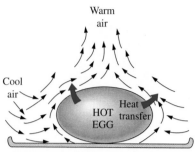

FIGURE 18-1

The cooling of a boiled egg in a cooler environment by natural convection.

FIGURE 18-2

The warming up of a cold drink in a warmer environment by natural convection.

18-1 ■ PHYSICAL MECHANISM OF NATURAL CONVECTION

Many familiar heat transfer applications involve *natural convection* as the *primary* mechanism of heat transfer. Some examples are cooling of electronic equipment such as power transistors, TVs, and VCRs; heat transfer from electric baseboard heaters or steam radiators; heat transfer from the refrigeration coils and power transmission lines; and heat transfer from the bodies of animals and human beings. Natural convection in gases is usually accompanied by radiation of comparable magnitude except for low-emissivity surfaces.

We know that a hot *boiled egg* (or a hot *baked potato*) on a plate eventually cools to the surrounding air temperature (Fig. 18-1). The egg is cooled by transferring heat by *convection* to the air and by radiation to the surrounding surfaces. Disregarding heat transfer by radiation, the physical mechanism of cooling a hot egg (or any hot object) in a cooler environment can be explained as follows:

As soon as the hot egg is exposed to cooler air, the temperature of the outer surface of the egg shell will drop somewhat, and the temperature of the air adjacent to the shell will rise as a result of heat conduction from the shell to the air. Consequently, the egg will soon be surrounded by a thin layer of warmer air, and heat will then be transferred from this warmer layer to the outer layers of air. The cooling process in this case would be rather *slow* since the egg would always be *blanketed* by warm air, and it would have no direct contact with the cooler air farther away. We may not notice any *air motion* in the vicinity of the egg, but careful measurements indicate otherwise.

The temperature of the air adjacent to the egg is higher, and thus its density is lower, since at constant pressure the *density* of a gas is inversely proportional to its *temperature*. Thus, we have a situation in which some low-density or "light" gas is surrounded by a high-density or "heavy" gas, and the natural laws dictate that *the light gas rise*. This is no different than the oil in a vinegar-and-oil salad dressing rising to the top (note that $\rho_{oil} < \rho_{vinegar}$). This phenomenon is characterized incorrectly by the phrase "heat rises," which is understood to mean *heated air rises*. The space vacated by the warmer air in the vicinity of the egg is replaced by the cooler air nearby, and the presence of cooler air in the vicinity of the egg speeds up the cooling process. The rise of warmer air and the flow of cooler air into its place continues until the egg is cooled to the temperature of the surrounding air. The motion that results from the continual replacement of the heated air in the vicinity of the egg by the cooler air nearby is called a **natural convection current,** and the heat transfer that is enhanced as a result of this natural convection current is called **natural convection heat transfer.** Note that in the absence of natural convection currents, heat transfer from the egg to the air surrounding it would be by *conduction* only, and the rate of heat transfer from the egg would be much *lower.*

Natural convection is just as effective in the heating of cold surfaces in a warmer environment as it is in the cooling of hot surfaces in a cooler environment, as shown in Fig. 18-2. Note that the direction of fluid motion is reversed in this case.

In a gravitational field, there seems to be a net force that pushes upward a light fluid placed in a heavier fluid. The upward force exerted by a fluid on a body completely or partially immersed in it is called the **buoyancy force.** The magnitude of the buoyancy force is equal to the weight of the *fluid displaced* by the body. That is,

$$F_{\text{buoyancy}} = \rho_{\text{fluid}} \, g V_{\text{body}} \qquad (18\text{-}1)$$

where ρ_{fluid} is the average density of the *fluid* (not the body), g is the gravitational acceleration, and V_{body} is the volume of the portion of the body immersed in the fluid (for bodies completely immersed in the fluid, it is the total volume of the body). In the absence of other forces, the net vertical force acting on a body is the difference between the weight of the body and the buoyancy force. That is,

$$\begin{aligned} F_{\text{net}} &= W - F_{\text{buoyancy}} \\ &= \rho_{\text{body}} \, g V_{\text{body}} - \rho_{\text{fluid}} \, g V_{\text{body}} \\ &= (\rho_{\text{body}} - \rho_{\text{fluid}}) \, g V_{\text{body}} \end{aligned} \qquad (18\text{-}2)$$

Note that this force is *proportional* to the difference in the *densities* of the fluid and the body immersed in it. Thus, a body immersed in a fluid will experience a "weight loss" in an amount equal to the weight of the fluid it displaces. This is known as *Archimedes' principle.*

To have a better understanding of the buoyancy effect, consider an egg dropped into water. If the average density of the egg is greater than the density of water (a sign of freshness), the egg will settle at the bottom of the container. Otherwise, it will rise to the top. When the density of the egg equals the density of water, the egg will settle somewhere in the water while remaining completely immersed, acting like a "weightless object" in space. This occurs when the upward buoyancy force acting on the egg equals the weight of the egg, which acts downward.

The *buoyancy effect* has far-reaching implications in life. For one thing, without buoyancy, heat transfer between a hot (or cold) surface and the fluid surrounding it would be by *conduction* instead of by *natural convection.* The natural convection currents encountered in the oceans, lakes, and the atmosphere owe their existence to buoyancy. Also, light boats as well as heavy warships made of steel float on water because of buoyancy (Fig. 18-3). Ships are designed on the basis of the principle that the entire weight of a ship and its contents is equal to the weight of the water that the submerged volume of the ship can contain. (Note that a larger portion of the hull of a ship will sink in fresh water than it does in salty water.) The "chimney effect" that induces the upward flow of hot combustion gases through a chimney is also due to the buoyancy effect, and the upward force acting on the gases in the chimney is proportional to the difference between the densities of the hot gases in the chimney and the cooler air outside. Note that there is *no gravity* in space, and thus there will be no natural convection heat transfer in a spacecraft, even if the spacecraft is filled with atmospheric air.

In heat transfer studies, the primary variable is *temperature,* and it is desirable to express the net buoyancy force (Eq. 18-2) in terms of temperature differences. But this requires expressing the density difference in terms of a temperature difference, which requires a knowledge of a property that represents the *variation of the density of a fluid with temperature at constant pressure. The property that provides that information is the* **volume expansion coefficient** β, defined as (Fig. 18-4)

$$\beta = \frac{1}{v}\left(\frac{\partial v}{\partial T}\right)_P = -\frac{1}{\rho}\left(\frac{\partial \rho}{\partial T}\right)_P \qquad (1/\text{K}) \qquad (18\text{-}3)$$

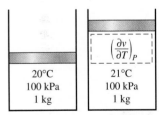

FIGURE 18-3

It is the buoyancy force that keeps the ships afloat in water ($W = F_{\text{buoyancy}}$ for floating objects).

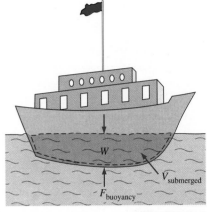

(a) A substance with a large β

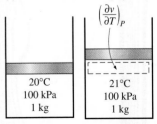

(b) A substance with a small β

FIGURE 18-4

The coefficient of volume expansion is a measure of the change in volume of a substance with temperature at constant pressure.

It can also be expressed approximately by replacing derivatives by differences as

$$\beta \approx -\frac{1}{\rho}\frac{\Delta\rho}{\Delta T} \quad \rightarrow \quad \Delta\rho \approx -\rho\beta\,\Delta T \qquad \text{(at constant } P\text{)}$$

We can show easily that the volume expansion coefficient β of an *ideal gas* $(P = \rho RT)$ at a temperature T is equivalent to the inverse of the temperature:

$$\beta_{\text{ideal gas}} = \frac{1}{T} \qquad (1/\text{K}) \tag{18-4}$$

where T is the *absolute* temperature. Note that a large value of β for a fluid means a large change in density with temperature, and that the product $\beta\,\Delta T$ represents the fraction of volume change of a fluid that corresponds to a temperature change ΔT at constant pressure. Also note that the buoyancy force is proportional to the *density difference,* which is proportional to the *temperature difference* at constant pressure. Therefore, the larger the temperature difference between the fluid adjacent to a hot (or cold) surface and the fluid away from it, the *larger* the buoyancy force and the *stronger* the natural convection currents, and thus the *higher* the heat transfer rate.

The magnitude of the natural convection heat transfer between a surface and a fluid is directly related to the *mass flow rate* of the fluid. The higher the mass flow rate, the higher the heat transfer rate. In fact, it is the very high flow rates that increase the heat transfer coefficient by orders of magnitude when forced convection is used. In natural convection, no blowers are used, and therefore the flow rate cannot be controlled externally. The flow rate in this case is established by the dynamic balance of *buoyancy* and *friction.*

As we have discussed earlier, the *buoyancy force* is caused by the density difference between the heated (or cooled) fluid adjacent to the surface and the fluid surrounding it, and is proportional to this density difference and the volume occupied by the warmer fluid. It is also well known that whenever two bodies in contact (solid–solid, solid–fluid, or fluid–fluid) move relative to each other, a *friction force* develops at the contact surface in the direction opposite to that of the motion. This opposing force slows down the fluid and thus reduces the flow rate of the fluid. Under steady conditions, the air flow rate driven by buoyancy is established at the point where these two effects *balance* each other. The friction force increases as more and more solid surfaces are introduced, seriously disrupting the fluid flow and heat transfer. For that reason, heat sinks with closely spaced fins are not suitable for natural convection cooling.

Most heat transfer correlations in natural convection are based on experimental measurements. The instrument used in natural convection experiments most often is the *Mach–Zehnder interferometer,* which gives a plot of isotherms in the fluid in the vicinity of a surface. The operation principle of interferometers is based on the fact that at low pressure, the lines of constant temperature for a gas correspond to the lines of constant density, and that the index of refraction of a gas is a function of its density. Therefore, the degree of refraction of light at some point in a gas is a measure of the temperature gradient at that point. An interferometer produces a map of interference fringes, which can be interpreted as lines of *constant temperature* as shown in Fig. 18-5. The smooth and parallel lines in (*a*) indicate that the flow is *laminar,* whereas the eddies and irregularities in (*b*) indicate that the flow is *turbulent.* Note that the lines are closest near the surface, indicating a *higher temperature gradient.*

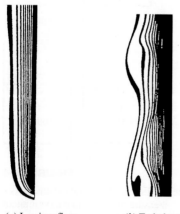

(*a*) Laminar flow   (*b*) Turbulent flow

FIGURE 18-5

Isotherms in natural convection over a hot plate in air.

The Grashof Number

We mentioned in the preceding chapter that the flow regime in forced convection is governed by the dimensionless *Reynolds number,* which represents the ratio of inertial forces to viscous forces acting on the fluid. The flow regime in natural convection is governed by another dimensionless number, called the **Grashof number,** which represents the ratio of the *buoyancy force* to the *viscous force* acting on the fluid. That is,

$$\text{Gr} = \frac{\text{Buoyancy forces}}{\text{Viscous forces}} = \frac{g\,\Delta\rho\,V}{\rho v^2} = \frac{g\beta\,\Delta T\,V}{v^2}$$

Since $\Delta\rho \approx \rho\beta\,\Delta T$, it is formally expressed as (Fig. 18-6)

$$\text{Gr} = \frac{g\beta(T_s - T_\infty)\delta^3}{v^2} \qquad (18\text{-}5)$$

where
    g = gravitational acceleration, m/s^2
    β = coefficient of volume expansion, 1/K ($\beta = 1/T$ for ideal gases)
    T_s = temperature of the surface, °C
    T_∞ = temperature of the fluid sufficiently far from the surface, °C
    δ = characteristic length of the geometry, m
    v = kinematic viscosity of the fluid, m^2/s

The role played by the *Reynolds number* in forced convection is played by the *Grashof number* in natural convection. As such, the Grashof number provides the main criterion in determining whether the fluid flow is laminar or turbulent in natural convection. For vertical plates, for example, the critical Grashof number is observed to be about 10^9. Therefore, the flow regime on a vertical plate becomes turbulent at Grashof numbers greater than 10^9.

The heat transfer rate in natural convection from a solid surface to the surrounding fluid is expressed by Newton's law of cooling as

$$\dot{Q}_{\text{conv}} = hA(T_s - T_\infty) \qquad (\text{W}) \qquad (18\text{-}6)$$

where A is the heat transfer surface area and h is the average heat transfer coefficient on the surface.

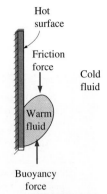

FIGURE 18-6

The Grashof number Gr is a measure of the relative magnitudes of the *buoyancy force* and the opposing *friction force* acting on the fluid.

18-2 ■ NATURAL CONVECTION OVER SURFACES

Natural convection heat transfer on a surface depends on the geometry of the surface as well as its orientation. It also depends on the variation of temperature on the surface and the thermophysical properties of the fluid involved.

The *velocity* and *temperature profiles* for natural convection over a vertical hot plate immersed in a quiescent fluid body are given in Fig. 18-7. As in forced convection, the thickness of the boundary layer increases in the flow direction. Unlike forced convection, however, the fluid velocity is *zero* at the *outer edge* of the velocity boundary layer as well as at the surface of the plate. This is expected since the fluid beyond the boundary layer is stationary. Thus, the fluid velocity increases with distance from the surface, reaches a maximum, and gradually decreases to zero at a distance sufficiently far from the surface. The *temperature* of the fluid will equal the plate temperature at the surface and gradually decrease to the temperature of the surrounding fluid at a distance sufficiently far from the surface, as shown in the figure. In the

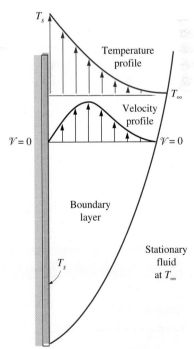

FIGURE 18-7

Typical velocity and temperature profiles for natural convection flow over a hot vertical plate at temperature T_s inserted in a fluid at temperature T_∞.

case of *cold surfaces,* the shape of the velocity and temperature profiles remains the same but their direction is reversed.

Natural Convection Correlations

Although we understand the mechanism of natural convection well, the complexities of fluid motion make it very difficult to obtain simple analytical relations for heat transfer by solving the governing equations of motion and energy. Some analytical solutions exist for natural convection, but such solutions lack generality since they are obtained for simple geometries under some simplifying assumptions. Therefore, with the exception of some simple cases, heat transfer relations in natural convection are based on experimental studies. Of the numerous such correlations of varying complexity and claimed accuracy available in the literature for any given geometry, we present below the *simpler* ones for two reasons: first, the accuracy of simpler relations is usually within the range of uncertainty associated with a problem, and second, we would like to keep the emphasis on the physics of the problems instead of formula manipulation.

The simple empirical correlations for the average *Nusselt number* Nu in natural convection are of the form (Fig. 18-8)

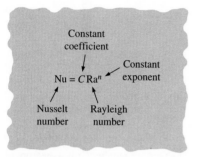

$$ \text{Nu} = \frac{h\delta}{k} = C(\text{Gr Pr})^n = C\,\text{Ra}^n \qquad (18\text{-}7) $$

where Ra is the **Rayleigh number,** which is the product of the Grashof and Prandtl numbers:

$$ \text{Ra} = \text{Gr Pr} = \frac{g\beta(T_s - T_\infty)\delta^3}{\nu^2}\,\text{Pr} \qquad (18\text{-}8) $$

FIGURE 18-8

Natural convection heat transfer correlations are usually expressed in terms of the Rayleigh number raised to a constant n multiplied by another constant C, both of which are determined experimentally.

The values of the constants C and n depend on the *geometry* of the surface and the *flow regime,* which is characterized by the range of the Rayleigh number. The value of n is usually $\frac{1}{4}$ for laminar flow and $\frac{1}{3}$ for turbulent flow. The value of the constant C is normally less than 1.

Simple relations for the average Nusselt number for various geometries are given in Table 18-1, together with sketches of the geometries. Also given in this table are the characteristic lengths of the geometries and the ranges of Rayleigh number in which the relation is applicable. All fluid properties are to be evaluated at the film temperature $T_f = \frac{1}{2}(T_s + T_\infty)$.

These relations have been obtained for the case of isothermal surfaces but could also be used approximately for the case of nonisothermal surfaces by assuming the surface temperature to be constant at some average value. The use of these relations is illustrated below with examples.

EXAMPLE 18-1   Heat Loss from Hot Water Pipes

A 6-m-long section of an 8-cm-diameter horizontal hot water pipe shown in Fig. 18-9 passes through a large room whose temperature is 18°C. If the outer surface temperature of the pipe is 70°C, determine the rate of heat loss from the pipe by natural convection.

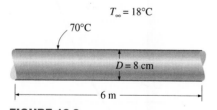

FIGURE 18-9

Schematic for Example 18-1.

Solution   A horizontal hot water pipe passes through a large room. The rate of heat loss from the pipe by natural convection is to be determined.

TABLE 18-1 789

Empirical correlations for the average Nusselt number for natural convection over surfaces

Geometry	Characteristic length δ	Range of Ra	Nu
Vertical plate	L	10^4–10^9 10^9–10^{13} Entire range	$Nu = 0.59Ra^{1/4}$ (18-9) $Nu = 0.1Ra^{1/3}$ (18-10) $Nu = \left\{0.825 + \dfrac{0.387Ra^{1/6}}{(1 + (0.492/Pr)^{9/16})^{8/27}}\right\}^2$ (18-11) (complex but more accurate)
Inclined plate	L		Use vertical plate equations as a first degree of approximation. Replace g by $g\cos\theta$ for $Ra < 10^9$
Horizontal plate (Surface area A and perimeter p) (a) Upper surface of a hot plate (or lower surface of a cold plate) (b) Lower surface of a hot plate (or upper surface of a cold plate) 	A/p	10^4–10^7 10^7–10^{11} 10^5–10^{11}	$Nu = 0.54Ra^{1/4}$ (18-12) $Nu = 0.15Ra^{1/3}$ (18-13) $Nu = 0.27Ra^{1/4}$ (18-14)
Vertical cylinder	L		A vertical cylinder can be treated as a vertical plate when $D \geq \dfrac{35L}{Gr^{1/4}}$ (18-15)
Horizontal cylinder	D	10^{-5}–10^{12}	$Nu = \left\{0.6 + \dfrac{0.387Ra^{1/6}}{(1 + (0.559/Pr)^{9/16})^{8/27}}\right\}^2$ (18-16)
Sphere	D	$Ra \leq 10^{11}$ $(Pr \geq 0.7)$	$Nu = 2 + \dfrac{0.589Ra^{1/4}}{(1 + (0.469/Pr)^{9/16})^{4/9}}$ (18-17)

Assumptions **1** Steady operating conditions exist. **2** Air is an ideal gas. **3** The local atmospheric pressure is 1 atm.

Properties The properties of air at the film temperature of $T_f = (T_s + T_\infty)/2 = (70 + 18)/2 = 44°C = 317$ K and 1 atm pressure are (Table A-18)

$$k = 0.0273 \text{ W/m} \cdot °C \qquad \text{Pr} = 0.710$$
$$\nu = 1.74 \times 10^{-5} \text{ m}^2/\text{s} \qquad \beta = \frac{1}{T_f} = \frac{1}{317 \text{ K}} = 0.00315 \text{ K}^{-1}$$

Analysis The characteristic length in this case is the outer diameter of the pipe, $\delta = D = 0.08$ m. Then the Rayleigh number becomes

$$\text{Ra} = \frac{g\beta(T_s - T_\infty)\delta^3}{\nu^2} \text{Pr}$$

$$= \frac{(9.81 \text{ m/s}^2)(0.00315 \text{ K}^{-1})[(70 - 18) \text{ K}](0.08 \text{ m})^3}{(1.74 \times 10^{-5} \text{ m}^2/\text{s})^2}(0.710) = 1.93 \times 10^6$$

Then the natural convection Nusselt number in this case can be determined from Eq. 18-16 to be

$$\text{Nu} = \left\{ 0.6 + \frac{0.387 \text{ Ra}^{1/6}}{[1 + (0.559/\text{Pr})^{9/16}]^{8/27}} \right\}^2$$

$$= \left\{ 0.6 + \frac{0.387(1.93 \times 10^6)^{1/6}}{[1 + (0.559/0.710)^{9/16}]^{8/27}} \right\}^2 = 17.2$$

Then

$$h = \frac{k}{D} \text{Nu} = \frac{0.0273 \text{ W/m} \cdot °C}{0.08 \text{ m}}(17.2) = 5.9 \text{ W/m}^2 \cdot °C$$

$$A = \pi DL = \pi(0.08 \text{ m})(6 \text{ m}) = 1.51 \text{ m}^2$$

and

$$\dot{Q} = hA(T_s - T_\infty) = (5.9 \text{ W/m}^2 \cdot °C)(1.51 \text{ m}^2)(70 - 18)°C = \textbf{463 W}$$

Therefore, the pipe will lose heat to the air in the room at a rate of 463 W by natural convection.

Discussion The pipe will lose heat to the surroundings by radiation as well as by natural convection. Assuming the outer surface of the pipe to be black (emissivity $\varepsilon = 1$) and the inner surfaces of the walls of the room to be at room temperature, the radiation heat transfer in this case is determined to be (Fig. 18-10)

$$\dot{Q}_{rad} = \varepsilon A\sigma(T_s^4 - T_\infty^4)$$
$$= (1)(1.51 \text{ m}^2)(5.67 \times 10^{-8} \text{ W/m}^2 \cdot \text{K}^4)[(70 + 273 \text{ K})^4 - (18 + 273 \text{ K})^4]$$
$$= 571 \text{ W}$$

which is as large as that for natural convection. The emissivity of a real surface is less than 1, and thus the radiation heat transfer for a real surface will be less. But radiation will still be significant for most systems cooled by natural convection. Therefore, a radiation analysis should normally accompany a natural convection analysis unless the emissivity of the surface is low.

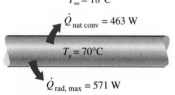

$T_\infty = 18°C$

$\dot{Q}_{\text{nat conv}} = 463$ W

$T_s = 70°C$

$\dot{Q}_{\text{rad, max}} = 571$ W

FIGURE 18-10

Radiation heat transfer is usually comparable to natural convection in magnitude and should be considered in heat transfer analysis.

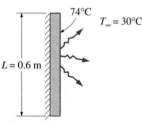

74°C

$T_\infty = 30°C$

$L = 0.6$ m

(*a*) Vertical

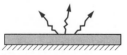

(*b*) Hot surface facing up

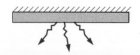

(*c*) Hot surface facing down

FIGURE 18-11

Schematic for Example 18-2.

EXAMPLE 18-2 Cooling of a Plate in Different Orientations

Consider a 0.6-m × 0.6-m thin square plate in a room at 30°C. One side of the plate is maintained at a temperature of 74°C, while the other side is insulated, as shown in Fig. 18-11. Determine the rate of heat transfer from the plate by natural convection if the plate is (*a*) vertical, (*b*) horizontal with hot surface facing up, and (*c*) horizontal with hot surface facing down.

Solution A hot plate with an insulated back is considered. The rate of heat loss by natural convection is to be determined for different orientations.

Assumptions **1** Steady operating conditions exist. **2** Air is an ideal gas. **3** The local atmospheric pressure is 1 atm.

Properties The properties of air at the film temperature of $T_f = (T_s + T_\infty)/2 = (74 + 30)/2 = 52°C = 325$ K and 1 atm pressure are (Table A-18)

$$k = 0.0279 \text{ W/m} \cdot °C \qquad \text{Pr} = 0.709$$

$$\nu = 1.815 \times 10^{-5} \text{ m}^2/\text{s} \qquad \beta = \frac{1}{T_f} = \frac{1}{325 \text{ K}} = 0.00308 \text{ K}^{-1}$$

Analysis (*a*) *Vertical.* The characteristic length in this case is the height of the plate, which is $\delta = 0.6$ m. The Rayleigh number is

$$\text{Ra} = \frac{g\beta(T_s - T_\infty)\delta^3}{\nu^2} \text{Pr}$$

$$= \frac{(9.81 \text{ m/s}^2)(0.00308 \text{ K}^{-1})[(74 - 30) \text{ K}](0.6 \text{ m})^3}{(1.815 \times 10^{-5} \text{ m}^2/\text{s})^2}(0.709) = 6.17 \times 10^8$$

Then the natural convection Nusselt number can be determined from Eq. 18-9 to be

$$\text{Nu} = 0.59 \text{ Ra}^{1/4} = 0.59(6.17 \times 10^8)^{1/4} = 93.0$$

Then

$$h = \frac{k}{\delta} \text{Nu} = \frac{0.0279 \text{ W/m} \cdot °C}{0.6 \text{ m}}(93.0) = 4.3 \text{ W/m}^2 \cdot °C$$

$$A = L^2 = (0.6 \text{ m})^2 = 0.36 \text{ m}^2$$

and

$$\dot{Q} = hA(T_s - T_\infty) = (4.3 \text{ W/m}^2 \cdot °C)(0.36 \text{ m}^2)(74 - 30)°C = \textbf{68.1 W}$$

(*b*) *Horizontal with hot surface facing up.* The characteristic length and the Rayleigh number in this case are

$$\delta = \frac{A}{p} = \frac{L^2}{4L} = \frac{L}{4} = \frac{0.6 \text{ m}}{4} = 0.15 \text{ m}$$

and

$$\text{Ra} = \frac{g\beta(T_s - T_\infty)\delta^3}{\nu^2} \text{Pr}$$

$$= \frac{(9.81 \text{ m/s}^2)(0.00308 \text{ K}^{-1})[(74 - 30) \text{ K}](0.15 \text{ m})^3}{(1.815 \times 10^{-5} \text{ m}^2/\text{s})^2}(0.709) = 9.65 \times 10^6$$

Then the natural convection Nusselt number can be determined from Eq. 18-12 to be

$$\text{Nu} = 0.54 \text{ Ra}^{1/4} = 0.54(9.65 \times 10^6)^{1/4} = 30.1$$

Therefore,

$$h = \frac{k}{\delta} \text{Nu} = \frac{0.0279 \text{ W/m} \cdot °C}{0.15 \text{ m}}(30.1) = 5.60 \text{ W/m}^2 \cdot °C$$

$$A = L^2 = (0.6 \text{ m})^2 = 0.36 \text{ m}^2$$

and

$$\dot{Q} = hA(T_s - T_\infty) = (5.60 \text{ W/m}^2 \cdot °C)(0.36 \text{ m}^2)(74 - 30)°C = \textbf{88.7 W}$$

(c) *Horizontal with hot surface facing down.* The characteristic length δ, the heat transfer surface area A, and the Rayleigh number in this case are the same as those determined in (b). But the natural convection Nusselt number is to be determined from Eq. 18-14.

$$Nu = 0.27 \, Ra^{1/4} = 0.27(9.66 \times 10^6)^{1/4} = 15.0$$

Then,

$$h = \frac{k}{\delta} Nu = \frac{0.0279 \, W/m \cdot °C}{0.15 \, m} (15.0) = 2.8 \, W/m^2 \cdot °C$$

and

$$\dot{Q} = hA(T_s - T_\infty) = (2.8 \, W/m^2 \cdot °C)(0.36 \, m^2)(74 - 30)°C = \textbf{44.4 W}$$

Note that the natural convection heat transfer is the lowest in the case of the hot surface facing down. This is not surprising, since the hot air is "trapped" under the plate in this case and cannot get away from the plate easily. As a result, the cooler air in the vicinity of the plate will have difficulty reaching the plate, which results in a reduced rate of heat transfer.

Discussion The plate will lose heat to the surroundings by radiation as well as by natural convection. Assuming the surface of the plate to be black (emissivity ε = 1) and the inner surfaces of the walls of the room to be at room temperature, the radiation heat transfer in this case is determined to be

$$\dot{Q}_{rad} = \varepsilon A\sigma(T_s^4 - T_\infty^4)$$
$$= (1)(0.36 \, m^2)(5.67 \times 10^{-8} \, W/m^2 \cdot K^4)[(74 + 273 \, K)^4 - (30 + 273 \, K)^4]$$
$$= 124 \, W$$

which is larger than that for natural convection heat transfer for each case. The emissivity of a real surface is less than 1, and thus the radiation heat transfer for a real surface will be less. But radiation can still be significant and needs to be considered in surfaces cooled by natural convection.

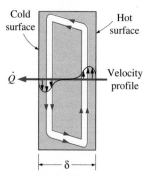

FIGURE 18-12

Convective currents in a vertical rectangular enclosure.

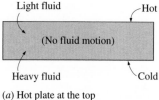

(a) Hot plate at the top

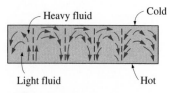

(b) Hot plate at the bottom

FIGURE 18-13

Convective currents in a horizontal enclosure with (a) hot plate at the top and (b) hot plate at the bottom.

18-3 ■ NATURAL CONVECTION INSIDE ENCLOSURES

A considerable portion of heat loss from a typical residence occurs through the windows. We certainly would insulate the windows, if we could, in order to conserve energy. The problem is finding an insulating material that is transparent. An examination of the thermal conductivities of the insulting materials reveals that *air* is a *better insulator* than most common insulating materials. Besides, it is transparent. Therefore, it makes sense to insulate the windows with a layer of air. Of course, we need to use another sheet of glass to trap the air. The result is an *enclosure,* which is known as a *double-pane window* in this case. Other examples of enclosures include wall cavities, solar collectors, and cryogenic chambers involving concentric cylinders or spheres.

Enclosures are frequently encountered in practice, and heat transfer through them is of practical interest. Heat transfer in enclosed spaces is complicated by the fact that the fluid in the enclosure, in general, does not remain stationary. In a vertical enclosure, the fluid adjacent to the hotter surface rises and the fluid adjacent to the cooler one falls, setting off a rotationary motion within the enclosure that enhances heat transfer through the enclosure. Typical flow patterns in vertical and horizontal rectangular enclosures are shown in Figs. 18-12 and 18-13.

The characteristics of heat transfer through a horizontal enclosure depend on whether the hotter plate is at the top or at the bottom, as shown in Fig.

18-13. When the *hotter plate* is at the *top,* no convection currents will develop in the enclosure, since the lighter fluid will always be on top of the heavier fluid. Heat transfer in this case will be by *pure conduction,* and we will have Nu = 1. When the *hotter plate* is at the *bottom,* the heavier fluid will be on top of the lighter fluid, and there will be a tendency for the lighter fluid to topple the heavier fluid and rise to the top, where it will come in contact with the cooler plate and cool down. Until that happens, however, the heat transfer is still by *pure conduction* and Nu = 1. When Ra > 1708, the buoyant force overcomes the fluid resistance and initiates natural convection currents, which are observed to be in the form of hexagonal cells called *Bénard cells.* For Ra > 3×10^5, the cells break down and the fluid motion becomes turbulent.

The Rayleigh number for an enclosure is determined from

$$Ra = \frac{g\beta(T_1 - T_2)\delta^3}{\nu^2} Pr \qquad (18\text{-}18)$$

where the characteristic length δ is the distance between the hot and cold surfaces, and T_1 and T_2 are the temperatures of the hot and cold surfaces, respectively. All fluid properties are to be evaluated at the average fluid temperature $T_{av} = \frac{1}{2}(T_1 + T_2)$.

Simple empirical correlations for the Nusselt number for various enclosures are given in Table 18-2. Once the Nusselt number is available, the heat transfer coefficient and the rate of heat transfer through the enclosure can be determined from

$$h = \frac{k}{\delta} Nu \qquad (18\text{-}19)$$

and

$$\dot{Q} = hA(T_1 - T_2) = k \, Nu \, A \, \frac{T_1 - T_2}{\delta} \qquad (18\text{-}20)$$

where

$$A = \begin{cases} HL & \text{rectangular enclosures} \\ \dfrac{\pi L(D_2 - D_1)}{\ln(D_2/D_1)} & \text{concentric cylinders} \\ \pi D_1 D_2 & \text{concentric spheres} \end{cases} \qquad (18\text{-}21)$$

For *inclined rectangular enclosures,* highly accurate but complex correlations are available in the literature. In the absence of such relations, the Nusselt number correlations for vertical enclosures can be used for inclined enclosures heated from below for inclination angles up to about $\theta = 20°$ from the vertical by replacing g in the Ra relation by $g \cos \theta$ (Fig. 18-14).

Effective Thermal Conductivity

You will recall from Chap. 3 that the rate of steady heat conduction across a layer of thickness δ, surface area A, and thermal conductivity k is

$$\dot{Q}_{cond} = kA \, \frac{T_1 - T_2}{\delta} \qquad (18\text{-}39)$$

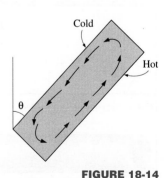

FIGURE 18-14

An inclined rectangular enclosure heated from below.

TABLE 18-2

Empirical correlations for the average Nusselt number for natural convection in enclosures (the characteristic length δ is as indicated on the respective diagram)

Geometry	Fluid	H/δ	Range of Pr	Range of Ra	Nusselt number	
Vertical rectangular enclosure (or vertical cylindrical enclosure)	Gas or liquid	—	—	Ra < 2000	Nu = 1	(18-22)
	Gas	11–42	0.5–2	2×10^3–2×10^5	$Nu = 0.197 Ra^{1/4} \left(\dfrac{H}{d}\right)^{-1/9}$	(18-23)
	Gas	11–42	0.5–2	2×10^5–10^7	$Nu = 0.073 Ra^{1/3} \left(\dfrac{H}{d}\right)^{-1/9}$	(18-24)
	Liquid	10–40	1–20,000	10^4–10^7	$Nu = 0.042 Pr^{0.012} Ra^{1/4} \left(\dfrac{H}{d}\right)^{-0.3}$	(18-25)
	Liquid	1–40	1–20	10^6–10^9	$Nu = 0.046 Ra^{1/3}$	(18-26)
Inclined rectangular enclosure					Use the correlations for vertical enclosures as a first-degree approximation for $\theta \le 20°$ by replacing g in the Ra relation by $g \cos \theta$	
Horizontal rectangular enclosure (hot surface at the top)	Gas or liquid	—	—	—	Nu = 1	(18-27)
Horizontal rectangular enclosure (hot surface at the bottom)	Gas or liquid	—	—	Ra < 1700	Nu = 1	(18-28)
	Gas	—	0.5–2	1.7×10^3–7×10^3	$Nu = 0.059 Ra^{0.4}$	(18-29)
	Gas	—	0.5–2	7×10^3–3.2×10^5	$Nu = 0.212 Ra^{1/4}$	(18-30)
	Gas	—	0.5–2	Ra > 3.2×10^5	$Nu = 0.061 Ra^{1/3}$	(18-31)
	Liquid	—	1–5000	1.7×10^3–6×10^3	$Nu = 0.012 Ra^{0.6}$	(18-32)
	Liquid	—	1–5000	6×10^3–3.7×10^4	$Nu = 0.375 Ra^{0.2}$	(18-33)
	Liquid	—	1–20	3.7×10^4–10^8	$Nu = 0.13 Ra^{0.3}$	(18-34)
	Liquid	—	1–20	Ra > 10^8	$Nu = 0.057 Ra^{1/3}$	(18-35)
Concentric rectangular cylinders	Gas or liquid	—	1–5000	6.3×10^3–10^6	$Nu = 0.11 Ra^{0.29}$	(18-36)
		—	1–5000	10^6–10^8	$Nu = 0.40 Ra^{0.20}$	(18-37)
Concentric spheres	Gas or liquid	—	0.7–4000	10^2–10^9	$Nu = 0.228 Ra^{0.226}$	(18-38)

where T_1 and T_2 are the temperatures on the two sides of the layer. A comparison of this relation with Eq. 18-20 reveals that the convection heat transfer in an enclosure is analogous to heat conduction across the fluid layer in the enclosure provided that the thermal conductivity k is replaced by kNu. That is, the fluid in an enclosure behaves like a fluid whose thermal conductivity is kNu as a result of convection currents. Therefore, the quantity kNu is called the **effective thermal conductivity** of the enclosure. That is,

$$k_{\text{eff}} = k\text{Nu} \qquad (18\text{-}40)$$

Note that for the special case of Nu = 1, the effective thermal conductivity of the enclosure becomes equal to the conductivity of the fluid. This is expected since this case corresponds to pure conduction (Fig. 18-15).

EXAMPLE 18-3   Heat Loss through a Double-Pane Window

The vertical 0.8-m-high, 2-m-wide double-pane window shown in Fig. 18-16 consists of two sheets of glass separated by a 2-cm air gap at atmospheric pressure. If the glass surface temperatures across the air gap are measured to be 12°C and 2°C, determine the rate of heat transfer through the window.

Solution   Two glasses of a double-pane window are maintained at specified temperatures. The rate of heat transfer through the window is to be determined.

Assumptions   **1** Steady operating conditions exist. **2** Air is an ideal gas.

Properties   The properties of air at the average temperature of $T_{ave} = (T_1 + T_2)/2 = (12 + 2)/2 = 7°C = 280$ K and 1 atm pressure are (Table A-18)

$$k = 0.0246 \text{ W/m} \cdot °C \qquad Pr = 0.717$$

$$\nu = 1.40 \times 10^{-5} \text{ m}^2/\text{s} \qquad \beta = \frac{1}{T_f} = \frac{1}{280 \text{ K}} = 0.00357 \text{ K}^{-1}$$

Analysis   We have a rectangular enclosure filled with air. The characteristic length in this case is the distance between the two glasses, $\delta = 0.02$ m. Then the Rayleigh number becomes

$$\text{Ra} = \frac{g\beta(T_1 - T_2)\delta^3}{\nu^2} Pr$$

$$= \frac{(9.81 \text{ m/s}^2)(0.00357 \text{ K}^{-1})[(12 - 2) \text{ K}](0.02 \text{ m})^3}{(1.40 \times 10^{-5} \text{ m}^2/\text{s})^2} (0.717) = 1.02 \times 10^4$$

Then the natural convection Nusselt number in this case can be determined from Eq. 18-23 to be

$$\text{Nu} = 0.197 \text{ Ra}^{1/4} \left(\frac{H}{\delta}\right)^{-1/9} = 0.197(1.02 \times 10^4)^{1/4} \left(\frac{0.8 \text{ m}}{0.02 \text{ m}}\right)^{-1/9} = 1.32$$

Then

$$A = H \times L = (0.8 \text{ m})(2 \text{ m}) = 1.6 \text{ m}^2$$

and

$$\dot{Q} = k \text{ Nu } A \frac{T_1 - T_2}{\delta}$$

$$= (0.0246 \text{ W/m} \cdot °C)(1.32)(1.6 \text{ m}^2) \frac{(12 - 2)°C}{0.02 \text{ m}} = \textbf{25.9 W}$$

Therefore, heat will be lost through the window at a rate of 25.9 W.

Discussion   Recall that a Nusselt number of Nu = 1 for an enclosure corresponds to pure conduction heat transfer through the enclosure. The air in the

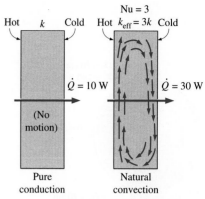

FIGURE 18-15

A Nusselt number of 3 for an enclosure indicates that heat transfer through the enclosure by *natural convection* is 3 times that by *pure conduction*.

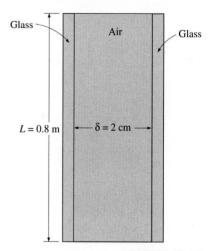

FIGURE 18-16

Schematic for Example 18-3.

enclosure in this case remains still, and no natural convection currents occur in the enclosure. The Nusselt number in our case is 1.32, which indicates that heat transfer through the enclosure is 1.32 times that by pure conduction. The increase in heat transfer is due to the natural convection currents that develop in the enclosure.

EXAMPLE 18-4   Heat Transfer through a Spherical Enclosure

The two concentric spheres of diameters $D_1 = 20$ cm and $D_2 = 30$ cm shown in Fig. 18-17 are separated by air at 1 atm pressure. The surface temperatures of the two spheres enclosing the air are $T_1 = 320$ K and $T_2 = 280$ K, respectively. Determine the rate of heat transfer from the inner sphere to the outer sphere by natural convection.

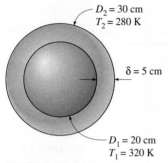

$D_2 = 30$ cm
$T_2 = 280$ K

$\delta = 5$ cm

$D_1 = 20$ cm
$T_1 = 320$ K

FIGURE 18-17
Schematic for Example 18-4.

Solution   Two surfaces of a spherical enclosure are maintained at specified temperatures. The rate of heat transfer through the enclosure is to be determined.

Assumptions   **1** Steady operating conditions exist. **2** Air is an ideal gas.

Properties   The properties of air at the average temperature of $T_{ave} = (T_1 + T_2)/2 = (320 + 280)/2 = 300$ K and 1 atm pressure are (Table A-18)

$$k = 0.0261 \text{ W/m} \cdot °C \qquad Pr = 0.712$$

$$\nu = 1.57 \times 10^{-5} \text{ m}^2/\text{s} \qquad \beta = \frac{1}{T_f} = \frac{1}{300 \text{ K}} = 0.00333 \text{ K}^{-1}$$

Analysis   We have a spherical enclosure filled with air. The characteristic length in this case is the distance between the two spheres, which is determined to be

$$\delta = \tfrac{1}{2}(D_2 - D_1) = \tfrac{1}{2}(0.3 - 0.2)\text{m} = 0.05 \text{ m}$$

Then the Rayleigh number becomes

$$Ra = \frac{g\beta(T_1 - T_2)\delta^3}{\nu^2} Pr$$

$$= \frac{(9.81 \text{ m/s}^2)(0.00333 \text{ K}^{-1})[(320 - 280) \text{ K}](0.05 \text{ m})^3}{(1.57 \times 10^{-5} \text{ m}^2/\text{s})^2}(0.712)$$

$$= 4.71 \times 10^5$$

Then the natural convection Nusselt number in this case can be determined from Eq. 18-38 to be

$$Nu = 0.228 \, Ra^{0.226} = 0.228(4.71 \times 10^5)^{0.226} = 4.37$$

That is, the air in the spherical enclosure will act like a stationary fluid whose thermal conductivity is 4.37 times that of air as a result of natural convection currents. Then,

$$A = \pi D_1 D_2 = \pi(0.2 \text{ m})(0.3 \text{ m}) = 0.188 \text{ m}^2$$

and

$$\dot{Q} = k \, Nu \, A \frac{T_1 - T_2}{\delta}$$

$$= (0.0261 \text{ W/m} \cdot °C)(4.37)(0.188 \text{ m}^2)\frac{(320 - 280) \text{ K}}{0.05 \text{ m}} = \textbf{17.2 W}$$

Therefore, heat will be lost from the inner sphere to the outer one at a rate of 17.2 W.

Discussion Assuming the surfaces of the spheres to be black (emissivity $\varepsilon = 1$), the rate of heat transfer between the two spheres by radiation is

$$\dot{Q}_{rad} = \varepsilon A_1 \sigma(T_1^4 - T_2^4)$$
$$= (1)\pi(0.2 \text{ m})^2(5.67 \times 10^{-8} \text{ W/m}^2 \cdot \text{K}^4)[(320 \text{ K})^4 - (280 \text{ K})^4]$$
$$= 30.9 \text{ W}$$

Thus, the maximum heat transfer by radiation is greater than the heat transfer by natural convection in this case. The emissivity of a real surface is less than 1, and thus the radiation heat transfer for a real enclosure will be less. But radiation can still be significant and needs to be considered.

EXAMPLE 18-5   Heating Water in a Tube by Solar Energy

A solar collector consists of a horizontal aluminum tube having an outer diameter of 2 in. enclosed in a concentric thin glass tube of 4-in.-diameter (Fig. 18-18). Water is heated as it flows through the tube, and the annular space between the aluminum and the glass tubes is filled with air at 1 atm pressure. The pump circulating the water fails during a clear day, and the water temperature in the tube starts rising. The aluminum tube absorbs solar radiation at a rate of 30 Btu/h per foot length, and the temperature of the ambient air outside is 70°F. Disregarding any heat loss by radiation, determine the temperature of the aluminum tube when steady operation is established (i.e., when the rate of heat loss from the tube equals the amount of solar energy gained by the tube).

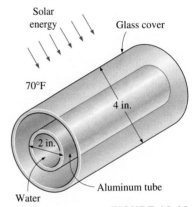

FIGURE 18-18
Schematic for Example 18-5.

Solution The circulating pump of a solar collector that consists of a horizontal tube and its glass cover fails one day. The temperature of the tube is determined when steady conditions are reached.

Assumptions 1 Steady operating conditions exist. 2 The tube and its cover are isothermal. 3 Air is an ideal gas. 4 Heat loss by radiation is negligible.

Properties The properties of air should be evaluated at the average temperature. But we do not know the exit temperature of the air in the duct, and thus we cannot determine the bulk fluid and glass cover temperatures at this point, and thus we cannot evaluate the average temperatures. Therefore, we will use the properties at an anticipated average temperature of 100°F (Table A-18E),

$$k = 0.0154 \text{ Btu/h} \cdot \text{ft} \cdot \text{°F} \qquad Pr = 0.72$$

$$\nu = 0.18 \times 10^{-3} \text{ ft}^2/\text{s} \qquad \beta = \frac{1}{T_f} = \frac{1}{(100 + 460) \text{ R}} = 0.001786 \text{ R}^{-1}$$

Analysis We have a horizontal cylindrical enclosure filled with air at 1 atm pressure. The problem involves heat transfer from the aluminum tube to the glass cover and from the outer surface of the glass cover to the surrounding ambient air. When steady operation is reached, these two heat transfer rates must equal the rate of heat gain. That is,

$$\dot{Q}_{tube-glass} = \dot{Q}_{glass-ambient} = \dot{Q}_{solar\ gain} = 30 \text{ Btu/h} \qquad \text{(per foot of tube)}$$

The heat transfer surface area of the glass cover is

$$A_2 = A_{glass} = (\pi DL)_{glass} = \pi\left(\frac{4}{12} \text{ ft}\right)(1 \text{ ft}) = 1.047 \text{ ft}^2 \qquad \text{(per foot of tube)}$$

To determine the Rayleigh number, we need to know the surface temperature of the glass, which is not available. Therefore, it is clear that the solution will require a trial-and-error approach. Assuming the glass cover temperature to be 100°F, the Rayleigh number, the Nusselt number, the convection heat transfer coefficient, and the rate of natural convection heat transfer from the glass cover to the ambient air are determined to be

$$\text{Ra} = \frac{g\beta(T_s - T_\infty)\delta^3}{\nu^2}\text{Pr}$$

$$= \frac{(32.2 \text{ ft/s}^2)(0.001786 \text{ R}^{-1})[(100 - 70)\text{ R}](\frac{4}{12}\text{ ft})^3}{(0.18 \times 10^{-3}\text{ ft}^2/\text{s})^2}(0.72) = 1.420 \times 10^6$$

$$\text{Nu} = \left\{0.6 + \frac{0.387 \text{ Ra}^{1/6}}{[1 + (0.559/\text{Pr})^{9/16}]^{8/27}}\right\}^2$$

$$= \left\{0.6 + \frac{0.387(1.420 \times 10^6)^{1/6}}{[1 + (0.559/0.72)^{9/16}]^{8/27}}\right\}^2 = 16.1$$

$$h = \frac{k}{D}\text{Nu} = \frac{0.0154 \text{ Btu/h} \cdot \text{ft} \cdot {}^\circ\text{F}}{\frac{4}{12}\text{ ft}}(16.1) = 0.743 \text{ Btu/h} \cdot \text{ft}^2 \cdot {}^\circ\text{F}$$

$$\dot{Q}_{\text{glass}} = hA_2(T_2 - T_\infty) = (0.743 \text{ Btu/h} \cdot \text{ft}^2 \cdot {}^\circ\text{F})(1.047 \text{ ft}^2)(100 - 70){}^\circ\text{F}$$
$$= 23.3 \text{ Btu/h}$$

which is less than 30 Btu/h. Therefore, the assumed temperature of 100°F for the glass cover is low. Repeating the calculations for a temperature of 110°F gives 33.8 Btu/h, which is high. Then the glass cover temperature corresponding to 30 Btu/h is determined by interpolation to be 106.4°F.

The temperature of the aluminum tube is determined in a similar manner using the natural convection relations for two horizontal concentric cylinders. The characteristic length in this case is the distance between the two cylinders, which is determined to be

$$\delta = \tfrac{1}{2}(D_2 - D_1) = \tfrac{1}{2}(4 - 2)\text{ in.} = 1\text{ in.}$$

Also,

$$A = \frac{\pi L(D_2 - D_1)}{\ln(D_2/D_1)} = \frac{\pi(1\text{ ft})(\frac{4}{12} - \frac{2}{12})\text{ ft}}{\ln(\frac{4}{2})} = 0.755 \text{ ft}^2$$

We start the calculations by assuming the tube temperature to be 200°F. This gives

$$\text{Ra} = \frac{g\beta(T_1 - T_2)\delta^3}{\nu^2}\text{Pr}$$

$$= \frac{(32.2 \text{ ft/s}^2)(0.001786 \text{ R}^{-1})[(200 - 107)\text{ R}](\frac{1}{12}\text{ ft})^3}{(0.18 \times 10^{-3}\text{ ft}^2/\text{s})^2}(0.72) = 6.88 \times 10^4$$

$$\text{Nu} = 0.11 \text{ Ra}^{0.29} = 0.11(6.88 \times 10^4)^{0.29} = 2.78$$

$$\dot{Q}_{\text{tube}} = k\text{ Nu }A\frac{T_1 - T_2}{\delta}$$

$$= (0.0154 \text{ Btu/h} \cdot \text{ft} \cdot {}^\circ\text{F})(2.78)(0.755 \text{ ft}^2)\frac{(200 - 107)\text{ R}}{\frac{1}{12}\text{ ft}} = 36.1 \text{ Btu/h}$$

which is more than 30 Btu/h. Therefore, the assumed temperature of 200°F for the tube is high. Repeating the calculations for a temperature of 180°F gives 26.4 Btu/h, which is low. Then the tube temperature corresponding to 30 Btu/h is determined by interpolation to be **188°F**. Therefore, the tube will reach an equilibrium temperature of 188°F when the pump fails.

This result above is obtained by using air properties at 100°F. It appears that this result can be improved by repeating the calculations above using air properties at the average temperature of 88.5°F for heat transfer from the glass cover to the ambient air and at 147.5°F for heat transfer from the tube to the glass cover. Also, we have not considered heat loss by radiation in the calculations, and thus the tube temperature determined above is probably too high. This problem is considered again in Chap. 19 by accounting for the effect of radiation heat transfer.

18-4 ■ COMBINED NATURAL AND FORCED CONVECTION

The presence of a temperature gradient in a fluid in a gravity field always gives rise to natural convection currents, and thus heat transfer by natural convection. Therefore, forced convection is always accompanied by natural convection.

We mentioned earlier that the convection heat transfer coefficient, natural or forced, is a strong function of the fluid velocity. Heat transfer coefficients encountered in forced convection are typically much higher than those encountered in natural convection because of the higher fluid velocities associated with forced convection. As a result, we tend to ignore natural convection in heat transfer analyses that involve forced convection, although we recognize that natural convection always accompanies forced convection. The error involved in ignoring natural convection is negligible at high velocities but may be considerable at low velocities associated with forced convection. Therefore, it is desirable to have a criterion to assess the relative magnitude of natural convection in the presence of forced convection.

For a given fluid, it is observed that the parameter Gr/Re^2 represents the importance of natural convection relative to forced convection. This is not surprising since the convection heat transfer coefficient is a strong function of the Reynolds number Re in forced convection and the Grashof number Gr in natural convection.

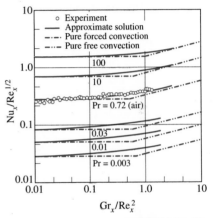

FIGURE 18-19

Variation of the local Nusselt number NU_x for combined natural and forced convection from a hot isothermal vertical plate (from Lloyd and Sparrow, Ref. 8).

A plot of the nondimensionalized heat transfer coefficient for combined natural and forced convection on a vertical plate is given in Fig. 18-19 for different fluids. We note from this figure that natural convection is negligible when $Gr/Re^2 < 0.1$, forced convection is negligible when $Gr/Re^2 > 10$, and neither is negligible when $0.1 < Gr/Re^2 < 10$. Therefore, both natural and forced convection must be considered in heat transfer calculations when the Gr and Re^2 are of the same order of magnitude (one is within a factor of 10 times the other). Note that forced convection is small relative to natural convection only in the rare case of extremely low forced flow velocities.

Natural convection may *help* or *hurt* forced convection heat transfer, depending on the relative directions of *buoyancy-induced* and the *forced convection* motions (Fig. 18-20):

1. In *assisting flow,* the buoyant motion is in the *same* direction as the forced motion. Therefore, natural convection assists forced convection and *enhances* heat transfer. An example is upward forced flow over a hot surface.

2. In *opposing flow,* the buoyant motion is in the *opposite* direction to the forced motion. Therefore, natural convection resists forced convection and *decreases* heat transfer. An example is upward forced flow over a cold surface.

3. In *transverse flow,* the buoyant motion is *perpendicular* to the forced motion. Transverse flow enhances fluid mixing and thus *enhances* heat transfer. An example is horizontal forced flow over a hot or cold cylinder or sphere.

When determining heat transfer under combined natural and forced convection conditions, it is tempting to add the contributions of natural and forced

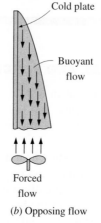

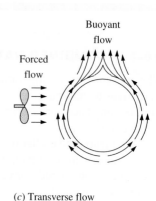

FIGURE 18-20

Natural convection can *enhance* or *inhibit* heat transfer, depending on the relative directions of *buoyancy-induced motion* and the *forced convection motion*.

(*a*) Assisting flow   (*b*) Opposing flow   (*c*) Transverse flow

convection in assisting flows and to subtract them in opposing flows. However, the evidence indicates differently. A review of experimental data suggests a correlation of the form

$$Nu_{combined} = (Nu_{forced}^n \pm Nu_{natural}^n)^{1/n} \qquad (18\text{-}41)$$

where Nu_{forced} and $Nu_{natural}$ are determined from the correlations for *pure forced* and *pure natural convection,* respectively. The plus sign is for *assisting* and *transverse* flows and the minus sign is for *opposing* flows. The value of the exponent n varies between 3 and 4, depending on the geometry involved. It is observed that $n = 3$ correlates experimental data for vertical surfaces well. Larger values of n are better suited for horizontal surfaces.

A question that frequently arises in the cooling of heat-generating equipment such as electronic components is whether to use a fan (or a pump if the cooling medium is a liquid)—that is, whether to utilize *natural* or *forced* convection in the cooling of the equipment. The answer depends on the maximum allowable operating temperature. Recall that the convection heat transfer rate from a surface at temperature T_s in a medium at T_∞ is given by

$$\dot{Q}_{conv} = hA(T_s - T_\infty)$$

where h is the convection heat transfer coefficient and A is the surface area. Note that for a fixed value of power dissipation and surface area, h and T_s are *inversely proportional.* Therefore, the device will operate at a *higher* temperature when h is low (typical of natural convection) and at a *lower* temperature when h is high (typical of forced convection).

Natural convection is the preferred mode of heat transfer since no blowers or pumps are needed and thus all the problems associated with these, such as noise, vibration, power consumption, and malfunctioning, are avoided. Natural convection is adequate for cooling *low-power-output* devices, especially when they are attached to extended surfaces such as heat sinks. For *high-power-output* devices, however, we have no choice but to use a blower or a pump to keep the operating temperature below the maximum allowable level. For *very-high-power-output* devices, even forced convection may not be sufficient to keep the surface temperature at the desirable levels. In such cases, we may have to use *boiling* and *condensation* to take advantage of the very high heat transfer coefficients associated with phase change processes.

18-5 ■ SUMMARY

In this chapter, we have considered *natural convection* heat transfer where any fluid motion occurs by natural means such as buoyancy. The fluid velocities associated with natural convection are low. Therefore, the heat transfer coefficients encountered in natural convection are usually much lower than those encountered in forced convection.

The upward force exerted by a fluid on a body completely or partially immersed in it is called the *buoyancy force*, whose magnitude is equal to the weight of the fluid displaced by the body. The *volume expansion coefficient* β of a substance represents the variation of the density of that substance with temperature at constant pressure and is defined as

$$\beta = \frac{1}{v}\left(\frac{\partial v}{\partial T}\right)_P = -\frac{1}{\rho}\left(\frac{\partial \rho}{\partial T}\right)_P \qquad (1/K)$$

For an ideal gas, it reduces to

$$\beta_{\text{ideal gas}} = \frac{1}{T} \qquad (1/K)$$

where T is the absolute temperature. The instrument used in natural convection experiments most often is the *Mach–Zehnder interferometer*, which gives a plot of isotherms in the fluid in the vicinity of a surface.

The flow regime in natural convection is governed by a dimensionless number called the *Grashof number*, which represents the ratio of the buoyancy force to the viscous force acting on the fluid and is expressed as

$$\text{Gr} = \frac{g\beta(T_s - T_\infty)\delta^3}{v^2}$$

where
g = gravitational acceleration, m/s²
β = coefficient of volume expansion, 1/K ($\beta = 1/T$ for ideal gases)
T_s = temperature of the surface, °C
T_∞ = temperature of the fluid sufficiently far from the surface, °C
δ = characteristic length of the geometry, m
v = kinematic viscosity of the fluid, m²/s

The Grashof number provides the main criterion in determining whether the fluid flow is laminar or turbulent in natural convection. The heat transfer rate in natural convection from a solid surface to the surrounding fluid is expressed by Newton's law of cooling as

$$\dot{Q}_{\text{conv}} = hA(T_s - T_\infty) \qquad (W)$$

where A is the heat transfer surface area and h is the average heat transfer coefficient on the surface.

Most heat transfer relations in natural convection are based on experimental studies, and the simple empirical correlations for the average *Nusselt number* Nu in natural convection are of the form

$$\text{Nu} = \frac{h\delta}{k} = C(\text{Gr Pr})^n = C\,\text{Ra}^n$$

where Ra is the *Rayleigh number*, which is the product of the Grashof and Prandtl numbers:

$$Ra = Gr\ Pr = \frac{g\beta(T_s - T_\infty)\delta^3}{\nu^2}\ Pr$$

The values of the constants C and n depend on the geometry of the surface and the flow regime, which is characterized by the range of the Rayleigh number. Simple relations for the average Nusselt number for various geometries are given in Table 18-1 together with a sketch of the geometry. All fluid properties are to be evaluated at the film temperature $T_f = \frac{1}{2}(T_s + T_\infty)$.

Simple empirical correlations for the Nusselt number for various enclosures are given in Table 18-2. Once the Nusselt number is available, the rate of heat transfer through the enclosure can be determined from

$$\dot{Q} = hA(T_1 - T_2) = k\ Nu\ A\ \frac{T_1 - T_2}{\delta}$$

where

$$A = \begin{cases} HL & \text{rectangular enclosures} \\ \dfrac{\pi L(D_2 - D_1)}{\ln(D_2/D_1)} & \text{concentric cylinders} \\ \pi D_1 D_2 & \text{concentric spheres} \end{cases}$$

The quantity $k\ Nu$ is called the *effective thermal conductivity* of the enclosure, since a fluid in an enclosure behaves like a quiescent fluid whose thermal conductivity is $k\ Nu$ as a result of convection currents.

For a given fluid, the parameter Gr/Re^2 represents the importance of natural convection relative to forced convection. Natural convection is negligible when $Gr/Re^2 < 0.1$, forced convection is negligible when $Gr/Re^2 > 10$, and neither is negligible when $0.1 < Gr/Re^2 < 10$.

REFERENCES AND SUGGESTED READING

1. S. W. Churchill. "Combined Free and Forced Convection around Immersed Bodies." *Heat Exchanger Design Handbook.* Section 2.5.9. New York: Hemisphere Publishing, 1986.

2. S. W. Churchill. "A Comprehensive Correlating Equation for Laminar Assisting Forced and Free Convection." *AIChE Journal* 23 (1977), pp. 10–16.

3. E. R. G. Eckerd and E. Soehngen. "Interferometric Studies on the Stability and Transition to Turbulence of a Free Convection Boundary Layer." *Proceedings of General Discussion, Heat Transfer ASME–IME,* London, 1951.

4. E. R. G. Eckerd and E. Soehngen. "Studies on Heat Transfer in Laminar Free Convection with Zehnder–Mach Interferometer." *USAF Technical Report 5747,* December 1948.

5. J. P. Holman. *Heat Transfer.* 7th ed. New York: McGraw-Hill, 1990.

6. F. P. Incropera and D. P. DeWitt. *Introduction to Heat Transfer.* 2nd ed. New York: John Wiley & Sons, 1990.

7. F. Kreith and M. S. Bohn. *Principles of Heat Transfer.* 5th ed. St. Paul, MN: West Publishing, 1993.

8. J. R. Lloyd and E. M. Sparrow. "Combined Forced and Free Convection Flow on Vertical Surfaces." *International Journal of Heat and Mass Transfer* 13 (1970), p. 434.

9. M. N. Özişik. *Heat Transfer—A Basic Approach.* New York: McGraw-Hill, 1985.

10. L. C. Thomas. *Heat Transfer.* Englewood Cliffs, NJ: Prentice Hall, 1992.

PROBLEMS*

Physical Mechanism of Natural Convection

18-1C   What is natural convection? How does it differ from forced convection? What force causes natural convection currents?

18-2C   In which mode of heat transfer is the convection heat transfer coefficient usually higher, natural convection or forced convection? Why?

18-3C   Consider a hot boiled egg in a spacecraft that is filled with air at atmospheric pressure and temperature at all times. Will the egg cool faster or slower when the spacecraft is in space instead of on the ground? Explain.

18-4C   What is buoyancy force? Compare the relative magnitudes of the buoyancy force acting on a body immersed in the following mediums: (*a*) air, (*b*) water, (*c*) mercury, and (*d*) an evacuated chamber.

18-5C   When will the hull of a ship sink in water deeper: when the ship is sailing in fresh water or in sea water? Why?

18-6C   A person weighs himself on a waterproof spring scale placed at the bottom of a 1-m-deep swimming pool. Will the person weigh more or less in water? Why?

18-7C   Consider two fluids, one with a large coefficient of volume expansion and the other with a small one. In what fluid will a hot surface initiate stronger natural convection currents? Why? Assume the viscosity of the fluids to be the same.

18-8C   Consider a fluid whose volume does not change with temperature at constant pressure. What can you say about natural convection heat transfer in this medium?

18-9C   What do the lines on an interferometer photograph represent? What do closely packed lines on the same photograph represent?

18-10C   Physically, what does the Grashof number represent? How does the Grashof number differ from the Reynolds number?

18-11   Show that the volume expansion coefficient of an ideal gas is $\beta = 1/T$, where T is the absolute temperature.

---

*Students are encouraged to answer *all* the concept "C" questions.

Natural Convection over Surfaces

18-12C How does the Rayleigh number differ from the Grashof number?

18-13C Under what conditions can the outer surface of a vertical cylinder be treated as a vertical plate in natural convection calculations?

18-14C Will a hot horizontal plate whose back side is insulated cool faster or slower when its hot surface is facing down instead of up?

18-15C Consider laminar natural convection from a vertical hot plate. Will the heat flux be higher at the top or at the bottom of the plate? Why?

18-16 An 8-m-long section of a 6-cm-diameter horizontal hot water pipe passes through a large room whose temperature is 22°C. If the temperature and the emissivity of the outer surface of the pipe are 65°C and 0.8, respectively, determine the rate of heat loss from the pipe by (*a*) natural convection and (*b*) radiation.

18-17 Consider a wall-mounted power transistor that dissipates 0.18 W of power in an environment at 35°C. The transistor is 0.45 cm long and has a diameter of 0.4 cm. The emissivity of the outer surface of the transistor is 0.1, and the average temperature of the surrounding surfaces is 25°C. Disregarding any heat transfer from the base surface, determine the surface temperature of the transistor. Use air properties at 100°C. *Answer:* 60°C

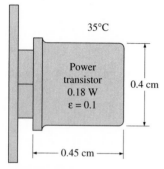

35°C

Power
transistor
0.18 W
$\varepsilon = 0.1$

0.4 cm

0.45 cm

FIGURE P18-17

18-18E Consider a 2-ft × 2-ft thin square plate in a room at 75°F. One side of the plate is maintained at a temperature of 130°F, while the other side is insulated. Determine the rate of heat transfer from the plate by natural convection if the plate is (*a*) vertical, (*b*) horizontal with hot surface facing up, and (*c*) horizontal with hot surface facing down.

18-19 A 500-W cylindrical resistance heater is 1 m long and 0.5 cm in diameter. The resistance wire is placed horizontally in a fluid at 20°C. Determine the outer surface temperature of the resistance wire in steady operation if the fluid is (*a*) air and (*b*) water. Ignore any heat transfer by radiation. Use properties at 1000 K for air and 40°C for water.

18-20 Water is boiling in a 12-cm-deep pan with an outer diameter of 25 cm that is placed on top of a stove. The ambient air and the surrounding surfaces are at a temperature of 25°C, and the emissivity of the outer surface of the pan is 0.95. Assuming the entire pan to be at an average temperature of 98°C, determine the rate of heat loss from the cylindrical side surface of the pan to the surroundings by (*a*) natural convection and (*b*) radiation. (*c*) If water is boiling at a rate of 2 kg/h at 100°C, determine the ratio of the heat lost from the side surfaces of the pan to that by the evaporation of water. The heat of vaporization of water at 100°C is 2257 kJ/kg.
Answers: 50 W, 56.1 W, 0.085

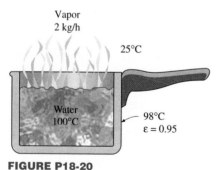

Vapor
2 kg/h

25°C

Water
100°C

98°C
$\varepsilon = 0.95$

FIGURE P18-20

18-21 Repeat Prob. 18-20 for a pan whose outer surface is polished and has an emissivity of 0.1.

18-22 In a plant that manufactures canned aerosol paints, the cans are temperature-tested in water baths at 55°C before they are shipped to ensure that they will withstand temperatures up to 55°C during transportation and shelving. The cans, moving on a conveyor, enter the open hot water bath, which is 0.5 m deep, 1 m wide, and 3.5 m long, and move slowly in the hot water toward the other end. Some of the cans fail the test and explode in the water

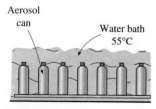

Aerosol
can

Water bath
55°C

FIGURE P18-22

bath. The water container is made of sheet metal, and the entire container is at about the same temperature as the hot water. The emissivity of the outer surface of the container is 0.7. If the temperature of the surrounding air and surfaces is 20°C, determine the rate of heat loss from the four side surfaces of the container (disregard the top surface, which is open).

The water is heated electrically by resistance heaters, and the cost of electricity is $0.085/kWh. If the plant operates 24 h a day 365 days a year and thus 8760 h a year, determine the annual cost of the heat losses from the container for this facility.

18-23 Reconsider Prob. 18-22. In order to reduce the heating cost of the hot water, it is proposed to insulate the side and bottom surfaces of the container with 5-cm-thick fiberglass insulation ($k = 0.035$ W/m · °C) and to wrap the insulation with aluminum foil ($\varepsilon = 0.1$) in order to minimize the heat loss by radiation. An estimate is obtained from a local insulation contractor, who proposes to do the insulation job for $350, including materials and labor. Would you support this proposal? How long will it take for the insulation to pay for itself from the energy it saves?

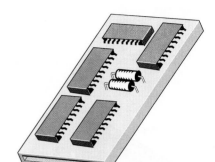

FIGURE P18-24

18-24 Consider a 15-cm × 20-cm printed circuit board (PCB) that has electronic components on one side. The board is placed in a room at 20°C. The heat loss from the back surface of the board is negligible. If the circuit board is dissipating 8 W of power in steady operation, determine the average temperature of the hot surface of the board, assuming the board is (a) vertical, (b) horizontal with hot surface facing up, and (c) horizontal with hot surface facing down. Take the emissivity of the surface of the board to be 0.8 and assume the surrounding surfaces to be at the same temperature as the air in the room.   *Answers:* (a) 46°C, (b) 42°C, (c) 50°C

18-25 A manufacturer makes absorber plates that are 1.2 m × 0.8 m in size for use in solar collectors. The back side of the plate is heavily insulated, while its front surface is coated with black chrome, which has an absorptivity of 0.87 for solar radiation and an emissivity of 0.09. Consider such a plate placed horizontally outdoors in calm air at 25°C. Solar radiation is incident on the plate at a rate of 700 W/m². Taking the effective sky temperature to be 10°C, determine the equilibrium temperature of the absorber plate. What would your answer be if the absorber plate is made of ordinary aluminum plate that has a solar absorptivity of 0.28 and an emissivity of 0.07?

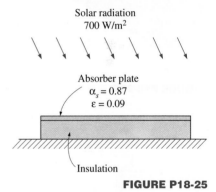

FIGURE P18-25

18-26 Repeat Prob. 18-25 for an aluminum plate painted flat black (solar absorptivity 0.98 and emissivity 0.98) and also for a plate painted white (solar absorptivity 0.26 and emissivity 0.90).

FIGURE P18-27

18-27 The following experiment is conducted to determine the natural convection heat transfer coefficient for a horizontal cylinder that is 50 cm long and 2 cm in diameter. A 50-cm-long resistance heater is placed along the centerline of the cylinder, and the surfaces of the cylinder are polished to minimize the radiation effect. The two circular side surfaces of the cylinder are well insulated. The resistance heater is turned on, and the power dissipation is maintained constant at 40 W. If the average surface temperature of the cylinder is measured to be 120°C in the 20°C room air when steady operation is reached, determine the natural convection heat transfer coefficient. If the emissivity of the outer surface of the cylinder is 0.1 and a 5 percent error is acceptable, do you think we need to do any correction for the radiation effect? Assume the surrounding surfaces to be at 20°C also.

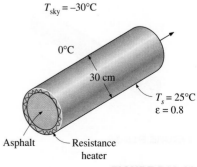

FIGURE P18-28

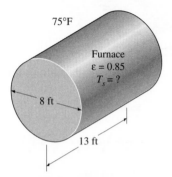

75°F

Furnace
ε = 0.85
T_s = ?

8 ft

13 ft

FIGURE P18-30E

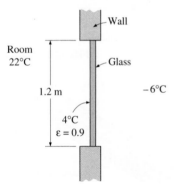

Wall

Room
22°C

Glass

1.2 m

−6°C

4°C
ε = 0.9

FIGURE P18-31

170°C
20°C   ε = 0.7

6.03 cm

60 m

Steam

FIGURE P18-33

18-28 Thick fluids such as asphalt and waxes and the pipes in which they flow are often heated in order to reduce the viscosity of the fluids and thus to reduce the pumping costs. Consider the flow of such a fluid through a 100-m-long pipe of outer diameter 30 cm in calm ambient air at 0°C. The pipe is heated electrically, and a thermostat keeps the outer surface temperature of the pipe constant at 25°C. The emissivity of the outer surface of the pipe is 0.8, and the effective sky temperature is −30°C, Determine the power rating of the electric resistance heater, in kW, that needs to be used. Also, determine the cost of electricity associated with heating the pipe during a 10-h period under the above conditions if the price of electricity is $0.09/kWh.
Answers: 29.2 kW, $26.3

18-29 Reconsider Prob. 18-28. To reduce the heating cost of the pipe, it is proposed to insulate it with sufficiently thick fiberglass insulation ($k = 0.035$ W/m · °C) wrapped with aluminum foil ($ε = 0.1$) to cut down the heat losses by 85 percent. Assuming the pipe temperature to remain constant at 25°C, determine the thickness of the insulation that needs to be used. How much money will the insulation save during this 10-h period?
Answers: 1.3 cm, $22.3

18-30E Consider an industrial furnace that resembles a 13-ft-long horizontal cylindrical enclosure 8 ft in diameter whose end surfaces are well insulated. The furnace burns natural gas at a rate of 48 therms/h (1 therm = 100,000 Btu). The combustion efficiency of the furnace is 82 percent (i.e., 18 percent of the chemical energy of the fuel is lost through the flue gases as a result of incomplete combustion and the flue gases leaving the furnace at high temperature). If the heat loss from the outer surfaces of the furnace by natural convection and radiation is not to exceed 1 percent of the heat generated inside, determine the highest allowable surface temperature of the furnace. Assume the air and wall surface temperature of the room to be 75°F, and take the emissivity of the outer surface of the furnace to be 0.85. If the cost of natural gas is $0.48/therm and the furnace operates 3000 h per year, determine the annual cost of this heat loss to the plant.

18-31 Consider a 1.2-m-high and 2-m-wide glass window with a thickness of 6 mm, thermal conductivity $k = 0.78$ W/m · °C, and emissivity $ε = 0.9$. The room and the walls that face the window are maintained at 22°C, and the average temperature of the inner surface of the window is measured to be 4°C. If the temperature of the outdoors is −6°C, determine (*a*) the convection heat transfer coefficient on the inner surface of the window, (*b*) the rate of total heat transfer through the window, and (*c*) the combined natural convection and radiation heat transfer coefficient on the outer surface of the window. Is it reasonable to neglect the thermal resistance of the glass in this case?

18-32 A 3-mm-diameter and 12-m-long electric wire is tightly wrapped with a 1.5-mm-thick plastic cover whose thermal conductivity and emissivity are $k = 0.15$ W/m · °C and $ε = 0.9$. Electrical measurements indicate that a current of 10 A passes through the wire and there is a voltage drop of 8 V along the wire. If the insulated wire is exposed to calm atmospheric air at $T_∞ = 30°C$, determine the temperature at the interface of the wire and the plastic cover in steady operation. Take the surrounding surfaces to be at about the same temperature as the air.

18-33 During a visit to a plastic sheeting plant, it was observed that a 60-m-long section of a 2-in. nominal (6.03-cm outer-diameter) steam pipe extended from one end of the plant to the other with no insulation on it. The temperature measurements at several locations revealed that the average temperature of the exposed surfaces of the steam pipe was 170°C, while the temperature of the surrounding air was 20°C. The outer surface of the pipe appeared to be oxidized, and its emissivity can be taken to be 0.7. Taking the temperature of the surrounding surfaces to be 20°C also, determine the rate of heat loss from the steam pipe.

Steam is generated in a gas furnace that has an efficiency of 78 percent, and the plant pays $0.538 per therm (1 therm = 105,500 kJ) of natural gas. The plant operates 24 h a day 365 days a year, and thus 8760 h a year. Determine the annual cost of the heat losses from the steam pipe for this facility.

18-34 Reconsider Prob. 18-33. In order to reduce heat losses, it is proposed to insulate the steam pipe with 5-cm-thick fiberglass insulation ($k = 0.038$ W/m · °C) and to wrap it with aluminum foil ($\varepsilon = 0.1$) in order to minimize the radiation losses. Also, an estimate is obtained from a local insulation contractor, who proposed to do the insulation job for $750, including materials and labor. Would you support this proposal? How long will it take for the insulation to pay for itself from the energy it saves? Assume the temperature of the steam pipe to remain constant at 170°C.

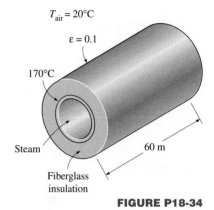

FIGURE P18-34

18-35 A 30-cm × 30-cm circuit board that contains 121 square chips on one side is to be cooled by combined natural convection and radiation by mounting it on a vertical surface in a room at 25°C. Each chip dissipates 0.05 W of power, and the emissivity of the chip surfaces is 0.7. Assuming the heat transfer from the back side of the circuit board to be negligible, and the temperature of the surrounding surfaces to be the same as the air temperature of the room, determine the surface temperature of the chips.

Answer: 33.4°C

18-36 Repeat Prob. 18-35 assuming the circuit board to be positioned horizontally with (*a*) chips facing up and (*b*) chips facing down.

18-37 The side surfaces of a 2-m-high cubic industrial furnace burning natural gas are not insulated, and the temperature at the outer surface of this section is measured to be 110°C. The temperature of the furnace room, including its surfaces, is 30°C, and the emissivity of the outer surface of the furnace is 0.7. It is proposed that this section of the furnace wall be insulated with glass wool insulation ($k = 0.038$ W/m · °C) wrapped by a reflective sheet ($\varepsilon = 0.2$) in order to reduce the heat loss by 90 percent. Assuming the outer surface temperature of the metal section still remains at about 110°C, determine the thickness of the insulation that needs to be used.

The furnace operates continuously throughout the year and has an efficiency of 78 percent. The price of the natural gas is $0.55/therm (1 therm = 105,500 kJ of energy content). If the installation of the insulation will cost $550 for materials and labor, determine how long it will take for the insulation to pay for itself from the energy it saves.

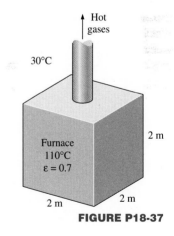

FIGURE P18-37

18-38 A 1.5-m-diameter, 5-m-long cylindrical propane tank is initially filled with liquid propane, whose density is 581 kg/m³. The tank is exposed to the ambient air at 25°C in calm weather. The outer surface of the tank is polished so that the radiation heat transfer is negligible. Now a crack develops at the top of the tank, and the pressure inside drops to 1 atm while the temperature drops to −42°C, which is the boiling temperature of propane at 1 atm. The heat of vaporization of propane at 1 atm is 425 kJ/kg. The propane is slowly

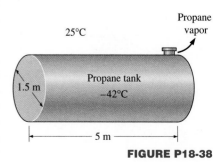

FIGURE P18-38

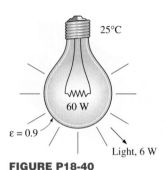

25°C

60 W

ε = 0.9

Light, 6 W

FIGURE P18-40

vaporized as a result of the heat transfer from the ambient air into the tank, and the propane vapor escapes the tank at −42°C through the crack. Assuming the propane tank to be at about the same temperature as the propane inside at all times, determine how long it will take for the tank to empty if it is not insulated.

18-39E An average person generates heat at a rate of 287 Btu/h while resting in a room at 77°F. Assuming one-quarter of this heat is lost from the head and taking the emissivity of the skin to be 0.9, determine the average surface temperature of the head when it is not covered. The head can be approximated as a 12-in.-diameter sphere, and the interior surfaces of the room can be assumed to be at the room temperature.

18-40 An incandescent light bulb is an inexpensive but highly inefficient device that converts electrical energy into light. It converts about 10 percent of the electrical energy it consumes into light while converting the remaining 90 percent into heat. The glass bulb of the lamp heats up very quickly as a result of absorbing all that heat and dissipating it to the surroundings by convection and radiation. Consider an 8-cm-diameter 60-W light bulb in a room at 25°C. The emissivity of the glass is 0.9. Assuming that 10 percent of the energy passes through the glass bulb as light with negligible absorption and the rest of the energy is absorbed and dissipated by the bulb itself by natural convection and radiation, determine the equilibrium temperature of the glass bulb. Assume the interior surfaces of the room to be at room temperature.
    Answer: 169°C

Natural Convection inside Enclosures

18-41C The upper and lower compartments of a well-insulated container are separated by two parallel sheets of glass with an air space between them. One of the compartments is to be filled with a hot fluid and the other with a cold fluid. If it is desired that heat transfer between the two compartments be minimal, would you recommend putting the hot fluid into the upper or the lower compartment of the container? Why?

18-42C Someone claims that the air space in a double-pane window enhances the heat transfer from a house because of the natural convection currents that occur in the air space and recommends that the double-pane window be replaced by a single sheet of glass whose thickness is equal to the sum of the thicknesses of the two glasses of the double-pane window to save energy. Do you agree with this claim?

18-43C Consider a double-pane window consisting of two glass sheets separated by a 1-cm-wide air space. Someone suggests inserting a thin vinyl sheet in the middle of the two glasses to form two 0.5-cm-wide compartments in the window in order to reduce natural convection heat transfer through the window. From a heat transfer point of view, would you be in favor of this idea to reduce heat losses through the window?

18-44C What does the effective conductivity of an enclosure represent? How is the ratio of the effective conductivity to thermal conductivity related to the Nusselt number?

18-45 Show that the thermal resistance of a rectangular enclosure can be expressed as $R = \delta/(Ak \, \text{Nu})$, where k is the thermal conductivity of the fluid in the enclosure.

18-46E A vertical 4-ft-high and 6-ft-wide double-pane window consists of two sheets of glass separated by a 1-in. air gap at atmospheric pressure. If the glass surface temperatures across the air gap are measured to be 65°F and 40°F, determine the rate of heat transfer through the window by (*a*) natural convection and (*b*) radiation. Also, determine the *R*-value of insulation of this window such that multiplying the inverse of the *R*-value by the surface area and the temperature difference gives the total rate of heat transfer through the window. The effective emissivity for use in radiation calculations between two large parallel glass plates can be taken to be 0.82.

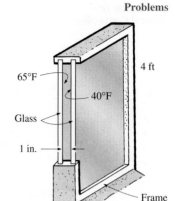

FIGURE P18-46E

18-47 Two concentric spheres of diameters 15 cm and 25 cm are separated by air at 1 atm pressure. The surface temperatures of the two spheres enclosing the air are $T_1 = 350$ K and $T_2 = 275$ K, respectively. Determine the rate of heat transfer from the inner sphere to the outer sphere by natural convection.

18-48 Flat-plate solar collectors are often tilted up toward the sun in order to intercept a greater amount of direct solar radiation. The tilt angle from the horizontal also affects the rate of heat loss from the collector. Consider a 2-m-high and 3-m-wide solar collector that is tilted at an angle θ from the horizontal. The back side of the absorber is heavily insulated. The absorber plate and the glass cover, which are spaced 2.5 cm from each other, are maintained at temperatures of 80°C and 32°C, respectively. Determine the rate of heat loss from the absorber plate by natural convection for $\theta = 0°$, 20°, and 90°.

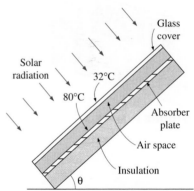

FIGURE P18-48

18-49 A simple solar collector is built by placing a 5-cm-diameter clear plastic tube around a garden hose whose outer diameter is 1.6 cm. The hose is painted black to maximize solar absorption, and some plastic rings are used to keep the spacing between the hose and the clear plastic cover constant. During a clear day, the temperature of the hose is measured to be 65°C, while the ambient air temperature is 26°C. Determine the rate of heat loss from the water in the hose per meter of its length by natural convection. Also, discuss how the performance of this solar collector can be improved.
    Answer: 6.5 W

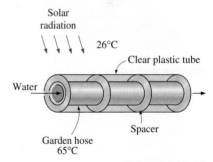

FIGURE P18-49

Combined Natural and Forced Convection

18-50C When is natural convection negligible and when is it not negligible in forced convection heat transfer?

18-51C Under what conditions does natural convection enhance forced convection, and under what conditions does it hurt forced convection?

18-52C When neither natural nor forced convection is negligible, is it correct to calculate each independently and add them to determine the total convection heat transfer?

18-53 Consider a 5-m-long vertical plate at 85°C in air at 30°C. Determine the forced motion velocity above which natural convection heat transfer from this plate is negligible.   *Answer:* 9.02 m/s

18-54 Consider a 3-m-long vertical plate at 60°C in water at 25°C. Determine the forced motion velocity above which natural convection heat transfer from this plate is negligible. Take $\beta = 0.0004$ K^{-1} for water.

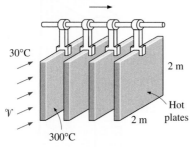

FIGURE P18-55

30°C

𝒱

300°C

2 m

2 m

Hot
plates

FIGURE P18-57E

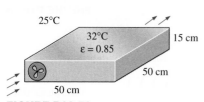

25°C

32°C
$\varepsilon = 0.85$

50 cm

50 cm

15 cm

FIGURE P18-59

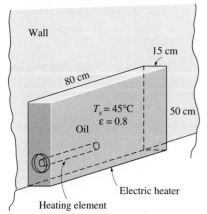

Wall

80 cm

$T_s = 45°C$
$\varepsilon = 0.8$

Oil

15 cm

50 cm

Electric heater

Heating element

FIGURE P18-60

18-55 In a production facility, thin square plates 2 m × 2 m in size coming out of the oven at 300°C are cooled by blowing ambient air at 30°C horizontally parallel to their surfaces. Determine the air velocity above which the natural convection effects on heat transfer are less than 10 percent and thus are negligible.

18-56 A 12-cm-high and 20-cm-wide circuit board houses 100 closely spaced logic chips on its surface, each dissipating 0.05 W. The board is cooled by a fan that blows air over the hot surface of the board at 35°C at a velocity of 0.5 m/s. The heat transfer from the back surface of the board is negligible. Determine the average temperature on the surface of the circuit board assuming the air flows vertically upwards along the 12-cm-long side by (a) ignoring natural convection and (b) considering the contribution of natural convection. Disregard any heat transfer by radiation.

Review Problems

18-57E A 0.1-W small cylindrical resistor mounted on a lower part of a vertical circuit board is 0.3 in. long and has a diameter of 0.2 in. The view of the resistor is largely blocked by another circuit board facing it, and the heat transfer through the connecting wires is negligible. The air is free to flow through the large parallel flow passages between the boards as a result of natural convection currents. If the air temperature at the vicinity of the resistor is 120°F, determine the approximate surface temperature of the resistor.
Answer: 212°F

18-58 An ice chest whose outer dimensions are 30 cm × 40 cm × 40 cm is made of 3-cm-thick styrofoam ($k = 0.033$ W/m · °C). Initially, the chest is filled with 40 kg of ice at 0°C, and the inner surface temperature of the ice chest can be taken to be 0°C at all times. The heat of fusion of water at 0°C is 333.7 kJ/kg, and the surrounding ambient air is at 20°C. Disregarding any heat transfer from the 40 cm × 40 cm base of the ice chest, determine how long it will take for the ice in the chest to melt completely if the ice chest is subjected to (a) calm air and (b) winds at 50 km/h. Assume the heat transfer coefficient on the front, back, and top surfaces to be the same as that on the side surfaces.

18-59 An electronic box that consumes 180 W of power is cooled by a fan blowing air into the box enclosure. The dimensions of the electronic box are 15 cm × 50 cm × 50 cm, and all surfaces of the box are exposed to the ambient except the base surface. Temperature measurements indicate that the box is at an average temperature of 32°C when the ambient temperature and the temperature of the surrounding walls are 25°C. If the emissivity of the outer surface of the box is 0.85, determine the fraction of the heat lost from the outer surfaces of the electronic box.

18-60 A 6-m internal-diameter spherical tank made of 1.5-cm-thick stainless steel ($k = 15$ W/m · °C) is used to store iced water at 0°C. The walls of the room are also at 20°C. The outer surface of the tank is black (emissivity $\varepsilon = 1$), and heat transfer between the outer surface of the tank and the surroundings is by natural convection and radiation. Assuming the entire steel tank to be at 0°C and thus the thermal resistance of the tank to be negligible, determine (a) the rate of heat transfer to the iced water in the tank and (b) the amount of ice at 0°C that melts during a 24-h period.
Answers: (a) 15.4 kW, (b) 3986 kg

18-61 Consider a 1.2-m-high and 2-m-wide double-pane window consisting of two 3-mm-thick layers of glass ($k = 0.78$ W/m · °C) separated by an 11-mm-wide air space. Determine the steady rate of heat transfer through this window and the temperature of its inner surface for a day during which the room is maintained at 24°C while the temperature of the outdoors is −5°C. Take the heat transfer coefficients on the inner and outer surfaces of the window to be $h_1 = 10$ W/m² · °C and $h_2 = 25$ W/m² · °C and disregard any heat transfer by radiation.

18-62 An electric resistance space heater is designed such that it resembles a rectangular box 50 cm high, 80 cm long, and 15 cm wide filled with 45 kg of oil. The heater is to be placed against a wall, and thus heat transfer from its back surface is negligible for safety considerations. The surface temperature of the heater is not to exceed 45°C in a room at 25°C. Disregarding heat transfer from the bottom and top surfaces of the heater in anticipation that the top surface will be used as a shelf, determine the power rating of the heater in W. Take the emissivity of the outer surface of the heater to be 0.8 and the average temperature of the ceiling and wall surfaces to be the same as the room air temperature.

Also, determine how long it will take for the heater to reach steady operation when it is first turned on (i.e., for the oil temperature to rise from 25°C to 45°C). State your assumptions in the calculations.

18-63 Skylights or "roof windows" are commonly used in homes and manufacturing facilities since they let natural light in during day time and thus reduce the lighting costs. However, they offer little resistance to heat transfer, and large amounts of energy are lost through them in winter unless they are equipped with a motorized insulating cover that can be used in cold weather and at nights to reduce heat losses. Consider a 1-m-wide and 2.5-m-long horizontal skylight on the roof of a house that is kept at 20°C. The glazing of the skylight is made of a single layer of 0.5-cm-thick glass ($k = 0.78$ W/m · °C and $\varepsilon = 0.9$). Determine the rate of heat loss through the skylight when the air temperature outside is −8°C and the effective sky temperature is −30°C. Compare your result with the rate of heat loss through an equivalent surface area of the roof that has a common R-5.34 construction in SI units (i.e., a thickness–to–effective-thermal-conductivity ratio of 5.34 m² · °C/W).

18-64 A solar collector consists of a horizontal copper tube of outer diameter 5 cm enclosed in a concentric thin glass tube of 9 cm diameter. Water is heated as it flows through the tube, and the annular space between the copper and glass tube is filled with air at 1 atm pressure. During a clear day, the temperatures of the tube surface and the glass cover are measured to be 60°C and 32°C, respectively. Determine the rate of heat loss from the collector by natural convection per meter length of the tube.   *Answer:* 14.9 W

18-65 A solar collector consists of a horizontal aluminum tube of outer diameter 4 cm enclosed in a concentric thin glass tube of 7 cm diameter. Water is heated as it flows through the aluminum tube, and the annular space between the aluminum and glass tubes is filled with air at 1 atm pressure. The pump circulating the water fails during a clear day, and the water temperature in the tube starts rising. The aluminum tube absorbs solar radiation at a rate of 20 W per meter length, and the temperature of the ambient air outside is 30°C. Approximating the surfaces of the tube and the glass cover as being black (emissivity $\varepsilon = 1$) in radiation calculations and taking the effective sky tem-

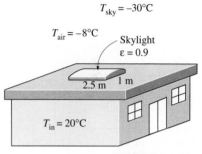

$T_{sky} = -30°C$

$T_{air} = -8°C$

Skylight
$\varepsilon = 0.9$

2.5 m   1 m

$T_{in} = 20°C$

FIGURE P18-63

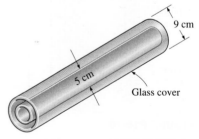

9 cm

5 cm

Glass cover

FIGURE P18-64

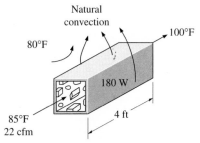

FIGURE P18-66E

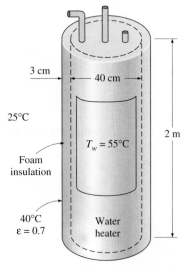

FIGURE P18-70

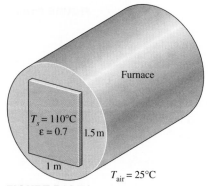

FIGURE P18-71

perature to be 10°C, determine the temperature of the aluminum tube when equilibrium is established (i.e., when the net heat loss from the tube by convection and radiation equals the amount of solar energy absorbed by the tube).

18-66E The components of an electronic system dissipating 180 W are located in a 4-ft-long horizontal duct whose cross-section is 6 in. × 6 in. The components in the duct are cooled by forced air, which enters at 85°F at a rate of 22 cfm and leaves at 100°F. The surfaces of the sheet metal duct are not painted, and thus radiation heat transfer from the outer surfaces is negligible. If the ambient air temperature is 80°F, determine (*a*) the heat transfer from the outer surfaces of the duct to the ambient air by natural convection and (*b*) the average temperature of the duct.

18-67E Repeat Prob. 18-66E for a circular horizontal duct of diameter 4 in.

18-68E Repeat Prob. 18-66E assuming the fan fails and thus the entire heat generated inside the duct must be rejected to the ambient air by natural convection through the outer surfaces of the duct.

18-69 Consider a cold aluminum canned drink that is initially at a uniform temperature of 5°C. The can is 12.5 cm high and has a diameter of 6 cm. The emissivity of the outer surface of the can is 0.6. Disregarding any heat transfer from the bottom surface of the can, determine how long it will take for the average temperature of the drink to rise to 7°C if the surrounding air and surfaces are at 25°C. *Answer:* 11.7 min

18-70 Consider a 2-m-high electric hot water heater that has a diameter of 40 cm and maintains the hot water at 55°C. The tank is located in a small room at 25°C whose walls and the ceiling are at about the same temperature. The tank is placed in a 46-cm-diameter sheet metal shell of negligible thickness, and the space between the tank and the shell is filled with foam insulation. The average temperature and emissivity of the outer surface of the shell are 40°C and 0.7, respectively. The price of electricity is $0.08/kWh. Hot water tank insulation kits large enough to wrap the entire tank are available on the market for about $30. If such an insulation is installed on this water tank by the home owner himself, how long will it take for this additional insulation to pay for itself? Disregard any heat loss from the top and bottom surfaces, and assume the insulation to reduce the heat losses by 80 percent.

18-71 During a plant visit, it was observed that a 1.5-m-high and 1-m-wide section of the vertical front section of a natural gas furnace wall was too hot to touch. The temperature measurements on the surface revealed that the average temperature of the exposed hot surface was 110°C, while the temperature of the surrounding air was 25°C. The surface appeared to be oxidized, and its emissivity can be taken to be 0.7. Taking the temperature of the surrounding surfaces to be 25°C also, determine the rate of heat loss from this furnace.

The furnace has an efficiency of 79 percent, and the plant pays $0.58 per therm (1 therm = 105,500 kJ) of natural gas. If the plant operates 10 h a day, 260 days a year, and thus 2600 h a year, determine the annual cost of the heat loss from this vertical hot surface on the front section of the furnace wall.

18-72 A group of 25 power transistors, dissipating 1.5 W each, are to be cooled by attaching them to a black-anodized square aluminum plate and mounting the plate on the wall of a room at 30°C. The emissivity of the transistor and the plate surfaces is 0.9. Assuming the heat transfer from the back side of the plate to be negligible and the temperature of the surrounding surfaces to be the same as the air temperature of the room, determine the size of the plate if the average surface temperature of the plate is not to exceed 50°C.   *Answer:* 43 cm × 43 cm

18-73 Repeat Prob. 18-72 assuming the plate to be positioned horizontally with (*a*) transistors facing up and (*b*) transistors facing down.

18-74E Hot water is flowing at an average velocity of 4 ft/s through a cast iron pipe ($k = 30$ Btu/h · ft · °F) whose inner and outer diameters are 1.0 in. and 1.2 in., respectively. The pipe passes through a 50-ft-long section of a basement whose temperature is 60°F. The emissivity of the outer surface of the pipe is 0.5, and the walls of the basement are also at about 60°F. If the inlet temperature of the water is 150°F and the heat transfer coefficient on the inner surface of the pipe is 30 Btu/h · ft^2 · °F, determine the temperature drop of water as it passes through the basement.

18-75 Consider a flat-plate solar collector placed horizontally on the flat roof of a house. The collector is 1.5 m wide and 6 m long, and the average temperature of the exposed surface of the collector is 42°C. Determine the rate of heat loss from the collector by natural convection during a calm day when the ambient air temperature is 15°C. Also, determine the heat loss by radiation by taking the emissivity of the collector surface to be 0.9 and the effective sky temperature to be −30°C.   *Answers:* 1295 W, 2921 W

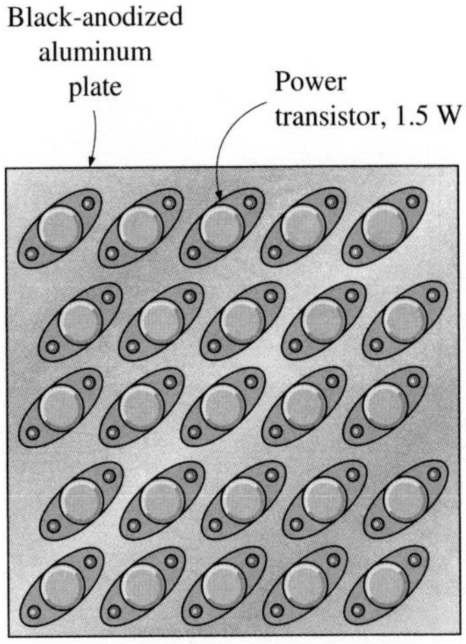

18-76 Solar radiation is incident on the glass cover of a solar collector at a rate of 650 W/m^2. The glass transmits 88 percent of the incident radiation and has an emissivity of 0.90. The hot water needs of a family in summer can be met completely by a collector 1.5 m high and 2 m wide, and tilted 40°C from the horizontal. The temperature of the glass cover is measured to be 35°C on a calm day when the surrounding air temperature is 23°C. The effective sky temperature for radiation exchange between the glass cover and the open sky is −40°C. Water enters the tubes attached to the absorber plate at a rate of 1 kg/min. Assuming the back surface of the absorber plate to be heavily insulated and the only heat loss occurs through the glass cover, determine (*a*) the total rate of heat loss from the collector, (*b*) the collector efficiency, which is the ratio of the amount of heat transferred to the water to the solar energy incident on the collector, and (*c*) the temperature rise of water as it flows through the collector.

Black-anodized aluminum plate

Power transistor, 1.5 W

FIGURE P18-72

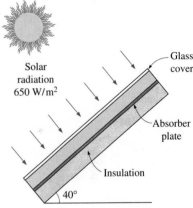

Solar radiation 650 W/m^2

Glass cover

Absorber plate

Insulation

40°

FIGURE P18-76

Computer, Design, and Essay Problems

18-77 Write a computer program to evaluate the variation of temperature with time of thin square metal plates that are removed from an oven at a specified temperature and placed vertically in a large room. The thickness, the size, the initial temperature, the emissivity, and the thermophysical properties of the plate as well as the room temperature are to be specified by the user. The program should evaluate the temperature of the plate at specified intervals and tabulate the results against time. The computer should list the assumptions made during calculations before printing the results.

For each step or time interval, assume the surface temperature to be constant and evaluate the heat loss during that time interval and the temperature drop of the plate as a result of this heat loss. This gives the temperature of the plate at the end of a time interval, which is to serve as the initial temperature of the plate for the beginning of the next time interval.

Try your program for 0.2-cm-thick vertical copper plates of 40 cm $\times$ 40 cm in size initially at 300°C cooled in a room at 25°C. Take the surface emissivity to be 0.9. Use a time interval of 1 s in calculations, but print the results at 10-s intervals for a total cooling period of 15 min.

18-78 Repeat Prob. 18-70 for a vertical slender cylindrical metal object instead of a square plate. The height and the diameter of the cylinder are to be specified by the user.

18-79 Write a computer program to optimize the spacing between the two glasses of a double-pane window. Assume the spacing is filled with dry air at atmospheric pressure. The program should evaluate the recommended practical value of the spacing to minimize the heat losses and list it when the size of the window (the height and the width) and the temperatures of the two glasses are specified.

18-80 Contact a manufacturer of aluminum heat sinks and obtain their product catalog for cooling electronic components by natural convection and radiation. Write an essay on how to select a suitable heat sink for an electronic component when its maximum power dissipation and maximum allowable surface temperature are specified.

18-81 The top surfaces of practically all flat-plate solar collectors are covered with glass in order to reduce the heat losses from the absorber plate underneath. Although the glass cover reflects or absorbs about 15 percent of the incident solar radiation, it saves much more from the potential heat losses from the absorber plate, and thus it is considered to be an essential part of a well-designed solar collector. Inspired by the energy efficiency of double-pane windows, someone proposes to use double glazing on solar collectors instead of a single glass. Investigate if this is a good idea for the town in which you live. Use local weather data and base your conclusion on heat transfer analysis and economic considerations.

Radiation Heat Transfer

So far, we have considered the conduction and convection modes of heat transfer, which are related to the nature of the materials involved and the presence of fluid motion, among other things. We now turn our attention to the third mechanism of heat transfer: *radiation,* which is characteristically different from the other two.

We start this chapter with a discussion of *electromagnetic waves* and the *electromagnetic spectrum,* with particular emphasis on *thermal radiation.* Then we introduce the idealized *blackbody, blackbody radiation,* and the *blackbody radiation function,* together with the *Stefan–Boltzmann law, Planck's distribution law,* and *Wien's displacement law.* This is followed by a discussion of radiation properties of materials such as *emissivity, absorptivity, reflectivity,* and *transmissivity* and their dependence on wavelength and temperature. The *greenhouse effect* is presented as an example of the consequences of the wavelength dependence of radiation properties. A separate section is devoted to the discussions of *atmospheric* and *solar radiation* because of their importance.

The second part of this chapter starts with a discussion of *view factors* and the rules associated with them. View factor *expressions* and *charts* for some common configurations are given, and the *crossed-strings method* is presented. We then discuss *radiation heat transfer,* first between black surfaces and then between nonblack surfaces. Finally, we consider *radiation shields* and discuss the *radiation effect* on temperature measurements and comfort.

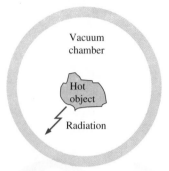

FIGURE 19-1

A hot object in a vacuum chamber loses heat by radiation only.

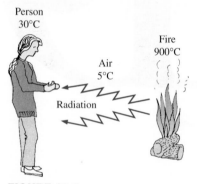

FIGURE 19-2

Unlike conduction and convection, heat transfer by radiation can occur between two bodies, even when they are separated by a medium colder than both of them.

19-1 ■ INTRODUCTION

Consider a hot object that is placed in an evacuated chamber whose walls are at room temperature (Fig. 19-1). Our experience tells us that the hot object will eventually cool down and reach thermal equilibrium with its surroundings. That is, it will lose heat until its temperature reaches the temperature of the walls of the chamber. Heat transfer between the object and the chamber could not have taken place by conduction or convection, because these two mechanisms cannot occur in a vacuum. Therefore, heat transfer must have occurred through another mechanism that involves the emission of the sensible internal energy of the object. This mechanism is *radiation.*

Radiation differs from the other two heat transfer mechanisms in that it does not require the presence of a material medium to take place. In fact, energy transfer by radiation is fastest (at the speed of light) and it suffers no attenuation in a *vacuum.* Also, radiation transfer occurs in solids as well as liquids and gases. In most practical applications, all three modes of heat transfer occur concurrently at varying degrees. But heat transfer through an evacuated space can occur only by radiation. For example, the energy of the sun reaches the earth by radiation.

You will recall that heat transfer by conduction or convection takes place in the direction of decreasing temperature; that is, from a high-temperature medium to a lower-temperature one. It is interesting that radiation heat transfer can occur between two bodies separated by a medium colder than both bodies (Fig. 19-2). For example, solar radiation reaches the surface of the earth after passing through extremely cold air layers at high altitudes. Also, the radiation-absorbing surfaces inside a greenhouse reach high temperatures even when its plastic or glass cover remains relatively cool.

The theoretical foundation of radiation was established in 1864 by physicist James Clerk Maxwell, who postulated that accelerated charges or changing electric currents give rise to electric and magnetic fields. These rapidly moving fields are called **electromagnetic waves** or **electromagnetic radiation,** and they represent the energy emitted by matter as a result of the changes in the electronic configurations of the atoms or molecules. In 1887, Heinrich Hertz experimentally demonstrated the existence of such waves. Electromagnetic waves transport energy just like other waves, and all electromagnetic waves travel at the *speed of light.* Electromagnetic waves are characterized by their *frequency* ν or *wavelength* λ. These two properties in a medium are related by

$$\lambda = \frac{c}{\nu} \qquad (19\text{-}1)$$

where c is the speed of light in that medium. In a vacuum, $c = c_0 = 2.998 \times 10^8$ m/s. The speed of light in a medium is related to the speed of light in a vacuum by $c = c_0/n$, where n is the *index of refraction* of that medium. The index of refraction is essentially unity for air and most gases and about 1.5 for water and glass. The commonly used unit of wavelength is the *micrometer* (μm), where 1 μm $= 10^{-6}$ m. Unlike the wavelength and the speed of propagation, the frequency of an electromagnetic wave depends only on the source and is independent of the medium through which the wave travels. The *frequency* (the number of oscillations per second) of an electromagnetic wave can range from a few cycles to millions of cycles and even higher per second, depending on the source. Note from Eq. 19-1 that the wavelength and the frequency of electromagnetic radiation are inversely proportional.

In radiation studies, it has proven useful to view electromagnetic radiation as the propagation of a collection of discrete packets of energy called **photons** or **quanta,** as proposed by Max Planck in 1900 in conjunction with his *quantum theory.* In this view, each photon of frequency ν is considered to have an energy of

$$e = h\nu = \frac{hc}{\lambda} \qquad (19\text{-}2)$$

where $h = 6.625 \times 10^{-34}$ J $\cdot$ s is *Planck's constant.* Note from the second part of Eq. 19-2 that h and c are constants, and thus the energy of a photon is inversely proportional to its wavelength. Therefore, shorter-wavelength radiation possesses larger photon energies. It is no wonder that we try to avoid very-short-wavelength radiation such as gamma rays and X-rays since they are highly destructive.

19-2 ■ THERMAL RADIATION

Although all electromagnetic waves have the same general features, waves of different wavelength differ significantly in their behavior. The electromagnetic radiation encountered in practice covers a wide range of wavelengths, varying from less than 10^{-10} μm for cosmic rays to more than 10^{10} μm for electrical power waves. The **electromagnetic spectrum** also includes gamma rays, X-rays, ultraviolet radiation, visible light, infrared radiation, thermal radiation, microwaves, and radio waves, as shown in Fig. 19-3.

Different types of electromagnetic radiation are produced differently through different mechanisms. For example, *gamma rays* are produced by nuclear reactions, *X-rays* by the bombardment of metals with high-energy electrons, *microwaves* by special types of electron tubes such as klystrons and magnetrons, and *radio waves* by the excitation of some crystals or by the flow of alternating current through electric conductors.

The short-wavelength gamma rays and X-rays are primarily of concern to nuclear engineers, while the long-wavelength microwaves and radio waves are of concern to electrical engineers. The type of electromagnetic radiation that is pertinent to heat transfer is the **thermal radiation** emitted as a result of vibrational and rotational motions of molecules, atoms, and electrons of a substance. Temperature is a measure of the strength of these activities at the microscopic level, and the rate of thermal radiation emission increases with increasing temperature. Thermal radiation is continuously emitted by all matter whose temperature is above absolute zero. That is, everything around us such as walls, furniture, and our friends constantly emits (and absorbs) radiation (Fig. 19-4). Thermal radiation is also defined as the portion of the electromagnetic spectrum that extends from about 0.1 to 100 μm, since the radiation emitted by bodies due to their temperature falls almost entirely into this wavelength range. Thus, thermal radiation includes the entire visible and infrared (IR) radiation as well as a portion of the ultraviolet (UV) radiation.

What we call **light** is simply the *visible* portion of the electromagnetic spectrum that lies between 0.40 and 0.76 μm. Light is characteristically no different than other electromagnetic radiation, except that it happens to trigger the sensation of seeing in the human eye. Light or the visible spectrum consists of narrow bands of color from violet (0.40–0.44 μm) to red (0.63–0.76 μm), as shown in Table 19-1. The color of a surface depends on its ability to *reflect* certain wavelengths. For example, a surface that reflects radiation in

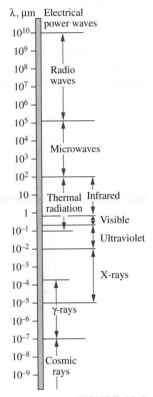

FIGURE 19-3

The electromagnetic wave spectrum.

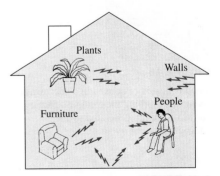

FIGURE 19-4

Everything around us constantly emits thermal radiation.

TABLE 19-1

The wavelength ranges of different colors

Color	Wavelength band
Violet	0.40-0.44 μm
Blue	0.44-0.49 μm
Green	0.49-0.54 μm
Yellow	0.54-0.60 μm
Orange	0.60-0.673 μm
Red	0.63-0.76 μm

FIGURE 19-5

Food is heated or cooked in a microwave oven by absorbing the electromagnetic radiation energy generated by the magnetron of the oven.

the wavelength range 0.63–0.76 μm while absorbing the rest of the visible radiation appears red to the eye. A surface that reflects all of the light appears *white,* while a surface that absorbs all of the light incident on it appears *black.* Then how do we see a black body?

A body that emits some radiation in the visible range is called a light source. The sun is obviously our primary light source. The electromagnetic radiation emitted by the sun is known as **solar radiation,** and nearly all of it falls into the wavelength band 0.3–3 μm. Almost *half* of solar radiation is light (i.e., it falls into the visible range), with the remaining being ultraviolet and infrared.

The radiation emitted by bodies at room temperature falls into the **infrared** region of the spectrum, which extends from 0.76 to 100 μm. Bodies start emitting noticeable visible radiation at temperatures above 800 K. The tungsten filament of a light bulb must be heated to temperatures above 2000 K before it can emit any significant amount of radiation in the visible range.

The **ultraviolet** radiation includes the low-wavelength end of the thermal radiation spectrum and lies between the wavelengths 0.01 and 0.40 μm. Ultraviolet rays are to be avoided since they can kill microorganisms and cause serious damage to humans and other living organisms. *About 12 percent of solar radiation is in the ultraviolet range,* and it would be devastating if it were to reach the surface of the earth. Fortunately, the ozone (O_3) layer in the atmosphere acts as a protective blanket and absorbs most of this ultraviolet radiation. The ultraviolet rays that remain in sunlight are still sufficient to cause serious sunburns to sun worshippers, and prolonged exposure to direct sunlight is the leading cause of skin cancer, which can be lethal. Recent discoveries of "holes" in the ozone layer have prompted the international community to ban the use of ozone-destroying chemicals such as the widely used refrigerant Freon-12 in order to save the earth. Ultraviolet radiation is also produced artificially in fluorescent lamps for use in medicine as a bacteria killer and in tanning parlors as an artificial tanner. The connection between skin cancer and ultraviolet rays has caused dermatologists to issue strong warnings against its use for tanning.

Microwave ovens utilize electromagnetic radiation in the **microwave** region of the spectrum generated by microwave tubes called *magnetrons.* Microwaves in the range of 10^2–10^5 μm are very suitable for use in cooking since they are *reflected* by metals, *transmitted* by glass and plastics, and *absorbed* by food (especially water) molecules. Thus, the electric energy converted to radiation in a microwave oven eventually becomes part of the internal energy of the food. The fast and efficient cooking of microwave ovens has made them some of the essential appliances in modern kitchens (Fig. 19-5).

Radars and cordless telephones also use electromagnetic radiation in the microwave region. The wavelength of the electromagnetic waves used in radio and TV broadcasting usually ranges between 1 and 1000 m in the **radio wave** region of the spectrum.

In heat transfer studies, we are interested in the energy emitted by bodies because of their temperature only. Therefore, we will limit our consideration to *thermal radiation,* which we will simply call *radiation.* The relations developed below are restricted to thermal radiation only and may not be applicable to other forms of electromagnetic radiation.

The electrons, atoms, and molecules of all solids, liquids, and gases above absolute zero temperature are constantly in motion, and thus radiation is constantly emitted, as well as being absorbed or transmitted throughout the entire volume of matter. That is, radiation is a **volumetric phenomenon.** However, for opaque (nontransparent) solids such as metals, wood, and rocks, radiation is considered to be a **surface phenomenon,** since the radiation emitted by the interior regions can never reach the surface, and the radiation incident on such bodies is usually absorbed within a few microns from the surface (Fig. 19-6). Note that the radiation characteristics of surfaces can be changed completely by applying thin layers of coatings on them.

19-3 ■ BLACKBODY RADIATION

A body at a temperature above absolute zero emits radiation in all directions over a wide range of wavelengths. The amount of radiation energy emitted from a surface at a given wavelength depends on the material of the body and the condition of its surface as well as the surface temperature. Therefore, different bodies may emit different amounts of radiation per unit surface area, even when they are at the same temperature. Thus, it is natural to be curious about the *maximum* amount of radiation that can be emitted by a surface at a given temperature. Satisfying this curiosity requires the definition of an idealized body, called a *blackbody,* to serve as a standard against which the radiative properties of real surfaces may be compared.

A **blackbody** is defined as *a perfect emitter and absorber of radiation.* At a specified temperature and wavelength, no surface can emit more energy than a blackbody. A blackbody absorbs *all* incident radiation, regardless of wavelength and direction. Also, a blackbody emits radiation energy uniformly in all directions (Fig. 19-7). That is, a blackbody is a *diffuse* emitter. The term *diffuse* means "independent of direction."

The radiation energy emitted by a blackbody per unit time and per unit surface area was determined experimentally by Joseph Stefan in 1879 and is expressed as

$$E_b = \sigma T^4 \qquad (\text{W/m}^2) \qquad (19\text{-}3)$$

where $\sigma = 5.67 \times 10^{-8}$ W/m² · K⁴ is the *Stefan–Boltzmann constant* and T is the absolute temperature of the surface in K. This relation was theoretically verified in 1884 by Ludwig Boltzmann. Equation 19-3 is known as the **Stefan–Boltzmann law** and E_b is called the **blackbody emissive power.** Note that the emission of thermal radiation is proportional to the *fourth power* of the absolute temperature.

Although a blackbody would appear *black* to the eye, a distinction should be made between the idealized blackbody and an ordinary black surface. Any surface that absorbs light (the visible portion of radiation) would appear black to the eye, and a surface that reflects it completely would appear white. Considering that visible radiation occupies a very narrow band of the spectrum from 0.4 to 0.76 μm, we cannot make any judgments about the blackness of a surface on the basis of visual observations. For example, snow and white paint reflect light and thus appear white. But they are essentially black for infrared radiation since they strongly absorb long-wavelength radiation. Surfaces coated with lampblack paint approach idealized blackbody behavior.

Another type of body that closely resembles a blackbody is a *large cavity with a small opening,* as shown in Fig. 19-8. Radiation coming in through the opening of area A will undergo multiple reflections, and thus it will have several chances to be absorbed by the interior surfaces of the cavity before any part of it can possibly escape. Also, if the surface of the cavity is isothermal at temperature T, the radiation emitted by the interior surfaces will stream

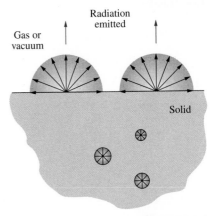

FIGURE 19-6

Radiation in opaque solids is considered a surface phenomenon since the radiation emitted only by the molecules at the surface can escape the solid.

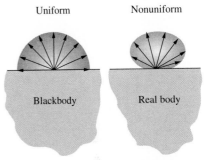

FIGURE 19-7

A blackbody is said to be a *diffuse* emitter since it emits radiation energy uniformly in all directions.

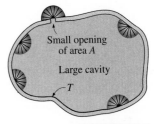

FIGURE 19-8

A large isothermal cavity at temperature T with a small opening of area A closely resembles a blackbody of surface area A at the same temperature.

through the opening after undergoing multiple reflections, and thus it will have a diffuse nature. Therefore, the cavity will act as a perfect absorber and perfect emitter, and the opening will resemble a blackbody of surface area A at temperature T, regardless of the actual radiative properties of the cavity.

The Stefan–Boltzmann law in Eq. 19-3 gives the *total* blackbody emissive power E_b, which is the sum of the radiation emitted over all wavelengths. Sometimes we need to know the **spectral blackbody emissive power,** which is *the amount of radiation energy emitted by a blackbody at an absolute temperature T per unit time, per unit surface area, and per unit wavelength about the wavelength* λ. For example, we are more interested in the amount of radiation an incandescent light bulb emits in the visible wavelength spectrum than we are in the total amount of radiation that the light bulb emits.

The relation for the spectral blackbody emissive power $E_{b\lambda}$ was developed by Max Planck in 1901 in conjunction with his famous quantum theory. This relation is known as **Planck's distribution law** and is expressed as

$$E_{b\lambda}(T) = \frac{C_1}{\lambda^5[\exp(C_2/\lambda T) - 1]} \qquad (\text{W/m}^2 \cdot \mu\text{m}) \qquad (19\text{-}4)$$

where

$$C_1 = 2\pi h c_0^2 = 3.742 \times 10^8 \text{ W} \cdot \mu\text{m}^4/\text{m}^2$$
$$C_2 = h c_0/k = 1.439 \times 10^4 \ \mu\text{m} \cdot \text{K}$$

Also, T is the absolute temperature of the surface, λ is the wavelength of the radiation emitted, and $k = 1.3805 \times 10^{-23}$ J/K is *Boltzmann's constant.* This relation is valid for a surface in a *vacuum* or a *gas.* For other mediums, it needs to be modified by replacing C_1 by C_1/n^2, where n is the index of refraction of the medium. Note that the term *spectral* indicates dependence on wavelength.

The variation of the blackbody emissive power with wavelength is plotted in Fig. 19-9 for selected temperatures. Several observations can be made from this figure:

1. The emitted radiation is a continuous function of *wavelength.* At any specified temperature, it increases with wavelength, reaches a peak, and then decreases with increasing wavelength.

2. At any wavelength, the amount of emitted radiation *increases* with increasing temperature.

3. As temperature increases, the curves get steeper and shift to the left to the shorter-wavelength region. Consequently, a larger fraction of the radiation is emitted at *shorter wavelengths* at higher temperatures.

4. The radiation emitted by the *sun,* which is considered to be a blackbody at 5762 K (or roughly at 5800 K), reaches its peak in the visible region of the spectrum. Therefore, the sun is in tune with our eyes. On the other hand, surfaces at $T \leq 800$ K emit almost entirely in the infrared region and thus are not visible to the eye unless they reflect light coming from other sources.

As the temperature increases, the peak of the curve in Fig. 19-9 shifts toward shorter wavelengths. The wavelength at which the peak occurs for a specified temperature is given by **Wien's displacement law** as

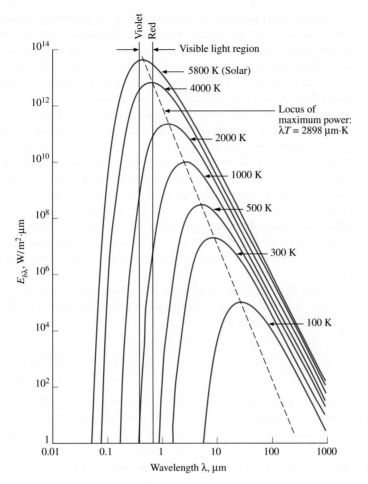

FIGURE 19-9

The variation of the blackbody emissive power with wavelength for several temperatures.

$$(\lambda T)_{\text{max power}} = 2897.8 \ \mu m \cdot K \qquad (19\text{-}5)$$

This relation was originally developed by Willy Wien in 1894 using classical thermodynamics, but it can also be obtained by differentiating Eq. 19-4 with respect to λ while holding T constant and setting the result equal to zero. A plot of Wien's displacement law, which is the locus of the peaks of the radiation emission curves, is also given in Fig. 19-9.

The peak of the solar radiation, for example, occurs at $\lambda = 2897.8/5762 = 0.50 \ \mu m$, which is near the middle of the visible range. The peak of the radiation emitted by a surface at room temperature ($T = 298$ K) occurs at $9.72 \ \mu m$, which is well into the infrared region of the spectrum.

An electrical resistance heater starts radiating heat soon after it is plugged in, and we can feel the emitted radiation energy by holding our hands against the heater. But this radiation is entirely in the infrared region and thus cannot be sensed by our eyes. The heater would appear dull red when its temperature reaches about 1000 K, since it will start emitting a detectable amount (about $1 \ \text{W/m}^2 \cdot \mu m$) of visible red radiation at that temperature. As the temperature rises even more, the heater appears bright red and is said to be *red hot*. When the temperature reaches about 1500 K, the heater emits enough radiation in the entire visible range of the spectrum to appear almost *white* to the eye, and it is called *white hot*.

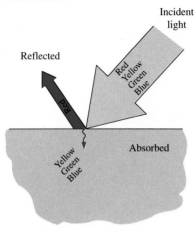

FIGURE 19-10

A surface that reflects red while absorbing the remaining parts of the incident light appears red to the eye.

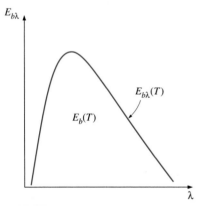

FIGURE 19-11

On an $E_{b\lambda}$–λ chart, the area under a curve for a given temperature represents the total radiation energy emitted by a blackbody at that temperature.

Although it cannot be sensed directly by the human eye, infrared radiation can be detected by infrared cameras, which transmit the information to microprocessors to display visual images of objects at night. *Rattlesnakes* can sense the infrared radiation or the "body heat" coming off warm-blooded animals, and thus they can see at night without using any instruments. Similarly, honeybees are sensitive to ultraviolet radiation.

It should be clear from the discussion above that the color of an object is not due to emission, which is primarily in the infrared region, unless the surface temperature of the object exceeds about 1000 K. Instead, the color of a surface depends on the absorption and reflection characteristics of the surface and is due to selective absorption and reflection of the incident visible radiation coming from a light source such as the sun or an incandescent light bulb. A piece of clothing containing a pigment that reflects red while absorbing the remaining parts of the incident light appears "red" to the eye (Fig. 19-10). Leaves appear "green" because their cells contain the pigment chlorophyll, which strongly reflects green while absorbing other colors.

It is left as an exercise to show that integration of the *spectral* blackbody emissive power $E_{b\lambda}$ over the entire wavelength spectrum gives the *total* blackbody emissive power E_b:

$$E_b(T) = \int_0^\infty E_{b\lambda}(T)\,d\lambda = \sigma T^4 \qquad (\text{W/m}^2) \qquad (19\text{-}6)$$

Thus, we obtained the Stefan–Boltzmann law (Eq. 19-3) by integrating Planck's distribution law (Eq. 19-4) over all wavelengths. Note that on an $E_{b\lambda}$–λ chart, $E_{b\lambda}$ corresponds to any value on the curve, whereas E_b corresponds to the area under the entire curve for a specified temperature (Fig. 19-11). Also, the term *total* means "integrated over all wavelengths."

EXAMPLE 19-1   Radiation Emission from a Black Ball

Consider a 20-cm-diameter spherical ball at 800 K suspended in the air as shown in Fig. 19-12. Assuming that the ball closely approximates a blackbody, determine (*a*) the total blackbody emissive power, (*b*) the total amount of radiation emitted by the ball in 5 min, and (*c*) the spectral blackbody emissive power at a wavelength of 3 μm.

Solution An isothermal sphere is suspended in the air. The total blackbody emissive power, the total radiation emitted in 5 minutes, and the spectral blackbody emissive power at 3 mm are to be determined.

Assumptions The ball behaves as a blackbody.

Analysis (*a*) The total blackbody emissive power is determined from the Stefan–Boltzmann law to be

$E_b = \sigma T^4 = (5.67 \times 10^{-8}\ \text{W/m}^2 \cdot \text{K}^4)(800\ \text{K})^4 = \textbf{23.2} \times \textbf{10}^3\ \textbf{W/m}^2 = \textbf{23.2 kW/m}^2$

That is, the ball emits 23.2 kJ of energy in the form of electromagnetic radiation per second per m² of the surface area of the ball.

(*b*) The total amount of radiation energy emitted from the entire ball in 5 min is determined by multiplying the blackbody emissive power obtained above by the total surface area of the ball and the given time interval:

$$A = \pi D^2 = \pi(0.2 \text{ m})^2 = 0.1257 \text{ m}^2$$

$$\Delta t = (5 \text{ min})\left(\frac{60 \text{ s}}{1 \text{ min}}\right) = 300 \text{ s}$$

$$Q_{rad} = E_b A \, \Delta t = (23.2 \text{ kW/m}^2)(0.1257 \text{ m}^2)(300 \text{ s})\left(\frac{1 \text{ kJ}}{1000 \text{ W} \cdot \text{s}}\right)$$

$$= \textbf{876 kJ}$$

That is, the ball loses 876 kJ of its internal energy in the form of electromagnetic waves to the surroundings in 5 min, which is enough energy to raise the temperature of 1 kg of water by 50°C. Note that the surface temperature of the ball cannot remain constant at 800 K unless there is an equal amount of energy flow to the surface from the surroundings or from the interior regions of the ball through some mechanisms such as chemical or nuclear reactions.

(c) The spectral blackbody emissive power at a wavelength of 3 μm is determined from Planck's distribution law to be

$$E_{b\lambda} = \frac{C_1}{\lambda^5\left[\exp\left(\dfrac{C_2}{\lambda T}\right) - 1\right]} = \frac{3.743 \times 10^8 \text{ W} \cdot \text{μm}^4/\text{m}^2}{(3 \text{ μm})^5\left[\exp\left(\dfrac{1.4387 \times 10^4 \text{ μm} \cdot \text{K}}{(3 \text{ μm})(800 \text{ K})}\right) - 1\right]}$$

$$= \textbf{3848 W/m}^2 \cdot \textbf{μm}$$

The Stefan–Boltzmann law $E_b(T) = \sigma T^4$ gives the *total* radiation emitted by a blackbody at all wavelengths from $\lambda = 0$ to $\lambda = \infty$. But we are often interested in the amount of radiation emitted over *some wavelength band*. For example, an incandescent light bulb is judged on the basis of the radiation it emits in the visible range rather than the radiation it emits at all wavelengths.

The radiation energy emitted by a blackbody per unit area over a wavelength band from $\lambda = 0$ to λ is determined from (Fig. 19-13)

$$E_{b, \, 0-\lambda}(T) = \int_0^\lambda E_{b\lambda}(T) \, d\lambda \qquad (\text{W/m}^2) \qquad (19\text{-}7)$$

It looks like we can determine $E_{b, \, 0-\lambda}$ by substituting the $E_{b\lambda}$ relation from Eq. 19-4 and performing this integration. But it turns out that this integration does not have a simple closed-form solution, and performing a numerical integration each time we need a value of $E_{b, \, 0-\lambda}$ is not practical. Therefore, we define a dimensionless quantity f_λ called the **blackbody radiation function** as

$$f_\lambda(T) = \frac{\displaystyle\int_0^\lambda E_{b\lambda}(T) \, d\lambda}{\sigma T^4} \qquad (19\text{-}8)$$

The function f_λ represents *the fraction of radiation emitted from a blackbody at temperature T in the wavelength band from $\lambda = 0$ to λ*. The values of f_λ are listed in Table 19-2 as a function of λT, where λ is in μm and T is in K.

The fraction of radiation energy emitted by a blackbody at temperature T over a finite wavelength band from $\lambda = \lambda_1$ to $\lambda = \lambda_2$ is determined from (Fig. 19-14)

$$f_{\lambda_1-\lambda_2}(T) = f_{\lambda_2}(T) - f_{\lambda_1}(T) \qquad (19\text{-}9)$$

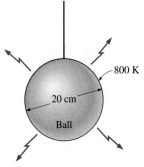

Blackbody Radiation

FIGURE 19-12

The spherical ball considered in Example 19-1.

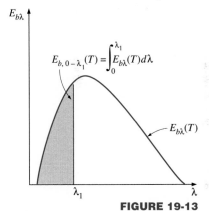

FIGURE 19-13

On an $E_{b\lambda} - \lambda$ chart, the area under the curve to the left of the $\lambda = \lambda_1$ line represents the radiation energy emitted by a blackbody in the wavelength range $0-\lambda_1$ for the given temperature.

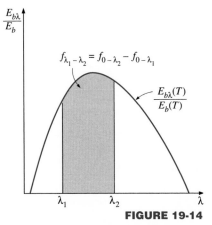

FIGURE 19-14

Graphical representation of the fraction of radiation emitted in the wavelength band from λ_1 to λ_2.

TABLE 19-2

Blackbody radiation functions f_λ

$\lambda T,$ $\mu m \cdot K$	f_λ	$\lambda T,$ $\mu m \cdot K$	f_λ
200	0.000000	6200	0.754140
400	0.000000	6400	0.769234
600	0.000000	6600	0.783199
800	0.000016	6800	0.796129
1000	0.000321	7000	0.808109
1200	0.002134	7200	0.819217
1400	0.007790	7400	0.829527
1600	0.019718	7600	0.839102
1800	0.039341	7800	0.848005
2000	0.066728	8000	0.856288
2200	0.100888	8500	0.874608
2400	0.140256	9000	0.890029
2600	0.183120	9500	0.903085
2800	0.227897	10,000	0.914199
3000	0.273232	10,500	0.923710
3200	0.318102	11,000	0.931890
3400	0.361735	11,500	0.939959
3600	0.403607	12,000	0.945098
3800	0.443382	13,000	0.955139
4000	0.480877	14,000	0.962898
4200	0.516014	15,000	0.969981
4400	0.548796	16,000	0.973814
4600	0.579280	18,000	0.980860
4800	0.607559	20,000	0.985602
5000	0.633747	25,000	0.992215
5200	0.658970	30,000	0.995340
5400	0.680360	40,000	0.997967
5600	0.701046	50,000	0.998953
5800	0.720158	75,000	0.999713
6000	0.737818	100,000	0.999905

where $f_{\lambda_1}(T)$ and $f_{\lambda_2}(T)$ are blackbody radiation functions corresponding to $\lambda_1 T$ and $\lambda_2 T$, respectively.

EXAMPLE 19-2   Emission of Radiation from an Incandescent Light Bulb

The temperature of the filament of an incandescent light bulb is 2500 K. Assuming the filament to be a blackbody, determine the fraction of the radiant energy emitted by the filament that falls in the visible range. Also, determine the wavelength at which the emission of radiation from the filament peaks.

Solution   The temperature of the filament of an incandescent light bulb is given. The fraction of visible radiation emitted by the filament and the wavelength at which the emission peaks are to be determined.

Assumptions   The filament behaves as a blackbody.

Analysis   The visible range of the electromagnetic spectrum extends from $\lambda_1 = 0.4$ μm to $\lambda_2 = 0.76$ μm. Noting that $T = 2500$ K, the blackbody radiation functions corresponding to $\lambda_1 T$ and $\lambda_2 T$ are determined from Table 19-2 to be

$$\lambda_1 T = (0.40 \ \mu m)(2500 \ K) = 1000 \ \mu m \cdot K \longrightarrow f_{\lambda_1} = 0.000321$$
$$\lambda_2 T = (0.76 \ \mu m)(2500 \ K) = 1900 \ \mu m \cdot K \longrightarrow f_{\lambda_2} = 0.053035$$

That is, 0.03 percent of the radiation is emitted at wavelengths less than 0.4 μm and 5.3 percent at wavelengths less than 0.76 μm. Then the fraction of radiation emitted between these two wavelengths is (Fig. 19-15)

$$f_{\lambda_1 - \lambda_2} = f_{\lambda_2} - f_{\lambda_1} = 0.053035 - 0.000321 = \mathbf{0.0527135}$$

That is, only about 5 percent of the radiation emitted by the filament of the light bulb falls in the visible range. The remaining 95 percent of the radiation appears in the infrared region in the form of radiant heat or "invisible light," as it used to be called. This is certainly not a very inefficient way of converting electrical energy to light and explains why fluorescent tubes are a wiser choice for lighting.

The wavelength at which the emission of radiation from the filament peaks is easily determined from Wien's displacement law (Eq. 19-5) to be

$$(\lambda T)_{max \ power} = 2897.8 \ \mu m \cdot K \quad \rightarrow \quad \lambda_{max \ power} = \frac{2897.8 \ \mu m \cdot K}{2500 \ K} = \mathbf{1.16 \ \mu m}$$

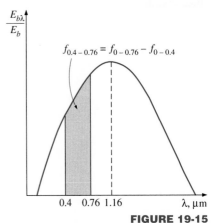

Discussion Note that the radiation emitted from the filament peaks in the infrared region.

FIGURE 19-15

Graphical representation of the fraction of radiation emitted in the visible range in Example 19-2.

19-4 ■ RADIATION PROPERTIES

Most materials encountered in practice, such as metals, wood, and bricks, are *opaque* to thermal radiation, and radiation is considered to be a *surface phenomenon* for such materials. That is, thermal radiation is emitted or absorbed within the first few microns of the surface, and thus we speak of radiation properties of *surfaces* for opaque materials.

Some other materials, such as glass and water, allow visible radiation to penetrate to considerable depths before any significant absorption takes place. Radiation through such *semitransparent* materials obviously cannot be considered to be a surface phenomenon since the entire volume of the material interacts with radiation. On the other hand, both glass and water are practically opaque to infrared radiation. Therefore, materials can exhibit different behavior at different wavelengths, and the dependence on wavelength is an important consideration in the study of radiation properties such as emissivity, absorptivity, reflectivity, and transmissivity of materials.

In the preceding section, we defined a *blackbody* as a perfect emitter and absorber of radiation and said that no body can emit more radiation than a blackbody at the same temperature. Therefore, a blackbody can serve as a convenient *reference* in describing the emission and absorption characteristics of real surfaces.

Emissivity

The **emissivity** of a surface is defined as *the ratio of the radiation emitted by the surface to the radiation emitted by a blackbody at the same temperature.* The emissivity of a surface is denoted by ε, and it varies between zero and one, $0 \le \varepsilon \le 1$. Emissivity is a measure of how closely a surface approximates a blackbody, for which $\varepsilon = 1$.

The emissivity of a real surface is not a constant. Rather, it varies with the *temperature* of the surface as well as the *wavelength* and the *direction* of the emitted radiation. Therefore, different emissivities can be defined for a surface, depending on the effects considered. For example, the emissivity of a surface at a specified wavelength is called *spectral emissivity* and is denoted by ε_λ. Likewise, the emissivity in a specified direction is called *directional emissivity,* denoted by ε_θ, where θ is the angle between the direction of radiation and the normal of the surface. The emissivity of a surface averaged over all directions is called the *hemispherical emissivity,* and the emissivity averaged over all wavelengths is called the *total emissivity.* Thus, the *total hemispherical emissivity* ε of a surface is simply the average emissivity over all directions and wavelengths and can be expressed as

$$\varepsilon(T) = \frac{E(T)}{E_b(T)} = \frac{E(T)}{\sigma T^4} \qquad (19\text{-}10)$$

where $E(T)$ is the total emissive power of the real surface. Equation 19-10 can be rearranged as

$$E(T) = \varepsilon(T)\sigma T^4 \qquad (\text{W/m}^2) \qquad (19\text{-}11)$$

Thus, the radiation emitted by the unit area of a real surface at temperature T is obtained by multiplying the radiation emitted by a blackbody at the same temperature by the emissivity of the surface.

Spectral emissivity is defined in a similar manner as

$$\varepsilon_\lambda(T) = \frac{E_\lambda(T)}{E_{b\lambda}(T)} \qquad (19\text{-}12)$$

where $E_\lambda(T)$ is the spectral emissive power of the real surface.

Radiation is a complex phenomenon as it is, and the consideration of wavelength and direction dependence of properties, assuming sufficient data exist, makes it even more complicated. Therefore, the *gray* and *diffuse* approximations are commonly utilized in radiation calculations. A surface is said to be *diffuse* if its properties are *independent of direction* and *gray* if its properties are *independent of wavelength.* Therefore, the emissivity of a gray, diffuse surface is simply the total hemispherical emissivity of that surface because of independence of direction and wavelength (Fig. 19-16).

A few comments about the validity of the diffuse approximation are in order. Although real surfaces do not emit radiation in a perfectly diffuse manner as a blackbody does, they usually come close. The variation of emissivity with direction for both conductors and nonconductors is given in Fig. 19-17. Here θ is the angle measured from the normal of the surface, and thus $\theta = 0$ for radiation emitted in a direction normal to the surface. Note that ε_θ remains nearly constant for about $\theta < 40°$ for conductors such as metals and for $\theta < 70°$ for nonconductors such as plastics. Therefore, the directional emissivity of a surface in the normal direction is representative of the hemispherical emissivity of the surface. In radiation analysis, it is common practice to assume the surfaces to be diffuse emitters with an emissivity equal to the value in the normal ($\theta = 0$) direction.

The effect of the gray approximation on emissivity and emissive power of a real surface is illustrated in Fig. 19-18. Note that the radiation emission from a real surface, in general, differs from the Planck distribution, and the emission curve may have several peaks and valleys.

A gray surface should emit as much radiation as the real surface it represents at the same temperature. Therefore, the areas under the emission curves

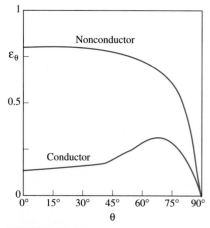

Real surface:
  $\varepsilon_\theta \neq$ constant
  $\varepsilon_\lambda \neq$ constant

Diffuse surface:
  $\varepsilon_\theta =$ constant

Gray surface:
  $\varepsilon_\lambda =$ constant

Diffuse, gray surface:
  $\varepsilon = \varepsilon_\lambda = \varepsilon_\theta =$ constant

FIGURE 19-16

The effect of diffuse and gray approximations on the emissivity of a surface.

FIGURE 19-17

Typical variations of emissivity with direction for electrical conductors and nonconductors.

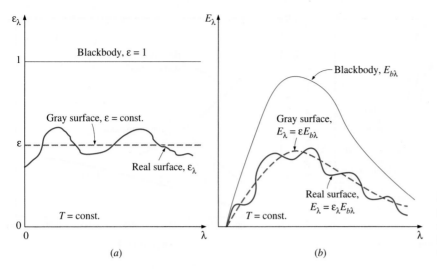

(a)

(b)

of the real and gray surfaces must be equal. That is, $\varepsilon(T)\sigma T^4 = \int_0^\infty \varepsilon_\lambda(T)E_{b\lambda}(T)\,d\lambda$. This requirement yields the following expression for the average emissivity:

$$\varepsilon(T) = \frac{\displaystyle\int_0^\infty \varepsilon_\lambda(T)E_{b\lambda}(T)\,d\lambda}{\sigma T^4} \qquad (19\text{-}13)$$

To perform this integration, we need to know the variation of spectral emissivity with wavelength at the specified temperature. The integrand is usually a complicated function, and the integration has to be performed numerically. However, the integration can be performed quite easily by dividing the spectrum into a sufficient number of *wavelength bands* and assuming the emissivity to remain constant over each band; that is, by expressing the function $\varepsilon(T)$ as a step function. This simplification offers great convenience for little sacrifice of accuracy, since it allows us to transform the integration into a summation in terms of blackbody emission functions.

As an example, consider the emissivity function plotted in Fig. 19-19. It seems like this function can be approximated reasonably well by a step function of the form

$$\varepsilon_\lambda = \begin{cases} \varepsilon_1 = \text{constant}, & 0 \le \lambda < \lambda_1 \\ \varepsilon_2 = \text{constant}, & \lambda_1 \le \lambda < \lambda_2 \\ \varepsilon_3 = \text{constant}, & \lambda_2 \le \lambda < \infty \end{cases} \qquad (19\text{-}14)$$

Then the average emissivity can be determined from Eq. 19-13 by breaking the integral into three parts and utilizing the definition of the blackbody radiation function as

$$\varepsilon(T) = \frac{\varepsilon_1 \displaystyle\int_0^{\lambda_1} E_{b\lambda}(T)\,d\lambda}{\sigma T^4} + \frac{\varepsilon_2 \displaystyle\int_{\lambda_1}^{\lambda_2} E_{b\lambda}(T)\,d\lambda}{\sigma T^4} + \frac{\varepsilon_3 \displaystyle\int_{\lambda_2}^{\infty} E_{b\lambda}(T)\,d\lambda}{\sigma T^4} \qquad (19\text{-}15)$$

$$= \varepsilon_1 f_{0-\lambda_1}(T) + \varepsilon_2 f_{\lambda_1-\lambda_2}(T) + \varepsilon_3 f_{\lambda_2-\infty}(T)$$

The emissivities of common materials are listed in Table A-24 in the appendix, and the variation of emissivity with wavelength and temperature is illustrated in Fig. 19-20. Typical ranges of emissivity of various materials are

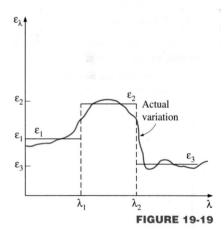

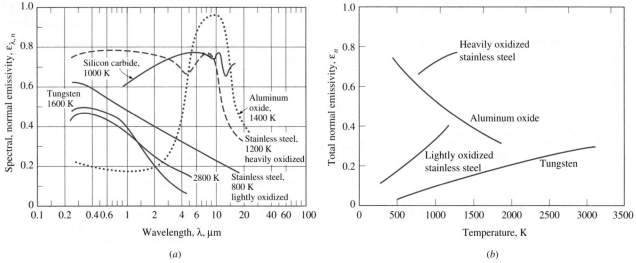

FIGURE 19-20

The variation of normal emissivity with (a) wavelength and (b) temperature for various materials.

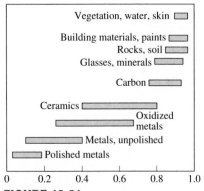

FIGURE 19-21

Typical ranges of emissivity for various materials.

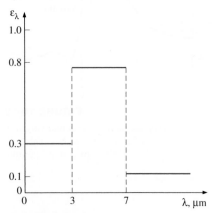

FIGURE 19-22

The spectral emissivity of the surface considered in Example 19-3.

given in Fig. 19-21. Note that metals generally have low emissivities, as low as 0.02 for polished surfaces, and nonmetals such as ceramics and organic materials have high ones. The emissivity of metals increases with temperature. Also, oxidation causes significant increases in the emissivity of metals. Heavily oxidized metals can have emissivities comparable to those of nonmetals.

Care should be exercised in the use and interpretation of radiation property data reported in the literature, since the properties strongly depend on the surface conditions such as oxidation, roughness, type of finish, and cleanliness. Consequently, there is considerable discrepancy and uncertainty in the reported values. This uncertainty is largely due to the difficulty in characterizing and describing the surface conditions precisely.

EXAMPLE 19-3 Average Emissivity of a Surface and Emissive Power

The spectral emissivity function of an opaque surface at 800 K is approximated as (Fig. 19-22)

$$\varepsilon_\lambda = \begin{cases} \varepsilon_1 = 0.3, & 0 \le \lambda < 3\ \mu m \\ \varepsilon_2 = 0.8, & 3\ \mu m \le \lambda < 7\ \mu m \\ \varepsilon_3 = 0.1, & 7\ \mu m \le \lambda < \infty \end{cases}$$

Determine the average emissivity of the surface and its emissive power.

Solution The variation of emissivity of a surface at a specified temperature with wavelength is given. The average emissivity of the surface and its emissive power are to be determined.

Analysis The variation of the emissivity of the surface with wavelength is given as a step function. Therefore, the average emissivity of the surface can be determined from Eq. 19-13 by breaking the integral into three parts,

$$\varepsilon(T) = \frac{\varepsilon_1 \int_0^{\lambda_1} E_{b\lambda}(T)\, d\lambda}{\sigma T^4} + \frac{\varepsilon_2 \int_{\lambda_1}^{\lambda_2} E_{b\lambda}(T)\, d\lambda}{\sigma T^4} + \frac{\varepsilon_3 \int_{\lambda_2}^{\infty} E_{b\lambda}(T)\, d\lambda}{\sigma T^4}$$

$$= \varepsilon_1\, f_{0-\lambda_1}(T) + \varepsilon_2\, f_{\lambda_1-\lambda_2}(T) + \varepsilon_3\, f_{\lambda_2-\infty}(T)$$

$$= \varepsilon_1\, f_{\lambda_1} + \varepsilon_2(f_{\lambda_2} - f_{\lambda_1}) + \varepsilon_3(1 - f_{\lambda_2})$$

where f_{λ_1} and f_{λ_2} are blackbody radiation functions corresponding to $\lambda_1 T$ and $\lambda_2 T$. These functions are determined from Table 19-2 to be

$$\lambda_1 T = (3 \ \mu m)(800 \ K) = 2400 \ \mu m \cdot K \quad \rightarrow \quad f_{\lambda_1} = 0.140256$$
$$\lambda_2 T = (7 \ \mu m)(800 \ K) = 5600 \ \mu m \cdot K \quad \rightarrow \quad f_{\lambda_2} = 0.701046$$

Note that $f_{0-\lambda_1} = f_{\lambda_1} - f_0 = f_{\lambda_1}$, since $f_0 = 0$, and $f_{\lambda_2-\infty} = f_\infty - f_{\lambda_2} = 1 - f_{\lambda_2}$, since $f_\infty = 1$. Substituting,

$$\varepsilon = 0.3 \times 0.140256 + 0.8(0.701046 - 0.140256) + 0.1(1 - 0.701046)$$
$$= \mathbf{0.521}$$

That is, the surface will emit as much radiation energy at 800 K as a gray surface having a constant emissivity $\varepsilon = 0.521$. The emissive power of the surface is

$$E = \varepsilon \sigma T^4 = 0.521(5.67 \times 10^{-8} \ W/m^2 \cdot K^4)(800 \ K)^4 = \mathbf{12,100 \ W/m^2}$$

Discussion Note that the surface emits 12.1 kJ of radiation energy per second per m^2 area of the surface.

Absorptivity, Reflectivity, and Transmissivity

Everything around us constantly emits radiation, and the emissivity represents the emission characteristics of those bodies. This means that every body, including our own, is constantly bombarded by radiation coming from all directions over a range of wavelengths. *The radiation energy incident on a surface per unit surface area per unit time* is called **irradiation** and is denoted by G.

When radiation strikes a surface, part of it is absorbed, part of it is reflected, and the remaining part, if any, is transmitted, as illustrated in Fig. 19-23. *The fraction of irradiation absorbed by the surface* is called the **absorptivity** α, *the fraction reflected by the surface* is called the **reflectivity** ρ, and *the fraction transmitted* is called the **transmissivity** τ. That is,

Absorptivity: $\alpha = \dfrac{\text{Absorbed radiation}}{\text{Incident radiation}} = \dfrac{G_{abs}}{G}, \quad 0 \le \alpha \le 1$ (19-16a)

Reflectivity: $\rho = \dfrac{\text{Reflected radiation}}{\text{Incident radiation}} = \dfrac{G_{ref}}{G}, \quad 0 \le \rho \le 1$ (19-16b)

Transmissivity: $\tau = \dfrac{\text{Transmitted radiation}}{\text{Incident radiation}} = \dfrac{G_{tr}}{G}, \quad 0 \le \tau \le 1$ (19-16c)

where G is the radiation energy incident on the surface, and G_{abs}, G_{ref}, and G_{tr} are the absorbed, reflected, and transmitted portions of it, respectively. The first law of thermodynamics requires that the sum of the absorbed, reflected, and transmitted radiation energy be equal to the incident radiation. That is,

$$G_{abs} + G_{ref} + G_{tr} = G$$

Dividing each term of this relation by G yields

$$\alpha + \rho + \tau = 1 \tag{19-17}$$

For opaque surfaces, $\tau = 0$, and thus

$$\alpha + \rho = 1 \tag{19-18}$$

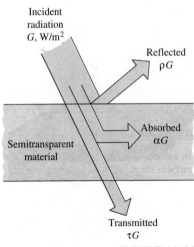

Incident radiation G, W/m^2

Reflected ρG

Semitransparent material

Absorbed αG

Transmitted τG

FIGURE 19-23

The absorption, reflection, and transmission of incident radiation by a semitransparent material.

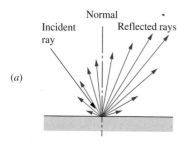

(a)

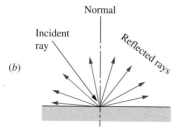

(b)

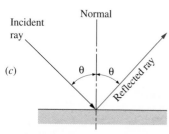

(c)

FIGURE 19-24

Different types of reflection from a surface: (a) actual or irregular, (b) diffuse, and (c) specular or mirrorlike.

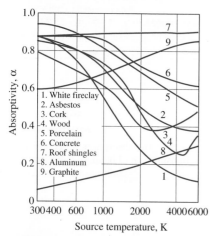

FIGURE 19-25

Variation of absorptivity with the temperature of the source of irradiation for various common materials at room temperature.

This is an important property relation since it allows us to determine both the absorptivity and reflectivity of an opaque surface by measuring either of these properties.

The definitions above are for *total hemispherical* properties, since G represents the radiation energy incident on the surface from all directions over the hemispherical space and over all wavelengths. Thus, α, ρ, and τ are the *average* properties of a medium for all directions and all wavelengths. However, like emissivity, these properties can also be defined for a specific wavelength or direction. For example, the *spectral* absorptivity, reflectivity, and transmissivity of a surface are defined in a similar manner as

$$\alpha_\lambda = \frac{G_{\lambda,\,abs}}{G_\lambda}, \qquad \rho_\lambda = \frac{G_{\lambda,\,ref}}{G_\lambda}, \qquad \tau_\lambda = \frac{G_{\lambda,\,tr}}{G_\lambda} \qquad (19\text{-}19)$$

where G_λ is the radiation energy incident at the wavelength λ and $G_{\lambda,\,abs}$, $G_{\lambda,\,ref}$, and $G_{\lambda,\,tr}$ are the absorbed, reflected, and transmitted portions of it, respectively. Similar definitions can be given for *directional* properties in direction θ by replacing all occurrences of the subscripts λ in Eq. 19-19 by θ.

The average absorptivity, reflectivity, and transmissivity of a surface can also be defined in terms of their spectral counterparts as

$$\alpha = \frac{\int_0^\infty \alpha_\lambda\, G_\lambda\, d\lambda}{\int_0^\infty G_\lambda\, d\lambda}, \qquad \rho = \frac{\int_0^\infty \rho_\lambda\, G_\lambda\, d\lambda}{\int_0^\infty G_\lambda\, d\lambda}, \qquad \tau = \frac{\int_0^\infty \tau_\lambda\, G_\lambda\, d\lambda}{\int_0^\infty G_\lambda\, d\lambda} \qquad (19\text{-}20)$$

The reflectivity differs somewhat from the other properties in that it is *bidirectional* in nature. That is, the value of the reflectivity of a surface depends not only on the direction of the incident radiation but also the direction of reflection. Therefore, the reflected rays of a radiation beam incident on a real surface in a specified direction will form an irregular shape, as shown in Fig. 19-24. Such detailed reflectivity data do not exist for most surfaces, and even if they did, they would be of little value in radiation calculations since this would usually add more complication to the analysis than it is worth.

In practice, for simplicity, surfaces are assumed to reflect in a perfectly *specular* or *diffuse* manner. In **specular** (or *mirrorlike*) **reflection,** *the angle of reflection equals the angle of incidence of the radiation beam.* In **diffuse reflection,** *radiation is reflected equally in all directions,* as shown in Fig. 19-24. Reflection from smooth and polished surfaces approximates specular reflection, whereas reflection from rough surfaces approximates diffuse reflection. In radiation analysis, smoothness is defined relative to wavelength. A surface is said to be *smooth* if the height of the surface roughness is much smaller than the wavelength of the incident radiation.

Unlike emissivity, the absorptivity of a material is practically independent of surface temperature. However, the absorptivity depends strongly on the temperature of the source at which the incident radiation is originating. This is also evident from Fig. 19-25, which shows the absorptivities of various materials at room temperature as functions of the temperature of the radiation source. For example, the absorptivity of the concrete roof of a house is about 0.6 for solar radiation (source temperature: 5762 K) and 0.9 for radiation originating from the surrounding trees and buildings (source temperature: 300 K), as illustrated in Fig. 19-26.

Notice that the absorptivity of aluminum increases with temperature, a characteristic for metals, and the absorptivity of electric nonconductors, in general, decreases with temperature. This decrease is most pronounced for surfaces that appear white to the eye. For example, the absorptivity of a white painted surface is low for solar radiation, although it is rather high for infrared radiation.

Kirchhoff's Law

Consider a small body of surface area A, emissivity ε, and absorptivity α at temperature T contained in a large isothermal enclosure at the same temperature, as shown in Fig. 19-27. Recall that a large isothermal enclosure forms a blackbody cavity regardless of the radiative properties of the enclosure surface, and the body in the enclosure is too small to interfere with the blackbody nature of the cavity. Therefore, the radiation incident on any part of the surface of the small body is equal to the radiation emitted by a blackbody at temperature T. That is, $G = E_b(T) = \sigma T^4$, and the radiation absorbed by the small body per unit of its surface area is

$$G_{abs} = \alpha G = \alpha \sigma T^4$$

The radiation emitted by the small body is (Eq. 19-3)

$$E_{emit} = \varepsilon \sigma T^4$$

Considering that the small body is in thermal equilibrium with the enclosure, the net rate of heat transfer to the body must be zero. Therefore, the radiation emitted by the body must be equal to the radiation absorbed by it:

$$A\varepsilon \sigma T^4 = A\alpha \sigma T^4$$

Thus, we conclude that

$$\varepsilon(T) = \alpha(T) \qquad (19\text{-}21)$$

That is, *the total hemispherical emissivity of a surface at temperature T is equal to its total hemispherical absorptivity for radiation coming from a blackbody at the same temperature.* This relation, which greatly simplifies the radiation analysis, was first developed by Gustav Kirchhoff in 1860 and is now called **Kirchhoff's law.** Note that this relation is derived under the condition that the surface temperature is equal to the temperature of the source of irradiation, and the reader is cautioned against using it when considerable difference (more than a few hundred degrees) exists between the surface temperature and the temperature of the source of irradiation.

The derivation above can also be repeated for radiation at a specified wavelength to obtain the *spectral* form of Kirchhoff's law:

$$\varepsilon_\lambda(T) = \alpha_\lambda(T) \qquad (19\text{-}22)$$

This relation is valid when the irradiation or the emitted radiation is independent of direction. The form of Kirchhoff's law that involves no restrictions is the *spectral directional* form expressed as $\varepsilon_{\lambda, \theta}(T) = \alpha_{\lambda, \theta}(T)$. That is, the emissivity of a surface at a specified wavelength, direction, and temperature is always equal to its absorptivity at the same wavelength, direction, and temperature.

It is very tempting to use Kirchhoff's law in radiation analysis since the relation $\varepsilon = \alpha$ together with $\rho = 1 - \alpha$ enables us to determine all three

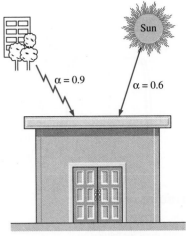

FIGURE 19-26

The absorptivity of a material may be quite different for radiation originating from sources at different temperatures.

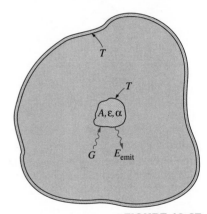

FIGURE 19-27

The small body contained in a large isothermal enclosure used in the development of Kirchhoff's law.

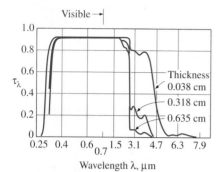

FIGURE 19-28

The spectral transmissivity of low-iron glass at room temperature for different thicknesses.

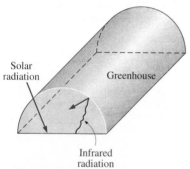

FIGURE 19-29

A greenhouse traps energy by allowing the solar radiation to come in but not allowing the infrared radiation to go out.

properties of an opaque surface from a knowledge of only *one* property. Although using Kirchhoff's law gives acceptable results in most cases, in practice, care should be exercised when there is considerable difference between the surface temperature and the temperature of the source of incident radiation.

The Greenhouse Effect

You have probably noticed that when you leave your car under direct sunlight on a sunny day, the interior of the car gets much warmer than the air outside, and you may have wondered why the car acts like a *heat trap*. The answer lies in the spectral transmissivity curve of the *glass*, which resembles an inverted U, as shown in Fig. 19-28. We observe from this figure that glass at thicknesses encountered in practice transmits over 90 percent of radiation in the visible range and is practically opaque (nontransparent) to radiation in the longer-wavelength infrared regions of the electromagnetic spectrum (roughly $\lambda > 3$ μm). Therefore, glass has a transparent window in the wavelength range 0.3 μm $< \lambda <$ 3 μm in which over 90 percent of solar radiation is emitted. On the other hand, the entire radiation emitted by surfaces at room temperature falls in the infrared region. Consequently, glass allows the solar radiation to enter but does not allow the infrared radiation from the interior surfaces to leave. This causes a rise in the interior temperature as a result of the energy build-up in the car. This heating effect, which is due to the nongray characteristic of glass (or clear plastics), is known as the **greenhouse effect,** since it is utilized primarily in greenhouses (Fig. 19-29).

The greenhouse effect is also experienced on a larger scale on earth. The surface of the earth, which warms up during the day as a result of the absorption of solar energy, cools down at night by radiating its energy into deep space as infrared radiation. The combustion gases such as CO_2 and water vapor in the atmosphere transmit the bulk of the solar radiation but absorb the infrared radiation emitted by the surface of the earth. Thus, there is concern that the energy trapped on earth will eventually cause global warming and thus drastic changes in weather patterns.

In *humid* places such as coastal areas, there is not a large change between the daytime and nighttime temperatures, because the humidity acts as a barrier on the path of the infrared radiation coming from the earth, and thus slows down the cooling process at night. In areas with clear skies such as deserts, there is a large swing between the daytime and nighttime temperatures because of the absence of such barriers for infrared radiation.

19-5 ■ ATMOSPHERIC AND SOLAR RADIATION

The sun is our primary source of energy. The energy coming off the sun, called *solar energy,* reaches us in the form of electromagnetic waves after experiencing considerable interactions with the atmosphere. The radiation energy emitted or reflected by the constituents of the atmosphere form the *atmospheric radiation.* Below we give an overview of the solar and atmospheric radiation because of their importance and relevance to daily life. Also, our familiarity with solar energy makes it an effective tool in developing a better understanding for some of the new concepts introduced earlier. Detailed treatment of this exciting subject can be found in numerous books devoted to this topic.

The *sun* is a nearly spherical body that has a diameter of $D \approx 1.39 \times 10^9$ m and a mass of $m \approx 2 \times 10^{30}$ kg and is located at a mean distance of $L = 1.50 \times 10^{11}$ m from the earth. It emits radiation energy continuously at a rate of $E_{sun} \approx 3.8 \times 10^{26}$ W. Less than a billionth of this energy (about 1.7×10^{17} W) strikes the earth, which is sufficient to keep the earth warm and to maintain life through the photosynthesis process. The energy of the sun is due to the continuous *fusion* reaction during which two hydrogen atoms fuse to form one atom of helium. Therefore, the sun is essentially a *nuclear reactor*, with temperatures as high as 40,000,000 K in its core region. The temperature drops to about 6000 K in the outer region of the sun, called the convective zone, as a result of the dissipation of this energy by radiation.

The solar energy reaching the earth's atmosphere is determined by a series of measurements taken in the late 1960s by using high-altitude aircraft, balloons, and spacecraft to be 1353 W/m². This quantity is called the *solar constant* G_s:

$$G_s = 1353 \text{ W/m}^2 \tag{19-23}$$

The **solar constant** represents *the rate at which solar energy is incident on a surface normal to the sun's rays at the outer edge of the atmosphere when the earth is at its mean distance from the sun* (Fig. 19-30). Owing to the ellipticity of the earth's orbit, the distance between the sun and the earth, and thus the actual value of the solar constant, changes throughout the year. It varies from a maximum of 1399 W/m² on December 21 to a minimum of 1310 W/m² on June 21. (Note that the earth is farthest away from the sun in summer in the northern hemisphere.) However, this variation, which remains within ±3.4 percent of the mean value, is considered negligible for most practical purposes, and G_s is taken to be a *constant* at its mean value of 1353 W/m².

The measured value of the solar constant can be used to estimate the effective surface temperature of the sun from the requirement that

$$(4\pi L^2)G_s = (4\pi r^2)\sigma T_{sun}^4 \tag{19-24}$$

where L is the mean distance between the sun's center and the earth and r is the radius of the sun. The left-hand side of this equation represents the total solar energy passing through a spherical surface whose radius is the mean earth–sun distance, and the right-hand side represents the total energy that leaves the sun's outer surface. The conservation of energy principle requires that these two quantities be equal to each other, since the solar energy experiences no attenuation (or enhancement) on its way through the vacuum (Fig. 19-31). The **effective surface temperature** of the sun is determined from Eq. 19-24 to be $T_{sun} = 5762$ K. That is, the sun can be treated as a blackbody at a temperature of 5762 K. This is also confirmed by the measurements of the spectral distribution of the solar radiation just outside the atmosphere plotted in Fig. 19-32, which shows only small deviations from the idealized blackbody behavior.

The spectral distribution of solar radiation on the ground plotted in Fig. 19-32 shows that the solar radiation undergoes considerable *attenuation* as it passes through the atmosphere as a result of *absorption* and *scattering*. About 99 percent of the atmosphere is contained within a distance of 30 km from the earth's surface. The several dips on the spectral distribution of radiation on the earth's surface are due to *absorption* by the gases O_2, O_3 (ozone), H_2O, and CO_2. Absorption by *oxygen* occurs in a narrow band about $\lambda = 0.76$ μm. The *ozone* absorbs *ultraviolet* radiation at wavelengths below 0.3 μm almost completely, and radiation in the range 0.3–0.4 μm

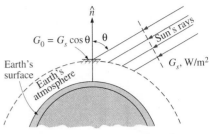

FIGURE 19-30

Solar radiation reaching the earth's atmosphere and the solar constant.

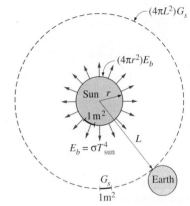

FIGURE 19-31

The total solar energy passing through concentric spheres remains constant, but the energy falling per unit area decreases with increasing radius.

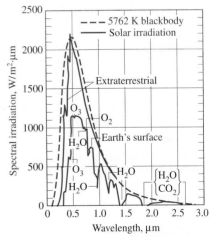

FIGURE 19-32

Spectral distribution of solar radiation just outside the atmosphere, at the surface of the earth on a typical day, and comparison with blackbody radiation at 5762 K.

considerably. Thus, the ozone layer in the upper regions of the atmosphere protects biological systems on earth from harmful ultraviolet radiation. In turn, we must protect the ozone layer from the destructive chemicals commonly used as refrigerants, cleaning agents, and propellants in aerosol cans. The use of these chemicals is now banned in many countries. The ozone gas also absorbs some radiation in the visible range. Absorption in the infrared region is dominated by *water vapor* and *carbon dioxide*. The dust particles and other pollutants in the atmosphere also absorb radiation at various wavelengths.

As a result of these absorptions, the solar energy reaching the *earth's surface* is weakened considerably, to about 950 W/m² on a clear day and much less on cloudy or smoggy days. Also, practically all of the solar radiation reaching the earth's surface falls in the wavelength band from 0.3 to 2.5 μm.

Another mechanism that attenuates solar radiation as it passes through the atmosphere is *scattering* or *reflection* by air molecules and the many other kinds of particles such as dust, smog, and water droplets suspended in the atmosphere. Scattering is mainly governed by the size of the particle relative to the wavelength of radiation. The oxygen and nitrogen molecules primarily scatter radiation at very short wavelengths, comparable to the size of the molecules themselves. Therefore, radiation at wavelengths corresponding to violet and blue colors is scattered the most. This molecular scattering in all directions is what gives the sky its bluish color. The same phenomenon is responsible for red sunrises and sunsets. Early in the morning and late in the afternoon, the sun's rays pass through a greater thickness of the atmosphere than they do at midday, when the sun is at the top. Therefore, the violet and blue colors of the light encounter a greater number of molecules by the time they reach the earth's surface, and thus a greater fraction of them are scattered (Fig. 19-33). Consequently, the light that reaches the earth's surface consists primarily of colors corresponding to longer wavelengths such as red, orange, and yellow. The clouds appear in reddish-orange color during sunrise and sunset because the light they reflect is reddish-orange at those times. For the same reason, a red traffic light is visible from a longer distance than is a green light under the same circumstances.

The solar energy incident on a surface on earth is considered to consist of *direct* and *diffuse* parts. The part of solar radiation that reaches the earth's surface without being scattered or absorbed by the atmosphere is called **direct solar radiation** G_D. The scattered radiation is assumed to reach the earth's surface uniformly from all directions and is called **diffuse solar radiation** G_d. Then the *total solar energy* incident on the unit area of a *horizontal surface* on the ground is (Fig. 19-34)

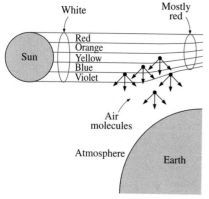

FIGURE 19-33

Air molecules scatter blue light much more than they do red light. At sunset, the light travels through a thicker layer of atmosphere, which removes much of the blue from the natural light, allowing the red to dominate.

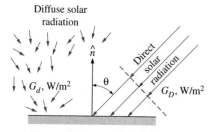

FIGURE 19-34

The direct and diffuse radiation incident on a horizontal surface at the earth's surface.

$$G_{\text{solar}} = G_D \cos \theta + G_d \qquad (\text{W/m}^2) \qquad (19\text{-}25)$$

where θ is the angle of incidence of direct solar radiation (the angle that the sun's rays make with the normal of the surface). The diffuse radiation varies from about 10 percent of the total radiation on a clear day to nearly 100 percent on a totally cloudy day.

The gas molecules and the suspended particles in the atmosphere *emit radiation* as well as absorbing it. The atmospheric emission is primarily due to the CO_2 and H_2O molecules and is concentrated in the regions from 5 to 8 μm and above 13 μm. Although this emission is far from resembling the distribution of radiation from a blackbody, it is found convenient in radiation calculations to treat the atmosphere as a blackbody at some lower fictitious temperature that emits an equivalent amount of radiation energy. This fictitious temperature is called the **effective sky temperature** T_{sky}. Then the radiation emission from the atmosphere to the earth's surface is expressed as

$$G_{sky} = \sigma T_{sky}^4 \qquad (W/m^2) \qquad (19\text{-}26)$$

The value of T_{sky} depends on the atmospheric conditions. It ranges from about 230 K for cold, clear-sky conditions to about 285 K for warm, cloudy-sky conditions.

Note that the effective sky temperature does not deviate much from the room temperature. Thus, in the light of Kirchhoff's law, we can take the absorptivity of a surface to be equal to its emissivity at room temperature, $\alpha = \varepsilon$. Then the sky radiation absorbed by a surface can be expressed as

$$E_{sky,\,absorbed} = \alpha G_{sky} = \alpha \sigma T_{sky}^4 = \varepsilon \sigma T_{sky}^4 \qquad (W/m^2) \qquad (19\text{-}27)$$

The net rate of radiation heat transfer to a surface exposed to solar and atmospheric radiation is determined from an energy balance (Fig. 19-35):

$$
\begin{aligned}
\dot{q}_{net,\,rad} &= \Sigma E_{absorbed} - \Sigma E_{emitted} \\
&= E_{solar,\,absorbed} + E_{sky,\,absorbed} - E_{emitted} \\
&= \alpha_s G_{solar} + \varepsilon \sigma T_{sky}^4 - \varepsilon \sigma T_s^4 \\
&= \alpha_s G_{solar} + \varepsilon \sigma (T_{sky}^4 - T_s^4) \qquad (W/m^2)
\end{aligned}
\qquad (19\text{-}28)
$$

where T_s is the temperature of the surface in K and ε is its emissivity at room temperature. A positive result for $\dot{q}_{net,\,rad}$ indicates a radiation heat gain by the surface and a negative result indicates a heat loss.

The absorption and emission of radiation by the *elementary gases* such as H_2, O_2, and N_2 at moderate temperatures are negligible, and a medium filled with these gases can be treated as a *vacuum* in radiation analysis. The absorption and emission of gases with *larger molecules* such as H_2O and CO_2, however, can be *significant* and may need to be considered when considerable amounts of such gases are present in a medium. For example, a 1-m-thick layer of water vapor at 1 atm pressure and 100°C emits more than 50 percent of the energy that a blackbody would emit at the same temperature.

In solar energy applications, the spectral distribution of incident solar radiation is very different than the spectral distribution of emitted radiation by the surfaces, since the former is concentrated in the short-wavelength region and the latter in the infrared region. Therefore, the radiation properties of surfaces will be quite different for the incident and emitted radiation, and the surfaces cannot be assumed to be gray. Instead, the surfaces are assumed to have two sets of properties: one for solar radiation and another for infrared radiation at room temperature. Table 19-3 lists the *emissivity* ε and the *solar absorptivity* α_s of the surfaces of some common materials. Surfaces that are intended to *collect solar energy,* such as the absorber surfaces of solar collectors, are desired to have high α_s but low ε values to maximize the absorption of solar radiation and to minimize the emission of radiation. Surfaces that are intended to *remain cool* under the sun, such as the outer surfaces of fuel tanks and refrigerator trucks, are desired to have just the opposite properties. Surfaces are often given the desired properties by coating them with thin layers of *selective* materials. A surface can be kept cool, for example, by simply painting it white.

We close this section by pointing out that what we call *renewable energy* is usually nothing more than the manifestation of solar energy in different

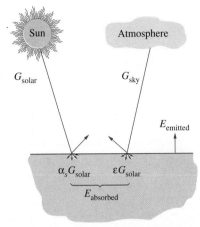

FIGURE 19-35

Radiation interactions of a surface exposed to solar and atmospheric radiation.

TABLE 19-3

Comparison of the solar absorptivity α_s of some surfaces with their emissivity ε at room temperature

Surface	α_s	ε
Aluminum		
Polished	0.09	0.03
Anodized	0.14	0.84
Foil	0.15	0.05
Copper		
Polished	0.18	0.03
Tarnished	0.65	0.75
Stainless steel		
Polished	0.37	0.60
Dull	0.50	0.21
Plated metals		
Black nickel oxide	0.92	0.08
Black chrome	0.87	0.09
Concrete	0.60	0.88
White marble	0.46	0.95
Red brick	0.63	0.93
Asphalt	0.90	0.90
Black paint	0.97	0.97
White paint	0.14	0.93
Snow	0.28	0.97
Human skin (caucasian)	0.62	0.97

forms. Such energy sources include wind energy, hydroelectric power, ocean thermal energy, ocean wave energy, and wood. For example, no hydroelectric power plant can generate electricity year after year unless the water evaporates by absorbing solar energy and comes back as a rainfall to replenish the water source (Fig. 19-36). Although solar energy is sufficient to meet the entire energy needs of the world, currently it is not economical to do so because of the low concentration of solar energy on earth and the high capital cost of harnessing it.

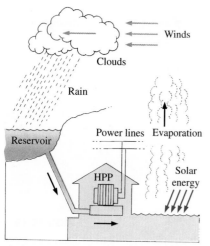

FIGURE 19-36

The cycle that water undergoes in a hydroelectric power plant.

EXAMPLE 19-4   Selective Absorber and Reflective Surfaces

Consider a surface exposed to solar radiation. At a given time, the direct and diffuse components of solar radiation are $G_D = 400$ and $G_d = 300$ W/m², and the direct radiation makes a 20° angle with the normal of the surface. The surface temperature is observed to be 320 K at that time. Assuming an effective sky temperature of 260 K, determine the net rate of radiation heat transfer for the following cases (Fig. 19-37):

(a) $\alpha_s = 0.9$ and $\varepsilon = 0.9$ (gray absorber surface)
(b) $\alpha_s = 0.1$ and $\varepsilon = 0.1$ (gray reflector surface)
(c) $\alpha_s = 0.9$ and $\varepsilon = 0.1$ (selective absorber surface)
(d) $\alpha_s = 0.1$ and $\varepsilon = 0.9$ (selective reflector surface)

Solution   A surface is exposed to solar and sky radiation. The net rate of radiation heat transfer is to be determined for four different combinations of emissivities and solar absorptivities.

Analysis   The total solar energy incident on the surface is

$$G_{solar} = G_D \cos \theta + G_d$$
$$= (400 \text{ W/m}^2) \cos 20° + (300 \text{ W/m}^2)$$
$$= 675.9 \text{ W/m}^2$$

Then the net rate of radiation heat transfer for each of the four cases is determined from:

$$\dot{q}_{net, \, rad} = \alpha_s G_{solar} + \varepsilon\sigma(T_{sky}^4 - T_s^4)$$

(a) $\alpha_s = 0.9$ and $\varepsilon = 0.9$ (gray absorber surface):

$$\dot{q}_{net, \, rad} = 0.9(675.9 \text{ W/m}^2) + 0.9(5.67 \times 10^{-8} \text{ W/m}^2 \cdot \text{K}^4)[(260 \text{ K})^4 - (320 \text{ K})^4]$$
$$= \textbf{306.5 W/m}^2$$

(b) $\alpha_s = 0.1$ and $\varepsilon = 0.1$ (gray reflector surface):

$$\dot{q}_{net, \, rad} = 0.1(675.9 \text{ W/m}^2) + 0.1(5.67 \times 10^{-8} \text{ W/m}^2 \cdot \text{K}^4)[(260 \text{ K})^4 - (320 \text{ K})^4]$$
$$= \textbf{34.1 W/m}^2$$

(c) $\alpha_s = 0.9$ and $\varepsilon = 0.1$ (selective absorber surface):

$$\dot{q}_{net, \, rad} = 0.9(675.9 \text{ W/m}^2) + 0.1(5.67 \times 10^{-8} \text{ W/m}^2 \cdot \text{K}^4)[(260 \text{ K})^4 - (320 \text{ K})^4]$$
$$= \textbf{574.8 W/m}^2$$

(d) $\alpha_s = 0.1$ and $\varepsilon = 0.9$ (selective reflector surface):

$$\dot{q}_{net, \, rad} = 0.1(675.9 \text{ W/m}^2) + 0.9(5.67 \times 10^{-8} \text{ W/m}^2 \cdot \text{K}^4)[(260 \text{ K})^4 - (320 \text{ K})^4]$$
$$= \textbf{-234.3 W/m}^2$$

Discussion   Note that the surface of an ordinary gray material of high absorptivity gains heat at a rate of 306.5 W/m². The amount of heat gain increases to 574.8 W/m² when the surface is coated with a selective material that has the same absorptivity for solar radiation but a low emissivity for infrared radiation. Also note that the surface of an ordinary gray material of high reflectivity still gains heat at a rate of 34.1 W/m². When the surface is coated with a selective material that has the same reflectivity for solar radiation but a high emissivity for infrared radiation, the surface loses 234.3 W/m² instead. Therefore, the temperature of the surface will decrease when a selective reflector surface is used.

19-6 ■ THE VIEW FACTOR

So far, we have considered the radiation properties of surfaces and the radiation interactions of a single surface. We are now in a position to consider *radiation heat transfer* between two or more surfaces, which is the primary quantity of interest in practice. Radiation heat transfer between surfaces depends on the *orientation* of the surfaces relative to each other as well as their radiation properties and temperatures, as illustrated in Fig. 19-38. For example, a camper will make the most use of a campfire on a cold night by standing as close to the fire as possible and by blocking as much of the radiation coming from the fire by turning her front to the fire instead of her side. Likewise, a person will maximize the amount of solar radiation incident on him by lying down on his back instead of standing up on his feet.

To account for the effects of orientation on radiation heat transfer between two surfaces, we define a new parameter called the *view factor,* which is a purely geometric quantity and is independent of the surface properties and temperature. It is also called the *shape factor, configuration factor,* and *angle factor.* The view factor based on the assumption that the surfaces are diffuse emitters and diffuse reflectors is called the *diffuse view factor,* and the view factor based on the assumption that the surfaces are diffuse emitters but specular reflectors is called the *specular view factor.* In this book, we will consider radiation exchange between diffuse surfaces only, and thus the term *view factor* will simply imply *diffuse view factor.*

The **view factor** from a surface i to a surface j is denoted by $F_{i \rightarrow j}$ and is defined as

> $F_{i \rightarrow j}$ = *the fraction of the radiation leaving surface i that strikes surface j directly*

Therefore, the view factor $F_{1 \rightarrow 2}$ represents the fraction of the radiation leaving surface 1 that strikes surface 2 and $F_{2 \rightarrow 1}$ represents the fraction of the radiation leaving surface 2 that strikes surface 1 directly. Note that the radiation that strikes a surface does not need to be absorbed by that surface. Also, radiation that strikes a surface after being reflected by other surfaces is not considered in the evaluation of the view factors. For the special case of $j = i$, we have

> $F_{i \rightarrow i}$ = *the fraction of radiation leaving surface i that strikes itself directly*

Noting that in the absence of strong electromagnetic fields radiation beams travel in straight paths, the view factor from a surface to itself will be zero unless the surface "sees" itself. Therefore, $F_{i \rightarrow i} = 0$ for *plane* or *convex* surfaces and $F_{i \rightarrow i} \neq 0$ for concave surfaces, as illustrated in Fig. 19-39.

The value of the view factor ranges between *zero* and *one.* The limiting case $F_{i \rightarrow j} = 0$ indicates that the two surfaces do not have a direct view of each other, and thus radiation leaving surface i cannot strike surface j directly. The

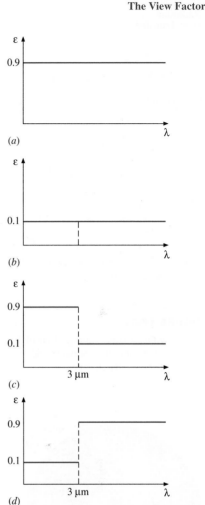

FIGURE 19-37

Graphical representation of the spectral emissivities of the four surfaces considered in Example 19-4.

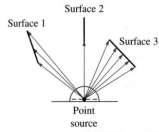

FIGURE 19-38

Radiation heat exchange between surfaces depends on the *orientation* of the surfaces relative to each other, and this dependence on orientation is accounted for by the *view factor.*

$$F_{1 \rightarrow 1} = 0$$

(a) Plane surface

$$F_{2 \rightarrow 2} = 0$$

(b) Convex surface

$$F_{3 \rightarrow 3} \neq 0$$

(c) Concave surface

FIGURE 19-39

The view factor from a surface to itself is *zero* for *plane* or *convex* surfaces and *nonzero* for *concave* surfaces.

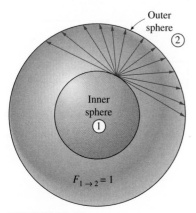

Outer
sphere
②

Inner
sphere
①

$$F_{1 \rightarrow 2} = 1$$

FIGURE 19-40

In a geometry that consists of two concentric spheres, the view factor $F_{1 \rightarrow 2} = 1$ since the entire radiation leaving the surface of the smaller sphere will be intercepted by the larger sphere.

other limiting case $F_{i \rightarrow j} = 1$ indicates that surface j completely surrounds surface i, so that the entire radiation leaving surface i is intercepted by surface j. For example, in a geometry consisting of two concentric spheres, the entire radiation leaving the surface of the smaller sphere (surface 1) will strike the larger sphere (surface 2), and thus $F_{1 \rightarrow 2} = 1$, as illustrated in Fig. 19-40.

The view factor has proven to be very useful in radiation analysis because it allows us to express the *fraction of radiation* leaving a surface that strikes another surface in terms of the orientation of these two surfaces relative to each other. The underlying assumption in this process is that the radiation a surface receives from a source is directly proportional to the angle the surface subtends when viewed from the source. This would be the case only if the radiation coming off the source is *uniform* in all directions throughout its surface and the medium between the surfaces does not *absorb, emit,* or *scatter* radiation. That is, it will be the case when the surfaces are *isothermal* and *diffuse* emitters and reflectors and the surfaces are separated by a *nonparticipating* medium such as a vacuum or air.

The view factor $F_{1 \rightarrow 2}$ between two surfaces A_1 and A_2 can be determined in a systematic manner first by expressing the view factor between two differential areas dA_1 and dA_2 in terms of the spatial variables and then by performing the necessary integrations. However, this approach is not practical, since, even for simple geometries, the resulting integrations are usually very complex and difficult to perform.

View factors for hundreds of common geometries are evaluated and the results are given in analytical, graphical, and tabular form in several publications. View factors for selected geometries are given in Tables 19-4 and 19-5 in *analytical* form and in Figs. 19-41 to 19-44 in *graphical* form. The view factors in Table 19-4 are for three-dimensional geometries. The view factors in Table 19-5, on the other hand, are for geometries that are *infinitely long* in the direction perpendicular to the plane of the paper and are therefore two-dimensional.

View Factor Relations

Radiation analysis on an enclosure consisting of N surfaces requires the evaluation of N^2 view factors, and this evaluation process is probably the most time-consuming part of a radiation analysis. However, it is neither practical nor necessary to evaluate all of the view factors directly. Once a sufficient number of view factors are available, the rest of them can be determined by utilizing some fundamental relations for view factors, as discussed below.

1 The Reciprocity Rule

The view factors $F_{i \rightarrow j}$ and $F_{j \rightarrow i}$ are *not* equal to each other unless the areas of the two surfaces are equal. That is,

$$F_{j \rightarrow i} = F_{i \rightarrow j} \qquad \text{when} \qquad A_i = A_j$$
$$F_{j \rightarrow i} \neq F_{i \rightarrow j} \qquad \text{when} \qquad A_i \neq A_j$$

Using the radiation intensity concept and going through some manipulations, it can be shown that the pair of view factors $F_{i \rightarrow j}$ and $F_{j \rightarrow i}$ are related to each other by

$$A_i F_{i \rightarrow j} = A_j F_{j \rightarrow i} \tag{19-29}$$

TABLE 19-4

View factor expressions for some common geometries of finite size (3D)

Geometry	Relation
Aligned parallel rectangles	$\overline{X} = X/L$ and $\overline{Y} = Y/L$ $$F_{i \to j} = \frac{2}{\pi \overline{X}\,\overline{Y}} \left\{ \ln \left[\frac{(1 + \overline{X}^2)(1 + \overline{Y}^2)}{1 + \overline{X}^2 + \overline{Y}^2} \right]^{1/2} \right.$$ $$+ \overline{X}(1 + \overline{Y}^2)^{1/2} \tan^{-1} \frac{\overline{X}}{(1 + \overline{Y}^2)^{1/2}}$$ $$+ \overline{Y}(1 + \overline{X}^2)^{1/2} \tan^{-1} \frac{\overline{Y}}{(1 + \overline{X}^2)^{1/2}}$$ $$\left. - \overline{X} \tan^{-1} \overline{X} - \overline{Y} \tan^{-1} \overline{Y} \right\}$$
Coaxial parallel disks	$R_i = r_i/L$ and $R_j = r_j/L$ $$S = 1 + \frac{1 + R_j^2}{R_i^2}$$ $$F_{i \to j} = \frac{1}{2} \left\{ S - \left[S^2 - 4 \left(\frac{r_j}{r_i} \right)^2 \right]^{1/2} \right\}$$
Perpendicular rectangles with a common edge	$H = Z/X$ and $W = Y/X$ $$F_{i \to j} = \frac{1}{\pi W} \left(W \tan^{-1} \frac{1}{W} + H \tan^{-1} \frac{1}{H} \right.$$ $$- (H^2 + W^2)^{1/2} \tan^{-1} \frac{1}{(H^2 + W^2)^{1/2}}$$ $$+ \frac{1}{4} \ln \left\{ \frac{(1 + W^2)(1 + H^2)}{1 + W^2 + H^2} \right.$$ $$\times \left[\frac{W^2(1 + W^2 + H^2)}{(1 + W^2)(W^2 + H^2)} \right]^{W^2}$$ $$\left. \left. \times \left[\frac{H^2(1 + H^2 + W^2)}{(1 + H^2)(H^2 + W^2)} \right]^{H^2} \right\} \right)$$

This relation is known as the **reciprocity rule,** and it enables us to determine the counterpart of a view factor from a knowledge of the view factor itself and the areas of the two surfaces. When determining the pair of view factors $F_{i \to j}$ and $F_{j \to i}$, it makes sense to evaluate first the easier one directly and then the harder one by applying the reciprocity rule.

2 The Summation Rule

The radiation analysis of a surface normally requires the consideration of the radiation coming in or going out in all directions. Therefore, most radiation problems encountered in practice involve enclosed spaces. When formulating a radiation problem, we usually form an *enclosure* consisting of the surfaces interacting radiatively. Even openings are treated as imaginary surfaces with radiation properties equivalent to those of the opening.

The conservation of energy principle requires that the entire radiation leaving any surface i of an enclosure be intercepted by the surfaces of

TABLE 19-5

View factor expressions for some infinitely long (2D) geometries

Geometry	Relation
Parallel plates with midlines connected by perpendicular line	$W_i = w_i/L$ and $W_j = w_j/L$ $$F_{i \to j} = \frac{[(W_i + W_j)^2 + 4]^{1/2} - (W_j - W_i)^2 + 4]^{1/2}}{2W_i}$$
Inclined plates of equal width and with a common edge	$$F_{i \to j} = 1 - \sin\frac{1}{2}\alpha$$
Perpendicular plates with a common edge	$$F_{i \to j} = \frac{1}{2}\left\{1 + \frac{w_j}{w_i} - \left[1 + \left(\frac{w_j}{w_i}\right)^2\right]^{1/2}\right\}$$
Three-sided enclosure	$$F_{i \to j} = \frac{w_i + w_j - w_k}{2w_i}$$
Infinite plane and row of cylinders	$$F_{i \to j} = 1 - \left[1 - \left(\frac{D}{s}\right)^2\right]^{1/2}$$ $$+ \frac{D}{s}\tan^{-1}\left(\frac{s^2 - D^2}{D^2}\right)^{1/2}$$

the enclosure. Therefore, *the sum of the view factors from surface i of an enclosure to all surfaces of the enclosure, including to itself, must equal unity.* This is known as the **summation rule** for an enclosure and is expressed as (Fig. 19-45)

$$\sum_{j=1}^{N} F_{i \to j} = 1 \qquad (19\text{-}30)$$

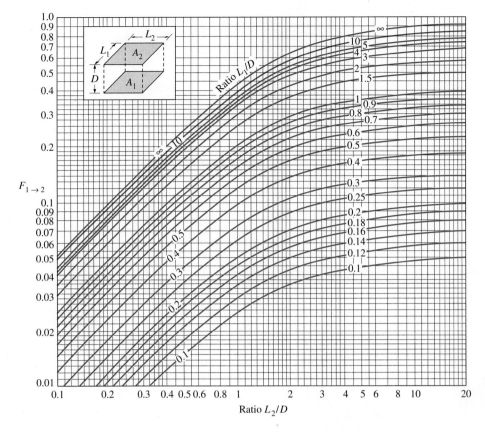

FIGURE 19-41

View factor between two aligned parallel rectangles of equal size.

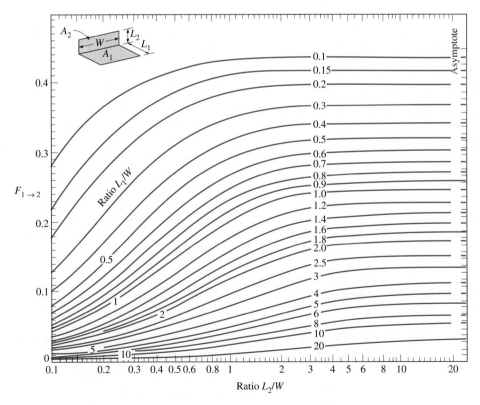

FIGURE 19-42

View factor between two perpendicular rectangles with a common edge.

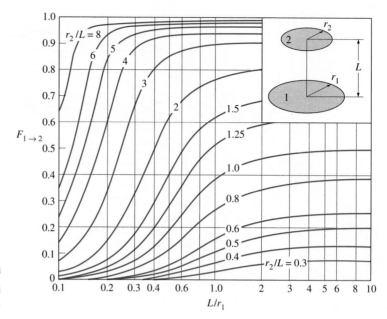

FIGURE 19-43

View factor between
two coaxial parallel disks.

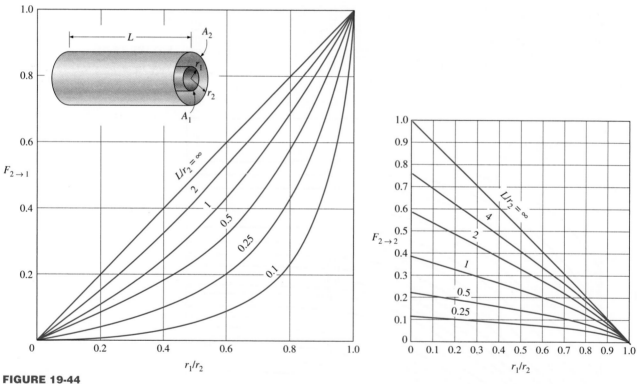

FIGURE 19-44

View factors for two concentric cylinders of finite length: (*a*) outer cylinder to inner cylinder; (*b*) outer cylinder to itself.

where N is the number of surfaces of the enclosure. For example, applying the summation rule to surface 1 of a three-surface enclosure yields

$$\sum_{j=1}^{3} F_{1 \rightarrow j} = F_{1 \rightarrow 1} + F_{1 \rightarrow 2} + F_{1 \rightarrow 3} = 1$$

The notation $F_{i \rightarrow j}$ is *instructive* for beginners, since it emphasizes that the view factor is for radiation that travels from surface i to surface j. However, this notation becomes rather awkward when it has to be used many times in a problem. In such cases, it is convenient to replace it by its *shorthand* version F_{ij}.

The summation rule can be applied to each surface of an enclosure by varying i from 1 to N. Therefore, the summation rule applied to each of the N surfaces of an enclosure gives N relations for the determination of the view factors. Also, the reciprocity rule gives $\frac{1}{2}N(N-1))$ additional relations. Then the total number of view factors that need to be evaluated directly for an N-surface enclosure becomes

$$N^2 - [N + \tfrac{1}{2}N(N-1)] = \tfrac{1}{2}N(N-1)$$

For example, for a six-surface enclosure, we need to determine only $\frac{1}{2} \times 6(6-1) = 15$ of the $6^2 = 36$ view factors directly. The remaining 21 view factors can be determined from the 21 equations that are obtained by applying the reciprocity and the summation rules.

EXAMPLE 19-5　View Factors Associated with Two Concentric Spheres

Determine the view factors associated with an enclosure formed by two spheres, shown in Fig. 19-46.

Solution　The view factors associated with two concentric spheres are to be determined.

Assumptions　The surfaces are diffuse emitters and reflectors.

Analysis　The outer surface of the smaller sphere (surface 1) and inner surface of the larger sphere (surface 2) form a two-surface enclosure. Therefore, $N = 2$ and this enclosure involves $N^2 = 2^2 = 4$ view factors, which are F_{11}, F_{12}, F_{21}, and F_{22}. In this two-surface enclosure, we need to determine only

$$\tfrac{1}{2}N(N-1) = \tfrac{1}{2} \times 2(2-1) = 1$$

view factor directly. The remaining three view factors can be determined by the application of the summation and reciprocity rules. But it turns out that we can determine not only one but *two* view factors directly in this case by a simple *inspection*:

$F_{11} = 0$,　　since no radiation leaving surface 1 strikes itself

$F_{12} = 1$,　　since all radiation leaving surface 1 strikes surface 2

Actually it would be sufficient to determine only one of these view factors by inspection, since we could always determine the other one from the summation rule applied to surface 1 as $F_{11} + F_{12} = 1$.

The view factor F_{21} is determined by applying the reciprocity rule to surfaces 1 and 2:

$$A_1 F_{12} = A_2 F_{21}$$

which yields

$$F_{21} = \frac{A_1}{A_2} F_{12} = \frac{4\pi r_1^2}{4\pi r_2^2} \times 1 = \left(\frac{r_1}{r_2}\right)^2$$

Finally, the view factor F_{22} is determined by applying the summation rule to surface 2:

$$F_{21} + F_{22} = 1$$

Surface i

FIGURE 19-45

Radiation leaving any surface i of an enclosure must be intercepted completely by the surfaces of the enclosure. Therefore, the sum of the view factors from surface i to each one of the surfaces of the enclosure must be unity.

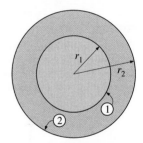

FIGURE 19-46

The geometry considered in Example 19-5.

and thus

$$F_{22} = 1 - F_{21} = 1 - \left(\frac{r_1}{r_2}\right)^2$$

Discussion Note that when the outer sphere is much larger than the inner sphere $(r_2 \gg r_1)$, F_{22} approaches one. This is expected, since the fraction of radiation leaving the outer sphere that is intercepted by the inner sphere will be negligible in that case. Also note that the two spheres considered above do not need to be concentric. However, the radiation analysis will be most accurate for the case of concentric spheres, since the radiation is most likely to be uniform on the surfaces in that case.

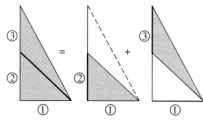

$$F_{1 \to (2, 3)} = F_{1 \to 2} + F_{1 \to 3}$$

FIGURE 19-47

The view factor from a surface to a composite surface is equal to the sum of the view factors from the surface to the parts of the composite surface.

3 The Superposition Rule

Sometimes the view factor associated with a given geometry is not available in standard tables and charts. In such cases, it is desirable to express the given geometry as the sum or difference of some geometries with known view factors, and then to apply the **superposition rule,** which can be expressed as follows: *the view factor from a surface i to a surface j is equal to the sum of the view factors from surface i to the parts of surface j.* Note that the reverse of this is not true. That is, the view factor from a surface *j* to a surface *i* is *not* equal to the sum of the view factors from the parts of surface *j* to surface *i*.

Consider the geometry in Fig. 19-47, which is infinitely long in the direction perpendicular to the plane of the paper. The radiation that leaves surface 1 and strikes the combined surfaces 2 and 3 is equal to the sum of the radiation that strikes surfaces 2 and 3. Therefore, the view factor from surface 1 to the combined surfaces of 2 and 3 is

$$F_{1 \to (2, 3)} = F_{1 \to 2} + F_{1 \to 3} \tag{19-31}$$

Suppose we need to find the view factor $F_{1 \to 3}$. A quick check of the view factor expressions and charts in this section will reveal that such a view factor cannot be evaluated directly. However, the view factor $F_{1 \to 3}$ can be determined from Eq. 19-31 after determining both $F_{1 \to 2}$ and $F_{1 \to (2, 3)}$ from the chart in Fig. 19-42. Therefore, it may be possible to determine some difficult view factors with relative ease by expressing one or both of the areas as the sum or differences of areas and then applying the superposition rule.

To obtain a relation for the view factor $F_{(2, 3) \to 1}$, we multiply Eq. 19-31 by A_1,

$$A_1 F_{1 \to (2, 3)} = A_1 F_{1 \to 2} + A_1 F_{1 \to 3}$$

and apply the reciprocity rule to each term to get

$$(A_2 + A_3)F_{(2, 3) \to 1} = A_2 F_{2 \to 1} + A_3 F_{3 \to 1}$$

or

$$F_{(2, 3) \to 1} = \frac{A_2 F_{2 \to 1} + A_3 F_{3 \to 1}}{A_2 + A_3} \tag{19-32}$$

Areas that are expressed as the sum of more than two parts can be handled in a similar manner.

FIGURE 19-48

The cylindrical enclosure considered in Example 19-6.

EXAMPLE 19-6 Fraction of Radiation Leaving through an Opening
Determine the fraction of the radiation leaving the base of the cylindrical enclosure shown in Fig. 19-48 that escapes through a coaxial ring opening at its top

surface. The radius and the length of the enclosure are r_1 = 10 cm and L = 10 cm, while the inner and outer radii of the ring are r_2 = 5 cm and r_3 = 8 cm, respectively.

Solution The fraction of radiation leaving the base of a cylindrical enclosure through a coaxial ring opening at its top surface is to be determined.

Assumptions The base surface is a diffuse emitter and reflector.

Analysis We are asked to determine the fraction of the radiation leaving the base of the enclosure that escapes through an opening at the top surface. Actually, what we are asked to determine is simply the *view factor* $F_{1 \to ring}$ from the base of the enclosure to the ring-shaped surface at the top.

We do not have an analytical expression or chart for view factors between a circular area and a coaxial ring, and so we cannot determine $F_{1 \to ring}$ directly. However, we do have a chart for view factors between two coaxial parallel disks, and we can always express a ring in terms of disks.

Let the base surface of radius r_1 = 10 cm be surface 1, the circular area of r_2 = 5 cm at the top be surface 2, and the circular area of r_3 = 8 cm be surface 3. Using the superposition rule, the view factor from surface 1 to surface 3 can be expressed as

$$F_{1 \to 3} = F_{1 \to 2} + F_{1 \to ring}$$

since surface 3 is the sum of surface 2 and the ring area. The view factors $F_{1 \to 2}$ and $F_{1 \to 3}$ are determined from the chart in Fig. 19-43 as follows:

$$\frac{L}{r_1} = \frac{10 \text{ cm}}{10 \text{ cm}} = 1 \quad \text{and} \quad \frac{r_2}{L} = \frac{5 \text{ cm}}{10 \text{ cm}} = 0.5 \xrightarrow{\text{(Fig. 19-43)}} F_{1 \to 2} = 0.11$$

$$\frac{L}{r_1} = \frac{10 \text{ cm}}{10 \text{ cm}} = 1 \quad \text{and} \quad \frac{r_3}{L} = \frac{8 \text{ cm}}{10 \text{ cm}} = 0.8 \xrightarrow{\text{(Fig. 19-43)}} F_{1 \to 3} = 0.28$$

Therefore,

$$F_{1 \to ring} = F_{1 \to 3} - F_{1 \to 2} = 0.28 - 0.11 = \mathbf{0.17}$$

which is the desired result. Note that $F_{1 \to 2}$ and $F_{1 \to 3}$ represent the fractions of radiation leaving the base that strike the circular surfaces 2 and 3, respectively, and their difference gives the fraction that strikes the ring area.

---

4 The Symmetry Rule

The determination of the view factors in a problem can be simplified further if the geometry involved possesses some sort of symmetry. Therefore, it is good practice to check for the presence of any *symmetry* in a problem before attempting to determine the view factors directly. The presence of symmetry can be determined *by inspection*, keeping the definition of the view factor in mind. Identical surfaces that are oriented in an identical manner with respect to another surface will intercept identical amounts of radiation leaving that surface. Therefore, the **symmetry rule** can be expressed as follows: *two (or more) surfaces that possess symmetry about a third surface will have identical view factors from that surface* (Fig. 19-49).

The symmetry rule can also be expressed as follows: *if the surfaces j and k are symmetric about the surface i then $F_{i \to j} = F_{i \to k}$.* Using the reciprocity rule, we can show that the relation $F_{j \to i} = F_{k \to i}$ is also true in this case.

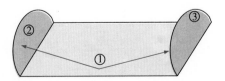

$$F_{1 \to 2} = F_{1 \to 3}$$
(Also, $F_{2 \to 1} = F_{3 \to 1}$)

FIGURE 19-49

Two surfaces that are symmetric about a third surface will have the same view factor from the third surface.

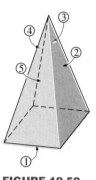

FIGURE 19-50
The pyramid considered in Example 19-7.

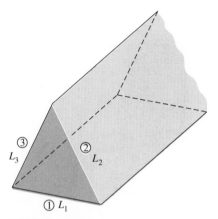

FIGURE 19-51
The infinitely long triangular duct
considered in Example 19-8.

EXAMPLE 19-7 View Factors Associated with a Tetragon
Determine the view factors from the base of the pyramid shown in Fig. 19-50 to each of its four side surfaces. The base of the pyramid is a square, and its side surfaces are isosceles triangles.

Solution The view factors from the base of a pyramid to each of its four side surfaces for the case of a square base are to be determined.

Assumptions The surfaces are diffuse emitters and reflectors.

Analysis The base of the pyramid (surface 1) and its four side surfaces (surfaces 2, 3, 4, and 5) form a five-surface enclosure. The first thing we notice about this enclosure is its symmetry. The four side surfaces are symmetric about the base surface. Then, from the *symmetry rule,* we have

$$F_{12} = F_{13} = F_{14} = F_{15}$$

Also, the *summation rule* applied to surface 1 yields

$$\sum_{j=1}^{5} F_{1j} = F_{11} + F_{12} + F_{13} + F_{14} + F_{15} = 1$$

However, $F_{11} = 0$, since the base is a *flat* surface. Then the two relations above yield

$$F_{12} = F_{13} = F_{14} = F_{15} = \mathbf{0.25}$$

Discussion Note that each of the four side surfaces of the pyramid receive one-fourth of the entire radiation leaving the base surface, as expected. Also note that the presence of symmetry greatly simplified the determination of the view factors.

EXAMPLE 19-8 View Factors Associated with a Long Triangular Duct
Determine the view factor from any one side to any other side of the infinitely long triangular duct whose cross-section is given in Fig. 19-51.

Solution The view factors associated with an infinitely long triangular duct are to be determined.

Assumptions The surfaces are diffuse emitters and reflectors.

Analysis The widths of the sides of the triangular cross-section of the duct are L_1, L_2, and L_3, and the surface areas corresponding to them are A_1, A_2, and A_3, respectively. Since the duct is infinitely long, the fraction of radiation leaving any surface that escapes through the ends of the duct is negligible. Therefore, the infinitely long duct can be considered to be a three-surface enclosure, $N = 3$.

This enclosure involves $N^2 = 3^2 = 9$ view factors, and we need to determine

$$\tfrac{1}{2}N(N-1) = \tfrac{1}{2} \times 3(3-1) = 3$$

of these view factors directly. Fortunately, we can determine all three of them by inspection to be

$$F_{11} = F_{22} = F_{33} = 0$$

since all three surfaces are flat. The remaining six view factors can be determined by the application of the summation and reciprocity rules.

Applying the summation rule to each of the three surfaces gives

$$F_{11} + F_{12} + F_{13} = 1$$
$$F_{21} + F_{22} + F_{23} = 1$$
$$F_{31} + F_{32} + F_{33} = 1$$

Noting that $F_{11} = F_{22} = F_{33} = 0$ and multiplying the first equation by A_1, the second by A_2, and the third by A_3 gives

$$A_1 F_{12} + A_1 F_{13} = A_1$$
$$A_2 F_{21} + A_2 F_{23} = A_2$$
$$A_3 F_{31} + A_3 F_{32} = A_3$$

Finally, applying the three reciprocity rules $A_1 F_{12} = A_2 F_{21}$, $A_1 F_{13} = A_3 F_{31}$, and $A_2 F_{23} = A_3 F_{32}$ gives

$$A_1 F_{12} + A_1 F_{13} = A_1$$
$$A_1 F_{12} + A_2 F_{23} = A_2$$
$$A_1 F_{13} + A_2 F_{23} = A_3$$

This is a set of three algebraic equations with three unknowns, which can be solved to obtain

$$F_{12} = \frac{A_1 + A_2 - A_3}{2A_1} = \frac{L_1 + L_2 - L_3}{2L_1}$$

$$F_{13} = \frac{A_1 + A_3 - A_2}{2A_1} = \frac{L_1 + L_3 - L_2}{2L_1} \qquad (19\text{-}33)$$

$$F_{23} = \frac{A_2 + A_3 - A_1}{2A_2} = \frac{L_2 + L_3 - L_1}{2L_2}$$

Discussion Note that we have replaced the areas of the side surfaces by their corresponding widths for simplicity, since $A = Ls$ and the length s can be factored out and canceled. We can generalize this result as follows: *the view factor from a surface of a very long triangular duct to another surface is equal to the sum of the widths of these two surfaces minus the width of the third surface, divided by twice the width of the first surface.*

View Factors between Infinitely Long Surfaces: The Crossed-Strings Method

Many problems encountered in practice involve geometries of constant cross-section such as channels and ducts that are *very long* in one direction relative to the other directions. Such geometries can conveniently be considered to be *two-dimensional,* since any radiation interaction through their end surfaces will be negligible. Then they can be modeled as being *infinitely long,* and the view factor between their surfaces can be determined by the amazingly simple *crossed-strings method* developed by H. C. Hottel in the 1950s. The surfaces of the geometry do not need to be flat; they can be convex, concave, or any irregular shape.

To demonstrate the method, consider the geometry shown in Fig. 19-52, and let us try to find the view factor $F_{1 \rightarrow 2}$ between surfaces 1 and 2. The first thing we do is identify the endpoints of the surfaces (the points A, B, C, and D) and connect them to each other with tightly stretched strings, which are indicated by dashed lines. Hottel has shown that the view factor $F_{1 \rightarrow 2}$ can be expressed in terms of the lengths of these stretched strings, which are straight lines, as

$$F_{1 \rightarrow 2} = \frac{(L_5 + L_6) - (L_3 + L_4)}{2L_1} \qquad (19\text{-}34)$$

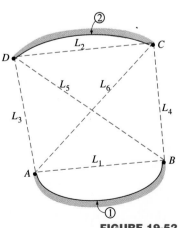

FIGURE 19-52

Determination of the view factor $F_{1 \rightarrow 2}$ by the application of the crossed-strings method.

Note that $L_5 + L_6$ is the sum of the lengths of the *crossed strings,* and $L_3 + L_4$ is the sum of the lengths of the *uncrossed strings* attached to the endpoints. Therefore, Hottel's crossed-strings method can be expressed verbally as

$$F_{i \to j} = \frac{\Sigma \, (\text{Crossed strings}) - \Sigma \, (\text{Uncrossed strings})}{2 \times (\text{String on surface } i)} \qquad (19\text{-}35)$$

The crossed-strings method is applicable even when the two surfaces considered share a common edge, as in a triangle. In such cases, the common edge can be treated as an imaginary string of zero length. The method can also be applied to surfaces that are partially blocked by other surfaces by allowing the strings to bend around the blocking surfaces.

EXAMPLE 19-9 The Crossed-Strings Method for View Factors

Two infinitely long parallel plates of widths $a = 12$ cm and $b = 5$ cm are located a distance $c = 6$ cm apart, as shown in Fig. 19-53. (*a*) Determine the view factor $F_{1 \to 2}$ from surface 1 to surface 2 by using the crossed-strings method. (*b*) Derive the crossed-strings formula by forming triangles on the given geometry and using Eq. 19-33 for view factors between the sides of triangles.

Solution The view factors between two infinitely long parallel plates are to be determined using the crossed-strings method, and the formula for the view factor is to be derived.

Assumptions The surfaces are diffuse emitters and reflectors.

Analysis (*a*) First we label the endpoints of both surfaces and draw straight dashed lines between the endpoints, as shown in Fig. 19-53. Then we identify the crossed and uncrossed strings and apply the crossed-strings method (Eq. 19-35) to determine the view factor $F_{1 \to 2}$:

$$F_{1 \to 2} = \frac{\Sigma \, (\text{Crossed strings}) - \Sigma \, (\text{Uncrossed strings})}{2 \times (\text{String on surface } 1)} = \frac{(L_5 + L_6) - (L_3 + L_4)}{2L_1}$$

where

$$
\begin{array}{ll}
L_1 = a = 12 \text{ cm} & L_4 = \sqrt{7^2 + 6^2} = 9.22 \text{ cm} \\
L_2 = b = 5 \text{ cm} & L_5 = \sqrt{5^2 + 6^2} = 7.81 \text{ cm} \\
L_3 = c = 6 \text{ cm} & L_6 = \sqrt{12^2 + 6^2} = 13.42 \text{ cm}
\end{array}
$$

Substituting,

$$F_{1 \to 2} = \frac{[(7.81 + 13.42) - (6 + 9.22)] \text{ cm}}{2 \times 12 \text{ cm}} = \boldsymbol{0.250}$$

(*b*) The geometry is infinitely long in the direction perpendicular to the plane of the paper, and thus the two plates (surfaces 1 and 2) and the two openings (imaginary surfaces 3 and 4) form a four-surface enclosure. Then applying the summation rule to surface 1 yields

$$F_{11} + F_{12} + F_{13} + F_{14} = 1$$

But $F_{11} = 0$ since it is a flat surface. Therefore,

$$F_{12} = 1 - F_{13} - F_{14}$$

where the view factors F_{13} and F_{14} can be determined by considering the triangles *ABC* and *ABD,* respectively, and applying Eq. 19-33 for view factors between the sides of triangles. We obtain

$$F_{13} = \frac{L_1 + L_3 - L_6}{2L_1}, \qquad F_{14} = \frac{L_1 + L_4 - L_5}{2L_1}$$

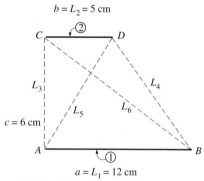

$b = L_2 = 5$ cm

$c = 6$ cm

$a = L_1 = 12$ cm

FIGURE 19-53

The two infinitely long parallel plates considered in Example 19-9.

Substituting,

$$F_{12} = 1 - \frac{L_1 + L_3 - L_6}{2L_1} - \frac{L_1 + L_4 - L_5}{2L_1}$$

$$= \frac{(L_5 + L_6) - (L_3 + L_4)}{2L_1}$$

which is the desired result. This is also a miniproof of the crossed-strings method for the case of two infinitely long plain parallel surfaces.

19-7 ■ RADIATION HEAT TRANSFER: BLACK SURFACES

So far, we have considered the nature of radiation, the radiation properties of materials, and the view factors, and we are now in a position to consider the rate of heat transfer between surfaces by radiation. The analysis of radiation exchange between surfaces, in general, is complicated because of reflection: a radiation beam leaving a surface may be reflected several times, with partial reflection occurring at each surface, before it is completely absorbed. The analysis is simplified greatly when the surfaces involved can be approximated as blackbodies because of the absence of reflection. In this section, we consider radiation exchange between *black surfaces* only; we will extend the analysis to reflecting surfaces in the next section.

Consider two black surfaces of arbitrary shape maintained at uniform temperatures T_1 and T_2, as shown in Fig. 19-54. Recognizing that radiation leaves a black surface at a rate of $E_b = \sigma T^4$ per unit surface area and that the view factor $F_{1 \rightarrow 2}$ represents the fraction of radiation leaving surface 1 that strikes surface 2, the *net* rate of radiation heat transfer from surface 1 to surface 2 can be expressed as

$$\dot{Q}_{1 \rightarrow 2} = \begin{pmatrix} \text{Radiation leaving} \\ \text{the entire surface 1} \\ \text{that strikes surface 2} \end{pmatrix} - \begin{pmatrix} \text{Radiation leaving} \\ \text{the entire surface 2} \\ \text{that strikes surface 1} \end{pmatrix} \quad (19\text{-}36)$$

$$= A_1 E_{b1} F_{1 \rightarrow 2} - A_2 E_{b2} F_{2 \rightarrow 1} \quad \text{(W)}$$

Applying the reciprocity rule $A_1 F_{1 \rightarrow 2} = A_2 F_{2 \rightarrow 1}$ yields

$$\dot{Q}_{1 \rightarrow 2} = A_1 F_{1 \rightarrow 2} \, \sigma(T_1^4 - T_2^4) \quad \text{(W)} \quad (19\text{-}37)$$

which is the desired relation. A negative value for $\dot{Q}_{1 \rightarrow 2}$ indicates that net radiation heat transfer is from surface 2 to surface 1.

Now consider an *enclosure* consisting of N *black* surfaces maintained at specified temperatures. The *net* radiation heat transfer *from* any surface i of this enclosure is determined by adding up the net radiation heat transfers from surface i to each of the surfaces of the enclosure:

$$\dot{Q}_i = \sum_{j=1}^{N} \dot{Q}_{i \rightarrow j} = \sum_{j=1}^{N} A_i F_{i \rightarrow j} \sigma(T_i^4 - T_j^4) \quad \text{(W)} \quad (19\text{-}38)$$

Again a negative value for $\dot{Q}$ indicates that net radiation heat transfer is *to* surface i (i.e., surface i *gains* radiation energy instead of losing). Also, the net heat transfer from a surface to itself is zero, regardless of the shape of the surface.

EXAMPLE 19-10 Radiation Heat Transfer in a Black Cubical Furnace

Consider the 5-m × 5-m × 5-m cubical furnace shown in Fig. 19-55, whose surfaces closely approximate black surfaces. The base, top, and side surfaces of

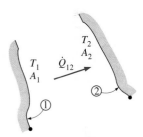

FIGURE 19-54

Two general black surfaces maintained at uniform temperatures T_1 and T_2.

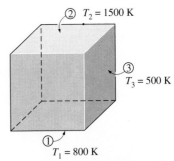

FIGURE 19-55

The cubical furnace of black surfaces considered in Example 19-10.

the furnace are maintained at uniform temperatures of 800 K, 1500 K, and 500 K, respectively. Determine (*a*) the net rate of radiation heat transfer between the base and the side surfaces, (*b*) the net rate of radiation heat transfer between the base and the top surface, and (*c*) the net radiation heat transfer from the base surface.

Solution   The surfaces of a cubical furnace are black and are maintained at uniform temperatures. The net rate of radiation heat transfer between the base and side surfaces, between the base and the top surface, from the base surface are to be determined.

Assumptions   The surfaces are black and isothermal.

Analysis   (*a*) Considering that the geometry involves six surfaces, we may be tempted at first to treat the furnace as a six-furnace enclosure. However, the four side surfaces possess the same properties, and thus we can treat them as a single side surface in radiation analysis. We consider the base surface to be surface 1, the top surface to be surface 2, and the side surfaces to be surface 3. Then the problem reduces to determining $\dot{Q}_{1 \to 3}$, $\dot{Q}_{1 \to 2}$, and $\dot{Q}_1$.

The net rate of radiation heat transfer $\dot{Q}_{1 \to 3}$ from surface 1 to surface 3 can be determined from Eq. 19-37, since both surfaces involved are black, by replacing the subscript 2 by 3:

$$\dot{Q}_{1 \to 3} = A_1 F_{1 \to 3} \sigma (T_1^4 - T_3^4)$$

But first we need to evaluate the view factor $F_{1 \to 3}$. After checking the view factor charts and tables, we realize that we cannot determine this view factor directly. However, we can determine the view factor $F_{1 \to 2}$ directly from Fig. 19-41 to be $F_{1 \to 2} = 0.2$, and we know that $F_{1 \to 1} = 0$ since surface 1 is a plane. Then applying the summation rule to surface 1 yields

$$F_{1 \to 1} + F_{1 \to 2} + F_{1 \to 3} = 1$$

or

$$F_{1 \to 3} = 1 - F_{1 \to 1} - F_{1 \to 2} = 1 - 0 - 0.2 = 0.8$$

Substituting,

$$\dot{Q}_{1 \to 3} = (25 \text{ m}^2)(0.8)(5.67 \times 10^{-8} \text{ W/m}^2 \cdot \text{K}^4)[(800 \text{ K})^4 - (500 \text{ K})^4]$$
$$= 393.6 \times 10^3 \text{ W} = \textbf{393.6 kW}$$

(*b*) The net rate of radiation heat transfer $\dot{Q}_{1 \to 2}$ from surface 1 to surface 2 is determined in a similar manner from Equation 19-37 to be

$$\dot{Q}_{1 \to 2} = A_1 F_{1 \to 2} \sigma (T_1^4 - T_2^4)$$
$$= (25 \text{ m}^2)(0.2)(5.67 \times 10^{-8} \text{ W/m}^2 \cdot \text{K}^4)[(800 \text{ K})^4 - (1500 \text{ K})^4]$$
$$= -1{,}319 \times 10^3 \text{ W} = \textbf{-1,319 kW}$$

The negative sign indicates that net radiation heat transfer is from surface 2 to surface 1.

(*c*) The net radiation heat transfer from the base surface $\dot{Q}_1$ is determined from Eq. 19-38 by replacing the subscript *i* by 1 and taking *N* = 3:

$$\dot{Q}_1 = \sum_{j=1}^{3} \dot{Q}_{1 \to j} = \dot{Q}_{1 \to 1} + \dot{Q}_{1 \to 2} + \dot{Q}_{1 \to 3}$$
$$= 0 + (-1{,}319{,}097 \text{ W}) + (393{,}611 \text{ W})$$
$$= -925.5 \times 10^3 \text{ W} = \textbf{-925.5 kW}$$

Again the negative sign indicates that net radiation heat transfer is *to* surface 1. That is, the base of the furnace is gaining net radiation at a rate of about 925 kW.

19-8 ■ RADIATION HEAT TRANSFER: DIFFUSE, GRAY SURFACES

851

Radiation
Heat Transfer:
Diffuse, Gray
Surfaces

The analysis of radiation transfer in enclosures consisting of black surfaces is relatively easy, as we have seen above, but most enclosures encountered in practice involve nonblack surfaces, which allow multiple reflections to occur. Radiation analysis of such enclosures becomes very complicated unless some simplifying assumptions are made.

To make a simple radiation analysis possible, it is common to assume the surfaces of an enclosure to be *opaque, diffuse,* and *gray.* That is, the surfaces are nontransparent, they are diffuse emitters and diffuse reflectors, and their radiation properties are independent of wavelength. Also, each surface of the enclosure is *isothermal,* and both the incoming and outgoing radiation are *uniform* over each surface. But first we introduce the concept of radiosity.

Radiosity

Surfaces emit radiation as well as reflect it, and thus the radiation leaving a surface consists of emitted and reflected parts. The calculation of radiation heat transfer between surfaces involves the *total* radiation energy streaming away from a surface, with no regard for its origin. Thus, we need to define a new quantity to represent the *total radiation energy leaving a surface per unit time and per unit area.* This quantity is called the **radiosity** and is denoted by J (Fig. 19-56).

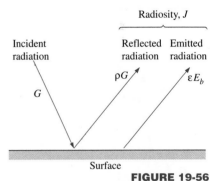

For a surface i that is *gray* and *opaque* ($\varepsilon_i = \alpha_i$ and $\alpha_i + \rho_i = 1$), the radiosity can be expressed as

$$J_i = \begin{pmatrix} \text{Radiation emitted} \\ \text{by surface } i \end{pmatrix} + \begin{pmatrix} \text{Radiation reflected} \\ \text{by surface } i \end{pmatrix}$$

$$= \varepsilon_i E_{bi} + \rho_i G_i \qquad (19\text{-}39)$$

$$= \varepsilon_i E_{bi} + (1 - \varepsilon_i)G_i \qquad (\text{W/m}^2)$$

FIGURE 19-56

Radiosity represents the sum of the radiation energy emitted and reflected by a surface.

where $E_{bi} = \sigma T_i^4$ is the blackbody emissive power of surface i and G_i is irradiation (i.e., the radiation energy incident on surface i per unit time per unit area).

For a surface that can be approximated as a *blackbody* ($\varepsilon_i = 1$), the radiosity relation reduces to

$$J_i = E_{bi} = \sigma T_i^4 \qquad (\text{blackbody}) \qquad (19\text{-}40)$$

That is, *the radiosity of a blackbody is equal to its emissive power.* This is expected, since a blackbody does not reflect any radiation, and thus radiation coming from a blackbody is due to emission only.

Net Radiation Heat Transfer to or from a Surface

During a radiation interaction, a surface *loses* energy by emitting radiation and *gains* energy by absorbing radiation emitted by other surfaces. A surface experiences a net gain or a net loss of energy, depending on which quantity is larger. The *net* rate of radiation heat transfer from a surface i of surface area A_i is denoted by $\dot{Q}_i$ and is expressed as

$$\dot{Q}_i = \begin{pmatrix} \text{Radiation leaving} \\ \text{entire surface } i \end{pmatrix} - \begin{pmatrix} \text{Radiation incident} \\ \text{on entire surface } i \end{pmatrix}$$

$$= A_i(J_i - G_i) \qquad (\text{W}) \qquad (19\text{-}41)$$

Solving for G_i from Eq. 19-39 and substituting into Eq. 19-41 yields

$$\dot{Q}_i = A_i\left(J_i - \frac{J_i - \varepsilon_i E_{bi}}{1 - \varepsilon_i}\right) = \frac{A_i\varepsilon_i}{1 - \varepsilon_i}(E_{bi} - J_i) \qquad \text{(W)} \qquad \text{(19-42)}$$

In an electrical analogy to Ohm's law, this equation can be rearranged as

$$\dot{Q}_i = \frac{E_{bi} - J_i}{R_i} \qquad \text{(W)} \qquad \text{(19-43)}$$

where

$$R_i = \frac{1 - \varepsilon_i}{A_i\varepsilon_i} \qquad \text{(19-44)}$$

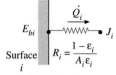

Surface i $R_i = \dfrac{1-\varepsilon_i}{A_i\varepsilon_i}$

FIGURE 19-57

Electrical analogy of surface resistance to radiation.

is the **surface resistance** to radiation. The quantity $E_{bi} - J_i$ corresponds to a *potential difference* and the net rate of radiation heat transfer corresponds to *current* in the electrical analogy, as illustrated in Fig. 19-57.

The direction of the net radiation heat transfer depends on the relative magnitudes of J_i (the radiosity) and E_{bi} (the emissive power of a blackbody at the temperature of the surface). It will be *from* the surface if $E_{bi} > J_i$ and *to* the surface if $J_i > E_{bi}$. A negative value for $\dot{Q}_i$ indicates that heat transfer is *to* the surface. All of this radiation energy gained must be removed from the other side of the surface through some mechanism if the surface temperature is to remain constant.

The surface resistance to radiation for a *blackbody* is *zero* since $\varepsilon_i = 1$ and $J_i = E_{bi}$. The net rate of radiation heat transfer in this case is determined directly from Eq. 19-41.

Some surfaces encountered in numerous practical heat transfer applications are modeled as being *adiabatic* since their back sides are well insulated and the net heat transfer through them is zero. When the convection effects on the front (heat transfer) side of such a surface is negligible and steady-state conditions are reached, the surface must lose as much radiation energy as it gains, and thus $\dot{Q}_i = 0$. In such cases, the surface is said to *reradiate* all the radiation energy it receives, and such a surface is called a **reradiating surface.** Setting $\dot{Q}_i = 0$ in Eq. 19-43 yields

$$J_i = E_{bi} = \sigma T_i^4 \qquad \text{(W/m}^2\text{)} \qquad \text{(19-45)}$$

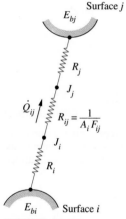

Surface j E_{bj}

R_j

J_j

$\dot{Q}_{ij}$ $R_{ij} = \dfrac{1}{A_i F_{ij}}$

J_i

R_i

E_{bi} Surface i

FIGURE 19-58

Electrical analogy of space resistance to radiation.

Therefore, the *temperature* of a reradiating surface under steady conditions can easily be determined from the equation above once its radiosity is known. Note that the temperature of a reradiating surface is *independent of its emissivity*. In radiation analysis, the surface resistance of a reradiating surface is disregarded since there is no net heat transfer through it. (This is like the fact that there is no need to consider a resistance in an electrical network if no current is flowing through it.)

Net Radiation Heat Transfer between Any Two Surfaces

Consider two diffuse, gray, and opaque surfaces of arbitrary shape maintained at uniform temperatures, as shown in Fig. 19-58. Recognizing that the radiosity J represents the rate of radiation leaving a surface per unit surface area and that the view factor $F_{i \rightarrow j}$ represents the fraction of radiation leaving surface i that strikes surface j, the *net* rate of radiation heat transfer from surface i to surface j can be expressed as

$$\dot{Q}_{i \to j} = \begin{pmatrix} \text{Radiation leaving} \\ \text{the entire surface } i \\ \text{that strikes surface } j \end{pmatrix} - \begin{pmatrix} \text{Radiation leaving} \\ \text{the entire surface } j \\ \text{that strikes surface } i \end{pmatrix} \quad (19\text{-}46)$$

$$= A_i J_i F_{i \to j} - A_j J_j F_{j \to i} \qquad \text{(W)}$$

Applying the reciprocity relation $A_i F_{i \to j} = A_j F_{j \to i}$ yields

$$\dot{Q}_{i \to j} = A_i F_{i \to j} (J_i - J_j) \qquad \text{(W)} \qquad (19\text{-}47)$$

Again in analogy to Ohm's law, this equation can be rearranged as

$$\dot{Q}_{i \to j} = \frac{J_i - J_j}{R_{i \to j}} \qquad \text{(W)} \qquad (19\text{-}48)$$

where

$$R_{i \to j} = \frac{1}{A_i R_{i \to j}} \qquad (19\text{-}49)$$

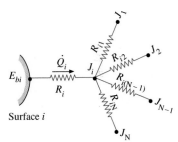

FIGURE 19-59

Network representation of net radiation heat transfer from surface i to the remaining surfaces of an N-surface enclosure.

is the **space resistance** to radiation. Again the quantity $J_i - J_j$ corresponds to a *potential difference,* and the net rate of heat transfer between two surfaces corresponds to *current* in the electrical analogy, as illustrated in Fig. 19-58.

The direction of the net radiation heat transfer between two surfaces depends on the relative magnitudes of J_i and J_j. A positive value for $\dot{Q}_{i \to j}$ indicates that net heat transfer is *from* surface i *to* surface j. A negative value indicates the opposite.

In an N-surface enclosure, the conservation of energy principle requires that the net heat transfer from surface i be equal to the sum of the net heat transfers from surface i to each of the N surfaces of the enclosure. That is,

$$\dot{Q}_i = \sum_{j=1}^{N} \dot{Q}_{i \to j} = \sum_{j=1}^{N} A_i F_{i \to j} (J_i - J_j) = \sum_{j=1}^{N} \frac{J_i - J_j}{R_{i \to j}} \qquad \text{(W)} \qquad (19\text{-}50)$$

The network representation of net radiation heat transfer from surface i to the remaining surfaces of an N-surface enclosure is given in Fig. 19-59. Note that $\dot{Q}_{i \to i}$ (the net rate of heat transfer from a surface to itself) is zero regardless of the shape of the surface. Combining Eqs. 19-43 and 19-50 gives

$$\frac{E_{bi} - J_i}{R_i} = \sum_{j=1}^{N} \frac{J_i - J_j}{R_{i \to j}} \qquad \text{(W)} \qquad (19\text{-}51)$$

which has the electrical analogy interpretation that *the net radiation flow from a surface through its surface resistance is equal to the sum of the radiation flows from that surface to all other surfaces through the corresponding space resistances.*

Methods of Solving Radiation Problems

In the radiation analysis of an enclosure, either the temperature or the net rate of heat transfer must be given for each of the surfaces to obtain a unique solution for the unknown surface temperatures and heat transfer rates. There are two methods commonly used to solve radiation problems. In the first method, Eqs. 19-50 (for surfaces with specified heat transfer rates) and 19-51 (for surfaces with specified temperatures) are simplified and rearranged as

| Surfaces with specified net heat transfer rate $\dot{Q}_i$ | $\dot{Q}_i = A_i \sum_{j=1}^{N} F_{i \to j}(J_i - J_j)$ | (19-52a) |
| Surfaces with specified temperature T_i | $\sigma T_i^4 = J_i + \dfrac{1 - \varepsilon_i}{\varepsilon_i} \sum_{j=1}^{N} F_{i \to j}(J_i - J_j)$ | (19-52b) |

Note that $\dot{Q}_i = 0$ for insulated (or reradiating) surfaces, and $\sigma T_i^4 = J_i$ for black surfaces since $\varepsilon_i = 1$ in that case. Also, the term corresponding to $j = i$ will drop out from either relation since $J_i - J_j = J_i - J_i = 0$ in that case.

The equations above give N linear algebraic equations for the determination of the N unknown radiosities for an N-surface enclosure. Once the radiosities $J_1, J_2, \ldots, J_N$ are available, the unknown heat transfer rates can be determined from Eq. 19-52a while the unknown surface temperatures can be determined from Eq. 19-52b. The temperatures of insulated or reradiating surfaces can be determined from $\sigma T_i^4 = J_i$. A positive value for $\dot{Q}_i$ indicates net radiation heat transfer *from* surface i to other surfaces in the enclosure while a negative value indicates net radiation heat transfer *to* the surface.

The systematic approach described above for solving radiation heat transfer problems is very suitable for use with today's popular equation solvers such as EES, Mathcad, and Matlab, especially when there are a large number of surfaces, and is known as the **direct method** (formerly, the *matrix method,* since it resulted in matrices and the solution required a knowledge of linear algebra). The second method described below, called the **network method,** is based on the electrical network analogy.

The network method was first introduced by A. K. Oppenheim in the 1950s and found widespread acceptance because of its simplicity and emphasis on the physics of the problem. The application of the method is straightforward: draw a surface resistance associated with each surface of an enclosure and connect them with space resistances. Then solve the radiation problem by treating it as an electrical network problem where the radiation heat transfer replaces the current and radiosity replaces the potential.

The network method is not practical for enclosures with more than three or four surfaces, however, because of the increased complexity of the network. Below we apply the method to solve radiation problems in two- and three-surface enclosures.

Radiation Heat Transfer in Two-Surface Enclosures

Consider an enclosure consisting of two opaque surfaces at specified temperatures T_1 and T_2, as shown in Fig. 19-60, and try to determine the net rate of radiation heat transfer between the two surfaces with the network method. Surfaces 1 and 2 have emissivities ε_1 and ε_2 and surface areas A_1 and A_2 and are maintained at uniform temperatures T_1 and T_2, respectively. There are only two surfaces in the enclosure, and thus we can write

$$\dot{Q}_{12} = \dot{Q}_1 = -\dot{Q}_2$$

That is, the net rate of radiation transfer from surface 1 to surface 2 must equal the net rate of radiation transfer *from* surface 1 and the net rate of radiation transfer *to* surface 2.

The radiation network of this two-surface enclosure consists of two surface resistances and one space resistance, as shown in Figure 19-60. In an

FIGURE 19-60

Schematic of a two-surface enclosure and the radiation network associated with it.

electrical network, the electric current flowing through these resistances connected in series would be determined by dividing the potential difference between points A and B by the total resistance between the same two points. The net rate of radiation transfer is determined in the same manner and is expressed as

$$\dot{Q}_{12} = \frac{E_{b1} - E_{b2}}{R_1 + R_{12} + R_2} = \dot{Q}_1 = -\dot{Q}_2$$

or

$$\dot{Q}_{12} = \frac{\sigma(T_1^4 - T_2^4)}{\dfrac{1 - \varepsilon_1}{A_1\,\varepsilon_1} + \dfrac{1}{A_1\,F_{12}} + \dfrac{1 - \varepsilon_2}{A_2\,\varepsilon_2}} \qquad \text{(W)} \qquad (19\text{-}53)$$

This important result is applicable to any two gray, diffuse, opaque surfaces that form an enclosure. The view factor F_{12} depends on the geometry and must be determined first. Simplified forms of Eq. 19-53 for some familiar arrangements that form a two-surface enclosure are given in Table 19-6. Note that $F_{12} = 1$ for all of these special cases.

EXAMPLE 19-11   Radiation Heat Transfer between Large Parallel Plates

Two very large parallel plates are maintained at uniform temperatures $T_1 = 800$ K and $T_2 = 500$ K and have emissivities $\varepsilon_1 = 0.2$ and $\varepsilon_2 = 0.7$, respectively, as shown in Fig. 19-61. Determine the net rate of radiation heat transfer between the two surfaces per unit surface area of the plates.

Solution   Two large parallel plates are maintained at uniform temperatures. The net rate of radiation heat transfer between the plates is to be determined.

Assumptions   Both surfaces are opaque, diffuse, and gray.

Analysis   The net rate of radiation heat transfer between the two plates per unit area is readily determined from Eq. 19-55 to be

$$\dot{q}_{12} = \frac{\dot{Q}_{12}}{A} = \frac{\sigma(T_1^4 - T_2^4)}{\dfrac{1}{\varepsilon_1} + \dfrac{1}{\varepsilon_2} - 1} = \frac{(5.67 \times 10^{-8}\ \text{W/m}^2 \cdot \text{K}^4)[(800\ \text{K})^4 - (500\ \text{K})^4]}{\dfrac{1}{0.2} + \dfrac{1}{0.7} - 1}$$

$$= \mathbf{3625\ W/m^2}$$

Discussion   Note that heat at a net rate of 3625 W is transferred from plate 1 to plate 2 by radiation per unit surface area of either plate.

Radiation Heat Transfer in Three-Surface Enclosures

We now consider an enclosure consisting of three opaque, diffuse, gray surfaces, as shown in Fig. 19-62. Surfaces 1, 2, and 3 have surface areas A_1, A_2, and A_3; emissivities ε_1, ε_2, and ε_3; and uniform temperatures T_1, T_2, and T_3, respectively. The radiation network of this geometry is constructed by following the standard procedure: draw a surface resistance associated with each of the three surfaces and connect these surface resistances with space resistances, as shown in the figure. Relations for the surface and space resistances are given by Eqs. 19-44 and 19-49. The three endpoint potentials E_{b1}, E_{b2}, and E_{b3} are considered known, since the surface temperatures are specified. Then all we need to find are the radiosities J_1, J_2, and J_3. The three equations for the determination of these three unknowns are obtained from the requirement that *the*

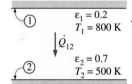

FIGURE 19-61
The two parallel plates considered in Example 19-11.

TABLE 19-6

Small object in a large cavity		
A_1, T_1, ε_1 A_2, T_2, ε_2	$\dfrac{A_1}{A_2} \approx 0$ $F_{12} = 1$	$\dot{Q}_{12} = A_1 \sigma \varepsilon_1 (T_1^4 - T_2^4)$ (19–54)
Infinitely large parallel plates		
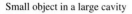 A_1, T_1, ε_1 A_2, T_2, ε_2	$A_1 = A_2 = A$ $F_{12} = 1$	$\dot{Q}_{12} = \dfrac{A\sigma(T_1^4 - T_2^4)}{\dfrac{1}{\varepsilon_1} + \dfrac{1}{\varepsilon_2} - 1}$ (19–55)
Infinitely long concentric cylinders		
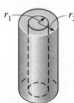 r_1 r_2	$\dfrac{A_1}{A_2} = \dfrac{r_1}{r_2}$ $F_{12} = 1$	$\dot{Q}_{12} = \dfrac{A_1 \sigma(T_1^4 - T_2^4)}{\dfrac{1}{\varepsilon_1} + \dfrac{1-\varepsilon_2}{\varepsilon_2}\left(\dfrac{r_1}{r_2}\right)}$ (19–56)
Concentric spheres		
r_1 r_2	$\dfrac{A_1}{A_2} = \left(\dfrac{r_1}{r_2}\right)^2$ $F_{12} = 1$	$\dot{Q}_{12} = \dfrac{A_1 \sigma(T_1^4 - T_2^4)}{\dfrac{1}{\varepsilon_1} + \dfrac{1-\varepsilon_2}{\varepsilon_2}\left(\dfrac{r_1}{r_2}\right)^2}$ (19–57)

algebraic sum of the currents (net radiation heat transfer) at each node must equal zero. That is,

$$\frac{E_{b1} - J_1}{R_1} + \frac{J_2 - J_1}{R_{12}} + \frac{J_3 - J_1}{R_{13}} = 0$$

$$\frac{J_1 - J_2}{R_{12}} + \frac{E_{b2} - J_2}{R_2} + \frac{J_3 - J_2}{R_{23}} = 0 \qquad (19\text{-}58)$$

$$\frac{J_1 - J_3}{R_{13}} + \frac{J_2 - J_3}{R_{23}} + \frac{E_{b3} - J_3}{R_3} = 0$$

Once the radiosities J_1, J_2, and J_3 are available, the net rate of radiation heat transfers at each surface can be determined from Eq. 19-50.

The set of equations above simplify further if one or more surfaces are "special" in some way. For example, $J_i = E_{bi} = \sigma T_i^4$ for a *black* or *reradiating*

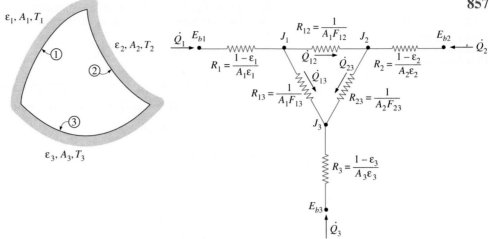

FIGURE 19-62

Schematic of a three-surface enclosure and the radiation network associated with it.

surface. Also, $\dot{Q}_i = 0$ for a reradiating surface. Finally, when the net rate of radiation heat transfer $\dot{Q}_i$ is specified at surface i instead of the temperature, the term $(E_{bi} - J_i)/R_i$ should be replaced by the specified $\dot{Q}_i$.

EXAMPLE 19-12 Radiation Heat Transfer in a Cylindrical Furnace

Consider a cylindrical furnace with $r_0 = H = 1$ m, as shown in Fig. 19-63. The top (surface 1) and the base (surface 2) of the furnace has emissivities $\varepsilon_1 = 0.8$ and $\varepsilon_2 = 0.4$, respectively, and are maintained at uniform temperatures $T_1 = 700$ K and $T_2 = 500$ K. The side surface closely approximates a blackbody and is maintained at a temperature of $T_3 = 400$ K. Determine the net rate of radiation heat transfer at each surface during steady operation and explain how these surfaces can be maintained at specified temperatures.

Solution The surfaces of a cylindrical furnace are maintained at uniform temperatures. The net rate of radiation heat transfer at each surface during steady operation is to be determined.

Assumptions **1** Steady operating conditions exist. **2** The surfaces are opaque, diffuse, and gray. **3** Convection heat transfer is not considered.

Analysis We will solve this problem systematically using the direct method to demonstrate its use. The cylindrical furnace can be considered to be a three-surface enclosure with surface areas of

$$A_1 = A_2 = \pi r_0^2 = \pi(1 \text{ m})^2 = 3.14 \text{ m}^2$$
$$A_3 = 2\pi r_0 H = 2\pi(1 \text{ m})(1 \text{ m}) = 6.28 \text{ m}^2$$

The view factor from the base to the top surface is, from Fig. 19-43, $F_{12} = 0.38$. Then the view factor from the base to the side surface is determined by applying the summation rule to be

$$F_{11} + F_{12} + F_{13} = 1 \quad \rightarrow \quad F_{13} = 1 - F_{11} - F_{12} = 1 - 0 - 0.38 = 0.62$$

since the base surface is flat and thus $F_{11} = 0$. Noting that the top and bottom surfaces are symmetric about the side surface, $F_{21} = F_{12} = 0.38$ and $F_{23} = F_{13} = 0.62$. The view factor F_{31} is determined from the reciprocity rule,

$$A_1 F_{13} = A_3 F_{31} \quad \rightarrow \quad F_{31} = F_{13}(A_1/A_3) = (0.62)(0.314/0.628) = 0.31$$

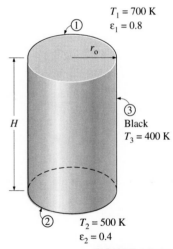

$T_1 = 700$ K
$\varepsilon_1 = 0.8$

r_0

H

③ Black
$T_3 = 400$ K

$T_2 = 500$ K
$\varepsilon_2 = 0.4$

FIGURE 19-63

The cylindrical furnace considered in Example 19-12.

Also, $F_{32} = F_{31} = 0.31$ because of symmetry. Now that all the view factors are available, we apply Eq. 19-52b to each surface to determine the radiosities:

Top surface ($i = 1$): $\qquad \sigma T_1^4 = J_1 + \dfrac{1 - \varepsilon_1}{\varepsilon_1}[F_{1\to2}(J_1 - J_2) + F_{1\to3}(J_1 - J_3)]$

Bottom surface ($i = 2$): $\quad \sigma T_2^4 = J_2 + \dfrac{1 - \varepsilon_2}{\varepsilon_2}[F_{2\to1}(J_2 - J_1) + F_{2\to3}(J_2 - J_3)]$

Side surface ($i = 3$): $\qquad \sigma T_3^4 = J_3 + 0$ (since surface 3 is black and thus $\varepsilon_3 = 1$)

Substituting the known quantities,

$$(5.67 \times 10^{-8} \text{ W/m}^2 \cdot \text{K}^4)(700 \text{ K})^4 = J_1 + \frac{1 - 0.8}{0.8}[0.38(J_1 - J_2) + 0.68(J_1 - J_3)]$$

$$(5.67 \times 10^{-8} \text{ W/m}^2 \cdot \text{K}^4)(500 \text{ K})^4 = J_2 + \frac{1 - 0.4}{0.4}[0.28(J_2 - J_1) + 0.68(J_2 - J_3)]$$

$$(5.67 \times 10^{-8} \text{ W/m}^2 \cdot \text{K}^4)(400 \text{ K})^4 = J_3$$

Solving the equations above for J_1, J_2, and J_3 gives

$$J_1 = 11{,}418 \text{ W/m}^2, \quad J_2 = 4562 \text{ W/m}^2, \quad \text{and} \quad J_3 = 1452 \text{ W/m}^2$$

Then the net rates of radiation heat transfer at the three surfaces are determined from Eq. 19-52a to be

$$\dot{Q}_1 = A_1[F_{1\to2}(J_1 - J_2) + F_{1\to3}(J_1 - J_3)]$$
$$= (3.14 \text{ m}^2)[0.38(11{,}418 - 4562) + 0.62(11{,}418 - 1452)] \text{ W/m}^2 = \mathbf{27{,}582 \text{ W}}$$

$$\dot{Q}_2 = A_2[F_{2\to1}(J_2 - J_1) + F_{2\to3}(J_2 - J_3)]$$
$$= (3.12 \text{ m}^2)[0.38(4562 - 11{,}418) + 0.62(4562 - 1452)] \text{ W/m}^2 = \mathbf{-2126 \text{ W}}$$

$$\dot{Q}_3 = A_3[F_{3\to1}(J_3 - J_1) + F_{3\to2}(J_3 - J_2)]$$
$$= (6.28 \text{ m}^2)[0.31(1452 - 11{,}418) + 0.31(1452 - 4562)] \text{ W/m}^2 = \mathbf{-25{,}456 \text{ W}}$$

Note that the direction of net radiation heat transfer is *from* the top surface *to* the base and side surfaces, and the algebraic sum of these three quantities must be equal to zero. That is,

$$\dot{Q}_1 + \dot{Q}_2 + \dot{Q}_3 = 27{,}582 + (-2126) + (-25{,}456) = 0$$

Discussion To maintain the surfaces at the specified temperatures, we must supply heat to the top surface continuously at a rate of 27,582 W while removing 2126 W from the base and 25,456 W from the side surfaces.

The direct method presented above is straightforward, and it does not require the evaluation of radiation resistances. Also, it can be applied to enclosures with any number of surfaces in the same manner.

EXAMPLE 19-13   Radiation Heat Transfer in a Triangular Furnace

A furnace is shaped like a long equilateral triangular duct, as shown in Fig. 19-64. The width of each side is 1 m. The base surface has an emissivity of 0.7 and is maintained at a uniform temperature of 600 K. The heated left-side surface closely approximates a blackbody at 1000 K. The right-side surface is well insulated. Determine the rate at which heat must be supplied to the heated side externally per unit length of the duct in order to maintain these operating conditions.

Solution   Two of the surfaces of a long equilateral triangular furnace are maintained at uniform temperatures while the third surface is insulated. The external rate of heat transfer to the heated side per unit length of the duct during steady operation is to be determined.

Assumptions   **1** Steady operating conditions exist. **2** The surfaces are opaque, diffuse, and gray. **3** Convection heat transfer is not considered.

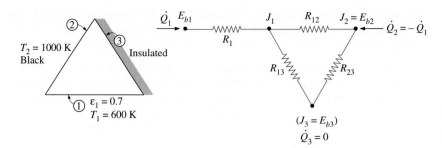

FIGURE 19-64
The triangular furnace considered in Example 19-13.

Analysis The furnace can be considered to be a three-surface enclosure with a radiation network as shown in the figure, since the duct is very long and thus the end effects are negligible. We observe that the view factor from any surface to any other surface in the enclosure is 0.5 because of symmetry. Surface 3 is a re-radiating surface since the net rate of heat transfer at that surface is zero. Then we must have $\dot{Q}_1 = -\dot{Q}_2$, since the entire heat lost by surface 1 must be gained by surface 2. The radiation network in this case is a simple series–parallel connection, and we can determine $\dot{Q}_1$ directly from

$$\dot{Q}_1 = \frac{E_{b1} - E_{b2}}{R_1 + \left(\dfrac{1}{R_{12}} + \dfrac{1}{R_{13} + R_{23}}\right)^{-1}} = \frac{E_{b1} - E_{b2}}{\dfrac{1 - \varepsilon_1}{A_1\,\varepsilon_1} + \left(\dfrac{1}{1/A_1\,F_{13} + 1/A_2\,F_{23}}\right)^{-1}}$$

where

$$A_1 = A_2 = A_3 = wL = 1\text{ m} \times 1\text{ m} = 1\text{ m}^2 \qquad \text{(per unit length of the duct)}$$
$$F_{12} = F_{13} = F_{23} = 0.5 \qquad \text{(symmetry)}$$
$$E_{b1} = \sigma T_1^4 = (5.67 \times 10^{-8}\text{ W/m}^2 \cdot \text{K}^4)(600\text{ K})^4 = 7348\text{ W/m}^2$$
$$E_{b2} = \sigma T_2^4 = (5.67 \times 10^{-8}\text{ W/m}^2 \cdot \text{K}^4)(1000\text{ K})^4 = 56{,}700\text{ W/m}^2$$

Substituting,

$$\dot{Q}_1 = \frac{(56{,}700 - 7348)\text{ W/m}^2}{\dfrac{1 - 0.7}{0.7 \times 1\text{ m}^2} + \left[(0.5 \times 1\text{ m}^2) + \dfrac{1}{1/(0.5 \times 1\text{ m}^2) + 1/(0.5 \times 1\text{ m}^2)}\right]^{-1}}$$
$$= 28{,}000\text{ W} = 28.0\text{ kW}$$

Therefore, heat at a rate of 28 kW must be supplied to the heated surface per unit length of the duct to maintain steady operation in the furnace.

EXAMPLE 19-14   Heat Transfer through Tubular Solar Collector
A solar collector consists of a horizontal aluminum tube having an outer diameter of 2 in. enclosed in a concentric thin glass tube of 4-in. diameter, as shown in Fig. 19-65. Water is heated as it flows through the tube, and the space between the aluminum and the glass tubes is filled with air at 1 atm pressure. The pump circulating the water fails during a clear day, and the water temperature in the tube starts rising. The aluminum tube absorbs solar radiation at a rate of 30 Btu/h per foot length, and the temperature of the ambient air outside is 70°F. The emissivities of the tube and the glass cover are 0.95 and 0.9, respectively. Taking the effective sky temperature to be 50°F, determine the temperature of the aluminum tube when steady operating conditions are established (i.e., when the rate of heat loss from the tube equals the amount of solar energy gained by the tube).

Solution The circulating pump of a solar collector that consists of a horizontal tube and its glass cover fails one day. The temperature of the tube is to be determined when steady conditions are reached.

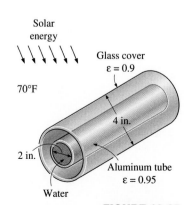

FIGURE 19-65
Schematic for Example 19-14.

Assumptions **1** Steady operating conditions exist. **2** The tube and its cover are isothermal. **3** Air is an ideal gas. **4** The surfaces are opaque, diffuse, and gray for infrared radiation. The outer surface is transparent to solar radiation.

Properties The properties of air should be evaluated at the average temperature. But we do not know the exit temperature of the air in the duct, and thus we cannot determine the bulk tube and glass cover temperatures at this point, and thus we cannot evaluate the average temperatures. Therefore, we will use the properties at an anticipated average temperature of 100°F (Table A-18),

$$k = 0.0154 \text{ Btu/h} \cdot \text{ft} \cdot \text{°F} \qquad Pr = 0.72$$
$$\nu = 0.18 \times 10^{-3} \text{ ft}^2/\text{s} \qquad \beta = \frac{1}{T_f} = \frac{1}{(100 + 460) \text{ R}} = 0.001786 \text{ R}^{-1}$$

The results obtained can then be refined for better accuracy, if necessary, using the evaluated surface temperatures.

Analysis This problem was solved in Chapter 18 by disregarding radiation heat transfer. Now we will repeat the solution by considering natural convection and radiation occurring simultaneously. We have a horizontal cylindrical enclosure filled with air at 1 atm pressure. The problem involves heat transfer from the aluminum tube to the glass cover and from the outer surface of the glass cover to the surrounding ambient air. When thermal equilibrium is established and steady operation is reached, these two heat transfer rates must equal the rate of heat gain. That is,

$$\dot{Q}_{\text{tube-glass}} = \dot{Q}_{\text{glass-ambient}} = \dot{Q}_{\text{solar gain}} = 30 \text{ Btu/h} \qquad \text{(per foot of tube)}$$

The heat transfer surface area of the glass cover is

$$A_2 = A_{\text{glass}} = (\pi DL)_{\text{glass}} = \pi(\tfrac{4}{12} \text{ ft})(1 \text{ ft}) = 1.047 \text{ ft}^2 \qquad \text{(per foot length of tube)}$$

To determine the Rayleigh number, we need to know the surface temperature of the glass, which is not available. Therefore, it is clear that the solution will require a trial-and-error approach. Assuming the glass cover temperature to be 80°F, the Raleigh number, the Nusselt number, the convection heat transfer coefficient, and the rate of natural convection heat transfer from the glass cover to the ambient air are determined to be

$$\text{Ra} = \frac{g\beta(T_s - T_\infty)\delta^3}{\nu^2} Pr$$

$$= \frac{(32.2 \text{ ft/s}^2)(0.001786 \text{ R}^{-1})[(80 - 70) \text{ K}](\tfrac{4}{12} \text{ ft})^3}{(0.18 \times 10^{-3} \text{ ft}^2/\text{s})^2} (0.72) = 4.733 \times 10^5$$

$$\text{Nu} = \left\{ 0.6 + \frac{0.387 \text{ Ra}^{1/6}}{[1 + (0.559/Pr)^{9/16}]^{8/27}} \right\}$$

$$= \left\{ 0.6 + \frac{0.387(4.733 \times 10^5)^{1/6}}{[1 + (0.559/0.72)^{9/16}]^{8/27}} \right\}^2 = 11.8$$

$$h = \frac{k}{D} \text{Nu} = \frac{0.0154 \text{ Btu/h} \cdot \text{ft} \cdot \text{°F}}{\tfrac{4}{12} \text{ ft}} (11.8) = 0.546 \text{ Btu/h} \cdot \text{ft}^2 \cdot \text{°F}$$

$$\dot{Q}_{2,\text{conv}} = hA_2(T_2 - T_\infty) = (0.546 \text{ Btu/h} \cdot \text{ft}^2 \cdot \text{°F})(1.047 \text{ ft}^2)(80 - 70)\text{°F}$$
$$= 5.7 \text{ Btu/h}$$

Also,

$$\dot{Q}_{2,\text{rad}} = \varepsilon\sigma A_2(T_2^4 - T_{\text{surr}}^4)$$
$$= (0.9)(0.1714 \times 10^{-8} \text{ Btu/h} \cdot \text{ft}^2 \cdot \text{R}^4)(1.047 \text{ ft}^2)[(540 \text{ R})^4 - (510 \text{ R})^4]$$
$$= 28.1 \text{ Btu/h}$$

and

$$\dot{Q}_{2,\text{total}} = \dot{Q}_{2,\text{conv}} + \dot{Q}_{2,\text{rad}} = 5.7 + 28.1 = 33.8 \text{ Btu/h}$$

which is more than 30 Btu/h. Therefore, the assumed temperature of 80°F for the glass cover is high. Repeating the calculations for a temperature of 75°F gives 25.5 Btu/h, which is low. Then the glass cover temperature corresponding to 30 Btu/h is determined by interpolation to be 78°F.

The temperature of the aluminum tube is determined in a similar manner using the natural convection radiation relations for two horizontal concentric cylinders. The characteristic length in this case is the distance between the two cylinders, which is determined to be

$$\delta = \tfrac{1}{2}(D_2 - D_1) = \tfrac{1}{2}(4 - 2) \text{ in.} = 1 \text{ in.}$$

Also,

$$A_1 = A_{\text{tube}} = (\pi DL)_{\text{tube}} = \pi(\tfrac{2}{12} \text{ ft})(1 \text{ ft}) = 0.524 \text{ ft}^2 \qquad \text{(per foot length of tube)}$$

$$A = \frac{\pi L(D_2 - D_1)}{\ln(D_2/D_1)} = \frac{\pi(1 \text{ ft})(\tfrac{4}{12} - \tfrac{2}{12}) \text{ ft}}{\ln(\tfrac{4}{2})} = 0.755 \text{ ft}^2$$

We start the calculations by assuming the tube temperature to be 120°F. This gives

$$\text{Ra} = \frac{g\beta(T_1 - T_2)\delta^3}{\nu^2} \text{Pr}$$

$$= \frac{(32.2 \text{ ft/s}^2)(0.001786 \text{ R}^{-1})[(120 - 78) \text{ R}](\tfrac{1}{12} \text{ ft})^3}{(0.18 \times 10^{-3} \text{ ft}^2/\text{s})^2}(0.72) = 3.11 \times 10^4$$

$$\text{Nu} = 0.11 \text{ Ra}^{0.29} = 0.11(3.11 \times 10^4)^{0.29} = 2.21$$

$$\dot{Q}_{1,\text{conv}} = k\text{Nu}A \frac{T_1 - T_2}{\delta}$$

$$= (0.0154 \text{ Btu/h} \cdot \text{ft} \cdot °\text{F})(2.21)(0.755 \text{ ft}^2) \frac{(120 - 78) \text{ R}}{\tfrac{1}{12} \text{ ft}} = 13.0 \text{ Btu/h}$$

Also,

$$\dot{Q}_{1,\text{rad}} = \frac{\sigma A_1(T_1^4 - T_2^4)}{\dfrac{1}{\varepsilon_1} + \dfrac{1 - \varepsilon_2}{\varepsilon_2}\left(\dfrac{D_1}{D_2}\right)}$$

$$= \frac{(0.1714 \times 10^{-8} \text{ Btu/h} \cdot \text{ft}^2 \cdot \text{R}^4)(0.524 \text{ ft}^2)[(580 \text{ R})^4 - (538 \text{ R})^4]}{\dfrac{1}{0.95} + \dfrac{1 - 0.9}{0.9}\left(\dfrac{2 \text{ in.}}{4 \text{ in.}}\right)}$$

$$= 23.8 \text{ Btu/h}$$

and

$$\dot{Q}_{1,\text{total}} = \dot{Q}_{1,\text{conv}} + \dot{Q}_{1,\text{rad}} = 13.0 + 23.8 = 36.8 \text{ Btu/h}$$

which is more than 30 Btu/h. Therefore, the assumed temperature of 120°F for the tube is high. Repeating the calculations for a temperature of 110°F gives 26.7 Btu/h, which is low. Then the tube temperature corresponding to 30 Btu/h is determined by interpolation to be **113°F**. Therefore, the tube will reach an equilibrium temperature of 113°F when the pump fails. Recall that disregarding radiation in the previous chapter gave a result of 188°F, which seems to be quite unrealistic.

Discussion Radiation should always be considered in systems that are heated or cooled by natural convection, unless the surfaces involved are polished and thus have emissivities close to zero.

19-9 ■ RADIATION SHIELDS AND THE RADIATION EFFECT

Radiation heat transfer between two surfaces can be reduced greatly by inserting a thin, high-reflectivity (low-emissivity) sheet of material between the two surfaces. Such highly reflective thin plates or shells are called **radiation shields.** Multilayer radiation shields constructed of about 20 sheets per cm thickness separated by evacuated space are commonly used in cryogenic and space applications. Radiation shields are also used in temperature measurements of fluids to reduce the error caused by the radiation effect when the temperature sensor is exposed to surfaces that are much hotter or colder than the fluid itself. The role of the radiation shield is to reduce the rate of radiation heat transfer by placing additional resistances in the path of radiation heat flow. The lower the emissivity of the shield, the higher the resistance.

Radiation heat transfer between two large parallel plates of emissivities ε_1 and ε_2 maintained at uniform temperatures T_1 and T_2 is given by Eq. 19-55:

$$\dot{Q}_{12,\,\text{no shield}} = \frac{A\sigma(T_1^4 - T_2^4)}{\dfrac{1}{\varepsilon_1} + \dfrac{1}{\varepsilon_2} - 1}$$

Now consider a radiation shield placed between these two plates, as shown in Fig. 19-66. Let the emissivities of the shield facing plates 1 and 2 be $\varepsilon_{3,\,1}$ and $\varepsilon_{3,\,2}$, respectively. Note that the emissivity of different surfaces of the shield may be different. The radiation network of this geometry is constructed, as usual, by drawing a surface resistance associated with each surface and connecting these surface resistances with space resistances, as shown in the figure. The resistances are connected in series, and thus the rate of radiation heat transfer is

$$\dot{Q}_{12,\,\text{one shield}} = \frac{E_{b1} - E_{b2}}{\dfrac{1 - \varepsilon_1}{A_1\varepsilon_1} + \dfrac{1}{A_1 F_{12}} + \dfrac{1 - \varepsilon_{3,\,1}}{A_3\varepsilon_{3,\,1}} + \dfrac{1 - \varepsilon_{3,\,2}}{A_3\varepsilon_{3,\,2}} + \dfrac{1}{A_3 F_{32}} + \dfrac{1 - \varepsilon_2}{A_2\varepsilon_2}}$$

(19-59)

Noting that $F_{13} = F_{23} = 1$ and $A_1 = A_2 = A_3 = A$ for parallel plates, Eq. 19-59 simplifies to

$$\dot{Q}_{12,\,\text{one shield}} = \frac{A\sigma(T_1^4 - T_2^4)}{\left(\dfrac{1}{\varepsilon_1} + \dfrac{1}{\varepsilon_2} - 1\right) + \left(\dfrac{1}{\varepsilon_{3,\,1}} + \dfrac{1}{\varepsilon_{3,\,2}} - 1\right)}$$

(19-60)

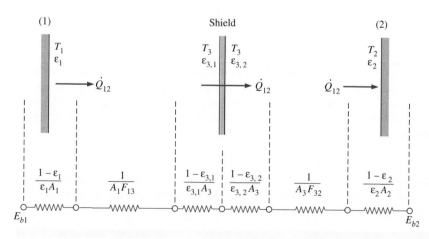

FIGURE 19-66

The radiation shield placed between two parallel plates and the radiation network associated with it.

where the terms in the second set of parentheses in the denominator represent the additional resistance to radiation introduced by the shield. The appearance of the equation above suggests that parallel plates involving multiple radiation shields can be handled by adding a group of terms like those in the second set of parentheses to the denominator for each radiation shield. Then the radiation heat transfer through large parallel plates separated by N radiation shields becomes

$$\dot{Q}_{12,\,N\,\text{shields}} = \frac{A\sigma(T_1^4 - T_2^4)}{\left(\dfrac{1}{\varepsilon_1} + \dfrac{1}{\varepsilon_2} - 1\right) + \left(\dfrac{1}{\varepsilon_{3,\,1}} + \dfrac{1}{\varepsilon_{3,\,2}} - 1\right) + \cdots + \left(\dfrac{1}{\varepsilon_{N,\,1}} + \dfrac{1}{\varepsilon_{N,\,2}} - 1\right)}$$

(19-61)

If the emissivities of all surfaces are equal, Eq. 19-61 reduces to

$$\dot{Q}_{12,\,N\,\text{shields}} = \frac{A\sigma(T_1^4 - T_2^4)}{(N+1)\left(\dfrac{1}{\varepsilon} + \dfrac{1}{\varepsilon} - 1\right)} = \frac{1}{N+1}\,\dot{Q}_{12,\,\text{no shields}} \qquad (19\text{-}62)$$

Therefore, when all emissivities are equal, 1 shield reduces the rate of radiation heat transfer to one-half, 9 shields reduce it to one-tenth, and 19 shields reduce it to one-twentieth (or 5 percent) of what it was when there were no shields.

The equilibrium temperature of the radiation shield T_3 in Fig. 19-66 can be determined by expressing Eq. 19-55 for $\dot{Q}_{13}$ or $\dot{Q}_{23}$ (which involves T_3) after evaluating $\dot{Q}_{12}$ from Eq. 19-60 and noting that $\dot{Q}_{12} = \dot{Q}_{13} = \dot{Q}_{23}$ when steady conditions are reached.

Radiation shields used to reduce the rate of radiation heat transfer between concentric cylinders and spheres can be handled in a similar manner. In case of one shield, Eq. 19-59 can be used by taking $F_{13} = F_{23} = 1$ for both cases and by replacing the A's by the proper area relations.

Radiation Effect on Temperature Measurements

A temperature measuring device indicates the temperature of its *sensor,* which is supposed to be, but is not necessarily, the temperature of the medium that the sensor is in. When a thermometer (or any other temperature measuring device such as a thermocouple) is placed in a medium, heat transfer takes place between the sensor of the thermometer and the medium by convection until the sensor reaches the temperature of the medium. But when the sensor is surrounded by surfaces that are at a different temperature than the fluid, radiation exchange will take place between the sensor and the surrounding surfaces. When the heat transfers by convection and radiation balance each other, the sensor will indicate a temperature that falls between the fluid and surface temperatures. Below we develop a procedure to account for the radiation effect and to determine the actual fluid temperature.

Consider a thermometer that is used to measure the temperature of a fluid flowing through a large channel whose walls are at a lower temperature than the fluid (Fig. 19-67). Equilibrium will be established and the reading of the thermometer will stabilize when heat gain by convection, as measured by the sensor, equals heat loss by radiation (or vice versa). That is, on a unit-area basis,

$$\dot{q}_{\text{conv, to sensor}} = \dot{q}_{\text{rad, from sensor}}$$
$$h(T_f - T_{\text{th}}) = \varepsilon_{\text{th}}\sigma(T_{\text{th}}^4 - T_w^4)$$

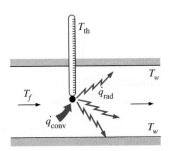

FIGURE 19-67

A thermometer used to measure the temperature of a fluid in a channel.

or

$$T_f = T_{th} + \frac{\varepsilon_{th}\,\sigma(T_{th}^4 - T_w^4)}{h} \qquad \text{(K)} \qquad \text{(19-63)}$$

where

T_f = actual temperature of the fluid, K
T_{th} = temperature value measured by the thermometer, K
T_w = temperature of the surrounding surfaces, K
h = convection heat transfer coefficient, W/m$^2 \cdot$ K or W/m$^2 \cdot$ °C
ε = emissivity of the sensor of the thermometer

The last term in Eq. 19-63 is due to the *radiation effect* and represents the radiation correction. Note that the radiation correction term is most significant when the convection heat transfer coefficient is small and the emissivity of the surface of the sensor is large. Therefore, the sensor should be coated with a material of high reflectivity (low emissivity) to reduce the radiation effect.

Placing the sensor in a radiation shield without interfering with the fluid flow also reduces the radiation effect. The sensors of temperature measurement devices used outdoors must be protected from direct sunlight since the radiation effect in that case is sure to reach unacceptable levels.

The radiation effect is also a significant factor in *human comfort* in heating and air-conditioning applications. A person who feels fine in a room at a specified temperature may feel chilly in another room at the same temperature as a result of the radiation effect if the walls of the second room are at a considerably lower temperature. For example, most people will feel comfortable in a room at 22°C if the walls of the room are also roughly at that temperature. When the wall temperature drops to 5°C for some reason, the interior temperature of the room must be raised to at least 27°C to maintain the same level of comfort. Therefore, well-insulated buildings conserve energy not only by reducing the heat loss or heat gain, but also by allowing the thermostats to be set at a lower temperature in winter and at a higher temperature in summer without compromising the comfort level.

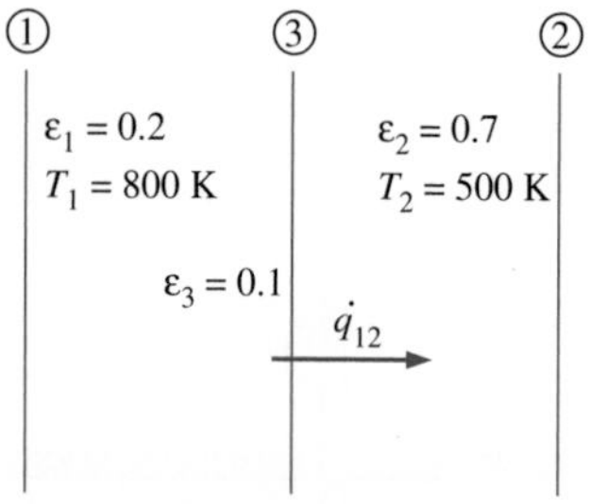

$\varepsilon_1 = 0.2$
$T_1 = 800$ K

$\varepsilon_2 = 0.7$
$T_2 = 500$ K

$\varepsilon_3 = 0.1$

$\dot{q}_{12}$

FIGURE 19-68
Schematic for Example 19-15.

EXAMPLE 19-15   Radiation Shields
A thin aluminum sheet with an emissivity of 0.1 on both sides is placed between two very large parallel plates that are maintained at uniform temperatures $T_1 =$ 800 K and $T_2 =$ 500 K and have emissivities $\varepsilon_1 =$ 0.2 and $\varepsilon_2 =$ 0.7, respectively, as shown in Fig. 19-68. Determine the net rate of radiation heat transfer between the two plates per unit surface area of the plates and compare the result to that without the shield.

Solution   A thin aluminum sheet is placed between two large parallel plates maintained at uniform temperatures. The net rates of radiation heat transfer between the two plates with and without the radiation shield are to be determined.

Assumptions   The surfaces are opaque, diffuse, and gray.

Analysis   The net rate of radiation heat transfer between these two plates without the shield was determined in Example 19-11 to be 3625 W/m^2. Heat transfer in the presence of one shield is determined from Eq. 19-60 to be

$$\dot{q}_{12,\text{ one shield}} = \frac{\dot{Q}_{12,\text{ one shield}}}{A} = \frac{\sigma(T_1^4 - T_2^4)}{\left(\dfrac{1}{\varepsilon_1} + \dfrac{1}{\varepsilon_2} - 1\right) + \left(\dfrac{1}{\varepsilon_{3,1}} + \dfrac{1}{\varepsilon_{3,2}} - 1\right)}$$

$$= \frac{(5.67 \times 10^{-8} \text{ W/m}^2 \cdot \text{K}^4)[(800 \text{ K})^4 - (500 \text{ K})^4]}{\left(\dfrac{1}{0.2} + \dfrac{1}{0.7} - 1\right) + \left(\dfrac{1}{0.1} + \dfrac{1}{0.1} - 1\right)}$$

$$= 805.6 \text{ W/m}^2$$

Discussion Note that the rate of radiation heat transfer reduces to about one-fourth of what it was as a result of placing a radiation shield between the two parallel plates.

EXAMPLE 19-16 Radiation Effect on Temperature Measurements

A thermocouple used to measure the temperature of hot air flowing in a duct whose walls are maintained at $T_w = 400$ K shows a temperature reading of $T_{th} = 650$ K (Fig. 19-69). Assuming the emissivity of the thermocouple junction to be $\varepsilon = 0.6$ and the convection heat transfer coefficient to be $h = 80$ W/m$^2 \cdot$ °C, determine the actual temperature of the air.

Solution The temperature of air in a duct is measured by a thermocouple that also exchanges heat by radiation with the surrounding surfaces. The radiation effect on the temperature measurement is to be quantified, and the actual air temperature is to be determined.

Assumptions The surfaces are opaque, diffuse, and gray.

Analysis The walls of the duct are at a considerably lower temperature than the air in it, and thus we expect the thermocouple to show a reading lower than the actual air temperature as a result of the radiation effect. The actual air temperature is determined from Eq. 19-63 to be

$$T_f = T_{th} + \frac{\varepsilon_{th}\,\sigma(T_{th}^4 - T_w^4)}{h}$$

$$= (650 \text{ K}) + \frac{0.6 \times (5.67 \times 10^{-8} \text{ W/m}^2 \cdot \text{K}^4)[(650 \text{ K})^4 - (400 \text{ K})^4]}{80 \text{ W/m}^2 \cdot \text{°C}}$$

$$= 715.0 \text{ K}$$

Note that the radiation effect causes a difference of 65°C (or 65 K since °C ≡ K for temperature differences) in temperature reading in this case.

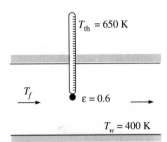

FIGURE 19-69

Schematic for Example 19-16.

19-10 ■ SUMMARY

Radiation propagates in the form of electromagnetic waves. The *frequency v* and *wavelength λ* of electromagnetic waves in a medium are related by $\lambda = c/v$, where c is the speed of light in that medium. All matter whose temperature is above absolute zero continuously emits *thermal radiation* as a result of vibrational and rotational motions of molecules, atoms, and electrons of a substance. Temperature is a measure of the strength of these activities at the microscopic level.

A *blackbody* is defined as a perfect emitter and absorber of radiation. At a specified temperature and wavelength, no surface can emit more energy than a blackbody. A blackbody absorbs all incident radiation, regardless of wavelength and direction. The radiation energy emitted by a blackbody per unit time and per unit surface area is called the *blackbody emissive power* and is expressed by the *Stefan–Boltzmann law* as

$$E_b = \sigma T^4 \qquad (\text{W/m}^2)$$

where $\sigma = 5.67 \times 10^{-8}$ W/m$^2 \cdot$ K^4 is the *Stefan–Boltzmann constant*, E_b is the blackbody emissive power, and T is the absolute temperature of the surface in K. At any specified temperature, the spectral blackbody emissive power $E_{b\lambda}$ increases with wavelength, reaches a peak, and then decreases with increasing wavelength. The wavelength at which the peak occurs for a specified temperature is given by *Wien's displacement law* as

$$(\lambda T)_{\text{max power}} = 2897.8 \ \mu\text{m} \cdot \text{K}$$

The *blackbody radiation function* f_λ represents the fraction of radiation emitted by a blackbody at temperature T in the wavelength band from $\lambda = 0$ to λ. The fraction of radiation energy emitted by a blackbody at temperature T over a finite wavelength band from $\lambda = \lambda_1$ to $\lambda = \lambda_2$ is determined from

$$f_{\lambda_1-\lambda_2}(T) = f_{\lambda_2}(T) - f_{\lambda_1}(T)$$

where $f_{\lambda_1}(T)$ and $f_{\lambda_2}(T)$ are the blackbody radiation functions corresponding to $\lambda_1 T$ and $\lambda_2 T$, respectively.

The *emissivity* of a surface is defined as the ratio of the radiation emitted by the surface to the radiation emitted by a blackbody at the same temperature. The *total hemispherical emissivity* ε of a surface is simply the average emissivity over all directions and wavelengths and is expressed as

$$\varepsilon(T) = \frac{E(T)}{E_b(T)} = \frac{E(T)}{\sigma T^4}$$

where $E(T)$ is the total emissive power of the surface.

The consideration of wavelength and direction dependence of properties makes radiation calculations very complicated. Therefore, the *gray* and *diffuse* approximations are commonly utilized in radiation calculations. A surface is said to be *diffuse* if its properties are *independent of direction* and *gray* if its properties are *independent of wavelength*.

The radiation energy incident on a surface per unit surface area per unit time is called *irradiation* and is denoted by G. When irradiation strikes a surface, part of it is absorbed, part of it is reflected, and the remaining part, if any, is transmitted. The fraction of irradiation absorbed by the surface is called the *absorptivity* α. The fraction of irradiation reflected by the surface is called the *reflectivity* ρ, and the fraction transmitted is called the *transmissivity* τ. The sum of the absorbed, reflected, and transmitted fractions of radiation energy must be equal to unity,

$$\alpha + \rho + \tau = 1$$

For *opaque* surfaces, $\tau = 0$, and thus

$$\alpha + \rho = 1$$

Surfaces are usually assumed to reflect in a perfectly *specular* or *diffuse* manner for simplicity. In *specular* (or *mirrorlike*) *reflection*, the angle of reflection equals the angle of incidence of the radiation beam. In *diffuse reflection*, radiation is reflected equally in all directions. Reflection from smooth and polished surfaces approximates specular reflection, whereas reflection from rough surfaces approximates diffuse reflection.

Kirchhoff's law of radiation is expressed as

$$\varepsilon(T) = \alpha(T)$$

That is, the total hemispherical emissivity of a surface at temperature T is equal to its total hemispherical absorptivity for radiation coming from a blackbody at the same temperature.

The *view factor* from a surface i to a surface j is denoted by $F_{i \to j}$ and is defined as the fraction of the radiation leaving surface i that strikes surface j directly. The view factor $F_{i \to i}$ represents the fraction of the radiation leaving surface i that strikes itself directly. $F_{i \to i} = 0$ for *plane* or *convex* surfaces and $F_{i \to i} \neq 0$ for *concave* surfaces. For view factors, the *reciprocity rule* is expressed as

$$A_i F_{i \to j} = A_j F_{j \to i}$$

The sum of the view factors from surface i of an enclosure to all surfaces of the enclosure, including to itself, must equal unity. This is known as the *summation rule* for an enclosure. The *superposition rule* is expressed as follows: the view factor from a surface i to a surface j is equal to the sum of the view factors from surface i to the parts of surface j. The symmetry rule is expressed as follows: if the surfaces j and k are symmetric about the surface i then $F_{i \to j} = F_{i \to k}$.

The rate of net radiation heat transfer between two *black* surfaces is determined from

$$\dot{Q}_{1 \to 2} = A_1 F_{1 \to 2} \sigma(T_1^4 - T_2^4) \qquad \text{(W)}$$

The *net* radiation heat transfer from any surface i of a *black* enclosure is determined by adding up the net radiation heat transfers from surface i to each of the surfaces of the enclosure:

$$\dot{Q}_i = \sum_{j=1}^{N} \dot{Q}_{i \to j} = \sum_{j=1}^{N} A_i F_{i \to j} \sigma(T_i^4 - T_j^4) \qquad \text{(W)}$$

The total radiation energy leaving a surface per unit time and per unit area is called the *radiosity* and is denoted by J. The *net* rate of radiation heat transfer from a surface i of surface area A_i is expressed as

$$\dot{Q}_i = \frac{E_{bi} - J_i}{R_i} \qquad \text{(W)}$$

where

$$R_i = \frac{1 - \varepsilon_i}{A_i \varepsilon_i}$$

is the *surface resistance* to radiation. The *net* rate of radiation heat transfer from surface i to surface j can be expressed as

$$\dot{Q}_{i \to j} = \frac{J_i - J_j}{R_{i \to j}} \qquad \text{(W)}$$

where

$$R_{i \to j} = \frac{1}{A_i F_{i \to j}}$$

is the *space resistance* to radiation. The *network method* is applied to radiation enclosure problems by drawing a surface resistance associated with each surface of an enclosure and connecting them with space resistances. Then the problem is solved by treating it as an electrical network problem where the radiation heat transfer replaces the current and the radiosity replaces the potential. The *direct method* is based on the following two equations:

Surfaces with specified net heat transfer rate $\dot{Q}_i$

$$\dot{Q}_i = A_i \sum_{j=1}^{N} F_{i \to j}(J_i - J_j)$$

Surfaces with specified temperature T_i

$$\sigma T_i^4 = J_i + \frac{1 - \varepsilon_i}{\varepsilon_i} \sum_{j=1}^{N} F_{i \to j}(J_i - J_j)$$

The first group (for surfaces with specified heat transfer rates) and the second group (for surfaces with specified temperatures) of equations give N linear algebraic equations for the determination of the N unknown radiosities for an N-surface enclosure. Once the radiosities $J_1, J_2, \ldots, J_N$ are available, the unknown surface temperatures and heat transfer rates can be determined from the equations above.

The net rate of radiation transfer between any two gray, diffuse, opaque surfaces that form an enclosure is given by

$$\dot{Q}_{12} = \frac{\sigma(T_1^4 - T_2^4)}{\dfrac{1 - \varepsilon_1}{A_1 \varepsilon_1} + \dfrac{1}{A_1 F_{12}} + \dfrac{1 - \varepsilon_2}{A_2 \varepsilon_2}} \qquad \text{(W)}$$

Radiation heat transfer between two surfaces can be reduced greatly by inserting between the two surfaces thin, high-reflectivity (low-emissivity) sheets of material called *radiation shields*. Radiation heat transfer between two large parallel plates separated by N radiation shields is

$$\dot{Q}_{12, \, N \, \text{shields}} = \frac{A\sigma(T_1^4 - T_2^4)}{\left(\dfrac{1}{\varepsilon_1} + \dfrac{1}{\varepsilon_2} - 1\right) + \left(\dfrac{1}{\varepsilon_{3,1}} + \dfrac{1}{\varepsilon_{3,2}} - 1\right) + \cdots + \left(\dfrac{1}{\varepsilon_{N,1}} + \dfrac{1}{\varepsilon_{N,2}} - 1\right)}$$

The radiation effect in temperature measurements can be properly accounted for by the relation

$$T_f = T_{\text{th}} + \frac{\varepsilon_{\text{th}} \, \sigma(T_{\text{th}}^4 - T_w^4)}{h} \qquad \text{(K)}$$

where T_f is the actual temperature of the fluid, T_{th} is the temperature value measured by the thermometer, and T_w is the temperature of the surrounding walls, all in K.

REFERENCES AND SUGGESTED READING

1. A. G. H. Dietz. "Diathermanous Materials and Properties of Surfaces." In *Space Heating with Solar Energy*, R. W. Hamilton. Cambridge, MA: MIT Press, 1954.

2. J. A. Duffy and W. A. Backman. *Solar Energy Thermal Process.* New York: John Wiley & Sons, 1974.

3. D. C. Hamilton and W. R. Morgan. "Radiation Interchange Configuration Factors." National Advisory Committee for Aeronautics, Technical Note 2836, 1952.

4. J. P. Holman. *Heat Transfer.* 7th ed. New York: McGraw-Hill, 1990.

5. H. C. Hottel. "Radiant Heat Transmission." In *Heat Transmission,* 3d ed., ed. W. H. McAdams. New York: McGraw-Hill, 1954.

6. J. R. Howell. *A Catalog of Radiation Configuration Factors.* New York: McGraw-Hill, 1982.

7. F. P. Incropera and D. P. DeWitt. *Introduction to Heat Transfer.* 2d ed. New York: John Wiley & Sons, 1990.

8. A. K. Oppenheim. "Radiation Analysis by the Network Method." Transactions of the ASME 78 (1956) pp. 725–35.

9. M. N. Özişik. *Heat Transfer—A Basic Approach.* New York: McGraw-Hill, 1985.

10. W. Sieber. *Zeitschrift für Technische Physics* 22 (1941), pp. 130–35.

11. L. C. Thomas. *Heat Transfer.* Englewood Cliffs, NJ: Prentice Hall, 1992.

12. Y. S. Touloukain and D. P. DeWitt. "Nonmetallic Solids." In *Thermal Radiative Properties.* Vol. 8. New York: IFI/Plenum, 1970.

13. Y. S. Touloukian and D. P. DeWitt. "Metallic Elements and Alloys." In Thermal Radiative Properties, Vol. 7. New York: IFI/Plenum, 1970.

PROBLEMS*

Electromagnetic and Thermal Radiation

19-1C   What is an electromagnetic wave? How does it differ from a sound wave?

19-2C   By what properties is an electromagnetic wave characterized? How are these properties related to each other?

19-3C   What is visible light? How does it differ from the other forms of electromagnetic radiation?

19-4C   How do ultraviolet and infrared radiation differ? Do you think your body emits any radiation in the ultraviolet range? Explain.

19-5C   What is thermal radiation? How does it differ from the other forms of electromagnetic radiation?

19-6C   What is the cause of color? Why do some objects appear blue to the eye while others appear red? Is the color of a surface at room temperature related to the radiation it emits?

19-7C   Why is radiation usually treated as a surface phenomenon?

19-8C   Why do skiers get sunburned so easily?

19-9C   How does microwave cooking differ from conventional cooking?

19-10   Electricity is generated and transmitted in power lines at a frequency of 60 Hz (1 Hz = 1 cycle per second). Determine the wavelength of the electromagnetic waves generated by the passage of electricity in power lines.

---

*Students are encouraged to answer *all* the concept "C" questions.

19-11 A microwave oven is designed to operate at a frequency of 2.8×10^9 Hz. Determine the wavelength of these microwaves and the energy of each microwave.

19-12 A radio station is broadcasting radio waves at a wavelength of 300 m. Determine the frequency of these waves. *Answer:* 1.0×10^6 Hz

19-13 A cordless telephone is designed to operate at a frequency of 8.5×10^8 Hz. Determine the wavelength of these telephone waves.

Blackbody Radiation

19-14C What is a blackbody? Does a blackbody actually exist?

19-15C Define the total and spectral blackbody emissive powers. How are they related to each other? How do they differ?

19-16C Why did we define the blackbody radiation function? What does it represent? For what is it used?

19-17C Consider two identical bodies, one at 1000 K and the other at 1500 K. Which body emits more radiation in the shorter-wavelength region? Which body emits more radiation at a wavelength of 20 μm?

19-18 Consider a 20-cm $\times$ 20-cm $\times$ 20-cm cubical body at 1000 K suspended in the air. Assuming the body closely approximates a blackbody, determine (*a*) the rate at which the cube emits radiation energy, in W, and (*b*) the spectral blackbody emissive power at a wavelength of 4 μm.

19-19E The sun can be treated as a blackbody at an effective surface temperature of 10,372 R. Determine the rate at which infrared radiation energy ($\lambda = 0.76$–100 μm) is emitted by the sun, in Btu/h $\cdot$ ft^2.

19-20 The temperature of the filament of an incandescent light bulb is 3200 K. Treating the filament as a blackbody, determine the fraction of the radiant energy emitted by the filament that falls in the visible range. Also, determine the wavelength at which the emission of radiation from the filament peaks.

19-21 An incandescent light bulb is desired to emit at least 20 percent of its energy at wavelengths shorter than 1 μm. Determine the minimum temperature to which the filament of the light bulb must be heated.

19-22 It is desired that the radiation energy emitted by a light source reach a maximum in the blue range ($\lambda = 0.47$ μm). Determine the temperature of this light source and the fraction of radiation it emits in the visible range ($\lambda = 0.40$–0.76 μm).

19-23 A 3-mm-thick glass window transmits 90 percent of the radiation between $\lambda = 0.3$ and 3.0 μm and is essentially opaque for radiation at other wavelengths. Determine the rate of radiation transmitted through a 2-m $\times$ 2-m glass window from blackbody sources at (*a*) 5800 K and (*b*) 1000 K.
Answers: (*a*) 218,400 kW, (*b*) 55.8 kW

Radiation Properties

19-24C Define the properties emissivity and absorptivity. When are these two properties equal to each other?

19-25C Define the properties reflectivity and transmissivity and discuss the different forms of reflection.

19-26C What is a graybody? How does it differ from a blackbody? What is a diffuse gray surface?

19-27C What is the greenhouse effect? Why is it a matter of great concern among atmospheric scientists?

19-28C We can see the inside of a microwave oven during operation through its glass door, which indicates that visible radiation is escaping the oven. Do you think that the harmful microwave radiation might also be escaping?

19-29 The spectral emissivity function of an opaque surface at 1000 K is approximated as

$$\varepsilon_\lambda = \begin{cases} \varepsilon_1 = 0.4, & 0 \le \lambda < 2 \ \mu m \\ \varepsilon_2 = 0.7, & 2 \ \mu m \le \lambda < 6 \ \mu m \\ \varepsilon_3 = 0.3, & 6 \ \mu m \le \lambda < \infty \end{cases}$$

Determine the average emissivity of the surface and the rate of radiation emission from the surface, in W/m^2.   *Answers:* 0.575, 32.6 kW/m^2

19-30 The reflectivity of aluminum coated with lead sulfate is 0.35 for radiation at wavelengths less than 3 μm and 0.95 for radiation greater than 3 μm. Determine the average reflectivity of this surface for solar radiation ($T \approx 5800$ K) and radiation coming from surfaces at room temperature ($T \approx 300$ K). Also, determine the emissivity and absorptivity of this surface at both temperatures. Do you think this material is suitable for use in solar collectors?

19-31 A furnace that has a 20-cm × 20-cm glass window can be considered to be a blackbody at 1200 K. If the transmissivity of the glass is 0.8 for radiation at wavelengths less than 3 μm and zero for radiation at wavelengths greater than 3 μm, determine the fraction and the rate of radiation coming from the furnace and transmitted through the window.

19-32 The emissivity of a tungsten filament can be approximated to be 0.5 for radiation at wavelengths less than 1 μm and 0.15 for radiation at greater than 1 μm. Determine the average emissivity of the filament at (*a*) 1500 K and (*b*) 3000 K. Also, determine the absorptivity and reflectivity of the filament at both temperatures.

19-33 The variations of the spectral emissivity of two surfaces are as given in Fig. P19-33. Determine the average emissivity of each surface at $T = 3000$ K. Also, determine the average absorptivity and reflectivity of each surface for radiation coming from a source at 3000 K. Which surface is more suitable to serve as a solar absorber?

19-34 The emissivity of a surface coated with aluminum oxide can be approximated to be 0.2 for radiation at wavelengths less than 5 μm and 0.9 for radiation at wavelengths greater than 5 μm. Determine the average emissivity of this surface at (*a*) 5800 K and (*b*) 300 K. What can you say about the absorptivity of this surface for radiation coming from sources at 5800 K and 300 K?   *Answers:* (*a*) 0.203, (*b*) 0.89

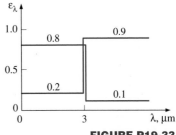
FIGURE P19-33

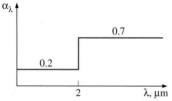

FIGURE P19-35

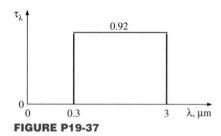

FIGURE P19-37

19-35 The variation of the spectral absorptivity of a surface is as given in Fig. P19-35. Determine the average absorptivity and reflectivity of the surface for radiation that originates from a source at $T = 2500$ K. Also, determine the average emissivity of this surface at 3000 K.

19-36E A 5-in.-diameter spherical ball is known to emit radiation at a rate of 120 Btu/h when its surface temperature is 800 R. Determine the average emissivity of the ball at this temperature.

19-37 The variation of the spectral transmissivity of a 0.6-cm-thick glass window is as given in Fig. P19-37. Determine the average transmissivity of this window for solar radiation ($T \approx 5800$ K) and radiation coming from surfaces at room temperature ($T \approx 300$ K). Also, determine the amount of solar radiation transmitted through the window for incident solar radiation of 650 W/m². *Answers:* 0.848, 0.00015, 551.1 W/m²

Atmospheric and Solar Radiation

19-38C What is the solar constant? How is it used to determine the effective surface temperature of the sun? How would the value of the solar constant change if the distance between the earth and the sun doubled?

19-39C What changes would you notice if the sun emitted radiation at an effective temperature of 2000 K instead of 5762 K?

19-40C Explain why the sky is blue and the sunset is yellow-orange.

19-41C When the earth is closest to the sun, we have winter in the northern hemisphere. Explain why. Also explain why we have summer in the northern hemisphere when the earth is farthest away from the sun.

19-42C What is the effective sky temperature?

19-43C You have probably noticed warning signs on the highways stating that bridges may be icy even when the roads are not. Explain how this can happen.

19-44C Unless you live in a warm southern state, you have probably had to scrape ice from the windshield and windows of your car many mornings. You may have noticed, with frustration, that the thickest layer of ice always forms on the windshield instead of the side windows. Explain why this is the case.

19-45C Explain why surfaces usually have quite different absorptivities for solar radiation and for radiation originating from the surrounding bodies.

19-46 A surface has an absorptivity of $\alpha_s = 0.85$ for solar radiation and an emissivity of $\varepsilon = 0.5$ at room temperature. The surface temperature is observed to be 350 K when the direct and the diffuse components of solar radiation are $G_D = 350$ and $G_d = 400$ W/m², respectively, and the direct radiation makes a 30° angle with the normal of the surface. Taking the effective sky temperature to be 280 K, determine the net rate of radiation heat transfer to the surface at that time.

19-47E Solar radiation is incident on the outer surface of a spaceship at a rate of 400 Btu/h · ft². The surface has an absorptivity of $\alpha_s = 0.10$ for solar radiation and an emissivity of $\varepsilon = 0.8$ at room temperature. The outer surface radiates heat into space at 0 R. If there is no net heat transfer into the spaceship, determine the equilibrium temperature of the surface.
Answer: 413.3 R

19-48 The air temperature on a clear night is observed to remain at about 4°C. Yet water is reported to have frozen that night due to radiation effect. Taking the convection heat transfer coefficient to be 10 W/m² · °C, determine the value of the effective sky temperature that night.

19-49 The absorber surface of a solar collector is made of aluminum coated with black chrome ($\alpha_s = 0.87$ and $\varepsilon = 0.09$). Solar radiation is incident on the surface at a rate of 600 W/m². The air and the effective sky temperatures are 25°C and 15°C, respectively, and the convection heat transfer coefficient is 10 W/m² · °C. For an absorber surface temperature of 70°C, determine the net rate of solar energy delivered by the absorber plate to the water circulating behind it.

19-50 Determine the equilibrium temperature of the absorber surface in Prob. 19-49 if the back side of the absorber is insulated.

The View Factor

19-51C What does the view factor represent? When is the view factor from a surface to itself not zero?

19-52C How can you determine the view factor F_{12} when the view factor F_{21} and the surface areas are available?

19-53C What are the summation rule and the superposition rule for view factors?

19-54C What is the crossed-strings method? For what kind of geometries is the crossed-strings method applicable?

19-55 Consider an enclosure consisting of seven surfaces. How many view factors does this geometry involve? How many of these view factors can be determined by the application of the reciprocity and the summation rules?

19-56 Consider an enclosure consisting of five surfaces. How many view factors does this geometry involve? How many of these view factors can be determined by the application of the reciprocity and summation rules?

19-57 Consider an enclosure consisting of 12 surfaces. How many view factors does this geometry involve? How many of these view factors can be determined by the application of the reciprocity and the summation rules?
Answers: 144, 78

19-58 Determine the view factors F_{13} and F_{23} between the rectangular surfaces shown in Fig. P19-58.

19-59 Consider a cylindrical enclosure whose height is twice the diameter of its base. Determine the view factor from the side surface of this cylindrical enclosure to its base surface.

19-60 Consider a hemispherical furnace with a flat circular base of diameter D. Determine the view factor from the dome of this furnace to its base.
Answer: 0.5

19-61 Determine the view factors F_{12} and F_{21} for the very long ducts shown in Fig. P19-61 without using any view factor tables or charts. Neglect end effects.

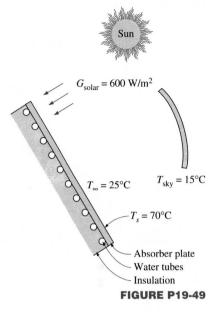

$G_{solar} = 600$ W/m²

$T_\infty = 25°C$ $T_{sky} = 15°C$

$T_s = 70°C$

Absorber plate
Water tubes
Insulation
FIGURE P19-49

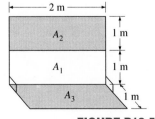

2 m

A_2 1 m

A_1 1 m

A_3 1 m

FIGURE P19-58

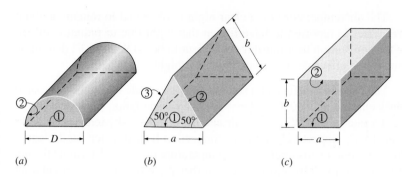

FIGURE P19-61

(a) Semicylindrical duct.
(b) Triangular duct.
(c) Rectangular duct.

(a)   (b)   (c)

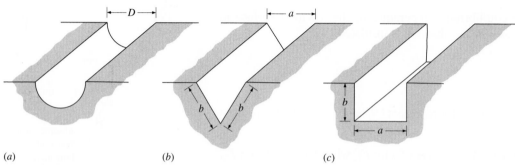

(a)   (b)   (c)

FIGURE P19-62

(a) Semicylindrical groove.
(b) Triangular groove.
(c) Rectangular groove.

FIGURE P19-64

19-62 Determine the view factors from the very long grooves shown in Fig. P19-62 to the surroundings without using any view factor tables or charts. Neglect end effects.

19-63 Determine the view factors from the base of a cube to each of the other five surfaces.   *Answer:* 0.2

19-64 Consider a conical enclosure of height h and base diameter D. Determine the view factor from the conical side surface to a hole of diameter d located at the center of the base.

19-65 Determine the four view factors associated with an enclosure formed by two very long concentric cylinders of radii r_1 and r_2. Neglect the end effects.

19-66 Determine the view factor F_{12} between the rectangular surfaces shown in Fig. P19-66.

19-67 Two infinitely long parallel cylinders of diameter D are located a distance s apart from each other. Determine the view factor F_{12} between these two cylinders.

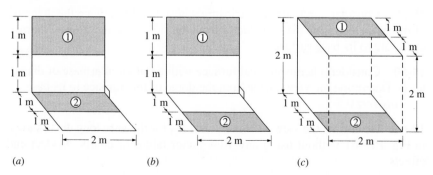

FIGURE P19-66   (a)   (b)   (c)

19-68 Three infinitely long parallel cylinders of diameter D are located a distance s apart from each other. Determine the view factor between the cylinder in the middle and the surroundings.

Radiation Heat Transfer between Surfaces

19-69C Why is the radiation analysis of enclosures that consist of black surfaces relatively easy? How is the rate of radiation heat transfer between two surfaces expressed in this case?

19-70C How does radiosity for a surface differ from the emitted energy? For what kind of surfaces are these two quantities identical?

19-71C What are the radiation surface and space resistances? How are they expressed? For what kind of surfaces is the radiation surface resistance zero?

19-72C What are the two methods used in radiation analysis? How do these two methods differ?

19-73C What is a reradiating surface? What simplifications does a reradiating surface offer in the radiation analysis?

19-74E Consider a 10-ft $\times$ 10-ft $\times$ 10-ft cubical furnace whose top and side surfaces closely approximate black surfaces and whose base surface has an emissivity $\varepsilon = 0.7$. The base, top, and side surfaces of the furnace are maintained at uniform temperatures of 800 R, 1600 R, and 2400 R, respectively. Determine the net rate of radiation heat transfer between (a) the base and the side surfaces and (b) the base and the top surfaces. Also, determine the net rate of radiation heat transfer to the base surface.

19-75 Two very large parallel plates are maintained at uniform temperatures of $T_1 = 600$ K and $T_2 = 400$ K and have emissivities $\varepsilon_1 = 0.5$ and $\varepsilon_2 = 0.9$, respectively. Determine the net rate of radiation heat transfer between the two surfaces per unit area of the plates.

19-76 A furnace is of cylindrical shape with $R = H = 2$ m. The base, top, and side surfaces of the furnace are all black and are maintained at uniform temperatures of 500, 700, and 800 K, respectively. Determine the net rate of radiation heat transfer to or from the top surface during steady operation.

19-77 Consider a hemispherical furnace of diameter $D = 5$ m with a flat base. The dome of the furnace is black, and the base has an emissivity of 0.7. The base and the dome of the furnace are maintained at uniform temperatures of 400 and 1000 K, respectively. Determine the net rate of radiation heat transfer from the dome to the base surface during steady operation.
Answer: 759 kW

19-78 Two very long concentric cylinders of diameters $D_1 = 0.2$ m and $D_2 = 0.5$ m are maintained at uniform temperatures of $T_1 = 800$ K and $T_2 = 500$ K and have emissivities $\varepsilon_1 = 1$ and $\varepsilon_2 = 0.7$, respectively. Determine the net rate of radiation heat transfer between the two cylinders per unit length of the cylinders.

19-79 The following experiment is conducted to determine the emissivity of a certain material. A long cylindrical rod of diameter $D_1 = 0.01$ m is coated with this new material and is placed in an evacuated long cylindrical enclosure of diameter $D_2 = 0.1$ m and emissivity $\varepsilon_2 = 0.95$, which is cooled externally and maintained at a temperature of 200 K at all times. The rod is heated by

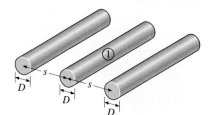

FIGURE P19-68

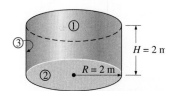

FIGURE P19-76

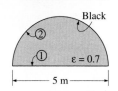

FIGURE P19-77

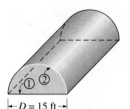

FIGURE P19-80E

passing electric current through it. When steady operating conditions are reached, it is observed that the rod is dissipating electric power at a rate of 8 W per unit of its length and its surface temperature is 500 K. Based on these measurements, determine the emissivity of the coating on the rod.

19-80E A furnace is shaped like a long semicylindrical duct of diameter $D = 15$ ft. The base and the dome of the furnace have emissivities of 0.5 and 0.9 and are maintained at uniform temperatures of 700 and 1800 R, respectively. Determine the net rate of radiation heat transfer from the dome to the base surface per unit length during steady operation.

19-81 Two parallel disks of diameter $D = 0.6$ m separated by $L = 0.4$ m are located directly on top of each other. Both disks are black and are maintained at a temperature of 700 K. The back sides of the disks are insulated, and the environment that the disks are in can be considered to be a blackbody at $T_\infty = 300$ K. Determine the net rate of radiation heat transfer from the disks to the environment.   *Answer:* 5505 W

19-82 A furnace is shaped like a long equilateral-triangular duct where the width of each side is 2 m. Heat is supplied from the base surface, whose emissivity is $\varepsilon_1 = 0.8$, at a rate of 800 W/m² while the side surfaces, whose emissivities are 0.5, are maintained at 500 K. Neglecting the end effects, determine the temperature of the base surface. Can you treat this geometry as a two-surface enclosure?

19-83 Consider a 4-m × 4-m × 4-m cubical furnace whose floor and ceiling are black and whose side surfaces are reradiating. The floor and the ceiling of the furnace are maintained at temperatures of 550 K and 1100 K, respectively. Determine the net rate of radiation heat transfer between the floor and the ceiling of the furnace.

19-84 Two concentric spheres of diameters $D_1 = 0.3$ m and $D_2 = 0.8$ m are maintained at uniform temperatures $T_1 = 700$ K and $T_2 = 400$ K and have emissivities $\varepsilon_1 = 0.5$ and $\varepsilon_2 = 0.7$, respectively. Determine the net rate of radiation heat transfer between the two spheres. Also, determine the convection heat transfer coefficient at the outer surface if both the surrounding medium and the surrounding surfaces are at 25°C. Assume the emissivity of the outer surface is 0.2.

FIGURE P19-85

19-85 A spherical tank of diameter $D = 2$ m that is filled with liquid nitrogen at 100 K is kept in an evacuated cubic enclosure whose sides are 3 m long. The emissivities of the spherical tank and the enclosure are $\varepsilon_1 = 0.1$ and $\varepsilon_2 = 0.8$, respectively. If the temperature of the cubic enclosure is measured to be 240 K, determine the net rate of radiation heat transfer to the liquid nitrogen.   *Answer:* 228 W

19-86 Repeat Prob. 19-85 by replacing the cubic enclosure by a spherical enclosure whose diameter is 3 m.

19-87 Consider a circular grill whose diameter is 0.3 m. The bottom of the grill is covered with hot coal bricks at 1100 K, while the wire mesh on top of the grill is covered with steaks initially at 5°C. The distance between the coal bricks and the steaks is 0.20 m. Treating both the steaks and the coal bricks as blackbodies, determine the initial rate of radiation heat transfer from the coal bricks to the steaks. Also, determine the initial rate of radiation heat transfer to the steaks if the side opening of the grill is covered by aluminum foil, which can be approximated as a reradiating surface.
     Answers: 1674 W, 3757 W

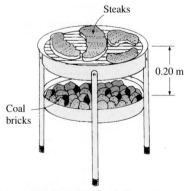

FIGURE P19-87

19-88E   A 19-ft-high room with a base area of 12 ft × 12 ft is to be heated by electric resistance heaters placed on the ceiling, which is maintained at a uniform temperature of 90°F at all times. The floor of the room is at 65°F and has an emissivity of 0.8. The side surfaces are well insulated. Treating the ceiling as a blackbody, determine the rate of heat loss from the room through the floor.

Radiation Shields and the Radiation Effect

19-89C   What is a radiation shield? Why is it used?

19-90C   What is the radiation effect? How does it influence the temperature measurements?

19-91C   Give examples of radiation effects that affect human comfort.

19-92   Consider a person whose exposed surface area is 1.7 m², emissivity is 0.7, and surface temperature is 32°C. Determine the rate of heat loss from that person by radiation in a large room whose walls are at a temperature of (a) 300 K and (b) 280 K.   *Answers:* (a) 37.4 W, (b) 169.2 W

19-93   A thin aluminum sheet with an emissivity of 0.15 on both sides is placed between two very large parallel plates, which are maintained at uniform temperatures $T_1 = 900$ K and $T_2 = 650$ K and have emissivities $\varepsilon_1 = 0.5$ and $\varepsilon_2 = 0.8$, respectively. Determine the net rate of radiation heat transfer between the two plates per unit surface area of the plates and compare the result with that without the shield.

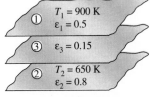

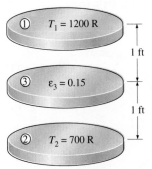

FIGURE P19-93

19-94   Two very large parallel plates are maintained at uniform temperatures of $T_1 = 1000$ K and $T_2 = 500$ K and have emissivities of $\varepsilon_1 = \varepsilon_2 = 0.2$, respectively. It is desired to reduce the net rate of radiation heat transfer between the two plates to one-fifth by placing thin aluminum sheets with an emissivity of 0.2 on both sides between the plates. Determine the number of sheets that need to be inserted.

19-95   Five identical thin aluminum sheets with emissivities of 0.1 on both sides are placed between two very large parallel plates, which are maintained at uniform temperatures of $T_1 = 800$ K and $T_2 = 450$ K and have emissivities of $\varepsilon_1 = \varepsilon_2 = 0.1$, respectively. Determine the net rate of radiation heat transfer between the two plates per unit surface area of the plates and compare the result to that without the shield.

19-96E   Two parallel disks of diameter $D = 3$ ft separated by $L = 2$ ft are located directly on top of each other. The disks are separated by a radiation shield whose emissivity is 0.15. Both disks are black and are maintained at temperatures of 1200 R and 700 R, respectively. The environment that the disks are in can be considered to be a blackbody at 540 R. Determine the net rate of radiation heat transfer through the shield under steady conditions.
      Answer: 866 Btu/h

FIGURE P19-96E

19-97   A radiation shield that has the same emissivity ε_3 on both sides is placed between two large parallel plates, which are maintained at uniform temperatures of $T_1 = 650$ K and $T_2 = 400$ K and have emissivities of $\varepsilon_1 = 0.6$ and $\varepsilon_2 = 0.9$, respectively. Determine the emissivity of the radiation shield if the radiation heat transfer between the plates is to be reduced to 15 percent of that without the radiation shield.

19-98 Two coaxial cylinders of diameters $D_1 = 0.10$ m and $D_2 = 0.30$ m and emissivities $\varepsilon_1 = 0.7$ and $\varepsilon_2 = 0.4$ are maintained at uniform temperatures of $T_1 = 750$ K and $T_2 = 500$ K, respectively. Now a coaxial radiation shield of diameter $D_3 = 0.20$ m and emissivity $\varepsilon_3 = 0.2$ is placed between the two cylinders. Determine the net rate of radiation heat transfer between the two cylinders per unit length of the cylinders and compare the result with that without the shield.

Review Problems

19-99 A thermocouple used to measure the temperature of hot air flowing in a duct whose walls are maintained at $T_w = 500$ K shows a temperature reading of $T_{th} = 850$ K. Assuming the emissivity of the thermocouple junction to be $\varepsilon = 0.6$ and the convection heat transfer coefficient to be $h = 60$ W/m² · °C, determine the actual temperature of air.
Answer: 1111 K

19-100 A thermocouple shielded by aluminum foil of emissivity 0.1 is used to measure the temperature of hot gases flowing in a duct whose walls are maintained at $T_w = 380$ K. The thermometer shows a temperature reading of $T_{th} = 530$ K. Assuming the emissivity of the thermocouple junction to be $\varepsilon = 0.8$ and the convection heat transfer coefficient to be $h = 120$ W/m² · °C, determine the actual temperature of the gas. What would the thermometer reading be if no radiation shield was used?

19-101 The spectral emissivity function of an opaque surface at 1200 K is approximated as

$$\varepsilon_\lambda = \begin{cases} \varepsilon_1 = 0.0, & 0 \le \lambda < 2 \ \mu m \\ \varepsilon_2 = 0.8, & 2 \ \mu m \le \lambda < 6 \ \mu m \\ \varepsilon_3 = 0.6, & 6 \ \mu m \le \lambda < \infty \end{cases}$$

Determine the average emissivity of the surface and the rate of radiation emission from the surface, in W/m².

19-102E Consider a sealed 8-in.-high electronic box whose base dimensions are 12 in. × 12 in. placed in a vacuum chamber. The emissivity of the outer surface of the box is 0.95. If the electronic components in the box dissipate a total of 100 W of power and the outer surface temperature of the box is not to exceed 130°F, determine the highest temperature at which the surrounding surfaces must be kept if this box is to be cooled by radiation alone. Assume the heat transfer from the bottom surface of the box to the stand to be negligible. *Answer:* 43°F

19-103 A 2-m-internal-diameter double-walled spherical tank is used to store iced water at 0°C. Each wall is 0.5 cm thick, and the 1.5-cm-thick air space between the two walls of the tank is evacuated in order to minimize heat transfer. The surfaces surrounding the evacuated space are polished so that each surface has an emissivity of 0.15. The temperature of the outer wall of the tank is measured to be 15°C. Assuming the inner wall of the steel tank to be at 0°C, determine (a) the rate of heat transfer to the iced water in the tank and (b) the amount of ice at 0°C that melts during a 24-h period.

19-104 Two concentric spheres of diameters $D_1 = 15$ cm and $D_2 = 25$ cm are separated by air at 1 atm pressure. The surface temperatures of the two spheres enclosing the air are $T_1 = 350$ K and $T_2 = 275$ K, respectively, and

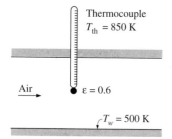

Thermocouple
$T_{th} = 850$ K

Air $\varepsilon = 0.6$

$T_w = 500$ K

FIGURE P19-99

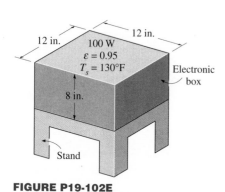

12 in.

12 in.

100 W
$\varepsilon = 0.95$
$T_s = 130°F$

Electronic box

8 in.

Stand

FIGURE P19-102E

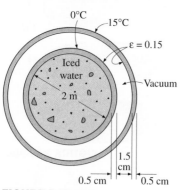

0°C
15°C
$\varepsilon = 0.15$

Iced water
2 m

Vacuum

1.5 cm
0.5 cm   0.5 cm

FIGURE P19-103

their emissivities are 0.5. Determine the rate of heat transfer from the inner sphere to the outer sphere by (a) natural convection and (b) radiation.

19-105 Consider a 1.5-m-high and 3-m-wide solar collector that is tilted at an angle 20° from the horizontal. The distance between the glass cover and the absorber plate is 3 cm, and the back side of the absorber is heavily insulated. The absorber plate and the glass cover are maintained at temperatures of 80°C and 32°C, respectively. The emissivity of the glass surface is 0.9 and that of the absorber plate is 0.8. Determine the rate of heat loss from the absorber plate by natural convection and radiation.

Answers: 713 W, 1289 W

19-106E A solar collector consists of a horizontal aluminum tube having an outer diameter of 2.5 in. enclosed in a concentric thin glass tube of diameter 5 in. Water is heated as it flows through the tube, and the annular space between the aluminum and the glass tube is filled with air at 0.5 atm pressure. The pump circulating the water fails during a clear day, and the water temperature in the tube starts rising. The aluminum tube absorbs solar radiation at a rate of 30 Btu/h per foot length, and the temperature of the ambient air outside is 75°F. The emissivities of the tube and the glass cover are 0.9. Taking the effective sky temperature to be 60°F, determine the temperature of the aluminum tube when thermal equilibrium is established (i.e., when the rate of heat loss from the tube equals the amount of solar energy gained by the tube).

19-107 A vertical 1.5-m-high and 3-m-wide double-pane window consists of two sheets of glass separated by a 1.5-cm-thick air gap. In order to reduce heat transfer through the window, the air space between the two glasses is partially evacuated to 0.3 atm pressure. The emissivities of the glass surfaces are 0.9. Taking the glass surface temperatures across the air gap to be 15°C and 5°C, determine the rate of heat transfer through the window by natural convection and radiation.

19-108 A simple solar collector is built by placing a 6-cm-diameter clear plastic tube around a garden hose whose outer diameter is 2 cm. The hose is painted black to maximize solar absorption, and some plastic rings are used to keep the spacing between the hose and the clear plastic cover constant. The emissivities of the hose surface and the glass cover are 0.9, and the effective sky temperature is estimated to be 15°C. The temperature of the plastic tube is measured to be 40°C, while the ambient air temperature is 25°C. Determine the rate of heat loss from the water in the hose by natural convection and radiation per meter of its length under steady conditions.

Answers: 6.0 W, 26.2 W

19-109 A solar collector consists of a horizontal copper tube of outer diameter 5 cm enclosed in a concentric thin glass tube of diameter 9 cm. Water is heated as it flows through the tube, and the annular space between the copper and the glass tubes is filled with air at 1 atm pressure. The emissivities of the tube surface and the glass cover are 0.85 and 0.9, respectively. During a clear day, the temperatures of the tube surface and the glass cover are measured to be 60°C and 32°C, respectively. Determine the rate of heat loss from the collector by natural convection and radiation per meter length of the tube.

Computer, Design, and Essay Problems

19-110 Consider an enclosure consisting of N diffuse and gray surfaces. The emissivity and temperature of each surface as well as all the view factors

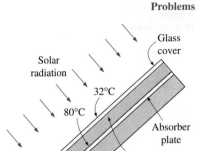

FIGURE P19-105

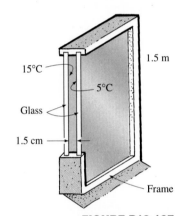

FIGURE P19-107

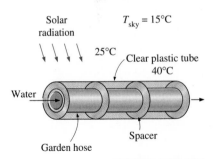

FIGURE P19-108

between the surfaces are specified. Write a program to determine the net rate of radiation heat transfer for each surface.

19-111 Radiation shields are commonly used in the design of superinsulations for use in space and cryogenic applications. Write an essay on superinsulations and how they are used in different applications.

19-112 Thermal comfort in a house is strongly affected by the so-called radiation effect, which is due to radiation heat transfer between the person and surrounding surfaces. A person feels much colder in the morning, for example, because of the lower surface temperature of the walls at that time, although the thermostat setting of the house is fixed. Write an essay on the radiation effect, how it affects human comfort, and how it is accounted for in heating and air-conditioning applications.

Heat Exchangers

Heat exchangers are devices that facilitate the *exchange of heat* between *two fluids* that are at different temperatures while keeping them from mixing with each other. Heat exchangers are commonly used in practice in a wide range of applications, from heating and air-conditioning systems in a household, to chemical processing and power production in large plants. Heat exchangers differ from mixing chambers in that they do not allow the two fluids involved to mix. In a car radiator, for example, heat is transferred from the hot water flowing through the radiator tubes to the air flowing through the closely spaced thin plates outside attached to the tubes.

Heat transfer in a heat exchanger usually involves *convection* in each fluid and *conduction* through the wall separating the two fluids. In the analysis of heat exchangers, it is convenient to work with an *overall heat transfer coefficient U* that accounts for the contribution of all these effects on heat transfer. The rate of heat transfer between the two fluids at a location in a heat exchanger depends on the magnitude of the temperature difference at that location, which varies along the heat exchanger. In the analysis of heat exchangers, it is usually convenient to work with the *logarithmic mean temperature difference LMTD,* which is an equivalent mean temperature difference between the two fluids for the entire heat exchanger.

Heat exchangers are manufactured in a variety of types, and thus we start this chapter with the *classification* of heat exchangers. We then discuss the determination of the overall heat transfer coefficient in heat exchangers, and the LMTD for some configurations. We then introduce the *correction factor F* to account for the deviation of the mean temperature difference from the LMTD in complex configurations. Next we discuss the effectiveness-NTU method, which enables us to analyze heat exchangers when the outlet temperatures of the fluids are not known. Finally, we discuss the selection of heat exchangers.

20-1 ■ TYPES OF HEAT EXCHANGERS

Different heat transfer applications require different types of hardware and different configurations of heat transfer equipment. The attempt to match the heat transfer hardware to the heat transfer requirements within the specified constraints has resulted in numerous types of innovative heat exchanger designs.

The simplest type of heat exchanger consists of two concentric pipes of different diameters, as shown in Fig. 20-1, called the **double-pipe** heat exchanger. One fluid in a double-pipe heat exchanger flows through the smaller pipe while the other fluid flows through the annular space between the two pipes. Two types of flow arrangement are possible in a double-pipe heat exchanger: in **parallel flow,** both the hot and cold fluids enter the heat exchanger at the same end and move in the *same* direction. In **counter flow,** on the other hand, the hot and cold fluids enter the heat exchanger at opposite ends and flow in *opposite* directions.

Another type of heat exchanger, which is specifically designed to realize a large heat transfer surface area per unit volume, is the **compact** heat exchanger. The ratio of the heat transfer surface area of a heat exchanger to its volume is called the *area density* β. A heat exchanger with $\beta > 700$ m^2/m^3 (or 200 ft^2/ft^3) is classified as being compact. Examples of compact heat exchangers are car radiators ($\beta \approx 1000$ m^2/m^3), glass ceramic gas turbine heat exchangers ($\beta \approx 6000$ m^2/m^3), the regenerator of a Stirling engine ($\beta \approx 15{,}000$ m^2/m^3), and the human lung ($\beta \approx 20{,}000$ m^2/m^3). Compact heat exchangers enable us to achieve high heat transfer rates between two fluids in

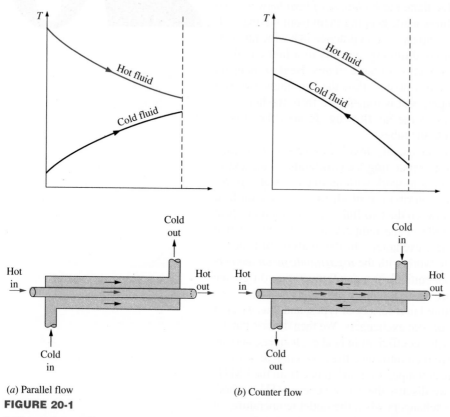

(*a*) Parallel flow

(*b*) Counter flow

FIGURE 20-1

Different flow regimes and associated temperature profiles in a double-pipe heat exchanger.

a small volume, and they are commonly used in applications with strict limitations on the weight and volume of heat exchangers (Fig. 20-2).

The large surface area in compact heat exchangers is obtained by attaching closely spaced *thin plate* or *corrugated fins* to the walls separating the two fluids. Compact heat exchangers are commonly used in gas-to-gas and gas-to-liquid (or liquid-to-gas) heat exchangers to counteract the low heat transfer coefficient associated with gas flow with increased surface area. In a car radiator, which is a water-to-air compact heat exchanger, for example, it is no surprise that fins are attached to the air side of the tube surface.

In compact heat exchangers, the two fluids usually move *perpendicular* to each other, and such flow configuration is called **cross-flow.** The cross-flow is further classified as *unmixed* and *mixed flow,* depending on the flow configuration, as shown in Fig. 20-3. In (*a*) the cross-flow is said to be *unmixed* since the plate fins force the fluid to flow through a particular interfin spacing and prevent it from moving in the transverse direction (i.e., parallel to the tubes). The cross-flow in (*b*) is said to be *mixed* since the fluid now is free to move in the transverse direction. Both fluids are unmixed in a car radiator. The presence of mixing in the fluid can have a significant effect on the heat transfer characteristics of the heat exchanger.

FIGURE 20-2

A gas-to-liquid compact heat exchanger for a residential air-conditioning system.

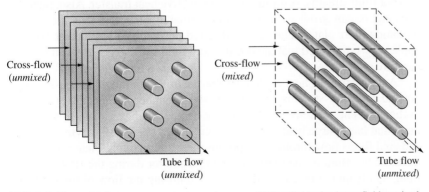

Cross-flow
(*unmixed*)

Tube flow
(*unmixed*)

(*a*) Both fluids unmixed

Cross-flow
(*mixed*)

Tube flow
(*unmixed*)

(*b*) One fluid mixed, one fluid unmixed

FIGURE 20-3

Different flow configurations in cross-flow heat exchangers.

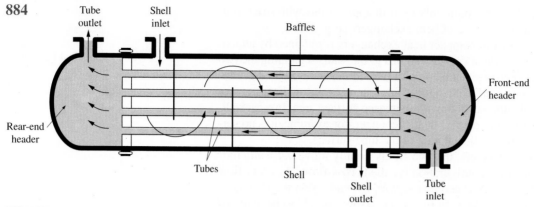

FIGURE 20-4

The schematic of a shell-and-tube heat exchanger (one-shell pass and one-tube pass).

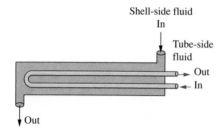

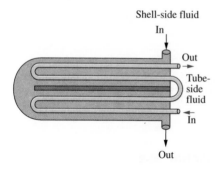

(a) One-shell pass and two-tube passes

(b) Two-shell passes and four-tube passes

FIGURE 20-5

Multipass flow arrangements in shell-and-tube heat exchangers.

Perhaps the most common type of heat exchanger in industrial applications is the **shell-and-tube** heat exchanger, shown in Fig. 20-4. Shell-and-tube heat exchangers contain a large number of tubes (sometimes several hundred) packed in a shell with their axes parallel to that of the shell. Heat transfer takes place as one fluid flows inside the tubes while the other fluid flows outside the tubes through the shell. *Baffles* are commonly placed in the shell to force the shell-side fluid to flow across the shell to enhance heat transfer and to maintain uniform spacing between the tubes. Despite their widespread use, shell-and-tube heat exchangers are not suitable for use in automotive, aircraft, and marine applications because of their relatively large size and weight. Note that the tubes in a shell-and-tube heat exchanger open to some large flow areas called *headers* at both ends of the shell, where the tube-side fluid accumulates before entering the tubes and after leaving them.

Shell-and-tube heat exchangers are further classified according to the number of shell and tube passes involved. Heat exchangers in which all the tubes make one U-turn in the shell, for example, are called *one-shell-pass and two-tube-passes* heat exchangers. Likewise, a heat exchanger that involves two passes in the shell and four passes in the tubes is called a *two-shell-passes and four-tube-passes* heat exchanger (Fig. 20-5).

An innovative type of heat exchanger that has found widespread use is the **plate and frame** (or just plate) heat exchanger, which consists of a series of plates with corrugated flat flow passages (Fig. 20-6). The hot and cold fluids flow in alternate passages, and thus each cold fluid stream is surrounded by two hot fluid streams, resulting in very effective heat transfer. Also, plate heat exchangers can grow with increasing demand for heat transfer by simply mounting more plates. They are well suited for liquid-to-liquid heat exchange applications, provided that the hot and cold fluid streams are at about the same pressure.

Another type of heat exchanger that involves the alternate passage of the hot and cold fluid streams through the same flow area is the **regenerative** heat exchanger. The *static*-type regenerative heat exchanger is basically a porous mass that has a large heat storage capacity, such as a ceramic wire mesh. Hot and cold fluids flow through this porous mass alternatively. Heat is transferred from the hot fluid to the matrix of the regenerator during the flow of the hot fluid, and from the matrix to the cold fluid during the flow of the cold fluid. Thus, the matrix serves as a temporary heat storage medium.

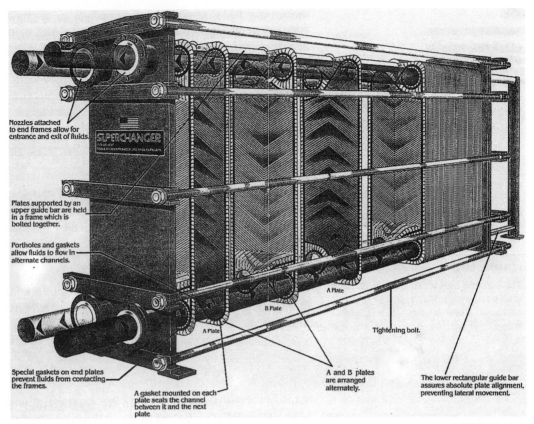

FIGURE 20-6

A plate-and-frame liquid-to-liquid heat exchanger (courtesy of Trante Corp.).

The *dynamic*-type regenerator involves a rotating drum and continuous flow of the hot and cold fluid through different portions of the drum so that any portion of the drum passes periodically through the hot stream, storing heat, and then through the cold stream, rejecting this stored heat. Again the drum serves as the medium to transport the heat from the hot to the cold fluid stream.

Heat exchangers are often given specific names to reflect the specific application for which they are used. For example, a *condenser* is a heat exchanger in which one of the fluids is cooled and condenses as it flows through the heat exchanger. A *boiler* is another heat exchanger in which one of the fluids absorbs heat and vaporizes. A *space radiator* is a heat exchanger that transfers heat from the hot fluid to the surrounding space by radiation.

20-2 ■ THE OVERALL HEAT TRANSFER COEFFICIENT

A heat exchanger typically involves two flowing fluids separated by a solid wall. Heat is first transferred from the hot fluid to the wall by *convection,* through the wall by *conduction,* and from the wall to the cold fluid again by *convection.* Any radiation effects are usually included in the convection heat transfer coefficients.

The thermal resistance network associated with this heat transfer process involves two convection and one conduction resistances, as shown in Fig. 20-7. Here the subscripts *i* and *o* represent the inner and outer surfaces of the

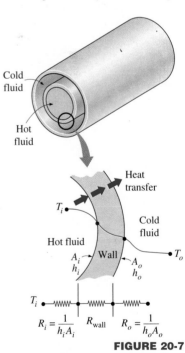

FIGURE 20-7

Thermal resistance network associated with heat transfer in a double-pipe heat exchanger.

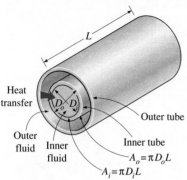

FIGURE 20-8

The two heat transfer surface areas associated with a double-pipe heat exchanger (for thin tubes, $D_i \approx D_o$ and thus $A_i \approx A_o$).

inner tube. For a double-pipe heat exchanger, we have $A_i = \pi D_i L$ and $A_o = \pi D_o L$, and the *thermal resistance* of the tube wall in this case is

$$R_{\text{wall}} = \frac{\ln (D_o/D_i)}{2\pi k L} \tag{20-1}$$

where k is the thermal conductivity of the wall material and L is the length of the tube. Then the *total thermal resistance* becomes

$$R = R_{\text{total}} = R_i + R_{\text{wall}} + R_o = \frac{1}{h_i A_i} + \frac{\ln (D_o/D_i)}{2\pi k L} + \frac{1}{h_o A_o} \tag{20-2}$$

The A_i is the area of the *inner surface* of the wall that separates the two fluids, and A_o is the area of the outer surface of the wall. In other words, A_i and A_o are surface areas of the separating wall wetted by the inner and the outer fluids, respectively. When one fluid flows inside a circular tube and the other outside of it, we have $A_i = \pi D_i L$ and $A_o = \pi D_o L$ (Fig. 20-8).

In the analysis of heat exchangers, it is convenient to combine all the thermal resistances in the path of heat flow from the hot fluid to the cold one into a single resistance R, and to express the rate of heat transfer between the two fluids as

$$\dot{Q} = \frac{\Delta T}{R} = UA\,\Delta T = U_i A_i\,\Delta T = U_o A_o\,\Delta T \tag{20-3}$$

where U is the **overall heat transfer coefficient**, whose unit is $W/m^2 \cdot {}^\circ C$, which is identical to the unit of the ordinary convection coefficient h. Canceling ΔT, Eq. 20-3 reduces to

$$\frac{1}{UA} = \frac{1}{U_i A_i} = \frac{1}{U_o A_o} = R = \frac{1}{h_i A_i} + R_{\text{wall}} + \frac{1}{h_o A_o} \tag{20-4}$$

Perhaps you are wondering why we have two overall heat transfer coefficients U_i and U_o for a heat exchanger. The reason is that every heat exchanger has two heat transfer surface areas A_i and A_o, which, in general, are not equal to each other.

Note that $U_i A_i = U_o A_o$, but $U_i \neq U_o$ unless $A_i = A_o$. Therefore, the overall heat transfer coefficient U of a heat exchanger is meaningless unless the area on which it is based is specified. This is especially the case when one side of the tube wall is finned and the other side is not, since the surface area of the finned side is several times that of the unfinned side.

When the wall thickness of the tube is small and the thermal conductivity of the tube material is high, as is usually the case, the thermal resistance of the tube is negligible ($R_{\text{wall}} \approx 0$) and the inner and outer surfaces of the tube are almost identical ($A_i \approx A_o \approx A$). Then Eq. 20-4 for the overall heat transfer coefficient simplifies to

$$\frac{1}{U} \approx \frac{1}{h_i} + \frac{1}{h_o} \tag{20-5}$$

where $U \approx U_i \approx U_o$. The individual convection heat transfer coefficients inside and outside the tube, h_i and h_o, are determined using the convection relations discussed in earlier chapters.

The overall heat transfer coefficient U in Eq. 20-5 is dominated by the *smaller* convection coefficient, since the inverse of a large number is small. When one of the convection coefficients is *much smaller* than the other (say, $h_i \ll h_o$), we have $1/h_i \gg 1/h_o$, and thus $U \approx h_i$. Therefore, the smaller heat transfer coefficient creates a *bottleneck* on the path of heat flow and seriously impedes heat transfer. This situation arises frequently when one of the fluids is a gas and the other is a liquid. In such cases, fins are commonly used on the gas side to enhance the product UA and thus the heat transfer on that side.

Representative values of the overall heat transfer coefficient U are given in Table 20-1. Note that the overall heat transfer coefficient ranges from about 10 W/m² · °C for gas-to-gas heat exchangers to about 10,000 W/m² · °C for heat exchangers that involve phase changes. This is not surprising, since gases have very low thermal conductivities, and phase-change processes involve very high heat transfer coefficients.

When the tube is *finned* on one side to enhance heat transfer, the total heat transfer surface area on the finned side becomes

$$A = A_{\text{total}} = A_{\text{fin}} + A_{\text{unfinned}} \qquad (20\text{-}6)$$

where A_{fin} is the surface area of the fins and A_{unfinned} is the area of the unfinned portion of the tube surface. For short fins of high thermal conductivity, we can use this total area in the convection resistance relation $R_{\text{conv}} = 1/hA$ since the fins in this case will be very nearly isothermal. Otherwise, we should determine the effective surface area A from

$$A = A_{\text{unfinned}} + \eta_{\text{fin}} A_{\text{fin}} \qquad (20\text{-}7)$$

TABLE 20-1

Representative values of the overall heat transfer coefficients in heat exchangers

Type of heat exchanger	U, W/m² · °C*
Water-to-water	850–1700
Water-to-oil	100–350
Water-to-gasoline or kerosene	300–1000
Feedwater heaters	1000–8500
Steam-to-light fuel oil	200–400
Steam-to-heavy fuel oil	50–200
Steam condenser	1000–6000
Freon condenser (water cooled)	300–1000
Ammonia condenser (water cooled)	800–1400
Alcohol condensers (water cooled)	250–700
Gas-to-gas	10–40
Water-to-air in finned tubes (water in tubes)	30–60[†]
	400–850[†]
Steam-to-air in finned tubes (steam in tubes)	30–300[†]
	400–4000[‡]

*Multiply the listed values by 0.176 to convert them to Btu/h · ft² · °F.

[†]Based on air-side surface area.

[‡]Based on water- or steam-side surface area.

where η_{fin} is the fin efficiency. This way, the temperature drop along the fins is accounted for. Note that $\eta_{fin} = 1$ for isothermal fins, and thus Equation 20-7 reduces to Eq. 20-6 in that case.

Fouling Factor

The performance of heat exchangers usually deteriorates with time as a result of accumulation of *deposits* on heat transfer surfaces. The layer of deposits represents *additional resistance* to heat transfer and causes the rate of heat transfer in a heat exchanger to decrease. The net effect of these accumulations on heat transfer is represented by a **fouling factor** R_f, which is a measure of the *thermal resistance* introduced by fouling.

The most common type of fouling is the *precipitation* of solid deposits in a fluid on the heat transfer surfaces. You can observe this type of fouling even in your house. If you check the inner surfaces of your teapot after prolonged use, you will probably notice a layer of calcium-based deposits on the surfaces at which boiling occurs. This is especially the case in areas where the water is hard. The scales of such deposits come off by scratching, and the surfaces can be cleaned of such deposits by chemical treatment. Now imagine those mineral deposits forming on the inner surfaces of fine tubes in a heat exchanger (Fig. 20-9) and the detrimental effect it may have on the flow passage area and the heat transfer. To avoid this potential problem, water in power and process plants is extensively treated and its solid contents are removed before it is allowed to circulate through the system. The solid ash particles in the flue gases accumulating on the surfaces of air preheaters create similar problems.

Another form of fouling, which is common in the chemical process industry, is *corrosion* and other *chemical fouling.* In this case, the surfaces are fouled by the accumulation of the products of chemical reactions on the surfaces. This form of fouling can be avoided by coating metal pipes with glass or using plastic pipes instead of metal ones. Heat exchangers may also be fouled by the growth of algae in warm fluids. This type of fouling is called *biological fouling* and can be prevented by chemical treatment.

In applications where it is likely to occur, fouling should be considered in the design and selection of heat exchangers. In such applications, it may be necessary to select a larger and thus more expensive heat exchanger to ensure that it meets the design heat transfer requirements even after fouling occurs.

FIGURE 20-9

Precipitation fouling of ash particles on superheater tubes (from *Steam, Its Generation, and Use,* Babcock and Wilcox Co., 1978).

The periodic cleaning of heat exchangers and the resulting down time are additional penalties associated with fouling.

The fouling factor is obviously zero for a new heat exchanger and increases with time as the solid deposits build up on the heat exchanger surface. The fouling factor depends on the *operating temperature* and the *velocity* of the fluids, as well as the length of service. Fouling increases with *increasing temperature* and *decreasing velocity*.

The overall heat transfer coefficient relation given above is valid for clean surfaces and needs to be modified to account for the effects of fouling on both the inner and the outer surfaces of the tube. For an unfinned shell-and-tube heat exchanger, it can be expressed as

$$\frac{1}{UA} = \frac{1}{U_i A_i} = \frac{1}{U_o A_o} = R = \frac{1}{h_i A_i} + \frac{R_{f,i}}{A_i} + \frac{\ln(D_o/D_i)}{2\pi kL} + \frac{R_{f,o}}{A_o} + \frac{1}{h_o A_o} \quad (20\text{-}8)$$

where $A_i = \pi D_i L$ and $A_o = \pi D_o L$ are the areas of inner and outer surfaces, and $R_{f,i}$ and $R_{f,o}$ are the fouling factors at those surfaces.

Representative values of fouling factors are given in Table 20-2. More comprehensive tables of fouling factors are available in handbooks. As you would expect, considerable uncertainty exists in these values, and they should be used as a guide in the selection and evaluation of heat exchangers to account for the effects of anticipated fouling on heat transfer. Note that most fouling factors in the table are of the order of 10^{-4} m$^2 \cdot$ °C/W, which is equivalent to the thermal resistance of a 0.2-mm-thick limestone layer ($k = 2.9$ W/m $\cdot$ °C) per unit surface area. Therefore, in the absence of specific data, we can assume the surfaces to be coated with 0.2 mm of limestone as a starting point to account for the effects of fouling.

EXAMPLE 20-1 Overall Heat Transfer Coefficient of a Heat Exchanger

Hot oil is to be cooled in a double-tube counter-flow heat exchanger. The copper inner tubes have a diameter of 2 cm and negligible thickness. The inner diameter of the outer tube (the shell) is 3 cm. Water flows through the tube at a rate of 0.5 kg/s, and the oil through the shell at a rate of 0.8 kg/s. Taking the average temperatures of the water and the oil to be 45°C and 80°C, respectively, determine the overall heat transfer coefficient of this heat exchanger.

Solution Hot oil is cooled by water in a double-tube counter-flow heat exchanger. The overall heat transfer coefficient is to be determined.

Assumptions **1** The thermal resistance of the inner tube is negligible since the tube material is highly conductive and its thickness is negligible. **2** Both the oil and water flow are fully developed. **3** Properties of the oil and water are constant.

Properties The properties of water at 45°C are (Table A-15)

$$\rho = 990 \text{ kg/m}^3 \qquad Pr = 3.91$$
$$k = 0.637 \text{ W/m} \cdot \text{°C} \qquad \nu = \mu/\rho = 0.602 \times 10^{-6} \text{ m}^2/\text{s}$$

The properties of oil at 80°C are (Table A-16).

$$\rho = 852 \text{ kg/m}^3 \qquad Pr = 490$$
$$k = 0.138 \text{ W/m} \cdot \text{°C} \qquad \nu = 37.5 \times 10^{-6} \text{ m}^2/\text{s}$$

Analysis The schematic of the heat exchanger is given in Fig. 20-10. The overall heat transfer coefficient U can be determined from Eq. 20-5:

$$\frac{1}{U} \approx \frac{1}{h_i} + \frac{1}{h_o}$$

TABLE 20-2

Representative fouling factors (thermal resistance due to fouling for a unit surface area)

Fluid	R_f, m$^2 \cdot$ °C/W
Distilled water, sea water, river water, boiler feedwater:	
Below 50°C	0.0001
Above 50°C	0.0002
Fuel oil	0.0009
Steam (oil-free)	0.0001
Refrigerants (liquid)	0.0002
Refrigerants (vapor)	0.0004
Alcohol vapors	0.0001
Air	0.0004

Source: Tubular Exchange Manufacturers Association.

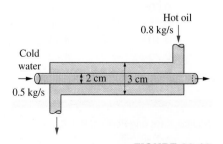

FIGURE 20-10

Schematic for Example 20-1.

where h_i and h_o are the convection heat transfer coefficients inside and outside the tube, respectively, which are to be determined using the forced convection relations.

The hydraulic diameter for a circular tube is the diameter of the tube itself, $D_h = D = 0.02$ m. The mean velocity of water in the tube and the Reynolds number are

$$\mathcal{V}_m = \frac{\dot{m}}{\rho A_c} = \frac{\dot{m}}{\rho(\frac{1}{4}\pi D^2)} = \frac{0.5 \text{ kg/s}}{(990 \text{ kg/m}^3)[\frac{1}{4}\pi(0.02 \text{ m})^2]} = 1.61 \text{ m/s}$$

and

$$\text{Re} = \frac{\mathcal{V}_m D_h}{\nu} = \frac{(1.61 \text{ m/s})(0.02 \text{ m})}{0.602 \times 10^{-6} \text{ m}^2/\text{s}} = 53,490$$

which is greater than 4000. Therefore, the flow of water is turbulent. Assuming the flow to be fully developed, the Nusselt number can be determined from

$$\text{Nu} = \frac{hD_h}{k} = 0.023 \text{ Re}^{0.8}\text{Pr}^{0.4} = 0.023(53,490)^{0.8}(3.91)^{0.4} = 240.6$$

Then,

$$h = \frac{k}{D_h}\text{Nu} = \frac{0.637 \text{ W/m} \cdot °\text{C}}{0.02 \text{ m}}(240.6) = 7663 \text{ W/m}^2 \cdot °\text{C}$$

Now we repeat the analysis above for oil. The properties of oil at 80°C are

$$\rho = 852 \text{ kg/m}^3 \qquad \nu = 37.5 \times 10^{-6} \text{ m}^2/\text{s}$$
$$k = 0.138 \text{ W/m} \cdot °\text{C} \qquad \text{Pr} = 490$$

The hydraulic diameter for the annular space is

$$D_h = D_o - D_i = 0.03 - 0.02 = 0.01 \text{ m}$$

The mean velocity and the Reynolds number in this case are

$$\mathcal{V}_m = \frac{\dot{m}}{\rho A_c} = \frac{\dot{m}}{\rho[\frac{1}{4}\pi(D_o^2 - D_i^2)]} = \frac{0.8 \text{ kg/s}}{(852 \text{ kg/m}^3)[\frac{1}{4}\pi(0.03^2 - 0.02^2)] \text{ m}^2} = 2.39 \text{ m/s}$$

and

$$\text{Re} = \frac{\mathcal{V}_m D_h}{\nu} = \frac{(2.39 \text{ m/s})(0.01 \text{ m})}{37.5 \times 10^{-6} \text{ m}^2/\text{s}} = 637$$

which is less than 4000. Therefore, the flow of oil is laminar. Assuming fully developed flow, the Nusselt number on the tube side of the annular space Nu_i corresponding to $D_i/D_o = 0.02/0.03 = 0.667$ can be determined from Table 20-3 by interpolation to be

$$\text{Nu} = 5.45$$

and

$$h_o = \frac{k}{D_h}\text{Nu} = \frac{0.138 \text{ W/m} \cdot °\text{C}}{0.01 \text{ m}}(5.45) = 75.2 \text{ W/m}^2 \cdot °\text{C}$$

Then the overall heat transfer coefficient for this heat exchanger becomes

$$U = \frac{1}{\frac{1}{h_i} + \frac{1}{h_o}} = \frac{1}{\frac{1}{7663 \text{ W/m}^2 \cdot °\text{C}} + \frac{1}{75.2 \text{ W/m}^2 \cdot °\text{C}}} = \textbf{74.5 W/m}^2 \cdot °\textbf{C}$$

Discussion Note that $U \approx h_o$ in this case, since $h_i \gg h_o$. This confirms our earlier statement that the overall heat transfer coefficient in a heat exchanger is dominated by the smaller heat transfer coefficient when the difference between the two values is large.

TABLE 20-3

Nusselt number for fully developed laminar flow in a circular annulus with one surface insulated and the other isothermal

D_i/D_o	Nu_i	Nu_o
0.00	—	3.66
0.05	17.46	4.06
0.10	11.56	4.11
0.25	7.37	4.23
0.50	5.74	4.43
1.00	4.86	4.86

Source: Kays and Perkins, Ref. 8.

To improve the overall heat transfer coefficient and thus the heat transfer in this heat exchanger, we must use some enhancement techniques on the oil side, such as a finned surface.

EXAMPLE 20-2   Effect of Fouling on the Overall Heat Transfer Coefficient

A double-pipe (shell-and-tube) heat exchanger is constructed of a stainless steel ($k = 15.1$ W/m · °C) inner tube of inner diameter $D_i = 1.5$ cm and outer diameter $D_o = 1.9$ cm and an outer shell of inner diameter 3.2 cm. The convection heat transfer coefficient is given to be $h_i = 800$ W/m² · °C on the inner surface of the tube and $h_o = 1200$ W/m² · °C on the outer surface. For a fouling factor of $R_{f,\,i} = 0.0004$ m² · °C/W on the tube side and $R_{f,\,o} = 0.0001$ m² · °C/W on the shell side, determine (a) the thermal resistance of the heat exchanger per unit length and (b) the overall heat transfer coefficients, U_i and U_o based on the inner and outer surface areas of the tube, respectively.

Solution   The heat transfer coefficients and the fouling factors on the tube and shell sides of a heat exchanger are given. The thermal resistance and the over-all heat transfer coefficients based on the inner and outer areas are to be determined.

Assumptions   The heat transfer coefficients and the fouling factors are constant and uniform.

Analysis   (a) The schematic of the heat exchanger is given in Fig. 20-11. The thermal resistance for an unfinned shell-and-tube heat exchanger with fouling on both heat transfer surfaces is given by Eq. 20-8 as

$$R = \frac{1}{UA} = \frac{1}{U_i A_i} = \frac{1}{U_o A_o} = \frac{1}{h_i A_i} + \frac{R_{f,\,i}}{A_i} + \frac{\ln (D_o/D_i)}{2\pi k L} + \frac{R_{f,\,o}}{A_o} + \frac{1}{h_o A_o}$$

where

$$A_i = \pi D_i L = \pi(0.015 \text{ m})(1 \text{ m}) = 0.0471 \text{ m}^2$$
$$A_o = \pi D_o L = \pi(0.019 \text{ m})(1 \text{ m}) = 0.0597 \text{ m}^2$$

Substituting, the total thermal resistance is determined to be

$$R = \frac{1}{(800 \text{ W/m}^2 \cdot \text{°C})(0.0471 \text{ m}^2)} + \frac{0.0004 \text{ m}^2 \cdot \text{°C/W}}{0.0471 \text{ m}^2}$$
$$+ \frac{\ln (0.019/0.015)}{2\pi(15.1 \text{ W/m} \cdot \text{°C})(1 \text{ m})}$$
$$+ \frac{0.0001 \text{ m}^2 \cdot \text{°C/W}}{0.0597 \text{ m}^2} + \frac{1}{(1200 \text{ W/m}^2 \cdot \text{°C})(0.0597 \text{ m}^2)}$$
$$= (0.02654 + 0.00849 + 0.0025 + 0.00168 + 0.01396)\text{°C/W}$$
$$= \mathbf{0.0532°C/W}$$

Note that about 19 percent of the total thermal resistance in this case is due to fouling and about 5 percent of it is due to the steel tube separating the two fluids. The rest (76 percent) is due to the convection resistances on the two sides of the inner tube.

(b) Knowing the total thermal resistance and the heat transfer surface areas, the overall heat transfer coefficient based on the inner and outer surfaces of the tube are determined again from Eq. 20-8 to be

$$U_i = \frac{1}{RA_i} = \frac{1}{(0.0532 \text{ °C/W})(0.0471 \text{ m}^2)} = \mathbf{399 \text{ W/m}^2 \cdot \text{°C}}$$

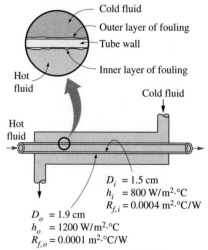

Cold fluid
Outer layer of fouling
Tube wall
Inner layer of fouling

Hot fluid

Cold fluid

Hot fluid

$D_i = 1.5$ cm
$h_i = 800$ W/m²·°C
$R_{f,i} = 0.0004$ m²·°C/W

$D_o = 1.9$ cm
$h_o = 1200$ W/m²·°C
$R_{f,o} = 0.0001$ m²·°C/W

FIGURE 20-11

Schematic for Example 20-2.

and

$$U_o = \frac{1}{RA_o} = \frac{1}{(0.0532 \ ^\circ\text{C/W})(0.0597 \ \text{m}^2)} = \textbf{315 W/m}^2 \cdot \ ^\circ\textbf{C}$$

Discussion Note that the two overall heat transfer coefficients differ significantly (by 27 percent) in this case because of the considerable difference between the heat transfer surface areas on the inner and the outer sides of the tube. For tubes of negligible thickness, the difference between the two overall heat transfer coefficients would be negligible.

20-3 ■ ANALYSIS OF HEAT EXCHANGERS

Heat exchangers are commonly used in practice, and an engineer often finds himself or herself in a position to *select a heat exchanger* that will achieve a *specified temperature change* in a fluid stream of known mass flow rate, or to *predict the outlet temperatures* of the hot and cold fluid streams in a *specified heat exchanger.*

In the following sections, we will discuss the two methods used in the analysis of heat exchangers. Of these, the *log mean temperature difference* (or LMTD) method is best suited for the first task and the *effectiveness-NTU* method for the second task stated above. But first we present some general considerations.

Heat exchangers usually operate for long periods of time with no change in their operating conditions. Therefore, they can be modeled as *steady-flow* devices. As such, the mass flow rate of each fluid remains constant, and the fluid properties such as temperature and velocity at any inlet or outlet remain the same. Also, the fluid streams experience little or no change in their velocities and elevations, and thus the *kinetic* and *potential energy changes* are negligible. The *specific heat* of a fluid, in general, changes with temperature. But, in a specified temperature range, it can be treated as a constant at some average value with little loss in accuracy. *Axial heat conduction* along the tube is usually insignificant and can be considered negligible. Finally, the outer surface of the heat exchanger is assumed to be *perfectly insulated,* so that there is no heat loss to the surrounding medium, and any heat transfer occurs between the two fluids only.

The idealizations stated above are closely approximated in practice, and they greatly simplify the analysis of a heat exchanger with little sacrifice of accuracy. Therefore, they are commonly used. Under these assumptions, the *first law of thermodynamics* requires that the rate of heat transfer from the hot fluid be equal to the rate of heat transfer to the cold one. That is,

$$\dot{Q} = \dot{m}_c C_{pc}(T_{c,\,\text{out}} - T_{c,\,\text{in}}) \tag{20-9}$$

and

$$\dot{Q} = \dot{m}_h C_{ph}(T_{h,\,\text{in}} - T_{h,\,\text{out}}) \tag{20-10}$$

where the subscripts c and h stand for *cold* and *hot* fluids, respectively, and

$$\dot{m}_c, \dot{m}_h = \text{mass flow rates}$$
$$C_{pc}, C_{ph} = \text{specific heats}$$
$$T_{c,\,\text{out}}, T_{h,\,\text{out}} = \text{outlet temperatures}$$
$$T_{c,\,\text{in}}, T_{h,\,\text{in}} = \text{inlet temperatures}$$

Note that the heat transfer rate $\dot{Q}$ is taken to be a positive quantity, and its direction is understood to be from the hot fluid to the cold one in accordance with the second law of thermodynamics.

In heat exchanger analysis, it is often convenient to combine the product of the *mass flow rate* and the *specific heat* of a fluid into a single quantity. This quantity is called the **heat capacity rate** and is defined as

$$C = \dot{m}C_p \tag{20-11}$$

The heat capacity rate of a fluid stream represents the rate of heat transfer needed to change the temperature of the fluid stream by 1°C as it flows through a heat exchanger. Note that in a heat exchanger, the fluid with a *large* heat capacity rate will experience a *small* temperature change, and the fluid with a *small* heat capacity rate will experience a *large* temperature change. Therefore, *doubling* the mass flow rate of a fluid while leaving everything else unchanged will *halve* the temperature change of that fluid.

With the definition of the heat capacity rate above, Eqs. 20-9 and 20-10 can also be expressed as

$$\dot{Q} = C_c(T_{c,\,\text{out}} - T_{c,\,\text{in}}) \tag{20-12}$$

and

$$\dot{Q} = C_h(T_{h,\,\text{in}} - T_{h,\,\text{out}}) \tag{20-13}$$

That is, the heat transfer rate in a heat exchanger is equal to the heat capacity rate of either fluid multiplied by the temperature change of that fluid. Note that *the only time the temperature rise of a cold fluid is equal to the temperature drop of the hot fluid is when the heat capacity rates of the two fluids are equal to each other* (Fig. 20-12).

Two special types of heat exchangers commonly used in practice are *condensers* and *boilers*. One of the fluids in a condenser or a boiler undergoes a phase-change process, and the rate of heat transfer is expressed as

$$\dot{Q} = \dot{m}h_{fg} \tag{20-14}$$

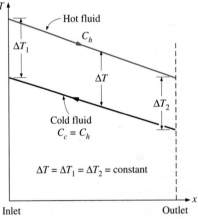

FIGURE 20-12

Two fluids that have the same mass flow rate and the same specific heat experience the same temperature change in a well-insulated heat exchanger.

where $\dot{m}$ is the rate of evaporation or condensation of the fluid and h_{fg} is the enthalpy of vaporization of the fluid at the specified temperature or pressure.

An ordinary fluid absorbs or releases a large amount of heat essentially at constant temperature during a phase-change process, as shown in Fig. 20-13. The heat capacity rate of a fluid during a phase-change process must approach infinity since the temperature change is practically zero. That is, $C = \dot{m}C_p \rightarrow \infty$ when $\Delta T \rightarrow 0$, so that the heat transfer rate $\dot{Q} = \dot{m}C_p\,\Delta T$ is a finite quantity. Therefore, in heat exchanger analysis, a condensing or boiling fluid is conveniently modeled as a fluid whose heat capacity rate is *infinity*.

The rate of heat transfer in a heat exchanger can also be expressed in an analogous manner to Newton's law of cooling as

$$\dot{Q} = UA\,\Delta T_m \tag{20-15}$$

where U is the overall heat transfer coefficient, A is the heat transfer area, and ΔT_m is an appropriate average temperature difference between the two fluids. Here the surface area A can be determined precisely using the dimensions of the heat exchanger. However, the overall heat transfer coefficient U and the

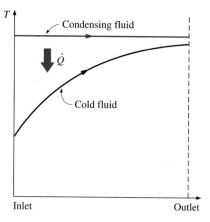

(a) Condenser ($C_h \rightarrow \infty$)

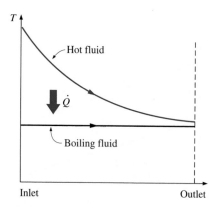

(b) Boiler ($C_c \rightarrow \infty$)

FIGURE 20-13

Variation of fluid temperatures in a heat exchanger when one of the fluids condenses or boils.

temperature difference ΔT between the hot and cold fluids, in general, are not constant and vary along the heat exchanger.

The average value of the overall heat transfer coefficient can be determined as described in the preceding section by using the average convection coefficients for each fluid. It turns out that the appropriate form of the mean temperature difference between the two fluids is *logarithmic* in nature, and its determination is presented in the next section.

20-4 ■ THE LOG MEAN TEMPERATURE DIFFERENCE METHOD

Earlier, we mentioned that the temperature difference between the hot and cold fluids varies along the heat exchanger, and it is convenient to have a *mean temperature difference* ΔT_m for use in the relation

$$\dot{Q} = UA\,\Delta T_m$$

In order to develop a relation for the equivalent average temperature difference between the two fluids, consider the *parallel-flow double-pipe* heat exchanger shown in Fig. 20-14. Note that the temperature difference ΔT between the hot and cold fluids is large at the inlet of the heat exchanger but decreases exponentially toward the outlet. As you would expect, the temperature of the hot fluid decreases and the temperature of the cold fluid increases along the heat exchanger, but the temperature of the cold fluid can never exceed that of the hot fluid no matter how long the heat exchanger is.

Assuming the outer surface of the heat exchanger to be well insulated so that any heat transfer occurs between the two fluids, and disregarding any changes in kinetic and potential energy, an energy balance on each fluid in a differential section of the heat exchanger can be expressed as

$$\delta \dot{Q} = -\dot{m}_h C_{ph}\,dT_h \tag{20-16}$$

and

$$\delta \dot{Q} = \dot{m}_c C_{pc}\,dT_c \tag{20-17}$$

That is, the rate of heat loss from the hot fluid at any section of a heat exchanger is equal to the rate of heat gain by the cold fluid in that section. The temperature change of the hot fluid is a *negative* quantity, and so a *negative sign* is added to Eq. 20-16 to make the heat transfer rate $\dot{Q}$ a positive quantity. Solving the equations above for dT_h and dT_c gives

$$dT_h = -\frac{\delta \dot{Q}}{\dot{m}_h C_{ph}} \tag{20-18}$$

and

$$dT_c = \frac{\delta \dot{Q}}{\dot{m}_c C_{pc}} \tag{20-19}$$

Taking their difference, we get

$$dT_h - dT_c = d(T_h - T_c) = -\delta \dot{Q}\left(\frac{1}{\dot{m}_h C_{ph}} + \frac{1}{\dot{m}_c C_{pc}}\right) \tag{20-20}$$

The rate of heat transfer in the differential section of the heat exchanger can also be expressed as

$$\delta \dot{Q} = U(T_h - T_c)\,dA \tag{20-21}$$

Substituting this equation into Eq. 20-20 and rearranging gives

$$\frac{d(T_h - T_c)}{T_h - T_c} = -U \, dA \left(\frac{1}{\dot{m}_h C_{ph}} + \frac{1}{\dot{m}_c C_{pc}} \right)$$ (20-22)

Integrating from the inlet of the heat exchanger to its outlet, we obtain

$$\ln \frac{T_{h,\text{out}} - T_{c,\text{out}}}{T_{h,\text{in}} - T_{c,\text{in}}} = -UA \left(\frac{1}{\dot{m}_h C_{ph}} + \frac{1}{\dot{m}_c C_{pc}} \right)$$ (20-23)

Finally, solving Eqs. 20-9 and 20-10 for $\dot{m}_c C_{pc}$ and $\dot{m}_h C_{ph}$ and substituting into Eq. 20-23 gives, after some rearrangement,

$$\dot{Q} = UA \, \Delta T_{\text{lm}}$$ (20-24)

where

$$\Delta T_{\text{lm}} = \frac{\Delta T_1 - \Delta T_2}{\ln (\Delta T_1/\Delta T_2)}$$ (20-25)

is the **log mean temperature difference**, which is the suitable form of the average temperature difference for use in the analysis of heat exchangers. Here ΔT_1 and ΔT_2 represent the temperature difference between the two fluids at the two ends (inlet and outlet) of the heat exchanger. It makes no difference which end of the heat exchanger is designated as the *inlet* or the *outlet* (Fig. 20-15).

The temperature difference between the two fluids decreases from ΔT_1 at the inlet to ΔT_2 at the outlet. Thus, it is tempting to use the arithmetic mean temperature $\Delta T_{\text{am}} = \frac{1}{2}(\Delta T_1 + \Delta T_2)$ as the average temperature difference. The logarithmic mean temperature difference ΔT_{lm} is obtained by tracing the actual temperature profile of the fluids along the heat exchanger and is an *exact* representation of the *average temperature difference* between the hot and cold fluids. It truly reflects the exponential decay of the local temperature difference.

Note that ΔT_{lm} is always less than ΔT_{am}. Therefore, using ΔT_{am} in calculations instead of ΔT_{lm} will overestimate the rate of heat transfer in a heat exchanger between the two fluids. When ΔT_1 differs from ΔT_2 by no more than 40 percent, the error in using the arithmetic mean temperature difference is less than 1 percent. But the error increases to undesirable levels when ΔT_1 differs from ΔT_2 by greater amounts. Therefore, we should always use the *logarithmic mean temperature difference* when determining the rate of heat transfer in a heat exchanger.

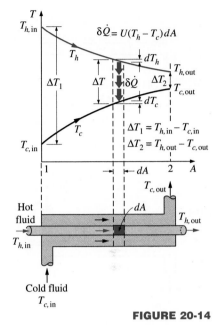

FIGURE 20-14

Variation of the fluid temperatures in a parallel-flow double-pipe heat exchanger.

Counter-Flow Heat Exchangers

The variation of temperatures of hot and cold fluids in a counter-flow heat exchanger is given in Fig. 20-16. Note that the hot and cold fluids enter the heat exchanger from opposite ends, and the outlet temperature of the *cold fluid* in this case may exceed the outlet temperature of the *hot fluid*. In the limiting case, the cold fluid will be heated to the inlet temperature of the hot fluid. However, the outlet temperature of the cold fluid can *never* exceed the inlet temperature of the hot fluid, since this would be a violation of the second law of thermodynamics.

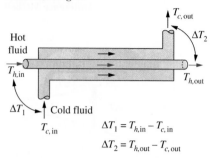

$$\Delta T_1 = T_{h,\text{in}} - T_{c,\text{in}}$$
$$\Delta T_2 = T_{h,\text{out}} - T_{c,\text{out}}$$

(a) Parallel-flow heat exchangers

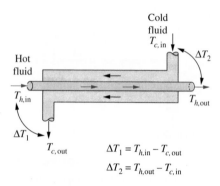

$$\Delta T_1 = T_{h,\text{in}} - T_{c,\text{out}}$$
$$\Delta T_2 = T_{h,\text{out}} - T_{c,\text{in}}$$

(b) Counter-flow heat exchangers

FIGURE 20-15

The ΔT_1 and ΔT_2 expressions in parallel-flow and counter-flow heat exchangers.

The relation above for the log mean temperature difference is developed using a parallel-flow heat exchanger, but we can show by repeating the analysis above for a counter-flow heat exchanger that is also applicable to counter-flow heat exchangers. But this time, ΔT_1 and ΔT_2 are expressed as shown in Fig. 20-15.

For specified inlet and outlet temperatures, the log mean temperature difference for a *counter-flow* heat exchanger is always *greater* than that for a parallel-flow heat exchanger. That is, $\Delta T_{\text{lm, }CF} > \Delta T_{\text{lm, }PF}$, and thus a smaller surface area (and thus a smaller heat exchanger) is needed to achieve a specified heat transfer rate in a counter-flow heat exchanger. Therefore, it is common practice to use counter-flow arrangements in heat exchangers.

In a counter-flow heat exchanger, the temperature difference between the hot and the cold fluids will remain constant along the heat exchanger when the *heat capacity rates* of the two fluids are *equal* (that is, ΔT = constant when $C_h = C_c$ or $\dot{m}_h C_{ph} = \dot{m}_c C_{pc}$). Then we have $\Delta T_1 = \Delta T_2$, and the log mean temperature difference relation above gives $\Delta T_{\text{lm}} = \frac{0}{0}$, which is indeterminate. It can be shown by the application of l'Hôpital's rule that in this case we have $\Delta T_{\text{lm}} = \Delta T_1 = \Delta T_2$, as expected.

A *condenser* or a *boiler* can be considered to be either a parallel- or counter-flow heat exchanger since both approaches give the same result.

Multipass and Cross-Flow Heat Exchangers: Use of a Correction Factor

The log mean temperature difference ΔT_{lm} relation developed earlier is limited to parallel-flow and counter-flow heat exchangers only. Similar relations are also developed for *cross-flow* and *multipass shell-and-tube* heat exchangers, but the resulting expressions are too complicated because of the complex flow conditions.

In such cases, it is convenient to relate the equivalent temperature difference to the log mean temperature difference relation for the counter-flow case as

$$\Delta T_{\text{lm}} = F \, \Delta T_{\text{lm, }CF} \qquad (20\text{-}26)$$

where F is the **correction factor,** which depends on the *geometry* of the heat exchanger and the inlet and outlet temperatures of the hot and cold fluid streams. The $\Delta T_{\text{lm, }CF}$ is the log mean temperature difference for the case of a *counter-flow* heat exchanger with the same inlet and outlet temperatures and is determined from Eq. 20-25 by taking $\Delta T_1 = T_{h,\text{in}} - T_{c,\text{out}}$ and $\Delta T_2 = T_{h,\text{out}} - T_{c,\text{in}}$ (Fig. 20-17).

The correction factor is less than unity for a cross-flow and multipass shell-and-tube heat exchanger. That is, $F \leq 1$. The limiting value of $F = 1$ corresponds to the counter-flow heat exchanger. Thus, the correction factor F for a heat exchanger is *a measure of deviation of the ΔT_{lm} from the corresponding values for the counter-flow case.*

The correction factor F for common cross-flow and shell-and-tube heat exchanger configurations is given in Fig. 20-18 versus two temperature ratios P and R defined as

$$P = \frac{t_2 - t_1}{T_1 - t_1} \qquad (20\text{-}27)$$

$$R = \frac{T_1 - T_2}{t_2 - t_1} = \frac{(\dot{m}C_p)_{\text{tube side}}}{(\dot{m}C_p)_{\text{shell side}}} \qquad (20\text{-}28)$$

where the subscripts 1 and 2 represent the *inlet* and *outlet,* respectively. Note that for a shell-and-tube heat exchanger, T and t represent the *shell-* and *tube-side* temperatures, respectively, as shown in the correction factor charts. It makes no difference whether the hot or the cold fluid flows through the shell or the tube. The determination of the correction factor F requires the availability of the *inlet* and the *outlet* temperatures for both the cold and hot fluids.

Note that the value of P ranges from 0 to 1. The value of R, on the other hand, ranges from 0 to infinity, with $R = 0$ corresponding to the phase-change (condensation or boiling) on the shell-side and $R \to \infty$ to phase-change on the tube side. The correction factor is $F = 1$ for both of these limiting cases. Therefore, the correction factor for a *condenser* or *boiler* is $F = 1$, regardless of the configuration of the heat exchanger.

EXAMPLE 20-3 The Condensation of Steam in a Condenser

Steam in the condenser of a power plant is to be condensed at a temperature of 30°C with cooling water from a nearby lake, which enters the tubes of the condenser at 14°C and leaves at 22°C. The surface area of the tubes is 45 m², and the overall heat transfer coefficient is 2100 W/m² · °C. Determine the mass flow rate of the cooling water needed and the rate of condensation of the steam in the condenser.

Solution Steam is condensed by cooling water in the condenser of a power plant. The mass flow rate of the cooling water and the rate of condensation are to be determined.

Assumptions **1** Steady operating conditions exist. **2** The heat exchanger is well insulated so that heat loss to the surroundings is negligible and thus heat transfer from the hot fluid is equal to the heat transfer to the cold fluid. **3** Changes in the kinetic and potential energies of fluid streams are negligible. **4** There is no fouling. **5** Fluid properties are constant.

Properties The heat of vaporization of water at 30°C is $h_{fg} = 2431$ kJ/kg and the specific heat of cold water at the average temperature of 18°C is $C_p = 4184$ J/kg · °C (Table A-15).

Analysis The schematic of the condenser is given in Fig. 20-19. The condenser can be treated as a counter-flow heat exchanger since the temperature of one of the fluids (the steam) remains constant.

The temperature difference between the steam and the cooling water at the two ends of the condenser is

$$\Delta T_1 = T_{h,\text{in}} - T_{c,\text{out}} = (30 - 22)°C = 8°C$$
$$\Delta T_2 = T_{h,\text{out}} - T_{c,\text{in}} = (30 - 14)°C = 16°C$$

That is, the temperature difference between the two fluids varies from 8°C at one end to 16°C at the other. The proper average temperature difference between the two fluids is the *logarithmic mean temperature difference* (not the arithmetic), which is determined from

$$\Delta T_{lm} = \frac{\Delta T_1 - \Delta T_2}{\ln(\Delta T_1/\Delta T_2)} = \frac{8 - 16}{\ln(8/16)} = 11.5°C$$

This is a little less than the arithmetic mean temperature difference of $\frac{1}{2}(8 + 16) = 12$°C. Then the heat transfer rate in the condenser is determined from

$$\dot{Q} = UA\,\Delta T_{lm} = (2100 \text{ W/m}^2 \cdot °C)(45 \text{ m}^2)(11.5°C) = \mathbf{1.087 \times 10^6 \text{ W} = 1087 \text{ kW}}$$

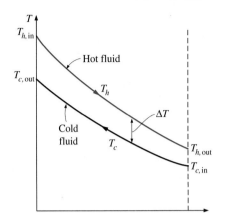

FIGURE 20-16

The variation of the fluid temperatures in a counter-flow double-pipe heat exchanger.

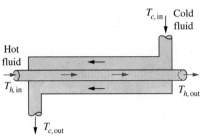

Heat transfer rate:
$$\dot{Q} = UAF\,\Delta T_{lm,CF}$$

where
$$\Delta T_{lm,CF} = \frac{\Delta T_1 - \Delta T_2}{\ln(\Delta T_1/\Delta T_2)}$$

$$\Delta T_1 = T_{h,\text{in}} - T_{c,\text{out}}$$
$$\Delta T_2 = T_{h,\text{out}} - T_{c,\text{in}}$$

and
$$F = \dots \text{ (Fig. 10–18)}$$

FIGURE 20-17

The determination of the heat transfer rate for cross-flow and multipass shell-and-tube heat exchangers using the correction factor.

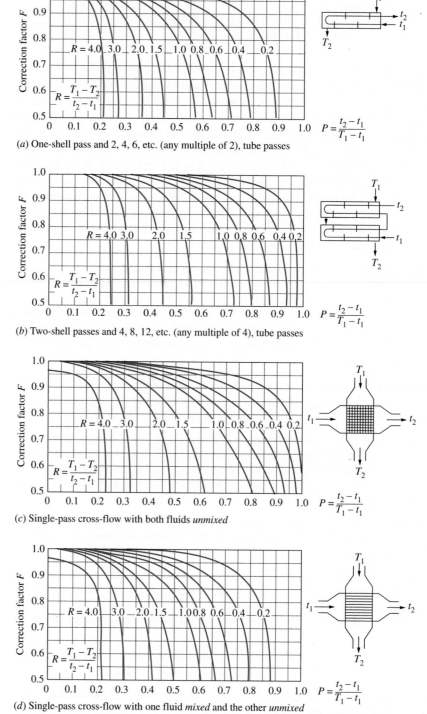

FIGURE 20-18

Correction factor F charts
for common shell-and-tube and
cross-flow heat exchangers (from
Bowman, Mueller, and Nagle, Ref. 2).

(a) One-shell pass and 2, 4, 6, etc. (any multiple of 2), tube passes

(b) Two-shell passes and 4, 8, 12, etc. (any multiple of 4), tube passes

(c) Single-pass cross-flow with both fluids *unmixed*

(d) Single-pass cross-flow with one fluid *mixed* and the other *unmixed*

Therefore, the steam will lose heat at a rate of 1,087 kW as it flows through the condenser, and the cooling water will gain practically all of it, since the condenser is well insulated.

The mass flow rate of the cooling water and the rate of the condensation of the steam are determined from $\dot{Q} = [\dot{m}C(T_{out} - T_{in})]_{cooling\ water} = (\dot{m}h_{fg})_{steam}$ to be

$$\dot{m}_{\text{cooling water}} = \frac{\dot{Q}}{C_p(T_{\text{out}} - T_{\text{in}})}$$

$$= \frac{1{,}087 \text{ kJ/s}}{(4.184 \text{ kJ/kg} \cdot {}^{\circ}\text{C})(22 - 14){}^{\circ}\text{C}} = \textbf{32.5 kg/s}$$

and

$$\dot{m}_{\text{steam}} = \frac{\dot{Q}}{h_{fg}} = \frac{1{,}087 \text{ kJ/s}}{2431 \text{ kJ/kg}} = \textbf{0.45 kg/s}$$

Therefore, we need to circulate about 72 kg of cooling water for each 1 kg of steam condensing to remove the heat released during the condensation process.

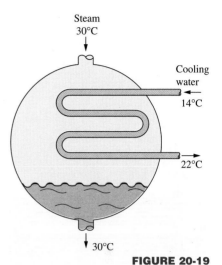

Cooling
water
14°C

22°C

30°C

FIGURE 20-19

Schematic for Example 20-3.

EXAMPLE 20-4   Heating Water in a Counter-Flow Heat Exchanger

A counter-flow double-pipe heat exchanger is to heat water from 20°C to 80°C at a rate of 1.2 kg/s. The heating is to be accomplished by geothermal water available at 160°C at a mass flow rate of 2 kg/s. The inner tube is thin-walled and has a diameter of 1.5 cm. If the overall heat transfer coefficient of the heat exchanger is 640 W/m² · °C, determine the length of the heat exchanger required to achieve the desired heating.

Solution   Water is heated in a counter-flow double-pipe heat exchanger by geothermal water. The required length of the heat exchanger is to be determined.

Assumptions   **1** Steady operating conditions exist. **2** The heat exchanger is well insulated so that heat loss to the surroundings is negligible and thus heat transfer from the hot fluid is equal to the heat transfer to the cold fluid. **3** Changes in the kinetic and potential energies of fluid streams are negligible. **4** There is no fouling. **5** Fluid properties are constant.

Properties   We take the specific heats of water and geothermal fluid to be 4.18 and 4.31 kJ/kg · °C, respectively.

Analysis   The schematic of the heat exchanger is given in Fig. 20-20. The rate of heat transfer in the heat exchanger can be determined from

$$\dot{Q} = [\dot{m}C_p(T_{\text{out}} - T_{\text{in}})]_{\text{water}} = (1.2 \text{ kg/s})(4.18 \text{ kJ/kg} \cdot {}^{\circ}\text{C})(80 - 20){}^{\circ}\text{C} = 301 \text{ kW}$$

Noting that all of this heat is supplied by the geothermal water, the outlet temperature of the geothermal water is determined to be

$$\dot{Q} = [\dot{m}C_p(T_{\text{in}} - T_{\text{out}})]_{\text{geothermal}} \longrightarrow T_{\text{out}} = T_{\text{in}} - \frac{\dot{Q}}{\dot{m}C_p}$$

$$= 160{}^{\circ}\text{C} - \frac{301 \text{ kW}}{(2 \text{ kg/s})(4.13 \text{ kJ/kg} \cdot {}^{\circ}\text{C})}$$

$$= 125{}^{\circ}\text{C}$$

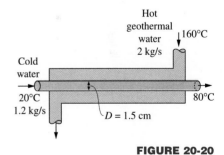

Hot
geothermal
water   160°C
2 kg/s

Cold
water
20°C   80°C
1.2 kg/s D = 1.5 cm

FIGURE 20-20

Schematic for Example 20-4.

Knowing the inlet and outlet temperatures of both fluids, the logarithmic mean temperature difference for this counter-flow heat exchanger becomes

$$\Delta T_1 = T_{h,\text{in}} - T_{c,\text{out}} = (160 - 80){}^{\circ}\text{C} = 80{}^{\circ}\text{C}$$
$$\Delta T_2 = T_{h,\text{out}} - T_{c,\text{in}} = (125 - 20){}^{\circ}\text{C} = 105{}^{\circ}\text{C}$$

and

$$\Delta T_{\text{lm}} = \frac{\Delta T_1 - \Delta T_2}{\ln(\Delta T_1/\Delta T_2)} = \frac{80 - 105}{\ln(80/105)} = 92.0{}^{\circ}\text{C}$$

Then the surface area of the heat exchanger is determined to be

$$\dot{Q} = UA \, \Delta T_{\text{lm}} \longrightarrow A = \frac{\dot{Q}}{U \, \Delta T_{\text{lm}}} = \frac{301{,}000 \text{ W}}{(640 \text{ W/m}^2 \cdot {}^{\circ}\text{C})(92.0{}^{\circ}\text{C})} = 5.11 \text{ m}^2$$

To provide this much heat transfer surface area, the length of the tube must be

$$A = \pi DL \longrightarrow L = \frac{A}{\pi D} = \frac{5.11 \text{ m}^2}{\pi(0.015 \text{ m})} = \textbf{108 m}$$

Discussion The inner tube of this counter-flow heat exchanger (and thus the heat exchanger itself) needs to be over 100 m long to achieve the desired heat transfer, which is impractical. In cases like this, we need to use a plate heat exchanger or a multipass shell-and-tube heat exchanger with multiple passes of tube bundles.

EXAMPLE 20-5 Heating of Glycerin in a Multipass Heat Exchanger

A 2-shell passes and 4-tube passes heat exchanger is used to heat glycerin from 20°C to 50°C by hot water, which enters the thin-walled 2-cm-diameter tubes at 80°C and leaves at 40°C (Fig. 20-21). The total length of the tubes in the heat exchanger is 60 m. The convection heat transfer coefficient is 25 W/m² · °C on the glycerin (shell) side and 160 W/m² · °C on the water (tube) side. Determine the rate of heat transfer in the heat exchanger (*a*) before any fouling occurs and (*b*) after fouling with a fouling factor of 0.0006 m² · °C/W occurs on the outer surfaces of the tubes.

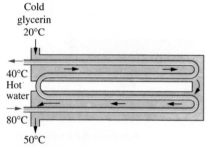

Cold
glycerin
20°C

40°C
Hot
water

80°C

50°C

FIGURE 20-21

Schematic for Example 20-5.

Solution Glycerin is heated in a 2-shell passes and 4-tube passes heat exchanger by hot water. The rate of heat transfer for the cases of fouling and no fouling are to be determined.

Assumptions **1** Steady operating conditions exist. **2** The heat exchanger is well insulated so that heat loss to the surroundings is negligible and thus heat transfer from the hot fluid is equal to heat transfer to the cold fluid. **3** Changes in the kinetic and potential energies of fluid streams are negligible. **4** Heat transfer coefficients and fouling factors are constant and uniform. **5** The thermal resistance of the inner tube is negligible since the tube is thin-walled and highly conductive.

Analysis The tubes are said to be thin-walled, and thus it is reasonable to assume the inner and outer surface areas of the tubes to be equal. Then the heat transfer surface area becomes

$$A = \pi DL = \pi(0.02 \text{ m})(60 \text{ m}) = 3.77 \text{ m}^2$$

The rate of heat transfer in this heat exchanger can be determined from

$$\dot{Q} = UAF \Delta T_{\text{lm, }CF}$$

where *F* is the correction factor and $\Delta T_{\text{lm, }CF}$ is the log mean temperature difference for the counter-flow arrangement. These two quantities are determined from

$$\Delta T_1 = T_{h,\text{ in}} - T_{c,\text{ out}} = (80 - 50)°C = 30°C$$

$$\Delta T_2 = T_{h,\text{ out}} - T_{c,\text{ in}} = (40 - 20)°C = 20°C$$

$$\Delta T_{\text{lm, }CF} = \frac{\Delta T_1 - \Delta T_2}{\ln(\Delta T_1/\Delta T_2)} = \frac{30 - 20}{\ln(30/20)} = 24.7°C$$

and

$$\left. \begin{array}{l} P = \dfrac{t_2 - t_1}{T_1 - t_1} = \dfrac{40 - 80}{20 - 80} = 0.67 \\[2mm] R = \dfrac{T_1 - T_2}{t_2 - t_1} = \dfrac{20 - 50}{40 - 80} = 0.75 \end{array} \right\} \quad F = 0.91 \qquad \text{(Fig. 20-18}b\text{)}$$

(*a*) In the case of no fouling, the overall heat transfer coefficient *U* is determined from

$$U = \cfrac{1}{\cfrac{1}{h_i} + \cfrac{1}{h_o}} = \cfrac{1}{\cfrac{1}{160 \text{ W/m}^2 \cdot {}^\circ\text{C}} + \cfrac{1}{25 \text{ W/m}^2 \cdot {}^\circ\text{C}}} = 21.6 \text{ W/m}^2 \cdot {}^\circ\text{C}$$

Then the rate of heat transfer becomes

$$\dot{Q} = UAF\,\Delta T_{\text{lm, }CF} = (21.6 \text{ W/m}^2 \cdot {}^\circ\text{C})(3.77 \text{m}^2)(0.91)(24.7 {}^\circ\text{C}) = \textbf{1730 W}$$

(*b*) When there is fouling on one of the surfaces, the overall heat transfer coefficient U is

$$U = \cfrac{1}{\cfrac{1}{h_i} + \cfrac{1}{h_o} + R_f} = \cfrac{1}{\cfrac{1}{160 \text{ W/m}^2 \cdot {}^\circ\text{C}} + \cfrac{1}{25 \text{ W/m}^2 \cdot {}^\circ\text{C}} + 0.0006 \text{ m}^2 \cdot {}^\circ\text{C/W}}$$

$$= 21.3 \text{ W/m}^2 \cdot {}^\circ\text{C}$$

The rate of heat transfer in this case becomes

$$\dot{Q} = UAF\,\Delta T_{\text{lm, }CF} = (21.3 \text{ W/m}^2 \cdot {}^\circ\text{C})(3.77 \text{ m}^2)(0.91)(24.7 {}^\circ\text{C}) = \textbf{1805 W}$$

Discussion   Note that the rate of heat transfer decreases as a result of fouling, as expected. The decrease is not dramatic, however, because of the relatively low convection heat transfer coefficients involved.

EXAMPLE 20-6   Cooling of an Automotive Radiator

A test is conducted to determine the overall heat transfer coefficient in an automotive radiator that is a compact cross-flow water-to-air heat exchanger with both fluids (air and water) unmixed (Fig. 20-22). The radiator has 40 tubes of internal diameter 0.5 cm and length 65 cm in a closely spaced plate-finned matrix. Hot water enters the tubes at 90°C at a rate of 0.6 kg/s and leaves at 65°C. Air flows across the radiator through the interfin spaces and is heated from 20°C to 40°C. Determine the overall heat transfer coefficient U_i of this radiator based on the inner surface area of the tubes.

Solution   During an experiment involving an automotive radiator, the inlet and exit temperatures of water and air and the mass flow rate of water are measured. The overall heat transfer coefficient based on the inner surface area is to be determined.

Assumptions   **1** Steady operating conditions exist. **2** Changes in the kinetic and potential energies of fluid streams are negligible. **3** Fluid properties are constant.

Properties   The specific heat of water at the average temperature of $(90 + 65)/2 = 77.5°C$ is 4.195 kJ/kg · °C.

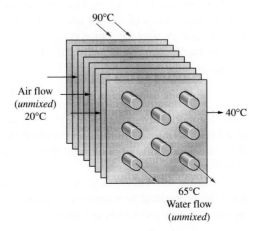

90°C

Air flow
(*unmixed*)
20°C

40°C

65°C
Water flow
(*unmixed*)

FIGURE 20-22

Schematic for Example 20-6.

Analysis The rate of heat transfer in this radiator from the hot water to the air is determined from an energy balance on water flow,

$$\dot{Q} = [\dot{m} C_p (T_{in} - T_{out})]_{water} = (0.6 \text{ kg/s})(4.195 \text{ kJ/kg} \cdot °C)(90 - 65)°C = 62.93 \text{ kW}$$

The tube-side heat transfer area is the total surface area of the tubes, and is determined from

$$A_i = n \pi D_i L = (40) \pi (0.005 \text{ m})(0.65 \text{ m}) = 0.408 \text{ m}^2$$

Knowing the rate of heat transfer and the surface area, the overall heat transfer coefficient can be determined from

$$\dot{Q} = U_i A_i F \Delta T_{lm, CF} \longrightarrow U_i = \frac{\dot{Q}}{A_i F \Delta T_{lm, CF}}$$

where F is the correction factor and $\Delta T_{lm, CF}$ is the log mean temperature difference for the counter-flow arrangement. These two quantities are found to be

$$\Delta T_1 = T_{h, in} - T_{c, out} = (90 - 40)°C = 50°C$$
$$\Delta T_2 = T_{h, out} - T_{c, in} = (65 - 20)°C = 45°C$$
$$\Delta T_{lm, CF} = \frac{\Delta T_1 - \Delta T_2}{\ln (\Delta T_1 / \Delta T_2)} = \frac{50 - 45}{\ln (50/45)} = 47.6°C$$

and

$$\left. \begin{array}{l} P = \dfrac{t_2 - t_1}{T_1 - t_1} = \dfrac{65 - 90}{20 - 90} = 0.36 \\[2mm] R = \dfrac{T_1 - T_2}{t_2 - t_1} = \dfrac{20 - 40}{65 - 90} = 0.80 \end{array} \right\} F = 0.97 \qquad \text{(Fig. 20-18c)}$$

Substituting, the overall heat transfer coefficient U_i is determined to be

$$U_i = \frac{\dot{Q}}{A_i F \Delta T_{lm, CF}} = \frac{62,930 \text{ W}}{(0.408 \text{ m}^2)(0.97)(47.6°C)} = 3341 \text{ W/m}^2 \cdot °C$$

Note that the overall heat transfer coefficient on the air side will be much lower because of the large surface area involved on that side.

20-5 ■ THE EFFECTIVENESS–NTU METHOD

The log mean temperature difference (LMTD) method discussed in the previous section is easy to use in heat exchanger analysis when the inlet and the outlet temperatures of the hot and cold fluids are known or can be determined from an energy balance. Once ΔT_{lm}, the mass flow rates, and the overall heat transfer coefficient are available, the heat transfer surface area of the heat exchanger can be determined from

$$\dot{Q} = UA \, \Delta T_{lm}$$

Therefore, the LMTD method is very suitable for determining the *size* of a heat exchanger to realize prescribed outlet temperatures when the mass flow rates and the inlet and outlet temperatures of the hot and cold fluids are specified.

With the LMTD method, the task is to *select* a heat exchanger that will meet the prescribed heat transfer requirements. The procedure to be followed by the selection process is as follows:

1. Select the type of heat exchanger suitable for the application.

2. Determine any unknown inlet or outlet temperature and the heat transfer rate using an energy balance.

3. Calculate the log mean temperature difference ΔT_{lm} and the correction factor F, if necessary.

4. Obtain (select or calculate) the value of the overall heat transfer coefficient U.

5 Calculate the heat transfer surface area A.

The task is completed by selecting a heat exchanger that has a heat transfer surface area equal to or larger than A.

A second kind of problem encountered in heat exchanger analysis is the determination of the *heat transfer rate* and the *outlet temperatures* of the hot and cold fluids for prescribed fluid mass flow rates and inlet temperatures when the *type* and *size* of the heat exchanger are specified. The heat transfer surface area A of the heat exchanger in this case is known, but the *outlet temperatures* are not. Here the task is to determine the heat transfer performance of a specified heat exchanger or to determine if a heat exchanger available in storage will do the job.

The LMTD method could still be used for this alternative problem, but the procedure would require tedious iterations, and thus it is not practical. In an attempt to eliminate the iterations from the solution of such problems, Kays and London came up with a method in 1955 called the **effectiveness-NTU method,** which greatly simplified heat exchanger analysis.

This method is based on a dimensionless parameter called the **heat transfer effectiveness ε,** defined as

$$\varepsilon = \frac{\dot{Q}}{Q_{\text{max}}} = \frac{\text{Actual heat transfer rate}}{\text{Maximum possible heat transfer rate}} \qquad (20\text{-}29)$$

The *actual* heat transfer rate in a heat exchanger can be determined from an energy balance on the hot or cold fluids and can be expressed as

$$\dot{Q} = C_c(T_{c,\,\text{out}} - T_{c,\,\text{in}}) = C_h(T_{h,\,\text{in}} - T_{h,\,\text{out}}) \qquad (20\text{-}30)$$

where $C_c = \dot{m}_c C_{pc}$ and $C_h = \dot{m}_c C_{ph}$ are the heat capacity rates of the cold and the hot fluids, respectively.

To determine the maximum possible heat transfer rate in a heat exchanger, we first recognize that the *maximum temperature difference* in a heat exchanger is the difference between the *inlet* temperatures of the hot and cold fluids. That is,

$$\Delta T_{\text{max}} = T_{h,\,\text{in}} - T_{c,\,\text{in}} \qquad (20\text{-}31)$$

The heat transfer in a heat exchanger will reach its maximum value when (1) the cold fluid is heated to the inlet temperature of the hot fluid or (2) the hot fluid is cooled to the inlet temperature of the cold fluid. These two limiting conditions will not be reached simultaneously unless the heat capacity rates of the hot and cold fluids are identical (i.e., $C_c = C_h$). When $C_c \neq C_h$, which is usually the case, the fluid with the *smaller* heat capacity rate will experience a larger temperature change, and thus it will be the first to experience the maximum temperature, at which point the heat transfer will come to a halt.

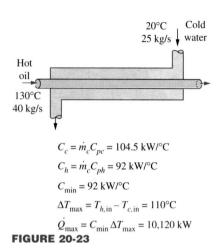

$$C_c = \dot{m}_c C_{pc} = 104.5 \text{ kW/°C}$$

$$C_h = \dot{m}_c C_{ph} = 92 \text{ kW/°C}$$

$$C_{min} = 92 \text{ kW/°C}$$

$$\Delta T_{max} = T_{h,in} - T_{c,in} = 110\text{°C}$$

$$\dot{Q}_{max} = C_{min} \Delta T_{max} = 10,120 \text{ kW}$$

FIGURE 20-23

The determination of the maximum rate of heat transfer in a heat exchanger.

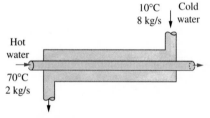

FIGURE 20-24

Schematic for Example 20-7.

Therefore, the maximum possible heat transfer rate in a heat exchanger is (Fig. 20-23)

$$\dot{Q}_{max} = C_{min}(T_{h,\,in} - T_{c,\,in}) \qquad (20\text{-}32)$$

where C_{min} is the smaller of $C_h = \dot{m}_h C_{ph}$ and $C_c = \dot{m}_c C_{pc}$. This is further clarified by the following example.

EXAMPLE 20-7   Upper Limit for Heat Transfer in a Heat Exchanger
Cold water enters a counter-flow heat exchanger at 10°C at a rate of 8 kg/s, where it is heated by a hot water stream that enters the heat exchanger at 70°C at a rate of 2 kg/s. Assuming the specific heat of water to remain constant at $C_p = 4.18$ kJ/ kg · °C, determine the maximum heat transfer rate and the outlet temperatures of the cold and the hot water streams for this limiting case.

Solution   Cold and hot water streams enter a heat exchanger at specified temperatures and flow rates. The maximum rate of heat transfer in the heat exchanger is to be determined.

Assumptions   **1** Steady operating conditions exist. **2** The heat exchanger is well insulated so that heat loss to the surroundings is negligible and thus heat transfer from the hot fluid is equal to heat transfer to the cold fluid. **3** Changes in the kinetic and potential energies of fluid streams are negligible. **4** Heat transfer coefficients and fouling factors are constant and uniform. **5** The thermal resistance of the inner tube is negligible since the tube is thin-walled and highly conductive.

Properties   The specific heat of water is given to be $C_p = 4.18$ kJ/kg · °C.

Analysis   A schematic of the heat exchanger is given in Fig. 20-24. The heat capacity rates of the hot and cold fluids are determined from

$$C_h = \dot{m}_h C_{ph} = (2 \text{ kg/s})(4.18 \text{ kJ/kg} \cdot \text{°C}) = 8.36 \text{ kW/°C}$$

and

$$C_c = \dot{m}_c C_{pc} = (8 \text{ kg/s})(4.18 \text{ kJ/kg} \cdot \text{°C}) = 33.4 \text{ kW/°C}$$

Therefore

$$C_{min} = C_h = 8.36 \text{ kW/°C}$$

which is the smaller of the two heat capacity rates. Then the maximum heat transfer rate is determined from Eq. 20-32 to be

$$\dot{Q}_{max} = C_{min}(T_{h,\,in} - T_{c,\,in})$$
$$= (8.36 \text{ kW/°C})(70 - 10)\text{°C}$$
$$= \textbf{502 kW}$$

That is, the maximum possible heat transfer rate in this heat exchanger is 502 kW. This value would be approached in a counter-flow heat exchanger with a *very large* heat transfer surface area.

   The maximum temperature difference in this heat exchanger is $\Delta T_{max} = T_{h,\,in} - T_{c,\,in} = (70 - 10)\text{°C} = 60\text{°C}$. Therefore, the hot water cannot be cooled by more than 60°C (to 10°C) in this heat exchanger, and the cold water cannot be heated by more than 60°C (to 70°C), no matter what we do. The outlet temperatures of the cold and the hot streams in this limiting case are determined to be

$$\dot{Q} = C_c(T_{c,\,out} - T_{c,\,in}) \longrightarrow T_{c,\,out} = T_{c,\,in} + \frac{\dot{Q}}{C_c} = 10\text{°C} + \frac{502 \text{ kW}}{33.4 \text{ kW/°C}} = \textbf{25°C}$$

$$\dot{Q} = C_h(T_{h,\,in} - T_{h,\,out}) \longrightarrow T_{h,\,out} = T_{h,\,in} - \frac{\dot{Q}}{C_h} = 70\text{°C} - \frac{502 \text{ kW}}{8.38 \text{ kW/°C}} = \textbf{10°C}$$

Discussion Note that the hot water is cooled to the limit of 10°C (the inlet temperature of the cold water stream), but the cold water is heated to 25°C only when maximum heat transfer occurs in the heat exchanger. This is not surprising, since the mass flow rate of the hot water is only one-fourth that of the cold water, and, as a result, the temperature of the cold water increases by 0.25°C for each 1°C drop in the temperature of the hot water.

You may be tempted to think that the cold water should be heated to 70°C in the limiting case of maximum heat transfer. But this will require the temperature of the hot water to drop to −170°C (below 10°C), which is impossible. Therefore, heat transfer in a heat exchanger reaches its maximum value when the fluid with the smaller heat capacity rate (or the smaller mass flow rate when both fluids have the same specific heat value) experiences the maximum temperature change. This example explains why we use C_{min} in the evaluation of $\dot{Q}_{max}$ instead of C_{max}.

We can show that the hot water will leave at the inlet temperature of the cold water and vice versa in the limiting case of maximum heat transfer when the mass flow rates of the hot and cold water streams are identical (Fig. 20-25). We can also show that the outlet temperature of the cold water will reach the 70°C limit when the mass flow rate of the hot water is greater than that of the cold water.

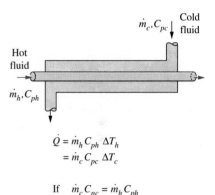

$$Q = \dot{m}_h C_{ph} \, \Delta T_h$$
$$= \dot{m}_c C_{pc} \, \Delta T_c$$

If $\dot{m}_c C_{pc} = \dot{m}_h C_{ph}$

then $\Delta T_h = \Delta T_c$

FIGURE 20-25

The temperature rise of the cold fluid in a heat exchanger will be equal to the temperature drop of the hot fluid when the mass flow rates and the specific heats of the hot and cold fluids are identical.

The determination of $\dot{Q}_{max}$ requires the availability of the *inlet temperature* of the hot and cold fluids and their *mass flow rates,* which are usually specified. Then, once the effectiveness of the heat exchanger is known, the actual heat transfer rate $\dot{Q}$ can be determined from

$$\dot{Q} = \varepsilon \dot{Q}_{max} = \varepsilon C_{min}(T_{h,\,in} - T_{c,\,in}) \qquad (20\text{-}33)$$

Therefore, the effectiveness of a heat exchanger enables us to determine the heat transfer rate without knowing the *outlet temperatures* of the fluids.

The effectiveness of a heat exchanger depends on the *geometry* of the heat exchanger as well as the *flow arrangement.* Therefore, different types of heat exchangers have different effectiveness relations. Below we illustrate the development of the effectiveness ε relation for the double-pipe *parallel-flow* heat exchanger.

Equation 20-23 developed in the previous section for a parallel-flow heat exchanger can be rearranged as

$$\ln \frac{T_{h,\,out} - T_{c,\,out}}{T_{h,\,in} - T_{c,\,in}} = -\frac{UA}{C_c}\left(1 + \frac{C_c}{C_h}\right) \qquad (20\text{-}34)$$

Also, solving Eq. 20-30 for $T_{h,\,out}$ gives

$$T_{h,\,out} = T_{h,\,in} - \frac{C_c}{C_h}(T_{c,\,out} - T_{c,\,in}) \qquad (20\text{-}35)$$

Substituting this relation into Eq. 20-34 after adding and subtracting $T_{c,\,in}$ gives

$$\ln \frac{T_{h,\,in} - T_{c,\,in} + T_{c,\,in} - T_{c,\,out} - \dfrac{C_c}{C_h}(T_{c,\,out} - T_{c,\,in})}{T_{h,\,in} - T_{c,\,in}} = -\frac{UA}{C_c}\left(1 + \frac{C_c}{C_h}\right)$$

which simplifies to

$$\ln \left[1 - \left(1 + \frac{C_c}{C_h}\right)\frac{T_{c,\,out} - T_{c,\,in}}{T_{h,\,in} - T_{c,\,in}}\right] = -\frac{UA}{C_c}\left(1 + \frac{C_c}{C_h}\right) \qquad (20\text{-}36)$$

We now manipulate the definition of effectiveness to obtain

$$\varepsilon = \frac{\dot{Q}}{\dot{Q}_{max}} = \frac{C_c(T_{c,\,out} - T_{c,\,in})}{C_{min}(T_{h,\,in} - T_{c,\,in})} \qquad \longrightarrow \qquad \frac{T_{c,\,out} - T_{c,\,in}}{T_{h,\,in} - T_{c,\,in}} = \varepsilon \frac{C_{min}}{C_c}$$

Substituting this result into Eq. 20-36 and solving for ε gives the following relation for the effectiveness of a *parallel-flow* heat exchanger:

$$\varepsilon_{parallel\ flow} = \frac{1 - \exp\left[-\dfrac{UA}{C_c}\left(1 + \dfrac{C_c}{C_h}\right)\right]}{\left(1 + \dfrac{C_c}{C_h}\right)\dfrac{C_{min}}{C_c}} \tag{20-37}$$

Taking either C_c or C_h to be C_{min} (both approaches give the same result), the relation above can be expressed more conveniently as

$$\varepsilon_{parallel\ flow} = \frac{1 - \exp\left[-\dfrac{UA}{C_{min}}\left(1 + \dfrac{C_{min}}{C_{max}}\right)\right]}{1 + \dfrac{C_{min}}{C_{max}}} \tag{20-38}$$

Again C_{min} is the *smaller* heat capacity ratio and C_{max} is the larger one, and it makes no difference whether C_{min} belongs to the hot or cold fluid.

Effectiveness relations of the heat exchangers typically involve the *dimensionless* group UA/C_{min}. This quantity is called the **number of transfer units NTU** and is expressed as

$$NTU = \frac{UA}{C_{min}} = \frac{UA}{(\dot{m}C_p)_{min}} \tag{20-39}$$

where U is the overall heat transfer coefficient and A is the heat transfer surface area of the heat exchanger. Note that NTU is proportional to A. Therefore, for specified values of U and C_{min}, the value of NTU *is a measure of the heat transfer surface area A*. Thus, the larger the NTU, the larger the heat exchanger.

In heat exchanger analysis, it is also convenient to define another dimensionless quantity called the **capacity ratio c** as

$$c = \frac{C_{min}}{C_{max}} \tag{20-40}$$

It can be shown that the effectiveness of a heat exchanger is a function of the number of transfer units NTU and the capacity ratio c. That is,

$$\varepsilon = \text{function } (UA/C_{min},\ C_{min}/C_{max}) = \text{function (NTU, } c)$$

Effectiveness relations have been developed for a large number of heat exchangers, and the results are given in Table 20-4. The effectivenesses of some common types of heat exchangers are also plotted in Fig. 20-26. More extensive effectiveness charts and relations are available in the literature. The dashed lines in Fig. 20-26f are for the case of C_{min} unmixed and C_{max} mixed, and the solid lines are for the opposite case. The analytic relations for the effectiveness give more accurate results than the charts, since reading errors in charts are unavoidable, and the relations are very suitable for computerized analysis of heat exchangers.

TABLE 20-4

Effectiveness relations for heat exchangers: NTU = UA/C_{min} and $c = C_{min}/C_{max} = (\dot{m}C_p)_{min}/(\dot{m}C_p)_{max}$

Heat exchanger type	Effectiveness relation
1 *Double pipe:*	
Parallel-flow	$\varepsilon = \dfrac{1 - \exp\left[-NTU(1 + c)\right]}{1 + c}$
Counter-flow	$\varepsilon = \dfrac{1 - \exp\left[-NTU(1 - c)\right]}{1 - c\exp\left[-NTU(1 - c)\right]}$
2 *Shell and tube:*	
One-shell pass 2, 4, ... tube passes	$\varepsilon = 2\left\{1 + c + \sqrt{1 + c^2}\ \dfrac{1 + \exp\left[-NTU\sqrt{1 + c^2}\right]}{1 - \exp\left[-NTU\sqrt{1 + c^2}\right]}\right\}^{-1}$
3 *Cross-flow (single-pass)*	
Both fluids unmixed	$\varepsilon = 1 - \exp\left\{\dfrac{NTU^{0.22}}{c}\left[\exp\left(-c\ NTU^{0.78}\right) - 1\right]\right\}$
C_{max} mixed, C_{min} unmixed	$\varepsilon = \dfrac{1}{c}\left(1 - \exp\{1 - c[1 - \exp(-NTU)]\}\right)$
C_{min} mixed, C_{max} unmixed	$\varepsilon = 1 - \exp\left\{-\dfrac{1}{c}[1 - \exp(-c\ NTU)]\right\}$
4 *All heat exchangers with c = 0*	$\varepsilon = 1 - \exp(-NTU)$

Source: Kays and London, Ref. 7.

We make the following observations from the effectiveness relations and charts given above:

1. The value of the effectiveness ranges from 0 to 1. It increases rapidly with NTU for small values (up to about NTU = 1.5) but rather slowly for larger values. Therefore, the use of a heat exchanger with a large NTU (usually larger than 3) and thus a large size cannot be justified economically, since a large increase in NTU in this case corresponds to a small increase in effectiveness. Thus, a heat exchanger with a very high effectiveness may be highly desirable from a heat transfer point of view but rather undesirable from an economical point of view.

2. For a given NTU and capacity ratio $c = C_{min}/C_{max}$, the *counter-flow* heat exchanger has the *highest* effectiveness, followed closely by the cross-flow heat exchangers with both fluids unmixed. As you might expect, the lowest effectiveness values are encountered in parallel-flow heat exchangers (Fig. 20-27).

3. The effectiveness of a heat exchanger is independent of the capacity ratio c for NTU values of less than about 0.3.

4. The value of the capacity ratio c ranges between 0 and 1. For a given NTU, the effectiveness becomes a *maximum* for $c = 0$ and a *minimum* for $c = 1$. The case $c = C_{min}/C_{max} \to 0$ corresponds to $C_{max} \to \infty$, which is

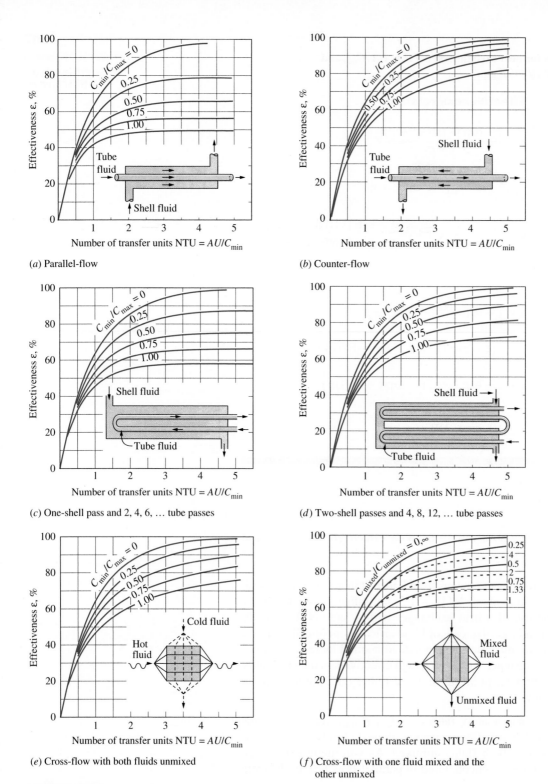

FIGURE 20-26

Effectiveness for heat exchangers (from Kays and London, Ref. 7).

realized during a phase-change process in a *condenser* or *boiler*. All effectiveness relations in this case reduce to

$$\varepsilon = \varepsilon_{max} = 1 - \exp(-NTU) \qquad (20\text{-}41)$$

regardless of the type of heat exchanger (Fig. 20-28). Note that the temperature of the condensing or boiling fluid remains constant in this case. The effectiveness is the *lowest* in the other limiting case of $cp = C_{min}/C_{max} = 1$, which is realized when the heat capacity rates of the two fluids are equal.

Once the quantities $c = C_{min}/C_{max}$ and $NTU = UA/C_{min}$ have been evaluated, the effectiveness ε can be determined from either the charts or (preferably) the effectiveness relation for the specified type of heat exchanger. Then the rate of heat transfer $\dot{Q}$ and the outlet temperatures $T_{h,\,out}$ and $T_{c,\,out}$ can be determined from Eqs. 20-33 and 20-30, respectively. Note that the analysis of heat exchangers with unknown outlet temperatures is a straightforward matter with the effectiveness-NTU method but requires rather tedious iterations with the LMTD method.

We mentioned earlier that when all the inlet and outlet temperatures are specified, the *size* of the heat exchanger can easily be determined using the LMTD method. Alternatively, it can also be determined from the effectiveness-NTU method by first evaluating the effectiveness ε from its definition (Eq. 20-29) and then the NTU from the appropriate NTU relation in Table 20-5.

Note that the relations in Table 20-5 are equivalent to those in Table 20-4. Both sets of relations are given for convenience. The relations in Table 20-4 give the effectiveness directly when NTU is known, and the relations in Table 20-5 give the NTU directly when the effectiveness ε is known.

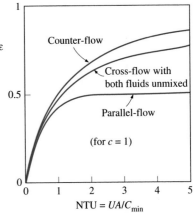

FIGURE 20-27

For a specified NTU and capacity ratio c, the counter-flow heat exchanger has the highest effectiveness and the parallel-flow the lowest.

EXAMPLE 20-8   Using the Effectiveness-NTU Method

Repeat Example 20-4, which was solved with the LMTD method, using the effectiveness-NTU method.

Solution   The schematic of the heat exchanger is redrawn in Fig. 20-29, and the same assumptions are utilized.

TABLE 20-5

NTU relations for heat exchangers $NTU = UA/C_{min}$ **and** $c = C_{min}/C_{max} = (\dot{m}C_p)_{min}/(\dot{m}C_p)_{max}$

Heat exchanger type	NTU relation
1 *Double-pipe:*	
Parallel-flow	$NTU = -\dfrac{\ln[1 - \varepsilon(1 + c)]}{1 + c}$
Counter-flow	$NTU = \dfrac{1}{c - 1}\ln\left(\dfrac{\varepsilon - 1}{\varepsilon c - 1}\right)$
2 *Shell and tube:* One-shell pass 2, 4, . . . tube passes	$NTU = -\dfrac{1}{\sqrt{1 + c^2}}\ln\left(\dfrac{2/\varepsilon - 1 - c - \sqrt{1 + c^2}}{2/\varepsilon - 1 - c + \sqrt{1 + c^2}}\right)$
3 *Cross-flow (single-pass)* C_{max} mixed, C_{min} unmixed	$NTU = -\ln\left[1 + \dfrac{\ln(1 - \varepsilon c)}{c}\right]$
C_{min} mixed, C_{max} unmixed	$NTU = -\dfrac{\ln[c\ln(1 - \varepsilon) + 1]}{c}$
4 *All heat exchangers with c = 0*	$NTU = -\ln(1 - \varepsilon)$

Source: Kays and London, Ref. 7.

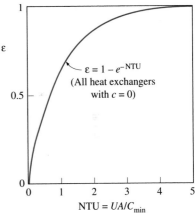

FIGURE 20-28

The effectiveness relation reduces to $\varepsilon = \varepsilon_{max} = 1 - \exp(-NTU)$ for all heat exchangers when the capacity ratio c = 0.

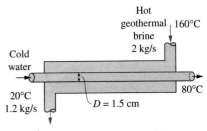

FIGURE 20-29

Schematic for Example 20-8.

Analysis In the effectiveness-NTU method, we first determine the heat capacity rates of the hot and cold fluids and identify the smaller one:

$$C_h = \dot{m}_h C_{ph} = (2 \text{ kg/s})(4.31 \text{ kJ/kg} \cdot °C) = 8.62 \text{ kW/°C}$$
$$C_c = \dot{m}_c C_{pc} = (1.2 \text{ kg/s})(4.18 \text{ kJ/kg} \cdot °C) = 5.02 \text{ kW/°C}$$

Therefore,

$$C_{min} = C_c = 5.02 \text{ kW/°C}$$

and

$$c = C_{min}/C_{max} = 5.02/8.62 = 0.583$$

Then the maximum heat transfer rate is determined from Eq. 20-32 to be

$$\dot{Q}_{max} = C_{min}(T_{h, \text{ in}} - T_{c, \text{ in}})$$
$$= (5.02 \text{ kW/°C})(160 - 20)°C$$
$$= 702.8 \text{ kW}$$

That is, the maximum possible heat transfer rate in this heat exchanger is 702.8 kW. The actual rate of heat transfer in the heat exchanger is

$$\dot{Q} = [\dot{m}C_p(T_{out} - T_{in})]_{water} = (1.2 \text{ kg/s})(4.18 \text{ kJ/kg} \cdot °C)(80 - 20)°C = 301.0 \text{ kW}$$

Thus, the effectiveness of the heat exchanger is

$$\varepsilon = \frac{\dot{Q}}{\dot{Q}_{max}} = \frac{301.0 \text{ kW}}{702.8 \text{ kW}} = 0.428$$

Knowing the effectiveness, the NTU of this counter-flow heat exchanger can be determined from Fig. 20-26b or the appropriate relation from Table 20-5. We choose the latter approach for greater accuracy:

$$NTU = \frac{1}{c - 1} \ln\left(\frac{\varepsilon - 1}{\varepsilon c - 1}\right) = \frac{1}{0.583 - 1} \ln\left(\frac{0.428 - 1}{0.428 \times 0.583 - 1}\right) = 0.651$$

Then the heat transfer surface area becomes

$$NTU = \frac{UA}{C_{min}} \longrightarrow A = \frac{NTU \, C_{min}}{U} = \frac{(0.651)(5020 \text{ W/°C})}{640 \text{ W/m}^2 \cdot °C} = 5.11 \text{ m}^2$$

To provide this much heat transfer surface area, the length of the tube must be

$$A = \pi D L \longrightarrow L = \frac{A}{\pi D} = \frac{5.11 \text{ m}^2}{\pi(0.015 \text{ m})} = \textbf{108 m}$$

Discussion Note that we obtained the same result with the effectiveness-NTU method in a systematic and straightforward manner.

EXAMPLE 20-9 Cooling Hot Oil by Water in a Multipass Heat Exchanger

Hot oil is to be cooled by water in a 1-shell-pass and 8-tube-passes heat exchanger. The tubes are thin-walled and are made of copper with an internal diameter of 1.4 cm. The length of each tube pass in the heat exchanger is 5 m, and the overall heat transfer coefficient is 310 W/m² · °C. Water flows through the tubes at a rate of 0.2 kg/s, and the oil through the shell at a rate of 0.3 kg/s. The water and the oil enter at temperatures of 20°C and 150°C, respectively. Determine the rate of heat transfer in the heat exchanger and the outlet temperatures of the water and the oil.

Solution Hot oil is to be cooled by water in a heat exchanger. The mass flow rates and the inlet temperatures are given. The rate of heat transfer and the outlet temperatures are to be determined.

Assumptions **1** Steady operating conditions exist. **2** The heat exchanger is well insulated so that heat loss to the surroundings is negligible and thus heat transfer from the hot fluid is equal to the heat transfer to the cold fluid. **3** The thickness of the tube is negligible since it is thin-walled. **4** Changes in the kinetic and potential energies of fluid streams are negligible. **5** The overall heat transfer coefficient is constant and uniform.

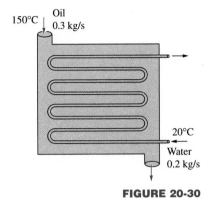

FIGURE 20-30
Schematic for Example 20-9.

Analysis The schematic of the heat exchanger is given in Fig. 20-30. The outlet temperatures are not specified, and they cannot be determined from an energy balance. The use of the LMTD method in this case will involve tedious iterations, and thus the ε-NTU method is indicated. The first step in the ε-NTU method is to determine the heat capacity rates of the hot and cold fluids and identify the smaller one:

$$C_h = \dot{m}_h C_{ph} = (0.3 \text{ kg/s})(2.13 \text{ kJ/kg} \cdot °C) = 0.639 \text{ kW/°C}$$
$$C_c = \dot{m}_c C_{pc} = (0.2 \text{ kg/s})(4.18 \text{ kJ/kg} \cdot °C) = 0.836 \text{ kW/°C}$$

Therefore,

$$C_{min} = C_h = 0.639 \text{ kW/°C}$$

and

$$c = \frac{C_{min}}{C_{max}} = \frac{0.639}{0.836} = 0.764$$

Then the maximum heat transfer rate is determined from Eq. 20-32 to be

$$\dot{Q}_{max} = C_{min}(T_{h, \text{ in}} - T_{c, \text{ in}})$$
$$= (0.639 \text{ kW/°C})(150 - 20)°C = 83.1 \text{ kW}$$

That is, the maximum possible heat transfer rate in this heat exchanger is 83.1 kW. The heat transfer surface area is

$$A = n(\pi DL) = 8\pi(0.014 \text{ m})(5 \text{ m}) = 1.76 \text{ m}^2$$

Then the NTU of this heat exchanger becomes

$$\text{NTU} = \frac{UA}{C_{min}} = \frac{(310 \text{ W/m}^2 \cdot °C)(1.76 \text{ m}^2)}{639 \text{ W/°C}} = 0.853$$

The effectiveness of this heat exchanger corresponding to $c = 0.764$ and NTU $= 0.853$ is determined from Fig. 20-26c to be

$$\varepsilon = 0.47$$

We could also determine the effectiveness from the third relation in Table 20-4 more accurately but with more labor. Then the actual rate of heat transfer becomes

$$\dot{Q} = \varepsilon \dot{Q}_{max} = (0.47)(83.1 \text{ kW}) = \textbf{39.1 kW}$$

Finally, the outlet temperatures of the cold and the hot fluid streams are determined to be

$$\dot{Q} = C_c(T_{c, \text{ out}} - T_{c, \text{ in}}) \longrightarrow T_{c, \text{ out}} = T_{c, \text{ in}} + \frac{\dot{Q}}{C_c}$$
$$= 20°C + \frac{39.1 \text{ kW}}{0.836 \text{ kW/°C}} = \textbf{66.8°C}$$

$$\dot{Q} = C_h(T_{h, \text{ in}} - T_{h, \text{ out}}) \longrightarrow T_{h, \text{ out}} = T_{h, \text{ in}} - \frac{\dot{Q}}{C_h}$$
$$= 150°C - \frac{39.1 \text{ kW}}{0.639 \text{ kW/°C}} = \textbf{88.8°C}$$

Therefore, the temperature of the cooling water will rise from 20°C to 66.8°C as it cools the hot oil from 150°C to 88.8°C in this heat exchanger.

20-6 ■ SELECTION OF HEAT EXCHANGERS

Heat exchangers are complicated devices, and the results obtained with the simplified approaches presented above should be used with care. For example, we assumed that the overall heat transfer coefficient U is constant throughout the heat exchanger and that the convection heat transfer coefficients can be predicted using the convection correlations. However, it should be kept in mind that the uncertainty in the predicted value of U can even exceed 30 percent. Thus, it is natural to tend to overdesign the heat exchangers in order to avoid unpleasant surprises.

Heat transfer enhancement in heat exchangers is usually accompanied by *increased pressure drop,* and thus *higher pumping power.* Therefore, any gain from the enhancement in heat transfer should be weighed against the cost of the accompanying pressure drop. Also, some thought should be given to which fluid should pass through the tube side and which through the shell side. Usually, the *more viscous fluid is more suitable for the shell side* (larger passage area and thus lower pressure drop) and *the fluid with the higher pressure for the tube* side.

Engineers in industry often find themselves in a position to select heat exchangers to accomplish certain heat transfer tasks. Usually, the goal is to heat or cool a certain fluid at a known mass flow rate and temperature to a desired temperature. Thus, the rate of heat transfer in the prospective heat exchanger is

$$\dot{Q}_{max} = \dot{m}C_p(T_{in} - T_{out})$$

which gives the heat transfer requirement of the heat exchanger before having any idea about the heat exchanger itself.

An engineer going through catalogs of heat exchanger manufacturers will be overwhelmed by the type and number of readily available off-the-shelf heat exchangers. The proper selection depends on several factors.

Heat Transfer Rate

This is the most important quantity in the selection of a heat exchanger. A heat exchanger should be capable of transferring heat at the specified rate in order to achieve the desired temperature change of the fluid at the specified mass flow rate.

Cost

Budgetary limitations usually play the most important role in the selection of heat exchangers, except for some specialized cases where "money is no object." An off-the-shelf heat exchanger has a definite cost advantage over those made to order. However, in some cases, none of the existing heat exchangers will do, and it may be necessary to undertake the expensive and time-consuming task of designing and manufacturing a heat exchanger from scratch to suit the needs. This is often the case when the heat exchanger is an integral part of the overall device to be manufactured.

The operation and maintenance costs of the heat exchanger are also important considerations in assessing the overall cost.

Pumping Power

In a heat exchanger, both fluids are usually forced to flow by pumps or fans that consume electrical power. The annual cost of electricity associated with the operation of the pumps and fans can be determined from

$$\text{Operating cost} = (\text{Pumping power, kW}) \times (\text{Hours of operation, h})$$
$$\times (\text{Price of electricity, \$/kWh})$$

where the pumping power is the total electrical power consumed by the motors of the pumps and fans. For example, a heat exchanger that involves a 1-hp pump and a $\frac{1}{3}$-hp fan (1 hp = 0.746 kW) operating 8 h a day and 5 days a week will consume 2017 kWh of electricity per year, which will cost \$161.4 at an electricity cost of 8 cents/kWh.

Minimizing the pressure drop and the mass flow rate of the fluids will *minimize* the operating cost of the heat exchanger, but it will *maximize* the size of the heat exchanger and thus the initial cost. As a rule of thumb, doubling the mass flow rate will reduce the initial cost by *half* but will increase the pumping power requirements by a factor of roughly *eight*.

Typically, fluid velocities encountered in heat exchangers range between 0.7 and 7 m/s for liquids and between 3 and 30 m/s for gases. Low velocities are helpful in avoiding erosion, tube vibrations, and noise as well as pressure drop.

Size and Weight

Normally, the *smaller* and the *lighter* the heat exchanger, the better it is. This is especially the case in the *automotive* and *aerospace* industries, where size and weight requirements are most stringent. Also, a larger heat exchanger normally carries a higher price tag. The space available for the heat exchanger in some cases limits the length of the tubes that can be used.

Type

The type of heat exchanger to be selected depends primarily on the type of *fluids* involved, the *size* and *weight* limitations, and the presence of any *phase-change* processes. For example, a heat exchanger is suitable to cool a liquid by a gas if the surface area on the gas side is many times that on the liquid side. On the other hand, a plate or shell-and-tube heat exchanger is very suitable for cooling a liquid by another liquid.

Materials

The materials used in the construction of the heat exchanger may be an important consideration in the selection of heat exchangers. For example, the thermal and structural *stress effects* need not be considered at pressures below 15 atm or temperatures below 150°C. But these effects are major considerations above 70 atm or 550°C and seriously limit the acceptable materials of the heat exchanger.

A temperature difference of 50°C or more between the tubes and the shell will probably pose *differential thermal expansion* problems and needs to be considered. In the case of corrosive fluids, we may have to select expensive *corrosion-resistant* materials such as stainless steel or even titanium if we are not willing to replace low-cost heat exchangers frequently.

Other Considerations

There are other considerations in the selection of heat exchangers that may or may not be important, depending on the application. For example, being *leak-tight* is an important consideration when *toxic* or *expensive* fluids are involved. Ease of servicing, low maintenance cost, and safety and reliability are some other important considerations in the selection process. Quietness is one of the primary considerations in the selection of liquid-to-air heat exchangers used in heating and air-conditioning applications.

EXAMPLE 20-10   Installing a Heat Exchanger to Save Energy and Money

In a dairy plant, milk is pasteurized by hot water supplied by a natural gas furnace. The hot water is then discharged to an open floor drain at 80°C at a rate of 15 kg/min. The plant operates 24 h a day and 365 days a year. The furnace has an efficiency of 80 percent, and the cost of the natural gas is $0.40 per therm (1 therm = 105,500 kJ). The average temperature of the cold water entering the furnace throughout the year is 15°C. The drained hot water cannot be returned to the furnace and recirculated, because it is contaminated during the process.

In order to save energy, installation of a water-to-water heat exchanger to preheat the incoming cold water by the drained hot water is proposed. Assuming that the heat exchanger will recover 75 percent of the available heat in the hot water, determine the heat transfer rating of the heat exchanger that needs to be purchased and suggest a suitable type. Also, determine the amount of money this heat exchanger will save the company per year from natural gas savings.

Solution   A water-to-water heat exchanger is to be installed to transfer energy from drained hot water to the incoming cold water to preheat it. The rate of heat transfer in the heat exchanger and the amount of energy and money saved per year are to be determined.

Assumptions   **1** Steady operating conditions exist. **2** The effectiveness of the heat exchanger remains constant.

Properties   We use the specific heat of water at room temperature, $C_p = 4.18$ kJ/kg · °C (Table A-15), and treat it as a constant.

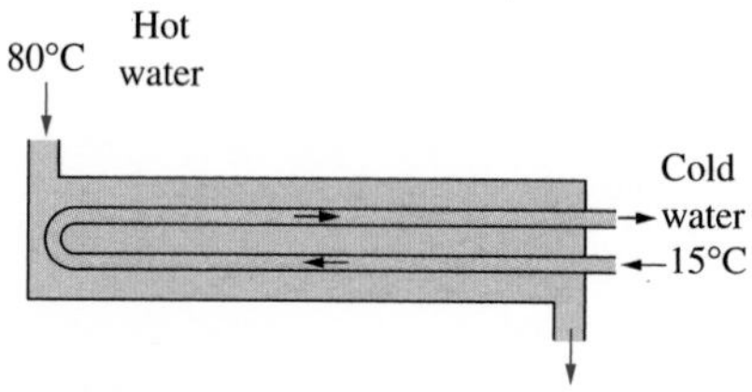

Analysis   A schematic of the prospective heat exchanger is given in Fig. 20-31. The heat recovery from the hot water will be a maximum when it leaves the heat exchanger at the inlet temperature of the cold water. Therefore,

$$\dot{Q}_{max} = \dot{m}_h C_p (T_{h,\,in} - T_{c,\,in})$$
$$= \left(\frac{15}{60}\text{ kg/s}\right)(4.18\text{ kJ/kg} \cdot °C)(80 - 15)°C$$
$$= 67.9\text{ kJ/s}$$

80°C Hot water

Cold water ←15°C

FIGURE 20-31
Schematic for Example 20-10.

That is, the existing hot water stream has the potential to supply heat at a rate of 67.9 kJ/s to the incoming cold water. This value would be approached in a counter-flow heat exchanger with a *very large* heat transfer surface area. A heat exchanger of reasonable size and cost can capture 75 percent of this heat transfer potential. Thus, the heat transfer rating of the prospective heat exchanger must be

$$\dot{Q} = \varepsilon \dot{Q}_{max} = (0.75)(67.9\text{ kJ/s}) = \textbf{50.9 kJ/s}$$

That is, the heat exchanger should be able to deliver heat at a rate of 50.9 kJ/s from the hot to the cold water. An ordinary plate or *shell-and-tube* heat exchanger should be adequate for this purpose, since both sides of the heat exchanger involve the same fluid at comparable flow rates and thus comparable heat transfer coefficients. (Note that if we were heating air with hot water, we would have to specify a heat exchanger that has a large surface area on the air side.)

The heat exchanger will operate 24 h a day and 365 days a year. Therefore, the annual operating hours are

Operating hours = (24 h/day)(365 days/year) = 8760 h/year

Noting that this heat exchanger saves 50.9 kJ of energy per second, the energy saved during an entire year will be

$$\text{Energy saved} = (\text{Heat transfer rate})(\text{Operation time})$$
$$= (50.9 \text{ kJ/s})(8760 \text{ h/year})(3600 \text{ s/h})$$
$$= 1.605 \times 10^9 \text{ kJ/year}$$

The furnace is said to be 80 percent efficient. That is, for each 80 units of heat supplied by the furnace, natural gas with an energy content of 100 units must be supplied to the furnace. Therefore, the energy savings determined above result in fuel savings in the amount of

$$\text{Fuel saved} = \frac{\text{Energy saved}}{\text{Furnace efficiency}} = \frac{1.605 \times 10^9 \text{ kJ/year}}{0.80}\left(\frac{1 \text{ therm}}{105,500 \text{ kJ}}\right)$$
$$= 19,020 \text{ therms/year}$$

Noting that the price of natural gas is $0.40 per therm, the amount of money saved becomes

$$\text{Money saved} = (\text{Fuel saved}) \times (\text{Price of fuel})$$
$$= (19,020 \text{ therms/year})(\$0.40/\text{therm})$$
$$= \mathbf{\$7607/year}$$

Therefore, the installation of the proposed heat exchanger will save the company $7607 a year, and the installation cost of the heat exchanger will probably be paid from the fuel savings in a short time.

20-7 ■ SUMMARY

Heat exchangers are devices that allow the exchange of heat between two fluids without allowing them to mix with each other. Heat exchangers are manufactured in a variety of types, the simplest being the *double-pipe* heat exchanger. In a *parallel-flow* type, both the hot and cold fluids enter the heat exchanger at the same end and move in the same direction, whereas in a *counter-flow* type, the hot and cold fluids enter the heat exchanger at opposite ends and flow in opposite directions. In *compact* heat exchangers, the two fluids move perpendicular to each other, and such a flow configuration is called *cross-flow.* Other common types of heat exchangers in industrial applications are the *plate* and the *shell-and-tube* heat exchangers.

Heat transfer in a heat exchanger usually involves convection in each fluid and conduction through the wall separating the two fluids. In the analysis of heat exchangers, it is convenient to work with an *overall heat transfer coefficient U* or a *total thermal resistance R*, expressed as

$$\frac{1}{UA} = \frac{1}{U_i A_i} = \frac{1}{U_o A_o} = R = \frac{1}{h_i A_i} + R_{\text{wall}} + \frac{1}{h_o A_o}$$

where the subscripts i and o stand for the inner and outer surfaces of the wall that separates the two fluids, respectively. When the wall thickness of the tube is small and the thermal conductivity of the tube material is high, the above relation simplifies to

$$\frac{1}{U} \approx \frac{1}{h_i} + \frac{1}{h_o}$$

where $U \approx U_i \approx U_o$. The effects of fouling on both the inner and the outer surfaces of the tubes of a heat exchanger can be accounted for by

$$\frac{1}{UA} = \frac{1}{U_i A_i} = \frac{1}{U_o A_o} = R = \frac{1}{h_i A_i} + \frac{R_{f,i}}{A_i} + \frac{\ln (D_o/D_i)}{2\pi k L} + \frac{R_{f,o}}{A_o} + \frac{1}{h_o A_o}$$

where $A_i = \pi D_i L$ and $A_o = \pi D_o L$ are the areas of the inner and outer surfaces and $R_{f,i}$ and $R_{f,o}$ are the fouling factors at those surfaces.

In a well-insulated heat exchanger, the rate of heat transfer from the hot fluid is equal to the rate of heat transfer to the cold one. That is,

$$\dot{Q} = \dot{m}_c C_{pc}(T_{c,\,out} - T_{c,\,in}) = C_c(T_{c,\,out} - T_{c,\,in})$$

and

$$\dot{Q} = \dot{m}_h C_{ph}(T_{h,\,in} - T_{h,\,out}) = C_h(T_{h,\,in} - T_{h,\,out})$$

where the subscripts c and h stand for the cold and hot fluids, respectively, and the product of the mass flow rate and the specific heat of a fluid $C = \dot{m} C_p$ is called the *heat capacity rate*.

Of the two methods used in the analysis of heat exchangers, the *log mean temperature difference* (or LMTD) method is best suited for determining the size of a heat exchanger when all the inlet and the outlet temperatures are known. The *effectiveness-NTU* method is best suited to predict the outlet temperatures of the hot and cold fluid streams in a specified heat exchanger. In the LMTD method, the rate of heat transfer is determined from

$$\dot{Q} = UA\,\Delta T_{lm}$$

where

$$\Delta T_{lm} = \frac{\Delta T_1 - \Delta T_2}{\ln (\Delta T_1/\Delta T_2)}$$

is the *log mean temperature difference,* which is the suitable form of the average temperature difference for use in the analysis of heat exchangers. Here ΔT_1 and ΔT_2 represent the temperature differences between the two fluids at the two ends (inlet and outlet) of the heat exchanger. For cross-flow and multipass shell-and-tube heat exchangers, the logarithmic mean temperature difference is related to the counter-flow one $\Delta T_{lm,\,CF}$ as

$$\Delta T_{lm} = F\,\Delta T_{lm,\,CF}$$

where F is the *correction factor,* which depends on the geometry of the heat exchanger and the inlet and outlet temperatures of the hot and cold fluid streams.

The *effectiveness* of a heat exchanger is defined as

$$\varepsilon = \frac{\dot{Q}}{\dot{Q}_{max}} = \frac{\text{Actual heat transfer rate}}{\text{Maximum possible heat transfer rate}}$$

where

$$\dot{Q}_{max} = C_{min}(T_{h,\,in} - T_{c,\,in})$$

and C_{min} is the smaller of $C_h = \dot{m}_h C_{ph}$ and $C_c = \dot{m}_c C_{pc}$. The effectiveness of heat exchangers can be determined from effectiveness relations or charts.

The selection or design of a heat exchanger depends on several factors such as the heat transfer rate, cost, pressure drop, size, weight, construction type, materials, and operating environment.

REFERENCES AND SUGGESTED READING

1. N. Afgan and E. U. Schlunder. *Heat Exchanger: Design and Theory Sourcebook.* Washington DC: McGraw-Hill/Scripta, 1974.

2. R. A. Bowman, A. C. Mueller, and W. M. Nagle. "Mean Temperature Difference in Design." *Transactions of the ASME* 62 (1940), p. 283.

3. A. P. Fraas. *Heat Exchanger Design.* 2d ed. New York: John Wiley & Sons, 1989.

4. K. A. Gardner. "Variable Heat Transfer Rate Correction in Multipass Exchangers, Shell Side Film Controlling." *Transactions of the ASME* 67 (1945), pp. 31–8.

5. J. P. Holman. *Heat Transfer.* 7th ed. New York: McGraw-Hill, 1990.

6. F. P. Incropera and D. P. DeWitt. *Introduction to Heat Transfer.* 2nd ed. New York: John Wiley & Sons, 1990.

7. W. M. Kays and A. L. London. *Compact Heat Exchangers.* 3rd ed. New York: McGraw-Hill, 1984.

8. W. M. Kays and H. C. Perkins. In *Handbook of Heat Transfer,* ed. W. M. Rohsenow and J. P. Hartnett. New York: McGraw-Hill, 1972, Chap. 7.

9. A. C. Mueller. "Heat Exchangers." In *Handbook of Heat Transfer,* ed. W. M. Rohsenow and J. P. Hartnett. New York: McGraw-Hill, 1972, Chap. 18.

10. M. N. Özişik. *Heat Transfer—A Basic Approach.* New York: McGraw-Hill, 1985.

11. E. U. Schlunder. *Heat Exchanger Design Handbook.* Washington, DC: Hemisphere, 1982.

12. *Standards of Tubular Exchanger Manufacturers Association.* New York: Tubular Exchanger Manufacturers Association, latest ed.

13. R. A. Stevens, J. Fernandes, and J. R. Woolf. "Mean Temperature Difference in One, Two, and Three Pass Crossflow Heat Exchangers." *Transactions of the ASME* 79 (1957), pp. 287–97.

14. J. Taborek, G. F. Hewitt, and N. Afgan. *Heat Exchangers: Theory and Practice.* New York: Hemisphere, 1983.

15. L. C. Thomas. *Heat Transfer.* Englewood Cliffs, NJ: Prentice Hall, 1992.

16. G. Walker. *Industrial Heat Exchangers.* Washington, DC: Hemisphere, 1982.

17. F. M. White. *Heat and Mass Transfer.* Reading, MA: Addison-Wesley, 1988.

PROBLEMS*

Types of Heat Exchangers

20-1C   Classify heat exchangers according to flow type and explain the characteristics of each type.

20-2C   Classify heat exchangers according to construction type and explain the characteristics of each type.

20-3C   When is a heat exchanger classified as being compact? Do you think a double-pipe heat exchanger can be classified as a compact heat exchanger?

20-4C   How does a cross-flow heat exchanger differ from a counter-flow one? What is the difference between mixed and unmixed fluids in cross-flow?

20-5C   What is the role of the baffles in a shell-and-tube heat exchanger? How does the presence of baffles affect the heat transfer and the pumping power requirements? Explain.

20-6C   Draw a 1-shell-pass and 6-tube-passes shell-and-tube heat exchanger. What are the advantages and disadvantages of using 6 tube passes instead of just 2 of the same diameter?

20-7C   Draw a 2-shell-passes and 8-tube-passes shell-and-tube heat exchanger. What is the primary reason for using so many tube passes?

20-8C   What is a regenerative heat exchanger? How does a static type of regenerative heat exchanger differ from a dynamic type?

The Overall Heat Transfer Coefficient

20-9C   What are the heat transfer mechanisms involved during heat transfer from the hot to the cold fluid?

20-10C   Under what conditions is the thermal resistance of the tube in a heat exchanger negligible?

20-11C   Consider a double-pipe parallel-flow heat exchanger of length L. The inner and outer diameters of the inner tube are D_1 and D_2, respectively, and the inner diameter of the outer tube is D_3. Explain how you would determine the two heat transfer surface areas A_i and A_o. When is it reasonable to assume $A_i \approx A_o \approx A$?

20-12C   Is the approximation $h_i \approx h_o \approx h$ for the convection heat transfer coefficient in a heat exchanger a reasonable one when the thickness of the tube wall is negligible?

20-13C   Under what conditions can the overall heat transfer coefficient of a heat exchanger be determined from $U = (1/h_i + 1/h_o)^{-1}$?

20-14C   What are the restrictions on the relation $UA = U_iA_i = U_oA_o$ for a heat exchanger? Here A is the heat transfer surface area and U is the overall heat transfer coefficient.

---

*Students are encouraged to answer *all* the concept "C" questions.

20-15C   In a thin-walled double-pipe heat exchanger, when is the approximation $U = h_i$ a reasonable one? Here U is the overall heat transfer coefficient and h_i is the convection heat transfer coefficient inside the tube.

20-16C   What are the common causes of fouling in a heat exchanger? How does fouling affect heat transfer and pressure drop?

20-17C   How is the thermal resistance due to fouling in a heat exchanger accounted for? How do the fluid velocity and temperature affect fouling?

20-18   A double-pipe heat exchanger is constructed of a copper ($k = 380$ W/m · °C) inner tube of internal diameter $D_i = 1.2$ cm and external diameter $D_o = 1.6$ cm and an outer tube of diameter 3.0 cm. The convection heat transfer coefficient is reported to be $h_i = 700$ W/m² · °C on the inner surface of the tube and $h_o = 1400$ W/m² · °C on its outer surface. For a fouling factor $R_{f, i} = 0.0005$ m² · °C/W on the tube side and $R_{f, o} = 0.0002$ m² · °C/W on the shell side, determine (*a*) the thermal resistance of the heat exchanger per unit length and (*b*) the overall heat transfer coefficients U_i and U_o based on the inner and outer surface areas of the tube, respectively.

20-19   Water at an average temperature of 107°C and an average velocity of 3.5 m/s flows through a 5-m-long stainless steel tube ($k = 14.2$ W/m · °C) in a boiler. The inner and outer diameters of the tube are $D_i = 1.0$ cm and $D_o = 1.4$ cm, respectively. If the convection heat transfer coefficient at the outer surface of the tube where boiling is taking place is $h_o = 8400$ W/m² · °C, determine the overall heat transfer coefficient U_i of this boiler based on the inner surface area of the tube.

20-20   Repeat Prob. 20-19, assuming a fouling factor $R_{f, i} = 0.0005$ m² · °C/W on the inner surface of the tube.

20-21   A long thin-walled double-pipe heat exchanger with tube and shell diameters of 1.0 cm and 2.5 cm, respectively, is used to condense refrigerant 134a at 20°C by water. The refrigerant flows through the tube, with a convection heat transfer coefficient of $h_i = 5000$ W/m² · °C. Water flows through the shell at a rate of 0.3 kg/s. Determine the overall heat transfer coefficient of this heat exchanger.   *Answer:* 2100 W/m² · °C

20-22   Repeat Prob. 20-21 by assuming a 2-mm-thick layer of limestone ($k = 1.3$ W/m · °C) forms on the outer surface of the inner tube.

20-23E   Water at an average temperature of 140°F and an average velocity of 8 ft/s flows through a thin-walled $\frac{3}{4}$-in.-diameter tube. The water is cooled by air that flows across the tube with a velocity of $U_\infty = 20$ ft/s at an average temperature of 80°F. Determine the overall heat transfer coefficient.

Analysis of Heat Exchangers

20-24C   What are the common approximations made in the analysis of heat exchangers?

20-25C   Under what conditions is the heat transfer relation

$$\dot{Q} = \dot{m}_c C_{pc}(T_{c, \text{ out}} - T_{c, \text{ in}}) = \dot{m}_h C_{ph} (T_{h, \text{ in}} - T_{h, \text{ out}})$$

valid for a heat exchanger?

20-26C   What is the heat capacity rate? What can you say about the temperature changes of the hot and cold fluids in a heat exchanger if both fluids have the same capacity rate? What does a heat capacity of infinity for a fluid in a heat exchanger mean?

20-27C   Consider a condenser in which steam at a specified temperature is condensed by rejecting heat to the cooling water. If the heat transfer rate in the condenser and the temperature rise of the cooling water is known, explain how the rate of condensation of the steam and the mass flow rate of the cooling water can be determined. Also, explain how the total thermal resistance R of this condenser can be evaluated in this case.

20-28C   Under what conditions will the temperature rise of the cold fluid in a heat exchanger be equal to the temperature drop of the hot fluid?

The Log Mean Temperature Difference Method

20-29C   In the heat transfer relation $\dot{Q} = UA\,\Delta T_{lm}$ for a heat exchanger, what is ΔT_{lm} called? How is it calculated for a parallel-flow and counter-flow heat exchanger?

20-30C   How does the log mean temperature difference for a heat exchanger differ from the arithmetic mean temperature difference (AMTD)? For specified inlet and outlet temperatures, which one of these two quantities is larger?

20-31C   The temperature difference between the hot and cold fluids in a heat exchanger is given to be ΔT_1 at one end and ΔT_2 at the other end. Can the logarithmic temperature difference ΔT_{lm} of this heat exchanger be greater than both ΔT_1 and ΔT_2? Explain.

20-32C   Can the logarithmic mean temperature difference ΔT_{lm} of a heat exchanger be a negative quantity? Explain.

20-33C   Can the outlet temperature of the cold fluid in a heat exchanger be higher than the outlet temperature of the hot fluid in a parallel-flow heat exchanger? How about in a counter-flow heat exchanger? Explain.

20-34C   For specified inlet and outlet temperatures, for what kind of heat exchanger will the ΔT_{lm} be greatest: double-pipe parallel-flow, double-pipe counter-flow, cross-flow, or multipass shell-and-tube heat exchanger?

20-35C   In the heat transfer relation $\dot{Q} = UAF\,\Delta T_{lm}$ for a heat exchanger, what is the quantity F called? What does it represent? Can F be greater than one?

20-36C   When the outlet temperatures of the fluids in a heat exchanger are not known, is it still practical to use the LMTD method? Explain.

20-37C   Explain how the LMTD method can be used to determine the heat transfer surface area of a multipass shell-and-tube heat exchanger when all the necessary information, including the outlet temperatures, is given.

20-38   Steam in the condenser of a steam power plant is to be condensed at a temperature of 50°C (h_{fg} = 2305 kJ/kg) with cooling water (C_p = 4180 J/kg · °C) from a nearby lake, which enters the tubes of the condenser at 18°C and leaves at 27°C. The surface area of the tubes is 58 m², and the overall heat transfer coefficient is 2400 W/m² · °C. Determine the mass flow rate of the cooling water needed and the rate of condensation of the steam in the condenser.   *Answers:* 101 kg/s, 1.65 kg/s

20-39 A double-pipe parallel-flow heat exchanger is to heat water (C_p = 4180 J/kg · °C) from 25°C to 60°C at a rate of 0.2 kg/s. The heating is to be accomplished by geothermal water (C_p = 4310 J/kg · °C) available at 140°C at a mass flow rate of 0.3 kg/s. The inner tube is thin-walled and has a diameter of 0.8 cm. If the overall heat transfer coefficient of the heat exchanger is 550 W/m² · °C, determine the length of the heat exchanger required to achieve the desired heating.

20-40E A 1-shell-pass and 8-tube-passes heat exchanger is used to heat glycerin (C_p = 0.60 Btu/lbm · °F) from 75°F to 140°F by hot water (C_p = 1.0 Btu/lbm · °F) that enters the thin-walled 0.5-in.-diameter tubes at 190°F and leaves at 120°F. The total length of the tubes in the heat exchanger is 500 ft. The convection heat transfer coefficient is 4 Btu/h · ft² · °F on the glycerin (shell) side and 50 Btu/h · ft² · °F on the water (tube) side. Determine the rate of heat transfer in the heat exchanger (*a*) before any fouling occurs and (*b*) after fouling with a fouling factor of 0.002 h · ft² · °F/Btu occurs on the outer surfaces of the tubes.

20-41 A test is conducted to determine the overall heat transfer coefficient in a shell-and-tube oil-to-water heat exchanger that has 24 tubes of internal diameter 1.2 cm and length 2 m in a single shell. Cold water (C_p = 4180 J/kg · °C) enters the tubes at 20°C at a rate of 5 kg/s and leaves at 55°C. Oil (C_p = 2150 J/kg · °C) flows through the shell and is cooled from 120°C to 45°C. Determine the overall heat transfer coefficient U_i of this heat exchanger based on the inner surface area of the tubes.   *Answer:* 13.9 kW/m² · °C

20-42 A double-pipe counter-flow heat exchanger is to cool ethylene glycol (C_p = 2560 J/kg · °C) flowing at a rate of 2 kg/s from 80°C to 40°C by water (C_p = 4180 J/kg · °C) that enters at 20°C and leaves at 55°C. The overall heat transfer coefficient based on the inner surface area of the tube is 250 W/m² · °C. Determine (*a*) the rate of heat transfer, (*b*) the mass flow rate of water, and (*c*) the heat transfer surface area on the inner side of the tube.

20-43 Water (C_p = 4180 J/kg · °C) enters the 2.5-cm-internal-diameter tube of a double-pipe counter-flow heat exchanger at 17°C at a rate of 3 kg/s. It is heated by steam condensing at 120°C (h_{fg} = 2203 kJ/kg) in the shell. If the overall heat transfer coefficient of the heat exchanger is 1500 W/m² · °C, determine the length of the tube required in order to heat the water to 80°C.

20-44 A thin-walled double-pipe counter-flow heat exchanger is to be used to cool oil (C_p = 2200 J/kg · °C) from 150°C to 40°C at a rate of 2 kg/s by water (C_p = 4180 J/kg · °C) that enters at 22°C at a rate of 1.5 kg/s. The diameter of the tube is 2.5 cm, and its length is 6 m. Determine the overall heat transfer coefficient of this heat exchanger.

20-45 Consider a water-to-water double-pipe heat exchanger whose flow arrangement is not known. The temperature measurements indicate that the cold water enters at 20°C and leaves at 50°C, while the hot water enters at 80°C and leaves at 45°C. Do you think this is a parallel-flow or counter-flow heat exchanger? Explain.

20-46 Cold water (C_p = 4180 J/kg · °C) leading to a shower enters a thin-walled double-pipe counter-flow heat exchanger at 15°C at a rate of 0.25 kg/s and is heated to 45°C by hot water (C_p = 4190 J/kg · °C) that enters at 100°C at a rate of 3 kg/s. If the overall heat transfer coefficient is 950 W/m² · °C, determine the rate of heat transfer and the heat transfer surface area of the heat exchanger.

20-47 Engine oil ($C_p = 2100$ J/kg · °C) is to be heated from 20°C to 60°C at a rate of 0.3 kg/s in a 2-cm-diameter thin-walled copper tube by condensing steam outside at a temperature of 130°C ($h_{fg} = 2174$ kJ/kg). For an overall heat transfer coefficient of 650 W/m² · °C, determine the rate of heat transfer and the length of the tube required to achieve it.   *Answers:* 25.2 kW, 7.0 m

20-48E Geothermal water ($C_p = 1.03$ Btu/lbm · °F) is to be used as the heat source to supply heat to the hydronic heating system of a house at a rate of 30 Btu/s in a double-pipe counter-flow heat exchanger. Water ($C_p = 1.0$ Btu/lbm · °F) is heated from 140°F to 200°F in the heat exchanger as the geothermal water is cooled from 250°F to 180°F. Determine the mass flow rate of each fluid and the total thermal resistance of this heat exchanger.

20-49 Glycerin ($C_p = 2400$ J/kg · °C) at 20°C and 0.3 kg/s is to be heated by ethylene glycol ($C_p = 2500$ J/kg · °C) at 60°C in a thin-walled double-pipe parallel-flow heat exchanger. The temperature difference between the two fluids is 15°C at the outlet of the heat exchanger. If the overall heat transfer coefficient is 240 W/m² · °C and the heat transfer surface area is 7.6 m², determine (*a*) the rate of heat transfer, (*b*) the outlet temperature of the glycerin, and (*c*) the mass flow rate of the ethylene glycol.

20-50 Air ($C_p = 1005$ J/kg · °C) is to be preheated by hot exhaust gases in a cross-flow heat exchanger before it enters the furnace. Air enters the heat exchanger at 95 kPa and 20°C at a rate of 0.8 m³/s. The combustion gases ($C_p = 1100$ J/kg · °C) enter at 180°C at a rate of 1.1 kg/s and leave at 95°C. The product of the overall heat transfer coefficient and the heat transfer surface area is $AU = 1200$ W/°C. Assuming both fluids to be unmixed, determine the rate of heat transfer and the outlet temperature of the air.

20-51 A shell-and-tube heat exchanger with 2-shell passes and 12-tube passes is used to heat water ($C_p = 4180$ J/kg · °C) in the tubes from 20°C to 70°C at a rate of 4.5 kg/s. Heat is supplied by hot oil ($C_p = 2300$ J/kg · °C) that enters the shell side at 170°C at a rate of 10 kg/s. For a tube-side overall heat transfer coefficient of 600 W/m² · °C, determine the heat transfer surface area on the tube side.   *Answer:* 15 m²

20-52 Repeat Prob. 20-51 for a mass flow rate of 2 kg/s for water.

20-53 A shell-and-tube heat exchanger with 2-shell passes and 8-tube passes is used to heat ethyl alcohol ($C_p = 2670$ J/kg · °C) in the tubes from 25°C to 70°C at a rate of 2.1 kg/s. The heating is to be done by water ($C_p = 4190$ J/kg · °C) that enters the shell side at 95°C and leaves at 45°C. If the overall heat transfer coefficient is 800 W/m² · °C, determine the heat transfer surface area of the heat exchanger.

20-54 A shell-and-tube heat exchanger with 2-shell passes and 12-tube passes is used to heat water ($C_p = 4180$ J/kg · °C) with ethylene glycol ($C_p = 2680$ J/kg · °C). Water enters the tubes at 22°C at a rate of 0.8 kg/s and leaves at 70°C. Ethylene glycol enters the shell at 110°C and leaves at 60°C. If the overall heat transfer coefficient based on the tube side is 280 W/m² · °C, determine the rate of heat transfer and the heat transfer surface area on the tube side.

20-55E Steam is to be condensed on the shell side of a 1-shell-pass and 8-tube-passes condenser, with 50 tubes in each pass at 90°F (h_{fg} = 1043 Btu/lbm) at a rate of 20 lbm/s. Cooling water (C_p = 1.0 Btu/lbm · °F) enters the tubes at 60°F and leaves at 73°F. The tubes are thin-walled and have a diameter of 3/4 in. and length of 5 ft per pass. If the overall heat transfer coefficient is 600 Btu/h · ft^2 · °F, determine (*a*) the rate of heat transfer, (*b*) the rate of condensation of steam, and (*c*) the mass flow rate of cold water.

20-56 A shell-and-tube heat exchanger with 1-shell pass and 20-tube passes is used to heat glycerin (C_p = 2480 J/kg · °C) in the shell, with hot water in the tubes. The tubes are thin-walled and have a diameter of 1.5 cm and length of 2 m per pass. The water enters the tubes at 100°C at a rate of 5 kg/s and leaves at 55°C. The glycerin enters the shell at 15°C and leaves at 55°C. Determine the mass flow rate of the glycerin and the overall heat transfer coefficient of the heat exchanger.

The Effectiveness–NTU Method

20-57C Under what conditions is the effectiveness-NTU method definitely preferred over the LMTD method in heat exchanger analysis?

20-58C What does the effectiveness of a heat exchanger represent? Can effectiveness be greater than one? On what factors does the effectiveness of a heat exchanger depend?

20-59C For a specified fluid pair, inlet temperatures, and mass flow rates, what kind of heat exchanger will have the highest effectiveness: double-pipe parallel-flow, double-pipe counter-flow, cross-flow, or multipass shell-and-tube heat exchanger?

20-60C Explain how you can evaluate the outlet temperatures of the cold and hot fluids in a heat exchanger after its effectiveness is determined.

20-61C Can the temperature of the hot fluid drop below the inlet temperature of the cold fluid at any location in a heat exchanger? Explain.

20-62C Can the temperature of the cold fluid rise above the inlet temperature of the hot fluid at any location in a heat exchanger? Explain.

20-63C Consider a heat exchanger in which both fluids have the same specific heats but different mass flow rates. Which fluid will experience a larger temperature change: the one with the lower or higher mass flow rate?

20-64C Explain how the maximum possible heat transfer rate $\dot{Q}_{max}$ in a heat exchanger can be determined when the mass flow rates, specific heats, and the inlet temperatures of the two fluids are specified. Does the value of $\dot{Q}_{max}$ depend on the type of the heat exchanger?

20-65C Consider two double-pipe counter-flow heat exchangers that are identical except that one is twice as long as the other one. Which heat exchanger is more likely to have a higher effectiveness?

20-66C Consider a double-pipe counter-flow heat exchanger. In order to enhance heat transfer, the length of the heat exchanger is now doubled. Do you think its effectiveness will also double?

20-67C Consider a shell-and-tube water-to-water heat exchanger with identical mass flow rates for both the hot and cold water streams. Now the mass flow rate of the cold water is reduced by half. Will the effectiveness of this heat exchanger increase, decrease, or remain the same as a result of this modification? Explain. Assume the overall heat transfer coefficient and the inlet temperatures remain the same.

20-68C Under what conditions can a counter-flow heat exchanger have an effectiveness of one? What would your answer be for a parallel-flow heat exchanger?

20-69C How is the NTU of a heat exchanger defined? What does it represent? Is a heat exchanger with a very large NTU (say, 10) necessarily a good one to buy?

20-70C Consider a heat exchanger that has an NTU of 4. Someone proposes to double the size of the heat exchanger and thus double the NTU to 8 in order to increase the effectiveness of the heat exchanger and thus save energy. Would you support this proposal?

20-71C Consider a heat exchanger that has an NTU of 0.1. Someone proposes to triple the size of the heat exchanger and thus triple the NTU to 0.3 in order to increase the effectiveness of the heat exchanger and thus save energy. Would you support this proposal?

20-72 Air ($C_p = 1005$ J/kg · °C) enters a cross-flow heat exchanger at 12°C at a rate of 3 kg/s, where it is heated by a hot water stream ($C_p = 4190$ J/kg · °C) that enters the heat exchanger at 90°C at a rate of 1 kg/s. Determine the maximum heat transfer rate and the outlet temperatures of the cold and the hot water streams for that case.

20-73 Hot oil ($C_p = 2200$ J/kg · °C) is to be cooled by water ($C_p = 4180$ J/kg · °C) in a 2-shell-passes and 12-tube-passes heat exchanger. The tubes are thin-walled and are made of copper with a diameter of 1.8 cm. The length of each tube pass in the heat exchanger is 3 m, and the overall heat transfer coefficient is 340 W/m² · °C. Water flows through the tubes at a total rate of 0.1 kg/s, and the oil through the shell at a rate of 0.2 kg/s. The water and the oil enter at temperatures 18°C and 160°C, respectively. Determine the rate of heat transfer in the heat exchanger and the outlet temperatures of the water and the oil.   *Answers:* 36.2 kW, 104.6°C, 77.7°C

20-74 Consider an oil-to-oil double-pipe heat exchanger whose flow arrangement is not known. The temperature measurements indicate that the cold oil enters at 20°C and leaves at 55°C, while the hot oil enters at 80°C and leaves at 45°C. Do you think this is a parallel-flow or counter-flow heat exchanger? Why? Assuming the mass flow rates of both fluids to be identical, determine the effectiveness of this heat exchanger.

20-75E Hot water enters a double-pipe counter-flow water-to-oil heat exchanger at 200°F and leaves at 100°F. Oil enters at 70°F and leaves at 130°F. Determine which fluid has the smaller heat capacity rate and calculate the effectiveness of this heat exchanger.

20-76 A thin-walled double-pipe parallel-flow heat exchanger is used to heat a chemical whose specific heat is 1800 J/kg · °C with hot water ($C_p = 4180$ J/kg · °C). The chemical enters at 20°C at a rate of 3 kg/s, while the water enters at 110°C at a rate of 2 kg/s. The heat transfer surface area of the heat exchanger is 7 m^2 and the overall heat transfer coefficient is 1200 W/m^2 · °C. Determine the outlet temperatures of the chemical and the water.

20-77 A cross-flow air-to-water heat exchanger with an effectiveness of 0.65 is used to heat water ($C_p = 4180$ J/kg · °C) with hot air ($C_p = 1010$ J/kg · °C). Water enters the heat exchanger at 20°C at a rate of 4 kg/s, while air enters at 100°C at a rate of 9 kg/s. If the overall heat transfer coefficient based on the water side is 260 W/m^2 · °C, determine the heat transfer surface area of the heat exchanger on the water side. Assume both fluids are unmixed.   *Answer:* 52.4 m^2

20-78 Water ($C_p = 4180$ J/kg · °C) enters the 2.5-cm-internal-diameter tube of a double-pipe counter-flow heat exchanger at 17°C at a rate of 3 kg/s. Water is heated by steam condensing at 120°C ($h_{fg} = 2203$ kJ/kg) in the shell. If the overall heat transfer coefficient of the heat exchanger is 900 W/m^2 · °C, determine the length of the tube required in order to heat the water to 80°C using (*a*) the LMTD method and (*b*) the ε-NTU method.

20-79 Ethanol is vaporized at 78°C ($h_{fg} = 846$ kJ/kg) in a double-pipe parallel-flow heat exchanger at a rate of 0.03 kg/s by hot oil ($C_p = 2200$ J/kg · °C) that enters at 120°C. If the heat transfer surface area and the overall heat transfer coefficients are 7.8 m^2 and 210 W/m^2 · °C, respectively, determine the outlet temperature and the mass flow rate of oil using (*a*) the LMTD method and (*b*) the ε-NTU method.

20-80 Water ($C_p = 4180$ J/kg · °C) is to be heated by solar-heated hot air ($C_p = 1010$ J/kg · °C) in a double-pipe counter-flow heat exchanger. Air enters the heat exchanger at 90°C at a rate of 0.3 kg/s, while water enters at 22°C at a rate of 0.1 kg/s. The overall heat transfer coefficient based on the inner side of the tube is given to be 80 W/m^2 · °C. The length of the tube is 12 m and the internal diameter of the tube is 1.2 cm. Determine the outlet temperatures of the water and the air.

20-81E A thin-walled double-pipe heat exchanger is to be used to cool oil ($C_p = 0.525$ Btu/lbm · °F) from 300°F to 105°F at a rate of 5 lbm/s by water ($C_p = 1.0$ Btu/lbm · °F) that enters at 70°F at a rate of 3 lbm/s. The diameter of the tube is 1 in. and its length is 20 ft. Determine the overall heat transfer coefficient of this heat exchanger using (*a*) the LMTD method and (*b*) the ε-NTU method.

20-82 Cold water ($C_p = 4180$ J/kg · °C) leading to a shower enters a thin-walled double-pipe counter-flow heat exchanger at 15°C at a rate of 0.25 kg/s and is heated to 45°C by hot water ($C_p = 4190$ J/kg · °C) that enters at 100°C at a rate of 3 kg/s. If the overall heat transfer coefficient is 950 W/m^2 · °C, determine the rate of heat transfer and the heat transfer surface area of the heat exchanger using the ε-NTU method.
    Answers: 31.35 kW, 0.482 m^2

20-83 Glycerin ($C_p = 2400$ J/kg · °C) at 20°C and 0.3 kg/s is to be heated by ethylene glycol ($C_p = 2500$ J/kg · °C) at 60°C and the same mass flow rate in a thin-walled double-pipe parallel-flow heat exchanger. If the overall heat transfer coefficient is 240 W/m^2 · °C and the heat transfer surface area is 7.6 m^2, determine (*a*) the rate of heat transfer and (*b*) the outlet temperatures of the glycerin and the glycol.

20-84 A cross-flow heat exchanger consists of 40 thin-walled tubes of 1-cm diameter located in a duct of 1 m × 1 m cross-section. There are no fins attached to the tubes. Cold water (C_p = 4180 J/kg · °C) enters the tubes at 18°C with an average velocity of 3 m/s, while hot air (C_p = 1010 J/kg · °C) enters the channel at 130°C and 105 kPa at an average velocity of 12 m/s. If the overall heat transfer coefficient is 80 W/m² · °C, determine the outlet temperatures of both fluids and the rate of heat transfer.

20-85 A shell-and-tube heat exchanger with 2-shell passes and 8-tube passes is used to heat ethyl alcohol (C_p = 2670 J/kg · °C) in the tubes from 25°C to 70°C at a rate of 2.1 kg/s. The heating is to be done by water (C_p = 4190 J/kg · °C) that enters the shell at 95°C and leaves at 45°C. If the overall heat transfer coefficient is 800 W/m² · °C, determine the heat transfer surface area of the heat exchanger using (*a*) the LMTD method and (*b*) the ε-NTU method. *Answer:* 17.4 m²

20-86 Steam is to be condensed on the shell side of a 1-shell-pass and 8-tube-passes condenser, with 50 tubes in each pass, at 30°C (h_{fg} = 2430 kJ/kg). Cooling water (C_p = 4180 J/kg · °C) enters the tubes at 15°C at a rate of 1800 kg/h. The tubes are thin-walled, and have a diameter of 1.5 cm and length of 2 m per pass. If the overall heat transfer coefficient is 3000 W/m² · °C, determine (*a*) the rate of heat transfer and (*b*) the rate of condensation of steam.

20-87 Cold water (C_p = 4180 J/kg · °C) enters the tubes of a heat exchanger with 2-shell-passes and 20-tube-passes at 20°C at a rate of 3 kg/s, while hot oil (C_p = 2200 J/kg · °C) enters the shell at 130°C at the same mass flow rate. The overall heat transfer coefficient based on the outer surface of the tube is 300 W/m² · °C and the heat transfer surface area on that side is 20 m². Determine the rate of heat transfer using (*a*) the LMTD method and (*b*) the ε-NTU method.

Selection of Heat Exchangers

20-88C A heat exchanger is to be selected to cool a hot liquid chemical at a specified rate to a specified temperature. Explain the steps involved in the selection process.

20-89C There are two heat exchangers that can meet the heat transfer requirements of a facility. One is smaller and cheaper but requires a larger pump, while the other is larger and more expensive but has a smaller pressure drop and thus requires a smaller pump. Both heat exchangers have the same life expectancy and meet all other requirements. Explain which heat exchanger you would choose under what conditions.

20-90C There are two heat exchangers that can meet the heat transfer requirements of a facility. Both have the same pumping power requirements, the same useful life, and the same price tag. But one is heavier and larger in size. Under what conditions would you choose the smaller one?

20-91 A heat exchanger is to cool oil (C_p = 2200 J/kg · °C) at a rate of 20 kg/s from 120°C to 50°C by air. Determine the heat transfer rating of the heat exchanger and propose a suitable type.

20-92 A shell-and-tube process heater is to be selected to heat water (C_p = 4190 J/kg · °C) from 20°C to 90°C by steam flowing on the shell side. The heat transfer load of the heater is 600 kW. If the inner diameter of the tubes is 1 cm and the velocity of water is not to exceed 3 m/s, determine how many tubes need to be used in the heat exchanger.

20-93 The condenser of a large power plant is to remove 500 MW of heat from steam condensing at 30°C (h_{fg} = 2430 kJ/kg). The cooling is to be accomplished by cooling water (C_p = 4180 J/kg · °C) from a nearby river, which enters the tubes at 18°C and leaves at 26°C. The tubes of the heat exchanger have an internal diameter of 2 cm, and the overall heat transfer coefficient is 3500 W/m² · °C. Determine the total length of the tubes required in the condenser. What type of heat exchanger is suitable for this task?
    Answer: 312.3 km

20-94 Repeat Prob. 20-93 for a heat transfer load of 300 MW.

Review Problems

20-95 Hot oil is to be cooled in a multipass shell-and-tube heat exchanger by water. The oil flows through the shell, with a heat transfer coefficient of h_o = 35 W/m² · °C, and the water flows through the tube with an average velocity of 3 m/s. The tube is made of brass (k = 110 W/m · °C) with internal and external diameters of 1.3 cm and 1.5 cm, respectively. Using water properties at 25°C, determine the overall heat transfer coefficient of this heat exchanger based on the inner surface.

20-96 Repeat Prob. 20-95 by assuming a fouling factor $R_{f,o}$ = 0.0004 m² · °C/W on the outer surface of the tube.

20-97 Cold water (C_p = 4180 J/kg · °C) enters the tubes of a heat exchanger with 2-shell passes and 20-tube passes at 20°C at a rate of 3 kg/s, while hot oil (C_p = 2200 J/kg · °C) enters the shell at 130°C at the same mass flow rate and leaves at 60°C. If the overall heat transfer coefficient based on the outer surface of the tube is 300 W/m² · °C, determine (*a*) the rate of heat transfer and (*b*) the heat transfer surface area on the outer side of the tube.
    Answers: (*a*) 462 kW, (*b*) 29.2 m²

20-98E Water (C_p = 1.0 Btu/lbm · °F) is to be heated by solar-heated hot air (C_p = 0.24 Btu/lbm · °F) in a double-pipe counter-flow heat exchanger. Air enters the heat exchanger at 190°F at a rate of 0.7 lbm/s and leaves at 120°F. Water enters at 70°F at a rate of 0.2 lbm/s. The overall heat transfer coefficient based on the inner side of the tube is given to be 20 Btu/h · ft² · °F. Determine the length of the tube required for a tube internal diameter of 0.5 in.

20-99 By taking the limit as $\Delta T_2 \longrightarrow \Delta T_1$, show that when $\Delta T_1 = \Delta T_2$ for a heat exchanger, the ΔT_{lm} relation reduces to $\Delta T_{lm} = \Delta T_1 = \Delta T_2$.

20-100 The condenser of a room air conditioner is designed to reject heat at a rate of 15,000 kJ/h from Refrigerant-134a as the refrigerant is condensed at a temperature of 40°C. Air (C_p = 1005 J/kg · °C) flows across the finned condenser coils, entering at 25°C and leaving at 35°C. If the overall heat transfer coefficient based on the refrigerant side is 150 W/m² · °C, determine the heat transfer area on the refrigerant side.   *Answer:* 3.05 m²

20-101 Air (C_p = 1005 J/kg · °C) is to be preheated by hot exhaust gases in a cross-flow heat exchanger before it enters the furnace. Air enters the heat exchanger at 95 kPa and 20°C at a rate of 0.8 m³/s. The combustion gases (C_p = 1100 J/kg · °C) enter at 180°C at a rate of 1.1 kg/s and leave at 95°C.

The product of the overall heat transfer coefficient and the heat transfer surface area is $AU = 1200$ W/°C. Assuming both fluids to be unmixed, determine the rate of heat transfer.

20-102 In a chemical plant, a certain chemical is heated by hot water supplied by a natural gas furnace. The hot water ($C_p = 4180$ J/kg · °C) is then discharged at 60°C at a rate of 8 kg/min. The plant operates 8 h a day, 5 days a week, 52 weeks a year. The furnace has an efficiency of 78 percent, and the cost of the natural gas is $0.54 per therm (1 therm = 100,000 Btu = 105,500 kJ). The average temperature of the cold water entering the furnace throughout the year is 14°C. In order to save energy, it is proposed to install a water-to-water heat exchanger to preheat the incoming cold water by the drained hot water. Assuming that the heat exchanger will recover 72 percent of the available heat in the hot water, determine the heat transfer rating of the heat exchanger that needs to be purchased and suggest a suitable type. Also, determine the amount of money this heat exchanger will save the company per year from natural gas savings.

Computer, Design, and Essay Problems

20-103 Write an interactive computer program that will give the effectiveness of a heat exchanger and the outlet temperatures of both the hot and cold fluids when the type of fluids, the inlet temperatures, the mass flow rates, the heat transfer surface area, the overall heat transfer coefficient, and the type of heat exchanger are specified. The program should allow the user to select from the fluids water, engine oil, glycerin, ethyl alcohol, and ammonia. Assume constant specific heats at about room temperature.

20-104 Repeat the problem above, accounting for the variation of specific heats with temperature.

20-105 Water flows through a shower head steadily at a rate of 8 kg/min. The water is heated in an electric water heater from 15°C to 45°C. In an attempt to conserve energy, it is proposed to pass the drained warm water at a temperature of 38°C through a heat exchanger to preheat the incoming cold water. Design a heat exchanger that is suitable for this task, and discuss the potential savings in energy and money for your area.

20-106 Open the engine compartment of your car and search for heat exchangers. How many do you have? What type are they? Why do you think those specific types are selected? If you were redesigning the car, would you use different kinds? Explain.

20-107 Write an essay on the static and dynamic types of regenerative heat exchangers and compile information about the manufacturers of such heat exchangers. Choose a few models by different manufacturers and compare their costs and performance.

Property Tables and Charts (SI Units)

TABLE A-1

Molar mass, gas constant, and critical-point properties

Substance	Formula	Molar mass, M kg/kmol	Gas constant, R kJ/kg · K*	Critical-point properties		
				Temperature, K	Pressure, MPa	Volume, m³/kmol
Air	—	28.97	0.2870	132.5	3.77	0.0883
Ammonia	NH_3	17.03	0.4882	405.5	11.28	0.0724
Argon	Ar	39.948	0.2081	151	4.86	0.0749
Benzene	C_6H_6	78.115	0.1064	562	4.92	0.2603
Bromine	Br_2	159.808	0.0520	584	10.34	0.1355
n-Butane	C_4H_{10}	58.124	0.1430	425.2	3.80	0.2547
Carbon dioxide	CO_2	44.01	0.1889	304.2	7.39	0.0943
Carbon monoxide	CO	28.011	0.2968	133	3.50	0.0930
Carbon tetrachloride	CCl_4	153.82	0.05405	556.4	4.56	0.2759
Chlorine	Cl_2	70.906	0.1173	417	7.71	0.1242
Chloroform	$CHCl_3$	119.38	0.06964	536.6	5.47	0.2403
Dichlorodifluoromethane (R-12)	CCl_2F_2	120.91	0.06876	384.7	4.01	0.2179
Dichlorofluoromethane (R-21)	$CHCl_2F$	102.92	0.08078	451.7	5.17	0.1973
Ethane	C_2H_6	30.070	0.2765	305.5	4.48	0.1480
Ethyl alcohol	C_2H_5OH	46.07	0.1805	516	6.38	0.1673
Ethylene	C_2H_4	28.054	0.2964	282.4	5.12	0.1242
Helium	He	4.003	2.0769	5.3	0.23	0.0578
n-Hexane	C_6H_{14}	86.179	0.09647	507.9	3.03	0.3677
Hydrogen (normal)	H_2	2.016	4.1240	33.3	1.30	0.0649
Krypton	Kr	83.80	0.09921	209.4	5.50	0.0924
Methane	CH_4	16.043	0.5182	191.1	4.64	0.0993
Methyl alcohol	CH_3OH	32.042	0.2595	513.2	7.95	0.1180
Methyl chloride	CH_3Cl	50.488	0.1647	416.3	6.68	0.1430
Neon	Ne	20.183	0.4119	44.5	2.73	0.0417
Nitrogen	N_2	28.013	0.2968	126.2	3.39	0.0899
Nitrous oxide	N_2O	44.013	0.1889	309.7	7.27	0.0961
Oxygen	O_2	31.999	0.2598	154.8	5.08	0.0780
Propane	C_3H_8	44.097	0.1885	370	4.26	0.1998
Propylene	C_3H_6	42.081	0.1976	365	4.62	0.1810
Sulfur dioxide	SO_2	64.063	0.1298	430.7	7.88	0.1217
Tetrafluoroethane (R-134a)	CF_3CH_2F	102.03	0.08149	374.3	4.067	0.1847
Trichlorofluoromethane (R-11)	CCl_3F	137.37	0.06052	471.2	4.38	0.2478
Water	H_2O	18.015	0.4615	647.3	22.09	0.0568
Xenon	Xe	131.30	0.06332	289.8	5.88	0.1186

*The unit kJ/kg · K is equivalent to kPa · m³/kg · K. The gas constant is calculated from $R = R_u/M$, where $R_u = 8.314$ kJ/kmol · K and M is the molar mass.

Source: K. A. Kobe and R. E. Lynn Jr., *Chemical Review* 52 (1953), pp. 117–236; and ASHRAE, *Handbook of Fundamentals* (Atlanta, GA: American Society of Heating, Refrigerating and Air-Conditioning Engineers, Inc., 1993), pp. 16.4 and 36.1.

TABLE A-2

Ideal-gas specific heats of various common gases
(*a*) At 300 K

Gas	Formula	Gas constant, R kJ/kg · K	C_p kJ/kg · K	C_v kJ/kg · K	k
Air	—	0.2870	1.005	0.718	1.400
Argon	Ar	0.2081	0.5203	0.3122	1.667
Butane	C_4H_{10}	0.1433	1.7164	1.5734	1.091
Carbon dioxide	CO_2	0.1889	0.846	0.657	1.289
Carbon monoxide	CO	0.2968	1.040	0.744	1.400
Ethane	C_2H_6	0.2765	1.7662	1.4897	1.186
Ethylene	C_2H_4	0.2964	1.5482	1.2518	1.237
Helium	He	2.0769	5.1926	3.1156	1.667
Hydrogen	H_2	4.1240	14.307	10.183	1.405
Methane	CH_4	0.5182	2.2537	1.7354	1.299
Neon	Ne	0.4119	1.0299	0.6179	1.667
Nitrogen	N_2	0.2968	1.039	0.743	1.400
Octane	C_8H_{18}	0.0729	1.7113	1.6385	1.044
Oxygen	O_2	0.2598	0.918	0.658	1.395
Propane	C_3H_8	0.1885	1.6794	1.4909	1.126
Steam	H_2O	0.4615	1.8723	1.4108	1.327

Note: The unit kJ/kg · K is equivalent to kJ/kg · °C.

Source: Gordon J. Van Wylen and Richard E. Sonntag, *Fundamentals of Classical Thermodynamics,* English/SI Version, 3rd ed. (New York: John Wiley & Sons, 1986), p. 687, Table A.8SI.

TABLE A-2

Ideal-gas specific heats of various common gases (*Continued*)
(*b*) At various temperatures

Temperature, K	C_p kJ/kg · K	C_v kJ/kg · K	k	C_p kJ/kg · K	C_v kJ/kg · K	k	C_p kJ/kg · K	C_v kJ/kg · K	k
	Air			**Carbon dioxide, CO_2**			**Carbon monoxide, CO**		
250	1.003	0.716	1.401	0.791	0.602	1.314	1.039	0.743	1.400
300	1.005	0.718	1.400	0.846	0.657	1.288	1.040	0.744	1.399
350	1.008	0.721	1.398	0.895	0.706	1.268	1.043	0.746	1.398
400	1.013	0.726	1.395	0.939	0.750	1.252	1.047	0.751	1.395
450	1.020	0.733	1.391	0.978	0.790	1.239	1.054	0.757	1.392
500	1.029	0.742	1.387	1.014	0.825	1.229	1.063	0.767	1.387
550	1.040	0.753	1.381	1.046	0.857	1.220	1.075	0.778	1.382
600	1.051	0.764	1.376	1.075	0.886	1.213	1.087	0.790	1.376
650	1.063	0.776	1.370	1.102	0.913	1.207	1.100	0.803	1.370
700	1.075	0.788	1.364	1.126	0.937	1.202	1.113	0.816	1.364
750	1.087	0.800	1.359	1.148	0.959	1.197	1.126	0.829	1.358
800	1.099	0.812	1.354	1.169	0.980	1.193	1.139	0.842	1.353
900	1.121	0.834	1.344	1.204	1.015	1.186	1.163	0.866	1.343
1000	1.142	0.855	1.336	1.234	1.045	1.181	1.185	0.888	1.335
	Hydrogen, H_2			**Nitrogen, N_2**			**Oxygen, O_2**		
250	14.051	9.927	1.416	1.039	0.742	1.400	0.913	0.653	1.398
300	14.307	10.183	1.405	1.039	0.743	1.400	0.918	0.658	1.395
350	14.427	10.302	1.400	1.041	0.744	1.399	0.928	0.668	1.389
400	14.476	10.352	1.398	1.044	0.747	1.397	0.941	0.681	1.382
450	14.501	10.377	1.398	1.049	0.752	1.395	0.956	0.696	1.373
500	14.513	10.389	1.397	1.056	0.759	1.391	0.972	0.712	1.365
550	14.530	10.405	1.396	1.065	0.768	1.387	0.988	0.728	1.358
600	14.546	10.422	1.396	1.075	0.778	1.382	1.003	0.743	1.350
650	14.571	10.447	1.395	1.086	0.789	1.376	1.017	0.758	1.343
700	14.604	10.480	1.394	1.098	0.801	1.371	1.031	0.771	1.337
750	14.645	10.521	1.392	1.110	0.813	1.365	1.043	0.783	1.332
800	14.695	10.570	1.390	1.121	0.825	1.360	1.054	0.794	1.327
900	14.822	10.698	1.385	1.145	0.849	1.349	1.074	0.814	1.319
1000	14.983	10.859	1.380	1.167	0.870	1.341	1.090	0.830	1.313

Source: Kenneth Wark, *Thermodynamics,* 4th ed. (New York: McGraw-Hill, 1983), p. 783, Table A-4M. Originally published in *Tables of Thermal Properties of Gases,* NBS Circular 564, 1955.

TABLE A-2

Ideal-gas specific heats of various common gases (*Concluded*)
(*c*) As a function of temperature

$$\bar{C}_p = a + bT + cT^2 + dT^3$$
(T in K, $\bar{C}_p$ in kJ/kmol · K)

Substance	Formula	a	b	c	d	Temperature range, K	% error Max.	% error Avg.
Nitrogen	N_2	28.90	-0.1571×10^{-2}	0.8081×10^{-5}	-2.873×10^{-9}	273–1800	0.59	0.34
Oxygen	O_2	25.48	1.520×10^{-2}	-0.7155×10^{-5}	1.312×10^{-9}	273–1800	1.19	0.28
Air	—	28.11	0.1967×10^{-2}	0.4802×10^{-5}	-1.966×10^{-9}	273–1800	0.72	0.33
Hydrogen	H_2	29.11	-0.1916×10^{-2}	0.4003×10^{-5}	-0.8704×10^{-9}	273–1800	1.01	0.26
Carbon monoxide	CO	28.16	0.1675×10^{-2}	0.5372×10^{-5}	-2.222×10^{-9}	273–1800	0.89	0.37
Carbon dioxide	CO_2	22.26	5.981×10^{-2}	-3.501×10^{-5}	7.469×10^{-9}	273–1800	0.67	0.22
Water vapor	H_2O	32.24	0.1923×10^{-2}	1.055×10^{-5}	-3.595×10^{-9}	273–1800	0.53	0.24
Nitric oxide	NO	29.34	-0.09395×10^{-2}	0.9747×10^{-5}	-4.187×10^{-9}	273–1500	0.97	0.36
Nitrous oxide	N_2O	24.11	5.8632×10^{-2}	-3.562×10^{-5}	10.58×10^{-9}	273–1500	0.59	0.26
Nitrogen dioxide	NO_2	22.9	5.715×10^{-2}	-3.52×10^{-5}	7.87×10^{-9}	273–1500	0.46	0.18
Ammonia	NH_3	27.568	2.5630×10^{-2}	0.99072×10^{-5}	-6.6909×10^{-9}	273–1500	0.91	0.36
Sulfur	S_2	27.21	2.218×10^{-2}	-1.628×10^{-5}	3.986×10^{-9}	273–1800	0.99	0.38
Sulfur dioxide	SO_2	25.78	5.795×10^{-2}	-3.812×10^{-5}	8.612×10^{-9}	273–1800	0.45	0.24
Sulfur trioxide	SO_3	16.40	14.58×10^{-2}	-11.20×10^{-5}	32.42×10^{-9}	273–1300	0.29	0.13
Acetylene	C_2H_2	21.8	9.2143×10^{-2}	-6.527×10^{-5}	18.21×10^{-9}	273–1500	1.46	0.59
Benzene	C_6H_6	-36.22	48.475×10^{-2}	-31.57×10^{-5}	77.62×10^{-9}	273–1500	0.34	0.20
Methanol	CH_4O	19.0	9.152×10^{-2}	-1.22×10^{-5}	-8.039×10^{-9}	273–1000	0.18	0.08
Ethanol	C_2H_6O	19.9	20.96×10^{-2}	-10.38×10^{-5}	20.05×10^{-9}	273–1500	0.40	0.22
Hydrogen chloride	HCl	30.33	-0.7620×10^{-2}	1.327×10^{-5}	-4.338×10^{-9}	273–1500	0.22	0.08
Methane	CH_4	19.89	5.024×10^{-2}	1.269×10^{-5}	-11.01×10^{-9}	273–1500	1.33	0.57
Ethane	C_2H_6	6.900	17.27×10^{-2}	-6.406×10^{-5}	7.285×10^{-9}	273–1500	0.83	0.28
Propane	C_3H_8	-4.04	30.48×10^{-2}	-15.72×10^{-5}	31.74×10^{-9}	273–1500	0.40	0.12
n-Butane	C_4H_{10}	3.96	37.15×10^{-2}	-18.34×10^{-5}	35.00×10^{-9}	273–1500	0.54	0.24
i-Butane	C_4H_{10}	-7.913	41.60×10^{-2}	-23.01×10^{-5}	49.91×10^{-9}	273–1500	0.25	0.13
n-Pentane	C_5H_{12}	6.774	45.43×10^{-2}	-22.46×10^{-5}	42.29×10^{-9}	273–1500	0.56	0.21
n-Hexane	C_6H_{14}	6.938	55.22×10^{-2}	-28.65×10^{-5}	57.69×10^{-9}	273–1500	0.72	0.20
Ethylene	C_2H_4	3.95	15.64×10^{-2}	-8.344×10^{-5}	17.67×10^{-9}	273–1500	0.54	0.13
Propylene	C_3H_6	3.15	23.83×10^{-2}	-12.18×10^{-5}	24.62×10^{-9}	273–1500	0.73	0.17

Source: B. G. Kyle, *Chemical and Process Thermodynamics* (Englewood Cliffs, N.J.: Prentice Hall, 1984). Used with permission.

TABLE A-3

Properties of common liquids, solids, and foods
(*a*) Liquids

Substance	Boiling data at 1 atm		Freezing data		Liquid properties		
	Normal boiling point, °C	Latent heat of vaporization, h_{fg} kJ/kg	Freezing point, °C	Latent heat of fusion, h_{if} kJ/kg	Temp., °C	Density, ρ kg/m^3	Specific heat, C_p kJ/kg · °C
Ammonia	−33.3	1357	−77.7	322.4	−33.3	682	4.43
					−20	665	4.52
					0	639	4.60
					25	602	4.80
Argon	−185.9	161.6	−189.3	28	−185.6	1394	1.14
Benzene	80.2	394	5.5	126	20	879	1.72
Brine (20% sodium chloride by mass)	103.9	—	−17.4	—	20	1150	3.11
n-Butane	−0.5	385.2	−138.5	80.3	−0.5	601	2.31
Carbon dioxide	−78.4*	230.5 (at 0°C)	−56.6		0	298	0.59
Ethanol	78.2	838.3	−114.2	109	25	783	2.46
Ethyl alcohol	78.6	855	−156	108	20	789	2.84
Ethylene glycol	198.1	800.1	−10.8	181.1	20	1109	2.84
Glycerine	179.9	974	18.9	200.6	20	1261	2.32
Helium	−268.9	22.8	—	—	−268.9	146.2	22.8
Hydrogen	−252.8	445.7	−259.2	59.5	−252.8	70.7	10.0
Isobutane	−11.7	367.1	−160	105.7	−11.7	593.8	2.28
Kerosene	204–293	251	−24.9	—	20	820	2.00
Mercury	356.7	294.7	−38.9	11.4	25	13560	0.139
Methane	−161.5	510.4	−182.2	58.4	−161.5	423	3.49
					−100	301	5.79
Methanol	64.5	1100	−97.7	99.2	25	787	2.55
Nitrogen	−195.8	198.6	−210	25.3	−195.8	809	2.06
					−160	596	2.97
Octane	124.8	306.3	−57.5	180.7	20	703	2.10
Oil (light)					25	910	1.80
Oxygen	−183	212.7	−218.8	13.7	−183	1141	1.71
Petroleum	—	230–384			20	640	2.0
Propane	−42.1	427.8	−187.7	80.0	−42.1	581	2.25
					0	529	2.53
					50	449	3.13
Refrigerant-134a	−26.1	216.8	−96.6	—	−50	1443	1.23
					−26.1	1374	1.27
					0	1294	1.34
					25	1206	1.42
Water	100	2257	0.0	333.7	0	1000	4.23
					25	997	4.18
					50	988	4.18
					75	975	4.19
					100	958	4.22

*Sublimation temperature. (At pressures below the triple-point pressure of 518 kPa, carbon dioxide exists as a solid or gas. Also, the freezing-point temperature of carbon dioxide is the triple-point temperature of −56.5°C.)

TABLE A-3

Properties of common liquids, solids, and foods (*Concluded*)

(*b*) Solids (values are for room temperature unless indicated otherwise)

Substance	Density, ρ kg/m³	Specific heat, C_p kJ/kg · °C	Substance	Density, ρ kg/m³	Specific heat, C_p kJ/kg · °C
Metals			**Nonmetals**		
Aluminum			Asphalt	2110	0.920
200 K		0.797	Brick, common	1922	0.79
250 K		0.859	Brick, fireclay (500°C)	2300	0.960
300 K	2,700	0.902	Concrete	2300	0.653
350 K		0.929	Clay	1000	0.920
400 K		0.949	Diamond	2420	0.616
450 K		0.973	Glass, window	2700	0.800
500 K		0.997	Glass, pyrex	2230	0.840
Bronze (76% Cu, 2% Zn, 2% Al)	8,280	0.400	Graphite	2500	0.711
			Granite	2700	1.017
Brass, yellow (65% Cu, 35% Zn)	8,310	0.400	Gypsum or plaster board	800	1.09
			Ice		
Copper			200 K		1.56
−173°C		0.254	220 K		1.71
−100°C		0.342	240 K		1.86
−50°C		0.367	260 K		2.01
0°C		0.381	273 K	921	2.11
27°C	8,900	0.386	Limestone	1650	0.909
100°C		0.393	Marble	2600	0.880
200°C		0.403	Plywood (Douglas Fir)	545	1.21
Iron	7,840	0.45	Rubber (soft)	1100	1.840
Lead	11,310	0.128	Rubber (hard)	1150	2.009
Magnesium	1,730	1.000	Sand	1520	0.800
Nickel	8,890	0.440	Stone	1500	0.800
Silver	10,470	0.235	Woods, hard (maple, oak, etc.)	721	1.26
Steel, mild	7,830	0.500	Woods, soft (fir, pine, etc.)	513	1.38
Tungsten	19,400	0.130			

(*c*) Foods

Food	Water content, % (mass)	Freezing point, °C	Specific heat, kJ/kg · °C — Above freezing	Specific heat, kJ/kg · °C — Below freezing	Latent heat of fusion, kJ/kg	Food	Water content, % (mass)	Freezing point, °C	Specific heat, kJ/kg · °C — Above freezing	Specific heat, kJ/kg · °C — Below freezing	Latent heat of fusion, kJ/kg
Apples	84	−1.1	3.65	1.90	281	Lettuce	95	−0.2	4.02	2.04	317
Bananas	75	−0.8	3.35	1.78	251	Milk, whole	88	−0.6	3.79	1.95	294
Beef round	67	—	3.08	1.68	224	Oranges	87	−0.8	3.75	1.94	291
Broccoli	90	−0.6	3.86	1.97	301	Potatoes	78	−0.6	3.45	1.82	261
Butter	16	—	—	1.04	53	Salmon fish	64	−2.2	2.98	1.65	214
Cheese, swiss	39	−10.0	2.15	1.33	130	Shrimp	83	−2.2	3.62	1.89	277
Cherries	80	−1.8	3.52	1.85	267	Spinach	93	−0.3	3.96	2.01	311
Chicken	74	−2.8	3.32	1.77	247	Strawberries	90	−0.8	3.86	1.97	301
Corn, sweet	74	−0.6	3.32	1.77	247	Tomatoes, ripe	94	−0.5	3.99	2.02	314
Eggs, whole	74	−0.6	3.32	1.77	247	Turkey	64	—	2.98	1.65	214
Ice cream	63	−5.6	2.95	1.63	210	Watermelon	93	−0.4	3.96	2.01	311

Source: Values are obtained from various handbooks and other sources or are calculated. Water content and freezing-point data of foods are from *ASHRAE, Handbook of Fundamentals,* SI version (Atlanta, GA: American Society of Heating, Refrigerating and Air-Conditioning Engineers, Inc., 1993), Chapter 30, Table 1. Freezing point is the temperature at which freezing starts for fruits and vegetables, and the average freezing temperature for other foods.

TABLE A-4

Saturated water—Temperature table

Temp., $T\ °C$	Sat. press., P_{sat} kPa	Specific volume, m³/kg		Internal energy, kJ/kg			Enthalpy, kJ/kg			Entropy, kJ/kg · K		
		Sat. liquid, v_f	Sat. vapor, v_g	Sat. liquid, u_f	Evap., u_{fg}	Sat. vapor, u_g	Sat. liquid, h_f	Evap., h_{fg}	Sat. vapor, h_g	Sat. liquid, s_f	Evap., s_{fg}	Sat. vapor, s_g
0.01	0.6113	0.001000	206.14	0.0	2375.3	2375.3	0.01	2501.3	2501.4	0.000	9.1562	9.1562
5	0.8721	0.001000	147.12	20.97	2361.3	2382.3	20.98	2489.6	2510.6	0.0761	8.9496	9.0257
10	1.2276	0.001000	106.38	42.00	2347.2	2389.2	42.01	2477.7	2519.8	0.1510	8.7498	8.9008
15	1.7051	0.001001	77.93	62.99	2333.1	2396.1	62.99	2465.9	2528.9	0.2245	8.5569	8.7814
20	2.339	0.001002	57.79	83.95	2319.0	2402.9	83.96	2454.1	2538.1	0.2966	8.3706	8.6672
25	3.169	0.001003	43.36	104.88	2304.9	2409.8	104.89	2442.3	2547.2	0.3674	8.1905	8.5580
30	4.246	0.001004	32.89	125.78	2290.8	2416.6	125.79	2430.5	2556.3	0.4369	8.0164	8.4533
35	5.628	0.001006	25.22	146.67	2276.7	2423.4	146.68	2418.6	2565.3	0.5053	7.8478	8.3531
40	7.384	0.001008	19.52	167.56	2262.6	2430.1	167.57	2406.7	2574.3	0.5725	7.6845	8.2570
45	9.593	0.001010	15.26	188.44	2248.4	2436.8	188.45	2394.8	2583.2	0.6387	7.5261	8.1648
50	12.349	0.001012	12.03	209.32	2234.2	2443.5	209.33	2382.7	2592.1	0.7038	7.3725	8.0763
55	15.758	0.001015	9.568	230.21	2219.9	2450.1	230.23	2370.7	2600.9	0.7679	7.2234	7.9913
60	19.940	0.001017	7.671	251.11	2205.5	2456.6	251.13	2358.5	2609.6	0.8312	7.0784	7.9096
65	25.03	0.001020	6.197	272.02	2191.1	2463.1	272.06	2346.2	2618.3	0.8935	6.9375	7.8310
70	31.19	0.001023	5.042	292.95	2176.6	2469.6	292.98	2333.8	2626.8	0.9549	6.8004	7.7553
75	38.58	0.001026	4.131	313.90	2162.0	2475.9	313.93	2321.4	2635.3	1.0155	6.6669	7.6824
80	47.39	0.001029	3.407	334.86	2147.4	2482.2	334.91	2308.8	2643.7	1.0753	6.5369	7.6122
85	57.83	0.001033	2.828	355.84	2132.6	2488.4	355.90	2296.0	2651.9	1.1343	6.4102	7.5445
90	70.14	0.001036	2.361	376.85	2117.7	2494.5	376.92	2283.2	2660.1	1.1925	6.2866	7.4791
95	84.55	0.001040	1.982	397.88	2102.7	2500.6	397.96	2270.2	2668.1	1.2500	6.1659	7.4159
	Sat. press., MPa											
100	0.10133	0.001044	1.6729	418.94	2087.6	2506.5	419.04	2257.0	2676.1	1.3069	6.0480	7.3549
105	0.12082	0.001048	1.4194	440.02	2072.3	2512.4	440.15	2243.7	2683.8	1.3630	5.9328	7.2958
110	0.14327	0.001052	1.2102	461.14	2057.0	2518.1	461.30	2230.2	2691.5	1.4185	5.8202	7.2387
115	0.16906	0.001056	1.0366	482.30	2041.4	2523.7	482.48	2216.5	2699.0	1.4734	5.7100	7.1833
120	0.19853	0.001060	0.8919	503.50	2025.8	2529.3	503.71	2202.6	2706.3	1.5276	5.6020	7.1296
125	0.2321	0.001065	0.7706	524.74	2009.9	2534.6	524.99	2188.5	2713.5	1.5813	5.4962	7.0775
130	0.2701	0.001070	0.6685	546.02	1993.9	2539.9	546.31	2174.2	2720.5	1.6344	5.3925	7.0269
135	0.3130	0.001075	0.5822	567.35	1977.7	2545.0	567.69	2159.6	2727.3	1.6870	5.2907	6.9777
140	0.3613	0.001080	0.5089	588.74	1961.3	2550.0	589.13	2144.7	2733.9	1.7391	5.1908	6.9299
145	0.4154	0.001085	0.4463	610.18	1944.7	2554.9	610.63	2129.6	2740.3	1.7907	5.0926	6.8833
150	0.4758	0.001091	0.3928	631.68	1927.9	2559.5	632.20	2114.3	2746.5	1.8418	4.9960	6.8379
155	0.5431	0.001096	0.3468	653.24	1910.8	2564.1	653.84	2098.6	2752.4	1.8925	4.9010	6.7935
160	0.6178	0.001102	0.3071	674.87	1893.5	2568.4	675.55	2082.6	2758.1	1.9427	4.8075	6.7502
165	0.7005	0.001108	0.2727	696.56	1876.0	2572.5	697.34	2066.2	2763.5	1.9925	4.7153	6.7078
170	0.7917	0.001114	0.2428	718.33	1858.1	2576.5	719.21	2049.5	2768.7	2.0419	4.6244	6.6663
175	0.8920	0.001121	0.2168	740.17	1840.0	2580.2	741.17	2032.4	2773.6	2.0909	4.5347	6.6256
180	1.0021	0.001127	0.19405	762.09	1821.6	2583.7	763.22	2015.0	2778.2	2.1396	4.4461	6.5857
185	1.1227	0.001134	0.17409	784.10	1802.9	2587.0	785.37	1997.1	2782.4	2.1879	4.3586	6.5465
190	1.2544	0.001141	0.15654	806.19	1783.8	2590.0	807.62	1978.8	2786.4	2.2359	4.2720	6.5079
195	1.3978	0.001149	0.14105	828.37	1764.4	2592.8	829.98	1960.0	2790.0	2.2835	4.1863	6.4698

Saturated water—Temperature table (*Concluded*)

		Specific volume, m³/kg		Internal energy, kJ/kg			Enthalpy, kJ/kg			Entropy, kJ/kg · K		
Temp., T °C	Sat. press., P_{sat} **MPa**	Sat. liquid, v_f	Sat. vapor, v_g	Sat. liquid, u_f	Evap., u_{fg}	Sat. vapor, u_g	Sat. liquid, h_f	Evap., h_{fg}	Sat. vapor, h_g	Sat. liquid, s_f	Evap., s_{fg}	Sat. vapor, s_g
200	1.5538	0.001157	0.12736	850.65	1744.7	2595.3	852.45	1940.7	2793.2	2.3309	4.1014	6.4323
205	1.7230	0.001164	0.11521	873.04	1724.5	2597.5	875.04	1921.0	2796.0	2.3780	4.0172	6.3952
210	1.9062	0.001173	0.10441	895.53	1703.9	2599.5	897.76	1900.7	2798.5	2.4248	3.9337	6.3585
215	2.104	0.001181	0.09479	918.14	1682.9	2601.1	920.62	1879.9	2800.5	2.4714	3.8507	6.3221
220	2.318	0.001190	0.08619	940.87	1661.5	2602.4	943.62	1858.5	2802.1	2.5178	3.7683	6.2861
225	2.548	0.001199	0.07849	963.73	1639.6	2603.3	966.78	1836.5	2803.3	2.5639	3.6863	6.2503
230	2.795	0.001209	0.07158	986.74	1617.2	2603.9	990.12	1813.8	2804.0	2.6099	3.6047	6.2146
235	3.060	0.001219	0.06537	1009.89	1594.2	2604.1	1013.62	1790.5	2804.2	2.6558	3.5233	6.1791
240	3.344	0.001229	0.05976	1033.21	1570.8	2604.0	1037.32	1766.5	2803.8	2.7015	3.4422	6.1437
245	3.648	0.001240	0.05471	1056.71	1546.7	2603.4	1061.23	1741.7	2803.0	2.7472	3.3612	6.1083
250	3.973	0.001251	0.05013	1080.39	1522.0	2602.4	1085.36	1716.2	2801.5	2.7927	3.2802	6.0730
255	4.319	0.001263	0.04598	1104.28	1596.7	2600.9	1109.73	1689.8	2799.5	2.8383	3.1992	6.0375
260	4.688	0.001276	0.04221	1128.39	1470.6	2599.0	1134.37	1662.5	2796.9	2.8838	3.1181	6.0019
265	5.081	0.001289	0.03877	1152.74	1443.9	2596.6	1159.28	1634.4	2793.6	2.9294	3.0368	5.9662
270	5.499	0.001302	0.03564	1177.36	1416.3	2593.7	1184.51	1605.2	2789.7	2.9751	2.9551	5.9301
275	5.942	0.001317	0.03279	1202.25	1387.9	2590.2	1210.07	1574.9	2785.0	3.0208	2.8730	5.8938
280	6.412	0.001332	0.03017	1227.46	1358.7	2586.1	1235.99	1543.6	2779.6	3.0668	2.7903	5.8571
285	6.909	0.001348	0.02777	1253.00	1328.4	2581.4	1262.31	1511.0	2773.3	3.1130	2.7070	5.8199
290	7.436	0.001366	0.02557	1278.92	1297.1	2576.0	1289.07	1477.1	2766.2	3.1594	2.6227	5.7821
295	7.993	0.001384	0.02354	1305.2	1264.7	2569.9	1316.3	1441.8	2758.1	3.2062	2.5375	5.7437
300	8.581	0.001404	0.02167	1332.0	1231.0	2563.0	1344.0	1404.9	2749.0	3.2534	2.4511	5.7045
305	9.202	0.001425	0.019948	1359.3	1195.9	2555.2	1372.4	1366.4	2738.7	3.3010	2.3633	5.6643
310	9.856	0.001447	0.018350	1387.1	1159.4	2546.4	1401.3	1326.0	2727.3	3.3493	2.2737	5.6230
315	10.547	0.001472	0.016867	1415.5	1121.1	2536.6	1431.0	1283.5	2714.5	3.3982	2.1821	5.5804
320	11.274	0.001499	0.015488	1444.6	1080.9	2525.5	1461.5	1238.6	2700.1	3.4480	2.0882	5.5362
330	12.845	0.001561	0.012996	1505.3	993.7	2498.9	1525.3	1140.6	2665.9	3.5507	1.8909	5.4417
340	14.586	0.001638	0.010797	1570.3	894.3	2464.6	1594.2	1027.9	2622.0	3.6594	1.6763	5.3357
350	16.513	0.001740	0.008813	1641.9	776.6	2418.4	1670.6	893.4	2563.9	3.7777	1.4335	5.2112
360	18.651	0.001893	0.006945	1725.2	626.3	2351.5	1760.5	720.3	2481.0	3.9147	1.1379	5.0526
370	21.03	0.002213	0.004925	1844.0	384.5	2228.5	1890.5	441.6	2332.1	4.1106	0.6865	4.7971
374.14	22.09	0.003155	0.003155	2029.6	0	2029.6	2099.3	0	2099.3	4.4298	0	4.4298

Source: Tables A-4 through A-8 are adapted from Gordon J. Van Wylen and Richard E. Sonntag, *Fundamentals of Classical Thermodynamics,* English/SI Version, 3rd ed. (New York: John Wiley & Sons, 1986), pp. 635–51. Originally published in Joseph H. Keenan, Frederick G. Keyes, Philip G. Hill, and Joan G. Moore, *Steam Tables,* SI Units (New York: John Wiley & Sons, 1978).

H₂O

TABLE A-5

Saturated water—Pressure table

H₂O

Press., P kPa	Sat. temp., T_{sat} °C	Specific volume, m³/kg		Internal energy, kJ/kg			Enthalpy, kJ/kg			Entropy, kJ/kg · K		
		Sat. liquid, v_f	Sat. vapor, v_g	Sat. liquid, u_f	Evap., u_{fg}	Sat. vapor, u_g	Sat. liquid, h_f	Evap., h_{fg}	Sat. vapor, h_g	Sat. liquid, s_f	Evap., s_{fg}	Sat. vapor, s_g
0.6113	0.01	0.001000	206.14	0.00	2375.3	2375.3	0.01	2501.3	2501.4	0.0000	9.1562	9.1562
1.0	6.98	0.001000	129.21	29.30	2355.7	2385.0	29.30	2484.9	2514.2	0.1059	8.8697	8.9756
1.5	13.03	0.001001	87.98	54.71	2338.6	2393.3	54.71	2470.6	2525.3	0.1957	8.6322	8.8279
2.0	17.50	0.001001	67.00	73.48	2326.0	2399.5	73.48	2460.0	2533.5	0.2607	8.4629	8.7237
2.5	21.08	0.001002	54.25	88.48	2315.9	2404.4	88.49	2451.6	2540.0	0.3120	8.3311	8.6432
3.0	24.08	0.001003	45.67	101.04	2307.5	2408.5	101.05	2444.5	2545.5	0.3545	8.2231	8.5776
4.0	28.96	0.001004	34.80	121.45	2293.7	2415.2	121.46	2432.9	2554.4	0.4226	8.0520	8.4746
5.0	32.88	0.001005	28.19	137.81	2282.7	2420.5	137.82	2423.7	2561.5	0.4764	7.9187	8.3951
7.5	40.29	0.001008	19.24	168.78	2261.7	2430.5	168.79	2406.0	2574.8	0.5764	7.6750	8.2515
10	45.81	0.001010	14.67	191.82	2246.1	2437.9	191.83	2392.8	2584.7	0.6493	7.5009	8.1502
15	53.97	0.001014	10.02	225.92	2222.8	2448.7	225.94	2373.1	2599.1	0.7549	7.2536	8.0085
20	60.06	0.001017	7.649	251.38	2205.4	2456.7	251.40	2358.3	2609.7	0.8320	7.0766	7.9085
25	64.97	0.001020	6.204	271.90	2191.2	2463.1	271.93	2346.3	2618.2	0.8931	6.9383	7.8314
30	69.10	0.001022	5.229	289.20	2179.2	2468.4	289.23	2336.1	2625.3	0.9439	6.8247	7.7686
40	75.87	0.001027	3.993	317.53	2159.5	2477.0	317.58	2319.2	2636.8	1.0259	6.6441	7.6700
50	81.33	0.001030	3.240	340.44	2143.4	2483.9	340.49	2305.4	2645.9	1.0910	6.5029	7.5939
75	91.78	0.001037	2.217	384.31	2112.4	2496.7	384.39	2278.6	2663.0	1.2130	6.2434	7.4564

Press., MPa												
0.100	99.63	0.001043	1.6940	417.36	2088.7	2506.1	417.46	2258.0	2675.5	1.3026	6.0568	7.3594
0.125	105.99	0.001048	1.3749	444.19	2069.3	2513.5	444.32	2241.0	2685.4	1.3740	5.9104	7.2844
0.150	111.37	0.001053	1.1593	466.94	2052.7	2519.7	467.11	2226.5	2693.6	1.4336	5.7897	7.2233
0.175	116.06	0.001057	1.0036	486.80	2038.1	2524.9	486.99	2213.6	2700.6	1.4849	5.6868	7.1717
0.200	120.23	0.001061	0.8857	504.49	2025.0	2529.5	504.70	2201.9	2706.7	1.5301	5.5970	7.1271
0.225	124.00	0.001064	0.7933	520.47	2013.1	2533.6	520.72	2191.3	2712.1	1.5706	5.5173	7.0878
0.250	127.44	0.001067	0.7187	535.10	2002.1	2537.2	535.37	2181.5	2716.9	1.6072	5.4455	7.0527
0.275	130.60	0.001070	0.6573	548.59	1991.9	2540.5	548.89	2172.4	2721.3	1.6408	5.3801	7.0209
0.300	133.55	0.001073	0.6058	561.15	1982.4	2543.6	561.47	2163.8	2725.3	1.6718	5.3201	6.9919
0.325	136.30	0.001076	0.5620	572.90	1973.5	2546.4	573.25	2155.8	2729.0	1.7006	5.2646	6.9652
0.350	138.88	0.001079	0.5243	583.95	1965.0	2548.9	584.33	2148.1	2732.4	1.7275	5.2130	6.9405
0.375	141.32	0.001081	0.4914	594.40	1956.9	2551.3	594.81	2140.8	2735.6	1.7528	5.1647	6.9175
0.40	143.63	0.001084	0.4625	604.31	1949.3	2553.6	604.74	2133.8	2738.6	1.7766	5.1193	6.8959
0.45	147.93	0.001088	0.4140	622.77	1934.9	2557.6	623.25	2120.7	2743.9	1.8207	5.0359	6.8565
0.50	151.86	0.001093	0.3749	639.68	1921.6	2561.2	640.23	2108.5	2748.7	1.8607	4.9606	6.8213
0.55	155.48	0.001097	0.3427	655.32	1909.2	2564.5	665.93	2097.0	2753.0	1.8973	4.8920	6.7893
0.60	158.85	0.001101	0.3157	669.90	1897.5	2567.4	670.56	2086.3	2756.8	1.9312	4.8288	6.7600
0.65	162.01	0.001104	0.2927	683.56	1886.5	2570.1	684.28	2076.0	2760.3	1.9627	4.7703	6.7331
0.70	164.97	0.001108	0.2729	696.44	1876.1	2572.5	697.22	2066.3	2763.5	1.9922	4.7158	6.7080
0.75	167.78	0.001112	0.2556	708.64	1866.1	2574.7	709.47	2057.0	2766.4	2.0200	4.6647	6.6847
0.80	170.43	0.001115	0.2404	720.22	1856.6	2576.8	721.11	2048.0	2769.1	2.0462	4.6166	6.6628
0.85	172.96	0.001118	0.2270	731.27	1847.4	2578.7	732.22	2039.4	2771.6	2.0710	4.5711	6.6421
0.90	175.38	0.001121	0.2150	741.83	1838.6	2580.5	742.83	2031.1	2773.9	2.0946	4.5280	6.6226
0.95	177.69	0.001124	0.2042	751.95	1830.2	2582.1	753.02	2023.1	2776.1	2.1172	4.4869	6.6041
1.00	179.91	0.001127	0.19444	761.68	1822.0	2583.6	762.81	2015.3	2778.1	2.1387	4.4478	6.5865
1.10	184.09	0.001133	0.17753	780.09	1806.3	2586.4	781.34	2000.4	2871.7	2.1792	4.3744	6.5536
1.20	187.99	0.001139	0.16333	797.29	1791.5	2588.8	798.65	1986.2	2784.8	2.2166	4.3067	6.5233
1.30	191.64	0.001144	0.15125	813.44	1777.5	2591.0	814.93	1972.7	2787.6	2.2515	4.2438	6.4953

TABLE A-5

Saturated water—Pressure table (*Concluded*)

Press., P **MPa**	Sat. temp., T_{sat} °C	Specific volume, m³/kg		Internal energy, kJ/kg			Enthalpy, kJ/kg			Entropy, kJ/kg · K		
		Sat. liquid, v_f	Sat. vapor, v_g	Sat. liquid, u_f	Evap., u_{fg}	Sat. vapor, u_g	Sat. liquid, h_f	Evap., h_{fg}	Sat. vapor, h_g	Sat. liquid, s_f	Evap., s_{fg}	Sat. vapor, s_g
1.40	195.07	0.001149	0.14084	828.70	1764.1	2592.8	830.30	1957.7	2790.0	2.2842	4.1850	6.4693
1.50	198.32	0.001154	0.13177	843.16	1751.3	2594.5	844.89	1947.3	2792.2	2.3150	4.1298	6.4448
1.75	205.76	0.001166	0.11349	876.46	1721.4	2597.8	878.50	1917.9	2796.4	2.3851	4.0044	6.3896
2.00	212.42	0.001177	0.09963	906.44	1693.8	2600.3	908.79	1890.7	2799.5	2.4474	3.8935	6.3409
2.25	218.45	0.001187	0.08875	933.83	1668.2	2602.0	936.49	1865.2	2801.7	2.5035	3.7937	6.2972
2.5	223.99	0.001197	0.07998	959.11	1644.0	2603.1	962.11	1841.0	2803.1	2.5547	3.7028	6.2575
3.0	233.90	0.001217	0.06668	1004.78	1599.3	2604.1	1008.42	1795.7	2804.2	2.6457	3.5412	6.1869
3.5	242.60	0.001235	0.05707	1045.43	1558.3	2603.7	1049.75	1753.7	2803.4	2.7253	3.4000	6.1253
4	250.40	0.001252	0.04978	1082.31	1520.0	2602.3	1087.31	1714.1	2801.4	2.7964	3.2737	6.0701
5	263.99	0.001286	0.03944	1147.81	1449.3	2597.1	1154.23	1640.1	2794.3	2.9202	3.0532	5.9734
6	275.64	0.001319	0.03244	1205.44	1384.3	2589.7	1213.35	1571.0	2784.3	3.0267	2.8625	5.8892
7	285.88	0.001351	0.02737	1257.55	1323.0	2580.5	1267.00	1505.1	2772.1	3.1211	2.6922	5.8133
8	295.06	0.001384	0.02352	1305.57	1264.2	2569.8	1316.64	1441.3	2758.0	3.2068	2.5364	5.7432
9	303.40	0.001418	0.02048	1350.51	1207.3	2557.8	1363.26	1378.9	2742.1	3.2858	2.3915	5.6722
10	311.06	0.001452	0.018026	1393.04	1151.4	2544.4	1407.56	1317.1	2724.7	3.3596	2.2544	5.6141
11	318.15	0.001489	0.015987	1433.7	1096.0	2529.8	1450.1	1255.5	2705.6	3.4295	2.1233	5.5527
12	324.75	0.001527	0.014263	1473.0	1040.7	2513.7	1491.3	1193.3	2684.9	3.4962	1.9962	5.4924
13	330.93	0.001567	0.012780	1511.1	985.0	2496.1	1531.5	1130.7	2662.2	3.5606	1.8718	5.4323
14	336.75	0.001611	0.011485	1548.6	928.2	2476.8	1571.1	1066.5	2637.6	3.6232	1.7485	5.3717
15	342.24	0.001658	0.010337	1585.6	869.8	2455.5	1610.5	1000.0	2610.5	3.6848	1.6249	5.3098
16	347.44	0.001711	0.009306	1622.7	809.0	2431.7	1650.1	930.6	2580.6	3.7461	1.4994	5.2455
17	352.37	0.001770	0.008364	1660.2	744.8	2405.0	1690.3	856.9	2547.2	3.8079	1.3698	5.1777
18	357.06	0.001840	0.007489	1698.9	675.4	2374.3	1732.0	777.1	2509.1	3.8715	1.2329	5.1044
19	361.54	0.001924	0.006657	1739.9	598.1	2338.1	1776.5	688.0	2464.5	3.9388	1.0839	5.0228
20	365.81	0.002036	0.005834	1785.6	507.5	2293.0	1826.3	583.4	2409.7	4.0139	0.9130	4.9269
21	369.89	0.002207	0.004952	1842.1	388.5	2230.6	1888.4	446.2	2334.6	4.1075	0.6938	4.8013
22	373.80	0.002742	0.003568	1961.9	125.2	2087.1	2022.2	143.4	2165.6	4.3110	0.2216	4.5327
22.09	374.14	0.003155	0.003155	2029.6	0	2029.6	2099.3	0	2099.3	4.4298	0	4.4298

TABLE A-6

Superheated water

T °C	v m³/kg	u kJ/kg	h kJ/kg	s kJ/kg·K	v m³/kg	u kJ/kg	h kJ/kg	s kJ/kg·K	v m³/kg	u kJ/kg	h kJ/kg	s kJ/kg·K
	P = 0.01 MPa (45.81°C)*				P = 0.05 MPa (81.33°C)				P = 0.10 MPa (99.63°C)			
Sat.†	14.674	2437.9	2584.7	8.1502	3.240	2483.9	2645.9	7.5939	1.6940	2506.1	2675.5	7.3594
50	14.869	2443.9	2592.6	8.1749								
100	17.196	2515.5	2687.5	8.4479	3.418	2511.6	2682.5	7.6947	1.6958	2506.7	2676.2	7.3614
150	19.512	2587.9	2783.0	8.6882	3.889	2585.6	2780.1	7.9401	1.9364	2582.8	2776.4	7.6134
200	21.825	2661.3	2879.5	8.9038	4.356	2659.9	2877.7	8.1580	2.172	2658.1	2875.3	7.8343
250	24.136	2736.0	2977.3	9.1002	4.820	2735.0	2976.0	8.3556	2.406	2733.7	2974.3	8.0333
300	26.445	2812.1	3076.5	9.2813	5.284	2811.3	3075.5	8.5373	2.639	2810.4	3074.3	8.2158
400	31.063	2968.9	3279.6	9.6077	6.209	2968.5	3278.9	8.8642	3.103	2967.9	3278.2	8.5435
500	35.679	3132.3	3489.1	9.8978	7.134	3132.0	3488.7	9.1546	3.565	3131.6	3488.1	8.8342
600	40.295	3302.5	3705.4	10.1608	8.057	3302.2	3705.1	9.4178	4.028	3301.9	3704.4	9.0976
700	44.911	3479.6	3928.7	10.4028	8.981	3479.4	3928.5	9.6599	4.490	3479.2	3928.2	9.3398
800	49.526	3663.8	4159.0	10.6281	9.904	3663.6	4158.9	9.8852	4.952	3663.5	4158.6	9.5652
900	54.141	3855.0	4396.4	10.8396	10.828	3854.9	4396.3	10.0967	5.414	3854.8	4396.1	9.7767
1000	58.757	4053.0	4640.6	11.0393	11.751	4052.9	4640.5	10.2964	5.875	4052.8	4640.3	9.9764
1100	63.372	4257.5	4891.2	11.2287	12.674	4257.4	4891.1	10.4859	6.337	4257.3	4891.0	10.1659
1200	67.987	4467.9	5147.8	11.4091	13.597	4467.8	5147.7	10.6662	6.799	4467.7	5147.6	10.3463
1300	72.602	4683.7	5409.7	11.5811	14.521	4683.6	5409.6	10.8382	7.260	4683.5	5409.5	10.5183
	P = 0.20 MPa (120.23°C)				P = 0.30 MPa (133.55°C)				P = 0.40 MPa (143.63°C)			
Sat.	0.8857	2529.5	2706.7	7.1272	0.6058	2543.6	2725.3	6.9919	0.4625	2553.6	2738.6	6.8959
150	0.9596	2576.9	2768.8	7.2795	0.6339	2570.8	2761.0	7.0778	0.4708	2564.5	2752.8	6.9299
200	1.0803	2654.4	2870.5	7.5066	0.7163	2650.7	2865.6	7.3115	0.5342	2646.8	2860.5	7.1706
250	1.1988	2731.2	2971.0	7.7086	0.7964	2728.7	2967.6	7.5166	0.5951	2726.1	2964.2	7.3789
300	1.3162	2808.6	3071.8	7.8926	0.8753	2806.7	3069.3	7.7022	0.6548	2804.8	3066.8	7.5662
400	1.5493	2966.7	3276.6	8.2218	1.0315	2965.6	3275.0	8.0330	0.7726	2964.4	3273.4	7.8985
500	1.7814	3130.8	3487.1	8.5133	1.1867	3130.0	3486.0	8.3251	0.8893	3129.2	3484.9	8.1913
600	2.013	3301.4	3704.0	8.7770	1.3414	3300.8	3703.2	8.5892	1.0055	3300.2	3702.4	8.4558
700	2.244	3478.8	3927.6	9.0194	1.4957	3478.4	3927.1	8.8319	1.1215	3477.9	3926.5	8.6987
800	2.475	3663.1	4158.2	9.2449	1.6499	3662.9	4157.8	9.0576	1.2372	3662.4	4157.3	8.9244
900	2.705	3854.5	4395.8	9.4566	1.8041	3854.2	4395.4	9.2692	1.3529	3853.9	4395.1	9.1362
1000	2.937	4052.5	4640.0	9.6563	1.9581	4052.3	4639.7	9.4690	1.4685	4052.0	4639.4	9.3360
1100	3.168	4257.0	4890.7	9.8458	2.1121	4256.8	4890.4	9.6585	1.5840	4256.5	4890.2	9.5256
1200	3.399	4467.5	5147.5	10.0262	2.2661	4467.2	5147.1	9.8389	1.6996	4467.0	5146.8	9.7060
1300	3.630	4683.2	5409.3	10.1982	2.4201	4683.0	5409.0	10.0110	1.8151	4682.8	5408.8	9.8780
	P = 0.50 MPa (151.86°C)				P = 0.60 MPa (158.85°C)				P = 0.80 MPa (170.43°C)			
Sat.	0.3749	2561.2	2748.7	6.8213	0.3157	2567.4	2756.8	6.7600	0.2404	2576.8	2769.1	6.6628
200	0.4249	2642.9	2855.4	7.0592	0.3520	2638.9	2850.1	6.9665	0.2608	2630.6	2839.3	6.8158
250	0.4744	2723.5	2960.7	7.2709	0.3938	2720.9	2957.2	7.1816	0.2931	2715.5	2950.0	7.0384
300	0.5226	2802.9	3064.2	7.4599	0.4344	2801.0	3061.6	7.3724	0.3241	2797.2	3056.5	7.2328
350	0.5701	2882.6	3167.7	7.6329	0.4742	2881.2	3165.7	7.5464	0.3544	2878.2	3161.7	7.4089
400	0.6173	2963.2	3271.9	7.7938	0.5137	2962.1	3270.3	7.7079	0.3843	2959.7	3267.1	7.5716
500	0.7109	3128.4	3483.9	8.0873	0.5920	3127.6	3482.8	8.0021	0.4433	3126.0	3480.6	7.8673
600	0.8041	3299.6	3701.7	8.3522	0.6697	3299.1	3700.9	8.2674	0.5018	3297.9	3699.4	8.1333
700	0.8969	3477.5	3925.9	8.5952	0.7472	3477.0	3925.3	8.5107	0.5601	3476.2	3924.2	8.3770
800	0.9896	3662.1	4156.9	8.8211	0.8245	3661.8	4156.5	8.7367	0.6181	3661.1	4155.6	8.6033
900	1.0822	3853.6	4394.7	9.0329	0.9017	3853.4	4394.4	8.9486	0.6761	3852.8	4393.7	8.8153
1000	1.1747	4051.8	4639.1	9.2328	0.9788	4051.5	4638.8	9.1485	0.7340	4051.0	4638.2	9.0153
1100	1.2672	4256.3	4889.9	9.4224	1.0559	4256.1	4889.6	9.3381	0.7919	4255.6	4889.1	9.2050
1200	1.3596	4466.8	5146.6	9.6029	1.1330	4466.5	5146.3	9.5185	0.8497	4466.1	5145.9	9.3855
1300	1.4521	4682.5	5408.6	9.7749	1.2101	4682.3	5408.3	9.6906	0.9076	4681.8	5407.9	9.5575

*The temperature in parentheses is the saturation temperature at the specified pressure.

†Properties of saturated vapor at the specified pressure.

TABLE A-6

Superheated water (*Continued*)

T °C	v m³/kg	u kJ/kg	h kJ/kg	s kJ/kg·K	v m³/kg	u kJ/kg	h kJ/kg	s kJ/kg·K	v m³/kg	u kJ/kg	h kJ/kg	s kJ/kg·K
	P = 1.00 MPa (179.91°C)				P = 1.20 MPa (187.99°C)				P = 1.40 MPa (195.07°C)			
Sat.	0.19444	2583.6	2778.1	6.5865	0.16333	2588.8	2784.8	6.5233	0.14084	2592.8	2790.0	6.4693
200	0.2060	2621.9	2827.9	6.6940	0.16930	2612.8	2815.9	6.5898	0.14302	2603.1	2803.3	6.4975
250	0.2327	2709.9	2942.6	6.9247	0.19234	2704.2	2935.0	6.8294	0.16350	2698.3	2927.2	6.7467
300	0.2579	2793.2	3051.2	7.1229	0.2138	2789.2	3045.8	7.0317	0.18228	2785.2	3040.4	6.9534
350	0.2825	2875.2	3157.7	7.3011	0.2345	2872.2	3153.6	7.2121	0.2003	2869.2	3149.5	7.1360
400	0.3066	2957.3	3263.9	7.4651	0.2548	2954.9	3260.7	7.3774	0.2178	2952.5	3257.5	7.3026
500	0.3541	3124.4	3478.5	7.7622	0.2946	3122.8	3476.3	7.6759	0.2521	3121.1	3474.1	7.6027
600	0.4011	3296.8	3697.9	8.0290	0.3339	3295.6	3696.3	7.9435	0.2860	3294.4	3694.8	7.8710
700	0.4478	3475.3	3923.1	8.2731	0.3729	3474.4	3922.0	8.1881	0.3195	3473.6	3920.8	8.1160
800	0.4943	3660.4	4154.7	8.4996	0.4118	3659.7	4153.8	8.4148	0.3528	3659.0	4153.0	8.3431
900	0.5407	3852.2	4392.9	8.7118	0.4505	3851.6	4392.2	8.6272	0.3861	3851.1	4391.5	8.5556
1000	0.5871	4050.5	4637.6	8.9119	0.4892	4050.0	4637.0	8.8274	0.4192	4049.5	4636.4	8.7559
1100	0.6335	4255.1	4888.6	9.1017	0.5278	4254.6	4888.0	9.0172	0.4524	4254.1	4887.5	8.9457
1200	0.6798	4465.6	5145.4	9.2822	0.5665	4465.1	5144.9	9.1977	0.4855	4464.7	5144.4	9.1262
1300	0.7261	4681.3	5407.4	9.4543	0.6051	4680.9	5407.0	9.3698	0.5186	4680.4	5406.5	9.2984
	P = 1.60 MPa (201.41°C)				P = 1.80 MPa (207.15°C)				P = 2.00 MPa (212.42°C)			
Sat.	0.12380	2596.0	2794.0	6.4218	0.11042	2598.4	2797.1	6.3794	0.09963	2600.3	2799.5	6.3409
225	0.13287	2644.7	2857.3	6.5518	0.11673	2636.6	2846.7	6.4808	0.10377	2628.3	2835.8	6.4147
250	0.14184	2692.3	2919.2	6.6732	0.12497	2686.0	2911.0	6.6066	0.11144	2679.6	2902.5	6.5453
300	0.15862	2781.1	3034.8	6.8844	0.14021	2776.9	3029.2	6.8226	0.12547	2772.6	3023.5	6.7664
350	0.17456	2866.1	3145.4	7.0694	0.15457	2863.0	3141.2	7.0100	0.13857	2859.8	3137.0	6.9563
400	0.19005	2950.1	3254.2	7.2374	0.16847	2947.7	3250.9	7.1794	0.15120	2945.2	3247.6	7.1271
500	0.2203	3119.5	3472.0	7.5390	0.19550	3117.9	3469.8	7.4825	0.17568	3116.2	3467.6	7.4317
600	0.2500	3293.3	3693.2	7.8080	0.2220	3292.1	3691.7	7.7523	0.19960	3290.9	3690.1	7.7024
700	0.2794	3472.7	3919.7	8.0535	0.2482	3471.8	3918.5	7.9983	0.2232	3470.9	3917.4	7.9487
800	0.3086	3658.3	4152.1	8.2808	0.2742	3657.6	4151.2	8.2258	0.2467	3657.0	4150.3	8.1765
900	0.3377	3850.5	4390.8	8.4935	0.3001	3849.9	4390.1	8.4386	0.2700	3849.3	4389.4	8.3895
1000	0.3668	4049.0	4635.8	8.6938	0.3260	4048.5	4635.2	8.6391	0.2933	4048.0	4634.6	8.5901
1100	0.3958	4253.7	4887.0	8.8837	0.3518	4253.2	4886.4	8.8290	0.3166	4252.7	4885.9	8.7800
1200	0.4248	4464.2	5143.9	9.0643	0.3776	4463.7	5143.4	9.0096	0.3398	4463.3	5142.9	8.9607
1300	0.4538	4679.9	5406.0	9.2364	0.4034	4679.5	5405.6	9.1818	0.3631	4679.0	5405.1	9.1329
	P = 2.50 MPa (223.99°C)				P = 3.00 MPa (233.90°C)				P = 3.50 MPa (242.60°C)			
Sat.	0.07998	2603.1	2803.1	6.2575	0.06668	2604.1	2804.2	6.1869	0.05707	2603.7	2803.4	6.1253
225	0.08027	2605.6	2806.3	6.2639								
250	0.08700	2662.6	2880.1	6.4085	0.07058	2644.0	2855.8	6.2872	0.05872	2623.7	2829.2	6.1749
300	0.09890	2761.6	3008.8	6.6438	0.08114	2750.1	2993.5	6.5390	0.06842	2738.0	2977.5	6.4461
350	0.10976	2851.9	3126.3	6.8403	0.09053	2843.7	3115.3	6.7428	0.07678	2835.3	3104.0	6.6579
400	0.12010	2939.1	3239.3	7.0148	0.09936	2932.8	3230.9	6.9212	0.08453	2926.4	3222.3	6.8405
450	0.13014	3025.5	3350.8	7.1746	0.10787	3020.4	3344.0	7.0834	0.09196	3015.3	3337.2	7.0052
500	0.13993	3112.1	3462.1	7.3234	0.11619	3108.0	3456.5	7.2338	0.09918	3103.0	3450.9	7.1572
600	0.15930	3288.0	3686.3	7.5960	0.13243	3285.0	3682.3	7.5085	0.11324	3282.1	3678.4	7.4339
700	0.17832	3468.7	3914.5	7.8435	0.14838	3466.5	3911.7	7.7571	0.12699	3464.3	3908.8	7.6837
800	0.19716	3655.3	4148.2	8.0720	0.16414	3653.5	4145.9	7.9862	0.14056	3651.8	4143.7	7.9134
900	0.21590	3847.9	4387.6	8.2853	0.17980	3846.5	4385.9	8.1999	0.15402	3845.0	4384.1	8.1276
1000	0.2346	4046.7	4633.1	8.4861	0.19541	4045.4	4631.6	8.4009	0.16743	4044.1	4630.1	8.3288
1100	0.2532	4251.5	4884.6	8.6762	0.21098	4250.3	4883.3	8.5912	0.18080	4249.2	4881.9	8.5192
1200	0.2718	4462.1	5141.7	8.8569	0.22652	4460.9	5140.5	8.7720	0.19415	4459.8	5139.3	8.7000
1300	0.2905	4677.8	5404.0	9.0291	0.24206	4676.6	5402.8	8.9442	0.20749	4675.5	5401.7	8.8723

TABLE A-6

Superheated water (*Continued*)

T °C	v m³/kg	u kJ/kg	h kJ/kg	s kJ/kg · K	v m³/kg	u kJ/kg	h kJ/kg	s kJ/kg · K	v m³/kg	u kJ/kg	h kJ/kg	s kJ/kg · K
	P = 4.0 MPa (250.40°C)				P = 4.5 MPa (257.49°C)				P = 5.0 MPa (263.99°C)			
Sat.	0.04978	2602.3	2801.4	6.0701	0.04406	2600.1	2798.3	6.0198	0.03944	2597.1	2794.3	5.9734
275	0.05457	2667.9	2886.2	6.2285	0.04730	2650.3	2863.2	6.1401	0.04141	2631.3	2838.3	6.0544
300	0.05884	2725.3	2960.7	6.3615	0.05135	2712.0	2943.1	6.2828	0.04532	2698.0	2924.5	6.2084
350	0.06645	2826.7	3092.5	6.5821	0.05840	2817.8	3080.6	6.5131	0.05194	2808.7	3068.4	6.4493
400	0.07341	2919.9	3213.6	6.7690	0.06475	2913.3	3204.7	6.7047	0.05781	2906.6	3195.7	6.6459
450	0.08002	3010.2	3330.3	6.9363	0.07074	3005.0	3323.3	6.8746	0.06330	2999.7	3316.2	6.8186
500	0.08643	3099.5	3445.3	7.0901	0.07651	3095.3	3439.6	7.0301	0.06857	3091.0	3433.8	6.9759
600	0.09885	3279.1	3674.4	7.3688	0.08765	3276.0	3670.5	7.3110	0.07869	3273.0	3666.5	7.2589
700	0.11095	3462.1	3905.9	7.6198	0.09847	3459.9	3903.0	7.5631	0.08849	3457.6	3900.1	7.5122
800	0.12287	3650.0	4141.5	7.8502	0.10911	3648.3	4139.3	7.7942	0.09811	3646.6	4137.1	7.7440
900	0.13469	3843.6	4382.3	8.0647	0.11965	3842.2	4380.6	8.0091	0.10762	3840.7	4378.8	7.9593
1000	0.14645	4042.9	4628.7	8.2662	0.13013	4041.6	4627.2	8.2108	0.11707	4040.4	4625.7	8.1612
1100	0.15817	4248.0	4880.6	8.4567	0.14056	4246.8	4879.3	8.4015	0.12648	4245.6	4878.0	8.3520
1200	0.16987	4458.6	5138.1	8.6376	0.15098	4457.5	5136.9	8.5825	0.13587	4456.3	5135.7	8.5331
1300	0.18156	4674.3	5400.5	8.8100	0.16139	4673.1	5399.4	8.7549	0.14526	4672.0	5398.2	8.7055
	P = 6.0 MPa (275.64°C)				P = 7.0 MPa (285.88°C)				P = 8.0 MPa (295.06°C)			
Sat.	0.03244	2589.7	2784.3	5.8892	0.02737	2580.5	2772.1	5.8133	0.02352	2569.8	2758.0	5.7432
300	0.03616	2667.2	2884.2	6.0674	0.02947	2632.2	2838.4	5.9305	0.02426	2590.9	2785.0	5.7906
350	0.04223	2789.6	3043.0	6.3335	0.03524	2769.4	3016.0	6.2283	0.02995	2747.7	2987.3	6.1301
400	0.04739	2892.9	3177.2	6.5408	0.03993	2878.6	3158.1	6.4478	0.03432	2863.8	3138.3	6.3634
450	0.05214	2988.9	3301.8	6.7193	0.04416	2978.0	3287.1	6.6327	0.03817	2966.7	3272.0	6.5551
500	0.05665	3082.2	3422.2	6.8803	0.04814	3073.4	3410.3	6.7975	0.04175	3064.3	3398.3	6.7240
550	0.06101	3174.6	3540.6	7.0288	0.05195	3167.2	3530.9	6.9486	0.04516	3159.8	3521.0	6.8778
600	0.06525	3266.9	3658.4	7.1677	0.05565	3260.7	3650.3	7.0894	0.04845	3254.4	3642.0	7.0206
700	0.07352	3453.1	3894.2	7.4234	0.06283	3448.5	3888.3	7.3476	0.05481	3443.9	3882.4	7.2812
800	0.08160	3643.1	4132.7	7.6566	0.06981	3639.5	4128.2	7.5822	0.06097	3636.0	4123.8	7.5173
900	0.08958	3837.8	4375.3	7.8727	0.07669	3835.0	4371.8	7.7991	0.06702	3832.1	4368.3	7.7351
1000	0.09749	4037.8	4622.7	8.0751	0.08350	4035.3	4619.8	8.0020	0.07301	4032.8	4616.9	7.9384
1100	0.10536	4243.3	4875.4	8.2661	0.09027	4240.9	4872.8	8.1933	0.07896	4238.6	4870.3	8.1300
1200	0.11321	4454.0	5133.3	8.4474	0.09703	4451.7	5130.9	8.3747	0.08489	4449.5	5128.5	8.3115
1300	0.12106	4669.6	5396.0	8.6199	0.10377	4667.3	5393.7	8.5475	0.09080	4665.0	5391.5	8.4842
	P = 9.0 MPa (303.40°C)				P = 10.0 MPa (311.06°C)				P = 12.5 MPa (327.89°C)			
Sat.	0.02048	2557.8	2742.1	5.6772	0.018026	2544.4	2724.7	5.6141	0.013495	2505.1	2673.8	5.4624
325	0.02327	2646.6	2856.0	5.8712	0.019861	2610.4	2809.1	5.7568				
350	0.02580	2724.4	2956.6	6.0361	0.02242	2699.2	2923.4	5.9443	0.016126	2624.6	2826.2	5.7118
400	0.02993	2848.4	3117.8	6.2854	0.02641	2832.4	3096.5	6.2120	0.02000	2789.3	3039.3	6.0417
450	0.03350	2955.2	3256.6	6.4844	0.02975	2943.4	3240.9	6.4190	0.02299	2912.5	3199.8	6.2719
500	0.03677	3055.2	3386.1	6.6576	0.03279	3045.8	3373.7	6.5966	0.02560	3021.7	3341.8	6.4618
550	0.03987	3152.2	3511.0	6.8142	0.03564	3144.6	3500.9	6.7561	0.02801	3125.0	3475.2	6.6290
600	0.04285	3248.1	3633.7	6.9589	0.03837	3241.7	3625.3	6.9029	0.03029	3225.4	3604.0	6.7810
650	0.04574	3343.6	3755.3	7.0943	0.04101	3338.2	3748.2	7.0398	0.03248	3324.4	3730.4	6.9218
700	0.04857	3439.3	3876.5	7.2221	0.04358	3434.7	3870.5	7.1687	0.03460	3422.9	3855.3	7.0536
800	0.05409	3632.5	4119.3	7.4596	0.04859	3628.9	4114.8	7.4077	0.03869	3620.0	4103.6	7.2965
900	0.05950	3829.2	4364.8	7.6783	0.05349	3826.3	4361.2	7.6272	0.04267	3819.1	4352.5	7.5182
1000	0.06485	4030.3	4614.0	7.8821	0.05832	4027.8	4611.0	7.8315	0.04658	4021.6	4603.8	7.7237
1100	0.07016	4236.3	4867.7	8.0740	0.06312	4234.0	4865.1	8.0237	0.05045	4228.2	4858.8	7.9165
1200	0.07544	4447.2	5126.2	8.2556	0.06789	4444.9	5123.8	8.2055	0.05430	4439.3	5118.0	8.0937
1300	0.08072	4662.7	5389.2	8.4284	0.07265	4460.5	5387.0	8.3783	0.05813	4654.8	5381.4	8.2717

TABLE A-6

Superheated water (*Concluded*)

T °C	v m³/kg	u kJ/kg	h kJ/kg	s kJ/kg · K	v m³/kg	u kJ/kg	h kJ/kg	s kJ/kg · K	v m³/kg	u kJ/kg	h kJ/kg	s kJ/kg · K
	P = 15.0 MPa (342.24°C)				P = 17.5 MPa (354.75°C)				P = 20.0 MPa (365.81°C)			
Sat.	0.010337	2455.5	2610.5	5.3098	0.007920	2390.2	2528.8	5.1419	0.005834	2293.0	2409.7	4.9269
350	0.011470	2520.4	2692.4	5.4421								
400	0.015649	2740.7	2975.5	5.8811	0.012447	2685.0	2902.9	5.7213	0.009942	2619.3	2818.1	5.5540
450	0.018445	2879.5	3156.2	6.1404	0.015174	2844.2	3109.7	6.0184	0.012695	2806.2	3060.1	5.9017
500	0.02080	2996.6	3308.6	6.3443	0.017358	2970.3	3274.1	6.2383	0.014768	2942.9	3238.2	6.1401
550	0.02293	3104.7	3448.6	6.5199	0.019288	3083.9	3421.4	6.4230	0.016555	3062.4	3393.5	6.3348
600	0.02491	3208.6	3582.3	6.6776	0.02106	3191.5	3560.1	6.5866	0.018178	3174.0	3537.6	6.5048
650	0.02680	3310.3	3712.3	6.8224	0.02274	3296.0	3693.9	6.7357	0.019693	3281.4	3675.3	6.6582
700	0.02861	3410.9	3840.1	6.9572	0.02434	3398.7	3824.6	6.8736	0.02113	3386.4	3809.0	6.7993
800	0.03210	3610.9	4092.4	7.2040	0.02738	3601.8	4081.1	7.1244	0.02385	3592.7	4069.7	7.0544
900	0.03546	3811.9	4343.8	7.4279	0.03031	3804.7	4335.1	7.3507	0.02645	3797.5	4326.4	7.2830
1000	0.03875	4015.4	4596.6	7.6348	0.03316	4009.3	4589.5	7.5589	0.02897	4003.1	4582.5	7.4925
1100	0.04200	4222.6	4852.6	7.8283	0.03597	4216.9	4846.4	7.7531	0.03145	4211.3	4840.2	7.6874
1200	0.04523	4433.8	5112.3	8.0108	0.03876	4428.3	5106.6	7.9360	0.03391	4422.8	5101.0	7.8707
1300	0.04845	4649.1	5376.0	8.1840	0.04154	4643.5	5370.5	8.1093	0.03636	4638.0	5365.1	8.0442
	P = 25.0 MPa				P = 30.0 MPa				P = 35.0 MPa			
375	0.0019731	1798.7	1848.0	4.0320	0.0017892	1737.8	1791.5	3.9305	0.0017003	1702.9	1762.4	3.8722
400	0.006004	2430.1	2580.2	5.1418	0.002790	2067.4	2151.1	4.4728	0.002100	1914.1	1987.6	4.2126
425	0.007881	2609.2	2806.3	5.4723	0.005303	2455.1	2614.2	5.1504	0.003428	2253.4	2373.4	4.7747
450	0.009162	2720.7	2949.7	5.6744	0.006735	2619.3	2821.4	5.4424	0.004961	2498.7	2672.4	5.1962
500	0.011123	2884.3	3162.4	5.9592	0.008678	2820.7	3081.1	5.7905	0.006927	2751.9	2994.4	5.6282
550	0.012724	3017.5	3335.6	6.1765	0.010168	2970.3	3275.4	6.0342	0.008345	2921.0	3213.0	5.9026
600	0.014137	3137.9	3491.4	6.3602	0.011446	3100.5	3443.9	6.2331	0.009527	3062.0	3395.5	6.1179
650	0.015433	3251.6	3637.4	6.5229	0.012596	3221.0	3598.9	6.4058	0.010575	3189.8	3559.9	6.3010
700	0.016646	3361.3	3777.5	6.6707	0.013661	3335.8	3745.6	6.5606	0.011533	3309.8	3713.5	6.4631
800	0.018912	3574.3	4047.1	6.9345	0.015623	3555.5	4024.2	6.8332	0.013278	3536.7	4001.5	6.7450
900	0.021045	3783.0	4309.1	7.1680	0.017448	3768.5	4291.9	7.0718	0.014883	3754.0	4274.9	6.9386
1000	0.02310	3990.9	4568.5	7.3802	0.019196	3978.8	4554.7	7.2867	0.016410	3966.7	4541.1	7.2064
1100	0.02512	4200.2	4828.2	7.5765	0.020903	4189.2	4816.3	7.4845	0.017895	4178.3	4804.6	7.4037
1200	0.02711	4412.0	5089.9	7.7605	0.022589	4401.3	5079.0	7.6692	0.019360	4390.7	5068.3	7.5910
1300	0.02910	4626.9	5354.4	7.9342	0.024266	4616.0	5344.0	7.8432	0.020815	4605.1	5333.6	7.7653
	P = 40.0 MPa				P = 50.0 MPa				P = 60.0 MPa			
375	0.0016407	1677.1	1742.8	3.8290	0.0015594	1638.6	1716.6	3.7639	0.0015028	1609.4	1699.5	3.7141
400	0.0019077	1854.6	1930.9	4.1135	0.0017309	1788.1	1874.6	4.0031	0.0016335	1745.4	1843.4	3.9318
425	0.002532	2096.9	2198.1	4.5029	0.002007	1959.7	2060.0	4.2734	0.0018165	1892.7	2001.7	4.1626
450	0.003693	2365.1	2512.8	4.9459	0.002486	2159.6	2284.0	4.5884	0.002085	2053.9	2179.0	4.4121
500	0.005622	2678.4	2903.3	5.4700	0.003892	2525.5	2720.1	5.1726	0.002956	2390.6	2567.9	4.9321
550	0.006984	2869.7	3149.1	5.7785	0.005118	2763.6	3019.5	5.5485	0.003956	2658.8	2896.2	5.3441
600	0.008094	3022.6	3346.4	6.0144	0.006112	2942.0	3247.6	5.8178	0.004834	2861.1	3151.2	5.6452
650	0.009063	3158.0	3520.6	6.2054	0.006966	3093.5	3441.8	6.0342	0.005595	3028.8	3364.5	5.8829
700	0.009941	3283.6	3681.2	6.3750	0.007727	3230.5	3616.8	6.2189	0.006272	3177.2	3553.5	6.0824
800	0.011523	3517.8	3978.7	6.6662	0.009076	3479.8	3933.6	6.5290	0.007459	3441.5	3889.1	6.4109
900	0.012962	3739.4	4257.9	6.9150	0.010283	3710.3	4224.4	6.7882	0.008508	3681.0	4191.5	6.6805
1000	0.014324	3954.6	4527.6	7.1356	0.011411	3930.5	4501.1	7.0146	0.009480	3906.4	4475.2	6.9127
1100	0.015642	4167.4	4793.1	7.3364	0.012496	4145.7	4770.5	7.2184	0.010409	4124.1	4748.6	7.1195
1200	0.016940	4380.1	5057.7	7.5224	0.013561	4359.1	5037.2	7.4058	0.011317	4338.2	5017.2	7.3083
1300	0.018229	4594.3	5323.5	7.6969	0.014616	4572.8	5303.6	7.5808	0.012215	4551.4	5284.3	7.4837

TABLE A-7

Compressed liquid water

T °C	v m³/kg	u kJ/kg	h kJ/kg	s kJ/kg · K	v m³/kg	u kJ/kg	h kJ/kg	s kJ/kg · K	v m³/kg	u kJ/kg	h kJ/kg	s kJ/kg · K
	P = 5 MPa (263.99°C)				P = 10 MPa (311.06°C)				P = 15 MPa (342.24°C)			
Sat.	0.0012859	1147.8	1154.2	2.9202	0.0014524	1393.0	1407.6	3.3596	0.0016581	1585.6	1610.5	3.6848
0	0.0009977	0.04	5.04	0.0001	0.0009952	0.09	10.04	0.0002	0.0009928	0.15	15.05	0.0004
20	0.0009995	83.65	88.65	0.2956	0.0009972	83.36	93.33	0.2945	0.0009950	83.06	97.99	0.2934
40	0.0010056	166.95	171.97	0.5705	0.0010034	166.35	176.38	0.5686	0.0010013	165.76	180.78	0.5666
60	0.0010149	250.23	255.30	0.8285	0.0010127	249.36	259.49	0.8258	0.0010105	248.51	263.67	0.8232
80	0.0010268	333.72	338.85	1.0720	0.0010245	332.59	342.83	1.0688	0.0010222	331.48	346.81	1.0656
100	0.0010410	417.52	422.72	1.3030	0.0010385	416.12	426.50	1.2992	0.0010361	414.74	430.28	1.2955
120	0.0010576	501.80	507.09	1.5233	0.0010549	500.08	510.64	1.5189	0.0010522	498.40	514.19	1.5145
140	0.0010768	586.76	592.15	1.7343	0.0010737	584.68	595.42	1.7292	0.0010707	582.66	598.72	1.7242
160	0.0010988	672.62	678.12	1.9375	0.0010953	670.13	681.08	1.9317	0.0010918	667.71	684.09	1.9260
180	0.0011240	759.63	765.25	2.1341	0.0011199	756.65	767.84	2.1275	0.0011159	753.76	770.50	2.1210
200	0.0011530	848.1	853.9	2.3255	0.0011480	844.5	856.0	2.3178	0.0011433	841.0	858.2	2.3104
220	0.0011866	938.4	944.4	2.5128	0.0011805	934.1	945.9	2.5039	0.0011748	929.9	947.5	2.4953
240	0.0012264	1031.4	1037.5	2.6979	0.0012187	1026.0	1038.1	2.6872	0.0012114	1020.8	1039.0	2.6771
260	0.0012749	1127.9	1134.3	2.8830	0.0012645	1121.1	1133.7	2.8699	0.0012550	1114.6	1133.4	2.8576
280					0.0013216	1220.9	1234.1	3.0548	0.0013084	1212.5	1232.1	3.0393
300					0.0013972	1328.4	1342.3	3.2469	0.0013770	1316.6	1337.3	3.2260
320									0.0014724	1431.1	1453.2	3.4247
340									0.0016311	1567.5	1591.9	3.6546
	P = 20 MPa (365.81°C)				P = 30 MPa				P = 50 MPa			
Sat.	0.002036	1785.6	1826.3	4.0139								
0	0.0009904	0.19	20.01	0.0004	0.0009856	0.25	29.82	0.0001	0.0009766	0.20	49.03	0.0014
20	0.0009928	82.77	102.62	0.2923	0.0009886	82.17	111.84	0.2899	0.0009804	81.00	130.02	0.2848
40	0.0009992	165.17	185.16	0.5646	0.0009951	164.04	193.89	0.5607	0.0009872	161.86	211.21	0.5527
60	0.0010084	247.68	267.85	0.8206	0.0010042	246.06	276.19	0.8154	0.0009962	242.98	292.79	0.8052
80	0.0010199	330.40	350.80	1.0624	0.0010156	328.30	358.77	1.0561	0.0010073	324.34	374.70	1.0440
100	0.0010337	413.39	434.06	1.2917	0.0010290	410.78	441.66	1.2844	0.0010201	405.88	456.89	1.2703
120	0.0010496	496.76	517.76	1.5102	0.0010445	493.59	524.93	1.5018	0.0010348	487.65	539.39	1.4857
140	0.0010678	580.69	602.04	1.7193	0.0010621	576.88	608.75	1.7098	0.0010515	569.77	622.35	1.6915
160	0.0010885	665.35	687.12	1.9204	0.0010821	660.82	693.28	1.9096	0.0010703	652.41	705.92	1.8891
180	0.0011120	750.95	773.20	2.1147	0.0011047	745.59	778.73	2.1024	0.0010912	735.69	790.25	2.0794
200	0.0011388	837.7	860.5	2.3031	0.0011302	831.4	865.3	2.2893	0.0011146	819.7	875.5	2.2634
220	0.0011695	925.9	949.3	2.4870	0.0011590	918.3	953.1	2.4711	0.0011408	904.7	961.7	2.4419
240	0.0012046	1016.0	1040.0	2.6674	0.0011920	1006.9	1042.6	2.6490	0.0011702	990.7	1049.2	2.6158
260	0.0012462	1108.6	1133.5	2.8459	0.0012303	1097.4	1134.3	2.8243	0.0012034	1078.1	1138.2	2.7860
280	0.0012965	1204.7	1230.6	3.0248	0.0012755	1190.7	1229.0	2.9986	0.0012415	1167.2	1229.3	2.9537
300	0.0013596	1306.1	1333.3	3.2071	0.0013304	1287.9	1327.8	3.1741	0.0012860	1258.7	1323.0	3.1200
320	0.0014437	1415.7	1444.6	3.3979	0.0013997	1390.7	1432.7	3.3539	0.0013388	1353.3	1420.2	3.2868
340	0.0015684	1539.7	1571.0	3.6075	0.0014920	1501.7	1546.5	3.5426	0.0014032	1452.0	1522.1	3.4557
360	0.0018226	1702.8	1739.3	3.8772	0.0016265	1626.6	1675.4	3.7494	0.0014838	1556.0	1630.2	3.6291
380					0.0018691	1781.4	1837.5	4.0012	0.0015884	1667.2	1746.6	3.8101

H₂O

TABLE A-8

Saturated ice—water vapor

Temp., T °C	Sat. press., P_{sat} kPa	Specific volume, m³/kg		Internal energy, kJ/kg			Enthalpy, kJ/kg			Entropy, kJ/kg · K		
		Sat. ice, $v_i \times 10^3$	Sat. vapor, v_g	Sat. ice, u_i	Subl., u_{ig}	Sat. vapor, u_g	Sat. ice, h_i	Subl., h_{ig}	Sat. vapor, h_g	Sat. ice, s_i	Subl., s_{ig}	Sat. vapor, s_g
0.01	0.6113	1.0908	206.1	−333.40	2708.7	2375.3	−333.40	2834.8	2501.4	−1.221	10.378	9.156
0	0.6108	1.0908	206.3	−333.43	2708.8	2375.3	−333.43	2834.8	2501.3	−1.221	10.378	9.157
−2	0.5176	1.0904	241.7	−337.62	2710.2	2372.6	−337.62	2835.3	2497.7	−1.237	10.456	9.219
−4	0.4375	1.0901	283.8	−341.78	2711.6	2369.8	−341.78	2835.7	2494.0	−1.253	10.536	9.283
−6	0.3689	1.0898	334.2	−345.91	2712.9	2367.0	−345.91	2836.2	2490.3	−1.268	10.616	9.348
−8	0.3102	1.0894	394.4	−350.02	2714.2	2364.2	−350.02	2836.6	2486.6	−1.284	10.698	9.414
−10	0.2602	1.0891	466.7	−354.09	2715.5	2361.4	−354.09	2837.0	2482.9	−1.299	10.781	9.481
−12	0.2176	1.0888	553.7	−358.14	2716.8	2358.7	−358.14	2837.3	2479.2	−1.315	10.865	9.550
−14	0.1815	1.0884	658.8	−362.15	2718.0	2355.9	−362.15	2837.6	2475.5	−1.331	10.950	9.619
−16	0.1510	1.0881	786.0	−366.14	2719.2	2353.1	−366.14	2837.9	2471.8	−1.346	11.036	9.690
−18	0.1252	1.0878	940.5	−370.10	2720.4	2350.3	−370.10	2838.2	2468.1	−1.362	11.123	9.762
−20	0.1035	1.0874	1128.6	−374.03	2721.6	2347.5	−374.03	2838.4	2464.3	−1.377	11.212	9.835
−22	0.0853	1.0871	1358.4	−377.93	2722.7	2344.7	−377.93	2838.6	2460.6	−1.393	11.302	9.909
−24	0.0701	1.0868	1640.1	−381.80	2723.7	2342.0	−381.80	2838.7	2456.9	−1.408	11.394	9.985
−26	0.0574	1.0864	1986.4	−385.64	2724.8	2339.2	−385.64	2838.9	2453.2	−1.424	11.486	10.062
−28	0.0469	1.0861	2413.7	−389.45	2725.8	2336.4	−389.45	2839.0	2449.5	−1.439	11.580	10.141
−30	0.0381	1.0858	2943	−393.23	2726.8	2333.6	−393.23	2839.0	2445.8	−1.455	11.676	10.221
−32	0.0309	1.0854	3600	−396.98	2727.8	2330.8	−396.98	2839.1	2442.1	−1.471	11.773	10.303
−34	0.0250	1.0851	4419	−400.71	2728.7	2328.0	−400.71	2839.1	2438.4	−1.486	11.872	10.386
−36	0.0201	1.0848	5444	−404.40	2729.6	2325.2	−404.40	2839.1	2434.7	−1.501	11.972	10.470
−38	0.0161	1.0844	6731	−408.06	2730.5	2322.4	−408.06	2839.0	2430.9	−1.517	12.073	10.556
−40	0.0129	1.0841	8354	−411.70	2731.3	2319.6	−411.70	2839.9	2427.2	−1.532	12.176	10.644

H₂O

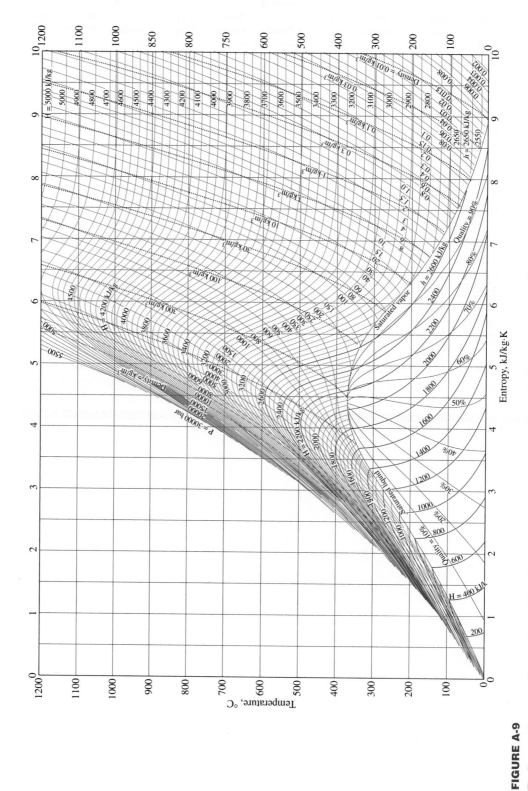

FIGURE A-9

T-s diagram for water. (*Source:* Lester Haar, John S. Gallagher, and George S. Kell, *NBS/NRC Steam Tables*, 1984. With permission from Hemisphere Publishing Corporation, New York.)

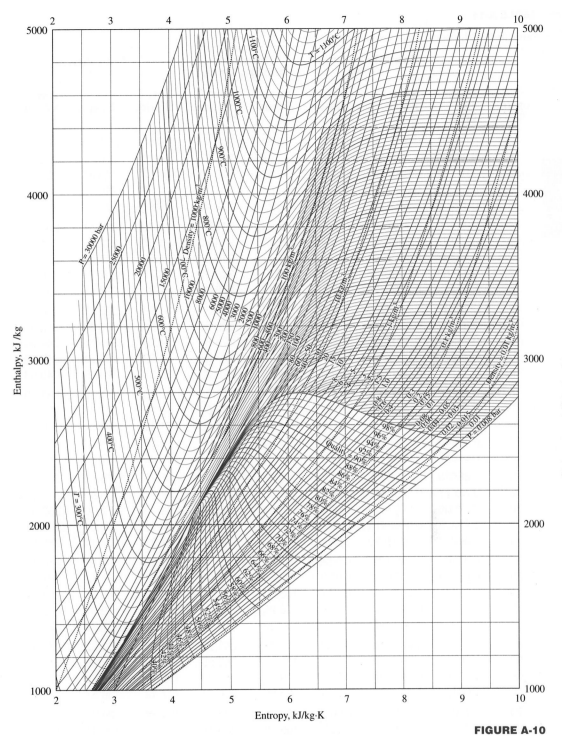

FIGURE A-10

Mollier diagram for water. (*Source:* Lester Haar, John S. Gallagher, and George S. Kell, *NBS/NRC Steam Tables,* 1984. With permission from Hemisphere Publishing Corporation, New York.)

TABLE A-11

Saturated refrigerant-134a—Temperature table

R-134a

Temp., T °C	Press., P_{sat} MPa	Specific volume, m³/kg		Internal energy, kJ/kg		Enthalpy, kJ/kg			Entropy, kJ/kg · K	
		Sat. liquid, v_f	Sat. vapor, v_g	Sat. liquid, u_f	Sat. vapor, u_g	Sat. liquid, h_f	Evap., h_{fg}	Sat. vapor, h_g	Sat. liquid, s_f	Sat. vapor, s_g
−40	0.05164	0.0007055	0.3569	−0.04	204.45	0.00	222.88	222.88	0.0000	0.9560
−36	0.06332	0.0007113	0.2947	4.68	206.73	4.73	220.67	225.40	0.0201	0.9506
−32	0.07704	0.0007172	0.2451	9.47	209.01	9.52	218.37	227.90	0.0401	0.9456
−28	0.09305	0.0007233	0.2052	14.31	211.29	14.37	216.01	230.38	0.0600	0.9411
−26	0.10199	0.0007265	0.1882	16.75	212.43	16.82	214.80	231.62	0.0699	0.9390
−24	0.11160	0.0007296	0.1728	19.21	213.57	19.29	213.57	232.85	0.0798	0.9370
−22	0.12192	0.0007328	0.1590	21.68	214.70	21.77	212.32	234.08	0.0897	0.9351
−20	0.13299	0.0007361	0.1464	24.17	215.84	24.26	211.05	235.31	0.0996	0.9332
−18	0.14483	0.0007395	0.1350	26.67	216.97	26.77	209.76	236.53	0.1094	0.9315
−16	0.15748	0.0007428	0.1247	29.18	218.10	29.30	208.45	237.74	0.1192	0.9298
−12	0.18540	0.0007498	0.1068	34.25	220.36	34.39	205.77	240.15	0.1388	0.9267
−8	0.21704	0.0007569	0.0919	39.38	222.60	39.54	203.00	242.54	0.1583	0.9239
−4	0.25274	0.0007644	0.0794	44.56	224.84	44.75	200.15	244.90	0.1777	0.9213
0	0.29282	0.0007721	0.0689	49.79	227.06	50.02	197.21	247.23	0.1970	0.9190
4	0.33765	0.0007801	0.0600	55.08	229.27	55.35	194.19	249.53	0.2162	0.9169
8	0.38756	0.0007884	0.0525	60.43	231.46	60.73	191.07	251.80	0.2354	0.9150
12	0.44294	0.0007971	0.0460	65.83	233.63	66.18	187.85	254.03	0.2545	0.9132
16	0.50416	0.0008062	0.0405	71.29	235.78	71.69	184.52	256.22	0.2735	0.9116
20	0.57160	0.0008157	0.0358	76.80	237.91	77.26	181.09	258.35	0.2924	0.9102
24	0.64566	0.0008257	0.0317	82.37	240.01	82.90	177.55	260.45	0.3113	0.9089
26	0.68530	0.0008309	0.0298	85.18	241.05	85.75	175.73	261.48	0.3208	0.9082
28	0.72675	0.0008362	0.0281	88.00	242.08	88.61	173.89	262.50	0.3302	0.9076
30	0.77006	0.0008417	0.0265	90.84	243.10	91.49	172.00	263.50	0.3396	0.9070
32	0.81528	0.0008473	0.0250	93.70	244.12	94.39	170.09	264.48	0.3490	0.9064
34	0.86247	0.0008530	0.0236	96.58	245.12	97.31	168.14	265.45	0.3584	0.9058
36	0.91168	0.0008590	0.0223	99.47	246.11	100.25	166.15	266.40	0.3678	0.9053
38	0.96298	0.0008651	0.0210	102.38	247.09	103.21	164.12	267.33	0.3772	0.9047
40	1.0164	0.0008714	0.0199	105.30	248.06	106.19	162.05	268.24	0.3866	0.9041
42	1.0720	0.0008780	0.0188	108.25	249.02	109.19	159.94	269.14	0.3960	0.9035
44	1.1299	0.0008847	0.0177	111.22	249.96	112.22	157.79	270.01	0.4054	0.9030
48	1.2526	0.0008989	0.0159	117.22	251.79	118.35	153.33	271.68	0.4243	0.9017
52	1.3851	0.0009142	0.0142	123.31	253.55	124.58	148.66	273.24	0.4432	0.9004
56	1.5278	0.0009308	0.0127	129.51	255.23	130.93	143.75	274.68	0.4622	0.8990
60	1.6813	0.0009488	0.0114	135.82	256.81	137.42	138.57	275.99	0.4814	0.8973
70	2.1162	0.0010027	0.0086	152.22	260.15	154.34	124.08	278.43	0.5302	0.8918
80	2.6324	0.0010766	0.0064	169.88	262.14	172.71	106.41	279.12	0.5814	0.8827
90	3.2435	0.0011949	0.0046	189.82	261.34	193.69	82.63	276.32	0.6380	0.8655
100	3.9742	0.0015443	0.0027	218.60	248.49	224.74	34.40	259.13	0.7196	0.8117

Source for Tables A-11 through A-13: M. J. Moran and H. N. Shapiro, *Fundamentals of Engineering Thermodynamics,* 2nd ed. (New York: John Wiley & Sons, 1992), pp. 710–15. Originally based on equations from D. P. Wilson and R. S. Basu, "Thermodynamic Properties of a New Stratospherically Safe Working Fluid—Refrigerant-134a," *ASHRAE Trans.* 94, Pt. 2 (1988), pp. 2095–118. Used with permission.

Saturated refrigerant-134a—Pressure table

Press., P **MPa**	Temp., T_{sat} **°C**	Specific volume, m³/kg		Internal energy, kJ/kg		Enthalpy, kJ/kg			Entropy, kJ/kg · K	
		Sat. liquid, v_f	Sat. vapor, v_g	Sat. liquid, u_f	Sat. vapor, u_g	Sat. liquid, h_f	Evap., h_{fg}	Sat. vapor, h_g	Sat. liquid, s_f	Sat. vapor, s_g
0.06	−37.07	0.0007097	0.3100	3.41	206.12	3.46	221.27	224.72	0.0147	0.9520
0.08	−31.21	0.0007184	0.2366	10.41	209.46	10.47	217.92	228.39	0.0440	0.9447
0.10	−26.43	0.0007258	0.1917	16.22	212.18	16.29	215.06	231.35	0.0678	0.9395
0.12	−22.36	0.0007323	0.1614	21.23	214.50	21.32	212.54	233.86	0.0879	0.9354
0.14	−18.80	0.0007381	0.1395	25.66	216.52	25.77	210.27	236.04	0.1055	0.9322
0.16	−15.62	0.0007435	0.1229	29.66	218.32	29.78	208.18	237.97	0.1211	0.9295
0.18	−12.73	0.0007485	0.1098	33.31	219.94	33.45	206.26	239.71	0.1352	0.9273
0.20	−10.09	0.0007532	0.0993	36.69	221.43	36.84	204.46	241.30	0.1481	0.9253
0.24	−5.37	0.0007618	0.0834	42.77	224.07	42.95	201.14	244.09	0.1710	0.9222
0.28	−1.23	0.0007697	0.0719	48.18	226.38	48.39	198.13	246.52	0.1911	0.9197
0.32	2.48	0.0007770	0.0632	53.06	228.43	53.31	195.35	248.66	0.2089	0.9177
0.36	5.84	0.0007839	0.0564	57.54	230.28	57.82	192.76	250.58	0.2251	0.9160
0.4	8.93	0.0007904	0.0509	61.69	231.97	62.00	190.32	252.32	0.2399	0.9145
0.5	15.74	0.0008056	0.0409	70.93	235.64	71.33	184.74	256.07	0.2723	0.9117
0.6	21.58	0.0008196	0.0341	78.99	238.74	79.48	179.71	259.19	0.2999	0.9097
0.7	26.72	0.0008328	0.0292	86.19	241.42	86.78	175.07	261.85	0.3242	0.9080
0.8	31.33	0.0008454	0.0255	92.75	243.78	93.42	170.73	264.15	0.3459	0.9066
0.9	35.53	0.0008576	0.0226	98.79	245.88	99.56	166.62	266.18	0.3656	0.9054
1.0	39.39	0.0008695	0.0202	104.42	247.77	105.29	162.68	267.97	0.3838	0.9043
1.2	46.32	0.0008928	0.0166	114.69	251.03	115.76	155.23	270.99	0.4164	0.9023
1.4	52.43	0.0009159	0.0140	123.98	253.74	125.26	148.14	273.40	0.4453	0.9003
1.6	57.92	0.0009392	0.0121	132.52	256.00	134.02	141.31	275.33	0.4714	0.8982
1.8	62.91	0.0009631	0.0105	140.49	257.88	142.22	134.60	276.83	0.4954	0.8959
2.0	67.49	0.0009878	0.0093	148.02	259.41	149.99	127.95	277.94	0.5178	0.8934
2.5	77.59	0.0010562	0.0069	165.48	261.84	168.12	111.06	279.17	0.5687	0.8854
3.0	86.22	0.0011416	0.0053	181.88	262.16	185.30	92.71	278.01	0.6156	0.8735

R-134a

TABLE A–13

Superheated refrigerant-134a

T °C	v m³/kg	u kJ/kg	h kJ/kg	s kJ/kg·K	v m³/kg	u kJ/kg	h kJ/kg	s kJ/kg·K	v m³/kg	u kJ/kg	h kJ/kg	s kJ/kg·K
	P = 0.06 MPa (T_{sat} = −37.07°C)				P = 0.10 MPa (T_{sat} = −26.43°C)				P = 0.14 MPa (T_{sat} = −18.80°C)			
Sat.	0.31003	206.12	224.72	0.9520	0.19170	212.18	231.35	0.9395	0.13945	216.52	236.04	0.9322
−20	0.33536	217.86	237.98	1.0062	0.19770	216.77	236.54	0.9602				
−10	0.34992	224.97	245.96	1.0371	0.20686	224.01	244.70	0.9918	0.14549	223.03	243.40	0.9606
0	0.36433	232.24	254.10	1.0675	0.21587	231.41	252.99	1.0227	0.15219	230.55	251.86	0.9922
10	0.37861	239.69	262.41	1.0973	0.22473	238.96	261.43	1.0531	0.15875	238.21	260.43	1.0230
20	0.39279	247.32	270.89	1.1267	0.23349	246.67	270.02	1.0829	0.16520	246.01	269.13	1.0532
30	0.40688	255.12	279.53	1.1557	0.24216	254.54	278.76	1.1122	0.17155	253.96	277.97	1.0828
40	0.42091	263.10	288.35	1.1844	0.25076	262.58	287.66	1.1411	0.17783	262.06	286.96	1.1120
50	0.43487	271.25	297.34	1.2126	0.25930	270.79	296.72	1.1696	0.18404	270.32	296.09	1.1407
60	0.44879	279.58	306.51	1.2405	0.26779	279.16	305.94	1.1977	0.19020	278.74	305.37	1.1690
70	0.46266	288.08	315.84	1.2681	0.27623	287.70	315.32	1.2254	0.19633	287.32	314.80	1.1969
80	0.47650	296.75	325.34	1.2954	0.28464	296.40	324.87	1.2528	0.20241	296.06	324.39	1.2244
90	0.49031	305.58	335.00	1.3224	0.29302	305.27	334.57	1.2799	0.20846	304.95	334.14	1.2516
100									0.21449	314.01	344.04	1.2785
	P = 0.18 MPa (T_{sat} = −12.73°C)				P = 0.20 MPa (T_{sat} = −10.09°C)				P = 0.24 MPa (T_{sat} = −5.37°C)			
Sat.	0.10983	219.94	239.71	0.9273	0.09933	221.43	241.30	0.9253	0.08343	224.07	244.09	0.9222
−10	0.11135	222.02	242.06	0.9362	0.09938	221.50	241.38	0.9256				
0	0.11678	229.67	250.69	0.9684	0.10438	229.23	250.10	0.9582	0.08574	228.31	248.89	0.9399
10	0.12207	237.44	259.41	0.9998	0.10922	237.05	258.89	0.9898	0.08993	236.26	257.84	0.9721
20	0.12723	245.33	268.23	1.0304	0.11394	244.99	267.78	1.0206	0.09339	244.30	266.85	1.0034
30	0.13230	253.36	277.17	1.0604	0.11856	253.06	276.77	1.0508	0.09794	252.45	275.95	1.0339
40	0.13730	261.53	286.24	1.0898	0.12311	261.26	285.88	1.0804	0.10181	260.72	285.16	1.0637
50	0.14222	269.85	295.45	1.1187	0.12758	269.61	295.12	1.1094	0.10562	269.12	294.47	1.0930
60	0.14710	278.31	304.79	1.1472	0.13201	278.10	304.50	1.1380	0.10937	277.67	303.91	1.1218
70	0.15193	286.93	314.28	1.1753	0.13639	286.74	314.02	1.1661	0.11307	286.35	313.49	1.1501
80	0.15672	295.71	323.92	1.2030	0.14073	295.53	323.68	1.1939	0.11674	295.18	323.19	1.1780
90	0.16148	304.63	333.70	1.2303	0.14504	304.47	333.48	1.2212	0.12037	304.15	333.04	1.2055
100	0.16622	313.72	343.63	1.2573	0.14932	313.57	343.43	1.2483	0.12398	313.27	343.03	1.2326
	P = 0.28 MPa (T_{sat} = −1.23°C)				P = 0.32 MPa (T_{sat} = 2.48°C)				P = 0.40 MPa (T_{sat} = 8.93°C)			
Sat.	0.07193	226.38	246.52	0.9197	0.06322	228.43	248.66	0.9177	0.05089	231.97	252.32	0.9145
0	0.07240	227.37	247.64	0.9238								
10	0.07613	235.44	256.76	0.9566	0.06576	234.61	255.65	0.9427	0.05119	232.87	253.35	0.9182
20	0.07972	243.59	265.91	0.9883	0.06901	242.87	264.95	0.9749	0.05397	241.37	262.96	0.9515
30	0.08320	251.83	275.12	1.0192	0.07214	251.19	274.28	1.0062	0.05662	249.89	272.54	0.9837
40	0.08660	260.17	284.42	1.0494	0.07518	259.61	283.67	1.0367	0.05917	258.47	282.14	1.0148
50	008992	268.64	293.81	1.0789	0.07815	268.14	293.15	1.0665	0.06164	267.13	291.79	1.0452
60	0.09319	277.23	303.32	1.1079	0.08106	276.79	302.72	1.0957	0.06405	275.89	301.51	1.0748
70	0.09641	285.96	312.95	1.1364	0.08392	285.56	312.41	1.1243	0.06641	284.75	311.32	1.1038
80	0.09960	294.82	322.71	1.1644	0.08674	294.46	322.22	1.1525	0.06873	293.73	321.23	1.1322
90	0.10275	303.83	332.60	1.1920	0.08953	303.50	332.15	1.1802	0.07102	302.84	331.25	1.1602
100	0.10587	312.98	342.62	1.2193	0.09229	312.68	342.21	1.1076	0.07327	312.07	341.38	1.1878
110	0.10897	322.27	352.78	1.2461	0.09503	322.00	352.40	1.2345	0.07550	321.44	351.64	1.2149
120	0.11205	331.71	363.08	1.2727	0.09774	331.45	362.73	1.2611	0.07771	330.94	362.03	1.2417
130									0.07991	340.58	372.54	1.2681
140									0.08208	350.35	383.18	1.2941

R-134a

TABLE A-13

Superheated refrigerant-134a (*Concluded*)

T °C	v m³/kg	u kJ/kg	h kJ/kg	s kJ/kg·K	v m³/kg	u kJ/kg	h kJ/kg	s kJ/kg·K	v m³/kg	u kJ/kg	h kJ/kg	s kJ/kg·K
	P = 0.50 MPa (T_{sat} = 15.74°C)				P = 0.60 MPa (T_{sat} = 21.58°C)				P = 0.70 MPa (T_{sat} = 26.72°C)			
Sat.	0.04086	253.64	256.07	0.9117	0.03408	238.74	259.19	0.9097	0.02918	241.42	261.85	0.9080
20	0.04188	239.40	260.34	0.9264								
30	0.04416	248.20	270.28	0.9597	0.03581	246.41	267.89	0.9388	0.02979	244.51	265.37	0.9197
40	0.04633	256.99	280.16	0.9918	0.03774	255.45	278.09	0.9719	0.03157	253.83	275.93	0.9539
50	0.04842	265.83	290.04	1.0229	0.03958	264.48	288.23	1.0037	0.03324	263.08	286.35	0.9867
60	0.05043	274.73	299.95	1.0531	0.04134	273.54	298.35	1.0346	0.03482	272.31	296.69	1.0182
70	0.05240	283.72	309.92	1.0825	0.04304	282.66	308.48	1.0645	0.03634	281.57	307.01	1.0487
80	0.05432	292.80	319.96	1.1114	0.04469	291.86	318.67	1.0938	0.03781	290.88	317.35	1.0784
90	0.05620	302.00	330.10	1.1397	0.04631	301.14	328.93	1.1225	0.03924	300.27	327.74	1.1074
100	0.05805	311.31	340.33	1.1675	0.04790	310.53	339.27	1.1505	0.04064	309.74	338.19	1.1358
110	0.05988	320.74	350.68	1.1949	0.04946	320.03	349.70	1.1781	0.04201	319.31	348.71	1.1637
120	0.06168	330.30	361.14	1.2218	0.05099	329.64	360.24	1.2053	0.04335	328.98	359.33	1.1910
130	0.06347	339.98	371.72	1.2484	0.05251	339.38	370.88	1.2320	0.04468	338.76	370.04	1.2179
140	0.06524	349.79	382.42	1.2746	0.05402	349.23	381.64	1.2584	0.04599	348.66	380.86	1.2444
150					0.05550	359.21	392.52	1.2844	0.04729	358.68	391.79	1.2706
160					0.05698	369.32	403.51	1.3100	0.04857	368.82	402.82	1.2963
	P = 0.80 MPa (T_{sat} = 31.33°C)				P = 0.90 MPa (T_{sat} = 35.53°C)				P = 1.00 MPa (T_{sat} = 39.39°C)			
Sat.	0.02547	243.78	264.15	0.9066	0.02255	245.88	266.18	0.9054	0.02020	247.77	267.97	0.9043
40	0.02691	252.13	273.66	0.9374	0.02325	250.32	271.25	0.9217	0.02029	248.39	268.68	0.9066
50	0.02846	261.62	284.39	0.9711	0.02472	260.09	282.34	0.9566	0.02171	258.48	280.19	0.9428
60	0.02992	271.04	294.98	1.0034	0.02609	269.72	293.21	0.9897	0.02301	268.35	291.36	0.9768
70	0.03131	280.45	305.50	1.0345	0.02738	279.30	303.94	1.0214	0.02423	278.11	302.34	1.0093
80	0.03264	289.89	316.00	1.0647	0.02861	288.87	314.62	1.0521	0.02538	287.82	313.20	1.0405
90	0.03393	299.37	326.52	1.0940	0.02980	298.46	325.28	1.0819	0.02649	297.53	324.01	1.0707
100	0.03519	308.93	337.08	1.1227	0.03095	308.11	335.96	1.1109	0.02755	307.27	334.82	1.1000
110	0.03642	318.57	347.71	1.1508	0.03207	317.82	346.68	1.1392	0.02858	317.06	345.65	1.1286
120	0.03762	328.31	358.40	1.1784	0.03316	327.62	357.47	1.1670	0.02959	326.93	356.52	1.1567
130	0.03881	338.14	369.19	1.2055	0.03423	337.52	368.33	1.1943	0.03058	336.88	367.46	1.1841
140	0.03997	348.09	380.07	1.2321	0.03529	347.51	379.27	1.2211	0.03154	346.92	378.46	1.2111
150	0.04113	358.15	391.05	1.2584	0.03633	357.61	390.31	1.2475	0.03250	357.06	389.56	1.2376
160	0.04227	368.32	402.14	1.2843	0.03736	367.82	401.44	1.2735	0.03344	367.31	400.74	1.2638
170	0.04340	378.61	413.33	1.3098	0.03838	378.14	412.68	1.2992	0.03436	377.66	412.02	1.2895
180	0.04452	389.02	424.63	1.3351	0.03939	388.57	424.02	1.3245	0.03528	388.12	423.40	1.3149
	P = 1.20 MPa (T_{sat} = 46.32°C)				P = 1.40 MPa (T_{sat} = 52.43°C)				P = 1.60 MPa (T_{sat} = 57.92°C)			
Sat.	0.01663	251.03	270.99	0.9023	0.01405	253.74	273.40	0.9003	0.01208	256.00	275.33	0.8982
50	0.01712	254.98	275.52	0.9164								
60	0.01835	265.42	287.44	0.9527	0.01495	262.17	283.10	0.9297	0.01233	258.48	278.20	0.9069
70	0.01947	275.59	298.96	0.9868	0.01603	272.87	295.31	0.9658	0.01340	269.89	291.33	0.9457
80	0.02051	285.62	310.24	1.0192	0.01701	283.29	307.10	0.9997	0.01435	280.78	303.74	0.9813
90	0.02150	295.59	321.39	1.0503	0.01792	293.55	318.63	1.0319	0.01521	291.39	315.72	1.0148
100	0.02244	305.54	332.47	1.0804	0.01878	303.73	330.02	1.0628	0.01601	301.84	327.46	1.0467
110	0.02335	315.50	343.52	1.1096	0.01960	313.88	341.32	1.0927	0.01677	312.20	339.04	1.0773
120	0.02423	325.51	354.58	1.1381	0.02039	324.05	352.59	1.1218	0.01750	322.53	350.53	1.1069
130	0.02508	335.58	365.68	1.1660	0.02115	334.25	363.86	1.1501	0.01820	332.87	361.99	1.1357
140	0.02592	345.73	376.83	1.1933	0.02189	344.50	375.15	1.1777	0.01887	343.24	373.44	1.1638
150	0.02674	355.95	388.04	1.2201	0.02262	354.82	386.49	1.2048	0.01953	353.66	384.91	1.1912
160	0.02754	366.27	399.33	1.2465	0.02333	365.22	397.89	1.2315	0.02017	364.15	396.43	1.2181
170	0.02834	376.69	410.70	1.2724	0.02403	375.71	409.36	1.2576	0.02080	374.71	407.99	1.2445
180	0.02912	387.21	422.16	1.2980	0.02472	386.29	420.90	1.2834	0.02142	385.35	419.62	1.2704
190					0.02541	396.96	432.53	1.3088	0.02203	396.08	431.33	1.2960
200					0.02608	407.73	444.24	1.3338	0.02263	406.90	443.11	1.3212

R-134a

951

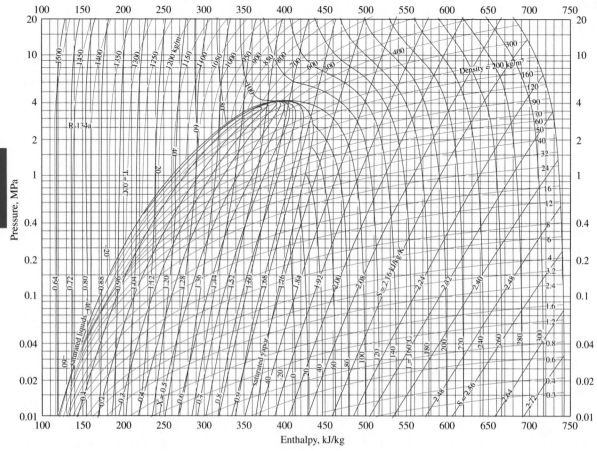

Note: The reference point used for the chart is different than that used in the R-134a tables. Therefore, problems should be solved using all property data either from the tables or from the chart, but not from both.

FIGURE A-14

P-h diagram for refrigerant-134a. (Reprinted by permission of American Society of Heating, Refrigerating, and Air-Conditioning Engineers, Inc., Atlanta, GA.)

TABLE A-15

Properties of saturated water

Temperature, T °C	Saturation pressure, P_{sat} kPa	Density, ρ kg/m³ Liquid	Density Vapor	Enthalpy of vaporization, h_{fg} kJ/kg	Specific heat, c_p J/kg·°C Liquid	Specific heat Vapor	Thermal conductivity, k W/m·°C Liquid	Thermal conductivity Vapor	Dynamic viscosity, μ kg/m·s Liquid	Dynamic viscosity Vapor	Prandtl number, Pr Liquid	Prandtl number Vapor	Volume expansion coefficient, β 1/K Liquid
0.01	0.6113	999.8	0.0048	2501	4217	1854	0.561	0.0171	1.792×10^{-3}	0.922×10^{-5}	13.5	1.00	-0.068×10^{-3}
5	0.8721	999.9	0.0068	2490	4205	1857	0.571	0.0173	1.519×10^{-3}	0.934×10^{-5}	11.2	1.00	0.015×10^{-3}
10	1.2276	999.7	0.0094	2478	4194	1862	0.580	0.0176	1.307×10^{-3}	0.946×10^{-5}	9.45	1.00	0.733×10^{-3}
15	1.7051	999.1	0.0128	2466	4186	1863	0.589	0.0179	1.138×10^{-3}	0.959×10^{-5}	8.09	1.00	0.138×10^{-3}
20	2.339	998.0	0.0173	2454	4182	1867	0.598	0.0182	1.002×10^{-3}	0.973×10^{-5}	7.01	1.00	0.195×10^{-3}
25	3.169	997.0	0.0231	2442	4180	1870	0.607	0.0186	0.891×10^{-3}	0.987×10^{-5}	6.14	1.00	0.247×10^{-3}
30	4.246	996.0	0.0304	2431	4178	1875	0.615	0.0189	0.798×10^{-3}	1.001×10^{-5}	5.42	1.00	0.294×10^{-3}
35	5.628	994.0	0.0397	2419	4178	1880	0.623	0.0192	0.720×10^{-3}	1.016×10^{-5}	4.83	1.00	0.337×10^{-3}
40	7.384	992.1	0.0512	2407	4179	1885	0.631	0.0196	0.653×10^{-3}	1.031×10^{-5}	4.32	1.00	0.377×10^{-3}
45	9.593	990.1	0.0655	2395	4180	1892	0.637	0.0200	0.596×10^{-3}	1.046×10^{-5}	3.91	1.00	0.415×10^{-3}
50	12.35	988.1	0.0831	2383	4181	1900	0.644	0.0204	0.547×10^{-3}	1.062×10^{-5}	3.55	1.00	0.451×10^{-3}
55	15.76	985.2	0.1045	2371	4183	1908	0.649	0.0208	0.504×10^{-3}	1.077×10^{-5}	3.25	1.00	0.484×10^{-3}
60	19.94	983.3	0.1304	2359	4185	1916	0.654	0.0212	0.467×10^{-3}	1.093×10^{-5}	2.99	1.00	0.517×10^{-3}
65	25.03	980.4	0.1614	2346	4187	1926	0.659	0.0216	0.433×10^{-3}	1.110×10^{-5}	2.75	1.00	0.548×10^{-3}
70	31.19	977.5	0.1983	2334	4190	1936	0.663	0.0221	0.404×10^{-3}	1.126×10^{-5}	2.55	1.00	0.578×10^{-3}
75	38.58	974.7	0.2421	2321	4193	1948	0.667	0.0225	0.378×10^{-3}	1.142×10^{-5}	2.38	1.00	0.607×10^{-3}
80	47.39	971.8	0.2935	2309	4197	1962	0.670	0.0230	0.355×10^{-3}	1.159×10^{-5}	2.22	1.00	0.653×10^{-3}
85	57.83	968.1	0.3536	2296	4201	1977	0.673	0.0235	0.333×10^{-3}	1.176×10^{-5}	2.08	1.00	0.670×10^{-3}
90	70.14	965.3	0.4235	2283	4206	1993	0.675	0.0240	0.315×10^{-3}	1.193×10^{-5}	1.96	1.00	0.702×10^{-3}
95	84.55	961.5	0.5045	2270	4212	2010	0.677	0.0246	0.297×10^{-3}	1.210×10^{-5}	1.85	1.00	0.716×10^{-3}
100	101.33	957.9	0.5978	2257	4217	2029	0.679	0.0251	0.282×10^{-3}	1.227×10^{-5}	1.75	1.00	0.750×10^{-3}

TABLE A-15

Properties of saturated water (*Continued*)

Temperature T °C	Saturation pressure P_{sat} kPa	Density ρ kg/m³ Liquid	Density ρ kg/m³ Vapor	Enthalpy of vaporization h_{fg} kJ/kg	Specific heat C_p J/kg·°C Liquid	Specific heat C_p J/kg·°C Vapor	Thermal conductivity k W/m·°C Liquid	Thermal conductivity k W/m·°C Vapor	Dynamic viscosity μ kg/m·s Liquid	Dynamic viscosity μ kg/m·s Vapor	Prandtl number, Pr Liquid	Prandtl number, Pr Vapor	Volume expansion coefficient, β 1/K Liquid
110	143.27	950.6	0.8263	2230	4229	2071	0.682	0.0262	0.255×10^{-3}	1.261×10^{-5}	1.58	1.00	0.798×10^{-3}
120	198.53	943.4	1.121	2203	4244	2120	0.683	0.0275	0.232×10^{-3}	1.296×10^{-5}	1.44	1.00	0.858×10^{-3}
130	270.1	934.6	1.496	2174	4263	2177	0.684	0.0288	0.213×10^{-3}	1.330×10^{-5}	1.33	1.01	0.913×10^{-3}
140	361.3	921.7	1.965	2145	4286	2244	0.683	0.0301	0.197×10^{-3}	1.365×10^{-5}	1.24	1.02	0.970×10^{-3}
150	475.8	916.6	2.546	2114	4311	2314	0.682	0.0316	0.183×10^{-3}	1.399×10^{-5}	1.16	1.02	1.025×10^{-3}
160	617.8	907.4	3.256	2083	4340	2420	0.680	0.0331	0.170×10^{-3}	1.434×10^{-5}	1.09	1.05	1.145×10^{-3}
170	791.7	897.7	4.119	2050	4370	2490	0.677	0.0347	0.160×10^{-3}	1.468×10^{-5}	1.03	1.05	1.178×10^{-3}
180	1002.1	887.3	5.153	2015	4410	2590	0.673	0.0364	0.150×10^{-3}	1.502×10^{-5}	0.983	1.07	1.210×10^{-3}
190	1254.4	876.4	6.388	1979	4460	2710	0.669	0.0382	0.142×10^{-3}	1.537×10^{-5}	0.947	1.09	1.280×10^{-3}
200	1553.8	864.3	7.852	1941	4500	2840	0.663	0.0401	0.134×10^{-3}	1.571×10^{-5}	0.910	1.11	1.350×10^{-3}
220	2318	840.3	11.60	1859	4610	3110	0.650	0.0442	0.122×10^{-3}	1.641×10^{-5}	0.865	1.15	1.520×10^{-3}
240	3344	813.7	16.73	1767	4760	3520	0.632	0.0487	0.111×10^{-3}	1.712×10^{-5}	0.836	1.24	1.720×10^{-3}
260	4688	783.7	23.69	1663	4970	4070	0.609	0.0540	0.102×10^{-3}	1.788×10^{-5}	0.832	1.35	2.000×10^{-3}
280	6412	750.8	33.15	1544	5280	4835	0.581	0.0605	0.094×10^{-3}	1.870×10^{-5}	0.854	1.49	2.380×10^{-3}
300	8581	713.8	46.15	1405	5750	5980	0.548	0.0695	0.086×10^{-3}	1.965×10^{-5}	0.902	1.69	2.950×10^{-3}
320	11,274	667.1	64.57	1239	6540	7900	0.509	0.0836	0.078×10^{-3}	2.084×10^{-5}	1.00	1.97	—
340	14,586	610.5	92.62	1028	8240	11,870	0.469	0.110	0.070×10^{-3}	2.255×10^{-5}	1.23	2.43	—
360	18,651	528.3	144.0	720	14,690	25,800	0.427	0.178	0.060×10^{-3}	2.571×10^{-5}	2.06	3.73	—
374.14	22,090	317.0	317.0	0	∞	∞	∞	∞	0.043×10^{-3}	4.313×10^{-5}	—	—	—

Note 1: Kinematic viscosity ν and thermal diffusivity α can be calculated from their definitions, $\nu = \mu/\rho$ and $\alpha = k/\rho C_p = \nu/Pr$. The temperatures 0.01°C, 100°C, and 374.14°C are the triple-, boiling-, and critical-point temperatures of water, respectively. The properties listed above (except the vapor density) can be used at any pressure with negligible error except at temperatures near the critical-point value.

Note 2: The unit kJ/kg · °C for specific heat is equivalent to kJ/kg · K, and the unit W/m · °C for thermal conductivity is equivalent to W/m · K.

Source: Viscosity and thermal conductivity data are from J. V. Sengers and J. T. R. Watson, *Journal of Physical and Chemical Reference Data* 15 (1986), pp. 1291–1322. Other data are obtained from various sources or calculated.

TABLE A-16

Properties of liquids

Tempera-ture, T °C	Density, ρ kg/m^3	Specific heat, C_p J/kg · °C	Thermal conductivity, k W/m · °C	Thermal diffusivity, α m^2/s	Dynamic viscosity, μ kg/m · s	Kinematic viscosity, ν m^2/s	Prandtl number, Pr
			Ammonia				
−40	692	4467	0.546	1.78×10^{-7}	2.81×10^{-4}	4.06×10^{-7}	2.28
−20	667	4509	0.546	1.82×10^{-7}	2.54×10^{-4}	3.81×10^{-7}	2.09
0	640	4635	0.540	1.82×10^{-7}	2.39×10^{-4}	3.73×10^{-7}	2.05
20	612	4798	0.521	1.78×10^{-7}	2.20×10^{-4}	3.59×10^{-7}	2.02
40	581	4999	0.493	1.70×10^{-7}	1.98×10^{-4}	3.40×10^{-7}	2.00
			Ethyl alcohol (C_2H_6O)				
−40	823	2037	0.186	1.11×10^{-7}	4.81×10^{-3}	5.84×10^{-6}	52.7
−20	815	2124	0.179	1.03×10^{-7}	2.83×10^{-3}	3.47×10^{-6}	33.6
0	806	2249	0.174	0.960×10^{-7}	1.77×10^{-3}	2.20×10^{-6}	22.9
20	789	2395	0.168	0.889×10^{-7}	1.20×10^{-3}	1.52×10^{-6}	17.0
40	772	2572	0.162	0.816×10^{-7}	0.834×10^{-3}	1.08×10^{-6}	13.2
60	755	2781	0.156	0.743×10^{-7}	0.592×10^{-3}	0.784×10^{-6}	10.6
80	738	3026	0.150	0.672×10^{-7}	0.430×10^{-3}	0.583×10^{-6}	8.7
			Ethylene glycol ($C_2H_6O_2$)				
0	1131	2295	0.254	9.79×10^{-8}	65.1×10^{-3}	57.5×10^{-6}	588
20	1117	2386	0.257	9.64×10^{-8}	21.4×10^{-3}	19.2×10^{-6}	199
40	1101	2476	0.259	9.50×10^{-8}	9.57×10^{-3}	8.69×10^{-6}	91
60	1088	2565	0.262	9.39×10^{-8}	5.17×10^{-3}	4.75×10^{-6}	51
80	1078	2656	0.265	9.26×10^{-8}	3.21×10^{-3}	2.98×10^{-6}	32
100	1059	2750	0.267	9.17×10^{-8}	2.15×10^{-3}	2.03×10^{-6}	22
			Freon-12 refrigerant (CCl_2F_2)				
−40	1515	885	0.069	5.14×10^{-8}	4.24×10^{-4}	2.80×10^{-7}	5.4
−20	1457	907	0.071	5.38×10^{-8}	3.43×10^{-4}	2.35×10^{-7}	4.4
0	1393	935	0.073	5.59×10^{-8}	2.98×10^{-4}	2.14×10^{-7}	3.8
20	1327	966	0.073	5.66×10^{-8}	2.62×10^{-4}	1.97×10^{-7}	3.5
40	1254	1002	0.069	5.46×10^{-8}	2.40×10^{-4}	1.91×10^{-7}	3.5
			Glycerin				
−20	1288	2143	0.282	1.02×10^{-7}	134	104×10^{-3}	1020×10^3
0	1276	2261	0.284	0.98×10^{-7}	12.1	9.5×10^{-3}	96×10^3
20	1264	2386	0.287	0.95×10^{-7}	1.49	1.2×10^{-3}	12.4×10^3
40	1252	2513	0.290	0.92×10^{-7}	0.27	0.2×10^{-3}	2.3×10^3
			Lead				
601*	10,588	161	15.5	0.91×10^{-5}	2.62×10^{-3}	2.47×10^{-7}	0.0272
700	10,476	157	17.4	1.06×10^{-5}	2.15×10^{-3}	2.05×10^{-7}	0.0194
800	10,359	153	19.0	1.20×10^{-5}	2.05×10^{-3}	1.98×10^{-7}	0.0165
900	10,237	149	20.3	1.33×10^{-5}	1.54×10^{-3}	1.50×10^{-7}	0.0113
1000	10,111	145	21.5	1.47×10^{-5}	1.32×10^{-3}	1.30×10^{-7}	0.0089
			Mercury				
234*	13,723	142	7.3	3.8×10^{-6}	2.00×10^{-3}	1.46×10^{-7}	0.0389
273	13,628	140	8.2	4.3×10^{-6}	1.69×10^{-3}	1.24×10^{-7}	0.0289
300	13,562	139	8.9	4.7×10^{-6}	1.51×10^{-3}	1.11×10^{-7}	0.0237
350	13,441	138	10.0	5.4×10^{-6}	1.31×10^{-3}	0.98×10^{-7}	0.0181
400	13,320	137	11.0	6.1×10^{-6}	1.18×10^{-3}	0.89×10^{-7}	0.0147
500	13,081	136	12.7	7.1×10^{-6}	1.02×10^{-3}	0.78×10^{-7}	0.0109
600	12,816	134	14.2	8.3×10^{-6}	0.84×10^{-3}	0.66×10^{-7}	0.0080
			Unused engine oil				
0	899	1796	0.147	9.11×10^{-8}	3850×10^{-3}	4280×10^{-6}	47,100
20	888	1880	0.145	8.72×10^{-8}	800×10^{-3}	901×10^{-6}	10,400
40	876	1964	0.144	8.34×10^{-8}	212×10^{-3}	242×10^{-6}	2870
60	864	2047	0.140	8.00×10^{-8}	72.5×10^{-3}	83.9×10^{-6}	1050
80	852	2131	0.138	7.69×10^{-8}	32.0×10^{-3}	37.5×10^{-6}	490
100	840	2219	0.137	7.38×10^{-8}	17.1×10^{-3}	20.3×10^{-6}	276
120	829	2307	0.135	7.10×10^{-8}	10.2×10^{-3}	12.4×10^{-6}	175
140	817	2395	0.133	6.86×10^{-8}	6.53×10^{-3}	8.0×10^{-6}	116
160	806	2483	0.132	6.63×10^{-8}	4.49×10^{-3}	5.6×10^{-6}	84

*Melting point.

Source: Tables A-16 and A-18 are adapted from Frank M. White, *Heat and Mass Transfer* (Reading, MA: Addison-Wesley, 1988), pp. 677–88 and 692–94. Originally compiled from various sources. Reprinted by permission of Addison-Wesley Longman Publishing Company, Inc.

TABLE A-17

Ideal-gas properties of air

T K	h kJ/kg	P_r	u kJ/kg	v_r	$s°$ kJ/kg · K	T K	h kJ/kg	P_r	u kJ/kg	v_r	$s°$ kJ/kg · K
200	199.97	0.3363	142.56	1707.0	1.29559	580	586.04	14.38	419.55	115.7	2.37348
210	209.97	0.3987	149.69	1512.0	1.34444	590	596.52	15.31	427.15	110.6	2.39140
220	219.97	0.4690	156.82	1346.0	1.39105	600	607.02	16.28	434.78	105.8	2.40902
230	230.02	0.5477	164.00	1205.0	1.43557	610	617.53	17.30	442.42	101.2	2.42644
240	240.02	0.6355	171.13	1084.0	1.47824	620	628.07	18.36	450.09	96.92	2.44356
250	250.05	0.7329	178.28	979.0	1.51917	630	638.63	19.84	457.78	92.84	2.46048
260	260.09	0.8405	185.45	887.8	1.55848	640	649.22	20.64	465.50	88.99	2.47716
270	270.11	0.9590	192.60	808.0	1.59634	650	659.84	21.86	473.25	85.34	2.49364
280	280.13	1.0889	199.75	738.0	1.63279	660	670.47	23.13	481.01	81.89	2.50985
285	285.14	1.1584	203.33	706.1	1.65055	670	681.14	24.46	488.81	78.61	2.52589
290	290.16	1.2311	206.91	676.1	1.66802	680	691.82	25.85	496.62	75.50	2.54175
295	295.17	1.3068	210.49	647.9	1.68515	690	702.52	27.29	504.45	72.56	2.55731
300	300.19	1.3860	214.07	621.2	1.70203	700	713.27	28.80	512.33	69.76	2.57277
305	305.22	1.4686	217.67	596.0	1.71865	710	724.04	30.38	520.23	67.07	2.58810
310	310.24	1.5546	221.25	572.3	1.73498	720	734.82	32.02	528.14	64.53	2.60319
315	315.27	1.6442	224.85	549.8	1.75106	730	745.62	33.72	536.07	62.13	2.61803
320	320.29	1.7375	228.42	528.6	1.76690	740	756.44	35.50	544.02	59.82	2.63280
325	325.31	1.8345	232.02	508.4	1.78249	750	767.29	37.35	551.99	57.63	2.64737
330	330.34	1.9352	235.61	489.4	1.79783	760	778.18	39.27	560.01	55.54	2.66176
340	340.42	2.149	242.82	454.1	1.82790	780	800.03	43.35	576.12	51.64	2.69013
350	350.49	2.379	250.02	422.2	1.85708	800	821.95	47.75	592.30	48.08	2.71787
360	360.58	2.626	257.24	393.4	1.88543	820	843.98	52.59	608.59	44.84	2.74504
370	370.67	2.892	264.46	367.2	1.91313	840	866.08	57.60	624.95	41.85	2.77170
380	380.77	3.176	271.69	343.4	1.94001	860	888.27	63.09	641.40	39.12	2.79783
390	390.88	3.481	278.93	321.5	1.96633	880	910.56	68.98	657.95	36.61	2.82344
400	400.98	3.806	286.16	301.6	1.99194	900	932.93	75.29	674.58	34.31	2.84856
410	411.12	4.153	293.43	283.3	2.01699	920	955.38	82.05	691.28	32.18	2.87324
420	421.26	4.522	300.69	266.6	2.04142	940	977.92	89.28	708.08	30.22	2.89748
430	431.43	4.915	307.99	251.1	2.06533	960	1000.55	97.00	725.02	28.40	2.92128
440	441.61	5.332	315.30	236.8	2.08870	980	1023.25	105.2	741.98	26.73	2.94468
450	451.80	5.775	322.62	223.6	2.11161	1000	1046.04	114.0	758.94	25.17	2.96770
460	462.02	6.245	329.97	211.4	2.13407	1020	1068.89	123.4	776.10	23.72	2.99034
470	472.24	6.742	337.32	200.1	2.15604	1040	1091.85	133.3	793.36	23.29	3.01260
480	482.49	7.268	344.70	189.5	2.17760	1060	1114.86	143.9	810.62	21.14	3.03449
490	492.74	7.824	352.08	179.7	2.19876	1080	1137.89	155.2	827.88	19.98	3.05608
500	503.02	8.411	359.49	170.6	2.21952	1100	1161.07	167.1	845.33	18.896	3.07732
510	513.32	9.031	366.92	162.1	2.23993	1120	1184.28	179.7	862.79	17.886	3.09825
520	523.63	9.684	374.36	154.1	2.25997	1140	1207.57	193.1	880.35	16.946	3.11883
530	533.98	10.37	381.84	146.7	2.27967	1160	1230.92	207.2	897.91	16.064	3.13916
540	544.35	11.10	389.34	139.7	2.29906	1180	1254.34	222.2	915.57	15.241	3.15916
550	555.74	11.86	396.86	133.1	2.31809	1200	1277.79	238.0	933.33	14.470	3.17888
560	565.17	12.66	404.42	127.0	2.33685	1220	1301.31	254.7	951.09	13.747	3.19834
570	575.59	13.50	411.97	121.2	2.35531	1240	1324.93	272.3	968.95	13.069	3.21751

TABLE A-17

Ideal-gas properties of air (*Concluded*)

T K	h kJ/kg	P_r	u kJ/kg	v_r	$s°$ kJ/kg · K	T K	h kJ/kg	P_r	u kJ/kg	v_r	$s°$ kJ/kg · K
1260	1348.55	290.8	986.90	12.435	3.23638	1600	1757.57	791.2	1298.30	5.804	3.52364
1280	1372.24	310.4	1004.76	11.835	3.25510	1620	1782.00	834.1	1316.96	5.574	3.53879
1300	1395.97	330.9	1022.82	11.275	3.27345	1640	1806.46	878.9	1335.72	5.355	3.55381
1320	1419.76	352.5	1040.88	10.747	3.29160	1660	1830.96	925.6	1354.48	5.147	3.56867
1340	1443.60	375.3	1058.94	10.247	3.30959	1680	1855.50	974.2	1373.24	4.949	3.58335
1360	1467.49	399.1	1077.10	9.780	3.32724	1700	1880.1	1025	1392.7	4.761	3.5979
1380	1491.44	424.2	1095.26	9.337	3.34474	1750	1941.6	1161	1439.8	4.328	3.6336
1400	1515.42	450.5	1113.52	8.919	3.36200	1800	2003.3	1310	1487.2	3.994	3.6684
1420	1539.44	478.0	1131.77	8.526	3.37901	1850	2065.3	1475	1534.9	3.601	3.7023
1440	1563.51	506.9	1150.13	8.153	3.39586	1900	2127.4	1655	1582.6	3.295	3.7354
1460	1587.63	537.1	1168.49	7.801	3.41247	1950	2189.7	1852	1630.6	3.022	3.7677
1480	1611.79	568.8	1186.95	7.468	3.42892	2000	2252.1	2068	1678.7	2.776	3.7994
1500	1635.97	601.9	1205.41	7.152	3.44516	2050	2314.6	2303	1726.8	2.555	3.8303
1520	1660.23	636.5	1223.87	6.854	3.46120	2100	2377.7	2559	1775.3	2.356	3.8605
1540	1684.51	672.8	1242.43	6.569	3.47712	2150	2440.3	2837	1823.8	2.175	3.8901
1560	1708.82	710.5	1260.99	6.301	3.49276	2200	2503.2	3138	1872.4	2.012	3.9191
1580	1733.17	750.0	1279.65	6.046	3.50829	2250	2566.4	3464	1921.3	1.864	3.9474

Note: The properties P_r (relative pressure) and v_r (relative specific volume) are dimensionless quantities used in the analysis of isentropic processes, and should not be confused with the properties pressure and specific volume.

Source: Kenneth Wark, *Thermodynamics,* 4th ed. (New York: McGraw-Hill, 1983), pp. 785–86, table A-5. Originally published in J. H. Keenan and J. Kaye, *Gas Tables* (New York: John Wiley & Sons, 1948).

Air

TABLE A-18

Properties of gases at 1 atm pressure

Temperature, T K	Density, ρ kg/m³	Specific heat, C_p J/kg · °C	Thermal conductivity, k W/m · °C	Thermal diffusivity, α m²/s	Dynamic viscosity, μ kg/m · s	Kinematic viscosity, ν m²/s	Prandtl number, Pr
\multicolumn{8}{c}{Air}
200	1.766	1003	0.0181	1.02×10^{-5}	1.34×10^{-5}	0.76×10^{-5}	0.740
250	1.413	1003	0.0223	1.57×10^{-5}	1.61×10^{-5}	1.14×10^{-5}	0.724
280	1.271	1004	0.0246	1.95×10^{-5}	1.75×10^{-5}	1.40×10^{-5}	0.717
290	1.224	1005	0.0253	2.08×10^{-5}	1.80×10^{-5}	1.48×10^{-5}	0.714
298	1.186	1005	0.0259	2.18×10^{-5}	1.84×10^{-5}	1.55×10^{-5}	0.712
300	1.177	1005	0.0261	2.21×10^{-5}	1.85×10^{-5}	1.57×10^{-5}	0.712
310	1.143	1006	0.0268	2.35×10^{-5}	1.90×10^{-5}	1.67×10^{-5}	0.711
320	1.110	1006	0.0275	2.49×10^{-5}	1.94×10^{-5}	1.77×10^{-5}	0.710
330	1.076	1007	0.0283	2.64×10^{-5}	1.99×10^{-5}	1.86×10^{-5}	0.708
340	1.043	1007	0.0290	2.78×10^{-5}	2.03×10^{-5}	1.96×10^{-5}	0.707
350	1.009	1008	0.0297	2.92×10^{-5}	2.08×10^{-5}	2.06×10^{-5}	0.706
400	0.883	1013	0.0331	3.70×10^{-5}	2.29×10^{-5}	2.60×10^{-5}	0.703
450	0.785	1020	0.0363	4.54×10^{-5}	2.49×10^{-5}	3.18×10^{-5}	0.700
500	0.706	1029	0.0395	5.44×10^{-5}	2.68×10^{-5}	3.80×10^{-5}	0.699
550	0.642	1039	0.0426	6.39×10^{-5}	2.86×10^{-5}	4.45×10^{-5}	0.698
600	0.589	1051	0.0456	7.37×10^{-5}	3.03×10^{-5}	5.15×10^{-5}	0.698
700	0.504	1075	0.0513	9.46×10^{-5}	3.35×10^{-5}	6.64×10^{-5}	0.702
800	0.441	1099	0.0569	11.7×10^{-5}	3.64×10^{-5}	8.25×10^{-5}	0.704
900	0.392	1120	0.0625	14.2×10^{-5}	3.92×10^{-5}	9.99×10^{-5}	0.705
1000	0.353	1141	0.0672	16.7×10^{-5}	4.18×10^{-5}	11.8×10^{-5}	0.709
1200	0.294	1175	0.0759	22.2×10^{-5}	4.65×10^{-5}	15.8×10^{-5}	0.720
1400	0.252	1201	0.0835	27.6×10^{-5}	5.09×10^{-5}	20.2×10^{-5}	0.732
1600	0.221	1240	0.0904	33.0×10^{-5}	5.49×10^{-5}	24.9×10^{-5}	0.753
1800	0.196	1276	0.0970	38.3×10^{-5}	5.87×10^{-5}	29.9×10^{-5}	0.772
2000	0.177	1327	0.1032	44.1×10^{-5}	6.23×10^{-5}	35.3×10^{-5}	0.801
\multicolumn{8}{c}{Ammonia (NH₃)}
200	1.038	2199	0.0153	0.67×10^{-5}	6.89×10^{-6}	0.66×10^{-5}	0.990
250	0.831	2248	0.0197	1.05×10^{-5}	8.53×10^{-6}	1.03×10^{-5}	0.973
300	0.692	2298	0.0246	1.55×10^{-5}	10.27×10^{-6}	1.48×10^{-5}	0.959
350	0.593	2349	0.0302	2.17×10^{-5}	12.06×10^{-6}	2.03×10^{-5}	0.938
400	0.519	2402	0.0364	2.92×10^{-5}	13.90×10^{-6}	2.68×10^{-5}	0.917
450	0.461	2455	0.0433	3.82×10^{-5}	15.76×10^{-6}	3.42×10^{-5}	0.894
500	0.415	2507	0.0506	4.86×10^{-5}	17.63×10^{-6}	4.25×10^{-5}	0.873
550	0.378	2559	0.0580	6.00×10^{-5}	19.5×10^{-6}	5.16×10^{-5}	0.860
600	0.346	2611	0.0656	7.26×10^{-5}	21.4×10^{-6}	6.18×10^{-5}	0.852
700	0.297	2710	0.0811	10.1×10^{-5}	25.1×10^{-6}	8.45×10^{-5}	0.839
800	0.260	2810	0.0977	13.4×10^{-5}	28.8×10^{-6}	11.1×10^{-5}	0.828
\multicolumn{8}{c}{Argon}
200	2.435	523.6	0.0124	0.98×10^{-5}	1.60×10^{-5}	0.66×10^{-5}	0.674
250	1.948	522.2	0.0152	1.49×10^{-5}	1.95×10^{-5}	1.00×10^{-5}	0.672
300	1.623	521.6	0.0177	2.09×10^{-5}	2.27×10^{-5}	1.40×10^{-5}	0.669
350	1.392	521.2	0.0201	2.78×10^{-5}	2.57×10^{-5}	1.85×10^{-5}	0.666
400	1.218	521.0	0.0223	3.52×10^{-5}	2.85×10^{-5}	2.34×10^{-5}	0.665
450	1.082	520.9	0.0244	4.33×10^{-5}	3.12×10^{-5}	2.88×10^{-5}	0.665

TABLE A-18

Properties of gases at 1 atm pressure (*Continued*)

Tempera-ture, T K	Density, ρ kg/m³	Specific heat, C_p J/kg · °C	Thermal conductivity, k W/m · °C	Thermal diffusivity, α m²/s	Dynamic viscosity, μ kg/m · s	Kinematic viscosity, ν m²/s	Prandtl number, Pr
500	0.974	520.8	0.0264	5.20×10^{-5}	3.37×10^{-5}	3.45×10^{-5}	0.664
550	0.886	520.7	0.0283	6.14×10^{-5}	3.60×10^{-5}	4.07×10^{-5}	0.662
600	0.812	520.6	0.0301	7.12×10^{-5}	3.83×10^{-5}	4.72×10^{-5}	0.662
700	0.696	520.6	0.0336	9.28×10^{-5}	4.25×10^{-5}	6.11×10^{-5}	0.658
800	0.609	520.5	0.0369	11.6×10^{-5}	4.64×10^{-5}	7.62×10^{-5}	0.655
900	0.541	520.5	0.0398	14.1×10^{-5}	5.01×10^{-5}	9.26×10^{-5}	0.654
1000	0.487	520.5	0.0427	16.8×10^{-5}	5.35×10^{-5}	11.0×10^{-5}	0.652
1200	0.406	520.5	0.0481	22.8×10^{-5}	5.99×10^{-5}	14.8×10^{-5}	0.648
1400	0.348	520.4	0.0535	29.6×10^{-5}	6.56×10^{-5}	18.9×10^{-5}	0.638
Carbon dioxide (CO₂)							
200	2.683	759	0.0095	0.47×10^{-5}	1.02×10^{-5}	0.38×10^{-5}	0.814
250	2.146	806	0.0129	0.75×10^{-5}	1.26×10^{-5}	0.59×10^{-5}	0.790
300	1.789	852	0.0166	1.09×10^{-5}	1.50×10^{-5}	0.84×10^{-5}	0.768
350	1.533	897	0.0205	1.49×10^{-5}	1.73×10^{-5}	1.13×10^{-5}	0.755
400	1.341	939	0.0244	1.94×10^{-5}	1.94×10^{-5}	1.45×10^{-5}	0.747
450	1.192	979	0.0283	2.43×10^{-5}	2.15×10^{-5}	1.80×10^{-5}	0.743
500	1.073	1017	0.0323	2.96×10^{-5}	2.35×10^{-5}	2.19×10^{-5}	0.740
550	0.976	1049	0.0363	3.55×10^{-5}	2.54×10^{-5}	2.60×10^{-5}	0.734
600	0.894	1077	0.0403	4.18×10^{-5}	2.72×10^{-5}	3.04×10^{-5}	0.727
700	0.767	1126	0.0487	5.64×10^{-5}	3.06×10^{-5}	3.99×10^{-5}	0.708
800	0.671	1169	0.0560	7.14×10^{-5}	3.39×10^{-5}	5.05×10^{-5}	0.708
900	0.596	1205	0.0621	8.65×10^{-5}	3.69×10^{-5}	6.19×10^{-5}	0.716
1000	0.537	1235	0.0680	10.25×10^{-5}	3.97×10^{-5}	7.40×10^{-5}	0.721
1200	0.447	1283	0.0780	13.6×10^{-5}	4.49×10^{-5}	10.04×10^{-5}	0.739
1400	0.383	1315	0.0867	17.2×10^{-5}	4.97×10^{-5}	13.0×10^{-5}	0.754
Carbon monoxide (CO)							
200	1.708	1045	0.0175	0.98×10^{-5}	1.27×10^{-5}	0.75×10^{-5}	0.763
250	1.366	1048	0.0214	1.50×10^{-5}	1.54×10^{-5}	1.13×10^{-5}	0.753
300	1.138	1051	0.0252	2.11×10^{-5}	1.78×10^{-5}	1.56×10^{-5}	0.743
350	0.976	1056	0.0288	2.80×10^{-5}	2.01×10^{-5}	2.05×10^{-5}	0.735
400	0.854	1060	0.0323	3.57×10^{-5}	2.21×10^{-5}	2.59×10^{-5}	0.727
450	0.759	1065	0.0355	4.39×10^{-5}	2.41×10^{-5}	3.18×10^{-5}	0.723
500	0.683	1071	0.0386	5.28×10^{-5}	2.60×10^{-5}	3.80×10^{-5}	0.720
550	0.621	1077	0.0416	6.22×10^{-5}	2.77×10^{-5}	4.46×10^{-5}	0.717
600	0.569	1084	0.0444	7.20×10^{-5}	2.94×10^{-5}	5.17×10^{-5}	0.718
700	0.488	1099	0.0497	9.27×10^{-5}	3.25×10^{-5}	6.66×10^{-5}	0.718
800	0.427	1114	0.0549	11.5×10^{-5}	3.54×10^{-5}	8.29×10^{-5}	0.718
900	0.379	1128	0.0596	13.9×10^{-5}	3.81×10^{-5}	10.04×10^{-5}	0.721
1000	0.342	1142	0.0644	16.5×10^{-5}	4.06×10^{-5}	11.9×10^{-5}	0.720
1100	0.310	1155	0.0692	19.3×10^{-5}	4.30×10^{-5}	13.9×10^{-5}	0.718
1200	0.285	1168	0.0738	22.2×10^{-5}	4.53×10^{-5}	15.9×10^{-5}	0.717
Helium							
200	0.2440	5197	0.115	0.91×10^{-4}	1.50×10^{-5}	0.61×10^{-4}	0.676
250	0.1952	5197	0.134	1.54×10^{-4}	1.75×10^{-5}	0.90×10^{-4}	0.680

TABLE A-18

Properties of gases at 1 atm pressure (*Continued*)

Temperature, T K	Density, ρ kg/m³	Specific heat, C_p J/kg · °C	Thermal conductivity, k W/m · °C	Thermal diffusivity, α m²/s	Dynamic viscosity, μ kg/m · s	Kinematic viscosity, ν m²/s	Prandtl number, Pr
300	0.1627	5197	0.150	1.77×10^{-4}	1.99×10^{-5}	1.22×10^{-4}	0.690
350	0.1394	5197	0.165	2.28×10^{-4}	2.21×10^{-5}	1.59×10^{-4}	0.698
400	0.1220	5197	0.180	2.83×10^{-4}	2.43×10^{-5}	1.99×10^{-4}	0.703
450	0.1085	5197	0.195	3.45×10^{-4}	2.63×10^{-5}	2.43×10^{-4}	0.702
500	0.0976	5197	0.211	4.17×10^{-4}	2.83×10^{-5}	2.90×10^{-4}	0.695
550	0.0887	5197	0.229	4.97×10^{-4}	3.02×10^{-5}	3.40×10^{-4}	0.684
600	0.0813	5197	0.247	5.84×10^{-4}	3.20×10^{-5}	3.93×10^{-4}	0.673
700	0.0697	5197	0.278	7.67×10^{-4}	3.55×10^{-5}	5.09×10^{-4}	0.663
800	0.0610	5197	0.307	9.68×10^{-4}	3.88×10^{-5}	6.37×10^{-4}	0.657
900	0.0542	5197	0.335	11.9×10^{-4}	4.20×10^{-5}	7.75×10^{-4}	0.652
1000	0.0488	5197	0.363	14.3×10^{-4}	4.50×10^{-5}	9.23×10^{-4}	0.645
1200	0.0407	5197	0.416	19.7×10^{-4}	5.08×10^{-5}	12.5×10^{-4}	0.635
1400	0.0349	5197	0.469	25.9×10^{-4}	5.61×10^{-5}	16.1×10^{-4}	0.622
1600	0.0305	5197	0.521	32.9×10^{-4}	6.10×10^{-5}	20.0×10^{-4}	0.608
1800	0.0271	5197	0.570	40.4×10^{-4}	6.57×10^{-5}	24.2×10^{-4}	0.599
2000	0.0244	5197	0.620	48.9×10^{-4}	7.00×10^{-5}	28.7×10^{-4}	0.587
Hydrogen							
200	0.1299	13,540	0.128	0.77×10^{-4}	0.68×10^{-5}	0.55×10^{-4}	0.717
250	0.0983	14,070	0.156	1.13×10^{-4}	0.79×10^{-5}	0.80×10^{-4}	0.713
300	0.0819	14,320	0.182	1.55×10^{-4}	0.89×10^{-5}	1.09×10^{-4}	0.705
350	0.0702	14,420	0.203	2.01×10^{-4}	0.99×10^{-5}	1.42×10^{-4}	0.705
400	0.0614	14,480	0.221	2.49×10^{-4}	1.09×10^{-5}	1.78×10^{-4}	0.714
450	0.0546	14,500	0.239	3.02×10^{-4}	1.18×10^{-5}	2.17×10^{-4}	0.719
500	0.0492	14,510	0.256	3.59×10^{-4}	1.27×10^{-5}	2.59×10^{-4}	0.721
550	0.0447	14,520	0.274	4.22×10^{-4}	1.36×10^{-5}	3.04×10^{-4}	0.722
600	0.0410	14,540	0.291	4.89×10^{-4}	1.45×10^{-5}	3.54×10^{-4}	0.724
700	0.0351	14,610	0.325	6.34×10^{-4}	1.61×10^{-5}	4.59×10^{-4}	0.724
800	0.0307	14,710	0.360	7.97×10^{-4}	1.77×10^{-5}	5.76×10^{-4}	0.723
900	0.0273	14,840	0.394	10.8×10^{-4}	1.92×10^{-5}	7.03×10^{-4}	0.723
1000	0.0246	14,990	0.428	11.6×10^{-4}	2.07×10^{-5}	8.42×10^{-4}	0.724
1200	0.0205	15,370	0.495	15.7×10^{-4}	2.36×10^{-5}	11.5×10^{-4}	0.733
Nitrogen							
200	1.708	1043	0.0183	1.02×10^{-5}	1.29×10^{-5}	0.75×10^{-5}	0.734
250	1.367	1042	0.0222	1.56×10^{-5}	1.55×10^{-5}	1.13×10^{-5}	0.725
300	1.139	1040	0.0260	2.19×10^{-5}	1.79×10^{-5}	1.57×10^{-5}	0.715
350	0.967	1041	0.0294	2.92×10^{-5}	2.01×10^{-5}	2.08×10^{-5}	0.711
400	0.854	1045	0.0325	3.64×10^{-5}	2.21×10^{-5}	2.59×10^{-5}	0.710
450	0.759	1050	0.0356	4.47×10^{-5}	2.41×10^{-5}	3.17×10^{-5}	0.709
500	0.683	1057	0.0387	5.36×10^{-5}	2.59×10^{-5}	3.79×10^{-5}	0.708
550	0.621	1065	0.0414	6.26×10^{-5}	2.76×10^{-5}	4.45×10^{-5}	0.711
600	0.569	1075	0.0441	7.20×10^{-5}	2.93×10^{-5}	5.14×10^{-5}	0.713
700	0.488	1098	0.0493	9.20×10^{-5}	3.24×10^{-5}	6.63×10^{-5}	0.720
800	0.427	1122	0.0541	11.3×10^{-5}	3.52×10^{-5}	8.24×10^{-5}	0.730
900	0.380	1146	0.0587	13.5×10^{-5}	3.79×10^{-5}	9.97×10^{-5}	0.739
1000	0.342	1168	0.0631	15.8×10^{-5}	4.04×10^{-5}	11.8×10^{-5}	0.747

TABLE A-18

Properties of gases at 1 atm pressure (*Concluded*)

Temperature, T K	Density, ρ kg/m³	Specific heat, C_p J/kg · °C	Thermal conductivity, k W/m · °C	Thermal diffusivity, α m²/s	Dynamic viscosity, μ kg/m · s	Kinematic viscosity, ν m²/s	Prandtl number, Pr
1200	0.285	1205	0.0713	20.8×10^{-5}	4.50×10^{-5}	15.8×10^{-5}	0.761
1400	0.244	1233	0.0797	26.5×10^{-5}	4.92×10^{-5}	20.2×10^{-5}	0.761
Oxygen							
200	1.951	906	0.0182	1.03×10^{-5}	1.47×10^{-5}	0.75×10^{-5}	0.728
250	1.561	914	0.0225	1.58×10^{-5}	1.78×10^{-5}	1.14×10^{-5}	0.721
300	1.301	920	0.0267	2.23×10^{-5}	2.07×10^{-5}	1.59×10^{-5}	0.711
350	1.115	929	0.0306	2.95×10^{-5}	2.34×10^{-5}	2.10×10^{-5}	0.710
400	0.976	942	0.0342	3.72×10^{-5}	2.59×10^{-5}	2.65×10^{-5}	0.713
450	0.867	956	0.0377	4.55×10^{-5}	2.83×10^{-5}	3.26×10^{-5}	0.717
500	0.780	971	0.0412	5.44×10^{-5}	3.05×10^{-5}	3.91×10^{-5}	0.720
550	0.709	987	0.0447	6.38×10^{-5}	3.27×10^{-5}	4.61×10^{-5}	0.722
600	0.650	1003	0.0480	7.36×10^{-5}	3.47×10^{-5}	5.34×10^{-5}	0.725
700	0.557	1032	0.0544	9.46×10^{-5}	3.85×10^{-5}	6.91×10^{-5}	0.730
800	0.488	1054	0.0603	11.7×10^{-5}	4.21×10^{-5}	8.63×10^{-5}	0.736
900	0.434	1074	0.0661	14.2×10^{-5}	4.54×10^{-5}	10.5×10^{-5}	0.738
1000	0.390	1091	0.0717	16.8×10^{-5}	4.85×10^{-5}	12.4×10^{-5}	0.738
1200	0.325	1116	0.0821	22.6×10^{-5}	5.42×10^{-5}	16.7×10^{-5}	0.737
1400	0.278	1136	0.0921	29.1×10^{-5}	5.95×10^{-5}	21.3×10^{-5}	0.734
Water vapor (steam)							
300	0.0253*	2041	0.0181	35.1×10^{-5}*	0.91×10^{-5}	36.1×10^{-5}*	1.03
350	0.258*	2037	0.0222	4.22×10^{-5}*	1.12×10^{-5}	4.33×10^{-5}*	1.02
400	0.555	2000	0.0264	2.38×10^{-5}	1.32×10^{-5}	2.38×10^{-5}	1.00
450	0.491	1968	0.0307	3.17×10^{-5}	1.52×10^{-5}	3.10×10^{-5}	0.98
500	0.441	1977	0.0357	4.09×10^{-5}	1.73×10^{-5}	3.92×10^{-5}	0.96
550	0.401	1994	0.0411	5.15×10^{-5}	1.93×10^{-5}	4.82×10^{-5}	0.94
600	0.367	2022	0.0464	6.25×10^{-5}	2.13×10^{-5}	5.82×10^{-5}	0.93
700	0.314	2083	0.0572	8.74×10^{-5}	2.54×10^{-5}	8.09×10^{-5}	0.93
800	0.275	2148	0.0686	11.6×10^{-5}	2.95×10^{-5}	10.7×10^{-5}	0.92
900	0.244	2217	0.078	14.4×10^{-5}	3.36×10^{-5}	13.7×10^{-5}	0.95
1000	0.220	2288	0.087	17.3×10^{-5}	3.76×10^{-5}	17.1×10^{-5}	0.99

*At saturation pressure (less than 1 atm).

For ideal gases, the properties C_p, k, μ, and Pr are independent of pressure. The properties ρ, ν, and α at a pressure P other than 1 atm are determined by multiplying the value of ρ at the given temperature by P and by dividing the values of ν and α at the given temperature by P, where P is in atm (1 atm = 101.325 kPa = 14.696 psi).

TABLE A-19

Properties of solid metals

Composition	Melting point, K	Properties at 300 K				Properties at various temperatures (K), k(W/m · K)/C_p(J/kg · K)					
		ρ kg/m³	C_p J/kg · K	k W/m · K	$\alpha \times 10^6$ m²/s	100	200	400	600	800	1000
Aluminum:											
Pure	933	2702	903	237	97.1	302	237	240	231	218	
						482	798	949	1033	1146	
Alloy 2024-T6 (4.5% Cu, 1.5% Mg, 0.6% Mn)	775	2770	875	177	73.0	65	163	186	186		
						473	787	925	1042		
Alloy 195, Cast (4.5% Cu)		2790	883	168	68.2			174	185		
Beryllium	1550	1850	1825	200	59.2	990	301	161	126	106	90.8
						203	1114	2191	2604	2823	3018
Bismuth	545	9780	122	7.86	6.59	16.5	9.69	7.04			
						112	120	127			
Boron	2573	2500	1107	27.0	9.76	190	55.5	16.8	10.6	9.60	9.85
						128	600	1463	1892	2160	2338
Cadmium	594	8650	231	96.8	48.4	203	99.3	94.7			
						198	222	242			
Chromium	2118	7160	449	93.7	29.1	159	111	90.9	80.7	71.3	65.4
						192	384	484	542	581	616
Cobalt	1769	8862	421	99.2	26.6	167	122	85.4	67.4	58.2	52.1
						236	379	450	503	550	628
Copper:											
Pure	1358	8933	385	401	117	482	413	393	379	366	352
						252	356	397	417	433	451
Commercial bronze (90% Cu, 10% Al)	1293	8800	420	52	14		42	52	59		
							785	160	545		
Phosphor gear bronze (89% Cu, 11% Sn)	1104	8780	355	54	17		41	65	74		
							—	—	—		
Cartridge brass (70% Cu, 30% Zn)	1188	8530	380	110	33.9	75	95	137	149		
							360	395	425		
Constantan (55% Cu, 45% Ni)	1493	8920	384	23	6.71	17	19				
						237	362				
Germanium	1211	5360	322	59.9	34.7	232	96.8	43.2	27.3	19.8	17.4
						190	290	337	348	357	375
Gold	1336	19,300	129	317	127	327	323	311	298	284	270
						109	124	131	135	140	145
Iridium	2720	22,500	130	147	50.3	172	153	144	138	132	126
						90	122	133	138	144	153
Iron:											
Pure	1810	7870	447	80.2	23.1	134	94.0	69.5	54.7	43.3	32.8
						216	384	490	574	680	975
Armco (99.75% pure)		7870	447	72.7	20.7	95.6	80.6	65.7	53.1	42.2	32.3
						215	384	490	574	680	975
Carbon steels:											
Plain carbon (Mn ≤ 1%, Si ≤ 0.1%)		7854	434	60.5	17.7			56.7	48.0	39.2	30.0
								487	559	685	1169

TABLE A-19

Properties of solid metals (*Continued*)

Composition	Melting point, K	ρ kg/m³	C_p J/kg·K	k W/m·K	$\alpha \times 10^6$ m²/s	100	200	400	600	800	1000
AISI 1010		7832	434	63.9	18.8			58.7	48.8	39.2	31.3
								487	559	685	1168
Carbon–silicon (Mn ≤ 1%, 0.1% < Si ≤ 0.6%)		7817	446	51.9	14.9			49.8	44.0	37.4	29.3
								501	582	699	971
Carbon–manganese–silicon (1% < Mn ≤ 1.65% 0.1% < Si ≤ 0.6%)		8131	434	41.0	11.6			42.2	39.7	35.0	27.6
								487	559	685	1090
Chromium (low) steels:											
$\frac{1}{2}$ Cr–$\frac{1}{4}$ Mo–Si (0.18% C, 0.65% Cr, 0.23% Mo, 0.6% Si)		7822	444	37.7	10.9			38.2	36.7	33.3	26.9
								492	575	688	969
1Cr–$\frac{1}{2}$ Mo (0.16% C, 1% Cr, 0.54% Mo, 0.39% Si)		7858	442	42.3	12.2			42.0	39.1	34.5	27.4
								492	575	688	969
1Cr–V (0.2% C, 1.02% Cr, 0.15% V)		7836	443	48.9	14.1			46.8	42.1	36.3	28.2
								492	575	688	969
Stainless steels:											
AISI 302		8055	480	15.1	3.91			17.3	20.0	22.8	25.4
								512	559	585	606
AISI 304	1670	7900	477	14.9	3.95	9.2	12.6	16.6	19.8	22.6	25.4
						272	402	515	557	582	611
AISI 316		8238	468	13.4	3.48			15.2	18.3	21.3	24.2
								504	550	576	602
AISI 347		7978	480	14.2	3.71			15.8	18.9	21.9	24.7
								513	559	585	606
Lead	601	11,340	129	35.3	24.1	39.7	36.7	34.0	31.4		
						118	125	132	142		
Magnesium	923	1740	1024	156	87.6	169	159	153	149	146	
						649	934	1074	1170	1267	
Molybdenum	2894	10,240	251	138	53.7	179	143	134	126	118	112
						141	224	261	275	285	295
Nickel:											
Pure	1728	8900	444	90.7	23.0	164	107	80.2	65.6	67.6	71.8
						232	383	485	592	530	562
Nichrome (80% Ni, 20% Cr)	1672	8400	420	12	3.4			14	16	21	
								480	525	545	
Inconel X-750 (73% Ni, 15% Cr, 6.7% Fe)	1665	8510	439	11.7	3.1	8.7	10.3	13.5	17.0	20.5	24.0
						—	372	473	510	546	626
Niobium	2741	8570	265	53.7	23.6	55.2	52.6	55.2	58.2	61.3	64.4
						188	249	274	283	292	301
Palladium	1827	12,020	244	71.8	24.5	76.5	71.6	73.6	79.7	86.9	94.2
						168	227	251	261	271	281
Platinum:											
Pure	2045	21,450	133	71.6	25.1	77.5	72.6	71.8	73.2	75.6	78.7
						100	125	136	141	146	152
Alloy 60Pt–40Rh (60% Pt, 40% Rh)	1800	16,630	162	47	17.4			52	59	65	69
								—	—	—	—

TABLE A-19

Properties of solid metals (*Concluded*)

Composition	Melting point, K	Properties at 300 K				Properties at various temperatures (K), k(W/m · K)/C_p(J/kg · K)					
		ρ kg/m³	C_p J/kg · K	k W/m · K	$\alpha \times 10^6$ m²/s	100	200	400	600	800	1000
Rhenium	3453	21,100	136	47.9	16.7	58.9	51.0	46.1	44.2	44.1	44.6
						97	127	139	145	151	156
Rhodium	2236	12,450	243	150	49.6	186	154	146	136	127	121
						147	220	253	274	293	311
Silicon	1685	2330	712	148	89.2	884	264	98.9	61.9	42.4	31.2
						259	556	790	867	913	946
Silver	1235	10,500	235	429	174	444	430	425	412	396	379
						187	225	239	250	262	277
Tantalum	3269	16,600	140	57.5	24.7	59.2	57.5	57.8	58.6	59.4	60.2
						110	133	144	146	149	152
Thorium	2023	11,700	118	54.0	39.1	59.8	54.6	54.5	55.8	56.9	56.9
						99	112	124	134	145	156
Tin	505	7310	227	66.6	40.1	85.2	73.3	62.2			
						188	215	243			
Titanium	1953	4500	522	21.9	9.32	30.5	24.5	20.4	19.4	19.7	20.7
						300	465	551	591	633	675
Tungsten	3660	19,300	132	174	68.3	208	186	159	137	125	118
						87	122	137	142	146	148
Uranium	1406	19,070	116	27.6	12.5	21.7	25.1	29.6	34.0	38.8	43.9
						94	108	125	146	176	180
Vanadium	2192	6100	489	30.7	10.3	35.8	31.3	31.3	33.3	35.7	38.2
						258	430	515	540	563	597
Zinc	693	7140	389	116	41.8	117	118	111	103		
						297	367	402	436		
Zirconium	2125	6570	278	22.7	12.4	33.2	25.2	21.6	20.7	21.6	23.7
						205	264	300	332	342	362

Source for Tables A-19 and A-20: Frank P. Incropera and David P. DeWitt, *Fundamentals of Heat and Mass Transfer*, 3rd ed. (New York: John Wiley & Sons, 1990), pp. A3–A8. Originally compiled from various sources. Reprinted by permission of John Wiley & Sons, Inc.

TABLE A-20

Properties of solid nonmetals

Composition	Melting point, K	ρ kg/m³	C_p J/kg·K	k W/m·K	$\alpha \times 10^6$ m²/s	100	200	400	600	800	1000
Aluminum oxide, sapphire	2323	3970	765	46	15.1	450 —	82 —	32.4 940	18.9 1110	13.0 1180	10.5 1225
Aluminum oxide, polycrystalline	2323	3970	765	36.0	11.9	133 —	55 —	26.4 940	15.8 1110	10.4 1180	7.85 1225
Beryllium oxide	2725	3000	1030	272	88.0			196 1350	111 1690	70 1865	47 1975
Boron	2573	2500	1105	27.6	9.99	190 —	52.5 —	18.7 1490	11.3 1880	8.1 2135	6.3 2350
Boron fiber epoxy (30% vol.) composite	590	2080									
k, ∥ to fibers				2.29		2.10	2.23	2.28			
k, ⊥ to fibers				0.59		0.37	0.49	0.60			
C_p			1122			364	757	1431			
Carbon											
Amorphous	1500	1950	—	1.60	—	0.67 —	1.18 —	1.89 —	21.9 —	2.37 —	2.53 —
Diamond, type IIa insulator	—	3500	509	2300		10,000 21	4000 194	1540 853			
Graphite, pyrolytic	2273	2210									
k, ∥ to layers				1950		4970	3230	1390	892	667	534
k, ⊥ to layers				5.70		16.8	9.23	4.09	2.68	2.01	1.60
C_p			709			136	411	992	1406	1650	1793
Graphite fiber epoxy (25% vol.) composite	450	1400									
k, heat flow ∥ to fibers				11.1		5.7	8.7	13.0			
k, heat flow ⊥ to fibers				0.87		0.46	0.68	1.1			
C_p			935			337	642	1216			
Pyroceram, Corning 9606	1623	2600	808	3.98	1.89	5.25 —	4.78 —	3.64 908	3.28 1038	3.08 1122	2.96 1197
Silicon carbide	3100	3160	675	490	230			— 880	— 1050	— 1135	87 1195
Silicon dioxide, crystalline (quartz)	1883	2650									
k, ∥ to c-axis				10.4		39	16.4	7.6	5.0	4.2	
k, ⊥ to c-axis				6.21		20.8	9.5	4.70	3.4	3.1	
C_p			745			—	—	885	1075	1250	
Silicon dioxide, polycrystalline (fused silica)	1883	2220	745	1.38	0.834	0.69 —	1.14 —	1.51 905	1.75 1040	2.17 1105	2.87 1155
Silicon nitride	2173	2400	691	16.0	9.65	— —	— 578	13.9 778	11.3 937	9.88 1063	8.76 1155
Sulfur	392	2070	708	0.206	0.141	0.165 403	0.185 606				
Thorium dioxide	3573	9110	235	13	6.1			10.2 255	6.6 274	4.7 285	3.68 295
Titanium dioxide, polycrystalline	2133	4157	710	8.4	2.8			7.01 805	5.02 880	3.94 910	3.46 930

TABLE A-21

Properties of building materials
(at a mean temperature of 24°C)

Material	Thickness, L mm	Density, ρ kg/m³	Thermal conductivity, k W/m · °C	Specific heat, C_p kJ/kg · °C	R-value (for listed thickness, L/k), °C · m²/W
Building Boards					
Asbestos–cement board	6 mm	1922	—	1.00	0.011
Gypsum of plaster board	10 mm	800	—	1.09	0.057
	13 mm	800	—	—	0.078
Plywood (Douglas fir)	—	545	0.12	1.21	—
	6 mm	545	—	1.21	0.055
	10 mm	545	—	1.21	0.083
	13 mm	545	—	1.21	0.110
	20 mm	545	—	1.21	0.165
Insulated board and sheating	13 mm	288	—	1.30	0.232
(regular density)	20 mm	288	—	1.30	0.359
Hardboard (high density, standard tempered)	—	1010	0.14	1.34	—
Particle board:					
Medium density	—	800	0.14	1.30	—
Underlayment	16 mm	640	—	1.21	0.144
Wood subfloor	20 mm	—	—	1.38	0.166
Building Membrane					
Vapor-permeable felt	—	—	—	—	0.011
Vapor-seal (2 layers of mopped 0.73 kg/m² felt)	—	—	—	—	0.021
Flooring Materials					
Carpet and fibrous pad	—	—	—	1.42	0.367
Carpet and rubber pad	—	—	—	1.38	0.217
Tile (asphalt, linoleum, vinyl)	—	—	—	1.26	0.009
Masonry Materials					
Masonry units:					
Brick, common		1922	0.72	—	—
Brick, face		2082	1.30	—	—
Brick, fire clay		2400	1.34	—	—
		1920	0.90	0.79	—
		1120	0.41	—	—
Concrete blocks (3 oval cores,	100 mm	—	0.77	—	0.13
sand and gravel aggregate)	200 mm	—	1.0	—	0.20
	300 mm	—	1.30	—	0.23
Concretes:					
Lightweight aggregates (including		1920	1.1	—	—
expanded shale, clay, or slate;		1600	0.79	0.84	—
expanded slags; cinders;		1280	0.54	0.84	—
pumice; and scoria)		960	0.33	—	—
		940	0.18	—	—

TABLE A-21

Properties of building materials (*Concluded*)
(at a mean temperature of 24°C)

Material	Thickness, L mm	Density, ρ kg/m³	Thermal conductivity, k W/m · °C	Specific heat, C_p kJ/kg · °C	R-value (for listed thickness, L/k), °C · m²/W
Cement/lime, mortar, and stucco		1920	1.40	—	—
		1280	0.65	—	—
Stucco		1857	0.72	—	—
Roofing					
Asbestos-cement shingles		1900	—	1.00	0.037
Asphalt roll roofing		1100	—	1.51	0.026
Asphalt shingles		1100	—	1.26	0.077
Built-in roofing	10 mm	1100	—	1.46	0.058
Slate	13 mm	—	—	1.26	0.009
Wood shingles (plain and plastic/film faced)		—	—	1.30	0.166
Plastering Materials					
Cement plaster, sand aggregate	19 mm	1860	0.72	0.84	0.026
Gypsum plaster:					
Lightweight aggregate	13 mm	720	—	—	0.055
Sand aggregate	13 mm	1680	0.81	0.84	0.016
Perlite aggregate	—	720	0.22	1.34	—
Siding Material (on flat surfaces)					
Asbestos-cement shingles	—	1900	—	—	0.037
Hardboard siding	11 mm	—	—	1.17	0.12
Wood (drop) siding	25 mm	—	—	1.30	0.139
Wood (plywood) siding, lapped	10 mm	—	—	1.21	0.111
Aluminum or steel siding (over sheeting):					
Hollow backed	10 mm	—	—	1.22	0.11
Insulating-board backed	10 mm	—	—	1.34	0.32
Architectural glass	—	2530	1.0	0.84	0.018
Woods					
Hardwoods (maple, oak, etc.)	—	721	0.159	1.26	—
Softwoods (fir, pine, etc.)	—	513	0.115	1.38	—
Metals					
Aluminum (1100)	—	2739	222	0.896	—
Steel, mild	—	7833	45.3	0.502	—
Steel, Stainless	—	7913	15.6	0.456	—

Source: Tables A-21 and A-22 are adapted from ASHRAE, *Handbook of Fundamentals* (Atlanta, GA: American Society of Heating, Refrigerating, and Air-Conditioning Engineers, 1993), Chap. 22, Table 4. Used with permission.

TABLE A-22

Properties of insulating materials
(at a mean temperature of 24°C)

Material	Thickness, L mm	Density, ρ kg/m³	Thermal conductivity, k W/m · °C	Specific heat, C_p kJ/kg · °C	R-value (for listed thickness, L/k), °C · m²/W
Blanket and Batt					
Mineral fiber (fibrous form	50 to 70 mm	4.8–32	—	0.71–0.96	1.23
processed from rock, slag,	75 to 90 mm	4.8–32	—	0.71–0.96	1.94
or glass)	135 to 165 mm	4.8–32	—	0.71–0.96	3.32
Board and Slab					
Cellular glass		136	0.055	1.0	—
Glass fiber (organic bonded)		64–144	0.036	0.96	—
Expanded polystyrene (molded beads)		16	0.040	1.2	—
Expanded polyurethane (R-11 expanded)		24	0.023	1.6	—
Expanded perlite (organic bonded)		16	0.052	1.26	—
Expanded rubber (rigid)		72	0.032	1.68	—
Mineral fiber with resin binder		240	0.042	0.71	—
Cork		120	0.039	1.80	—
Sprayed or Formed in Place					
Polyurethane foam		24–40	0.023–0.026	—	—
Glass fiber		56–72	0.038–0.039	—	—
Urethane, two-part mixture (rigid foam)		70	0.026	1.045	—
Mineral wool granules with asbestos/inorganic binders (sprayed)		190	0.046	—	—
Loose Fill					
Mineral fiber (rock, slag, or glass)	~ 75 to 125 mm	9.6–32	—	0.71	1.94
	~165 to 222 mm	9.6–32	—	0.71	3.35
	~191 to 254 mm	—	—	0.71	3.87
	~185 mm	—	—	0.71	5.28
Silica aerogel		122	0.025	—	—
Vermiculite (expanded)		122	0.068	—	—
Perlite, expanded		32–66	0.039–0.045	1.09	—
Sawdust or shavings		128–240	0.065	1.38	—
Cellulosic insulation (milled paper or wood pulp)		37–51	0.039–0.046	—	—
Roof Insulation					
Cellular glass	—	144	0.058	1.0	—
Preformed, for use above deck	13 mm	—	—	1.0	0.24
	25 mm	—	—	2.1	0.49
	50 mm	—	—	3.9	0.93
Reflective Insulation					
Silica powder (evacuated)		160	0.0017	—	—
Aluminum foil separating fluffy glass mats; 10–12 layers (evacuated); for cryogenic applications (150 K)		40	0.00016	—	—
Aluminum foil and glass paper laminate; 75–150 layers (evacuated); for cryogenic applications (150 K)		120	0.000017	—	—

TABLE A-23

Properties of miscellaneous materials
(Values are at 300 K unless indicated otherwise)

Material	Density, ρ kg/m³	Thermal conductivity, k, W/m · K	Specific heat, C_p J/kg · K
Asphalt	2115	0.062	920
Bakelite	1300	1.4	1465
Brick, refractory			
Chrome brick			
473 K	3010	2.3	835
823 K	—	2.5	—
1173 K	—	2.0	—
Fire clay, burnt 1600 K			
773 K	2050	1.0	960
1073 K	—	1.1	—
1373 K	—	1.1	—
Fire clay, burnt 1725 K			
773 K	2325	1.3	960
1073 K	—	1.4	—
1373 K	—	1.4	—
Fire clay brick			
478 K	2645	1.0	960
922 K	—	1.5	—
1478 K	—	1.8	—
Magnesite			
478 K	—	3.8	1130
922 K	—	2.8	—
1478 K	—	1.9	—
Chicken meat, white (74.4% water content)			
198 K	—	1.60	—
233 K	—	1.49	—
253 K	—	1.35	—
273 K	—	0.48	—
293 K	—	0.49	—
Clay, dry	1550	0.930	—
Clay, wet	1495	1.675	—
Coal, anthracite	1350	0.26	1260
Concrete (stone mix)	2300	1.4	880
Cork	86	0.048	2030
Cotton	80	0.06	1300
Fat	—	0.17	—
Glass			
Window	2800	0.7	750
Pyrex	2225	1–1.4	835
Crown	2500	1.05	—
Lead	3400	0.85	—

Material	Density, ρ kg/m³	Thermal conductivity, k, W/m · K	Specific heat, C_p J/kg · K
Ice			
273 K	920	1.88	2040
253 K	922	2.03	1945
173 K	928	3.49	1460
Leather, sole	998	0.159	—
Linoleum	535	0.081	—
	1180	0.186	—
Mica	2900	0.523	—
Paper	930	0.180	1340
Plastics			
Plexiglass	1190	0.19	1465
Teflon			
300 K	2200	0.35	1050
400 K	—	0.45	—
Lexan	1200	0.19	1260
Nylon	1145	0.29	—
Polypropylene	910	0.12	1925
Polyester	1395	0.15	1170
PVC, vinyl	1470	0.1	840
Porcelain	2300	1.5	—
Rubber, natural	1150	0.28	—
Rubber, vulcanized			
Soft	1100	0.13	2010
Hard	1190	0.16	—
Sand	1515	0.2–1.0	800
Snow, fresh	100	0.60	—
Snow, 273 K	500	2.2	—
Soil, dry	1500	1.0	1900
Soil, wet	1900	2.0	2200
Sugar	1600	0.58	—
Tissue, human			
Skin	—	0.37	—
Fat layer	—	0.2	—
Muscle	—	0.41	—
Vaseline	—	0.17	—
Wood, cross-grain			
Balsa	140	0.055	—
Fir	415	0.11	2720
Oak	545	0.17	2385
White pine	435	0.11	—
Yellow pine	640	0.15	2805
Wood, radial			
Oak	545	0.19	2385
Fir	420	0.14	2720
Wool, ship	145	0.05	—

Source: Compiled from various sources.

TABLE A-24

Emissivities of surfaces

(*a*) Metals

Material	Temperature, K	Emissivity, ε	Material	Temperature, K	Emissivity, ε
Aluminum			Magnesium, polished	300–500	0.07–0.13
Polished	300–900	0.04–0.06	Mercury	300–400	0.09–0.12
Commercial sheet	400	0.09	Molybdenum		
Heavily oxidized	400–800	0.20–0.33	Polished	300–2000	0.05–0.21
Anodized	300	0.8	Oxidized	600–800	0.80–0.82
Bismuth, bright	350	0.34	Nickel		
Brass			Polished	500–1200	0.07–0.17
Highly polished	500–650	0.03–0.04	Oxidized	450–1000	0.37–0.57
Polished	350	0.09	Platinum, polished	500–1500	0.06–0.18
Dull plate	300–600	0.22	Silver, polished	300–1000	0.02–0.07
Oxidized	450–800	0.6	Stainless steel		
Chromium, polished	300–1400	0.08–0.40	Polished	300–1000	0.17–0.30
Copper			Lightly oxidized	600–1000	0.30–0.40
Highly polished	300	0.02	Highly oxidized	600–1000	0.70–0.80
Polished	300–500	0.04–0.05	Steel		
Commercial sheet	300	0.15	Polished sheet	300–500	0.08–0.14
Oxidized	600–1000	0.5–0.8	Commercial sheet	500–1200	0.20–0.32
Black oxidized	300	0.78	Heavily oxidized	300	0.81
Gold			Tin, polished	300	0.05
Highly polished	300–1000	0.03–0.06	Tungsten		
Bright foil	300	0.07	Polished	300–2500	0.03–0.29
Iron			Filament	3500	0.39
Highly polished	300–500	0.05–0.07	Zinc		
Case iron	300	0.44	Polished	300–800	0.02–0.05
Wrought iron	300–500	0.28	Oxidized	300	0.25
Rusted	300	0.61			
Oxidized	500–900	0.64–0.78			
Lead					
Polished	300–500	0.06–0.08			
Unoxidized, rough	300	0.43			
Oxidized	300	0.63			

TABLE A-24

Emissivities of surfaces (*Concluded*)

(*b*) Nonmetals

Material	Temperature, K	Emissivity, ε	Material	Temperature, K	Emissivity, ε
Alumina	800–1400	0.65–0.45	Paper, white	300	0.90
Aluminum oxide	600–1500	0.69–0.41	Plaster, white	300	0.93
Asbestos	300	0.96	Porcelain, glazed	300	0.92
Asphalt pavement	300	0.85–0.93	Quartz, rough, fused	300	0.93
Brick			Rubber		
Common	300	0.93–0.96	Hard	300	0.93
Fireclay	1200	0.75	Soft	300	0.86
Carbon filament	2000	0.53	Sand	300	0.90
Cloth	300	0.75–0.90	Silicon carbide	600–1500	0.87–0.85
Concrete	300	0.88–0.94	Skin, human	300	0.95
Glass			Snow	273	0.80–0.90
Window	300	0.90–0.95	Soil, earth	300	0.93–0.96
Pyrex	300–1200	0.82–0.62	Soot	300–500	0.95
Pyroceram	300–1500	0.85–0.57	Teflon	300–500	0.85–0.92
Ice	273	0.95–0.99	Water, deep	273–373	0.95–0.96
Magnesium oxide	400–800	0.69–0.55	Wood		
Masonry	300	0.80	Beech	300	0.94
Paints			Oak	300	0.90
Aluminum	300	0.40–0.50			
Black, lacquer, shiny	300	0.88			
Oils, all colors	300	0.92–0.96			
Red primer	300	0.93			
White acrylic	300	0.90			
White enamel	300	0.90			

TABLE A-25

Solar radiative properties of materials

Description/Composition	Solar absorptivity, α_s	Emissivity, ε, at 300 K	Ratio, α_s/ε	Solar transmissivity, τ_s
Aluminum				
Polished	0.09	0.03	3.0	
Anodized	0.14	0.84	0.17	
Quartz-overcoated	0.11	0.37	0.30	
Foil	0.15	0.05	3.0	
Brick, red (Purdue)	0.63	0.93	0.68	
Concrete	0.60	0.88	0.68	
Galvanized sheet metal				
Clean, new	0.65	0.13	5.0	
Oxidized, weathered	0.80	0.28	2.9	
Glass, 3.2-mm thickness				
Float or tempered				0.79
Low iron oxide type				0.88
Marble, slightly off-white (nonreflective)	0.40	0.88	0.45	
Metal, plated				
Black sulfide	0.92	0.10	9.2	
Black cobalt oxide	0.93	0.30	3.1	
Black nickel oxide	0.92	0.08	11	
Black chrome	0.87	0.09	9.7	
Mylar, 0.13-mm thickness				0.87
Paints				
Black (Parsons)	0.98	0.98	1.0	
White, acrylic	0.26	0.90	0.29	
White, zinc oxide	0.16	0.93	0.17	
Paper, white	0.27	0.83	0.32	
Plexiglas, 3.2-mm thickness				0.90
Porcelain tiles, white (reflective glazed surface)	0.26	0.85	0.30	
Roofing tiles, bright red				
Dry surface	0.65	0.85	0.76	
Wet surface	0.88	0.91	0.96	
Sand, dry				
Off-white	0.52	0.82	0.63	
Dull red	0.73	0.86	0.82	
Snow				
Fine particles, fresh	0.13	0.82	0.16	
Ice granules	0.33	0.89	0.37	
Steel				
Mirror-finish	0.41	0.05	8.2	
Heavily rusted	0.89	0.92	0.96	
Stone (light pink)	0.65	0.87	0.74	
Tedlar, 0.10-mm thickness				0.92
Teflon, 0.13-mm thickness				0.92
Wood	0.59	0.90	0.66	

Source: V. C. Sharma and A. Sharma, "Solar Properties of Some Building Elements," *Energy* 14 (1989), pp. 805–10, and other sources.

TABLE A-26

Properties of the atmosphere at high altitude

Altitude, m	Temperature, °C	Pressure, kPa	Gravity, g m/s²	Speed of sound, m/s	Density, kg/m³	Viscosity, μ kg/m · s	Thermal conductivity, W/m · °C
0	15.00	101.33	9.807	340.3	1.225	1.789×10^{-5}	0.0253
200	13.70	98.95	9.806	339.5	1.202	1.783×10^{-5}	0.0252
400	12.40	96.61	9.805	338.8	1.179	1.777×10^{-5}	0.0252
600	11.10	94.32	9.805	338.0	1.156	1.771×10^{-5}	0.0251
800	9.80	92.08	9.804	337.2	1.134	1.764×10^{-5}	0.0250
1000	8.50	89.88	9.804	336.4	1.112	1.758×10^{-5}	0.0249
1200	7.20	87.72	9.803	335.7	1.090	1.752×10^{-5}	0.0248
1400	5.90	85.60	9.802	334.9	1.069	1.745×10^{-5}	0.0247
1600	4.60	83.53	9.802	334.1	1.048	1.739×10^{-5}	0.0245
1800	3.30	81.49	9.801	333.3	1.027	1.732×10^{-5}	0.0244
2000	2.00	79.50	9.800	332.5	1.007	1.726×10^{-5}	0.0243
2200	0.70	77.55	9.800	331.7	0.987	1.720×10^{-5}	0.0242
2400	−0.59	75.63	9.799	331.0	0.967	1.713×10^{-5}	0.0241
2600	−1.89	73.76	9.799	330.2	0.947	1.707×10^{-5}	0.0240
2800	−3.19	71.92	9.798	329.4	0.928	1.700×10^{-5}	0.0239
3000	−4.49	70.12	9.797	328.6	0.909	1.694×10^{-5}	0.0238
3200	−5.79	68.36	9.797	327.8	0.891	1.687×10^{-5}	0.0237
3400	−7.09	66.63	9.796	327.0	0.872	1.681×10^{-5}	0.0236
3600	−8.39	64.94	9.796	326.2	0.854	1.674×10^{-5}	0.0235
3800	−9.69	63.28	9.795	325.4	0.837	1.668×10^{-5}	0.0234
4000	−10.98	61.66	9.794	324.6	0.819	1.661×10^{-5}	0.0233
4200	−12.3	60.07	9.794	323.8	0.802	1.655×10^{-5}	0.0232
4400	−13.6	58.52	9.793	323.0	0.785	1.648×10^{-5}	0.0231
4600	−14.9	57.00	9.793	322.2	0.769	1.642×10^{-5}	0.0230
4800	−16.2	55.51	9.792	321.4	0.752	1.635×10^{-5}	0.0229
5000	−17.5	54.05	9.791	320.5	0.736	1.628×10^{-5}	0.0228
5200	−18.8	52.62	9.791	319.7	0.721	1.622×10^{-5}	0.0227
5400	−20.1	51.23	9.790	318.9	0.705	1.615×10^{-5}	0.0226
5600	−21.4	49.86	9.789	318.1	0.690	1.608×10^{-5}	0.0224
5800	−22.7	48.52	9.785	317.3	0.675	1.602×10^{-5}	0.0223
6000	−24.0	47.22	9.788	316.5	0.660	1.595×10^{-5}	0.0222
6200	−25.3	45.94	9.788	315.6	0.646	1.588×10^{-5}	0.0221
6400	−26.6	44.69	9.787	314.8	0.631	1.582×10^{-5}	0.0220
6600	−27.9	43.47	9.786	314.0	0.617	1.575×10^{-5}	0.0219
6800	−29.2	42.27	9.785	313.1	0.604	1.568×10^{-5}	0.0218
7000	−30.5	41.11	9.785	312.3	0.590	1.561×10^{-5}	0.0217
8000	−36.9	35.65	9.782	308.1	0.526	1.527×10^{-5}	0.0212
9000	−43.4	30.80	9.779	303.8	0.467	1.493×10^{-5}	0.0206
10,000	−49.9	26.50	9.776	299.5	0.414	1.458×10^{-5}	0.0201
12,000	−56.5	19.40	9.770	295.1	0.312	1.422×10^{-5}	0.0195
14,000	−56.5	14.17	9.764	295.1	0.228	1.422×10^{-5}	0.0195
16,000	−56.5	10.53	9.758	295.1	0.166	1.422×10^{-5}	0.0195
18,000	−56.5	7.57	9.751	295.1	0.122	1.422×10^{-5}	0.0195

Source: U.S. Standard Atmosphere Supplements, U.S. Government Printing Office, 1966. Based on year-round mean conditions at 45° latitude and varies with the time of the year and the weather patterns. The conditions at sea level ($z = 0$) are taken to be $P = 101.325$ kPa, $T = 15$°C, $\rho = 1.2250$ kg/m³, $g = 9.80665$ m²/s.

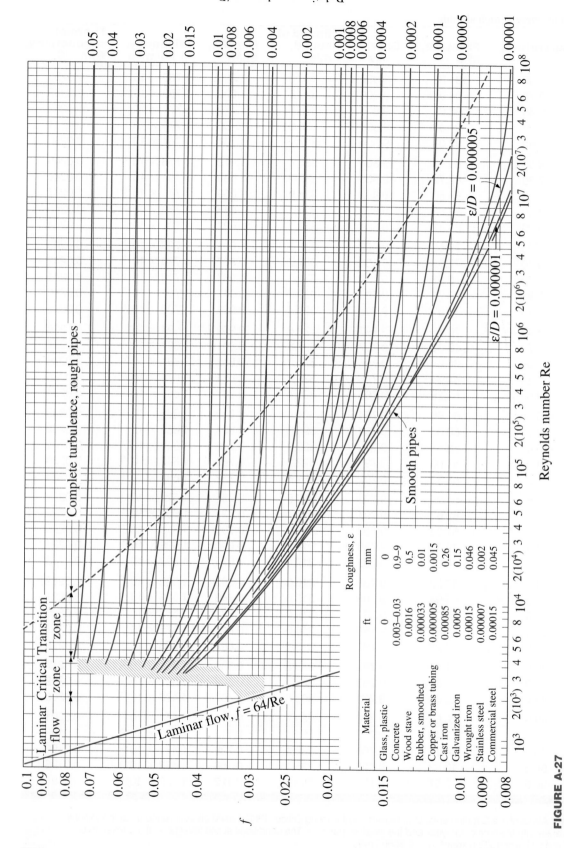

FIGURE A-27

The Moody chart for the friction factor for fully developed flow in circular tubes.

Material	Roughness, ε	
	ft	mm
Glass, plastic	0	0
Concrete	0.003–0.03	0.9–9
Wood stave	0.0016	0.5
Rubber, smoothed	0.000033	0.01
Copper or brass tubing	0.000005	0.0015
Cast iron	0.00085	0.26
Galvanized iron	0.0005	0.15
Wrought iron	0.00015	0.046
Stainless steel	0.000007	0.002
Commercial steel	0.00015	0.045

Property Tables and Charts (English Units)

2

TABLE A-1E

Molar mass, gas constant, and critical-point properties

Substance	Formula	Molar mass, M lbm/lbmol	Gas constant, R Btu/ lbm · R*	Gas constant, R psia · ft³/ lbm · R*	Critical-point properties Tempera- ture, R	Critical-point properties Pres- sure, psia	Critical-point properties Volume, ft³/lbmol
Air	—	28.97	0.06855	0.3704	238.5	547	1.41
Ammonia	NH_3	17.03	0.1166	0.6301	729.8	1636	1.16
Argon	Ar	39.948	0.04971	0.2686	272	705	1.20
Benzene	C_6H_6	78.115	0.02542	0.1374	1012	714	4.17
Bromine	Br_2	159.808	0.01243	0.06714	1052	1500	2.17
n-Butane	C_4H_{10}	58.124	0.03417	0.1846	765.2	551	4.08
Carbon dioxide	CO_2	44.01	0.04513	0.2438	547.5	1071	1.51
Carbon monoxide	CO	28.011	0.07090	0.3831	240	507	1.49
Carbon tetrachloride	CCl_4	153.82	0.01291	0.06976	1001.5	661	4.42
Chlorine	Cl_2	70.906	0.02801	0.1517	751	1120	1.99
Chloroform	$CHCl_3$	119.38	0.01664	0.08988	965.8	794	3.85
Dichlorodifluoromethane (R-12)	CCl_2F_2	120.91	0.01643	0.08874	692.4	582	3.49
Dichlorofluoromethane (R-21)	$CHCl_2F$	102.92	0.01930	0.1043	813.0	749	3.16
Ethane	C_2H_6	30.020	0.06616	0.3574	549.8	708	2.37
Ethyl alcohol	C_2H_5OH	46.07	0.04311	0.2329	929.0	926	2.68
Ethylene	C_2H_4	28.054	0.07079	0.3825	508.3	742	1.99
Helium	He	4.003	0.4961	2.6805	9.5	33.2	0.926
n-Hexane	C_6H_{14}	86.178	0.02305	0.1245	914.2	439	5.89
Hydrogen (normal)	H_2	2.016	0.9851	5.3224	59.9	188.1	1.04
Krypton	Kr	83.80	0.02370	0.1280	376.9	798	1.48
Methane	CH_4	16.043	0.1238	0.6688	343.9	673	1.59
Methyl alcohol	CH_3OH	32.042	0.06198	0.3349	923.7	1154	1.89
Methyl chloride	CH_3Cl	50.488	0.03934	0.2125	749.3	968	2.29
Neon	Ne	20.183	0.09840	0.5316	80.1	395	0.668
Nitrogen	N_2	28.013	0.07090	0.3830	227.1	492	1.44
Nitrous oxide	N_2O	44.013	0.04512	0.2438	557.4	1054	1.54
Oxygen	O_2	31.999	0.06206	0.3353	278.6	736	1.25
Propane	C_3H_8	44.097	0.04504	0.2433	665.9	617	3.20
Propylene	C_3H_6	42.081	0.04719	0.2550	656.9	670	2.90
Sulfur dioxide	SO_2	64.063	0.03100	1.1675	775.2	1143	1.95
Tetrafluoroethane (R-134a)	CF_3CH_2F	102.03	0.01946	0.1052	673.7	589.9	2.96
Trichlorofluoromethane (R-11)	CCl_3F	137.37	0.01446	0.07811	848.1	635	3.97
Water	H_2O	18.015	0.1102	0.5956	1165.3	3204	0.90
Xenon	Xe	131.30	0.01513	0.08172	521.55	852	1.90

*Calculated from $R = R_u/M$, where R_u = 1.986 Btu/lbmol · R = 10.73 psia · ft³/lbmol · R and M is the molar mass.

Source: K. A. Kobe and R. E. Lynn Jr., Chemical Review 52 (1953), pp. 117–236, and ASHRAE, Handbook of Fundamentals (Atlanta, GA: American Society of Heating, Refrigerating, and Air-Conditioning Engineers, Inc., 1993), pp. 16.4 and 36.1.

TABLE A-2E

Ideal-gas specific heats of various common gases

(*a*) At 80°F

Gas	Formula	Gas constant, R Btu/lbm · R	C_p Btu/lbm · R	C_v Btu/lbm · R	k
Air	—	0.06855	0.240	0.171	1.400
Argon	Ar	0.04971	0.1253	0.0756	1.667
Butane	C_4H_{10}	0.03424	0.415	0.381	1.09
Carbon dioxide	CO_2	0.04513	0.203	0.158	1.285
Carbon monoxide	CO	0.07090	0.249	0.178	1.399
Ethane	C_2H_6	0.06616	0.427	0.361	1.183
Ethylene	C_2H_4	0.07079	0.411	0.340	1.208
Helium	He	0.4961	1.25	0.753	1.667
Hydrogen	H_2	0.9851	3.43	2.44	1.404
Methane	CH_4	0.1238	0.532	0.403	1.32
Neon	Ne	0.09840	0.246	0.1477	1.667
Nitrogen	N_2	0.07090	0.248	0.177	1.400
Octane	O_8H_{18}	0.01742	0.409	0.392	1.044
Oxygen	O_2	0.06206	0.219	0.157	1.395
Propane	C_3H_8	0.04504	0.407	0.362	1.124
Steam	H_2O	0.1102	0.445	0.335	1.329

Source: Gordon J. Van Wylen and Richard E. Sonntag, *Fundamentals of Classical Thermodynamics,* English/SI Version, 3rd ed. (New York: John Wiley & Sons, 1986), p. 687, Table A-8E.

TABLE A-2E

Ideal-gas specific heats of various common gases (*Continued*)
(*b*) At various temperatures

Temp., °F	C_p Btu/lbm · R	C_v Btu/lbm · R	k	C_p Btu/lbm · R	C_v Btu/lbm · R	k	C_p Btu/lbm · R	C_v Btu/lbm · R	k
	Air			**Carbon dioxide, CO_2**			**Carbon monoxide, CO**		
40	0.240	0.171	1.401	0.195	0.150	1.300	0.248	0.177	1.400
100	0.240	0.172	1.400	0.205	0.160	1.283	0.249	0.178	1.399
200	0.241	0.173	1.397	0.217	0.172	1.262	0.249	0.179	1.397
300	0.243	0.174	1.394	0.229	0.184	1.246	0.251	0.180	1.394
400	0.245	0.176	1.389	0.239	0.193	1.233	0.253	0.182	1.389
500	0.248	0.179	1.383	0.247	0.202	1.223	0.256	0.185	1.384
600	0.250	0.182	1.377	0.255	0.210	1.215	0.259	0.188	1.377
700	0.254	0.185	1.371	0.262	0.217	1.208	0.262	0.191	1.371
800	0.257	0.188	1.365	0.269	0.224	1.202	0.266	0.195	1.364
900	0.259	0.191	1.358	0.275	0.230	1.197	0.269	0.198	1.357
1000	0.263	0.195	1.353	0.280	0.235	1.192	0.273	0.202	1.351
1500	0.276	0.208	1.330	0.298	0.253	1.178	0.287	0.216	1.328
2000	0.286	0.217	1.312	0.312	0.267	1.169	0.297	0.226	1.314
	Hydrogen, H_2			**Nitrogen, N_2**			**Oxygen, O_2**		
40	3.397	2.412	1.409	0.248	0.177	1.400	0.219	0.156	1.397
100	3.426	2.441	1.404	0.248	0.178	1.399	0.220	0.158	1.394
200	3.451	2.466	1.399	0.249	0.178	1.398	0.223	0.161	1.387
300	3.461	2.476	1.398	0.250	0.179	1.396	0.226	0.164	1.378
400	3.466	2.480	1.397	0.251	0.180	1.393	0.230	0.168	1.368
500	3.469	2.484	1.397	0.254	0.183	1.388	0.235	0.173	1.360
600	3.473	2.488	1.396	0.256	0.185	1.383	0.239	0.177	1.352
700	3.477	2.492	1.395	0.260	0.189	1.377	0.242	0.181	1.344
800	3.494	2.509	1.393	0.262	0.191	1.371	0.246	0.184	1.337
900	3.502	2.519	1.392	0.265	0.194	1.364	0.249	0.187	1.331
1000	3.513	2.528	1.390	0.269	0.198	1.359	0.252	0.190	1.326
1500	3.618	2.633	1.374	0.283	0.212	1.334	0.263	0.201	1.309
2000	3.758	2.773	1.355	0.293	0.222	1.319	0.270	0.208	1.298

Note: The unit Btu/lbm · R is equivalent to Btu/ lbm · °F.

Source: Kenneth Wark, *Thermodynamics,* 4th ed. (New York: McGraw-Hill, 1983), p. 830, Table A-4. Originally published in *Tables of Thermal Properties of Gases,* NBS Circular 564, 1955.

TABLE A-2E

Ideal-gas specific heats of various common gases (*Concluded*)
(*c*) As a function of temperature

$$\bar{C}_p = a + bT + cT^2 + dT^3$$
(*T* in R, $\bar{C}_p$ in Btu/lbmol · R)

Substance	Formula	a	b	c	d	Temperature range, R	% error Max.	% error Avg.
Nitrogen	N_2	6.903	-0.02085×10^{-2}	0.05957×10^{-5}	-0.1176×10^{-9}	491-3240	0.59	0.34
Oxygen	O_2	6.085	0.2017×10^{-2}	-0.05275×10^{-5}	0.05372×10^{-9}	491-3240	1.19	0.28
Air	—	6.713	0.02609×10^{-2}	0.03540×10^{-5}	-0.08052×10^{-9}	491-3240	0.72	0.33
Hydrogen	H_2	6.952	-0.02542×10^{-2}	0.02952×10^{-5}	-0.03565×10^{-9}	491-3240	1.02	0.26
Carbon monoxide	CO	6.726	0.02222×10^{-2}	0.03960×10^{-5}	-0.09100×10^{-9}	491-3240	0.89	0.37
Carbon dioxide	CO_2	5.316	0.79361×10^{-2}	-0.2581×10^{-5}	0.3059×10^{-9}	491-3240	0.67	0.22
Water vapor	H_2O	7.700	0.02552×10^{-2}	0.07781×10^{-5}	-0.1472×10^{-9}	491-3240	0.53	0.24
Nitric oxide	NO	7.008	-0.01247×10^{-2}	0.07185×10^{-5}	-0.1715×10^{-9}	491-2700	0.97	0.36
Nitrous oxide	N_2O	5.758	0.7780×10^{-2}	-0.2596×10^{-5}	0.4331×10^{-9}	491-2700	0.59	0.26
Nitrogen dioxide	NO_2	5.48	0.7583×10^{-2}	-0.260×10^{-5}	0.322×10^{-9}	491-2700	0.46	0.18
Ammonia	NH_3	6.5846	0.34028×10^{-2}	0.073034×10^{-5}	-0.27402×10^{-9}	491-2700	0.91	0.36
Sulfur	S_2	6.499	0.2943×10^{-2}	-0.1200×10^{-5}	0.1632×10^{-9}	491-3240	0.99	0.38
Sulfur dioxide	SO_2	6.157	0.7689×10^{-2}	-0.2810×10^{-5}	0.3527×10^{-9}	491-3240	0.45	0.24
Sulfur trioxide	SO_3	3.918	1.935×10^{-2}	-0.8256×10^{-5}	1.328×10^{-9}	491-2340	0.29	0.13
Acetylene	C_2H_2	5.21	1.2227×10^{-2}	-0.4812×10^{-5}	0.7457×10^{-9}	491-2700	1.46	0.59
Benzene	C_6H_6	-8.650	6.4322×10^{-2}	-2.327×10^{-5}	3.179×10^{-9}	491-2700	0.34	0.20
Methanol	CH_4O	4.55	1.214×10^{-2}	-0.0898×10^{-5}	-0.329×10^{-9}	491-1800	0.18	0.08
Ethanol	C_2H_6O	4.75	2.781×10^{-2}	-0.7651×10^{-5}	0.821×10^{-9}	491-2700	0.40	0.22
Hydrogen chloride	HCl	7.244	-0.1011×10^{-2}	0.09783×10^{-5}	-0.1776×10^{-9}	491-2740	0.22	0.08
Methane	CH_4	4.750	0.6666×10^{-2}	0.09352×10^{-5}	-0.4510×10^{-9}	491-2740	1.33	0.57
Ethane	C_2H_6	1.648	2.291×10^{-2}	-0.4722×10^{-5}	0.2984×10^{-9}	491-2740	0.83	0.28
Propane	C_3H_8	-0.966	4.044×10^{-2}	-1.159×10^{-5}	1.300×10^{-9}	491-2740	0.40	0.12
n-Butane	C_4H_{10}	0.945	4.929×10^{-2}	-1.352×10^{-5}	1.433×10^{-9}	491-2740	0.54	0.24
i-Butane	C_4H_{10}	-1.890	5.520×10^{-2}	-1.696×10^{-5}	2.044×10^{-9}	491-2740	0.25	0.13
n-Pentane	C_5H_{12}	1.618	6.028×10^{-2}	-1.656×10^{-5}	1.732×10^{-9}	491-2740	0.56	0.21
n-Hexane	C_6H_{14}	1.657	7.328×10^{-2}	-2.112×10^{-5}	2.363×10^{-9}	491-2740	0.72	0.20
Ethylene	C_2H_4	0.944	2.075×10^{-2}	-0.6151×10^{-5}	0.7326×10^{-9}	491-2740	0.54	0.13
Propylene	C_3H_6	0.753	3.162×10^{-2}	-0.8981×10^{-5}	1.008×10^{-9}	491-2740	0.73	0.17

Source: B. G. Kyle, *Chemical and Process Thermodynamics* (Englewood Cliffs, N.J.: Prentice Hall, 1984). Used with permission.

TABLE A-3E

Properties of common liquids, solids, and foods
(*a*) Liquids

Substance	Boiling data at 1 atm		Freezing data		Liquid properties		
	Normal boiling point, °F	Latent heat of vaporization, h_{fg} Btu/lbm	Freezing point, °F	Latent heat of fusion, h_{if} Btu/lbm	Temperature, °F	Density, ρ lbm/ft³	Specific heat, C_p Btu/lbm · °F
Ammonia	−27.9	24.54	−107.9	138.6	−27.9	42.6	1.06
					0	41.3	1.083
					40	39.5	1.103
					80	37.5	1.135
Argon	−302.6	69.5	−308.7	12.0	−302.6	87.0	0.272
Benzene	176.4	169.4	41.9	54.2	68	54.9	0.411
Brine (20% sodium chloride by mass)	219.0		0.7	—	68	71.8	0.743
n-Butane	31.1	165.6	−217.3	34.5	31.1	37.5	0.552
Carbon dioxide	−109.2*	99.6 (at 32°F)	−69.8	—	32	57.8	0.583
Ethanol	172.8	360.5	−173.6	46.9	77	48.9	0.588
Ethyl alcohol	173.5	368	−248.8	46.4	68	49.3	0.678
Ethylene glycol	388.6	344.0	12.6	77.9	68	69.2	0.678
Glycerine	355.8	419	66.0	86.3	68	78.7	0.554
Helium	−452.1	9.80	—	—	−452.1	9.13	5.45
Hydrogen	−423.0	191.7	−434.5	25.6	−423.0	4.41	2.39
Isobutane	10.9	157.8	−255.5	45.5	10.9	37.1	0.545
Kerosene	399–559	108	−12.8	—	68	51.2	0.478
Mercury	674.1	126.7	−38.0	4.90	77	847	0.033
Methane	−258.7	219.6	296.0	25.1	−258.7	26.4	0.834
					−160	20.0	1.074
Methanol	148.1	473	−143.9	42.7	77	49.1	0.609
Nitrogen	−320.4	85.4	−346.0	10.9	−320.4	50.5	0.492
					−260	38.2	0.643
Octane	256.6	131.7	−71.5	77.9	68	43.9	0.502
Oil (light)	—				77	56.8	0.430
Oxygen	−297.3	91.5	−361.8	5.9	−297.3	71.2	0.408
Petroleum	—	99–165			68	40.0	0.478
Propane	−43.7	184.0	−305.8	34.4	−43.7	36.3	0.538
					32	33.0	0.604
					100	29.4	0.673
Refrigerant-134a	−15.0	93.2	−141.9	—	−40	88.5	0.283
					−15	86.0	0.294
					32	80.9	0.318
					90	73.6	0.348
Water	212	970.5	32	143.5	32	62.4	1.01
					90	62.1	1.00
					150	61.2	1.00
					212	59.8	1.01

*Sublimation temperature. (At pressures below the triple-point pressure of 75.1 psia, carbon dioxide exists as a solid or gas. Also, the freezing-point temperature of carbon dioxide is the triple-point temperature of −69.8°F.)

TABLE A-3E

Properties of common liquids, solids, and foods (*Concluded*)

(b) Solids (values are for room temperature unless indicated otherwise)

Substance	Density, ρ lbm/ft^3	Specific heat, C_p Btu/lbm · °F	Substance	Density, ρ lbm/ft^3	Specific heat, C_p Btu/lbm · °F
Metals			**Nonmetals**		
Aluminum			Asphalt	132	0.220
−100°F		0.192	Brick, common	120	0.189
32°F		0.212	Brick, fireclay (500°C)	144	0.229
100°F	170	0.218	Concrete	144	0.156
200°F		0.224	Clay	62.4	0.220
300°F		0.229	Diamond	151	0.147
400°F		0.235	Glass, window	169	0.191
500°F		0.240	Glass, pyrex	139	0.200
Bronze (76% Cu, 2% Zn, 2% Al)	517	0.0955	Graphite	156	0.170
			Granite	169	0.243
Brass, yellow (65% Cu, 35% Zn)	519	0.0955	Gypsum or plaster board	50	0.260
			Ice		
Copper			−100°F		0.375
−240°F		0.0674	−50°F		0.424
−150°F		0.0784	0°F		0.471
−60°F		0.0862	20°F		0.491
0°F		0.0893	32°F	57.5	0.502
100°F	555	0.0925	Limestone	103	0.217
200°F		0.0938	Marble	162	0.210
390°F		0.0963	Plywood (Douglas fir)	34.0	
Iron	490	0.107	Rubber (soft)	68.7	
Lead	705	0.030	Rubber (hard)	71.8	
Magnesium	108	0.239	Sand	94.9	
Nickel	555	0.105	Stone	93.6	
Silver	655	0.056	Woods, hard (maple, oak, etc.)	45.0	
Steel, mild	489	0.119	Woods, soft (fir, pine, etc.)	32.0	
Tungsten	1211	0.031			

(c) Foods

Food	Water content, % (mass)	Freezing point, °F	Specific heat, Btu/lbm · °F Above freezing	Specific heat, Btu/lbm · °F Below freezing	Latent heat of fusion, Btu/lbm	Food	Water content, % (mass)	Freezing point, °F	Specific heat, Btu/lbm · °F Above freezing	Specific heat, Btu/lbm · °F Below freezing	Latent heat of fusion, Btu/lbm
Apples	84	30	0.873	0.453	121	Lettuce	95	32	0.961	0.487	136
Bananas	75	31	0.801	0.426	108	Milk, whole	88	31	0.905	0.465	126
Beef round	67	—	0.737	0.402	96	Oranges	87	31	0.897	0.462	125
Broccoli	90	31	0.921	0.471	129	Potatoes	78	31	0.825	0.435	112
Butter	16	—	—	0.249	23	Salmon fish	64	28	0.713	0.393	92
Cheese, swiss	39	14	0.513	0.318	56	Shrimp	83	28	0.865	0.450	119
Cherries	80	29	0.841	0.441	115	Spinach	93	31	0.945	0.481	134
Chicken	74	27	0.793	0.423	106	Strawberries	90	31	0.921	0.471	129
Corn, sweet	74	31	0.793	0.423	106	Tomatoes, ripe	94	31	0.953	0.484	135
Eggs, whole	74	31	0.793	0.423	106	Turkey	64	—	0.713	0.393	92
Ice cream	63	22	0.705	0.390	90	Watermelon	93	31	0.945	0.481	134

Source: Values are obtained from various handbooks and other sources or are calculated. Water content and freezing-point data of foods are from ASHRAE, *Handbook of Fundamentals,* I-P version (Atlanta, GA: American Society of Heating, Refrigerating, and Air-Conditioning Engineers, Inc., 1993), Chap. 30, Table 1. Freezing point is the temperature at which freezing starts for fruits and vegetables, and the average freezing temperature for other foods.

TABLE A-4E

Saturated water—Temperature table

Temp., T °F	Sat. press., P_{sat} **psia**	Specific volume, ft³/lbm Sat. liquid, v_f	Sat. vapor, v_g	Internal energy, Btu/lbm Sat. liquid, u_f	Evap., u_{fg}	Sat. vapor, u_g	Enthalpy, Btu/lbm Sat. liquid, h_f	Evap., h_{fg}	Sat. vapor, h_g	Entropy, Btu/lbm · R Sat. liquid, s_f	Evap., s_{fg}	Sat. vapor, s_g
32.018	0.08866	0.016022	3302	0.00	1021.2	1021.2	0.01	1075.4	1075.4	0.00000	2.1869	2.1869
35	0.09992	0.016021	2948	2.99	1019.2	1022.2	3.00	1073.7	1076.7	0.00607	2.1704	2.1764
40	0.12166	0.016020	2445	8.02	1015.8	1023.9	8.02	1070.9	1078.9	0.01617	2.1430	2.1592
45	0.14748	0.016021	2037	13.04	1012.5	1025.5	13.04	1068.1	1081.1	0.02618	2.1162	2.1423
50	0.17803	0.016024	1704.2	18.06	1009.1	1027.2	18.06	1065.2	1083.3	0.03607	2.0899	2.1259
60	0.2563	0.016035	1206.9	28.08	1002.4	1030.4	28.08	1059.6	1087.7	0.05555	2.0388	2.0943
70	0.3632	0.016051	867.7	38.09	995.6	1033.7	38.09	1054.0	1092.0	0.07463	1.9896	2.0642
80	0.5073	0.016073	632.8	48.08	988.9	1037.0	48.09	1048.3	1096.4	0.09332	1.9423	2.0356
90	0.6988	0.016099	467.7	58.07	982.2	1040.2	58.07	1042.7	1100.7	0.11165	1.8966	2.0083
100	0.9503	0.016130	350.0	68.04	975.4	1043.5	68.05	1037.0	1105.0	0.12963	1.8526	1.9822
110	1.2763	0.016166	265.1	78.02	968.7	1046.7	78.02	1031.3	1109.3	0.14730	1.8101	1.9574
120	1.6945	0.016205	203.0	87.99	961.9	1049.8	88.00	1025.5	1113.5	0.16465	1.7690	1.9336
130	2.225	0.016247	157.17	97.97	955.1	1053.0	97.98	1019.8	1117.8	0.18172	1.7292	1.9109
140	2.892	0.016293	122.88	107.95	948.2	1056.2	107.96	1014.0	1121.9	0.19851	1.6907	1.8892
150	3.722	0.016343	96.99	117.95	941.3	1059.3	117.96	1008.1	1126.1	0.21503	1.6533	1.8684
160	4.745	0.016395	77.23	127.94	934.4	1062.3	127.96	1002.2	1130.1	0.23130	1.6171	1.8484
170	5.996	0.016450	62.02	137.95	927.4	1065.4	137.97	996.2	1134.2	0.24732	1.5819	1.8293
180	7.515	0.016509	50.20	147.97	920.4	1068.3	147.99	990.2	1138.2	0.26311	1.5478	1.8109
190	9.343	0.016570	40.95	158.00	913.3	1071.3	158.03	984.1	1142.1	0.27866	1.5146	1.7932
200	11.529	0.016634	33.63	168.04	906.2	1074.2	168.07	977.9	1145.9	0.29400	1.4822	1.7762
210	14.125	0.016702	27.82	178.10	898.9	1077.0	178.14	971.6	1149.7	0.30913	1.4508	1.7599
212	14.698	0.016716	26.80	180.11	897.5	1077.6	180.16	970.3	1150.5	0.31213	1.4446	1.7567
220	17.188	0.016772	23.15	188.17	891.7	1079.8	188.22	965.3	1153.5	0.32406	1.4201	1.7441
230	20.78	0.016845	19.386	198.26	884.3	1082.6	198.32	958.8	1157.1	0.33880	1.3901	1.7289
240	24.97	0.016922	16.327	208.36	876.9	1085.3	208.44	952.3	1160.7	0.35335	1.3609	1.7143
250	29.82	0.017001	13.826	218.49	869.4	1087.9	218.59	945.6	1164.2	0.36772	1.3324	1.7001
260	35.42	0.017084	11.768	228.64	861.8	1090.5	228.76	938.8	1167.6	0.38193	1.3044	1.6864
270	41.85	0.017170	10.066	238.82	854.1	1093.0	238.95	932.0	1170.9	0.39597	1.2771	1.6731
280	49.18	0.017259	8.650	249.02	846.3	1095.4	249.18	924.9	1174.1	0.40986	1.2504	1.6602
290	57.53	0.017352	7.467	259.25	838.5	1097.7	259.44	917.8	1177.2	0.42360	1.2241	1.6477
300	66.98	0.017448	6.472	269.52	830.5	1100.0	269.73	910.4	1180.2	0.43720	1.1984	1.6356
310	77.64	0.017548	5.632	279.81	822.3	1102.1	280.06	903.0	1183.0	0.45067	1.1731	1.6238
320	89.60	0.017652	4.919	290.14	814.1	1104.2	290.43	895.3	1185.8	0.46400	1.1483	1.6123
330	103.00	0.017760	4.312	300.51	805.7	1106.2	300.43	887.5	1188.4	0.47722	1.1238	1.6010
340	117.93	0.017872	3.792	310.91	797.1	1108.0	311.30	879.5	1190.8	0.49031	1.0997	1.5901
350	134.53	0.017988	3.346	321.35	788.4	1109.8	321.80	871.3	1193.1	0.50329	1.0760	1.5793
360	152.92	0.018108	2.961	331.84	779.6	1111.4	332.35	862.9	1195.2	0.51617	1.0526	1.5688
370	173.23	0.018233	2.628	342.37	770.6	1112.9	342.96	854.2	1197.2	0.52894	1.0295	1.5585
380	195.60	0.018363	2.339	352.95	761.4	1114.3	353.62	845.4	1199.0	0.54163	1.0067	1.5483
390	220.2	0.018498	2.087	363.58	752.0	1115.6	364.34	836.2	1200.6	0.55422	0.9841	1.5383
400	247.1	0.018638	1.8661	374.27	742.4	1116.6	375.12	826.8	1202.0	0.56672	0.9617	1.5284
410	276.5	0.018784	1.6726	385.01	732.6	1117.6	385.97	817.2	1203.1	0.57916	0.9395	1.5187
420	308.5	0.018936	1.5024	395.81	722.5	1118.3	396.89	807.2	1204.1	0.59152	0.9175	1.5091
430	343.3	0.019094	1.3521	406.68	712.2	1118.9	407.89	796.9	1204.8	0.60381	0.8957	1.4995
440	381.2	0.019260	1.2192	417.62	701.7	1119.3	418.98	786.3	1205.3	0.61605	0.8740	1.4900
450	422.1	0.019433	1.1011	428.6	690.9	1119.5	430.2	775.4	1205.6	0.6282	0.8523	1.4806
460	466.3	0.019614	0.9961	439.7	679.8	1119.6	441.4	764.1	1205.5	0.6404	0.8308	1.4712
470	514.1	0.019803	0.9025	450.9	668.4	1119.4	452.8	752.4	1205.2	0.6525	0.8093	1.4618
480	565.5	0.020002	0.8187	462.2	656.7	1118.9	464.3	740.3	1204.6	0.6646	0.7878	1.4524
490	620.7	0.020211	0.7436	473.6	644.7	1118.3	475.9	727.8	1203.7	0.6767	0.7663	1.4430

H₂O

982

TABLE A-4E

Saturated water—Temperature table (*Concluded*)

Temp., T °**F**	Sat. press., P_{sat} **psia**	Specific volume, ft³/lbm Sat. liquid, v_f	Sat. vapor, v_g	Internal energy, Btu/lbm Sat. liquid, u_f	Evap., u_{fg}	Sat. vapor, u_g	Enthalpy, Btu/lbm Sat. liquid, h_f	Evap., h_{fg}	Sat. vapor, h_g	Entropy, Btu/lbm · R Sat. liquid, s_f	Evap., s_{fg}	Sat. vapor, s_g
500	680.0	0.02043	0.6761	485.1	632.3	1117.4	487.7	714.8	1202.5	0.6888	0.7448	1.4335
520	811.4	0.02091	0.5605	508.5	606.2	1114.8	511.7	687.3	1198.9	0.7130	0.7015	1.4145
540	961.5	0.02145	0.4658	532.6	578.4	1111.0	536.4	657.5	1193.8	0.7374	0.6576	1.3950
560	1131.8	0.02207	0.3877	557.4	548.4	1105.8	562.0	625.0	1187.0	0.7620	0.6129	1.3749
580	1324.3	0.02278	0.3225	583.1	515.9	1098.9	588.6	589.3	1178.0	0.7872	0.5668	1.3540
600	1541.0	0.02363	0.2677	609.9	480.1	1090.0	616.7	549.7	1166.4	0.8130	0.5187	1.3317
620	1784.4	0.02465	0.2209	638.3	440.2	1078.5	646.4	505.0	1151.4	0.8398	0.4677	1.3075
640	2057.1	0.02593	0.1805	668.7	394.5	1063.2	678.6	453.4	1131.9	0.8681	0.4122	1.2803
660	2362	0.02767	0.14459	702.3	340.0	1042.3	714.4	391.1	1105.5	0.8990	0.3493	1.2483
680	2705	0.03032	0.11127	741.7	269.3	1011.0	756.9	309.8	1066.7	0.9350	0.2718	1.2068
700	3090	0.03666	0.07438	801.7	145.9	947.7	822.7	167.5	990.2	0.9902	0.1444	1.1346
705.44	3204	0.05053	0.05053	872.6	0	872.6	902.5	0	902.5	1.0580	0	1.0580

Source: Tables A-4E through A-8E are adapted from Gordon J. Van Wylen and Richard E. Sonntag, *Fundamentals of Classical Thermodynamics,* English/SI Version, 3rd ed. (New York: John Wiley & Sons, 1986), pp. 619–33. Originally published in Joseph H. Keenan, Frederick G. Keyes, Philip G. Hill, and Joan G. Moore, *Steam Tables* (New York: John Wiley & Sons, 1969).

TABLE A-5E

Saturated water—Pressure table

Press., P **psia**	Sat. temp., T_{sat} °**F**	Specific volume, ft³/lbm Sat. liquid, v_f	Sat. vapor, v_g	Internal energy, Btu/lbm Sat. liquid, u_f	Evap., u_{fg}	Sat. vapor, u_g	Enthalpy, Btu/lbm Sat. liquid, h_f	Evap., h_{fg}	Sat. vapor, h_g	Entropy, Btu/lbm · R Sat. liquid, s_f	Evap., s_{fg}	Sat. vapor, s_g
1.0	101.70	0.016136	333.6	69.74	974.3	1044.0	69.74	1036.0	1105.8	0.13266	1.8453	1.9779
2.0	126.04	0.016230	173.75	94.02	957.8	1051.8	94.02	1022.1	1116.1	0.17499	1.7448	1.9198
3.0	141.43	0.016300	118.72	109.38	947.2	1056.6	109.39	1013.1	1122.5	0.20089	1.6852	1.8861
4.0	152.93	0.016358	90.64	120.88	939.3	1060.2	120.89	1006.4	1127.3	0.21983	1.6426	1.8624
5.0	162.21	0.016407	73.53	130.15	932.9	1063.0	130.17	1000.9	1131.0	0.23486	1.6093	1.8441
6.0	170.03	0.016451	61.98	137.98	927.4	1065.4	138.00	996.2	1134.2	0.24736	1.5819	1.8292
8.0	182.84	0.016526	47.35	150.81	918.4	1069.2	150.84	988.4	1139.3	0.26754	1.5383	1.8058
10	193.19	0.016590	38.42	161.20	911.0	1072.2	161.23	982.1	1143.3	0.28358	1.5041	1.7877
14.696	211.99	0.016715	26.80	180.10	897.5	1077.6	180.15	970.4	1150.5	0.31212	1.4446	1.7567
15	213.03	0.016723	26.29	181.14	896.8	1077.9	181.19	969.7	1150.9	0.31367	1.4414	1.7551
20	227.96	0.016830	20.09	196.19	885.8	1082.0	196.26	960.1	1156.4	0.33580	1.3962	1.7320

TABLE A-5E

Saturated water—Pressure table (*Concluded*)

Press., P psia	Sat. temp., T_{sat} °F	Specific volume, ft³/lbm		Internal energy, Btu/lbm			Enthalpy, Btu/lbm			Entropy, Btu/lbm · R		
		Sat. liquid, v_f	Sat. vapor, v_g	Sat. liquid, u_f	Evap., u_{fg}	Sat. vapor, u_g	Sat. liquid, h_f	Evap., h_{fg}	Sat. vapor, h_g	Sat. liquid, s_f	Evap., s_{fg}	Sat. vapor, s_g
25	240.08	0.016922	16.306	208.44	876.9	1085.3	208.52	952.2	1160.7	0.35345	1.3607	1.7142
30	250.34	0.017004	13.748	218.84	869.2	1088.0	218.93	945.4	1164.3	0.36821	1.3314	1.6996
35	259.30	0.017073	11.900	227.93	862.4	1090.3	228.04	939.3	1167.4	0.38093	1.3064	1.6873
40	267.26	0.017146	10.501	236.03	856.2	1092.3	236.16	933.8	1170.0	0.39214	1.2845	1.6767
45	274.46	0.017209	9.403	243.37	850.7	1094.0	243.51	928.8	1172.3	0.40218	1.2651	1.6673
50	281.03	0.017269	8.518	250.08	845.5	1095.6	250.24	924.2	1174.4	0.41129	1.2476	1.6589
55	287.10	0.017325	7.789	256.28	840.8	1097.0	256.46	919.9	1176.3	0.41963	1.2317	1.6513
60	292.73	0.017378	7.177	262.06	836.3	1098.3	262.25	915.8	1178.0	0.42733	1.2170	1.6444
65	298.00	0.017429	6.657	267.46	832.1	1099.5	267.67	911.9	1179.6	0.43450	1.2035	1.6380
70	302.96	0.017478	6.209	272.56	828.1	1100.6	272.79	908.3	1181.0	0.44120	1.1909	1.6321
75	307.63	0.017524	5.818	277.37	824.3	1101.6	277.61	904.8	1182.4	0.44749	1.1790	1.6265
80	312.07	0.017570	5.474	281.95	820.6	1102.6	282.21	901.4	1183.6	0.45344	1.1679	1.6214
85	316.29	0.017613	5.170	286.30	817.1	1103.5	286.58	898.2	1184.8	0.45907	1.1574	1.6165
90	320.31	0.017655	4.898	290.46	813.8	1104.3	290.76	895.1	1185.9	0.46442	1.1475	1.6119
95	324.16	0.017696	4.654	294.45	810.6	1105.0	294.76	892.1	1186.9	0.46952	1.1380	1.6076
100	327.86	0.017736	4.434	298.28	807.5	1105.8	298.61	889.2	1187.8	0.47439	1.1290	1.6034
110	334.82	0.017813	4.051	305.52	801.6	1107.1	305.88	883.7	1189.6	0.48355	1.1122	1.5957
120	341.30	0.017886	3.730	312.27	796.0	1108.3	312.67	878.5	1191.1	0.49201	1.0966	1.5886
130	347.37	0.017957	3.457	318.61	790.7	1109.4	319.04	873.5	1192.5	0.49989	1.0822	1.5821
140	353.08	0.018024	3.221	324.58	785.7	1110.3	325.05	868.7	1193.8	0.50727	1.0688	1.5761
150	358.48	0.018089	3.016	330.24	781.0	1111.2	330.75	864.2	1194.9	0.51422	1.0562	1.5704
160	363.60	0.018152	2.836	335.63	776.4	1112.0	336.16	859.8	1196.0	0.52078	1.0443	1.5651
170	368.47	0.018214	2.676	340.76	772.0	1112.7	341.33	855.6	1196.9	0.52700	1.0330	1.5600
180	373.13	0.018273	2.533	345.68	767.7	1113.4	346.29	851.5	1197.8	0.53292	1.0223	1.5553
190	377.59	0.018331	2.405	350.39	763.6	1114.0	351.04	847.5	1198.6	0.53857	1.0122	1.5507
200	381.86	0.018387	2.289	354.9	759.6	1114.6	355.6	843.7	1199.3	0.5440	1.0025	1.5464
250	401.04	0.018653	1.8448	375.4	741.4	1116.7	376.2	825.8	1202.1	0.5680	0.9594	1.5274
300	417.43	0.018896	1.5442	393.0	725.1	1118.2	394.1	809.8	1203.9	0.5883	0.9232	1.5115
350	431.82	0.019124	1.3267	408.7	710.3	1119.0	409.9	795.0	1204.9	0.6060	0.8917	1.4978
400	444.70	0.019340	1.1620	422.8	696.7	1119.5	424.2	781.2	1205.5	0.6218	0.8638	1.4856
450	456.39	0.019547	1.0326	435.7	683.9	1119.6	437.4	768.2	1205.6	0.6360	0.8385	1.4746
500	467.13	0.019748	0.9283	447.7	671.7	1119.4	449.5	755.8	1205.3	0.6490	0.8154	1.4645
550	477.07	0.019943	0.8423	458.9	660.2	1119.1	460.9	743.9	1204.8	0.6611	0.7941	1.4551
600	486.33	0.02013	0.7702	469.4	649.1	1118.6	471.7	732.4	1204.1	0.6723	0.7742	1.4464
700	503.23	0.02051	0.6558	488.9	628.2	1117.0	491.5	710.5	1202.0	0.6927	0.7378	1.4305
800	518.36	0.02087	0.5691	506.6	608.4	1115.0	509.7	689.6	1199.3	0.7110	0.7050	1.4160
900	532.12	0.02123	0.5009	523.0	589.6	1112.6	526.6	669.5	1196.0	0.7277	0.6750	1.4027
1000	544.75	0.02159	0.4459	538.4	571.5	1109.9	542.4	650.0	1192.4	0.7432	0.6471	1.3903
1200	567.37	0.02232	0.3623	566.7	536.8	1103.5	571.7	612.3	1183.9	0.7712	0.5961	1.3673
1400	587.25	0.02307	0.3016	592.7	503.3	1096.0	598.6	575.5	1174.1	0.7964	0.5497	1.3461
1600	605.06	0.02386	0.2552	616.9	470.5	1087.4	624.0	538.9	1162.9	0.8196	0.5062	1.3258
1800	621.21	0.02472	0.2183	640.0	437.6	1077.7	648.3	502.1	1150.4	0.8414	0.4645	1.3060
2000	636.00	0.02565	0.18813	662.4	404.2	1066.6	671.9	464.4	1136.3	0.8623	0.4238	1.2861
2500	668.31	0.02860	0.13059	717.7	313.4	1031.0	730.9	360.5	1091.4	0.9131	0.3196	1.2327
3000	695.52	0.03431	0.08404	783.4	185.4	968.8	802.5	213.0	1015.5	0.9732	0.1843	1.1575
3203.6	705.44	0.05053	0.05053	872.6	0	872.6	902.5	0	902.5	1.0580	0	1.0580

TABLE A-6E

Superheated water

T °F	v ft³/lbm	u Btu/lbm	h Btu/lbm	s Btu/lbm·R	v ft³/lbm	u Btu/lbm	h Btu/lbm	s Btu/lbm·R	v ft³/lbm	u Btu/lbm	h Btu/lbm	s Btu/lbm·R
	P = 1.0 psia (101.70°F)*				*P* = 5.0 psia (162.21°F)				*P* = 10.0 psia (193.19°F)			
Sat.†	333.6	1044.0	1105.8	1.9779	73.53	1063.0	1131.0	1.8441	38.42	1072.2	1143.3	1.7877
200	392.5	1077.5	1150.1	2.0508	78.15	1076.3	1148.6	1.8715	38.85	1074.7	1146.6	1.7927
240	416.4	1091.2	1168.3	2.0775	83.00	1090.3	1167.1	1.8987	41.32	1089.0	1165.5	1.8205
280	440.3	1105.0	1186.5	2.1028	87.83	1104.3	1185.5	1.9244	43.77	1103.3	1184.3	1.8467
320	464.2	1118.9	1204.8	2.1269	92.64	1118.3	1204.0	1.9487	46.20	1117.6	1203.1	1.8714
360	488.1	1132.9	1223.2	2.1500	97.45	1132.4	1222.6	1.9719	48.62	1131.8	1221.8	1.8948
400	511.9	1147.0	1241.8	2.1720	102.24	1146.6	1241.2	1.9941	51.03	1146.1	1240.5	1.9171
440	535.8	1161.2	1260.4	2.1932	107.03	1160.9	1259.9	2.0154	53.44	1160.5	1259.3	1.9385
500	571.5	1182.8	1288.5	2.2235	114.20	1182.5	1288.2	2.0458	57.04	1182.2	1287.7	1.9690
600	631.1	1219.3	1336.1	2.2706	126.15	1219.1	1335.8	2.0930	63.03	1218.9	1335.5	2.0164
700	690.7	1256.7	1384.5	2.3142	138.08	1256.5	1384.3	2.1367	69.01	1256.3	1384.0	2.0601
800	750.3	1294.9	1433.7	2.3550	150.01	1294.7	1433.5	2.1775	74.98	1294.6	1433.3	2.1009
1000	869.5	1373.9	1534.8	2.4294	173.86	1373.9	1534.7	2.2520	86.91	1373.8	1534.6	2.1755
1200	988.6	1456.7	1639.6	2.4967	197.70	1456.6	1639.5	2.3192	98.84	1456.5	1639.4	2.2428
1400	1107.7	1543.1	1748.1	2.5584	221.54	1543.1	1748.1	2.3810	110.76	1543.0	1748.0	2.3045
	P = 14.696 psia (211.99°F)				*P* = 20 psia (227.96°F)				*P* = 40 psia (267.26°F)			
Sat.	26.80	1077.6	1150.5	1.7567	20.09	1082.0	1156.4	1.7320	10.501	1092.3	1170.0	1.6767
240	28.00	1087.9	1164.0	1.7764	20.47	1086.5	1162.3	1.7405				
280	29.69	1102.4	1183.1	1.8030	21.73	1101.4	1181.8	1.7676	10.711	1097.3	1176.6	1.6857
320	31.36	1116.8	1202.1	1.8280	22.98	1116.0	1201.0	1.7930	11.360	1112.8	1196.9	1.7124
360	33.02	1131.2	1221.0	1.8516	24.21	1130.6	1220.1	1.8168	11.996	1128.0	1216.8	1.7373
400	34.67	1145.6	1239.9	1.8741	25.43	1145.1	1239.2	1.8395	12.623	1143.0	1236.4	1.7606
440	36.31	1160.1	1258.8	1.8956	26.64	1159.6	1258.2	1.8611	13.243	1157.8	1255.8	1.7828
500	38.77	1181.8	1287.3	1.9263	28.46	1181.5	1286.8	1.8919	14.164	1180.1	1284.9	1.8140
600	42.86	1218.6	1335.2	1.9737	31.47	1218.4	1334.8	1.9395	15.685	1217.3	1333.4	1.8621
700	46.93	1256.1	1383.8	2.0175	34.47	1255.9	1383.5	1.9834	17.196	1255.1	1382.4	1.9063
800	51.00	1294.4	1433.1	2.0584	37.46	1294.3	1432.9	2.0243	18.701	1293.7	1432.1	1.9474
1000	59.13	1373.7	1534.5	2.1330	43.44	1373.5	1534.3	2.0989	21.70	1373.1	1533.8	2.0223
1200	67.25	1456.5	1639.3	2.2003	49.41	1456.4	1639.2	2.1663	24.69	1456.1	1638.9	2.0897
1400	75.36	1543.0	1747.9	2.2621	55.37	1542.9	1747.9	2.2281	27.68	1542.7	1747.6	2.1515
1600	83.47	1633.2	1860.2	2.3194	61.33	1633.2	1860.1	2.2854	30.66	1633.0	1859.9	2.2089
	P = 60 psia (292.73°F)				*P* = 80 psia (312.07°F)				*P* = 100 psia (327.86°F)			
Sat.	7.177	1098.3	1178.0	1.6444	5.474	1102.6	1183.6	1.6214	4.434	1105.8	1187.8	1.6034
320	7.485	1109.5	1192.6	1.6634	5.544	1106.0	1188.0	1.6271				
360	7.924	1125.3	1213.3	1.6893	5.886	1122.5	1209.7	1.6541	4.662	1119.7	1205.9	1.6259
400	8.353	1140.8	1233.5	1.7134	6.217	1138.5	1230.6	1.6790	4.934	1136.2	1227.5	1.6517
440	8.775	1156.0	1253.4	1.7360	6.541	1154.2	1251.0	1.7022	5.199	1152.3	1248.5	1.6755
500	9.399	1178.6	1283.0	1.7678	7.017	1177.2	1281.1	1.7346	5.587	1175.7	1279.1	1.7085
600	10.425	1216.3	1332.1	1.8165	7.794	1215.3	1330.7	1.7838	6.216	1214.2	1329.3	1.7582
700	11.440	1254.4	1381.4	1.8609	8.561	1253.6	1380.3	1.8285	6.834	1252.8	1379.2	1.8033
800	12.448	1293.0	1431.2	1.9022	9.321	1292.4	1430.4	1.8700	7.445	1291.8	1429.6	1.8449
1000	14.454	1372.7	1533.2	1.9773	10.831	1372.3	1532.6	1.9453	8.657	1371.9	1532.1	1.9204
1200	16.452	1455.8	1638.5	2.0448	12.333	1455.5	1638.1	2.0130	9.861	1455.2	1637.7	1.9882
1400	18.445	1542.5	1747.3	2.1067	13.830	1542.3	1747.0	2.0749	11.060	1542.0	1746.7	2.0502
1600	20.44	1632.8	1859.7	2.1641	15.324	1632.6	1859.5	2.1323	12.257	1632.4	1859.3	2.1076
1800	22.43	1726.7	1975.7	2.2179	16.818	1726.5	1975.5	2.1861	13.452	1726.4	1975.3	2.1614
2000	24.41	1824.0	2095.1	2.2685	18.310	1823.9	2094.9	2.2367	14.647	1823.7	2094.8	2.2121

*The temperature in parentheses is the saturation temperature at the specified pressure.

†Properties of saturated vapor at the specified pressure.

TABLE A-6E

Superheated water (*Continued*)

T °F	v ft³/ lbm	u Btu/ lbm	h Btu/ lbm	s Btu/ lbm·R	v ft³/ lbm	u Btu/ lbm	h Btu/ lbm	s Btu/ lbm·R	v ft³/ lbm	u Btu/ lbm	h Btu/ lbm	s Btu/ lbm·R
	P = 120 psia (341.30°F)*				**P = 140 psia (353.08°F)**				**P = 160 psia (363.60°F)**			
Sat.	3.730	1108.3	1191.1	1.5886	3.221	1110.3	1193.8	1.5761	2.836	1112.0	1196.0	1.5651
360	3.844	1116.7	1202.0	1.6021	3.259	1113.5	1198.0	1.5812				
400	4.079	1133.8	1224.4	1.6288	3.466	1131.4	1221.2	1.6088	3.007	1128.8	1217.8	1.5911
450	4.360	1154.3	1251.2	1.6590	3.713	1152.4	1248.6	1.6399	3.228	1150.5	1246.1	1.6230
500	4.633	1174.2	1277.1	1.6868	3.952	1172.7	1275.1	1.6682	3.440	1171.2	1273.0	1.6518
550	4.900	1193.8	1302.6	1.7127	4.184	1192.6	1300.9	1.6944	3.646	1191.3	1299.2	1.6784
600	5.164	1213.2	1327.8	1.7371	4.412	1212.1	1326.4	1.7191	3.848	1211.1	1325.0	1.7034
700	5.682	1252.0	1378.2	1.7825	4.860	1251.2	1377.1	1.7648	4.243	1250.4	1376.0	1.7494
800	6.195	1291.2	1428.7	1.8243	5.301	1290.5	1427.9	1.8068	4.631	1289.9	1427.0	1.7916
1000	7.208	1371.5	1531.5	1.9000	6.173	1371.0	1531.0	1.8827	5.397	1370.6	1530.4	1.8677
1200	8.213	1454.9	1637.3	1.9679	7.036	1454.6	1636.9	1.9507	6.154	1454.3	1636.5	1.9358
1400	9.214	1541.8	1746.4	2.0300	7.895	1541.6	1746.1	2.0129	6.906	1541.4	1745.9	1.9980
1600	10.212	1632.3	1859.0	2.0875	8.752	1632.1	1858.8	2.0704	7.656	1631.9	1858.6	2.0556
1800	11.209	1726.2	1975.1	2.1413	9.607	1726.1	1975.0	2.1242	8.405	1725.9	1974.8	2.1094
2000	12.205	1823.6	2094.6	2.1919	10.461	1823.5	2094.5	2.1749	9.153	1823.3	2094.3	2.1601
	P = 180 psia (373.13°F)				**P = 200 psia (381.86°F)**				**P = 225 psia (391.87°F)**			
Sat.	2.533	1113.4	1197.8	1.5553	2.289	1114.6	1199.3	1.5464	2.043	1115.8	1200.8	1.5365
400	2.648	1126.2	1214.4	1.5749	2.361	1123.5	1210.8	1.5600	2.073	1119.9	1206.2	1.5427
450	2.850	1148.5	1243.4	1.6078	2.548	1146.4	1240.7	1.5938	2.245	1143.8	1237.3	1.5779
500	3.042	1169.6	1270.9	1.6372	2.724	1168.0	1268.8	1.6239	2.405	1165.9	1266.1	1.6087
550	3.228	1190.0	1297.5	1.6642	2.893	1188.7	1295.7	1.6512	2.588	1187.0	1293.5	1.6366
600	3.409	1210.0	1323.5	1.6893	3.058	1208.9	1322.1	1.6767	2.707	1207.5	1320.2	1.6624
700	3.763	1249.6	1374.9	1.7357	3.379	1248.8	1373.8	1.7234	2.995	1247.7	1372.4	1.7095
800	4.110	1289.3	1426.2	1.7781	3.693	1288.6	1425.3	1.7660	3.276	1287.8	1424.2	1.7523
900	4.453	1329.4	1477.7	1.8175	4.003	1328.9	1477.1	1.8055	3.553	1328.3	1476.2	1.7920
1000	4.793	1370.2	1529.8	1.8545	4.310	1369.8	1529.3	1.8425	3.827	1369.3	1528.6	1.8292
1200	5.467	1454.0	1636.1	1.9227	4.918	1453.7	1635.7	1.9109	4.369	1453.4	1635.3	1.8977
1400	6.137	1541.2	1745.6	1.9849	5.521	1540.9	1745.3	1.9732	4.906	1540.7	1744.9	1.9600
1600	6.804	1631.7	1858.4	2.0425	6.123	1631.6	1858.2	2.0308	5.441	1631.3	1857.9	2.0177
1800	7.470	1725.8	1974.6	2.0964	6.722	1725.6	1974.4	2.0847	5.975	1725.4	1974.2	2.0716
2000	8.135	1823.2	2094.2	2.1470	7.321	1823.0	2094.0	2.1354	6.507	1822.9	2093.8	2.1223
	P = 250 psia (401.04°F)				**P = 275 psia (409.52°F)**				**P = 300 psia (417.43°F)**			
Sat.	1.8448	1116.7	1202.1	1.5274	1.6813	1117.5	1203.1	1.5192	1.5442	1118.2	1203.9	1.5115
450	2.002	1141.1	1233.7	1.5632	1.8026	1138.3	1230.0	1.5495	1.6361	1135.4	1226.2	1.5365
500	2.150	1163.8	1263.3	1.5948	1.9407	1161.7	1260.4	1.5820	1.7662	1159.5	1257.5	1.5701
550	2.290	1185.3	1291.3	1.6233	2.071	1183.6	1289.0	1.6110	1.8878	1181.9	1286.7	1.5997
600	2.426	1206.1	1318.3	1.6494	2.196	1204.7	1316.4	1.6376	2.004	1203.2	1314.5	1.6266
650	2.558	1226.5	1344.9	1.6739	2.317	1225.3	1343.2	1.6623	2.117	1224.1	1341.6	1.6516
700	2.688	1246.7	1371.1	1.6970	2.436	1245.7	1369.7	1.6856	2.227	1244.6	1368.3	1.6751
800	2.943	1287.0	1423.2	1.7401	2.670	1286.2	1422.1	1.7289	2.442	1285.4	1421.0	1.7187
900	3.193	1327.6	1475.3	1.7799	2.898	1327.0	1474.5	1.7689	2.653	1326.3	1473.6	1.7589
1000	3.440	1368.7	1527.9	1.8172	3.124	1368.2	1527.2	1.8064	2.860	1367.7	1526.5	1.7964
1200	3.929	1453.0	1634.8	1.8858	3.570	1452.6	1634.3	1.8751	3.270	1452.2	1633.8	1.8653
1400	4.414	1540.4	1744.6	1.9483	4.011	1540.1	1744.2	1.9376	3.675	1539.8	1743.8	1.9279
1600	4.896	1631.1	1857.6	2.0060	4.450	1630.9	1857.3	1.9954	4.078	1630.7	1857.0	1.9857
1800	5.376	1725.2	1974.0	2.0599	4.887	1725.0	1973.7	2.0493	4.479	1724.9	1973.5	2.0396
2000	5.856	1822.7	2093.6	2.1106	5.323	1822.5	2093.4	2.1000	4.879	1822.3	2093.2	2.0904

TABLE A-6E

Superheated water (*Continued*)

T °F	v ft³/lbm	u Btu/lbm	h Btu/lbm	s Btu/lbm · R	v ft³/lbm	u Btu/lbm	h Btu/lbm	s Btu/lbm · R	v ft³/lbm	u Btu/lbm	h Btu/lbm	s Btu/lbm · R
	P = 350 psia (431.82°F)				*P* = 400 psia (444.70°F)				*P* = 450 psia (456.39°F)			
Sat.	1.3267	1119.0	1204.9	1.4978	1.1620	1119.5	1205.5	1.4856	1.0326	1119.6	1205.6	1.4746
450	1.3733	1129.2	1218.2	1.5125	1.1745	1122.6	1209.6	1.4901				
500	1.4913	1154.9	1251.5	1.5482	1.2843	1150.1	1245.2	1.5282	1.1226	1145.1	1238.5	1.5097
550	1.5998	1178.3	1281.9	1.5790	1.3833	1174.6	1277.0	1.5605	1.2146	1170.7	1271.9	1.5436
600	1.7025	1200.3	1310.6	1.6068	1.4760	1197.3	1306.6	1.5892	1.2996	1194.3	1302.5	1.5732
650	1.8013	1221.6	1338.3	1.6323	1.5645	1219.1	1334.9	1.6153	1.3803	1216.6	1331.5	1.6000
700	1.8975	1242.5	1365.4	1.6562	1.6503	1240.4	1362.5	1.6397	1.4580	1238.2	1359.6	1.6248
800	2.085	1283.8	1418.8	1.7004	1.8163	1282.1	1416.6	1.6844	1.6077	1280.5	1414.4	1.6701
900	2.267	1325.0	1471.8	1.7409	1.9776	1323.7	1470.1	1.7252	1.7524	1322.4	1468.3	1.7113
1000	2.446	1366.6	1525.0	1.7787	2.136	1365.5	1523.6	1.7632	1.8941	1364.4	1522.2	1.7495
1200	2.799	1451.5	1632.8	1.8478	2.446	1450.7	1631.8	1.8327	2.172	1450.0	1630.8	1.8192
1400	3.148	1539.3	1743.1	1.9106	2.752	1538.7	1742.4	1.8956	2.444	1538.1	1741.7	1.8823
1600	3.494	1630.2	1856.5	1.9685	3.055	1629.8	1855.9	1.9535	2.715	1629.3	1855.4	1.9403
1800	3.838	1724.5	1973.1	2.0225	3.357	1724.1	1972.6	2.0076	2.983	1723.7	1972.1	1.9944
2000	4.182	1822.0	2092.8	2.0733	3.658	1821.6	2092.4	2.0584	3.251	1821.3	2092.0	2.0453
	P = 500 psia (467.13°F)				*P* = 600 psia (486.33°F)				*P* = 700 psia (503.23°F)			
Sat.	0.9283	1119.4	1205.3	1.4645	0.7702	1118.6	1204.1	1.4464	0.6558	1117.0	1202.0	1.4305
500	0.9924	1139.7	1231.5	1.4923	0.7947	1128.0	1216.2	1.4592				
550	1.0792	1166.7	1266.6	1.5279	0.8749	1158.2	1255.4	1.4990	0.7275	1149.0	1243.2	1.4723
600	1.1583	1191.1	1298.3	1.5585	0.9456	1184.5	1289.5	1.5320	0.7929	1177.5	1280.2	1.5081
650	1.2327	1214.0	1328.0	1.5860	1.0109	1208.6	1320.9	1.5609	0.8520	1203.1	1313.4	1.5387
700	1.3040	1236.0	1356.7	1.6112	1.0727	1231.5	1350.6	1.5872	0.9073	1226.9	1344.4	1.5661
800	1.4407	1278.8	1412.1	1.6571	1.1900	1275.4	1407.6	1.6343	1.0109	1272.0	1402.9	1.6145
900	1.5723	1321.0	1466.5	1.6987	1.3021	1318.4	1462.9	1.6766	1.1089	1315.6	1459.3	1.6576
1000	1.7008	1363.3	1520.7	1.7371	1.4108	1361.2	1517.8	1.7155	1.2036	1358.9	1514.9	1.6970
1100	1.8271	1406.0	1575.1	1.7731	1.5173	1404.2	1572.7	1.7519	1.2960	1402.4	1570.2	1.7337
1200	1.9518	1449.2	1629.8	1.8072	1.6222	1447.7	1627.8	1.7861	1.3868	1446.2	1625.8	1.7682
1400	2.198	1537.6	1741.0	1.8704	1.8289	1536.5	1739.5	1.8497	1.5652	1535.3	1738.1	1.8321
1600	2.442	1628.9	1854.8	1.9285	2.033	1628.0	1853.7	1.9080	1.7409	1627.1	1852.6	1.8906
1800	2.684	1723.3	1971.7	1.9827	2.236	1722.6	1970.8	1.9622	1.9152	1721.8	1969.9	1.9449
2000	2.926	1820.9	2091.6	2.0335	2.438	1820.2	2090.8	2.0131	2.0887	1819.5	2090.1	1.9958
	P = 800 psia (518.36°F)				*P* = 1000 psia (544.75°F)				*P* = 1250 psia (572.56°F)			
Sat.	0.5691	1115.0	1199.3	1.4160	0.4459	1109.9	1192.4	1.3903	0.3454	1101.7	1181.6	1.3619
550	0.6154	1138.8	1229.9	1.4469	0.4534	1114.8	1198.7	1.3966				
600	0.6776	1170.1	1270.4	1.4861	0.5140	1153.7	1248.8	1.4450	0.3786	1129.0	1216.6	1.3954
650	0.7324	1197.2	1305.6	1.5186	0.5637	1184.7	1289.1	1.4822	0.4267	1167.2	1266.0	1.4410
700	0.7829	1222.1	1338.0	1.5471	0.6080	1212.0	1324.6	1.5135	0.4670	1198.4	1306.4	1.4767
750	0.8306	1245.7	1368.6	1.5730	0.6490	1237.2	1357.3	1.5412	0.5030	1226.1	1342.4	1.5070
800	0.8764	1268.5	1398.2	1.5969	0.6878	1261.2	1388.5	1.5664	0.5364	1251.8	1375.8	1.5341
900	0.9640	1312.9	1455.6	1.6408	0.7610	1307.3	1448.1	1.6120	0.5984	1300.0	1438.4	1.5820
1000	1.0482	1356.7	1511.9	1.6807	0.8305	1352.2	1505.9	1.6530	0.6563	1346.4	1498.2	1.6244
1100	1.1300	1400.5	1567.8	1.7178	0.8976	1396.8	1562.9	1.6908	0.7116	1392.0	1556.6	1.6631
1200	1.2102	1444.6	1623.8	1.7526	0.9630	1441.5	1619.7	1.7261	0.7652	1437.5	1614.5	1.6991
1400	1.3674	1534.2	1736.6	1.8167	1.0905	1531.9	1733.7	1.7909	0.8689	1529.0	1730.0	1.7648
1600	1.5218	1626.2	1851.5	1.8754	1.2152	1624.4	1849.3	1.8499	0.9699	1622.2	1846.5	1.8243
1800	1.6749	1721.0	1969.0	1.9298	1.3384	1719.5	1967.2	1.9046	1.0693	1717.6	1965.0	1.8791
2000	1.8271	1818.8	2089.3	1.9808	1.4608	1817.4	2087.7	1.9557	1.1678	1815.7	2085.8	1.9304

H₂O

TABLE A-6E

Superheated water (*Concluded*)

T °F	v ft³/ lbm	u Btu/ lbm	h Btu/ lbm	s Btu/ lbm · R	v ft³/ lbm	u Btu/ lbm	h Btu/ lbm	s Btu/ lbm · R	v ft³/ lbm	u Btu/ lbm	h Btu/ lbm	s Btu/ lbm · R
	P = 1500 psia (596.39°F)				*P* = 1750 psia (617.31°F)				*P* = 2000 psia (636.00°F)			
Sat.	0.2769	1091.8	1168.7	1.3359	0.2268	1080.2	1153.7	1.3109	0.18813	1066.6	1136.3	1.2861
600	0.2816	1096.6	1174.8	1.3416								
650	0.3329	1147.0	1239.4	1.4012	0.2627	1122.5	1207.6	1.3603	0.2057	1091.1	1167.2	1.3141
700	0.3716	1183.4	1286.6	1.4429	0.3022	1166.7	1264.6	1.4106	0.2487	1147.7	1239.8	1.3782
750	0.4049	1214.1	1326.5	1.4767	0.3341	1201.3	1309.5	1.4485	0.2803	1187.3	1291.1	1.4216
800	0.4350	1241.8	1362.5	1.5058	0.3622	1231.3	1348.6	1.4802	0.3071	1220.1	1333.8	1.4562
850	0.4631	1267.7	1396.2	1.5320	0.3878	1258.8	1384.4	1.5081	0.3312	1249.5	1372.0	1.4860
900	0.4897	1292.5	1428.5	1.5562	0.4119	1284.8	1418.2	1.5334	0.3534	1276.8	1407.6	1.5126
1000	0.5400	1340.4	1490.3	1.6001	0.4569	1334.3	1482.3	1.5789	0.3945	1328.1	1474.1	1.5598
1100	0.5876	1387.2	1550.3	1.6399	0.4990	1382.2	1543.8	1.6197	0.4325	1377.2	1537.2	1.6017
1200	0.6334	1433.5	1609.3	1.6765	0.5392	1429.4	1604.0	1.6571	0.4685	1425.2	1598.6	1.6398
1400	0.7213	1526.1	1726.3	1.7431	0.6158	1523.1	1722.6	1.7245	0.5368	1520.2	1718.8	1.7082
1600	0.8064	1619.9	1843.7	1.8031	0.6896	1617.6	1841.0	1.7850	0.6020	1615.4	1838.2	1.7692
1800	0.8899	1715.7	1962.7	1.8582	0.7617	1713.9	1960.5	1.8404	0.6656	1712.0	1958.3	1.8249
2000	0.9725	1814.0	2083.9	1.9096	0.8330	1812.3	2082.0	1.8919	0.7284	1810.6	2080.2	1.8765
	P = 2500 psia (668.31°F)				*P* = 3000 psia (695.52°F)				*P* = 3500 psia			
Sat.	0.13059	1031.0	1091.4	1.2327	0.08404	968.8	1015.5	1.1575				
650									0.02491	663.5	679.7	0.8630
700	0.16839	1098.7	1176.6	1.3073	0.09771	1003.9	1058.1	1.1944	0.03058	759.5	779.3	0.9506
750	0.2030	1155.2	1249.1	1.3686	0.14831	1114.7	1197.1	1.3122	0.10460	1058.4	1126.1	1.2440
800	0.2291	1195.7	1301.7	1.4112	0.17572	1167.6	1265.2	1.3675	0.13626	1134.7	1223.0	1.3226
850	0.2513	1229.5	1345.8	1.4456	0.19731	1207.7	1317.2	1.4080	0.15818	1183.4	1285.9	1.3716
900	0.2712	1259.5	1385.4	1.4752	0.2160	1241.8	1361.7	1.4414	0.17625	1222.4	1336.5	1.4096
950	0.2896	1288.2	1422.2	1.5018	0.2328	1272.7	1402.0	1.4705	0.19214	1256.4	1380.8	1.4416
1000	0.3069	1315.2	1457.2	1.5262	0.2485	1301.7	1439.6	1.4967	0.2066	1287.6	1421.4	1.4699
1100	0.3393	1366.8	1523.8	1.5704	0.2772	1356.2	1510.1	1.5434	0.2328	1345.2	1496.0	1.5193
1200	0.3696	1416.7	1587.7	1.6101	0.3036	1408.0	1576.6	1.5848	0.2566	1399.2	1565.3	1.5624
1400	0.4261	1514.2	1711.3	1.6804	0.3524	1508.1	1703.7	1.6571	0.2997	1501.9	1696.1	1.6368
1600	0.4795	1610.2	1832.6	1.7424	0.3978	1606.3	1827.1	1.7201	0.3395	1601.7	1821.6	1.7010
1800	0.5312	1708.2	1954.0	1.7986	0.4416	1704.5	1949.6	1.7769	0.3776	1700.8	1945.4	1.7583
2000	0.5820	1807.2	2076.4	1.8506	0.4844	1803.9	2072.8	1.8291	0.4147	1800.6	2069.2	1.8108
	P = 4000 psia				*P* = 5000 psia				*P* = 6000 psia			
650	0.02447	657.7	675.8	0.8574	0.02377	648.0	670.0	0.8482	0.01222	640.0	665.8	0.8405
700	0.02867	742.1	763.4	0.9345	0.02676	721.8	746.6	0.9156	0.02563	708.1	736.5	0.9028
750	0.06331	960.7	1007.5	1.1395	0.03364	821.4	852.6	1.0049	0.02978	788.6	821.7	0.9746
800	0.10522	1095.0	1172.9	1.2740	0.05932	987.2	1042.1	1.1583	0.03942	896.9	940.7	1.0708
850	0.12833	1156.5	1251.5	1.3352	0.08556	1092.7	1171.9	1.2596	0.05818	1018.8	1083.4	1.1820
900	0.14622	1201.5	1309.7	1.3789	0.10385	1155.1	1251.1	1.3190	0.07588	1102.9	1187.2	1.2599
950	0.16151	1239.2	1358.8	1.4144	0.11853	1202.2	1311.9	1.3629	0.09008	1162.0	1262.0	1.3140
1000	0.17520	1272.9	1402.6	1.4449	0.13120	1242.0	1363.4	1.3988	0.10207	1209.1	1322.4	1.3561
1100	0.19954	1333.9	1481.6	1.4973	0.15302	1310.6	1452.2	1.4577	0.12218	1286.4	1422.1	1.4222
1200	0.2213	1390.1	1553.9	1.5423	0.17199	1371.6	1530.8	1.5066	0.13927	1352.7	1507.3	1.4752
1300	0.2414	1443.7	1622.4	1.5823	0.18918	1428.6	1603.7	1.5493	0.15453	1413.3	1584.9	1.5206
1400	0.2603	1495.7	1688.4	1.6188	0.20517	1483.2	1673.0	1.5876	0.16854	1470.5	1657.6	1.5608
1600	0.2959	1597.1	1816.1	1.6841	0.2348	1587.9	1805.2	1.6551	0.19420	1578.7	1794.3	1.6307
1800	0.3296	1697.1	1941.1	1.7420	0.2626	1689.8	1932.7	1.7142	0.21801	1682.4	1924.5	1.6910
2000	0.3625	1797.3	2065.6	1.7948	0.2895	1790.8	2058.6	1.7676	0.24087	1784.3	2051.7	1.7450

TABLE A-7E

Compressed liquid water

T °F	v ft³/lbm	u Btu/lbm	h Btu/lbm	s Btu/lbm·R	v ft³/lbm	u Btu/lbm	h Btu/lbm	s Btu/lbm·R	v ft³/lbm	u Btu/lbm	h Btu/lbm	s Btu/lbm·R
	P = 500 psia (467.13°F)				P = 1000 psia (544.75°F)				P = 1500 psia (596.39°F)			
Sat.	0.019748	447.70	449.53	0.649.04	0.021591	538.39	542.38	0.74320	0.023461	604.97	611.48	0.80824
32	0.015994	0.00	1.49	0.00000	0.015967	0.03	2.99	0.00005	0.015939	0.05	4.47	0.00007
50	0.015998	18.02	19.50	0.03599	0.015972	17.99	20.94	0.03592	0.015946	17.95	22.38	0.03584
100	0.016106	67.87	69.36	0.12932	0.016082	67.70	70.68	0.12901	0.016058	67.53	71.99	0.12870
150	0.016318	117.66	119.17	0.21457	0.016293	117.38	120.40	0.21410	0.016268	117.10	121.62	0.21364
200	0.016608	167.65	169.19	0.29341	0.016580	167.26	170.32	0.29281	0.016554	166.87	171.46	0.29221
250	0.016972	217.99	219.56	0.36702	0.016941	217.47	220.61	0.36628	0.016910	216.96	221.65	0.36554
300	0.017416	268.92	270.53	0.43641	0.017379	268.24	271.46	0.43552	0.017343	267.58	272.39	0.43463
350	0.017954	320.71	322.37	0.50249	0.017909	319.83	323.15	0.50140	0.017865	318.98	323.94	0.50034
400	0.018608	373.68	375.40	0.56604	0.018550	372.55	375.98	0.56472	0.018493	371.45	376.59	0.56343
450	0.019420	428.40	430.19	0.62798	0.019340	426.89	430.47	0.62632	0.019264	425.44	430.79	0.62470
500					0.02036	483.8	487.5	0.6874	0.02024	481.8	487.4	0.6853
550									0.02158	542.1	548.1	0.7469
	P = 2000 psia (636.00°F)				P = 3000 psia (695.52°F)				P = 5000 psia			
Sat.	0.025649	662.40	671.89	0.86227	0.034310	783.45	802.50	0.97320				
32	0.015912	0.06	5.95	0.00008	0.015859	0.09	8.90	0.00009	0.015755	0.11	14.70	−0.00001
50	0.015920	17.91	23.81	0.03575	0.015870	17.84	26.65	0.03555	0.015773	17.67	32.26	0.03508
100	0.016034	67.37	73.30	0.12839	0.015987	67.04	75.91	0.12777	0.015897	66.40	81.11	0.12651
200	0.016527	166.49	172.60	0.29162	0.016476	165.74	174.89	0.29046	0.016376	164.32	179.47	0.28818
300	0.017308	266.93	273.33	0.43376	0.017240	265.66	275.23	0.43205	0.017110	263.25	279.08	0.42875
400	0.018439	370.38	377.21	0.56216	0.018334	368.32	378.50	0.55970	0.018141	364.47	381.25	0.55506
450	0.019191	424.04	431.14	0.62313	0.019053	421.36	431.93	0.62011	0.018803	416.44	433.84	0.61451
500	0.02014	479.8	487.3	0.6832	0.019944	476.2	487.3	0.6794	0.019603	469.8	487.9	0.6724
560	0.02172	551.8	559.8	0.7565	0.021382	546.2	558.0	0.7508	0.020835	536.7	556.0	0.7411
600	0.02330	605.4	614.0	0.8086	0.02274	597.0	609.6	0.8004	0.02191	584.0	604.2	0.7876
640					0.02475	654.3	668.0	0.8545	0.02334	634.6	656.2	0.8357
680					0.02879	728.4	744.3	0.9226	0.02535	690.6	714.1	0.8873
700									0.02676	721.8	746.6	0.9156

TABLE A-8E

Saturated ice—water vapor

Temp., T °F	Sat. press., P_{sat} psia	Specific volume, ft³/lbm		Internal energy, Btu/lbm			Enthalpy, Btu/lbm			Entropy, Btu/lbm · R		
		Sat. ice, v_i	Sat. vapor, $v_g \times 10^{-3}$	Sat. ice, u_i	Subl., u_{ig}	Sat. vapor, u_g	Sat. ice, h_i	Subl., h_{ig}	Sat. vapor, h_g	Sat. ice, s_i	Subl., s_{ig}	Sat. vapor, s_g
32.018	0.0887	0.01747	3.302	−143.34	1164.6	1021.2	−143.34	1218.7	1075.4	−0.292	2.479	2.187
32	0.0886	0.01747	3.305	−143.35	1164.6	1021.2	−143.35	1218.7	1075.4	−0.292	2.479	2.187
30	0.0808	0.01747	3.607	−144.35	1164.9	1020.5	−144.35	1218.9	1074.5	−0.294	2.489	2.195
25	0.0641	0.01746	4.506	−146.84	1165.7	1018.9	−146.84	1219.1	1072.3	−0.299	2.515	2.216
20	0.0505	0.01745	5.655	−149.31	1166.5	1017.2	−149.31	1219.4	1070.1	−0.304	2.542	2.238
15	0.0396	0.01745	7.13	−151.75	1167.3	1015.5	−151.75	1219.7	1067.9	−0.309	2.569	2.260
10	0.0309	0.01744	9.04	−154.17	1168.1	1013.9	−154.17	1219.9	1065.7	−0.314	2.597	2.283
5	0.0240	0.01743	11.52	−156.56	1168.8	1012.2	−156.56	1220.1	1063.5	−0.320	2.626	2.306
0	0.0185	0.01743	14.77	−158.93	1169.5	1010.6	−158.93	1220.2	1061.2	−0.325	2.655	2.330
−5	0.0142	0.01742	19.03	−161.27	1170.2	1008.9	−161.27	1220.3	1059.0	−0.330	2.684	2.354
−10	0.0109	0.01741	24.66	−163.59	1170.9	1007.3	−163.59	1220.4	1056.8	−0.335	2.714	2.379
−15	0.0082	0.01740	32.2	−165.89	1171.5	1005.6	−165.89	1220.5	1054.6	−0.340	2.745	2.405
−20	0.0062	0.01740	42.2	−168.16	1172.1	1003.9	−168.16	1220.6	1052.4	−0.345	2.776	2.431
−25	0.0046	0.01739	55.7	−170.40	1172.7	1002.3	−170.40	1220.6	1050.2	−0.351	2.808	2.457
−30	0.0035	0.01738	74.1	−172.63	1173.2	1000.6	−172.63	1220.6	1048.0	−0.356	2.841	2.485
−35	0.0026	0.01737	99.2	−174.82	1173.8	988.9	−174.82	1220.6	1045.8	−0.361	2.874	2.513
−40	0.0019	0.01737	133.8	−177.00	1174.3	997.3	−177.00	1220.6	1043.6	−0.366	2.908	2.542

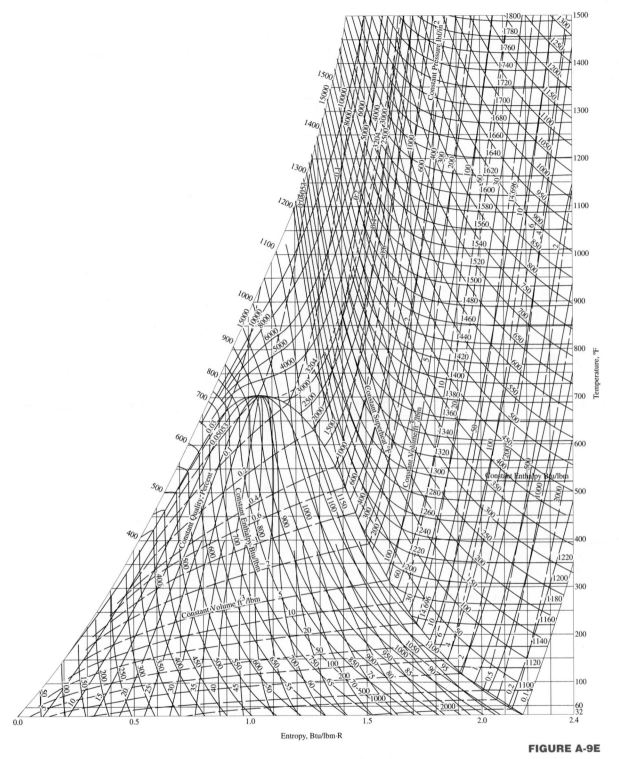

FIGURE A-9E

T-s diagram for water. [*Source:* Joseph H. Keenan, Frederick G. Keyes, Philip G. Hill, and Joan G. Moore, *Steam Tables* (New York: John Wiley & Sons, 1969).]

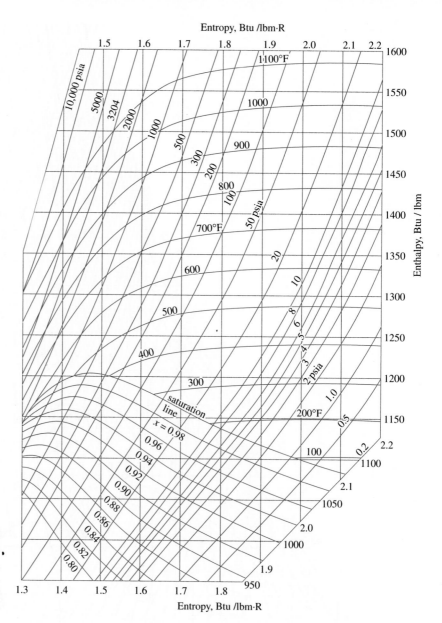

FIGURE A-10E

Mollier diagram for water. [*Source:* Joseph H. Keenan, Frederick G. Keyes, Philip G. Hill, and Joan G. Moore, *Steam Tables* (New York: John Wiley & Sons, 1969).]

TABLE A-11E

Saturated refrigerant-134a—Temperature table

Temp., T °F	Press., P_{sat} psia	Specific volume, ft³/lbm		Internal energy, Btu/lbm		Enthalpy, Btu/lbm			Entropy, Btu/lbm · R	
		Sat. liquid, v_f	Sat. vapor, v_g	Sat. liquid, u_f	Sat. vapor, u_g	Sat. liquid, h_f	Evap., h_{fg}	Sat. vapor, h_g	Sat. liquid, s_f	Sat. vapor, s_g
−40	7.490	0.01130	5.7173	−0.02	87.90	0.00	95.82	95.82	0.0000	0.2283
−30	9.920	0.01143	4.3911	2.81	89.26	2.83	94.49	97.32	0.0067	0.2266
−20	12.949	0.01156	3.4173	5.69	90.62	5.71	93.10	98.81	0.0133	0.2250
−15	14.718	0.01163	3.0286	7.14	91.30	7.17	92.38	99.55	0.0166	0.2243
−10	16.674	0.01170	2.6918	8.61	91.98	8.65	91.64	100.29	0.0199	0.2236
−5	18.831	0.01178	2.3992	10.09	92.66	10.13	90.89	101.02	0.0231	0.2230
0	21.203	0.01185	2.1440	11.58	93.33	11.63	90.12	101.75	0.0264	0.2224
5	23.805	0.01193	1.9208	13.09	94.01	13.14	89.33	102.47	0.0296	0.2219
10	26.651	0.01200	1.7251	14.60	94.68	14.66	88.53	103.19	0.0329	0.2214
15	29.756	0.01208	1.5529	16.13	95.35	16.20	87.71	103.90	0.0361	0.2209
20	33.137	0.01216	1.4009	17.67	96.02	17.74	86.87	104.61	0.0393	0.2205
25	36.809	0.01225	1.2666	19.22	96.69	19.30	86.02	105.32	0.0426	0.2200
30	40.788	0.01233	1.1474	20.78	97.35	20.87	85.14	106.01	0.0458	0.2196
40	49.738	0.01251	0.9470	23.94	98.67	24.05	83.34	107.39	0.0522	0.2189
50	60.125	0.01270	0.7871	27.14	99.98	27.28	81.46	108.74	0.0585	0.2183
60	72.092	0.01290	0.6584	30.39	101.27	30.56	79.49	110.05	0.0648	0.2178
70	85.788	0.01311	0.5538	33.68	102.54	33.89	77.44	111.33	0.0711	0.2173
80	101.37	0.01334	0.4682	37.02	103.78	37.27	75.29	112.56	0.0774	0.2169
85	109.92	0.01346	0.4312	38.72	104.39	38.99	74.17	113.16	0.0805	0.2167
90	118.99	0.01358	0.3975	40.42	105.00	40.72	73.03	113.75	0.0836	0.2165
95	128.62	0.01371	0.3668	42.14	105.60	42.47	71.86	114.33	0.0867	0.2163
100	138.83	0.01385	0.3388	43.87	106.18	44.23	70.66	114.89	0.0898	0.2161
105	149.63	0.01399	0.3131	45.62	106.76	46.01	69.42	115.43	0.0930	0.2159
110	161.04	0.01414	0.2896	47.39	107.33	47.81	68.15	115.96	0.0961	0.2157
115	173.10	0.01429	0.2680	49.17	107.88	49.63	66.84	116.47	0.0992	0.2155
120	185.82	0.01445	0.2481	50.97	108.42	51.47	65.48	116.95	0.1023	0.2153
140	243.86	0.01520	0.1827	58.39	110.41	59.08	59.57	118.65	0.1150	0.2143
160	314.63	0.01617	0.1341	66.26	111.97	67.20	52.58	119.78	0.1280	0.2128
180	400.22	0.01758	0.0964	74.83	112.77	76.13	43.78	119.91	0.1417	0.2101
200	503.52	0.02014	0.0647	84.90	111.66	86.77	30.92	117.69	0.1575	0.2044
210	563.51	0.02329	0.0476	91.84	108.48	94.27	19.18	113.45	0.1684	0.1971

Source for Tables A-11E through A-13E: M. J. Moran and H. N. Shapiro, *Fundamentals of Engineering Thermodynamics,* 2nd ed. (New York: John Wiley & Sons, 1992), pp. 754–58. Originally based on equations from D. P. Wilson and R. S. Basu, "Thermodynamic Properties of a New Stratospherically Safe Working Fluid—Refrigerant 134a," *ASHRAE Trans.* 94, Pt. 2 (1988), pp. 2095–118. Used with permission.

TABLE A-12E

Saturated refrigerant-134a—Pressure table

Press., P **psia**	Temp., T_{sat} **°F**	Specific volume, ft³/lbm		Internal energy, Btu/lbm		Enthalpy, Btu/lbm			Entropy, Btu/lbm · R	
		Sat. liquid, v_f	Sat. vapor, v_g	Sat. liquid, u_f	Sat. vapor, u_g	Sat. liquid, h_f	Evap., h_{fg}	Sat. vapor, h_g	Sat. liquid, s_f	Sat. vapor, s_g
5	−53.48	0.01113	8.3508	−3.74	86.07	−3.73	97.53	93.79	−0.0090	0.2311
10	−29.71	0.01143	4.3581	2.89	89.30	2.91	94.45	97.37	0.0068	0.2265
15	−14.25	0.01164	2.9747	7.36	91.40	7.40	92.27	99.66	0.0171	0.2242
20	−2.48	0.01181	2.2661	10.84	93.00	10.89	90.50	101.39	0.0248	0.2227
30	15.38	0.01209	1.5408	16.24	95.40	16.31	87.65	103.96	0.0364	0.2209
40	29.04	0.01232	1.1692	20.48	97.23	20.57	85.31	105.88	0.0452	0.2197
50	40.27	0.01252	0.9422	24.02	98.71	24.14	83.29	107.43	0.0523	0.2189
60	49.89	0.01270	0.7887	27.10	99.96	27.24	81.48	108.72	0.0584	0.2183
70	58.35	0.01286	0.6778	29.85	101.05	30.01	79.82	109.83	0.0638	0.2179
80	65.93	0.01302	0.5938	32.33	102.02	32.53	78.28	110.81	0.0686	0.2175
90	72.83	0.01317	0.5278	34.62	102.89	34.84	76.84	111.68	0.0729	0.2172
100	79.17	0.01332	0.4747	36.75	103.68	36.99	75.47	112.46	0.0768	0.2169
120	90.54	0.01360	0.3941	40.61	105.06	40.91	72.91	113.82	0.0839	0.2165
140	100.56	0.01386	0.3358	44.07	106.25	44.43	70.52	114.95	0.0902	0.2161
160	109.56	0.01412	0.2916	47.23	107.28	47.65	68.26	115.91	0.0958	0.2157
180	117.74	0.01438	0.2569	50.16	108.18	50.64	66.10	116.74	0.1009	0.2154
200	125.28	0.01463	0.2288	52.90	108.98	53.44	64.01	117.44	0.1057	0.2151
220	132.27	0.01489	0.2056	55.48	109.68	56.09	61.96	118.05	0.1101	0.2147
240	138.79	0.01515	0.1861	57.93	110.30	58.61	59.96	118.56	0.1142	0.2144
260	144.92	0.01541	0.1695	60.28	110.84	61.02	57.97	118.99	0.1181	0.2140
280	150.70	0.01568	0.1550	62.53	111.31	63.34	56.00	119.35	0.1219	0.2136
300	156.17	0.01596	0.1424	64.71	111.72	65.59	54.03	119.62	0.1254	0.2132
350	168.72	0.01671	0.1166	69.88	112.45	70.97	49.03	120.00	0.1338	0.2118
400	179.95	0.01758	0.0965	74.81	112.77	76.11	43.80	119.91	0.1417	0.2102
450	190.12	0.01863	0.0800	79.63	112.60	81.18	38.08	119.26	0.1493	0.2079
500	199.38	0.02002	0.0657	84.54	111.76	86.39	31.44	117.83	0.1570	0.2047

R-134a

TABLE A-13E

Superheated refrigerant-134a

T °F	v ft³/lbm	u Btu/lbm	h Btu/lbm	s Btu/lbm · R	v ft³/lbm	u Btu/lbm	h Btu/lbm	s Btu/lbm · R	v ft³/lbm	u Btu/lbm	h Btu/lbm	s Btu/lbm · R
	P = 10 psia (T_{sat} = −29.71°F)				P = 15 psia (T_{sat} = −14.25°F)				P = 20 psia (T_{sat} = −2.48°F)			
Sat.	4.3581	89.30	97.37	0.2265	2.9747	91.40	99.66	0.2242	2.2661	93.00	101.39	0.2227
−20	4.4718	90.89	99.17	0.2307								
0	4.7026	94.24	102.94	0.2391	3.0893	93.84	102.42	0.2303	2.2816	93.43	101.88	0.2238
20	4.9297	97.67	106.79	0.2472	3.2468	97.33	106.34	0.2386	2.4046	96.98	105.88	0.2323
40	5.1539	101.19	110.72	0.2553	3.4012	100.89	110.33	0.2468	2.5244	100.59	109.94	0.2406
60	5.3758	104.80	114.74	0.2632	3.5533	104.54	114.40	0.2548	2.6416	104.28	114.06	0.2487
80	5.5959	108.50	118.85	0.2709	3.7034	108.28	118.56	0.2626	2.7569	108.05	118.25	0.2566
100	5.8145	112.29	123.05	0.2786	3.8520	112.10	122.79	0.2703	2.8705	111.90	122.52	0.2644
120	6.0318	116.18	127.34	0.2861	3.9993	116.01	127.11	0.2779	2.9829	115.83	126.87	0.2720
140	6.2482	120.16	131.72	0.2935	4.1456	120.00	131.51	0.2854	3.0942	119.85	131.30	0.2795
160	6.4638	124.23	136.19	0.3009	4.2911	124.09	136.00	0.2927	3.2047	123.95	135.81	0.2869
180	6.6786	128.38	140.74	0.3081	4.4359	128.26	140.57	0.3000	3.3144	128.13	140.40	0.2922
200	6.8929	132.63	145.39	0.3152	4.5801	132.52	145.23	0.3072	3.4236	132.40	145.07	0.3014
220									3.5323	136.76	149.83	0.3085
	P = 30 psia (T_{sat} = 15.38°F)				P = 40 psia (T_{sat} = 29.04°F)				P = 50 psia (T_{sat} = 40.27°F)			
Sat.	1.5408	95.40	103.96	0.2209	1.1692	97.23	105.88	0.2197	0.9422	98.71	107.43	0.2189
20	1.5611	96.26	104.92	0.2229								
40	1.6465	99.98	109.12	0.2315	1.2065	99.33	108.26	0.2245				
60	1.7293	103.75	113.35	0.2398	1.2723	103.20	112.62	0.2331	0.9974	102.62	111.85	0.2276
80	1.8098	107.59	117.63	0.2478	1.3357	107.11	117.00	0.2414	1.0508	106.62	116.34	0.2361
100	1.8887	111.49	121.98	0.2558	1.3973	111.08	121.42	0.2494	1.1022	110.65	120.85	0.2443
120	1.9662	115.47	126.39	0.2635	1.4575	115.11	125.90	0.2573	1.1520	114.74	125.39	0.2523
140	2.0426	119.53	130.87	0.2711	1.5165	119.21	130.43	0.2650	1.2007	118.88	129.99	0.2601
160	2.1181	123.66	135.42	0.2786	1.5746	123.38	135.03	0.2725	1.2484	123.08	134.64	0.2677
180	2.1929	127.88	140.05	0.2859	1.6319	127.62	139.70	0.2799	1.2953	127.36	139.34	0.2752
200	2.2671	132.17	144.76	0.2932	1.6887	131.94	144.44	0.2872	1.3415	131.71	144.12	0.2825
220	2.3407	136.55	149.54	0.3003	1.7449	136.34	149.25	0.2944	1.3873	136.12	148.96	0.2897
240					1.8006	140.81	154.14	0.3015	1.4326	140.61	153.87	0.2969
260					1.8561	145.36	159.10	0.3085	1.4775	145.18	158.85	0.3039
280					1.9112	149.98	164.13	0.3154	1.5221	149.82	163.90	0.3108
	P = 60 psia (T_{sat} = 49.89°F)				P = 70 psia (T_{sat} = 58.35°F)				P = 80 psia (T_{sat} = 65.93°F)			
Sat.	0.7887	99.96	108.72	0.2183	0.6778	101.05	109.83	0.2179	0.5938	102.02	110.81	0.2175
60	0.8135	102.03	111.06	0.2229	0.6814	101.40	110.23	0.2186				
80	0.8604	106.11	115.66	0.2316	0.7239	105.58	114.96	0.2276	0.6211	105.03	114.23	0.2239
100	0.9051	110.21	120.26	0.2399	0.7640	109.76	119.66	0.2361	0.6579	109.30	119.04	0.2327
120	0.9482	114.35	124.88	0.2480	0.8023	113.96	124.36	0.2444	0.6927	113.56	123.82	0.2411
140	0.9900	118.54	129.53	0.2559	0.8393	118.20	129.07	0.2524	0.7261	117.85	128.60	0.2492
160	1.0308	122.79	134.23	0.2636	0.8752	122.49	133.82	0.2601	0.7584	122.18	133.41	0.2570
180	1.0707	127.10	138.98	0.2712	0.9103	126.83	138.62	0.2678	0.7898	126.55	138.25	0.2647
200	1.1100	131.47	143.79	0.2786	0.9446	131.23	143.46	0.2752	0.8205	130.98	143.13	0.2722
220	1.1488	135.91	148.66	0.2859	0.9784	135.69	148.36	0.2825	0.8506	135.47	148.06	0.2796
240	1.1871	140.42	153.60	0.2930	1.0118	140.22	153.33	0.2897	0.8803	140.02	153.05	0.2868
260	1.2251	145.00	158.60	0.3001	1.0448	144.82	158.35	0.2968	0.9095	144.63	158.10	0.2940
280	1.2627	149.65	163.67	0.3070	1.0774	149.48	163.44	0.3038	0.9384	149.32	163.21	0.3010
300	1.3001	154.38	168.81	0.3139	1.1098	154.22	168.60	0.3107	0.9671	154.06	168.38	0.3079
320									0.9955	158.88	173.62	0.3147

R-134a

TABLE A-13E

Superheated refrigerant-134a (*Concluded*)

T °F	v ft³/lbm	u Btu/lbm	h Btu/lbm	s Btu/lbm · R	v ft³/lbm	u Btu/lbm	h Btu/lbm	s Btu/lbm · R	v ft³/lbm	u Btu/lbm	h Btu/lbm	s Btu/lbm · R
	P = 90 psia (T_{sat} = 72.83°F)				**P = 100 psia (T_{sat} = 79.17°F)**				**P = 120 psia (T_{sat} = 90.54°F)**			
Sat.	0.5278	102.89	111.68	0.2172	0.4747	103.68	112.46	0.2169	0.3941	105.06	113.82	0.2165
80	0.5408	104.46	113.47	0.2205	0.4761	103.87	112.68	0.2173				
100	0.5751	108.82	118.39	0.2295	0.5086	108.32	117.73	0.2265	0.4080	107.26	116.32	0.2210
120	0.6073	113.15	123.27	0.2380	0.5388	112.73	122.70	0.2352	0.4355	111.84	121.52	0.2301
140	0.6380	117.50	128.12	0.2463	0.5674	117.13	127.63	0.2436	0.4610	116.37	126.61	0.2387
160	0.6675	121.87	132.98	0.2542	0.5947	121.55	132.55	0.2517	0.4852	120.89	131.66	0.2470
180	0.6961	126.28	137.87	0.2620	0.6210	125.99	137.49	0.2595	0.5082	125.42	136.70	0.2550
200	0.7239	130.73	142.79	0.2696	0.6466	130.48	142.45	0.2671	0.5305	129.97	141.75	0.2628
220	0.7512	135.25	147.76	0.2770	0.6716	135.02	147.45	0.2746	0.5520	134.56	146.82	0.2704
240	0.7779	139.82	152.77	0.2843	0.6960	139.61	152.49	0.2819	0.5731	139.20	151.92	0.2778
260	0.8043	144.45	157.84	0.2914	0.7201	144.26	157.59	0.2891	0.5937	143.89	157.07	0.2850
280	0.8303	149.15	162.97	0.2984	0.7438	148.98	162.74	0.2962	0.6140	148.63	162.26	0.2921
300	0.8561	153.91	168.16	0.3054	0.7672	153.75	167.95	0.3031	0.6339	153.43	167.51	0.2991
320	0.8816	158.73	173.42	0.3122	0.7904	158.59	173.21	0.3099	0.6537	158.29	172.81	0.3060
	P = 140 psia (T_{sat} = 100.56°F)				**P = 160 psia (T_{sat} = 109.55°F)**				**P = 180 psia (T_{sat} = 117.74°F)**			
Sat.	0.3358	106.25	114.95	0.2161	0.2916	107.28	115.91	0.2157	0.2569	108.18	116.74	0.2154
120	0.3610	110.90	120.25	0.2254	0.3044	109.88	118.89	0.2209	0.2595	108.77	117.41	0.2166
140	0.3846	115.58	125.24	0.2344	0.3269	114.73	124.41	0.2303	0.2814	113.83	123.21	0.2264
160	0.4066	120.21	130.74	0.2429	0.3474	119.49	129.78	0.2391	0.3011	118.74	128.77	0.2355
180	0.4274	124.82	135.89	0.2511	0.3666	124.20	135.06	0.2475	0.3191	123.56	134.19	0.2441
200	0.4474	129.44	141.03	0.2590	0.3849	128.90	140.29	0.2555	0.3361	128.34	139.53	0.2524
220	0.4666	134.09	146.18	0.2667	0.4023	133.61	145.52	0.2633	0.3523	133.11	144.84	0.2603
240	0.4852	138.77	151.34	0.2742	0.4192	138.34	150.75	0.2709	0.3678	137.90	150.15	0.2680
260	0.5034	143.50	156.54	0.2815	0.4356	143.11	156.00	0.2783	0.3828	142.71	155.46	0.2755
280	0.5212	148.28	161.78	0.2887	0.4516	147.92	161.29	0.2856	0.3974	147.55	160.79	0.2828
300	0.5387	153.11	167.06	0.2957	0.4672	152.78	166.61	0.2927	0.4116	152.44	166.15	0.2899
320	0.5559	157.99	172.39	0.3026	0.4826	157.69	171.98	0.2996	0.4256	157.38	171.55	0.2969
340	0.5730	162.93	177.78	0.3094	0.4978	162.65	177.39	0.3065	0.4393	162.36	177.00	0.3038
360	0.5898	167.93	183.21	0.3162	0.5128	167.67	182.85	0.3132	0.4529	167.40	182.49	0.3106
	P = 200 psia (T_{sat} = 125.28°F)				**P = 300 psia (T_{sat} = 156.17°F)**				**P = 400 psia (T_{sat} = 179.95°F)**			
Sat.	0.2288	108.98	117.44	0.2151	0.1424	111.72	119.62	0.2132	0.0965	112.77	119.91	0.2102
140	0.2446	112.87	121.92	0.2226								
160	0.2636	117.94	127.70	0.2321	0.1462	112.95	121.07	0.2155				
180	0.2809	122.88	133.28	0.2410	0.1633	118.93	128.00	0.2265	0.0965	112.79	119.93	0.2102
200	0.2970	127.76	138.75	0.2494	0.1777	124.47	134.34	0.2363	0.1143	120.14	128.60	0.2235
220	0.3121	132.60	144.15	0.2575	0.1905	129.79	140.36	0.2453	0.1275	126.35	135.79	0.2343
240	0.3266	137.44	149.53	0.2653	0.2021	134.99	146.21	0.2537	0.1386	132.12	142.38	0.2438
260	0.3405	142.30	154.90	0.2728	0.2130	140.12	151.95	0.2618	0.1484	137.65	148.64	0.2527
280	0.3540	147.18	160.28	0.2802	0.2234	145.23	157.63	0.2696	0.1575	143.06	154.72	0.2610
300	0.3671	152.10	165.69	0.2874	0.2333	150.33	163.28	0.2772	0.1660	148.39	160.67	0.2689
320	0.3799	157.07	171.13	0.2945	0.2428	155.44	168.92	0.2845	0.1740	153.69	166.57	0.2766
340	0.3926	162.07	176.60	0.3014	0.2521	160.57	174.56	0.2916	0.1816	158.97	172.42	0.2840
360	0.4050	167.13	182.12	0.3082	0.2611	165.74	180.23	0.2986	0.1890	164.26	178.26	0.2912
380					0.2699	170.94	185.92	0.3055	0.1962	169.57	184.09	0.2983
400					0.2786	176.18	191.64	0.3122	0.2032	174.90	189.94	0.3051

R-134a

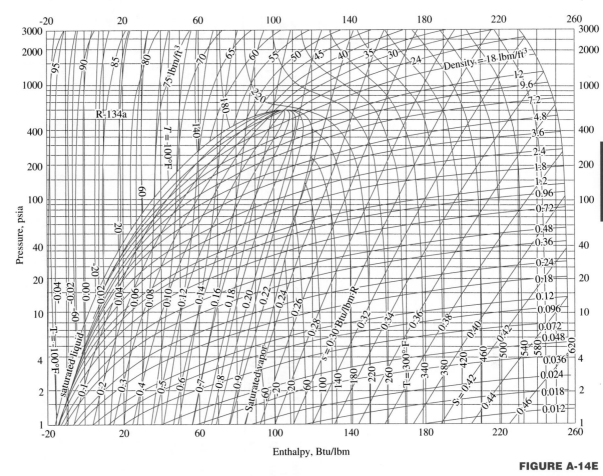

FIGURE A-14E

P-h diagram for refrigerant-134a. (Reprinted by permission of the American Society of Heating, Refrigerating, and Air-Conditioning Engineers, Inc., Atlanta, GA.)

TABLE A-15E

Properties of saturated water

Temperature, T °F	Saturation pressure, P_{sat} psia	Density, ρ lbm/ft³ Liquid	Density, ρ lbm/ft³ Vapor	Enthalpy of vaporization, h_{fg} Btu/lbm	Specific heat C_p Btu/lbm · °F Liquid	Specific heat C_p Btu/lbm · °F Vapor	Thermal conductivity, k Btu/h · ft · °F Liquid	Thermal conductivity, k Btu/h · ft · °F Vapor	Dynamic viscosity, μ lbm/ft · h Liquid	Dynamic viscosity, μ lbm/ft · h Vapor	Prandtl number, Pr Liquid	Prandtl number, Pr Vapor	Volume expansion coefficient, β 1/R Liquid
32.02	0.0887	62.41	0.00030	1075	1.010	0.446	0.324	0.0099	4.336	0.0223	13.5	1.00	-0.038×10^{-3}
40	0.1217	62.42	0.00034	1071	1.004	0.447	0.329	0.0100	3.740	0.0226	11.4	1.01	0.003×10^{-3}
50	0.1780	62.41	0.00059	1065	1.000	0.448	0.335	0.0102	3.161	0.0229	9.44	1.01	0.047×10^{-3}
60	0.2563	62.36	0.00083	1060	0.999	0.449	0.341	0.0104	2.713	0.0232	7.95	1.00	0.080×10^{-3}
70	0.3632	62.30	0.00115	1054	0.999	0.450	0.347	0.0106	2.360	0.0236	6.79	1.00	0.115×10^{-3}
80	0.5073	62.22	0.00158	1048	0.999	0.451	0.352	0.0108	2.075	0.0240	5.89	1.00	0.145×10^{-3}
90	0.6988	62.12	0.00214	1043	0.999	0.453	0.358	0.0110	1.842	0.0244	5.14	1.00	0.174×10^{-3}
100	0.9503	62.00	0.00286	1037	0.999	0.454	0.363	0.0112	1.648	0.0248	4.54	1.01	0.200×10^{-3}
110	1.2763	61.86	0.00377	1031	0.999	0.456	0.367	0.0115	1.486	0.0252	4.05	1.00	0.224×10^{-3}
120	1.6945	61.71	0.00493	1026	0.999	0.458	0.371	0.0117	1.348	0.0256	3.63	1.00	0.246×10^{-3}
130	2.225	61.55	0.00636	1020	0.999	0.460	0.375	0.0120	1.230	0.0260	3.28	1.00	0.267×10^{-3}
140	2.892	61.38	0.00814	1014	0.999	0.463	0.378	0.0122	1.129	0.0264	2.98	1.00	0.287×10^{-3}
150	3.722	61.19	0.0103	1008	1.000	0.465	0.381	0.0125	1.040	0.0269	2.73	1.00	0.306×10^{-3}
160	4.745	60.99	0.0129	1002	1.000	0.468	0.384	0.0128	0.963	0.0273	2.51	1.00	0.325×10^{-3}
170	5.996	60.79	0.0161	996	1.001	0.472	0.386	0.0131	0.894	0.0278	2.90	1.00	0.346×10^{-3}
180	7.515	60.57	0.0199	990	1.002	0.475	0.388	0.0134	0.834	0.0282	2.15	1.00	0.367×10^{-3}
190	9.343	60.35	0.0244	984	1.004	0.479	0.390	0.0137	0.781	0.0287	2.01	1.00	0.382×10^{-3}
200	11.53	60.12	0.0297	978	1.005	0.483	0.391	0.0141	0.733	0.0291	1.88	1.00	0.395×10^{-3}

210	14.125	59.87	0.0359	972	1.007	0.487	0.392	0.0144	0.690	0.0296	1.77	1.00	0.412×10^{-3}
212	14.698	59.82	0.0373	970	1.007	0.488	0.392	0.0145	0.682	0.0297	1.75	1.00	0.417×10^{-3}
220	17.19	59.62	0.0432	965	1.009	0.492	0.393	0.0148	0.651	0.0300	1.67	1.00	0.429×10^{-3}
230	20.78	59.36	0.0516	959	1.011	0.497	0.394	0.0152	0.616	0.0305	1.58	1.00	0.443×10^{-3}
240	24.97	59.09	0.0612	952	1.013	0.503	0.394	0.0156	0.585	0.0310	1.50	1.00	0.462×10^{-3}
250	29.82	58.82	0.0723	946	1.015	0.509	0.395	0.0160	0.556	0.0310	1.43	1.00	0.480×10^{-3}
260	35.42	58.53	0.0850	939	1.018	0.516	0.395	0.0164	0.530	0.0319	1.37	1.00	0.497×10^{-3}
270	41.85	58.24	0.0993	932	1.020	0.523	0.395	0.0168	0.506	0.0324	1.31	1.01	0.514×10^{-3}
280	49.18	57.94	0.1156	925	1.023	0.530	0.395	0.0172	0.484	0.0328	1.25	1.01	0.532×10^{-3}
290	57.53	57.63	0.3390	918	1.026	0.538	0.395	0.0177	0.464	0.0333	1.21	1.01	0.549×10^{-3}
300	66.98	57.31	0.1545	910	1.029	0.547	0.394	0.0182	0.445	0.0338	1.16	1.02	0.566×10^{-3}
320	89.60	56.65	0.2033	895	1.036	0.567	0.393	0.0191	0.412	0.0347	1.09	1.03	0.636×10^{-3}
340	117.93	55.95	0.2637	880	1.044	0.590	0.391	0.0202	0.383	0.0356	1.02	1.04	0.656×10^{-3}
360	152.92	55.22	0.3377	863	1.054	0.617	0.389	0.0213	0.359	0.0365	0.973	1.06	0.681×10^{-3}
380	195.60	54.46	0.4275	845	1.065	0.647	0.385	0.0224	0.337	0.0375	0.932	1.08	0.720×10^{-3}
400	241.1	53.65	0.5359	827	1.078	0.683	0.382	0.0237	0.318	0.0384	0.893	1.11	0.771×10^{-3}
450	422.1	51.46	0.9082	775	1.121	0.799	0.370	0.0271	0.278	0.0407	0.842	1.20	0.912×10^{-3}
500	680.0	48.95	1.479	715	1.188	0.972	0.352	0.0312	0.246	0.0432	0.830	1.35	1.111×10^{-3}
550	1046.7	45.96	4.268	641	1.298	1.247	0.329	0.0368	0.219	0.0461	0.864	1.56	1.445×10^{-3}
600	1541	42.32	3.736	550	1.509	1.759	0.299	0.0461	0.194	0.0497	0.979	1.90	1.885×10^{-3}
650	2210	37.31	6.152	422	2.086	3.103	0.267	0.0677	0.167	0.0555	1.30	2.54	—
700	3090	27.28	13.44	168	13.80	25.90	0.254	0.1964	0.123	0.0736	6.68	9.71	—
705.44	3204	19.79	19.79	0	∞	∞	∞	∞	0.104	0.1043	—	—	—

Note 1: Kinematic viscosity ν and thermal diffusivity α can be calculated from their definitions, $\nu = \mu/\rho$ and $\alpha = k/\rho C_p = \nu/Pr$. The temperatures 32.02°F, 212°F, and 705.44°F are the triple-, boiling-, and critical-point temperatures of water, respectively. All properties listed above (except the vapor density) can be used at any pressure with negligible error except at temperatures near the critical-point value.

Note 2: The unit Btu/lbm · °F for specific heat is equivalent to Btu/lbm · R, and the unit Btu/h · ft · °F for thermal conductivity is equivalent to Btu/h · ft · R.

Source: Viscosity and thermal conductivity data are from J. V. Sengers and J. T. R. Watson, *Journal of Physical and Chemical Reference Data* 15 (1986), pp. 1291–322. Other data are obtained from various sources or calculated.

TABLE A-16E

Properties of liquids

Tempera-ture, T °F	Density, ρ lbm/ft³	Specific heat, C_p Btu/lbm · °F	Thermal conduc-tivity, k Btu/h · ft · °F	Dynamic viscosity, μ lbm/ft · s	Kinematic viscosity, ν ft²/s	Thermal diffusiv-ity, α ft²/h	Volume expansiv-ity, β 1/R	Prandtl number, Pr
Ammonia								
−20	42.4	1.07	0.317	17.6×10^{-5}	0.417×10^{-5}	6.94×10^{-3}		2.15
0	41.6	1.08	0.316	17.1×10^{-5}	0.410×10^{-5}	7.04×10^{-3}		2.09
10	40.8	1.09	0.314	16.6×10^{-5}	0.407×10^{-5}	7.08×10^{-3}		2.07
32	40.0	1.11	0.312	16.1×10^{-5}	0.402×10^{-5}	7.03×10^{-3}	1.2×10^{-3}	2.05
50	39.1	1.13	0.307	15.5×10^{-5}	0.396×10^{-5}	6.95×10^{-3}	1.3×10^{-3}	2.04
80	37.2	1.17	0.293	14.5×10^{-5}	0.386×10^{-5}	6.73×10^{-3}		2.01
120	35.2	1.22	0.275	13.0×10^{-5}	0.355×10^{-5}	6.40×10^{-3}		1.99
Benzene								
60	55.1	0.40	0.093	46.0×10^{-5}	0.835×10^{-5}	4.22×10^{-3}	0.60×10^{-3}	7.2
80	54.6	0.42	0.092	39.6×10^{-5}	0.725×10^{-5}	4.01×10^{-3}		6.5
100	54.0	0.44	0.087	35.1×10^{-5}	0.650×10^{-5}	3.53×10^{-3}		5.1
150	53.5	0.46		26.0×10^{-5}	0.480×10^{-5}			4.5
200				20.3×10^{-5}				4.0
n-Butyl alcohol								
60	50.5	0.55	0.097	226×10^{-5}	4.48×10^{-5}	3.49×10^{-3}		46.6
100	49.7	0.61	0.096	129×10^{-5}	2.60×10^{-5}	3.16×10^{-3}	0.45×10^{-3}	29.5
150	48.5	0.68	0.095	67.5×10^{-5}	1.39×10^{-5}	2.88×10^{-3}	0.48×10^{-3}	17.4
200	47.2	0.77	0.094	38.6×10^{-5}	0.815×10^{-5}	2.58×10^{-3}		11.3
Glycerin								
50	79.3	0.554	0.165	2.56	0.0323	3.76×10^{-3}		31,000
70	78.9	0.570	0.165	1.0	0.0127	3.67×10^{-3}	0.28×10^{-3}	12,500
85	78.5	0.584	0.164	0.424	0.0054	3.58×10^{-3}	0.30×10^{-3}	5,400
100	78.2	0.600	0.163	0.188	0.0024	3.45×10^{-3}		2,500
120	77.7	0.617		0.124	0.0016			1,600
Light oil								
60	57.0	0.43	0.077	5820×10^{-5}	102×10^{-5}	3.14×10^{-3}	0.38×10^{-3}	1170
80	56.8	0.44	0.077	2780×10^{-5}	49×10^{-5}	3.09×10^{-3}	0.38×10^{-3}	570
100	56.0	0.46	0.076	1530×10^{-5}	27.4×10^{-5}	2.95×10^{-3}	0.39×10^{-3}	340
150	54.3	0.48	0.075	530×10^{-5}	9.8×10^{-5}	2.88×10^{-3}	0.40×10^{-3}	122
200	54.0	0.51	0.074	250×10^{-5}	4.6×10^{-5}	2.69×10^{-3}	0.42×10^{-3}	62
250	53.0	0.52	0.074	139×10^{-5}	2.6×10^{-5}	2.67×10^{-3}	0.44×10^{-3}	35
300	51.8	0.54	0.073	830×10^{-5}	1.6×10^{-5}	2.62×10^{-3}	0.45×10^{-3}	22
Mercury								
50	847	0.033	4.7	1.07×10^{-3}	1.2×10^{-6}	0.17	1.01×10^{-4}	0.027
200	834	0.033	6.0	0.84×10^{-3}	1.0×10^{-6}	0.22	1.01×10^{-6}	0.016
300	826	0.033	6.7	0.74×10^{-3}	0.90×10^{-6}	0.25	1.01×10^{-6}	0.012
400	817	0.032	7.2	0.67×10^{-3}	0.82×10^{-6}	0.27	1.01×10^{-6}	0.011
600	802	0.032	8.1	0.58×10^{-3}	0.72×10^{-6}	0.31	1.03×10^{-6}	0.0084
Sodium								
200	58.0	0.33	49.8	0.47×10^{-3}	8.1×10^{-6}	2.6		0.011
400	56.3	0.32	46.4	0.29×10^{-3}	5.1×10^{-6}	2.6		0.0072
700	53.7	0.31	41.8	0.19×10^{-3}	3.5×10^{-6}	2.5		0.0050
1000	51.2	0.30	37.8	0.14×10^{-3}	2.7×10^{-6}	2.4		0.0040
1300	48.6	0.30	34.5	0.12×10^{-3}	2.5×10^{-6}	2.4		0.0038

Source: F. Kreith, *Principles of Heat Transfer,* 3rd ed. (New York: Addison-Wesley Educational Publishers, Inc., 1973). Originally compiled from various sources. Reprinted by permission of Addison-Wesley Educational Publishers, Inc.

TABLE A-17E

Ideal-gas properties of air

T R	h Btu/lbm	P_r	u Btu/lbm	v_r	$s°$ Btu/lbm · R	T R	h Btu/lbm	P_r	u Btu/lbm	v_r	$s°$ Btu/lbm · R
360	85.97	0.3363	61.29	396.6	0.50369	1600	395.74	71.13	286.06	8.263	0.87130
380	90.75	0.4061	64.70	346.6	0.51663	1650	409.13	80.89	296.03	7.556	0.87954
400	95.53	0.4858	68.11	305.0	0.52890	1700	422.59	90.95	306.06	6.924	0.88758
420	100.32	0.5760	71.52	270.1	0.54058	1750	436.12	101.98	316.16	6.357	0.89542
440	105.11	0.6776	74.93	240.6	0.55172						
						1800	449.71	114.0	326.32	5.847	0.90308
460	109.90	0.7913	78.36	215.33	0.56235	1850	463.37	127.2	336.55	5.388	0.91056
480	114.69	0.9182	81.77	193.65	0.57255	1900	477.09	141.5	346.85	4.974	0.91788
500	119.48	1.0590	85.20	174.90	0.58233	1950	490.88	157.1	357.20	4.598	0.92504
520	124.27	1.2147	88.62	158.58	0.59173	2000	504.71	174.0	367.61	4.258	0.93205
537	128.10	1.3593	91.53	146.34	0.59945						
540	129.06	1.3860	92.04	144.32	0.60078	2050	518.71	192.3	378.08	3.949	0.93891
						2100	532.55	212.1	388.60	3.667	0.94564
560	133.86	1.5742	95.47	131.78	0.60950	2150	546.54	223.5	399.17	3.410	0.95222
580	138.66	1.7800	98.90	120.70	0.61793	2200	560.59	256.6	409.78	3.176	0.95919
600	143.47	2.005	102.34	110.88	0.62607	2250	574.69	281.4	420.46	2.961	0.96501
620	148.28	2.249	105.78	102.12	0.63395						
640	153.09	2.514	109.21	94.30	0.64159	2300	588.82	308.1	431.16	2.765	0.97123
						2350	603.00	336.8	441.91	2.585	0.97732
660	157.92	2.801	112.67	87.27	0.64902	2400	617.22	367.6	452.70	2.419	0.98331
680	162.73	3.111	116.12	80.96	0.65621	2450	631.48	400.5	463.54	2.266	0.98919
700	167.56	3.446	119.58	75.25	0.66321	2500	645.78	435.7	474.40	2.125	0.99497
720	172.39	3.806	123.04	70.07	0.67002						
740	177.23	4.193	126.51	65.38	0.67665	2550	660.12	473.3	485.31	1.996	1.00064
						2600	674.49	513.5	496.26	1.876	1.00623
760	182.08	4.607	129.99	61.10	0.68312	2650	688.90	556.3	507.25	1.765	1.01172
780	186.94	5.051	133.47	57.20	0.68942	2700	703.35	601.9	518.26	1.662	1.01712
800	191.81	5.526	136.97	53.63	0.69558	2750	717.83	650.4	529.31	1.566	1.02244
820	196.69	6.033	140.47	50.35	0.70160						
840	201.56	6.573	143.98	47.34	0.70747	2800	732.33	702.0	540.40	1.478	1.02767
						2850	746.88	756.7	551.52	1.395	1.03282
860	206.46	7.149	147.50	44.57	0.71323	2900	761.45	814.8	562.66	1.318	1.03788
880	211.35	7.761	151.02	42.01	0.71886	2950	776.05	876.4	573.84	1.247	1.04288
900	216.26	8.411	154.57	39.64	0.72438	3000	790.68	941.4	585.04	1.180	1.04779
920	221.18	9.102	158.12	37.44	0.72979						
940	226.11	9.834	161.68	35.41	0.73509	3050	805.34	1011	596.28	1.118	1.05264
						3100	820.03	1083	607.53	1.060	1.05741
960	231.06	10.61	165.26	33.52	0.74030	3150	834.75	1161	618.82	1.006	1.06212
980	236.02	11.43	168.83	31.76	0.74540	3200	849.48	1242	630.12	0.955	1.06676
1000	240.98	12.30	172.43	30.12	0.75042	3250	864.24	1328	641.46	0.907	1.07134
1040	250.95	14.18	179.66	27.17	0.76019						
1080	260.97	16.28	186.93	24.58	0.76964	3300	879.02	1418	652.81	0.8621	1.07585
						3350	893.83	1513	664.20	0.8202	1.08031
1120	271.03	18.60	194.25	22.30	0.77880	3400	908.66	1613	675.60	0.7807	1.08470
1160	281.14	21.18	201.63	20.29	0.78767	3450	923.52	1719	687.04	0.7436	1.08904
1200	291.30	24.01	209.05	18.51	0.79628	3500	938.40	1829	698.48	0.7087	1.09332
1240	301.52	27.13	216.53	16.93	0.80466						
1280	311.79	30.55	224.05	15.52	0.81280	3550	953.30	1946	709.95	0.6759	1.09755
						3600	968.21	2068	721.44	0.6449	1.10172
1320	322.11	34.31	231.63	14.25	0.82075	3650	983.15	2196	732.95	0.6157	1.10584
1360	332.48	38.41	239.25	13.12	0.82848	3700	998.11	2330	744.48	0.5882	1.10991
1400	342.90	42.88	246.93	12.10	0.83604	3750	1013.1	2471	756.04	0.5621	1.11393
1440	353.37	47.75	254.66	11.17	0.84341						
1480	363.89	53.04	262.44	10.34	0.85062	3800	1028.1	2618	767.60	0.5376	1.11791
						3850	1043.1	2773	779.19	0.5143	1.12183
1520	374.47	58.78	270.26	9.578	0.85767	3900	1058.1	2934	790.80	0.4923	1.12571
1560	385.08	65.00	278.13	8.890	0.86456	3950	1073.2	3103	802.43	0.4715	1.12955

Air

1001

Ideal-gas properties of air (*Concluded*)

T R	h Btu/lbm	P_r	u Btu/lbm	v_r	$s°$ Btu/lbm · R	T R	h Btu/lbm	P_r	u Btu/lbm	v_r	$s°$ Btu/lbm · R
4000	1088.3	3280	814.06	0.4518	1.13334	4600	1270.4	6089	955.04	0.2799	1.17575
						4700	1300.9	6701	978.73	0.2598	1.18232
4050	1103.4	3464	825.72	0.4331	1.13709	4800	1331.5	7362	1002.5	0.2415	1.18876
4100	1118.5	3656	837.40	0.4154	1.14079						
4150	1133.6	3858	849.09	0.3985	1.14446	4900	1362.2	8073	1026.3	0.2248	1.19508
4200	1148.7	4067	860.81	0.3826	1.14809	5000	1392.9	8837	1050.1	0.2096	1.20129
4300	1179.0	4513	884.28	0.3529	1.15522	5100	1423.6	9658	1074.0	0.1956	1.20738
						5200	1454.4	10,539	1098.0	0.1828	1.21336
4400	1209.4	4997	907.81	0.3262	1.16221	5300	1485.3	11,481	1122.0	0.1710	1.21923
4500	1239.9	5521	931.39	0.3019	1.16905						

Note: The properties P_r (relative pressure) and v_r (relative specific volume) are dimensionless quantities used in the analysis of isentropic processes, and should not be confused with the properties pressure and specific volume.

Source: Kenneth Wark, *Thermodynamics,* 4th ed. (New York: McGraw-Hill, 1983), pp. 832–33, Table A-5. Originally published in J. H. Keenan and J. Kaye, *Gas Tables* (New York: John Wiley & Sons, 1948).

Air

TABLE A-18E

Properties of gases at 1 atm pressure*

Temperature, T °F	Density, ρ lbm/ft³	Specific heat, C_p Btu/lbm · °F	Thermal conductivity, k Btu/h · ft · °F	Dynamic viscosity, μ lbm/ft · s	Kinematic viscosity, ν ft²/s	Thermal diffusivity, α ft²/h	Volume expansivity, β 1/R	Prandtl number, Pr
**Air**								
0	0.086	0.239	0.0133	1.110×10^{-5}	0.130×10^{-3}	0.646	2.18×10^{-3}	0.73
32	0.081	0.240	0.0140	1.165×10^{-5}	0.145×10^{-3}	0.720	2.03×10^{-3}	0.72
60	0.077	0.240	0.0146	1.214×10^{-5}	0.159×10^{-3}	0.796	1.93×10^{-3}	0.72
80	0.074	0.240	0.0150	1.250×10^{-5}	0.170×10^{-3}	0.851	1.86×10^{-3}	0.72
100	0.071	0.240	0.0154	1.285×10^{-5}	0.180×10^{-3}	0.905	1.79×10^{-3}	0.72
120	0.069	0.240	0.0158	1.316×10^{-5}	0.192×10^{-3}	0.964	1.74×10^{-3}	0.72
140	0.067	0.241	0.0162	1.347×10^{-5}	0.204×10^{-3}	1.023	1.68×10^{-3}	0.72
160	0.064	0.241	0.0166	1.378×10^{-5}	0.215×10^{-3}	1.082	1.63×10^{-3}	0.72
180	0.062	0.241	0.0170	1.409×10^{-5}	0.227×10^{-3}	1.141	1.57×10^{-3}	0.72
200	0.060	0.241	0.0174	1.440×10^{-5}	0.239×10^{-3}	1.20	1.52×10^{-3}	0.72
300	0.052	0.243	0.0193	1.610×10^{-5}	0.306×10^{-3}	1.53	1.32×10^{-3}	0.71
400	0.046	0.245	0.0212	1.750×10^{-5}	0.378×10^{-3}	1.88	1.16×10^{-3}	0.689
500	0.0412	0.247	0.0231	1.890×10^{-5}	0.455×10^{-3}	2.27	1.04×10^{-3}	0.683
600	0.0373	0.250	0.0250	2.000×10^{-5}	0.540×10^{-3}	2.68	0.943×10^{-3}	0.685
700	0.0341	0.253	0.0268	2.14×10^{-5}	0.625×10^{-3}	3.10	0.862×10^{-3}	0.690
800	0.0314	0.256	0.0286	2.25×10^{-5}	0.717×10^{-3}	3.56	0.794×10^{-3}	0.697
900	0.0291	0.259	0.0303	2.36×10^{-5}	0.815×10^{-3}	4.02	0.735×10^{-3}	0.705
1000	0.0271	0.262	0.0319	2.47×10^{-5}	0.917×10^{-3}	4.50	0.685×10^{-3}	0.713
1500	0.0202	0.276	0.0400	3.00×10^{-5}	1.47×10^{-3}	7.19	0.510×10^{-3}	0.739
2000	0.0161	0.286	0.0471	3.45×10^{-5}	2.14×10^{-3}	10.2	0.406×10^{-3}	0.753
**Carbon dioxide**								
0	0.132	0.184	0.0076	0.88×10^{-5}	0.067×10^{-3}	0.313	2.18×10^{-3}	0.77
100	0.108	0.203	0.0100	1.05×10^{-5}	0.098×10^{-3}	0.455	1.79×10^{-3}	0.77
200	0.092	0.216	0.0125	1.22×10^{-5}	0.133×10^{-3}	0.63	1.52×10^{-3}	0.76
500	0.063	0.247	0.0198	1.67×10^{-5}	0.266×10^{-3}	1.27	1.04×10^{-3}	0.75
1000	0.0414	0.280	0.0318	2.30×10^{-5}	0.558×10^{-3}	2.75	0.685×10^{-3}	0.73
1500	0.0308	0.298	0.0420	2.86×10^{-5}	0.925×10^{-3}	4.58	0.510×10^{-3}	0.73
2000	0.0247	0.309	0.050	3.30×10^{-5}	1.34×10^{-3}	6.55	0.406×10^{-3}	0.735
3000	0.0175	0.322	0.061	3.92×10^{-5}	2.25×10^{-3}	10.8	0.289×10^{-3}	0.745
**Carbon monoxide**								
0	0.0835	0.2482	0.0129	1.065×10^{-5}	0.128×10^{-3}	0.621	2.18×10^{-3}	0.75
200	0.0582	0.2496	0.0169	1.390×10^{-5}	0.239×10^{-3}	1.16	1.52×10^{-3}	0.74
400	0.0446	0.2532	0.0208	1.670×10^{-5}	0.374×10^{-3}	1.84	1.16×10^{-3}	0.73
600	0.0362	0.2592	0.0246	1.910×10^{-5}	0.527×10^{-3}	2.62	0.943×10^{-3}	0.725
800	0.0305	0.2662	0.0285	2.134×10^{-5}	0.700×10^{-3}	3.50	0.794×10^{-3}	0.72
1000	0.0263	0.2730	0.0322	2.336×10^{-5}	0.887×10^{-3}	4.50	0.685×10^{-3}	0.71
1500	0.0196	0.2878	0.0414	2.783×10^{-5}	1.420×10^{-3}	7.33	0.510×10^{-3}	0.70
**Helium**								
0	0.012	1.24	0.078	1.140×10^{-5}	0.950×10^{-3}	5.25	2.18×10^{-3}	0.67
200	0.00835	1.24	0.097	1.480×10^{-5}	1.77×10^{-3}	9.36	1.52×10^{-3}	0.686
400	0.0064	1.24	0.115	1.780×10^{-5}	2.78×10^{-3}	14.5	1.16×10^{-3}	0.70
600	0.0052	1.24	0.129	2.02×10^{-5}	3.89×10^{-3}	20.0	0.943×10^{-3}	0.715
800	0.00436	1.24	0.138	2.285×10^{-5}	5.24×10^{-3}	25.5	0.794×10^{-3}	0.73
1000	0.00377	1.24	—	2.520×10^{-5}	6.69×10^{-3}	—	0.685×10^{-3}	—
1500	0.0028	1.24	—	3.160×10^{-5}	11.10×10^{-3}	—	0.510×10^{-3}	—

TABLE A-18E

Properties of gases at 1 atm pressure* (*Concluded*)

Tempera-ture, T °F	Density, ρ lbm/ft³	Specific heat, C_p Btu/lbm · °F	Thermal conduc-tivity, k Btu/h · ft · °F	Dynamic viscosity, μ lbm/ft · s	Kinematic viscosity, ν ft²/s	Thermal diffusiv-ity, α ft²/h	Volume expansiv-ity, β 1/R	Prandtl number, Pr
Hydrogen
0	0.0060	3.39	0.094	0.540×10^{-5}	0.89×10^{-3}	4.62	2.18×10^{-3}	0.70
100	0.0049	3.42	0.110	0.620×10^{-5}	1.26×10^{-3}	6.56	1.79×10^{-3}	0.695
200	0.0042	3.44	0.122	0.692×10^{-5}	1.65×10^{-3}	8.45	1.52×10^{-3}	0.69
500	0.0028	3.47	0.160	0.884×10^{-5}	3.12×10^{-3}	16.5	1.04×10^{-3}	0.69
1000	0.0019	3.51	0.208	1.160×10^{-5}	6.2×10^{-3}	31.2	0.685×10^{-3}	0.705
1500	0.0014	3.62	0.260	1.415×10^{-5}	10.2×10^{-3}	51.4	0.510×10^{-3}	0.71
2000	0.0011	3.76	0.307	1.64×10^{-5}	14.4×10^{-3}	74.2	0.406×10^{-3}	0.72
3000	0.0008	4.02	0.380	1.72×10^{-5}	24.2×10^{-3}	118.0	0.289×10^{-3}	0.66
Nitrogen
0	0.0840	0.2478	0.0132	1.055×10^{-5}	0.125×10^{-3}	0.635	2.18×10^{-3}	0.713
100	0.0690	0.2484	0.0154	1.222×10^{-5}	0.177×10^{-3}	0.898	1.79×10^{-3}	0.71
200	0.0585	0.2490	0.0174	1.380×10^{-5}	0.236×10^{-3}	1.20	1.52×10^{-3}	0.71
400	0.0449	0.2515	0.0212	1.660×10^{-5}	0.370×10^{-3}	1.88	1.16×10^{-3}	0.71
600	0.0364	0.2564	0.0252	1.915×10^{-5}	0.526×10^{-3}	2.70	0.943×10^{-3}	0.70
800	0.0306	0.2623	0.0291	2.145×10^{-5}	0.702×10^{-3}	3.62	0.794×10^{-3}	0.70
1000	0.0264	0.2689	0.0330	2.355×10^{-5}	0.891×10^{-3}	4.65	0.685×10^{-3}	0.69
1500	0.0197	0.2835	0.0423	2.800×10^{-5}	1.420×10^{-3}	7.58	0.500×10^{-3}	0.676
Oxygen
0	0.0955	0.2185	0.0131	1.215×10^{-5}	0.127×10^{-3}	0.627	2.18×10^{-3}	0.73
100	0.0785	0.2200	0.0159	1.420×10^{-5}	0.181×10^{-3}	0.880	1.79×10^{-3}	0.71
200	0.0666	0.2228	0.0179	1.610×10^{-5}	0.242×10^{-3}	1.20	1.52×10^{-3}	0.722
400	0.0511	0.2305	0.0228	1.955×10^{-5}	0.382×10^{-3}	1.94	1.16×10^{-3}	0.710
600	0.0415	0.2390	0.0277	2.26×10^{-5}	0.545×10^{-3}	2.79	0.943×10^{-3}	0.704
800	0.0349	0.2465	0.0324	2.53×10^{-5}	0.725×10^{-3}	3.76	0.794×10^{-3}	0.695
1000	0.0301	0.2528	0.0366	2.78×10^{-5}	0.924×10^{-3}	4.80	0.685×10^{-3}	0.690
1500	0.0224	0.2635	0.0465	3.32×10^{-5}	1.480×10^{-3}	7.88	0.510×10^{-3}	0.677
Water vapor
212	0.0372	0.451	0.0145	0.870×10^{-5}	0.234×10^{-3}	0.864	1.49×10^{-3}	0.96
300	0.0328	0.456	0.0171	1.000×10^{-5}	0.303×10^{-3}	1.14	1.32×10^{-3}	0.95
400	0.0288	0.462	0.0200	1.130×10^{-5}	0.395×10^{-3}	1.50	1.16×10^{-3}	0.94
500	0.0258	0.470	0.0228	1.265×10^{-5}	0.490×10^{-3}	1.88	1.04×10^{-3}	0.94
600	0.0233	0.477	0.0257	1.420×10^{-5}	0.610×10^{-3}	2.31	0.943×10^{-3}	0.94
700	0.0213	0.485	0.0288	1.555×10^{-5}	0.725×10^{-3}	2.79	0.862×10^{-3}	0.93
800	0.0196	0.494	0.0321	1.700×10^{-5}	0.855×10^{-3}	3.32	0.794×10^{-3}	0.92
900	0.0181	0.50	0.0355	1.810×10^{-5}	0.987×10^{-3}	3.93	0.735×10^{-3}	0.91
1000	0.0169	0.51	0.0388	1.920×10^{-5}	1.13×10^{-3}	4.50	0.685×10^{-3}	0.91
1200	0.0149	0.53	0.0457	2.14×10^{-5}	1.44×10^{-3}	5.80	0.603×10^{-3}	0.88
1400	0.0133	0.55	0.053	2.36×10^{-5}	1.78×10^{-3}	7.25	0.537×10^{-3}	0.87
1600	0.0120	0.56	0.061	2.58×10^{-5}	2.14×10^{-3}	9.07	0.485×10^{-3}	0.87
1800	0.0109	0.58	0.068	2.81×10^{-5}	2.58×10^{-3}	10.8	0.442×10^{-3}	0.87
2000	0.0100	0.60	0.076	3.03×10^{-5}	3.03×10^{-3}	12.7	0.406×10^{-3}	0.86

*For ideal gases, the properties C_p, k, μ, and Pr are independent of pressure. The properties ρ, ν, and α at a pressure P other than 1 atm are determined by multiplying the value of ρ at the given temperature by P and by dividing the values of ν and α at the given temperature by P, where P is in atm (1 atm = 101.325 kPa = 14.696 psi).

Source: F. Kreith, *Principles of Heat Transfer*, 3rd ed. (New York: Addison-Wesley Educational Publishers, Inc., 1973). Originally compiled from various sources. Reprinted by permission of Addison-Wesley Educational Publishers, Inc.

TABLE A-19E

Properties of solid metals

Composition	Melting point, R	ρ lbm/ft³	C_p Btu/ lbm · R	k Btu/ h · ft · R	$\alpha \times 10^6$ ft²/s	180	360	720	1080	1440	1800
Aluminum: Pure	1679	168	0.216	137	1045	174.5	137	138.6	133.4	126	
						0.115	0.191	0.226	0.246	0.273	
Alloy 2024-T6 (4.5% Cu, 1.5% Mg, 0.6% Mn)	1395	173	0.209	102.3	785.8	37.6	94.2	107.5	107.5		
							0.113	0.188	0.22	0.249	
Alloy 195, cast (4.5% Cu)		174.2	0.211	97	734			100.5	106.9		
								—	—		
Beryllium	2790	115.5	0.436	115.6	637.2	572	174	93	72.8	61.3	52.5
						0.048	0.266	0.523	0.621	0.624	0.72
Bismuth	981	610.5	0.029	4.6	71	9.5	5.6	4.06			
						0.026	0.028	0.03			
Boron	4631	156	0.264	15.6	105	109.7	32.06	9.7	6.1	5.5	5.7
						0.03	0.143	0.349	0.451	0.515	0.558
Cadmium	1069	540	0.055	55.6	521	117.3	57.4	54.7			
						0.047	0.053	0.057			
Chromium	3812	447	0.107	54.1	313.2	91.9	64.1	52.5	46.6	41.2	37.8
						0.045	0.091	0.115	0.129	0.138	0.147
Cobalt	3184	553.2	0.101	57.3	286.3	96.5	70.5	49.3	39	33.6	30.1
						0.056	0.09	0.107	0.12	0.131	0.145
Copper: Pure	2445	559	0.092	231.7	1259.3	278.5	238.6	227.07	219	212	203.4
						0.06	0.085	0.094	0.01	0.103	0.107
Commercial bronze (90% Cu, 10% Al)	2328	550	0.1	30	150.7		24.3	30	34		
							0.187	0.109	0.130		
Phosphor gear bronze (89% Cu, 11% Sn)	1987	548.1	0.084	31.2	183		23.7	37.6	42.8		
							—	—	—		
Cartridge brass (70% Cu, 30% Zn)	2139	532.5	0.09	63.6	364.9	43.3	54.9	79.2	86.0		
							0.09	0.09	0.101		
Constantan (55% Cu, 45% Ni)	2687	557	0.092	13.3	72.3	9.8	1.1				
						0.06	0.09				
Germanium	2180	334.6	0.08	34.6	373.5	134	56	25	15.7	11.4	10.05
						0.045	0.069	0.08	0.083	0.085	0.089
Gold	2180	334.6	0.08	34.6	373.5	189	186.6	179.7	172.2	164.09	156
	2405	1205	0.03	183.2	1367	0.026	0.029	0.031	0.032	0.033	0.034
Iridium	4896	1404.6	0.031	85	541.4	99.4	88.4	83.2	79.7	76.3	72.8
						0.021	0.029	0.031	0.032	0.034	0.036
Iron: Pure	3258	491.3	0.106	46.4	248.6	77.4	54.3	40.2	31.6	25.01	19
						0.051	0.091	0.117	0.137	0.162	0.232
Armco (99.75% pure)		491.3	0.106	42	222.8	55.2	46.6	38	30.7	24.4	18.7
						0.051	0.091	0.117	0.137	0.162	0.233
Carbon steels: Plain carbon (Mn ≤ 1%, Si ≤ 0.1%)		490.3	0.103	35	190.6			32.8	27.7	22.7	17.4
								0.116	0.113	0.163	0.279
AISI 1010		489	0.103	37	202.4			33.9	28.2	22.7	18
								0.116	0.133	0.163	0.278
Carbon–silicon (Mn ≤ 1%, 0.1% < Si ≤ 0.6%)		488	0.106	30	160.4			28.8	25.4	21.6	17
								0.119	0.139	0.166	0.231

TABLE A-19E

Properties of solid metals (*Continued*)

Composition	Melting point, R	ρ lbm/ft³	C_p Btu/lbm·R	k Btu/h·ft·R	$\alpha \times 10^6$ ft²/s	180	360	720	1080	1440	1800
Carbon–manganese–silicon (1% < Mn ≤ 1.65%, 0.1% < Si ≤ 0.6%)		508	0.104	23.7	125			24.4 0.116	23 0.133	20.2 0.163	16 0.260
Chromium (low) steels: $\frac{1}{2}$Cr–$\frac{1}{4}$ Mo–Si (0.18% C, 0.65% Cr, 0.23% Mo, 0.6% Si)		488.3	0.106	21.8	117.4			22 0.117	21.2 0.137	19.3 0.164	15.6 0.231
1 Cr–$\frac{1}{2}$ Mo (0.16% C, 1% Cr, 0.54% Mo, 0.39% Si)		490.6	0.106	24.5	131.3			24.3 0.117	22.6 0.137	20 0.164	15.8 0.231
1 Cr–V (0.2% C, 1.02% Cr, 0.15% V)		489.2	0.106	28.3	151.8			27.0 0.117	24.3 0.137	21 0.164	16.3 0.231
Stainless steels: AISI 302		503	0.114	8.7	42			10 0.122	11.6 0.133	13.2 0.140	14.7 0.144
AISI 304	3006	493.2	0.114	8.6	42.5	5.31 0.064	7.3 0.096	9.6 0.123	11.5 0.133	13 0.139	14.7 0.145
AISI 316		514.3	0.111	7.8	37.5			8.8 0.12	10.6 0.131	12.3 0.137	14 0.143
AISI 347		498	0.114	8.2	40			9.1 0.122	1.1 0.133	12.7 0.14	14.3 0.144
Lead	1082	708	0.03	20.4	259.4	23 0.028	21.2 0.029	19.7 0.031	18.1 0.034		
Magnesium	1661	109	0.245	90.2	943	87.9 0.155	91.9 0.223	88.4 0.256	86.0 0.279	84.4 0.302	
Molybdenum	5209	639.3	0.06	79.7	578	1034 0.033	82.6 0.053	77.4 0.062	72.8 0.065	68.2 0.068	64.7 0.070
Nickel: Pure	3110	555.6	0.106	52.4	247.6	94.8 0.055	61.8 0.091	46.3 0.115	37.9 0.141	39 0.126	41.4 0.134
Nichrome (80% Ni, 20% Cr)	3010	524.4	0.1	6.9	36.6			8.0 0.114	9.3 0.125	12.2 0.130	
Inconel X-750 (73% Ni, 15% Cr, 6.7% Fe)	2997	531.3	0.104	6.8	33.4	5 —	5.9 0.088	7.8 0.112	9.8 0.121	11.8 0.13	13.9 0.149
Niobium	4934	535	0.063	31	254	31.9 0.044	30.4 0.059	32 0.065	33.6 0.067	35.4 0.069	32.2 0.071
Palladium	3289	750.4	0.058	41.5	263.7	44.2 0.04	41.4 0.054	42.5 0.059	46 0.062	50 0.064	54.4 0.067
Platinum: Pure	3681	1339	0.031	41.4	270	44.7 0.024	42 0.03	41.5 0.032	42.3 0.034	43.7 0.035	45.5 0.036
Alloy 60Pt–40Rh (60% Pt, 40% Rh)	3240	1038.2	0.038	27.2	187.3			30 —	34 —	37.5 —	40 —
Rhenium	6215	1317.2	0.032	27.7	180	34 0.023	30 0.03	26.6 0.033	25.5 0.034	25.4 0.036	25.8 0.037
Rhodium	4025	777.2	0.058	86.7	534	107.5 0.035	89 0.052	84.3 0.06	78.5 0.065	73.4 0.069	70 0.074
Silicon	3033	145.5	0.17	85.5	960.2	510.8 0.061	152.5 0.132	57.2 0.189	35.8 0.207	24.4 0.218	18.0 0.226

TABLE A-19E

Properties of solid metals (*Concluded*)

Composition	Melting point, R	Properties at 540 R ρ lbm/ft³	C_p Btu/ lbm · R	k Btu/ h · ft · R	α × 10⁶ ft²/s	Properties at various temperatures (R), k(Btu/h · ft · R)/C_p(Btu/lbm · R) 180	360	720	1080	1440	1800
Silver	2223	656	0.056	248	1873	257	248.4	245.5	238	228.8	219
						0.044	0.053	0.057	0.059	0.062	0.066
Tantalum	5884	1036.3	0.033	33.2	266	34.2	33.2	33.4	34	34.3	34.8
						0.026	0.031	0.034	0.035	0.036	0.036
Thorium	3641	730.4	0.028	31.2	420.9	34.6	31.5	31.4	32.2	32.9	32.9
						0.024	0.027	0.029	0.032	0.035	0.037
Tin	909	456.3	0.054	38.5	431.6	49.2	42.4	35.9			
						0.044	0.051	0.058			
Titanium	3515	281	0.013	12.7	100.3	17.6	14.2	11.8	11.2	11.4	12
						0.071	0.111	0.131	0.141	0.151	0.161
Tungsten	6588	1204.9	0.031	100.5	735.2	120.2	107.5	92	79.2	72.2	68.2
						0.020	0.029	0.032	0.033	0.034	0.035
Uranium	2531	1190.5	0.027	16	134.5	12.5	14.5	17.1	19.6	22.4	25.4
						0.022	0.026	0.029	0.035	0.042	0.043
Vanadium	3946	381	0.117	17.7	110.9	20.7	18	18	19.3	20.6	22.0
						0.061	0.102	0.123	0.128	0.134	0.142
Zinc	1247	445.7	0.093	67	450	67.6	68.2	64.1	59.5		
						0.07	0.087	0.096	0.104		
Zirconium	3825	410.2	0.067	13.1	133.5	19.2	14.6	12.5	12	12.5	13.7
						0.049	0.063	0.072	0.77	0.082	0.087

Source: Tables A-19E and A-20E are obtained from the respective tables in SI units in Appendix 1 using proper conversion factors.

TABLE A-20E

Properties of solid nonmetals

Composition	Melting point, R	Properties at 540 R				Properties at various temperatures (R), k(Btu/h·ft·R)/C_p(Btu/lbm·R)					
		ρ lbm/ft³	C_p Btu/lbm·R	k Btu/h·ft·R	α × 10⁶ ft²/s	180	360	720	1080	1440	1800
Aluminum oxide, sapphire	4181	247.8	0.182	26.6	162.5	260	47.4	18.7	11	7.5	6
						—	—	0.224	0.265	0.281	0.293
Aluminum oxide, polycrystalline	4181	247.8	0.182	20.8	128	76.8	31.7	15.3	9.3	6	4.5
						—	—	0.244	0.265	0.281	0.293
Beryllium oxide	4905	187.3	0.246	157.2	947.3			113.2	64.2	40.4	27.2
								0.322	0.40	0.44	0.459
Boron	4631	156	0.264	16	107.5	109.8	30.3	10.8	6.5	4.6	3.6
								0.355	0.445	0.509	0.561
Boron fiber epoxy (30% vol.) composite	1062	130									
k, ∥ to fibers				1.3		1.2	1.3	1.31			
k, ⊥ to fibers				0.34		0.21	0.28	0.34			
C_p			0.268			0.086	0.18	0.34			
Carbon Amorphous	2700	121.7	—	0.92	—	0.38	0.68	1.09	1.26	1.36	1.46
Diamond, type IIa insulator	—	219	0.121	1329	—	5778	2311.2	889.8			
								0.005	0.046	0.203	
Graphite, pyrolytic	4091	138									
k, ∥ to layers				1126.7		2871.6	1866.3	803.2	515.4	385.4	308.5
k, ⊥ to layers				3.3		9.7	5.3	2.4	1.5	1.16	0.92
C_p			0.169			0.032	0.098	0.236	0.335	0.394	0.428
Graphite fiber epoxy (25% vol.) composite	810	87.4									
k, heat flow ∥ to fibers				6.4		3.3	5.0	7.5			
k, heat flow ⊥ to fibers				0.5	5	0.4	0.63				
C_p			0.223			0.08	0.153	0.29			
Pyroceram, Corning 9606	2921	162.3	0.193	2.3	20.3	3.0	2.3	2.1	1.9	1.7	1.7
Silicon carbide	5580	197.3	0.161	283.1	2475.7			—	—	—	50.3
								0.210	0.25	0.27	0.285
Silicon dioxide, crystalline (quartz)	3389	165.4									
k, ∥ to c-axis				6		22.5	9.5	4.4	2.9	2.4	
k, ⊥ to c-axis				3.6		12.0	5.9	2.7	2	1.8	
C_p			0.177			—	—	0.211	0.256	0.298	
Silicon dioxide, polycrystalline (fused silica)	3389	138.6	0.177	0.79	9	0.4	0.65	0.87	1.01	1.25	1.65
						—	—	0.216	0.248	0.264	0.276
Silicon nitride	3911	150	0.165	9.2	104	—	—	8.0	6.5	5.7	5.0
						—	0.138	0.185	0.223	0.253	0.275
Sulfur	706	130	0.169	0.1	1.51	0.095	0.1				
						0.962	0.144				
Thorium dioxide	6431	568.7	0.561	7.5	65.7			5.9	3.8	2.7	2.12
								0.609	0.654	0.680	0.704
Titanium dioxide, polycrystalline	3840	259.5	0.170	4.9	30.1			4.0	2.9	2.3	2
								0.192	0.210	0.217	0.222

TABLE A-21E

Properties of building materials
(at a mean temperature of 75°F)

Material	Thickness, L in.	Density, ρ lbm/ft^3	Thermal conductivity, k Btu-in./ft · °F	Specific heat, C_p Btu/lbm · °F	R-value (for listed thickness, L/k), °F · h · ft^2/Btu
Building Boards					
Asbestos–cement board	$\frac{1}{4}$ in.	120	—	0.24	0.06
Gypsum of plaster board	$\frac{3}{8}$ in.	50	—	0.26	0.32
	$\frac{1}{2}$ in.	50	—	—	0.45
Plywood (Douglas fir)	—	34	0.80	0.29	—
	$\frac{1}{4}$ in.	34	—	0.29	0.31
	$\frac{3}{8}$ in.	34	—	0.29	0.47
	$\frac{1}{2}$ in.	34	—	0.29	0.62
	$\frac{3}{4}$ in.	34	—	0.29	0.93
Insulated board and sheating	$\frac{1}{2}$ in.	18	—	0.31	1.32
(regular density)	$\frac{25}{32}$ in.	18	—	0.31	2.06
Hardboard (high density, standard tempered)	—	63	1.00	0.32	—
Particle board:					
Medium density	—	50	0.94	0.31	—
Underlayment	$\frac{5}{8}$ in.	40	—	0.29	0.82
Wood subfloor	$\frac{3}{4}$ in.	—	—	0.33	0.94
Building Membranes					
Vapor-permeable felt	—	—	—	—	0.06
Vapor-seal (2 layers of mopped 17.3 lbm/ft^2 felt)	—	—	—	—	0.12
Flooring Materials					
Carpet and fibrous pad	—	—	—	0.34	2.08
Carpet and rubber pad	—	—	—	0.33	1.23
Tile (asphalt, linoleum, vinyl)	—	—	—	0.30	0.05
Masonry Materials					
Masonry units:					
Brick, common		120	5.0	—	—
Brick, face		130	9.0	—	—
Brick, fire clay		150	9.3	—	—
		120	6.2	0.19	—
		70	2.8	—	—
Concrete blocks (3 oval cores, sand and gravel aggregate)	4 in.	—	5.34	—	0.71
	8 in.	—	6.94	—	1.11
	12 in.	—	9.02	—	1.28
Concretes:					
Lightweight aggregates		120	5.2	—	—
(including expanded shale,		100	3.6	0.2	—
clay, or slate; expanded slags;		80	2.5	0.2	—
cinders; pumice; and scoria)		60	1.7	—	—
		40	1.15	—	—

TABLE A-21E

Properties of building materials (*Concluded*)
(at a mean temperature of 75°F)

Material	Thickness, L in.	Density, ρ lbm/ft³	Thermal conductivity, k Btu-in./ft · °F	Specific heat, C_p Btu/lbm · °F	R-value (for listed thickness, L/k), °F · h · ft²/Btu
Cement/lime, mortar, and stucco		120	9.7	—	—
		80	4.5	—	—
Stucco		116	5.0	—	—
Roofing					
Asbestos-cement shingles		120	—	0.24	0.21
Asphalt roll roofing		70	—	0.36	0.15
Asphalt shingles		70	—	0.30	0.44
Built-in roofing	$\frac{3}{8}$ in.	70	—	0.35	0.33
Slate	$\frac{1}{2}$ in.	—	—	0.30	0.05
Wood shingles (plain and plastic film faced)		—	—	0.31	0.94
Plastering Materials					
Cement plaster, sand aggregate	$\frac{3}{4}$ in.	116	5.0	0.20	0.15
Gypsum plaster:					
Lightweight aggregate	$\frac{1}{2}$ in.	45	—	—	0.32
Sand aggregate	$\frac{1}{2}$ in.	105	5.6	0.20	0.09
Perlite aggregate	—	45	1.5	0.32	—
Siding Material (on flat surfaces)					
Asbestos-cement shingles	—	120	—	—	0.21
Hardboard siding	$\frac{7}{16}$ in.	—	—	0.28	0.67
Wood (drop) siding	1 in.	—	—	0.31	0.79
Wood (plywood) siding, lapped	$\frac{3}{8}$ in.	—	—	0.29	0.59
Aluminum or steel siding (over sheeting):					
Hollow backed	$\frac{3}{8}$ in.	—	—	0.29	0.61
Insulating-board backed	$\frac{3}{8}$ in.	—	—	0.32	1.82
Architectural glass	—	158	6.9	0.21	0.10
Woods					
Hardwoods (maple, oak, etc.)	—	45	1.10	0.30	—
Softwoods (fir, pine, etc.)	—	32	0.80	0.33	—
Metals					
Aluminum (1100)	—	171	1536	0.214	—
Steel, mild	—	489	314	0.120	—
Steel, Stainless	—	494	108	0.109	—

Source: Tables A-21E and A-22E are adapted from ASHRAE, *Handbook of Fundamentals* (Atlanta, GA: American Society of Heating, Refrigerating, and Air-Conditioning Engineers, 1993), Chap. 22, Table 4. Used with permission.

TABLE A-22E

Properties of insulating materials
(at a mean temperature of 75°F)

Material	Thickness, L in.	Density, ρ lbm/ft^3	Thermal conductivity, k Btu-in./ft · °F	Specific heat, C_p Btu/lbm · °F	R-value (for listed thickness, L/k) °F · h · ft^2/Btu
Blanket and Batt					
Mineral fiber (fibrous form	~2 to $2\frac{3}{4}$ in.	0.3–2.0	—	0.17–0.23	7
processed from rock, slag,	~3 to $3\frac{1}{2}$ in.	0.3–2.0	—	0.17–0.23	11
or glass)	~$5\frac{1}{4}$ to $6\frac{1}{2}$ in.	0.3–2.0	—	0.17–0.23	19
Board and Slab					
Cellular glass		8.5	0.38	0.24	—
Glass fiber (organic bonded)		4–9	0.25	0.23	—
Expanded polystyrene (molded beads)		1.0	0.28	0.29	—
Expanded polyurethane (R-11 expanded)		1.5	0.16	0.38	—
Expanded perlite (organic bonded)		1.0	0.36	0.30	—
Expanded rubber (rigid)		4.5	0.22	0.40	—
Mineral fiber with resin binder		15	0.29	0.17	—
Cork		7.5	0.27	0.43	—
Sprayed or Formed in Place					
Polyurethane foam		1.5–2.5	0.16–0.18	—	—
Glass fiber		3.5–4.5	0.26–0.27	—	—
Urethane, two-part mixture (rigid foam)		4.4	0.18	0.25	—
Mineral wool granules with asbestos/inorganic binders (sprayed)		12	0.32	—	—
Loose Fill					
Mineral fiber (rock, slag,	~3.75 to 5 in.	0.6–0.20	—	0.17	11
or glass)	~6.5 to 8.75 in.	0.6–0.20	—	0.17	19
	~7.5 to 10 in.	—	—	0.17	22
	~7.25 in.	—	—	0.17	30
Silica aerogel		7.6	0.17	—	—
Vermiculite (expanded)		7–8	0.47	—	—
Perlite, expanded		2–4.1	0.27–0.31	—	—
Sawdust or shavings		8–15	0.45	—	—
Cellulosic insulation (milled paper or wood pulp)		0.3–3.2	0.27–0.32	—	—
Cork, granulated		10	0.31	—	—
Roof Insulation					
Cellular glass	—	9	0.4	0.24	—
Preformed, for use above deck	$\frac{1}{2}$ in.	—	—	0.24	1.39
	1 in.	—	—	0.50	2.78
	2 in.	—	—	0.94	5.56
Reflective Insulation					
Silica powder (evacuated)		10	0.0118	—	—
Aluminum foil separating fluffy glass mats; 10–12 layers (evacuated); for cryogenic applications (270 R)		2.5	0.0011	—	—
Aluminum foil and glass paper laminate; 75–150 layers (evacuated); for cryogenic applications (270 R)		7.5	0.00012	—	—

TABLE A-23E

Properties of miscellaneous materials
(Values are at 540 R unless indicated otherwise)

Material	Density, ρ lbm/ft^3	Thermal conductivity, k Btu/h · ft · R	Specific heat, C_p Btu/lbm · R	Material	Density, ρ lbm/ft^3	Thermal conductivity, k Btu/h · ft · R	Specific heat, C_p Btu/lbm · R
Asphalt	132.0	0.036	0.220	Ice			
Bakelite	81.2	0.81	0.350	492 R	57.4	1.09	0.487
Brick, refractory				455 R	57.6	1.17	0.465
Chrome brick				311 R	57.9	2.02	0.349
851 R	187.9	1.33	0.199	Leather, sole	62.3	0.092	—
1481 R	—	1.44	—	Linoleum	33.4	0.047	—
2111 R	—	1.16	—		73.7	0.11	—
Fire clay, burnt				Mica	181.0	0.30	—
2880 R				Paper	58.1	0.10	0.320
1391 R	128.0	0.58	0.229	Plastics			
1931 R	—	0.64	—	Plexiglass	74.3	0.11	0.350
2471 R	—	0.64	—	Teflon			
Fire clay, burnt				540 R	137.3	0.20	0.251
3105 R				720 R	—	0.26	—
1391 R	145.1	0.75	0.229	Lexan	74.9	0.11	0.301
1931 R	—	0.81	—	Nylon	71.5	0.17	—
2471 R	—	0.81	—	Polypropylene	56.8	0.069	0.388
Fire clay brick				Polyester	87.1	0.087	0.279
860 R	165.1	0.58	0.229	PVC, vinyl	91.8	0.058	0.201
1660 R	—	0.87	—	Porcelain	143.6	0.87	—
2660 R	—	1.04	—	Rubber, natural	71.8	0.16	—
Magnesite				Rubber, vulcanized			
860 R	—	2.20	0.270	Soft	68.7	0.075	0.480
1660 R	—	1.62	—	Hard	74.3	0.092	—
2660 R	—	1.10	—	Sand	94.6	0.1–0.6	0.191
Chicken meat, white (74.4% water content)				Snow, fresh	6.24	0.35	—
				Snow, 492 R	31.2	1.27	—
356 R	—	0.92	—	Soil, dry	93.6	0.58	0.454
419 R	—	0.86	—	Soil, wet	118.6	1.16	0.525
455 R	—	0.78	—	Sugar	99.9	0.34	—
492 R	—	0.28	—	Tissue, human			
527 R	—	0.28	—	Skin	—	0.21	—
Clay, dry	96.8	0.54	—	Fat layer	—	0.12	—
Clay, wet	93.3	0.97	—	Muscle	—	0.24	—
Coal, anthracite	84.3	0.15	0.301	Vaseline	—	0.098	—
Concrete (stone mix)	143.6	0.81	0.210	Wood, cross-grain			
Cork	5.37	0.028	0.485	Balsa	8.74	0.032	—
Cotton	5.0	0.035	0.311	Fir	25.9	0.064	0.650
Fat	—	0.10	—	Oak	34.0	0.098	0.570
Glass				White pine	27.2	0.064	—
Window	174.8	0.40	0.179	Yellow pine	40.0	0.087	0.670
Pyrex	138.9	0.6–0.8	0.199	Wood, radial			
Crown	156.1	0.61	—	Oak	34.0	0.11	0.570
Lead	212.2	0.49	—	Fir	26.2	0.081	0.650
				Wool, ship	9.05	0.029	—

TABLE A-26E

Properties of the atmosphere at high altitude

Altitude, ft	Temperature, °F	Pressure, psia	Gravity, g ft/s²	Speed of sound, ft/s	Density, lbm/ft³	Viscosity, μ lbm/ft · s	Thermal conductivity, Btu/h · ft · °F
0	59.00	14.7	32.174	1116	0.07647	1.202×10^{-5}	0.0146
500	57.22	14.4	32.173	1115	0.07536	1.199×10^{-5}	0.0146
1000	55.43	14.2	32.171	1113	0.07426	1.196×10^{-5}	0.0146
1500	53.65	13.9	32.169	1111	0.07317	1.193×10^{-5}	0.0145
2000	51.87	13.7	32.168	1109	0.07210	1.190×10^{-5}	0.0145
2500	50.09	13.4	32.166	1107	0.07104	1.186×10^{-5}	0.0144
3000	48.30	13.2	32.165	1105	0.06998	1.183×10^{-5}	0.0144
3500	46.52	12.9	32.163	1103	0.06985	1.180×10^{-5}	0.0143
4000	44.74	12.7	32.162	1101	0.06792	1.177×10^{-5}	0.0143
4500	42.96	12.5	32.160	1099	0.06690	1.173×10^{-5}	0.0142
5000	41.17	12.2	32.159	1097	0.06590	1.170×10^{-5}	0.0142
5500	39.39	12.0	32.157	1095	0.06491	1.167×10^{-5}	0.0141
6000	37.61	11.8	32.156	1093	0.06393	1.164×10^{-5}	0.0141
6500	35.83	11.6	32.154	1091	0.06296	1.160×10^{-5}	0.0141
7000	34.05	11.3	32.152	1089	0.06200	1.157×10^{-5}	0.0140
7500	32.26	11.1	32.151	1087	0.06105	1.154×10^{-5}	0.0140
8000	30.48	10.9	32.149	1085	0.06012	1.150×10^{-5}	0.0139
8500	28.70	10.7	32.148	1083	0.05919	1.147×10^{-5}	0.0139
9000	26.92	10.5	32.146	1081	0.05828	1.144×10^{-5}	0.0138
9500	25.14	10.3	32.145	1079	0.05738	1.140×10^{-5}	0.0138
10,000	23.36	10.1	32.145	1077	0.05648	1.137×10^{-5}	0.0137
11,000	19.79	9.72	32.140	1073	0.05473	1.130×10^{-5}	0.0136
12,000	16.23	9.34	32.137	1069	0.05302	1.124×10^{-5}	0.0136
13,000	12.67	8.99	32.134	1065	0.05135	1.117×10^{-5}	0.0135
14,000	9.12	8.63	32.131	1061	0.04973	1.110×10^{-5}	0.0134
15,000	5.55	8.29	32.128	1057	0.04814	1.104×10^{-5}	0.0133
16,000	+1.99	7.97	32.125	1053	0.04659	1.097×10^{-5}	0.0132
17,000	−1.58	7.65	32.122	1049	0.04508	1.090×10^{-5}	0.0132
18,000	−5.14	7.34	32.119	1045	0.04361	1.083×10^{-5}	0.0130
19,000	−8.70	7.05	32.115	1041	0.04217	1.076×10^{-5}	0.0129
20,000	−12.2	6.76	32.112	1037	0.04077	1.070×10^{-5}	0.0128
22,000	−19.4	6.21	32.106	1029	0.03808	1.056×10^{-5}	0.0126
24,000	−26.5	5.70	32.100	1020	0.03553	1.042×10^{-5}	0.0124
26,000	−33.6	5.22	32.094	1012	0.03311	1.028×10^{-5}	0.0122
28,000	−40.7	4.78	32.088	1003	0.03082	1.014×10^{-5}	0.0121
30,000	−47.8	4.37	32.082	995	0.02866	1.000×10^{-5}	0.0119
32,000	−54.9	3.99	32.08	987	0.02661	0.986×10^{-5}	0.0117
34,000	−62.0	3.63	32.07	978	0.02468	0.971×10^{-5}	0.0115
36,000	−69.2	3.30	32.06	969	0.02285	0.956×10^{-5}	0.0113
38,000	−69.7	3.05	32.06	968	0.02079	0.955×10^{-5}	0.0113
40,000	−69.7	2.73	32.05	968	0.01890	0.955×10^{-5}	0.0113
45,000	−69.7	2.148	32.04	968	0.01487	0.955×10^{-5}	0.0113
50,000	−69.7	1.691	32.02	968	0.01171	0.955×10^{-5}	0.0113
55,000	−69.7	1.332	32.00	968	0.00922	0.955×10^{-5}	0.0113
60,000	−69.7	1.048	31.99	968	0.00726	0.955×10^{-5}	0.0113

Source: U.S. Standard Atmosphere Supplements, U.S. Government Printing Office, 1966. Based on year-round mean conditions at 45° latitude and varies with the time of the year and the weather patterns. The conditions at sea level ($z = 0$) are taken to be $P = 14.696$ psia, $T = 59$°F, $\rho = 0.076474$ lbm/ft³, $g = 32.1741$ ft²/s.

Introduction to EES

OVERVIEW

EES (pronounced "ease") is an acronym for Engineering Equation Solver. The basic function provided by EES is the numerical solution of nonlinear algebraic and differential equations. In addition, EES provides built-in thermodynamic and transport property functions for many fluids, including water, dry and moist air, refrigerants, combustion gases, and others. Additional property data can be added by the user. The combination of equation-solving capability and engineering property data makes EES a very powerful tool.

A license for EES is provided by McGraw-Hill to departments of educational institutions that adopt this text. If you need more information, contact your local McGraw-Hill representative, call 1-800-338-3987, or visit our website at www.mhhe.com/engcs/mech/ees. A commercial version of EES can be obtained from:

F-Chart Software
4406 Fox Bluff Road
Middleton, WI 53562
Phone: (608) 836-8531
Fax: (608) 836-8536
http://www.fchart.com

BACKGROUND INFORMATION

The EES program is probably installed on your departmental computer. In addition, the license agreement for EES allows students and faculty in a participating educational department to copy the program for educational use on their personal computer systems. Ask your instructor for details.

To start EES from the Windows File Manager or Explorer, double-click on the EES program icon or on any file created by EES. You can also start EES from the Windows Run command in the Start menu. EES begins by displaying a dialog window that shows registration information, the version number, and other information. Click the OK button to dismiss the dialog window.

Detailed help is available at any point in EES. Pressing the F1 key will bring up a Help window relating to the foremost window. Clicking the Contents button will present the Help index shown in Fig. 1. Clicking on an underlined word (shown in green on color monitors) will provide help relating to that subject.

EES commands are distributed among nine pull-down menus. A brief summary of their functions follows. (A tenth pull-down menu, which is made visible with the Load Textbook command described below, provides access to problems from this text.)

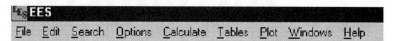

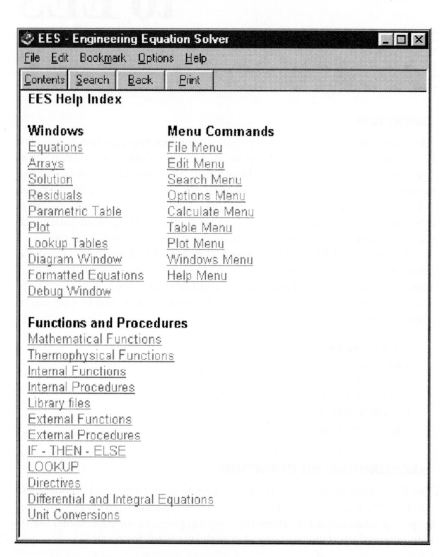

FIGURE 1
EES Help Index.

The System menu appears above the File menu. The System menu is not part of EES, but rather a feature of the Windows Operating System. It holds commands that allow window moving, resizing, and switching to other applications.

The File menu provides commands for loading, merging, and saving work files and libraries, and printing. The Load Textbook command in this menu reads the problem disk developed for this text and creates a new menu to the right of the Help menu for easy access to EES problems accompanying this text.

The Edit menu provides the editing commands to cut, copy, and paste information.

The Search menu provides Find and Replace commands for use in the Equations window.

The Options menu provides commands for setting the guess values and bounds of variables, the unit system, default information, and program preferences. A command is also provided for displaying information on built-in and user-supplied functions.

The Calculate menu contains the commands to check, format, and solve the equation set.

The Tables menu contains commands to set up and alter the contents of the Parametric and Lookup Tables and to do linear regression on the data in these tables. The Parametric Table, which is similar to a spreadsheet, allows the equation set to be solved repeatedly while varying the values of one or more variables. The Lookup Table holds user-supplied data that can be interpolated and used in the solution of the equation set.

The Plot menu provides commands to modify an existing plot or prepare a new plot of data in the Parametric, Lookup, or Array tables. Curve-fitting capability is also provided.

The Windows menu provides a convenient method of bringing any of the EES windows to the front or to organize the windows.

The Help menu provides commands for accessing the online help documentation.

A basic capability provided by EES is the solution of a set of nonlinear algebraic equations. To demonstrate this capability, start EES and enter this simple example problem in the Equations window.

Text is entered in the same manner as for any word processor. Formatting rules are as follows:

1. Upper- and lowercase letters are not distinguished. EES will (optionally) change the case of all variables to match the manner in which they first appear.

2. Blank lines and spaces may be entered as desired since they are ignored.

3. Comments must be enclosed within braces { } or within quote marks " ". Comments may span as many lines as needed. Comments within braces may be nested, in which case only the outermost set of braces is recognized. Comments within quotes will also be displayed in the Formatted Equations window.

4. Variable names must start with a letter and consist of any keyboard characters except ( ) ' | * / + - ^ { } : " or ;. Array variables are identified with square brackets around the array index or indices, for example, X[5,3]. The maximum variable length is 30 characters.

5. Multiple equations may be entered on one line if they are separated by a semicolon (;). The maximum line length is 255 characters.

6. The caret symbol (^) or ** is used to indicate raising to a power.

7. The order in which the equations are entered does not matter.

8. The position of knowns and unknowns in the equation does not matter.

If you wish, you may view the equations in mathematical notation by selecting the **Formatted Equations** command from the **Windows** menu.

Select the **Solve** command from the **Calculate** menu. A dialog window

will appear indicating the progress of the solution. When the calculations are completed, the button will change from Abort to Continue (Fig. 2). Click the Continue button. The solution to this equation set will then be displayed.

A HEAT TRANSFER EXAMPLE PROBLEM

In this section, Prob. 14-58 from the text is worked from start to finish illustrating some of the capabilities of the EES program. EES is particularly ap-

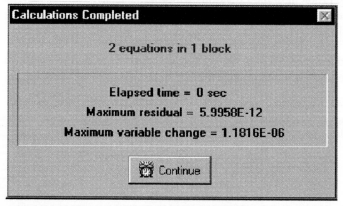

FIGURE 2
Calculations Completed window.

propriate for this problem since a trial-and-error solution would be required if the problem were done by hand. The problem to be solved is stated as follows.

The roof of a house consists of a 15-cm-thick concrete slab (k = 2 W/m · °C) that is 15 m wide and 20 m long. The emissivity of the outer surface of the roof is 0.9 and the convection heat transfer coefficient on that surface is estimated to be 15 W/m² · °C. The inner surface of the roof is maintained at 15°C. On a clear winter night, the ambient air is reported to be at 10°C while the night sky temperature for radiation heat transfer is 255 K. Considering both radiation and convection heat transfer, determine the outer surface temperature and the rate of heat transfer through the roof.

If the house is heated by a furnace burning natural gas with an efficiency of 85 percent and the unit cost of natural gas is $0.60/therm, determine the money lost through the roof that night during a 14-hour period.

This problem involves conduction, convection, and radiation. The problem is facilitated by the energy-flow diagram in Fig. 3.

To solve this problem, it is necessary to equate the conduction heat transfer rate through the concrete roof to the combined convective and radiative heat transfer rates from the roof surface.

$$\dot{Q}_{\text{cond}} = k\,A(T_{\text{inner}} - T_{\text{surface}})/L \qquad (1)$$

$$\dot{Q}_{\text{conv}} = h\,A(T_{\text{surface}} - T_{\text{ambient}}) \qquad (2)$$

$$\dot{Q}_{\text{rad}} = \sigma\,\varepsilon\,A(T^4_{\text{surface}} - T^4_{\text{sky}}) \qquad (3)$$

$$\dot{Q}_{\text{cond}} = \dot{Q}_{\text{conv}} + \dot{Q}_{\text{rad}} \qquad (4)$$

where

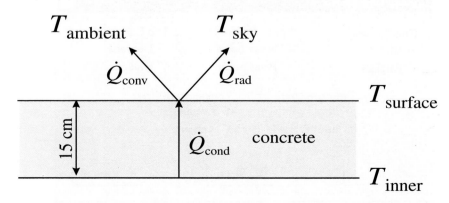

FIGURE 3
Energy-flow diagram for example problem.

A = roof area (m^2)
k = thermal conductivity (W/m · K)
h = convection coefficient (W/m^2 · K)
T_{inner} = temperature of the inner surface of the roof
$T_{surface}$ = temperature of the outer surface of the roof
$T_{ambient}$ = temperature of the outdoor air
T_{sky} = effective radiative temperature of the sky
σ = Stefan–Boltzman constant (5.67E–8 W/m^2 · K^4)
ε = emissivity of the roof surface
L = thickness of the concrete slab

There are a total of 13 variables in this problem. Of these, 9 (A, k, h, T_{inner}, $T_{ambient}$, T_{sky}, σ, ε, L) are specified in the problem statement, leaving 4 unknown variables. There are four equations relating these variables, so the problem is completely defined. However, the equations are nonlinear and, as a result, cumbersome to solve. This is where EES can help.

Start EES or select the New command from the File menu if you have already been using the program. A blank Equations window will appear. Since this problem does not require any of the thermophysical property functions built into EES, it is not necessary to specify the unit system. However, it is good practice to set the unit system at the start of a problem. To view or change the unit system, select Unit System from the Options menu (Fig. 4).

After entering the equations for this problem and (optionally) checking the syntax using the Check/Format command in the Calculate menu, the Equations window will appear as shown in Fig. 5. Comments are normally displayed in blue on a color monitor. Other formatting options are set with the Preferences command in the Options menu.

Many of the variable names include an underscore. The underscore is a special formatting character that indicates the start of a subscript. Using underscores is optional but it improves the display of the Formatted Equations and Solutions windows.

Only one built-in function has been used in this problem: the Convert function, which is used to convert Wh into therms. The Convert function should be able to provide any unit conversions you encounter. There are many other useful functions in EES, including functions that provide thermodynamic and transport property data. Select the Function Information command in the Options menu. This command will bring up the Function Information dialog window. Click on the radio buttons at the top of the dialog

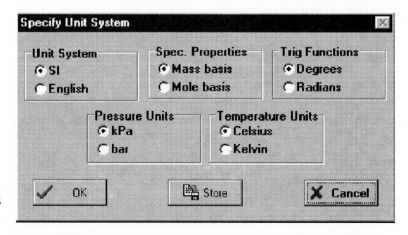

FIGURE 4
Unit System dialog.

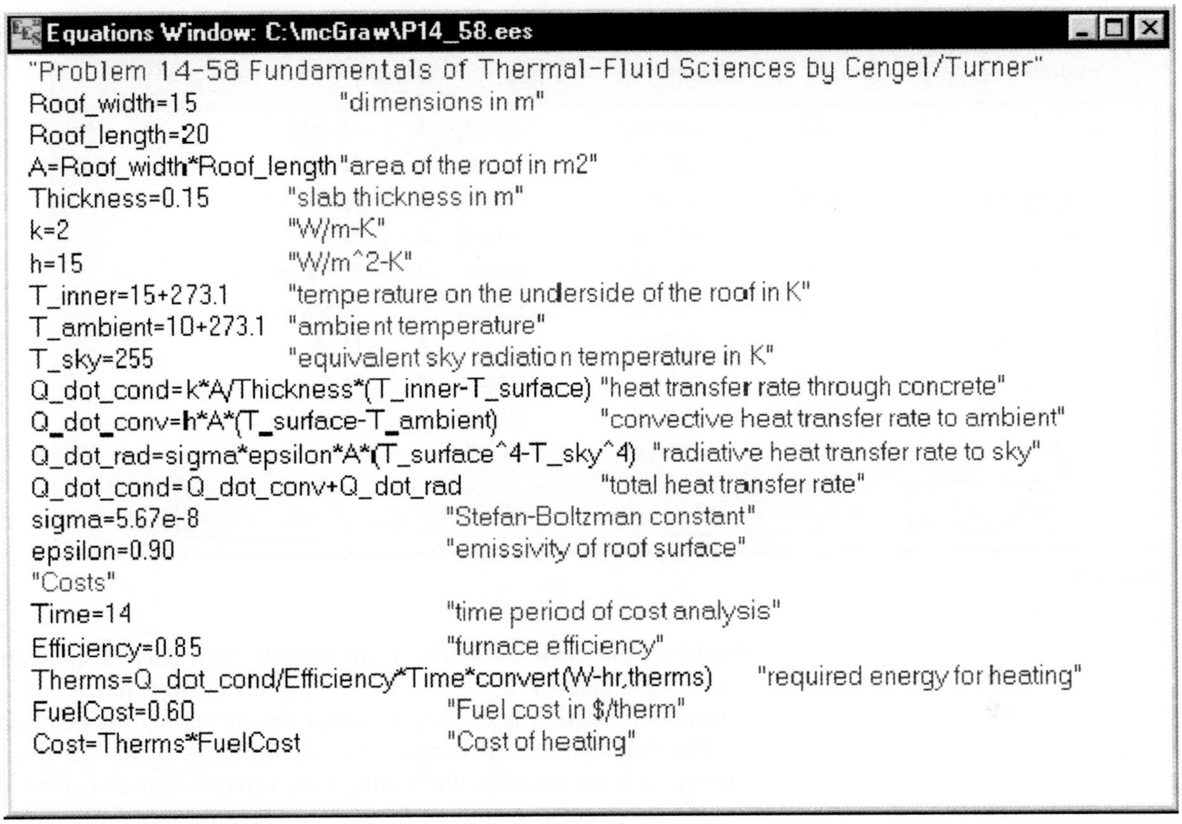

```
"Problem 14-58 Fundamentals of Thermal-Fluid Sciences by Cengel/Turner"
Roof_width=15              "dimensions in m"
Roof_length=20
A=Roof_width*Roof_length   "area of the roof in m2"
Thickness=0.15             "slab thickness in m"
k=2                        "W/m-K"
h=15                       "W/m^2-K"
T_inner=15+273.1           "temperature on the underside of the roof in K"
T_ambient=10+273.1         "ambient temperature"
T_sky=255                  "equivalent sky radiation temperature in K"
Q_dot_cond=k*A/Thickness*(T_inner-T_surface)  "heat transfer rate through concrete"
Q_dot_conv=h*A*(T_surface-T_ambient)          "convective heat transfer rate to ambient"
Q_dot_rad=sigma*epsilon*A*(T_surface^4-T_sky^4)  "radiative heat transfer rate to sky"
Q_dot_cond=Q_dot_conv+Q_dot_rad               "total heat transfer rate"
sigma=5.67e-8                        "Stefan-Boltzman constant"
epsilon=0.90                         "emissivity of roof surface"
"Costs"
Time=14                              "time period of cost analysis"
Efficiency=0.85                      "furnace efficiency"
Therms=Q_dot_cond/Efficiency*Time*convert(W-hr,therms)   "required energy for heating"
FuelCost=0.60                        "Fuel cost in $/therm"
Cost=Therms*FuelCost                 "Cost of heating"
```

FIGURE 5

Equations window.

Only one built-in function has been used in this problem: the Convert function, which is used to convert Wh into therms. The Convert function should be able to provide any unit conversions you encounter. There are many other useful functions in EES, including functions that provide thermodynamic and transport property data. Select the Function Information command in the Options menu. This command will bring up the Function Information dialog window. Click on the radio buttons at the top of the dialog window to view the available functions. The Info button in this dialog provides additional information relating to the selected function.

It is usually a good idea to set the guess values and (possibly) the lower and upper bounds for the variables before attempting to solve the equations. This is done with the Variable Information command in the Options menu. Before displaying the Variable Information dialog, EES checks syntax and compiles newly entered and/or changed equations, and then solves all equations with one unknown. The Variable Information dialog will then appear.

The Variable Information dialog (Fig. 6) contains a line for each variable appearing in the Equations window. By default, each variable has a guess value of 1.0 with lower and upper bounds of negative and positive infinity. (The lower and upper bounds are shown in italics if EES has previously calculated the value of the variable. In this case, the Guess value column displays the calculated value. The italicized values may still be edited.)

The A in the Display options column indicates that EES will automatically determine the display format for the numerical value of the variable when it is displayed in the Solution window. In this case, EES will select an appropriate number of digits, so the digits column to the right of the A is

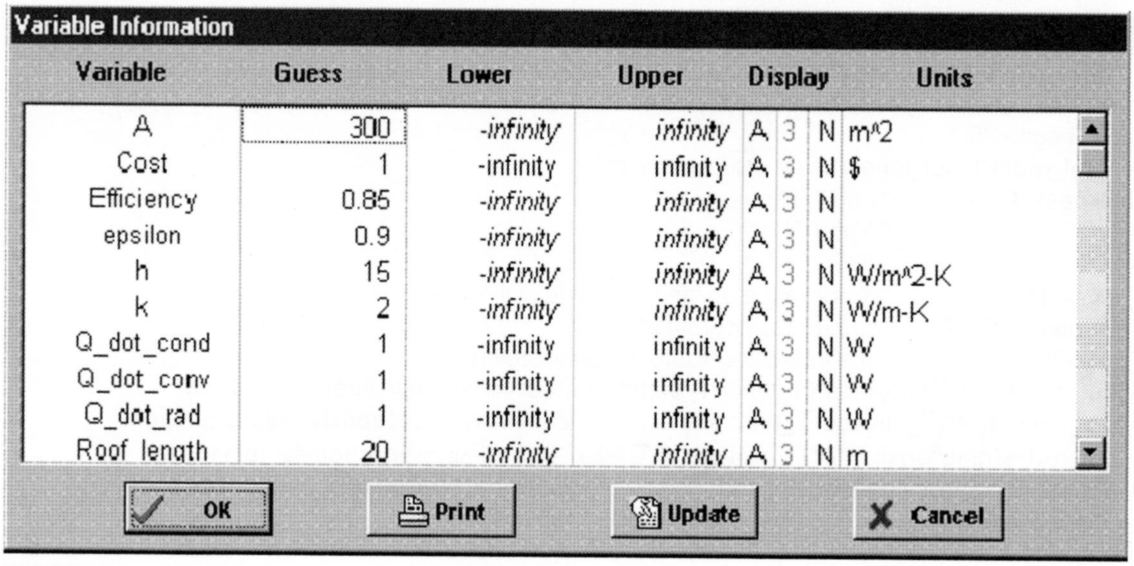

FIGURE 6

Variable Information dialog.

disabled. Automatic formatting is the default. Alternative display options are F (for fixed number of digits to the right of the decimal point) and E (for exponential format). The display and other defaults can easily be changed with the Default Information command in the Options menu. The third Display options column controls the highlighting effects such as normal (default), bold, and boxed. The units of the variables can be specified, if desired. The units will be displayed with the variable in the Solution window and/or in the Parametric Table. The units can also be set directly in the Solution window. EES does not automatically do unit conversions but it can provide unit conversions using the Convert function. The units information entered here is only for display purposes.

With nonlinear equations, it is sometimes necessary to provide reasonable guess values and bounds in order to determine the desired solution. (It is not necessary for this problem.) The bounds of some variables are known from the physics of the problem. In the example problem, the roof surface temperature should be somewhere between 255 and 285 K. It would be good practice to set the guess value for T_surface to something reasonable, for example, 270 K.

To solve the equation set, select the Solve command from the Calculate menu. An information dialog will appear indicating the elapsed time, the maximum residual (i.e., the difference between the left-hand side and the right-hand side of an equation), and the maximum change in the values of the variables since the last iteration. When the calculations are completed, EES displays the total number of equations in the problem and the number of blocks. A block is a subset of equations that can be solved independently. EES automatically blocks the equation set, whenever possible, to improve the calculation efficiency. When the calculations are completed, the button will change from Abort to Continue.

By default, the calculations are stopped when 100 iterations have occurred, the elapsed time exceeds 3600 sec, the maximum residual is less than 10^{-6}, or the maximum variable change is less than 10^{-9}. These defaults can be changed with the Stop Criteria command in the Options menu. If the maximum residual is larger than the value set for the stopping criteria, the

1022

Solution
A = 300 [m2]
Cost = 8.601 [$]
Efficiency = 0.85
ε = 0.9
FuelCost = 0.6
h = 15 [W/m2-K]
k = 2 [W/m-K]
Q̇cond = 25508 [W]
Q̇conv = -6197 [W]
Q̇rad = 31705 [W]
Rooflength = 20 [m]
Roofwidth = 15 [m]
σ = 5.670E-08 [W/m2-K4]
Therms = 14.34 [therms]
Thickness = 0.15 [m]
Time = 14 [hr]
Tambient = 283.1 [K]
Tinner = 288.1 [K]
Tsky = 255 [K]
Tsurface = 281.7 [K]

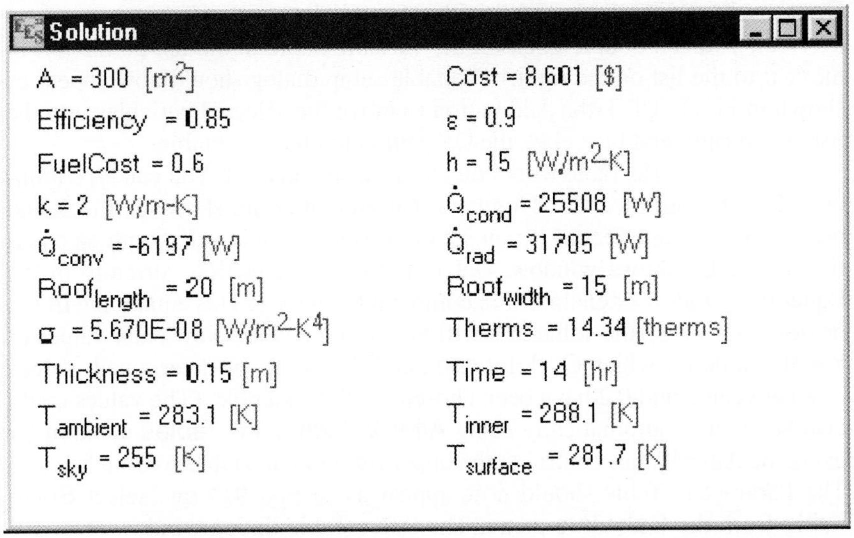

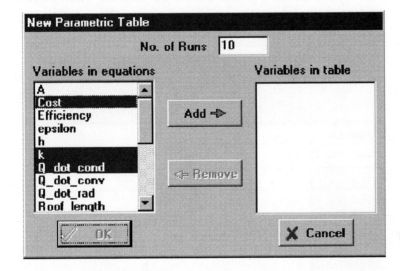

FIGURE 7
Solution window.

New Parametric Table
No. of Runs
10
Variables in equations
Variables in table
A
Cost
Efficiency
epsilon
h
k
Q_dot_cond
Q_dot_conv
Q_dot_rad
Roof_length
Add
Remove
OK
Cancel

FIGURE 8
New Parametric Table dialog.

equations were not correctly solved, possibly because the bounds on one or more variables constrained the solution. Clicking the Continue button will remove the information dialog and display the Solution window (Fig. 7). The problem is now completed since the values of T_surface, Q_dot_cond, Q_dot_conv, and Q_dot_rad are determined.

An interesting and unexpected result is evident in the solution. Because of the very low sky temperature, the radiant losses cause the roof surface temperature to be lower than the ambient temperature and thus the roof is actually warmed by the convection from the air.

One of the most useful features of EES is its ability to provide parametric studies. For example, in this problem, it may be of interest to see how the heating cost varies with the thermal conductivity of the roof material. A series of calculations can be automated and plotted in EES.

Select the New Table command in the Tables menu. A dialog will be displayed listing the variables appearing in the Equations window. In this case, we will construct a table containing the variables k, Q_dot_cond, and Cost. Click on k from the variable list on the left. This will cause k to be highlighted and the Add button will become active (Fig. 8). Repeat for Q_dot_cond and

Cost, using the scroll bar to bring the variable into view if necessary. (As a shortcut, you can double-click on the variable name in the list on the left to move it to the list on the right.) The table setup dialog should now appear as shown in Fig. 8. Click the Add button to move the selected variables into the list on the right and then click the OK button to create the table.

The Parametric Table works much like a spreadsheet. You can type numbers directly into the cells. Numbers that you enter are shown in black and produce the same effect as if you set the variable to that value with an equation in the Equations window. Delete the k = 2 equation currently in the Equations window or enclose it in comment braces { }. This equation will not be needed because the value of k will be set in the table. Now enter values of k in the table for which Q_dot_cond and Cost are to be determined. Values of k between 2 and 0.2 have been chosen for this example. (The values could also be entered automatically using Alter Values in the Tables menu or by using the Alter Values control at the upper right of each table column header.) The Parametric Table should now appear as in Fig. 9. Now, select Solve Table from the Calculate menu. The Solve Table dialog window (Fig. 10) will appear, allowing you to choose the runs for which calculations will be done.

Parametric Table

	1 $\dot{Q}_{cond}$ [W]	2 k [W/m-K]	3 Cost [$]
Run 1		2	
Run 2		1.8	
Run 3		1.6	
Run 4		1.4	
Run 5		1.2	
Run 6		1	
Run 7		0.8	
Run 8		0.6	
Run 9		0.4	
Run 10		0.2	

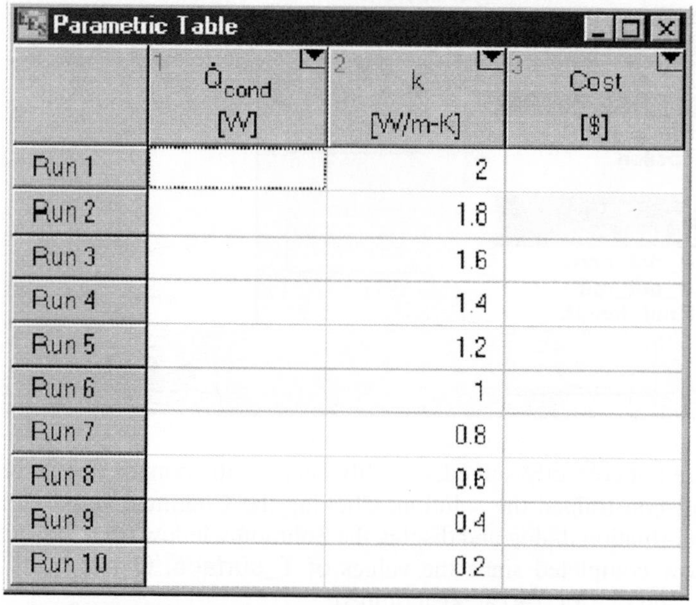

FIGURE 9
Parametric Table window.

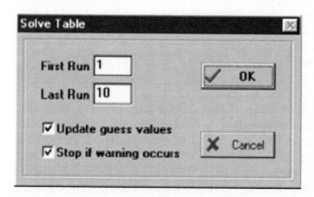

FIGURE 10
Solve Table dialog.

When the Update Guess Values control is selected, as shown, the solution for the last run will provide guess values for the current run. Click the OK button. A status window will be displayed, indicating the progress of the solution. When the calculations are completed (Fig. 11), the calculated values of Q_dot_cond and Cost will be entered into the table. The values calculated by EES will be displayed in blue, bold, or italic type depending on the setting made in the Screen Display tab of the Preferences command in the Options menu.

The relationship between variables such as Cost and k is now apparent, but it can be seen more clearly with a plot. Select New Plot Window from the Plot menu. The New Plot Window dialog (Fig. 12) will appear. Choose k to

Parametric Table

	$\dot{Q}_{cond}$ [W]	k [W/m-K]	Cost [$]
Run 1	25508	2	8.601
Run 2	23927	1.8	8.068
Run 3	22207	1.6	7.488
Run 4	20329	1.4	6.855
Run 5	18269	1.2	6.16
Run 6	16000	1	5.395
Run 7	13488	0.8	4.548
Run 8	10691	0.6	3.605
Run 9	7558	0.4	2.548
Run 10	4022	0.2	1.356

FIGURE 11

Parametric Table window after calculations have been completed.

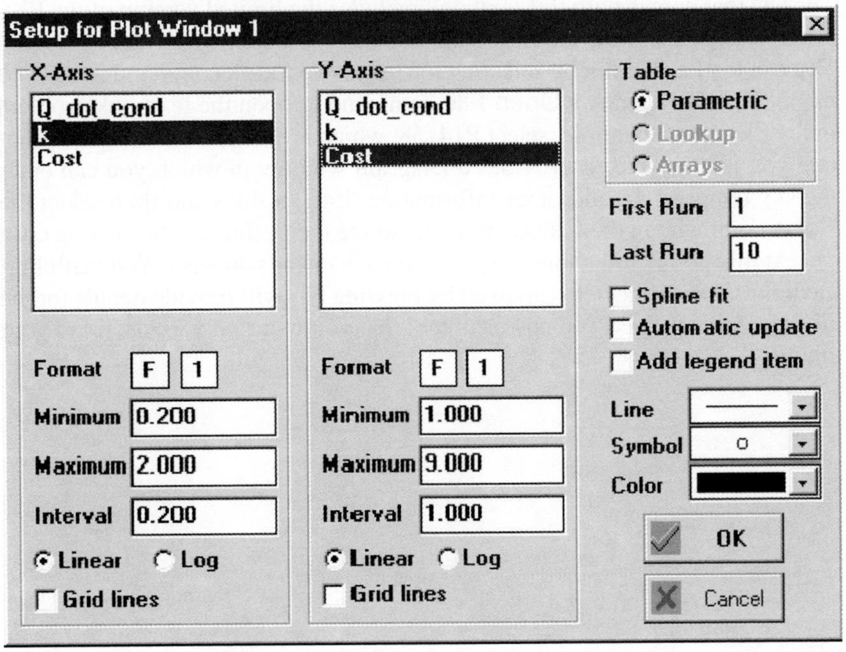

FIGURE 12

New Plot Window dialog.

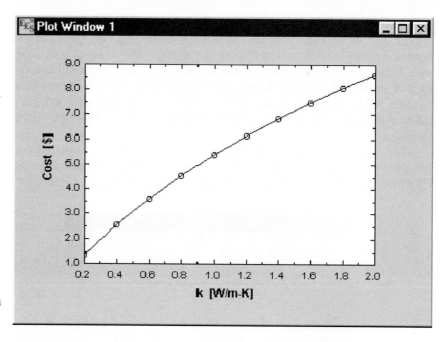

FIGURE 13

Plot window.

be the *x*-axis by clicking on k in the X-Axis list. Click on Cost in the Y-Axis list. You may wish to adjust the scale limits or add grid lines. When you click the OK button, the plot will be constructed and the plot window will appear (Fig. 13). The plot shows how the Cost varies with the thermal conductivity, indicating why most people do not have roofs made of concrete.

Once created, there are a variety of ways in which the appearance of the plot can be changed. Double-click the mouse in the plot rectangle or on the plot axis to see some of these options.

LOADING A TEXTBOOK FILE

The CD that comes with this textbook includes the limited version of the EES program together with the EES solutions of 57 problems from the textbook. You can load a solution by selecting the Load Textbook command in the File menu. Use the Windows Open File command to open the textbook problem index file. As an example, select P14_58, which is a modification of the problem you just entered. It provides a Diagram window in which you can enter the sky temperature and other information. Enter values and then select the Solve command in the Calculate menu to see their effect on the heating cost.

At this point, you should explore. Try whatever you wish. You can't hurt anything. The online help (invoked by pressing F1) will provide details for the EES commands. EES is a powerful tool that you will find very useful in your studies.

Index

Index

1029

Conversion Factors

DIMENSION	METRIC	METRIC/ENGLISH
Acceleration	$1 \text{ m/s}^2 = 100 \text{ cm/s}^2$	$1 \text{ m/s}^2 = 3.2808 \text{ ft/s}^2$ $1 \text{ ft/s}^2 = 0.3048^* \text{ m/s}^2$
Area	$1 \text{ m}^2 = 10^4 \text{ cm}^2 = 10^6 \text{ mm}^2$ $= 10^{-6} \text{ km}^2$	$1 \text{ m}^2 = 1550 \text{ in}^2 = 10.764 \text{ ft}^2$ $1 \text{ ft}^2 = 144 \text{ in}^2 = 0.09290304^* \text{ m}^2$
Density	$1 \text{ g/cm}^3 = 1 \text{ kg/L} = 1000 \text{ kg/m}^3$	$1 \text{ g/cm}^3 = 62.428 \text{ lbm/ft}^3 = 0.036127 \text{ lbm/in}^3$ $1 \text{ lbm/in}^3 = 1728 \text{ lbm/ft}^3$ $1 \text{ kg/m}^3 = 0.062428 \text{ lbm/ft}^3$
Energy, heat, work, internal energy, enthalpy	$1 \text{ kJ} = 1000 \text{ J} = 1000 \text{ Nm} = 1 \text{ kPa} \cdot \text{m}^3$ $1 \text{ kJ/kg} = 1000 \text{ m}^2/\text{s}^2$ $1 \text{ kWh} = 3600 \text{ kJ}$ $1 \text{ cal}^\dagger = 4.184 \text{ J}$ $1 \text{ IT cal}^\dagger = 4.1868 \text{ J}$ $1 \text{ Cal}^\dagger = 4.1868 \text{ kJ}$	$1 \text{ kJ} = 0.94782 \text{ Btu}$ $1 \text{ Btu} = 1.055056 \text{ kJ}$ $\quad = 5.40395 \text{ psia} \cdot \text{ft}^3 = 778.169 \text{ lbf} \cdot \text{ft}$ $1 \text{ Btu/lbm} = 25.037 \text{ ft}^2/\text{s}^2 = 2.326^* \text{ kJ/kg}$ $1 \text{ kJ/kg} = 0.430 \text{ Btu/lbm}$ $1 \text{ kWh} = 3412.14 \text{ Btu}$ $1 \text{ therm} = 10^5 \text{ Btu} = 1.055 \times 10^5 \text{ kJ}$ (natural gas)
Force	$1 \text{ N} = 1 \text{ kg} \cdot \text{m/s}^2 = 10^5 \text{ dyne}$ $1 \text{ kgf} = 9.80665 \text{ N}$	$1 \text{ lbf} = 32.174 \text{ lbm} \cdot \text{ft/s}^2 = 4.44822 \text{ N}$ $1 \text{ N} = 0.22481 \text{ lbf}$
Heat flux	$1 \text{ W/cm}^2 = 10^4 \text{ W/m}^2$	$1 \text{ W/m}^2 = 0.3171 \text{ Btu/h} \cdot \text{ft}^2$
Heat generation rate	$1 \text{ W/cm}^3 = 10^6 \text{ W/m}^3$	$1 \text{ W/m}^3 = 0.09665 \text{ Btu/h} \cdot \text{ft}^3$
Heat transfer coefficient	$1 \text{ W/m}^2 \cdot {}^\circ\text{C} = 1 \text{ W/m}^2 \cdot \text{K}$	$1 \text{ W/m}^2 \cdot {}^\circ\text{C} = 0.17612 \text{ Btu/h} \cdot \text{ft}^2 \cdot {}^\circ\text{F}$
Length	$1 \text{ m} = 100 \text{ cm} = 1000 \text{ mm}$ $1 \text{ km} = 1000 \text{ m}$	$1 \text{ m} = 39.370 \text{ in} = 3.2808 \text{ ft} = 1.0926 \text{ yd}$ $1 \text{ ft} = 12 \text{ in} = 0.3048^* \text{ m}$ $1 \text{ mile} = 5280 \text{ ft} = 1.6093 \text{ km}$ $1 \text{ in} = 2.54^* \text{ cm}$
Mass	$1 \text{ kg} = 1000 \text{ g}$ $1 \text{ metric ton} = 1000 \text{ kg}$	$1 \text{ kg} = 2.2046226 \text{ lbm}$ $1 \text{ lbm} = 0.45359237^* \text{ kg}$ $1 \text{ ounce} = 28.3495 \text{ g}$ $1 \text{ slug} = 32.174 \text{ lbm} = 14.5939 \text{ kg}$ $1 \text{ short ton} = 2000 \text{ lbm} = 907.1847 \text{ kg}$

*Exact conversion factor between metric and English units.

†Calorie is originally defined as the amount of heat needed to raise the temperature of 1 g of water by 1°C, but it varies with temperature. The international steam table (IT) calorie (generally preferred by engineers) is exactly 4.1868 J by definition and corresponds to the specific heat of water at 15°C. The thermochemical calorie (generally preferred by physicists) is exactly 4.184 J by definition and corresponds to the specific heat of water at room temperature. The difference between the two is about 0.06 percent, which is negligible. The capitalized Calorie used by nutritionists is actually a kilocalorie (1000 IT calories).

DIMENSION	METRIC	METRIC/ENGLISH
Power, heat transfer rate	$1 \text{ W} = 1 \text{ J/s}$ $1 \text{ kW} = 1000 \text{ W} = 1.341 \text{ hp}$ $1 \text{ hp}^{\ddagger} = 745.7 \text{ W}$	$1 \text{ kW} = 3412.14 \text{ Btu/h}$ $\quad = 737.56 \text{ lbf} \cdot \text{ft/s}$ $1 \text{ hp} = 550 \text{ lbf} \cdot \text{ft/s} = 0.7068 \text{ Btu/s}$ $\quad = 42.41 \text{ Btu/min} = 2544.5 \text{ Btu/h}$ $\quad = 0.74570 \text{ kW}$ $1 \text{ boiler hp} = 33,475 \text{ Btu/h}$ $1 \text{ Btu/h} = 1.055056 \text{ kJ/h}$ $1 \text{ ton of refrigeration} = 200 \text{ Btu/min}$
Pressure	$1 \text{ Pa} = 1 \text{ N/m}^2$ $1 \text{ kPa} = 10^3 \text{ Pa} = 10^{-3} \text{ MPa}$ $1 \text{ atm} = 101.325 \text{ kPa} = 1.01325 \text{ bars}$ $\quad = 760 \text{ mmHg at } 0°C$ $\quad = 1.03323 \text{ kgf/cm}^2$ $1 \text{ mmHg} = 0.1333 \text{ kPa}$	$1 \text{ Pa} = 1.4504 \times 10^{-4} \text{ psia}$ $\quad = 0.020886 \text{ lbf/ft}^2$ $1 \text{ psia} = 144 \text{ lbf/ft}^2 = 6.894757 \text{ kPa}$ $1 \text{ atm} = 14.696 \text{ psia} = 29.92 \text{ inHg at } 30°F$ $1 \text{ inHg} = 3.387 \text{ kPa}$
Specific heat	$1 \text{ kJ/kg} \cdot °C = 1 \text{ kJ/kg} \cdot K$ $\quad = 1 \text{ J/g} \cdot °C$	$1 \text{ Btu/lbm} \cdot °F = 4.1868 \text{ kJ/kg} \cdot °C$ $1 \text{ Btu/lbmol} \cdot R = 4.1868 \text{ kJ/kmol} \cdot K$ $1 \text{ kJ/kg} \cdot °C = 0.23885 \text{ Btu/lbm} \cdot °F$ $\quad = 0.23885 \text{ Btu/lbm} \cdot R$
Specific volume	$1 \text{ m}^3/\text{kg} = 1000 \text{ L/kg}$ $\quad = 1000 \text{ cm}^3/\text{g}$	$1 \text{ m}^3/\text{kg} = 16.02 \text{ ft}^3/\text{lbm}$ $1 \text{ ft}^3/\text{lbm} = 0.062428 \text{ m}^3/\text{kg}$
Temperature	$T(\text{K}) = T(°C) + 273.15$ $\Delta T(\text{K}) = \Delta T(°C)$	$T(\text{R}) = T(°F) + 459.67 = 1.8T(\text{K})$ $T(°F) = 1.8 \, T(°C) + 32$ $\Delta T(°F) = \Delta T(\text{R}) = 1.8^* \, \Delta T(\text{K})$
Thermal conductivity	$1 \text{ W/m} \cdot °C = 1 \text{ W/m} \cdot K$	$1 \text{ W/m} \cdot °C = 0.57782 \text{ Btu/h} \cdot \text{ft} \cdot °F$
Thermal resistance	$1°C/\text{W} = 1 \text{ K/W}$	$1 \text{ K/W} = 0.52750 °F/\text{h} \cdot \text{Btu}$
Velocity	$1 \text{ m/s} = 3.60 \text{ km/h}$	$1 \text{ m/s} = 3.2808 \text{ ft/s} = 2.237 \text{ mi/h}$ $1 \text{ mi/h} = 1.46667 \text{ ft/s}$ $1 \text{ mi/h} = 1.609 \text{ km/h}$
Viscosity, dynamic	$1 \text{ kg/m} \cdot \text{s} = 1 \text{ N} \cdot \text{s/m}^2 = 1 \text{ Pa} \cdot \text{s} = 10 \text{ poise}$	$1 \text{ kg/m} \cdot \text{s} = 2419.1 \text{ lbf/ft} \cdot \text{h}$ $\quad = 0.020886 \text{ lbf} \cdot \text{s/ft}^2$ $\quad = 5.8016 \times 10^{-6} \text{ lbf} \cdot \text{h/ft}^2$
Viscosity, kinematic	$1 \text{ m}^2/\text{s} = 10^4 \text{ cm}^2/\text{s}$ $1 \text{ stoke} = 1 \text{ cm}^2/\text{s} = 10^{-4} \text{ m}^2/\text{s}$	$1 \text{ m}^2/\text{s} = 10.764 \text{ ft}^2/\text{s} = 3.875 \times 10^4 \text{ ft}^2/\text{h}$ $1 \text{ m}^2/\text{s} = 10.764 \text{ ft}^2/\text{s}$
Volume	$1 \text{ m}^3 = 1000 \text{ L} = 10^6 \text{ cm}^3 \text{ (cc)}$	$1 \text{ m}^3 = 6.1024 \times 10^4 \text{ in}^3 = 35.315 \text{ ft}^3$ $\quad = 264.17 \text{ gal (U.S.)}$ $1 \text{ U.S. gallon} = 231 \text{ in}^3 = 3.7854 \text{ L}$ $1 \text{ fl ounce} = 29.5735 \text{ cm}^3 = 0.0295735 \text{ L}$ $1 \text{ U.S. gallon} = 128 \text{ fl ounces}$

*Exact conversion factor between metric and English units.

‡Mechanical horsepower. The electrical horsepower is taken to be exactly 746 W.

Some Physical Constants

Universal gas constant	R_u = 8.31434 kJ/kmol · K
	= 8.31434 kPa · m³/kmol · K
	= 0.0831434 bar · m³/kmol · K
	= 82.05 L · atm/kmol · K
	= 1.9858 Btu/lbmol · R
	= 1545.35 ft · lbf/lbmol · R
	= 10.73 psia · ft³/lbmol · R
Standard acceleration of gravity	g = 9.80665 m/s²
	= 32.174 ft/s²
Standard atmospheric pressure	1 atm = 101.325 kPa
	= 1.01325 bar
	= 14.696 psia
	= 760 mmHg (0°C)
	= 29.9213 inHg (32°F)
	= 10.3323 mH₂O (4°C)
Stefan–Boltzmann constant	σ = 5.66961 × 10⁻⁸ W/m² · K⁴
	= 0.1714 × 10⁻⁸ Btu/h · ft² · R⁴
Boltzmann's constant	k = 1.380622 × 10⁻²³ J/K
Speed of light in vacuum	c = 2.9979 × 10⁸ m/s
	= 9.836 × 10⁸ ft/s
Speed of sound in dry air at 0°C and 1 atm	C = 331.36 m/s
	= 1089 ft/s
Heat of fusion of water at 1 atm	h_{if} = 333.7 kJ/kg
	= 143.5 Btu/lbm
Heat of vaporization of water at 1 atm	h_{fg} = 2257.1 kJ/kg
	= 970.4 Btu/lbm